Genetics

SECOND EDITION

Peter J. Russell

REED COLLEGE

Scott, Foresman and Company

GLENVIEW, ILLINOIS
BOSTON
LONDON

Library of Congress Cataloging-in-Publication Data

Russell, Peter J.
    Genetics.

    Bibliography: p.
    Includes index.
    1. Genetics.   I. Title.
QH430.R87   1989   575.1   88–26440
ISBN 0–673–39843–9

    2 3 4 5 6 7 8 9 10—VHJ—95 94 93 92 91 90

Printed in the United States of America

Design and illustration development: J/B Woolsey
    Associates, Bala Cynwyd PA
Designer: Wladislawa Finne
Illustrators: Bette Woolsey, John D. Woolsey, Patrick
    Lane

Credits and Acknowledgments appear on page 893.

Cover illustration: This computer-generated graphic
was created by Arthur J. Olson of the Research Institute of Scripps Clinic, La Jolla, California, using a
Convex C1 minisupercomputer and a Silicon Graphics 4D Graphics Display, and the molecular surface
rendering software MCS developed by Michael L.
Connolly. Copyright © 1988.

Preface

Genetics has long been one of the key areas in biology, and in the twenty-five years since the first unravelings of the DNA mysteries, we have seen an unparalleled explosion of knowledge and understanding about genetics and its related disciplines. Not only is knowledge of genetics accumulating rapidly, but its many applications are affecting our daily lives and bringing benefit to humanity. In writing *Genetics,* my goal is to provide today's students with a necessary new synthesis of this vital field.

It is most gratifying when a project that has demanded a great deal of time and effort is successful. The first edition of *Genetics* has achieved such a positive response from its users that the publisher and I have invested our time and effort to improve that book so that we have a superior new edition. The second edition retains the overall approach, logical progression of ideas, organization, and pedagogical features (such as "Analytical Approaches for Solving Genetics Problems," "Keynotes") that made the first edition readily accessible to students. It retains too a format that allows you to use the chapters out of sequence to accommodate your own teaching strategies. Some significant changes, however, appear in this second edition:

1. All the molecular aspects of genetics are updated, so that the book continues to reflect our current understanding of genes at the molecular level.
2. Molecular approaches applied to the traditional, classical areas of genetics (such as diagnosis of human disease) have been developed in more detail.
3. Because recombinant DNA technology has revolutionized many areas of biology, a chapter is added, concentrating on recombinant DNA technology and the manipulation of DNA. This addition is significant in that it shows how, by manipulating DNA, many applications have become available for use in biological research, commercial ventures, and medical diagnoses. The first edition had some coverage of this topic, but I have expanded and updated it to reflect the most recent uses of this technology.

4. A new chapter on genetic recombination shows models for generating genetic change by crossover events.

5. The chapter on gene regulation in eukaryotes is significantly rewritten and expanded.

6. The chapters on genetics of populations and quantitative genetics are expanded and significantly strengthened by additional examples and by more comprehensive treatment of all topics.

7. The art program is completely revised. Many figures have been added, and all the original illustrations have been redrawn for greater clarity. The innovative new art program emphasizes key genetic processes, making genetics more accessible to students and pedagogically sound for instructors.

8. Many questions and problems are added to the chapters. All answers have been checked for accuracy by independent reviewers.

9. Many references have been added to the "Suggested Readings" that accompany each chapter.

10. Many of the first-edition "boxed" special topics have been integrated into the text so that these topics receive as much attention as all the other topics.

Organization and Coverage

The second edition's twenty-four chapters contrast to the first edition's eighteen. This change reflects added chapters as well as reorganized material. Dividing some of the chapters in two allows for greater clarity and depth of coverage.

The first seven chapters deal with transmission of the genetic material. In Chapter 1 we review the structure of viruses and of prokaryotic and eukaryotic cells, and asexual and sexual reproduction. Mitosis and meiosis are discussed in the context of both animal and plant life cycles. Chapter 2 is focused on the contributions Mendel made to our understanding of the principles of heredity, and Chapters 2 and 3 both present the basic principles of genetics in rela-

tion to Mendel's laws. Chapter 3 covers experimental evidence for the relationship between genes and chromosomes, methods of sex determination, and Mendelian genetics in human beings.

The exceptions to and extensions of Mendelian analysis (such as the existence of multiple alleles, lethal alleles, and modifications of dominance relationships) are described in Chapter 4. In Chapter 5, we describe how the order of and distance between the genes on eukaryotic chromosomes are determined in genetic experiments designed to quantify the crossovers that occur during meiosis. In the next two chapters, we consider advanced mapping analysis in eukaryotes, and genetic mapping in prokaryotes. (This material had been combined in one chapter in the first edition.) Chapter 6 is focused on tetrad analysis, primarily in fungal systems, on using mitotic analysis in mapping eukaryotic genes, and on procedures for mapping genes to human chromosomes by means of somatic cell hybrids. In Chapter 7, we discuss the ways of mapping genes in bacteriophages and in bacteria (transformation, conjugation, and transduction).

In Chapters 8 through 15, the "molecular core" of the book, we present the current state of knowledge on the molecular aspects of genetics. In Chapter 8, we cover the structure of DNA, and present the classical experiments revealing that DNA and RNA are genetic material, and the double helix model for the structure of DNA. In this edition, the details of DNA structure and the organization of DNA in chromosomes in prokaryotes are exposed in Chapter 9. We cover DNA replication in prokaryotes and eukaryotes in Chapter 10.

After thoroughly explaining the nature of the gene and its relationship to chromosome structure, we discuss the first step in the expression of a gene transcription. In this edition, we describe the general process of transcription in Chapter 11 and we present the currently understood details of the structure, transcription, and processing of messenger RNA, transfer RNA, and ribosomal RNA molecules for both prokaryotes and eukaryotes in Chapter 12. In Chapter 13, we describe the structure of proteins, the evidence for the nature of the genetic code, and a detailed expression of our current knowledge on translation in both prokaryotes and eukaryotes. In Chapter 14, we examine genetic fine structure and some aspects of gene function, such as genetic control of protein and enzyme structure and function, and involvement of particular sets of genes in directing and controlling biochemical pathways. In the new Chapter 15, on recombinant DNA technology, we discuss in detail the modern tools used in molecular genetics. Recombinant DNA technology for cloning and characterizing genes and manipulating DNA, are described followed by the commercial applications of recombinant DNA technology.

In the next four chapters, we describe the ways in which genetic material can change or be changed or both. The processes of gene mutation and the procedures which screen for potential mutagens and carcinogens (Ames test) and which select for particular classes of mutants from a heterogeneous population are discussed in Chapter 16. Chromosome aberrations—that is, changes in normal chromosome structure or chromosome number—are described in Chapter 17, and the structures and movement of transposable genetic elements in prokaryotes and eukaryotes are presented in Chapter 18. Models for generating genetic change by genetic recombination events are discussed in still another addition, Chapter 19.

The next two chapters are focused on regulation of gene expression in prokaryotes (Chapter 20) and eukaryotes (Chapter 21). In Chapter 20, we discuss the operon as a unit of gene regulation, the current molecular details in the regulation of gene expression in bacterial operons, and regulation of genes in bacteriophages. In Chapter 21, we explain how gene expression is regulated in eukaryotes by focusing on the molecular changes that accompany gene regulation, short-term gene regulation in lower and higher eukaryotes, gene regulation in development and differentiation, genetic defects in human development, and oncogenes and their relationship to cancer.

In Chapter 22, we address the organization and genetics of extranuclear genomes of mito-

chondria and chloroplasts. We cover the classical genetic experiments, which established that a gene is extranuclear. In this chapter we also discuss current molecular information about the organization of genes within the extranuclear genomes.

In Chapters 23 and 24, we describe quantitative genetics and the genetics of populations, respectively. In Chapter 23, we develop the concept that some heritable traits, the quantitative traits, show continuous variation over a range of phenotypes. In Chapter 23 we also discuss heritability: the relative extent to which a characteristic is determined by genes or by the environment. In Chapter 24, we present the basic principles in the genetics of populations and extend our treatment of the gene from the individual organism to a population of organisms.

Pedagogical Features

Because the study of genetics is so complex and difficult, I have incorporated a number of special pedagogical features to help students and enhance their understanding and appreciation of genetic principles:

1. Most chapters have a section on "Analytical Approaches for Solving Genetics Problems." The principles of genetics have always been best taught with a problem-solving approach. Often, however, beginning students do not acquire experience with the basic concepts that would enable them to methodically attack assigned problems. In the "Analytical Approaches" sections, typical genetic problems are "talked through" in step-by-step detail to help students understand how to tackle a genetics problem by applying fundamental principles.

2. More than 400 questions and problems in the problem sets closing each chapter give students further practice in solving genetics problems. For each chapter, the problems are chosen to represent a range of difficulty and topics. The answers to selected questions (indicated by *) are included at the back of the

book, and answers to all questions are available in a separate supplement.

3. Keynote summaries throughout each chapter emphasize important ideas and critical points.

4. Important terms and concepts are clearly defined where they are introduced in the text and are collected in a Glossary at the back of the book for easy reference.

5. Comprehensive and up-to-date bibliographies for each chapter are included at the back of the book.

6. Some chapters include boxes covering special topics related to chapter coverage. Some of these boxed topics are: *What Organisms Are Suitable for Use in Genetic Experiments* (Chapter 2); and *Equilibrium Density Gradient Centrifugation* (Chapter 10).

Acknowledgments

I would like to extend special thanks to Ben Pierce (Baylor University) for his creative rewritings of Chapter 23 and 24, "Quantitative Genetics" and "Genetics of Populations." He did an excellent job in expanding two very difficult topics, making a significant contribution to the overall quality of this book.

Also making an invaluable contribution to this text are John and Bette Woolsey of J/B Woolsey Associates, Inc., who developed and executed an art program that is innovative and exciting.

I want to again acknowledge my wife, Jenny, and my children, Steven and Kristie, for enduring the time I invested in this book. I am also grateful to my students of genetics here at Reed for their encouragement in this endeavor. I would also like to thank all the reviewers involved in this project, including: Fred W. Allendorf (University of Montana), Phillip T. Barnes (Connecticut College), John Belote (Syracuse University), Jean A. Bowles (Metropolitan State College), L. Herbert Bruneau (Oklahoma State University), Scott K. Davis (Texas A&M University), Max P. Dunford (New Mexico State University), Paul Goldstein (University of Texas, El Paso), J.L. Hamrick (University of Georgia), Duane L. Johnson (Colorado State University), Elliot Krause (Seton Hall University), Peter Kuempel (University of Colorado), E.N. Larter (University of Manatoba), Thomas J. Leonard (University of Wisconsin, Madison), Mark F. Sanders (University of California, Davis), Bruce A. Voyles (Grinnell College), William Wissinger, Ph.D. (St. Bonaventure University), Harry Wistrand (Agnes Scott College). For their time and effort in reviewing our new art program, I give special thanks to: Joseph O. Falkinham, III (Virginia Polytechnic Institute), Clint Magill (Texas A&M University), Jim Price (Texas A&M University) for their thorough examination of the material. I am grateful to the Literary Executor of the late Sir Ronald A. Fisher, F.R.S., to Dr. Frank Yates, F.R.S., and to Longman Group Ltd. London, for permission to reprint Table IV from their book *Statistical Table for Biological, Agricultural and Medical Research* (6th edition, 1974).

Finally, I thank everyone in Scott, Foresman/Little, Brown's College Division who contributed to this effort. They include Bonnie Roesch, Maxine Effenson, Jane Tufts, Janice Friedman, Cynthia Frank, Maryann Zoll, Julia Slater, and Jane Betz.

Contents

Viruses, Cells, and Cellular Reproduction

Genetics is the science of heredity and involves studying the structure, function, and transmission of genes (the units of heredity) from one generation to the next. The differences between organisms are the result of differences in the genes they carry, which have resulted from the evolutionary processes of mutation (a heritable change in the genetic material), recombination (exchange of genetic material between chromosomes), and selection (the favoring of particular combinations of genes in a given environment). In comparison with evolutionary time, very little time has been spent studying genetics. The principles of heredity were recognized by Gregor Mendel in the 1860s, the development of the subject began about 1900, and the key modern discoveries have been made in the past three or four decades.

The study of genetics has examined two aspects of the genetic material of living organisms: how physical traits are coded for and expressed in the organism, and how these traits are inherited (copied and passed on) from one generation to the next (Figure 1.1).

At present, we have a good understanding of the nature of genetic material, how it is replicated and passed on from generation to generation, how it is expressed in the cell, and how that expression is regulated. Clearly, gene activity is of central importance to cell growth and function and to the development and differentiation of organisms. In these respects the study of genetics interfaces directly with cellular biology, embryology, and developmental biology.

In the past decade genetics has received attention above and beyond its basic contributions to our understanding of life processes and of genetic diseases. This interest has come about from public attention given to the development and application of particular molecular techniques known collectively as *recombinant DNA technology.* Research in this area has led to the development of industries dealing in what is called *biotechnology,* or *genetic engineering.* While most of this research was undertaken to study fundamental genetic processes without regard to practical applications, there has been, and will continue to be, a lot of applied work

a) Expression of physical traits (phenotype)

b) Heredity of genetic code

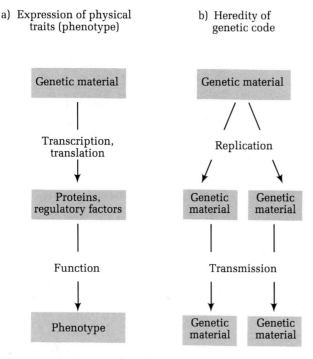

Figure 1.1 *The two roles of genetic material: (a) coding for the physical traits (phenotype) of an organism through the processes of transcription, translation, and protein manufacture; (b) ensuring, through replication and division, that the organism's genetic code is accurately transmitted from one generation to the next.*

done that is of direct benefit to humankind. In the area of plant breeding, genetic engineering is being applied to improve crop yields and to develop plants resistant to pests and plant diseases. In the area of animal breeding, genetic engineering is being employed in the beef, dairy cattle, and poultry industries to improve yields and to develop better strains. In medicine the potential is equally impressive. For example, recombinant DNA technology is proving very important in the production of antibiotics, hormones, and other medically important agents and in the diagnosis and treatment of certain human genetic diseases (such as Huntington's disease and Tay-Sachs disease). In short, the science of genetics is currently in a dramatic growth phase. By understanding the basic principles of genetics, as described in this text, you

will have a greater capacity to understand, foster, and perhaps contribute to the new and exciting applications of this subject.

Viruses

A **virus** is a noncellular organism that can reproduce only within a host cell. It contains genetic material, either **DNA (deoxyribonucleic acid)** or **RNA (ribonucleic acid),** within a membrane or protein coat. Every type of cellular organism can be infected by viruses. Thus there are plant viruses, animal viruses, and bacterial viruses. The latter are also called bacteriophages ("bacteria eaters"), or simply phages.

Viruses vary considerably in size, shape, and composition (Figure 1.2a-g) and have evolved specific mechanisms to infect a host cell or cells. We will discuss some of these mechanisms in other chapters, but for now, remember that the genetic material of the virus, once it is within the host cell, directs the cellular machinery to produce progeny viruses.

Cells—The Units of Life

After many years of experiments the **cell theory** was put forward in the early nineteenth century. There are three basic tenets of the cell theory:

1. The **cell** is the smallest building unit of a multicellular organism and, as a unit, is itself an elementary organism.
2. Each cell in a multicellular organism has a specific role.
3. A cell can only be produced from another cell by cell division.

Beyond these three basic ideas our knowledge and understanding of the structure of organisms and cells has been advanced through the use of microscopes. Under the microscope we can see that while there is a great diversity of cell types, they all have some features in common. We will start our study of the genetics of various life forms by examining cell structure.

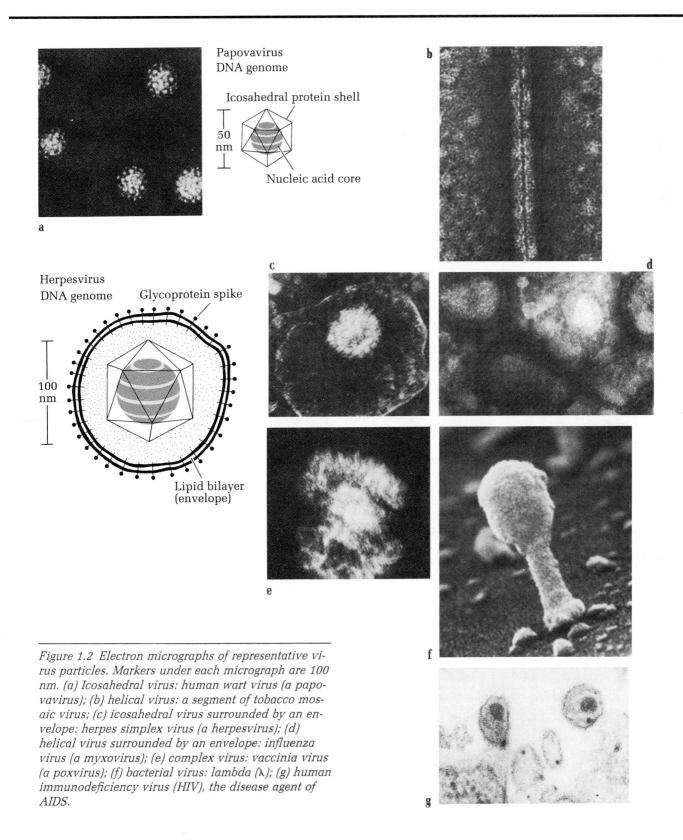

Figure 1.2 *Electron micrographs of representative virus particles. Markers under each micrograph are 100 nm. (a) Icosahedral virus: human wart virus (a papovavirus); (b) helical virus: a segment of tobacco mosaic virus; (c) icosahedral virus surrounded by an envelope: herpes simplex virus (a herpesvirus); (d) helical virus surrounded by an envelope: influenza virus (a myxovirus); (e) complex virus: vaccinia virus (a poxvirus); (f) bacterial virus: lambda (λ); (g) human immunodeficiency virus (HIV), the disease agent of AIDS.*

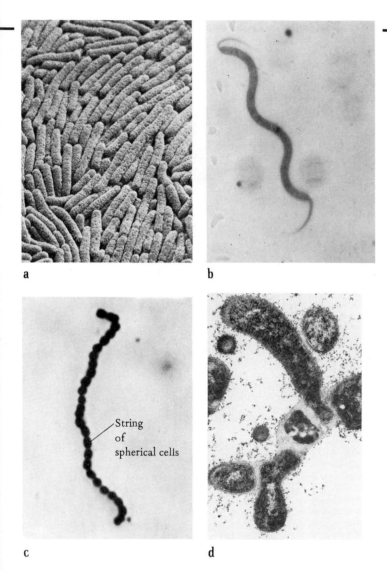

Figure 1.3 Four prokaryotes: (a) Escherichia coli, *a rod-shaped bacterium common in human intestines; (b)* Spirillum volutans, *a helically coiled bacterium; (c)* Streptococcus lactis, *a spherical bacterium common in milk; (d)* Actinomycete, *a prokaryote with branching, multicellular filaments.*

Prokaryotes: Bacteria and Cyanobacteria

Prokaryotes (meaning "prenuclear") are the simplest cellular organisms. Within this group are all the bacteria (Figure 1.3).

Bacteria are spherical, rod-shaped, or spiral-shaped single-celled or multicellular, filamentous organisms (Figure 1.3) that vary in size.

Some are only 100 nm (1 nm $= 10^{-9}$ m) in diameter, while very large ones are 6 μm (1 μm $= 10^{-6}$ m) in diameter and 60 μm long. The genetic material of bacteria is found in a single, circular DNA molecule that is associated with few proteins. In bacteria this genetic material is not separated from the rest of the cytoplasm by a nuclear membrane such as is found in eukaryotic ("true nuclear") organisms, which is a major distinguishing feature of prokaryotes. In their cytoplasm bacteria also contain a large number of ribosomes, the organelles on which protein synthesis takes place. The shape of the bacterium is maintained by a rigid cell wall located outside the cell membrane.

The most intensely studied bacterium in genetics is *Escherichia coli* (Figure 1.3a), a rod-shaped bacterium common in human intestines. Studies of this bacterium have resulted in significant advances in our understanding of the regulation of gene expression and in the development of the whole field of molecular biology. Today *E. coli* is used extensively in recombinant DNA experiments.

Certain bacteria, the cyanobacteria, are unicellular or multicellular, filamentous photosynthetic organisms (Figure 1.4). Like other bacteria, they have a cell wall, their genetic material is DNA that is not associated with protein or bounded by a nuclear membrane, and they contain large numbers of ribosomes. Their cells are approximately 6–10 μm in diameter.

Bacteria and cyanobacteria reproduce by **binary fission;** that is, after their genetic material replicates, each cell simply divides in two (Figure 1.5).

Eukaryotes

Eukaryotes (means "true nucleus") are organisms that have cells in which the genetic material is located in the **nucleus,** a discrete structure within the cell that is bounded by a nuclear membrane. Eukaryotes can be unicellular or multicellular and are often grouped by the terms *lower* and *higher.* Lower eukaryotes have relatively simple cellular organization, and while their DNA is significantly more complex than

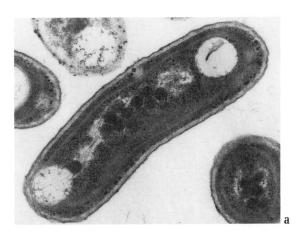
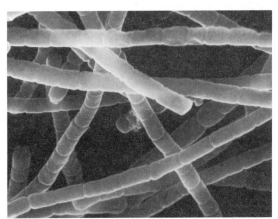

Figure 1.4 Unicellular (a) and filamentous (b) cyano-bacteria.

that of prokaryotes, it is less complex than that of the higher eukaryotes. Higher eukaryotes include genetically more complex multicellular organisms.

The following eukaryotic organisms are commonly used in genetic research, and you will see a number of them frequently throughout this text: *Drosophila melanogaster* (fruit fly), *Pisum sativum* (garden pea), *Zea mays* (corn), *Chlamydomonas reinhardi* (a green alga), *Neurospora crassa* (common bread mold), *Saccharomyces cerevisiae* (baker's yeast), *Mus musculus* (mouse), and *Homo sapiens* (humans) (Figure 1.6). In the above list, *Chlamydomonas* and *Saccharomyces* are unicellular organisms and the rest are multicellular. *Neurospora, Saccharomyces,* and *Chlamydomonas* are lower eukaryotes, while those remaining are higher eukaryotes.

Cell structure. The structure of the eukaryotic cell is very different from that of a bacterial cell. Figure 1.7a is an electron micrograph of a cross section through an animal cell, Figure 1.7b shows a diagrammatic representation of a thin section through a generalized eukaryotic (animal) cell, Figure 1.7c is an electron micrograph of a cross section through a higher plant cell, and Figure 1.7d is a diagrammatic representation of a thin section through a generalized higher plant cell.

Nucleus. The nucleus is separated from the remainder of the cell, the cytoplasm, by a double membrane called the nuclear membrane. An electron micrograph of a cross section of an animal cell through the nucleus is shown in Figure 1.8a (see p. 8) and a diagram of an animal cell nucleus is shown in Figure 1.8b. The membrane is selectively permeable so that material can move between the nucleus and the cytoplasm. Some cellular compounds can pass through the membrane itself, whereas others move through holes called nuclear pores, which are about 20–80 nm in diameter, and are each surrounded by a thickened, electron-dense ring, or annulus.

The nucleus contains most of the genetic material of the cell. The genetic material is complexed with protein and is organized into a

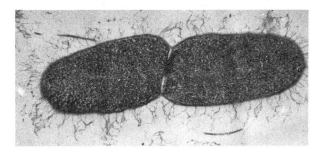

Figure 1.5 Binary fission of E. coli.

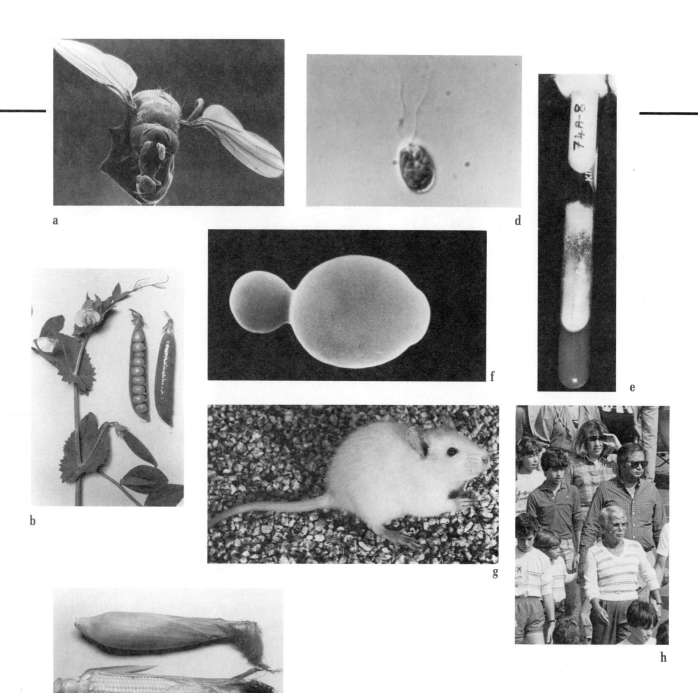

Figure 1.6 *Eukaryotic organisms commonly used in genetics research: (a)* Drosophila melanogaster *(fruit fly); (b)* Pisum sativum *(garden pea); (c)* Zea mays *(corn) (d)* Chlamydomonas reinhardi *(a green alga); (e)* Neurospora crassa *(a bread mold); (f)* Saccharomyces cerevisiae *(a budding yeast); (g)* Mus musculus *(mouse); (h)* Homo sapiens *(humans).*

Figure 1.7 *Eukaryotic cell: (a) Electron micrograph of a thin section through a eukaryotic (animal) cell. LD: lipid droplet; Ly: lysosome; M: mitochondrion; MV: microvilli; N: nucleus; RER, SER: rough and smooth endoplasmic reticulum, respectively. (b) Section drawing of a thin section through an animal cell showing the main organizational features and the principal organelles. (c) Thin section of a root tip cell from* Elodea canadensis. *Easily seen are the cell wall, nucleus, vacuoles, mitochondria, endoplasmic reticulum, Golgi apparatus, and ribosomes. (d) Section drawing of a thin section through a plant cell. (Note that not all cell parts in the diagrams are readily visible in the electron micrographs.)*

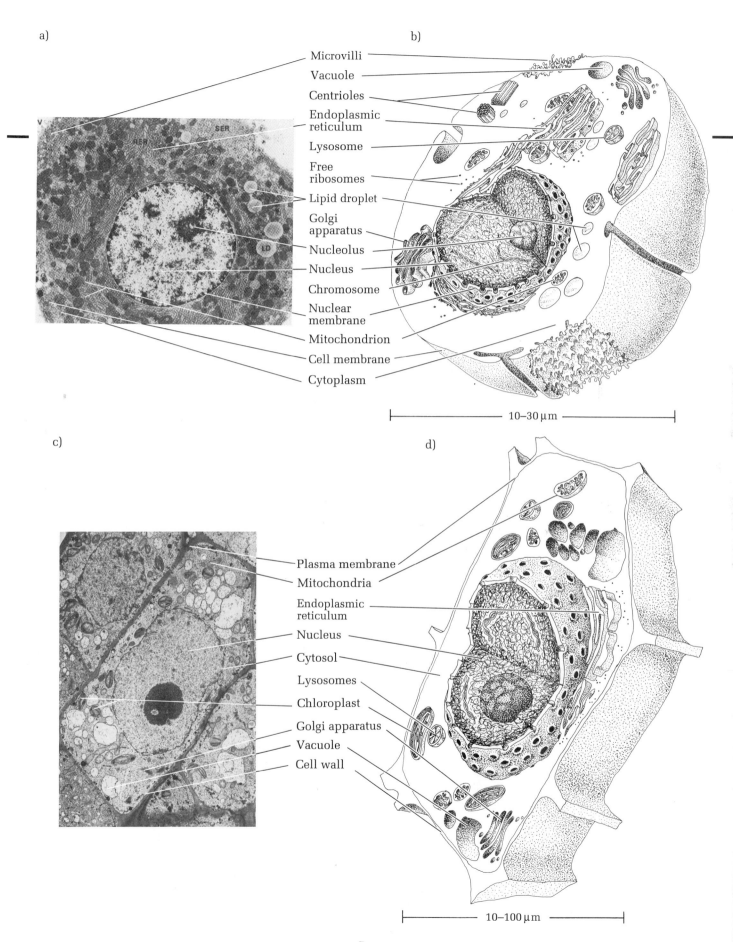

a)

b)

Microvilli
Vacuole
Centrioles
Endoplasmic reticulum
Lysosome
Free ribosomes
Lipid droplet
Golgi apparatus
Nucleolus
Nucleus
Chromosome
Nuclear membrane
Mitochondrion
Cell membrane
Cytoplasm

10–30 μm

c)

d)

Plasma membrane
Mitochondria
Endoplasmic reticulum
Nucleus
Cytosol
Lysosomes
Chloroplast
Golgi apparatus
Vacuole
Cell wall

10–100 μm

a)

b)

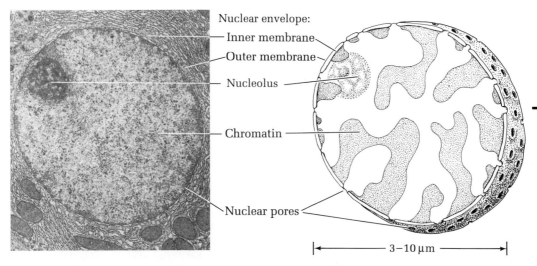

Nuclear envelope:
Inner membrane
Outer membrane
Nucleolus
Chromatin
Nuclear pores

3–10 μm

Figure 1.8 (a) Electron micrograph of a cross section of an animal cell through the nucleus. (b) Diagram of an animal cell nucleus.

a) Rough endoplasmic reticulum

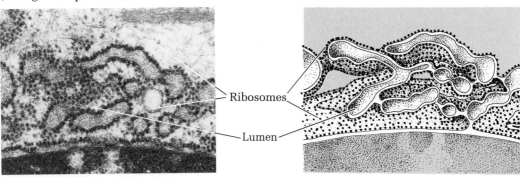

Ribosomes
Lumen

b) Smooth endoplasmic reticulum

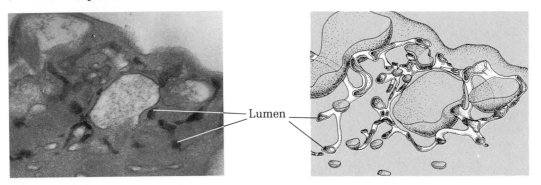

Lumen

Figure 1.9 (a) Rough and (b) smooth endoplasmic reticulum.

number of linear structures called **chromosomes.** *Chromosome* means "colored body" and is so named because these threadlike structures are visible under the light microscope only after they are stained with dyes. However, in a nondi-viding cell such as the one shown in Figure 1.7, the chromosomes are in an extended state and are relatively difficult to see even when stained. Also within the nucleus is the *nucleolus,* a structure within which the ribosomes are assembled.

Cytoplasm. The cytoplasm contains a vast amount of different materials and organelles. Of special interest for geneticists are the ribosomes, mitochondria, chloroplasts, centrioles, endoplasmic reticulum (ER), and Golgi apparatus.

The endoplasmic reticulum is a double-membrane system that runs through the cell (Figure 1.9). The double membrane is continuous with both the nuclear membrane and the plasma membrane (which forms the perimeter of the cell). Figure 1.10 presents a photograph and diagrams the relationship between the ER and the nuclear membranes. Under the electron microscope we can see two types of ER: rough (Figure 1.9a) and smooth (Figure 1.9b). The former has ribosomes attached to it, giving it a rough appearance, and the latter does not. Ribosomes bound to the endoplasmic reticulum synthesize proteins to be secreted by the cell. These proteins are moved into the space between the two membranes and are then translocated to the Golgi apparatus (Figure 1.11), where they are packaged into vesicles. The vesicles are thought to carry material between the Golgi apparatus and different compartments of the cell. In secretory cells, some of the vesicles migrate to the cell surface and, by fusion with the outer membrane, release their contents to the outside of the cell. The synthesis of all other kinds of proteins, such as enzymes and cellular structural proteins, is performed by ribosomes that are free in the cytoplasm.

Eukaryotic cells typically contain mitochondria (Figure 1.12). These large organelles are surrounded by a double membrane, the inner one of which is highly convoluted. The mitochondria play an extremely important role in energy processing for the cell. They also contain genetic material in the form of a circular molecule of DNA. As in bacteria, probably no structural proteins are associated with the genetic material of mitochondria.

Last, many plant cells contain chloroplasts, the large, chlorophyll-containing organelles involved in photosynthesis (Figure 1.13). Chloroplasts also contain genetic material, and as in mitochondria, its form is a circular DNA molecule. The plant cell also has a cell wall.

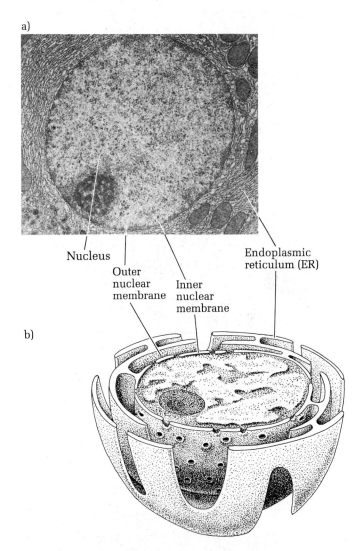

Figure 1.10 Electron micrograph (a) and schematic diagram (b) illustrating the relationship between the nuclear membranes and the endoplasmic reticulum.

Table 1.1 summarizes the key similarities and differences between prokaryotic and eukaryotic cells. Notably prokaryotes do not undergo meiosis, although they have developed unique ways to incorporate new genetic materials into their ultimate progeny (e.g., conjugation, transformation).

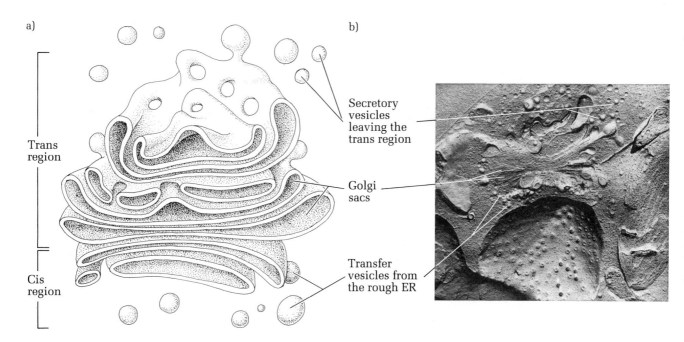

a)

Trans
region

Cis
region

b)

Secretory
vesicles
leaving the
trans region

Golgi
sacs

Transfer
vesicles from
the rough ER

Figure 1.11 Schematic diagram (a) and electron micrograph (b) of the Golgi apparatus. Cis region is near the ER. Trans region is towards the cell surface.

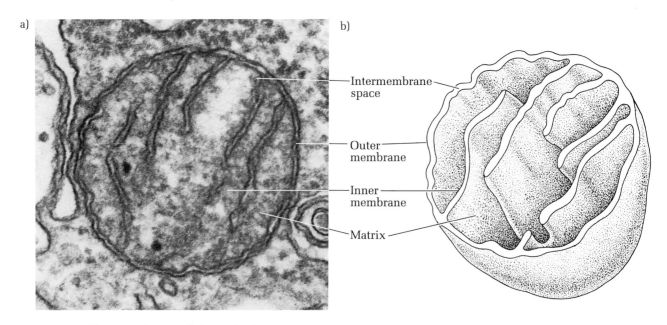

a)

b)

Intermembrane
space

Outer
membrane

Inner
membrane

Matrix

Figure 1.12 Electron micrograph (a) and schematic diagram (b) of a mitochondrion.

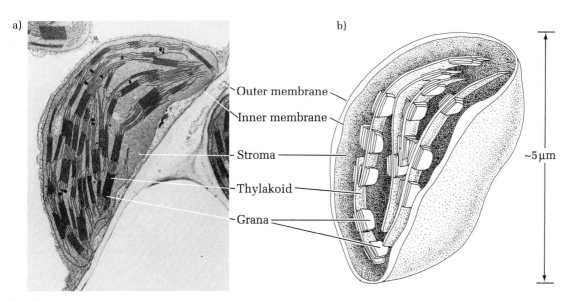

Figure 1.13 (a) Electron micrograph of a section through a plant leaf showing chloroplasts; (b) Diagram of a chloroplast.

Table 1.1 Key Differences among the Cells of Prokaryotes, Higher Plants, and Animals

| Feature | Prokaryotic Cell | Eukaryotes | |
		Higher Plant Cell	Animal Cell
Cell membrane	External only (two separated by periplasmic space)	External and internal	External and internal
Supporting structure	Polysaccharide cell wall (murein)	Polysaccharide cell wall (cellulose)	Protein cytoskeleton
Nuclear membrane	Absent	Present	Present
Chromosomes	Single, circular DNA only	Multiple, linear, complexed with protein	Multiple, linear, complexed with protein
Membrane-bound organelles	Absent	Many, including mitochondria, large vacuoles, and chloroplasts	Many, including mitochondria, lysosomes
Endoplasmic reticulum	Absent	Present	Present
Ribosomes	Smaller, free	Larger, some membrane-bound	Large, some membrane-bound
Cell division	Fission	Mitosis	Mitosis
Sexual recombination	Only by plasmid-mediated transfer	Meiosis and fertilization	Meiosis and fertilization
Centrioles	None	Usually absent	Usually present

Cellular Reproduction in Eukaryotes

The cell theory states that all living things are composed of cells, and all cells arise from previously existing cells. We now turn to the subject of cell division in eukaryotes.

Chromosome Complement of Eukaryotes

The genetic material of eukaryotes is distributed among multiple chromosomes; the number of chromosomes is characteristic of the organism. For example, the nuclei of human cells have 46 chromosomes, those of the fruit fly *Drosophila melanogaster* have 8, and those of the bread mold *Neurospora crassa* have 7. Many eukaryotes have two copies of each type of chromosome in their nuclei and are said to be **diploid,** or 2N, in terms of their chromosome complement. They are produced by the fusion of two gametes, one from the female parent and one from the male parent, to produce a diploid zygote, which then divides. Each gamete has only one set of chromosomes and is said to be **haploid** (N). The complete haploid complement of genetic information is called the **genome.** Of the three organisms just listed, humans and fruit flies are diploid (2N) and the bread mold is haploid (N). Thus humans have 23 pairs of chromosomes, *Drosophila melanogaster* has 4 pairs, and *Neurospora crassa* has 7 single chromosomes.

The members of a chromosome pair are called **homologous chromosomes;** each individual member of a pair is called a **homolog.** Homologous chromosomes are identical with respect to the arrangement of genes they contain and with respect to their visible structure. Chromosomes from different pairs are called **nonhomologous chromosomes.** Figure 1.14 illustrates the chromosomal organization of haploid and diploid organisms.

In animals and in some plants there are differences in the chromosome complement of male and female cells. One sex has a matched pair of **sex chromosomes,** chromosomes related to the sex of the organism; the other sex has an un-

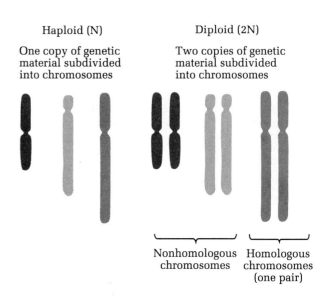

Haploid (N)

One copy of genetic material subdivided into chromosomes

Diploid (2N)

Two copies of genetic material subdivided into chromosomes

Nonhomologous chromosomes

Homologous chromosomes (one pair)

Figure 1.14 Chromosomal organization of haploid and diploid organisms.

matched pair or an unpaired chromosome. For example, human females have two X chromosomes (XX), while human males have one X and one Y (XY). Chromosomes other than sex chromosomes are called **autosomes.**

Keynote Diploid eukaryotic cells have a double set of chromosomes, one set coming from each parent. The members of a pair of chromosomes, one from each parent, are called homologous chromosomes. Haploid eukaryotic cells have only one set of chromosomes.

Under the microscope we see that chromosomes differ in their size and morphology within and between species. Each chromosome has a specialized region somewhere along its length that is often seen as a constriction under the microscope. This constriction, called a **centromere** (or **kinetochore**), is important in the activities of the chromosomes during cellular division and can be located in one of four general positions in the chromosome (see Figure 1.15). A **metacentric** chromosome has the centromere

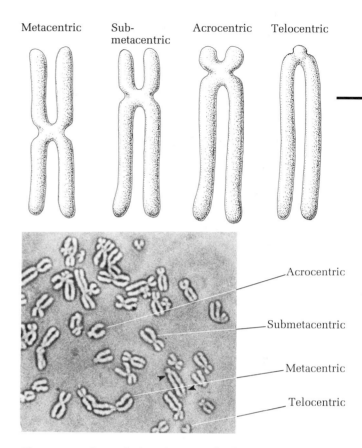

in approximately the center of the chromosome, a **submetacentric** chromosome has the centromere slightly displaced from the middle of the chromosome, an **acrocentric** chromosome has the centromere near the end of the chromosome, and a **telocentric** chromosome has the centromere at the end. As a result of these locations, a metacentric chromosome has two approximately equal-length arms, an acrocentric has one long arm and one short arm, and a telocentric has only one arm. Chromosomes also vary in relative size. Chromosomes of mice, for example, are all similar in length, whereas those of humans show a wide range of relative lengths.

Asexual and Sexual Reproduction

Eukaryotes can reproduce by asexual or sexual reproduction. In **asexual reproduction** a new individual develops from either a single cell (agamogony) or from a group of cells (vegetative reproduction) in the absence of any sexual process. It may be the only mode of reproduction of a species, or it may occur within the sexually reproductive organism as an essential or nonessential part of its life cycle. Asexual reproduction is found in both unicellular and multicellular eukaryotes. Single-celled eukaryotes (such as yeast) grow, double their genetic material, and generate two progeny cells, each of which contains an exact copy of the genetic material found in the parental cell. This process repeats as long as there are sufficient nutrients in the growth medium. In multicellular organisms asexual reproduction is sometimes referred to as **vegetative reproduction.** Multicellular fungi, for example, can be propagated vegetatively by taking a small piece of the growth mass and transferring it to new medium. Many higher plants, such as roses and fruit trees, are commercially maintained by grafting, which is an asexual means of propagation.

Sexual reproduction is the fusion of two haploid gametes (sex cells) to produce a single diploid zygote cell from which a new multicellular individual develops by mitotic division under programmed control from genetic material. Sexual reproduction involves the alternation of dip-

Figure 1.15 *General classification of eukaryotic chromosomes into metacentric, submetacentric, acrocentric, and telocentric types, based on the position of the centromere.*

loid and haploid phases. The main biological significance of sexual reproduction is that it achieves genetic recombination; that is, it generates gene combinations in the offspring that are distinct from those in the parents. With the exception of self-fertilizing organisms (such as many plants), the two gametes are from different parents, and it is during the production of the gametes that genetic recombination takes place. Figure 1.16 shows the cycle of growth and sexual reproduction in a higher eukaryote and also shows how the number of chromosomes characteristic of an organism is kept constant from generation to generation.

Sexually reproducing organisms have two sorts of cells: somatic (body) cells and germ cells. Somatic cells, which are diploid, reproduce by a process called mitosis. Germ cells (or gametes), which are haploid, are produced by meiosis. Figure 1.17 illustrates the differences between asexual and sexual reproduction.

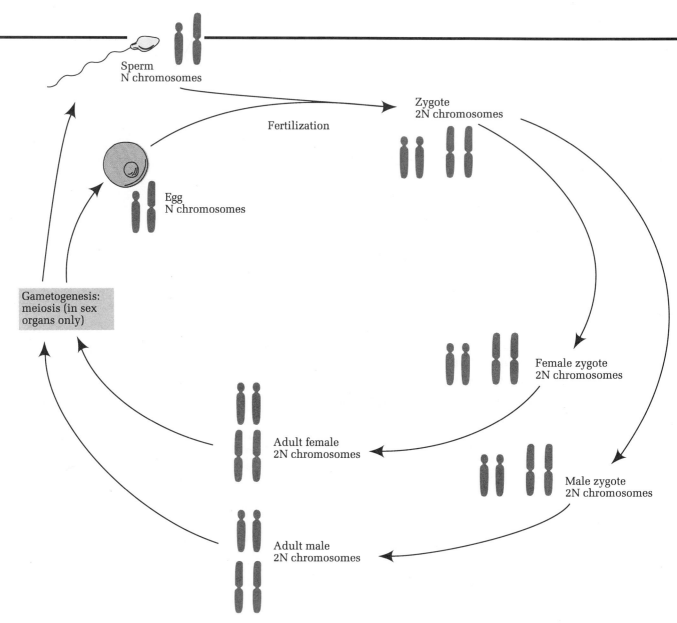

Figure 1.16 Cycle of growth and reproduction in a sexually reproducing higher eukaryote. Sexual reproduction involves the alternation of diploid (somatic) and haploid (gametic) phases.

Mitosis: Nuclear Division

In both unicellular and multicellular eukaryotes, cellular reproduction is a cyclical process of growth, **mitosis (nuclear division** or **karyokinesis)** and (usually) **cell division (cytokinesis).** This process is called the **cell cycle.** In prolifer-

ating somatic cells the cell cycle consists of four phases (Figure 1.18): the mitotic (or dividing) phase (M) and an interphase between divisions that consists of three stages, G_1 (gap 1), S, and G_2 (gap 2). During G_1, the cell prepares for DNA and chromosome replication, which take place in the S phase. In G_2, the cell prepares for cell division, or the M phase. (The molecular events that occur in each cell cycle phase will be described in a later chapter.) Most of the cell cycle is spent in interphase, although there is great variation among cell types in the relative time spent in each of the four stages.

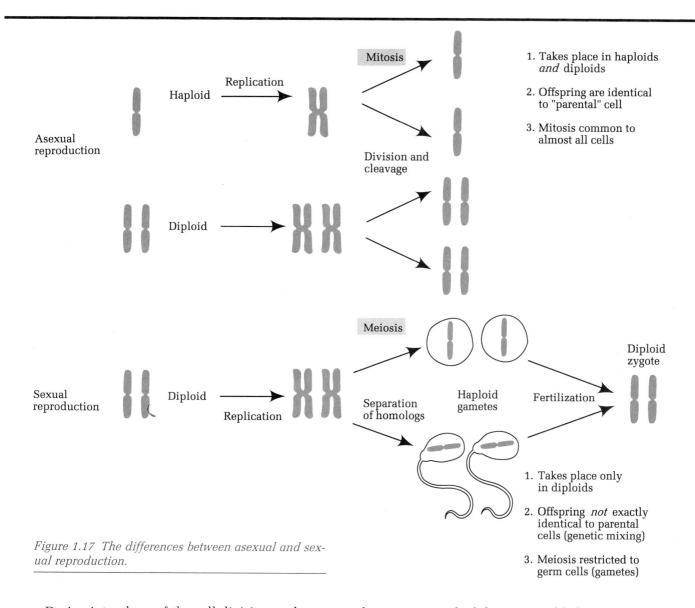

Figure 1.17 The differences between asexual and sexual reproduction.

During interphase of the cell division cycle, the chromosomes are extended and are difficult to see under the light microscope. During interphase (that is, before mitosis itself begins), each chromosome replicates, and the DNA of the centromere also replicates, although only one centromere structure is seen under the microscope. The product of the replication of a chromosome is two exact copies called **sister chromatids,** or simply chromatids, that are held together by the replicated but unseparated centromere. More precisely, a **chromatid** is one of the two visibly distinct, longitudinal subunits of all replicated

chromosomes, which become visible between early prophase and metaphase of mitosis (and between diplonema and the second metaphase of meiosis, discussed later). After these stages the chromatids are known as **daughter chromosomes.** The chromosomes are defined by their respective centromeres.

Cell division in eukaryotic cells involves two processes that may or may not occur together: mitosis (division of the nucleus—karyokinesis), which involves the precise replication of chromosomes and the distribution of a complete chromosome set to each progeny cell; and cyto-

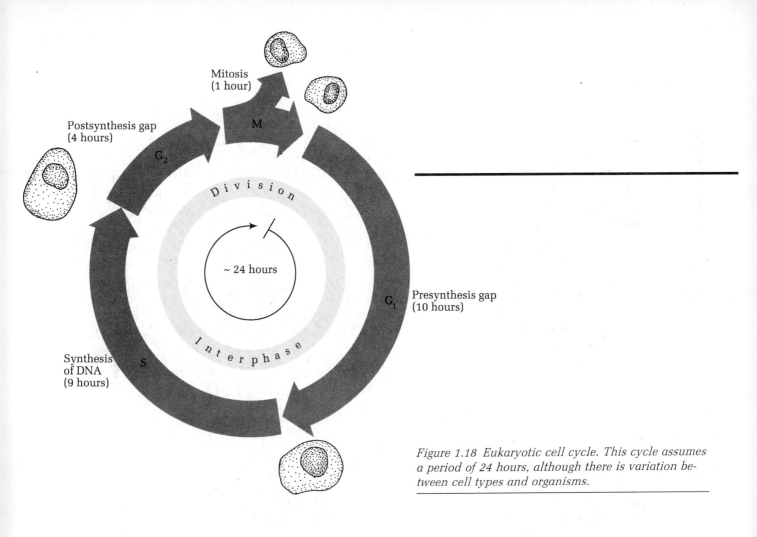

Figure 1.18 *Eukaryotic cell cycle. This cycle assumes a period of 24 hours, although there is variation between cell types and organisms.*

Mitosis
(1 hour)

Postsynthesis gap
(4 hours)

G_2

M

Division

~ 24 hours

Interphase

G_1

Presynthesis gap
(10 hours)

Synthesis
of DNA
(9 hours)

S

kinesis, the division of the cytoplasm.

Mitosis occurs in both the haploid and diploid cells. It is a continuous process, but for purposes of discussion it is usually divided into four cytologically distinguishable stages called prophase, metaphase, anaphase, and telophase. Figure 1.19 presents photographs showing the typical chromosome morphology in interphase and in the four stages of mitosis in plant (onion root tip) and animal (newt lung epithelium) cells. Figure 1.20 shows the four stages in simplified diagrams.

Prophase. At the beginning of **prophase** (Figure 1.20a) the chromatids are very elongated and cannot be seen under the light microscope. In preparation for mitosis they begin to coil tightly so that they appear shorter and fatter under the microscope. By late prophase each chromosome, which was duplicated during the preceding S phase, is seen to consist of two sister chromatids.

During prophase, the mitotic spindle (spindle apparatus) assembles outside the nucleus. Each of the spindle fibers in the bipolar mitotic spin-

dle is approximately 25 nm in diameter and consists of microtubules made of special proteins called *tubulins.* In most animals cells, the foci for spindle assembly are the centrioles (see Figure 1.7). Higher-plant cells usually do not have centrioles, but they do have a mitotic spindle. Prior to the S phase, the cell's pair of centrioles have replicated and each new centriole pair becomes the focus for a radial array of microtubules called the *aster.* Early in prophase the two asters are adjacent to one another close to the nuclear membrane. By late prophase the two asters are far apart along the outside of the nucleus and are spanned by the microtubular spindle fibers.

Near the end of prophase the nuclear membrane breaks down and the nucleolus or nucleoli disappear, allowing the spindle to enter the nuclear area. Specialized structures called *kinetochores* form on either face of the centromeres of each chromosome and become attached to special microtubules called *kinetochore microtubules* (Figure 1.21). These microtubules radiate in opposite directions from each side of each chromosome and interact with the spindle microtubules.

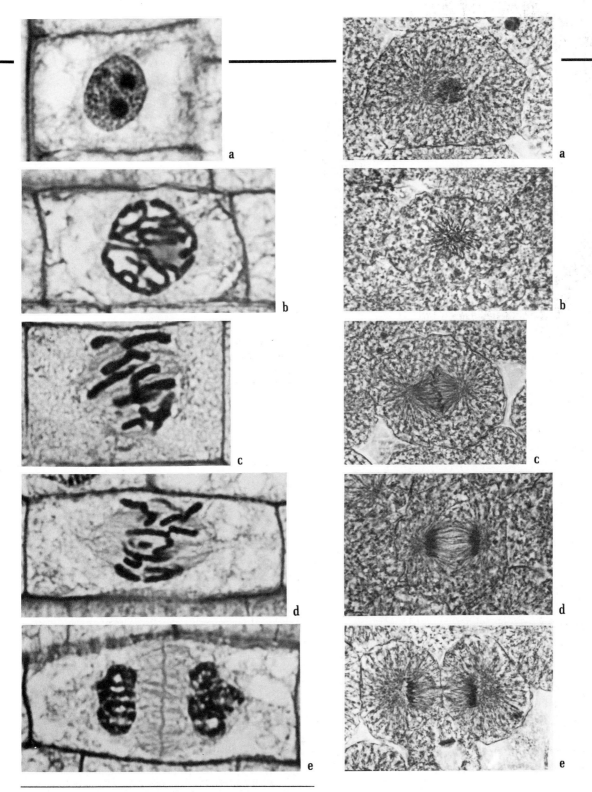

Figure 1.19 Stages of postmeiotic (haploid) mitosis in onion root tip (left) and newt lung epithelium (right): (a) interphase; (b) prophase; (c) metaphase; (d) anaphase; (e) telophase.

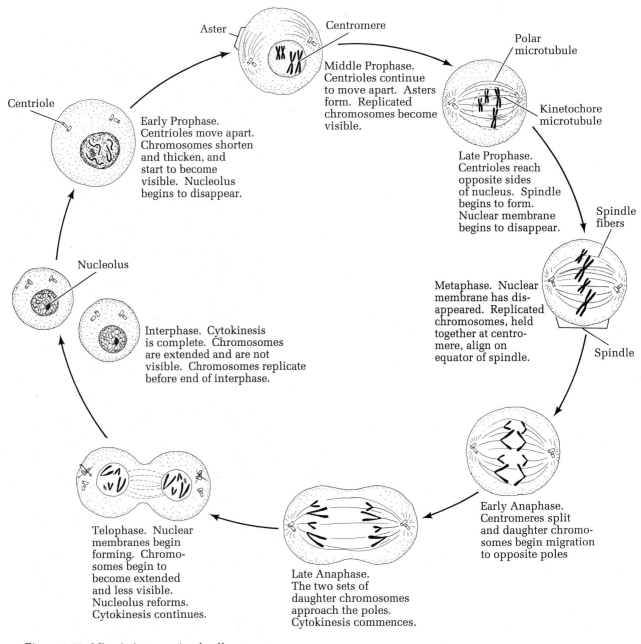

Aster

Centromere

Centriole

Early Prophase. Centrioles move apart. Chromosomes shorten and thicken, and start to become visible. Nucleolus begins to disappear.

Middle Prophase. Centrioles continue to move apart. Asters form. Replicated chromosomes become visible.

Polar microtubule

Kinetochore microtubule

Late Prophase. Centrioles reach opposite sides of nucleus. Spindle begins to form. Nuclear membrane begins to disappear.

Spindle fibers

Nucleolus

Metaphase. Nuclear membrane has disappeared. Replicated chromosomes, held together at centromere, align on equator of spindle.

Spindle

Interphase. Cytokinesis is complete. Chromosomes are extended and are not visible. Chromosomes replicate before end of interphase.

Telophase. Nuclear membranes begin forming. Chromosomes begin to become extended and less visible. Nucleolus reforms. Cytokinesis continues.

Late Anaphase. The two sets of daughter chromosomes approach the poles. Cytokinesis commences.

Early Anaphase. Centromeres split and daughter chromosomes begin migration to opposite poles

Figure 1.20 Mitosis in an animal cell.

Metaphase. Metaphase (Figure 1.20) begins when the nuclear envelope has completely disappeared and the chromosomes and nucleoplasm have mixed with the cytoplasm. During metaphase the chromosomes become arranged so that their centromeres become aligned in one plane halfway between the two spindle poles and with the long axes of the chromosomes at 90 degrees to the spindle axis. The kinetochore microtubules are responsible for this chromosome alignment event. The plane where the chromosomes become aligned is called the **metaphase plate.** Electron micrographs of human chromosomes at this stage of the cell cycle are shown in Figure 1.22. Note the highly condensed state of the sister chromatids. Figure 1.22c shows an electron micrograph (EM) of a human chromosome from which much of the protein has been removed. In the center is a dense framework of protein that retains the form of the chromosome. It is surrounded by a halo of DNA filaments that have uncoiled and spread outward.

Anaphase. Anaphase (Figure 1.20c) is initiated by a splitting of each chromosome caused by the breaking apart of the sister chromatids at their junction point at the centromere. Once the paired kinetochores on each chromosome separate, the sister chromatid pairs undergo **disjunc-**

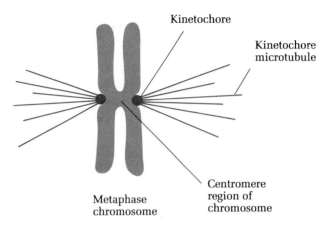

Figure 1.21 Kinetochores and kinetochore microtubules.

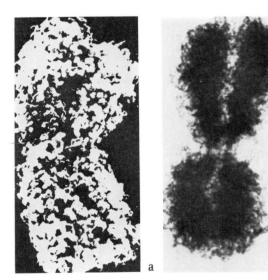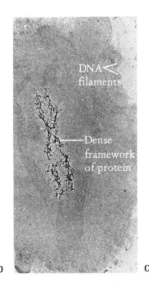

Figure 1.22 Human metaphase chromosome: (a) intact chromosome from scanning electron micrograph; (b) intact chromosome from transmission electron micrograph; (c) transmission electron micrograph of a chromosome from which much of the protein has been removed; the DNA filaments have uncoiled.

Figure 1.23 Mitotic apparatus at anaphase in an animal cell. A centriole pair is located at each pole, with spindle microtubules spanning the cell. The separation of the sister chromatids depends on spindle microtubules attached to the centromere of each chromatid.

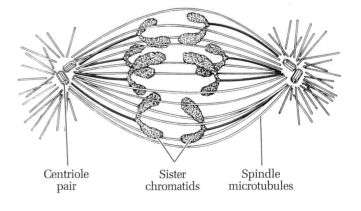

a) Animal cell   b) Plant cell

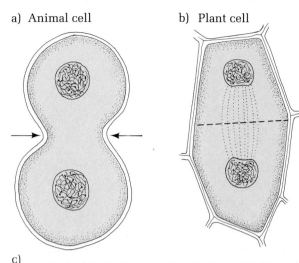

c)

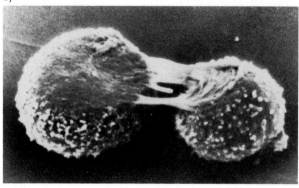

Figure 1.24 Cytokinesis (cell division): (a) diagram of cytokinesis in an animal cell; (b) diagram of cytokinesis in a higher plant cell; (c) scanning electron micrograph of an animal cell in tissue culture shortly after its division.

tion (separation), and the daughter chromosomes (as each sister chromatid is now called) move toward the poles. In anaphase the two centromeres of sister chromatids migrate toward the opposite poles of the cell. Now the chromosomes assume the characteristic shapes related to the location of the centromere along the length of the chromosome. For example, a metacentric chromosome will appear as a V as the two roughly equal-length chromosome arms trail the centromere in its migration toward the pole. Figure 1.23 gives an interpretation of the mitotic apparatus in anaphase in an animal cell.

Telophase. During **telophase** (Figure 1.20d), the migration of the daughter chromosomes to the two poles is completed. The two sets of progeny chromosomes are assembled into two groups at opposite ends of the cell. The chromosomes begin to uncoil and assume the extended state characteristic of interphase. A nuclear membrane forms around each chromosome group, the spindle microtubules disappear, and the nucleolus or nucleoli re-form. The cell has two nuclei at this point.

Cytokinesis. **Cytokinesis** refers to division of the cytoplasm. In most cases the telophase stage of mitosis is accompanied by cytokinesis. Cytokinesis compartmentalizes the two new nuclei into separate daughter cells and completes the mitotic cell division process. In animal cells cytokinesis occurs by the formation of a constriction in the middle of the cell until two daughter cells are produced (Figure 1.24a). However, most plant cells do not divide by the formation of a constriction. Instead, a new cell membrane and cell wall are assembled between the two new nuclei to form a cell plate (Figure 1.24b). Cell wall material coats each side of the cell plate, and the result is two progeny cells. Figure 1.24c is a scanning electron micrograph of an animal cell in tissue culture shortly after its division.

Genetic significance of mitosis. The process of mitosis provides continuity to and maintenance of the genetic content of a cell from generation to generation. Mitosis occurs after DNA and

chromosome duplication has taken place. It is a highly ordered process in which a duplicated chromosome set is partitioned equally into the two daughter cells.

Keynote Mitosis is the process of nuclear division in eukaryotes. It is one part of the cell cycle (i.e., G_1, S, G_2, and M) and results in the production of daughter nuclei that contain identical chromosome numbers and are genetically identical to one another and to the parent nucleus from which they arose. Prior to mitosis, DNA synthesis occurs to double the chromosome number. Mitosis is a genetically controlled process usually followed by cytokinesis. Both haploid and diploid cells can undergo mitosis.

Meiosis

Meiosis is the term applied to the two successive divisions of a diploid nucleus following only one DNA replication cycle. That results in the formation of haploid gametes **(gametogenesis)** or of meiospores (in sporogenesis). During meiosis, homologous chromosomes pair, replicate once, and undergo assortment so that each of the four cells resulting from the two meiotic divisions receives one chromosome of each chromosome set. The two nuclear divisions of a normal meiosis are called **meiosis I** and **meiosis II.** The first meiotic division results in reduction in the number of chromosomes (reductional division), and the second division results in separation of the chromatids (equational division). In most cases the divisions are accompanied by cytokinesis, so the result of the meiosis of a single diploid cell is four haploid cells.

Meiosis I: The first meiotic division. Meiosis I, in which the chromosome number is reduced from diploid to haploid, consists of four cytologically distinguishable stages: *prophase I, metaphase I, anaphase I,* and *telophase I.* Figure 1.25 presents photographs of the stages of meiosis I, and diagrams of the stages.

Prophase I. In brief, in prophase I the chromosomes become shorter and thicker, crossing-over occurs, the spindle apparatus forms, and the nuclear membrane and nucleolus/nucleoli disappear (Figure 1.25). Prophase I is divided into a number of parts. As it begins, the chromosomes have already replicated. In **leptonema** (early prophase) the chromosomes have begun to coil and are now visible. The threadlike chromosomes seen at this time are the sister chromatids held together at the centromere. The number of chromosomes present at this stage is the same as the number in the diploid cell. Once a cell enters leptonema, it's committed to the meiotic process.

In the **zygonema** (early/mid prophase) stage homologous chromosomes begin to pair in a highly specific way and twist around one another. This chromosome pairing is called **synapsis.** Each synapsed set of homologous chromosomes consists of four chromatids and is referred to as a **bivalent** or tetrad.

Following zygonema is **pachynema** (midprophase). The word *pachynema* means "thick strand," and at this stage the chromosomes become much shorter and thicker. They are also more intimately synapsed. During pachynema a most significant event takes place: **crossing-over,** the physical exchange of chromosome regions between homologous chromosomes. A crossover is the place on the chromatids where crossing-over has occurred. If there are genetic differences between the homologs, crossing-over can produce new gene combinations in a chromatid. There is usually no loss or addition of genetic material to either chromosome, since crossing-over involves reciprocal exchanges.

The process of crossing-over is facilitated by the tight synapse (alignment side by side) of the four chromatids and involves the formation of a zipperlike structure along the length of the chromatids called the **synaptinemal complex** (Figure 1.26, see p. 24.) Although we know that the structure consists mainly of protein with some RNA attached, we do not know its role in crossing-over. In some organisms, such as male *Drosophila,* the synaptinemal complex is not formed and no crossing-over occurs in that case.

A chromosome that emerges from meiosis with a combination of genes that differs from the combination with which it started is called a **recombinant chromosome.** Therefore crossing-over is a mechanism that can give rise to **genetic recombination,** a concept we will examine more fully in later chapters.

The next stage in prophase I is **diplonema** (mid/late prophase). Here the chromosomes begin to move apart. The results of crossing-over become visible during diplonema as a cross-shaped structure called a **chiasma** (plural, chiasmata). Since all four chromatids may be involved in crossing-over events along the length of the homologs, the chiasma pattern at this stage may be very complex.

Diplonema is followed rapidly in most organisms by the remaining stages of meiosis. However, in many animals the oocytes can remain in diplonema for very long periods. For example, in the human female, oocytes in the ovary go through meiosis up to diplonema by the seventh month of fetal development and then remain arrested in this stage for many years. At the onset of puberty and until menopause, one oocyte per menstrual cycle is matured to a haploid ovum (egg) and is released into the fallopian tubes.

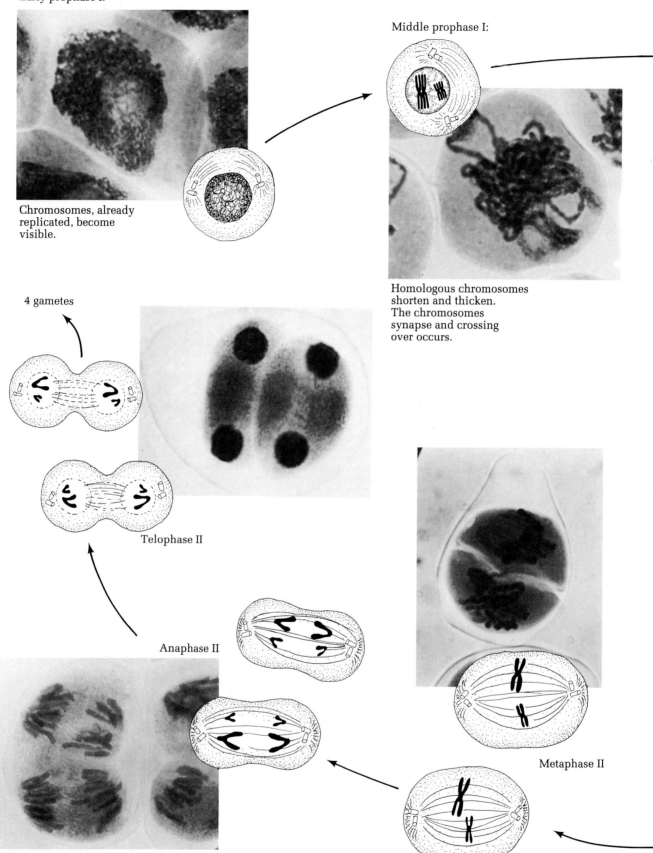

Early prophase I:

Chromosomes, already replicated, become visible.

Middle prophase I:

Homologous chromosomes shorten and thicken. The chromosomes synapse and crossing over occurs.

4 gametes

Telophase II

Anaphase II

Metaphase II

Figure 1.25 Meiosis. The diagrams show the stages of meiosis in an animal cell, and the photographs show the same stages of meiosis in the plant, chives.

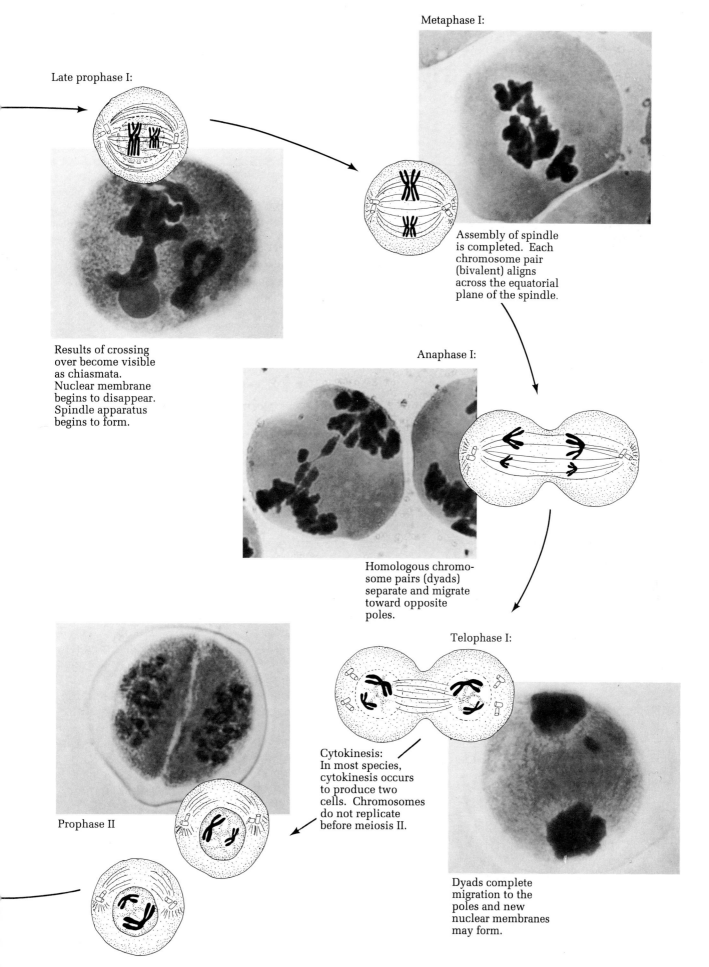

Late prophase I:

Results of crossing over become visible as chiasmata. Nuclear membrane begins to disappear. Spindle apparatus begins to form.

Metaphase I:

Assembly of spindle is completed. Each chromosome pair (bivalent) aligns across the equatorial plane of the spindle.

Anaphase I:

Homologous chromosome pairs (dyads) separate and migrate toward opposite poles.

Telophase I:

Cytokinesis: In most species, cytokinesis occurs to produce two cells. Chromosomes do not replicate before meiosis II.

Dyads complete migration to the poles and new nuclear membranes may form.

Prophase II

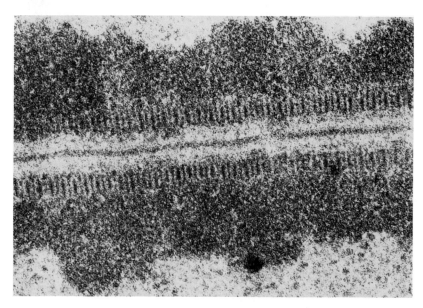

Figure 1.26 Electron micrograph of the synaptinemal complex separating two homologous chromatid pairs during the zygonema stage of prophase I.

Diakinesis (late prophase) follows diplonema. During this stage, the four chromatids of each tetrad are even more condensed, and the chiasmata often *terminalize;* that is, they move down the chromatids to the ends. As a result, the chromatids at this stage appear to be attached together near the tips. New crossing-over events have not occurred, however; rather, the chromosomes are sliding past each other so that the chiasmata, in a sense, delay the separation of the chromatids. In addition, the nucleoli disappear, and the nuclear membrane begins to break down. The chromosomes can be most easily counted at this stage of meiosis.

Metaphase I. By the beginning of metaphase I (Figure 1.25), the nuclear membrane has completely broken down, and the tetrads become aligned across the equatorial plane of the cell. The spindle apparatus is completely formed now, and the microtubules are attached to the centromere regions of the homologs. Note particularly that it is the synapsed pairs of homologs (the tetrads) that are found at the metaphase plate. In contrast, in mitosis replicated homolo-

gous chromosomes (sister chromatid pairs) align independently at the metaphase plate.

Anaphase I. In anaphase I (Figure 1.25) the chromosomes in each tetrad separate, so homologous pairs (the bivalents) disjoin and migrate toward the opposite poles. In this way the maternally derived and paternally derived homologs are segregated (except for the parts of chromosomes exchanged during the crossing-over process). The sister chromatid pairs (dyads) that have segregated remain attached at their respective centromere regions.

Telophase I. Telophase I (Figure 1.25) varies considerably in length among species. In telophase I the dyads complete their migration to the opposite poles of the cell, and new nuclear membranes may form. In most species, cytokinesis follows to produce two cells, each with one nucleus. In each nucleus the number of chromosomes is reduced to the haploid level.

Meiosis II: The second meiotic division. The second meiotic division is very similar to a mitotic division. Figure 1.25 presents photographs and diagrammatic representations of prophase II, metaphase II, anaphase II, and telophase II of meiosis II.

Prophase II. (Figure 1.25) is a stage of chromosome contraction that does not occur in all organisms.

In **metaphase II** (Figure 1.25) each of the two cells of a dyad organizes a spindle apparatus that attaches to the now-divided centromeres. The centromeres line up on the equator of the second-division spindles.

During **anaphase II** (Figure 1.25) the centromeres (and therefore the chromatids) are pulled to the opposite poles of the spindle: One sister chromatid of each pair goes to one pole, while the other goes to the opposite pole.

In the last stage, **telophase II** (Figure 1.25), a nuclear membrane forms around each set of chromosomes and cytokinesis takes place. After telophase II, the chromosomes become more extended and again are invisible under the light microscope.

Since there is *no chromosome replication between meiosis I and meiosis II,* the end products of the two meiotic divisions are four haploid cells from one original diploid cell. Each of the four progeny cells has one chromosome from each homologous pair of chromosomes. Remember, however, that these chromosomes are not exact copies of the original chromosomes because of the crossing-over that occurs between chromosomes during pachynema of meiosis I.

Figure 1.27 presents a diagrammatic comparison of mitosis and meiosis.

Genetic Significance of Meiosis

Meiosis has three significant results:

1. Meiosis generates cells with half the number of chromosomes found in the diploid cell that entered the process because two division cycles follow only one cycle of replication (S period). Fusion of the haploid nuclei (called fertilization or syngamy) restores the diploid number. Therefore through a cycle of meiosis and fertilization, the chromosome number is maintained in sexually reproducing organisms.

2. In metaphase I of meiosis, each maternally derived and paternally derived chromosome has an equal chance of aligning on one or the other side of the equatorial metaphase plate. As a result, each nucleus generated by meiosis will have a combination of maternal and paternal chromosomes.

   The number of possible chromosome combinations is large, especially when the number of chromosomes in an organism is large. Consider a hypothetical organism with two pairs of chromosomes in a diploid cell entering meiosis. Figure 1.28 shows all the possible combinations (four) of maternal and paternal chromosomes that can occur at the metaphase plate.

   The general formula is that the number of possible chromosome arrangements is 2^n, where n is the number of chromosome pairs. In *Drosophila,* which has four pairs of chromosomes, the number of possible arrangements is 2^4, or 16; in humans, which have 23 chromosome pairs, over 8 million metaphase arrangements are possible. Therefore since there are many gene differences between the maternally derived and paternally derived chromosomes, the nuclei produced by meiosis will be genetically quite different from the parental cell and from each other.

3. The crossing-over that occurs between maternal and paternal chromatid pairs during meiosis I results in still more variation in the final combinations. Crossing-over occurs during every meiosis, and because the sites of crossing-over vary from one meiosis to another, the number of different kinds of progeny nuclei produced by meiosis is extremely large.

Keynote Meiosis occurs in all sexually reproducing eukaryotes. It is a process in which a diploid (2N) cell or cell nucleus is transformed into four haploid (N) cells or nuclei through one round of chromosome replication and two rounds of nuclear division, the first of

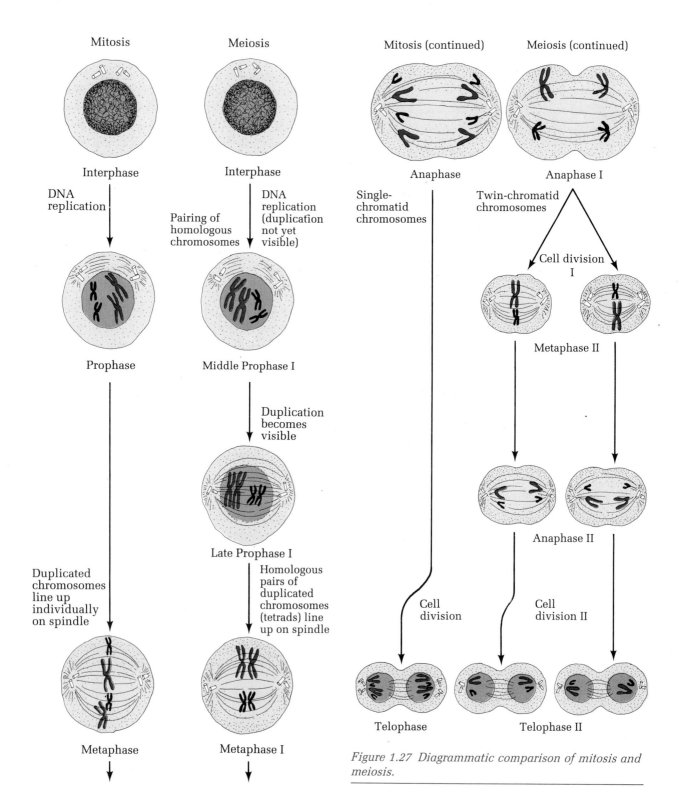

Figure 1.27 Diagrammatic comparison of mitosis and meiosis.

which involves the pairing (synapsis) of homologous chromosomes. The meiotic process results in the conservation of the number of chromosomes from generation to generation. It also generates genetic variability through the various ways in which maternal and paternal chromosomes are combined in the progeny nuclei and by crossing-over (the physical exchange of genes between maternally derived and paternally derived homologs).

Locations of Meiosis in the Life Cycle

Meiosis in animals. Most multicellular animals are diploid through most of their life cycle. Meiosis in such animals produces haploid gametes, which produce a diploid zygote when their nuclei fuse in the fertilization process. The zygote then divides mitotically to produce the new diploid organism. Thus the gametes are the only haploid stages of the life cycle. The gametes are only formed in specialized cells. In the male the gamete is the sperm, produced through a process called **spermatogenesis.** The female gamete is the egg, produced by **oogenesis.** Spermatogenesis and oogenesis are illustrated in Figure 1.29.

In male animals the **sperm cells** (also called **spermatozoa**) are produced by the testes. The testes contain the primordial germ cells *(primary spermatogonia),* which, through mitotic division, produce *secondary spermatogonia.* Spermatogonia transform into *primary spermatocytes (meiocytes),* each of which undergoes meiosis I and gives rise to two *secondary spermatocytes,* and each spermatocyte undergoes meiosis II. As a result of these two divisions, four haploid *spermatids* arise, which eventually differentiate into the mature male gametes, the spermatozoa.

In female animals the ovary contains the primordial germ cells *(primary oogonia),* which, by mitosis, give rise to *secondary oogonia.* These cells transform into *primary oocytes,* which grow until the end of oogenesis. The diploid, primary oocyte goes through meiosis I and unequal cytokinesis to give two cells: a large one called the **secondary oocyte** and a very small

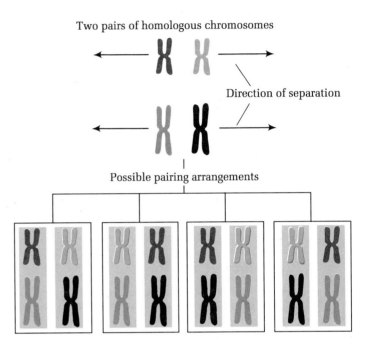

Two pairs of homologous chromosomes

Direction of separation

Possible pairing arrangements

Figure 1.28 Four possible arrangements of two pairs of homologous chromosomes on the metaphase plate of the first meiotic division. Paternal chromosomes are shown in black and grey and maternal chromosomes in color.

one called the *first polar body.* In the second meiotic division the secondary oocyte produces two haploid cells, one a very small one called a *second polar body* and the other a large cell that rapidly matures into the mature egg cell, or **ovum.** The first polar body may or may not divide. Thus in the female animal only one mature gamete (the ovum) is produced by meiosis of a diploid cell. As we have intimated, the egg has a very large supply of cytoplasm necessary for the early stages of embryo growth.

Meiosis in plants. The life cycle of sexually reproducing plants typically has two phases, the **gametophyte** or haploid stage in which gametes are produced, and the **sporophyte** or diploid stage in which haploid spores are produced by meiosis.

In angiosperms, the flowering plants, the flower is the structure in which sexual reproduction occurs. Figure 1.30 shows a generalized

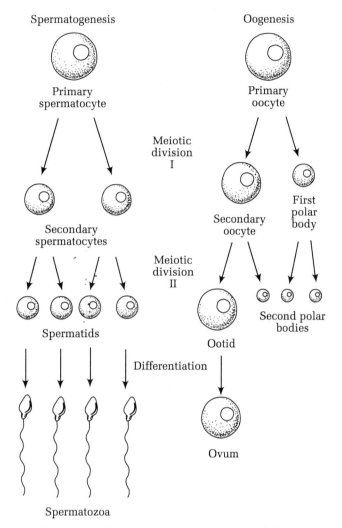

Figure 1.29 *Spermatogenesis and oogenesis in an animal.*

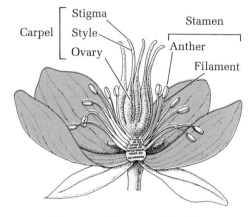

Figure 1.30 *Structure of a generalized flower.*

flower that contains both male and female reproductive organs, the **stamens** and **carpels,** respectively. Each stamen consists of a single stalk, the filament, on the top of which is an anther. From the anther are released the pollen grains, which are immature male gametophytes. Each carpel contains the female gametophytes and typically consists of the stigma, a sticky surface specialized to receive the pollen; the style, a thin stalk down which the pollen tube grows; and at the base of the structure, the ovary, within which are the ovules. Each ovule encloses a female ga-

metophyte with a single egg cell. When the egg cell is fertilized, the ovule develops into a seed.

Meiosis occurs in specific ways in the female and male parts of the flower. Meiosis in the female part of the flower is shown in Figure 1.31. Within the ovary of the flower, each ovule is a single large 2N cell called the megaspore mother cell. This cell undergoes meiosis, producing four haploid products, only one of which remains viable—the megaspore, or haploid female gametophyte. The production of the megaspore is called **megasporogenesis.**

Meiosis in the megaspore mother cell involves two divisions as usual, resulting in four haploid cells. Three of the cells disintegrate leaving a single large cell with one haploid nucleus. By three successive nuclear mitotic divisions, a megagametophyte cell with eight identical haploid nuclei is produced. The eight nuclei migrate in two sets of four to the two ends of the cell, and then two of the nuclei—one from each end—migrate back to the cell center. Cell walls are produced, resulting in unequal division of the cytoplasm. The larger cell forms around the two central nuclei: this binucleate cell, when it is fertilized, gives rise to the **endosperm** of the seed; hence it is called the endosperm mother cell. Each of the other six haploid nuclei is now found in its own cell. The entire seven-celled megagametophyte structure is called the *embryo sac.* One of the six haploid cells is the egg cell which is the single female gamete of the ovule.

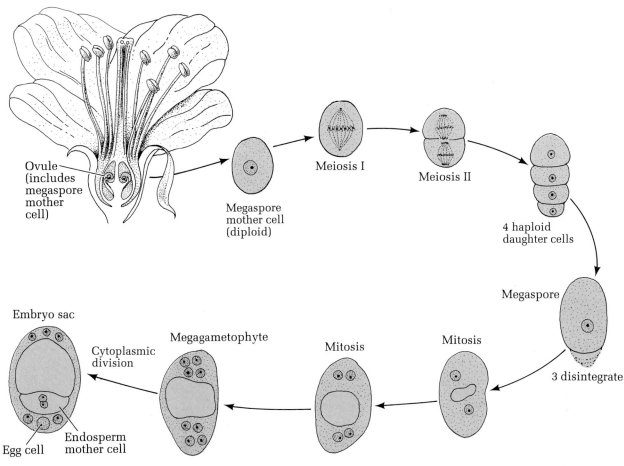

Ovule
(includes
megaspore
mother
cell)

Megaspore
mother cell
(diploid)

Meiosis I

Meiosis II

4 haploid
daughter cells

Megaspore

3 disintegrate

Mitosis

Mitosis

Embryo sac

Cytoplasmic
division

Megagametophyte

Egg cell

Endosperm
mother cell

*Figure 1.31 Megasporogenesis: Meiosis in the female
part of the flower and the production of the embryo
sac.*

Meiosis in the anthers occurs by similar, although less complex, events to produce the pollen grains on male gametophytes (Figure 1.32). This series of events is called **microsporogenesis.** Anthers contain four pollen sacs within which are many diploid microspore mother cells that undergo meiosis to produce four haploid microspores each. The haploid nucleus of each microspore divides once mitotically, producing a haploid generative nucleus and a haploid tube nucleus. Each microspore then generates a tough coat and is then a pollen grain.

During pollination, the pollen grains are deposited on the stigma and each "germinates,"

producing a pollen tube (Figure 1.33). The pollen grows down through the stigma and into the style, producing enzymes to digest the tissues ahead of its growth. As the pollen tube grows down the style, the nuclei stay close to the growing tip: the haploid generative nucleus produces two sperm nuclei (the gametes) by mitosis, while the tube nucleus does not divide at all. The pollen tube eventually penetrates the tissues of the ovule at a very small pore called the micropyle, allowing the two sperm nuclei to enter the embryo sac. One sperm nucleus fertilizes the haploid egg cell, producing the diploid zygote. The other sperm nucleus fuses with the

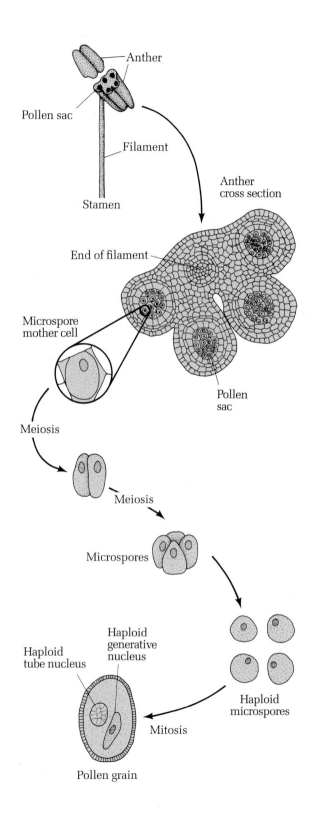

Anther

Pollen sac

Filament

Stamen

Anther
cross section

End of filament

Microspore
mother cell

Pollen
sac

Meiosis

Meiosis

Microspores

Haploid
generative
nucleus

Haploid
tube nucleus

Haploid
microspores

Mitosis

Pollen grain

two haploid nuclei in the endosperm mother cell, producing a triploid nucleus, thereby completing fertilization. This **double fertilization** is found only in the life cycle of flowering plants.

After fertilization, the diploid zygote, the triploid endosperm cell, and the other tissues begin the cell divisions necessary to produce the plant embryo within a seed. In this, the embryos derive from the zygote, and the stored food within the seed derives from the triploid endosperm cell.

Among living organisms only plants produce gametes from special bodies called gametophytes. Thus plants have two distinct reproductive phases, called the **alternation of generations** (Figure 1.34). Meiosis and fertilization are at the interfaces of these stages. The haploid gametophyte generation begins with spores that are produced by meiosis. Meiosis in flowering plants occurs in anthers and ovules and produces, respectively, pollen and a tiny female gametophyte located deep within the ovule tissue. These two structures produce sperm and egg, respectively, and fertilization initiates the diploid sporophyte generation. The diploid sporophyte generation produces specialized cells called spores, which, in turn, produce gametophytes. Thus the sporophyte is the second alternate generation.

Figure 1.34 also indicates the relationship of each phase of the alternation of generations to meiosis. The products of meiosis in plants are the spores. The spores of ferns and mosses, the cells in gymnosperms and angiosperms that produce male gametophytes, and the cells in the ovules of flowers that produce the female gametophyte are all produced by meiosis. Thus in all green plants the alternation of generations involves an alternation between stages of haploid cells and diploid cells: The gametophyte cells are haploid, and the sporophyte cells are diploid.

Figure 1.32 Microsporogenesis: Meiosis in the male part of the flower and the production of the pollen grain.

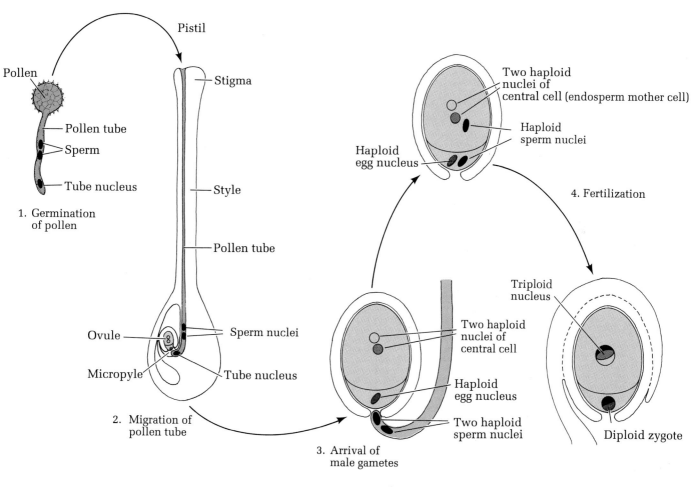

Figure 1.33 Pollination and fertilization in a flowering plant.

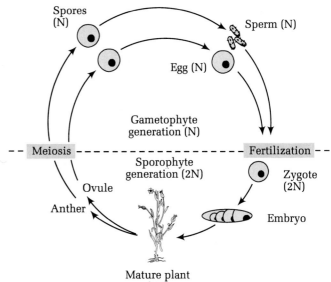

Figure 1.34 Alternation of gametophyte and sporophyte generations in green plants.

Analytical
Approaches
for Solving
Genetics
Problems

Because this chapter is primarily descriptive, there are no quantitative problems for this material.

Questions
and
Problems

In 1.1 through 1.3, select the correct answer.

*1.1 Interphase is a period corresponding to the cell cycle phases of

a. mitosis.
b. S.
c. $G_1 + S + G_2$.
d. $G_1 + S + G_2 + M$.

1.2 Chromatids joined together by a replicated but unseparated centromere are called

a. sister chromatids.
b. homologs.
c. alleles.
d. tetrads.

*1.3 Mitosis and meiosis always differ in regard to the presence of

a. chromatids.
b. homologs.
c. bivalents.
d. centromeres.
e. spindles.

1.4 State whether each of the following statements is true or false. Explain your choice.

a. The chromosomes in a somatic cell of any organism are all morphologically alike.
b. During mitosis the chromosomes divide and the resulting sister chromatids separate at anaphase, ending up in two nuclei, each of which has the same number of chromosomes as the parental cell.
c. Any chromosome may synapse with any other chromosome in the same cell at zygonema.

*1.5 a. Can meiosis occur in haploid species?
b. Can meiosis occur in a haploid individual?

In 1.6 through 1.8, select the correct answer.

1.6 The general life cycle of a eukaryotic organism has the sequence

   a. 1N → meiosis → 2N → fertilization → 1N.
   b. 2N → meiosis → 1N → fertilization → 2N.
   c. 1N → mitosis → 2N → fertilization → 1N.
   d. 2N → mitosis → 1N → fertilization → 2N.

*1.7 Which statement is true?

   a. Gametes are 2N; zygotes are 1N.
   b. Gametes and zygotes are 2N.
   c. The number of chromosomes can be the same in gamete cells and in somatic cells.
   d. The zygotic and the somatic chromosome numbers cannot be the same.
   e. Haploid organisms have haploid zygotes.

1.8 All of the following happen in prophase I of meiosis *except*

   a. chromosome condensation.
   b. pairing of homologs.
   c. chiasma formation.
   d. terminalization.
   e. segregation.

*1.9 Give the name of the stages of mitosis or meiosis at which the following events occur:

   a. Chromosomes are located in a plane at the center of the spindle.
   b. The chromosomes move away from the spindle equator to the poles.

1.10 Given the diploid, meiotic mother cell in the figure, diagram the chromosomes as they would appear (a) in late pachynema; (b) in a nucleus at prophase of the second meiotic division; (c) in one of the three polar bodies resulting from oogenesis in an animal.

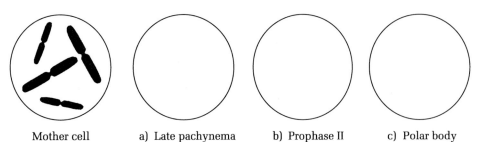

Mother cell   a) Late pachynema   b) Prophase II   c) Polar body

1.11 Does mitosis or meiosis have greater significance in genetics (i.e., the study of heredity)? Explain your choice.

*1.12 Consider a diploid organism that has three pairs of chromosomes. Assume that the organism receives chromosomes A, B, and C from the female parent and A′, B′, and C′ from the male parent. Assume in the following questions that no crossing-over occurs.

a. What proportion of the gametes of this organism would be expected to contain all the chromosomes of maternal origin?
b. What proportion of the gametes would be expected to contain some chromosomes from both maternal and paternal origin?

* Solutions to Questions and Problems marked with an asterisk are found on pages 877–892.

Mendelian Genetics

2

By simple observation it is evident that a lot of variation occurs among individuals of a given species. Among dogs, for example, are many breeds, including dachshunds, Chihuahuas, bassett hounds, Dalmatians, spaniels, and so on (Figure 2.1). These breeds are clearly distinguishable by their size, shape, and color, yet they are all representatives of the same species: *Canis.* Similarly, differences among individual humans include eye color, height, and hair color even though they all belong to the species *Homo sapiens.* Within a species each generation perpetuates individual differences, yet the species remains clearly identifiable. The differences among individuals within and between species are the result of differences in their genes (DNA segments), which determine the structure, function, and development of the cell or organism. Thus the genetic information coded in DNA is responsible for species and individual variation.

Breeding experiments on plants and animals have been carried out for many centuries in an attempt to understand how characteristics are transmitted from one generation to the next. The understanding of how genes are transmitted from parent to offspring began with the work of Gregor Johann Mendel (1822–1884), an Austrian monk. In this chapter we will learn the basic principles of transmission genetics by examining Mendel's experiments and their results. Throughout this chapter we must remember that the segregation of genes is directly related to the behavior of chromosomes. However, while Mendel analyzed the patterns of segregation of hereditary traits that he called characters, he did not know that genes (i.e., DNA segments) controlled the characters, or that genes are located in chromosomes. Indeed, he did not know of the existence of chromosomes.

Genotype and Phenotype

Before we begin our study of Mendel's work, we must distinguish between the nature of the genetic material each individual organism possesses and the physical characteristics this genetic material gives rise to.

a

b

c

The characteristics of an individual that are transmitted from one generation to another are sometimes called **hereditary traits** (Mendel called them **characters**). These traits are under the control of DNA segments called **genes.** The genetic constitution of an organism is called its **genotype,** while the physical manifestation of a genetic trait is called a **phenotype.** Different forms of the gene that controls eye color in humans, for example, produce a number of phenotypes: brown eyes, blue eyes, green eyes, and so on.

The genes we carry are essentially like a blueprint that can be interpreted in several ways. Genes give only the potential for the development of a particular phenotypic characteristic; the extent to which that potential is realized depends not only on interactions with other genes and their products but also on environmental influences (Figure 2.2). A person's height, for example, is controlled by many genes, which, in turn, are significantly affected by internal and external environmental influences. Notable among these influences are nutrition and hormonal effects during puberty. While we can say that a child is "tall like her father," there is no simple genetic explanation for that statement. She may be tall because of excellent nutritional habits interacting with the genetic potential for height.

Figure 2.1 Variation among dogs: (a) dachshund; (b) Chihuahua; (c) bassett-hound; (d) dalmation; (e) spaniel.

d

e

Because of the effects of environment, then, individuals may have identical genotypes (as in identical twins) but very different phenotypes. Studies have shown, for instance, that identical twins raised apart and exposed to different environmental effects tend to vary more in their appearance than identical twins raised together in the same environment (Figure 2.3). Similarly, individuals may have virtually identical phenotypes but very different genotypes. These situations reinforce the points that genes are merely a starting position for determining the structure and function of an organism and that the road to the mature phenotypic state is highly complex and involves a myriad of interacting biochemical pathways. We will develop the relationship between genotype and phenotype in more detail as the text proceeds, but for our current task of examining Mendel's experiments, the important aspects mentioned above will suffice.

Keynote *The genotype is the genetic constitution of an organism. The phenotype is the physical manifestation of the genetic traits. The genes give the potential for the development of characteristics; this potential is affected by interactions with other genes and by environment. Thus individuals with the same genotype can have different phenotypes, and individuals with the same phenotypes may have different genotypes.*

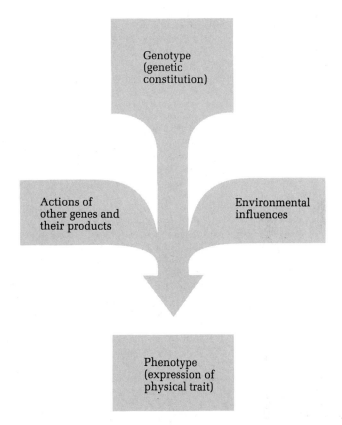

Figure 2.2 *Influences on the physical manifestation (phenotype) of the genetic blueprint (genotype): interactions with other genes and their products (such as hormones) and environment (such as nutrition).*

Mendel's Experiments

Mendel's Experimental Design

The first person to obtain some understanding of the principles of heredity was Gregor Johann Mendel (Figure 2.4), whose work is considered to be the foundation of modern genetics. Mendel left his monastery at Brunn, Austria, for a short time to study a variety of sciences at the University of Vienna. This education evidently gave him an excellent background for designing experiments and analyzing experimental data.

Mendel returned to the monastery in 1853, and in 1854 he obtained and grew 34 strains of peas and tested them for constancy of phenotype in the following year. In 1856 he began a series of experiments with the garden pea, *Pisum sativum L,* in an attempt to learn something about the mechanisms of heredity. It is not certain why Mendel was curious about heredity. He was interested in a number of scientific areas at the time, particularly in honeybees (he was an active member of the local beekeepers' society). It is possible that a paper published about this time on segregation of traits in bees may have stimulated Mendel to begin his genetic experiments.

From the results of crossbreeding pea plants that exhibited differences in characteristics such as height, flower color, and seed shape, Mendel

Figure 2.3 Identical twins reared apart (a) and reared together (b).

Figure 2.4 Gregor Johann Mendel, the father of the science of genetics.

developed a simple theory to explain the transmission of hereditary characteristics or traits from generation to generation. (Note that Mendel had no knowledge of mitosis and meiosis. Now, of course, we know that the segregation of genes conforms to the behavior of chromosomes.) Figure 2.5 outlines the general procedure of a genetic cross. Although his conclusions were reported in 1865, it wasn't until the late 1800s and early 1900s that their significance was realized. Today Mendel's methods of analysis continue to be used in genetic studies. In view of his significant contributions, Mendel is regarded as the father of genetics.

Compared with earlier researchers, Mendel used several new procedures in his experiments. In the paper he wrote on his pea-breeding results, Mendel said, "Among all the numerous experiments made [by my predecessors], not one has been carried out to an extent and in such a way as to make it possible to determine the number of different forms under which the offspring of hybrids appear, or to arrange these forms with certainty according to their separate generations, or definitely to ascertain their statistical relations." It was because of his clear conception of these three primary necessities that Mendel's work was so successful; this quantitative approach was new in Mendel's time. Contributing to his success was the fact that the characters he chose to study were stable.

Mendel's new experimental approach was effective because he also developed a simple interpretation of the ratios he obtained and then carried out direct and convincing experiments to test this hypothesis. In his initial breeding experiments he reasoned that he should study one characteristic at a time to see how it was inherited. He made carefully controlled matings (crosses) between strains of peas that exhibited differences in heritable characteristics, and most importantly, he kept very careful records of the outcome of the crosses and the number of each type of pea produced. The numerical data he obtained enabled him to do a rigorous analysis of the hereditary transmission of characteristics.

Mendel performed all his significant genetic experiments with the garden pea, a good choice

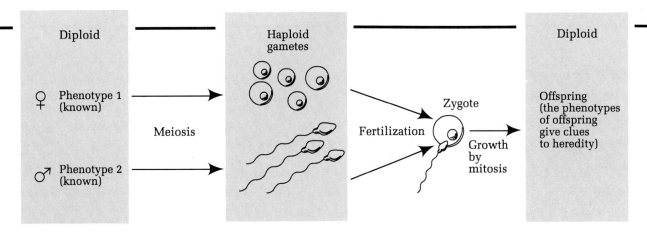

Figure 2.5 General procedure of a genetic cross.

since the pea fits many of the criteria that make organisms suitable for use in genetic experiments (Box 2.1). Furthermore, the pea plant is easy to grow, and it produces a large number of seeds, the progeny produced by sexually reproducing plants. When the seeds are planted, they germinate to produce new pea plants.

Figure 2.6 presents a cross section of a flower of the garden pea, showing the stamens and the pistils (the male and the female reproductive organs, respectively). The pea normally reproduces by **self-fertilization;** that is, pollen produced from the stamens lands on the pistil within the same flower and fertilizes the plant. This process is also called **selfing.** Fortunately, it is a relatively simple procedure to prevent self-fertilization of the pea: remove the stamens from a developing flower bud before they produce any mature pollen. The pistils of that flower can then be pollinated with pollen taken from the stamens of another flower and dusted onto the stigma of the pistil of the emasculated one.

Box 2.1 What Organisms Are Suitable for Use in Genetic Experiments?

There are several qualities that make an organism well suited for genetic experimentation. (1) If an experimenter wants to test a genetic hypothesis, the genetic history of the organism involved must be known well. Therefore the genetic background of the parents used in the experimental crosses must be known. (2) The organism must have a relatively short life cycle so that a large number of generations occur within a relatively short time. In this way data over many generations can be rapidly obtained. (3) A consideration is the number of offspring produced from a mating, since much genetic information can be obtained if there are numerous progeny to study. (4) The organism should be easy to handle. Hundreds of fruit flies can easily be kept in half-pint milk bottles for experimental purposes, but hundreds of elephants would certainly be more difficult and much more expensive to maintain. (5) Most importantly, individuals in the population must differ in a number of ways. If there are no discernible differences among the individual organisms under study, it is impossible to study the inheritance of traits. The more marked the differences, the easier the genetic analysis.

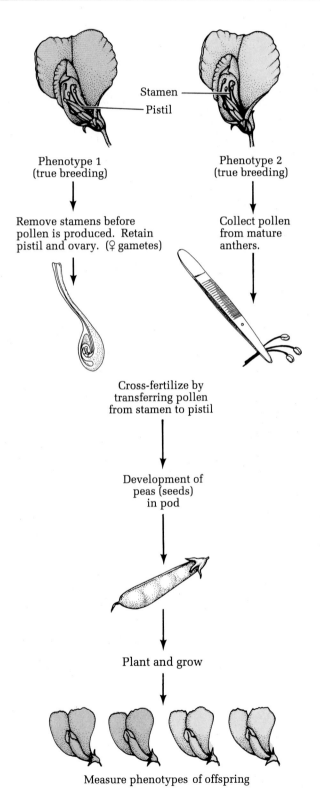

Stamen

Pistil

Phenotype 1
(true breeding)

Phenotype 2
(true breeding)

Remove stamens before
pollen is produced. Retain
pistil and ovary. (♀ gametes)

Collect pollen
from mature
anthers.

Cross-fertilize by
transferring pollen
from stamen to pistil

Development of
peas (seeds)
in pod

Plant and grow

Measure phenotypes of offspring

Cross-fertilization, or more simply a **cross,** is the term used for the fusion of male and female gametes (pollen and egg in this case) from different individuals (Figure 2.6). Once cross-fertilization has occurred, the zygote develops into seeds (peas), which are then planted. The phenotypes of the plants that grow from the seeds are then analyzed.

Mendel knew enough about the reproductive cycle of the garden pea to plan the crosses carefully. For his experiments he obtained 34 strains of pea plants that differed in a number of traits. He allowed each strain to self-fertilize for many generations to ensure that the traits he wanted to study were inherited and to remove from consideration those strains that produced progeny with traits different from the parental type. Thus he only worked with pea strains in which the trait under investigation remained unchanged from parent to offspring for many generations. Such strains are called **true-breeding** or **pure-breeding strains.**

Next, Mendel selected seven traits to study in breeding experiments. Each trait had two easily distinguishable, alternative appearances (phenotypes), as shown in Figure 2.7. These traits affect the appearance of most parts of the pea plant, including: (1) flower and seed coat color (grey versus white seed coats, and purple versus white flowers [Note: A single gene controls these particular color properties of seed coats and flowers.]); (2) seed color (yellow versus green); (3) seed shape (smooth versus wrinkled); (4) pod color (green versus yellow); (5) pod shape (inflated versus pinched); (6) stem height (tall versus short); and (7) flower position (axial versus terminal).

Monohybrid Crosses and Mendel's Principle of Segregation

Before we discuss Mendel's experiment, we will clarify the terminology encountered in breeding experiments. The parental generation is called

Figure 2.6 Procedure for crossing pea plants.

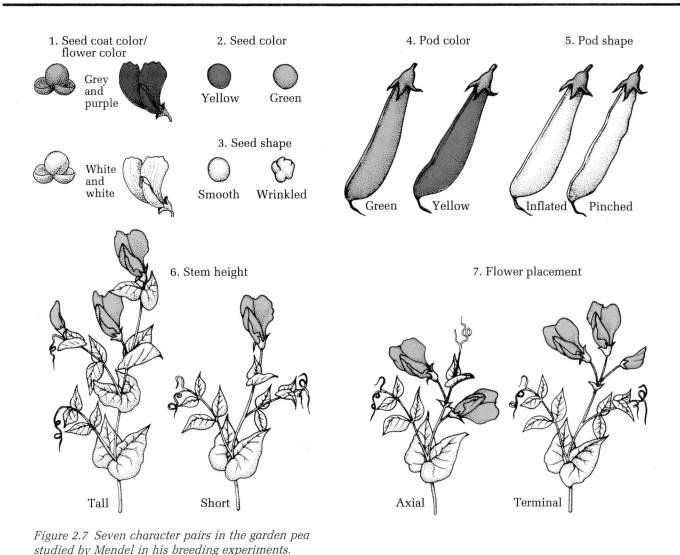

1. Seed coat color/ flower color

Grey and purple

White and white

2. Seed color

Yellow   Green

3. Seed shape

Smooth   Wrinkled

4. Pod color

Green   Yellow

5. Pod shape

Inflated   Pinched

6. Stem height

Tall   Short

7. Flower placement

Axial   Terminal

Figure 2.7 Seven character pairs in the garden pea studied by Mendel in his breeding experiments.

the **P generation.** The progeny produced from the P mating is called the **first filial generation, or F$_1$.** The subsequent generation produced by breeding together the F$_1$ offspring is termed the **F$_2$ generation.** Interbreeding the offspring of each generation results in F$_3$, F$_4$, F$_5$, generations, and so on.

To begin testing his hypothesis about the nature of segregation of traits, Mendel first performed crosses between true-breeding strains of peas that differed in a single trait. Such crosses are called **monohybrid crosses.** For example, he pollinated pea plants that gave rise only to

smooth seeds with pollen from a true-breeding variety that produced wrinkled seeds.* As Figure 2.8 shows, the outcome of this monohybrid cross was all smooth seeds. The same result was obtained when the parental types were reversed; that is, when the pollen from a smooth-seeded plant was used to pollinate a pea plant that gave wrinkled seeds. [Matings that are done both

* Seeds are the diploid progeny of sexual reproduction. If a phenotype concerns the seed itself, it can be directly seen by inspection of the seeds. If a phenotype concerns a part of the mature plant, such as stem size or flower color, the seeds must be germinated before that phenotype can be observed.

P generation

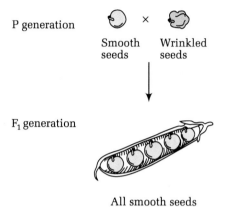

Smooth   Wrinkled
seeds   seeds

F₁ generation

All smooth seeds

Figure 2.8 Results of one of Mendel's breeding crosses. In the parental generation he crossed a true-breeding strain of peas that produced smooth seeds with a true-breeding strain that produced wrinkled peas. All the F₁ progeny seeds were smooth.

ways, smooth female (♀) × wrinkled male (♂) and wrinkled female × smooth male, are called **reciprocal crosses.** Conventionally, the female is given first in crosses.]

The significant point was that all the F₁ progeny seeds of the smooth × wrinkled reciprocal

crosses were smooth; that is, they exactly resembled only one of the parents in this character.

In the next stage of the experiment Mendel planted the seeds and allowed the F₁ plants to produce the F₂ seed by self-fertilization. Both smooth and wrinkled seeds appeared in the F₂ generation, and both types could be found within the same pod. Then, typical of his analytical approach to the experiments, he counted the number of each type. He found that 5474 were smooth and 1850 were wrinkled (Figure 2.9). The calculated ratio of smooth to wrinkled seeds was 2.96:1, which is a statistical deviation from a 3:1 ratio.

Using this same quantitative approach, Mendel analyzed the behavior of the six other pairs of traits. Qualitatively and quantitatively, the same results were obtained (Table 2.1). From the seven sets of crosses he made the following generalizations about his data:

1. The results of reciprocal crosses were always the same.
2. All F₁ progeny resembled one of the parental strains.
3. In the F₂ generation the parental trait that had disappeared in the F₁ generation reap-

Table 2.1 Mendel's Results in Crosses Between Plants Differing in One of Seven Characters

Character[a]	*F₁*	*F₂ (Number)*			*F₂ (Percent)*	
		Dominant	*Recessive*	*Total*	*Dominant*	*Recessive*
Seeds: smooth versus wrinkled	All smooth	5,474	1850	7,324	74.7	25.3
Seeds: yellow versus green	All yellow	6,022	2001	8,023	75.1	24.9
Seed coats: grey versus white	All grey	705	224	929	75.9	24.1
Flowers: purple versus white	All purple					
Flowers: axial versus terminal	All axial	651	207	858	75.9	24.1
Pods: inflated versus pinched	All inflated	882	299	1,181	74.7	25.3
Pods: green versus yellow	All green	428	152	580	73.8	26.2
Stem: tall versus short	All tall	787	277	1,064	74.0	26.0
Total or average		14,949	5010	19,959	74.9	25.1

[a] The dominant trait is always written first.

Source: From F. J. Ayala and J. A. Kiger, Jr., *Modern genetics.* 2nd ed. Copyright © 1984 and 1980 by The Benjamin/Cummings Publishing Company, Inc. Reprinted by permission.

peared. Further, the trait seen in the F_1 was always found in the F_2 at about three times the frequency of the other trait.

The key question for Mendel was how a trait present in the P generation could disappear in the F_1 and then appear with full expression in the F_2. Mendel observed that while the F_1 progeny resembled only one of the parents in their phenotype, they did not breed true as the parent they resembled did. The F_1 progeny possessed the potential to produce F_2 progeny that expressed the P phenotypic characteristic that had disappeared in the F_1 generation. Mendel concluded that the alternative traits in the crosses—for example, smoothness or wrinkledness of the seeds—were determined by **particulate factors.** He reasoned that these factors, which were transmitted from parents to progeny through the gametes, carried hereditary information. We now know these factors by another name, *genes.* Since Mendel was examining pairs of traits (e.g., wrinkled/smooth), each factor was considered to exist in alternative forms (which we now call **alleles**), each of which specified one of the traits. For the gene that controls the shape of the pea seed, for example, there is one form, or allele, that results in the production of a smooth seed and another allele that results in a wrinkled seed.

Mendel reasoned further that for pure-breeding strains of the peas, both egg and pollen must have carried or transmitted an identical form of the factor. Since both traits were seen in the F_2, whereas only one appeared in the F_1, then each F_1 individual must have contained both factors, one for each of the alternative traits. Furthermore, since only one of the characters was seen in the F_1's, the expression of the other trait must somehow have been masked by the other factor, a feature called *dominance.* For the smooth × wrinkled example the F_1 seeds were all smooth. Thus the allele for smoothness is **dominant** to the allele for wrinkledness. Conversely, wrinkled is said to be **recessive** to smooth. Similar conclusions can be made for the other six pairs of traits. The dominant and the recessive forms for each pair of traits are indicated in Table 2.1.

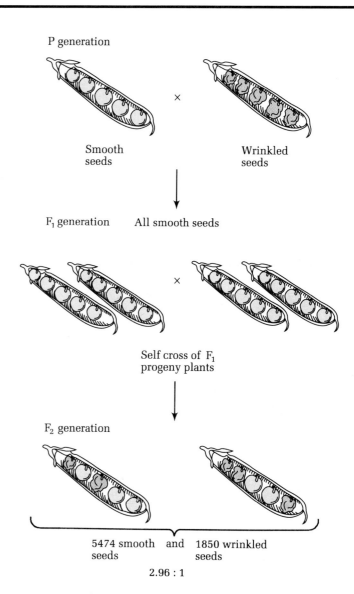

P generation

Smooth seeds × Wrinkled seeds

F_1 generation   All smooth seeds

Self cross of F_1 progeny plants

F_2 generation

5474 smooth and 1850 wrinkled
seeds   seeds

2.96 : 1

Figure 2.9 The F_2 progeny of the cross shown in Figure 2.8. When the plants grown from the F_1 seeds were self-pollinated, both smooth and wrinkled F_2 progeny seeds were produced. Commonly, both seed types were found in the same pod. In Mendel's actual experiments he counted 5474 smooth and 1850 wrinkled F_2 progeny seeds, for a ratio of 2.96:1.

We can see the crosses more easily if we use some symbols for the alleles. For the smooth × wrinkled cross we can give the symbol *S* to the factor for smoothness and the symbol *s* to the

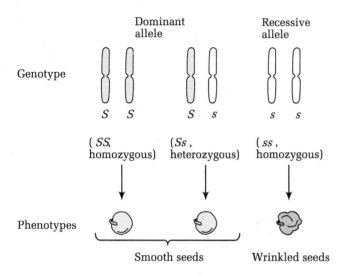

Figure 2.10 Dominant and recessive alleles of a gene for seed shape in peas.

factor for wrinkledness. The convention is that the dominant allele is given the uppercase (capital) letter and the recessive allele the lowercase (small) letter (see Figure 2.10). Using these symbols, we can diagram the cross as shown in Figure 2.11: the production of the F_1 generation is shown in Figure 2.11a, and the F_2 generation in Figure 2.11b. (In both Figure 2.10 and 2.11 the genes are located on chromosomes. To remind you, gene segregation follows the behavior of chromosomes although the existence of chromosomes was unknown to Mendel.) Since each parent is true breeding, each must contain two copies of the same allele. Thus the genotype of the parental plant grown from the smooth seeds is *SS* and that of the wrinkled parent is *ss*. True-breeding individuals that contain only one specific allele are **homozygous.**

When these plants produce gametes by meiosis, each gamete contains only one copy of the gene (i.e., one allele); the plants from smooth seeds produce *S*-bearing gametes, and the plants from wrinkled seeds produce *s*-bearing gametes. When the gametes fuse during the fertilization process, the resulting zygote has one *S* and one *s* factor, a genotype of *Ss*. Plants that have two different alleles at a specific gene lo-

cus are called **heterozygous.** Because of the dominance of the smooth *S* allele, only smooth seeds develop from the F_1 zygotes.

The plants derived from the F_1 seeds differ from the smooth parent in that they produce two types of gametes in equal numbers: *S*-bearing and *s*-bearing. All the possible F_1 gametic fusions are represented in the matrix in Figure 2.11, called a **Punnett square** after its originator, R. Punnett. These fusions give rise to the zygotes that produce the F_2 generation.

Three types of zygotes are produced: *SS, Ss,* and *ss*. As a result of the random fusing of gametes, the relative proportion of these zygotes is 1:2:1, respectively. However, since the *S* factor is dominant to the *s* factor, both the *SS* and *Ss* seeds are smooth, and the F_2 generation seeds exhibit a phenotypic ratio of 3:1, smooth:wrinkled seeds. The results are the same for crosses involving the other six character pairs.

The Principle of Segregation

From the sort of data we have discussed, Mendel proposed his **first law,** the **principle of segregation,** which states that the two members of a gene pair (alleles) segregate (separate) from each other in the formation of gametes. As a result, half the gametes carry one allele and the other half carry the other allele. Thus each gamete carries only one allele at each gene locus. The progeny are produced by the random combination of gametes from the two parents. In proposing the principle of segregation, Mendel had clearly differentiated between the factors (genes) that determined the traits (the genotype) and the traits themselves (the phenotype). From a modern perspective this law means that at the gene level the members of a pair of alleles of a gene on a chromosome segregate during meiosis so that any offspring receives only one member of a pair from each parent. Thus **gene segregation** parallels the separation of homologous pairs of chromosomes at anaphase I in meiosis.

Since the genetics concepts and terms we have learned so far in this chapter are often difficult to assimilate, you may wish to take a moment to review them (see Box 2.2). A thorough

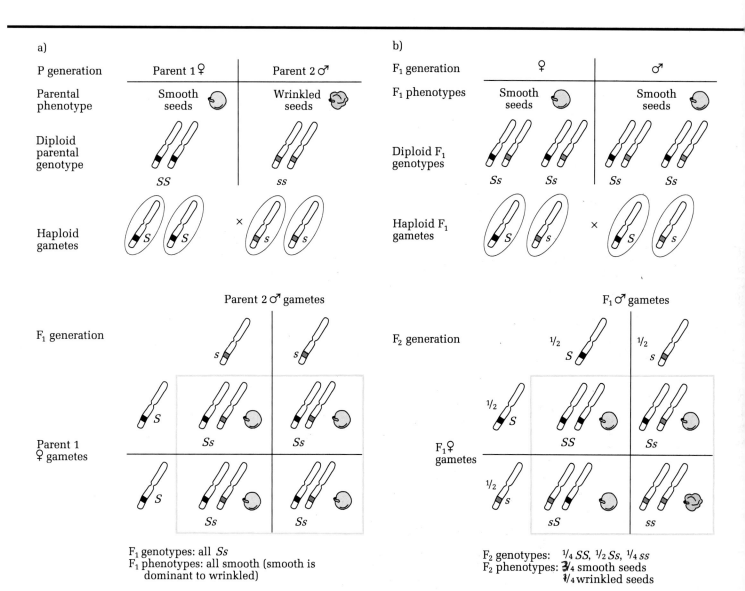

a)

b)

F₁ genotypes: all *Ss*
F₁ phenotypes: all smooth (smooth is
      dominant to wrinkled)

F₂ genotypes: ¼ *SS*, ½ *Ss*, ¼ *ss*
F₂ phenotypes: ¾ smooth seeds
      ¼ wrinkled seeds

*Figure 2.11 The same cross as in Figures 2.8 and 2.9
but with genetic symbols used to illustrate the principle of segregation of Mendelian factors. (a) Production of F₁ generation; (b) Production of F₂ generation.*

understanding of these terms is essential to your
study of genetics.

Keynote   *Mendel's first law, the principle of
segregation, states that the two
members of a gene pair (allele) segregate (separate) from each other in the formation of gametes; half the gametes carry one allele, and the
other half carry the other allele.*

Representing Crosses with a Branch Diagram

The use of a Punnett square to consider the pairing of all possible gamete types from the two
parents (shown in Figure 2.11) is an acceptable
way of predicting the relative frequencies of genotypes in the next generation. In this section,
though, we'll describe an alternative method:
the branch diagram. (Box 2.3 discusses some el-

ementary principles of probability that will help you understand this approach.) Figure 2.12 illustrates the application of the branch diagram approach for the analysis of the $F_1 \times F_1$ self of the smooth × wrinkled cross diagrammed in Figure 2.11.

The F_1 seeds from the cross have the genotype *Ss*. Both eggs and pollen are produced in the flowers of the plants grown from these seeds. In the meiotic process an equal number of *S* and *s* gametes are expected to be produced, so we can say that half the gametes are *S* and the other half are *s*. Thus 1/2 is the predicted frequency of these two types. But just as tossing a coin many times does not always give exactly half heads

and half tails, the gamete frequency may not be exactly realized. However, the more chances (e.g., tosses) the more likely you will obtain the true frequency.

From the rules of probability the relative expected frequencies of the three possible genotypes in the F_2 generation can be predicted. To produce an *SS* plant, an *S* egg must pair with an *S* pollen grain. The frequency of *S* eggs in the population of eggs is 1/2, and the frequency of *S* pollen grains in the pollen population is also 1/2. Therefore the relative expected proportion of *SS* plants in the F_2 is 1/2 × 1/2 = 1/4. A similar proportion is predicted for the wrinkled, or *ss*, progeny.

Box 2.2 *Genetic Terminology*

Cross: a mating between two individuals leading to the fusion of gametes.

Zygote: the cell produced by the fusion of the male and female gametes.

Gene (Mendelian factor): the determinant of a characteristic of an organism.

Locus (gene locus): the specific place on a chromosome where a gene is located.

Alleles: alternative forms of a gene. For example, *S* and *s* alleles represent the smoothness and wrinkledness of the pea seed. The rationale for the symbols used in our discussion of Mendel's work is still applicable in plant genetics: The dominant allele is assigned the uppercase letter and the recessive is assigned the lowercase letter. The actual letter used is based on the dominant phenotype, for instance, *S* for smooth, *s* for wrinkled, *Y* for yellow, and *y* for green. Later we will see a different way of assigning gene symbols that is more generally used in genetics today.

Genotype: the genetic constitution of an organism. An organism that has the same allele for a given gene locus on both members of a homologous pair of chromosomes in a diploid organism is said to be **homozygous** for that allele. Homozygotes produce only one gametic type. Thus true-breeding smooth individuals have the genotype *SS,* and true-breeding wrinkled individuals have the genotype *ss;* both are homozygous. The smooth parent is **homozygous dominant;** the wrinkled parent is **homozygous recessive.**

Diploid organisms that have two different alleles at a specific gene locus on homologous chromosomes are said to be **heterozygous.** So F_1 hybrid plants from the cross of *SS* and *ss* parents have one *S* allele and one *s* allele. Individuals heterozygous for two allelic forms of a gene produce only two kinds of gametes (*S* and *s*).

Phenotype: the physical manifestation of a genetic trait that results from a specific genotype and its interaction with the environment. In our example the *S* allele was dominant to the *s* allele, so in the heterozygous condition the seed is smooth. Therefore both the homozygous dominant *SS* and the heterozygous *Ss* seeds have the same phenotype (smooth), though they differ in genotype.

What about the *Ss* progeny? Again, the frequency of *S* in one gametic type is 1/2, and the frequency of *s* in the other gametic type is also 1/2. However, there are two ways in which *Ss* progeny can be obtained. The first involves the fusion of an *S* egg with *s* pollen, and the second is a fusion of an *s* egg with *S* pollen. Using the product rule (Box 2.3), the probability of each of these events occurring is 1/2 × 1/2 = 1/4. So, using the sum rule (Box 2.3), the probability of one or the other occurring is the sum of the individual probabilities, or 1/4 + 1/4 = 1/2.

From this discussion the prediction is that a fourth of the F_2 progeny will be *SS*, half will be *Ss*, and a fourth will be *ss*, exactly as was found with the method shown in Figure 2.11. Either method, then, is applicable to any cross.

Confirming the Principle of Segregation: The Use of Testcrosses

Mendel did a number of tests to ensure the validity of his results when formulating his principle of segregation. He continued the self-fertilizations to the F_6 generation and found that in every generation both the dominant and recessive characters appeared. He concluded that the principle of segregation was valid no matter how many generations were carried out.

Box 2.3 Elementary Principles of Probability

A **probability** is the ratio of the number of times a particular event occurs to the number of trials during which the event could have happened.

For example, the probability of picking a heart (from 13 hearts) from a deck of cards (52 cards) is p(heart) = 13/52 = 1/4. That is, we would expect, on the average, to pick a heart from a deck of cards once in every four trials.

Probabilities and the *laws of chance* are involved in the transmission of genes. As a simple example, let's consider a couple and the chance that their child will be a boy or a girl. Assume that an equal number of boys and girls are born (which is not actually true, but we can assume it to be so for the sake of discussion). The probability that the child will be a boy is 1/2, or 0.5. Similarly, the probability that the child will be a girl is 1/2.

Now a rule of probability can be introduced: the **product rule.** This rule states that the probability of two independent events occurring simultaneously is the product of each of their probabilities. Thus the probability that a family will have two girls in a row is 1/4. That is, the probability of the first child being a girl is 1/2, the probability of the second being a girl is also 1/2, and by the product rule the probability of the first and second being girls is 1/2 × 1/2, or 1/4. Similarly, the probability of having three boys in a row is 1/2 × 1/2 × 1/2 = 1/8.

Another rule of probability is the **sum rule,** which states that the probability that one of two mutually exclusive events will occur is the sum of their individual probabilities. For example, if two dice are thrown, what is the probability of getting two sixes or two ones? The probability of getting two sixes is found by using the product rule. The probability of getting one six, p(one six), is 1/6, since there are six faces to a die. Therefore the probability of getting two sixes, p(two sixes), when two dice are thrown is 1/6 × 1/6 = 1/36. Similarly, p(two ones) = 1/36. To roll two sixes or two ones involves mutually exclusive events, so the sum rule is used to find the probability. The answer is 1/36 + 1/36 = 1/18.

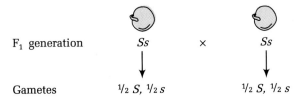

| F₁ generation | Ss | × | Ss |

| Gametes | ½ S, ½ s | | ½ S, ½ s |

Random combination of gametes results in:

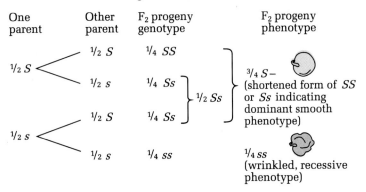

| One parent | Other parent | F₂ progeny genotype | F₂ progeny phenotype |

Figure 2.12 *Calculating the ratios of phenotypes in the F₂ generation of the cross of Figure 2.11 by using the branch diagram approach.*

Another important test concerned the F₂ plants. As shown in Figure 2.11, a ratio of 1:2:1 occurs for the genotypes *SS, Ss,* and *ss* for the smooth × wrinkled example. Phenotypically, the ratio of smooth to wrinkled is 3:1. At the time of his experiments Mendel knew only the phenotypic ratio, although he hypothesized the presence of segregating factors that were responsible for the smooth and wrinkled phenotypes. To be sure of his factor hypothesis, Mendel tested the F₂ plants by allowing them to self-pollinate. As he expected, the plants produced from wrinkled seeds bred true, supporting his contention that they were pure for the *s* factor (gene).

Selfing the plants derived from the F₂ smooth seeds produced two different results. One-third produced only smooth seeds, whereas the other two-thirds produced both smooth and wrinkled seeds in each pod (Figure 2.13). In fact, the ratio of the two types of seeds obtained from the plants was the same as that of all the F₂ progeny. These results completely support the prin-

ciple of segregation of genes. The random combination of gametes that form the zygotes of the F₂ produces two genotypes that give rise to the smooth phenotype (e.g., Figures 2.11 and 2.12). Moreover, the relative proportion of the two genotypes *SS* and *Ss* is 1:2. The *SS* seeds give rise to true-breeding plants, whereas the *Ss* seeds give rise to plants that behave exactly like the F₁ plants when they are self-pollinated. *Mendel explained these results by proposing that each plant had two factors, while each gamete had only one. He also proposed that the random combination of the gametes generated the progeny in the proportions he found. Mendel obtained the same results in all seven sets of crosses.* The segregation pattern seen, as we now know, results from the behavior of chromosomes.

The self-fertilization test of the F₂'s proved to be a useful way of confirming the genotype of a plant with a given phenotype. A more common test to find out the genotype of an organism is to perform a **testcross,** a cross of an individual of unknown genotype, usually expressing the dominant phenotype, with a homozygous recessive individual in order to determine the genotype of the individual.

In a testcross involving plants grown from the F₁ seeds (Figure 2.14), the plants produced by the F₁ smooth seeds are *Ss* and produce both *S* and *s* gametes in equal proportion. The other

Figure 2.13 *Determination of the genotypes of the F₂ smooth progeny of Figure 2.11 by selfing the plants grown from the smooth seeds.*

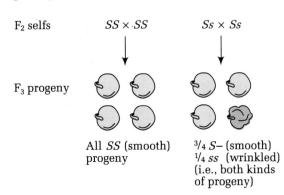

| F₂ selfs | $SS \times SS$ | $Ss \times Ss$ |

| F₃ progeny | All *SS* (smooth) progeny | ¾ *S*– (smooth) ¼ *ss* (wrinkled) (i.e., both kinds of progeny) |

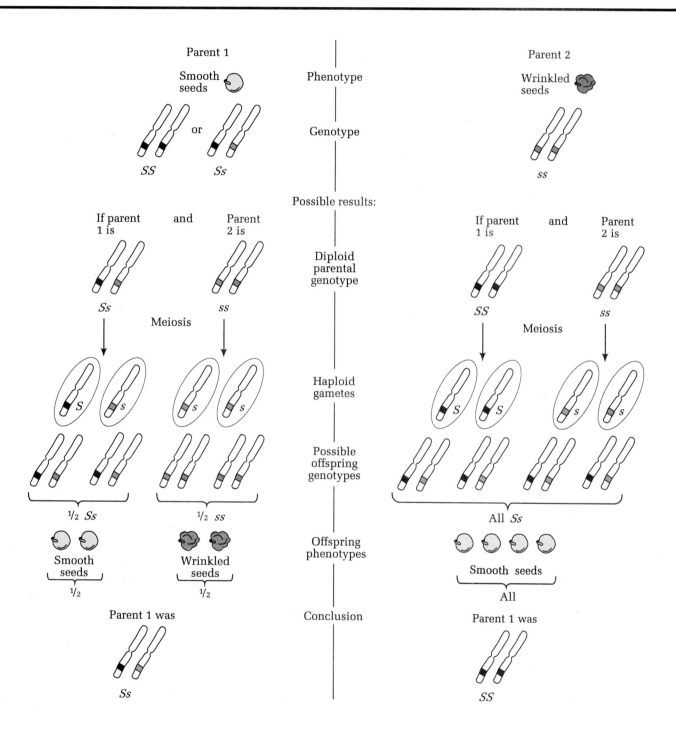

Figure 2.14 Determination of the genotypes of the F₁ and F₂ generation seeds of Figure 2.11 by testcrossing plants grown from the seeds with a homozygous recessive strain.

parent in the testcross is the homozygous recessive *ss* plant grown from wrinkled seeds, which produces only *s* gametes. As a result of random combination of gametes, half the progeny of the testcross are *Ss* heterozygotes and have a smooth phenotype owing to the dominance of the *S* allele, and the other half are *ss* homozygotes and have a wrinkled phenotype.

In actual practice the observation of a 1:1 ratio of dominant to recessive phenotypes would indicate that the F_1 smooth seed used in the testcross is heterozygous. We can similarly predict the outcome of a testcross of the F_2's showing the dominant phenotype. For those F_2's that are homozygous *SS,* the result of a testcross with an *ss* plant will be all smooth seeds. As Figure 2.14 shows, the *SS*'s produce only *S* gametes. Thus all zygotes are *Ss* and all the seeds are smooth in phenotype. To review the logic in the other direction, if an organism in a testcross gives only progeny with the dominant phenotype, then the testcrossed plant must have been homozygous dominant. For the *Ss* F_2 plants a 1:1 ratio of dominant to recessive phenotypes is expected, as was true of the F_1 testcross.

In summary, testcrosses of Mendel's plants grown from F_2 seeds with the dominant phenotype would give the following results: A third of the plants would give rise only to progeny seeds expressing the dominant phenotype. These F_2 seeds would be homozygous for the dominant gene. The other two-thirds would give a 1:1 ratio of heterozygous and homozygous recessive seeds, indicating that the F_2 seeds must be heterozygous.

Keynote   *A testcross is a cross of an individual of unknown genotype, usually expressing the dominant phenotype, with a homozygous recessive in order to determine the genotype of the unknown individual. The phenotypes of the progeny of the testcross indicate the genotype of the individual tested. If the progeny all show the dominant phenotype, the individual was homozygous dominant. If there is an approximately 1:1 ratio of progeny with dominant and recessive phenotypes, the individual was heterozygous.*

Dihybrid Crosses and the Mendelian Principle of Independent Assortment

The next logical extension of Mendel's experiments was to determine what happens when two pairs of characters were simultaneously involved in the cross. Mendel did a number of these crosses, and in each case he obtained the same results. From these experiments he proposed his **second law,** the **principle of independent assortment,** which states that the factors (genes) for different traits assort independently of one another. In other words, genes on different chromosomes behave independently in the production of gametes. We can turn this around the other way. That is, since genes are on chromosomes, and since maternal and paternal chromosomes assort randomly and independently at metaphase, then if genes assort independently they are on separate chromosomes.

Consider an example involving smooth *(S)*/wrinkled *(s)* and yellow *(Y)*/green *(y)* seed traits (yellow is dominant to green). Mendel made crosses between true-breeding smooth-yellow plants *(SS YY)* and wrinkled-green plants *(ss yy),* with the results shown in Figure 2.15. All the F_1 seeds from this cross were smooth and yellow, as predicted from the results of the monohybrid crosses. As Figure 2.15a shows, the smooth-yellow parent produces only *S Y* gametes, which give rise to *Ss Yy* zygotes upon fusion with the *s y* gametes from the other parent. Because of the dominance of the smooth and the yellow traits, all F_1 seeds are smooth and yellow.

The seeds of the F_1 generation will grow into plants that are heterozygous for two pairs of alleles at two different loci. Such plants are called dihybrid plants, and a cross between two dihybrids of the same type is called a **dihybrid cross.**

When Mendel self-pollinated the F_1 plants to give rise to the F_2 generation, (Figure 2.15b) he considered two possible outcomes. One was that the genes for the traits from the original parents would be transmitted to the progeny together. In

51
*Dihybrid
Crosses and
the Mendelian
Principle of
Independent
Assortment*

this case a phenotypic ratio of 3:1 smooth-yellow: wrinkled-green would be predicted. The other possibility was that the traits would be inherited independently of one another. The dihybrid plants grown from the F_1 seeds produce four types of gametes: *S Y, S y, s Y,* and *s y*. Because of the independence of the two pairs of genes, there is an equal frequency of each gametic type. In the $F_1 \times F_2$ selfing these four types of gametes fuse randomly in all possible combinations to give rise to the zygotes and, hence, the progeny seeds. All the possible gametic fusions are represented in the large Punnett square in Figure 2.15. In a dihybrid cross there are 16 possible gametic fusions. The result is 9 different genotypes but, because of dominance, only 4 phenotypes:

1 *SS YY*, 2 *Ss YY*,
2 *SS Yy*, 4 *Ss Yy* = 9 smooth-yellow

1 *SS yy*, 2 *Ss yy* = 3 smooth-green

1 *ss YY*, 2 *ss Yy* = 3 wrinkled-yellow

1 *ss yy* = 1 wrinkled-green

If pairs of characters are inherited independently in a dihybrid cross, then the F_2 from an $F_1 \times F_1$ selfing will give a 9:3:3:1 ratio of the four possible phenotypic classes. This ratio arises as a result of the independent assortment of the two gene pairs into the gametes and of the random fusion of the gametes. The 9:3:3:1 ratio is the product of two 3:1 ratios; that is, $(3:1)^2 =$ 9:3:3:1. This prediction was met in all dihybrid crosses that Mendel performed. In every case the F_2 ratio was close to 9:3:3:1. For our example he counted 315 smooth-yellow, 108 smooth-green, 101 wrinkled-yellow, and 32 wrinkled-green seeds—very close to the predicted ratio. To Mendel this result meant that the factors (genes) determining different character pairs were transmitted independently. We now know that this statement is true because genes are located on chromosomes, which are distributed to gametes by the process of meiosis.

a)

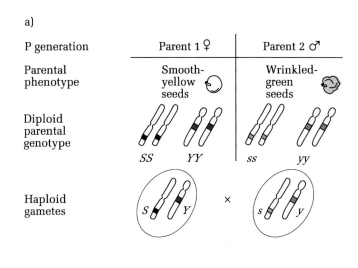

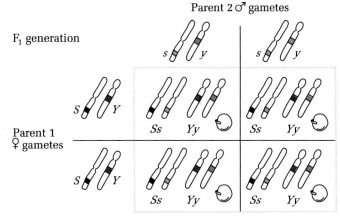

F_1 genotypes: all *Ss Yy*
F_1 phenotypes: all smooth-yellow seeds

Figure 2.15 Illustration of the principle of independent assortment in a dihybrid cross. This cross, actually done by Mendel, involves the smooth/wrinkled and yellow/green character pairs of the garden pea. (a) Production of F_1 generation.

Keynote   Mendel's second law, the principle of independent assortment, states that genes on different chromosomes behave independently in the production of gametes.

Branch Diagram of Dihybrid Cross

As perhaps is already apparent, it is quite tedious and slow to construct a Punnett square of gamete combinations and then to count up the

b)

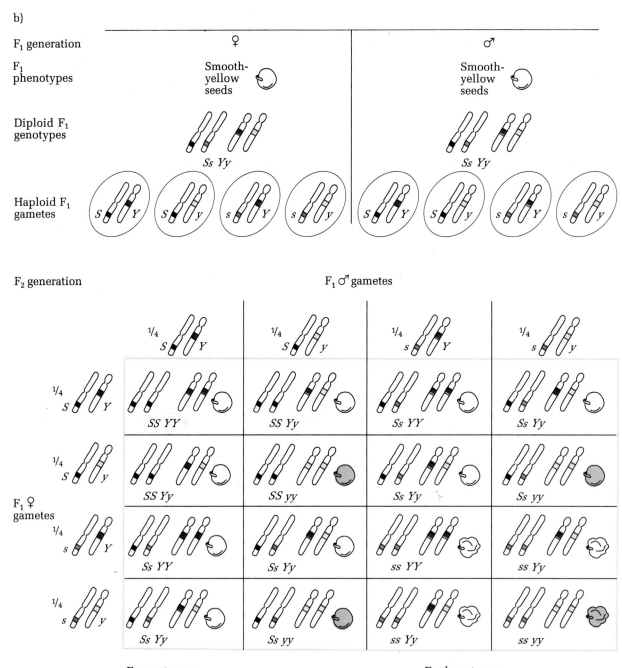

F₁ generation ♀ ♂

F₁ phenotypes — Smooth-yellow seeds — Smooth-yellow seeds

Diploid F₁ genotypes — *Ss Yy* — *Ss Yy*

Haploid F₁ gametes — *S Y*, *S y*, *s Y*, *s y* — *S Y*, *S y*, *s Y*, *s y*

F₂ generation — F₁ ♂ gametes

F₁ ♀ gametes

F₂ genotypes:

$^1/_{16}$ (*SS YY*) + $^2/_{16}$ (*Ss YY*) + $^2/_{16}$ (*SS Yy*) + $^4/_{16}$ (*Ss Yy*) = $^9/_{16}$ smooth-yellow seeds
$^1/_{16}$ (*SS yy*) + $^2/_{16}$ (*Ss yy*) = $^3/_{16}$ smooth-green seeds
$^1/_{16}$ (*ss YY*) + $^2/_{16}$ (*ss Yy*) = $^3/_{16}$ wrinkled-yellow seeds
$^1/_{16}$ (*ss yy*) = $^1/_{16}$ wrinkled-green seeds

F₂ phenotypes:

53
*Dihybrid
Crosses and
the Mendelian
Principle of
Independent
Assortment*

Figure 2.15 (continued) (b) The F_2 genotypes and 9:3:3:1 phenotypic ratio of smooth-yellow:smooth-green:wrinkled-yellow: wrinkled-green peas are derived by using the Punnett square.

numbers of each phenotypic class from all the genotypes produced. This chore is not too difficult for a dihybrid cross, but for more than two gene pairs it becomes rather complex. It is easier to get into the habit of calculating the expected ratios of phenotypic classes by using a branch diagram.

Using the same example in which the two gene pairs assort independently into the gametes, we will consider each gene pair in turn. We showed earlier that an F_1 self of an *Ss* heterozygote gave rise to progeny of which three-fourths were smooth and a fourth were wrinkled. Genotypically, the former class had at least one dominant *S* allele; that is, they were *SS* or *Ss*. A convenient way to signify this situation is to use a dash to indicate an allele that has no effect on the phenotype. Thus *S–* means that phenotypically the seeds are smooth and genotypically they are either *SS* or *Ss*.

Now consider the F_2 produced from a selfing of *Yy* heterozygotes; again, a 3:1 ratio is seen, with three-fourths of the seeds being yellow and a fourth of the seeds being green. Since this segregation occurs independently of the segregation of the smooth/wrinkled pair, we can consider all possible combinations of the phenotypic classes. For example, the expected proportion of F_2 seeds that are smooth and yellow is the product of the probability that an F_2 seed will be smooth and the probability that it will be yellow, or 3/4 × 3/4 = 9/16. Similarly, the expected proportion of F_2's that are wrinkled and yellow is 3/4 × 1/4 = 3/16. Extending this calculation to all possible phenotypes, as shown in Figure 2.16, we obtain the 9:3:3:1 ratio.

The testcross can be used to check the genotypes of the F_1's and the F_2's from a dihybrid cross. The F_1 is a double heterozygote, *Ss Yy* in the example. This F_1 produces four types of gametes in equal proportions, as was the case in Figure 2.15: *S Y, S y, s Y,* and *s y.* In a testcross

with a doubly homozygous recessive plant, in this case *ss yy,* the phenotypic ratio of the progeny is a direct reflection of the ratio of gametic types produced by the F_1 parent. In a testcross like this one, then, there will be a 1:1:1:1 ratio in the offspring of *Ss Yy:Ss yy:ss Yy:ss yy* genotypes, which means a 1:1:1:1 ratio of smooth-yellow:smooth-green:wrinkled-yellow:wrinkled-green phenotypes. This 1:1:1:1 phenotypic ratio is diagnostic of testcrosses in which the "unknown" parent is a double heterozygote.

In the F_2 of a dihybrid cross there were nine different genotypic classes but only four phenotypic classes. The genotypes can be ascertained by testcrossing as we've shown. Table 2.2 (p. 57) lists the expected ratios of progeny phenotypes from such testcrosses. No two patterns are the same, so the testcross is truly a diagnostic approach here to confirm genotypic type.

Trihybrid Crosses

Mendel also confirmed his laws for three characters segregating in other garden pea crosses. Such crosses are called **trihybrid crosses.** Here the proportions of F_2 genotypes and phenotypes are predicted with precisely the same logic used before, with each character pair considered in-

Figure 2.16 The same example as in Figure 2.15 but here the F_2 phenotypic ratio is calculated by using the branch diagram approach.

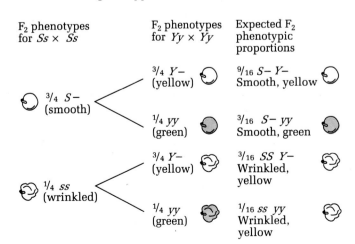

F_2 phenotypes for *Ss* × *Ss*	F_2 phenotypes for *Yy* × *Yy*	Expected F_2 phenotypic proportions
3/4 *S–* (smooth)	3/4 *Y–* (yellow)	9/16 *S– Y–* Smooth, yellow
	1/4 *yy* (green)	3/16 *S– yy* Smooth, green
1/4 *ss* (wrinkled)	3/4 *Y–* (yellow)	3/16 *SS Y–* Wrinkled, yellow
	1/4 *yy* (green)	1/16 *ss yy* Wrinkled, yellow

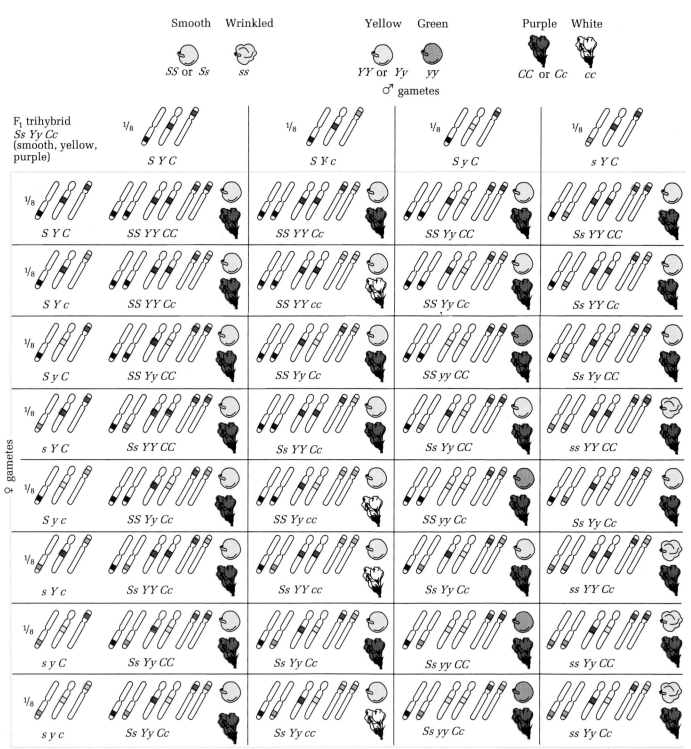

Smooth Wrinkled Yellow Green Purple White

SS or Ss ss YY or Yy yy CC or Cc cc

♂ gametes

F₂ genotypes and phenotypes:

$\frac{1}{64}$ ($SS\,YY\,Cc$) + $\frac{2}{64}$ ($SS\,Yy\,CC$) + $\frac{2}{64}$ ($SS\,Yy\,Cc$) + $\frac{4}{64}$ ($Ss\,YY\,Cc$) + $\frac{4}{64}$ ($SS\,Yy\,CC$) + $\frac{4}{64}$ ($Ss\,YY\,Cc$) + $\frac{4}{64}$ ($Ss\,Yy\,CC$)

$\frac{1}{64}$ ($SS\,YY\,cc$) + $\frac{2}{64}$ ($SS\,Yy\,cc$)

$\frac{1}{64}$ ($SS\,yy\,CC$) + $\frac{2}{64}$ ($SS\,yy\,Cc$)

$\frac{1}{64}$ ($ss\,YY\,CC$) + $\frac{2}{64}$ ($ss\,Yy\,CC$)

Figure 2.17 Genotypes and phenotypes produced in
the progeny of trihybrid pea plants allowed to self-
fertilize: (a) derivation using the Punnett square.

55
Dihybrid
Crosses and
the Mendelian
Principle of
Independent
Assortment

♂ gametes

$\frac{1}{8}$	$\frac{1}{8}$	$\frac{1}{8}$	$\frac{1}{8}$
$S\,y\,c$	$s\,Y\,c$	$s\,y\,C$	$s\,y\,c$
$SS\,Yy\,Cc$	$Ss\,YY\,Cc$	$Ss\,Yy\,CC$	$Ss\,Yy\,Cc$
$SS\,Yy\,cc$	$Ss\,YY\,cc$	$Ss\,Yy\,Cc$	$Ss\,Yy\,cc$
$SS\,yy\,Cc$	$Ss\,Yy\,Cc$	$Ss\,yy\,CC$	$Ss\,yy\,Cc$
$Ss\,Yy\,Cc$	$ss\,YY\,Cc$	$ss\,Yy\,CC$	$ss\,Yy\,Cc$
$SS\,yy\,cc$	$Ss\,Yy\,cc$	$Ss\,yy\,Cc$	$Ss\,yy\,cc$
$Ss\,Yy\,cc$	$ss\,YY\,cc$	$ss\,Yy\,Cc$	$ss\,Yy\,cc$
$Ss\,yy\,Cc$	$ss\,Yy\,Cc$	$ss\,yy\,CC$	$ss\,yy\,Cc$
$Ss\,yy\,cc$	$ss\,Yy\,cc$	$ss\,yy\,Cc$	$ss\,yy\,cc$

$+ \;{}^{4}/_{64}\,(\,Ss\,Yy\,CC\,) + \;{}^{8}/_{64}\,(\,Ss\,Yy\,Cc\,) = {}^{27}/_{64}$ smooth, yellow, purple

$+ \;{}^{2}/_{64}\,(\,Ss\,YY\,cc\,) + \;{}^{4}/_{64}\,(\,Ss\,Yy\,cc\,) = {}^{9}/_{64}$ smooth, yellow, white

$+ \;{}^{2}/_{64}\,(\,Ss\,yy\,CC\,) + \;{}^{4}/_{64}\,(\,Ss\,yy\,Cc\,) = {}^{9}/_{64}$ smooth, green, purple

$+ \;{}^{2}/_{64}\,(\,ss\,YY\,Cc\,) + \;{}^{4}/_{64}\,(\,ss\,Yy\,Cc\,) = {}^{9}/_{64}$ wrinkled, yellow, purple

$\;{}^{1}/_{64}\,(\,SS\,yy\,cc\,) + \;{}^{2}/_{64}\,(\,Ss\,yy\,cc\,) = {}^{3}/_{64}$ smooth, green, white

$\;{}^{1}/_{64}\,(\,ss\,YY\,cc\,) + \;{}^{2}/_{64}\,(\,ss\,Yy\,cc\,) = {}^{3}/_{64}$ wrinkled, yellow, white

$\;{}^{1}/_{64}\,(\,ss\,yy\,CC\,) + \;{}^{2}/_{64}\,(\,ss\,yy\,Cc\,) = {}^{3}/_{64}$ wrinkled, green, purple

$\;{}^{1}/_{64}\,(\,ss\,yy\,cc\,) = {}^{1}/_{64}$ wrinkled, green, white

b)

| Expected F$_2$ phenotypes for $Ss \times Ss$ | Expected F$_2$ phenotypes for $Yy \times Yy$ | Expected F$_2$ phenotypes for $Cc \times Cc$ | Expected F$_2$ phenotypic proportions |

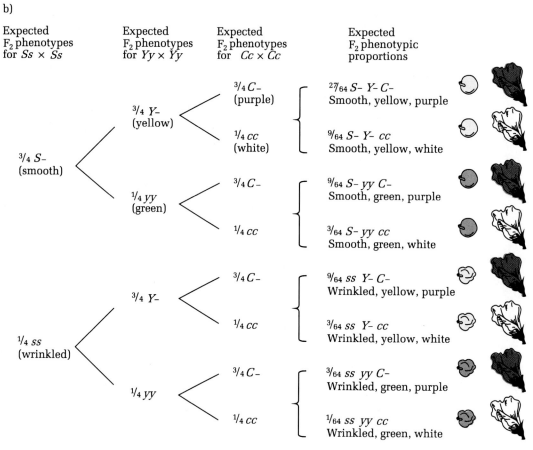

$^3/_4$ $S-$
(smooth)

$^1/_4$ ss
(wrinkled)

$^3/_4$ $Y-$
(yellow)

$^1/_4$ yy
(green)

$^3/_4$ $Y-$

$^1/_4$ yy

$^3/_4$ $C-$
(purple)

$^1/_4$ cc
(white)

$^3/_4$ $C-$

$^1/_4$ cc

$^3/_4$ $C-$

$^1/_4$ cc

$^3/_4$ $C-$

$^1/_4$ cc

$^{27}/_{64}$ $S- Y- C-$
Smooth, yellow, purple

$^9/_{64}$ $S- Y- cc$
Smooth, yellow, white

$^9/_{64}$ $S- yy C-$
Smooth, green, purple

$^3/_{64}$ $S- yy cc$
Smooth, green, white

$^9/_{64}$ $ss Y- C-$
Wrinkled, yellow, purple

$^3/_{64}$ $ss Y- cc$
Wrinkled, yellow, white

$^3/_{64}$ $ss yy C-$
Wrinkled, green, purple

$^1/_{64}$ $ss yy cc$
Wrinkled, green, white

Figure 2.17 (continued) (b) Branch diagram derivation of the relative frequencies of the eight phenotypic classes.

dependently. Although we will not derive the results in detail, Figure 2.17 shows a very large Punnett square (Figure 2.17a) and an alternative branch diagram (Figure 2.17b). The independently assorting character pairs in the cross are smooth versus wrinkled seeds, yellow versus green seeds, and purple versus white flowers. There are 64 combinations of 8 maternal and 8 paternal gametes. Combination of these gametes gives rise to 27 different genotypes and 8 different phenotypes.

Now that enough examples have been considered, we can make some generalizations about phenotypic and genotypic classes. In each of the previous examples, the F$_1$ is heterozygous for each gene involved in the cross, and the F$_2$ is generated by selfing where that is possible or, in organisms where it is not possible (e.g., humans, mice, and *Drosophila*), by allowing the F$_1$'s to interbreed. In monohybrid crosses there were 2 phenotypic classes in the F$_2$, in dihybrid crosses there were 4, and in trihybrid crosses there were 8. The general rule is that there will be 2^n phenotypic classes in the F$_2$, where n is the number of independently assorting, heterozygous gene pairs (Table 2.3). (This rule holds only when a

57
*Dihybrid
Crosses and
the Mendelian
Principle of
Independent
Assortment*

Table 2.2 Proportions of Phenotypic Classes Expected from Testcrosses of Strains with Various Genotypes for Two Gene Pairs

	Proportion of Phenotypic Classes			
Testcrosses	$A-B-$	$A-bb$	$aa\,B-$	$aa\,bb$
$AA\,BB \times aa\,bb$	1	0	0	0
$Aa\,BB \times aa\,bb$	1/2	0	1/2	0
$AA\,Bb \times aa\,bb$	1/2	1/2	0	0
$Aa\,Bb \times aa\,bb$	1/4	1/4	1/4	1/4
$AA\,bb \times aa\,bb$	0	1	0	0
$Aa\,bb \times aa\,bb$	0	1/2	0	1/2
$aa\,BB \times aa\,bb$	0	0	1	0
$aa\,Bb \times aa\,bb$	0	0	1/2	1/2
$aa\,bb \times aa\,bb$	0	0	0	1

Table 2.3 Number of Phenotypic and Genotypic Classes Expected from Self Crosses of Heterozygotes in Which All Genes Show Complete Dominance

Number of Segregating Genes	Number of Phenotypic Classes	Number of Genotypic Classes
1 [a]	2	3
2	4	9
3	8	27
4	16	81
n	2^n	3^n

[a] For example, from $Aa \times Aa$, two phenotypic classes are expected, with genotypic classes of AA, Aa, and aa.

dominant-recessive relation holds for each of the gene pairs.) Furthermore, we saw that there were 3 genotypic classes in the F_2 of monohybrid crosses, 9 in dihybrid crosses, and 27 in trihybrid crosses. A simple rule is that the number of genotypic classes will be 3^n, where n is the number of heterozygous gene pairs.

Incidentally, the phenotypic rule (2^n) can also be used to predict the number of classes that will come from a multiple heterozygous F_1 used in a testcross. Here the number of genotypes in the next generation will be the same as the number of phenotypes.

Rediscovery of Mendel's Principles

Although the science of genetics formally started in 1866 when Mendel published the results of his pea-breeding experiments his impact on the scientific community at the time was minuscule since other scientists of his time did not recognize the validity and importance of his conclusions. It wasn't until 35 years after he presented his work that the significance of his research was realized.

At about the turn of the century three researchers working independently on breeding experiments came to precisely the same conclusions as had Mendel, and in a sense they rediscovered Mendelism. The three men were Carl Correns, Hugo de Vries, and Erich von Tschermak. Correns concentrated mostly on maize (corn) and peas, de Vries worked with a number of different plant species, and von Tschermak studied peas.

The first demonstration that Mendelism applied to animals came in 1902 from the work of William Bateson, who experimented with fowl. Bateson also coined the terms *genetics, zygote,* F_1, F_2, and **allelomorph** (literally, "alternative form," meaning one of an array of different forms of a gene). The last was shortened by others to **allele**. The term *gene* as a replacement for Mendelian factor was introduced by W. L. Johannsen in 1909. In the years subsequent to Bateson's work a number of investigators showed the general applicability of Mendelian principles to all eukaryotic organisms.

Analytical Approaches for Solving Genetics Problems

The most practical way to reinforce Mendelian principles is to solve genetics problems. These problems generally present familiar and unfamiliar examples and pose questions designed to get problem solvers to work the material forward and backward so that they become totally familiar with it. In this and many other chapters we will discuss how to approach genetics problems by presenting examples of such problems and discussing their answers.

Q.1 A purple-flowered pea plant is crossed with a white-flowered pea plant. All the F_1 plants produced purple flowers. When the F_1 plants are allowed to self-pollinate, 401 of the F_2's have purple flowers and 131 have white flowers. What are the genotypes of the parental and F_1 generation plants?

A.1 The ratio of plant phenotypes in the F_2 is very close to the 3:1 ratio expected of a monohybrid cross. More specifically, this ratio is expected to result from an $F_1 \times F_1$ cross in which both are identically heterozygous for a specific gene pair. In addition, since only one phenotypic class appeared in the F_1, we can deduce that both parental plants were true breeding. Further, since the F_1 phenotype exactly resembled one of the parental phenotypes, we can say that purple is dominant to white flowers. Assigning the symbol *P* to the gene that determines purpleness of flowers and the symbol *p* to the alternative form of the gene that determines whiteness, we can write the genotypes:

P generation: *PP,* for the purple-flowered plant
  pp, for the white-flowered plant
F_1 generation: *Pp,* which, because of dominance, is purple-flowered

We could further deduce that the F_2 plants have an approximately 1:2:1 ratio or *PP:Pp:pp.*

Q.2 In chickens the white plumage of the leghorn breed is dominant over colored plumage, feathered shanks are dominant over clean shanks, and pea comb is dominant over single comb. Each of the gene pairs segregates independently. If a homozygous white, feathered, pea-combed chicken is crossed with a homozygous colored, clean, single-comb chicken, and the F_1's are allowed to interbreed, what proportion of the white, feathered, pea-combed birds in the F_2 will produce only white, feathered, pea-combed progeny if mated to colored, clean-shanked, single-combed birds?

A.2 This example is typical of a question that presents the unfamiliar in an attempt to get at the familiar. The best approach to such questions is to reduce them to their simple parts and, wherever possible, to assign gene symbols for each character. We are told which character is dominant for each of the three gene pairs, so we can use *W* for white and *w* for colored, *F* for feathered and *f* for clean shanks, and *P* for pea comb and *p* for single comb. The cross involves true-breeding strains and can be written as follows:

P generation: *WW FF PP* × *ww ff pp*

F_1 generation: *Ww Ff Pp*

Now the question asks the proportion of the white, feathered, pea-combed birds in the F_2 that will produce only white, feathered, pea-combed progeny if mated to colored, clean-shanked, single-combed birds. The latter are homozygous recessive for all three genes; that is, *ww ff pp,* as in the parental generation. For the result listed the white, feathered, pea-combed birds must be homozygous for the dominant alleles of the respective genes. What we are seeking, then, is the proportion of the F_2 chickens that are *WW FF PP* in genotype. We know that each gene pair segregates independently, so the answer can be calculated by using simple probability rules. We consider each gene pair in turn. For the white/colored case the $F_1 \times F_1$ is *Ww × Ww,* and we know from Mendelian principles that the relative proportion of F_2 genotypes will be 1 *WW:*2 *Ww:*1 *ww.* Therefore the proportion of the F_2's that will be *WW* is 1/4. The same relationship holds for the other two pairs of genes. Since the segregation of the three gene pairs is independent, we must multiply the probabilities of each occurrence to calculate the probability for *WW FF PP* individuals. The answer is 1/4 × 1/4 × 1/4 = 1/64.

Q.3 Consider three gene pairs *Aa, Bb,* and *Cc,* each of which affects a different character. In each case the uppercase letter signifies the dominant allele and the lowercase letter the recessive allele. These three gene pairs assort independently of each other. Calculate the probability of obtaining:
(a) an *Aa BB Cc* zygote from a cross *Aa Bb Cc × Aa Bb Cc;*
(b) an *Aa BB cc* zygote from a cross *aa BB cc × AA bb CC;*
(c) an *A B C* phenotype from a cross *Aa Bb CC × Aa Bb cc;*
(d) an *a b c* phenotype from a cross *Aa Bb Cc × aa Bb cc.*

A.3 Again, we must reduce the question into simple parts so that basic Mendelian principles can be applied. The key is that the genes assort independently, so we must multiply the probabilities of the individual occurrences to obtain the final answer.

a. First, we must consider the *Aa* gene pair. The cross is *Aa × Aa,* so the probability of the zygote being *Aa* is 2/4 since the expected distribution of genotypes is 1 *AA:*2 *Aa:*1 *aa,* as we have discussed. Then the probability of *BB* from *Bb × Bb* is 1/4, and that of *Cc* from *Cc × Cc* is 2/4, following the same sort of logic. The probability of an *Aa BB Cc* zygote is 1/2 × 1/4 × 1/2 = 1/16.

b. Similar logic is needed here, although we must be sure of the genotypes of the parental types since they differ from one gene pair to another. For the *Aa* pair the probability of getting *Aa* from *AA × aa* has to be 1. Next, the probability of getting *BB* from *BB × bb* is 0, so on these grounds alone we cannot get the zygote asked for from the cross given.

c. This question and the next ask for the probability of getting a particular phenotype, so we must start thinking of dominance. Again, we break up the question and consider each character pair in turn. The probability of an *A* phenotype from *Aa × Aa* is 3/4, from basic Mendelian principles. Similarly, the probability of a *B* phenotype from *Bb × Bb* is 3/4. Lastly, the probability of a *C* phenotype from *CC × cc* is 1. Overall, the probability of an *A B C* phenotype is 3/4 × 3/4 × 1 = 9/16.

d. An *a b c* phenotype from *Aa Bb Cc × aa Bb cc* is 1/2 × 1/4 × 1/2 = 1/16.

*Questions
and
Problems*

*2.1 In tomatoes, red fruit color is dominant to yellow. Suppose a tomato plant homozygous for red is crossed with one homozygous for yellow. Determine the appearance of (a) the F_1; (b) the F_2; (c) the offspring of a cross of the F_1 back to the red parent; (d) the offspring of a cross of the F_1 back to the yellow parent.

2.2 A red-fruited tomato plant, when crossed with a yellow-fruited one, produces progeny about half of which are red-fruited and half of which are yellow-fruited. What are the genotypes of the parents?

2.3 In maize, a dominant gene A is necessary for seed color as opposed to colorless (a). A recessive gene wx results in waxy starch as opposed to normal starch (Wx). The two genes segregate independently. Give phenotypes and relative frequencies for offspring resulting when a plant of genetic constitution $Aa\ Wx\ Wx$ is testcrossed.

*2.4 F_2 plants segregate 3/4 colored:1/4 colorless. If a colored plant is picked at random and selfed, what is the probability that more than one type will segregate among a large number of its progeny?

*2.5 In guinea pigs rough coat *(R)* is dominant over smooth coat *(r)*. A rough-coated guinea pig is bred to a smooth one, giving eight rough and seven smooth progeny in the F_1.

a. What are the genotypes of the parents and their offspring?
b. If one of the rough F_1 animals is mated to its rough parent, what progeny would you expect?

2.6 In cattle the polled (hornless) condition *(P)* is dominant over the horned *(p)* phenotype. A particular polled bull is bred to three cows. With cow A, which is horned, a horned calf is produced; with a polled cow B a horned calf is produced; and with horned cow C a polled calf is produced. What are the genotypes of the bull and the three cows, and what phenotypic ratios do you expect in the offspring of these three matings?

*2.7 In the Jimsonweed purple flowers are dominant to white. When a particular purple-flowered Jimsonweed is self-fertilized, there are 28 purple-flowered and 10 white-flowered progeny. What proportion of the purple-flowered progeny will breed true?

*2.8 Two black female mice are crossed with the same brown male. In a number of litters female X produced 9 blacks and 7 browns and female Y produced 14 blacks. What is the mode of inheritance of black and brown coat color in mice? What are the genotypes of the parents?

2.9 Bean plants may differ in their symptoms when infected with a virus. Some show local lesions that do not seriously harm the plant. Other plants show general systemic infection. The following genetic analysis was made:

P   local lesions × systemic lesions

F_1 all local lesions

F_2 785 local lesions:269 systemic lesions

What is probably the genetic basis of this difference in beans? Assign gene symbols to all the genotypes occurring in the above experiment. Design a testcross to verify your assumptions.

2.10 In Jimsonweed purple flower *(P)* is dominant to white *(p)*, and spiny pods *(S)* are dominant to smooth *(s)*. In a cross between a Jimsonweed homozygous for white flowers and spiny pods and one homozygous for purple flowers and smooth pods, determine the phenotype of (a) the F_1; (b) the F_2; (c) the progeny of a cross of the F_1 back to the white, spiny parent; (d) the progeny of a cross of the F_1 back to the purple, smooth parent.

2.11 What progeny would you expect from the following Jimsonweed crosses?

a. *PP ss* × *pp SS*
b. *Pp SS* × *pp ss*
c. *Pp Ss* × *Pp SS*
d. *Pp Ss* × *Pp Ss*
e. *Pp Ss* × *Pp ss*
f. *Pp Ss* × *pp ss*

*2.12 In summer squash white fruit *(W)* is dominant over yellow *(w)*, and disk-shaped fruit *(D)* is dominant over sphere-shaped fruit *(d)*. In the following problems the appearances of the parents and their progeny are given. Determine the genotypes of the parents in each case.

a. White, disk × yellow, sphere gives 1/2 white, disk and 1/2 white, sphere.
b. White, sphere × white, sphere gives 3/4 white, sphere and 1/4 yellow, sphere.
c. Yellow, disk × white, sphere gives all white, disk progeny.
d. White, disk × yellow, sphere gives 1/4 white, disk, 1/4 white, sphere, 1/4 yellow, disk, and 1/4 yellow, sphere.
e. White, disk × white, sphere gives 3/8 white, disk, 3/8 white, sphere, 1/8 yellow, disk, and 1/8 yellow, sphere.

*2.13 Genes *a, b,* and *c* assort independently and are recessive to their respective alleles *A, B,* and *C.* Two triply heterozygous *(Aa Bb Cc)* individuals are crossed.

a. What is the probability that a given offspring will be phenotypically *ABC*, that is, exhibit all three dominant traits?
b. What is the probability that a given offspring will be genotypically homozygous for all three dominant alleles?

2.14 In garden peas tall stem *(T)* is dominant over short stem *(t)*, green pods *(G)* are dominant over yellow pods *(g)*, and smooth seeds *(S)* are dominant over wrinkled seeds *(s)*. Suppose a homozygous short, green, wrinkled pea plant is crossed with a homozygous tall, yellow, smooth one.

a. What will be the appearance of the F_1?
b. What will be the appearance of the F_2?
c. What will be the appearance of the offspring of a cross of the F_1 back to its short, green, wrinkled parent?
d. What will be the appearance of the offspring of a cross of the F_1 back to its tall, yellow, smooth parent?

2.15 Two homozygous strains of corn are hybridized. They are distinguished by six different pairs of genes, all of which assort independently and produce an independent phenotypic effect. The F_1 hybrid is selfed to give an F_2.

a. What is the number of possible genotypes in the F_2?
b. How many of these genotypes will be homozygous at all six gene loci?
c. If all gene pairs act in a dominant-recessive fashion, what proportion of the F_2 will be homozygous for all dominants?
d. What proportion of the F_2 will show all dominant phenotypes?

*2.16 The coat color of mice is controlled by several genes. The agouti pattern, characterized by a yellow band of pigment near the tip of the hairs, is produced by the dominant allele *A;* homozygous *aa* mice do not have the band and are nonagouti. The dominant allele *B* determines black hairs, and the recessive allele *b* determines brown. Homozygous $c^h c^h$ individuals allow pigments to be deposited only at the extremities (e.g., feet, nose, and ears) in a pattern called Himalayan. The genotype *C–* allows pigment to be distributed over the entire body.

a. If a true-breeding black mouse is crossed with a true-breeding brown, agouti, Himalayan mouse, what will be the phenotypes of the F_1 and F_2?
b. What proportion of the black agouti F_2 will be of genotype-*Aa BB Cc^h*?
c. What proportion of the Himalayan mice in the F_2 are expected to show brown pigment?
d. What proportion of all agoutis in the F_2 are expected to show black pigment?

2.17 In cocker spaniels, solid coat color is dominant over spotted coat. Suppose a true-breeding solid-colored dog is crossed with a spotted dog, and the F_1 dogs are interbred.

a. What is the probability that the first puppy born will have a spotted coat?
b. What is the probability that if four puppies are born, all of them will have a solid coat?

3

Chromosomal Basis of Inheritance, Sex Linkage, and Mendelian Genetics in Humans

Soon after the beginning of the twentieth century, scientists realized that Mendelian laws applied to a large variety of organisms and that the principles of Mendelian analysis could be used to predict the outcome of crosses in these organisms. On Mendel's foundation early geneticists began to build genetic hypotheses that could be tested by appropriate crosses and began to investigate the nature of Mendelian factors. In this chapter we will examine the evidence showing that Mendelian factors are what we now know as genes and that genes are located in chromosomes. We will learn about various mechanisms of sex determination, some of which involve special chromosomes that are related to the sex of the organism, called sex chromosomes. We will also learn about the genetic segregation patterns of genes located in the sex chromosomes and in the other chromosomes (the autosomes) of the cell.

Chromosome Theory of Inheritance

By the end of the nineteenth century, several important discoveries had been made in the field of cytology, the scientific study of cells. Through the research of biologists, notably K. Nageli, M. J. Schleiden, and T. Schwann, many details of cell structure and cell division were known. R. Brown named and described the cell nucleus in 1833, and in the latter part of the century O. Hertwig showed that the nucleus was involved directly in sea urchin fertilization. In plants a similar fusion of sex cell nuclei was described by E. Strasburger. Strasburger also coined the terms **nucleoplasm** and **cytoplasm** for the semisolid matrix of cell material in the nucleus and in the surrounding cell body, respectively.

Toward the end of the century W. C. Schneider and W. Flemming independently showed that chromosomes within the nucleus divided longitudinally during cell division. E. van Beneden subsequently showed that the divided chro-

64
Chromosomal
Basis of
Inheritance,
Sex Linkage,
and Mendelian
Genetics
in Humans

mosomes were distributed in equal numbers to the two daughter cells.

Cytologists also firmly established that the total number of chromosomes remains constant in all cells of an organism (there are a few exceptions) except during gamete formation, when the number of chromosomes is reduced by half. They showed that the 2N chromosome number is maintained in the next generation when gametes fuse during fertilization to produce the zygote cell from which the mature organism develops. While the number of chromosomes in a cell is constant in individuals of the same species (with few exceptions), the chromosome number varies considerably between species (Table 3.1).

In 1900 Mendel's work on the nature of hereditary factors was rediscovered. Within a short time Mendel's experiments became known and appreciated throughout the scientific world. In 1903 Walter Sutton and Theodor Boveri independently recognized that the transmission of chromosomes from one generation to the next closely paralleled the pattern of transmission of genes from one generation to the next. To explain this correlation, they put forward the hypothesis that genes are on chromosomes, a proposal known as the **chromosome theory of heredity.** In the subsequent ten years many experiments by cytologists and geneticists provided irrefutable evidence for the relationship between genes and chromosomes.

Keynote *The chromosome theory of heredity states that the chromosomes are the carriers of the genes. The first clear formulation of the chromosome theory was made in 1903 by Sutton and by Boveri, who independently recognized that the transmission of chromosomes from one generation to the next closely paralleled the pattern of transmission of genes from one generation to the next.*

Sex Chromosomes

The proof of the chromosome theory of heredity came from experiments that related the hereditary behavior of particular genes to the transmis-

Table 3.1 Chromosome Number in Various Organisms[a]

Organism	Total Chromosome Number
Human	46
Chimpanzee	48
Rhesus monkey	42
Dog	78
Cat	72
Horse	64
Mallard	80
Chicken	78
Alligator	32
Cobra	38
Bullfrog	26
Toad	36
Goldfish	94
Starfish	36
Fruit fly	8
Housefly	12
Mosquito	6
Australian ant (myrecia pilosula)	♂ 1, ♀ 2
Nematode	11 ♂, 12 ♀
Neurospora (haploid)	7
Sphagnum moss (haploid)	23
Field horsetail	216
Giant sequoia	22
Tobacco	48
Cotton	52
Kidney bean	22
Broad bean	12
Onion	16
Potato	48
Tomato	24
Bread wheat	42
Rice	24
Baker's yeast	34

[a] Except as noted, all chromosome numbers are for diploid cells.

sion of the **sex chromosome,** the chromosome in eukaryotic organisms that is represented differently in the two sexes. The discovery of sex chromosomes will be reviewed here as a prelude to discussing the experimental proof of the Sutton-Boveri hypothesis.

In the early 1900s, C. E. McClung, N. Stevens, and E. B. Wilson, all of whom worked with insects, independently obtained evidence that particular chromosomes determined the sex of an organism. In 1901 McClung proposed that a particular chromosome, which he called the accessory chromosome, in grasshoppers and other members of the order Orthoptera was involved in determining each insect's sex at fertilization. In 1905 Stevens and Wilson provided more information about McClung's proposal. They found that in some Orthoptera the female has an even number of chromosomes while the male has an odd number. One of the chromosomes is present in two copies in the female but in only one copy in the male. It was this extra chromosome in the female that McClung had hypothesized as the accessory chromosome. Stevens called the extra chromosome an **X chromosome.** Since there is a relationship between this chromosome and the sex of the organism, the X chromosome is an example of a sex chromosome.

When female grasshoppers form eggs by meiosis, each egg receives one of each chromosome, including an X chromosome. Half the male's sperm cells produced by meiosis receive an X chromosome and the other half do not. Consequently, the sex of the progeny grasshoppers is determined by the chromosome composition of the sperm that fertilizes the egg. If the sperm carries an X chromosome, then the resulting fertilized egg will have a pair of X's, and the individual that develops from the zygote will be female. If the sperm does not have an X, the fertilized egg will have an unpaired X and the individual will be a male.

Not all insects have an unpaired sex chromosome as do grasshoppers. For example, Stevens found that in the common mealworm, *Tenebrio molitor,* the male has a partner chromosome for the X chromosome that is much smaller and

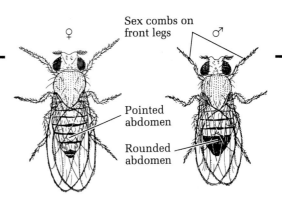

Figure 3.1 Female (left) *and male* (right) Drosophila melanogaster *(fruit fly), an organism used extensively in experiments of geneticists.*

clearly distinguishable from the X chromosome. Stevens called the partner chromosome the **Y chromosome,** and like the X chromosome, it is a sex chromosome. The sperm cells of the mealworm contain either an X or a Y chromosome, and the sex of the offspring is determined by the sperm type that fertilizes the X chromosome-bearing egg: XX mealworms are female and XY mealworms are male.

Stevens and Wilson observed similar X–Y sex chromosome complements in other organisms. Perhaps the most important of these organisms to geneticists is the fruit fly, *Drosophila melanogaster.* Figure 3.1 shows the appearance of male and female *Drosophila,* Figure 3.2 shows the chromosome complements of the two sexes, and

Figure 3.2 Chromosomes of D. melanogaster *diagramed to show their morphological differences. A female has four pairs of chromosomes in her somatic cells, including a pair of X chromosomes. The only difference in the male is an XY pair of sex chromosomes instead of two X's.*

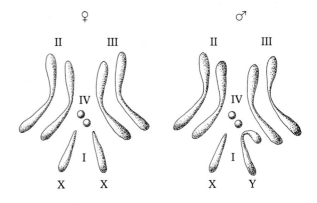

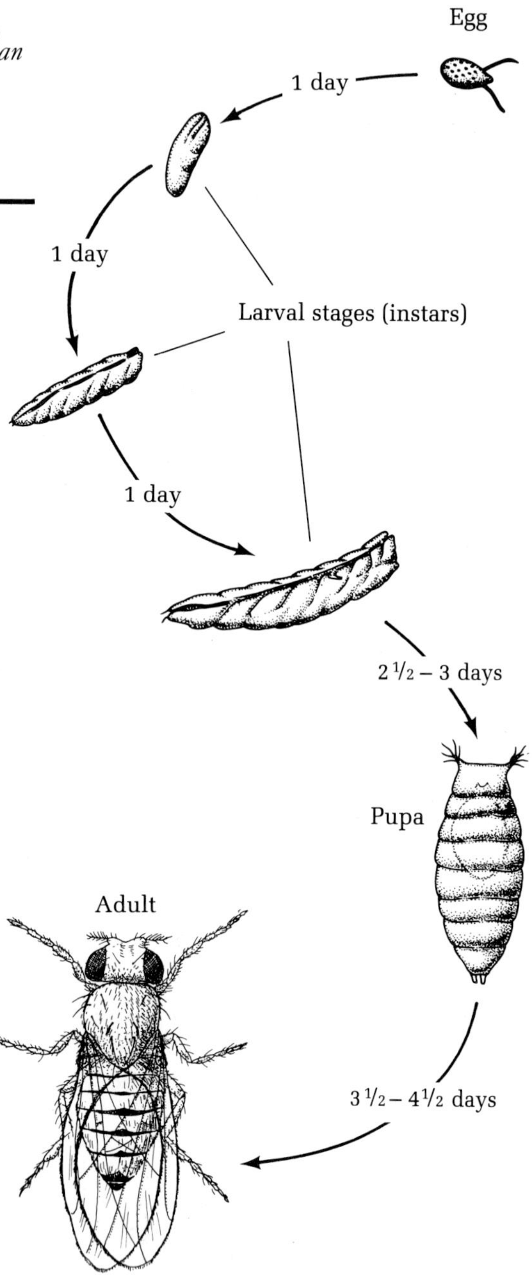

66
*Chromosomal
Basis of
Inheritance,
Sex Linkage,
and Mendelian
Genetics
in Humans*

Egg

1 day

1 day

Larval stages (instars)

1 day

$2^{1}/_{2}$ – 3 days

Pupa

Adult

$3^{1}/_{2}$ – $4^{1}/_{2}$ days

Figure 3.3 Life cycle of D. melanogaster.

Figure 3.3 shows the life cycle. Because the male produces two kinds of gametes with respect to sex chromosome content (X or Y) and the female produces only one gametic type (X), the male is called the **heterogametic sex** and the female is called the **homogametic sex.** The X and Y chromosomes are similar in size but their shapes are different.

As shown in Figure 3.4, the pattern of transmission of the X and the Y chromosomes from generation to generation is very straightforward. The production of the F_1 generation is shown in

Figure 3.4a, and the F_2 generation in Figure 3.4b. In this figure the X is indicated by a vertical straight line and the Y by a wavy line. (*Note:* In some organisms the male is the homogametic sex and the female is the heterogametic sex. This reversal will be discussed later in the chapter.)

Keynote *A sex chromosome is a chromosome in eukaryotic organisms that is represented differently in the two sexes. In many of the organisms encountered in genetic studies, one sex possesses a pair of identical chromosomes (the X chromosomes). The opposite sex possesses a pair of visibly different chromosomes: one is an X chromosome, and the other is structurally and functionally different and is called the Y chromosome. Commonly, the XX sex is female and the XY sex is male. The XX and XY sexes are called the homogametic and heterogametic sexes, respectively.*

Sex Linkage

The conclusive evidence to support the chromosome theory of heredity was obtained by using *Drosophila* (fruit flies) in genetic experiments. In 1910 Thomas Hunt Morgan reported his results of crossing experiments with the fruit fly.

Morgan found a male fly in one of his true-breeding stocks that had white eyes instead of the bright red eyes that are characteristic of the **wild type.** The term *wild type* refers to a strain, organism, or gene that is predominant in the wild population. For example, a *Drosophila* strain with all wild-type genes will have bright red eyes. Variants of a wild-type strain arise from mutational changes of the wild-type genes that produce **mutant alleles;** the result is strains with mutant characteristics. Mutant alleles may be recessive or dominant to the wild-type allele; for example, the mutant allele that causes white eyes in *Drosophila* is recessive to the wild-type (red-eye) allele.

When Morgan crossed the white-eyed male with a red-eyed female from the same stock, he found that all the F_1 flies were red-eyed and concluded that the white-eyed trait was reces-

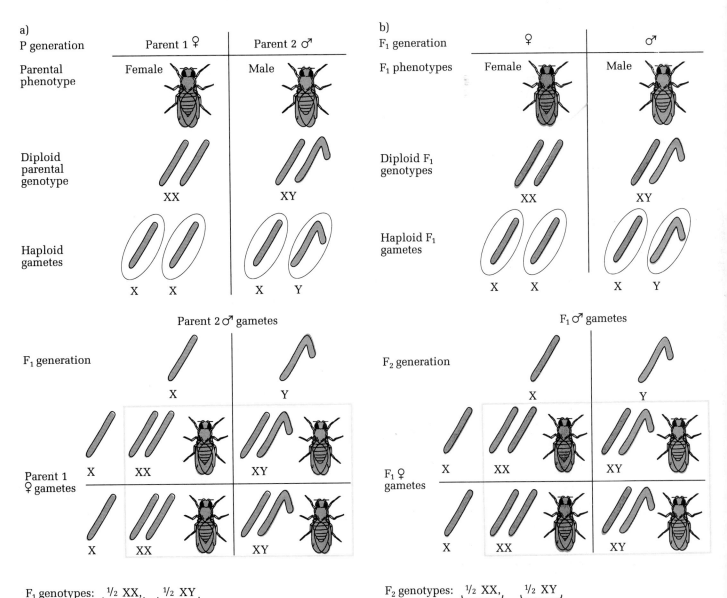

a)

P generation | Parent 1 ♀ | Parent 2 ♂

Parental phenotype: Female — Male

Diploid parental genotype: XX — XY

Haploid gametes: X X — X Y

Parent 2 ♂ gametes

F₁ generation

Parent 1 ♀ gametes

	X	Y
X	XX	XY
X	XX	XY

F₁ genotypes: ½ XX, ½ XY
F₁ phenotypes: ½ female, ½ male

b)

F₁ generation | ♀ | ♂

F₁ phenotypes: Female — Male

Diploid F₁ genotypes: XX — XY

Haploid F₁ gametes: X X — X Y

F₁ ♂ gametes

F₂ generation

F₁ ♀ gametes

	X	Y
X	XX	XY
X	XX	XY

F₂ genotypes: ½ XX, ½ XY
F₂ phenotypes: ½ female, ½ male

Figure 3.4 Inheritance pattern of X and Y chromosomes in organisms where the female is XX and the male is XY: (a) Production of F₁ generation; (b) production of F₂ generation.

sive. Next, he allowed the F₁ progeny to interbreed and counted the phenotypic classes in the F₂ generation. There were 3470 red-eyed and 782 white-eyed flies. The number of individuals with the recessive phenotype was too small to fit the Mendelian 3:1 ratio. In addition, Morgan noticed that all the white-eyed flies were male. (Later, he determined that the deficiency of flies

68
*Chromosomal
Basis of
Inheritance,
Sex Linkage,
and Mendelian
Genetics
in Humans*

with the recessive phenotype occurs because flies with white eyes are not as viable as those with red eyes.)

Next, Morgan crossed a white-eyed male fly with one of its red-eyed daughters in order to find out more about the inheritance of the white trait. The result was 129 red-eyed females, 132 red-eyed males, 88 white-eyed females, and 86 white-eyed males. The closest ratio to the observed results is 1:1:1:1, with the deviation from this ratio again being the result of decreased viability of white-eyed flies. The results showed that females could have white eyes and that approximately equal numbers of red- and white-eyed flies occur in both sexes.

Figure 3.5 diagrams the first set of crosses, which produced the 3:1 ratio of red-eyed:white-eyed flies in the F_2. The *Drosophila* gene symbolism used is described in Box 3.1. Note as we go through the discussions the mother-son inheritance pattern that we shall see is the result of the segregation of genes located on a sex chromosome.

In interpreting the data from his two crosses, Morgan proposed that the gene for the eye color variant is located on the X chromosome and that there is no equivalent gene on the Y chromosome. The condition of X-linked genes in males is said to be **hemizygous** since they have no allele on the Y and only half the zygotes receive an X chromosome. For example, the white-eyed *Drosophila* males have an X chromosome with a white allele and no other allele of that gene in the gene complement: These males are said to be hemizygous for the white allele. Since the white allele of the gene is recessive, the original white-eyed male must have had the recessive allele for white eyes (designated w; see Box 3.1) in his X chromosome. The red-eyed female came from a true-breeding stock, so both of her X chromosomes must have carried the dominant allele for red eyes, w^+.

The F_1's are produced in the following way (Figure 3.5a): The males receive their only X chromosome from their mother and hence have the w^+ allele and are red-eyed. The females receive a dominant w^+ allele from their mother and a recessive w allele from their father. As a

consequence, the F_1 females are also red-eyed.

To produce the F_2 flies, Morgan interbred F_1 red-eyed males and red-eyed females (Figure 3.5b). In the F_2 the males that received an X chromosome with the w allele from their mothers are white-eyed; those that received an X chromosome with the w^+ allele are red-eyed.

From these results we can interpret Morgan's second cross, the cross of a white-eyed male with its red-eyed daughter that gave an approximately 1:1:1:1 ratio in the progeny of red-eyed females, red-eyed males, white-eyed females, and white-eyed males (Figure 3.6, p. 72). The white-eyed male produces sperm that either is Y bearing or contains an X chromosome with the recessive w allele. As a result, the eye phenotype of the progeny is determined by the genetic makeup of the egg. This gene transmission from a male parent to a female child to a male grandchild is called **crisscross inheritance.**

Morgan did one more cross in this series, crossing a red-eyed male with a true-breeding white-eyed female (Figure 3.7, p. 73). (This cross is the *reciprocal cross* of the first cross Morgan performed—white male × red female.) The parental male here is hemizygous for the w^+ allele, and the parental female is homozygous for the w allele. All the F_1 females receive a w^+-bearing X from their father and a w-bearing X from their mother (Figure 3.7a). Consequently, they are heterozygous w^+/w and have red eyes. All the F_1 males receive a w-bearing X from their mother and a Y from their father, and so they have white eyes (Figure 3.7a). This result is distinct from that of the first cross diagramed in Figure 3.5. In addition, note that these results are different from the normal results of a reciprocal cross because of the sex linkage.

Interbreeding of the F_1's (Figure 3.7b) follows the same pattern as the testcross diagramed in Figure 3.6 since the male is w/Y and the female is w^+/w. Thus an approximately equal number of male and female red- and white-eyed flies appear in the F_2. This ratio differs from the results obtained in the first cross, where an approximately 3:1 ratio of red-eyed:white-eyed flies was obtained and where none of the females and approximately half the males exhibited the white-

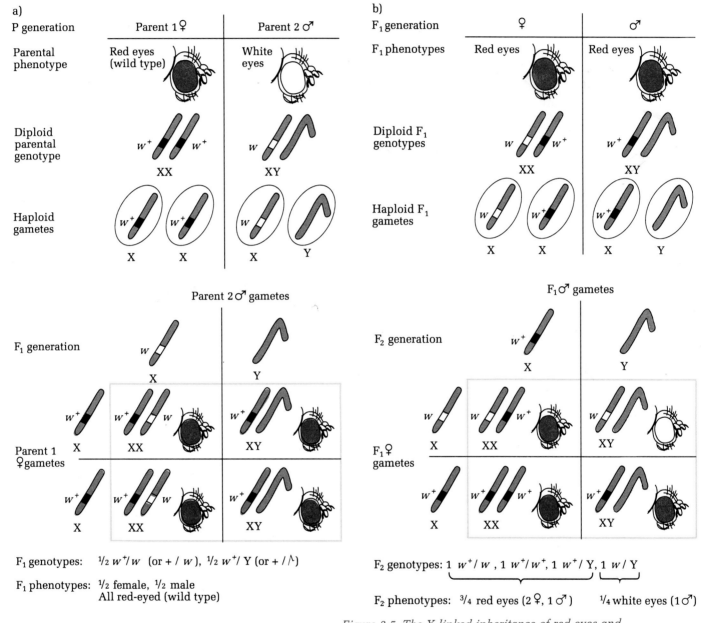

a)

	Parent 1 ♀	Parent 2 ♂
P generation		
Parental phenotype	Red eyes (wild type)	White eyes
Diploid parental genotype	w^+ w^+ XX	w XY
Haploid gametes	w^+ X w^+ X	w X Y

Parent 2 ♂ gametes

F₁ generation

F₁ genotypes: ½ w^+/w (or $+/w$), ½ $w^+/$ Y (or $+/$ ∧)

F₁ phenotypes: ½ female, ½ male
All red-eyed (wild type)

b)

	♀	♂
F₁ generation		
F₁ phenotypes	Red eyes	Red eyes
Diploid F₁ genotypes	w w^+ XX	w^+ XY
Haploid F₁ gametes	w X w^+ X	w^+ X Y

F₁♂ gametes

F₂ generation

F₂ genotypes: 1 w^+/w, 1 w^+/w^+, 1 $w^+/$ Y, 1 $w/$ Y

F₂ phenotypes: ¾ red eyes (2 ♀, 1 ♂)   ¼ white eyes (1 ♂)

Figure 3.5 The X-linked inheritance of red eyes and white eyes in D. melanogaster. *The symbols* w *and* w⁺ *indicate the white- and red-eyed alleles, respectively. Here a red-eyed female is crossed with a white-eyed male (a), and the F₁ flies are interbred to produce the F₂'s (b).*

eyed phenotype. The difference in phenotypic ratios in the two sets of crosses reflects the transmission patterns of sex chromosomes and the genes they contain.

Morgan's crosses of *Drosophila* involved eye color characteristics that we now know are

70
*Chromosomal
Basis of
Inheritance,
Sex Linkage,
and Mendelian
Genetics
in Humans*

*Box 3.1
Genetic
Symbols
Revisited*

The gene symbols used for *Drosophila* are different from those used for peas in Chapter 2. This *Drosophila* symbolism is commonly used in genetics today. In this system the symbol + indicates a wild-type allele of a gene; that is, the allele most frequently found in natural populations of the organism. A lowercase letter designates mutant alleles of a gene that are recessive to the wild-type allele, and an uppercase letter is used for alleles that are dominant to the wild-type allele. The letters are chosen on the basis of the phenotype of the organism expressing the mutant allele. For example, a variant strain of *Drosophila* has bright scarlet eyes instead of the usual brick red. The allele in-

volved is recessive to the wild-type brick red allele, and because the eye color is close to vermilion in tint, the allele is designated v and is called the vermilion allele. The wild-type allele of v is v^+, but when there is no chance of confusing it with other genes in the cross, it is often shortened to $+$.

A conventional way to represent the chromosomes (instead of the way we have been using in the figures) is to use the slash (/). Thus v^+/v or $+/v$ indicates that there are *two* homologous chromosomes, one with the wild-type allele (v^+ or $+$) and the other with the recessive allele (v). The Y chromosome is usually symbolized as a Y or a bent slash ($\rightarrow$). Thus Morgan's cross of a true-breeding red-eyed female fly with a white-eyed male could be written $w^+/w^+ \times w/Y$ or $+/+ \times w/\rightarrow$.

The same rules apply when alleles that are dominant to the wild-type allele are involved. For instance, some *Drosophila* called *bar* mutants have bar-shaped eyes instead of the normal round eyes (Box Figure 1), because the shape of the eye is related to the number of facets, and the bar mutation results in a reduced number of facets compared with the wild type. In the heterozygote the allele for bar-shaped eye is dominant although not completely so, such that females homozygous for *bar* have much smaller eyes than females heterozygous for *bar*. The symbol for this allele is $B,$ and the wild-type allele is B^+, or $+$ in the shorthand version. The *bar* gene is located on the X chromosome, and there is no allele on the Y. Note that because fewer facets are produced, the eye is much smaller in the hemizygous males and homozygous B/B females than in the B/B^+ (i.e., $B/+$) heterozygous females. Box Figure 2 illustrates the inheritance of the bar-shaped eye characteristic in reciprocal crosses. These crosses show that the bar-eye phenotype is determined by an X-linked gene and that no equivalent allele is present in the Y chromosome.

a) Round   b) Bar-eye
   eye   female

c) Bar-eye   d) Wide
   male   bar-eye
                 (female)

Box Figure 1 Eye of Drosophila melanogaster, *a typical insect compound eye made up of a large number of facets: (a) round eye of wild-type flies; (b) bar-eye (homozygous B/B) female; (c) bar-eye male; (d) intermediate wide-bar-eye phenotype in heterozygous (B/B $^+$) females.*

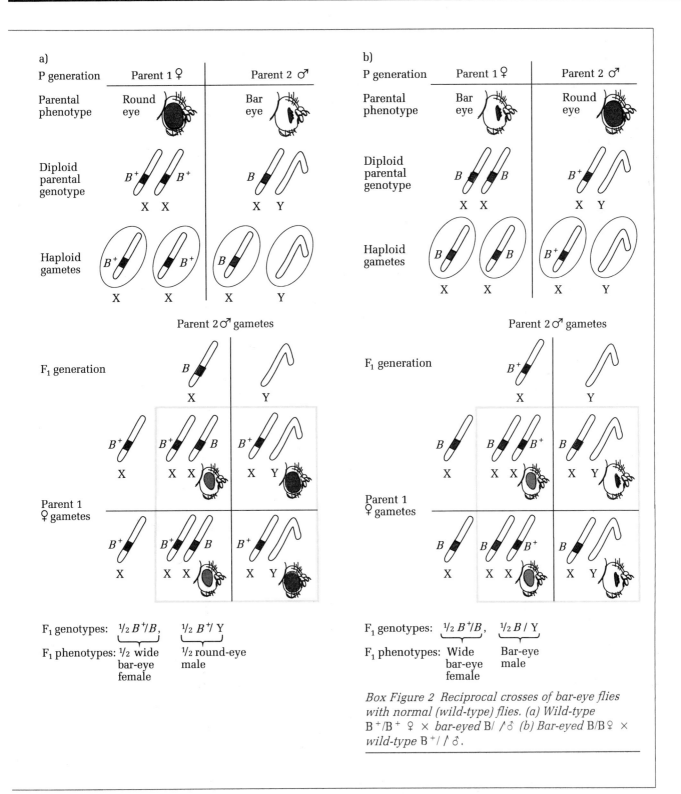

Box Figure 2 *Reciprocal crosses of bar-eye flies
with normal (wild-type) flies. (a) Wild-type
B⁺/B⁺ ♀ × bar-eyed B/ /♂ (b) Bar-eyed B/B♀ ×
wild-type B⁺/ / ♂.*

72
*Chromosomal
Basis of
Inheritance,
Sex Linkage,
and Mendelian
Genetics
in Humans*

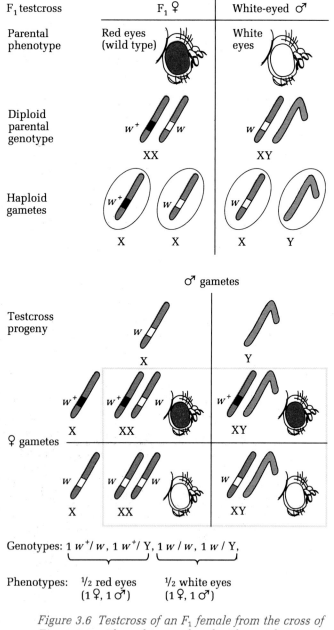

Genotypes: 1 w^+/w, 1 $w^+/$ Y, 1 w/w, 1 $w/$ Y,

Phenotypes: ½ red eyes   ½ white eyes
(1♀, 1♂)   (1♀, 1♂)

Figure 3.6 Testcross of an F_1 female from the cross of Figure 3.5 with a white-eyed male. The result is a 1:1 ratio of red-eyed:white-eyed flies in each sex.

coded for by genes found on the X chromosome. These characteristics and the genes that give rise to them are referred to as **sex-linked,** or, more correctly, **X-linked.** *Sex-linked inheritance is the* term used for the pattern of hereditary transmis-

sion of sex-linked genes. When the results of reciprocal crosses are not the same, sex-linked characteristics may well be involved. By comparison, the results of reciprocal crosses are the same when they involve genes located on the **autosomes** (chromosomes other than the sex chromosomes). Most importantly, Morgan's results strongly supported the hypothesis that genes were located on chromosomes, although his data only showed a strong correlation between the behavior of genes and chromosomes and did not prove the Sutton-Boveri hypothesis. Morgan found many other examples of genes on the X chromosome in *Drosophila* and in other organisms, thereby showing that his observations were not confined to a single species.

Keynote **Sex linkage** *is the association of genes with the sex-determining chromosomes of eukaryotes. Such genes, as well as the phenotypic characteristics these genes control, are referred to as sex-linked. Morgan's pioneering work with the inheritance of sex-linked genes of* Drosophila *strongly supported, but did not prove, the chromosome theory of heredity.*

Nondisjunction of X Chromosomes

While Morgan's results strongly supported the chromosome theory of inheritance, the actual proof for the theory came from the work of Morgan's student, Calvin Bridges, who was carrying out basic genetic and cytological studies with *Drosophila.* Morgan's work showed that all the F_1 males from a cross of a white-eyed female (w/w ♀) with a red-eyed male ($w^+/$Y ♂) should be white-eyed and all the females should be red-eyed. Bridges found that there are rare exceptions to this result: About 1 in 2000 of the F_1 flies from such a cross are either white-eyed females or red-eyed males.

To explain these data, Bridges hypothesized that occasionally the two X chromosomes fail to separate in meiosis, so eggs can be produced either with two X chromosomes or with no X chromosomes instead of the usual one. This phenomenon is called **X chromosome nondis-**

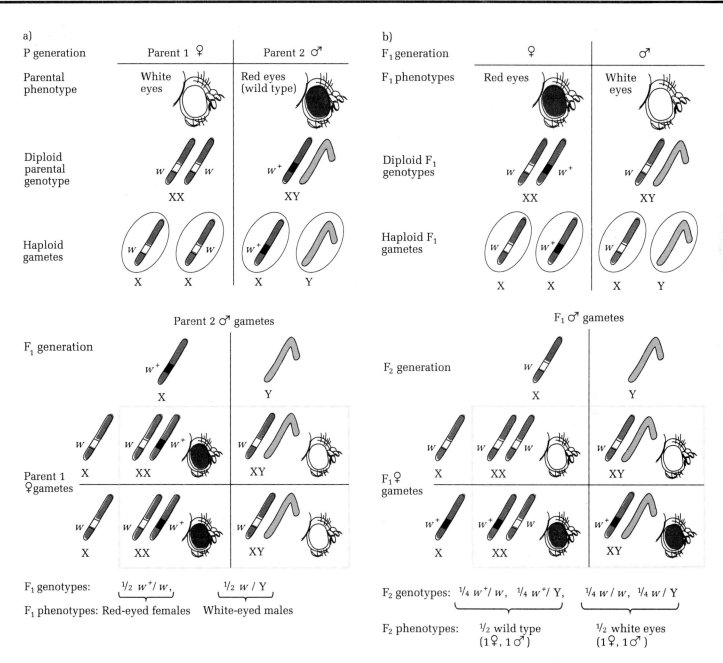

a)

| P generation | Parent 1 ♀ | Parent 2 ♂ |

Parental phenotype: White eyes / Red eyes (wild type)

Diploid parental genotype: w w XX / w^+ XY

Haploid gametes: w X, w X / w^+ X, Y

F₁ generation — Parent 2 ♂ gametes: w^+ X, Y

Parent 1 ♀ gametes: w X, w X

F₁ genotypes: ½ w^+/w, ½ w/Y
F₁ phenotypes: Red-eyed females   White-eyed males

b)

| F₁ generation | ♀ | ♂ |

F₁ phenotypes: Red eyes / White eyes

Diploid F₁ genotypes: w w^+ XX / w XY

Haploid F₁ gametes: w X, w^+ X / w X, Y

F₂ generation — F₁ ♂ gametes: w X, Y

F₁ ♀ gametes: w X, w^+ X

F₂ genotypes: ¼ w^+/w, ¼ w^+/Y,   ¼ w/w, ¼ w/Y
F₂ phenotypes: ½ wild type (1♀, 1♂)   ½ white eyes (1♀, 1♂)

junction (Figure 3.8), and when it occurs in an individual with a normal set of chromosomes, it is called **primary nondisjunction.** Figure 3.9 shows how primary nondisjunction of the X chromosomes can explain the exceptional flies in Bridges's first experimental cross.

When primary nondisjunction occurs in the *w/w* female (Figure 3.9) two classes of excep-

Figure 3.7 Reciprocal cross of that shown in Figure 3.5. A homozygous white-eyed female is crossed with a red-eyed (wild-type) male (a), and the F₁ flies are interbred to produce the F₂'s (b). This cross gives different results from the cross shown in Figure 3.5 because of the way the sex chromosomes segregate in crosses.

74
*Chromosomal
Basis of
Inheritance,
Sex Linkage,
and Mendelian
Genetics
in Humans*

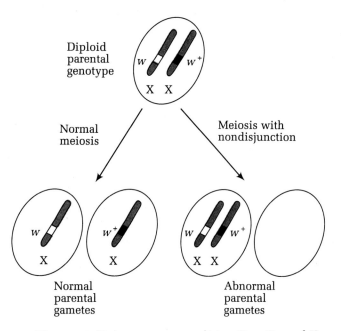

Diploid
parental
genotype

Normal
meiosis

Meiosis with
nondisjunction

Normal
parental
gametes

Abnormal
parental
gametes

Figure 3.8 X chromosome nondisjunction. Normal X chromosome segregation is shown on the left.

tional eggs result with equal frequency: those with two X chromosomes and those with no X chromosomes. The male is w^+/Y and produces equal numbers of w^+- and Y-bearing sperm. When they are fertilized by the two types of sperm, four types of zygotes are produced. Two of these types usually do not survive: the Y0 class, which has a Y chromosome but no X chromosome, and the triplo-X (XXX) class, which has three X chromosomes and, in this case, the genotype $w/w/w^+$. The former die because they lack the X chromosome and its genes that code for essential cell functions. The latter often die because the flies apparently cannot function with three doses of each of the X chromosome genes, at least under standard culture conditions. The surviving classes are the red-eyed X0 males, with no Y chromosome and a w^+ allele on the X, and the white-eyed XXY females, with a w allele on each X. The males are red-eyed because they received their X chromosome from their fathers, and the females have white eyes because their two X chromosomes came from their mothers. This result is unusual

since sons normally get their X from their mothers, and daughters get one X from each parent.

Bridges's hypothesis was very plausible and readily testable by examining the chromosome composition of the exceptional flies. This cytological examination showed that, indeed, the white-eyed females were XXY and the red-eyed males were X0.

Bridges found that XXY *Drosophila* females are normal and fertile, and to test his hypothesis further, he crossed the exceptional white-eyed XXY females obtained in the F_1 of the above cross with normal red-eyed XY males (Figure 3.10). The XXY female is homozygous for the w allele on her two X's. The male has the w^+ allele on his X. Both parents have no equivalent eye color allele on their Y chromosomes. Again, flies with unexpected phenotypes resulted from this cross: About 4 percent of the male progeny had red eyes and about 4 percent of the female progeny had white eyes.

To account for these unusual phenotypes, Bridges hypothesized that segregation of chromosomes in the meiosis of an XXY female can occur in two ways. Most commonly, the two X chromosomes separate and migrate to opposite poles with one of them accompanied by the Y, producing equal numbers of X- and XY-bearing eggs. This result, normal disjunction, is what should happen with respect to the X's during meiosis; it occurs 96 percent of the time (Figure 3.10a). In the second way, which takes place the remaining 4 percent of the time, there is nondisjunction of the X's (Figure 3.10b). Bridges called this segregation **secondary nondisjunction** because it occurred in the progenies of females that were produced by primary nondisjunction. Secondary nondisjunction results in the two X's migrating together to one pole and the Y migrating to the other; the eggs are XX and Y. When these eggs are fertilized by the two classes of sperm (X and Y), the two surviving classes are the exceptional red-eyed males and white-eyed females. Bridges verified his secondary nondisjunction hypothesis by microscopic examination of the chromosomes of the flies collected from the cross.

In conclusion, Bridges's experiments showed

without any doubt that a specific gene was associated with a specific chromosome. These results were taken as proof of the chromosome theory of heredity.

Keynote *Bridges observed an unexpected pattern of inheritance of an X-linked mutant gene in* Drosophila. *He correlated this pattern directly with a rare event during meiosis, called nondisjunction, in which members of a homologous pair of chromosomes do not segregate to the opposite poles. This association of a specific gene with a specific chromosome was taken as definite proof for the chromosome theory of heredity.*

Sex Determination and Sex Linkage in Eukaryotic Systems

In this section some of the mechanisms for sex determination will be discussed. In *genotypic sex determination systems,* sex is governed by the genotype of the zygote or spores; in *phenotypic* (or *environmental*) *sex determination systems,* sex is governed by internal and external environmental conditions.

Genotypic Sex Determination Systems

Genotypic sex determination, in which the sex chromosomes play a decisive role in the inheritance and determination of sex, may occur in one of two ways. In the **X chromosome–autosome balance system** (seen in *Drosophila*), the main factor in sex determination is the ratio between the number of X chromosomes and the number of autosomes. Sex is determined at the time of fertilization, and sex differences are assumed to be due to the action during development of two sets of genes located on the X chromosomes and on the autosomes. In this system the Y chromosome has no effect on sex determination.

In the other system (seen in humans), the Y chromosome of the heterogametic sex is active

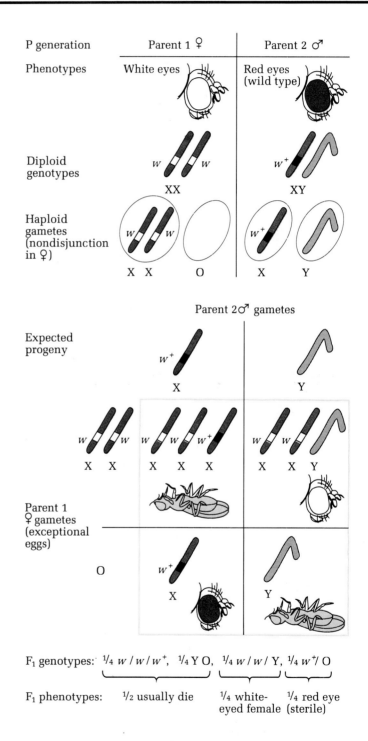

Figure 3.9 Primary nondisjunction during meiosis in a white-eyed female D. melanogaster.

76
*Chromosomal
Basis of
Inheritance,
Sex Linkage,
and Mendelian
Genetics
in Humans*

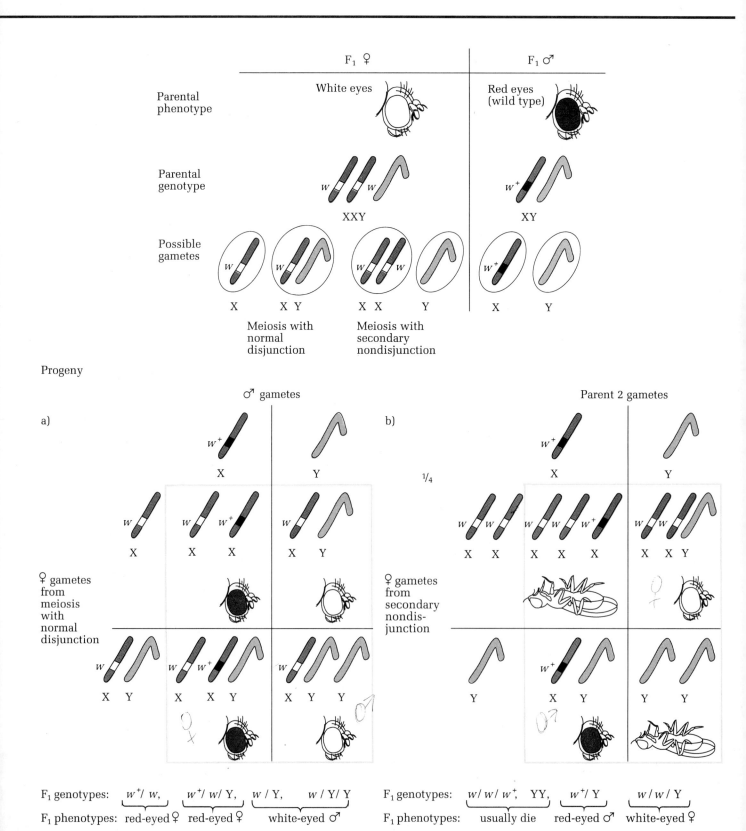

F₁ ♀   F₁ ♂

Parental phenotype — White eyes / Red eyes (wild type)

Parental genotype — w w / w^+

XXY / XY

Possible gametes:

X   X Y   X X   Y / X   Y

Meiosis with normal disjunction   Meiosis with secondary nondisjunction

Progeny

♂ gametes   Parent 2 gametes

a)   b)

w^+ X   Y   ¼ w^+ X   Y

♀ gametes from meiosis with normal disjunction

♀ gametes from secondary nondis-junction

F₁ genotypes: w^+/w, $w^+/w/Y$, w/Y, $w/Y/Y$

F₁ phenotypes: red-eyed ♀ red-eyed ♀ white-eyed ♂

F₁ genotypes: $w/w/w^+$, YY, w^+/Y $w/w/Y$

F₁ phenotypes: usually die red-eyed ♂ white-eyed ♀

in determining the sex of an individual. Individuals carrying the Y chromosome are of a definite sex, irrespective of their autosomal genotype. In short, the sex of the embryo is determined at the time of fertilization by the sex chromosomes contributed by the sperm and the ovum.

Sex determination in *Drosophila*. *Drosophila melanogaster* has four pairs of chromosomes: three pairs of autosomes and one pair of sex chromosomes. In this organism the homogametic sex is the female (XX) and the heterogametic sex is the male (XY). However, the sex of the fly is not the consequence of the presence or absence of a Y chromosome: An XXY fly is female and an X0 fly is male. In fact, the sex of the fly, as many studies have shown, is determined by the ratio of the number of X chromosomes (X) to the number of sets of autosomes (A). *Drosophila* is the prototype of the *X chromosome-autosome balance system of sex determination.*

Table 3.2 presents some chromosome complements and the sex of the resulting flies. In a normal female there are two X's and two sets of autosomes; hence the X:A ratio is 1.00. A normal male has a ratio of 0.50. If the X:A ratio is greater than or equal to 1.00, the fly will be female; if the X:A ratio is less than or equal to 0.50, the fly will be male. If the ratio is between 0.50 and 1.00, the fly is neither male nor female; it is an intersex. Intersex flies are variable in appearance, having, in general, complex mixtures of male and female attributes for the internal sex organs and external genitalia. Such flies are sterile.

Apparently, in *Drosophila* and other organisms with this mechanism of sex determination, there are sex-determining genes on the X chromosome and on the autosomes that operate in a counteracting way during development. The two sets of genes are not equally effective; maleness

genes are more effective on the autosomes (individuals with one X and two autosome sets are male), while femaleness genes are more numerous or more effective on the X chromosomes (individuals with two X chromosomes and two autosome sets are female). The Y chromosome has no influence on sex determination, although it is required for directing the fertility of the sperm.

Sex determination in mammals. In contrast, in humans and other mammals the presence of the Y chromosome determines maleness, and its absence determines femaleness. This mode of sex determination is called the *chromosomal mechanism of sex determination.* Evidence for this mechanism came from studies of humans and other mammals that have an abnormal sex chromosome complement as a result of meiotic nondisjunction. While sex determination in these cases follows directly from the sex chromosomes present, those individuals who have unusual chromosome complements display many unusual characteristics.

Table 3.2 Sex Balance Theory of Sex Determination in Drosophila melanogaster

Sex Chromosome Complement	Autosome Complement (A)	X:A Ratio[a]	Sex of Flies
XX	AA	1.00	♀
XY	AA	0.50	♂
XXX	AA	1.50	Metafemale (sterile)
XXY	AA	1.00	♀
XXX	AAAA	0.75	Intersex
XXXX	AAA	1.33	Metafemale (sterile)
XX	AAA	0.67	Intersex
X	AA	0.50	♂ (sterile)

[a] If the X chromosome:autosome ratio is greater than or equal to 1.00 (X:A ≥ 1.00), the fly will be a female. If the X chromosome:autosome ratio is less than or equal to 0.50 (X:A ≤ 0.50), the fly will be male. Between these two ratios, the fly will be an intersex.

Figure 3.10 Results of a cross between the exceptional white-eyed XXY female of Figure 3.9 with a normal red-eyed XY male. (a) Normal disjunction of the X chromosomes in the XXY female. (b) Secondary nondisjunction of the X chromosomes in the XXY female.

78

*Chromosomal
Basis of
Inheritance,
Sex Linkage,
and Mendelian
Genetics
in Humans*

Nondisjunction can produce, for example, exceptional X0 individuals that have an X but no Y chromosome. In *Drosophila,* X0 flies are male if there is a normal set of autosomes, but in humans these individuals are female and sterile, and they exhibit **Turner's syndrome.** These females have few noticeable major defects until puberty, when they fail to develop secondary sexual characteristics. They tend to be shorter than average, and they have weblike necks, poorly developed breasts, and immature internal sexual organs. Typically, they exhibit mental deficiencies and are usually infertile. Nondisjunction can also result in XXY humans, which are male and are said to have **Klinefelter's syndrome.** Many of the affected males are mentally deficient, have underdeveloped testes, and are taller than average.

Table 3.3 summarizes the consequences of exceptional X and Y chromosome complements in humans. In each case two sets of autosomes are associated with the sex chromosomes. With rare exceptions mammals cannot tolerate any variation in the number of autosomes; mammals with an unusual number of autosomes usually die. Furthermore, a mechanism in mammals com-

pensates for X chromosomes in excess of the normal complement but not for extra autosomes. The excess X chromosomes are observed to be cytologically condensed and are not active in guiding the formation of phenotypic characteristics. This so-called **lyonization** of the excess X chromosomes (named after Mary Lyon, who first described this mechanism) will be explained more fully in Chapter 17.

Proof that the Y chromosome plays an active role in the determination of sex in humans and other mammals was obtained in 1955 when E. J. Eichwald and C. R. Silmser set out to see whether they could detect differences between males and females that could be related to sex chromosomes. The background for their experiments is as follows: Individuals that are genetically very similar can exchange skin grafts or other organ grafts without having antibodies formed to counter the grafts. (An **antibody** is a protein molecule that recognizes and binds to a foreign substance introduced into the organism. Any large molecule that stimulates the production of specific antibodies or that binds specifically to an antibody is called an **antigen.**) The grafts are not recognized as foreign since the cell surfaces have the same antigens as the cells of the recipient organism. Strains of experimental organisms have been produced by inbreeding them (making brother-sister matings) for so many generations that the individuals have essentially the same genotype. These strains are called **isogenic.**

Eichwald and Silmser reasoned that the only genetic differences between isogenic brothers and sisters would be the result of the sex chromosome differences between the two sexes. They transplanted skin from females of an inbred mouse strain to male mice of the same strain and vice versa. In the former there was no rejection of the transplant, but in the latter the skin transplanted to the female mice was eventually rejected. They concluded that there must have been a functional gene (or genes) in males that produced an antigen unique to males and to which the female made antibodies. Since the only genetic difference in males and females of the strain was the presence of the Y chromo-

Table 3.3 Consequences of Various Numbers of X and Y Chromosome Abnormalities in Humans, Showing Role of the Y in Sex Determination

Chromosome Constitution[a]	*Designation of Individual*
46, XX	Normal ♀
46, XY	Normal ♂
45, X	Turner's syndrome ♀
47, XXX	Triplo-X ♀
48, XXXX	Tetra-X ♀
47, XXY	Klinefelter's syndrome ♂
48, XXXY	Klinefelter's syndrome ♂
48, XXYY	Klinefelter's syndrome ♂
47, XYY	XYY ♂

[a] The first number indicates the total number of chromosomes in the nucleus, and the X's and Y's indicate the sex chromosome complement.

some in the males, they reasoned that the antigen was coded for by Y-linked genes.

The acceptance of tissue grafts is called **histocompatibility,** and the Y-linked gene is said to have a histocompatibility locus, named the *H–Y locus.* The product of this locus is the **H–Y antigen.** Years after Eichwald and Silmser did their experiments, E. Goldberg discovered that the H–Y antigen, as defined by antisera (blood serum containing antibodies specific to some particular antigen) rather than by transplantation, is conserved in evolution and that it is present at the earliest stages of embryo development, even in an eight-cell mouse embryo, for example. In 1975, S. Ohno proposed the H–Y hypothesis, that the H–Y antigen is the diffusible substance that results in testes formation and Wolffian duct development in mammals.

There are some data, however, that cast doubt on the hypothesized role of the H–Y antigen. A. McLaren found a mouse strain that had two X chromosomes but was male because part of a Y chromosome had broken off and had become attached (translocated) to one of the X chromosomes. These mice developed normal testes. She found that these mice did not make the H–Y antigen and concluded that instead of determining testes formation the H–Y antigen might instead play a role in spermatogenesis since the mice did not produce sperm.

More recently D. Page looked for situations in which the chromosomes and the gonads did not match or in which there was an abnormal Y chromosome. By studying patients who were attending infertility clinics he found the human equivalent of McLaren's mice. Among the patients he analyzed were 6 females who were XY instead of the normal XX and 20 normal-looking males who had two X chromosomes instead of the normal XY. The unusual females were found to have deletions of small segments of the short arm of Y chromosome (because of its centromere position, the Y chromosome is divided into two uneven portions, a short arm and a long arm). The Y chromosome deletions in these females share a common overlapping region which Page concludes may be essential for male determination. The unusual males have a piece of the short arm of the Y chromosome—including the segment deleted in the XY females—translocated onto one of their X chromosomes. Like McLaren's mice, these males do not produce sperm, because two X chromosomes are incompatible with male fertility. In addition, the XX males contain only part of the Y chromosome, lacking the long arm. Possibly the long arm of the Y chromosome plays an important role in spermatogenesis.

The gene for the H–Y antigen is near the centromere of the Y chromosome, possibly on the long arm, far away from the segment of the chromosome deleted in the XY females. Therefore, the role, if any, of the H–Y antigen in mammalian sex determination remains to be discovered.

Sex chromosomes in other organisms. Not all organisms have an X–Y sex chromosome makeup like that found in mammals and in *Drosophila.* In birds, butterflies, moths, and some fish, the sex chromosome composition is the opposite of that in mammals: The male in these organisms is the homogametic sex and the female is the heterogametic sex. So that we do not confuse the situation here with the X and Y chromosome convention, we designate the chromosomes in these organisms as Z and W. Thus the males are ZZ and the females are ZW. Genes on the Z chromosome behave just like X-linked genes in the examples discussed previously except that hemizygosity is found only in females. All the daughters of a male homozygous for a Z-linked recessive gene express the recessive trait, and so on. As an example, the inheritance of the sex-linked (Z-linked), dominant barred plumage *(B)* in poultry is illustrated in Figure 3.11.

Several sex determination mechanisms are encountered among the insects. In bees, for example, sex is determined by the number of sets of chromosomes a bee has. Thus fertilized eggs produce diploid females, while unfertilized eggs divide asexually to give rise to haploid males. The latter are fertile, producing sperm by mitosis rather than by meiosis. The life cycle of the bee proceeds when a queen bee uses the sperm to produce females and lays unfertilized eggs to produce males.

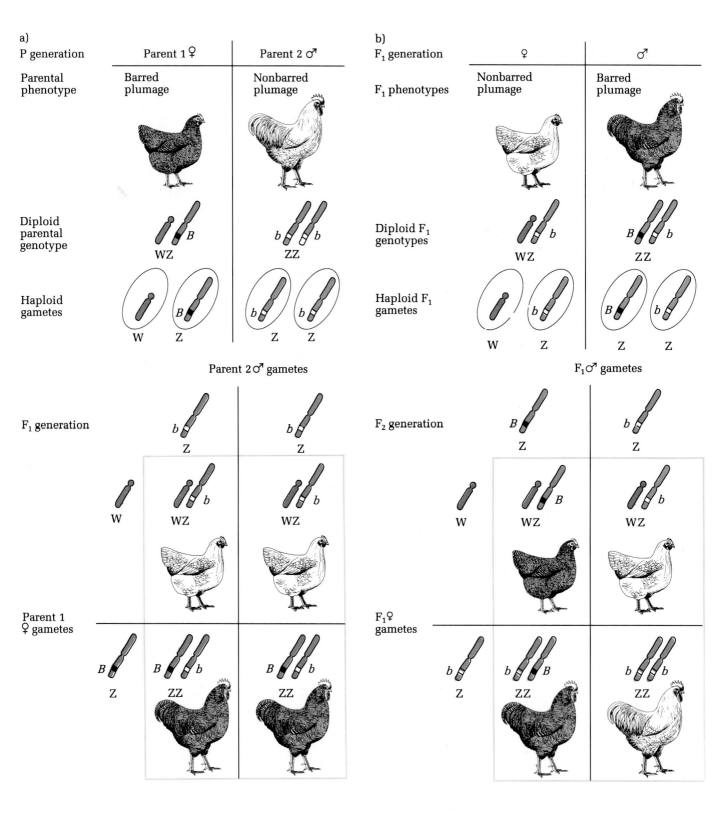

a)

	Parent 1 ♀	Parent 2 ♂
P generation		
Parental phenotype	Barred plumage	Nonbarred plumage

Diploid parental genotype: WZ (B) / ZZ (b b)

Haploid gametes: W, Z (B) / Z (b), Z (b)

Parent 2 ♂ gametes

F₁ generation: b Z / b Z

Parent 1 ♀ gametes:

	W		
W		WZ	WZ
B Z	B Z	ZZ	B b ZZ

F₁ genotypes: ½ b/W, ½ B/b

F₁ phenotypes: ½ nonbarred females, ½ barred males

b)

	♀	♂
F₁ generation		
F₁ phenotypes	Nonbarred plumage	Barred plumage

Diploid F₁ genotypes: WZ (b) / ZZ (B b)

Haploid F₁ gametes: W, Z (b) / Z (B), Z (b)

F₁ ♂ gametes

F₂ generation: B Z / b Z

F₁ ♀ gametes:

	W		
W		WZ B	WZ b
b Z	b Z	ZZ b B	ZZ b b

F₂ genotypes: ¼ B/W, ¼ B/b, ¼ b/W, ¼ b/b

F₂ phenotypes: ½ barred (1♀, 1♂), ½ nonbarred (1♀, 1♂)

Higher plants exhibit quite a variety of sexual situations. Some species have plants of separate sexes, with male plants producing flowers that contain only stamens and female plants producing flowers that contain only pistils. These species are called **dioecious.** Other species have both male and female sex organs on the same plant. If the sex organs are in the same flower, the plant is said to be **hermaphroditic** (e.g., the rose and the buttercup), and the flower is said to be a *perfect flower.* If the sex organs are in different flowers on the same plant, it is said to be **monoecious** (e.g., corn), and the flower is said to be an *imperfect flower.*

Some dioecious plants have sex chromosomes that differ between the sexes, and a large proportion of these plants have an X–Y system. Only in a few instances has sex determination in such plants been studied with anything like the rigor applied to the corresponding problem in *Drosophila.* These studies have shown that an X chromosome-autosome balance system like that found in *Drosophila* determines the sex of the individual. Many other sex determination systems are seen in the dioecious plants. For example, some of these plants have visibly different (heteromorphic) chromosomes in the two sexes, but the constitution of the different sexes is more complex than that of the *Drosophila* X–Y type. Other dioecious plants have no visibly heteromorphic pair of chromosomes, and the mechanism of sex determination is not known in these instances.

Many species, particularly eukaryotic microorganisms, do not have sex chromosomes but instead rely on a *genic system* for the determination of sex; that is, a system in which the sexes are specified by simple allelic differences at a small number of gene loci. For example, the orange bread mold *Neurospora crassa* is a haploid

Figure 3.12 Photograph of boat shell snail.

fungus that has two sexes referred to as **mating types.** The sexes are morphologically indistinguishable and are determined by the *A* and *a* alleles of a single gene locus in chromosome I of this organism. How these alleles function in sexual reproduction is not known. If two strains of *Neurospora* are placed together on the appropriate medium, they will not mate if each carries the same allele. However, if one is mating type *A* and the other is mating type *a,* the two strains will fuse, a diploid nucleus will be produced by fusion of two haploid nuclei, and meiosis will be initiated. Thus in some way the mating-type alleles function in a compatibility system connected with the sexual reproduction cycle.

Phenotypic sex determination systems. The environment plays a major role in systems of phenotypic sex determination. In boat shell snails (Figure 3.12), for example, each individual typically goes through a developmental sequence in which an early asexual stage is followed by a male phase, then a transitional phase, then finally a female phase. But when individuals in the male phase are suitably mated and sedentary, their transformation to females is deferred. Conversely, nonsedentary, unmated males quickly convert to the female phase. In other words, the transformation of sexes appears to be strongly affected by the environment.

A similar phenomenon is found in the marine worm *Bonellia* (Figure 3.13). In this organism the free-swimming larval forms are sexually undifferentiated. If an individual settles down alone, it becomes female. If a larva attaches to the body of an adult female, it will differentiate into a male. Thus sex differentiation in *Bonellia*

Figure 3.11 Sex-linked inheritance in chickens. A barred-feather female is crossed with a nonbarred male in (a). The barred-feather (B) phenotype is caused by a gene that is dominant to the allele for nonbarred (b) plumage. In (b), an F₁ nonbarred female is crossed with an F₁ barred male.

82
*Chromosomal
Basis of
Inheritance,
Sex Linkage,
and Mendelian
Genetics
in Humans*

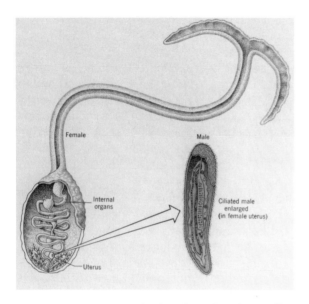

Figure 3.13 Drawing of a female and male Bonellia, *a marine worm.*

Keynote *Many eukaryotic organisms have sex chromosomes that are represented differentially in the two sexes; in humans and many other eukaryotes the male is XY and the female is XX. In many cases sex determination is related to the sex chromosomes. For humans and many other mammals, for instance, the presence of the Y chromosome confers maleness, and its absence results in femaleness.* Drosophila *has an X chromosome–autosome balance system of sex determination: The sex of the individual is related to the ratio of the number of X chromosomes to the number of sets of autosomes, and the Y chromosome has no effect. Several other sex-determining systems are known in the eukaryotes, including genic systems, found particularly in the lower eukaryotes, and phenotypic (environmental) systems.*

is not determined at fertilization by some genetic component but is instead directed by environmental factors related either to the association or lack of association with other members of the species.

a   b

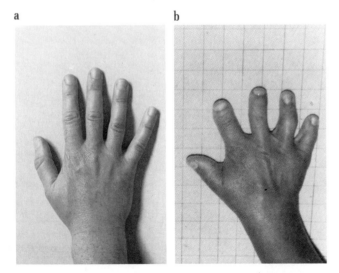

Figure 3.14 Photograph of normal human hand (a) alongside human hand with brachydactyly (b).

Mendelian Genetics in Humans

After the rediscovery of Mendel's laws in 1900, geneticists found that the inheritance of chromosomes and the processes of chromosome and gene segregation are the same in all sexually reproducing eukaryotes, including humans. W. Farabee in 1905 was the first to document a Mendelian trait in humans, brachydactyly (abnormally broad and short fingers: Figure 3.14). By analyzing pedigrees, Farabee learned that brachydactyly is inherited. The pattern of transmission of the abnormality over several generations led to the conclusion that the trait is a simple dominant trait. In this section we will examine some of the methods used to determine the mechanism of hereditary transmission, and we will learn about some inherited human traits.

Pedigree Analysis

The study of human genetics is complicated because controlled matings of humans cannot be made. This does not mean that genetic analysis of humans is impossible, but it does mean that

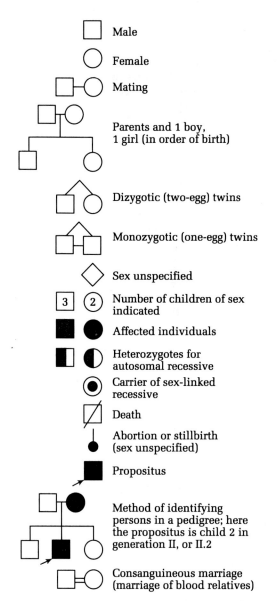

Figure 3.15 Symbols used in human pedigree analysis.

compilation of phenotypic records of the family over several generations. The individual through which the pedigree is discovered is called the **propositus.** The more information there is, the more likely the investigator will be able to make some conclusions about the mechanism of inheritance of the gene (or genes) responsible for the trait being studied.

One of the modern applications of pedigree analysis is called **genetic counseling:** A geneticist makes predictions about the probabilities of particular traits (deleterious or not) occurring among children of a couple. In most cases the couple comes to the counselor because there is some possibility of a heritable trait in one or both families. As we might expect, pedigree analysis is most useful for traits that are the result of a single gene difference.

Because it is difficult to assign genotypes to the individuals in a family with respect to a trait, it is not possible to diagram crosses in the way we did for *Drosophila* crosses. Therefore, pedigree analysis has its own set of symbols. Figure 3.15 summarizes the basic symbols that will be used here and elsewhere in the text, and Figure 3.16 presents a hypothetical pedigree to show how the symbols are assigned to the family tree.

geneticists must study crosses that happen by *chance* rather than by control. The inheritance patterns of human traits are usually established by examining the occurrences of the trait in the family trees of individuals who clearly exhibit the trait. The family tree investigation is called **pedigree analysis,** and it involves the careful

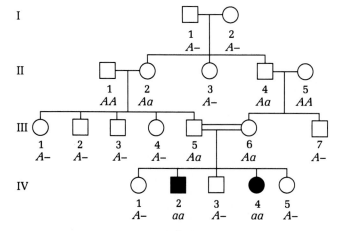

Figure 3.16 Example of a human pedigree, showing the use of pedigree symbols.

84
Chromosomal
Basis of
Inheritance,
Sex Linkage,
and Mendelian
Genetics
in Humans

The trait exemplified in Figure 3.16 is determined by a recessive allele *a*. Gene symbols are included here to show the deductive reasoning possible with such pedigrees; normally, such symbols would not be present. Note that generations are numbered with roman numerals while individuals are numbered with arabic numerals, a system that facilitates reference to particular people in the pedigree. The occurrence of the trait in this pedigree resulted from homozygosity for the rare allele, brought about by cousins marrying. Since cousins share a fair proportion of their genes, it can be expected that a number of genes will become homozygous, and in this case one of the genes resulted in an identifiable genetic trait. Since a number of genes in the homozygous condition give rise to serious diseases, marriage between cousins is often prohibited by law.

Mechanisms of Gene Inheritance

In humans the gene responsible for a trait can be inherited in five different ways: autosomal dominant, autosomal recessive, sex-linked dominant, sex-linked recessive, and Y-linked. In this section we will examine traits whose genes are inherited in these ways in order to learn something about the analytical thinking that is required to determine the mechanism of inheritance.

Autosomal dominant inheritance. A trait due to a dominant mutant gene carried in an autosome is called an **autosomal dominant** trait. Over 400 human traits are known to be inherited in an autosomal dominant fashion. Figure 3.17 illustrates one such trait called woolly hair, in which an individual's hair is very tightly kinked, is very brittle, and breaks before it can grow very long. The best examples of pedigrees for this trait come from Norwegian families; one of these pedigrees is presented in Figure 3.18. Since it is a fairly rare trait and since not all children of an affected parent show the trait, it can be assumed that most woolly-haired individuals are heterozygous for the autosomal dominant allele involved.

Figure 3.17 Members of a Norwegian family, some exhibiting the trait of woolly hair.

Generation:

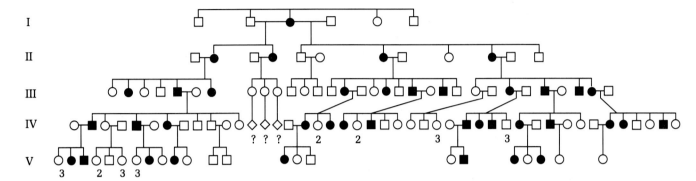

Figure 3.18 Part of a pedigree showing the transmission of the autosomal dominant trait of woolly hair.

Several generalizations can be made about autosomal dominant traits. Dominant mutant alleles are expressed in a heterozygote when they are in combination with the wild-type allele. Because many dominant mutant alleles that give rise to recognizable traits are rare, it is unusual to find individuals homozygous for the dominant allele. Thus an affected person in a pedigree is likely to be a heterozygote, and most marriages are between a heterozygote and a homozygous recessive (wild type). Most dominant mutant genes that are clinically significant (resulting in medical problems) fall into this category. Also, most known autosomal dominant traits produce less severe clinical effects than do recessive traits. If the dominant trait has serious effects, natural selection tends to decrease the frequency of that allele or to remove that allele from the population. That is, mutant individuals may be unable to reproduce or may not survive.

Another very common feature of autosomal dominant traits is that they can vary quite widely in expression from one individual to another. **Expressivity** refers to the degree to which an individual manifests a phenotype corresponding to its genotype. Variations in expressivity presumably reflect the biochemical effects of the dominant allele, the activity of modifier genes, or the sensitivity of the individual to environmental conditions.

For a pedigree to indicate that a trait is caused by an autosomal dominant mutant gene, it must satisfy four basic requirements (refer to Figure 3.18):

1. Every affected person in the pedigree must have at least one affected parent.
2. Approximately equal numbers of males and females in the pedigree must express the trait, and those individuals should transmit the mutant gene to their progeny.
3. The pedigree must have male and female individuals in each generation who express the mutant gene. In other words, the trait must not skip generations, nor should it appear in males only in one generation, in females only in the next generation, in males only in the generation after that, and so on. That transmission pattern is characteristic of sex-linked inheritance. Father-to-son and mother-to-daughter transmission is expected to occur as frequently as father-to-daughter and mother-to-son transmission in autosomal dominant inheritance.
4. An affected heterozygous individual must transmit the mutant gene to half his or her progeny. Suppose the dominant mutant allele is designated A, and its wild-type allele is A^+. Then most crosses will be $A/A^+ \times A^+/A^+$. From basic Mendelian principles half the

86
*Chromosomal
Basis of
Inheritance,
Sex Linkage,
and Mendelian
Genetics
in Humans*

progeny will be A^+/A^+ and the other half will be A/A^+, showing the trait.

Autosomal recessive inheritance. A trait due to a homozygous recessive mutant gene carried in an autosome is called an **autosomal recessive** trait. More than 500 human traits are known to be caused by autosomal recessive genes. Individuals expressing the autosomal recessive trait of albinism (deficient pigmentation) are shown in Figure 3.19a, and a pedigree for this trait is shown in Figure 3.19b.

For the traits to be expressed, the allele must be homozygous. Unlike dominant mutant alleles, recessive mutant alleles are commonly associated with serious abnormalities or diseases. Individuals with albinism, for instance, do not produce the melanin pigment that protects the skin from harmful ultraviolet radiation. As a consequence, albinos have considerable skin and eye sensitivity to sunlight. Frequencies of recessive mutant alleles are usually higher than frequencies of dominant mutant alleles because heterozygotes are not at a significant selective disadvantage. Nonetheless, most autosomal recessive traits are rare. In the United States approximately 1 in 38,000 of the white population and 1 in 10,000 of the black population are albinos.

It is sometimes difficult to prove that a trait is caused by an autosomal recessive mutant allele, especially if the trait is a relatively rare one. For example, environmental factors could lead to some families having more than one affected sibling, thereby giving the false impression that the trait might be inherited. The following characteristics of autosomal recessive inheritance for a relatively rare trait should appear in pedigrees (refer to Figure 3.19):

1. Most affected individuals have two normal parents, so the trait tends to skip generations. In other words, the trait is not shown in the generation represented by the normal parents because they are both heterozygous. The trait appears in the F_1 since a quarter of the progeny are expected to be homozygous for the autosomal recessive allele. If the trait is rare or relatively rare, then an individual expressing the trait is likely to mate with a homozygous normal individual; thus the next generation is represented by heterozygotes who do not express the trait.

2. The trait should be expressed in both sexes and transmitted by either sex to both male and female offspring in approximately equal proportions. This inheritance pattern reflects the meiotic segregation patterns of chromosomes other than the sex chromosomes.

3. Matings between normal heterozygotes should yield both normal progeny and progeny exhibiting the trait, in a ratio of about 3:1. The problem here is that it is difficult to obtain enough numbers to make the data statistically significant, especially when a biochemical test is necessary to confirm the presence of the trait. In this case only the living members of a family can be assayed. For autosomal recessive genes that have less deleterious effects, the allele can reach significant frequencies in the population. An example of such a recessive trait is attached ear lobes. Significant numbers of heterozygotes and homozygous recessives for this trait occur in the population. As a result, there is a strong possibility of $a^+/a \times a/a$ matings from which half the progeny will have the trait. Recall that this ratio of 1:1 is also characteristic for a rare autosomal dominant trait; it is therefore important to obtain enough other data so that we don't jump to the wrong conclusion.

4. Particularly with rare traits, the recessive allele carried by each parent has typically originated with a common ancestor, and the heterozygous, phenotypically normal parents of an individual expressing the trait may be related to each other.

5. When both parents are affected, all their progeny will exhibit the trait.

X-linked recessive inheritance. A trait due to a recessive mutant gene carried on the X chromosome is called an **X-linked recessive** trait. At least 100 traits are known whose inheritance has been traced to the X chromosome. Most of the traits involve X-linked recessive genes. The most well known X-linked recessive pedigree is that

a)

b) Pedigree

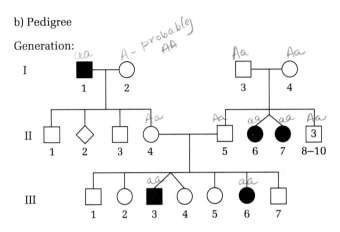

Figure 3.19 (a) Human albinos and (b) a pedigree showing the transmission of the autosomal recessive trait of albinism.

of hemophilia A in Queen Victoria's family (Figure 3.20). Hemophilia (the bleeder's disease) is a serious ailment in which the blood lacks a clotting factor. A cut or even a serious bruise can be fatal to a hemophiliac. In Queen Victoria's pedigree the first instance of hemophilia was in one of her sons, so she was a carrier for this trait.

In X-linked recessive traits the female must be homozygous for the recessive allele in order to express the mutant trait. The trait is expressed in the male who possesses but one copy of the mutant allele on the X chromosome (except in the rare instance of a homologous gene on the Y chromosome). Therefore affected males normally transmit the mutant gene to all their daughters but to none of their sons. The instance of a father-to-son inheritance of a trait in a pedigree would tend to rule out X-linked recessive inheritance. As mentioned earlier in this chapter, the passage of sex-linked traits from mother to son and from father to daughter is called crisscross inheritance.

Other characteristics of X-linked recessive inheritance are as follows (refer to Figure 3.20):

1. For rare X-linked recessive genes many more males than females should exhibit the trait,

owing to the different number of X chromosomes in the two sexes. Most females will be heterozygous under these conditions.

2. The sons of heterozygous (carrier) mothers should show an approximately 1:1 ratio of normal individuals to individuals expressing the trait. That is, $a^+/a \times a^+/Y$ gives half a^+/Y and half a/Y sons. If only unaffected sons are produced in a small family, the trait will skip a generation.

3. All daughters from a mating of a carrier female with a normal male will be normal, but half will be carriers. That is, $a^+/a \times a^+/Y$ gives half a^+/a^+ and half a^+/a females. In turn, half the sons of the F_1 carrier females will exhibit the trait.

4. A male expressing the trait, when mated with a homozygous normal female, will produce all normal children, but all the female progeny will be carriers. That is, $a^+/a^+ \times a/Y$ gives a^+/a females and a^+/Y (normal) males.

X-linked dominant inheritance. A trait due to a dominant mutant gene carried on the X chromosome is called an **X-linked dominant** trait. Only a few dominant X-linked traits have been identified.

Generation:

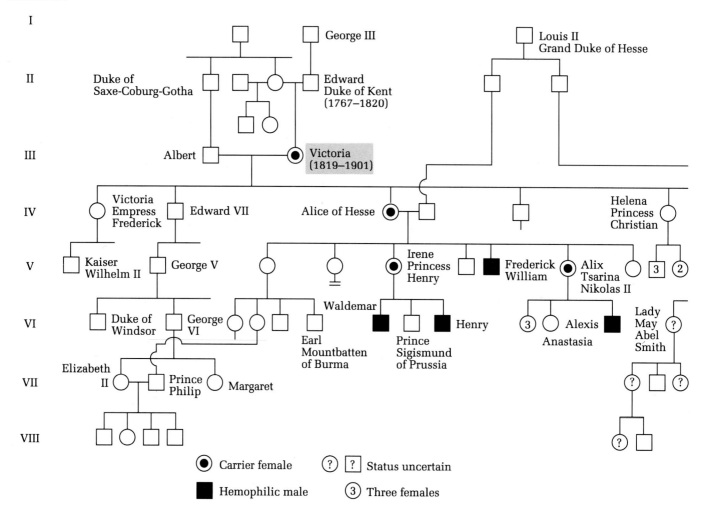

Carrier female • ? Status uncertain

■ Hemophilic male • ③ Three females

An example of an X-linked dominant trait that causes faulty tooth enamel and dental discoloration is shown in Figure 3.21a and a pedigree for this trait is shown in Figure 3.21b. Note that all the daughters and none of the sons of an affected father (III.1) are affected, and that heterozygous mothers (IV.3) transmit the trait to half their sons and half their daughters. Webbing to the tips of the toes in an Irish family in South Dakota (studied in the 1930s) has also been ascribed to an X-linked dominant mutant allele, as has a severe bleeding anomaly called constitutional thrombopathy, studied in two Finnish families on an archipelago in the Baltic Sea. In this case (also studied in the 1930s) bleeding is not due to the absence of a clotting factor (as in hemophilia) but to interference in the formation of blood platelets, which are needed for blood clotting.

In general, X-linked dominant traits tend to be milder in the female than in the male. The X-linked dominant traits follow the same sort of inheritance rules as the X-linked recessives, except that heterozygous females express the trait. Since females have twice the number of X chromosomes as males, X-linked dominant traits are more frequent in females than in males. If the trait is a rare one, we expect two-thirds of the affected persons to be female. For rare traits most females in a pedigree would be heterozygous. These females should pass on the trait to half their progeny, regardless of their sex. As in X-linked recessive traits, males who have an X-linked dominant trait transmit the allele to their daughters but not to their sons. As a result, we see father-to-daughter inheritance of the trait.

Y-linked inheritance. A trait due to a mutant

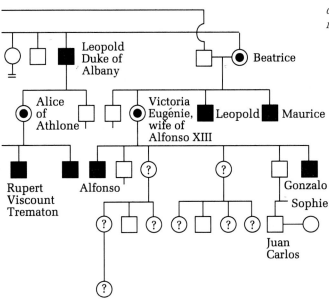

Figure 3.20 Pedigree of Queen Victoria (III.2) and her descendants, showing the inheritance of hemophilia. Either Queen Victoria was heterozygous for the sex-linked recessive hemophilia allele, or the trait arose as a mutation in her germ cells (the cells that give rise to the gametes).

gene carried on the Y chromosome but with no counterpart on the X is called a **Y-linked,** or **holandric** ("wholly male"), trait. Such traits should be readily recognizable since every son of an affected male should have the trait but no females should ever express it. Several traits with Y-linked inheritance have been suggested. In most cases the genetic evidence for such inheritance is poor or nonexistent. In fact, there is no clear evidence for genetic loci on the Y chromosome other than those involved with sex determination in the male.

A putative example of Y-linked inheritance is the hairy ears, or hairy pinnae, trait in which bristly hairs of atypical length grow from the ears (Figure 3.22). This trait is common in parts of India, although some white populations also exhibit it. While this trait shows father-to-son inheritance, there is no doubt that it is a complex phenotype, and many of the collected pedigrees can be interpreted in other ways, such as autosomal inheritance. Another putative Y-linked trait is bilateral radio-ulnar synostosis. In this trait the radius and ulna bones of both arms are fused.

In sum, collecting reliable human pedigree data is a difficult task. In many cases the accuracy of record keeping within the families involved is open to question. Also, particularly with small families, there may not be enough affected people to allow an unambiguous determination of the mechanism of inheritance involved, especially when rare traits are involved. Moreover, the degree to which a trait is expressed may vary so that some individuals may be erroneously classified as being normal. It is also possible for a mutant phenotype to be produced by alleles of different genes, and therefore different pedigrees may, quite correctly, indicate mechanisms of inheritance of the "same" trait.

90
Chromosomal
Basis of
Inheritance,
Sex Linkage,
and Mendelian
Genetics
in Humans

a)

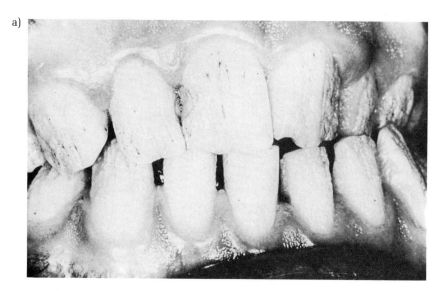

b) Pedigree

Generation:

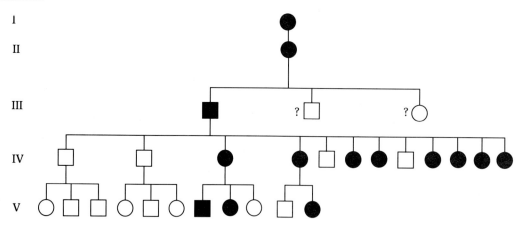

Figure 3.21 (a) A person with the X-linked dominant trait of faulty enamel and (b) a pedigree showing the transmission of the faulty enamel trait. This pedigree illustrates a shorthand convention in which parents who do not exhibit the trait are omitted. Thus it is a given that the female in generation I paired with a male who did not exhibit the trait.

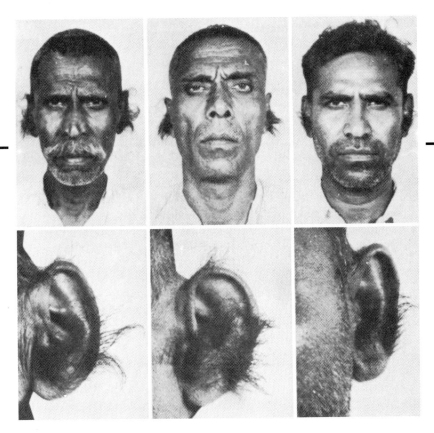

Figure 3.22 *Individuals exhibiting the putatively Y-linked trait of hairy ears.*

Keynote *Controlled matings cannot be made in humans, so analysis of the inheritance of genes in humans must rely on pedigree analysis, which involves the careful study of the phenotypic records of the family extending over several generations. The data obtained from pedigree analysis enable geneticists to make judgments, with varying degrees of confidence, about whether a mutant gene is inherited as an autosomal recessive, an autosomal dominant, an X-linked recessive, an X-linked dominant, or a Y-linked allele. Information obtained from pedigree analysis can be used in genetic counseling to advise prospective parents of the chance of having a child with a genetic defect.*

Analytical Approaches for Solving Genetics Problems

The concepts introduced in this chapter may be reinforced by solving genetics problems. The types of problems are similar to those introduced in Chapter 2. When sex linkage is involved, remember that one sex has two kinds of sex chromosomes, whereas the other sex has only one; this feature alters the inheritance patterns slightly. Most sample problems presented in this section center on interpreting data and predicting the outcome of particular crosses.

Q.1 A female from a pure-breeding strain of *Drosophila* with vermilion-colored eyes is crossed with a male from a pure-breeding wild-type, red-eyed strain. All the F_1 males have vermilion-colored eyes, and all the females have wild-type red eyes. What conclusions can you draw about the mode of inheritance of the vermilion trait, and how could you test them?

92
Chromosomal
Basis of
Inheritance,
Sex Linkage,
and Mendelian
Genetics
in Humans

A.1 The observation is the classic one that suggests sex linkage of the trait involved. Since none of the F_1 daughters have the trait and all the F_1 males do, the trait is presumably X-linked recessive. The results fit this hypothesis because the F_1 males receive the X chromosome from their homozygous v/v mother, with the v gene on the X chromosome. Further, the F_1 females are v^+/v since they receive a v^+-bearing X chromosome from the wild-type male parent and a v-bearing X chromosome from the female parent. If the trait were autosomal recessive, all the F_1 flies would have had wild-type eyes; and if it were autosomal dominant, both the F_1 males and females would have had vermilion-colored eyes. If the trait were X-linked dominant, all the F_1 flies would have had vermilion eyes.

The easiest way to verify this hypothesis is to let the F_1 flies interbreed. This cross is v^+/v ♀ $\times$ v/Y ♂, and the expectation is that there will be a 1:1 ratio of wild:vermilion eyes in both sexes in the F_2. That is, half the females are v^+/v and half are v/v; half the males are v^+/Y and half are v/Y. This ratio is certainly not the 3:1 ratio that would result from an $F_1 \times F_1$ cross for an autosomal gene.

Q.2 In humans hemophilia A or B is caused by an X-linked recessive gene. A woman who is a nonbleeder had a father who was a hemophiliac. (Actually, hemophiliac fathers are relatively rare, but we can at least consider one for this question.) She marries a nonbleeder, and they plan to have children. Calculate the probability of hemophilia in the female and male offspring.

A.2 Since hemophilia is an X-linked trait, and since her father was a hemophiliac, the woman must be heterozygous for this recessive gene. If we assign the symbol h to this recessive mutation and h^+ to the wild-type allele, she must be h^+/h. The man she marries is normal with regard to blood clotting and hence must be hemizygous for h^+, that is, h^+/Y. All their daughters receive an X chromosome from the father, and so each must have an h^+ gene. In fact, half the daughters are h^+/h^+ and the other half are h^+/h. Since the wild-type allele is dominant, none of the daughters are hemophiliacs. However, all the sons of the marriage receive their X chromosome from their mother. Therefore they have a probability of 1/2 that they will receive the chromosome carrying the h allele, which means they will be hemophiliacs. Thus the probability of hemophilia among daughters of this marriage is 0, and among sons it is 1/2.

Q.3 Tribbles are hypothetical animals that have an X–Y sex determination mechanism like that of humans. The trait bald (b) is X-linked and recessive to furry (b^+), and the trait long leg (l) is autosomal and recessive to short leg (l^+). You make reciprocal crosses between true-breeding bald, long-legged tribbles and true-bleeding furry, short-legged tribbles. Do you expect a 9:3:3:1 ratio in the F_2 of either or both of these crosses? Explain your answer.

A.3 This question focuses on the fundamentals of X chromosome and autosome segregation during a genetic cross and tests whether or not we have grasped the principles involved in gene segregation. The figure on page 93 diagrams

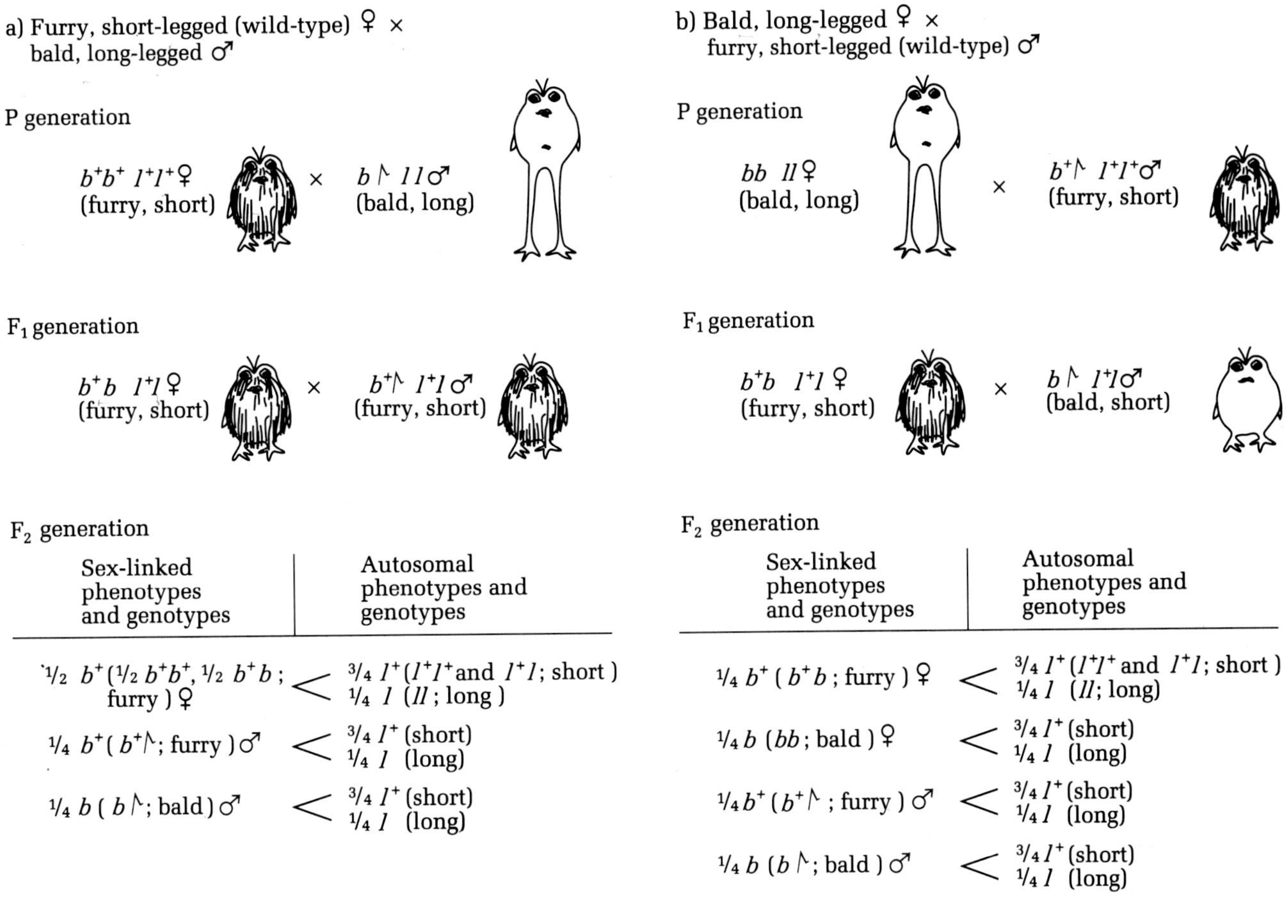

a) Furry, short-legged (wild-type) ♀ ×
   bald, long-legged ♂

P generation

b^+b^+ l^+l^+♀ × $b\mathrel{\mkern-2mu}$ ll♂
(furry, short)   (bald, long)

F₁ generation

b^+b l^+l♀ × $b^+\mathrel{\mkern-2mu}$ l^+l♂
(furry, short)   (furry, short)

F₂ generation

Sex-linked phenotypes and genotypes	Autosomal phenotypes and genotypes
½ b^+ (½ b^+b^+, ½ b^+b; furry) ♀	¾ l^+ (l^+l^+ and l^+l; short) ¼ l (ll; long)
¼ b^+ ($b^+\mathrel{\mkern-2mu}$; furry) ♂	¾ l^+ (short) ¼ l (long)
¼ b ($b\mathrel{\mkern-2mu}$; bald) ♂	¾ l^+ (short) ¼ l (long)

Phenotypic ratios:

	Furry-short		Furry-long		Bald-short		Bald-long
	b^+l^+		b^+l		$b l^+$		$b l$
♀	6	:	2	:	0	:	0
♂	3	:	1	:	3	:	1
Total	9	:	3	:	3	:	1

b) Bald, long-legged ♀ ×
   furry, short-legged (wild-type) ♂

P generation

bb ll♀  × $b^+\mathrel{\mkern-2mu}$ l^+l^+♂
(bald, long)   (furry, short)

F₁ generation

b^+b l^+l ♀ × $b\mathrel{\mkern-2mu}$ l^+l♂
(furry, short)   (bald, short)

F₂ generation

Sex-linked phenotypes and genotypes	Autosomal phenotypes and genotypes
¼ b^+ (b^+b ; furry) ♀	¾ l^+ (l^+l^+ and l^+l; short) ¼ l (ll; long)
¼ b (bb; bald) ♀	¾ l^+ (short) ¼ l (long)
¼ b^+ ($b^+\mathrel{\mkern-2mu}$; furry) ♂	¾ l^+ (short) ¼ l (long)
¼ b ($b\mathrel{\mkern-2mu}$; bald) ♂	¾ l^+ (short) ¼ l (long)

Phenotypic ratios:

	Furry-short		Furry-long		Bald-short		Bald-long
	b^+l^+		b^+l		$b l^+$		$b l$
♀	3	:	1	:	3	:	1
♂	3	:	1	:	3	:	1
Total	6	:	2	:	6	:	2

the two crosses involved, and we can discuss the answer by referring to it.

Let us first consider the cross of a wild-type female tribble (b^+/b^+, l^+/l^+) with a male double-mutant tribble (b/Y, l/l). Part (a) of the figure diagrams this cross. The F₁'s of this cross are all normal—that is, with furry bodies and short legs—because for the autosomal character both sexes are heterozygous, and for the X-linked character the female is heterozygous and the male is hemizygous for the b^+ allele donated by the normal mother. With the production of the F₂ progeny the best approach is to treat the X-linked and

94
*Chromosomal
Basis of
Inheritance,
Sex Linkage,
and Mendelian
Genetics
in Humans*

autosomal traits separately. For the X-linked trait random combination of the gametes produced gives a 1:1:1:1 genotypic ratio of b^+/b^+ (furry female): b^+/b (furry female):b^+/Y (furry male):b/Y (bald male) progeny. Collecting by phenotypes, we see that all the females are furry, and half the males are furry and half are bald. For the autosomal leg trait the $F_1 \times F_1$ is a cross of two heterozygotes, so we expect a 3:1 phenotypic ratio of short-legged:long-legged tribbles in the F_2. Since the segregation of the autosome is independent of the inheritance of the X chromosome, we can multiply the probabilities of occurrence of the X-linked and autosomal traits to calculate their relative frequencies. The calculations are presented in part (a) of the figure, from which we see that the ratio of the four possible phenotypic classes differs in females and males.

This cross, then, has a 9:3:3:1 ratio of the four possible phenotypes in the F_2. However, note that the ratio in each sex is not 9:3:3:1 owing to the inheritance pattern of the X chromosome. This result contrasts markedly with the situation of two autosomal genes segregating independently, where the 9:3:3:1 ratio is found for both sexes.

The reciprocal cross is diagrammed in part (b) of the figure. Since the parental female is homozygous for the sex-linked trait, all the F_1 males are bald. Genotypically, the F_1 males and females are different from those in the first cross with respect to the sex chromosome but just the same with respect to the autosome. Again, considering the X chromosome first as we go to the F_2, we find a 1:1:1:1 genotypic ratio of furry females:bald females:furry males: bald males. In this case, then, half of both males and females are furry and half are bald, in contrast to the results of the first cross in which no bald females were produced in the F_2. For the autosomal trait we expect a 3:1 ratio of short:long in the F_2, as before. Putting the two traits together, we get the calculations presented in part (b) of the figure. (*Note:* We use the total 6:2:6:2 here rather than 3:1:3:1 because the numbers add to 16, as does 9 + 3 + 3 + 1.) So in this case we do not get a 9:3:3:1 ratio, and moreover, the ratio is the same in both sexes.

This question has forced us to think through the segregation of two types of chromosomes and has shown that we must be careful about predicting the outcomes of crosses in which sex chromosomes are involved. Nonetheless, the basic principles for the analysis were no different from those used before: Reduce the questions to their basic parts and then put the puzzle together step by step.

Questions and Problems

3.1 At the time of synapsis preceding the reduction division in meiosis, the homologous chromosomes align in pairs and one member of each pair passes to each of the daughter nuclei (see Chapter 1). In an animal with five pairs of chromosomes, assume that chromosomes 1, 2, 3, 4, and 5 have come from the father, and 1', 2', 3', 4', and 5' have come from the mother. In what proportion of the germ cells of this animal will all the paternal chromosomes be present together?

*3.2 In a male *Homo sapiens,* from which grandparent could each sex chromosome have been derived? (Indicate "Yes" or "No" for each option.)

	Mother's		Father's	
	Mother	Father	Mother	Father
X chromosome	_____	_____	_____	_____
Y chromosome	_____	_____	_____	_____

3.3 In *Drosophila,* white eyes are a sex-linked character. The mutant allele for white eyes (w) is recessive to the wild-type allele for brick red eye color (w^+).

a. A white-eyed female is crossed with a red-eyed male, and an F_1 female from this cross is mated with her father and an F_1 male is mated with his mother. What will be the appearance of the offspring of these last two crosses with respect to eye color?

b. A white-eyed female is crossed with a red-eyed male, and the F_2 from this cross is interbred. What will be the appearance of the F_3 with respect to eye color?

*3.4 One form of color blindness (c) in humans is caused by a sex-linked recessive mutant gene. A woman with normal color vision (c^+) whose father was color-blind marries a man of normal vision whose father was also color-blind. What proportion of their offspring will be color-blind? (Give your answer separately for males and females.)

3.5 In humans, red-green color blindness is due to an X-linked recessive gene. A color-blind daughter is born to a woman with normal color vision whose husband is color-blind. What is the mother's genotype with respect to the alleles concerned?

*3.6 In humans, red-green color blindness is recessive and X-linked, while albinism is recessive and autosomal. What types of children can be produced as the result of marriages between two homozygous parents, a normal-visioned albino woman and a color-blind, normally pigmented man?

*3.7 In *Drosophila,* vestigial wings (vg) are recessive to normal long wings (vg^+), and the gene for this trait is autosomal. The gene for the white eye traits is on the X chromosome. Suppose a homozygous white, long-winged female fly is crossed with a homozygous red, vestigial-winged male.

a. What will be the appearance of the F_1?
b. What will be the appearance of the F_2?
c. What will be the appearance of the offspring of a cross of the F_1 back to each parent?

3.8 In *Drosophila,* two red-eyed, long-winged flies are bred together and produce the following offspring: Females are 3/4 red, long and 1/4 red, vestigial;

96
*Chromosomal
Basis of
Inheritance,
Sex Linkage,
and Mendelian
Genetics
in Humans*

males are 3/8 red, long, 3/8 white, long, 1/8 red, vestigial, and 1/8 white, vestigial. What are the genotypes of the parents?

3.9 In poultry a dominant sex-linked gene (B) produces barred feathers, and the recessive allele (b), when homozygous, produces nonbarred feathers. Suppose a nonbarred cock is crossed with a barred hen.

a. What will be the appearance of the F_1 birds?
b. If an F_1 female is mated with her father, what will be the appearance of the offspring?
c. If an F_1 male is mated with his mother, what will be the appearance of the offspring?

*3.10 A man (A) suffering from defective tooth enamel, which results in brown-colored teeth, marries a normal woman. All their daughters have brown teeth, but the sons are normal. The sons of man A marry normal women, and all their children are normal. The daughters of man A marry normal men, and 50 percent of their children have brown teeth. Explain these facts.

3.11 In humans, differences in the ability to taste phenylthiourea are due to a pair of autosomal alleles. Inability to taste is recessive to ability to taste. A child who is a nontaster is born to a couple who can both taste the substance. What is the probability that their next child will be a taster?

3.12 Huntington's disease is a human disease inherited as a Mendelian autosomal dominant. The disease results in choreic (uncontrolled) movements, progressive mental deterioration, and eventually death. The disease affects the carriers of the trait anytime between 15 and 65 years of age. The American folksinger Woody Guthrie died of Huntington's disease, as did one of his parents. Marjorie Mazia, Woody's wife, had no history of this disease in her family. The Guthries had three children. What is the probability that a particular Guthrie child will die of Huntington's disease?

3.13 Suppose gene A is on the X chromosome, and genes B, C, and D are on three different autosomes. Thus $A-$ signifies the dominant phenotype in the male or female. An equivalent situation holds for $B-$, $C-$, and $D-$. The cross $AA\ BB\ CC\ DD$ females $\times$ $aY\ bb\ cc\ dd$ males is made.

a. What is the probability of obtaining an $A-$ individual in the F_1?
b. What is the probability of obtaining an a male in the F_1?
c. What is the probability of obtaining an $A-\ B-\ C-\ D-$ female in the F_1?
d. How many different F_2 genotypes will there be?
e. What proportion of F_2's will be heterozygous for the four genes?
f. Determine the probabilities of obtaining each of the following types in the F_2: (1) $A-\ bb\ CC\ dd$ (female); (2) $aY\ BB\ Cc\ Dd$ (male); (3) $AY\ bb\ CC\ dd$ (male); (4) $aa\ bb\ Cc\ Dd$ (female).

*3.14 As a famous mad scientist, you have cleverly devised a method to isolate *Drosophila* ova that have undergone primary nondisjunction of the sex chromosomes. In one experiment you used females homozygous for the sex-linked recessive mutation causing white eyes (w) as your source of nondisjunction ova. The ova were collected and fertilized with sperm from red-eyed males. The progeny of this "engineered" cross were then backcrossed separately to the two parental strains. What classes of progeny (genotype and phenotype) would you expect to result from these backcrosses? (The genotype of the original parents may be denoted as ww for the females and w^+Y for the males.)

3.15 In *Drosophila* the bobbed gene (bb^+) is located on the X chromosome. Unlike most X-linked genes, however, the Y chromosome also carries a bobbed gene. The mutant allele bb is recessive to bb^+. If a wild-type F$_1$ female that resulted from primary nondisjunction in oogenesis in a cross of a bobbed female with a wild-type male is mated to a bobbed male, what will be the phenotypes and their frequencies in the offspring? List males and females separately in your answer. (*Hint:* Refer to the text for information about the frequency of nondisjunction in *Drosophila*.)

*3.16 In human genetics the pedigree is used for analysis of inheritance patterns. The female is represented by a circle and the male by a square. The figure below presents three, two-generation family pedigrees for a trait in humans. Normal individuals are represented by unshaded symbols and people with the trait by shaded symbols. For each pedigree (A, B, and C), state, by answering yes or no in the appropriate blank space, whether transmission of the trait can be accounted for on the basis of each of the listed simple modes of inheritance:

	Pedigree A	Pedigree B	Pedigree C
Autosomal recessive	_____	_____	yes
Autosomal dominant	yes	yes	_____
X-linked recessive	_____	_____	_____
X-linked dominant	_____	yes	_____

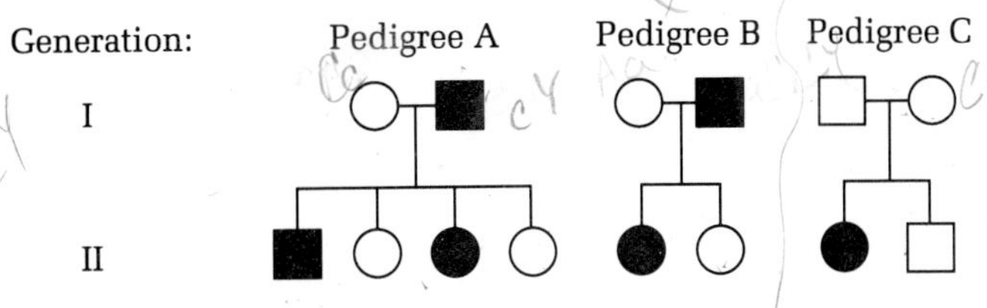

3.17 Looking at the pedigree in the following figure, in which shaded symbols represent a "trait," which of the progeny (as designated by numbers) eliminate X-linked recessiveness as a mode of inheritance for the trait?

98
*Chromosomal
Basis of
Inheritance,
Sex Linkage,
and Mendelian
Genetics
in Humans*

a. 1 and 2   c. 5
b. 4   d. 2 and 4

Generation:

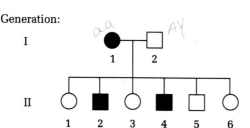

*3.18 When constructing human pedigrees, geneticists often refer to particular persons by a number. For example, in the accompanying pedigree the female with the asterisk would be I.2. Using this means to designate specific individuals in the pedigree, determine the probable mode of inheritance for the trait shown in the affected individuals by answering the following questions. Affected individuals are those denoted by shaded symbols. Assume the condition is caused by a single gene.

a. Y-linked inheritance can be excluded at a glance. What two other mechanisms of inheritance can be definitely excluded? Why can these be excluded?

b. Of the remaining mechanisms of inheritance which is the most likely? Why?

Generation:

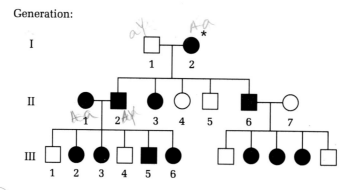

3.19 Phenylketonuria (PKU) is an inborn error in the metabolism of the amino acid phenylalanine. The characteristic feature of PKU is severe mental retardation; many untreated patients with this trait have IQs below 20. The three-generation pedigree shown in the figure on page 99 is of an affected family. The generations are labeled by roman numerals and the individuals in each generation by arabic numerals.

a. What is the mode of inheritance of PKU?

b. Which persons in the pedigree are known to be heterozygous for PKU?

c. What is the probability that III.2 is a carrier (heterozygous)?

d. If III.3 and III.4 marry, what is the probability that their first child will have PKU?

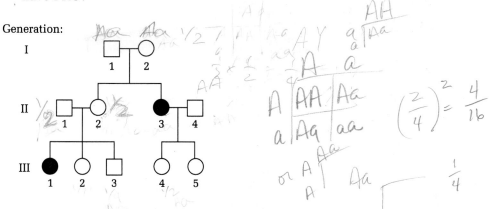

3.20 For the more complex pedigrees shown in the next figure, indicate the probable mode of inheritance: autosomal recessive, autosomal dominant, X-linked recessive, X-linked dominant, Y-linked.

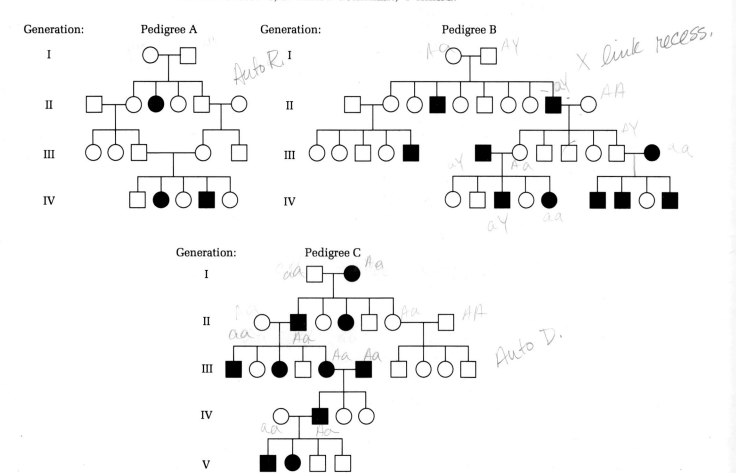

100
Chromosomal
Basis of
Inheritance,
Sex Linkage,
and Mendelian
Genetics
in Humans

*3.21 If a rare genetic disease is inherited on the basis of an X-linked dominant gene, one would expect to find that:

    a. Affected fathers have 100% affected sons.
    b. Affected mothers have 100% affected daughters
    c. Affected fathers have 100% affected daughters.
    d. Affected mothers have 100% affected sons.

3.22 If a genetic disease is inherited on the basis of an autosomal dominant gene, one would expect to find that:

    a. Affected fathers have only affected children.
    b. Affected mothers never have affected sons.
    c. If both parents are affected, all of their offspring have the disease.
    d. If a child has the disease, one of his or her grandparents also had it.

*3.23 If a genetic disease is inherited as an autosomal recessive, one would expect to find that:

    a. Two affected individuals never have an unaffected child.
    b. Two affected individuals have affected male offspring but no affected female children.
    c. If a child has the disease, one of his or her grandparents will have had it.
    d. In a marriage between an affected individual and an unaffected one, all the children are unaffected.

3.24 Which of the following statements is *not* true for a disease that is inherited as a rare X-linked dominant?

    a. All daughters of an affected male will inherit the disease.
    b. Sons will inherit the disease only if their mothers have the disease.
    c. Both affected males and affected females will pass the trait to half their children.
    d. Daughters will inherit the disease only if their father has the disease.

*3.25 Women who were known to be carriers of the X-linked, recessive, hemophilia gene were studied in order to determine the amount of time required for the blood clotting reaction. It was found that the time required for clotting was extremely variable from individual to individual. The values obtained ranged from normal clotting time at one extreme all the way to clinical hemophilia at the other extreme. What is the most probable explanation for these findings?

Extensions of Mendelian Genetic Analysis

It was originally thought that Mendel's laws apply to all eukaryotic organisms and that they form the foundation for predicting the outcome of crosses in which segregation and independent assortment might be occurring. As more and more geneticists did experiments, though, they found that there are exceptions and extensions to Mendel's laws. Several of these cases will be discussed in this chapter. We will examine examples of genes that have many different alleles rather than just two, cases in which one allele may not be completely dominant to another allele at the gene locus, situations in which products of different genes interact to produce modified Mendelian ratios, lethal genes, and the effects of the environment on gene expression. This discussion will give us a broader knowledge of genetic analysis and particularly how genes relate to the phenotypes of an organism.

Multiple Alleles

So far in our genetic analyses we have considered only pairs of alleles controlling characters, such as smooth versus wrinkled peas, red versus white eyes in *Drosophila,* and unattached versus attached ear lobes in humans. The allele found in the standard laboratory strain of the organism is the **wild-type allele,** and the alternative allele is the variant or **mutant allele.** In a population of individuals, however, there can be many alleles of a given gene, not just two. Such genes are said to have **multiple alleles,** and the alleles are said to constitute a multiple allelic series (Figure 4.1). Although a gene may have multiple alleles in a given population of individuals, a single diploid individual can possess only two of these alleles. We will turn now to the historical evidence for multiple alleles of a gene.

Alleles are alternative forms of a gene, each of which may affect a character differently. The w^+ and w alleles in *Drosophila,* for example, are involved in the development of the eye color character. The w^+ allele results in red eyes, and the w allele, when homozygous or hemizygous, results in white eyes.

It was Morgan's work with a white-eyed var-

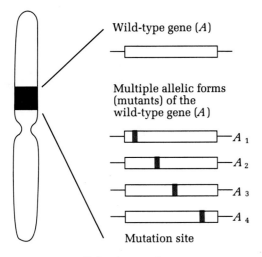

Wild-type gene (*A*)

Multiple allelic forms
(mutants) of the
wild-type gene (*A*)

A_1

A_2

A_3

A_4

Mutation site

Figure 4.1 Allelic forms of a gene.

iant of *Drosophila* that indicated the presence of genes on the X chromosome. Soon after that discovery he found evidence for other distinct genes in the *Drosophila* X chromosome (such as the bar-eye shape and the vermilion eye color genes). Morgan crossed strains that differed with respect to the eye color genes. When he crossed a strain homozygous for the recessive white-eye trait (*w/w*) with a strain homozygous for the recessive vermilion-eye trait (*v/v*), he had an unexpected result: The F$_1$ females were all red-eyed. The simplest explanation is that the white and vermilion eye color traits are specified by *nonallelic genes*—both X chromosomes of the white-eyed parental strain have the genetic makeup wv^+, while the vermilion-eyed parental strain has X chromosomes of genetic type w^+v. The F$_1$ females from this pairing are doubly heterozygous wv^+/w^+v and therefore show the dominant effect resulting from the presence of the wild-type allele for each allelic pair; namely, the brick red eyes characteristic of wild-type flies. (*Note:* In the absence of any other mutant genes, a w^+ or v^+ allele alone will give wild-type brick red eyes.)

In 1912 Morgan obtained data for X-linked eye color genes that could not be explained so simply. The eye color variants were white and eosin (reddish orange). Like white, eosin is recessive

to the wild-type eye color. However, when a female from an eosin-eyed strain is crossed with a male from a white-eyed strain, all F$_1$ females have eosin eyes. In 1913, Alfred Sturtevant concluded that red (wild-type) eye color is dominant to eosin and to white and that eosin is recessive to wild type but dominant to white. The explanation is that eosin and white are both mutant alleles of a single gene; in other words, there are multiple alleles of the white gene.

This concept becomes clearer if we assign symbols to the alleles. A lowercase letter (or letters) designates the gene, and superscripts designate the different alleles, as in the following example: w^+ is the wild-type allele of the white-eye gene, w is the recessive white allele, and w^e is the eosin allele. In this notation the cross of an eosin-eyed female with a white-eyed male is as diagramed in Figure 4.2a. The F$_1$ females are w^e/w and have eosin eyes because w^e is dominant over w. If these F$_1$ females are crossed with red-eyed males (Figure 4.2b), all the female progeny are heterozygous and red-eyed since they contain the w^+ allele; they are either w^+/w^e or w^+/w. Half the male progeny are eosin-eyed (w^e/Y), and the other half are white (w/Y).

The number of possible allelic forms of a gene is not restricted to three, and indeed many hundreds of alleles are known for some genes. The number of possible genotypes in a multiple allelic series depends on the number of alleles involved (Table 4.1). With one allele, only one genotype is possible. With two alleles, A^1 and A^2, three genotypes are possible: A^1A^1 and A^2A^2 homozygotes and the A^1A^2 heterozygote. With n alleles $n(n + 1)/2$ genotypes are possible, of which n are homozygous and $n(n - 1)/2$ are heterozygotes.

Multiple allelic series exist for all types of genes, not just X-linked ones. But while considering the white-eyed gene of *Drosophila,* let us expand on the example a little. There are many known alleles of the white gene that are distinguishable because different eye colors result when the alleles are homozygous or hemizygous. The eye color phenotype is related to the amount of pigment deposited in the eye cells. It is possible to extract the pigments from the eyes

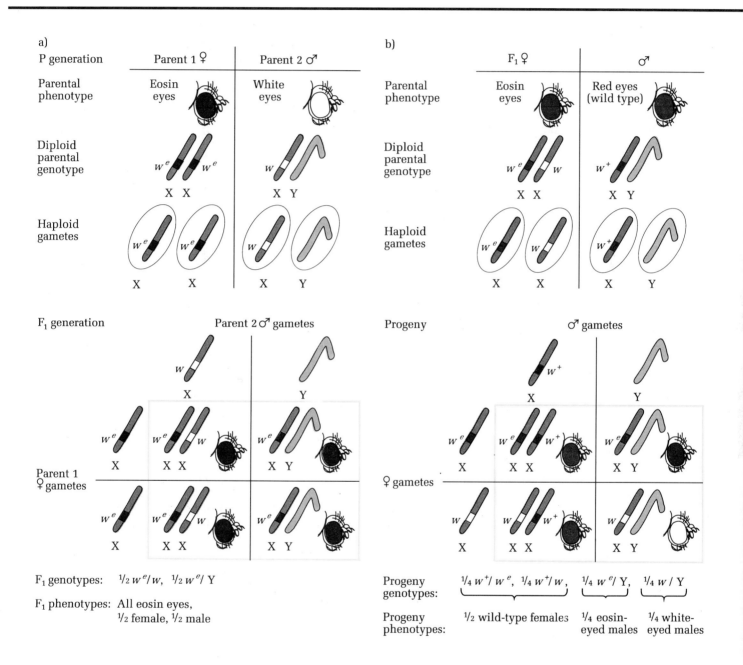

a)

P generation — Parent 1 ♀ | Parent 2 ♂

Parental phenotype: Eosin eyes | White eyes

Diploid parental genotype: w^e w^e — X X | w — X Y

Haploid gametes: w^e X | w^e X ‖ w X | Y

F₁ generation — Parent 2 ♂ gametes

Parent 1 ♀ gametes: w^e X | w^e X

	w (X)	(Y)
w^e X	w^e w X X	w^e X Y
w^e X	w^e w X X	w^e X Y

F₁ genotypes: ½ w^e/w, ½ $w^e/$Y

F₁ phenotypes: All eosin eyes, ½ female, ½ male

b)

F₁ ♀ | **♂**

Parental phenotype: Eosin eyes | Red eyes (wild type)

Diploid parental genotype: w^e w — X X | w^+ — X Y

Haploid gametes: w^e X | w X ‖ w^+ X | Y

Progeny — ♂ gametes

♀ gametes: w^e X | w X

	w^+ (X)	(Y)
w^e X	w^e w^+ X X	w^e X Y
w X	w w^+ X X	w X Y

Progeny genotypes: ¼ w^+/w^e, ¼ w^+/w, ¼ $w^e/$Y, ¼ $w/$Y

Progeny phenotypes: ½ wild-type females | ¼ eosin-eyed males | ¼ white-eyed males

and to quantify the amount of pigment present in wild type and in strains with different mutant alleles of the white locus (Table 4.2). As expected, the original mutant allele *w* has the least amount of pigment. Then a range of pigment amounts occurs, all the way up to the wild-type value.

Figure 4.2 Results of crosses of Drosophila melanogaster *involving two mutant alleles of the same locus, white* (w) *and white-eosin* (wᵉ). *(a) white-eosin eyed* (wᵉ/wᵉ) *♀ × white eyed* (w/Y) *♂. (b) F₁* (wᵉ/w) *♀ × red-eyed (wild type)* w⁺/Y *♂.*

Table 4.1 Genotype Number in Multiple Alleles

Number of Alleles	Kinds of Genotypes	Kinds of Homozygotes	Kinds of Heterozygotes
1	1	1	0
2	3	2	1
3	6	3	3
4	10	4	6
5	15	5	10
n	$\dfrac{n(n+1)}{2}$	n	$\dfrac{n(n-1)}{2}$

Table 4.2 Eye Pigment Quantification

Genotypes	Relative Amount of Total Pigment
w^+/w^+ (wild type)	1.0000
w/w (white)	0.0044
w^t/w^t (tinged)	0.0062
w^a/w^a (apricot)	0.0197
w^{bl}/w^{bl} (blood)	0.0310
w^e/w^e (eosin)	0.0324
w^{ch}/w^{ch} (cherry)	0.0410
w^{a3}/w^{a3} (apricot-3)	0.0632
w^w/w^w (wine)	0.0650
w^{co}/w^{co} (coral)	0.0798
w^{sat}/w^{sat} (satsuma)	0.1404
w^{col}/w^{col} (colored)	0.1636

Source: From spectrophotometric data of D. J. Nolte, 1959. The eye-pigmentary system of *Drosophila. Heredity* 13:219–281.

Thus far two main points have emerged from our analysis:

1. The existence of multiple alleles of a gene (the white gene in our example) can be determined by genetic experiments.
2. The range of eye colors represents a range of phenotypic expressions of the same gene and reflects the varying extents to which the biological activity of the white gene product has been altered.

Another example of multiple alleles of a gene is found in the human ABO blood group series. Since certain blood groups are incompatible, these alleles are of particular importance when blood transfusions are contemplated.

Four blood group phenotypes occur in the ABO system; O, A, B, and AB; Table 4.3 gives their possible genotypes. The six genotypes involved with the four phenotypes represent various combinations of three ABO blood group alleles, I^A, I^B, and i. People homozygous for the recessive i allele are of blood group O. Both I^A and I^B are dominant to i, and thus you will express blood group A if you are either I^AI^A or I^Ai, and you will express blood group B if you are either I^BI^B or I^Bi. Since heterozygous I^AI^B individuals express blood group AB, and I^A and I^B alleles are considered to exhibit codominance (see p. 108).

The genetics of this system follows basic Mendelian principles. An individual who expresses blood group O, for example, must be ii in geno-

type. The parents of this person could both be O ($ii \times ii$), they could both be A ($I^Ai \times I^Ai$ to give one-fourth ii progeny), they could both be B ($I^Bi \times I^Bi$), or one could be A and one could be B ($I^Ai \times I^Bi$). Simply put, each parent would have to be either homozygous for i or heterozygous, with i being one of the two alleles.

Blood-typing (the determination of an individual's blood group) and the analysis of the inheritance of blood groups are sometimes used in legal medicine in cases of disputed paternity or

Table 4.3 ABO Blood Groups in Humans Determined by the Alleles I^A, I^B, *and* i

Phenotype (Blood Group)	Genotype
O	ii
A	I^AI^A or I^Ai
B	I^BI^B or I^Bi
AB	I^AI^B

maternity or in cases of an inadvertent baby switch in a hospital. In such cases genetic data may indicate that a person could be the parent, but it cannot prove that he or she is, indeed, the parent. Genetic analysis can only be used to show that an individual is not the parent of a particular child; for example, a child of phenotype AB (genotype $I^A I^B$) could not be the child of a parent of phenotype O (genotype *ii*). (*Note:* Blood-type data alone are usually not sufficient for a legal decision according to the laws in most states.)

When giving blood transfusions, doctors must match carefully the blood groups of donors and recipients because the blood group alleles specify molecular groups, called cellular antigens, that attach to the outside of the red blood cells. The I^A allele specifies the A antigen, and in people with blood group A ($I^A I^A$ or $I^A i$) the blood serum contains naturally occurring antibodies for the B antigen (called anti-B antibodies). Conversely, people of the B blood group have serum

containing naturally occurring anti-A antibodies. For the AB blood type both A and B antigens are found on the blood cells, and neither anti-A nor anti-B antigens occur in the serum of people with this blood type. In people with blood type O the blood cells have neither A nor B antigen, and therefore the serum contains both anti-A and anti-B antibodies. These antigen-antibody relationships are shown in Figure 4.3. This figure also shows the reaction that occurs when blood cells from the four groups are added to serum from each of the four groups. In a number of instances the blood cells clump, or agglutinate. Agglutination may lead to organ failure

Figure 4.3 Antigenic reactions that characterize the human ABO blood groups. Blood serum from each of the four blood groups was mixed with blood cells from the four types in all possible combinations. In some cases, such as a mix of B serum with A cells, the cells become clumped.

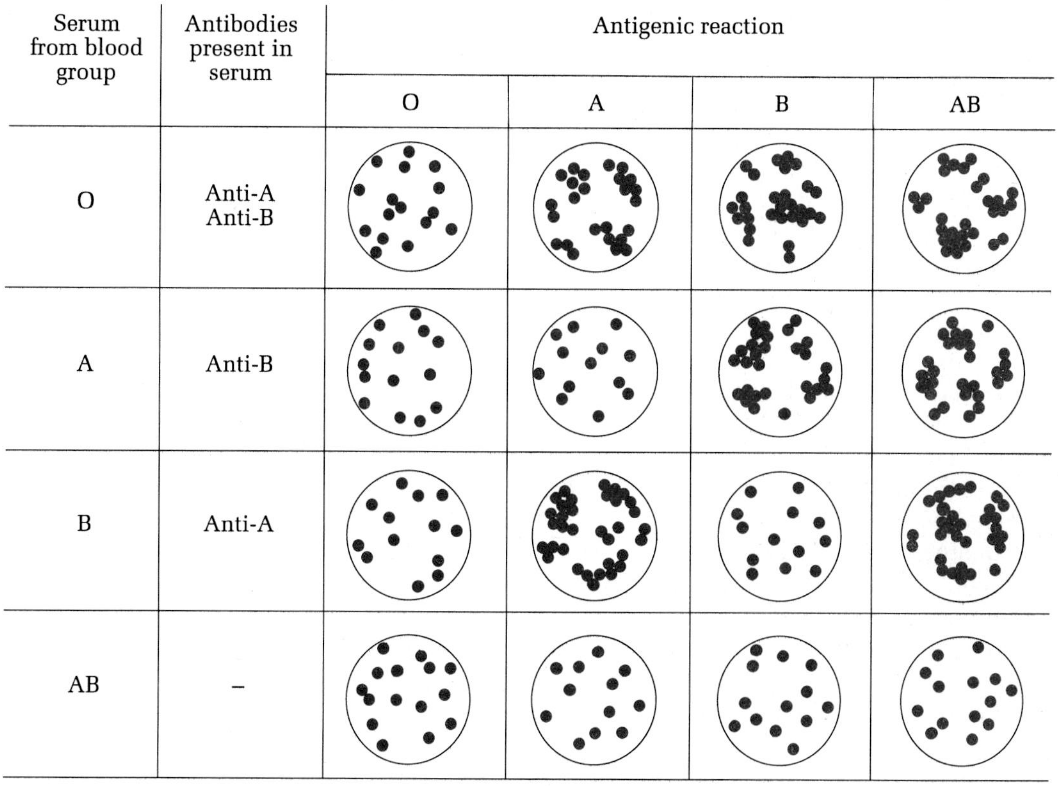

Serum from blood group	Antibodies present in serum	Antigenic reaction			
		O	A	B	AB
O	Anti-A Anti-B				
A	Anti-B				
B	Anti-A				
AB	–				

and possibly death since the clumped cells cannot move through the fine capillaries. Note that individuals with the AB blood group can receive blood of any group in transfusions because they have no antibodies in their serum—they are universal acceptors. Similarly, individuals with O blood group have no antigens in their serum, so they can give blood to any group; they are universal donors.

Because of their medical significance, the ABO blood groups have been studied in a wide variety of human populations. The data indicate a large variation in the relative distribution of O, A, B, and AB blood groups in different populations (Table 4.4) and have served for the study of the genetics of populations (Chapter 24).

Keynote *Many allelic forms of a gene can exist in a population. When they do, the gene is said to show multiple allelism, and the alleles involved are said to constitute a multiple allelic series. Any given diploid individual, however, can possess only two different alleles. Multiple alleles obey the same rule of transmission as alleles of which there are only two kinds, although the dominance relationships among multiple alleles vary from one group to another.*

Modifications of Dominance Relationships

Most of the genetic examples discussed so far have been cases in which one allele is dominant to the other so that at the phenotypic level the heterozygote is essentially indistinguishable from the homozygous dominant, a phenomenon called **complete dominance.** In **complete recessiveness,** the allele is phenotypically expressed only when it is homozygous. Complete dominance and complete recessiveness are the two extremes of a range of dominance relationships. For example, in *Drosophila* the white (*w*) allele is completely recessive to the wild-type (*w*$^+$) allele, which, therefore, is completely dominant to

the *w* allele. Similar dominance/recessiveness relationships are the case for the allelic pairs controlling all the character pairs studied by Mendel. Many allelic pairs, however, do not exhibit this dominance relationship.

Incomplete Dominance

In many cases one allele is not completely dominant to another allele, a phenomenon called **incomplete,** or **partial, dominance.** In incomplete dominance the heterozygote's phenotype is between that of individuals homozygous for either individual allele involved.

In chickens, for example, there is a mutant gene called frizzle that, when homozygous, results in the animal's having abnormal feathers (Figure 4.4). Instead of being closely interwebbed, as are feathers of the wild type, frizzled feathers are weak and stringy. By mechanical abrasion the feathers break, and the chickens be-

Table 4.4 Relative Frequencies of ABO Blood Groups in Some Human Populations

Population	Blood Group			
	O	*A*	*B*	*AB*
Armenians	.289	.499	.132	.080
Austrians	.427	.391	.115	.066
Bolivian Indians	.931	.053	.016	.001
Chinese	.439	.270	.233	.058
Danes	.423	.434	.101	.042
Eskimos (Greenland)	.472	.452	.059	.017
French	.417	.453	.091	.039
Irish	.542	.323	.106	.029
Nigerians	.515	.214	.232	.039
U.S. whites (St. Louis)	.453	.413	.099	.035
U.S. blacks (Iowa)	.491	.265	.201	.043

Source: From A. E. Mourant, A. C. Kopec, and K. Domaniewska-Sobczak, 1976. *The Distribution of the Human Blood Groups.* 2nd ed., London, UK: Oxford University Press.

Figure 4.4 Frizzled fowl. (a) Phenotype of frizzled fowl; (b) feather of frizzled fowl showing the lack of the closely interwoven structure of a normal feather (c). As a result, the bird has an unusual, sometimes naked appearance.

come virtually naked. When a frizzled fowl is crossed with a true-breeding normal bird, the F_1 heterozygotes exhibit a phenotype intermediate between the two extremes: They are mildly frizzled chickens (Figure 4.5). When the F_1's are interbred, instead of seeing a 3:1 ratio characteristic of complete dominance, we see a 1:2:1 ratio of normal:mildly frizzled:frizzled chickens. Thus frizzle is an incompletely dominant phenotype. This phenotype ratio exactly parallels the genotypic ratio and is characteristic of incomplete dominance. Testcrosses of the mildly frizzled birds show that they are heterozygous, since a 1:1 ratio of mildly frizzled to totally frizzled fowl is seen in the progeny.

Initially, commercial interest was shown in frizzled fowl since it seemed that their feather phenotype would facilitate plucking for marketing purposes. However, the feather phenotype is only the most obvious phenotypic expression of the mutant frizzle gene. Frizzled fowl have many other problems, including an enlarged

heart, functional abnormalities of other organs, and, because of the feather defect, great difficulty in maintaining body heat. Suffice it to say that frizzled fowl are not used commercially.

This story illustrates an important point: The mutant allele of a gene may have more than one phenotypic consequence in the organism. **Pleiotropy** refers to the multiple phenotypic effects of a single mutant allele. Pleiotropy is a reasonable thing to expect if we recognize that the phenotypic expression of a gene is determined at the biochemical level. Often, a gene product may interact with or participate in many biochemical reactions.

Another example of incomplete dominance involves plumage color in chickens. Crosses between a true-breeding black strain and a true-breeding white strain give F_1 birds with grey plumage (Figure 4.6a), called Andalusian blues by chicken breeders. An Andalusian blue cannot be true breeding because it is heterozygous, so in Andalusian × Andalusian crosses (Figure

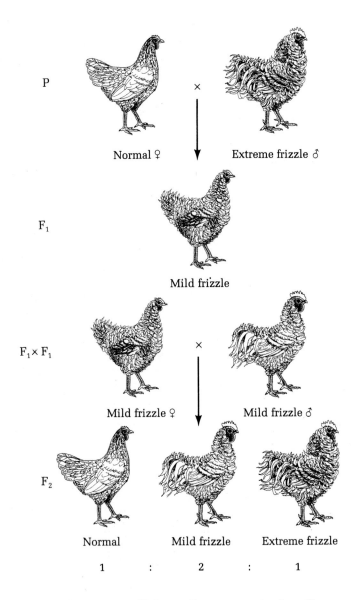

P

Normal ♀   ×   Extreme frizzle ♂

F₁

Mild frizzle

F₁ × F₁

Mild frizzle ♀   ×   Mild frizzle ♂

F₂

Normal   Mild frizzle   Extreme frizzle

1   :   2   :   1

Figure 4.5 Results of a cross between a normal chicken and an extremely frizzled fowl. The F₁ is heterozygous and, because of incomplete dominance, has a mildly frizzled condition. The F₂ produced by interbreeding the F₁ generation shows a 1:2:1 ratio of normal:mildly frizzled:extremely frizzled birds.

red:pink:white (Figure 4.7b). Thus the red and the white flowers are determined by homozygosity for the respective color alleles involved and the pink flowers by heterozygosity for the two alleles.

One example of incomplete dominance has already been discussed, although not specifically in those terms: the dominant X-linked bar eyes in *Drosophila* (Chapter 3). Recall that in females the heterozygous B/B^+ flies had a less extreme intermediate bar eye (called wide bar) than did B/B females. Wide bar results from incomplete dominance of the mutant B allele.

Codominance

Codominance is another modification of the dominance relationship. Codominance is a situation in which the heterozygote exhibits the phenotypes of both homozygotes. It differs from incomplete dominance, in which the heterozygote exhibits a phenotype intermediate between the two homozygotes.

A good example of codominance is found in the human ABO blood series, described earlier in this chapter (pp. 104–106). Heterozygous $I^A I^B$ individuals are blood group AB because both the A antigen (product of the I^A allele) and the B antigen (product of the I^B allele) are produced. Thus, the I^A and I^B alleles are considered to be codominant.

The M–N human blood group system is another example of codominance. This system is of less clinical importance than the ABO system in terms of transfusion incompatibility. In the M–N system three blood types occur: M, N, and MN. They are determined by the genotypes $L^M L^M$, $L^N L^N$, and $L^M L^N$, respectively. As in the ABO system, the M–N alleles result in the formation of antigens on the red blood cell surface. The heterozygote in this case has both the

4.6b) the two alleles will segregate in the offspring and produce black, Andalusian blue, and white fowl in a ratio of 1:2:1. The most efficient way to produce Andalusian blues is to cross black × white, since all progeny of this mating are Andalusian blues.

There are many examples of incomplete dominance in the plant kingdom, such as flower color in the snapdragon (Figure 4.7, p. 111). A cross of a red-flowered variety with a white-flowered variety produces F₁ plants with pink flowers (Figure 4.7a). The F₂ shows a 1:2:1 ratio of

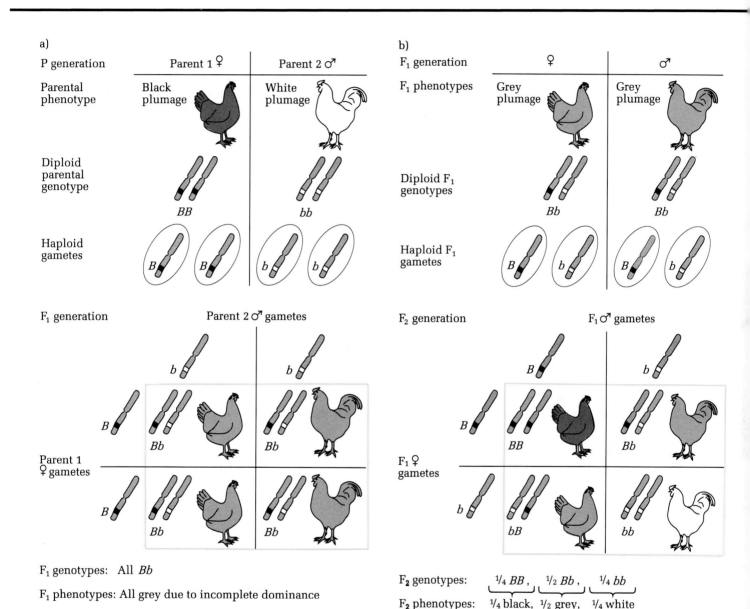

a)

| P generation | Parent 1 ♀ | Parent 2 ♂ |

Parental phenotype: Black plumage — White plumage

Diploid parental genotype: *BB* — *bb*

Haploid gametes: *B* *B* — *b* *b*

F₁ generation — Parent 2 ♂ gametes

Parent 1 ♀ gametes

Bb *Bb*
Bb *Bb*

F₁ genotypes: All *Bb*

F₁ phenotypes: All grey due to incomplete dominance

b)

| F₁ generation | ♀ | ♂ |

F₁ phenotypes: Grey plumage — Grey plumage

Diploid F₁ genotypes: *Bb* — *Bb*

Haploid F₁ gametes: *B* *b* — *B* *b*

F₂ generation — F₁ ♂ gametes

F₁ ♀ gametes

BB *Bb*
bB *bb*

F₂ genotypes: ¼ *BB*, ½ *Bb*, ¼ *bb*

F₂ phenotypes: ¼ black, ½ grey, ¼ white

Figure 4.6 Incomplete dominance in Andalusian fowls. (a) A cross between a white and a black bird produces F₁'s of intermediate grey color, called Andalusian blues. (b) The F₂ generation shows the 1:2:1 phenotype ratio characteristic of incomplete dominance.

M and the N antigens and shows the phenotypes of both homozygotes.

There is no distinct dividing line between codominance and incomplete dominance, and in many situations it is not easy to distinguish

which is involved. To make such a distinction requires the ability to detect, in the heterozygote, the distinct substances specified by each member of an allelic pair. Thus the molecular basis of a gene's expression must be understood for accurate description of the basis for each phenotype.

Keynote *In complete dominance the same phenotype results whether an allele is heterozygous or homozygous. In complete recessiveness the allele is phenotypically expressed only when it is homozygous; the recessive allele has no effect on the phenotype of the heterozygote. Complete dominance and complete recessiveness are two extremes between which all transitional degrees of dominance are possible. In incomplete dominance, for example, the phenotype of the heterozygote is intermediate between those of the two homozygotes, while in codominance the heterozygote exhibits the phenotypes of both homozygotes (essentially, no dominance is involved).*

Gene Interactions and Modified Mendelian Ratios

No gene acts by itself in determining the phenotype of an individual; the phenotype is the result of highly complex and integrated patterns of molecular reactions that are under direct gene control. All the genetic examples we have discussed and will discuss have discrete biochemical bases, and in a number of cases the complex interactions between genes can be detected by genetic analysis. Some of these examples will be discussed in this section.

In Chapter 2 we discussed Mendel's principle of independent assortment, which states that genes on different chromosomes behave independently in the production of gametes. Let's assume that there are two independently assorting gene pairs, each with two alleles; A and a, and B and b. The outcome of a cross between individuals each of whom is doubly heterozygous ($A/a\ B/b \times A/a\ B/b$) will be nine genotypes in the following proportions: 1/16 $A/A\ B/B$:2/16 $A/A\ B/b$:2/16 $A/a\ B/B$:4/16 $A/a\ B/b$:1/16 $A/A\ b/b$:2/16 $A/a\ b/b$:1/16 $a/a\ B/B$:2/16 $a/a\ B/b$:1/16 $a/a\ b/b$. If the phenotypes determined by the two allelic pairs are distinct—for example, smooth

versus wrinkled peas, long versus short stems—then we get the familiar dihybrid phenotypic ratio of 9:3:3:1. That is, 9/16 of the progeny show both dominant phenotypes, 3/16 show one dominant phenotype and the other gene pair's recessive phenotype, 3/16 show the first gene pair's recessive phenotype and the other pair's dominant phenotype, and 1/16 show both recessive phenotypes. Any alteration in this standard 9:3:3:1 ratio indicates that the phenotype is the product of the interaction of two or more genes. This genotypic distribution will be referred to as each modified Mendelian ratio is discussed.

For the 9:3:3:1 phenotypic ratio the genotypes for these phenotypes can be represented in a shorthand way as, respectively, $A-\ B-$, $A-\ bb$, $aa\ B-$, $aa\ bb$. The dash indicates that the phenotype is the same whether the gene is homozygous dominant or heterozygous (e.g., $A-$ means AA and/or Aa). This system cannot be used when incomplete dominance or codominance is involved.

In the following sections, the main processes that result in modified Mendelian ratios will be discussed. In the first section, examples of interactions between nonallelic genes that control the same general phenotypic attribute will be described. In the second section, examples of interactions of nonallelic genes in which one gene masks the expression of another gene will be discussed. This second type of interaction is called **epistasis.** In both cases the discussions will be confined to dihybrid crosses in which the two pairs of alleles assort independently. In the "real world" there are many more complex examples of gene interactions involving more than two pairs of alleles and/or genes that do not assort independently.

Gene Interactions That Produce New Phenotypes

In all the previous examples of dihybrid crosses, the two character pairs have acted independently with respect to the phenotypes followed. For example, the allelic pair for round/wrinkled peas had no effect on the allelic pair for long/short stem. If, however, the two allelic pairs af-

111
Gene
Interactions
and Modified
Mendelian
Ratios

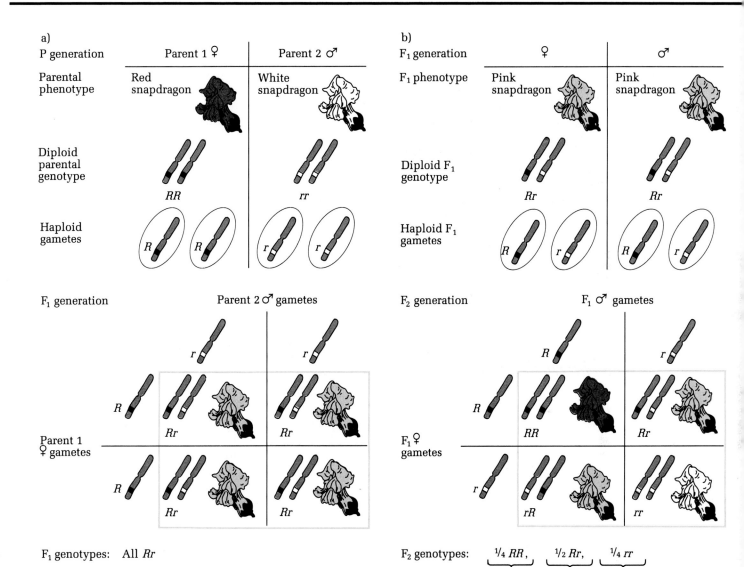

a)

| P generation | Parent 1 ♀ | Parent 2 ♂ |

| Parental phenotype | Red snapdragon | White snapdragon |

Diploid parental genotype *RR* / *rr*

Haploid gametes: *R* *R* / *r* *r*

F₁ generation — Parent 2 ♂ gametes: *r* *r*

Parent 1 ♀ gametes: *R* *R*

Rr *Rr* / *Rr* *Rr*

F₁ genotypes: All *Rr*

F₁ phenotypes: All pink due to incomplete dominance

b)

| F₁ generation | ♀ | ♂ |

| F₁ phenotype | Pink snapdragon | Pink snapdragon |

Diploid F₁ genotype *Rr* / *Rr*

Haploid F₁ gametes: *R* *r* / *R* *r*

F₂ generation — F₁ ♂ gametes: *R* *r*

F₁ ♀ gametes: *R* *r*

RR *Rr* / *rR* *rr*

F₂ genotypes: ¼ *RR*, ½ *Rr*, ¼ *rr*

F₂ phenotypes: ¼ red, ½ pink, ¼ white

Figure 4.7 Incomplete dominance in snapdragons. (a) A cross of a true-breeding red-flowered variety with a white-flowered variety produces F₁'s with pink flowers. (b) A 1:2:1 ratio of red:pink:white is seen in the F₂.

fect the same phenotypic characteristic, there is a chance of gene product interaction, and a variety of phenotypic ratios can result, depending on the type and the extent of interaction between the products of the nonallelic genes.

Comb shape in chickens. A classical example of such gene interactions is comb shape in chickens; new comb shape phenotypes are produced as a consequence of the interactions between two allelic pairs. Figure 4.8 shows the four comb phenotypes that result from the interaction of the alleles of two gene loci. Each of these types can be bred true.

Crosses made between true-breeding rose-combed and single-combed varieties showed

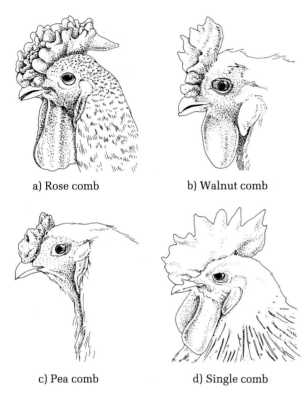

a) Rose comb

b) Walnut comb

c) Pea comb

d) Single comb

Figure 4.8 Four distinct comb shape phenotypes in chickens that result from all possible combinations of a dominant and a recessive allele at each of two gene loci: (a) rose comb (R− pp); (b) walnut comb (R− P−); (c) pea comb (rr P−); (d) single comb (rr pp).

that rose was dominant over single. When the F_1 rose-combed birds were bred together, there was a clear segregation into 3 rose:1 single in the F_2. Similarly, pea comb was found to be dominant over single, with a 3 pea:1 single ratio in the F_2. When true-breeding rose and pea varieties were crossed, however, a new and interesting result was obtained (Figure 4.9). Instead of showing either rose or pea combs, all birds in the F_1 showed a new comb form different from either the rose or the pea comb (Figure 4.9a). This new form was called walnut comb since it resembles half a walnut meat.

When the F_1 walnut-combed birds were bred together, another fascinating result was observed in the F_2 (Figure 4.9b): Not only did walnut-, rose-, and pea-combed birds appear, but so did

single-combed birds. These four comb types occurred in a ratio of 9 walnut:3 rose:3 pea:1 single. Such a ratio is characteristic in F_2 progeny from a cross of parents differing in two genes. The doubly dominant class in the F_2 was walnut, while the proportion of the singles indicated that this class contained both recessive alleles involved.

The overall explanation of the result is as follows: The walnut comb depends on the presence of two dominant alleles, *R* and *P,* of two independently assorting gene loci. In the presence of at least one *R* allele, and with homozygosity for the recessive *p* allele, a rose comb results. Birds with at least one *P* allele, and homozygous for the recessive *r* allele, have a pea comb. Doubly homozygous *rr pp* birds have a single comb. These interpretations are reflected in the Punnett square of Figure 4.9.

Thus the interaction of both dominant alleles, each of which individually produces a different phenotype, produces a new phenotype. The biochemical basis for the four comb types is not known. At a very general level we can propose that the single-comb phenotype results from the activities of a number of genes other than the *R* and *P* genes. Thus *rr pp* birds do not produce any functional gene product that influences the comb phenotype beyond the basic single appearance. The dominant *R* allele might produce a gene product that interacts with the products of those genes controlling the single-comb phenotype such that a rose-shaped comb results. Similarly, the dominant *P* allele might produce a gene product that interacts with the products of the single-comb genes to produce a pea-shaped comb. When the products of both the *R* and *P* alleles are present, they interact so that the comb develops differently, resulting in the walnut comb.

Fruit shape in summer squash (9:6:1 ratio). In the comb shape example each dominant allele alone produces a different phenotype, whereas both dominant alleles together produce a new phenotype. And when both recessive alleles are homozygous, yet another comb phenotype results. Similarly, fruit shape in summer squashes

113
*Gene
Interactions
and Modified
Mendelian
Ratios*

shows complete dominance at both gene pairs, and interaction between both dominants results in a new phenotype. In this case, though, each dominant allele alone results in the same phenotype, so this example differs from the comb shape example discussed above.

In summer squash two varieties are spherical fruit and long fruit (Figure 4.10, p. 116). The long-fruit varieties are always true breeding. However, in some crosses between different true-breeding spherical varieties, the F_1 fruit is disk-shaped (Figure 4.10), and in the F_2 fruit approximately 9/16 is disk-shaped, 6/16 is sphere-shaped, and 1/16 is long-shaped. The explanation is as follows: Either dominant allele alone (*A– bb* or *aa B–*) specifies spherical fruit, while the two nonallelic dominant alleles (*A– B–*) interact together to produce a new phenotype, namely, disk-shaped fruit. The doubly homozygous recessive (*aa bb*) gives a long fruit shape. Thus in the cross above the sphere-shaped parentals are *AA bb* and *aa BB*. The F_1's are disk-shaped and doubly heterozygous *Aa Bb*. The F_2 disk-shaped fruits are *A– B–*, the spherical fruits are *A– bb* and *aa B–*, and the long-shaped fruits are *aa bb,* thereby giving the 9:6:1 ratio.

Again, the precise biochemical bases for the different shapes of squash fruit are not known. In the absence of *A–* and *B–* allele products, in the *aa bb* squash, the functions of other genes determine the long fruit shape. If either the *A–* or the *B–* allele product, *but not both,* is present, the basic long fruit shape is modified into a sphere shape. The disk fruit shape presumably occurs through a modification of the sphere shape as a result of the interaction of the *A–* and *B–* allele products together.

Epistasis

Epistasis is the interaction of nonallelic genes in which one gene masks the expression of another. Genes whose expression is masked by nonallelic genes are said to be *hypostatic.* Epistasis may be caused by the presence of homozygous recessives of one gene pair so that *aa* masks the effect of the *B–* gene, if we think of our general listing of F_2 genotypes presented at the beginning of this section. Or epistasis may result from the presence of one dominant allele in a gene pair. For example, *A* might mask the effect of the *B* gene. Moreover, the epistatic effect need not be just in one direction, as in the examples we just mentioned. Epistasis can occur in both directions between two gene pairs. With all these possibilities there can be quite a number of modifications of the 9:3:3:1 ratio.

Coat color in rodents (the 9:3:4 ratio). In one type of epistasis *aa B–* and *aa bb* individuals have the same phenotype, so the overall phenotypic ratio is 9:3:4. An example is coat color in rodents. The ancestral coat color of mice, for example, is the greyish color seen in ordinary wild mice. This color is due to the presence of two pigments in the fur. Individual hairs are mostly black with narrow yellow bands near the tip (Figure 4.11, p. 116). This coloration, the agouti pattern, has a camouflage function and is found in many wild rodents, including the wild rabbit, the guinea pig, the grey squirrel and wild mice.

Several variations in the agouti coat coloration have been preserved under domestication. The most familiar example is the albino, in which the coat is white and the eyes are pink because of the complete lack of pigment in the iris. Albinos are true breeding, and this variation behaves as a complete recessive to any other color. Another variant has a black coat color, which is presumably produced as a result of the absence of yellow pigment from the agouti pattern. Black also breeds true and is recessive to agouti. When black mice are crossed with albinos, the F_1 progeny are all agouti. When these F_1 agoutis are interbred, the F_2 progeny consist of approximately 9/16 agouti animals, 3/16 black, and 4/16 albino. The explanation of the results is that the parents differ in a gene necessary for the development of any color, which the black mice have but the albinos do not, and in a gene for the agouti pattern, which results in a banding of the black hairs with yellow.

It is now known that three gene loci are involved in the color phenotypes discussed above. At one locus the dominant *C* allele specifies a product that is necessary for the production of

a)

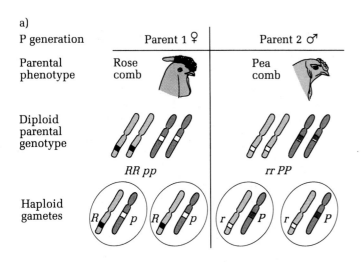

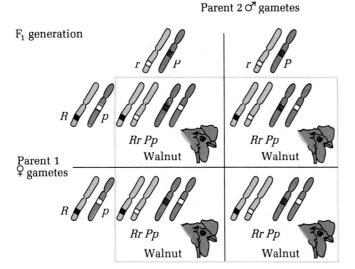

F₁ genotypes: All *Rr Pp*

F₁ phenotypes: All walnut comb

Figure 4.9 Results of genetic crosses showing the interaction of genes for comb shape in fowl. (a) The cross of a true-breeding rose-combed bird with a true-breeding pea-combed bird gives all walnut-combed offspring in the F₁. (b) When the F birds are interbred, a 9:3:3:1 ratio of walnut:rose:pea:single occurs in the F₂.

that governs the synthesis of black pigment. The recessive allele *b*, when homozygous, results in brown pigment. This latter locus is relevant to the example only insofar as the basic hair color is black; therefore all the mice involved must have at least one *B* allele.

The results in the example can be interpreted, then, on the assumptions that the parental black was *aa CC* and the parental albino was *AA cc.* This parentage gives F₁ progeny with the genotype *Aa Cc,* and these progeny are agoutis since the *A* and *C* products for agouti pattern and color production, respectively, are both synthesized. In the F₂ animals the *A– C–* are agouti, the *aa C–* are black, and the *A– cc* and *aa cc* are albino, giving a 9:3:4 ratio.

Flower color in sweet peas (the 9:7 ratio). The sweet pea occurs in a number of true-breeding varieties, most of which have descended from a purple-flowered wild sweet pea of Sicily. Purple flower color is dominant to white and gives a typical 3:1 ratio in the F₂. The white-flowered varieties breed true, and crosses between different white varieties usually produce white-flowered progeny. In some cases, however, crosses of two true-breeding white varieties give only purple-flowered F₁ plants. When these F₁ hybrids are self-fertilized, they produce an F₂ generation consisting of about 9/16 purple-flowered sweet peas and 7/16 white-flowered. All the F₂ white-flowered plants breed true when self-fertilized. A ninth of the purple-flowered F₂ plants breed true.

These results may be explained by considering the interaction of two nonallelic genes. In contrast to the results in the example of comb shape in chickens, no new traits appear in the F₁ or F₂. The occurrence of purple flowers in

any pigment in the coat; the recessive *c* allele, when homozygous, prevents pigment formation and hence the mice are albino. At a second locus the dominant allele *A* specifies a product that determines the agouti factor. Its recessive allele *a* is present in the homozygous state in all nonagouti mice, such as blacks. The dominant allele *B* of the third locus specifies a product

115
*Gene
Interactions
and Modified
Mendelian
Ratios*

b)

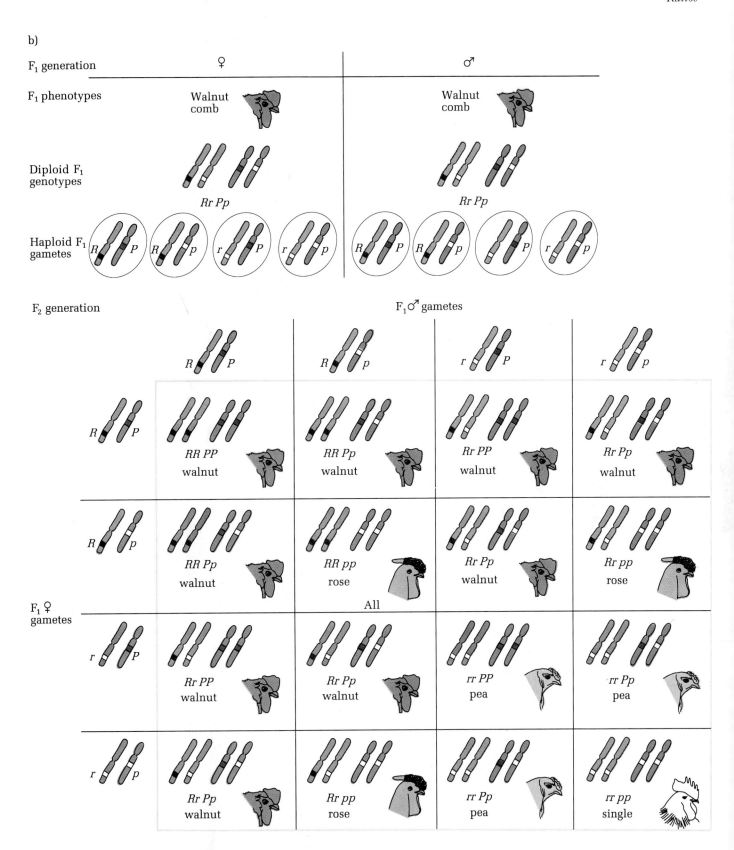

F₁ generation ♀ ♂

F₁ phenotypes — Walnut comb — Walnut comb

Diploid F₁ genotypes — *Rr Pp* — *Rr Pp*

Haploid F₁ gametes — *R P* *R p* *r P* *r p* — *R P* *R p* *r P* *r p*

F₂ generation — F₁♂ gametes

F₁♀ gametes

	R P	*R p*	*r P*	*r p*
R P	*RR PP* walnut	*RR Pp* walnut	*Rr PP* walnut	*Rr Pp* walnut
R p	*RR Pp* walnut	*RR pp* rose (All)	*Rr Pp* walnut	*Rr pp* rose
r P	*Rr PP* walnut	*Rr Pp* walnut	*rr PP* pea	*rr Pp* pea
r p	*Rr Pp* walnut	*Rr pp* rose	*rr Pp* pea	*rr pp* single

F₂ phenotypes: ⁹⁄₁₆ walnut, ³⁄₁₆ rose, ³⁄₁₆ pea, ¹⁄₁₆ single comb (*rr pp*)

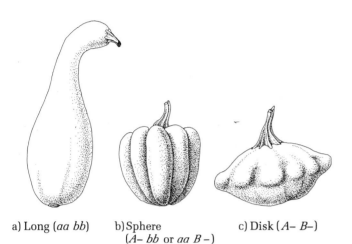

a) Long (*aa bb*)   b) Sphere   c) Disk (*A– B–*)
                      (*A– bb* or *aa B –*)

Figure 4.10 Three true-breeding fruit shapes of summer squash: (a) long; (b) sphere; (c) disk.

about 9/16 of the F_2 plants suggests that colored flowers appear only when two independent dominant factors are present together and that the color purple results from some interaction between them. White flower color would then be due to the absence of either or both of these factors. Thus we propose that gene pair *C/c* specifies whether or not the flower can be colored, and gene pair *P/p* specifies whether or not purple flower color will result.

Figure 4.12 presents a hypothetical pathway for the production of purple pigment. In this pathway a colorless precursor compound is converted through several steps (via compounds 1, 2, and 3) to a purple end product. Each step is controlled by a functional gene product. To explain the F_2 ratio in the sweet pea example, a hypothesis can be put forward that genes *C* and *P* control different steps early in the pathway; *C* controls the conversion of white compound 1 to white compound 2, and *P* controls the conversion of compound 2 to compound 3 (Figure 4.12a). Therefore, homozygosity for the recessive allele of either or both of the *C* and *P* genes will result in a block in the pathway, so only white pigment rather than purple pigment will accumulate. That is, *C– pp*, *cc P–*, and *cc pp* genotypes will all be white (Figure 4.12b). The only

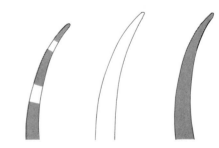

a) Agouti   b) Albino   c) Black
 (*A– C–*)   (*A– cc* or *aa cc*)   (*aa C–*)

Figure 4.11 Pigment patterns in rodent fur: (a) agouti; (b) albino; and (c) black.

117
*Gene
Interactions
and Modified
Mendelian
Ratios*

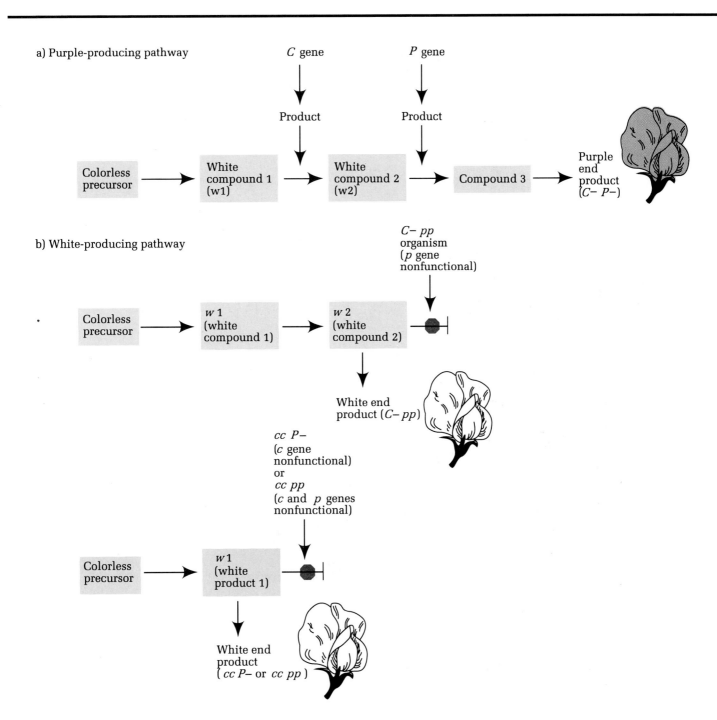

*Figure 4.12 Hypothetical pathway for the production
of purple pigment (a) and white pigment (b) in sweet
peas to explain the 9 purple:7 white ratio in the F_2 of
a dihybrid self.*

plants that produce purple flowers will be those in which the two steps are completed so that the rest of the pathway leading to the colored pigment can be carried out. This situation occurs only in *C– P–* plants. (The hypothetical pathway proposed for this sweet pea example is applicable to other 9:7 ratios, with modification, of course, for the particular phenotypes encountered.)

In the cross described, the two white parentals were *CC pp* and *cc PP,* and the F_1 plants were purple and doubly heterozygous. Interbreeding the F_1 gives a 9:7 ratio of purple:white in the F_2. The true-breeding purple F_2 plants were *CC PP.* In this case we have epistasis occurring in both directions between two gene pairs. The consequence of this gene interaction is that the same phenotype (white) is exhibited whenever one or the other gene pair is homozygous recessive.

Fruit color in summer squash (the 12:3:1 ratio). Summer squash has three common fruit colors: white, yellow, and green. In crosses between white and yellow and between white and green, white is always expressed. In crosses between yellow and green, yellow is expressed. Yellow thus acts as a recessive in relation to white but as a dominant in relation to green. The simplest explanation is that there is an allele for white that is epistatic to those for yellow and green.

Consider two gene pairs, *W/w* and *Y/y.* In squashes that are *W–* in genotype, the fruit is white no matter what the genotype is at the other locus. In *ww* plants the fruit will be yellow if a dominant allele of the other locus is present and green if it is absent. That is, *W– Y–* and *W– yy* plants have white fruits, *ww Y–* plants have yellow fruits, and *ww yy* plants have green fruits. The F_2 progeny of an F_1 self of doubly heterozygous individuals shows a 12:3:1 ratio of white:yellow:green fruits in the plants. This example shows complete dominance at both gene pairs, with one gene (the white one here), when dominant, epistatic to the other.

A hypothetical biochemical pathway to explain the 12:3:1 ratio of squash color is shown in Figure 4.13. The assumption is that a white

substance is converted to a yellow end product via a green intermediate. Both steps are controlled by genes, the green to yellow step being controlled by the *Y* gene. The F_2 phenotypic ratio will be generated if the dominant *Y* allele is needed for the conversion of the green substance to yellow and if the dominant *W* allele specifies a product that acts as an inhibitor of the white-to-green conversion step. Thus all plants that have at least one *W* allele will be white-fruited no matter which alleles are present at the *Y* locus, since no green substance can be made in this case (Figure 4.13a). As a result, 12/16 of the F_2 squashes are white-fruited and have the genotypes *W– Y–* and *W– yy.* Also, 3/16 of the F_2's are yellow-fruited and are the *ww Y–* individuals (Figure 4.13b). In this case green substance is made since there is no inhibition of the white-to-green step. Further, functional *Y* allele product is present and catalyzes the conversion of green substance to yellow substance. Lastly, 1/16 of the F_2's are green-fruited and are the *ww yy* individuals (Figure 4.13c). Again, green substance is produced because there is no inhibition of the white-to-green step, but in the absence of a dominant *Y* allele the green substance cannot be converted to yellow substance.

In summary, many types of interaction are possible between the products of gene pairs (Table 4.5). Geneticists detect such interactions when they observe modifications of the expected phenotypic ratios in crosses. We have discussed some examples in which two nonallelic genes assort independently and in which complete dominance is exhibited in each gene pair. The ratios we discussed would necessarily be modified further if the genes did not assort independently or if incomplete dominance or codominance prevailed.

Keynote *In many instances in plants and animals, nonallelic genes do not function independently in determining phenotypic characteristics. In some cases interaction between gene products results in new phenotypes without modification of typical Mendelian ratios. In another type of gene interaction, called epistasis, interaction between gene products*

119
Gene
Interactions
and Modified
Mendelian
Ratios

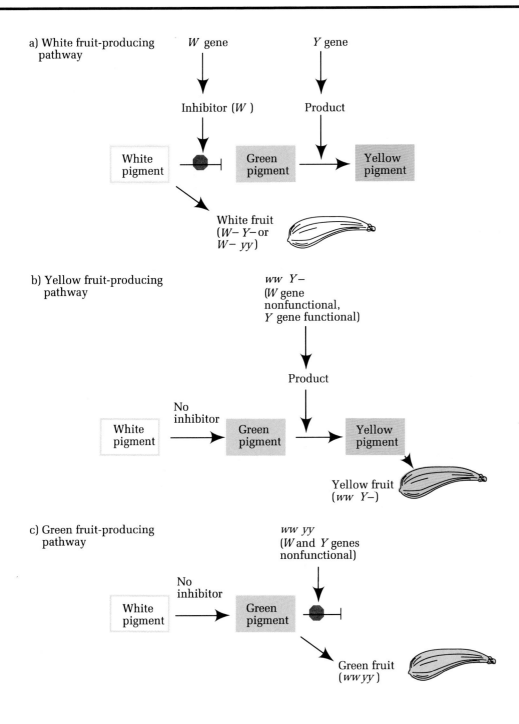

a) White fruit-producing pathway

b) Yellow fruit-producing pathway

c) Green fruit-producing pathway

Figure 4.13 Hypothetical pathway to explain the F₂ ratio of 12 white:3 yellow:1 green color in squashes. (a) Production of white color; (b) Production of yellow color; (c) Production of green color.

causes modifications of Mendelian ratios because one gene product interferes with the phenotypic expression of another nonallelic gene (or genes). The phenotype is controlled largely by the former gene and not the latter when both genes occur together in the genotype. Epistasis is complicated further when one or both gene pairs involve incomplete dominance or codominance or when gene pairs do not assort independently.

Lethal Genes

For a few years after the rediscovery of Mendelism, geneticists believed that mutations only

changed the appearance of a living organism, but then they discovered that a mutant allele could cause the death of an organism. In a sense this mutation is still a change in phenotype, with the new phenotype being lethality. An allele that results in the death of an organism is called a **lethal allele.**

In 1905 Lucien Cuenot reported the results of his studies of a mouse body color gene. Wild-type mice have a dark body color, and Cuenot was interested in the inheritance of a gene that causes yellow body color. The yellow variety resembles Andalusian fowl in that it never breeds true. Matings between two yellows produce progeny of which about 2/3 are yellow and 1/3 are of another color (black, brown, or grey).

Table 4.5 Summary of Dominance Modifications

		AA BB	AA Bb	Aa BB	Aa Bb	AA bb	Aa bb	aa BB	aa Bb	aa bb
~~More~~ Fewer than four phenotypic classes	A and B both incompletely dominant	1	2	2	4	1	2	1	2	1
	A incompletely dominant; B completely dominant	3			6	1	2	3		1
Four phenotypic classes	A and B both completely dominant (classic ratio)	9				3		3		1
~~Fewer~~ More than four phenotypic classes	aa epistatic to B and b; recessive epistasis	9				3		4		
	A epistatic to B and b; dominant epistasis	12						3		1
	A epistatic to B and b; bb epistatic to A and a; dominant and recessive epistasis	13[a]						3		
	aa epistatic to B and b; bb epistatic to A and a; duplicate recessive epistasis	9						7		
	A epistatic to B and b; B epistatic to A and a; duplicate dominant epistasis	15								1
	Duplicate interaction	9					6			1

[a] The 13 is composed of the 12 classes immediately above plus the one *aa bb* from the last column.

When yellows were bred to nonyellows, about half the young produced were yellow and half were nonyellow. The latter ratio is what is expected from the mating of a heterozygote with a recessive, which suggested that yellow mice were heterozygous. When heterozygotes are interbred, however, we would expect 1/4 to be pure yellow, 1/2 to be heterozygous yellow, and 1/4 to be nonyellow.

Cuenot's data were correctly interpreted in 1910 by W. Castle and C. Little, who proposed that the yellow homozygotes were aborted in utero. In other words, the yellow allele has a dominant effect with regard to coat color, but a recessive allele, when homozygous, leads to death. The correctness of the Castle-Little hypothesis was provided by the examination of pregnant mice. The yellow × yellow cross is shown in Figure 4.14. With the symbol Y assigned to the yellow allele, the cross is $Yy \times Yy$. We expect a genotypic ratio of 1 YY : 2 Yy : 1 yy among the progeny. The 1/4 YY's die before birth, leaving 2/4 Yy and 1/4 yy survivors, which gives a 2:1 phenotypic ratio of yellow (Yy):nonyellow (yy) mice in the offspring. An allele that causes lethality in the homozygous state, as Y does, is called a **recessive lethal.** Characteristically, recessive lethal alleles are recognized by a 2:1 ratio of progeny types when two heterozygotes are crossed.

Numerous examples of lethal genes are found in all diploid organisms. For example, the recessive lethal Dexter gene in cattle has properties very similar to those of the yellow gene in the mouse. Heterozygous Dexter cattle are viable but have short limbs. (In fact, someone once suggested that Dexter cattle would be a useful strain to raise since they would be less likely to escape from fenced pastures.) In the homozygous state the Dexter gene produces calves with extremely short limbs and a bulldoglike appearance. These calves are usually aborted.

In humans there are also many known recessive lethal alleles. One particular autosomal recessive mutant allele, when homozygous, causes *Tay-Sachs disease.* Homozygotes appear normal at birth, but before about one year of age they begin to show symptoms of central nervous sys-

tem deterioration. Progressive mental retardation, blindness, and loss of neuromuscular control follow. The afflicted children usually die at three to four years of age. The genetic defect results in a deficiency in an enzyme called *hexosaminidase A* (hex A), which is required for the synthesis of complex chemical structures called *sphingolipids,* structures necessary for proper nerve function.

Sex-linked lethal genes as well as autosomal lethal genes exist, and dominant lethal genes as well as recessive lethals. By definition, dominant lethals exert their effect in heterozygotes so that, in general, the organism dies at conception or at quite a young age. Dominant lethals cannot be studied genetically unless death occurs after the organism has reached reproductive age. One example is the autosomal dominant trait Huntington's disease. This disease results in involuntary movements, progressive central nervous system degeneration, and eventually death. The onset of the symptoms may not occur until the early thirties or so, and death occurs usually when the afflicted persons are in their forties or fifties. As a result, the individuals may have passed on the gene to their offspring before knowing whether they are afflicted with the disease. The American folksinger Woody Guthrie died from Huntington's disease, as mentioned earlier.

Keynote   *A lethal gene is one that, when expressed, is fatal to the individual. Recessive lethal and dominant lethal genes exist, and they can be sex-linked or autosomal.*

The Environment and Gene Expression

The genetic constitution of the zygote only specifies the potential for the organism to develop and function. Many things can influence gene expression as the organism develops and differentiates, and one such influence is the organism's environment.

A gene is a segment of DNA, and an organ-

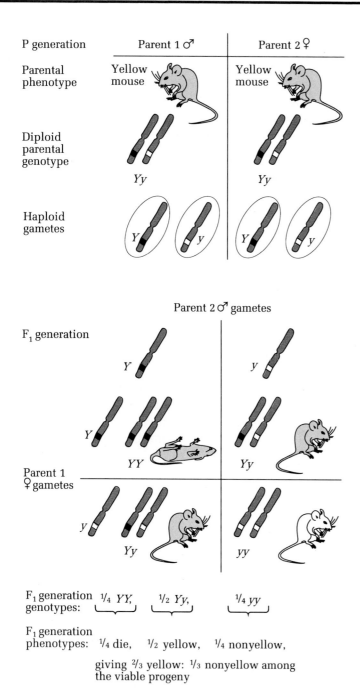

F₁ generation
genotypes: ¼ *YY,*   ½ *Yy,*   ¼ *yy*

F₁ generation
phenotypes: ¼ die,   ½ yellow,   ¼ nonyellow,

giving ⅔ yellow: ⅓ nonyellow among
the viable progeny

Figure 4.14 Inheritance of a lethal gene Y *in mice. A
mating of yellow by yellow gives 1/4 nonyellow
(black) mice, 1/2 yellow mice, and 1/4 dead embryos.
The viable yellow mice are heterozygous* Yy, *and the
dead individuals are homozygous* YY.

ism's set of genes (its genome) is found on its chromosomes. The *development* of an organism from a zygote is a process of regulated *growth* and *differentiation* that results from the interactions of the genome with the internal cellular environment and with the external environment. Development is a programmed series of phenotypic changes that are under temporal, spatial, and quantitative control. Development is essentially irreversible under normal environmental conditions. Four major processes interact with one another to constitute the complex process of development: replication of the genetic material, growth, differentiation of the various cell types, and the aggregation of differentiated cells into defined tissues and organs.

We must think of development as a series of intertwined, complex biochemical pathways whose steps are under gene control. Any of these pathways are susceptible to environmental influences if the products of the genes controlling the pathways are affected by the internal or external environmental parameters. Though genes influence development, no gene by itself totally determines a particular phenotype. In other words, a phenotype is the result of closely interwoven interactions among many factors during development. Therefore the same genetic constitution does not necessarily mean the same phenotypes will result for all individuals with that constitution. This phenomenon is most readily studied in experimental organisms where the genotype is unequivocally known. The extent to which the gene manifests its effects under varying environmental conditions can then be seen.

In some cases not all individuals who are known to have a particular gene show the phenotype specified by that gene. The frequency with which a dominant or homozygous recessive gene manifests itself in individuals in a population is called the **penetrance** of the gene. For example, an organism may be genotypically *A–* or *aa* but may not display the phenotype typically associated with that genotype. If 80 percent, say, of the individuals carrying a particular gene show the corresponding phenotype, we say that there is 80 percent penetrance. Pen-

etrance depends on both the genotype (e.g., the presence of epistatic or other genes) and the environment. Penetrance is complete (100 percent) when all the homozygous recessives show one phenotype, when all the homozygous dominants show another phenotype, and when all the heterozygotes are alike. Many genes show complete penetrance: All the seven gene pairs in Mendel's experiments and the alleles in the human ABO blood group system are two examples. If less than 100 percent of the carriers of a particular genotype exhibit the phenotype expected, penetrance is incomplete. Figure 4.15 illustrates the concept of penetrance.

An example of incomplete penetrance is retinoblastoma (tumor of the eye) disease in humans. This disease is inherited as an autosomal dominant, and the mutant gene shows incomplete penetrance since not all the individuals who have the gene for retinoblastoma express that trait. Another human gene that shows incomplete penetrance is the autosomal dominant gene that determines polydactyly (extra fingers or toes).

At a different level we can also determine the degree to which a gene influences a phenotype. **Expressivity** refers to the kind of or degree to which a penetrant gene or genotype is phenotypically expressed. Expressivity may be described in either qualitative or quantitative terms (Figure 4.16). Like penetrance, expressivity depends on both the genotype and the external environment, and it may be constant or variable. A classical example of variation in expression is found in the human condition osteogenesis imperfecta. The three main features of this disease are blue sclerae (the whites of the eyes), very fragile bones, and deafness. This condition is inherited as an autosomal dominant with almost 100 percent penetrance; that is, 100 percent of individuals with the genotype have the condition. However, the trait shows variable expressivity: A person with the gene may have any one or any combination of the three traits. Moreover, the fragility of the bones for those who exhibit this condition is also very variable. Therefore, we see that in medical genetics it is important to recognize that a gene may vary

Known genotype

Expected phenotype
produced

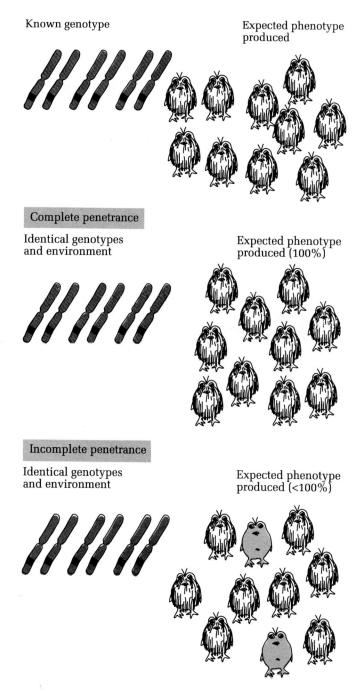

Complete penetrance

Identical genotypes
and environment

Expected phenotype
produced (100%)

Incomplete penetrance

Identical genotypes
and environment

Expected phenotype
produced (<100%)

Figure 4.15 Complete and incomplete penetrance.

widely in its expression, a qualification that makes the task of genetic counseling that much more difficult.

Keynote *Penetrance is the frequency with which a dominant or homozygous recessive gene manifests itself in the phenotype of an individual. Expressivity is the kind or degree of phenotypic manifestation of a penetrant gene or genotype.*

Effects of the Internal Environment

Complex biochemical responses occur in cells and in the organism as a result of certain stimuli from the internal environment. Not enough is known about cell biochemistry as it relates to development and adult function, though, to describe these environmental effects at the molecular level.

Age. The age of the organism is a relevant internal environmental change that can affect gene function. All genes do not continually function; instead, programmed activation and deactivation of genes occur as the organism develops and functions with time. The Huntington's disease gene is one example of a gene whose effects are not manifested until later in the organism's existence. Somehow, the gene recognizes the age of the individual, and the phenotypic effects, devastating as they are, commence. Numerous other age-dependent genetic diseases occur in humans; pattern baldness (i.e., the bald spot creeping forward from the crown of the head) appears in males between 20 and 30 years, and Duchenne severe muscular dystrophy appears in children between 2 and 5 years. While such interactions with age can be described, and while in some cases we know the defective gene product involved, in most cases the nature of the age dependency is not understood.

Sex. The expression of particular genes may be influenced by the sex of the individual. In the case of sex-linked genes, as mentioned earlier, differences in the phenotypes of the two sexes are related to different complements of genes on the sex chromosomes. However, in some cases genes that are not located on the sex chromosomes affect a particular character that appears in one sex but not the other. Traits of this kind are called **sex-limited traits.**

One such trait is the feathering pattern of chickens (Figure 4.17). A common difference between male and female fowls is in the structure of many of the feathers. Cocks often have long, narrow, pointed feathers on their hackles and saddle, pointed feathers on the cape, back, and wing, and long, curving, pointed sickle feathers on the tail. This pattern is called cock feathering. Hens usually show the contrasting feather characteristics of short, broad, blunt, and straight feathers. This pattern is called hen feathering. If F_1 heterozygous (h^+h) hen-feathered males and females are interbred, all F_2 females are hen-feathered, whereas F_2 males exhibit a 3:1 ratio of hen-feathered:cock-feathered. The interpretation is that the trait is controlled by an autosomal gene, with hen feathering (h^+) dominant to cock feathering (h). Assigning gene symbols, we can conclude that all three possible F_2 genotypes in the females, h^+h^+, h^+h, and hh, give rise to hen feathering of the tail, but in the F_2 males the h^+h^+ and h^+h chickens are hen-feathered while the hh chickens are cock-feathered. In other words, the cock-feathered trait is male-limited and shows that sex as an internal environmental parameter can affect gene expression. Several experiments have indicated that differences in sex hormones between the two sexes in association with the genetic constitution of the skin (in which the feathers originate) determine the state of feathering in the fowl.

A slightly different situation is found in **sex-influenced traits.** Such traits appear in both sexes, but either the frequency of occurrence in the two sexes is different or the relationship between genotype and phenotype is different. For example, the human trait spina bifida, in which the spinal cord does not completely close as the embryo develops, is found much more frequently in females than in males.

Another straightforward example of a sex-influenced trait is pattern baldness in humans

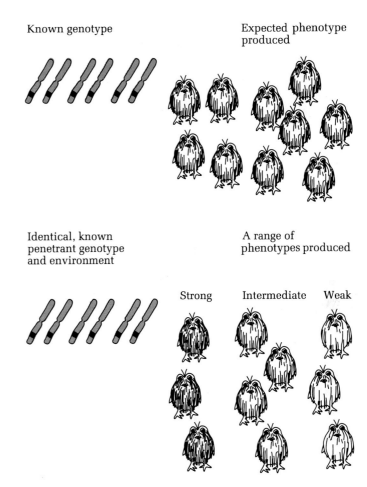

Known genotype

Expected phenotype produced

Identical, known penetrant genotype and environment

A range of phenotypes produced

Strong   Intermediate   Weak

Figure 4.16 Expressivity of a gene: mild, intermediate, or strong.

(Figure 4.18). One pair of alleles of an autosomal gene is involved in this trait, b^+ and b. The bb genotype specifes pattern baldness in both males and females, and the b^+b^+ genotype gives a nonbald phenotype. The difference lies in the heterozygote: In males it leads to the bald phenotype, and in females it leads to the nonbald phenotype. In other words, the b allele acts as a dominant in males but as a recessive in females; its expression is influenced by the sex hormones of the individual. This pattern of inheritance and gene expression explains why pattern baldness is far more frequent among men than among women. From a large sampling of

Hen-feathered cock (♂)

Cock (♂)   Hen (♀)

| Genotype | Phenotype | |
	♀	♂
$h^+ h^+$	Hen-feathered	Hen-feathered
$h^+ h$	Hen-feathered	Hen-feathered
$h h$	Hen-feathered	Cock-feathered

Figure 4.17 Sex-limited inheritance of feathering in chickens. From a cross between heterozygous F_1 parents the F_2 progenies show a 3:1 ratio of hen-feathered:cock-feathered male birds, while all females are hen-feathered. The gene controlling the phenotype is autosomal.

the progeny of matings between two heterozygotes, 3/4 of the daughters are nonbald and 1/4 are bald, and 3/4 of the sons are bald and 1/4 are nonbald.

Effects of the External Environment

Although many factors in the external environment influence gene expression (nutrition, light, chemicals, infectious agents, etc.), we will only discuss two factors here, temperature and chemicals (nutrition).

Temperature. The rate at which a chemical reaction occurs depends in large part on temperature, as is also true, within limits, for biochemical reactions taking place within organisms. Therefore it is reasonable to expect that temperature might affect gene expression, with significant effects on development. A good example is fur color in Himalayan rabbits (Figure 4.19). Certain genotypes of this white rabbit cause dark fur to develop at the extremities (ears, nose, and paws), where the temperature is lower. Since all the body cells develop from a single zygote, this distinct fur pattern cannot be the result of a genotypic difference of the cells in those areas. It must result from the external environmental influence.

This hypothesis can be tested by rearing Himalayan rabbits under different temperature conditions (Figure 4.20, p. 129). When a rabbit is reared at a temperature above 30°C, all its fur, including the ears, nose, and paws, is white (Figure 4.20a). If a rabbit is raised at a temperature of approximately 25°C, the typical Himalayan phenotype results (Figure 4.20b). Lastly, if a rabbit is raised at 25°C while its left rear flank is artificially cooled to a temperature below 25°C, the rabbit develops the Himalayan coat phenotype, with additional dark fur on the cooled flank area (Figure 4.20c).

Chemicals

Phenylketonuria. Ultimately, all the interactions we have discussed occur at the chemical level. The human disease phenylketonuria (PKU), for example, is a disorder controlled by an autosomal recessive gene. In individuals homozygous for the recessive allele, a variety of symptoms appear, most notably mental retardation at an early age. The cause of the phenotypes is a defective gene product in a biochemical pathway that is used for the metabolism of phenylalanine, an amino acid. (An amino acid is a building block of a protein.) The manifestation of the symptoms depends on the dietary intake of protein containing phenylalanine, such as the protein of mother's milk, which may be considered as an external environment effect.

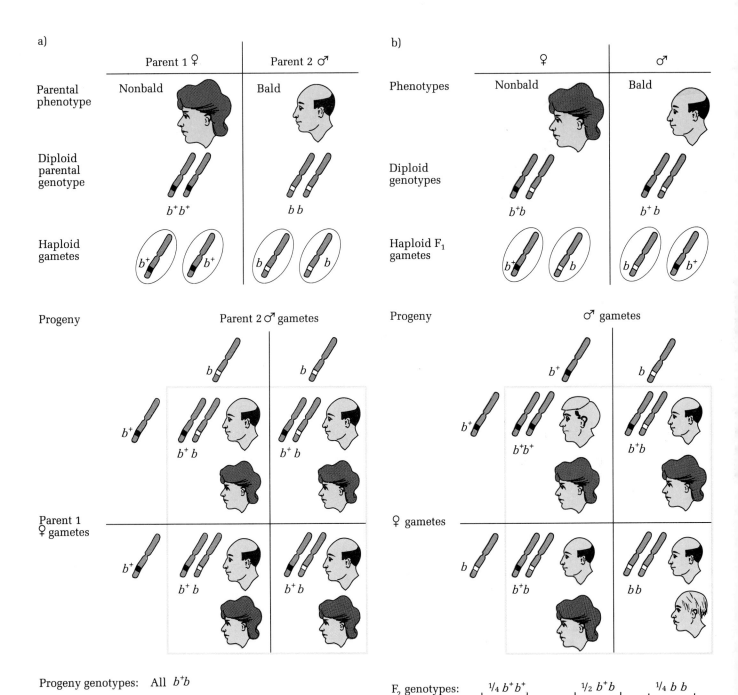

Figure 4.18 Sex-influenced inheritance of pattern baldness in humans. The allele is recessive in one sex and dominant in the other, so a cross between two heterozygotes produces a 3:1 ratio of bald:nonbald in males and 1:3 ratio of bald:nonbald in females. (a) Production of F₁ generation; (b) Production of F₂ generation.

Figure 4.19 Himalayan rabbit.

Phenylketonuria can be diagnosed soon after birth by a simple blood test and then treated by restricting the amount of phenylalanine in the diet.

Phenocopies induced by chemicals. Changes in the environment can influence the expression of one or more genes. The most sensitive area is development, since relatively small changes that take place early in development can result in great changes later. When administered to a developing embryo, certain drugs, chemicals, and viruses have been shown to produce effects that mimic those of known specific mutant alleles. The abnormal individual resulting from this treatment is called a **phenocopy** (phenotypic copies). A phenocopy is defined as a nonhereditary, phenotypic modification (caused by special environmental conditions) that mimics a similar phenotype caused by a gene mutation. The agent that produces a phenocopy is called a *phenocopying agent.* There are many examples of phenocopies, and studies of phenocopies have, in some instances, provided useful information about the actual molecular defects caused by the mutant counterpart. Remember that the abnormalities seen in these situations occur in a genotypically normal individual without any alteration of the genetic constitution.

Phenocopies occur in all sorts of organisms,

from the simple to the complex. In the bread mold *Neurospora crassa,* for example, the normal-growth phenotype is a many-branched, web-like mass in which individual filaments extend fairly rapidly. The result is an orange-colored, fluffy growth similar to the grey or green mold that sometimes grows on bread. In some mutants of *Neurospora* the branching is much more frequent and growth is slower, resulting in what is called a colonial-like growth habit since it extends only slowly from the point of inoculation and looks like a button on a solid medium. Figure 4.21 shows these two growth phenotypes. The colonial growth is a result of a defect in cell wall structure, which makes it weaker. Consequently, the internal pressure of the cytoplasm in the filaments induces many more branches than normal. This same colonial phenotype can be phenocopied by adding chemicals to the growth medium that affect normal cell wall development. In wild-type *Neurospora* the addition of the unusual sugar sorbose brings about colonial growth. If a piece of the colonial culture is now placed onto a normal medium, the wild-type growth habit is reestablished. At the biochemical level cell wall abnormalities similar to those found in some colonial mutants can be detected in the sorbose-induced phenocopies.

There are also many examples of phenocopies

a) White extremities,
   reared at >30° C

b) Normal
   Himalayan pattern,
   reared at 25° C

c) Himalayan pattern
   with dark patch
   on flank, reared at
   25° C, flank cooled
   to below 25° C

*Figure 4.20 Phenotypic appearance of Himalayan
rabbits raised under different temperature conditions.*

for which there is a phenocopy is phocomelia, a suppression of the development of the long bones of the limbs caused by a rare dominant allele with variable expressivity. Between 1959 and 1961 similar phenotypes were produced by the sedative thalidomide when taken by expectant mothers during the 35th to 50th day of gestation. The drug was removed from the market when the devastating effects were discovered.

In conclusion, at least for the more complex organisms, phenocopies are produced when an organism is exposed to a phenocopying agent during a sharply defined period of its development, and the same effect can often be produced by very different agents.

Twin Studies

We are left with the nature/nurture question: How do we decide to what extent a character is controlled by genes and to what extent it is controlled by the environment? One solution is to study genetically identical individuals in varying environments. For laboratory-reared animals these studies remove the uncertainty of the variation in genotype on the experimental results. Typically, the experimental subjects are lines of mice or rats that have been inbred (brother × sister matings) over a large number of generations. As a consequence of this inbreeding, most genes are homozygous and therefore have, as nearly as is reasonably possible, a genetically homogeneous population of organisms. Such inbred (isogenic) lines are used routinely in studies of the effects of unknown agents, such as new pharmaceutical agents, food additives, food colorings, and dyes, where the animal is used to assess the potential risk to humans.

Although we cannot produce inbred lines of humans for such studies, we can study twins. Two types of twins are possible: **fraternal (dizygotic) twins** and **identical (monozygotic) twins.** Fraternal twins result from the fertilization of two eggs released by the ovaries at the same time. Two eggs mean two distinct meioses, and since meiosis is the time when genetic exchange and chromosome segregation take place, the two eggs are likely to be very different genetically

in humans. Cataracts, deafness, and defects of the heart are sometimes produced when an individual is homozygous for rare recessive alleles and may also result if the mother is infected with rubella (German measles) virus during the first 12 weeks of pregnancy. Another human trait

because most humans are heterozygous for many of their genes. Note that we expect a roughly 1:2:1 ratio of boy-boy:boy-girl:girl-girl pairs among the two-egg twins. Identical twins result from the fertilization of a single egg. After the first mitotic division of the fertilized egg, the two cells separate, and each can give rise to a separate individual. These two individuals are genetically identical since they come from the same fertilized egg: No boy-girl identical twins are expected here.

In human populations, then, it is possible to study whether a trait is inherited by examining its occurrence in twins since twins, if they are raised together, are exposed to essentially the same environmental influences. In brief, an experimenter collects data on the frequency with which both members of a set of twins possess a trait. If both have the trait or if both do not have the trait, they are said to be **concordant.** Conversely, if one twin has the trait and the other does not, they are **discordant.** By determining the concordance-discordance frequencies for a particular trait, the experimenter can make some conclusions about the relative contributions of the genetic and nongenetic environments to that trait. More specifically, the results for identical

and fraternal twins can be compared; the conclusions depend on the results. If there is a high degree of concordance for the identical twins but a great discordance for the fraternal twins, the trait in question can be said to have a strong hereditary component. On the other hand, nearly equal concordance and discordance ratios for identical and fraternal twins would suggest that the trait was determined more by environmental factors than by hereditary ones.

Let us briefly consider some examples with the understanding that these examples involve complex genotypic environmental components. Lots of children catch measles, and therefore there is a high concordance frequency in both monozygotic and dizygotic twins. Thus whether an individual gets measles is largely determined by environmental influences. There is also about equal concordance for handedness (i.e., whether you are left- or right-handed), for the age at which a baby sits up, and for coffee drinking. However, marked differences show up for such things as eye color and hair color, as we might expect since these characters are known to involve genes. There is also a marked difference in concordance in the two types of twins, with a high concordance in monozygotic twins, for

Figure 4.21 The fungus Neurospora crassa: *the wild-type-growth phenotype* (left) *and a colonial-growth phenocopy* (right). *The phenocopy is produced by growing the wild-type strain on a medium containing sorbose. The colonial morphology is very similar to that seen in genetic mutants of this fungus.*

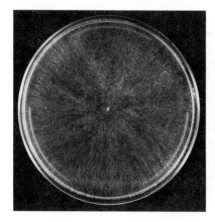
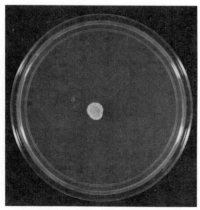

131
Analytical
Approaches
for Solving
Genetics
Problems

traits such as pulse rate, blood pressure, schizo-phrenia, and (believe it or not) antisocial behav-ior. It would do little good to catalog those traits that are or are not controlled mainly by genes. Instead, we can conclude that by studying twins, we can obtain some evidence for the in-volvement or noninvolvement of genes in var-ious traits. But we must be careful in interpret-ing the results of such a study since the variation in penetrance can complicate the re-sults.

Keynote *The phenotypic expression of a gene depends on several factors, in-cluding its dominance relations, the genetic constitution of the rest of the genome (e.g., the presence of epistatic or modifier genes), and the influences of the internal and external environ-ments. In some cases special environmental conditions can cause a phenocopy, a nonheredi-tary and phenotypic modification that mimics a similar phenotype caused by a gene mutation.*

Analytical Approaches for Solving Genetics Problems

Q.1 In snapdragons red flower color (R) is incompletely dominant to white flower color (r); the heterozygote has pink flowers. Also, normal broad leaves (B) are incompletely dominant to narrow, grasslike leaves (b); the heterozy-gote has an intermediate leaf breadth. If a red-flowered, narrow-leaved snap-dragon is crossed with a white-flowered, broad-leaved one, what will be the phenotypes of the F_1 and F_2 generations, and what will be the frequencies of the different classes?

A.1 This basic question on gene segregation includes the issue of incomplete dominance. In the case of incomplete dominance, remember that the geno-type can be directly determined from the phenotype. Therefore we do not need to ask whether or not a strain is true breeding.

  The best approach here is to assign genotypes to the parental snapdragons. Thus the red, narrow plant is *RR bb,* and the white, broad plant is *rr BB.* The F_1 plants from this cross will all be double heterozygotes, *Rr Bb.* Owing to the incomplete dominance, these plants are pink-flowered and have leaves of intermediate breadth. Interbreeding the F_1's gives the F_2 generation, but it does not have the usual 9:3:3:1 ratio. Instead, there is a different phenotype for each genotype. These genotypes and phenotypes and their relative fre-quencies are shown in the figure on page 132.

Q.2 In snapdragons red flower color is incompletely dominant to white, with the heterozygote being pink; normal flowers are completely dominant to peloric-shaped ones; and tallness is completely dominant to dwarfness. The three gene pairs segregate independently. If a homozygous red, tall, normal-flowered plant is crossed with a homozygous white, dwarf, peloric-flowered one, what proportion of the F_2 will resemble the F_1 in appearance?

A.2 Let us assign symbols: R = red and r = white; N = normal flowers and n = peloric; T = tall and t = dwarf. Then the initial cross becomes *RR TT NN* × *rr tt nn.* From this cross we see that all the F_1 plants are triple heterozygotes with the genotype *Rr Tt Nn* and the phenotype pink, tall, normal-flowered. Interbreeding the F_1 generation will produce 27 different genotypes in the F_2; this answer follows from the rule that the number of genotypes is 3^n, where n is the number of heterozygous gene pairs involved in the $F_1 \times F_1$

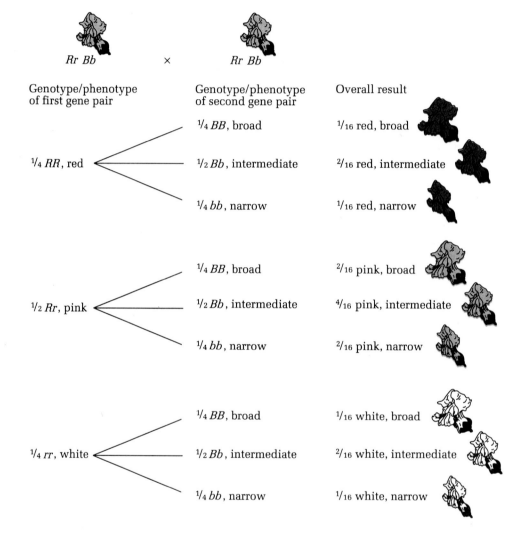

Rr Bb × *Rr Bb*

Genotype/phenotype of first gene pair	Genotype/phenotype of second gene pair	Overall result

¼ *RR*, red

¼ *BB*, broad — 1/16 red, broad

½ *Bb*, intermediate — 2/16 red, intermediate

¼ *bb*, narrow — 1/16 red, narrow

½ *Rr*, pink

¼ *BB*, broad — 2/16 pink, broad

½ *Bb*, intermediate — 4/16 pink, intermediate

¼ *bb*, narrow — 2/16 pink, narrow

¼ *rr*, white

¼ *BB*, broad — 1/16 white, broad

½ *Bb*, intermediate — 2/16 white, intermediate

¼ *bb*, narrow — 1/16 white, narrow

cross. We are asked specifically for the proportion of F_2 progeny that resemble the F_1 in appearance. We can calculate this proportion in a direct way that does not involve displaying all the possible genotypes and collecting those classes with the appropriate phenotype. First, we calculate the frequency of pink-flowered plants in the F_2, and then we determine the proportion of these plants that have the other two attributes. From an $Rr \times Rr$ cross we calculate that half of the progeny will be heterozygous Rr and therefore pink. Next, we determine the proportion of F_2 plants that are phenotypically like the F_1 with respect to height (i.e., tall). Either TT or Tt plants will be tall, and so 3/4 of the F_2 will be tall. Similarly, 3/4 of the F_2 plants will be normal-flowered like the F_1's. To obtain the probability of occurrence of all three of these phenotypes together (i.e., pink, tall, normal), we must multiply the individual probabilities since the gene pairs segregate independently. The answer is $1/2 \times 3/4 \times 3/4$, or 9/32.

133
Analytical
Approaches
for Solving
Genetics
Problems

Q.3 a. An $F_1 \times F_1$ self gives a 9:7 genotypic ratio in the F_2. What phenotypic ratio would you expect if you testcrossed the F_1?

  b. Answer the same question for an $F_1 \times F_1$ cross that gives a 9:3:4 ratio.

  c. Answer the same question for a 15:1 ratio.

A.3 This question deals with epistatic effects. In answering the question, we must consider the interaction between the different genotypes in order to proceed with the testcross part. Let us set up the general genotypes that we will deal with throughout. The simplest are allelic pairs a^+ and a, and b^+ and b, where the wild-type alleles are completely dominant to the other member of the pair.

  a. A 9:7 ratio in the F_2 implies that both members of the F_1 are double heterozygotes and that epistasis is involved. Essentially, any genotype with a homozygous recessive condition has the same phenotype, so the 3, 3, and 1 parts of a 9:3:3:1 ratio are phenotypically combined into one class. Genotypically, 9/16 are $a^+ - b^+ -$ types, and the other 7/16 are $a^+ - bb$, aa $b^+ -$, and aa bb. (As always, the use of the $-$ after a wild-type allele signifies that the same phenotype results whether the missing allele is a wild type or a mutant.) Now the testcross asked for is a^+a $b^+b \times aa$ bb. Following the same logic used before in questions like this one, we can predict a 1:1:1:1 ratio of a^+a b^+b:a^+a bb:aa b^+b:aa bb. The first genotype will have the same phenotype as the 9/16 class of the F_2, but because of epistasis, the other three genotypes will have the same phenotype as the 7/16 class of the F_2. In sum, the answer is a phenotypic ratio of 1:3 in the progeny of a testcross of the F_1.

  b. We are asked to answer the same question for a 9:3:4 ratio in the F_2. Again, this question involves a modified dihybrid ratio where two classes of the 9:3:3:1 have the same phenotype. Complete dominance for each of the two gene pairs occurs here also, so the F_1's are a^+a b^+b. Perhaps both the $a^+ - bb$ and aa bb classes in the F_2 will have the same phenotype, while the $a^+ - b^+ -$ and aa $b^+ -$ classes will have phenotypes distinct from each other and from the interaction class. The genotypic ratio of a testcross of the F_1 is the same as in part a. Considering them in the same order as we did in part a, the second and fourth classes would have the same phenotype owing to epistasis. So there are only three possible phenotypic classes instead of the four found in the testcross of a dihybrid F_1 where there is complete dominance and no interaction. The phenotypic ratio here is 1:1:2, where these phenotypes are listed in the same relative order as in the 9:3:4.

  c. This question is yet another example of epistasis. Since $15 + 1$ is 16, this number gives the outcome of an F_1 self of a dihybrid where there is complete dominance for each gene pair and interaction between the dominant alleles. In this case the $a^+ - b^+ b$, $a^+ - bb$, and aa $b^+ -$ classes have one phenotype and include 15/16 of the F_2 progeny, and the aa bb class has the other phenotype and 1/16 of the F_2. The genotypic results of a testcross of the F_1 are the same as in parts a and b, that is, a 1:1:1:1 ratio of a^+a b^+b:a^+a bb:aa b^+b:aa bb. The first three classes have the same phenotype, which is the same as that of the 15/16 of the F_2's, and the last class has the other phenotype. The answer, then, is a 3:1 phenotypic ratio.

*Questions
and
Problems*

4.1 In rabbits, C = agouti coat color, c^{ch} = chinchilla, c^h = Himalayan, and c = albino. The four alleles constitute a multiple allelic series. The agouti C is dominant to the three other alleles, c is recessive to all three other alleles, and chinchilla is dominant to Himalayan. Determine the phenotypes of progeny from the following crosses:

a. $CC \times cc$ d. $Cc^h \times c^h c$ g. $c^h c \times cc$
b. $Cc^{ch} \times Cc$ e. $Cc^h \times cc$ h. $Cc^h \times Cc$
c. $Cc \times Cc$ f. $c^{ch} c^h \times c^h c$ i. $Cc^h \times Cc^{ch}$

*4.2 If a given population of diploid organisms contains three, and only three, alleles of a particular gene (say *w, w1,* and *w2*), how many different diploid genotypes are possible in the populations? List all possible genotypes of diploids (considering only these three alleles).

4.3 The genetic basis of the ABO blood types seems most likely to be

a. multiple alleles.
b. polyexpressive hemizygotes.
c. allelically excluded alternates.
d. three independently assorting genes.

4.4 In humans the three alleles I^A, I^B, and i constitute a multiple allelic series that determine the ABO blood group system, as we described in this chapter. For the following problems, state whether the child mentioned can actually be produced from the marriage. Explain your answer.

a. An O child from the marriage of two A individuals.
b. An O child from the marriage of an A to a B.
c. An AB child from the marriage of an A to an O.
d. An O child from the marriage of an AB to an A.
e. An A child from the marriage of an AB to a B.

4.5 A man is blood type O,M. A woman is blood type A,M and her child is type A,MN. The aforesaid man cannot be the father of the child because:

a. O men cannot have type A children.
b. O men cannot have MN children.
c. An O man and an A woman cannot have an A child.
d. An M man and an M woman cannot have an MN child.

*4.6 A woman of blood group AB marries a man of blood group A whose father was group O. What is the probability that

a. their two children will both be group A?
b. one child will be group B and the other group O?

c. the first child will be a son of group AB, and their second child a son of group B?

4.7 If a mother and her child belong to blood group O, what blood group could the father *not* belong to?

*4.8 A man of what blood group could not be a father to a child of blood type AB?

4.9 In snapdragons, red flower color (R) is incompletely dominant to white (r); the Rr heterozygotes are pink. A red-flowered snapdragon is crossed with a white-flowered one. Determine the flower color of (a) the F_1; (b) the F_2; (c) the progeny of a cross of the F_1 to the red parent; (d) the progeny of a cross of the F_1 to the white parent.

*4.10 In Shorthorn cattle the heterozygous condition of the alleles for red coat color (R) and white coat color (r) is roan coat color. If two roan cattle are mated, what proportion of the progeny will resemble their parents in coat color?

4.11 What progeny will a roan Shorthorn have if bred to (a) red; (b) roan; (c) white?

*4.12 In peaches, fuzzy skin (F) is completely dominant to smooth (nectarine) skin (f), and the heterozygous conditions of oval glands at the base of the leaves (O) and no glands (o) give round glands. A homozygous fuzzy, no-gland peach variety is bred to a smooth, oval-gland variety.

a. What will be the appearance of the F_1?
b. What will be the appearance of the F_2?
c. What will be the appearance of the offspring of a cross of the F_1 back to the smooth, oval-glanded parent?

4.13 In guinea pigs, short hair (L) is dominant to long hair (l), and the heterozygous conditions of yellow coat (W) and white coat (w) give cream coat. A short-haired, cream guinea pig is bred to a long-haired, white guinea pig, and a long-haired, cream baby guinea pig is produced. When the baby grows up, it is bred back to the short-haired, cream parent. What phenotypic classes and in what proportions are expected among the offspring?

4.14 The shape of radishes may be long (LL), oval (Ll), or round (ll), and the color of radishes may be red (RR), purple (Rr), or white (rr). If a long, red radish plant is crossed with a round, white plant, what will be the appearance of the F_1 and the F_2?

4.15 In poultry the genes for rose comb (R) and pea comb (P), if present together, give walnut comb. The recessive alleles of each gene, when present together in a homozygous state, give single comb. What will be the comb characters of the offspring of the following crosses?

a. $RR\ Pp \times rr\ Pp$ c. $Rr\ pp \times rr\ Pp$ e. $Rr\ pp \times Rr\ pp$

b. $rr\ PP \times Rr\ Pp$ d. $Rr\ Pp \times Rr\ Pp$

4.16 For the following crosses involving the comb character in poultry, determine the genotypes of the two parents:

a. A walnut crossed with a single produces offspring 1/4 of which are walnut, 1/4 rose, 1/4 pea, and 1/4 single.
b. A rose crossed with a walnut produces offspring 3/8 of which are walnut, 3/8 rose, 1/8 pea, and 1/8 single.
c. A rose crossed with a pea produces five walnut and six rose offspring.
d. A walnut crossed with a walnut produces one rose, two walnut, and one single offspring.

4.17 In poultry feathered shanks (F) are dominant to clean (f), and white plumage of white leghorns (I) is dominant to black (i).

a. A feathered-shanked, white, rose-combed bird crossed with a clean-shanked, white, walnut-combed bird produces these offspring: two feathered, white, rose; four clean, white, walnut; three feathered, black, pea; one clean, black, single; one feathered white, single; two clean, white, rose. What are the genotypes of the parents?
b. A feathered-shanked, white, walnut-combed bird crossed with a clean-shanked, white, pea-combed bird produces a single offspring, which is clean-shanked, black, and single-combed. In further offspring from this cross, what proportion may be expected to resemble each parent, respectively?

*4.18 F_2 plants segregate 9/16 colored:7/16 colorless. If a colored plant from the F_2 is chosen at random and selfed, what is the probability that there will be no segregation of the two phenotypes among its progeny?

4.19 The gene l in *Drosophila* is recessive, sex-linked, and lethal when homozygous or hemizygous (the condition in the male). If a female of genotype Ll is crossed with a normal male, what is the probability that the first two surviving progeny to be observed will be males?

*4.20 A locus in mice is involved with pigment production; when parents heterozygous at this locus are mated, 3/4 of the progeny are colored and 1/4 are albino. Another phenotype concerns the coat color produced in the mice; when two yellow mice are mated, 2/3 of the progeny are yellow and 1/3 are agouti. The albino mice cannot express whatever alleles they may have at the independently assorting agouti locus.

a. When yellow mice are crossed with albino, they produce an F_1 consisting of 1/2 albino, 1/3 yellow, and 1/6 agouti. What are the probable genotypes of the parents?
b. If yellow F_1 mice are crossed among themselves, what phenotypic ratio would you expect among the progeny? What proportion of the yellow progeny produced here would be expected to be true breeding?

4.21 In *Drosophila melanogaster,* a recessive autosomal gene, ebony (*e*), produces an ebony body color when homozygous, and an independently assorting autosomal gene, black (*bl*), produces a black color when homozygous. Flies with genotypes *ee bl$^+$ −*, *e$^+$ − bl bl*, and *ee bl bl* are phenotypically identical with respect to body color. If true-breeding ebony flies are crossed with true-breeding black flies,

   a. what will be the phenotype of the F$_1$'s?
   b. what phenotypes and what proportions would occur in the F$_2$ generation?
   c. what phenotypic ratios would you expect to find in the progeny of these backcrosses: (1) F$_1$ × true-breeding ebony and (2) F$_1$ × true-breeding black?

*4.22 In four-o'clocks two genes, *Y* and *R*, affect flower color. Neither is completely dominant, and the two interact on each other to produce seven different flower colors:

YY RR = crimson	*YY Rr* = orange-red	*YY rr* = yellow
Yy RR = magenta	*Yy Rr* = magenta-rose	*Yy rr* = pale yellow
yy RR, *yy Rr*, and *yy rr* = white		

   a. In a cross of a crimson-flowered plant with a white one (*yy rr*), what will be the appearances of the F$_1$, the F$_2$, and the offspring of the F$_1$ backcrossed to the crimson parent?
   b. What will be the flower colors in the offspring of a cross of orange-red × pale yellow?
   c. What will be the flower colors in the offspring of a cross of a yellow with a *yy Rr* white?

4.23 Two four-o'clock plants were crossed and gave the following offspring: 1/8 crimson, 1/8 orange-red, 1/4 magenta, 1/4 magenta-rose, and 1/4 white. Unfortunately, the person who made the crosses was color-blind and could not record the flower colors of the parents. From the results of the cross, deduce the genotypes and flower colors of the two parents.

*4.24 Genes *A*, *B*, and *C* are independently assorting and control production of a black pigment.

   a. Assume that *A*, *B*, and *C* act in a pathway as follows:

$$\text{colorless} \xrightarrow{\ A\ } \xrightarrow{\ B\ } \xrightarrow{\ C\ } \text{black}$$

   The alternative alleles that give abnormal functioning of these genes are designated *a*, *b*, and *c*, respectively. A black *AA BB CC* is crossed by a colorless *aa bb cc* to give a black F$_1$. The F$_1$ is selfed. What proportion of the F$_2$ is colorless? (Assume that the products of each step except the last are colorless, so only colorless and black phenotypes are observed.)
   b. Assume that *C* produces an inhibitor that prevents the formation of black by destroying the ability of *B* to carry out its function, as follows:

$$\text{colorless} \xrightarrow{A} \xrightarrow{B} \text{black}$$
$$\uparrow$$
$$C \text{ (inhibitor)}$$

A colorless *AA BB CC* is crossed with a colorless *aa bb cc,* giving a colorless F_1. The F_1 is selfed to give an F_2. What is the ratio of colorless to black in the F_2? (Only colorless and black phenotypes are observed, as in part a.)

4.25 In *Drosophila* a mutant strain has plum-colored eyes. A cross between a plum-eyed male and a plum-eyed female gives 2/3 plum-eyed and 1/3 red-eyed (wild-type) progeny flies. A second mutant strain of *Drosophila,* called stubble, has short bristles instead of the normal long bristles. A cross between a stubble female and a stubble male gives 2/3 stubble and 1/3 normal-bristled flies in the offspring. Assuming that the plum gene assorts independently from the stubble gene, what will be the phenotypes and their relative proportions in the progeny of a cross between two plum-eyed, stubble-bristled flies? (Both genes are autosomal.)

*4.26 In sheep, white fleece (*W*) is dominant over black (*w*), and horned (*H*) is dominant over hornless (*h*) in males but recessive in females. If a homozygous horned white ram is bred to a homozygous hornless black ewe, what will be the appearance of the F_1 and the F_2?

4.27 A horned black ram bred to a hornless white ewe has the following offspring: Of the males 1/4 are horned, white; 1/4 are horned, black; 1/4 are hornless, white; and 1/4 are hornless, black. Of the females 1/2 are hornless and black, and 1/2 are hornless and white. What are the genotypes of the parents?

*4.28 A horned white ram is bred to the following four ewes and has one offspring by the first three and two by the fourth: Ewe A is hornless and black; the offspring is a horned white female. Ewe B is hornless and white; the offspring is a hornless black female. Ewe C is horned and black; the offspring is a horned white female. Ewe D is hornless and white; the offspring are one hornless black male and one horned white female. What are the genotypes of the five parents?

*4.29 Common pattern baldness is more frequent in males than in females. This appreciable difference in frequency is assumed to be due to

  a. Y-linkage of this trait.
  b. X-linked recessive mode of inheritance for the trait in question.
  c. sex-influenced autosomal inheritance.
  d. excessive beer-drinking in males, consumption of gin being approximately equal between the sexes.

5

Linkage, Crossing-Over, and Genetic Mapping in Eukaryotes

As we have seen throughout Chapters 1–4, the independent assortment of genes during meiosis is a fundamental principle of genetics. In many instances, however, certain genes (and hence the phenotypic characters they control) are inherited together because they are located on the same chromosome. Genes on the same chromosome are called linked genes and are said to belong to a linkage group (Figure 5.1). The number of linkage groups in an organism equals the haploid number of chromosomes. Through testcrosses we can determine which genes are linked to each other and can then construct a linkage map or genetic map of each chromosome.

Genetic mapping provides information that is useful in many aspects of genetic analysis. Genetic mapping can tell us, for example, whether genes that work together to produce a given phenotype are located together on the chromosome. Such information can be used to construct models that explain the regulation of gene expression in an organism. Also, the knowledge of the locations of genes on chromosomes has been useful in recombinant DNA research and in understanding the DNA sequences in and around genes and at parts of the chromosomes (such as the centromeres) that are of particular functional interest.

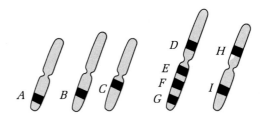

A, B, C: Unlinked genes

D–E–F–G: Linked genes

H–I: Linked genes

[A], [B], [C], [D–E–F–G], [H–I]: Five linkage groups

Figure 5.1 Linked and unlinked genes. Genes located on the same chromosome are called linked genes and belong to linkage groups.

140
Linkage,
Crossing-Over,
and Genetic
Mapping in
Eukaryotes

In this chapter we will learn about the discovery of gene linkage, about the evidence that gene recombination is the result of a physical exchange of chromosome parts (crossing-over), and about how a genetic map of a chromosome is constructed.

Discovery of Genetic Linkage

Partial Linkage in the Sweet Pea

In 1905, W. Bateson, E. R. Saunders, and R. C. Punnett, in their studies of heredity in the sweet pea, *Lathyrus odoratus L,* discovered the first exception to the law of independent assortment of gene pairs. They crossed two true-breeding strains, one that produced purple flowers and long pollen and another that produced red flowers and round pollen. All F_1 plants had purple flowers and long pollen, indicating that purple flower color was dominant to red flower color and that long pollen was dominant to round pollen. The 381 plants of the F_2 consisted of 305 plants with purple petals and 76 plants with red petals. (This is not a good fit to the 3:1 ratio predicted by the law of segregation for a cross of two heterozygotes so presumably there was a difference in viability for the two phenotypic classes.) For the pollen shape trait the F_2 plants segregated into 305 with long pollen and 76 with round pollen. These results indicated that both the flower color trait and the pollen shape trait are controlled by a single pair of alleles at their respective genes.

A natural extension of the data analysis was to see whether the two character pairs assorted independently in the cross. From the $F_1 \times F_1$ cross the F_2 progeny displayed the following phenotypes:

284 (74.6 percent) purple flowers, long pollen
21 (5.5 percent) purple flowers, round pollen
21 (5.5 percent) red flowers, long pollen
55 (14.4 percent) red flowers, round pollen

More precise frequencies determined by Punnett in 1917 and based on 6952 F_2 plants derived from the same type of purple and long × red and round cross gave these results:

4831 (69.5 percent) purple, long
390 (5.6 percent) purple, round
393 (5.6 percent) red, long
1338 (19.3 percent) red, round

If the genes assorted independently, the expected proportions would be 9/16:3/16:3/16:1/16 for the four phenotypes in the order given, or in percentages, 56.25 percent:18.75 percent:18.75 percent:6.25 percent, respectively. The observed values differ significantly from those expected according to the independent assortment hypothesis. (What constitutes a "significant" difference between observed results and results expected on the basis of a particular hypothesis may be determined by a **chi-square test,** a statistical procedure described on pp. 151–153.)

Bateson, Saunders, and Punnett recognized that the dominant purple and long characters occurred together more often than predicted by Mendel's second law but could not explain why this result occurred. They tried to explain their exceptional results in terms of modified Mendelian ratios but without success. We know now that the genes they were studying are located on the same chromosome. Genes located on the same chromosome are said to exhibit **linkage** and are considered to be **linked genes.** Thus Bateson and colleagues' results can be explained as follows: If we symbolize the purple allele as *P,* the red allele as *p,* the long allele as *L,* and the round pollen allele as *l,* the cross is *PP LL* × *pp ll.* The F_1 double heterozygote is *Pp Ll.* If independent assortment were occurring (Figure 5.2), equal proportions of *P L, P l, p L,* and *p l* gametes would be produced by the F_1 plants, and the resulting F_2 generation would exhibit the expected 9:3:3:1 ratio.

Instead, the observed ratio of F_2 phenotypes can best be explained if 44 percent of the gametes are *P L,* 44 percent are *p l,* and 6 percent each are *P l* and *p L* (Figure 5.3). At the time of pollen and egg formation in the F_1 sweet peas, there was a tendency for the parental *P* and *L* alleles to stay together and for the parental *p* and *l* alleles to stay together. Since two *nonpa-*

F₁ generation
(diploid)

Genotype

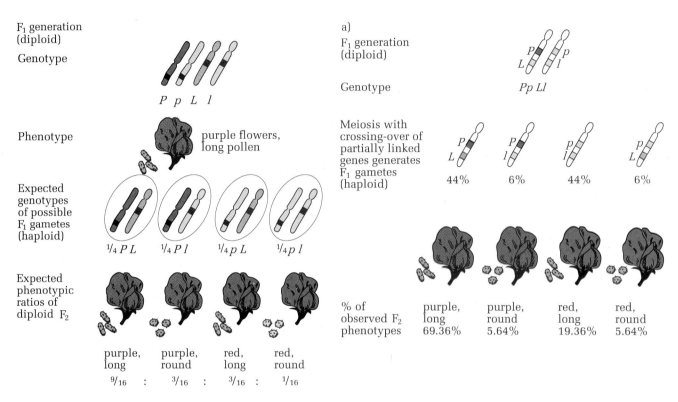

P p L l

Phenotype

purple flowers,
long pollen

Expected
genotypes
of possible
F₁ gametes
(haploid)

¼ *P L*   ¼ *P l*   ¼ *p L*   ¼ *p l*

Expected
phenotypic
ratios of
diploid F₂

purple,   purple,   red,   red,
long   round   long   round
⁹/₁₆   :   ³/₁₆   :   ³/₁₆   :   ¹/₁₆

*Figure 5.2 Expected results of F₁ × F₁ cross if genes
for flower color and pollen size assort independently.*

rental combinations of phenotypes were produced (purple and round, and red and long), al-
beit at low frequency, *p L* and *P l* gametes must
also have been generated. Thus the linkage between the two alleles in these crosses was not
complete. This **partial linkage** occurs when homologous chromosomes exchange corresponding
parts during meiosis through the process called
crossing-over (see Chapter 1 and later discussion
in this chapter). Figure 5.4 shows what the observed results would be if the linkage between

*Figure 5.3 Observed results indicating partial linkage
of genes: (a) under assumption that P L and p l gametes (pollen and eggs) each occur with a frequency
of 44 percent and the P l and p L gametes each occur
with a frequency of 6 percent; (b) production of gametes from F₁ double heterozygote; (c) combination of
gametes to produce the F₂ classes.*

a)
F₁ generation
(diploid)

Genotype   *Pp Ll*

Meiosis with
crossing-over of
partially linked
genes generates
F₁ gametes
(haploid)   44%   6%   44%   6%

% of   purple,   purple,   red,   red,
observed F₂   long   round   long   round
phenotypes   69.36%   5.64%   19.36%   5.64%

b) Random union of F₁ gametes to produce F₂

		Pollen			
Gametes from F₁ heterozygote (%)		*P L* 44	*P l* 6	*p l* 44	*p L* 6
		Genotype frequencies in F₂ (%)			
	P L 44	19.36	2.64	19.36	2.64
	P l 6	2.64	0.36	2.64	0.36
Eggs	*p l* 44	19.36	2.64	19.36	2.64
	p L 6	2.64	0.36	2.64	0.36

c) Phenotype frequencies in F₂ (%)

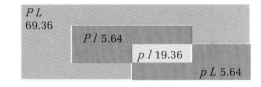

F₂ phenotype
totals (%)

P L 69.36   *P l* 5.64   *p l* 19.36   *p L* 5.64

142
Linkage,
Crossing-Over,
and Genetic
Mapping in
Eukaryotes

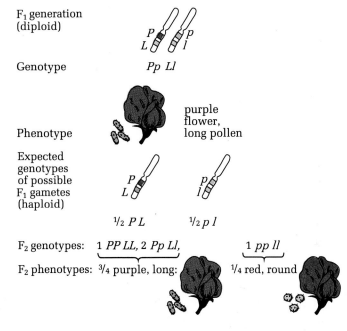

F₁ generation
(diploid)

Genotype   *Pp Ll*

Phenotype   purple
                  flower,
                  long pollen

Expected
genotypes
of possible
F₁ gametes
(haploid)

½ *P L*   ½ *p l*

F₂ genotypes: 1 *PP LL*, 2 *Pp Ll*,   1 *pp ll*

F₂ phenotypes: ¾ purple, long: ¼ red, round

Figure 5.4 Expected results of F₁ × F₁ cross if genes are completely linked.

the two alleles was complete. In the F₁ all are purple, long, and in the F₂ a 3:1 ratio is observed of purple, long:red, round

Morgan's Linkage Experiments with *Drosophila*

Our modern understanding of genetic linkage stems from the work of T. H. Morgan and his colleagues with linkage in *Drosophila melanogaster,* which was done around 1911. We described some of Morgan's work on sex linkage in *Drosophila* in Chapter 3. By 1911 Morgan had accumulated a number of spontaneously arising sex-linked mutants. He found that two recessive genes *w* (white eye) and *m* (miniature wing) were located on the X chromosome. Strains carrying these mutant genes were used in experiments to see whether or not genetic exchange could occur between the linked genes on the chromosomes.

Morgan crossed a white miniature (*w m/w m*) female with a wild-type male (*w⁺ m⁺/Y*) (Figure 5.5). As expected, all the F₁ males were white-eyed and had miniature wings, while all females were wild type for both the eye color and wing size traits. The F₁ flies were interbred and the 2441 F₂ flies were analyzed. In results similar to those of Bateson, Saunders, and Punnett, the most frequent phenotypic classes in both sexes were the grandparental phenotypes of mutant white eyes plus miniature wings or wild-type red eyes plus large wings. Conventionally, we refer to the original genotypes of the two chromosomes as **parental genotypes, parental classes,** or, more simply, **parentals.** The term is also used to describe phenotypes, so the original white miniature females and wild-type males in the crosses under discussion are defined as the parentals.

Morgan was surprised to find that significant numbers of flies had the nonparental combinations of the eye and wing phenotypes of white eyes plus normal wings and red eyes plus miniature wings. The two nonparental combinations of linked genes are called **recombinants.** In all, 900 of 2441 F₂ flies, or 36.9 percent, exhibited recombinant phenotypes. To explain the recombinants, Morgan proposed that in 36.9 percent of the meioses an exchange of genes had occurred between the two X chromosomes of the F₁ females. (No such exchange needs to be postulated for sperm production since the males are hemizygous and no genetic exchange occurs between the nonhomologous X and Y chromosomes.)

Next, Morgan extended the work to include another sex-linked character, which was yellow body color. He crossed a true-breeding white-eyed, yellow-bodied mutant female with a wild-type red-eyed, grey-bodied male (where red eyes and a grey body are dominant traits). As expected, the F₁ progeny consisted of wild-type females and white-eyed, yellow-bodied males; that is, the two parental phenotypic classes. Among the 2205 F₂ flies obtained, the original grandparental combinations of white eyes plus yellow body and red eyes plus grey body were the most frequent phenotypic classes, while only 1.3 per-

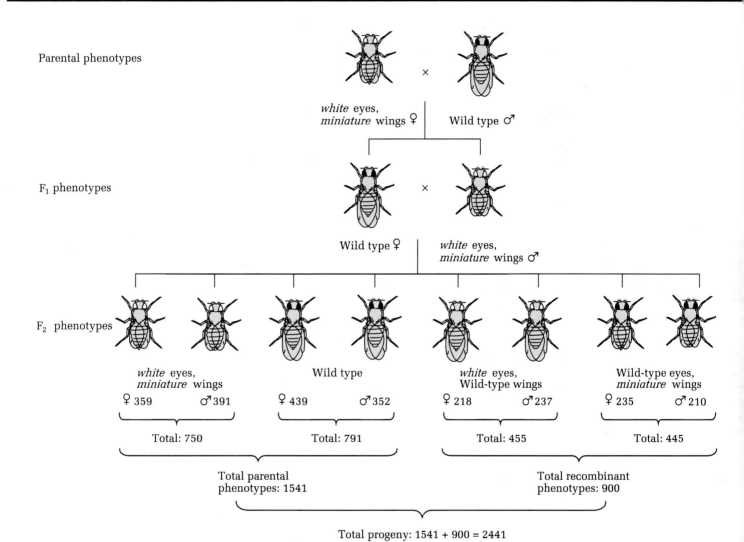

Figure 5.5 Morgan's experimental crosses of white-eye and miniature-wing variants of D. melanogaster, *showing evidence of linkage and recombination in the X chromosome.*

cent of the flies (29 of 2205) showed recombinant phenotypes. Morgan concluded that the genes that controlled the eye color and the body color traits were more closely linked than were the genes for eye color and wing size.

A large number of other crosses of this type were done and the conclusions were always the same. *In each case the parental phenotypic classes were the most frequent while the recom-* *binant classes occurred much less frequently.* Approximately equal numbers of each of the two parental classes were obtained, and similar results were obtained for the recombinant classes. Morgan's general conclusion was that *during segregation of alleles at meiosis, certain ones tend to remain together because they lie near each other on the same chromosome.* To take this conclusion one step further, characters

144
Linkage,
Crossing-Over,
and Genetic
Mapping in
Eukaryotes

remain together to a greater extent the closer the corresponding genes are on the chromosome.

Morgan was also able to relate the production of the phenotypic classes in the F_2 progeny of the cross to chromosomal segregation. Earlier, in 1909, F. Janssens had described *chiasmata* (singular = chiasma; cytologically observable physical exchange between homologous chromosomes) during prophase I of meiosis in the salamander. Janssens thought (but could not prove) that the chiasmata might be sites of physical exchange between maternal and paternal homologues (a concept already discussed in Chapter 1). Morgan hypothesized that the partial linkage occurred when two genes on the same chromosome were separated physically from one another by chiasma during meiosis. In 1912 Morgan and E. Cattell introduced the term **crossing-over** to describe the process of chromosomal interchange by which recombinants (new combinations of linked genes) arise.

To clarify the terminology: A chiasma is the place on a homologous pair of chromosomes at which a physical exchange is occurring; that is, it is the site of crossing-over. Crossing-over is the actual process of exchange of chromosome segments of nonsister chromatids of homologous chromosomes by symmetrical breakage and crosswise rejoining. Figure 5.6 presents a simplified diagram of this highly complex process. Therefore, the reason that 36.9 percent of the *Drosophila* progeny were recombinant for *w* and *m*, and only 1.3 percent were recombinant for *w* and *y*, is that *w* and *m* are relatively far apart on the X chromosome while *w* and *y* are quite close. Thus the probability of crossing-over occurring between *w* and *m* is much higher than the probability of crossing-over taking place between *w* and *y*.

Keynote *The production of genetic recombinants results from physical exchanges between homologous chromosomes that have become tightly aligned during the meiotic prophase. A chiasma is the site of crossing-over. Crossing-over is the exchange of chromosome parts of nonsister chromatids of homologous chromosomes by symmetrical breakage and crosswise rejoining.*

Gene Recombination and the Role of Chromosomal Exchange

Morgan's hypothesis that physical exchange between chromosomes during meiosis leads to genetic recombination is now universally accepted. At the time, however, it was only a hypothesis, and convincing evidence that the appearance of recombinants was associated with the process of crossing-over was not obtained until the 1930s.

Corn Experiments

The first evidence for the association of gene recombination with chromosomal exchange came from the work of Harriet B. Creighton and Barbara McClintock in 1931 with corn, *Zea mays.* They performed an experiment in which the two chromosomes under study differed cytologically on both sides of the chromosomal segment in which the crossing-over event occurs.

Crosses were made of a strain of corn that was heterozygous for two genes on chromosome 9 (Figure 5.7a). One of the genes determines colored (*C*) or colorless (*c*) seeds. The other gene determines the forms of starch synthesized by the plants; standard-type plants (*Wx*) produced two forms of starch, amylose and amylopectin, while waxy plants (*wx*) produce amylopectin but no amylose. One of the chromosomes had a normal appearance and carried *c* and *Wx* genes. Its homolog carried *C* and *wx*, possessed a large, darkly staining knob at the end nearer *C*, and was longer than the *c Wx* chromosome because it had a piece of another chromosome attached to the end nearer *wx*. Cytologically distinguishable features such as these are called **cytological**

145
*Gene Recombination
and the
Role of
Chromosomal
Exchange*

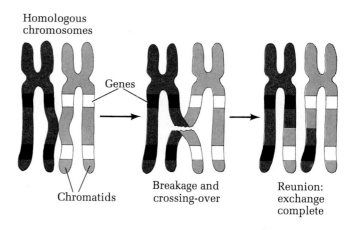

Figure 5.6 Mechanism of crossing-over: a highly simplified diagram of a crossover between two nonsister chromatids during meiotic prophase, giving rise to recombinant (nonparental) combinations of linked genes.

markers; the genes involved are called **genetic markers** or **gene markers.**

During meiosis, crossing-over occurs between the two loci (Figure 5.7b). When Creighton and McClintock examined the two classes of recombinants, *c wx* and *C Wx,* that occurred in the progeny, they found that whenever the genes had recombined, the cytological features (the knob and the extra piece) had also recombined (Figure 5.7c). No such exchange was evident in the parental (nonrecombinant) classes of progeny. These data provided strong evidence that genetic recombination is associated with physical exchange of parts between homologous chromosomes.

Drosophila Experiments

Within a few weeks of the publication of Creighton and McClintock's results, C. Stern reported similar results for experiments done with *Drosophila melanogaster.* His strategy was similar: He used genetic markers and cytological markers; the cross he analyzed is diagramed in Figure 5.8.

The male parent carried normal X and Y chromosomes with the recessive allele *car* (carnation-colored eye) and the wild-type allele of the *B* (bar-eye) gene. The *car* gene is located toward the end of the X chromosome relative to the *B* gene. Since the male parent is hemizygous, these flies had wild-type-shaped, carnation-

colored eyes. The female parent had two abnormal and cytologically distinct X chromosomes. One X chromosome had the wild-type alleles of both the *car* and the *B* genes and had a portion of the Y chromosome attached to it. The other X

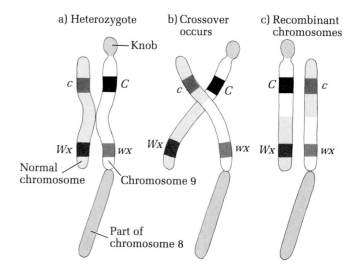

Figure 5.7 Evidence for association of gene recombination with chromosomal exchange in corn. (a) Physical and genetic constitutions of the two chromosomes in the heterozygote; (b) crossover occurs; and (c) physical and genetic constitutions of the recombinant chromosomes.

146
Linkage,
Crossing-Over,
and Genetic
Mapping in
Eukaryotes

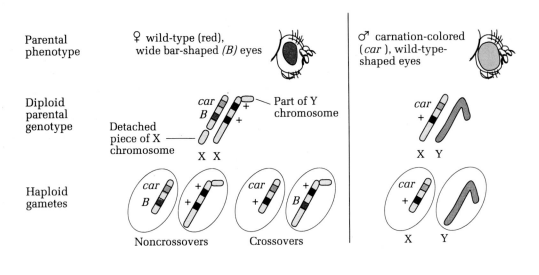

Parental phenotype	♀ wild-type (red), wide bar-shaped *(B)* eyes	♂ carnation-colored *(car)*, wild-type-shaped eyes
Diploid parental genotype	Detached piece of X chromosome — *car B* +/+ Part of Y chromosome — X X	*car* +/ X Y
Haploid gametes	*car B* + + / *car* + B Noncrossovers Crossovers	*car* + / X Y

Parent 2 ♂ gametes

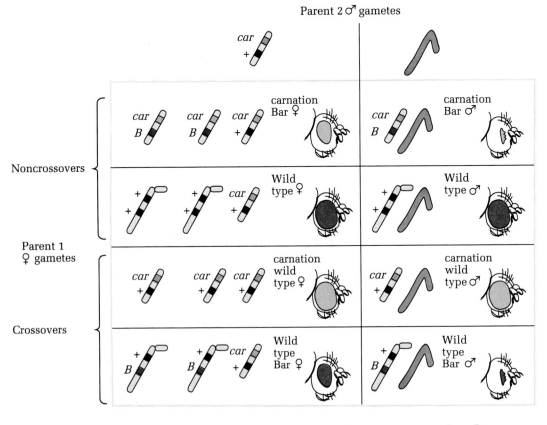

Parent 1 ♀ gametes

Noncrossovers

Crossovers

F₁ genotypes: *car B*/*car* +; *car B*/ Y
 + +/*car* +; + +/ Y
 car +/*car* +; *car* +/ Y
 + *B*/*car* +; + *B*/ Y

F₁ phenotypes: carnation, Bar; 1♀, 1♂
Wild type (red), nonbar; 1♀, 1♂
carnation, wild type; 1♀, 1♂
Wild type, (red), Bar; 1♀, 1♂

147
*Gene Recombination
and the
Role of
Chromosomal
Exchange*

chromosome had the recessive *car* allele and the dominant *B* allele. Phenotypically, the parental females had a wide-bar eye since they were $B/+$, and the eye was wild-type in color since they were heterozygous $+/car$. In addition, the latter X chromosome was distinctly shorter than normal X since part of it had broken off and was attached to the small chromosome 4 (see the unattached piece in Figure 5.8).

In gamete formation only two classes of sperm were produced: the Y bearing and the X bearing. The Y sperm had no genetic markers of relevance here, and the X sperm carried the *car* and B^+ alleles. However, four classes of eggs were produced. Two resulted from meioses in which no crossing-over had occurred between the chromosomes, and the other two were recombinant gametes produced by such a crossing-over event. As with the corn experiments, every case where genetic recombination had occurred was accompanied by an exchange of identifiable chromosome segments. Where no recombination had occurred, the two phenotypic classes of flies produced were wild-type and carnation plus bar. No exchanges of chromosome parts were evident among these nonrecombinants. The two classes of recombinants were carnation-colored, round eyes, and wild-type colored, bar eyes. The carnation flies had a complete X chromosome, and the bar flies had a shorter-than-normal X chromosome to which a piece of the Y chromosome was attached, while the rest of the X chromosome was attached to chromosome 4. This chromosomal makeup could only have resulted from physical exchanges of homologous chromosome parts. There is no doubt, therefore, *that genetic recombination results from physical crossing-over between chromosomes.* (Note that as with Bridges's work on nondisjunction, the application of a combined genetic and cytological approach generated an important and fundamental piece of information.)

Figure 5.8 Experiment of Stern to demonstrate the relationship between genetic recombination and chromosomal exchange in D. melanogaster.

Keynote *The proof that genetic recombination occurs when crossing-over takes place during meiosis came from breeding experiments in which the parents differed with respect to both genetic and cytological markers. Results of these experiments showed that when recombinant phenotypes occurred, the cytological markers indicated that crossing-over had occurred.*

Crossing-Over at the Tetrad Stage of Meiosis

In Chapter 1 we found that crossing-over occurs in prophase I of meiosis; further studies have shown that crossing-over occurs at the tetrad stage in prophase I. The organism *Neurospora crassa* (the orange bread mold) is used here to show how evidence was obtained to support this conclusion and, therefore, to disprove the alternative hypothesis that crossing-over occurred before meiosis in interphase.

Life cycle of *Neurospora crassa*. To understand the proof for crossing-over at the tetrad stage, we must understand the life cycle of *Neurospora crassa* (Figure 5.9). Then we can appreciate how suitable this organism is for the experiment.

Neurospora crassa is a mycelial-form fungus, meaning that it spreads over its growth medium in a weblike pattern (see Figure 4.21). It produces a furry mycelial mat when it grows on bread, and its designation as an orange bread mold comes from the color of the asexual spores (called conidia) it produces when it grows. *Neurospora* has several properties that make it useful for genetic and biochemical studies: It is a haploid (N) organism, so the effects of mutations may be seen directly, and its short life cycle facilitates the study of the segregation of genetic defects.

Neurospora can be propagated vegetatively (asexually) so that unlimited populations of strains with particular genotypes can be produced for study. For asexual propagation of *Neurospora*, pieces of the mycelial growth or the asexual spores (conidia) can be used to inoculate a suitable growth medium to give rise to a new mycelium. This growth involves only mi-

148
Linkage,
Crossing-Over,
and Genetic
Mapping in
Eukaryotes

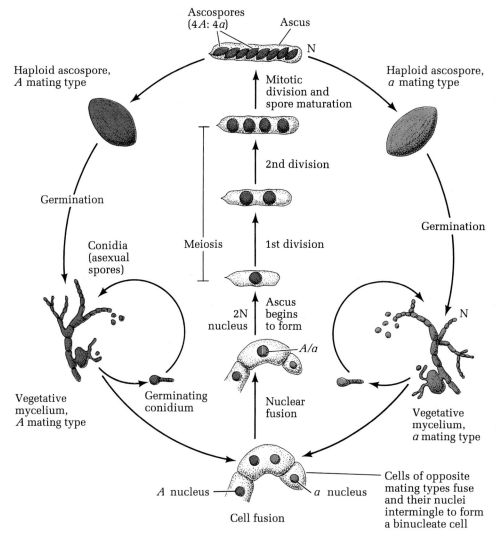

Ascospores
(4*A*: 4*a*)   Ascus

N

Haploid ascospore,
A mating type

Mitotic
division and
spore maturation

Haploid ascospore,
a mating type

2nd division

Germination

Meiosis   1st division

Germination

Conidia
(asexual
spores)

N

2N
nucleus

Ascus
begins
to form

A/a

Vegetative
mycelium,
A mating type

Germinating
conidium

Nuclear
fusion

Vegetative
mycelium,
a mating type

A nucleus

a nucleus

Cells of opposite
mating types fuse
and their nuclei
intermingle to form
a binucleate cell

Cell fusion

*Figure 5.9 Life cycle of the haploid, mycelial-form
fungus* Neurospora crassa.

totic division of the nuclei and the production of new cell mass.

Neurospora crassa can also reproduce by sexual means. There are two mating types (sexes, in a loose sense), called *A* and *a*. Both mating types look identical and can only be distinguished by the fact that strains of the *A* mating type do not mate with other *A* strains, and *a* strains do not mate with other *a* strains. (Thus the original designation of mating types as *A* or

a was arbitrary.) Under normal conditions *N. crassa* reproduces asexually. However, if nutrients (particularly nitrogen sources) become seriously depleted in the growth medium and if strains of the two mating types are present, the sexual cycle is initiated. Cells of the two mating types fuse, followed by fusion of two haploid nuclei to produce an *A/a* diploid nucleus, which is the only diploid stage of the life cycle. The diploid nucleus immediately undergoes

149
*Gene Recombination
and the
Role of
Chromosomal
Exchange*

meiosis and produces four haploid nuclei (two *A* and two *a*) within an elongating sac called an ascus. A subsequent mitotic division results in a linear arrangement of eight haploid nuclei around which spore walls form to produce eight sexual ascospores (four *A* and four *a*). Each ascus, then, contains all the products of a single meiosis. When the ascus is ripe, the ascospores (sexual spores) are shot out of the ascus and out of the fruiting body to be dispersed by wind currents. A particularly important feature is that all the products of a single meiosis are found together in the ascus and can be isolated and cultured for analysis. Moreover, the order of the four spore pairs within an ascus reflects exactly the orientation of the four chromatids of each tetrad at the metaphase plate in the first meiotic division: These tetrads are called **ordered tetrads.**

Recombination at the tetrad stage in *Neurospora*. For a determination of whether crossing-over occurs at the tetrad stage of prophase I, or prior to meiosis in interphase, crosses are made between haploid strains of *Neurospora* that differ genetically. A theoretical example is shown in Figure 5.10. The genetic constitution of the two strains is as follows: One of them, of mating type *A*, is unable to synthesize the amino acid methionine, and that compound must be provided in the growth medium in order for the strain to grow. This *Met⁻* strain is wild-type for all other genes. The other strain, of mating type *a*, is wild-type for its ability to synthesize methionine but is unable to synthesize another amino acid, histidine. The latter strain is referred to as the *His⁻* strain. Genotypically, the former strain is *met his⁺*, while the latter is *met⁺ his*. Both genes are in the same linkage group; that is, on the same chromosome.

The diploid nucleus produced from the mating is heterozygous for the two genes. We can now make some theoretical predictions. In meiosis, crossing-over will occur between the two genes at a frequency that is a function of the distance between them on the chromosome. Recombinant progeny will be produced with the genotypes *met⁺ his⁺* and *met his.* The former is wild-type for both genes, and the latter is a double mutant strain that requires both methionine and histidine in the growth medium in order to grow.

The progeny strains may be tested for their phenotypes (which are equivalent to their genotypes since the organism is haploid) by determining whether they can grow on an unsupplemented medium, a medium with methionine, one with histidine, or one with methionine and histidine. If no crossing-over occurs, the resulting ascus will contain four spores of each parental type. Figure 5.10a shows the consequences of crossovers occurring at the two-strand stage, that is, prior to meiosis in interphase: The resulting asci will contain only recombinant spores (four of each type). If, however, crossing-over occurs at the four-strand stage, the results would be as diagramed in Figure 5.10b. The crossing-over occurs between just two of the four chromatids and involves one from each parental strain. Those chromatids become recombinant for the gene markers and give rise to four recombinant spores in the ascus. The other two chromatids not involved in the crossover give rise to four parental spores in the same ascus. The overall result is an ascus with four parental and four recombinant progeny spores, representing all four possible phenotypic classes. Asci of this last type are found routinely in such experiments. These results provide unequivocal proof that crossing-over occurs at the tetrad stage of prophase I.

Keynote *Crossing-over occurs at the four-strand (tetrad) stage in prophase I in meiosis. The evidence for this principle came from the analysis of genetic experiments with the haploid, mycelial fungus* Neurospora crassa *in which all the products of a single meiotic event are contained in a structure called the ascus. The crucial result was that a single ascus contained two spores with the parental genotypes and two spores with the recombinant genotypes.*

a) Production of progeny spores if crossing-over occurs at two-strand stage of meiosis

b) Production of progeny spores if crossing-over occurs at four-strand stage of meiosis

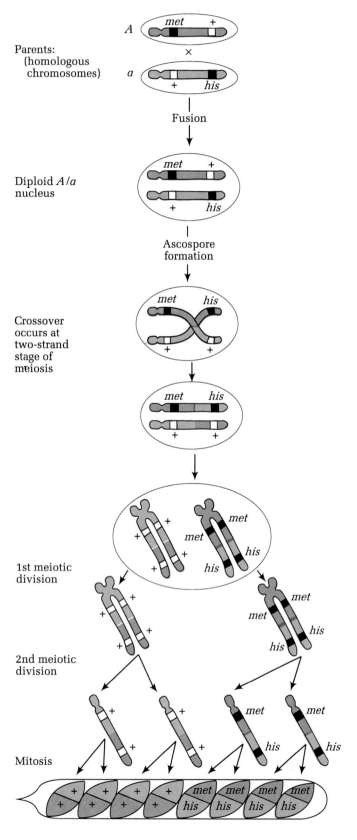

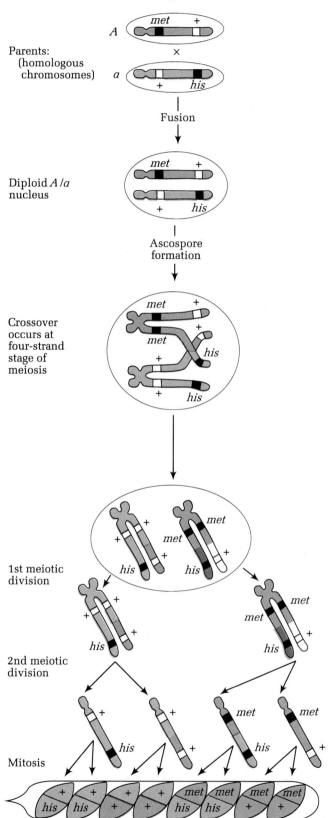

151
*Locating
Genes on
Chromosomes:
Mapping
Techniques*

Locating Genes on Chromosomes: Mapping Techniques

So far in the chapter we have learned two significant concepts. The first, derived principally from Morgan's experiments with *Drosophila,* is that genetic recombinants result from crossing-over between homologous chromosomes. More specifically, the number of genetic recombinants produced is characteristic of the two linked genes involved. The second concept, derived from genetic studies with the fungus *Neurospora crassa,* is that crossing-over takes place at the tetrad stage in prophase I of meiosis. We will now examine how genetic experiments can be used to locate genes on chromosomes relative to one another in eukaryotic organisms. This process is called **genetic mapping.**

Detecting Linkage through Testcrosses

Before beginning any experiments to construct a **genetic map** (also called a **linkage map**), geneticists must show that the genes under consideration are linked (located on the same chromosome); if they are not linked, then the genes are assorting independently. Therefore the simplest way to test for linkage is to analyze the results of crosses to see whether the data deviate significantly from those expected by independent assortment.

*Figure 5.10 Experiment showing that crossing-over occurs at the four-strand stage of meiosis; two haploid strains with different genetic markers (*met *and* his*) are crossed: (a) If a single crossover occurred at the two-strand stage (i.e., in interphase before chromosome duplication), an ascus would be produced that contains only the two types of recombinant ascospores, a result differing from experimental findings that asci resulting from a single crossover contain both types of parental ascospores and both types of recombinant ascospores. (b) If a single crossover occurred at the four-strand stage, an ascus would be produced that contains both types of parental ascospores and both types of recombinant ascospores, an outcome matching observed results.*

In Chapter 1 we showed that in diploid organisms, if two allelic pairs assort independently and if there is complete dominance of one allele of each pair, then the progeny from a self will show a 9:3:3:1 phenotypic ratio. This cross is not the best one to use to check for linkage, however. The testcross (a cross of an individual with an individual homozygous recessive for all genes involved) is better because the distribution of phenotypes is the result of segregation events in only one of the parents; the other parent contributes only recessive alleles to the progeny. We saw in Chapter 2 that a testcross between $a^+a\ b^+b$ and $aa\ bb$, where genes a and b are unlinked, gives a 1:1:1:1 ratio of the four possible phenotypic classes $a^+b^+{:}a^+b{:}a\ b^+{:}a\ b$. Since any *significant* deviation from this ratio in the direction of too many parental types and too few recombinant types would suggest that the two genes are linked, it is important to know how large a deviation must be in order to be considered "significant." The chi-square test can be used to make such a decision.

The chi-square test. The observed phenotypic ratios among progeny of a cross rarely match the expected or predicted ratios even though the hypothesis on which the expected ratios are based is correct. For example, none of the F_2's of Mendel's monohybrid crosses exhibited an exact 3:1 ratio, although the observed ratios were close enough so that he was not bothered by the discrepancy. Observed results may not match the expected results for several reasons, such as inadequate sample size, sampling error, or decreased viability of some of the genotypes involved.

Since genetic analysis often depends on comparing observed data with predictions based on a given hypothesis, we must decide when the deviation between the observed and the expected result is large enough to indicate that the hypothesis might be wrong. A common way to decide is to use a statistical test called the **chi-square** (χ^2) **test,** which, as used in the analysis of genetic data, is, essentially, a *goodness-of-fit test.*

The simplest way to illustrate the use of the

152
*Linkage,
Crossing-Over,
and Genetic
Mapping in
Eukaryotes*

chi-square test is to analyze an example from the text. Let us consider the F_2 data of one of Mendel's dihybrid crosses involving the smooth/ wrinkled and yellow/green gene pairs in the garden pea (see Figure 2.10). The dominant alleles are those for the smooth and yellow phenotypes. The F_2 data are as follows:

> 315 smooth, yellow
> 108 smooth, green
> 101 wrinkled, yellow
> 32 wrinkled, green

If the two genes are unlinked, we hypothesize a 9:3:3:1 ratio of the above, respective phenotypes. To test this hypothesis, we employ the chi-square test, as shown in Table 5.1.

First, we list the four phenotypes expected in the F_2 of the cross (column 1). Then we list the observed (o) numbers for each phenotype, using actual numbers and not percentages or proportions (column 2). Next, we calculate the expected (e) number for each phenotypic class from the total number of progeny (556) and from the hypothesis under evaluation, in this case 9:3:3:1 (column 3); thus we list $9/16 \times 556 = 312.75$, and so on. (Fractional numbers are used for accuracy in the later computations.) Now we subtract the expected number (e) from the observed number (o) for each class to find the deviation value d. The sum of the d values is always zero (column 4).

In column (5) the deviation squared (d^2) is

computed by multiplying each deviation value in column (4) by itself. In column (6) the deviation squared is then divided by the expected number (e). The chi-square value, χ^2 (item 7 in the table), is the total of the four values in column (6). In our example χ^2 is 0.4700. The general formula is

$$\chi^2 = \Sigma \frac{d^2}{e}$$

The last value in the table, item (8), is the degrees of freedom (df) for the set of data. The degrees of freedom in a test involving n classes are usually equal to $n - 1$. That is, if the total number of progeny (556 in our example) is divided among n classes (four phenotypic classes in the example), then once the expected numbers have been computed for $n - 1$ classes (three in our example), the expected number of the last class is set. Thus in our example there are only three degrees of freedom in the analysis.

The χ^2 value and the degrees of freedom are next used with a table to determine the probability P that the deviation of the observed values from the expected values is due to chance. Generally, if the probability of obtaining the observed χ^2 values is greater than 5 in 100 ($P > 0.05$), the deviation is considered not statistically significant and could have occurred by chance alone. Thus the hypothesis being tested could apply to the data being analyzed. If $P \leq 0.05$, we consider the deviation from the expected values to be statistically significant and

Table 5.1 Chi-Square Test

(1) Phenotypes	(2) Observed Number (o)	(3) Expected Number (e)	(4) d	(5) d^2	(6) d^2/e
Smooth, yellow	315	312.75	2.25	5.0625	0.0162
Smooth, green	108	104.25	3.75	14.0625	0.1349
Wrinkled, yellow	101	104.25	-3.25	10.5625	0.1013
Wrinkled, green	32	34.75	-2.75	7.5625	0.2176
Total	556	556.00	0.0		0.4700

(7) $\chi^2 = 0.4700$ (8) df = 3

153
*Locating
Genes on
Chromosomes:
Mapping
Techniques*

not due to chance alone; the hypothesis may well be inapplicable, If $P \leq 0.01$, the deviation is highly significant and some nonchance factor is involved; the hypothesis is almost certainly invalid.

The P value for a set of data is obtained from tables of χ^2 values for various degrees of freedom. Table 5.2 presents part of a table of chi-square probabilities. For our example, where $\chi^2 = 0.4700$ with three degrees of freedom, the P value is greater than 0.90 and less than 0.95 $(0.90 < P < 0.95)$. Thus from 90 to 95 times out of 100 we could expect chance deviations from the expected values of this magnitude. This P value, being greater than 0.05, indicates that the hypothesis that the two pairs of genes are assorting independently is valid. Had the P value been highly significant, then the independent assortment hypothesis would have been considered invalid and we would have had to think of alternative hypotheses, such as that linked genes might be involved.

Table 5.2 Chi-Square Probabilities

df	\multicolumn{10}{c}{Probabilities}
	0.95	*0.90*	*0.70*	*0.50*	*0.30*	*0.20*	*0.10*	*0.05*	*0.01*	*0.001*
1	0.004	0.016	0.15	0.46	1.07	1.64	2.71	3.84	6.64	10.83
2	0.10	0.21	0.71	1.39	2.41	3.22	4.61	5.99	9.21	13.82
3	0.35	0.58	1.42	2.37	3.67	4.64	6.25	7.82	11.35	16.27
4	0.71	1.06	2.20	3.36	4.88	5.99	7.78	9.49	13.28	18.47
5	1.15	1.61	3.00	4.35	6.06	7.29	9.24	11.07	15.09	20.52
6	1.64	2.20	3.83	5.35	7.23	8.56	10.65	12.59	16.81	22.46
7	2.17	2.83	4.67	6.35	8.38	9.80	12.02	14.07	18.48	24.32
8	2.73	3.49	5.53	7.34	9.52	11.03	13.36	15.51	20.09	26.13
9	3.33	4.17	6.39	8.34	10.66	12.24	14.68	16.92	21.67	27.88
10	3.94	4.87	7.27	9.34	11.78	13.44	15.99	18.31	23.21	29.59
11	4.58	5.58	8.15	10.34	12.90	14.63	17.28	19.68	24.73	31.26
12	5.23	6.30	9.03	11.34	14.01	15.81	18.55	21.03	26.22	32.91
13	5.89	7.04	9.93	12.34	15.12	16.99	19.81	22.36	27.69	34.53
14	6.57	7.79	10.82	13.34	16.22	18.15	21.06	23.69	29.14	36.12
15	7.26	8.55	11.72	14.34	17.32	19.31	22.31	25.00	30.58	37.70
20	10.85	12.44	16.27	19.34	22.78	25.04	28.41	31.41	37.57	45.32
25	14.61	16.47	20.87	24.34	28.17	30.68	34.38	37.65	44.31	52.62
30	18.49	20.60	25.51	29.34	33.53	36.25	40.26	43.77	50.89	59.70
50	34.76	37.69	44.31	49.34	54.72	58.16	63.17	67.51	76.15	86.66

←——— Accept | Reject ———→
at 0.05 level

Source: Taken from Table IV of Fisher and Yates, *Statistical Tables for Biological, Agricultural and Medical Research,* published by Oliver & Boyd, Edinburgh, and by permission of the authors and publishers.

154
Linkage,
Crossing-Over,
and Genetic
Mapping in
Eukaryotes

The concept of a genetic map. The data obtained by Morgan from *Drosophila* crosses indicated that the frequency of crossing-over (and hence of recombinants) for linked genes is characteristic of the gene pairs involved: For the X-linked genes white (w) and miniature (m) the frequency of crossing-over was 36.9 percent, and for the white and yellow (y) genes it was 1.3 percent. Moreover, the frequency of recombinants for two linked genes is approximately constant regardless of how the alleles involved are arranged on the chromosomes. In an individual doubly heterozygous for the w and m alleles, for example, the alleles can be arranged in two ways:

$$\frac{w^+\, m^+}{w\ \ m} \quad \text{or} \quad \frac{w^+\, m}{w\ \ m^+}$$

In the arrangement on the left the two wild-type alleles are on one homozygous chromosome and the two recessive mutant alleles are on the other: This arrangement is called **coupling.** Crossing-over between the two genes produces $w^+\, m$ and $w\, m^+$ recombinants. In the arrangement on the right each homologous chromosome carries the wild-type allele of one gene and the mutant allele of the other gene: This arrangement is called **repulsion.** Crossing-over between the two genes produces $w^+\, m^+$ and $w\, m$ recombinants. While the actual phenotypes of the recombinant classes are different for the two arrangements, the percentage of recombinants among the total progeny will be 36.9 percent in each case.

Morgan thought that the characteristic crossover frequencies for linked genes might be related to the physical distances separating the genes on the chromosome. In 1913 a student of Morgan's, Alfred Sturtevant, devised the testcross method to analyze the linkage relationships between genes. He suggested that the percentage of recombinants (which are produced by crossovers in gamete production) could be used as a quantitative measure of the distance between two gene pairs on a genetic map. This distance is measured in **map units** (mu) (Figure 5.11). A crossover frequency of 1 percent between two genes equals 1 map unit and means

that in every 100 gametes formed, one crossover will occur between these two genes. The map unit is sometimes called a **centi-Morgan** (cM) in honor of Morgan.

Sturtevant's concept of genetic map distance led him to the important discovery that the distances between a series of linked genes are additive. Thus the genes on a chromosome can be represented by a linear genetic map that shows the genes belonging to the chromosome in linear order. The position of a gene on a genetic map is called its **locus** (plural, loci). Crossover and recombination values are used to show a linear order of the genes on a chromosome and to provide information about the relative distance between any two genes. The greater the crossover frequency between two genes, the farther apart they are (Figure 5.12).

The first genetic map was based on recombination frequencies from crosses involving the sex-linked genes w, m, and y discussed earlier, where w gives white eyes, m gives miniature wings, and y gives yellow body. From these mapping experiments, the recombination frequencies for the $w \times m$, $w \times y$, and $m \times y$ crosses were established as 32.6, 1.3, and 33.9 percent, respectively. The percentages are quantitative measures of the distance between the genes involved.

The logic used in constructing a genetic map of these three genes is as follows: The genes must be arranged at different points on a line that best accommodate the data. The recombination frequencies show that w and y are closely linked and that m is quite a distance away from the other two genes. Since the w–m distance is less than the y–m distance (as shown by the smaller percentage of the recombinants in the $w \times m$ cross), the order of genes must be y–w–m. Thus the three genes are ordered and spaced as follows, with the y gene assigned an arbitrary value of 0 and the other map units indicating genetic distance from y:

155
Locating
Genes on
Chromosomes:
Mapping
Techniques

Recombination frequencies are used to construct genetic maps in all gene-mapping experiments, whether the organism is a eukaryote or a prokaryote.

Gene Mapping by Using Two-Point Testcrosses

From Sturtevant's work we have seen that the percentage of recombinants resulting from crossing-over is used as a measurement of the genetic distance between two linked genes. Quite simply, by carrying out two-point testcrosses such as those shown in Figure 5.13, we can determine the relative numbers of parental and recombinant classes in the progeny. For autosomal recessives a double heterozygote is crossed with a doubly homozygous recessive mutant strain (Figure 5.13a). When the double heterozygous $a^+ b^+/a b$ F_1 progeny from a cross of $a^+ b^+/$ $a^+ b^+$ with $a b/a b$ are testcrossed with $a b/a b$, four phenotypic classes are found among the F_2 progeny. Two of these classes have the parental phenotypes $a^+ b^+$ and $a b$ and derive from zygotes in which no crossover had occurred. Since this same no-crossover event produced both classes, approximately equal numbers of these two types are expected.

The other two F_2 phenotypic classes have recombinant phenotypes $a^+ b$ and $a b^+$, which derive from zygotes in which a single crossover occurred between the chromosomes. Again, we expect approximately equal numbers of these two recombinant classes. Because a single crossover event occurs more rarely than no crossing over, an excess of parental phenotypes over recombinant phenotypes in the progeny of a testcross indicates linkage between the genes involved. For autosomal dominants a double heterozygote is crossed with a wild-type strain since the recessive allele of a dominant mutant is the wild-type allele (Figure 5.13b). When the doubly heterozygous $A B/A^+ B^+$ F_1 progeny from a cross of $A B/A B$ with $A^+ B^+/A^+ B^+$ are testcrossed with $A^+ B^+/A^+ B^+$, four phenotypic classes are found among the F_2 progeny. Two of these classes have the parental phenotypes $A B$ and $A^+ B^+$ and derive from zygotes in which no crossover had occurred. The other

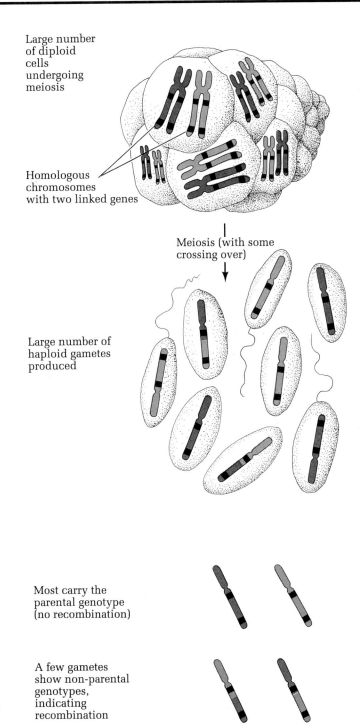

Large number of diploid cells undergoing meiosis

Homologous chromosomes with two linked genes

Meiosis (with some crossing over)

Large number of haploid gametes produced

Most carry the parental genotype (no recombination)

A few gametes show non-parental genotypes, indicating recombination

Figure 5.11 Percentage of recombinants as a quantitative measure of the distance (map unit) between two genes pairs on a genetic map.

156
Linkage,
Crossing-Over,
and Genetic
Mapping in
Eukaryotes

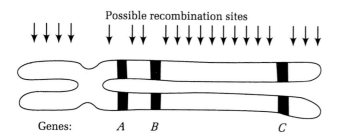

Possible recombination sites

Genes: *A*   *B*   *C*

Possible sites between genes *A* and *B* = 2
Possible sites between genes *B* and *C* = 10
Possible sites between genes *A* and *C* = 12

Figure 5.12 Recombination between linked genes and map distance. The possible sites for recombination between genes A, B, *and* C *are as follows: 2 sites between genes* A *and* B; *10 sites between genes* B *and* C; *12 sites between genes* A *and* C. *Thus there are more chances that recombination will take place between genes that are far apart (like* A *and* C) *than between genes that are close together (like* A *and* B). *The percentage of recombinants can provide information about the relative distance between two linked genes.*

two F_2 phenotypic classes have recombinant phenotypes *A B^+* and *A^+ B,* and these derive from zygotes in which a single crossover occurred between the chromosomes. As for the autosomal recessives, an excess of parental phenotypes over recombinant phenotypes is expected.

Two-point testcrosses are essentially the same when sex-linked genes are involved. For X-linked recessives a double heterozygous female is crossed with a homozygous male carrying the recessive alleles

$$\frac{+\,+}{a\ \ b} \times \frac{a\ \ b}{\longrightarrow}$$

For X-linked dominants a doubly heterozygous female is crossed with a male carrying wild-type alleles on the X chromosome:

$$\frac{A\ \ B}{+\,+} \times \frac{+\,+}{\longrightarrow}$$

In all cases a two-point testcross should yield a pair of parental types that occur with about

equal frequencies and a pair of recombinant types that also occur with equal frequencies. The phenotypes will, of course, depend on the configuration of the two allelic pairs in the homologous chromosomes, that is, whether they are in coupling or repulsion. A count of the representatives of each progeny class gives the percentage of recombinant types when the following formula is used:

$$\frac{\text{number of recombinants}}{\text{total number of testcross progeny}} \times 100 = \begin{array}{l}\text{percent}\\\text{recombinants}\end{array}$$

The value for the percentage of recombinants is usually directly converted into map units, so, for example, 20 percent recombination between genes *a* and *b* indicates that the two genes are 20 mu apart.

The two-point method of mapping is most accurate when the two genes examined are close together; when genes are far apart, other factors can influence the production of recombinant progeny. When we are drawing a map of the genes on a chromosome, these factors must be taken into consideration. Large numbers of progeny must also be counted (scored) to ensure a high degree of accuracy. We have already seen from Sturtevant's work the logic behind constructing an actual linkage map from recombination frequencies involving all possible pairwise crosses of the genes under study. From mapping experiments carried out in all types of organisms, we know that genes are linearly arranged

Figure 5.13 Testcross to show that two genes are linked: (a) Genes a *and* b *are recessive mutant alleles linked on the same autosome. A homozygous* a^+b^+/a^+b^+ *individual is crossed with a homozygous recessive* a b/a b *individual, and the doubly heterozygous* F_1 *progeny* (a^+b^+/a b) *are testcrossed with homozygous* a b/a b *individuals. (b) Genes* A *and* B *are dominant mutant alleles linked on the same autosome. A homozygous dominant individual* (A B/A B) *is crossed with a wild-type* A^+ B^+/A^+B^+ *(homozygous recessive) individual, and doubly heterozygous* F_1 *progeny* A B/A^+ B^+ *are testcrossed with the wild-type* A^+ B^+/A^+ B^+.

157
Locating
Genes on
Chromosomes:
Mapping
Techniques

in linkage groups. There is a one-to-one correspondence of linkage groups and chromosomes, so the sequence of genes on the linkage group reflects the sequence of genes on the chromosome.

Generating a Genetic Map

A genetic map is generated from counting the number of times a crossover event occurred in a particular segment of the chromosome out of all meioses examined (in the form of progeny organisms). Since in many cases the probability of a crossing-over event is not uniform along a chromosome, we must be cautious about how far we extrapolate the genetic map (which derives from data produced by genetic crosses) to the chromosomal map (which derives from cyto-

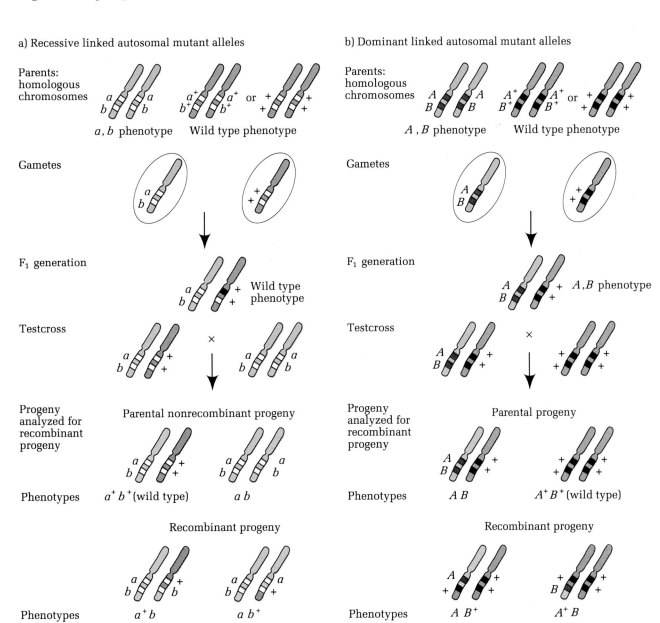

158
*Linkage,
Crossing-Over,
and Genetic
Mapping in
Eukaryotes*

genetic determinations of the locations of genes along the chromosome itself). Nonetheless, a good working hypothesis is to consider crossovers as being randomly distributed along the chromosome.

The recombination frequencies observed between genes may also be used to predict the outcome of genetic crosses. For example, a recombination frequency of 20 percent between genes indicates that the probability of a crossover occurring in that region of the chromosome in a meiosis is 0.2. Since crossing-over is a random process, more than one crossover may occur in that region in a meiosis. The probabilities of these so-called **multiple crossovers** can readily be computed. That is, the probability of two crossovers (called a **double crossover**) is

$$0.2 \times 0.2 = 0.04$$

The probability of three crossovers (a triple crossover) is

$$0.2 \times 0.2 \times 0.2 = 0.008$$

and so on.

For any testcross the percentage of recombinants in the progeny cannot exceed 50 percent. That is, if the genes are assorting independently, an equal number of recombinants and parentals are expected in the progeny, so the frequency of recombinants is 50 percent. With such a value in hand we state that the two genes are unlinked. Genes may be unlinked in two ways. One way results when the genes are on different chromosomes, a case we discussed before. The second way occurs when the genes are on the same chromosome but are so far apart that many different crossovers occur between them. As Figure 5.14 shows, the consequence of an even number of crossovers is parental progeny, whereas an odd number of crossovers gives rise to recombinant progeny. When genes are far apart, crossovers occur so often that the frequency of even-numbered crossovers is about equal to the frequency of odd-numbered crossovers. In this case we get 50 percent recombinants.

The point is, if two genes are unlinked, and therefore show 50 percent recombination, it does not necessarily mean that they are on different chromosomes. More data would be needed to determine whether the genes are on the same chromosome or different chromosomes. One way to find out is to map a number of other genes in the linkage group. For example, if *a* and *m* show 50 percent recombination, perhaps we will find that *a* shows 27 percent recombination with *e,* and *e* shows 26 percent recombination with *m.* This result would indicate that *a* and *m* are in the same linkage group.

Double Crossovers

Data for the F₂ progeny of a cross between a yellow-bodied, white-eyed, miniature-winged fe-

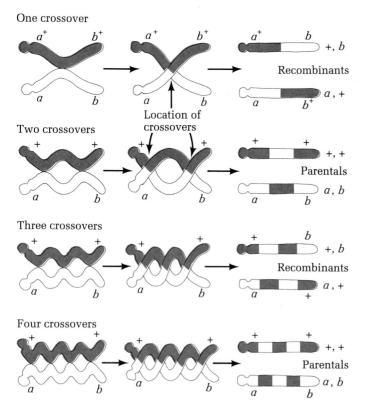

Figure 5.14 Consequences of multiple crossovers between two genes. An even number of crossovers results in parental progeny; an odd number of crossovers results in recombinant progeny. (In this figure, only the two chromatids involved in crossing over are shown.)

159
*Locating
Genes on
Chromosomes:
Mapping
Techniques*

Table 5.3 Data for the F_2 Progeny of a Cross between a Yellow-bodied, White-eyed, Miniature-winged (y w m/y w m) Mutant Female and a Wild-Type (+ + +/Y) Male Drosophila

Phenotypes				Crossovers		
Body Color	Eye Color	Wing Size	Number	Body Color and Eye Color	Eye Color and Wing Size	Body Color and Wing Size
Normal	Red (normal)	Long (normal)	758	—	—	—
Yellow	White	Miniature	700	—	—	—
Normal	Red	Miniature	401	—	401	401
Yellow	White	Long	317	—	317	317
Normal	White	Miniature	16	16	—	16
Yellow	Red	Long	12	12	—	12
Normal	White	Long	1	1	1	—
Yellow	Red	Miniature	0	0	0	—
		Total	2205	29	719	746
		Percentage	100	1.3	32.6	33.8

male and a normal male *Drosophila* are shown in Table 5.3. We have already shown that the gene order is *y–w–m*. The data show 29 exchanges between the genes for body color and eye color and 719 exchanges between genes for eye color and wing size, giving a total of 748. However, only 746 exchanges are apparent between the body color and wing size genes. The discrepancy is due to the single normal-bodied, white-eyed, normal-winged fly. In the production of the egg from which this fly developed, two exchanges (a double crossover) must have occurred between the X chromosomes, with the result that the body color and wing size genes (normal body color and normal-length wings) appeared together as in the grandparent, although the intervening gene (white eyes) was derived from the other grandparent. Thus large crossover frequencies give an inaccurate map, since unless intervening genes are available, there is no way of detecting double crossovers. Various empirical formulas have been devised to allow for the effects of double crossing-over by correcting recombination frequencies to give map units.

To see mechanistically how double crossovers

are a significant factor in genetic-mapping experiments, consider a hypothetical case of two allelic pairs a^+/a and b^+/b in coupling and separated by quite a distance on the same chromosome. Figure 5.15a shows that double-crossover chromatids do not show recombination of the allelic pairs involved, so only parental progeny result. Exactly the same result would occur if no crossing-over took place between the two chromatids in this region. However, it is the percentage of crossing-over between genes that is a measure of the distance between them. Therefore since the double crossover in Figure 5.15a did not generate recombinants, the two crossover events will be uncounted, and the estimate of map distance between genes a and b will be low. How low the estimate will be will depend on how many double crossovers take place per meiosis in the region being examined.

Three Point Cross

How can we avoid this pitfall? Double crossing-over occurs quite rarely within distances of 5 mu (map units) or less, and it occurs only some-

160
*Linkage,
Crossing-Over,
~and Genetic
Mapping in
Eukaryotes*

a) Single crossover

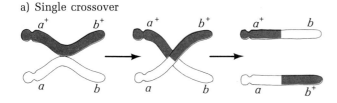

b) Double crossover

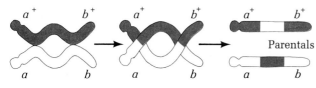

Parentals

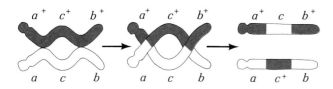

Gametes are recombinant for the c^+/c gene pair relative to the a^+/a and b^+/b gene pairs

Figure 5.15 (a) Progeny of single and double crossovers. A single crossover between linked genes generates recombinant gametes, and a double crossover between linked genes gives parental gametes. Thus inaccurate map distances between genes result, since not all crossovers can be accounted for. (b) A possible solution to the double-crossover problem. The presence of a third allelic pair between the two allelic pairs of Figure 5.15a enables us to detect the double-crossover event. In a double crossover the middle gene will change positions relative to the outside genes. (In this figure, only the two chromatids involved in crossing over are shown.)

times in greater distances for certain chromosomal regions. Consequently, one way to get accurate map distances is to study closely linked genes. Another efficient way to obtain such data is to use a **three-point testcross** involving three genes within a relatively short section of a chromosome. (The evidence for the closely linked

nature of three genes would derive from separate two-point testcrosses and the drawing of a preliminary genetic map.)

The potential advantage of the three-point testcross can be seen in the theoretical case if we have a third allelic pair c^+/c between the a^+/a and b^+/b allelic pairs of Figure 5.15a, as diagramed in Figure 5.15b. A double-crossover event between a and b might be detected by the recombination of the c^+/c allelic pair relative to the other two allelic pairs. Of course, we can now ask: What if the double crossover occurred between genes a and c? Nonetheless, the principle holds: Accurate mapping data can be obtained when the genes are relatively close together. In fact, three-point testcrosses are routinely used as an efficient way of mapping genes both for their order in the chromosome and for the distances between them, as we will see in the next section.

Keynote *The map distance between two genes is based on the frequency of recombinants between the two genes (an approximation of the frequency of crossovers between the two genes). By using appropriate genetic markers in crosses, geneticists are able to compute the recombination frequency between genes in chromosomes as the percentage of progeny showing the reciprocal recombinant phenotypes. The more closely the recombination frequency parallels the crossover frequency, the closer the genes are. But when the genetic distance increases between genes, the incidence of multiple crossovers, and particularly double crossovers, causes the recombination frequency to be an underestimate of the crossover frequency and hence of the map distance.*

Mapping Chromosomes by Using Three-Point Testcrosses

A three-point testcross is preferable to a two-point testcross for mapping purposes because it aids in the detection of double crossovers and because the relative order of the genes on the chromosome can be established from a single

161
*Locating
Genes on
Chromosomes:
Mapping
Techniques*

cross. In diploid organisms the three-point test-cross is a cross of a triple heterozygote with a triply homozygous recessive. If the mutant genes in the cross are all recessive, a typical three-point testcross might be

$$\frac{+\ \ +\ \ +}{a\ \ b\ \ c} \times \frac{a\ b\ c}{a\ b\ c}$$

If any of the mutant genes are dominant, the homozygous parent in the testcross will carry the wild-type allele of these genes. For example, a testcross of a strain heterozygous for two recessive mutations (*a* and *c*) and one dominant mutation (*B*) might be

$$\frac{+\ \ +\ \ +}{a\ \ B\ \ c} \times \frac{a\ +\ c}{a\ +\ c}$$

In a testcross involving sex-linked genes, the female is the heterozygous strain (assuming that the female is the homogametic sex) and the male is hemizygous for the recessive markers.

Three-point mapping is also done in haploid eukaryotic organisms, but in this case a testcross is not needed. Thus in *Neurospora crassa* we might cross two haploid strains to generate the triply heterozygous diploid cell. A cross of an *a b c* strain with a + + + strain, for instance, would give a + + +/*a b c* diploid cell. That cell goes through meiosis to generate haploid spores from which progeny *Neurospora* cultures grow. The phenotypes of the progeny are determined directly by the genotypes since the cells are haploid.

Let us suppose we have a (hypothetical) plant in which there are three linked genes. A recessive allele *p* of the first gene determines purple flower color in contrast to the white color of the wild type. A recessive allele *r* of the second gene results in a round fruit shape as compared with an elongated fruit in the wild type, and a recessive allele *j* of the third gene gives a juicy fruit instead of the wild-type dry fruit. The task before us is to determine the order of the genes on the chromosome and to determine the map distance between the genes. To do so, we must carry out the appropriate testcross of a triple heterozygote (+ + +/*p r j*) with a triply homozygous recessive (*p r j/p r j*) and then count the different phenotypic classes in the progeny (Figure 5.16).

Three genes are heterozygous in the cross. For each, two different phenotypes occur in the testcross progeny. Therefore for the three genes a total of $(2)^3 = 8$ phenotypic classes will appear in the progeny, representing all possible combinations of phenotypes. In an actual experiment it may happen that not all the phenotypic classes are generated. Nevertheless, the absence of a phenotypic class is important information, and the experimenter should enter a 0 in the class for which no progeny are found.

Establishing the order of the genes. The first step in finding the genetic map distances of the three genes is to determine the order in which the genes are located on the chromosome; that is, which gene is in the middle? Each progeny type is generated from a zygote formed by the fusion of the gametes from each parent. One parent carries the recessive alleles for all three genes, and the other is heterozygous for all three genes. Therefore the phenotype of the resulting individual is determined by the alleles in the gamete from the triply heterozygous parent; the gamete from the other parent will carry only recessive alleles. We know from the genotypes of the original parents that all three genes are in coupling. Since the heterozygous parent in the test-cross was + + +/*p r j*, classes 1 and 2 in Figure 5.16 are parental progeny: Class 1 is produced by the fusion of a + + + gamete with a *p r j* gamete from the triply homozygous recessive parent, and class 2 is produced by the fusion of a *p r j* gamete from the triply heterozygous parent and a *p r j* gamete from the triply homozygous recessive parent. These classes are generated from meioses in which no crossing-over occurs on the region of the chromosome in which the three genes are located.

The double-crossover progeny classes can often be found by inspection of the numbers of each phenotypic class. Since the frequency of a double crossover in a region is expected to be lower than the frequency of a single crossover, *double-crossover gametes are the least frequent to be produced.* Thus to identify the double-

162
Linkage,
Crossing-Over,
and Genetic
Mapping in
Eukaryotes

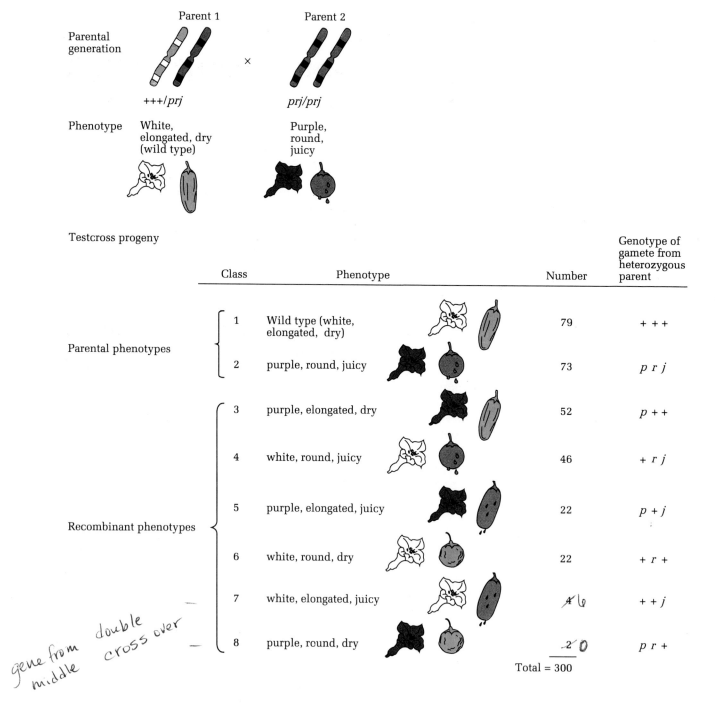

Figure 5.16 *Theoretical analysis of a three-point mapping, showing the testcross used and the resultant progeny.*

163
Locating
Genes on
Chromosomes:
Mapping
Techniques

crossover progeny, we can examine the progeny to find the pair of classes that have the lowest number of representatives. In Figure 5.16 classes 7 and 8 are such a pair. The genotypes of the gametes from the heterozygous parent that give rise to these phenotypes are $+ + j$ and $p\,r\,+$.

Referring now to Figure 5.15b, we see that a double crossover changes the orientation of the gene in the center of the sequence (c^+/c in that case) with respect to the two flanking allelic pairs. Therefore genes p, r, and j must be arranged in such a way that the center gene switches to give classes 7 and 8. To determine the arrangement, we must check the relative organization of the genes in the parental heterozygote so that it is clear which genes are in coupling and which are in repulsion. In this example the parental (noncrossover) gametes are $+ + +$ and $p\,r\,j$, so all are in coupling. The double-crossover gametes are $+ + j$ and $p\,r\,+$, so the only possible gene order that is compatible with the data is $p\,j\,r$, with the genotype of the heterozygous parent being $+ + +/p\,j\,r$. Figure 5.17 illustrates the generation of the double-crossover gametes from that parent.

Calculating the map distances between genes. Now the data can be rewritten as shown in Figure 5.18 to reflect the newly determined gene order. For convenience in the analysis the region between genes p and j is called region I and that between genes j and r is called region II.

Map distances can now be calculated in the way described previously. The frequency of crossing-over (genetic recombination) is computed between two genes at a time. For the p–j distance all the crossovers that occurred in region I must be added. Thus we must consider the recombinant progeny resulting from a single crossover in that region (classes 3 and 4) and the recombinant progeny produced by a double crossover (classes 7 and 8). The latter must be included since each double crossover includes a crossover in region I and therefore is a recombination event between genes p and j. From Figure 5.16 there are 98 recombinant progeny in classes 3 and 4, and 6 in classes 7 and 8, for a total of 104 progeny that resulted from a recombination

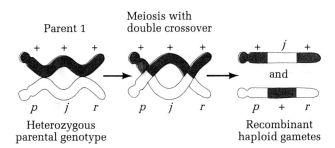

Figure 5.17 Rearrangement of the three genes of Figure 5.16 to p j r. *The evidence is that a double crossover (as it is shown in this figure) generates the least frequent pair of recombinant phenotypes (in this case class 7 with six progeny and class 8 with zero progeny).*

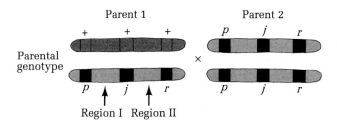

Testcross progeny

Class	Genotype of gamete from heterozygous parent	Number	Origin
1	$+ + +$	79	Parentals, no crossover
2	$p\,j\,r$	73	
3	$p + +$	52	Recombinants, single crossover, region I
4	$+ j\,r$	46	
5	$p\,j +$	22	Recombinants, single crossover, region II
6	$+ + r$	22	
7	$+ j +$	4	Recombinants, double crossover
8	$p + r$	2	

Total = 300

Figure 5.18 Rewritten form of the testcross and testcross progeny of Figure 5.16 based on the actual gene order p j r.

164
*Linkage,
Crossing-Over,
and Genetic
Mapping in
Eukaryotes*

event in region I. Since there are a total of 300 progeny, the percentage of progeny generated by crossing-over in region I is 34.66 percent. In other words, the percentage of recombination between p and j is determined as follows:

$$\text{percentage} = \frac{\begin{array}{c}\text{number of single cross-}\\\text{overs in region I }(p - j) +\\\text{number of double crossovers}\end{array}}{\text{total progeny}} \times 100$$

$$= \frac{98 + 6}{300} \times 100$$

$$= 34.66\%$$

In other words, the distance between genes p and j is 34.66 mu. This map distance, which is quite large, is chosen mainly for illustration. If this cross were an actual cross, we would have to be cautious about that map distance value since it is in the range where double crossovers might occur at a significant frequency and might go undetected. In an actual cross the value of 34.66 mu would probably underestimate the true distance.

The same method is used to obtain the map distance between genes j and r. That is, we calculate the frequency of crossovers in the cross that gave rise to progeny recombinant for genes j and r and directly relate that frequency to map distance. In this case all the crossovers that occurred in region II (see Figure 5.18) must be added (classes 5, 6, 7, and 8). The percentage of crossovers is calculated as follows:

$$\text{percentage} = \frac{\begin{array}{c}\text{number of single cross-}\\\text{overs in region II }(j - r) +\\\text{number of double crossovers}\end{array}}{\text{total progeny}} \times 100$$

$$= \frac{44 + 6}{300} \times 100$$

$$= 16.66\%$$

Thus the map distance between genes j and r is 16.66 map units.

In summary, a genetic map of the three genes

in the example has been generated (Figure 5.19). The example has illustrated that the three-point testcross is an effective way of establishing the order of genes and of calculating map distances. In a calculation of map distances from three-point testcross data, the double-crossover figure must be added to each of the single-crossover figures since in each case a double crossover represents single-crossover events occurring in the same meiosis simultaneously in regions I and II.

To compute the map distance between the two outside genes, we simply add the two map distances. Thus in the example the $p–r$ distance is 34.66 + 16.66 = 51.33 mu. This map distance can be computed directly from the data by combining the two formulas discussed previously. So the $p–r$ distance is calculated as follows:

$$\text{distance} = \frac{\begin{array}{c}\text{(number of single crossovers}\\\text{in region I + number of dou-}\\\text{ble crossovers) + (number of}\\\text{single crossovers in region II}\\\text{+ number of double cross-}\\\text{overs)}\end{array}}{\text{total progeny}} \times 100$$

$$= \frac{\begin{array}{c}\text{number of single crossovers}\\\text{in region I + number of sin-}\\\text{gle crossovers in region II +}\\\text{2(number of double cross-}\\\text{overs)}\end{array}}{\text{total progeny}} \times 100$$

$$= \frac{98 + 44 + 2(6)}{300} \times 100$$

$$= 51.33 \text{ map units}$$

Keynote *The map distance between genes is calculated from the results of test-crosses between strains carrying appropriate genetic markers. The most accurate testcross is the three-point testcross using a diploid organism in which a triple heterozygote is crossed with a strain that is homozygous recessive for all three genes. The most accurate map distances are obtained when the three genes are reasonably close to one another, with, say, 5–10 map units*

165
Locating
Genes on
Chromosomes:
Mapping
Techniques

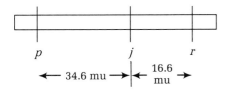

Figure 5.19 Genetic linkage map of the p-j-r *region of the chromosome as computed from the recombination data of Figure 5.18.*

between each locus. The unit of genetic distance is the map unit (mu), which is the distance between gene pairs for which 1 product out of 100 is recombinant, or, in other words, a recombinant frequency of 1 percent is 1 mu. For illustration, the genetic maps for Drosophila melanogaster *and* Zea mays *are shown in Figures 5.20 and 5.21.*

Interference and coincidence. The map distances determined by three-point mapping are primarily useful in elaborating the overall organization of genes on a chromosome. In addition, the map distances obtained are useful in telling us a little about the recombination mechanisms themselves. For example, the map distance of 34.66 mu between gene *p* and *j* means that 34.66 percent of the gametes should result from crossing-over between the two gene loci. Similarly, the map distance of 16.66 mu between *j* and *r* indicates that 16.66 percent of the gametes should result from crossing-over between these two gene loci.

If in the example of Figure 5.18 crossing-over in region I is independent of crossing-over in region II, then the probability of simultaneous crossing-over (i.e., a double crossover) in the two regions is equal to the product of the probabilities of the two events occurring separately. Thus if crossing-over in the two regions is independent, we have

$$0.3466 \times 0.1666 = 0.0577$$

or 5.77 percent double crossovers are expected to occur. However, only 6/300 = 2% double crossovers occurred in this cross (in classes 7 and 8).

This discrepancy is not just a simple experimental error due to small sample size, misscored progeny, and so on, associated with this cross. It is characteristic of such mapping crosses that double-crossover progeny typically do not appear as often as the map distances between the genes lead us to expect. Thus once a crossing-over event has occurred in one part of the meiotic tetrad, the probability of another crossing-over event occurring nearby is reduced, most probably by physical interference caused by the breaking and rejoining chromatids. This phenomenon is called **chiasma interference** (also **chromosomal interference**).

There is no way to predict the extent of interference since it tends to vary from chromosome to chromosome and even between segments of the same chromosome. But for the gene maps to be useful in predicting the outcome of crosses, we should know the magnitude of the interference throughout the map. The usual way to express the extent of interference is as a **coefficient of coincidence,** which is the ratio

$$\frac{\text{observed double-crossover frequency}}{\text{expected double-crossover frequency}}$$

And

$$\text{interference} = 1 - \text{coefficient of coincidence}$$

For the portion of the map in our example the coefficient of coincidence is

$$0.020/0.0577 = 0.347$$

The coefficient of coincidence normally varies between zero and one and may be interpreted as follows: A coincidence of one means that all expected double crossovers for a given region occur; there is no interference and the interference value is zero. A coefficient of coincidence of zero means that none of the expected double crossovers occur. Here there is total interference, and one crossover completely prevents a second crossover from occurring in the region under examination. The interference value is one. Thus the coincidence values and the interference values are inversely related. In the example the coefficient of coincidence of 0.347 means that the interference value is 0.653: Only 34.7 per-

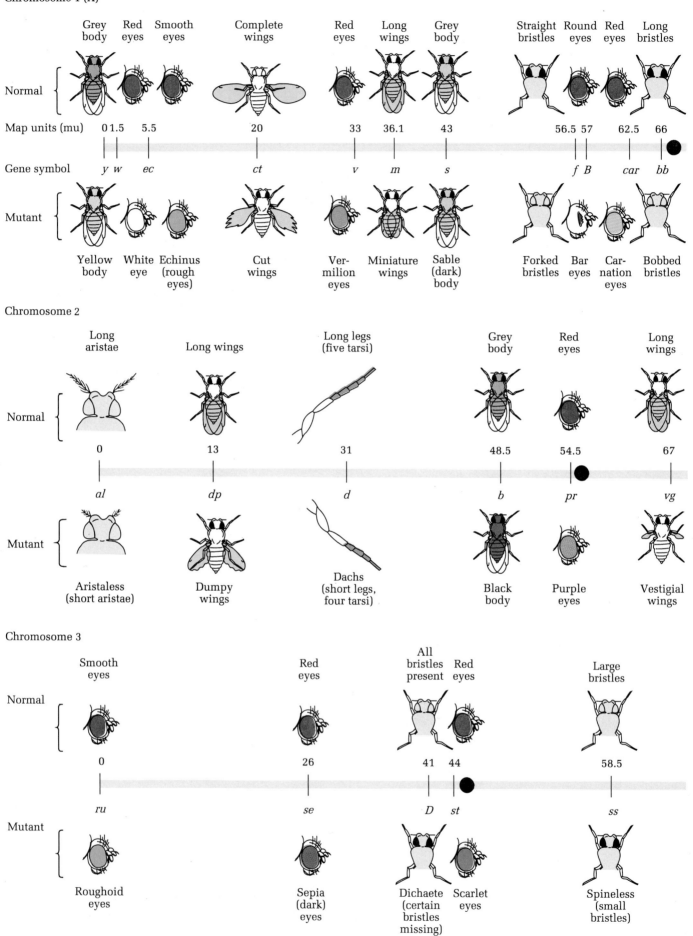

Chromosome 1 (X)

| | Grey body | Red eyes | Smooth eyes | Complete wings | | Red eyes | Long wings | Grey body | | Straight bristles | Round eyes | Red eyes | Long bristles |

Normal

Map units (mu) 0 1.5 5.5 20 33 36.1 43 56.5 57 62.5 66

Gene symbol *y* *w* *ec* *ct* *v* *m* *s* *f* *B* *car* *bb*

Mutant

Yellow body · White eye · Echinus (rough eyes) · Cut wings · Vermilion eyes · Miniature wings · Sable (dark) body · Forked bristles · Bar eyes · Carnation eyes · Bobbed bristles

Chromosome 2

| | Long aristae | Long wings | Long legs (five tarsi) | Grey body | Red eyes | Long wings |

Normal

0 13 31 48.5 54.5 67

al *dp* *d* *b* *pr* *vg*

Mutant

Aristaless (short aristae) · Dumpy wings · Dachs (short legs, four tarsi) · Black body · Purple eyes · Vestigial wings

Chromosome 3

| | Smooth eyes | Red eyes | All bristles present | Red eyes | Large bristles |

Normal

0 26 41 44 58.5

ru *se* *D* *st* *ss*

Mutant

Roughoid eyes · Sepia (dark) eyes · Dichaete (certain bristles missing) · Scarlet eyes · Spineless (small bristles)

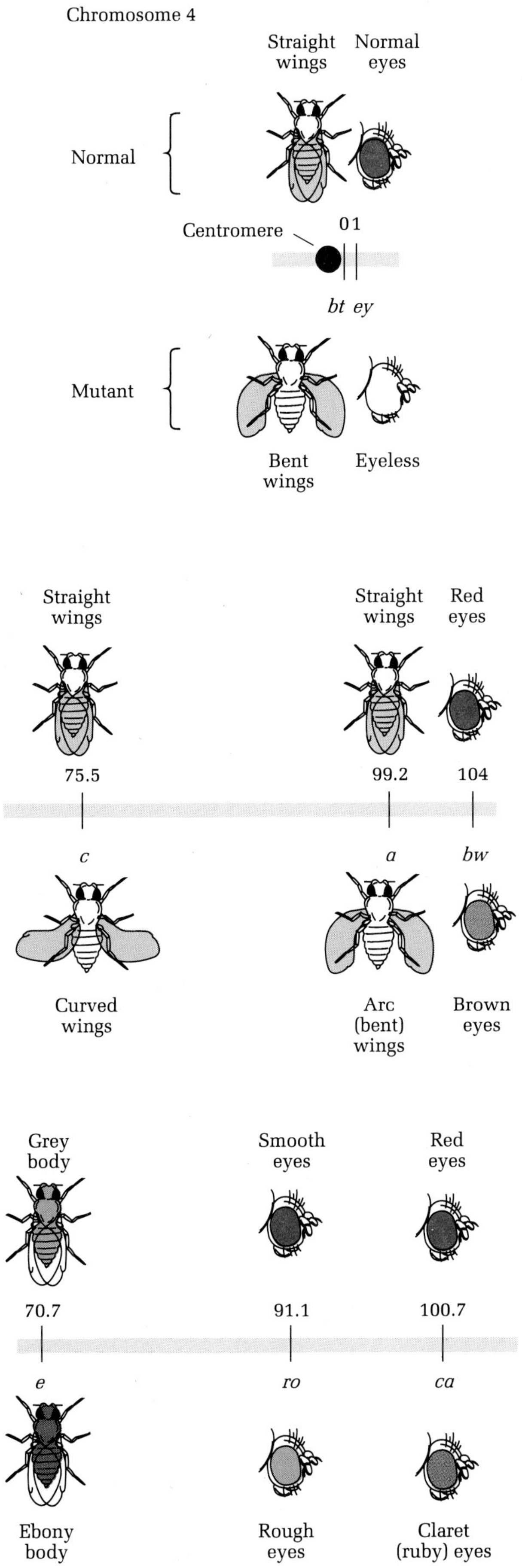

Figure 5.20 *Part of the genetic map of* Drosophila melanogaster *showing (pictorially) several commonly used genetic markers.*

cent of the expected double crossovers took place in the cross.

Keynote The occurrence of a chiasma between two chromatids may physically impede the occurrence of a second chiasma nearby, a phenomenon called chiasma interference. The extent of interference is expressed by the coefficient of coincidence and is calculated by dividing the number of observed double crossovers by the number of expected double crossovers. The coefficient of coincidence ranges from zero to one, and the extent of interference is measured as 1 − coefficient of coincidence.

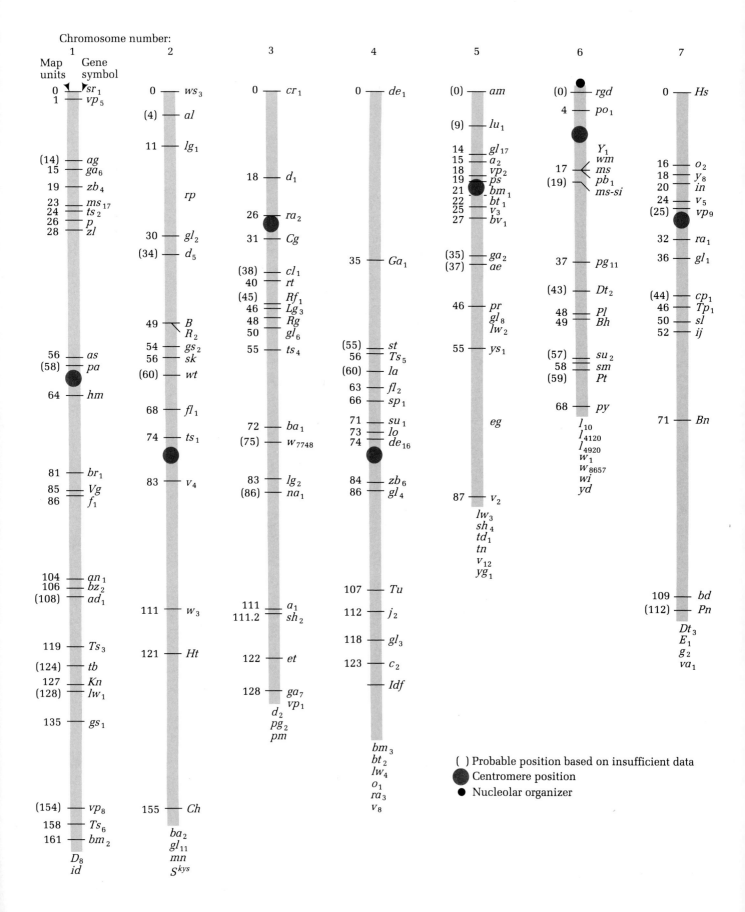

Chromosome number:

| 1 | 2 | 3 | 4 | 5 | 6 | 7 |

Map units — Gene symbol

Chromosome 1:
- 0 — sr_1
- 1 — vp_5
- (14) — ag
- 15 — ga_6
- 19 — zb_4
- 23 — ms_{17}
- 24 — ts_2
- 26 — p
- 28 — zl
- 56 — as
- (58) — pa
- 64 — hm
- 81 — br_1
- 85 — Vg
- 86 — f_1
- 104 — an_1
- 106 — bz_2
- (108) — ad_1
- 119 — Ts_3
- (124) — tb
- 127 — Kn
- (128) — lw_1
- 135 — gs_1
- (154) — vp_8
- 158 — Ts_6
- 161 — bm_2
- D_8
- id

Chromosome 2:
- 0 — ws_3
- (4) — al
- 11 — lg_1
- rp
- 30 — gl_2
- (34) — d_5
- 49 — B / R_2
- 54 — gs_2
- 56 — sk
- (60) — wt
- 68 — fl_1
- 74 — ts_1
- 83 — v_4
- 111 — w_3
- 121 — Ht
- 155 — Ch
- ba_2
- gl_{11}
- mn
- S^{kys}

Chromosome 3:
- 0 — cr_1
- 18 — d_1
- 26 — ra_2
- 31 — Cg
- (38) — cl_1
- 40 — rt
- (45) — Rf_1
- 46 — Lg_3
- 48 — Rg
- 50 — gl_6
- 55 — ts_4
- 72 — ba_1
- (75) — w_{7748}
- 83 — lg_2
- (86) — na_1
- 111 — a_1
- 111.2 — sh_2
- 122 — et
- 128 — ga_7
- vp_1
- d_2
- pg_2
- pm

Chromosome 4:
- 0 — de_1
- 35 — Ga_1
- (55) — st
- 56 — Ts_5
- (60) — la
- 63 — fl_2
- 66 — sp_1
- 71 — su_1
- 73 — lo
- 74 — de_{16}
- 84 — zb_6
- 86 — gl_4
- 107 — Tu
- 112 — j_2
- 118 — gl_3
- 123 — c_2
- Idf
- bm_3
- bt_2
- lw_4
- o_1
- ra_3
- v_8

Chromosome 5:
- (0) — am
- (9) — lu_1
- 14 — gl_{17}
- 15 — a_2
- 18 — vp_2
- 19 — ps
- 21 — bm_1
- 22 — bt_1
- 25 — v_3
- 27 — bv_1
- (35) — ga_2
- (37) — ae
- 46 — pr
- gl_8
- lw_2
- 55 — ys_1
- eg
- 87 — v_2
- lw_3
- sh_4
- td_1
- tn
- v_{12}
- yg_1

Chromosome 6:
- (0) — rgd
- 4 — po_1
- Y_1
- wm
- 17 — ms
- pb_1
- (19) — $ms\text{-}si$
- 37 — pg_{11}
- (43) — Dt_2
- 48 — Pl
- 49 — Bh
- (57) — su_2
- 58 — sm
- (59) — Pt
- 68 — py
- l_{10}
- l_{4120}
- l_{4920}
- w_1
- w_{8657}
- wi
- yd

Chromosome 7:
- 0 — Hs
- 16 — o_2
- 18 — y_8
- 20 — in
- 24 — v_5
- (25) — vp_9
- 32 — ra_1
- 36 — gl_1
- (44) — cp_1
- 46 — Tp_1
- 50 — sl
- 52 — ij
- 71 — Bn
- 109 — bd
- (112) — Pn
- Dt_3
- E_1
- g_2
- va_1

( ) Probable position based on insufficient data
● Centromere position
• Nucleolar organizer

169
*Locating
Genes on
Chromosomes:
Mapping
Techniques*

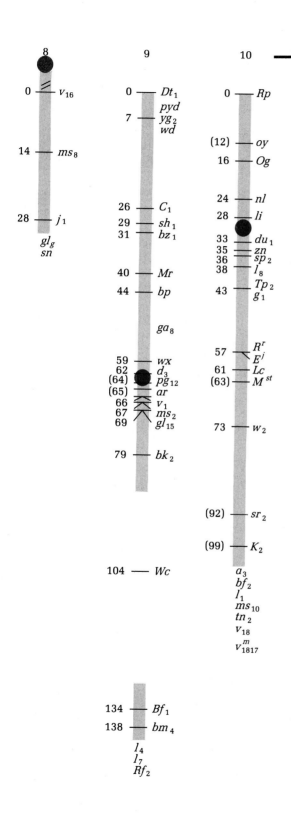

Figure 5.21 Genetic map of Zea mays.

170
Linkage,
Crossing-Over,
and Genetic
Mapping in
Eukaryotes

Analytical Approaches for Solving Genetics Problems

Q.1 In corn the gene for colored (C) seeds is completely dominant to the gene for colorless (c) seeds. Similarly, for the character of the endosperm (the part of the seed that contains the food stored for the embryo), a single gene pair controls whether the endosperm is full or shrunken. Full (Sh) is dominant to shrunken (sh). A true-breeding colored, full-seeded plant was crossed with a colorless, shrunken-seeded one. The F_1 colored, full plants were test-crossed to the doubly recessive type, that is, colorless and shrunken. The result was as follows:

colored, full	4032
colored, shrunken	149
colorless, full	152
colorless, shrunken	4035
Total	8368

Is there evidence that the gene for color and the gene for endosperm shape are linked? If so, what is the map distance between the two loci?

A.1 The best approach is to diagram the cross, using gene symbols:

P: colored and full   ×   colorless and shrunken
     CC SS   *cc ss*
               ↓

F_1:   colored and full
                Cc Ss

Testcross: colored and full   ×   colorless and shrunken
        Cc Ss   *cc ss*

If the genes were unlinked, a 1:1:1:1 ratio of colored and full:colored and shrunken:colorless and full:colorless and shrunken would result in the progeny of this testcross. By inspection, we can see that the actual progeny deviate a great deal from this ratio, showing a 27:1:1:27 ratio. If we did a chi-square test (using the actual numbers, not the percentages or ratios), we would see immediately that the hypothesis that the genes are unlinked is invalid, and we must consider the two genes to be linked in coupling. More specifically, the parental combinations (colored plus full and colorless plus shrunken) are more numerous than expected, while the recombinant types (colorless plus full and colored plus shrunken) are correspondingly less numerous than expected. This result comes directly from the inequality of the four gamete types produced by meiosis in the colored and full F_1 parent.

To calculate the map distance between the two genes, we need to compute the frequency of crossovers in that region of the chromosome during meiosis. We cannot do that directly, but we can compute the percentage of recombinant progeny that must have resulted from such crossovers:

Parental types:	colored, full	4032
	colorless, shrunken	4035
		8067

Recombinant types:	colored, shrunken	149
	colorless, full	152
		301

171
Analytical
Approaches
for Solving
Genetics
Problems

This calculation gives about 3.6 percent recombinant types (i.e. $\frac{301}{8368} \times 100$) and about 96.4 percent parental types (i.e. $\frac{8067}{8368} \times 100$). Since the recombination frequency can be used directly as an indication of map distance, especially when the distance is small, we can conclude that the distance between the two genes is 3.6 mu (3.6 cM).

We would get approximately the same result if the two genes were in repulsion rather than in coupling. That is, the crossovers are occurring between homologous chromosomes regardless of whether or not there are genetic differences in the two homologs that we, as experimenters, use as markers in genetic crosses. This same cross in repulsion would be as follows:

P: colorless and full × colored and shrunken

$$\frac{c \ S}{c \ S} \qquad\qquad \frac{C \ s}{C \ s}$$

$$\downarrow$$

F$_1$: colored and full

$$\frac{C \ s}{c \ S}$$

The testcross of the F$_1$ with colorless and shrunken (*cc ss*) gives 638 colored and full (recombinant):21,379 colored and shrunken (parental):21,906 colorless and full (parental):672 colorless and full (recombinant) in a total of 44,595 progeny. Thus 2.94 percent are recombinants, for a map distance between the two genes of 2.94 mu, which is reasonably close to the results of the cross made in coupling.

Q.2 In the Chinese primrose slate-colored flower (*s*) is recessive to blue flower (*S*); red stigma (*r*) is recessive to green stigma (*R*); and long style (*l*) is recessive to short style (*L*). The F$_1$ of a cross between two true-breeding strains, when testcrossed, gave the following progeny:

Phenotype	Number of Progeny
slate flower, green stigma, short style	27
slate flower, red stigma, short style	85
blue flower, red stigma, short style	402
slate flower, red stigma, long style	977
slate flower, green stigma, long style	427
blue flower, green stigma, long style	95
blue flower, green stigma, short style	960
blue flower, red stigma, long style	27
Total	3000

a. What were the genotypes of the parents in the cross of the two true-breeding strains?
b. Make a map of these genes, showing gene order and distance between them.
c. Derive the coefficient of coincidence for interference between these genes.

A.2 a. With three gene pairs eight phenotypic classes are expected, and eight are observed. The reciprocal pairs of classes with the most representatives are

172
Linkage,
Crossing-Over,
and Genetic
Mapping in
Eukaryotes

those resulting from no crossovers, and these pairs can tell us the genotypes of the original parents. The two classes are slate plus red plus long and blue plus green plus short. Thus the F_1 triply heterozygous parent of this generation must have been $S\ R\ L/s\ r\ l$, so the true-breeding parents were $S\ R\ L/S\ R\ L$ (blue, green, short) and $s\ r\ l/s\ r\ l$ (slate, red, long).

b. The order of the genes can be determined by inspecting the reciprocal pairs of phenotypic classes that represent the results of double crossing-over. These classes have the least numerous representatives, so the double-crossover classes are slate plus green plus short ($s\ R\ L$) and blue plus red plus long ($S\ r\ l$). The gene pair that has changed its position relative to the other two pairs of alleles is the central gene, S/s in this case. Therefore the order of genes is $R\ S\ L$ (or $L\ S\ R$). We can diagram the F_1 testcross as follows:

$$\frac{R\ S\ L}{r\ s\ l} \times \frac{r\ s\ l}{r\ s\ l}$$

A single crossover between the R and S genes gives the green plus slate plus long ($R\ s\ l$) and red plus blue plus short ($r\ S\ L$) classes, which have 427 and 402 members, respectively, for a total of 829. The double-crossover classes have already been defined, and they yield 54 progeny. The map distance between R and S is given by the crossover frequency in that region, which is the sum of the single crossovers and double crossovers divided by the total number of progeny, times 100 percent. Thus

$$\frac{829 + 54}{3000} \times 100\% = \frac{883}{3000} \times 100\%$$

$$= 29.43\% \text{ or } 29.43 \text{ map units}$$

With similar logic the distance between S and L is given by the crossover frequency in that region, which is the sum of the single-crossover and double-crossover progeny classes. The single-crossover progeny classes are green plus blue plus long ($R\ S\ l$) and red plus slate plus short ($r\ s\ L$), which have 95 and 85 members, respectively, for a total of 180. The map distance is given by

$$180 + \frac{54}{3000} \times 100\% = \frac{234}{3000} \times 100\%$$

$$= 7.8\% = 7.8 \text{ map units}$$

The data we have derived give us the following map:

c. The coefficient of coincidence is given by

$$\frac{\text{frequency of observed double crossovers}}{\text{frequency of expected double crossovers}}$$

The frequency of observed double crossovers is $(54/3000) \times 100\% = 1.8\%$.

The expected frequency of double crossovers is the product of the map distances between r and s and between s and l, that is, 29.4 × 7.8% = 2.3%. The coefficient of coincidence, therefore, is 1.8/2.3 = 0.78. In other words, 78% of the expected double crossovers did indeed take place; there was only 22% interference.

Questions and Problems

5.1 A cross $AA\ BB \times aa\ bb$ results in an F_1 of phenotype $A\ B$; the following numbers are obtained in the F_2 (phenotypes):

$A\ B$	110
$A\ b$	16
$a\ B$	19
$a\ b$	15
Total	160

Are genes at the a and b loci linked or independent? What F_2 numbers would otherwise be expected?

5.2 In the F_2 of his cross of red-flowered × white-flowered *Pisum*, Mendel obtained 705 plants with red flowers and 224 with white.

a. Is this result consistent with his hypothesis of factor segregation, from which a 3:1 ratio would be predicted?
b. In how many similar experiments would a deviation as great as or greater than this one be expected? (Calculate χ^2 and obtain the approximate value of P from the table.)

*5.3 In corn a dihybrid for the recessive a and b is testcrossed. The distribution of the phenotypes was as follows:

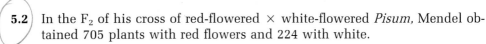

$A-\ B-$	122	
$A-\ bb$	118	400
$aa\ B-$	81	
$aa\ bb$	79	

Are the genes assorting independently? Test the hypothesis with a χ^2 test. Explain tentatively any deviation from expectation, and tell how you would test your explanation.

5.4 Why are crosses used for chromosome mapping in *Drosophila* set up with females heterozygous and males homozygous recessive?

5.5 The F_1 from a cross of $A\ B/A\ B \times a\ b/a\ b$ is testcrossed, resulting in the following phenotypic ratios:

$A\ B$	308
$A\ b$	190
$a\ b$	292
$a\ B$	210

What is the frequency of recombination between genes a and b?

174
Linkage,
Crossing-Over,
and Genetic
Mapping in
Eukaryotes

*5.6 In rabbits the English type of coat (white-spotted) is dominant over non-English (unspotted), and short hair is dominant over long hair (Angora). When homozygous English, short-haired rabbits were crossed with non-English Angoras and the F_1 crossed back to non-English Angoras, the following offspring were obtained: 72 English and short-haired; 69 non-English and Angora; 11 English and Angora; and 6 non-English and short-haired. What is the map distance between the genes for coat color and hair length?

5.7 In *Drosophila* the mutant black (*b*) has a black body, and the wild type has a grey body; the mutant vestigial (*vg*) has wings that are much shortened and crumpled compared with the long wings of the wild type. In the following cross, the true-breeding parents are given together with the counts of offspring of F_1 females × black and vestigial males:

P black and normal × grey and vestigial
F_1 females × black and vestigial males

grey, normal	283
grey, vestigial	1294
black, normal	1418
black, vestigial	241

From these data, calculate the map distance between the black and vestigial genes.

*5.8 Use the following two-point recombination data to map the genes concerned. Show the order and the length of the shortest intervals.

Gene loci	% recombination	Gene loci	% recombination
a,b	50	b,d	13
a,c	15	b,e	50
a,d	38	c,d	50
a,e	8	c,e	7
b,c	50	d,e	45

5.9 Use the following two-point recombination data to map the genes concerned. Show the order and the length of the shortest intervals.

Loci	% recombination	Loci	% recombination
a,b	50	c,d	50
a,c	17	c,e	50
a,d	50	c,f	7
a,e	50	c,g	19
a,f	12	d,e	7
a,g	3	d,f	50
b,c	50	d,g	50
b,d	2	e,f	50
b,e	5	e,g	50
b,f	50	f,g	15
b,g	50		

5.10 The following data are from Bridges and Morgan's work on the recombination between the genes black, curved, purple, speck, and vestigial in chromosome 2 of *Drosophila*. On the basis of the data, map the chromosome for these five genes as accurately as possible. Remember that determinations for short distances are more accurate than those for long ones.

Genes in cross	Total progeny	Number of recombinants
black, curved	62,679	14,237
black, purple	48,931	3,026
black, speck	685	326
black, vestigial	20,153	3,578
curved, purple	51,136	10,205
curved, speck	10,042	3,037
curved, vestigial	1,720	141
purple, speck	11,985	5,474
purple, vestigial	13,601	1,609
speck, vestigial	2,054	738

*5.11 Genes *a* and *b* are linked, with 10 percent recombination. What would be the phenotypes, and the probability of each, among progeny of the following cross?

$$\frac{a \ b^+}{a^+ b} \times \frac{a \ b}{a \ b}$$

*5.12 Genes *A* and *B* are sex-linked and are located 7 mu apart in the X chromosome of *Drosophila*. A female of genotype *A b/a B* is mated with a wild type (*A B*) male.

a. What is the probability that one of her sons will be either *A B* or *a B* in phenotype?
b. What is the probability that one of her daughters will be *A B* in phenotype?

5.13 *A* and *B* are linked autosomal genes in *Drosophila* whose recombination frequency in females is 5 percent. *C* and *D* are X-linked genes, located 10 map units apart. A homozygous dominant female is mated to a recessive male and the daughters are testcrossed. Which of the following would you expect to observe in the testcross progeny?

a. Different ratios in males and females.
b. Nearly equal frequency of *AbCD, Abcd, aBCD,* and *aBcd* classes.
c. Independent segregation of some genes with respect to others involved in the cross.
d. Double-crossover classes less frequent than expected because of interference between the two marked regions.

5.14 In maize the dominant genes *A* and *C* are both necessary for colored seeds. Homozygous recessive plants give colorless seed regardless of the genes at the second locus. Genes *A* and *C* show independent segregation, while the

176
Linkage,
Crossing-Over,
and Genetic
Mapping in
Eukaryotes

recessive mutant gene waxy endosperm (wx) is linked with C (20 percent recombination). The dominant Wx allele results in starchy endosperm.

a. What phenotypic ratios would be expected when a plant of constitution c Wx/C wx A/A is testcrossed?
b. What phenotypic ratios would be expected when a plant of constitution c Wx/C wx A/a is testcrossed?

*5.15 Assume that genes A and B are linked and show 20 percent crossing-over.

a. If a homozygous A B/A B individual is crossed with an a b/a b individual, what will be the genotype of the F_1? What gametes will the F_1 produce and in what proportions? If the F_1 is testcrossed with a doubly homozygous recessive individual, what will be the proportions and genotypes of the offspring?
b. If, instead, the original cross is A b/A b × a B/a B, what will be the genotype of the F_1? What gametes will the F_1 produce and in what proportions? If the F_1 is testcrossed with a doubly homozygous recessive, what will be the proportions and genotypes of the offspring?

5.16 In tomatoes tall vine is dominant over dwarf, and spherical fruit shape is dominant over pear shape. Vine height and fruit shape are linked, with a crossover percentage of 20. A certain tall, spherical-fruited tomato plant crossed with a dwarf, pear-fruited one produces 81 tall plus spherical; 79 dwarf plus pear; 22 tall plus pear; and 17 dwarf plus spherical. Another tall and spherical plant crossed with a dwarf and pear plant produces 21 tall plus pear; 18 dwarf plus spherical; 5 tall plus spherical; and 4 dwarf plus pear. What are the genotypes of the two tall and spherical plants? If they were crossed, what would their offspring be?

*5.17 Genes A and B are in one chromosome, 20 mu apart; C and D are in another chromosome, 10 mu apart. Genes E and F are in yet another chromosome and are 30 mu apart. Cross a homozygous $ABCDEF$ individual with an $abcdef$ one, and cross the F_1 back to an $abcdef$ individual. What are the chances of getting individuals of the following phenotypes in the progeny?

a. $ABCDEF$
b. $ABCdeF$
c. $AbcDEf$
d. $aBCdef$
e. $abcDeF$

*5.18 Genes d and p occupy loci 5 map units apart in the same autosomal linkage group. Gene h is a separate autosomal linkage group and therefore segregates independently of the other two. What types of offspring are expected, and what is the probability of each, when individuals of the following genotypes are testcrossed:

a. $\dfrac{D\,P}{d\,p}\;\dfrac{h}{h}$

b. $\dfrac{d \; P \; H}{D \; p \; h}$

5.19 A hairy-winged (h) *Drosophila* female is mated with a yellow-bodied (y), white-eyed (w) male. The F_1 are all normal. The F_1 progeny are then crossed, and the F_2 that emerge are as follows:

Females:	wild type	757
	hairy	243
Males:	wild type	390
	hairy	130
	yellow	4
	white	3
	hairy, yellow	1
	hairy, white	2
	yellow, white	360
	hairy, yellow, white	110

Give genotypes of the parents and the F_1, and note the linkage relations and distances, where appropriate.

5.20 In the Maltese bippy amiable (A) is dominant to nasty (a), benign (B) is dominant to active (b), and crazy (C) is dominant to sane (c). A true-breeding amiable, active, crazy bippy was mated, with some difficulty, to a true-breeding nasty, benign, sane bippy. An F_1 individual from this cross was then used in a testcross (to a nasty, active, sane bippy) and produced, in typical prolific bippy fashion, 4000 offspring. From an ancient manuscript entitled *The Genetics of the Bippy, Maltese and Other,* you discover that all three genes are autosomal, A is linked to B but not to C, and the map distance between A and B is 20 mu.

a. Predict all the expected phenotypes and the numbers of each type from this cross.
b. Which phenotypic classes would be missing had A and B shown complete linkage?
c. Which phenotypic classes would be missing if A and B were unlinked?
d. Again, assuming A and B to be unlinked, predict all the expected phenotypes of nasty bippies and the frequencies of each type resulting from a self-cross of the F_1.

5.21 How many possible linear orders are there for three linked genes? How many are there if the left versus the right end of the chromosome is taken into consideration?

5.22 Fill in the blanks. Continuous bars indicate linkage, and the order of linked genes is correct as shown. If all types of gametes are equally probable, write "none" in the right column headed "Least frequent classes." In the right column, show two gamete genotypes, unless all types are equally frequent, in which case write "none."

178
*Linkage,
Crossing-Over,
and Genetic
Mapping in
Eukaryotes*

Parent genotypes	Number of different possible gamete genotypes	Least frequent classes	
$\dfrac{A\ b\ C}{a\ B\ c}$	_____	_____	_____
$\dfrac{A\ b\ C}{a\ B\ c}$	_____	_____	_____
$\dfrac{A\ b\ C\ D}{a\ B\ c\ d}$	_____	_____	_____
$\dfrac{A\ b\ C\ D\ e\ f}{A\ B\ C\ d\ e\ f}$	_____	_____	_____
$\dfrac{b\ D}{B\ d}$	_____	_____	_____

5.23 For each of the following tabulations of testcross progeny phenotypes and numbers, state which locus is in the middle, and reconstruct the genotype of the tested triple heterozygotes.

a.		b.		c.	
ABC	191	*CDE*	9	*FGH*	110
abc	180	*cde*	11	*fgh*	114
Abc	5	*Cde*	35	*Fgh*	37
aBC	5	*cDE*	27	*fGH*	33
ABc	21	*CDe*	78	*FGh*	202
abC	31	*cdE*	81	*fgH*	185
AbC	104	*CdE*	275	*FgH*	4
aBc	109	*cDe*	256	*fGh*	0

*5.24 Genes at loci *f, m,* and *w* are linked, but their order is unknown. The F_1 heterozygotes from a cross of *FF MM WW* × *ff mm ww* are testcrossed. The most frequent phenotypes in testcross progeny will be *FMW* and *fmw* regardless of what the gene order turns out to be.

a. What classes of testcross progeny (phenotypes) would be least frequent if locus *m* is in the middle?
b. What classes would be least frequent if locus *f* is in the middle?
c. What classes would be least frequent if locus *w* is in the middle?

5.25 The following numbers were obtained for testcross progeny in *Drosophila* (phenotypes):

+ *m* +	218
w + *f*	236
+ + *f*	168
w m +	178
+ *m f*	95
w + +	101
+ + +	3
w m f	1
Total	1000

Construct a genetic map.

*5.26 Three of the many recessive mutations in *Drosophila melanogaster* that affect body color, wing shape, or bristle morphology are black (*b*) body versus grey in the wild type, dumpy (*dp*), obliquely truncated wings versus long wings in the wild type, and hooked (*hk*) bristles at the tip versus not hooked in the wild type. From a cross of a dumpy female with a black and hooked male, all the F_1 were wild type for all three characters. The testcross of an F_1 female with a dumpy, black, hooked male gave the following results:

wild type	169
black	19
black, hooked	301
dumpy, hooked	21
hooked	8
hooked, dumpy, black	172
dumpy, black	6
dumpy	305
Total	1000

a. Construct a genetic map of the linkage group (or groups) these genes occupy. If applicable, show the order and give the map distances between the genes.

b. (1) Determine the coefficient of coincidence for the portion of the chromosome involved in the cross. (2) How much interference is there?

5.27 In corn colorless aleurone (*c*) is recessive to colored (*C*), shrunken endosperm (*sh*) is recessive to full (*Sh*), and waxy endosperm (*wx*) is recessive to starchy (*Wx*). The F_1 plants from the cross of true-breeding colored, shrunken, and starchy × true-breeding colorless, full, and waxy were crossed with colorless, shrunken, and waxy plants, and the following progeny were generated:

colored, shrunken, starchy	2538
colorless, fully, waxy	2708
colored, full, waxy	116
colorless, shrunken, starchy	113
colored, shrunken, waxy	601
colorless, full, starchy	626
colored, full, starchy	4
colorless, shrunken, waxy	2

Map the positions of the *c, sh,* and *wx* genes in the chromosome.

5.28 In Chinese primroses long style (*l*) is recessive to short (*L*), red flower (*r*) is recessive to magenta (*R*), and red stigma (*rs*) is recessive to green (*Rs*). From a cross of homozygous short, magenta flower, and green stigma with long, red flower, and red stigma, the F_1 was crossed back to long, red flower, and red stigma. The following offspring were obtained:

180
Linkage,
Crossing-Over,
and Genetic
Mapping in
Eukaryotes

Style	Flower	Stigma	Number
short	magenta	green	1063
long	red	red	1032
short	magenta	red	634
long	red	green	526
short	red	red	156
long	magenta	green	180
short	red	green	39
long	magenta	red	54

Map the genes involved.

5.29 The frequencies of gametes of different genotypes, determined by testcrossing a triple heterozygote, are as follows:

Gamete genotype	%
+ + +	12.9
a b c	13.5
+ + c	6.9
a b +	6.5
+ b c	26.4
a + +	27.2
a + c	3.1
+ b +	3.5
Total	100.0

a. Which gametes are known to have been involved in double crossovers?
b. Which gamete types have not been involved in any exchanges?
c. The order shown is not necessarily correct. Which gene locus is in the middle?

*5.30 Genes *a*, *b*, and *c* are recessive. Females heterozygous at these three loci are crossed to phenotypically wild-type males. The progeny are phenotypically as follows:

Daughters: all + + +

Sons:			
+ + +	23
a b c	26
+ + c	45
a b +	54
+ b c	427
a + +	424
a + c	1
+ b +	0
Total	1000

a. What is known of the genotype of the females' parents with respect to these three loci? Give gene order and the arrangement in the homologs.
b. What is known of the genotype of the male parents?
c. Map the three genes.

5.31 Two normal-looking *Drosophila* are crossed and yield the following phenotypes among the progeny:

Females:	+	+	+	2000
Males:	+	+	+	3
	a	b	c	1
	+	b	c	839
	a	+	+	825
	a	b	+	86
	+	+	c	90
	a	+	c	81
	+	b	+	75
			Total	4000

Give parental genotypes, gene arrangement in the female parent, map distance, and the coefficient of coincidence.

5.32 Three different semidominant mutations affect the tail of mice. They are linked genes, and all three are lethal in the embryo when homozygous. Fused-tail (*Fu*) and kinky-tail (*Ki*) mice have kinky-appearing tails, while brachyury (*T*) mice have short tails. A fourth gene, histocompatibility-2 (*H–2*), is linked to the three tail genes and is concerned with tissue transplantation. Mice that are *H–2/+* will accept tissue grafts, whereas *+/+* mice will not. In the following crosses the normal allele is represented by a +. The phenotypes of the progeny are given for four crosses.

(1) $\dfrac{Fu\ +}{+\ Ki} \times \dfrac{+\ +}{+\ +}$

Fused tail	106
Kinky tail	92
Normal tail	1
Fused-kinky tail	1

(2) $\dfrac{Fu\ H\text{–}2}{+\ +} \times \dfrac{+\ +}{+\ +}$

Fused tail, accepts graft	88
Normal tail, rejects graft	104
Normal tail, accepts graft	5
Fused tail, rejects graft	3

(3) $\dfrac{T\ H\text{–}2}{+\ +} \times \dfrac{+\ +}{+\ +}$

Brachy tail, accepts graft	1048
Normal tail, rejects graft	1152
Brachy tail, rejects graft	138
Normal tail, accepts graft	162

(4) $\dfrac{Fu\ +}{+\ T} \times \dfrac{+\ +}{+\ +}$

Fused tail	146
Brachy tail	130
Normal tail	14
Fused-brachy tail	10

Make a map of the four genes involved in these crosses, giving gene order and map distances between the genes.

*5.33 The cross in *Drosophila* of

$$\frac{a^+\ b^+\ c\ d\ e}{a\ b\ c^+\ d^+\ e^+} \times \frac{a\ b\ c\ d\ e}{a\ b\ c\ d\ e}$$

182
Linkage,
Crossing-Over,
and Genetic
Mapping in
Eukaryotes

gave 1000 progeny of the following 16 phenotypes:

Genotype	Number		Genotype	Number
(1) a^+ b^+ c d e	220	(9) a b^+ c^+ d e^+	14
(2) a^+ b^+ c d e^+	230	(10) a b^+ c^+ d e	13
(3) a b c^+ d^+ e	210	(11) a^+ b c d^+ e^+	8
(4) a b c^+ d^+ e^+	215	(12) a^+ b c d^+ e	8
(5) a b^+ c^+ d^+ e	12	(13) a^+ b^+ c^+ d e^+	7
(6) a b^+ c^+ d^+ e^+	13	(14) a^+ b^+ c^+ d e	7
(7) a^+ b c d e^+	16	(15) a b c d^+ e^+	6
(8) a^+ b c d e	14	(16) a b c d^+ e	7

a. Draw a genetic map of the chromosome, indicating the linkage of the five genes and the number of map units separating each.

b. From the single-crossover frequencies, what would be the expected frequency of a^+ b^+ c^+ d^+ e^+ flies?

5.34 Complete the diagrams in the figure on the preceding page by correctly showing the centromeres, chromosome strands, and alleles on each strand.

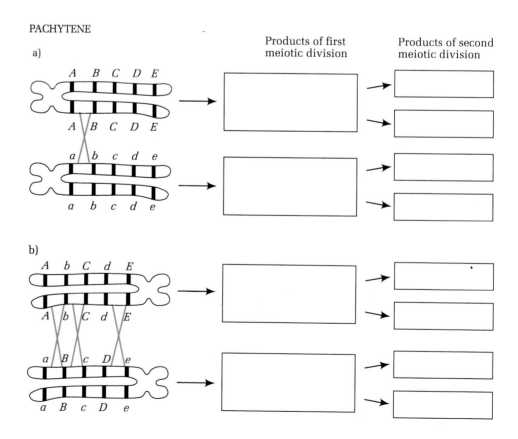

c)

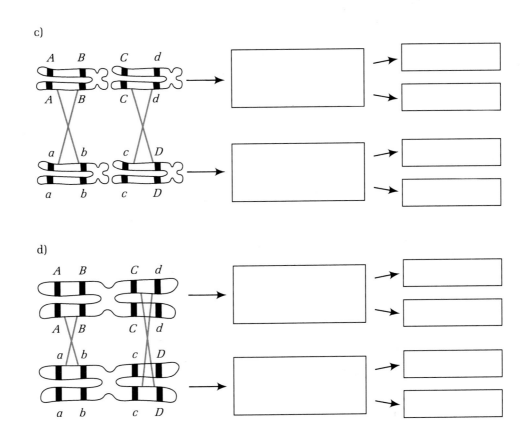

d)

6

Advanced Genetic Mapping in Eukaryotes

In the previous chapter we considered the classical experiments involving gene linkage in eukaryotes. We saw that the outcome of crosses can be used to determine map distances and that the map distances obtained are useful in predicting the outcome of crosses. In this chapter we will see how the basic mapping principles are applied and extended in determining map distance by tetrad analysis in certain amenable haploid organisms, by nonmeiotic means, and in the mapping of human chromosomes.

Tetrad Analysis

Tetrad analysis is a mapping technique that can only be used to map the genes of certain eukaryotic organisms in which the products of a single meiosis, the meiotic tetrad, are contained within a single structure. The eukaryotic organisms in which this phenomenon occurs are either fungi or single-celled algae and all are haploid. *Neurospora crassa* and yeast (both fungi) and *Chlamydomonas reinhardi* (a single-celled alga) are frequently used in tetrad analysis.

By analyzing the phenotypes of the meiotic tetrads, geneticists can directly infer the genotypes of each member of the tetrad. Haploid organisms have no genetic dominance since there is only one copy of each gene; hence the genotype is expressed directly in the phenotype. Much valuable information about how and when genetic recombination (crossing-over) occurs and about gene segregation has been discovered through tetrad analysis of various organisms, and we will outline this technique in the following sections.

Since tetrad analysis is done primarily on the three microorganisms mentioned above, it will be helpful to begin by outlining the life cycles of yeast and *Chlamydomonas reinhardi.* The life cycle of *Neurospora crassa* has already been described (p. 148).

Life Cycle of Yeast

Figure 6.1 diagrams the life cycle of baker's yeast, *Saccharomyces cerevisiae* (the yeast re-

a)

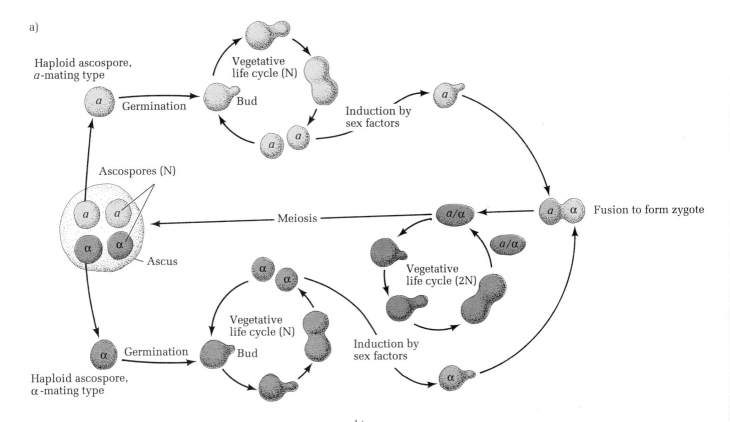

Haploid ascospore,
a-mating type

Germination

Vegetative
life cycle (N)

Bud

Induction by
sex factors

Ascospores (N)

Ascus

Meiosis

Fusion to form zygote

Vegetative
life cycle (2N)

Vegetative
life cycle (N)

Bud

Induction by
sex factors

Haploid ascospore,
α-mating type

Germination

b)

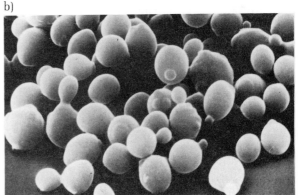

Figure 6.1 (a) Life cycle of the yeast Saccharomyces
cerevisiae. *(b) A scanning electron micrograph of
budding yeast cells.*

sponsible for making bread rise). Two mating
types occur in yeast, *a* and α. The haploid vegetative cells of this organism reproduce mitotically, with the new cell arising from the parental
cell by budding. Fusion of haploid *a* and α cells
produces a diploid cell that is stable and also
reproduces by budding. The nuclei of these diploid cells have twice the volume of the haploid
cell nuclei.

As we saw in *Neurospora crassa,* under conditions of nitrogen starvation diploid *a*/α cells
sporulate; that is, they go through meiosis. The
four haploid meiotic products of a diploid cell,
the ascospores, are contained within a roughly
spherical ascus. Two of these ascospores are of
mating type *a* and two are of mating type α.
When the ascus is ripe, the ascospores are released, and they germinate to produce haploid
vegetative cells. On a solid medium each ascospore develops into a discrete colony. In yeast
the four ascospores are arranged randomly
within the ascus so that only unordered tetrads

can be isolated from this organism (in contrast
to *Neurospora crassa,* where the ascospores are
linearly arranged in the ascus in a way that reflects the orientation of the four chromatids of
the meiotic tetrad at metaphase I).

Life Cycle of *Chlamydomonas reinhardi*

Figure 6.2 diagrams the life cycle of *Chlamydomonas reinhardi.* Like yeast, *Chlamydomonas reinhardi* has haploid vegetative cells. Each is a single green algal cell that can swim freely as a result of the motion of its two flagella. When nitrogen is limited, the cells change morphologically to become gametes so that mating is possible. As in yeast and *Neurospora,* there are two mating types, here designated plus (+) and minus (−). Also as in *Neurospora* and yeast, only gametes of opposite mating types (+ and   in this case) can fuse to produce diploid zygotes. No fusion occurs between gametes of like mating types. After a maturation process, the zygote enters meiosis. The four haploid meiotic products are contained within a sac as an unordered tetrad, and there are two + and two − cells. When these cells are released, they are free swimming, and by mitosis they give rise to clones of those meiotic products. Any one of the haploid vegetative cells can become differentiated into a gamete under nitrogen-poor conditions.

Using Random-Spore Analysis to Map Genes in Haploid Eukaryotes

In the three haploid eukaryotes we have discussed, the meiotic products can be collected after they have been released from the ascus or sac. In the case of the fungi (*Neurospora crassa* and yeast), spores can be induced to germinate and the resulting cultures can be analyzed. The free-swimming meiotic products of *Chlamydomonas* can be analyzed directly. The haploid nature of the mature stages of all three organisms, in fact, simplifies the analysis, since this stage is exactly equivalent to that of the gametes produced after meiosis in a diploid eukaryote. Thus we can make three-point crosses to map genes on a chromosome, with essentially the same approach as the one we used for a diploid eukaryote.

Figure 6.3 shows a three-point-mapping cross in a haploid eukaryote to illustrate the similarities with three-point testcrosses in diploids. Here a wild-type (haploid) strain is crossed with a strain that carries three mutant genes in the same chromosome. A diploid zygote is produced that is triply heterozygous. When this zygote undergoes meiosis to produce the haploid progeny organisms, eight genotypic (and hence phenotypic) classes potentially can be produced in a fashion identical to that in three-point testcrosses involving diploid eukaryotes; the only difference is that the parents and the progeny are haploid and not diploid. Thus random-spore analysis can be used with haploid eukaryotes to obtain data for drawing a genetic map of the chromosomes in these organisms. This information is important for understanding the structural and functional organization of the genome in organisms such as these.

Using Tetrad Analysis to Map Two Linked Genes

Random-spore analysis to map genes in *N. crassa,* yeast, and *C. reinhardi* is formally little different from analysis of the progeny of testcrosses in diploid eukaryotes. As we have mentioned, one unique property of these organisms is that the products of a meiosis, the meiotic tetrad, are contained within a single structure that can be isolated and analyzed by the geneticist. Such analysis produces data for drawing genetic maps, as we will now see.

Let us consider how we can determine the map distance between two linked genes by analyzing a number of complete meiotic tetrads. The analysis to be described requires only unordered tetrads, and hence is applicable to *N. crassa,* yeast, and *C. reinhardi.* A diploid zygote is constructed that is heterozygous for both genes, and after meiosis, the resulting tetrads are analyzed. For a cross of *a b* × + + three possible tetrad types result, called **parental-ditype (PD), tetratype (T),** and **nonparental-ditype (NPD)** tetrads (Figure 6.4, p. 189).

The PD asci contain only two types of spores, both of which are of the parental type *a b* and + + (hence the name parental ditype). The T asci contain all four possible types of spores, that is, the two parental types and the two recombinant types *a* + and + *b.* The NPD asci

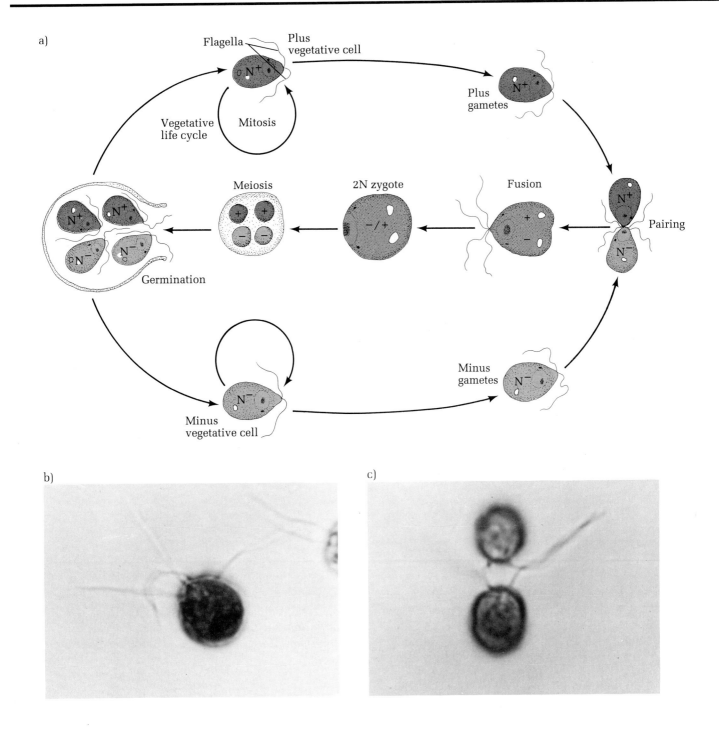

Figure 6.2 (a) Life cycle of the unicellular green alga Chlamydomonas reinhardi. *Light micrographs of* Chlamydomonas *are shown in (b) and (c).*

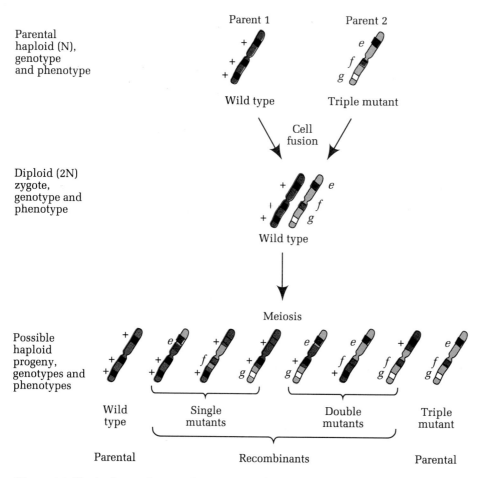

Parent 1   Parent 2

Parental
haploid (N),
genotype
and phenotype

Wild type   Triple mutant

Cell
fusion

Diploid (2N)
zygote,
genotype and
phenotype

Wild type

Meiosis

Possible
haploid
progeny,
genotypes and
phenotypes

Wild   Single   Double   Triple
type   mutants   mutants   mutant

Parental   Recombinants   Parental

*Figure 6.3 Typical genetic cross for mapping three
genes in a haploid organism such as yeast or* Neurospora crassa.

contain two types of spores, both of which are
of the nonparental (recombinant) types $a +$ and
$+ b$.

Keynote *In organisms in which the products
of a meiosis (the meiotic tetrad) are
contained within a single structure, three types
of tetrads are possible when two genes are segregating in a cross. The parental-ditype (PD) tetrad contains four nuclei, all of parental genotypes: two of one parent and two of the other
parent. The nonparental-ditype (NPD) tetrad
contains four nuclei, all of recombinant (non-*
*parental) genotypes, that is, two of each possi-
ble type. The tetratype (T) tetrad contains two
parental and two recombinant nuclei (one of
each parental type and one of each recombinant
type).*

Tetrad analysis helps to determine whether or
not two genes are linked. This determination is
based on the ways each tetrad type is produced
when genes are unlinked and when genes are
linked.

Figure 6.5 shows how PD, NPD, and T tetrads
are produced when two genes are on different

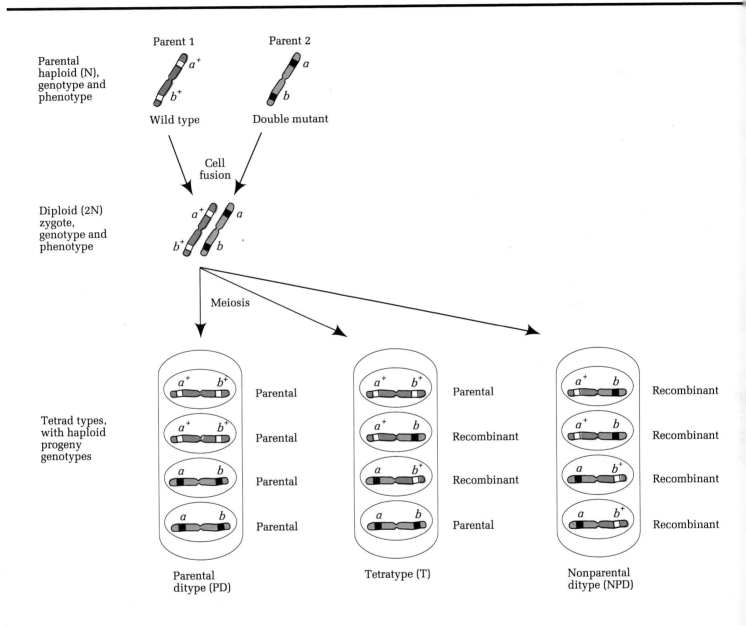

Figure 6.4 *Three types of tetrads produced from a cross of* a b × + + : *parental-ditype, tetratype, and nonparental-ditype tetrads.*

chromosomes. In this case the PD and NPD tetrads result from events in which no crossovers are involved; the relative metaphase plate orientation of the four chromatids for the two chromosomes determines whether a PD or an NPD tetrad results. Since the two sets of four chromatids align at the metaphase plate independently, the PD and NPD orientations shown occur with approximately equal frequency. Thus if two genes are unlinked, the frequency of PD tetrads

will equal the frequency of NPD tetrads. For linked genes the results are different: T tetrads are produced when there is, for example, a single crossover between one of the genes and the centromere on that chromosome. The frequency

One metaphase alignment

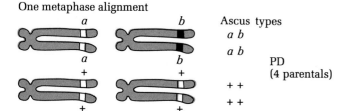

Alternative metaphase alignment

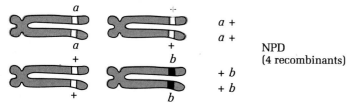

Single crossover between *a* gene and its centromere

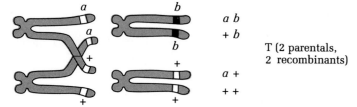

Figure 6.5 Origin of tetrad types for a cross a b ×
*+ + in which the two genes are located on different
chromosomes that assort independently.*

of these T tetrads would depend on how far the
genes are from their respective centromeres.

Figure 6.6 shows the origins of each tetrad
type when two genes are linked on the same
chromosome. If no crossing-over (Figure 6.6a)
occurs between the genes, then a PD ascus results. A single crossover (Figure 6.6b) produces
two parental and two recombinant chromatids
and hence a T ascus. In double crossovers (Figure 6.6c) the chromatid strands involved must
be taken into account. In a two-strand double
crossover (Figure 6.6c) the two crossover events
between the two genes take place between the
same two chromatids. This crossover results in a
PD ascus since no recombinant progeny are produced. Three-strand double crossovers (Figure

6.6c) include three of the four chromatids, and
there are two possible ways in which this event
can happen. In each case two recombinant and
two parental progeny types are produced in
each ascus, which is a T ascus. Lastly, in four-
strand double crossovers (Figure 6.6c) each
crossover event involves two distinct chromatids, so all four chromatids of the tetrad are involved. This crossover is the only way in which
NPD tetrads are produced. Therefore since PD
tetrads are produced either when there is no
crossing-over or when there is a two-strand double crossover, and since NPD tetrads are only
produced by four-strand double crossovers
(which are rare), we can state that two genes are
linked if the frequency of PD tetrads is far
greater than the frequency of NPD tetrads (i.e.,
PD ≫ NPD).

Once we know that two genes are linked and
we have data on the relative numbers of each
type of meiotic tetrad, the distance between the
two genes can be computed by using the basic
mapping formula:

$$\frac{\text{number of recombinants}}{\text{total number of progeny}} \times 100$$

In tetrad analysis, however, we analyze types of
tetrads rather than individual progeny. By examining the nature of the four products in each
tetrad, we see that a PD tetrad has all parental
spores, a T tetrad has two parental and two recombinant spores, and an NPD tetrad has four
recombinant spores. Then the basic mapping
formula can be converted into tetrad terms so
that the recombination frequency between genes
a and *b* is

$$\frac{1/2\ \text{T} + \text{NPD}}{\text{total tetrads}} \times 100$$

In this formula the 1/2 T and the NPD represent the only recombinants; the other 1/2 T and the
PD represent the nonrecombinants (the parentals). Thus the formula does indeed compute the
percentage of recombinants. So if there are 200
asci with 140 PD, 48 T, and 12 NPD, the recombination frequency between the genes is

$$\frac{1/2(48) + 12}{200} \times 100 = 18\%$$

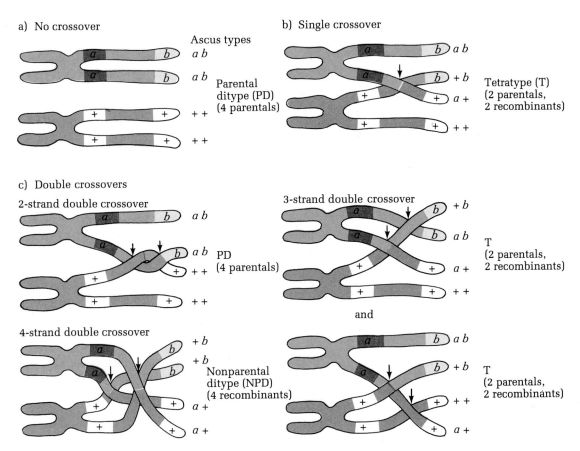

Figure 6.6 Origin of tetrad types for a cross a b ×
+ + *in which both genes are located on the same
chromosome. The arrows indicate the sites at which
crossovers occurred. (a) No crossover; (b) single crossover; (c) three types of double crossovers.*

In sum, we have derived a formula for calculating the map distance between two linked
genes by using the frequencies of the three possible types of tetrads rather than by analyzing
individual progeny. Where more than two genes
are linked in a cross, the data may best be analyzed by considering two genes at a time and
classifying each tetrad into PD, NPD, and T for
each pair.

Keynote *In organisms in which all products
of meiosis are contained within a
single structure, the analysis of the relative proportion of tetrad types provides another way to
compute the map distance between genes. The
general formula when two linked genes are
being mapped is*

$$\frac{1/2 \text{ T} + \text{NPD}}{\text{total tetrads}} \times 100$$

Calculating Gene-Centromere Distance in Organisms with Linear Tetrads

In *Neurospora crassa* and some other haploid microorganisms, the meiotic products are arranged linearly within the ascus in what is called an ordered tetrad. By contrast, in organisms such as yeast and *Chlamydomonas reinhardi*, the meiotic products are arranged randomly within the ascus or sac in what is called an unordered tetrad. The tetrad analysis we have discussed so far is applicable to both types of tetrad arrangement.

Neurospora crassa actually has eight spores, since the four meiotic products undergo one more mitotic division before the ascospores are produced. For the purposes of our genetic discussions the eight spores can be considered as four pairs. The most interesting thing about them is that their genetic content directly reflects the orientation of the four chromatids of

each chromosome pair in the diploid zygote nucleus at metaphase I. This fact allows us to map the distance between genes and their centromere, which is not possible using unordered tetrads. Locating the centromeres on the genetic maps of chromosomes makes the maps more complete. In higher organisms centromeres are located primarily through cytological studies (which are usually not possible in lower eukaryotes because their chromosomes are too small).

In this example we will map the position of the mating-type locus of *N. crassa* relative to its centromere. Mating type is a function of which allele, *A* or *a*, is present at a locus in linkage group I. If an *N. crassa* strain of mating type *A* is crossed with one of mating type *a*, a diploid zygote of genotype *A/a* results. Figure 6.7 shows how this zygote might give rise to the four meiotic products, the ascospores. (To simplify things, we will ignore the last mitotic division that produces eight spores.) For the purposes of illustration a • will be used to indicate the centromeres of the *A* parent and a ○ will be used to indicate the centromere of the *a* parent. (In reality there is no difference between the two.) At the zygote stage of the life cycle the chromosomes have divided but the centromeres have not divided.

If no crossing-over occurs between the mating-type locus and the centromere, the resulting ascospores have the genotypes shown in Figure

Figure 6.7 Gene-centromere distance of the mating-type locus in N. crassa. *The ascus develops from a diploid zygote in which no crossing-over occurred between the centromere and the mating-type locus, and it shows first-division segregation for the mating-type alleles. (In* N. crassa *a mitotic division after the second meiotic division produces eight spores, thereby doubling each progeny type; i.e., the 2:2 ratio is actually 4:4. The mitotic division is not considered here for simplicity's sake.)*

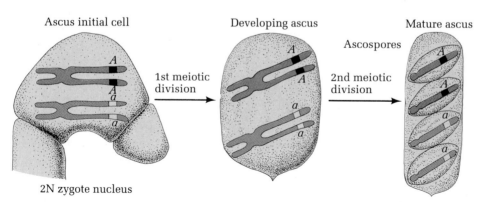

6.7. An important point here is that the centromeres do not separate until just before the second meiotic division. Therefore the spores in the top half of the ascus always have the centromere from one parent (the • centromere in this case), and the spores in the other half of the ascus always have the centromere from the other parent (∘ here). Since the two types of centromere segregate to different nuclear areas after the first meiotic division, we say that they show first-division segregation. And since no crossing-over occurs between the mating-type locus and the centromere, each allelic pair also shows *first-division segregation.* Note particularly that all four spores in an ascus showing first-division segregation of alleles are parental types: The *A* allele is on the chromosome with a • centromere and the *a* allele is on the chromosome with the ∘ centromere. Furthermore, since it is equally likely that the four chromatids in the diploid zygote will be rotated 180°, equal numbers of first-division segregation asci are expected in which the • *A* spores are in the bottom half and the ∘ *a* spores are in the top half.

To determine the map distance between a gene and its centromere, we must measure the crossover frequency between the two chromosomal sites. Thus we need to predict the consequences of a single crossover between the mating-type locus and the centromere. Figure 6.8 shows that four ascus types are produced. These four types are generated in equal frequencies since they reflect the four possible different orientations of the four chromatids in the diploid zygote at metaphase I. Each has first-division segregation of the centromere. By contrast, *A* and *a* do not segregate into separate nuclei until the second division. This situation is called *second-division segregation* for the gene, and different patterns of gene segregation are produced depending on which chromatids are involved in the crossover event. The 1:1:1:1 (*A a A a*, *a A a A*) and 1:2:1 (*A a a A*, *a A A a*) second-division segregation patterns are readily distinguishable from the 2:2 (*A A a a*, *a a A A*) first-division segregation pattern.

By analyzing ordered tetrads, we can count the number of asci that show second-division segregation for a particular gene marker. For the mating-type locus about 14 percent of the asci show second-division segregation. This value must be converted to a map distance, and in this case it is not a direct conversion. So the diagrammatic distinction of the two centromere types comes in useful here. If we consider the centromere to be a gene marker (and in a sense it is), then the parental types are • *A* and ∘ *a,* and the recombinant types are • *a* and ∘ *A.* In a first-division segregation ascus (Figure 6.7) all the spores are parentals, while in a second-division segregation ascus (Figure 6.8) half are parentals and half are recombinants. Therefore to convert the tetrad data to crossover or recombination data, we divide the percentage of second-division segregation asci (14 percent) by 2. Thus the mating-type gene is 7 mu from the centromere of linkage group I.

In essence, the computation of gene-centromere distance is a special case of mapping the distance between two genes. If ordered tetrads are isolated, then not only can genes be mapped one to another, but each can be mapped to its centromere.

Keynote *In some microorganisms the products of a meiosis are arranged in a specialized structure in a way that reflects the orientation of the four chromatids of each homologous pair of chromosomes at metaphase I. The ordered tetrads allow us to map the distance of a gene from its centromere. Where no crossover occurs between the gene and the centromere, a first-division segregation tetrad results in which one parental type is found in half the ordered tetrad and the other parental type is found in the other half to give a 2:2 segregation pattern. When a single crossover occurs between the gene and its centromere, several different tetrad segregation patterns are found, and each exemplifies second-division segregation. The gene-centromere map distance is computed as the percentage of second-division tetrads divided by 2.*

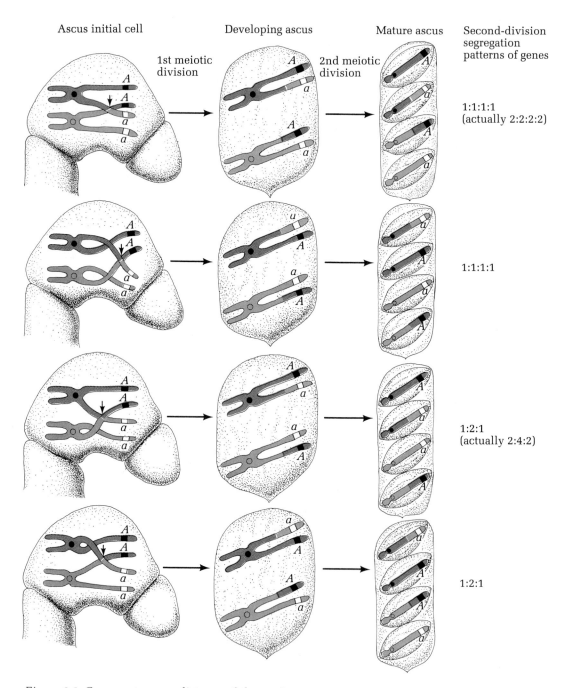

Ascus initial cell   Developing ascus   Mature ascus   Second-division segregation patterns of genes

1st meiotic division

2nd meiotic division

1:1:1:1
(actually 2:2:2:2)

1:1:1:1

1:2:1
(actually 2:4:2)

1:2:1

Figure 6.8 Gene-centromere distance of the mating-type locus in N. crassa. Production of asci after a single crossover occurs between the mating-type locus and its centromere. The arrows show the points at which the crossovers occurred. The asci show second-division segregation for the mating-type locus, and the four types of asci are produced in equal proportions.

Mitotic Recombination

Discovery of Mitotic Recombination in *Drosophila*

Experimental evidence shows that crossing-over can occur during mitosis as well as during meiosis in some organisms. The first demonstration of **mitotic crossing-over** was observed by Curt Stern in 1936 in crosses involving *Drosophila* strains that carried recessive sex-linked mutations causing yellow body color (*y*) instead of the normal grey body color and short, twisty bristles (singed, *sn*) instead of the normal long, curved bristles. In flies with a wild-type grey body color, all bristles are black; in yellow-bodied (mutant) flies the bristles are yellow.

From a cross of homozygous $y^+ \, sn/y^+ \, sn$ females (grey bodies, singed bristles) with $y \, sn^+ /\!/$ males (yellow bodies, normal bristles), Stern found, as expected, that the female progeny were mostly wild type in appearance: They had grey bodies and normal bristles. However, instead of being all one color, some females had sectors of yellow and/or singed bristles. The origin of these flies could be explained by chromosome nondisjunction or by chromosomal loss. (These changes are somatic rather than germline changes.)

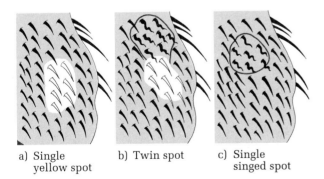

a) Single yellow spot   b) Twin spot   c) Single singed spot

Figure 6.9 Body surface phenotype segregation in a Drosophila *strain* y⁺ sn/y sn⁺. *The* sn *allele causes short, twisted (singed) bristles, and the* y *allele results in a yellow body coloration: (a) single yellow spot in normal–body color background; (b) twin spot of yellow color and singed bristles; (c) single singed-bristle spot in normal-bristle phenotype background.*

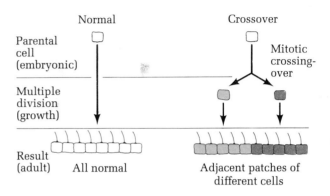

Figure 6.10 Production of twin spots by mitotic crossing-over followed by normal development into an adult.

Other females had **twin spots,** two adjacent regions of bristles, one showing the yellow phenotype and the other showing the singed phenotype, that is, a *mosaic* phenotype. Surrounding the twin spot, and constituting essentially the rest of the bristles, the phenotype was wild type. Figure 6.9 diagrams this twin spot and contrasts it with the single-phenotype spots. Stern reasoned that since the two parts of a twin spot were always adjacent, the twin spots must be the reciprocal products of the same genetic event. The best explanation was that they were generated by a mitotic crossing-over event, an event that has been found in a number of organisms, although it occurs rarely. It was deemed the most likely explanation since the tissue involved is produced by mitotic events after the *Drosophila* embryo has differentiated. The production of twin spots by mitotic crossing-over is shown schematically in Figure 6.10.

Mechanism of Mitotic Crossing-over

Mitotic crossing-over (mitotic recombination) can only be studied in diploid cells. It is a process during mitosis that produces a progeny cell with a combination of genes that differs from that of the diploid parental cell that entered the mitotic cycle. Mitotic crossing-over occurs at a stage similar to the four-strand stage of meiosis. Recall that during mitosis each pair of homologous chromosomes replicates and that the two pairs of chromatids then align independently at the metaphase plate. When the chromatids separate, each progeny cell receives one copy of each parental homolog so that each cell has the same genetic constitution as the parental cell. Figure 6.11 shows the pattern of gene segregation for a theoretical cell heterozygous for all the genes on its one pair of homologous chromosomes. Both the parental and progeny cells are wild type in phenotype.

Under rare circumstances, after each chromosome has replicated and prior to metaphase, the two pairs of maternally derived and paternally derived chromatids come together to form a tetrad that is analogous to the four-strand stage in meiosis. It is during this stage that crossing-over can occur. Figure 6.12 diagrams the consequences of mitotic crossing-over between genes c and d in a cell of the same genotype as the cell in Figure 6.11. After the crossover has occurred, the two pairs of chromatids separate and migrate independently to the metaphase plate, resulting in two possible orientations (A and B in Figure 6.12).

When type A completes mitosis, the two diploid cells 1 and 2 are produced. Progeny cell 1 is homozygous $d\ e/d\ e$ and hence expresses the phenotypes associated with these two mutant alleles. Progeny cell 2 is homozygous $d^+\ e^+/d^+\ e^+$ and therefore has a wild-type phenotype indistinguishable from that of the parental cell. For both progeny cell types all other genes are heterozygous, and thus the phenotypes expressed are those of the dominant, wild-type alleles of those genes: Progeny cell 2 is phenotypically like the parental cell. In general, then, mitotic crossing-over can make all those genes distal to the crossover point homozygous if the chromatid pairs align appropriately at the metaphase plate. This phenomenon applies only to those genes on the same chromosome arm as the crossover; that is, to those genes from the centromere outward. Therefore the crossover be-

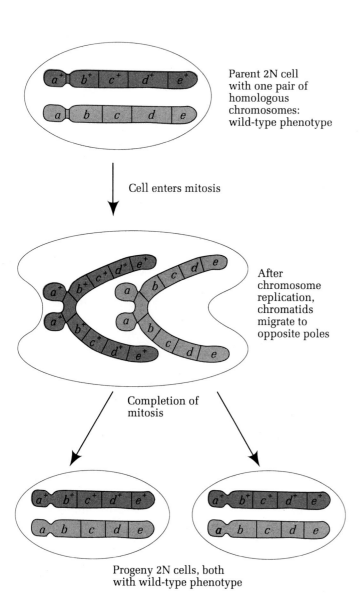

Parent 2N cell with one pair of homologous chromosomes: wild-type phenotype

Cell enters mitosis

After chromosome replication, chromatids migrate to opposite poles

Completion of mitosis

Progeny 2N cells, both with wild-type phenotype

Figure 6.11 Normal mitotic segregation of genes in a theoretical diploid cell with one homologous pair of chromosomes. The cell is heterozygous for all five genes on the chromosome. During mitosis each homolog replicates, and the two pairs of chromatids align independently on the metaphase plate. Each progeny cell receives one chromatid of each pair and hence is heterozygous for all the five genes; it is genetically identical to the parental cell and phenotypically wild type.

Figure 6.12 Result of a mitosis of the same cell type as the cell in Figure 6.10 but in which a rare mitotic crossing-over event occurs.

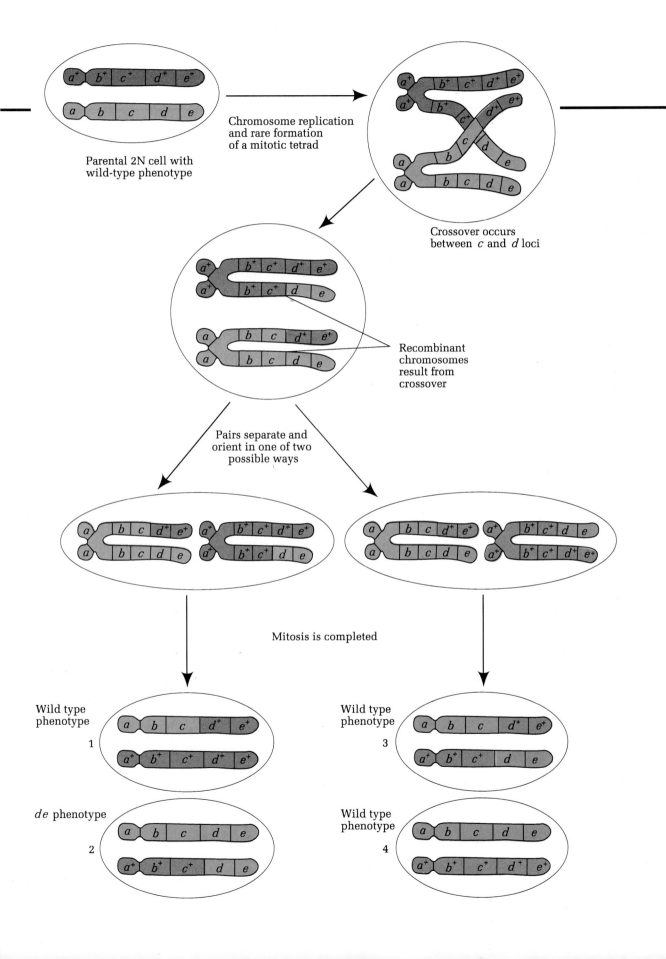

Parental 2N cell with
wild-type phenotype

Chromosome replication
and rare formation
of a mitotic tetrad

Crossover occurs
between *c* and *d* loci

Recombinant
chromosomes
result from
crossover

Pairs separate and
orient in one of two
possible ways

Mitosis is completed

Wild type
phenotype
1

de phenotype
2

Wild type
phenotype
3

Wild type
phenotype
4

tween *c* and *d* has no effect on gene *a,* which is in the other arm of this chromosome.

When type B completes mitosis, the two diploid cells 3 and 4 are produced. In both progeny cells, as a result of the orientation of the chromatid pairs at metaphase, all genes are heterozygous, and the cells are phenotypically wild type. Nonetheless, inspection of Figure 6.12 shows that genetic recombination has occurred since all the genes are not in coupling, as they were in the parental cell. Since genetic-mapping studies depend on the detection and counting of progeny with phenotypes that are recombined from those found in the parents, the progeny cells from type A metaphase are the ones to analyze in mitotic recombination studies.

Keynote Crossing-over can occur during mitosis as well as during meiosis, although it occurs much more rarely during mitosis. As in meiosis, mitotic crossing-over occurs at a four-strand stage. Single crossovers during mitosis can be detected in a heterozygote because loci distal to the crossover and on the same chromosome arm may become homozygous.

Mitotic Recombination in Fungal Systems

Mitotic recombination has been studied most extensively in fungi. Here we will look at some classic experiments done since the 1950s with the fungus *Aspergillus nidulans,* which showed that mitotic recombination analysis can be used to construct genetic maps.

Life cycle of *Aspergillus nidulans.* The fungus *Aspergillus nidulans* is a mycelial-form fungus like *Neurospora,* and its colonies are greenish. Its use in genetic studies is summarized in Figure 6.13. *Aspergillus* is a suitable organism for genetic studies using mitotic recombination since (1) the asexual spores of this fungus (cf. the conidia of *Neurospora*), instead of having many nuclei, as is the case for many fungi, have only a single nucleus—that is, they are uninucleate; (2) the phenotype of each asexual spore is controlled by the genotype of the nucleus it

carries; and (3) two haploid strains can be fused by mixing them together. The result of fusion is a mycelium (the typical growth habit of this organism) in which the two nuclear types from the two original haploid strains (distinguished by shading in Figure 6.13) coexist and divide mitotically within the same cytoplasm. A cell—or a collection of cells, as in a mycelium—that possesses any number of genetically different nuclei in a common cytoplasm is called a **heterokaryon.** As we shall see, the formation of heterokaryons in *Aspergillus* enables geneticists to construct heterozygous strains that can be used in mitotic recombination studies.

When a heterokaryon of *Aspergillus* produces asexual spores, each spore contains only one of the two haploid nuclear types since the spores are uninucleate (Figure 6.13). Rarely, two haploid nuclei in a heterokaryon will fuse to produce a diploid nucleus in a process called *diploidization.* Each uninucleate, asexual spore produced by a diploid strain of *Aspergillus* will have a diploid nucleus. Diploid cells can undergo mitotic crossing-over, thus producing mitotic genetic recombinants. These diploid nuclei (which are quite unstable) divide by mitosis to produce haploid progeny nuclei (haploid segregants), which may contain combinations of alleles that differ from those of the two parental haploid nuclei. This formation of haploid nuclei from a diploid nucleus is called **haploidization.**

In haploidization the two homologs of each chromosome independently assort. As with the independent assortment that occurs during meiosis, which homolog ends up in which haploid segregant is the result of a random event. Thus the result of haploidization is similar to that of independent assortment during meiosis— the independent segregation of blocks of genes during haploidization will indicate on which chromosomes the genes belong.

Systems that achieve genetic recombination by means other than the regular alternation of meiosis and fertilization are called **parasexual systems.** In fungi (such as *Aspergillus*) the parasexual cycle consists of the following sequence of events: the formation of a heterokaryon in a multinucleate mycelium, the rare fusion of haploid nuclei differing in genotype within the heterokaryon, mitotic crossing-over within diploid fusion nuclei that multiply side by side with the haploid nuclei, and the subsequent haploidization of the diploid fusion nuclei without a meiotic process. The parasexual cycle has the same effect as a regular meiotic sexual cycle in that it results in genetic recombination.

Mitotic recombination analysis in *Aspergillus nidulans.*

As in meiotic recombination analysis, in mitotic recombination analysis it is necessary to construct a strain that is heterozygous for the genes that are to be studied. In *Aspergillus* a heterokaryon is constructed between two strains that differ genetically, and then the rare diploid nuclei that form are selected. Usually, the strains to be fused have genotypes that are conducive to the formation of heterokaryons, as we shall see.

For example, consider the following two haploid strains:

Strain 1: w ad^+ pro $paba^+$ y^+ bi
Strain 2: w^+ ad pro^+ $paba$ y bi^+

The alleles *ad, pro, paba,* and *bi* are recessive to their respective wild-type alleles, and they specify that adenine, proline, para-aminobenzoic acid, and biotin, respectively, must be added to the growth medium in order for the strain carrying those mutant alleles to survive. Thus either strain alone cannot grow without the appropriate growth supplements. However, a heterokaryon resulting from the fusion of the two strains requires no growth supplements since all four genes are then heterozygous; this procedure is the typical way of directing the formation of, or forcing, a heterokaryon for experimental analysis.

The w and y alleles control the color of the asexual spores and hence the overall color of the colony (Figure 6.14). The w allele is recessive to the wild-type green allele w^+ and results in white spores. The y allele is also recessive to the wild-type green allele y^+ and results in yellow spores. A strain with genotype w^+ y^+ is green, a w y^+ strain is white, a w^+ y strain is yellow, and a w y strain is white because of epistatic effects (see pp. 113–118). The heterokaryon of

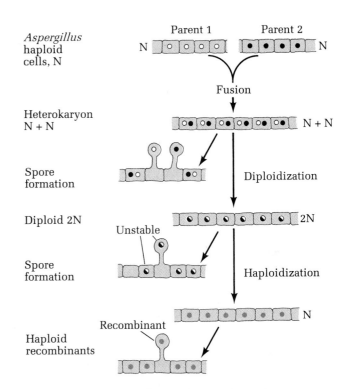

Figure 6.13 Parasexual reproduction: heterokaryons and diploids in Aspergillus nidulans.

strain 1 and strain 2, however, is not green, as we would expect from the heterozygosity of both genes. Since the spores are uninucleate, their color (and hence the colony coloration) is controlled by the genotype of the nucleus contained in each spore. Therefore the heterokaryon has a mixture of mostly yellow and white spores and has a mottled appearance.

Some of the spores on the heterokaryotic colony will be green as a result of the rare diploidization process. With respect to the spore coloration the diploid nucleus genotype is w^+ y/w y^+. The diploid spores (detectably larger than haploid spores) may then be isolated and cultured for study. The diploid cultures derived from these spores require no growth supplements since wild-type alleles for all the nutritional-requirement genes are present: w^+ ad^+ pro $paba^+$ y^+ bi^+/w ad pro^+ $paba$ y bi^+.

When a diploid *Aspergillus* spore with the genotype above is used to inoculate a solid

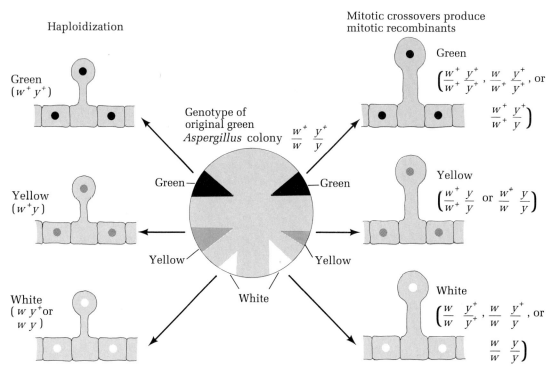

Haploidization

Mitotic crossovers produce mitotic recombinants

Green
$(w^+ y^+)$

Genotype of original green *Aspergillus* colony $\dfrac{w^+}{w}\ \dfrac{y^+}{y}$

Green

Green — Green

Yellow
$(w^+ y)$

Yellow — Yellow

White
$(w\ y^+$ or $w\ y)$

White

Green
$\left(\dfrac{w^+}{w^+}\ \dfrac{y^+}{y^+},\ \dfrac{w}{w^+}\ \dfrac{y^+}{y^+},\ \text{or}\ \dfrac{w^+}{w^+}\ \dfrac{y^+}{y}\right)$

Yellow
$\left(\dfrac{w^+}{w^+}\ \dfrac{y}{y}\ \text{or}\ \dfrac{w^+}{w}\ \dfrac{y}{y}\right)$

White
$\left(\dfrac{w}{w}\ \dfrac{y^+}{y^+},\ \dfrac{w}{w}\ \dfrac{y^+}{y},\ \text{or}\ \dfrac{w}{w}\ \dfrac{y}{y}\right)$

Figure 6.14 Stylized diagram of a green Aspergillus *colony with the starting diploid genotype* w⁺ y⁺/w y. *Through haploidization and mitotic crossovers new haploid and diploid genotypes are produced and are detected as colored sectors: green (not detectable in the green colony background), yellow, and white.*

growth medium, a colony will be produced that radiates from the point of inoculation. As long as the spores produced are diploid and have the same spore coloration genotype as the inoculating spore (i.e., $w^+ y/w y^+$), the colony will be green. During the growth of the green colony, either haploidization, diploidization, or both may occur. When spore coloration genes are involved, the cells and their offspring give rise to sectors of different colors as the colony continues to grow (Figure 6.14). Three types of colored sectors are produced with genotypes that differ from the diploid spore used to inoculate the medium: green (indistinguishable from the parental colony phenotype), yellow, and white. With respect to the spore coloration genes the haploid

and diploid genotypes for the three types of sectors are as follows: green, $w^+ y^+$ (haploid), $w^+/w^+\ y^+/y^+$, $w^+/w\ y^+/y^+$, or $w^+/w^+\ y^+/y$ (diploid); yellow, $w^+\ y$ (haploid), $w^+/w\ y/y$ or $w^+/w^+\ y/y$ (diploid); white, $w\ y^+$ or $w\ y$ (both haploid), $w/w\ y^+/y^+$, $w/w\ y^+/y$, or $w/w\ y/y$ (diploid). Diploid sectors may be distinguished from haploid sectors on the basis of the larger size of the spores.

When diploid *Aspergillus* cells reproduce to yield colonies, the colonies are predominantly green, with white and/or yellow sectors occurring rarely (Figure 6.14). Both haploid and diploid white and yellow sectors may be produced. Whether the sector is haploid or diploid can be determined by examining the size of the spores. All haploid sectors are produced by haploidization. As an example, let us consider the haploid white sectors. About half the white haploid sectors have the genotype $w\ ad^+\ pro\ paba^+\ y^+\ bi$, and half have the genotype $w\ ad\ pro^+\ paba\ y\ bi^+$. With the exception of the common white allele, these two genotypes are reciprocals.

The 50:50 segregation of the two sets of five

alleles indicates that they are located on a different chromosome from that carrying the *w* gene. Thus in some haploid white sectors a chromosome with *ad pro⁺ paba y bi⁺* has segregated, while in others its homolog with *ad⁺ pro paba⁺ y⁺ bi* alleles is segregated. The interpretation of the haploid white sector data is that the six gene loci are located on two nonhomologous chromosomes; the white gene is on one chromosome and the other five genes are on the other. It is not possible to determine gene order by this analysis, although the correct gene order is given in the following diagram:

$$w \qquad\qquad + \quad pro \quad + \quad + \quad bi$$
$$+ \qquad ad \quad + \quad paba \quad y \quad +$$

To put the above analysis in context, we note that the study of haploid segregants provides essential information about which genes are linked on which chromosomes. Once that information is known, the next stage of the mitotic analysis is to establish gene order and determine map distances between genes on the same chromosome.

To determine gene order and establish map distance, we must study a second type of segregant, the diploid segregants. From the diploid cell we constructed, the easily discernible diploid segregants are those resulting in white or yellow sectors. That such sectors are diploid would be established by examining spore size. The selected diploid segregants may then be analyzed phenotypically to determine the genotypes for the other genes on the chromosomes.

The diploid segregants are produced by mitotic crossing-over. Since this event is very rare, for all intents and purposes, only single crossovers need to be considered in each chromosome arm. As was discussed earlier, a crossing-over event in mitosis makes all those genes distal to the crossing-over point (on the same chromosome arm) homozygous. Thus any recessive alleles that are heterozygous in the diploid cell may become homozygous recessive as a result of the crossover, and a new phenotype, the recessive one, will be manifested.

One way in which a diploid yellow sector can arise is diagramed in Figure 6.15. A mitotic crossover between the *pro* and *paba* genes has produced a segregant that is homozygous for the *y* allele and hence is yellow. The same crossover produces a twin spot that is homozygous y^+/y^+, but since it is green, it is not detected in the overall green color of the colony. The crossover diagramed has made all genes distal to that point homozygous. Therefore the yellow segregant is also *paba/paba* and requires para-aminobenzoic acid in order to grow. The homozygosity for the *bi⁺* allele goes undetected.

One other crossover that could produce a diploid yellow sector is a crossover between the *paba* and *y* loci (Figure 6.16). In this case the yellow sector is still heterozygous for *paba* since mitotic crossing-over produces homozygosity only for genes distal to the crossover point. These facts give us a way to determine the gene order for each chromosome arm. A gene marker that is far away from the centromere is chosen, and then diploid segregants homozygous for that gene that result from mitotic recombination are isolated. With a number of genes marking the chromosome from the centromere out to the distal marker, it is then a simple matter to see which sets of genes become homozygous in the various segregants. In the example, *y* and *paba y* genotypes were found in the various yellow sectors. From the mechanics of mitotic recombination the order of genes is centromere–*paba*–*y*. A position cannot be assigned for those genes that become homozygous wild type, so their relative positions must be determined from other mitotic recombination experiments.

After diploid segregants, which are produced by mitotic recombination, are obtained, they can be counted. These quantitative data can be used to compute map distance between the genes; that is, the mitotic recombination frequency between genes can be computed by using the same formula used in meiotic recombination studies. The map distance between the *paba* and *y* genes, for example, is given by the percentage of yellow segregants that result from crossing-over in the *paba–y* region. Those particular segregants are still heterozygous for *paba* and *pro*.

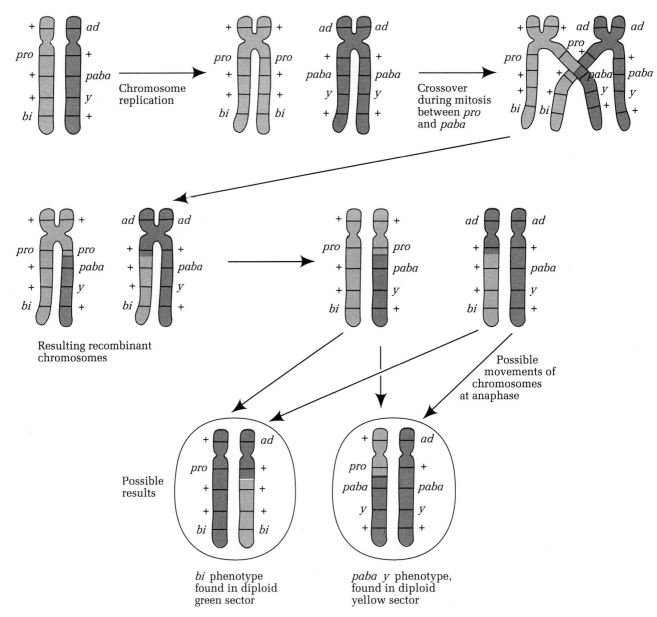

Figure 6.15 Possible mitotic crossing-over event between the pro *and* paba *loci that can give rise to a diploid yellow sector in the green diploid* Aspergillus *strain of Figure 6.13.*

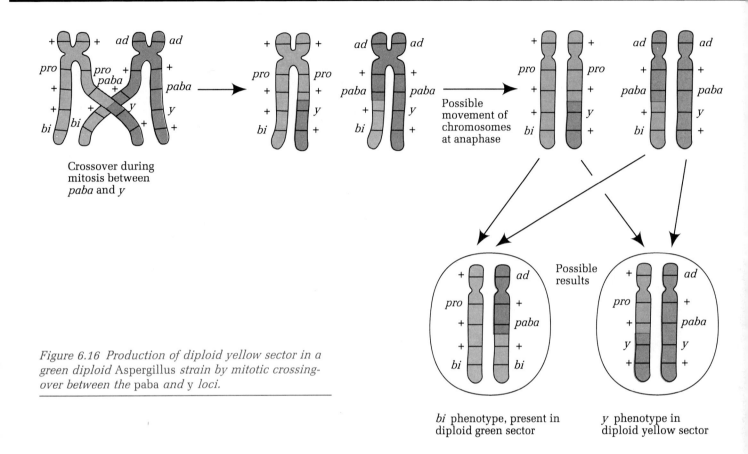

Figure 6.16 Production of diploid yellow sector in a green diploid Aspergillus *strain by mitotic crossing-over between the* paba *and* y *loci.*

The parasexual cycle describes genetic systems that achieve genetic recombination by means other than the regular alternation of meiosis and fertilization. The parasexual cycle in fungi such as Aspergillus consists of (1) the formation of a heterokaryon by mycelial fusion and then fusion of the two haploid nuclei to give a diploid nucleus; (2) mitotic crossing-over within the diploid nucleus; (3) haploidization of the diploid nuclei without meiosis, a process that produces haploid nuclei into which one or the other of each parental chromosome has segregated randomly.

Mapping Genes in Human Chromosomes

For practical reasons, with humans it is not possible to do genetic mapping experiments of the kind discussed for other organisms. Nonetheless, we have a strong interest in mapping genes in human chromosomes since there are so many known diseases and traits that have a genetic basis. In Chapter 3 we saw that pedigree analysis could be used to determine the mode by which a particular genetic trait is inherited. In this way many genes have been localized to the X chromosome. Pedigree analysis cannot show on which chromosome a particular autosomal gene is located, though.

Although modern statistical and computer analysis has made it possible to obtain some linkage data for autosomal genes, many of the more than 1200 known autosomal genes have not been located on one of the 23 autosomes. Moreover, probably between 10,000 and 50,000 genes have not even been identified yet, much less located on all the chromosomes. One technique has been developed that allows geneticists to map both sex-linked and autosomal genes in humans. This technique, which involves the fusion of human and rodent cultured cells in a process called **somatic cell hybridization,** will be discussed after a description of mapping human X-linked genes by recombination analysis.

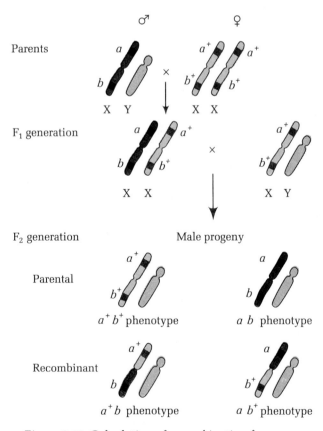

Figure 6.17 Calculation of recombination frequency for two X-linked human genes by analyzing the male progeny of women doubly heterozygous for the two genes.

Mapping Human Genes by Recombination Analysis

As we have seen, it is not possible to set up appropriate testcrosses for genetic mapping in humans by recombination analysis. In a very few cases certain pedigrees have proved to have individuals with appropriate genotypes to permit analysis of linkage between autosomal genes. Recombination analysis in humans has been carried out more for X-linked genes, however, because the hemizygosity of the X chromosome in males has provided a richer source of useful genotypic pairings in pedigrees. Consider the following theoretical example (Figure 6.17). A male with two rare X-linked recessive alleles a and b marries a woman who expresses neither of the traits involved. Since the traits are rare it is likely that the woman is homozygous for the wild-type allele of each gene, that is, she is $a^+ b^+/a^+ b^+$. A female offspring from these parents would be doubly heterozygous $a^+ b^+/a b.$ Recombinant gametes from this female would be produced by crossing-over between the two genes at a frequency related to the genetic distance they are apart. If this female pairs with a normal $a^+ b^+/Y$ male, all female progeny will be $a^+ b^+$ in phenotype because of the $a^+ b^+$ chromosome transmitted from the father. The male progeny, however, will express all four possible phenotype classes, that is, the parental $a^+ b^+$ and $a b,$ and the recombinant $a^+ b$ and $a b^+,$ because of the hemizygosity of the X chromosome. Thus, analysis of the male progeny from pairings such as this ($a^+ b^+/a b \times a^+ b^+/Y$) in a large number of pedigrees will produce a value for the frequency of recombination between the two loci involved, and an estimate of genetic map distance can be obtained.

Using this approach a number of genes have been mapped along the human X chromosome. For example, it was found that the distance between the green weakness gene, g (the recessive allele of which is responsible for a form of color blindness) and the hemophilia A gene, $h,$ is 8 map units.

Mapping Human Genes by Somatic Cell Hybridization Techniques

Formation of somatic cell hybrids. Somatic cells can be isolated from a multicellular eukaryote and under appropriate conditions can be cultured in vitro to give rise to what is known as a cell culture line. If the cell used to initiate the line is taken from mature somatic tissue, the cell line tends to have a finite life time during which there is no change in the chromosomal constitution of progeny cells relative to that found in the parental cell. More useful, though, are established cell lines in which the cells have been selected to grow and divide essentially indefinitely in cell culture. Often the cell lines are derived from malignant tissues, and often the chromosomal constitution differs from that characteristic of the wild-type parental organism.

Figure 6.18 diagrams the formation of a somatic cell hybrid of a human cell and an established cell line from the mouse. To cause the cells to fuse, geneticists mix suspensions of the two cell types with polyethylene glycol (PEG), which causes the membranes to fuse. Once the cells have fused, the nuclei in the then binucleate heterokaryon often fuse to form a single nucleus with both sets of chromosomes present. Each cell of this uninucleate cell line is called a **synkaryon.** By mitosis the cell line reproduces so that each original synkaryotic cell gives rise to a colony of synkaryotic cells that can then be studied.

Using the HAT technique to locate hybrid cells. One of the problems in producing a hybrid cell is finding it among the parental cells. Fortunately, a convenient selective procedure prevents the parental cells from growing but allows the hybrid cells to reproduce. This procedure is the *HAT technique* (hypoxanthine-aminopterin-thymidine). The HAT technique uses conditions under which only hybrid cells are able to synthesize DNA and thus divide and increase in number. The nonhybrid cells are unable to synthesize DNA and hence they cannot divide.

To understand how the HAT technique can work, we must understand a little about the bio-

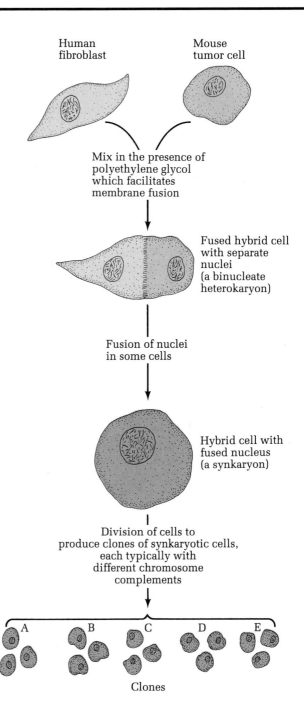

Figure 6.18 Technique for producing a human/mouse hybrid somatic cell. The cell fusion results in cells that contain all the mouse chromosomes and an incomplete set of human chromosomes. A fibroblast cell is a somatic cell from fibrous connective tissue.

chemistry of DNA synthesis. In the cell, DNA synthesis requires the presence of molecules called purine and pyrimidine triphosphates, which are made from their respective monophosphates. (The structure and synthesis of DNA will be developed more fully in later chapters.) The purine and pyrimidine monophosphates can be generated either by their synthesis from simpler precursor molecules or by what are called salvage pathways in which enzymes recycle purines and pyrimidines produced by degradation of DNA or RNA. If the drug aminopterin is present in the culture medium (as it is in the HAT technique), the synthesis of new purines and pyrimidines from precursors is inhibited. As a consequence, the cells become completely dependent on the salvage pathways for the production of purine and pyrimidine triphosphates needed for the synthesis of DNA.

Figure 6.19 illustrates the use of the HAT technique to produce a human-mouse hybrid cell. In this technique two cells are fused, cells that have different genetic defects in the purine and pyrimidine salvage pathways. In general, aminopterin will inhibit the new synthesis of purines and pyrimidines from precursors so that the cells have to rely on the salvage pathways for purine and pyrimidine production. The genetic defects facilitate the isolation of hybrid cells: One cell line will lack an enzyme needed in the purine salvage pathway, and the other cell line will lack an enzyme needed in the pyrimidine salvage pathway. In the example the mouse cell line is defective in the activity of hypoxanthine phosphoribosyl transferase (HGPRT), an enzyme needed for the purine salvage pathway. These cells have the enzyme thymidine kinase (TK), which is needed for the pyrimidine salvage pathway. The human cell line has normal HGPRT activity but is deficient in TK activity. In the presence of aminopterin neither human nor mouse cells can grow since the human cells cannot make pyrimidine monophosphates, and the mouse cells cannot make the purine monophosphates. However, any hybrid cells formed by fusion of the two cell types will grow, because the human chromosomes contribute a normal HGPRT gene and the mouse chromosomes con-

tribute a normal TK gene. Therefore both enzymes can be made, both salvage pathways can operate, and DNA synthesis can occur. The HAT technique is an excellent example of using gene defects in a powerful selection scheme.

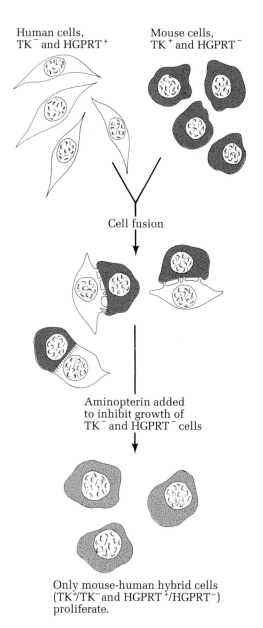

Human cells,
TK$^-$ and HGPRT$^+$

Mouse cells,
TK$^+$ and HGPRT$^-$

Cell fusion

Aminopterin added
to inhibit growth of
TK$^-$ and HGPRT$^-$ cells

Only mouse-human hybrid cells
(TK$^+$/TK$^-$ and HGPRT$^+$/HGPRT$^-$)
proliferate.

Figure 6.19 HAT technique for selecting fused, hybrid mouse-human cells.

Keynote *Geneticists can make a somatic cell hybrid between a human cell and a mouse cell line. The two cell types are fused by the presence of a suitable agent such as polyethylene glycol. Once the cells have fused, the nuclei often fuse to form a uninucleate cell line called a synkaryon. Within the nucleus the chromosomes of the two cell types are easily distinguished.*

Using hybrid somatic cells to locate genes. Two interesting properties of human/rodent hybrid cells make them particularly useful for somatic cell genetics. First, the human and rodent chromosomes are readily distinguishable under the microscope, particularly with the use of fluorescent dyes, which produce characteristic banding patterns on the chromosomes. The 46 human chromosomes vary extensively in size, while the 40 mouse chromosomes are quite small and uniform. Second, as the hybrid cell reproduces, the number of chromosomes decreases since some chromosomes are discarded; for unknown reasons human chromosomes are preferentially lost. On the selective medium with aminopterin just described, the eventual surviving, stable hybrid cell line is usually one containing the complete set of mouse chromosomes plus a small number of human chromosomes, which vary in number and type from cell line to cell line. These lines must contain the human chromosome that supplies a function that is deficient in the rodent genome.

To map human genes by using hybrid somatic cell lines, we start with a human cell that carries one or more genetic markers. Generally, we use markers that can readily be detected by analyzing tissue culture cells. Examples of markers include genes that control resistance to antibiotics and to other drugs, that code for enzymes, that determine nutritional requirements, and that specify cell surface antigens that are detectable by the use of fluorescence-labeled antibodies.

Once the human genetic markers are chosen, various stable hybrid cell lines (i.e., cultures no longer losing chromosomes) are analyzed for the presence or absence of the genetic markers.

These data are correlated for a number of somatic cell lines with the presence or absence of particular chromosomes. Given enough cell lines, we can show that a particular marker is present only when one particular chromosome is present, and it is absent when that chromosome is absent. For example, if a drug resistance phenotype is detected in several different cell lines that have in common human chromosome 14, then we would conclude that the gene controlling resistance to the drug is on chromosome 14. In this way genes amenable to analysis can be localized to individual chromosomes. Genes that are localized to a particular chromosome through this experimental approach are called **syntenic** ("together thread," i.e., another term for *linked*).

With this experimental technique many genes have been localized to individual chromosomes in the human genome. Further localization of genes to particular regions of chromosomes has also been possible in some instances. In these experiments the cells used have chromosomal abnormalities. For example, in some cells a part of a chromosome may spontaneously break off and be lost (**deleted**) or become attached (**translocated**) to another chromosome. If either of these events occurs, we can reinvestigate the location of genes known to be on the affected chromosome or chromosomes by forming hybrid somatic cell lines between the human cell that contains the chromosome and a deletion or a translocated chromosome. The loss or addition of various pieces of a particular chromosome can then be correlated with the presence or absence of specific genetic markers. This concept, involving a deletion of part of a chromosome, is depicted in Figure 6.20. In this example the enzyme is detectable in cell line A containing both chromosomes 6 and 12 (Figure 6.20a), thereby indicating that the gene for the enzyme is on either chromosome 6 or 12. Obviously a cell line with only one of the two chromosomes would indicate which chromosome carried the gene by the presence or absence of the enzyme activity. In cell line B (Figure 6.20b) both chromosomes are present, but chromosome 6 has suffered a deletion of most of the short arm of that chro-

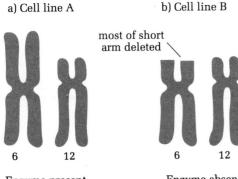

a) Cell line A   b) Cell line B

most of short arm deleted

6   12   6   12

Enzyme present   Enzyme absent

Figure 6.20 Locating a human gene coding for an enzyme to the short arm of chromosome 6 by somatic cell hybridization. In cell line A (a) the two human chromosomes remaining are 6 and 12 and the enzyme can be detected. In cell line B (b) both chromosomes 6 and 12 are present but chromosome 6 has suffered a deletion. Since the enzyme is not detectable, the gene for the enzyme must have been on the deleted segment of chromosome 6.

mosome. The absence of enzyme activity in this cell line indicates that the gene for the enzyme must be on the short arm of chromosome 6. While this method gives more specific information about the location of genes on human chromosomes, it is still not as precise a method as conventional genetic-mapping methods used with laboratory organisms.

As a summary, Figure 6.21 shows a genetic map of human chromosomes with some of the details of the banding patterns indicated. Table 6.1 gives the key to the symbols used in the figure. Despite the wealth of information in the figure and the table, hundreds of known human genes still remain to be localized.

Keynote *Hybrid somatic cell lines help in localizing human genes to particular chromosomes. Suitable genes for mapping are those that can be detected by analyzing tissue culture cells, such as genes that affect biochemical requirements or enzyme activities. As a somatic cell hybrid reproduces, a preferential and random loss of human chromosome occurs. Eventually, stable hybrid lines are established with various subsets of the human chromosomes. With enough cell lines of this kind a particular trait can be identified with a particular human chromosome.*

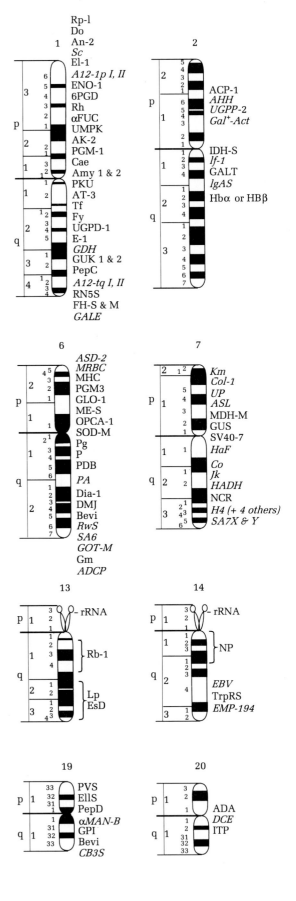

Figure 6.21 Genetic map of human chromosomes. Table 6.1 gives the key to the symbols.

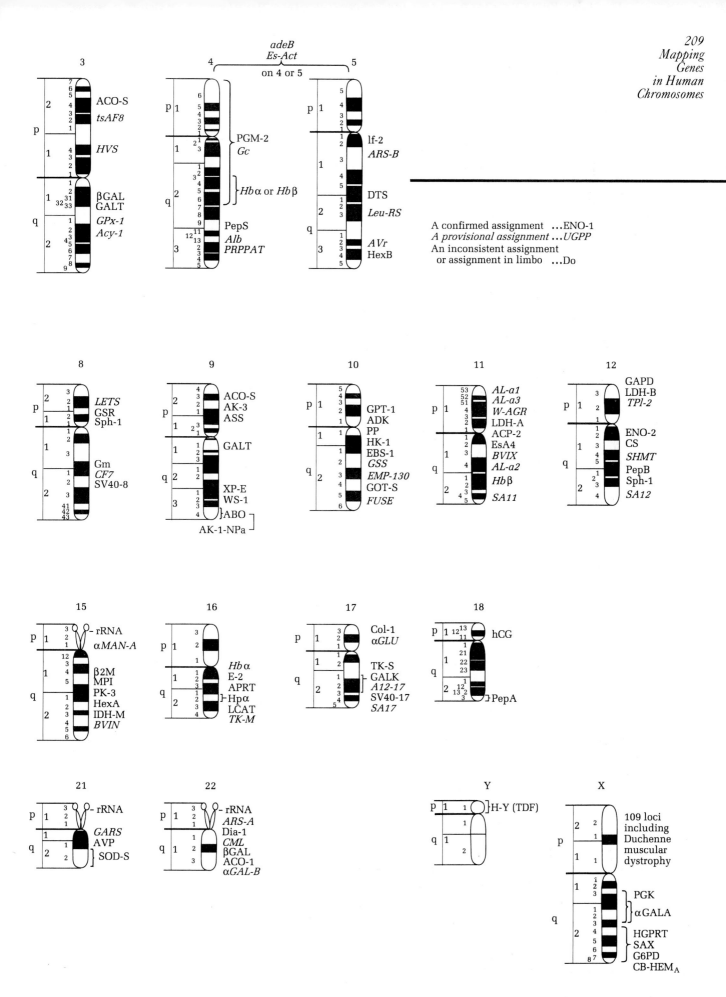

A confirmed assignment ...ENO-1
A provisional assignment ...*UGPP*
An inconsistent assignment
or assignment in limbo ...Do

Concluding Remarks

In this chapter we have learned how to map genes in certain microorganisms by using tetrad analysis. In a subset of those microorganisms the meiotic tetrads are ordered in a way that reflects the orientation of the four chromatids of each homologous pair of chromosomes at metaphase I. Ordered tetrads make it possible to map a gene's location relative to its centromere. In this chapter we also discovered how to map genes by using nonmeiotic means. The meiotic gene-mapping methods described in the previous chapter cannot be used on certain eukaryotes, so the discovery of nonmeiotic mapping procedures has been beneficial in the sense of allowing us to map genes in those eukaryotes. For example, genes have been located on human chromosomes by using somatic cell hybrids between human and rodent cells, which preferentially discard human chromosomes in a random way.

Table 6.1 Key to Gene Symbols in Figure 6.21

ABO	= ABO blood group (chr. 9)
ACO–M	= Aconitase, mitochondrial (chr. 22)
ACO–S	= Aconitase, soluble (chr. 9)
ACP–1	= Acid phosphatase-1 (chr. 2)
#ACP–2	= Acid phosphatase-2 (chr. 11)
Acy–1	= Aminoacylase-1 (chr. 3)
adeB	= Formylglycinamide ribotide amidotransferase (chr. 4 or 5)
#ADA	= Adenosine deaminase (chr. 20)
#ADCP	= Adenosine deaminase complexing protein (chr. 6)
ADK	= Adenosine kinase (chr. 10)
A12–1pI,II	= Adenovirus-12 chromosome modification site-1p I and II (chr. 1 p)
A12–1qI,II	= Adenovirus-12 chromosome modification site-1q I and II (chr. 1 q)
A12–17	= Adenovirus-12 chromosome modification site-17 (chr. 17)
#AH–3	= Adrenal hyperplasia III (21-hydroxylase deficiency) (chr. 6)
AHH	= Aryl hydrocarbon hydroxylase (chr. 2)
AK–1	= Adenylate kinase-1 (chr. 9)
AK–2	= Adenylate kinase-2 (chr. 1)
AK–3	= Adenylate kinase-3 (chr. 9)
AL	= Lethal antigen: 3 loci (a1, a2, a3) (chr. 11)
#Alb	= Albumin (chr. 4)
Amy–1	= Amylase, salivary (chr. 1)
Amy–2	= Amylase, pancreatic (chr. 1)
#An–2	= Aniridia, type II Baltimore (chr. 1)
#ARS–A	= Arylsulfatase A (chr. 22)
#ARS–B	= Arylsulfatase B (chr. 5)
#ARPT	= Adenine phosphoribosyltransferase (chr. 16)
#ASD–2	= Atrial septal defect, secundum type (chr. 6)
#ASL	= Argininosuccinate lyase (chr. 7)
#ASS	= Argininosuccinate synthetase (chr. 9)
#AT–3	= Antithrombin III (chr. 1)
AVP	= Antiviral protein (chr. 21)
AVr	= Antiviral state regulator (chr. 5)
Bevi	= Baboon M7 virus infection (chr. 6 or 19)

Bf	=	Properdin factor B (chr. 6)
β2M	=	β2–microglobulin (chr. 15)
BVIN	=	BALB virus induction, N-tropic (chr. 15)
BVIX	=	BALB virus induction xeno-tropic (chr. 11)
#C2	=	Complement component-2 (chr. 6)
C4F	=	Complement component-4 fast (chr. 6)
C4S	=	Complement component-4 slow (chr. 6)
C6	=	Complement component-6 (chr. 6)
#C8	=	Complement component-8 (chr. 6)
#Cae	=	Cataract, zonular pulverulent (chr. 1)
CB	=	Color blindness (deutan and protan) (X chr.)
CB3S	=	Coxsackie B3 virus susceptibility (chr. 19)
#CF7	=	Clotting factor VII (chr. 8)
Ch	=	Chido blood group (chr. 6)—same as C4S
#CML	=	Chronic myeloid leukemia (chr. 22)
Co	=	Colton blood group (chr. 7)
#Col–1	=	Collagen I (α–1 and α–2) (chr. 7 and 17)
CS	=	Citrate synthase, mitochondrial (chr. 12)
#Dia–1	=	NADH-diaphorase (chr. 6 or 22)
#DMJ	=	Juvenile diabetes mellitus (chr. 6)
Do	=	Dombrock blood group (chr. 1)
DCE	=	Desmosterol-to-cholesterol enzyme (chr. 20)
#DTS	=	Diphtheria toxin sensitivity (chr. 5)
#EBS–1	=	Epidermolysis bullosa, Ogna type (chr. 10)
#EBV	=	Epstein-Barr virus integration site (chr. 14)
#E–1	=	Pseudocholinesterase-1 (chr. 1)
#E–2	=	Pseudocholinesterase-2 (chr. 16)
E11–S	=	Echo 11 sensitivity (chr. 19)
E1–1	=	Elliptocytosis-1 (chr. 1)

EMP–130	=	External membrane protein-130 (chr. 10)
EMP–195	=	External membrane protein-195 (chr. 14)
ENO–1	=	Enolase-1 (chr. 1)
ENO–2	=	Enolase-2 (chr. 12)
Es–Act	=	Esterase activator (chr. 4 or 5)
EsA4	=	Esterase-A4 (chr. 11)
EsD	=	Esterase D (chr. 13)
FH–M	=	Fumarate hydratase, mitochondrial (chr. 1)
FH–S	=	Fumarate hydratase, soluble (chr. 1)
#αFUC	=	Alpha-L-fucosidase (chr. 1)
FUSE	=	Polykaryocytosis inducer (chr. 10)
#Fy	=	Duffy blood group (chr. 1)
Gal+-Act	=	Galactose + activator (chr. 2)
#αGAL A	=	α-galactosidase A (Fabry disease) (X chr.)
αGAL B	=	α-galactosidase B (chr. 22)
#βGAL	=	β-galactosidase (chr. 3) (see also chr. 12 and 22)
#GALK	=	Galactokinase (chr. 17)
#GALT	=	Galactose-1-phosphate uridyltransferase (chr. 2, 3, or 9)
#GALE	=	Galactose-4-epimerase (chr. 1)
#αGLU	=	α-glucosidase (chr. 17)
GAPD	=	Glyceraldehyde-3-phosphate dehydrogenase (chr. 12)
GAPS	=	Phosphoribosyl glycinamide synthetase (chr. 21)
Gc	=	Group-specific component (chr. 4)
GDH	=	Glucose dehydrogenase (chr. 1)
GLO–1	=	Glyoxylase I (chr. 6)
Gm	=	Immunoglobulin heavy chain (chr. 8)
GOT–M	=	Glutamate oxaloacetate transaminase, mitochondrial (chr. 6)
GOT–S	=	Glutamate oxaloacetate transaminase, soluble (chr. 10)
#GPI	=	Glucosephosphate isomerase (chr. 19)
GPT–1	=	Glutamate pyruvate transaminase, soluble (chr. 10)
#GPx–1	=	Glutathione peroxidase-1 (chr. 3)

Table 6.1 *continued*

#G6PD	=	Glucose-6-phosphate dehydrogenase (X chr.)
#GSR	=	Glutathione reductase (chr. 8)
GSS	=	Glutamate-gamma-semialdehyde synthetase (chr. 10)
GARS	=	Glycinamide ribonucleotide synthetase (chr. 21)
GUK–1 & 2	=	Guanylate kinase-1 & 2 (chr. 1)
#GUS	=	Beta-glucuronidase (chr. 7)
H4	=	Histone H4 and 4 other histone genes (chr. 7)
HADH	=	Hydroxyacyl–CoA dehydrogenase (chr. 7)
HaF	=	Hageman factor (chr. 7)
#Hbα	=	Hemoglobin alpha chain (chr. 2, 4, 5 or 16)
#Hbβ	=	Hemoglobin beta chain (chr. 2, 4, 5 or 11)
hCG	=	Human chorionic gonadotropin (chr. 18)
#Hch	=	Hemochromatosis (chr. 6)
#HEM–A	=	Classic hemophilia (X chr.)
#HexA	=	Hexosaminidase A (chr. 15)
#HexB	=	Hexosaminidase B (chr. 5)
#HGPRT	=	Hypoxanthine-guanine phosphoribosyltransferase (X chr.)
HK–1	=	Hexokinase-1 (chr. 10)
HLA (A–D)	=	Human leukocyte antigens (chr. 6)
HLA–DR	=	Human leukocyte antigen, D-related (chr. 6)
Hpα	=	Haptoglobin, alpha (chr. 16)
#HVS	=	Herpes virus sensitivity (chr. 6)
H–Y	=	Y histocompatibility antigen (Y chr.)
IgAS	=	Immunoglobulin heavy chains attachment site (chr. 2)
If–1	=	Interferon-1 (chr. 2)
If–2	=	Interferon-2 (chr. 5)
IDH–M	=	Isocitrate dehydrogenase, mitochondrial (chr. 15)
IDH–S	=	Isocitrate dehydrogenase, soluble (chr. 2)
#ITP	=	Inosine triphosphatase (chr. 20)
Jk	=	Kidd blood group (chr. 7)

Km	=	Kappa immunoglobulin light chains, Inv (chr. 7)
#LAP	=	Laryngeal adductor paralysis (chr. 6)
#LCAT	=	Lecithin-cholesterol acyltransferase (chr. 16)
LDH–A	=	Lactate dehydrogenase A (chr. 11)
LDH–B	=	Lactate dehydrogenase B (chr. 12)
LETS	=	Large, external, transformation-sensitive protein (chr. 8)
LeuRS	=	Leucyl-tRNA synthetase (chr. 5)
Lp	=	Lipoprotein β–Lp (chr. 13)
αMAN–A	=	Cytoplasmic α–D-mannosidase (chr. 15)
#αMAN–B	=	Lysosomal α–D-mannosidase (chr. 19)
#MDH–M	=	Malate dehydrogenase, mitochondrial (chr. 7)
MDH–S	=	Malate dehydrogenase, soluble (chr. 2)
ME–S	=	Malic enzyme, soluble (chr. 6)
MHC	=	Major histocompatibility complex (chr. 6)
MLC–W	=	Mixed lymphocyte culture, weak (chr. 6)
MPI	=	Mannosephosphate isomerase (chr. 15)
MRBC	=	Monkey red blood cell receptor (chr. 6)
MTR	=	5-Methyltetrahydrofolate: L-homocysteine S–S-methyltransferase (chr. 1)
NCR	=	Neutrophil chemotactic response (chr. 7)
NDF	=	Neutrophil differentiation factor (chr. 6)
#NP	=	Nucleoside phosphorylase (chr. 14)
#NPa	=	Nail-patella syndrome (chr. 9)
#OPCA–1	=	Olivopontocerebellar atrophy I (chr. 6)
P	=	P blood group (chr. 6)
PA	=	Plasminogen activator (chr. 6)

#PDB	= Paget disease of bone (chr. 6)		SOD–M	= Superoxide dismutase, mitochondrial (chr. 6)
PepA	= Peptidase A (chr. 18)		SOD–S	= Superoxide dismutase, soluble (chr. 21)
PepB	= Peptidase B (chr. 12)		Sph–1	= Sherocytosis, Denver type (chr. 8 or 12)
PepC	= Peptidase C (chr. 1)		SV40–7	= SV40-integration site-7 (chr. 7)
PepD	= Peptidase D (chr. 19)		SV40–8	= SV40-integration site-8 (chr. 8)
PepS	= Peptidase S (chr. 4)		SV40–17	= SV40-integration site-17 (chr. 17)
Pg	= Pepsinogen (chr. 6)			
PGK	= Phosphoglycerate kinase (X chr.)		TDF	= Testis-determining factor (Y chr.)—probably same as H–Y
PGM–1	= Phosphoglucomutase-1 (chr. 1)		Tf	= Transferrin (chr. 1)
PGM–2	= Phosphoglucomutase-2 (chr. 4)		TK–M	= Thymidine kinase, mitochondrial (chr. 16)
PGM–3	= Phosphoglucomutase-3 (chr. 6)		TK–S	= Thymidine kinase, soluble (chr. 17)
6PGD	= 6-phosphogluconate dehydrogenase (chr. 1)		#TPI–1 & 2	= Triosephosphate isomerase-1 and -2 (chr. 12)
PRPPAT	= Phosphoribosylpyrophosphate amidotransferase (chr. 4)		TrpRS	= Tryptophanyl-tRNA synthetase (chr. 14)
PK–3	= Pyruvate kinase-3 (chr. 15)		tsAF8	= Temperature-sensitive (AF8) complement (chr. 3)
#PKU	= Phenylketonuria (chr. 1)			
PP	= Inorganic pyrophosphatase (chr. 10)		UGPP–1	= Uridyl diphosphate glucose pyrophosphorylase-1 (chr. 1)
#PVS	= Polio sensitivity (chr. 19)		UGPP–2	= Uridyl diphosphate glucose pyrophosphorylase-2 (chr. 2)
#RB–1	= Retinoblastoma-1 (chr. 13)		UMPK	= Uridine monophosphate kinase (chr. 1)
rC3b	= Receptor for C3b (chr. 6)		UP	= Uridine phosphorylase (chr. 7)
rC3d	= Receptor for C3d (chr. 6)			
Rg	= Rodgers blood group (chr. 6) same as C4F		#WS–1	= Waardenburg syndrome-1 (chr. 9)
#Rh	= Rhesus blood group (chr. 1)		#W–AGR	= Wilms tumor—aniridia/ambiguous genitalia/mental retardation (chr. 11)
RN5S	= 5S RNA gene(s) (chr. 1)			
#RP–1	= Retinitis pigmentosa-1 (chr. 1)			
rRNA	= Ribosomal RNA (chr. 13, 14, 15, 21, 22)		#XP–E	= Xeroderma pigmentosum, Egyptian (chr. 9)
#RwS	= Ragweed sensitivity (chr. 6)			
SA6	= Surface antigen 6 (chr. 6)			
SA7	= Surface antigen 7 (chr. 7)			
SA11	= Surface antigen 11 (chr. 11)			
SA12	= Surface antigen 12 (chr. 12)			
SA17	= Surface antigen 17 (chr. 17)			
SAX	= X-linked species (or surface) antigen (X chr.)			
Sc	= Scianna blood group (chr. 1)			
SHMT	= Serine hydroxymethyltransferase (chr. 12)			

Analytical Approaches for Solving Genetics Problems

Q.1 A *Neurospora* strain that required both adenine (*ad*) and tryptophan (*trp*) for growth was mated to a wild-type strain (*ad*⁺ *trp*⁺), and this cross produced the following ordered tetrads:

Spore pair 1 : *ad trp*	*ad* +	*ad trp*	*ad trp*
Spore pair 2 : *ad trp*	*ad* +	*ad* +	+ *trp*
Spore pair 3 : + +	+ *trp*	+ *trp*	*ad* +
Spore pair 4 : + +	+ *trp*	+ +	+ +
(1) 63	(2) 3	(3) 15	(4) 9

Spore pair 1 : *ad trp*	*ad* +	+ *trp*
Spore pair 2 : + +	+ *trp*	+ +
Spore pair 3 : *ad trp*	*ad* +	*ad trp*
Spore pair 4 : + +	+ *trp*	*ad* +
(5) 3	(6) 1	(7) 6

a. Determine the gene-centromere distance for the two genes.

b. From the data given, calculate the map distance between the two genes.

A.1 a. The gene-centromere distance is given by the formula

$$\frac{\text{percent second-division tetrads}}{2}$$

For the *ad* gene tetrads 4, 5, and 6 show second-division segregation; the total number of such tetrads is 13. There are 100 tetrads, so the *ad* gene is (13/2)% = 6.5 mu from its centromere. For the *trp* gene tetrads, 3, 5, 6, and 7 show second-division segregation, and the total number of such tetrads is 25, indicating the *trp* gene is 12.5 mu from its centromere.

b. The linkage relationship between the genes can be determined by analyzing the relative number of parental-ditype (PD), nonparental-ditype (NPD), and tetratype (T) tetrads. Tetrads 1 and 5 are PD, 2 and 6 are NPD, and 3, 4, and 7 are T. If two genes are unlinked, the frequency of PD tetrads will approximately equal the frequency of NPD tetrads. If two genes are linked, the frequency of PD tetrads will greatly exceed that of the NPD tetrads. Here the latter case prevails, so the two genes must be linked. The map distance between two genes is given by the general formula

$$\frac{1/2\ \text{T} + \text{NPD}}{\text{total}} \times 100$$

For this example the number of T tetrads is 30, and the number of NPD tetrads is 4. Thus the map distance between *ad* and *trp* is

$$\frac{(1/2 \times 30) + 4}{100} \times 100 = 19\ \text{mu}$$

Thus we have the following map, with • indicating the centromere:

215
Analytical
Approaches
for Solving
Genetics
Problems

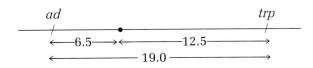

Q.2 In *Aspergillus,* forced diploids were constructed between a wild-type strain and a strain containing the mutant genes *y* (yellow), *w* (white), *pro* (proline requirement), *met* (methionine requirement), and *ad* (adenine requirement). All these genes are known to be in a single chromosome.

Homozygous yellow and homozygous white segregants were isolated and analyzed for the presence of the other gene markers. The following phenotypic results were obtained:

y/y segregants:	w^+ pro^+ met^+ ad^+	15
	w^+ pro met^+ ad^+	28
w/w segregants:	y^+ pro^+ met ad	6
	y^+ pro^+ met^+ ad	12

Draw a map of the chromosome, giving the order of the genes and the position of the centromere.

A.2 The segregants are all diploid. In mitotic recombination a single crossover renders all gene loci distal to that point homozygous. In this regard the crossover events in one chromosome arm are independent of those in the other chromosome arm. Therefore we must inspect the data with these concepts in mind.

There are two classes of *y/y* segregants: wild-type segregants and proline-requiring segregants, which are homozygous for the *pro* gene. Thus of the four loci other than yellow, only *pro* is in the same chromosome arm as *y.* Further, since not all the *y/y* segregants are *pro* in phenotype, the *pro* locus must be closer to the centromere than the *y* locus, as shown in the following map:

A single crossover between *pro* and *y* will give *y/y* segregants that are wild-type for all other genes, whereas a single crossover between the centromere and *pro* will give homozygosity for both *pro* and *y*—hence, the *pro* requirement.

Similar logic can be applied to the *w/w* segregants. Again, there are two classes. Both classes are also phenotypically *ad,* indicating that the *ad* locus is further from the centromere than the *w* locus. Hence every time *w* becomes homozygous, so does *ad.* The remaining gene to be located is *met.* Some of the *w/w* segregants are *met*$^+$ and some are *met,* so the *met* gene is closer to the centromere than the *w* gene. The reasoning here is analogous to that for the placement of the *pro* gene in the other arm. Taking all the conclusions together, we have the following gene order:

y	*pro*	*met*	*w*	*ad*

*Questions
and
Problems*

6.1 In *Saccharomyces, Neurospora,* and *Chlamydomonas,* what meiotic events give rise to PD, NPD, and T tetrads?

6.2 What important item of information regarding crossing-over can be obtained from tetrad analysis (as in *Neurospora*) but not from single-strand analysis (as in *Drosophila*)?

6.3 A cross was made between a pantothenate-requiring (*pan*) strain and a lysine-requiring (*lys*) strain of *Neurospora crassa,* and 750 random ascospores were plated on a minimal medium (a medium lacking pantothenate and lysine). Thirty colonies subsequently grew. Map the *pan* and *lys* loci.

*6.4 In *Neurospora* the following crosses yielded the progeny as shown:

$$a^+ \, b \times a \, b^+ \rightarrow$$

981	$a^+ \, b$
1000	$a \quad b^+$
10	$a^+ \, b^+$
9	$a \quad b$
2000	

$$a^+ \, c \times a \, c^+ \rightarrow$$

850	$a^+ \, c$
833	$a \quad c^+$
169	$a^+ \, c^+$
148	$a \quad c$
2000	

$$b^+ \, c \times b \, c^+ \rightarrow$$

850	$b^+ \, c$
850	$b \quad c^+$
140	$b^+ \, c^+$
160	$b \quad c$
2000	

What is the probable gene order, and what are the approximate map distances between adjacent genes?

6.5 Four different albino strains of *Neurospora* were each crossed to the wild type. All crosses resulted in half wild-type and half albino progeny. Crosses were made between the first strain and the other three with the following results:

1 × 2:	975 albino,	25 wild type
1 × 3:	1000 albino	
1 × 4:	750 albino,	250 wild type

Which mutations represent different genes, and which genes are linked? How did you arrive at your conclusions?

*6.6 Genes *met* and *thi* are linked in *Neurospora crassa;* we wish to locate *arg* with respect to *met* and *thi*. From the cross *arg*+ + × + *thi met* the following random ascospore isolates were obtained. Map these three genes.

arg thi met	26		*arg* + +	51
arg thi +	17		+ *thi* +	4
arg + *met*	3		+ + *met*	14
+ *thi met*	56		+ + +	29

6.7 Given a *Neurospora* zygote of the constitution shown in the following figure, diagram the significant events producing an ascus where the *A* alleles segregate at the first division and the *B* alleles segregate at the second division.

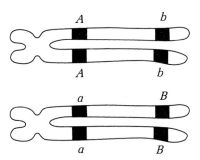

6.8 Double exchanges between two loci can be of several types, called two-strand, three-strand, and four-strand doubles.

a. Four recombination gametes would be produced from a tetrad in which the first of two exchanges was as depicted in the first figure below. Draw in the second exchange.

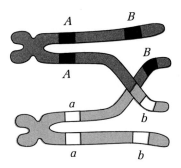

b. Draw in the second exchange so that four nonrecombination gametes would result from the tetrad shown in the next figure.

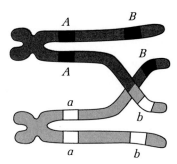

c. Other possible double-crossover types would result in two recombination and two parental gametes. If all possible multiple-crossover types occur at random, the frequency of recombination between genes at two loci will never exceed a certain percentage regardless of how far apart they are on the chromosome. What is that percentage? Explain the reason why the percentage cannot theoretically exceed that value.

*6.9 A cross between a pink (p^-) yeast strain of mating type mt^+ and a grey strain (p^+) of mating type mt^- produced the following tetrads:

Number of Tetrads	Kind of Tetrad
18	$p^+\ mt^+, p^+\ mt^+, p^-\ mt^-, p^-\ mt^-$
8	$p^+\ mt^+, p^-\ mt^+, p^+\ mt^-, p^-\ mt^-$
20	$p^+\ mt^-, p^+\ mt^-, p^-\ mt^+, p^-\ mt^+$

On the basis of these results, are the p and mt genes on separate chromosomes?

6.10 In *Neurospora* the peach gene (*pe*) is on one chromosome and the colonial gene (*col*) is on another. Disregarding the occurrence of chiasmata, what kinds of tetrads (asci) would you expect and in what proportions if these two strains were crossed?

6.11 The following unordered asci were obtained from the cross *leu* + × + *rib* in yeast. Draw the linkage map and determine the map distance.

30	45	41	4	39	2	39
leu +	leu rib	leu +	+ +	leu +	leu rib	leu +
+ rib	leu +	leu +	leu rib	+ rib	+ +	leu +
leu +	+ +	+ rib	leu rib	+ +	leu rib	+ rib
+ rib	+ rib	+ rib	+ +	leu rib	+ +	+ rib

*6.12 The genes *a*, *b*, and *c* are on the same chromosome arm in *Neurospora crassa*. The following ordered asci were obtained from the cross *a b* + × + + *c*:

45	5	146	1	10	20	15	58
a b +	a b +	a b +	a b +	a b +	a b +	a b +	a b +
+ b c	a + +	a b +	+ + +	a + c	+ + c	a b c	+ b +
a + +	+ b c	+ + c	a b c	+ b +	a b +	+ + +	a + c
+ + c	+ + c	+ + c	+ + c	+ + c	+ + c	+ + c	+ + c

a. Determine the correct gene order.
b. Determine all gene-gene and gene-centromere distances.

*6.13 If 10 percent of the asci analyzed in a particular two-point cross of *Neurospora crassa* show that an exchange has occurred between two loci (i.e., exhibit second-division segregation), what is the map distance between the two loci?

6.14 A diploid strain of *Aspergillus nidulans* (forced between wild type and a multiple mutant) that was heterozygous for the recessive mutations *y* (yellow), *w* (white), *ad* (adenine), *sm* (small), *phe* (phenylalanine), and *pu* (putrescine) produced haploid segregants. Forty-one haploid yellow segregants were tested and found to have the following genotypes and numbers:

$$
\begin{array}{llllll}
y & w & pu & ad & sm & phe & 7 \\
y & w & pu & ad & + & + & 11 \\
y & + & + & + & sm & phe & 16 \\
y & + & + & + & + & + & 7
\end{array}
$$

What are the linkage relationships of these genes?

6.15 A heterokaryon was established in the fungus *Aspergillus nidulans* between a *met⁻*, *trp⁻* auxotroph and a *leu⁻*, *nic⁻* auxotroph. A diploid strain was selected from this heterokaryon. From this diploid strain, the following eight haploid strains were obtained from conidial isolates, via the parasexual cycle, in approximately equal frequencies:

1. nic^+ leu^+ met^- trp^-
2. met^+ leu^+ nic^+ trp^+
3. trp^- met^- leu^+ nic^-
4. leu^- nic^- trp^+ met^+
5. leu^+ nic^- met^+ trp^+
6. met^- nic^- leu^- trp^-
7. trp^+ leu^- met^+ nic^+
8. nic^+ met^- trp^- leu^-

Which, if any, of these four marker genes are linked, and which are unlinked?

*6.16 A (green) diploid of *Aspergillus nidulans* is heterozygous for *each* of the following recessive mutant genes: *sm, pu, phe, bi, w* (white), *y* (yellow), and *ad*. Analysis of white and yellow haploid segregants from this diploid indicated several classes with the following genotypes:

Genotype

sm	pu	phe	bi	w	y	ad
sm	pu	phe	+	w	y	ad
+	pu	+	+	w	y	ad
+	pu	+	bi	w	+	ad
sm	+	phe	+	+	y	+
+	+	+	+	+	y	+
sm	pu	phe	bi	w	+	ad

How many linkage groups are involved, and which genes are on which linkage group?

6.17 A (green) diploid of *Aspergillus nidulans* is homozygous for the recessive mutant gene *ad* and heterozygous for the following recessive mutant genes: *paba, ribo, y* (yellow), *an, bi, pro,* and *su-ad.* Those recessive alleles that are on the same chromosome are in coupling. The *su-ad* allele is a recessive suppressor of the *ad* allele: The +/*su-ad* genotype does not suppress the adenine requirement of the *ad/ad* diploid, whereas the *su-ad/su-ad* genotype does suppress that requirement. Therefore the parental diploid requires adenine for growth. From this diploid two classes of segregants were selected: yellow and adenine-independent. The accompanying table lists the types of segregants obtained.

Segregant Type Selected	Phenotype
Adenine-independent	+
	ribo
	ribo an
	ribo an pro paba y bi
Yellow	*y ad bi*
	paba y ad bi
	pro paba y ad bi
	ribo an pro paba y bi

Analyze these results as completely as possible to determine the location of the centromere and the relative locations of the genes.

6.18 The accompanying table shows the only human chromosomes present in stable human-mouse somatic cell hybrid lines.

	Human Chromosomes			
	2	4	10	19
A	−	+	+	−
B	+	−	+	+
C	−	+	+	+
D	+	+	−	−

The presence of four enzymes, I, II, III, and IV, was investigated: I was present in *A, B,* and *C* but absent in *D*; II was in *B* and *D* but absent in *A* and *C*; III was in *A, C,* and *D* but not in *B*; and IV was in *B* and *C* but not in *A* and *D*. On what chromosomes are the genes for the four enzymes?

7

Genetic Mapping in Prokaryotes

In the previous two chapters we considered the principles of genetic mapping in eukaryotic organisms. To map genes in prokaryotes, geneticists use essentially the same experimental strategies. Crosses are made between strains that differ in genetic markers, and recombinants—the evidence of the exchange of genetic material—are detected and counted. The principles applied in analyzing the data obtained from such prokaryotic crosses are also the same; that is, the frequency with which exchanges occur between two sets of genes gives us an idea of the physical distances between the genes' loci.

Mapping Genes in Bacteria

The most commonly used bacterium for genetic and molecular analysis is *Escherichia coli* (*E. coli*), which is found most commonly in animal (including human) intestines. It is a good subject for study since it can be grown on a simple, defined medium and can be handled with simple microbiological techniques.

As we learned in Chapter 1, bacterial cells are relatively small compared with eukaryotic cells. Bacterium *E. coli* is a cylindrical organism about

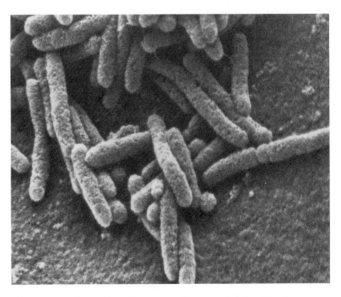

Figure 7.1 Scanning electron micrograph of cells of the bacterium Escherichia coli.

1 μm long and 0.5 μm in diameter (Figure 7.1). Its cytoplasm is full of ribosomes (the organelles responsible for protein synthesis), while the central nucleoid region contains the genetic material, a single circular chromosome consisting of DNA. As in all prokaryotes, there is no membrane between the bacterium's nucleoid region and the rest of the cell.

Like other bacteria, *E. coli* can be grown both in liquid medium and on the surface of growth medium solidified with agar. Genetic analysis of bacteria typically is done by plating cells out on agar-solidified media. Wherever a single bacterium lands on the agar surface it will grow and divide repeatedly, resulting in the formation of a visible cluster of genetically identical cells called a *colony* (Figure 7.2). The concentration of bacterial cells in a suspension (the *titer*) can be determined by spreading known volumes of the suspension or of a known dilution of the suspension on the agar surface, incubating the plates, and then counting the number of resulting colonies. By varying the nutrient composition of the media it is possible to detect and quantify the various genotypic classes in a genetic analysis.

Figure 7.2 Bacterial colonies growing on a nutrient medium in a Petri dish.

Genetic material can be transferred between bacteria by three processes. In each case: (1) transfer is unidirectional; (2) there is no true zygote; and (3) the cells are essentially always haploid. Thus, these genetic transfer processes are different from those in eukaryotes.

1. **Transformation** is a process in which there is a transfer of genetic information by means of extracellular pieces of DNA in bacteria. The DNA fragments are derived from donor bacteria and may be taken up by another living bacterium, the recipient. If the donor and the recipient bacteria differ genetically, then genetic recombinants are produced by crossover events involving the DNA fragments and the recipient's chromosome. The genetic recombinants generated by the transformation process are called **transformants.**

2. **Conjugation** is a process in which there is a unidirectional transfer of genetic information through direct cellular contact between a donor and a recipient bacterial cell. The contact is followed by the formation of a cellular bridge physically connecting both cells. Then a segment (rarely all) of the donor's chromosome may be transferred into the recipient and may undergo genetic recombination with a homologous chromosome segment of the recipient cell. The recipients inheriting donor DNA are called **transconjugants.**

3. **Transduction** is a process by which bacteriophages mediate the transfer of bacterial genetic information from one bacterium (the donor) to another (the recipient). Since the amount of DNA a phage can carry is limited, the amount of genetic material that can be transferred is usually less than 1 percent of that in the bacterial chromosome. Once the donor genetic material has been introduced into the recipient, it may undergo genetic recombination with a homologous chromosome region. The recipients inheriting donor DNA in this way are called **transductants.**

It is possible to map bacterial genes by using any one of these methods. However, not all methods can be used on each species of bacterium.

Bacterial Transformation

Transformation of bacteria was first observed by F. Griffith in 1928, and in 1944 O. Avery and his colleagues showed that DNA was responsible for the genetic change that was observed (see Chapter 8). In transformation, fragments of free DNA are taken up by a living bacterium. This process is generally used to map genes in bacterial species in which conjugation or transduction does not occur. In transformation mapping experiments, DNA from a donor strain is extracted and purified. The extraction process breaks the DNA into small fragments, and this genetic material is then added to a suspension of recipient bacteria with a different genotype. The donor DNA that is taken up by the recipient cell may undergo genetic recombination with the homologous parts of the recipient's chromosome to produce a recombinant chromosome. Those recipients whose phenotypes are changed by a recombination event are called **transformants.**

While most bacterial species are probably capable of the recombination event of the transformation process, there is a great deal of variability in their ability to take up DNA. Wild-type *E. coli,* for example, is not readily transformable because the bacterium rapidly degrades incoming DNA. Other bacteria, such as *Bacillus subtilis* and *Streptococcus pneumoniae,* can be easily transformed in a test tube, and mapping has been done on these bacteria by using this technique. Even in a bacterial species that is capable of being transformed, only a small population will actually take up DNA. Over the years, though, scientists have devised experimental conditions for a number of bacterial species that produce a population of cells, called *competent cells,* most of which are able to take up DNA because they are leaky, meaning extraneous DNA is readily passed through the membrane of competent cells.

Once a donor DNA fragment has been taken up by a recipient bacterium with a different genetic constitution, we are in a position to detect the transformants produced by genetic exchange between the transforming DNA pieces and the recipient cell's chromosome. Since the bacterial chromosome is circular and the DNA fragment is linear (Figures 7.3a and 7.3b), a single crossover (or any odd number of crossovers) (Figure 7.3c) will open up the chromosome and add the transforming DNA to make a long, linear chro-

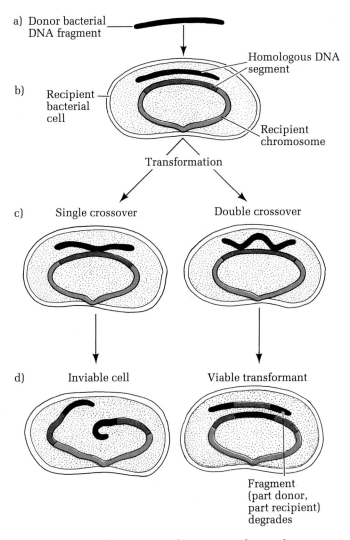

a) Donor bacterial DNA fragment

b) Recipient bacterial cell — Homologous DNA segment — Recipient chromosome

Transformation

c) Single crossover — Double crossover

d) Inviable cell — Viable transformant

Fragment (part donor, part recipient) degrades

Figure 7.3 Transformation in bacteria: (a) linear donor bacterial DNA fragment; (b) donor transforming DNA fragment finds homologous region on the bacterial chromosome; (c) single crossover opens chromosome to produce linear molecule resulting in an inviable cell; (d) double crossover results in exchange of homologous DNA segments and in a viable transformant.

mosome. A linear chromosome is unable to replicate in the bacterium and will often degrade, leading to the death of the cell. Thus the integration of part or all of the transforming DNA pieces requires an even number of crossovers, for example, a double crossover (Figure 7.3d). First, the donor DNA migrates to the homologous region in the recipient chromosome, and then the crossovers take place. These crossovers leave an intact, circular, recombinant bacterial chromosome and a residual linear piece of DNA (part donor and part recipient) that degrades. Given highly competent recipient cells and an excess of DNA fragments, transformation of most genes will occur at a frequency of about 10^{-3} (i.e., one cell per 10^3 cells).

Using transformation it is possible to determine gene linkage, gene order, and map distance. The principles are as follows. If two genes, x^+ and y^+, are far apart on the donor chromosome, they will always be found on different DNA fragments. Thus, given an $x^+ y^+$ donor and an $x y$ recipient, the probability of simultaneous transformation (*cotransformation*) of the recipient to $x^+ y^+$ (from the product rule) is the square of the probability of transformation with one gene alone. If transformation occurred at a frequency of 1 in 10^3 cells per gene (see above), $x^+ y^+$ transformants would be expected to appear at a frequency of 1 in 10^6 recipient cells ($10^{-3} \times 10^{-3}$). Therefore, if two genes are close enough that they often are carried on the same DNA fragment, the cotransformation frequency would be close to the frequency of transformation by a single gene. To turn this argument around to the way it is used practically, if cotransformation occurs at a frequency that is substantially higher than the product of the two single-gene transformations, the two genes must be close together.

Gene order can also be determined from cotransformation data. For example, if genes p and q are often cotransformed, and genes q and o are often cotransformed, but genes o and p are never cotransformed, the gene order must be p-q-o (Figure 7.4).

Since it is possible to control the size of DNA fragments used in transformation experiments,

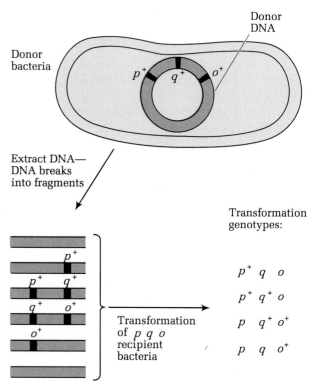

Figure 7.4 Demonstration of gene order by cotransformation.

the probability of cotransformation of two genes can be related to the average molecular size of the transforming DNA. Using this approach of relating cotransformation frequency to average size of the transforming DNA, a physical map of the genes (i.e., a map of the relative physical locations of genes along the DNA) can be derived.

Keynote *Transformation is the transfer of genetic material between members of the same or related species by means of extracellular DNA. In transformation, DNA is extracted from a donor strain and added to recipient cells. Some DNA fragments are taken up by the recipient cells and may associate with the homologous region of the recipient's genome. By a genetic recombination event, part of the transforming DNA molecule can exchange with part of the recipient's chromosomal DNA. Frequent cotransformation of donor genes indicates close physical linkage of those genes. Analysis of co-*

transformants can be used to determine gene order and map distance between genes. Transformation has been used to construct genetic maps for bacterial species for which conjugation or transduction analyses are not possible.

Conjugation in Bacteria

Discovery of Conjugation in *E. coli*

In 1946 another method of genetic transfer between bacteria, conjugation, was discovered by Joshua Lederberg and Edward Tatum. They knew that meiosis does not occur in bacteria, but they wanted to find out if any genetic exchange between bacterial cells occurred and if it was similar to the exchange found in sexually reproducing eukaryotic organisms. They studied two strains of *E. coli* that differed in their nutritional requirements. Strain A had the genotype *met bio thr⁺ leu⁺ thi⁺*, and strain B had the genotype *met⁺ bio⁺ thr leu thi*. Strains carrying any of the mutant genes require nutritional supplements in their growth medium in order to grow, while strains carrying wild-type genes require no supplements. Thus strain A can only grow on a medium supplemented with the amino acid methionine (*met*) and the vitamin biotin (*bio*) but does not need the amino acids threonine (*thr*) or leucine (*leu*), or the vitamin thiamine (*thi*). In contrast, strain B can only grow on a medium supplemented with *thr, leu,* and *thi* but does not require *met* or *bio*. Strains that must have additives in the medium in order to grow are called **auxotrophs.** A strain that is wild-type for all nutritional-requirement genes and thus requires no supplements in its growth medium is called a **prototroph.**

In their experiment Lederberg and Tatum mixed strains A and B together and spread (plated) them onto plates of minimal medium, that is, a medium with no growth additives (Figure 7.5). In control experiments the two strains were plated separately. As expected, neither strain could grow without the appropriate nutritional supplement. The mixed cells, however, gave rise to some prototrophic colonies. Geno-

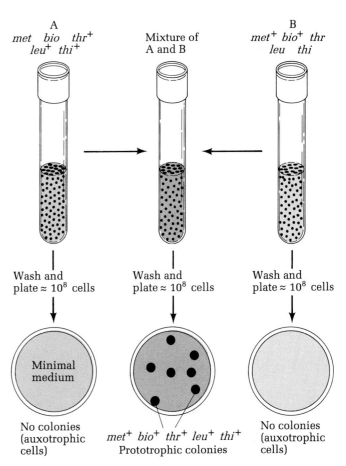

Figure 7.5 Lederberg and Tatum experiment showing that sexual recombination occurs between cells of E. coli. *After the cells have been mixed and the mixture plated, a few colonies grow on the minimal medium, indicating that they can now make the essential constituents. These colonies are recombinants produced by an exchange of genetic material between the strains.*

typically, these prototrophic strains were *met⁺ bio⁺ thr⁺ leu⁺ thi⁺*, and they arose at a frequency of about 1 in 10 million cells (1×10^{-7}). Since no colonies appeared on the control plates, mutation was ruled out as the origin of the prototrophic colonies. The results suggested that some type of genetic exchange had occurred between the two strains.

To determine whether cell-to-cell contact was needed for the genetic exchange to occur, the researchers performed an experiment using a U tube apparatus made by Bernard Davis (Figure 7.6). Strains A and B were placed in a liquid medium on either side of the U separated by a fine filter whose pores were too small to allow bacteria to move through it. After several hours of incubation during which the medium was moved between compartments by alternating suction and pressure, the cells were plated on a minimal medium to check for the appearance of prototrophic colonies. No prototrophic colonies appeared. The interpretation was that cell-cell contact was required in order for the previously observed genetic exchange to occur. In other words, *E. coli* has a type of mating system, called **conjugation**, in which genetic material can be transferred from one bacterium to another through a tube transiently linking the two cells. Figure 7.7 shows two bacteria conjugating.

The Sex Factor *F*

In 1953 William Hayes obtained evidence that the genetic exchange in *E. coli* is not reciprocal.

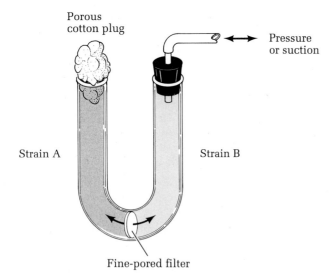

Figure 7.6 Davis U tube experiment showing that physical contact between the two strains of bacteria used in the experiment of Figure 7.6 was needed in order for genetic exchange to occur.

Instead, as he discovered, one cell acts as a donor and the other cell acts as a recipient. Hayes proposed that the transfer of genetic material between the strains is mediated by a fertility factor called *F* that the donor cell possesses (F^+) and the recipient cell lacks (F^-). The *F* factor is found only in *E. coli* and is an example of a **plasmid,** a self-replicating genetic element found in the cytoplasm of bacteria. A wide range of different plasmids occur among bacterial species.

In crosses between F^+ and F^- cells of *E. coli,* recombinants for chromosomal (nonplasmid) gene markers are rare, but the *F* factor is transferred frequently during the conjugation process. The *F* factor itself is a circular, double-stranded DNA molecule that is distinct from the host cell chromosome. About a fortieth of the size of the host chromosome, the *F* factor contains a region of DNA, called the **origin,** that is required for its replication within *E. coli* and contains fertility genes, which are required if a cell is to be a donor cell.

When F^+ and F^- cells are mixed, they conjugate. No conjugation occurs between two cells of the same mating type (i.e., two F^+ bacteria or two F^- bacteria). One strand of the *F* factor is nicked at the origin, and DNA replication proceeds from that point. As replication proceeds, a copy of the *F* factor is transferred to the F^- cell as shown in (Figure 7.8a). The complementary strand is synthesized in the F^- cell. There is a polarity of transfer; that is, the origin of replication is always transferred first, followed by the rest of the *F* factor in one particular orientation.

When the complete *F* factor has been transferred, the F^- cell becomes an F^+ cell as a result of the genes specific to the *F* factor. These genes include genes that specify hairlike cell surface components called **F-pili** or sex-pili, which allow the physical union of F^+ and F^- cells to take place. In a population of cells, only some of which are F^+ to start with, the end result of the conjugation of F^+ and F^- cells and the subsequent transfer of the *F* factor is an epidemic conversion of the population to the F^+ state.

Keynote Some E. coli *bacteria possess a plasmid, called the* F *factor, that is required for mating. The* E. coli *cells with the* F *factor are* F^+*, and those without it are* F^-*. The* F^+ *cells (donors) can mate with* F^- *cells (recipients) in a process called conjugation, which leads to the one-way transfer of the* F *factor from donor to recipient during replication of the* F *factor. As a result, both donor and recipient are* F^+*. None of the bacterial chromosome is transferred during* $F^+ \times F^-$ *conjugation.*

High-Frequency Recombination Strains

Although geneticists established that F^- cells can be converted to the F^+ state by the transfer of the F factor, they did not know how conjugation resulted in the formation of recombinants for chromosomal genes. This puzzle was solved when W. Hayes and L. Cavalli-Sforza separately isolated a different type of donor strain from an F^+ stock. In a cross of this donor strain with an F^- strain that differed with respect to chromosomal gene markers, the frequency of recombinants for the chromosomal genes was about a thousand times greater than that normally seen for chromosomal genes in $F^+ \times F^-$ crosses. This new, highly recombinant type of F^+ donor strain is called an ***Hfr,*** or **high-frequency recombination,** strain. The particular strains were later called *Hfr H* and *Hfr C* after their discoverers. The *Hfr* strains are the strains generally used in mapping *E. coli* genes through conjugation.

The *Hfr* strains originate as a result of a rare event in which the F factor actually integrates into the bacterial chromosome by a single crossover event (Figure 7.8b). When the F factor is integrated, it no longer replicates independently; instead, it is replicated as part of the host chromosome. Since the F factor genes are still functional, *Hfr* cells are able to mate with F^- cells. When mating happens, the same events occur as in the $F^+ \times F^-$ matings. The integrated F factor is nicked in one of the two DNA strands and replication begins. During replication part of the factor first moves into the recipient cell. With time some of the bacterial chromosome is also transferred into the recipient; so if there are ge-

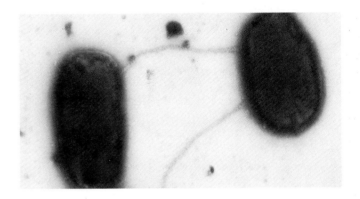

Figure 7.7 Electron micrograph of bacterial conjugation between an F^+ *donor and* F^- *recipient* E. coli *bacterium. Spherical donor specific RNA phage MS-2 can be seen attached to the* F *pilus bridging the two bacteria. Magnification 30,000X.*

netic marker differences between donor and recipient, recombinants can be isolated. In these *Hfr* $\times$ F^- matings the F^- cell almost never acquires the F^+ phenotype. That is, for the recipient cell to become F^+, it must receive a complete copy of the F factor. However, only part of the F factor is transferred at the beginning of conjugation; the rest of the F factor is at the end of the donor chromosome. All of the donor chromosome would have to be transferred in order for a complete functional F factor to be found in the recipient. This is an extremely rare event because all the while the bacteria are conjugating, they are "jiggling" around by Brownian motion and there is a very high probability that the mating pair will break apart long before the second F factor part is transferred.

The low-frequency recombination of chromosomal gene markers in $F^+ \times F^-$ crosses can be understood when we consider that only about 1 in 10,000 F^+ cells in a population become *Hfr* cells by integration of the F factor. The reverse process, excision of the F factor, also occurs spontaneously and at low frequency, producing an F^+ cell from an *Hfr* cell. In excision the F factor loops out of the *Hfr* chromosome, and by a single crossing-over event just like the integration event, a circular host chromosome and a

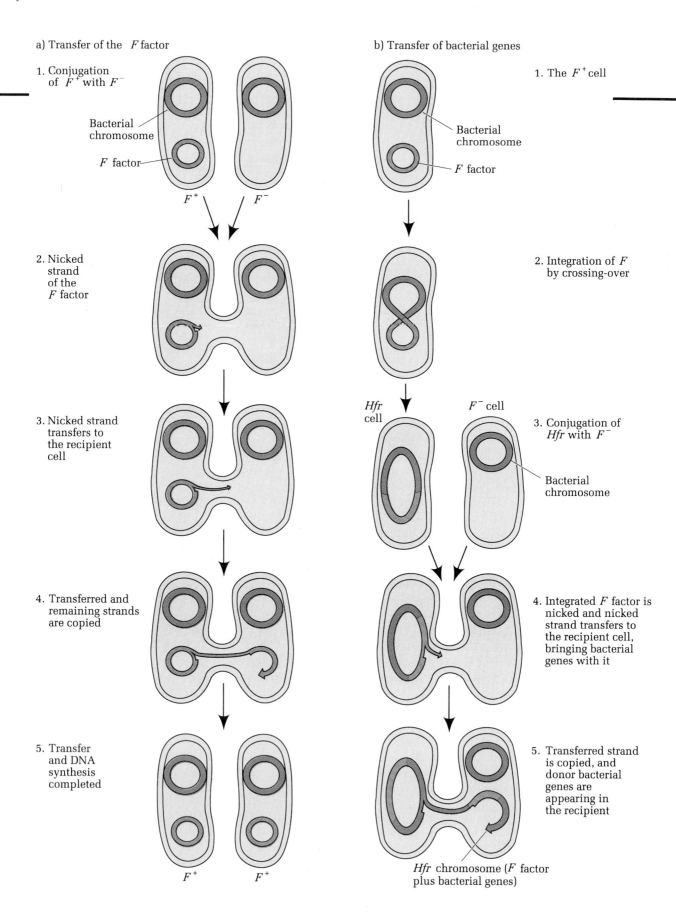

a) Transfer of the *F* factor

1. Conjugation of *F*⁺ with *F*⁻

   Bacterial chromosome

   F factor

   F⁺   *F*⁻

2. Nicked strand of the *F* factor

3. Nicked strand transfers to the recipient cell

4. Transferred and remaining strands are copied

5. Transfer and DNA synthesis completed

   F⁺   *F*⁺

b) Transfer of bacterial genes

1. The *F*⁺ cell

   Bacterial chromosome

   F factor

2. Integration of *F* by crossing-over

   Hfr cell   *F*⁻ cell

3. Conjugation of *Hfr* with *F*⁻

   Bacterial chromosome

4. Integrated *F* factor is nicked and nicked strand transfers to the recipient cell, bringing bacterial genes with it

5. Transferred strand is copied, and donor bacterial genes are appearing in the recipient

   Hfr chromosome (*F* factor plus bacterial genes)

circular extrachromosomal *F* factor are generated.

F' Factors

Occasionally, the excision of the *F* factor from the host chromosome is not precise. As a result, the excised *F* factor may contain a relatively small part of the host chromosome that was adjacent to it when it was integrated. Consider an *E. coli* strain in which the *F* factor has integrated next to the *lac*⁺ region, a set of genes required for the breakdown of lactose (Figure 7.9a). In the excision process, if the looping out is not precise, then the adjacent host chromosomal genes, in this case *lac*⁺, may be included in the loop (Figure 7.9b). Then by single crossover the looped-out DNA will be separated from the host chromosome (Figure 7.9c). The newly generated plasmid contains, in addition to the the *F* factor, a piece of the host chromosome containing *lac*⁺. The *F* factors containing bacterial genes are called *F'* (F prime) factors, and they are named for the genes they have picked up. For example, an *F'* with the *lac* region is called *F' (lac)*.

Cells with *F'* factors can conjugate with *F*⁻ cells since all the *F* factor functions are present. As in regular conjugation, a copy of the *F'* factor is transferred to the *F*⁻ cell, which then becomes *F*⁺ in phenotype since it now has an *F* factor. The recipient also receives a copy of the bacterial gene on the *F* factor. Typically, the recipient has its own copy of that gene, so the resulting cell line will be partially diploid, having two copies of one or a few genes and only one copy of all the others. This particular type of conjugation is called *F* **duction** or **sexduction.**

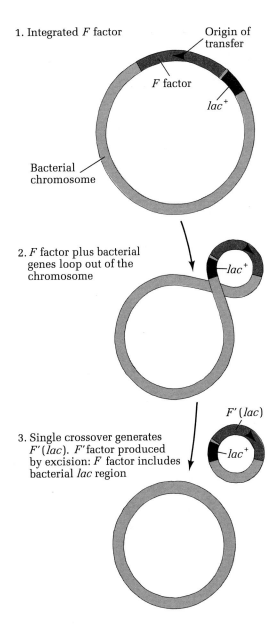

1. Integrated *F* factor

Origin of transfer

F factor

lac⁺

Bacterial chromosome

2. *F* factor plus bacterial genes loop out of the chromosome

lac⁺

3. Single crossover generates *F' (lac)*. *F'* factor produced by excision: *F* factor includes bacterial *lac* region

F' (lac)

lac⁺

Figure 7.9 Production of an F' factor: (1) region of bacterial chromosome into which the F factor has integrated; (2) the F factor looping out incorrectly, so it includes a piece of bacterial chromosome, the lac⁺ genes; (3) excision, in which a single crossover between the looped-out DNA segment and the rest of the bacterial chromosome (i.e., the reverse of integration) results in an F' factor, called F⁺(lac).

Figure 7.8 Transfer of genetic material during conjugation in E. coli: (a) transfer of the F factor from donor to recipient cell during F⁺ × F⁻ matings; (b) production of Hfr strain by integration of F factor and transfer of bacterial genes from donor to recipient cell during Hfr × F⁻ matings.

a) Progressive transfer of donor genes to recipient during *Hfr* × *F* ⁻ conjugation

b) Appearance of donor genetic markers in recipient as a function of time

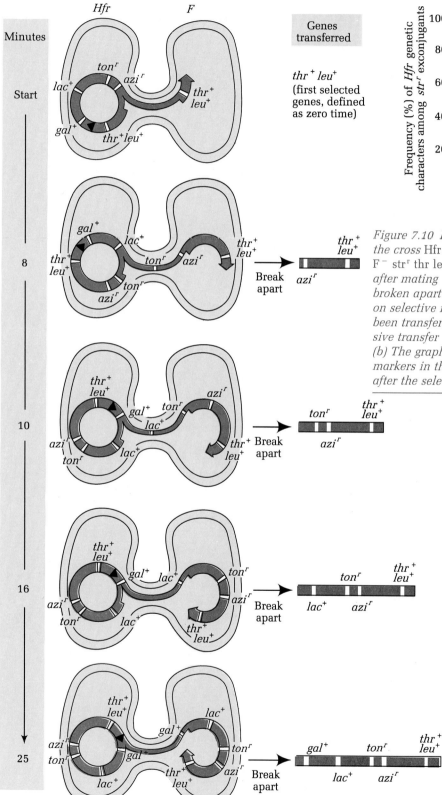

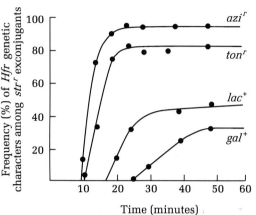

Genes transferred

thr ⁺ *leu* ⁺ (first selected genes, defined as zero time)

Figure 7.10 Interrupted-mating experiment involving the cross Hfr H str ⁺ thr ⁺ leu ⁺ azi ʳ ton ʳ lac ⁺ gal ⁺ × F ⁻ str ʳ thr leu azi ˢ ton ˢ lac gal: *(a) At various times after mating commences, the conjugating pairs are broken apart and the transconjugant cells are plated on selective media to determine which genes have been transferred from the* Hfr *to the* F ⁻· *The progressive transfer of donor genes with time is illustrated. (b) The graph shows the appearance of donor genetic markers in the* F ⁻ *cell as a function of time; that is, after the selected* thr ⁺ *and* leu ⁺ *genes have entered.*

Using Conjugation and Interrupted Mating to Map Bacterial Genes

In the late 1950s François Jacob and Elie Wollman studied the transfer of chromosomal genes from *Hfr* strains to F^- cells. Their experimental design was to make an *Hfr* $\times$ F^- mating and, at various times after conjugation commenced, to break apart the conjugating pairs by mixing the cells in a kitchen blender (Figure 7.10a). The separated cells were then analyzed to determine the relative times at which donor genes entered the recipients and produced genetic recombinants. That is, if gene *a* entered the recipient after 9 minutes, and gene *b* after 15 minutes, genes *a* and *b* would be considered to be 6 minutes apart on the genetic map. In this way the distance between genes was measured as time units (minutes) between the two genes in terms of their appearance as recombinants in the recipient. The genetic map of *E. coli* was constructed in just this way, with map units in minutes, timed from an arbitrarily located origin.

The use of interrupted mating to map bacterial genes may be illustrated by considering one of Jacob and Wollman's experiments, which involved the following cross:

Donor:

   Hfr H str^s thr$^+$ leu$^+$ azi^r ton^r lac$^+$ gal$^+$

Recipient:

   F^- *str^r thr   leu$^+$ azi^s ton^s lac   gal*

The *Hfr H* strain is an *Hfr* strain of *E. coli* distinct from that found by Cavalli-Sforza, and it was isolated by William Hayes, as already discussed. *Hfr H* is prototrophic and is sensitive to the antibiotic streptomycin. The F^- cell carries a streptomycin resistance gene and also a number of mutant genes. These genes cause the F^- to be auxotrophic for threonine (*thr*) and leucine (*leu*), to be sensitive to sodium azide (*azi^s*) and to infection by bacteriophage T1 (*ton^s*), and to be unable to ferment lactose (*lac*) or galactose (*gal*) as a carbon source.

In a conjugation experiment the two cell types are mixed together in a nutrient medium at 37°C, the normal growth temperature for *E. coli*. After a few minutes have passed so that the *Hfr H* and F^- cells may pair, the culture is diluted to prevent the formation of new mating pairs. This procedure ensures some synchrony in the chromosome transfer between cells. Samples are removed from the mating mixture at various times and are then blended to break the pairs apart. These **transconjugants** are plated on a selective medium designed to allow recombinant types to grow and divide while killing both the *Hfr H* and F^- parental types.

For this particular cross the medium contains streptomycin to kill the *Hfr H* and lacks threonine and leucine so that the parental F^- dies. In this cross the threonine (*thr$^+$*) and leucine (*leu$^+$*) genes are the first donor chromosomal genes to be transferred to the F^-, so recombinants formed by the exchange of those genes with the *thr leu* genes of the F^- recipient are produced and will grow on the selective medium. This exchange process occurs in the same way described for transformation; that is, by double crossovers. Appropriate media can be used to test for the appearance of other (unselected) marker genes among the selected transconjugants.

Figure 7.10b shows an example of the results. The transconjugants here are *thr$^+$*, *leu$^+$*, and *str^r*. The first unselected marker gene to be transferred is *azi^r*, and recombinants for this gene are seen at about 9 minutes. The second gene transferred is *ton^r* at 10 minutes, followed by *lac$^+$* at about 17 minutes and *gal$^+$* at about 25 minutes. Figure 7.10b also shows the rates of appearance and the maximum frequencies obtained for each recombinant type. It is apparent that with increasing time after conjugation commences, both the rates of appearance and the maximum frequencies of recombinants decrease. The rate of appearance of recombinants decreases because the population of mating cells become asynchronous with respect to the transfer of genetic material; that is, the rate of transfer from mating couple to mating couple is not constant. The maximum frequency decreases because with time there is an increasing chance that mating pairs will break apart.

In this experiment, then, each gene from an *Hfr* appears in recombinants at a different but

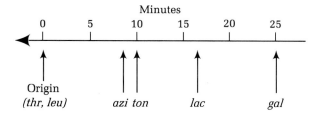

Figure 7.11 Chromosome map of the genes in the interrupted-mating experiment of Figure 7.11. The marker positions represent the time of entry of the genes into the recipient during the experiment. Thus the map distance is given in minutes. The azi marker, for example, appeared about 8 minutes after conjugation began, whereas the gal marker appeared after about 25 minutes.

reproducible time after mating commences, and the time intervals between gene appearances are used as a measure of genetic distance. Also, the maximal frequency of recombinants becomes smaller the later the gene enters the recipient. From such data we can conclude that an *Hfr* chromosome is transferred into the F^- cell in a linear way. The transfer starts at a specific point within the integrated *F* factor called the origin, or *O*. Genes located far from the origin tend not to be transferred because of the high probability that the conjugating pairs will be broken apart by that time. Thus from the experiment described (*Hfr H × F$^-$*) and from the time intervals, the map shown in Figure 7.11 may be constructed.

Circularity of the *E. coli* Map

It would take about 100 minutes to transfer the entire chromosome plus *F* factor from an *Hfr* cell to an *F$^-$* cell. Normally, however, the conjugating pairs break apart well before such a complete transfer. Then how can we construct a complete genetic map of the *E. coli* chromosome?

Further experiments showed that different *Hfr* strains transfer donor genes into the recipient strain with different starting points and polarities of transfer. That is, since the *F* factor can integrate into the bacterial chromosome at a

number of places, and since the *F* factor itself determines the direction of gene transfer, chromosomal transfer may be in different directions and from different places in the map. Figure 7.12a diagrams the order of chromosomal gene transfer for *Hfr* strains *H, 1, 2,* and *3*. In *Hfr H*, the genes are transferred in the order *O–thr–pro–lac–pur–gal* . . .; in *Hfr 1* the order is *O–thr–thi–gly–his–gal* . . .; in *Hfr 2* it is *O–pro–thr–thi–gly–his* . . .; and in *Hfr 3* it is *O–pur–lac–pro–thr–thi.* . . . Moreover, the genetic distance in time units between a particular pair of genes is constant, within experimental error, no matter which *Hfr* strain is used as donor; that is, the genetic distance between *thr* and *pro*, for example, is the same in *H, 1, 2,* and *3*.

From the data in Figure 7.12a, the best way to construct a genetic map of the chromosome is to try to align the genes transferred by each *Hfr* as shown in Figure 7.12b. In view of the overlap of the genes, the simplest map that can be drawn from these data is a circular one, as shown in Figure 7.12c. This was a truly significant finding since all previous genetic maps of eukaryotic chromosomes were linear.

Keynote *The F factor can integrate into the bacterial chromosome by a single crossover event. Strains in which this integration has happened can conjugate with F$^-$ strains, and transfer of the bacterial chromosome occurs. The strains with the integrated F factor are Hfr (high-frequency recombination) strains. In Hfr × F$^-$ matings the chromosome is transferred in a one-way fashion from the Hfr to the F$^-$, beginning at a specific site called the origin (O). The farther a gene is from O, the later it is transferred to the F$^-$, and this result forms the basis for mapping genes by their times of entry into the F$^-$.*

Transduction in Bacteria

Transduction is the process in which genetic material is transferred between bacterial strains by bacterial viruses called bacteriophages (phages, for short), which are viruses that infect

a) Orders of gene transfer

    Hfr strains:

    H   origin – *thr* – *pro* – *lac* – *pur* – *gal* – *his* – *gly* – *thi*

    1   origin – *thr* – *thi* – *gly* – *his* – *gal* – *pur* – *lac* – *pro*

    2   origin – *pro* – *thr* – *thi* – *gly* – *his* – *gal* – *pur* – *lac*

    3   origin – *pur* – *lac* – *pro* – *thr* – *thi* – *gly* – *his* – *gal*

b) Alignments of gene transfer for the *Hfr* strains

    H   *thr* – *pro* – *lac* – *pur* – *gal* – *his* – *gly* – *thi*

    1   *pro* – *lac* – *pur* – *gal* – *his* – *gly* – *thi* – *thr*

    2   *lac* – *pur* – *gal* – *his* – *gly* – *thi* – *thr* – *pro*

    3   *gal* – *his* – *gly* – *thi* – *thr* – *pro* – *lac* – *pur*

c) Circular *E. coli* chromosome map derived from *Hfr* gene transfer data

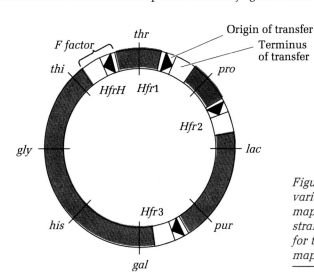

Figure 7.12 Interrupted-mating experiments with a variety of Hfr *strains showing that the* E. coli *linkage map is circular: (a) orders of gene transfer for the* Hfr *strains H, 1, 2, and 3; (b) alignment of gene transfer for the* Hfr *strains; (c) circular* E. coli *chromosome map derived from the* Hfr *gene transfer data.*

bacteria. Before discussing how transduction occurs and can be used to map bacterial genes, we will discuss the genetics of bacteriophages.

Genetics of Some Bacteriophages

Most bacterial strains have phages that are specific for them. For example, *E. coli* is susceptible to infection by DNA-containing phages such as T2, T4, T5, and λ, among others. As a result of extensive study, most of the genes located on the commonly used phages have been identified and mapped. Moreover, the specific functions of many of the genes have been identified.

The structure of all phages is simple. A phage contains its genetic material (either DNA or RNA) in a single chromosome surrounded by a coat of protein molecules. Variation in the num-

ber and organization of the proteins gives the phages their characteristic appearances. Figure 7.13 presents electron micrographs and diagrams of two phages of great genetic significance, T4 (Figure 7.13a) and λ (Figure 7.13b). Phage T4 is one of a series of similar phages (T2, T4, and T6), collectively called T-even phages. It has a number of distinct structural components: a head (which contains the genetic material DNA), a core, a sheath, a base plate, and tail fibers (the latter two structures enable the phage to attach itself to a bacterium). The head and all other structural components consist of proteins.

The T4 and λ phages have different life cycles. T4 is shown in Figure 7.14; the life cycle of λ is illustrated in Figure 7.15. Several stages occur in the T4 life cycle. First, the phage particle attaches to the surface of the bacterial cell.

a) T4 phage

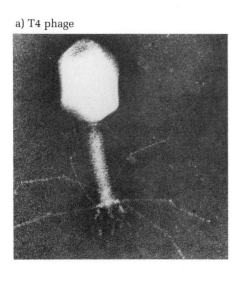
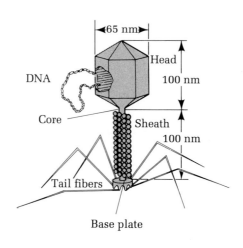

b) λ phage

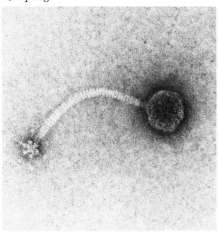
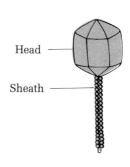

Figure 7.13 Electron micrographs and diagrams of two bacteriophages (1 nm = 10^{-9}m): (a) T4 phage, which is representative of T-even phages; (b) λ phage, a temperate phage.

Then the phage chromosome is injected into the bacterium in a series of events that includes a springlike contraction of the phage sheath. Finally, progeny phages are produced within the bacteria by the action of the phage genes. The progeny phages are eventually released from the bacteria as the cells are broken open (lysed). The suspension of released progeny phages is called a **phage lysate.** In essence, the phage takes over the bacterium, breaks down the bacterial chro-

mosome, and directs its growth and reproductive activities to express the phage's genes and to produce progeny phages. This type of phage life cycle is called the **lytic cycle,** and phages like T4 that always follow that cycle when they infect bacteria are called **virulent phages.**

Phage λ has a structure similar to that of T4. When its DNA is injected into *E. coli,* there are two paths that the phage can follow (Figure 7.15). One is a lytic cycle exactly like that of the

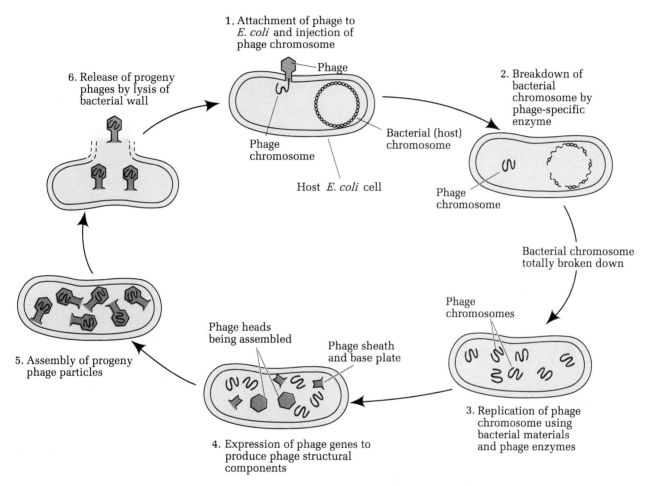

Figure 7.14 Life cycle of a virulent phage, such as T2 or T4.

T phages. The other is the **lysogenic pathway.** In the lysogenic pathway the λ chromosome does not replicate; instead, it inserts itself physically into a specific region of the host cell's chromosome in a way that is essentially the same as *F* factor integration. In this integrated state the phage chromosome is called a **prophage.** Thus every time the host cell chromosome replicates, the integrated λ chromosome is replicated. The bacterium that contains a phage in the prophage state is said to be **lysogenic** for that phage; the phenomenon of the insertion of a phage chromosome into a bacterial chromosome is called **lyso-**

geny. Phages that have a choice between lytic and lysogenic pathways are called *temperate phages.* Occasionally, the mechanism that maintains the prophage state (which is a function of phage genes) breaks down. The lytic cycle is then initiated, and progeny λ phages are released.

Mapping Phage Genes

The same principles used to map eukaryotic genes are used to map phage genes. Crosses are made between phage strains that differ in ge-

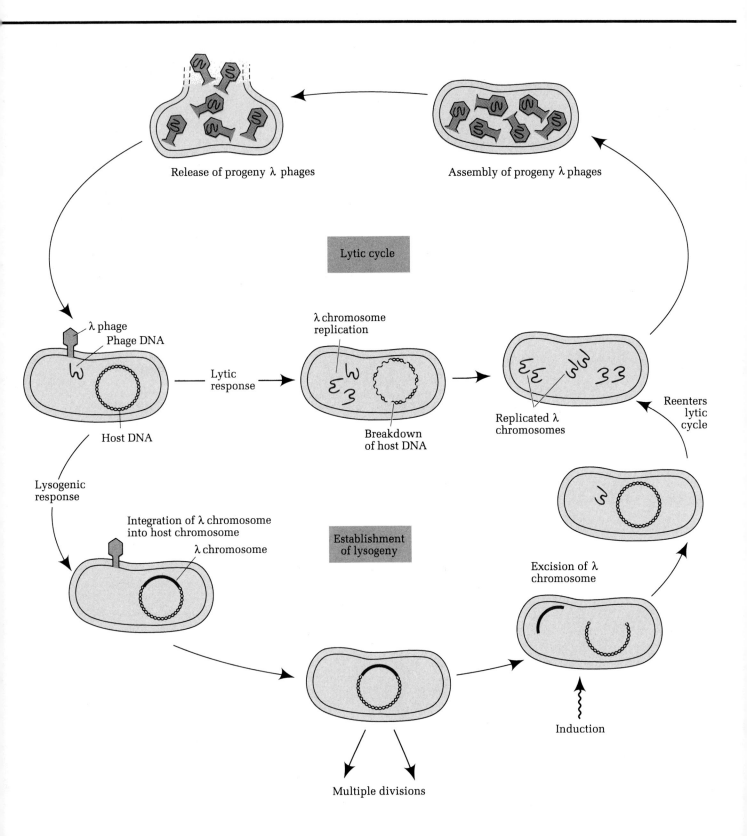

Release of progeny λ phages

Assembly of progeny λ phages

Lytic cycle

λ phage
Phage DNA

λ chromosome
replication

Lytic
response

Host DNA

Breakdown
of host DNA

Replicated λ
chromosomes

Reenters
lytic
cycle

Lysogenic
response

Integration of λ chromosome
into host chromosome

λ chromosome

Establishment
of lysogeny

Excision of λ
chromosome

Induction

Multiple divisions

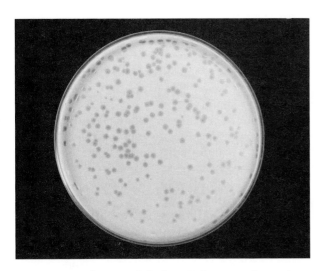

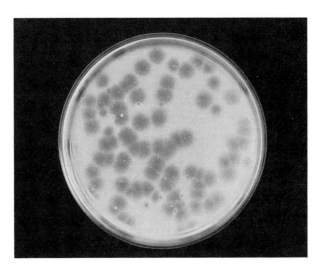

Figure 7.16 Plaques of the E. coli *bacteriophages (a)
T4 and (b)* λ

netic markers, and the proportion of recombinants among the total progeny is determined. Thus the basic procedure of mapping genes in two-, three-, or four-gene crosses involves the mixed infection of bacteria with phages of different genotypes.

In phages this basic experimental design must be adapted to the phage life cycle. First, we must be able to count phage types. We do so by plating a mixture of phages and bacteria on a solid medium. The concentration of bacteria is chosen so that with growth an entire "lawn" of bacteria forms. Phages are present in much lower concentrations. Each phage infects a bacterium and goes through the lytic cycle. The released progeny phages infect neighboring bacteria, and the lytic cycle is repeated. The result is a cleared patch in the lawn of bacteria. Each clearing is called a **plaque,** and one plaque derives from one of the original bacteriophages that was plated. Figure 7.16 shows plaques of the *E. coli* phages T4 and λ: note the clear T4

*Figure 7.15 Life cycle of a temperate phage, such as
λ. When a temperate phage infects a cell, it may go
through either the lytic or lysogenic cycle.*

plaque and the turbid center of the λ plaque.

Second, we must be able to distinguish phage phenotypes easily. Since individual phages are visible only under the electron microscope, mutants that affect phage morphology cannot be used. However, several mutants affect the phage life cycle and hence give rise to differences in the appearances of plaques on a bacterial lawn. For example, one of the earliest studies of phage recombination was done in the late 1940s by Alfred Hershey and R. Rotman. They used strains of T2 that differed in two phenotypes: plaque morphology and host range (i.e., which bacterial strain the phage can lyse). One phage strain, of the genotype $h^+ \, r$, was wild-type for the host range gene (h^+, able to lyse only the *B* strain of *E. coli*) and mutant for the plaque morphology gene (*r*, producing large plaques with distinct borders). The other phage strain, $h \, r^+$, was mutant for the host range (able to lyse both the *B* and the *B/2* strains of *E. coli*) and wild-type for the plaque morphology gene (r^+, producing small plaques with fuzzy borders). In addition, when plated on a lawn containing both the *B* and *B/2* strain, any phage carrying the mutant host range allele *h* (infects both *B* and *B/2*) results in clear plaques, while phages carrying the wild-type h^+ allele result in cloudy plaques, since they can only infect the *B* bacteria, leaving a background cloudiness of uninfected *B/2* bacteria.

To map these two genes, we must make a genetic cross. The cross is achieved by adding both types of phages ($h^+ r$ and $h r^+$) to a culture of bacteria, making sure that a high enough concentration of each phage type is present to ensure that a high proportion of bacteria are infected simultaneously with both phages.

After the life cycle is completed, the progeny

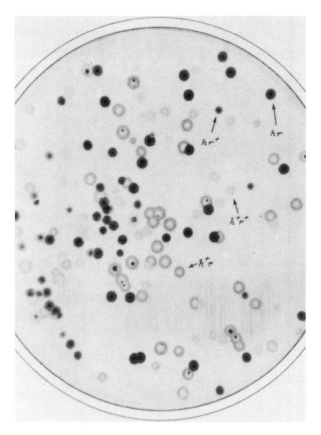

Figure 7.17 Plaques produced by progeny of a cross of T2 strains h r$^+$ × h$^+$ r. *Four plaque phenotypes, representing both parental types and the two recombinants, may be discerned. The parental* h r$^+$ *type produces clear, small plaques, and the parental* h$^+$ r *type produces cloudy, large plaques. The recombinant* h r *type produces a clear, large plaque, and the recombinant* h$^+$ r$^+$ *type produces a cloudy, small plaque.*

phages are plated onto a bacterial lawn containing a mixture of *E. coli* strains *B* and *B/2*. Four plaque phenotypes are found from this experiment (Figure 7.17), two parental types and two recombinant types. The parental type $h r^+$ gives a clear and small plaque; the other parental $h^+ r$ gives a cloudy and large plaque. The reciprocal recombinant types give recombined phenotypes: The $h^+ r^+$ plaques are cloudy and small, and the $h r$ plaques are clear and large.

Since the four progeny types can be readily analyzed on the solid medium, we can calculate the recombination frequency. In this case the frequency is given by

$$\frac{(h^+ r^+) + (h r) \text{ plaques}}{\text{total plaques}} \times 100$$

With the same assumption used in mapping eukaryotic genes, namely, that crossovers occur randomly along the chromosome, the recombination frequency can be considered to be the map distance (in map units) between the phage genes. In addition, experiments have shown that the recombination frequency is the same when the parental strains are $h^+ r^+$ and $h r$, supporting this conclusion. Soon after these first experiments were performed, a number of other T2 genes were mapped in much the same way.

Transduction Mapping of Bacterial Chromosomes

Transduction, the process whereby pieces of bacterial DNA are carried between bacterial strains by a phage (called a **phage vector**), is a useful mechanism for mapping bacterial genes. Of necessity, the useful phages are temperate phages, that is, those phages able to lysogenize the bacteria they infect. Examples are phages P1 and λ for *E. coli* and phage P22 for *Salmonella typhimurium*. Two types of transduction occur: In **generalized transduction** any gene may be transferred between bacteria; in **specialized transduction** only specific genes are transferred.

Generalized transduction. The discovery of generalized transduction is credited to Joshua and Esther Lederberg and Norton Zinder. In 1952 these researchers were testing to see whether a conjugation mechanism occurred in the bacterial species *Salmonella typhimurium,* which causes typhus in the mouse. Their experiment was essentially similar to the one that showed conjugation existed in *E. coli.* They mixed together two multiple auxotrophic strains and looked for the appearance of prototrophs. One strain was *phe$^+$ trp$^+$ tyr$^+$ met his,* which required methionine and histidine, and the other strain was *phe trp tyr met$^+$ his$^+$,* which required phenylalanine, tryptophan, and tyrosine. When they crossed these two auxotrophic strains, they found prototrophic recombinants *phe$^+$ trp$^+$ tyr$^+$ met$^+$ his$^+$* at a low frequency.

When Zinder and the Lederbergs used the U tube apparatus (which showed the existence of conjugation in *E. coli*), they still found prototrophs. This result indicated that recombinants were being produced by a mechanism other than conjugation, since cell-to-cell contact was not necessary. The interpretation was that some noncellular agent was responsible for the formation of recombinants. They called this agent a *filterable agent* since it could pass through a filter with pores small enough to block bacterial movement. In this particular case the filterable agent was identified as the temperate phage P22.

The mechanism for generalized transduction by P22 is as follows: Normally, the P22 phage enters the lysogenic state when it infects a bacterium, and it then exists in the prophage state in these cells (i.e., the phages genes required for taking the phage through the lytic cycle are suppressed). As we have said, the mechanism that maintains this state breaks down occasionally, and the phage genome goes through the lytic cycle and produces progeny phages. During the lytic cycle, the bacterial DNA is degraded, and by mistake, instead of the phage DNA being packaged into the phage head, rarely a piece of bacterial DNA becomes so packaged (Figure 7.18). Since this event is rare, only a very small proportion of the progeny phages carry bacterial genetic material. These phages are called **transducing phages** since they are the vehicles by which genetic material is shuttled between bacteria. The population of phages in the phage lysate, consisting mostly of normal phages but with some transducing phages present ($1/10^5$ particles), can now be used to lysogenize a new population of bacteria. Recipient bacteria that show genetic recombination of the transferred donor genes are called *transductants.*

An experiment like this one is typically designed so that the donor cell type and the recipient cell type differ in genetic markers; then the transduction events can be followed. The strategy behind a transduction experiment is similar to the strategy behind a transformation experiment—and, in fact, behind all genetic-mapping experiments. In the example we are discussing, the donor cell is *thr$^+$* and the recipient cell is *thr.* Prototrophic transductants are then detectable since the cell no longer requires threonine in order to grow. By doing the experiment in this way, researchers are able to pick out the extremely low frequency of transductants among all the cells present by testing for those cells that are able to do something the nontransduced cells cannot, namely, grow on a minimal medium.

The process we've just described is called generalized transduction because the piece of bacterial DNA that the phage erroneously picks up and shuttles to a recipient bacterium is a random piece of the fragmented bacterial chromosome. Thus any genes can be transduced: All that are needed are a temperate phage and bacterial strains carrying different genetic markers. To map genes by using generalized transduction, researchers usually carry out two-factor or three-factor transduction experiments. In a typical transduction experiment, transductants for one of two or more donor markers are selected and these transductants are then analyzed for the presence or absence of other donor markers. Transduction from an *a$^+$b$^+$* donor to an *a b* recipient produces various transductants for *a$^+$* and *b$^+$*, namely *a$^+$b, ab$^+$*, and *a$^+$b$^+$*. If we select for one or other donor markers, linkage information for the two genes can be determined. If we select for *a$^+$* transductants, linkage between genes *a* and *b* is given by:

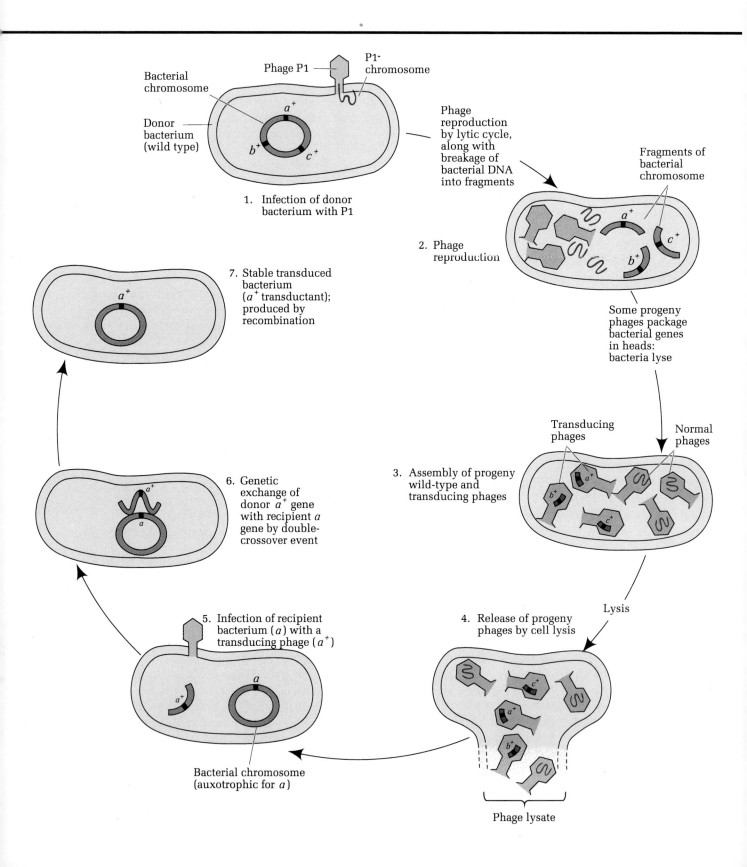

Phage P1

P1-chromosome

Bacterial chromosome

Donor bacterium (wild type)

a^+

b^+ c^+

1. Infection of donor bacterium with P1

Phage reproduction by lytic cycle, along with breakage of bacterial DNA into fragments

Fragments of bacterial chromosome

a^+

c^+

b^+

2. Phage reproduction

Some progeny phages package bacterial genes in heads: bacteria lyse

7. Stable transduced bacterium (a^+ transductant); produced by recombination

a^+

Transducing phages

Normal phages

a^+

b^+

c^+

3. Assembly of progeny wild-type and transducing phages

6. Genetic exchange of donor a^+ gene with recipient a gene by double-crossover event

a^+

a

Lysis

5. Infection of recipient bacterium (a) with a transducing phage (a^+)

a

a^+

4. Release of progeny phages by cell lysis

c^+

a^+

b^+

Bacterial chromosome (auxotrophic for a)

Phage lysate

$$\frac{\text{number of single transductants}}{\text{number of total transductants}} \times 100\% =$$

$$\frac{(a^+ b)}{(a^+ b) + (a^+ b^+)} \times 100\%$$

If we select for b^+ transductants, linkage between a and b is given by

$$\frac{(a \, b^+)}{(a \, b^+) + (a^+ b^+)} \times 100\%$$

Since the production of a transducing phage is a rare event, and since the amount of bacterial DNA the phage can carry is limited by the capacity of the phage head, the chance of either a^+ or b^+ being included individually in the phage head is proportional to the distance between them. If the genes are close enough to be included in the transducing phage, an $a^+ b^+$ transductant can be produced which is called a *cotransductant*. Thus, the **cotransduction** of two or more genes is a good indication that the genes are closely linked.

The map distances between cotransduced genes may be determined by using transduction, and it is by this procedure that linkage maps of bacterial chromosomes have been constructed. For example, consider the mapping of some genes in *E. coli* by using transduction with the temperate phage P1. The donor *E. coli* strain is $leu^+ \, thr^+ \, azi^r$ (able to grow on a minimal medium and resistant to the metabolic poison sodium azide), and the recipient cell is $leu \, thr \, azi^s$ (needs leucine and threonine as supplements in the medium and sensitive to azide). The P1

Figure 7.18 Generalized transduction between strains of E. coli: *(1) wild-type donor cell of* E. coli *infected with the temperate bacteriophage P1; (2) the host cell DNA is broken up during the lytic cycle; (3) during assembly of progeny phages, some pieces of the bacterial chromosome are incorporated into some of the progeny phages to produce transducing phages; (4) following cell lysis, a low frequency of transducing phages is found in the phage lysate; (5) the transducing phage infecting an auxotrophic recipient bacterium; (6) a double-crossover event results in the exchange of the donor* a$^+$ *gene with the recipient mutant* a *gene; (7) stable* a$^+$ *transductant, with all descendants of that cell having the same genotype.*

Table 7.1 Transduction Data for Deducing Gene Order

Selected Marker	Unselected Markers
leu^+	$50\% = azi^r$
	$2\% = thr^+$
thr^+	$3\% = leu^+$
	$0\% = azi^r$

phages are grown on the bacterial donor cells, and the phage lysate is used to transduce the recipient bacterial cells. Transductants are selected for any one of the donor markers and then analyzed for the presence of the other two (in this case) unselected markers. Typical data from such an experiment are shown in Table 7.1.

Consider the leu^+ selected transductants. Of these transductants 50 percent are also transduced for azi^r and 2 percent for thr^+. For the thr^+ transductants 3 percent are leu^+ and 0 percent are azi^r. The simplest interpretation is that the *leu* gene is closer to the *thr* gene than the *thr* gene is to the *azi* gene. The order of genes then is

The transductants are produced by crossing-over between the piece of donor bacterial chromosome brought in by the infecting phage and the homologous region on the recipient bacterial chromosome. The infected donor DNA "finds" the region of the recipient chromosome to which it is homologous, and the exchange of parts is accomplished by double (or other even-numbered) crossovers. Map distance can be obtained from these two- or three-factor transduction experiments by determining relative cotransduction frequencies.

Specialized transduction. Some temperate bacteriophages can transduce only certain sections of the bacterial chromosome, in contrast to generalized transducing phages, which can carry any part of the bacterial chromosome. An example of such a **specialized transducing phage** is λ, which infects *E. coli*. Figure 7.19 shows the

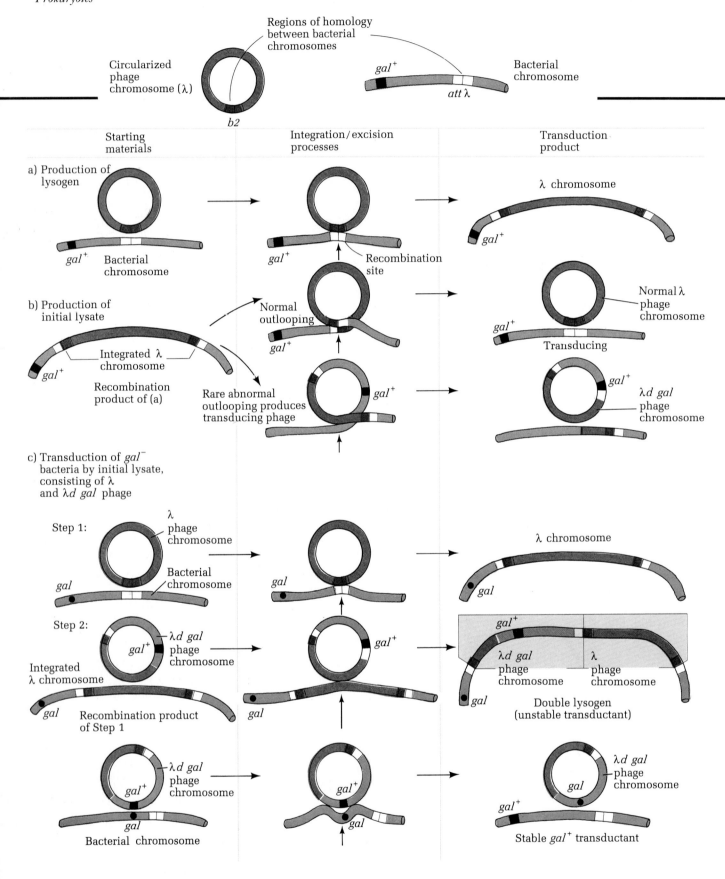

Circularized
phage
chromosome (λ)

b2

Regions of homology
between bacterial
chromosomes

gal⁺

att λ

Bacterial
chromosome

Starting
materials

Integration/excision
processes

Transduction
product

a) Production of
lysogen

gal⁺   Bacterial
chromosome

gal⁺

λ chromosome

gal⁺

Recombination
site

b) Production of
initial lysate

Integrated λ
chromosome

Recombination
product of (a)

Normal
outlooping

gal⁺

Rare abnormal
outlooping produces
transducing phage

gal⁺

gal⁺

Normal λ
phage
chromosome

Transducing

gal⁺

λ*d gal*
phage
chromosome

c) Transduction of *gal⁻*
bacteria by initial lysate,
consisting of λ
and λ*d gal* phage

Step 1:

λ
phage
chromosome

gal   Bacterial
chromosome

gal

λ chromosome

gal

Step 2:

Integrated
λ chromosome

gal   Recombination product
of Step 1

λ*d gal*
phage
chromosome

gal⁺

gal

gal⁺

gal

gal⁺

λ*d gal*
phage
chromosome

λ
phage
chromosome

gal   Double lysogen
(unstable transductant)

λ*d gal*
phage
chromosome

gal⁺

gal   Bacterial chromosome

gal⁺

gal

gal⁺

gal

λ*d gal*
phage
chromosome

Stable *gal⁺* transductant

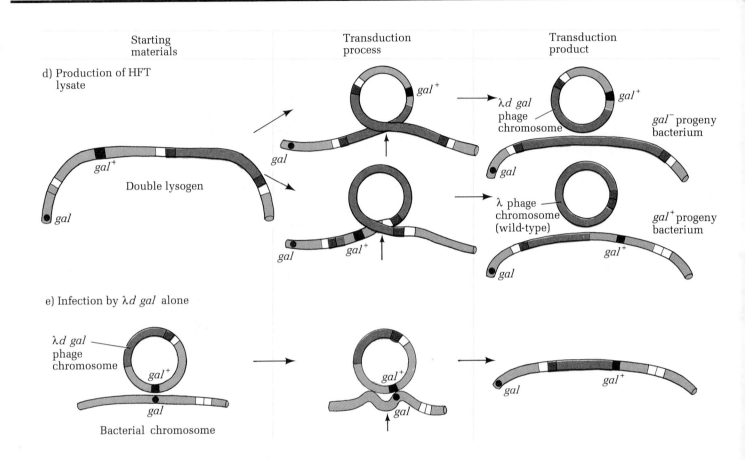

Starting materials | Transduction process | Transduction product

d) Production of HFT lysate

Double lysogen

gal⁺

gal

gal⁺

gal

λ*d gal* phage chromosome

gal⁺

gal⁻ progeny bacterium

λ phage chromosome (wild-type)

gal⁺ progeny bacterium

gal

gal⁺

gal

e) Infection by λ*d gal* alone

λ*d gal* phage chromosome

gal⁺

gal

Bacterial chromosome

gal⁺

gal

gal

gal⁺

Figure 7.19 Specialized transduction by bacteriophage λ; (a) production of a lysogenic bacterial strain by crossing-over in the region of homology between the circular bacterial chromosome (att λ) and the circularized phage chromosome (b2); (b) production of initial lysate, in which induction of the lysogenic bacterium causes out-looping, which produces many normal λ and a low frequency of a transducing phage, λd gal, which carries the bacterial gal gene as a result of incorrect out-looping; (c) transduction of gal⁻ bacteria by the initial lysate, producing either unstable transductants by integration of both λ and λd gal (resulting in a double lysogen) or stable transductants (single lysogens) by crossing-over around the gal region; (d) the unstable double lysogen containing λ and λd gal producing about equal numbers of λd gal and wild-type λ phages, thus giving rise to a high-frequency transducing (HFT) lysate; (e) low concentrations of phage from the HFT lysate used to infect bacteria so that a cell can be infected by λd gal alone.

series of events by which phage λ can transduce bacterial genes.

The life cycle of λ was described earlier (see Figure 7.15). The genetic material circularizes soon after it is injected into the cell, and then it integrates into the bacterial chromosome at a specific site, producing what is called a lysogen (Figure 7.19a). That site is called *att* λ, and it is homologous with a site called *b2* in the λ DNA. By a single-crossover event the λ chromosome integrates. In the integrated state the phage, now called a prophage, is maintained by the action of specific gene products, which prevent expression of the λ genes essential for the lytic cycle. The point of integration for this particular phage in the *E. coli* chromosome is between the *gal* (galactose fermentation) and the *bio* (biotin biosynthesis) genes.

The particular *E. coli* strain that λ lysogenizes is *E. coli K12*, and when it contains the λ prophage, it is designated *E. coli K12*(λ). Let us focus just on the *gal* gene and assume that the particular *K12* strain that λ lysogenized is *gal*$^+$, that is, it can ferment galactose as a carbon source. This phenotype is readily detectable by plating the cells on a solid medium containing galactose as a carbon source and containing a dye that changes color in response to the products of galactose fermentation. The *gal*$^+$ colonies are pink, while *gal* colonies are white on this medium. If we induce the prophage—that is, reverse the inhibition of phage functions—then the lytic cycle is initiated. Induction may be accomplished by any procedure that destroys the protein repressor molecules responsible for the inhibition (ultraviolet light irradiation is one commonly used procedure).

After the lytic cycle is initiated, the phage chromosome loops out, and a separate circular λ chromosome is generated by a single-crossover event at the *att* λ/*b2* sites (Figure 7.19b). This sequence of events is exactly the reverse of the integration process and parallels the events of *F* factor integration and excision in conjugation. In most cases the excision of the phage chromosome is precise so that the complete λ chromosome and only the λ chromosome is produced. In rare cases the excision events are not precise so that by crossing-over between the phage chromosome and the bacterial chromosome at sites other than the homologous recognition sites, an abnormal circular DNA product results. In the case diagramed a piece of λ chromosome has been left in the bacterial chromosome, while a piece of bacterial chromosome, including the *gal*$^+$ gene, has been added to the rest of the λ chromosome. When a bacterial gene (or genes) is included in a progeny phage, we have a transducing phage, called, in this case, λ*d gal*. The *d* stands for "defective" since not all phage genes are present, and the *gal* indicates that the bacterial host cell *gal* gene has been picked up. This event is similar to *F'* production by defective excision of the *F* factor.

Since the abnormal looping-out phenomenon is a rare event, the phage lysate produced from the initial infection diagramed contains mostly normal phages and a few transducing phages (1/10^5). The phage lysate can be used to infect nonlysogenic bacteria (bacteria that do not contain phages) that are *gal*. Nonlysogenic bacteria must be used because once a bacterium contains a prophage, it is immune to infection by a second phage of the same type. The phage lysate contains mostly wild-type phages and relatively few (about one in a million) *gal*$^+$ transducing phages. Because of the small proportion of transducing phages, the lysate is called a *low-frequency transducing (LFT) lysate*.

Infection of the *gal* bacterial cells with the LFT lysate produces two types of transductants (Figure 7.19c). One type is unstable: The wild-type λ integrates at its normal *att* λ site, and then the λ*d gal*$^+$ phage integrates by a crossing-over event within the common λ sequences. In this case both types of phages are integrated in the bacterial chromosome, and the bacterium is heterozygous *gal*$^+$/*gal* and hence can ferment galactose. This transductant is unstable because by induction phage growth can be initiated. (This event will be described a little later.) The second type is stable: These transductants are produced when only a λ*d gal*$^+$ phage is present (Figure 7.19c). The *gal*$^+$ gene carried by the phage may be exchanged for the bacterial *gal* gene by a double-crossover event. Such a transductant is stable because the bacterial chromosome contains only one type of *gal* gene and no phage genes are integrated.

A *high-frequency transducing (HFT) lysate* may be obtained with this system (Figure 7.19d). The starting point is the doubly lysogenic, unstable transductant of Figure 7.19c in which both a wild-type λ and a λ*d gal*$^+$ genome are integrated side by side in the bacterial chromosome. If this strain is induced, λ*d gal*$^+$ phage are produced at high frequency. When the HFT lysate is used to infect new nonlysogenic *gal* bacteria so that only one phage infects each bacterium (Figure 7.19e), the defective-phage genome is integrated by a single crossover in the *gal* gene region. This integration produces a *gal*$^+$/*gal* transductant that is stable since it lacks sufficient λ genes to lyse the cells. Thus by manipulating the number of phages that infect a bacterium, and using appropriate selection events, we can detect specialized transduction and cause marker genes to be transduced at high frequency. Then we can make a detailed analysis of the genes transduced in this way.

Keynote *Transduction is the process by which bacteriophages mediate the transfer of genetic information from one bacterium (the donor) to another (the recipient). The capacity of the phage particle is limited, so the amount of DNA transferred is usually less than 1 percent of that in the bacterial chromosome. In generalized transduction any bacterial gene can be accidentally incorporated into the transducing phage during the phage life cycle and subsequently transferred to a recipient bacterium. Specialized transduction is mediated by temperate phages (such as λ) whose prophages associate with only one site of the bacterial chromosome. In this case the transducing phage is generated by abnormal excision of the prophage from the host chromosome so that the prophage includes both bacterial and phage genes.*

Concluding Remarks

In this chapter, we have seen how genetic mapping can be accomplished in prokaryotes such as bacteria and bacteriophages. Essentially the same experimental strategy is used for all gene mapping; that is, genetic material is exchanged between strains differing in genetic markers and recombinants are detected and counted. Depending on the particular bacterium, the mechanism of gene transfer may be transformation, conjugation, or transduction. Bacteriophage chromosomes may also be mapped by infecting bacteria simultaneously with two phage strains.

As a summary, two linkage maps of prokaryotes of genetic significance are shown: *E. coli* (Figure 7.20) and bacteriophage T4 (Figure 7.21).

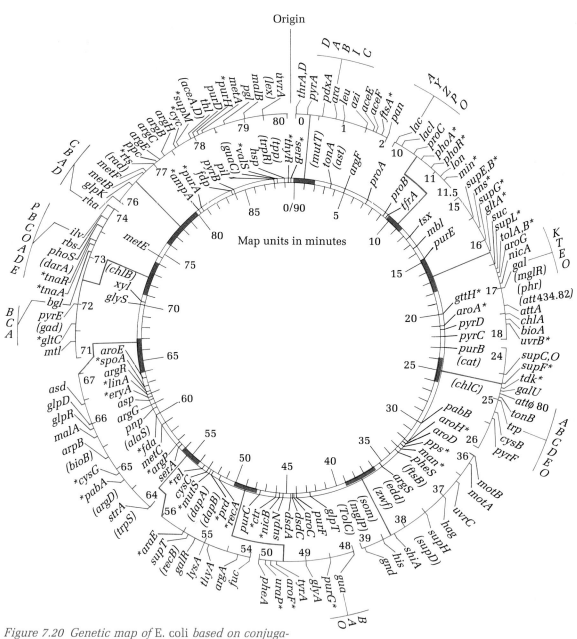

Figure 7.20 Genetic map of E. coli *based on conjugation experiments. Units are in minutes timed from an arbitrary origin at 12 o'clock.*

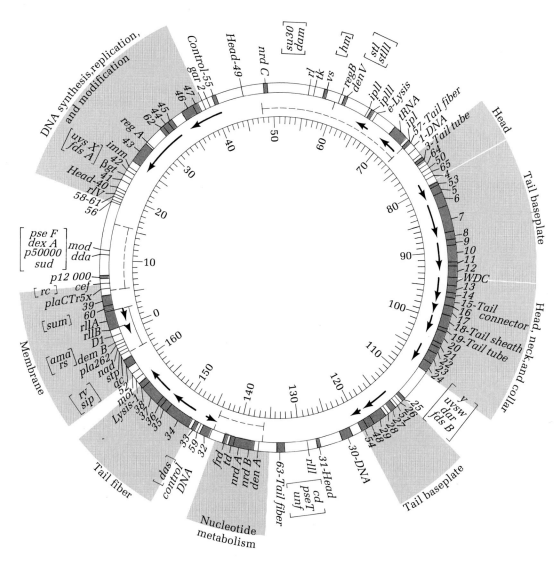

Figure 7.21 Genetic map of the bacteriophage T4.
Note the clustering of genes of related function
around the perimeter of the map.

*Analytical
Approaches
for Solving
Genetics
Problems*

Q.1 In *E. coli* the following *Hfr* strains donate the markers shown in the order given:

Hfr Strain	Order of Gene Transfer
1	*G E B D N A*
2	*P Y L G E B*
3	*X T J F P Y*
4	*B E G L Y P*

All the *Hfr* strains were derived from the same F^+ strain. What is the order of genes in the original F^+ chromosome?

A.1 This question is an exercise in piecing together various segments of the circumference of a circle. The best approach is to draw a circle and label it with the genes transferred from one *Hfr* and then to see which of the other *Hfr*'s transfers an overlapping set. For example, *Hfr 1* transfers *E*, then *B*, then *D*, and so on; and *Hfr 4* transfers *B*, then *E*, and so forth. Now we can juxtapose the two sets of genes transferred by the two *Hfr*'s and deduce that the polarities of transfer are opposite:

Hfr 1	$\quad$ *G E B D N A*
Hfr 4	*P Y L G E B*

Extending this reasoning to the other *Hfr*'s, we can draw an unambiguous map (see the figure below), with the arrowheads indicating the order of transfer.

$\quad$ The same logic would be used if the question gave the relative time units of entry of each of the genes. In that case we would expect that the time "distance" between any two genes would be approximately the same regardless of the order of transfer or how far the genes were from the origin.

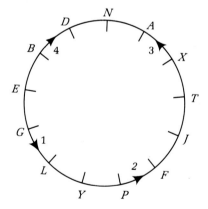

7.1 Distinguish among F^-, F^+, F', and *Hfr* strains of *E. coli*.

*7.2 In $F^+ \times F^-$ crosses the F^- recipient is converted to a donor with very high frequency. However, it is rare for a recipient to become a donor in *Hfr* $\times$ F^- crosses. Explain why.

7.3 Distinguish between the lysogenic and lytic cycles.

7.4 Distinguish between generalized and specialized transduction.

*7.5 If an *E. coli* auxotroph A could only grow on a medium containing thymine, and an auxotroph B could only grow on a medium containing leucine, how would you test whether DNA from A could transform B?

*7.6 With the technique of interrupted mating four *Hfr* strains were tested for the sequence in which they transmitted a number of different genes to an F^- strain. Each *Hfr* strain was found to transmit its genes in a unique sequence, as shown in the accompanying table (only the first six genes transmitted were scored for each strain).

Order of Transmission	*Hfr* Strain			
	1	2	3	4
First	O	R	E	O
	F	H	M	G
	B	M	H	X
	A	E	R	C
	E	A	C	R
Last	M	B	X	H

What is the gene sequence in the original strain from which these *Hfr* strains derive? Indicate on your diagram the origin and polarity of each of the four *Hfr*'s.

7.7 When *Hfr* donors conjugate with F^- recipients that are lysogenic for phage λ, the recipients (the transconjugant zygotes) usually survive. However, when *Hfr* donors that are lysogenic for λ conjugate with F^- cells that are nonlysogens, the transconjugant zygotes produced from matings that have lasted at least 100 minutes usually lyse, releasing mature λ phage particles. This event is called zygotic induction of λ.

a. Explain zygotic induction.
b. Explain how the locus of the integrated λ prophage can be determined.

7.8 At time zero an *Hfr* strain (strain 1) was mixed with an F^- strain, and at various times after mixing, samples were removed and agitated to separate conjugating cells. The cross may be written as:

$$Hfr\ 1: a^+\ b^+\ c^+\ d^+\ e^+\ f^+\ g^+\ h^+\ str^s$$
$$F^-: a^-\ b^-\ c^-\ d^-\ e^-\ f^-\ g^-\ h^-\ str^r$$

(No order is implied in listing the markers.)

The samples were then plated onto selective media to measure the frequency of h^+ str^r recombinants that had received certain genes from the *Hfr* cell. The graph of the number of recombinants against time is shown in the figure below.

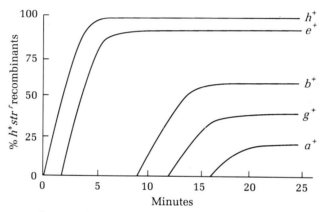

a. Indicate whether each of the following statements is true or false.
   i. All F^- cells that received a^+ from the *Hfr* in the chromosome transfer process must also have received b^+.
   ii. The order of gene transfer from *Hfr* to F^- was a^+ (first), then g^+, then b^+, then e^+, then h^+.
   iii. Most e^+ str^r recombinants are likely to be *Hfr* cells.
   iv. None of the b^+ str^r recombinants plated at 15 minutes are also a^+.
b. Draw a linear map of the *Hfr* chromosome indicating the (i) point of insertion, or origin; (ii) the order of the genes a^+, b^+, e^+, g^+, and h^+; (iii) the shortest distance between consecutive genes on the chromosomes.

7.9 Construct a map from the following two factor phage cross data (show map distance):

	% recombination
$r_1 \times r_2$	0.10
$r_1 \times r_3$	0.05
$r_1 \times r_4$	0.19
$r_2 \times r_3$	0.15
$r_2 \times r_4$	0.10
$r_3 \times r_4$	0.23

*7.10 The following two-factor crosses were made to analyze the genetic linkage between four genes in phage λ: *c, mi, s,* and *co:*

Parents	Progeny
$c +$ × $+ mi$	1213 $c +$, 1205 $+ mi$, 84 $+ +$, 75 $c mi$
$c +$ × $+ s$	566 $c +$, 808 $+ s$, 19 $+ +$, 20 $c s$
$co +$ × $+ mi$	5162 $co +$, 6510 $+ mi$, 311 $+ +$, 341 $co mi$
$mi +$ × $+ s$	502 $mi +$, 647 $+ s$, 65 $+ +$, 56 $mi s$

Construct a genetic map of the four genes.

7.11 Three gene loci in T4 that affect plaque morphology in easily distinguishable ways are *r*(rapid lysis), *m*(minute), and *tu*(turbid). A culture of *E. coli* is mixedly infected with two types of phage: *r m tu* and $r^+ m^+ tu^+$. Progeny phage are collected and the following genotype classes are found:

r^+	m^+	tu^+	3,729
r^+	m^+	tu	965
r^+	m	tu^+	520
r	m^+	tu^+	172
r^+	m	tu	162
r	m^+	tu	474
r	m	tu^+	853
r	m	tu	3,467
			10,342

Construct a map of the three genes. What is the coefficient of coincidence, and what does the value suggest?

*7.12 A stock of T4 phage is diluted by a factor of 10^{-8} and 0.1 mL of it is mixed with 0.1 mL of 10^8 *E. coli B*/mL and 2.5 mL melted agar, and poured on the surface of an agar petri dish. The next day 20 plaques are visible. What is the concentration of T4 phages in the original T4 stock?

7.13 Wild-type phage T4 grows on both *E. coli B* and *E. coli K12(λ)*, producing turbid plaques. The *rII* mutants of T4 grow on *E. coli B,* producing clear plaques, but do not grow on *E. coli K12(λ)*. This host range property permits the detection of a very low number of r^+ phages among a large number of *rII* phages. With this sensitive system it is possible to determine the genetic distance between two mutations within the same gene, in this case the *rII* locus. Suppose *E. coli B* is mixedly infected with *rIIx* and *rIIy,* two separate mutants in the *rII* locus. Suitable dilutions of progeny phages are plated on *E. coli B* and *E. coli K12(λ)*. A 0.1-mL sample of a thousandfold dilution plated on *E. coli B* showed 672 plaques. A 0.2-mL sample of undiluted phage plated on *E. coli K12(λ)* showed 470 turbid plaques. What is the genetic distance between the two *rII* mutations?

*7.14 The *rII* mutants of bacteriophage T4 grow in *E. coli B* but not in *E. coli K12(λ)*. The *E. coli* strain *B* is doubly infected with two *rII* mutants. A 6×10^7 dilution of the lysate is plated on *E. coli B*. A 2×10^5 dilution is plated on *E. coli K12(λ)*. Twelve plaques appeared on strain *K12(λ)* and 16 on strain *B*. Calculate the amount of recombination between these two mutants.

7.15 Wild-type (r^+) strains of T4 produce turbid plaques, whereas *rII* mutant strains produce larger, clearer plaques. Five *rII* mutations *(a–e)* in the *A* cistron of the *rII* region of T4 give the following percentages of wild-type recombinants in two-point crosses:

$a \times b$	0.2 percent
$a \times c$	0.9 percent
$a \times d$	0.4 percent
$b \times c$	0.7 percent
$e \times a$	0.3 percent
$e \times d$	0.7 percent
$e \times c$	1.2 percent
$e \times b$	0.5 percent
$b \times d$	0.2 percent
$d \times c$	0.5 percent

What is the order of the mutational sites and what are the map distances between the sites?

7.16 In a transduction cross with the donor $aceF^+$ dhl and the recipient $aceF$ dhl^+, selection was for $aceF^+$. Of the unselected markers, 88 percent were dhl and 12 percent were dhl^+. When the donor was $aceF^+$ leu and the recipient was $aceF$ leu^+, selection was for $aceF^+$. Of the unselected markers, 34 percent were leu and 66 percent were leu^+. Is dhl or leu closer to $aceF$?

*7.17 In P1 transduction the donor is $cysB^+$ $trpE$ and the recipient is $cysB$ $trpE^+$. Selection was for cys^+. Of the $cysB^+$ colonies, 37 percent were $trpE$ as well. When the same experiment was done with a $cysB^+$ $trpB$ donor and a $cysB$ $trpB^+$ recipient, 53 percent of the $cysB^+$ colonies were $trpB^+$ as well. Is $trpE$ or $trpB$ closer to $cysB$?

7.18 Consider the following data with P1 transduction:

Donor		Recipient		Selected Marker	Unselected Marker	%
$aroA$	$pyrD^+$	$aroA^+$	$pyrD$	$pyrD^+$	$aroA$	5
$aroA^+$	$cmlB$	$aroA$	$cmlB^+$	$aroA^+$	$cmlB$	26
$cmlB$	$pyrD^+$	$cmlB^+$	$pyrD$	$pyrD^+$	$cmlB$	54

Choose the correct order:

a. $aroA - cmlB - pyrD$
b. $aroA - pyrD - cmlB$
c. $cmlB - aroA - pyrD$

7.19 Order the mutants trp, $pyrF$, and qts on the basis of the following three-factor transduction cross:

Donor	trp^+ pyr^+ qts
Recipient	trp pyr qts^+
Selected Marker	trp^+

Unselected Markers	Number
pyr^+ qts^+	22
pyr^+ qts	10
pyr qts^+	68
pyr qts	0

*7.20 Order *cheA, cheB, eda,* and *supD* from the following data:

Markers	% cotransduction
cheA-eda	15
cheA-supD	5
cheB-eda	28
cheB-supD	2.7
eda-supD	0

8

The Structure of Genetic Material

In the previous seven chapters we have learned much about how traits are inherited from one generation to another, how variations arise (through recombination), and how the genes governing certain traits have been located on given chromosomes (by establishing linkage and by gene mapping). In all our discussions, however, we have taken as given the existence and the function of the genetic material—genes and the chromosomes on which they are located. In the next several chapters we will explore the molecular structure and function of genetic material—either DNA (deoxyribonucleic acid) or RNA (ribonucleic acid)—and we will examine in detail the mechanisms by which genetic information is transmitted from generation to generation. In addition, we will see exactly what the genetic message is, and we will learn how DNA directs the manufacture of proteins, including enzymes.

In this chapter we begin our study of the molecular aspects of genetics by discussing the qualities that the genetic material possesses and the evidence that DNA and RNA are genetic material. In addition, we will examine the structure of DNA and RNA molecules.

The Nature of Genetic Material: DNA and RNA

Long before DNA and RNA were proved to carry genetic information, geneticists recognized that special molecules must fulfill that function. They postulated that the material responsible for inheritable traits would have to have three principal characteristics:

1. It must contain *all the information* for an organism's cell structure, function, development, and reproduction in a stable form.
2. It must *replicate accurately* so that progeny cells have the same genetic information as the parental cell.
3. It must be capable of *variation*. Without variation (such as through mutation and recombination), organisms would be incapable of

change and adaptation, and evolution could not occur.

From experiments in the mid to late 1800s and the early 1900s, scientists suspected that genetic material was composed of protein. Chromosomes were first seen under the microscope in the latter part of the nineteenth century. Chemical analysis over the first 40 years of this century revealed that chromosomes contained a unique molecular constituent, deoxyribonucleic acid (DNA). It was not understood, however, that DNA was the chemical constituent of genes. Rather, DNA was considered to serve as a molecular framework for a theoretical class of special proteins that carried genetic information. In the 1940s, G. Beadle and E. Tatum obtained evidence from experiments with the orange bread mold, *Neurospora crassa,* that genes function by controlling the synthesis of specific enzymes (this was called the "one gene—one enzyme" hypothesis; see Chapter 14, pp. 451–456, for a detailed discussion), thereby proving a hypothesis proposed by A. Garrod in 1902 (see Chapter 14, pp. 446–447). Evidence gathered from experiments in the first half of this century showed unequivocally that the genetic material of living organisms consisted of one of two types of nucleic acids: DNA or RNA.

The Discovery of DNA as Genetic Material

One of the first studies (done in 1928) to suggest that DNA is the genetic material involved the bacterium *Diplococcus pneumoniae* (also called pneumococcus) (Figure 8.1). Two strains of pneumococcus are known. The smooth (*S*) strain is infectious (virulent) and results in the death of the infected animal. Each bacterial cell of the *S* type is surrounded by a polysaccharide coat that gives the strain its infectious properties and results in the smooth, shiny appearance of *S* colonies. The rough (*R*) strain is noninfectious (avirulent) since *R* cells lack the polysaccharide coat; colonies of this strain are nonshiny. Figure 8.2 shows the colony phenotypes of the two strains.

The *S* strains can be classified further into *IIS*

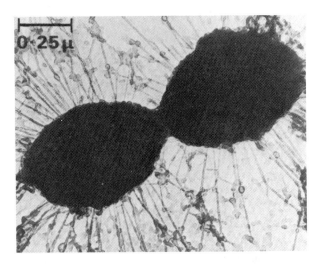

Figure 8.1 EM of the bacterium Diplococcus pneumoniae. *(The cell is in the process of dividing.)*

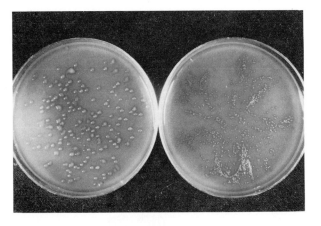

Figure 8.2 Colonies of smooth (S) *and rough* (R) *strains of* Diplococcus pneumoniae.

and *IIS*, which have clearly defined differences in the chemical composition of the polysaccharide coat. The differences between *IIS* and *IIIS* are genetically determined, as are the *S* and *R* states. In about one *S* cell out of 10^7 there will be a mutational change from *S* to *R*. Rarely will the *R* colony that results give rise to an *S* cell by a second mutation. The *S* colony that develops has the same capsule type as the *S* strain from which the *R* mutant was originally derived. That

is, if the original was *IIS*, then the reversion of the *R* strain that derives from it will also be *IIS*, not *IIIS*.

In 1928 F. Griffith injected mice with *R* bacteria, produced by mutation of type *IIS* bacteria (Figure 8.3). Genetically these bacteria lacked the ability to make the polysaccharide coat and, *if* a coat could be made at all, had the genes for making a type II polysaccharide coat. The *R* bacteria did not affect the mice, and after a while the bacteria disappeared from the animals' bloodstreams. Griffith also injected mice with living type *IIIS* bacteria. Those animals died, and living type *IIIS* bacteria could be isolated from their blood. If the type *IIIS* bacteria were heat-killed before injecting them, however, the mice survived. These two experiments showed that the bacteria had to be alive and had to pos-

Figure 8.3 Transformation experiment of Griffith. Mice injected with type IIIS *pneumococcus died, whereas mice injected with either type* IIR *or heat-killed type* IIIS *bacteria survived. An injected mixture of living type* IIR *and heat-killed type* IIIS *bacteria, however, caused the mice to die.*

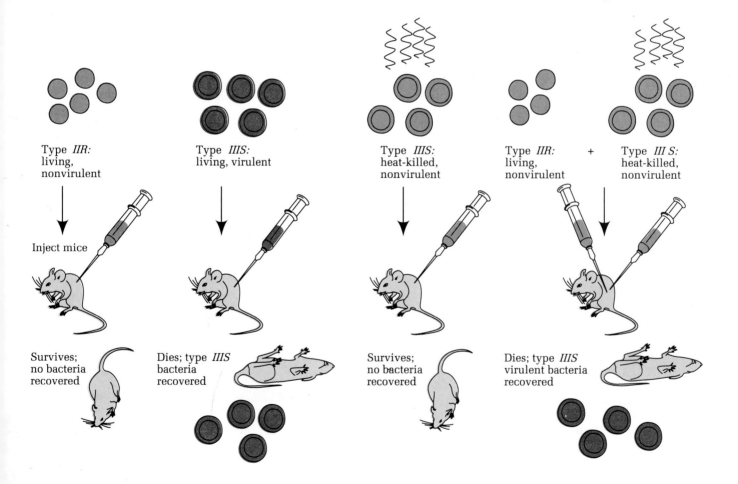

sess the polysaccharide coat in order for them to be infectious.

Lastly, Griffith infected mice with a mixture of living *R* bacteria (derived from *IIS*) and heat-killed type *IIIS* bacteria. In this case the mice died. Moreover, living *S* bacteria were present in the blood. These living cells were all of type *IIIS* and therefore could not have arisen by mutation of the *R* bacteria since that would have produced *IIS* colonies. Rather, Griffith concluded that some *R* bacteria had somehow been transformed into smooth, infectious type *IIIS* cells by interaction with the dead type *IIIS* cells. The transformed *IIIS* cells retained their infectious properties and capsule type in successive generations, indicating that the transformation was stable. Griffith believed the unknown agent responsible for the change in genetic material was a protein. He referred to the agent as the *transforming principle.*

From a modern molecular perspective, the data indicate that the genetic material of the virulent bacteria was not destroyed by the heat treatment and that it was released from the heat-killed cells and entered the living avirulent cells where recombination occurred to generate a virulent transformant (Figure 8.4). Subsequent research has verified this genetic hypothesis. That is, the *S* gene codes for a key enzyme needed for the synthesis of the carbohydrate-containing capsule. The *R* allele of the gene produces no active enzyme and the capsule is not made.

The nature of the transforming principle. In 1944, O. T. Avery, C. M. MacLeod, and M. McCarty showed that the transforming principle was not protein but was instead a molecule of DNA. They used a system in which the transformation of bacteria from *R* to *S* type was done in

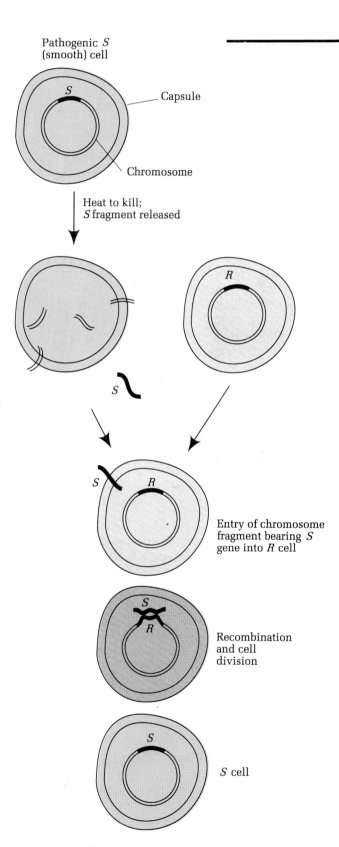

Figure 8.4 Transformation of a genetic characteristic of a bacterial cell (Diplococcus pneumoniae) *by addition of heat-killed cells of a genetically different strain. In the example, an* R *cell receives a chromosome fragment containing the* S *gene. Since most* R *cells receive other chromosomal fragments, the efficiency of transformation for a given gene is usually less than 1 percent.*

the test tube. In their experiments they lysed type *IIIS* cells and separated the lysate into the various cellular macromolecular components such as lipids, polysaccharides, proteins, and the nucleic acids RNA and DNA (Figure 8.5). They tested each component to see whether it was the transforming principle by checking whether or not it could transform living *R* bacteria derived from *IIS*. The nucleic acids (at this point not separated into the two types) were the only components of the type *IIIS* cells that could transform the *R* cells into *IIIS*.

Avery and his colleagues then used specific **nucleases**—enzymes that degrade nucleic acids—to determine whether DNA or RNA was the transforming principle. When they treated the nucleic acids with **ribonuclease** (RNase), which degrades RNA, the transforming activity was still present. However, when they used **deoxyribonuclease** (DNase), which degrades DNA, no transformation resulted. Thus according to the results, DNA was the genetic material. Although Avery's work was important, it was criticized because the nucleic acids isolated from the bacteria were not completely pure and, in fact, contained some protein contamination.

Constancy of chromosomal DNA amount. Support for the hypothesis that DNA was the genetic material came from the results of experiments that showed that DNA was located almost exclusively in the nucleus of eukaryotic cells and never in cell locations where chromosomes, the carriers of genetic information, were absent. In addition, A. Mirsky and H. Ris in 1949 demonstrated by cytochemical analysis that the amount of DNA per diploid set of chromosomes was constant for a given organism. Moreover, the diploid DNA amount was shown to be equal to twice the amount of DNA found in the haploid sperm. Different organisms were shown to have different amounts of DNA. We now know that the DNA content of certain cells of an organism can vary depending on the tissue of origin. In general, though, with the exception of spontaneous chromosome loss or breakage, the DNA content is usually a multiple of the DNA content of a zygote cell. For example, the DNA

content of human cells may be from two to four times that of a diploid cell, and the DNA content of the root nodule cells of leguminous plants such as the pea is characteristically double that of the cells of the rest of the plant.

Three other aspects of DNA that strongly supported its role as the genetic material were: (1) DNA is metabolically stable, that is, it is not rapidly synthesized and degraded like many other cellular molecules. (2) The amount of DNA per cell is approximately related to the complexity of the organism. Thus cells of higher organisms contain considerably more DNA than do bacterial cells. (3) The amount of DNA in gametes is approximately one-half that found in 2N cells of the same species.

The Hershey and Chase bacteriophage experiments. In 1953, Alfred D. Hershey and Martha Chase reported experimental results conclusively showing that DNA is genetic material. They were studying the replication of bacteriophage T2 to see which phage components were needed to complete the life cycle. (See Figure 7.15 for the life cycle of this bacteriophage.) It was known at the time that T2 infected *Escherichia coli* by first attaching itself with its tail fibers and then injecting into the bacterium some genetic material that somehow controlled the production of progeny T2 particles. Ultimately, the bacterium lysed, releasing 100–200 phages that could infect other bacteria. However, the nature of T2's genetic material was not known.

Hershey and Chase grew cells of *E. coli* in media containing either a radioactive isotope of phosphorus (^{32}P) or a radioactive isotope of sulfur (^{35}S) (Figure 8.6a). They used these isotopes since DNA contains phosphorus but no sulfur, and protein contains sulfur but no phosphorus. They infected the bacteria with T2 and collected the progeny phages that were produced. At this point, then, Hershey and Chase had two batches of T2: One had the proteins radioactively la-

Figure 8.5 Chemical method used in the original isolation of a chemically pure transforming agent.

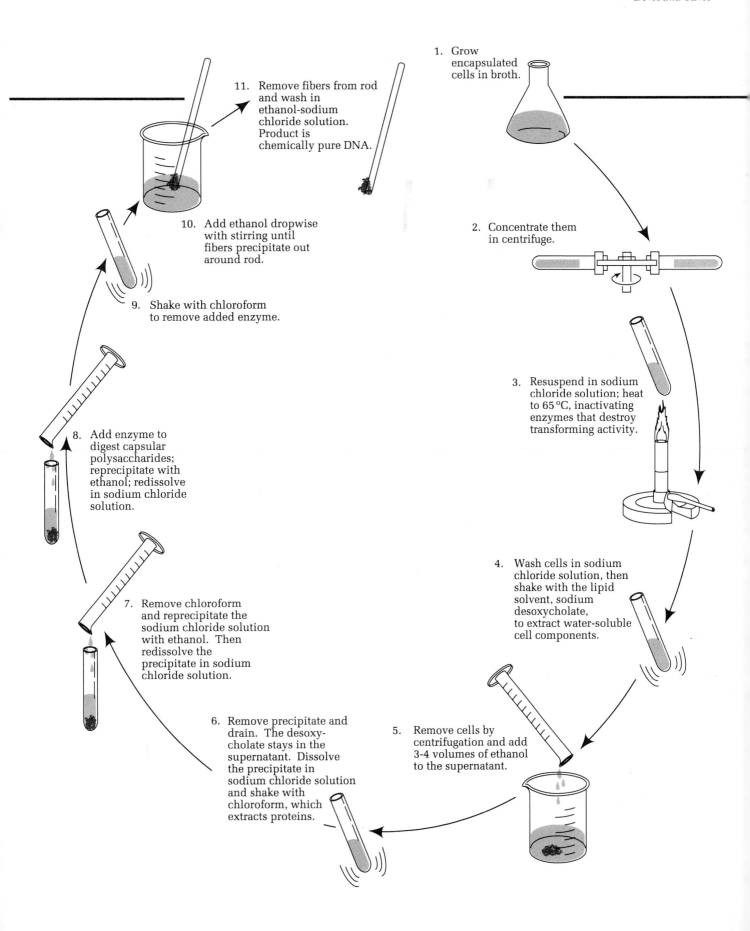

1. Grow encapsulated cells in broth.

2. Concentrate them in centrifuge.

3. Resuspend in sodium chloride solution; heat to 65 °C, inactivating enzymes that destroy transforming activity.

4. Wash cells in sodium chloride solution, then shake with the lipid solvent, sodium desoxycholate, to extract water-soluble cell components.

5. Remove cells by centrifugation and add 3-4 volumes of ethanol to the supernatant.

6. Remove precipitate and drain. The desoxycholate stays in the supernatant. Dissolve the precipitate in sodium chloride solution and shake with chloroform, which extracts proteins.

7. Remove chloroform and reprecipitate the sodium chloride solution with ethanol. Then redissolve the precipitate in sodium chloride solution.

8. Add enzyme to digest capsular polysaccharides; reprecipitate with ethanol; redissolve in sodium chloride solution.

9. Shake with chloroform to remove added enzyme.

10. Add ethanol dropwise with stirring until fibers precipitate out around rod.

11. Remove fibers from rod and wash in ethanol-sodium chloride solution. Product is chemically pure DNA.

a) Preparation of radioactively labeled T2 bacteriophage

(1)

Protein
coat

DNA

T2 phage

Infect *E. coli*
and grow in
^{32}P-containing
medium

E. coli

Lysis

Progeny
phages with
^{32}P-labeled DNA

(2)

Infect *E. coli*
and grow in
^{35}S-containing
medium

Lysis

Progeny
phages with
^{35}S-labeled
protein

b) Experiment that showed DNA to be the genetic material of T2

(1)

^{32}P DNA

Blend
briefly

Phage ghosts

Radioactivity
recovered in
host and
passed on to
phage progeny

(2)

^{35}S protein

Blend
briefly

Radioactivity
recovered in
phage ghosts

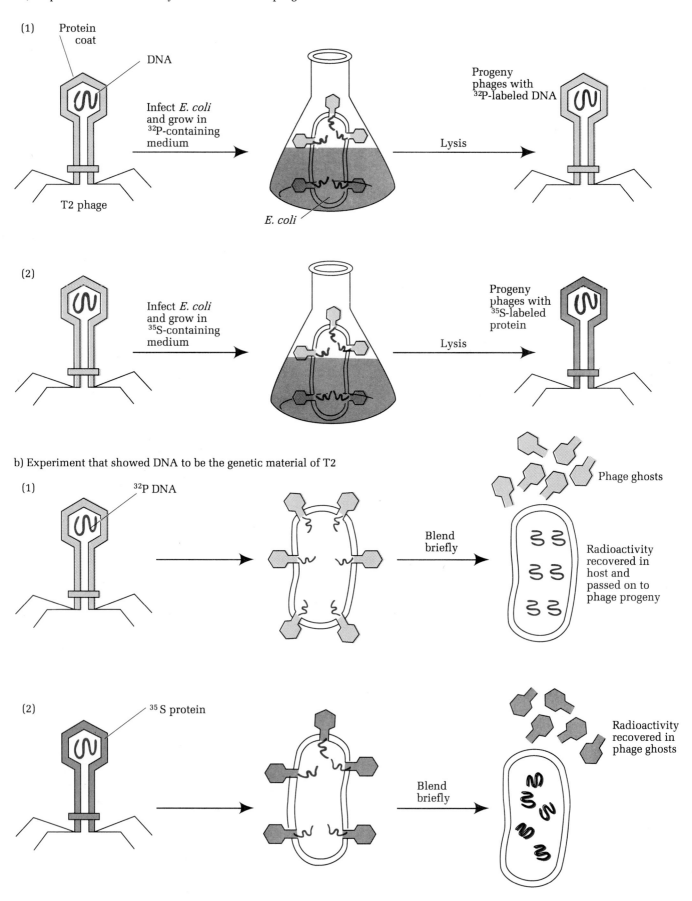

beled with ^{35}S, and the other had the DNA labeled with ^{32}P.

Since they knew that each T2 phage consisted of approximately half DNA and half protein, they knew that one of these two classes of molecules must be the genetic material. To determine which class it was, they next infected unlabeled *E. coli* with the two types of radioactively labeled T2 and obtained the interesting results shown in Figure 8.6b. When the infecting phage was ^{32}P-labeled, most of the radioactivity could be found within the bacteria soon after infection. Very little could be found in protein parts of the phage (the phage ghosts) released from the cell surface by mixing the cells in a kitchen blender. After lysis, some of the ^{32}P was found in the progeny phages. In contrast, after infecting *E. coli* with ^{35}S-labeled T2, virtually none of the radioactivity appeared within the cell, and none was found in the progeny phage particles. Most of the radioactivity could be found in the phage ghosts released after treating the cultures in the kitchen blender. Since genes serve as the blueprint for making the progeny virus particles, it was also presumed that the blueprint must get into the bacterial cell in order for new phage particles to be built. Therefore, since it was DNA and not protein that entered the cell, as evidenced by the presence of ^{32}P, Hershey and Chase reasoned that DNA must be the material responsible for the function and reproduction of phage T2. The protein, they hypothesized, provided a structural framework to contain the DNA and the specialized structures required to inject the DNA into the bacterial cell.

Similarly, other experiments showed that the DNA of mouse polyoma virus is the genetic material. In these experiments DNA was purified from the virus particles (which consist of DNA and proteins) and used to infect mouse cells in culture. The DNA was able to direct the synthesis of thousands of new polyoma particles, thereby proving that DNA is the genetic material and indicating that the proteins function as a protective framework for the DNA in its movement between cells.

The Discovery of RNA as Genetic Material

Most of the organisms discussed in this book (such as humans, *Drosophila,* yeast, *E. coli,* and phage T2) have DNA as their genetic material. Some bacterial viruses and some animal and plant viruses, however, have RNA as their genetic material. The following classical experiment showed that RNA is the genetic material of the *tobacco mosaic virus* (TMV).

Like phage T2, TMV contains two chemical components, in this case RNA and protein in a spiral (helical) configuration. The protein surrounds the RNA core, protects it from attack by nucleases (Figure 8.7), and along with the RNA core, functions in the infection of the plant cells. Many varieties of TMV are known; they differ in the plants they infect and by the extent to which they affect these plants.

In 1956, A. Gierer and G. Schramm showed that when tobacco plants were inoculated with the purified RNA of TMV (i.e., without the protein coat), they developed typical virus-induced

Figure 8.6 Hershey-Chase experiment: (a) the production of T2 phages either with (1) ^{32}P-labeled DNA or with (2) ^{35}S-labeled protein; (b) the experimental evidence showing that DNA is the genetic material in T2: (1) The ^{32}P is found within the bacteria and appears in progeny phages, (2) while the ^{35}S is not found within the bacteria and is released with the phage ghosts.

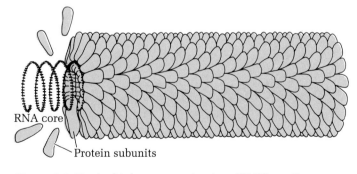

RNA core

Protein subunits

Figure 8.7 Typical tobacco mosaic virus (TMV) particle. The helical RNA core is surrounded by a helical arrangement of protein subunits.

lesions. This result indicated strongly that RNA was the genetic material of TMV, a conclusion that was supported by the observation that no lesions were produced when the RNA had been degraded by treatment with ribonuclease and then injected into the plant.

In an experiment in 1957, H. Fraenkel-Conrat and B. Singer confirmed Gierer and Schramm's conclusions. They isolated the RNA and protein components of two distinct TMV strains and reconstituted the RNA of one type with the protein of the other type, and vice versa (Figure 8.8). They then infected the tobacco leaves with the two hybrid viruses. The progeny viruses isolated from the resulting lesions were of the type specified by the RNA and not by the protein. Thus they conclusively showed that RNA is the genetic material of TMV.

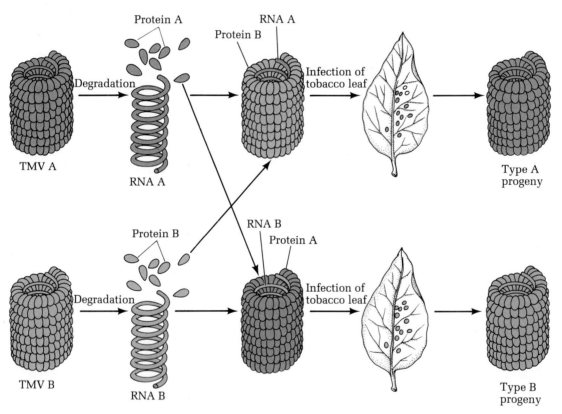

Figure 8.8 Demonstration that RNA is the genetic material in tobacco mosaic virus (TMV). Hybrid particles were made from the protein subunits of one TMV strain and the RNA of a different TMV strain. Tobacco leaves were infected with the reconstituted hybrid viruses, and the progeny viruses isolated from the resulting leaf lesions were analyzed. The progeny viruses always had protein subunits specified by the RNA component; that is, the character of the protein coat cannot be transmitted from the hybrid particles in their progeny.

Keynote Genetic material must contain all the information for the cell structure and function of an organism and must replicate accurately so that progeny cells have the same genetic information as the parental cell. In addition, genetic material must be capable of variation, one of the bases for evolutionary change. A series of experiments proved that the genetic material of organisms consists of one of two types of nucleic acids, DNA or RNA. Of the two, DNA is most common.

The Chemical Composition of DNA and RNA

Both DNA and RNA are **macromolecules,** which means that they have a molecular weight of at least a few thousand daltons (1 dalton is equivalent to a twelfth of the mass of the carbon 12 atom, or 1.67×10^{-24} g). In this respect they differ markedly from many other molecules important to cell function, such as sugars and amino acids, which weigh from a hundred to a few hundred daltons.

Both DNA and RNA are polymeric molecules made up of monomeric molecules called **nucleotides.** Each nucleotide consists of three distinct parts: (1) a pentose (5-carbon) sugar, (2) a **nitrogenous** (nitrogen-containing) **base,** and (3) a **phosphate group** (PO_4^{2-}). Because the phosphate groups are acidic, these macromolecules are called nucleic acids.

For RNA the pentose sugar is **ribose,** and for DNA the sugar is **deoxyribose** (Figure 8.9). The two sugars differ by the chemical groups attached to the 2′ carbon: a hydroxyl group (OH) in ribose, a hydrogen group (H) in deoxyribose. (The carbon atoms in the pentose sugars are numbered 1′ to 5′ to distinguish them from the carbon and nitrogen atoms in the rings of the bases.)

The nitrogenous bases fall into two classes, the **purines** and the **pyrimidines.** In DNA the most commonly found purines are **adenine** (A) and **guanine** (G), and the most common pyrimidines are **thymine** (T) and **cytosine** (C). The RNA molecule also contains adenine, guanine, and cytosine, but thymine is replaced by **uracil** (U). The chemical structures of these five bases are given in Figure 8.10. Note that thymine contains a methyl group (CH_3) not found in uracil.

In DNA and RNA, bases are always attached to the 1′ carbon of the pentose sugar by a covalent bond. The purine bases are bonded at the 9 nitrogen, while the pyrimidines bond at the 1 nitrogen. The phosphate group (PO_4^{2-}) is attached to the 5′ carbon of the sugar in both DNA and RNA. Examples of a DNA nucleotide (a **deoxyribonucleotide**) and an RNA nucleotide (a

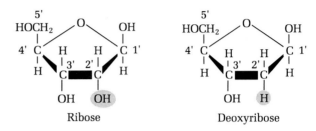

Figure 8.9 *Structures of ribose and deoxyribose, the pentose sugars of RNA and DNA, respectively.*

ribonucleotide) are shown in Figure 8.11a. For future discussions, remember that a *nucleotide,* the basic building block of the DNA and RNA molecules, consists of the sugar, a base, and a phosphate group. The sugar plus base only is called a nucleoside, so a nucleotide is also called a *nucleoside* phosphate. A complete listing of the names for the bases, nucleosides, and nucleotides is presented in Table 8.1 (page 266).

Because DNA contains the pentose sugar deoxyribose and the pyrimidine thymine, while RNA contains the pentose sugar ribose and the pyrimidine uracil, these two nucleic acids have different chemical and biological properties. The presence of the 2′ OH, for example, means that RNA can be degraded with alkali, while DNA is resistant to that treatment. Also, the cellular enzymes that catalyze nucleic acid synthesis (polymerases) and nucleic acid degradation (nucleases) are usually DNA-specific or RNA-specific. On the practical side the differences between the two molecules permit them to be separated and purified for study in the laboratory.

In DNA and RNA the nucleotides are linked together to form polynucleotide chains by a covalent bond between the phosphate group (which is attached to the 5′ carbon of the sugar ring) of one nucleotide and the 3′ carbon of the pentose sugar of another nucleotide. These 5′–3′ phosphate linkages are called **phosphodiester bonds.** A short polynucleotide chain is diagramed in Figure 8.11b. (An RNA polynucleotide is similar.) The phosphodiester bonds are relatively strong, and as a consequence, the repeated sugar-phosphate-sugar-phosphate backbone of DNA and RNA is a stable structure.

To understand how a polynucleotide chain is synthesized (which we will study in a later chapter), we must be aware of one more feature of the chain: It has polarity; that is, the 5′ phosphate (5′–P) end and the 3′–OH end of the polynucleotide chain can be clearly identified, as they are in Figure 8.11b.

Keynote *DNA and RNA usually occur in nature as macromolecules composed of smaller building block molecules called nucleotides. Each nucleotide consists of a 5-carbon sugar (deoxyribose in DNA, ribose in RNA) to which is attached a phosphate group and one of four nitrogenous bases, adenine, guanine, cytosine, and thymine (in DNA) or uracil (in RNA).*

The Physical Structure of DNA: The Double Helix

In 1953 James D. Watson and Francis H. C. Crick

published a paper in which they proposed a model for the physical and chemical structure of the DNA molecule. According to their model, most DNA consists of two polynucleotide chains wound around each other in a right-handed (clockwise) helix. In generating their model, Watson and Crick used three main pieces of evidence:

1. The DNA molecule was known to be composed of bases, sugars, and phosphate groups linked together as a **polynucleotide** (deoxyribonucleotide) chain.
2. Erwin Chargaff had hydrolyzed the DNA of a number of organisms by chemical treatment and had quantified the purines and pyrimidines released. His studies showed that in all the DNAs the amount of the purines was equal to the amount of the pyrimidines. More importantly, the amount of adenine (A) was equal to that of thymine (T), and the amount

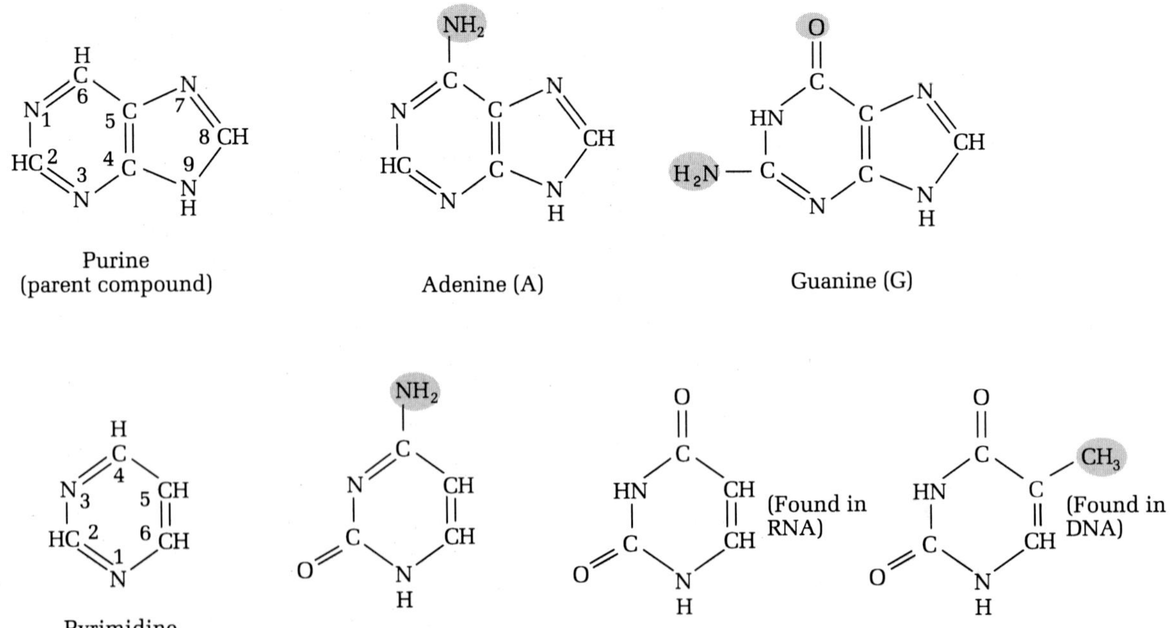

Figure 8.10 Structures of the nitrogenous bases in DNA and RNA. The parent compounds shown at the top, purine and pyrimidine, give the numbering system for these ring compounds.

a) DNA and RNA nucleotides

b) DNA polynucleotide chain

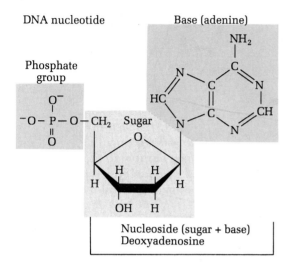

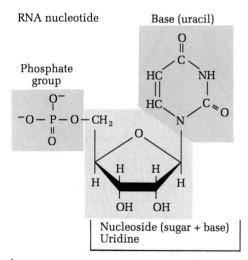

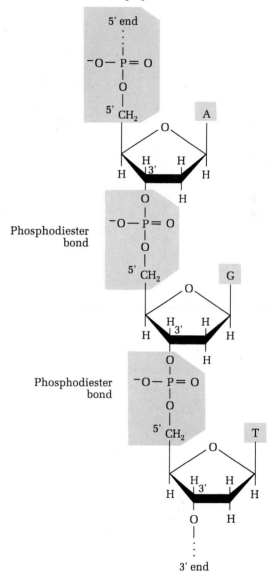

Figure 8.11 Chemical structure of DNA and RNA: (a) basic structures of DNA and RNA nucleosides (sugar plus base) and nucleotides (sugar plus base plus phosphate group), the basic building blocks of DNA and RNA molecules; (b) a segment of a polynucleotide chain, in this case a single strand of DNA. The deoxyribose sugars are linked by phosphodiester bonds between the 3' carbon of one sugar and the 5' carbon of the next sugar.

Table 8.1 Names of the Base, Nucleoside, and Nucleotide Components Found in DNA and RNA

	Base: Purines (Pu)		Base: Pyrimidines (Py)		
	Adenine (A)	Guanine (G)	Cytosine (C)	Thymine (T) (deoxyribose only)	Uracil (U) (ribose only)
Nucleoside: pentose sugar [(1) deoxyribose or (2) ribose] + base	1. Deoxyadenosine (dA)	1. Deoxyguanosine (dG)	1. Deoxycytidine (dC)	1. Thymidine (dT)	
	2. Adenosine (rA)	2. Guanosine (rG)	2. Cytidine (rC)		2. Uridine (rU)
Nucleotide: pentose sugar [(1) or (2)] + base + phosphate group	1. Deoxyadenylic acid or deoxyadenosine monophosphate (dAMP)	1. Deoxyguanylic acid or deoxyguanosine monophosphate (dGMP)	1. Deoxycytidylic acid or deoxycytidine monophosphate (dCMP)	1. Thymidylic acid or thymidine monophosphate (TMP)	
	2. Adenylic acid or adenosine monophosphate (AMP)	2. Guanylic acid or guanosine monophosphate (GMP)	2. Cytidylic acid or cytidine monophosphate (CMP)		2. Uridylic acid or uridine monophosphate (UMP)

of guanine (G) was equal to that of cytosine (C). These equivalences have become known as Chargaff's rules. In comparisons of DNAs from different organisms, the A/T and G/C ratios are always the same, although the (A + T)/(G + C) ratio (typically presented as %GC) varies (see Table 8.2).

3. Rosalind Franklin, working with Maurice H. F. Wilkins, studied isolated fibers of DNA by using the X-ray diffraction technique, a procedure in which a beam of parallel X rays is directed on a regular, repeating array of atoms. The beam is diffracted by the atoms in a pattern that is characteristic of the atomic weight and spatial arrangement of the atoms. The diffracted X rays are recorded on a photographic plate. By analyzing the photograph, Franklin could propose the spatial arrangement of the atoms that gave rise to the diffraction pattern. The analysis of X-ray diffraction patterns is extremely complicated since the X rays cannot be focused. Thus given diffraction patterns can usually be interpreted in more than one way, and models built of the analyzed molecules may not be accurate. Moreover, since the experiments usually use molecules in a crystalline or fiber formation, the structures proposed may not precisely describe the molecules as they exist in the aqueous environment of the cell.

The diffraction patterns obtained by directing X rays along the length of drawn-out fibers of DNA indicated that the molecule is organized in a highly ordered, helical structure. An example of DNA's X-ray diffraction pattern and the method by which it was obtained are illustrated in Figure 8.12. Franklin interpreted the data to mean that DNA consisted of two polynucleotide chains wound around each other.

Watson and Crick considered all the evidence just described and began to build three-dimensional models for the structure of DNA. The model they devised, which fit all the known data on the composition of the DNA molecule,

Table 8.2 Base Compositions of DNAs from Various Organisms

DNA Origin	Percentage of Base in DNA				Ratios		
	A	T	G	C	A/T	G/C	(A + T)/(G + C)
Human (sperm)	31.0	31.5	19.1	18.4	0.98	1.03	1.67
Corn (*Zea*)	25.6	25.3	24.5	24.6	1.01	1.00	1.04
Drosophila	27.3	27.6	22.5	22.5	0.99	1.00	1.22
Euglena nucleus	22.6	24.4	27.7	25.8	0.93	1.07	0.88
Escherichia coli	26.1	23.9	24.9	25.1	1.09	0.99	1.00

was the now-famous double-helix model for DNA. Figure 8.13a shows a three-dimensional model of the DNA molecule, and Figure 8.13b is a diagram of the DNA molecule showing the sugar-phosphate backbone and base pairs in a stylized way. As indicated in these representations, the DNA molecule consists of two polynucleotide chains wound around each other in a right-handed double helix; that is, viewed on end, the two strands wind around each other in a clockwise (right-handed) fashion. The sugar-phosphate backbones are on the outsides of the double helix, and the bases are oriented toward the central axis and arranged like a stack of pennies. The two bonds that attach a base pair to its sugar rings are not directly opposite each other. Because of this the two sugar-phosphate backbones of the double helix are not equally spaced along the helical axis, and the grooves that form between the backbones are not of equal size, resulting in what are called the *major groove* and the *minor groove* (see Figure 8.13).

As an important consequence of their model, Watson and Crick realized that for the structure of DNA to be stable, the adenine of one chain must pair with the thymine of the other chain, and guanine must pair with cytosine. Pairings between a purine and a pyrimidine are the only ones possible because two purines would take up too much space and two pyrimidines would take up too little space to permit the formation of a regular helix. Mispairing between purines and pyrimidines (e.g., A with C, or T with G) is inherently unstable owing to failure of proper hydrogen bond formation. The bonds involved

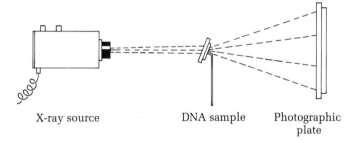

X-ray source   DNA sample   Photographic plate

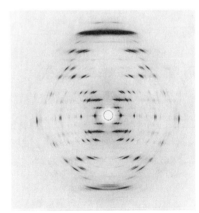

X-ray diffraction pattern

Figure 8.12 X-ray diffraction analysis of DNA: The X-ray diffraction pattern of DNA that Watson and Crick used in developing their double-helix model. The dark areas that form an X shape in the center of the photograph indicate the helical nature of the DNA. The dark crescents at the top and bottom of the photograph indicate the 0.34-nm distance between the base pairs.

a) Molecular model

b) Stylized diagram

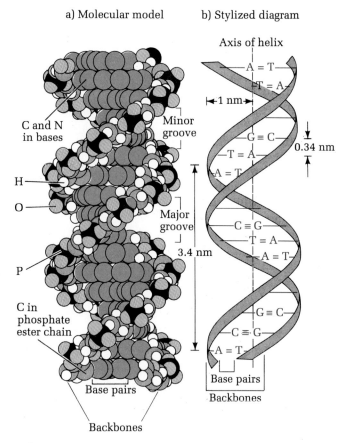

Figure 8.13 Molecular structure of DNA: (a) three-dimensional molecular model of the Watson and Crick DNA double helix; (b) stylized representation of the DNA double helix.

a) Adenine-thymine base (Double hydrogen bond)

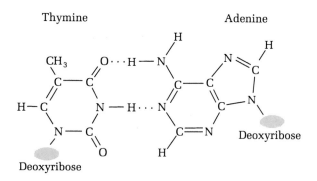

b) Guanine-cytosine base (Triple hydrogen bond)

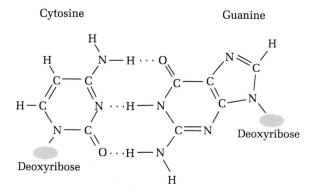

Figure 8.14 Structures of the complementary-base pairs found in DNA. In both cases a purine pairs with a pyrimidine: (a) the adenine-thymine bases, which pair through two hydrogen bonds; (b) the guanine-cytosine bases, which pair through three hydrogen bonds.

in the two pairings are relatively weak hydrogen bonds (two between adenine and thymine, three between guanine and cytosine) (Figure 8.14). These weak bonds make it relatively easy to separate the two strands of the DNA. The adenine-thymine and guanine-cytosine pairs are the only ones that can fit the physical dimensions of the helical model, and they are totally in accord with Chargaff's rules. The specific A–T and G–C pairs are called **complementary-base pairs,** and thus the nucleotide sequence in one strand dictates the nucleotide sequence of the other.* For instance, if we have the sequence 5′-TATTCCGA-3′ on one chain, the opposite

chain must bear the sequence 3′-ATAAGGCT-5′; the two strands are antiparallel (see below).

---

*As a result of their work, Watson, Crick, and Wilkins received the Nobel Prize. Franklin, who worked for Wilkins and is now deceased, has only recently been recognized for her contributions to the formulation of the double-helix model. Any Nobel Prize can be shared among only a maximum of three people, so the three recipients in this case were appropriately chosen. In addition, Nobel Prizes are never awarded posthumously.

Three important features of the DNA double helix were deduced from the early X-ray diffraction and other data:

1. The diameter of the helix is 2 nm, and the base pairs are 0.34 nm apart [1 nanometer (nm) = $1/10^9$ m = 10 angstroms (Å)].
2. A complete (360°) turn of the helix takes 3.4 nm; therefore there are 10 base pairs per turn.
3. The two strands are aligned in an **antiparallel** fashion; that is, they have **opposite polarity.** One strand is oriented in the 5′–3′ way, while the other is oriented 3′–5′.

A Preview of DNA Replication

With great foresight Watson and Crick realized that the double-helical model of DNA they had proposed suggested how the DNA molecule might replicate to produce two identical daughter DNA molecules in what has become known as the **semiconservative replication model.** They reasoned that if the ladderlike DNA molecule unwound and separated into two strands (like a zipper unfastening), each could be a template for the synthesis of the complementary strand. Thus they hypothesized that as the helix unwound from one end, the base sequence of the new strand would be determined by the nucleotide sequence of the template strand because only A–T and G–C pairs could be formed (Figure 8.15). By following the complementary-base-pairing rules, then, the genetic material would be replicated accurately (one of our three requirements for genetic material). Watson and Crick's hypothesis about the method of DNA replication proved correct; we will examine the DNA replication process in the next chapter.

Keynote *The DNA molecule usually consists of two polynucleotide chains joined by hydrogen bonds between pairs of bases (A and T, G and C) in a double helix. The diameter of the helix is 2 nm, and there are 10 base pairs in each complete turn (3.4 nm). The double-helix model was proposed by Watson and Crick from chemical and physical analyses of DNA.*

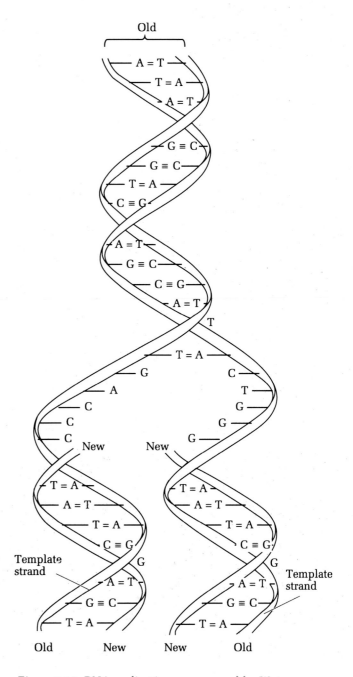

Figure 8.15 DNA replication as proposed by Watson and Crick. The two complementary strands unwind, and each is a template for the synthesis of a complementary strand.

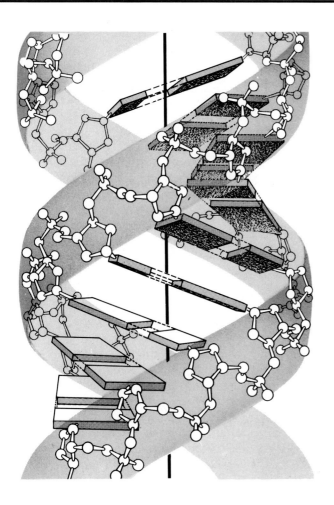
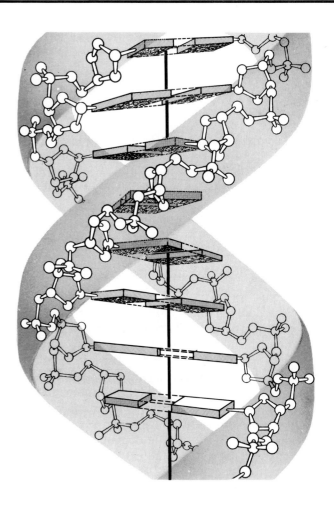

Figure 8.16 Schematic drawings of A-DNA (left) and B-DNA (right) based on analyses of DNA fibers. The sugar-phosphate backbones of the double helix are represented as ribbons and the runglike base pairs connecting them as planks. In A-DNA, the base pairs are tilted and are pulled away from the axis of the double helix. In B-DNA, the base pairs sit astride the helix axis and are perpendicular to it. (© Irving Geiss.)

Other DNA Structures

A more thorough analysis of the early X-ray diffraction patterns of DNA fibers revealed that DNA can exist in different forms depending on the conditions. When humidity is high, DNA is in what is called the B form (= the double-helical form deduced by Watson and Crick). When the humidity is relatively low, DNA is found in the A form. Figure 8.16 shows schematic drawings of A-DNA and B-DNA based on analyses of DNA fibers. Both of these DNA forms involve right-handed helices. Note that in A-DNA the base pairs are tilted and pulled away from the axis of the double helix. In contrast, in B-DNA the helix axis passes through the base pairs which are, themselves, oriented perpendicular to that axis.

An inherent problem in analyzing DNA fibers by X-ray diffraction is that the DNA strands are generally not regularly ordered around and along the fiber axis. Thus the DNA models generated from the resulting data show averaged helical structures and cannot show local structural variations that might occur because of the presence of particular base sequences. This limi-

b

c

Figure 8.17 Space-filling models of a) B-DNA, b) A-DNA, and c) Z-DNA.

tation has been overcome in recent years because methods have been developed to synthesize short DNA molecules (called **oligomers** [oligo = few]) of defined sequences. The pure DNA oligomers can be crystallized and analyzed by single-crystal X-ray diffraction. Some sequences produced the B helix while others produced the A helix. Unexpectedly certain sequences produced a third DNA form called Z-DNA, which has a left-handed helix and a zigzag sugar-phosphate backbone. (The latter property gave this DNA form its "Z" designation.) Figure 8.17 shows space-filling models and Figure 8.18 shows perspective drawings of A-, B-, and Z-DNA based on single-crystal studies. Some of the key structural features of each DNA type as deduced from the single-crystal studies are given in Table 8.3 and are discussed in the following two subsections.

Features of A-DNA and B-DNA. A-DNA and B-DNA are right-handed double helices with 10.9 and 10.0 base pairs per 360-degree turn of the helix, respectively. These values correspond

Table 8.3 Properties of A-DNA, B-DNA, and Z-DNA

Property	A-DNA	B-DNA	Z-DNA
Helix direction	Right-handed	Right-handed	Left-handed
Base pairs per helix turn	10.9	10.0	12.0
Overall morphology	Short and broad	Longer and thinner	Elongated and thin
Major groove	Extremely narrow and very deep	Wide and of intermediate depth	Flattened out on helix surface
Minor groove	Very broad and shallow	Narrow and of intermediate depth	Extremely narrow and very deep
Helix axis location	Major groove	Through base pairs	Minor groove
Base inclination from helix axis (degrees)	13.0	2.0	8.8
Mean base pair propeller twist (degrees)	15.4	11.7	4.4
Sugar-base bond conformation*	*anti*	*anti*	*anti* at C; *syn* at G

* The words *anti* and *syn* refer to the two conformations of the C-N bond connecting each base to its sugar ring (see Figure 8.20).

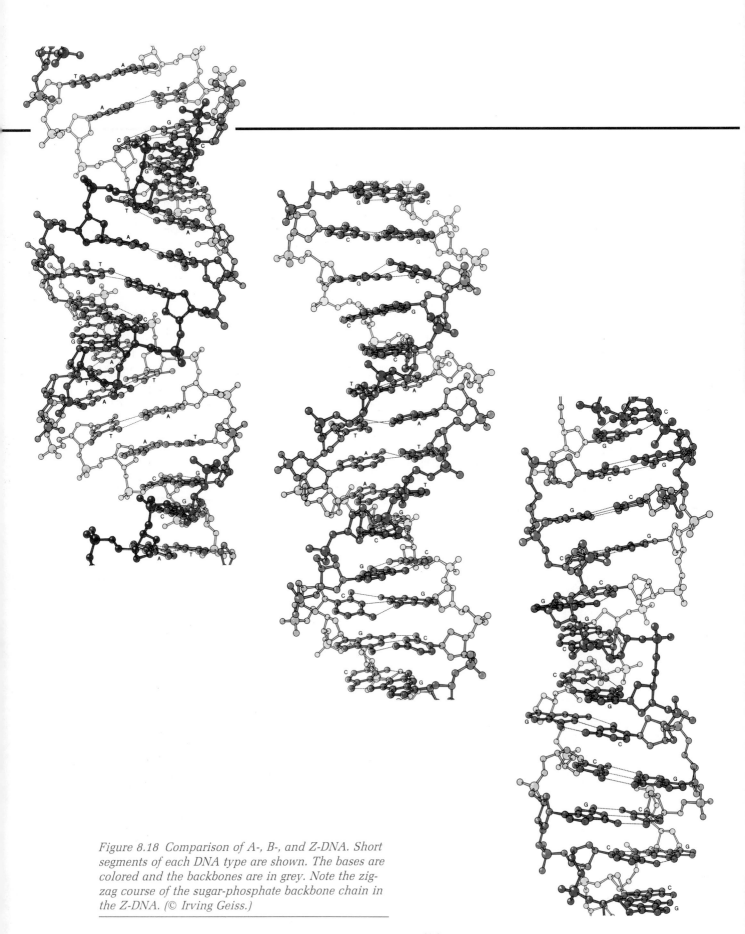

Figure 8.18 Comparison of A-, B-, and Z-DNA. Short segments of each DNA type are shown. The bases are colored and the backbones are in grey. Note the zig-zag course of the sugar-phosphate backbone chain in the Z-DNA. (© Irving Geiss.)

closely with the data derived from DNA fiber studies. The bases are inclined away from the perpendicular from the helix axis (which would be 0 degrees) by 13 degrees in A-DNA and 2 degrees in B-DNA.

The single crystal studies showed a significant deviation from the fiber studies' predictions in a property of individual base pairs called **propeller twist** (Figure 8.19). Propeller twist is a rotation of the two bases of a base pair in opposite directions about their long axis. Fiber studies did not predict any propeller twist and the resulting models showed the flat purine and pyrimidine rings aligned relative to each other in the same plane. It is now known that propeller twist exists and that it serves to stabilize the helix. The mean values for propeller twist are 15 degrees for A-DNA and 12 degrees for B-DNA. Individual propeller twist values for specific base pairs in the helix vary considerably in order to produce the most stable double helix structure; the range is 3 to 25 degrees.

The single crystal studies confirmed the fiber studies' conclusion that the A-DNA double helix is short and broad with a deep major groove and a broad and shallow minor groove. The B-DNA double helix is thinner and taller for the same number of base pairs than A-DNA, with a wide major groove and a narrow minor groove; both grooves are of similar depths.

Features of Z-DNA. Z-DNA was discovered when single crystals of an alternating cytosine-guanine DNA oligomer were subjected to X-ray diffraction analysis. DNA oligomers containing A-T base pairs do not form Z-DNA. Z-DNA is a left-handed helix generated because of a different bond conformation between one of its bases and the deoxyribose sugar (this bond is called a glycosidic bond) than is found in A- and B-DNA. Figure 8.20 shows the two conformations of the C-N glycosidic bond connecting each base to its sugar. The *anti* conformation occurs for all bases in A- and B-DNA and at cytosines in Z-DNA. The *syn* conformation is found at the guanines in Z-DNA. The alternation of a purine (G) and a pyrimidine (C) along a Z-DNA double strand produces the *anti-syn* alternation that

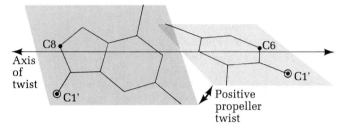

Figure 8.19 Propeller twist is defined by a schematic drawing of a purine-pyrimidine base pair. A clockwise rotation of the nearer base as one sights down the long axis (color) of the base pair in either direction is considered positive propeller twist.

gives the backbone of Z-DNA a zigzag appearance. Schematic molecular drawings of Z-DNA and B-DNA contrasting the zigzag backbone of the former with the regular helical backbone of the latter are presented in Figure 8.21, and the key structural features of Z-DNA are given in Table 8.3.

In Z-DNA there are 12.0 base pairs per complete helical turn. The bases are inclined away

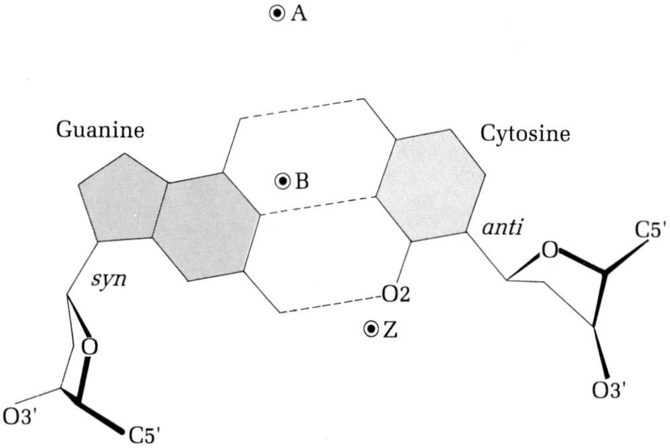

Figure 8.20 Two conformations of the C-N bond connecting each base to its sugar ring are shown. The anti *conformation (right) appears in all A- and B-DNA and at the cytosines in Z-DNA; the* syn *conformation (left) appears at the guanines in Z-DNA. The three points labeled A, B, and Z indicate for each type of helix the location of the helix axis with respect to a base pair. The axis is in the major groove in A-DNA, passes through the base pair in B-DNA, and is in the minor groove in Z-DNA.*

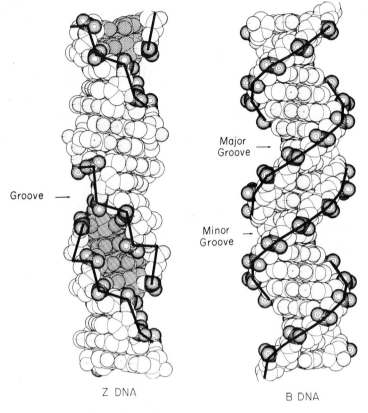

Groove →

Major
Groove →

Minor
Groove →

Z DNA

B DNA

Figure 8.21 Molecular models of Z-DNA and B-DNA. The irregularity of the Z-DNA backbone is illustrated by the heavy lines that go from phosphate to phosphate residue along the chain. The groove in Z-DNA is quite deep, extending to the axis of the double helix. In contrast, B-DNA has a smooth line connecting the phosphate groups and two grooves.

from the perpendicular from the helix axis by 8.8 degrees. The base pairs have smaller mean propeller twist values than A- or B-DNA, that is, 4.4 degrees. The Z-DNA helix is thin and elongated with a deep, minor groove. The major groove is pushed to the surface of the helix so that it is not really evident as a distinct groove. The helix axis passes through the minor groove in Z-DNA, through the major groove in A-DNA, and through the base pairs in B-DNA (see Figure 8.20).

DNA in the Cell

In solution, DNA usually is found in the B form and hence B-DNA is the form typically found in cells. Since A-DNA is found only under relatively low humidity conditions, it is unlikely that any lengthy sections of A-DNA exist within cells. There is evidence, however, that Z-DNA exists in at least some cells. This evidence has come from the use of antibodies made against Z-DNA molecules. That is, Z-DNA is much more immmunogenic (capable of inducing antibodies against it) than B-DNA. Experiments have been done in which anti-Z-DNA antibodies have been added to fixed cells and chromosomes. These antibodies will bind to any region of the chromosomes that contain stretches of Z-DNA. The locations where these antibodies bound could not be seen directly. To do that, antibodies were prepared against the anti-Z-DNA antibodies and fluorescent molecules were attached to these new antibodies. When the new antibodies were added to the fixed cells they bound specifically to the anti-Z-DNA antibodies and the fluorescent tags they carried served to pinpoint the Z-DNA regions. Figure 8.22 shows fluorescent areas of various plant nuclei resulting from this experimental approach. The fluorescence is evidence for the presence of Z-DNA in the chromosomes.

Despite evidence for its presence in chromosomes, however, it is not clear what role Z-DNA has in cellular function. It has been postulated that Z-DNA may be involved in gene regulation, although it has not been rigorously proved that any proteins bind specifically to Z-DNA, which presumably would be a facet of gene regulatory events.

Keynote X-ray diffraction analysis of single crystals of DNA oligomers of known sequence has produced detailed molecular models of three DNA types: A-DNA, B-DNA, and Z-DNA. A-DNA and B-DNA are both right-handed double helices while Z-DNA is a left-handed double helix. The number of base pairs per helical turn are 10.9, 10.0, and 12.0 for A-, B-, and Z-DNA, respectively. A-DNA is a short and

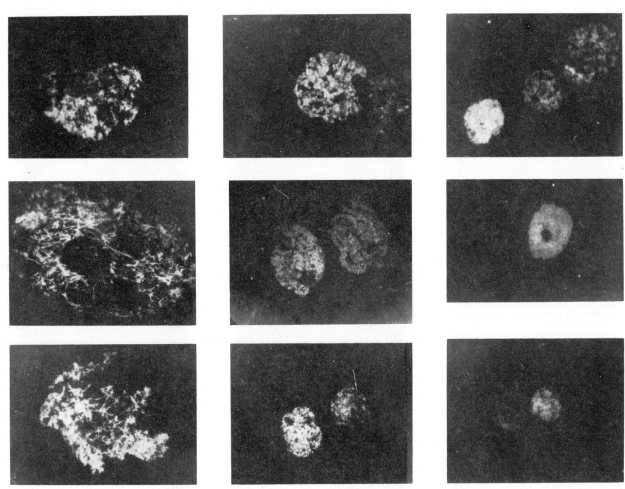

broad molecule, B-DNA is a longer and thinner
molecule, and Z-DNA is elongated and thin.
The three DNA forms differ in a number of molecular attributes, including dimensions of the
major and minor grooves and location of the helix axis. B-DNA is the form found predominantly in cells. Z-DNA has also been found in
the chromosomes of some organisms but its role
in cellular function, if any, is not clear. Since
A-DNA forms only under conditions of low humidity, it appears unlikely that this DNA type
exists in a stable form in cells.

*Figure 8.22 Fluorescence micrographs showing Z-
DNA in various plant nuclei: rye (left column), broad
bean (center column), pea (right column, top and
middle), tobacco (right column, bottom).*

Bends in DNA

Even though the DNA molecule is drawn in
textbooks as a very straight, helical molecule,
implying that it is a rigid rodlike molecule, this

is actually not so. In solution, B-DNA is quite a
flexible molecule, undergoing shape changes as
it responds to the local environmental ionic and
temperature conditions. It is possible that particular bends in DNA might favor the binding
of specific protein molecules, and this could
be part of the gene regulation process. While
specific information on this point is only just
coming to light, it is known that the ability of
B-DNA to bend slightly is enhanced by the
presence of specific base-pair sequences. The
sequence CAAAAAT, for example, particularly
if it is repeated three or four times separated

by 10 or so base pairs, has been shown to cause DNA to bend.

Concluding Remarks

In this chapter we have learned about the structure of DNA and RNA and have focused on DNA as the carrier of genetic information and on the various structures that DNA can assume. Not all genetic material exists as double-stranded DNA, as our TMV example illustrated. That is, in some organisms the genetic material is double-stranded or single-stranded RNA, or single-stranded DNA. Double-helical RNA forms a right-handed, double-helical structure much like double-stranded DNA. Single-stranded DNA and RNA are generally elongated molecules, but different parts of a single-stranded molecule (if it is not complexed with proteins) may form sections of double-helical structures if there are sufficient regions of bases that can form base pairs; these localized double-helical sections are like double-helical DNA or RNA, but they may not be very stable, especially if the region is relatively short.

Analytical Approaches for Solving Genetics Problems

Q.1 The linear chromosome of phage T2 is 52 μm long. The chromosome consists of double-stranded DNA, with 0.34 nm between each base pair. The average weight of a base pair is 660 daltons. What is the molecular weight of the T2 molecule?

A.1 This question involves the careful conversion of different units of measurement. The first step is to put the lengths in the same units: 52 μm is 52 millionths of a meter, or $52,000 \times 10^{-9}$ m, or 52,000 nm. One base occupies 0.34 nm in the double helix, so the number of base pairs in this chromosome is 52,000 divided by 0.34, or 152,941 base pairs. Each base pair, on the average, weighs 660 daltons, and therefore the molecular weight of the chromosome is $152,941 \times 660 = 1.01 \times 10^8$ daltons, or 101 million daltons.

The average length of the double helix in a human chromosome is 3.8 cm, which is 3.8 hundredths of a meter, or 38 million nm—substantially longer than the T2 chromosome! There are over 111.7 million base pairs in the average human chromosome.

Q.2 The accompanying table lists the relative percentages of bases of nucleic acids isolated from different species. For each one, what type of nucleic acid is involved? Is it double- or single-stranded? Explain your answer.

Species	Adenine	Guanine	Thymine	Cytosine	Uracil
(i)	21	29	21	29	0
(ii)	29	21	29	21	0
(iii)	21	21	29	29	0
(iv)	21	29	0	29	21
(v)	21	29	0	21	29

A.2 This question focuses on the base-pairing rules and the difference between DNA and RNA. In analyzing the data, we should determine first whether the nucleic acid is RNA or DNA, and then whether it is double- or single-stranded. If the nucleic acid has thymine, it is DNA; if it has uracil, it is

RNA. Thus species (i), (ii), and (iii) must have DNA as their genetic material, and species (iv) and (v) must have RNA as their genetic material.

Next, the data must be analyzed for strandedness. Double-stranded DNA must have equal percentages of A and T and of G and C. Similarly, double-stranded RNA must have equal percentages of A and U and of G and C. Hence species (i) and (ii) have double-stranded DNA, while species (iii) must have single-stranded DNA since the base-pairing rules are violated, with A = G and T = C but A ≠ T and G ≠ C. As for the RNA-containing species, (iv) contains double-stranded RNA since A = U and G = C, and (v) must contain single-stranded RNA.

*Questions
and
Problems*

8.1 In the 1920s while working with *Diplococcus pneumoniae,* the agent that causes pneumonia, Griffith discovered an interesting phenomenon. In the experiments mice were injected with different types of bacteria. For each of the following bacteria type(s) injected, indicate whether the mice lived or died:

a. type *IIR*;   c. heat-killed *IIIS*;
b. type *IIIS*;   d. type *IIR* + heat-killed *IIIS*.

8.2 Several years after Griffith described the transforming principle, Avery, MacLeod, and McCarty investigated the same phenomenon.

a. Describe their experiments.
b. What did their experiments demonstrate beyond Griffith's?
c. How were enzymes used as a control in their experiments?

*8.3 By differentially labeling the coat protein and the DNA of phage T2, Hershey and Chase demonstrated that (choose the correct answer)

a. only the protein enters the infected cell.
b. the entire virus enters the infected cell.
c. a metaphase chromosome is composed of two chromatids, each containing a single DNA molecule.
d. the phage genetic material is most probably DNA.
e. the phage coat protein directs synthesis of new progeny phage.

8.4 What is the evidence that the genetic material of TMV (tobacco mosaic virus) is RNA?

*8.5 In DNA and RNA, which carbon atoms of the sugar molecule are connected by a phosphodiester bond?

8.6 Which base is unique to DNA and which base is unique to RNA?

8.7 How do nucleosides and nucleotides differ?

8.8 What chemical group is found at the 5′ end of a DNA chain? At the 3′ end of a DNA chain?

*8.9 What is the base sequence of the DNA strand that would be complementary to the following single-stranded DNA molecules:

a. 5' AGTTACCTGATCGTA 3'
b. 5' TTCTCAAGAATTCCA 3'

*8.10 Is an adenine-thymine or guanine-cytosine base pair harder to break apart? Explain your answer.

8.11 The double-helix model of DNA, as suggested by Watson and Crick, was based on a variety of lines of evidence gathered on DNA by other researchers. The facts fell into the following two general categories; give three examples of each:

a. chemical composition;
b. physical structure.

*8.12 For double-stranded DNA, which of the following base ratios always equals 1?

a. $(A + T)/(G + C)$ d. $(G + T)/(A + C)$
b. $(A + G)/(C + T)$ e. A/G
c. C/G

8.13 If the ratio of $(A + T)$ to $(G + C)$ in a particular DNA is 1.00, does this result indicate that the DNA is most likely constituted of two complementary strands of DNA or a single strand of DNA, or is more information necessary?

8.14 Explain whether the $(A + T)/(G + C)$ ratio in double-stranded DNA is expected to be the same as the $(A + C)/(G + T)$ ratio.

8.15 What is a DNA oligomer?

8.16 The genetic material of bacteriophage ΦX174 is single-stranded DNA. What base equalities or inequalities might we expect for single-stranded DNA?

*8.17 A double-stranded DNA molecule is 100,000 base pairs (100 kilobases) long.

a. How many nucleotides does it contain?
b. How many complete turns are there in the molecule?
c. How long is the DNA molecule?

9

The Organization of DNA in Chromosomes

In the last chapter we discussed the structures of DNA. In this chapter we will examine how the DNA is organized into chromosomes. As would be expected, the chromosomes of eukaryotes are much more complex than the chromosomes of prokaryotes. Eukaryotic chromosomes consist of a highly ordered complex of DNA and proteins, with special regions—centromeres and telomeres—that are of special importance for chromosome function. Ultimately, geneticists will need an even greater understanding of chromosome structure than we have now so that a complete picture can be obtained of how their expression is regulated.

The Structural Characteristics of Prokaryotic Chromosomes

The Watson and Crick double-helix model of DNA structure does not by itself describe the structural characteristics of chromosomes. In the following section we will examine the organization of DNA molecules into the chromosomes found in the bacteria and viruses that have been studied extensively by geneticists and molecular biologists. Since an organism's genes are discrete segments of DNA (perhaps a thousand or so base pairs), it is important to understand how DNA—and hence the genes—are organized in chromosomes.

Bacterial Chromosomes

In Chapter 7 we learned that the DNA of the bacterium *Escherichia coli* is located in a central region called the nucleoid region (see Figure 7.1). If an *E. coli* cell is lysed gently, the DNA is released in a highly folded state, presumably close to the natural state found within the cell. The DNA is present as a single chromosome, approximately 1100 μm long [4×10^3 kbp (1 kbp = 1 kilobase pairs = 1000 base pairs)], which contains all the genes necessary for the bacterium to grow and survive in a variety of natural and laboratory environments.

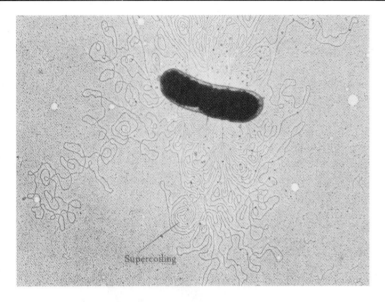

Figure 9.1 Electron micrograph of the E. coli *chromosome, showing the extensive supercoiling.*

Supercoiled DNA. Like the DNA in the Watson-Crick model, *E. coli's* DNA is double-stranded, but instead of being linear (as are the chromosomes found in the nuclei of eukaryotes), the *E. coli* chromosome is circular. So that all the DNA is packed into the nucleoid region, the chromosome is supercoiled. In other words, the DNA double helix is helically wound around itself to form what are called supercoils or superhelical twists. The supercoiled structure of the *E. coli* DNA is shown in Figure 9.1. Electron micrographs of the circular DNA molecule of bacteriophage PM2 show the difference between nonsupercoiled (= relaxed) DNA (Figure 9.2a) and supercoiled DNA (Figure 9.2b).

What are the constraints on circular DNA compared with linear DNA? In a linear DNA molecule the ends are free, allowing the molecule to rotate freely. This enables the molecule to exhibit a range in the number of times the two strands of the double helix twist around each other. This is called the **linkage number.** In a given circular DNA molecule, however, the number of times the two chains twist around each other is a defined value. When changes occur in the average number of base pairs per helix turn, this results in the generation of an ap-

propriate number of supercoils in the opposite direction. Thus, untwisting of the helix produces negative supercoiling (left-handed direction) and overtwisting results in positive supercoiling (right-handed direction).

The *E. coli* chromosome is negatively supercoiled. Negative supercoiling apparently makes it easier for processes to occur that require the helix to unwind (e.g., DNA replication, gene expression, and genetic crossing-over processes). Supercoiled DNA is less stable than nonsupercoiled DNA. Therefore, if a break (nick) is made in the backbone of one of the two strands the tension is relieved and DNA converts to its non-supercoiled (relaxed) state by the rotation of the free ends around the other strand. You can simulate this by twisting a double strand of rope until it is tightly supercoiled and then letting go; the rope will spin until it is completely relaxed. As DNA molecules (and ropes) become increasingly supercoiled they become increasingly compacted. As a result, supercoiled DNA molecules will sediment more rapidly in the centrifuge than will relaxed DNA molecules of the same type.

The supercoiling of the DNA is controlled by a class of enzymes called **topoisomerases.** Topoisomerases are enzymes that convert one topological form of DNA into another. They do this by altering the linkage number of the DNA molecule. Two topoisomerases have been identified and characterized in both prokaryotes and eukaryotes. In prokaryotes these two enzymes, called topoisomerase I and II, have been best characterized in *E. coli*. Topoisomerase II (also called DNA gyrase) untwists DNA in the left-handed direction to produce negative supercoils. Topoisomerase I does the opposite; that is, it converts negatively supercoiled DNA to the relaxed state. In the presence of both enzymes, then, DNA can be interchanged between the negatively supercoiled and relaxed states. Both types of topoisomerases must be present and functional in cells for DNA to move between supercoiled and relaxed states.

Both topoisomerases function by making a nick or nicks in the DNA chain. The broken ends produced are held together by the topoisomerase enzyme itself so that the ends are unable to rotate freely about the other continuous strand. If that were to happen, only relaxed DNA molecules would result from topoisomerase action. Topoisomerase II functions to introduce two negative turns in the linkage number by passing DNA chains through transient double-strand breaks (Figure 9.3). Topoisomerase I introduces one positive turn in the linkage number by passing DNA chains through transient single-strand breaks (Figure 9.4). Most topoisomerases relax the supertwisting of supercoiled DNA, but only topoisomerase II catalyzes the formation of negatively supercoiled DNA. The DNA gyrases have been found only in bacteria, while the relaxing topoisomerases are found in both bacteria and eukaryotes.

Proteins complexed to bacterial chromosomes.
As we will see, eukaryotic DNA is complexed with a number of discrete proteins called histones, which serve to compact the DNA into the chromosome structures characteristic of eukaryotic nuclei. Until recently it was believed that no analogous proteins existed in bacteria and that, therefore, bacterial chromosomes consisted of naked DNA. Now, however, compacted sections of *E. coli* chromosomes have been isolated using very gentle techniques. These sections contain basic proteins called DNA-binding proteins that appear to be functionally analogous to eukaryotic histones. Regions of the proteins rich in basic (positively charged) amino acids apparently are in a configuration suitable for interacting with the negatively charged phosphates of one turn of the DNA backbone.

Plasmids in bacteria. In addition to the main chromosome, bacterial cells often contain DNA in plasmids, double-stranded DNA circles that are much smaller than the circular chromosome. In natural populations of bacteria the amount of plasmid DNA may approach 1 or 2 percent of the cellular DNA amount. These plasmids are supercoiled and they replicate autonomously, that is, they replicate independently of the main chromosome. For this replication to occur, the plasmids must contain genes that control their

a

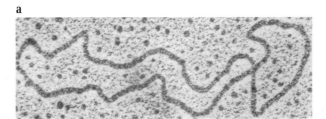

b

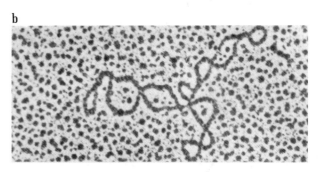

Figure 9.2 Electron micrographs of bacteriophage PM2 DNA: (a) relaxed (nonsupercoiled) DNA; (b) supercoiled DNA.

replication and maintain a balance in the ratio of the number of plasmids to the main chromosome.

T-Even Phage Chromosomes

The T2, T4, and T6 bacteriophages infect *E. coli,* and we have discussed some facets of their genetics in Chapter 7. (To review their structures, refer to Figure 7.14.) The characteristics of these phages, along with features of other viruses we will discuss, are shown in Table 9.1. Like all viruses, T2, T4, and T6 (collectively called the T-even phages) contain a single chromosome surrounded by a protein coat. The T-even phages are virulent, and infection by them always results in the lysis of the infected bacterial cell.

From the results of genetic-mapping experiments, geneticists suggested that the genetic maps of phages T2 and T4 are circular and that the phage chromosomes might have ends at different points around the circular maps, with the ends overlapping. The phage DNA itself was

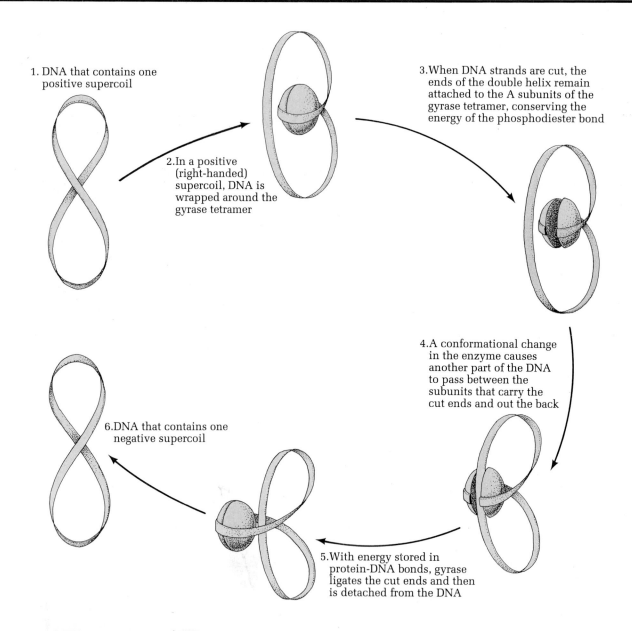

1. DNA that contains one positive supercoil

2. In a positive (right-handed) supercoil, DNA is wrapped around the gyrase tetramer

3. When DNA strands are cut, the ends of the double helix remain attached to the A subunits of the gyrase tetramer, conserving the energy of the phosphodiester bond

4. A conformational change in the enzyme causes another part of the DNA to pass between the subunits that carry the cut ends and out the back

5. With energy stored in protein-DNA bonds, gyrase ligates the cut ends and then is detached from the DNA

6. DNA that contains one negative supercoil

Figure 9.3 Model for the action of topoisomerase II (DNA gyrase). Starting with DNA, which has one positive supercoil, this enzyme introduces two negative turns resulting in DNA with one negative supercoil.

proposed to be a population of linear chromosomes with ends at different places around the circle, an arrangement referred to as *circularly permuted* chromosomes (Figure 9.5a). The experimental demonstration of the existence of circularly permuted chromosomes is shown in Figure 9.5b.

Here is the logic behind the experiment. If a double-stranded DNA molecule is heated, it denatures into single-stranded molecules be-

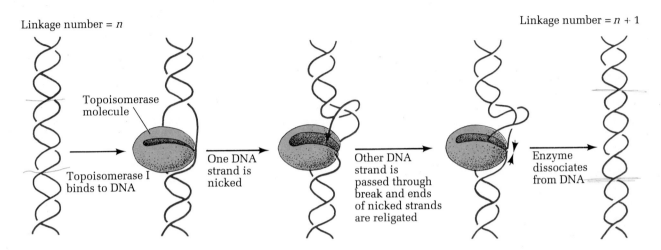

Figure 9.4 *Model for the action of topoisomerase I.
As a result of the action of this enzyme, the DNA has
one less negative supercoil.*

Table 9.1 *Characteristics of the Chromosomes of Certain Viruses*

Virus	*Host*	*Structure and Type of Genetic Material*	*Chromosome Description*	*Length of Chromosome (μm)*	*%GC*	
T-even phages	*E. coli*	Double-stranded DNA	Linear; circularly permuted	60	35	
T7	*E. coli*	Double-stranded DNA	Linear; unique sequence	12	48	
λ	*E. coli*	Double-stranded DNA	Linear; single-stranded "sticky ends"	16	49	
P22	*Salmonella typhimurium*	Double-stranded DNA	Linear; unique sequence	14	48	
ΦX174	*E. coli*	Single-stranded DNA	Circular	1.8	A25 T33	G24 C18
Qβ	*E. coli*	Single-stranded RNA		1.4	A22 U29	G24 C25
Reovirus	Mammals	Double-stranded RNA	Several pieces	8.3	A38 U28	G17 C17
SV40	Human	Double-stranded DNA	Supercoiled ring	1.7	41	
Murine leukemia	Mouse	Single-stranded RNA	Several pieces		A25 U23	G25 C27
Tobacco mosaic virus (TMV)	Tobacco	Single-stranded RNA	Linear		A30 U26	G25 C19

a) Circular permutations of complementary DNA strands

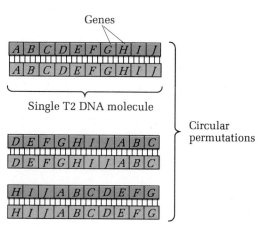

Figure 9.5 Circularly permuted DNA molecules: (a) examples of double-stranded DNA molecules with sequences that are circular permutations of each other; (b) experimental procedure to test whether a population of chromosomes is circularly permuted. The double-stranded DNA is denatured by heat into single strands. When they cool, random reassociation of double-stranded DNA will occur and, because of the circular permutation, molecules will be produced with complementary single-stranded ends, forming circles, visible by electron microscopy.

b) Experimental procedure

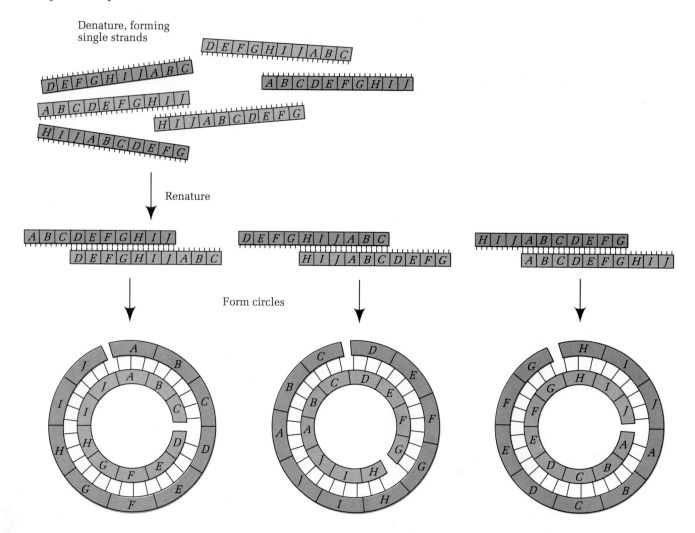

cause the hydrogen bonds between the bases are broken. If the heated solution is cooled, complementary regions of DNA strands come together again (renature or reanneal) to form double-stranded molecules held together by complementary-base pairing. If the experiment involves a circularly permuted population of DNA molecules (such as those extracted from the phage progeny in a bacterium that has been infected with a single phage), the terminal region of one strand will be complementary to the middle part of another strand. Therefore renaturation of these two strands will produce a molecule that is double-stranded in the middle and single-stranded on each end. The single strands have complementary sequences of bases. The complementary single-stranded ends will renature to form a circle of double-stranded DNA whose circumference is the same as the length of the linear T2 chromosomes. Since circles can be seen under the electron microscope when this experiment is performed with T2 and T4 chromosomes, we can conclude that the chromosomes of these phages are circularly permuted. This fact correlates very well with the fact that the genetic maps of T2 and T4 are circular, as we saw in Chapter 7.

There is also evidence that the linear DNA released from the T2 and T4 phages has the same sequence of nucleotides at each end of the molecule. This *terminal redundancy,* as it is called, is shown diagrammatically in Figure 9.6a. Terminal redundancy in a linear molecule can be shown experimentally as diagramed in Figure 9.6b. A chromosome is treated with the enzyme exonuclease III (an exonuclease is an enzyme that digests DNA from free ends), which removes nucleotides from the 3′ ends of each strand of the double-helical DNA. If the molecule is terminally redundant, then the enzyme treatment leaves complementary single-stranded ends, which can pair to form circles.

How are the native, circularly permuted and terminally redundant T2 and T4 chromosomes generated from a single parental phage? The answer is found in the mechanism used to package DNA into the phage heads (Figure 9.7). After the phage chromosome is injected into the host bacterium, it replicates several times. This replication produces a number of chromosomes, all of which have the same terminally redundant sequence as the parental DNA. The next event that occurs is molecular recombination between the DNA molecules at the terminally redundant ends, a splicing process that produces very long molecules, called *concatamers,* that repeat the base sequence of the original unit phage chromosome in a tandem fashion. The concatamers then undergo several rounds of replication, and late in infection the DNA is packaged into the phage heads that have been produced concurrently. The packaging is done by the *headful;* the length of DNA that is put into each head is determined by the volume of the head. Since the head can hold a little more than a genome's worth of DNA, we see how each phage chromosome contains a terminally redundant region. In addition, the successive clipping of the concatamer molecules into headful lengths leads to the circular permutation of the population of progeny phage chromosomes. In other words, each phage contains the same amount of DNA, with the same sequences represented and with terminal redundancy. In a population of phage chromosomes, individual chromosomes will have different terminal sequences as a result of different permutations of the same sequence.

Bacteriophage ΦX174 Chromosome

The virulent phage ΦX174 is a DNA phage that infects *E. coli.* It has attracted the attention of geneticists because of its small size relative to other phages. An electron micrograph of ΦX174 and diagrammatic representations of its structure is shown in Figure 9.8, p. 288. The ΦX174 phage is an icosahedron consisting of protein subunits surrounding the genetic material. At each vertex of the protein coat, there is a spike which is involved in the infection process. Unlike the T-even pahges and the λ phage, ΦX174 does not have a contractile sheath, baseplate, or tail fibers. The study of this phage has provided valuable information about the molecular biology of prokaryotic DNA replication.

In 1959 R. Sinsheimer found that the DNA of

a) Terminal redundancy and circular permutation

ABCDEFGHIJAB
ABCDEFGHIJAB
Permutation 1

DEFGHIJABCDE
DEFGHIJABCDE
Permutation 2

GHIJABCDEFGH
GHIJABCDEFGH
Permutation 3

Figure 9.6 Terminally redundant DNA molecules: (a) examples of terminally redundant double-stranded DNA molecules; (b) experimental procedure to test whether a linear, double-stranded DNA molecule is terminally redundant. The molecules were digested to a limited extent with exonuclease III, which removes deoxyribonucleotides from the 3' ends. If the molecules are terminally redundant, complementary single-stranded ends are exposed which form circles when paired, visible through an electron microscope.

b) Experimental procedure

Exonuclease III
digestion

Short-term digestion of 1

Moderate-term digestion of 2

Long-term digestion of 3

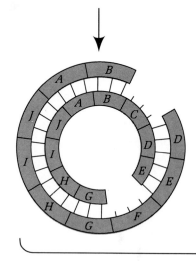

Form circles by renaturation of complementary ends

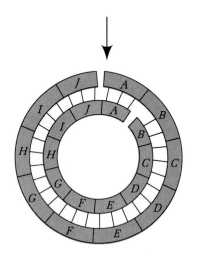

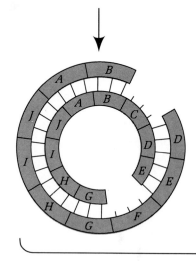

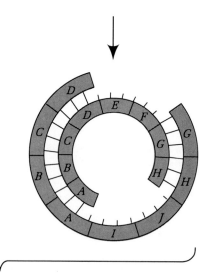

Circles with gaps because of amount of exonuclease III digestion

ΦX174 has a base composition that does not fit the complementary-base-pairing rules. The ΦX174 chromosome is not double-stranded but single-stranded. In addition, if the chromosome is treated with an exonuclease, the DNA is not affected. The simplest explanation is that the ΦX174 chromosome is circular, a fact that has been corroborated by electron microscopy. Further, the chromosome has been sequenced in its entirety, so it is known to contain 5386 nucleotides. How the 11 ΦX174 genes are arranged in the ΦX174 chromosome is an illustration of effi-

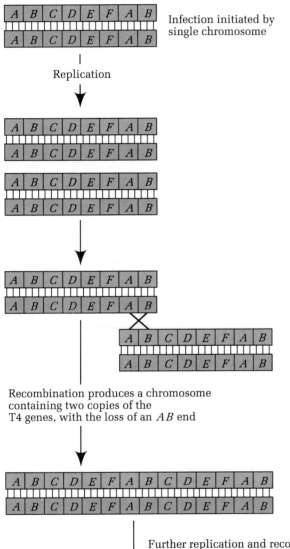

Infection initiated by
single chromosome

Replication

Recombination produces a chromosome
containing two copies of the
T4 genes, with the loss of an *AB* end

Further replication and recombination
produce concatameric T4 DNA

Packaging of DNA into pieces large enough to fill the
phage head results in terminally redundant
and circularly permuted chromosomes.

cient use of DNA, since in some cases parts or even all of one gene is found within the sequence of another gene.

Bacteriophage λ Chromosome

The structure and life cycle of the temperate bacteriophage λ has been described in Chapter 7 (pp. 234–235). This phage has been studied extensively for a number of years, so many of its gene functions are well understood. More recently, specially constructed derivatives of phage λ have been used in recombinant DNA experiments.

Like the Watson-Crick model, the phage λ chromosome is double-stranded DNA, is linear, and has no proteins associated with it. The two ends of the λ DNA molecule are single-stranded, and just like the T2 and T4 chromosomes after treatment with exonuclease III, the native, single-stranded termini of phage λ DNA are complementary. The single-stranded terminus is 12 nucleotides long; its sequence is shown in Figure 9.9a.

Figure 9.7 Circularly permuted and terminally redundant T4 progeny chromosomes produced experimentally by making concatameric molecules and cleaving them into pieces large enough to fill the phage heads.

a)   b)   c)

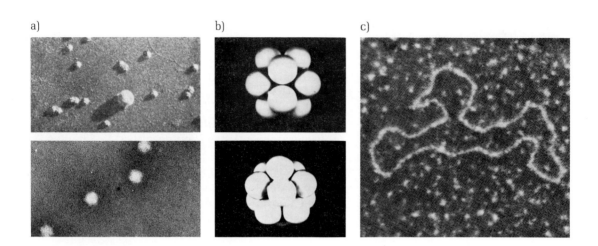

*Figure 9.8 Bacteriophage ΦX174: (a) electron micro-
graphs of ΦX174 phage particles; (b) models of the
ΦX174 phage particle: top—view along an axis of 2-
fold symmetry; bottom—view along an axis of 5-fold
symmetry; (c) single-stranded circular DNA chromosome of ΦX174.*

Regardless of whether λ goes through the lytic
or the lysogenic cycle, the first step after the λ
DNA is injected into the host cell is the conversion of the linear molecule into a circular molecule (Figure 9.9a). The "sticky ends" pair and
the single-stranded gaps are bonded in an
enzyme-catalyzed reaction. In the lysogenic
cycle the circular DNA finds a particular site in
the *E. coli* chromosome, and by a crossing-over
event the DNA is integrated into the main chromosome (see p. 236).

In the lytic cycle the DNA replicates and produces a long concatameric molecule similar to
the one for T2 and T4. It is from this concatameric structure that progeny phage λ chromosomes are generated. Generation is accomplished by the following process: The phage λ
chromosome has a gene called *ter* (for "termi-
nus-generating activity"), the product of which
is a DNA endonuclease (an enzyme that digests
a nucleic acid chain from somewhere along its
length rather than at the termini). The endonuclease recognizes a sequence of nucleotides in
the concatamer known as *cos* (Figure 9.9b).
Once it is aligned on the DNA at the recognition

site, the endonuclease makes a staggered cut
such that linear λ chromosomes with complementary, 12-base-long, single-stranded ends are
produced. The chromosomes are then packaged
in the assembled phage heads, and progeny λ
phages are assembled and then released from
the cell when it lyses.

Keynote   *In prokaryotes in which the genetic
material is DNA, the chromosome
consists of DNA with little or no protein associated with it. In bacteria the chromosome is a
circular, double-stranded DNA molecule that is
compacted by twisting and supertwisting of the
helix. In bacteriophages the genetic material
may be double-stranded or single-stranded DNA
or RNA. Phage chromosomes can be linear or
circular.*

The Structural Characteristics of Eukaryotic Chromosomes

So far we have concentrated on describing the
genetic material in prokaryotic organisms whose
genetic material is either DNA or RNA, which
can be either single-stranded or double-stranded,
circular or linear.

Now we will examine the general structure of
eukaryotic chromosomes, which differs from the
structures of prokaryotic chromosomes. In addi-

a) Linear λ chromosome (~48,000 base pairs) forms circular λ chromosome

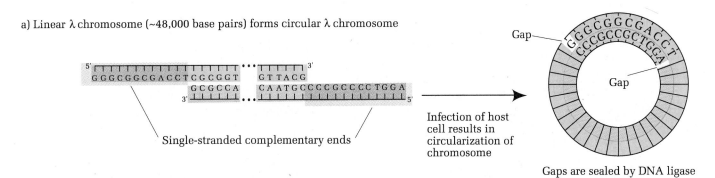

Single-stranded complementary ends

Infection of host cell results in circularization of chromosome

Gap

Gap

Gaps are sealed by DNA ligase

b) Production of progeny, linear λ chromosomes from concatamers (multiple copies linked end-to-end at complementary ends)

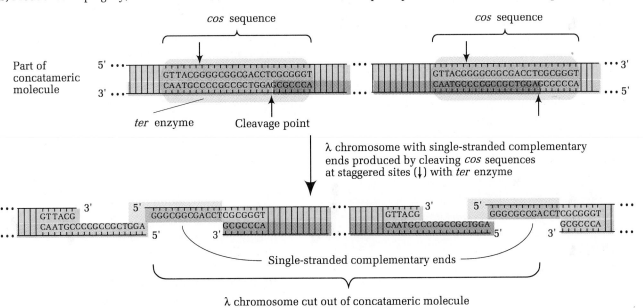

cos sequence

cos sequence

Part of concatameric molecule

ter enzyme   Cleavage point

λ chromosome with single-stranded complementary ends produced by cleaving *cos* sequences at staggered sites (↓) with *ter* enzyme

Single-stranded complementary ends

λ chromosome cut out of concatameric molecule

Figure 9.9 λ chromosome structure varies at stages of lytic infection of E. coli: (a) parts of the λ chromosome showing the nucleotide sequence of the two single-stranded, complementary ("sticky") ends, and the chromosome circularizing after infection by pairing of the ends, with the single-stranded gaps filled in to produce a covalently closed circle; (b) generation of the "sticky" ends of the λ DNA during the lytic cycle. During replication of the λ chromosome, a giant concatameric DNA molecule is produced; it contains tandem repeats of the λ genome. The diagram shows the "join" between two adjacent λ chromosomes and the extent of the cos *sequence. The* cos *sequence is recognized by the* ter *gene product, an endonuclease that makes two cuts at the sites shown by the arrows. These cuts produce two λ chromosomes (shaded and unshaded in the diagram) from the concatamer.*

tion, we will study how the chromosome is organized at a fundamental molecular level in order to understand how the chromosomes replicate, recombine, condense, and become extended again as they go through the cell cycle. This information is related to the regulation of gene expression in the eukaryotic chromosome.

In the following sections some of the interesting molecular aspects of the organization of the eukaryotic chromosome will be described. As you proceed, think about how the information might relate to chromosome mechanics during mitosis and to gene function.

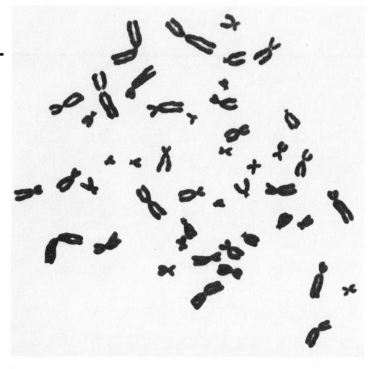

Figure 9.10 Mitotic metaphase chromosomes of a human male.

The Eukaryotic Cell Cycle

A fundamental difference between prokaryotes and eukaryotes is that prokaryotes have but one chromosome, while eukaryotes have several chromosomes, and these are usually diploid in number. When a eukaryotic cell divides, then, it is essential that each daughter cell receive one copy of each chromosome. As we saw in Chapter 1 (pp. 14–20), this distribution occurs in the process of mitosis.

The eukaryotic cell cycle consists of four sequential phases: G_1–S–G_2–M (see Figure 1.18). The first three phases constitute the interphase stage, or the part of the cell cycle in which cell division (M) is not occurring. The G_1 and G_2 phases are gaps with respect to DNA synthesis and the mitotic processes. The DNA synthesis (replication) occurs in the S phase, and the progeny chromosomes are segregated into the daughter cells in the mitotic (M) phase. Thus there is a discrete gap between the time the chromosomes replicate and the time they are segregated into the progeny cells. The relative time a dividing cell spends in each phase varies from cell type to cell type and from species to species.

The Eukaryotic Chromosome Complement

Although chromosomes in the interphase nucleus are quite difficult to see, they become highly condensed during mitosis, and when stained, they are readily visible under the light microscope. In metaphase, before the chromosomes separate into the two progeny cells, each chromosome has replicated into two sister chromatids and is in its most highly condensed state. The sister chromatids are attached at one point along their length, the centromere. Almost all cytogenetic analysis of chromosomes has been done with condensed metaphase chromosomes. A photograph of a stained set of human male chromosomes in metaphase is shown in Figure 9.10. There are 46 chromosomes (each with an X shape representing a pair of sister chromatids joined at the centromere), which vary in size and in the position of the centromere.

The Karyotype

A complete set of all the metaphase chromosomes in a cell is called its **karyotype** (literally, "nucleus type"). For most organisms, all cells have the same karyotype. However, the karyotype is species-specific so that a wide range of number, size, and shape of metaphase chromosomes is seen among eukaryotic organisms. Moreover, there is no simple relationship between the karyotype and evolutionary complexity; that is, closely related organisms can have quite different karyotypes.

Figure 9.11 shows the karyotype for the cell of a normal human male. It is customary, particularly with human chromosomes, to arrange chromosomes in order according to size. Figure 9.11 shows 46 chromosomes: 2 pairs of each of the 22 autosomes and 1 pair of each of the 2 sex chromosomes. (In a karyotype of a human female there are also 46 chromosomes: 2 pairs of each of the 22 autosomes and 2 pairs of the X chromosome.) The number of pairs is 46 because humans are diploid (2N) organisms possessing one haploid (N) set of chromosomes (23

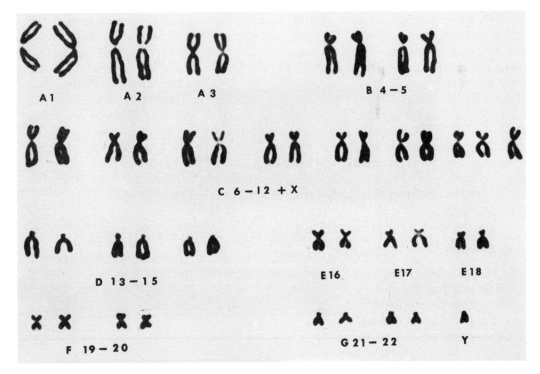

Figure 9.11 Human male metaphase chromosomes arranged as a karyotype. Note that groups of chromosomes with similar morphologies have letter designations (A through G). This arrangement is based on the size of the condensed chromosomes. In the male all chromosomes except the X and Y sex chromosomes are present in pairs.

chromosomes) from the egg and another haploid set from the sperm. (In general, the gametes of sexually reproducing organisms are haploid in their chromosome content.)

A knowledge of the size, overall morphology, and banding patterns (see p. 292) of chromosomes permits geneticists to identify certain chromosome aberrations that correlate with congenital abnormalities or dysfunctions. For example, occasionally chromosomes undergo changes in morphology by gaining or losing a portion of a chromosome or by exchanging pieces with a nonhomologous chromosome. In addition, changes in the number of chromosomes in the karyotype may occur because of an error in cell division. In humans, some congenital syndromes (such as Down syndrome or cri-du-chat syndrome) result in characteristic features that correlate with an extra chromosome 21 (trisomy 21), in the case of Down syndrome, or with the loss of the distal portion of chromosome 5, in the case of cri-du-chat syndrome. Extra or missing chromosomes and large additions and deletions to chromosomes can be observed under the light microscope.

For each chromosome in a human karyotype, the chromosomes are numbered for easy identification. Conventionally, the largest pair of homologous chromosomes is designated 1, the next largest 2, and so on. In humans chromosomes 1 (largest) through 22 (smallest) are called the autosomes to distinguish them from the pair of sex chromosomes. Formally, the sex chromosomes constitute pair 23 even though, in humans at least, they do not fit properly in the size scale. As shown in Figure 9.11, the X chromosome is a large metacentric chromosome, and the Y chromosome is the smallest chromosome in the display. In nonhuman karyotypes, however, the Y chromosome is not always the smallest chromosome.

When karyotypes are listed for several eukaryotic organisms, we immediately see that the number of chromosomes varies from species to species (Table 9.2). The diploid number of chromosomes in humans is 46, in chimpanzees it is 48, in cats it is 38, and in garden peas (which began the science of genetics in Mendel's garden) it is 14. In fruit flies, *Drosophila melanogaster,* it is 8. In haploid organisms the number of chromosomes is, for example, 16 in the green alga *Chlamydomonas reinhardi,* 17 in baker's yeast *Saccharomyces cerevisiae,* and 7 in the pink bread mold *Neurospora crassa.*

Chromosomal Banding Patterns

The human metaphase chromosomes shown in Figure 9.11 were seen after staining the chromosomes with Feulgen stain. This material stains the chromosomes uniformly, making it difficult to distinguish chromosomes that are similar in size and general morphology. For this reason the human chromosomes in the figure are grouped into subsets, and within each subset one cannot determine which chromosome is which on the basis of size alone.

Fortunately, a number of techniques have been developed to stain certain regions or *bands* of the chromosomes more intensely than other regions. The banding patterns are specific for each chromosome, thereby enabling each chromosome in the karyotype to be distinguished clearly. This makes it easier to distinguish chromosomes of similar sizes and shapes such as those in the C group in Figure 9.11.

One of these staining techniques is called **Q banding.** In it, metaphase chromosomes are stained with quinacrine mustard, producing what are known as Q bands on the chromosomes. The Q bands can be discerned by their fluorescence under ultraviolet light. A metaphase karyotype of human chromosomes showing the Q bands is presented in Figure 9.12. The Q bands are unique for each chromosome in the human set. Although Q banding is a useful diagnostic tool, it is limited by the fact that the fluorescent bands are not permanent.

A more useful technique involves first treating the chromosomes with mild heat or proteolytic enzymes (enzymes that digest proteins) and then staining with Giemsa stain (a permanent DNA dye) to produce a *G banding* pattern (Figure 9.13). G and Q bands have the same locations and are presumed to reveal the same underlying chromosome structures. The G bands are stable and are even visible in scanning electron micrographs of chromosomes as constrictions at the banding regions. The drawings and map of human chromosomes in Figure 6.10 are derived from analysis of G-banded chromosomes.

Cellular DNA Content and Phylogeny

As discussed earlier, the haploid complement of chromosomes in an organism constitutes that organism's genome. In prokaryotes and in the viruses that infect them, the genome equals

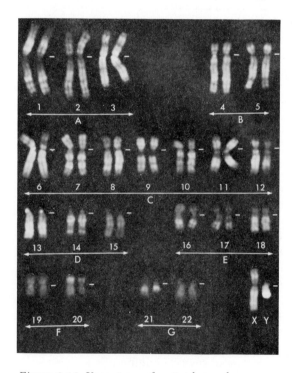

Figure 9.12 Karyotype of metaphase chromosomes from a normal human male showing Q banding. The Q-banding patterns were produced by staining of the chromosomes with the fluorescent stain quinacrine.

Table 9.2 Haploid Chromosome Numbers in Eukaryotic Organisms

Scientific Name	Common Name	Number of Chromosomes*
DIPLOID ORGANISMS		
ANIMALS		
Homo sapiens	Human	46
Pan troglodytes	Chimpanzee	48
Equus caballus	Horse	64
Bos taurus	Cattle	60
Canis familiaris	Dog	78
Felis domesticus	Cat	38
Oryctolagus cuniculus	Rabbit	44
Rattus norvegicus	Rat	42
Mus musculus	House mouse	40
Cavia cobaya	Guinea pig	64
Meleagris gallopavo	Turkey	82
Gallus domesticus	Chicken	≈78
Rana pipiens	Frog	26
Xenopus laevis	Toad	36
Planaria torva	Flatworm	16
Caenorhabditis elegans	Nematode	11♂/12♀
Musca domestica	Housefly	12
Drosophila melanogaster	Fruit fly	8
PLANTS		
Pinus ponderosa	Ponderosa pine	24
Brassica oleracea	Cabbage	18
Pisum sativum	Garden pea	14
Lycopersicon esculentum	Tomato	24
Solanum tuberosum	Potato	48
Nicotiana tabacum	Tobacco	48
Triticum aestivum	Bread wheat	42
Oryza sativa	Rice	42
Zea mays	Corn	20
Vicia faba	Broad bean	12
HAPLOID ORGANISMS		
Chlamydomonas reinhardi	A unicellular alga	16
Neurospora crassa	Orange bread mold	7
Aspergillus nidulans	A green bread mold	8
Saccharomyces cerevisiae	Brewer's yeast	≈17

* For the diploid organisms, the diploid number of chromosomes is given.

Source: Data adapted from P. L. Altman and D. S. Dittmer (eds.), 1972. *Biology data book,* 2nd ed., vol. 1. Bethesda, MD: Federation of American Societies for Experimental Biology and from other sources.

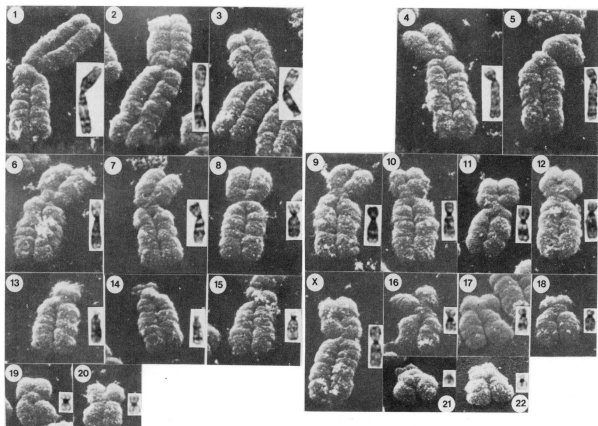

Figure 9.13 *Morphologies of human chromsomes after G banding. Shown are autosomes 1 through 22 and X chromosome. The scanning electron micrographs exhibit constriction at the site where the G bands are located, and the light micrographs (insets) show the G banding.*

the total amount of genetic material present in the one chromosome they possess. In eukaryotes, though, the genome size is defined as the amount (often called the **C value**) of DNA found in the haploid set of chromosomes. Table 9.3 gives the haploid amount of DNA for some prokaryotes and eukaryotes. The length of the DNA is also presented as calculated from the number of base pairs.

The table shows that the amount of DNA (measured in daltons or base pairs) found in the different species varies considerably. Even within a given species the *C* value may vary 18 to 20 percent. The table shows, for example, that phage ΦX174 contains 1.6×10^6 daltons (5386 nucleotides) of single-stranded DNA, while at the other end of the range, the *C* value of a plant in the lily family, *Lilium longiflorum,* is 2 $\times$

10^{14} daltons (3×10^{11} base pairs). In the lower eukaryotes the yeast *Saccharomyces cerevisiae* has only about four times the amount of DNA as *E. coli,* and *Neurospora crassa* has about ten times the amount of DNA as *E. coli.* Since their genomes are relatively small, these eukaryotic organisms are often used in research experiments. Nonetheless the data presented in Table 9.3 do show somewhat of a trend in terms of the amount of DNA and the complexity of the organism. Yeasts, *Drosophila melanogaster,* sea urchins, and humans clearly are increasingly complex and have successively larger amounts of DNA in their haploid genomes. Of the vertebrates, however, amphibians have the greatest amount of DNA per haploid genome, and they are significantly less complex than humans. And, as we have already noted, lilies have about 100 times the amount of DNA per haploid genome as do humans.

In a more detailed view of this so-called *C value paradox* (the lack of a direct relationship between the *C* value and phylogenetic complexity), Figure 9.14 shows the *C* values for a variety of animals. Note that some animal groups, such as the mammals, have a narrow range of genome size, while others, such as the insects and the amphibians, have a very broad range.

Keynote *In eukaryotes the complete set of metaphase chromatid pairs in a cell is called its karyotype. The karyotype is species-specific. The haploid (N) amount of DNA in a eukaryotic cell is called the* C *value. The amount of genetic material varies greatly among prokaryotes and eukaryotes. There is not a direct relationship between the* C *value and phylogenetic complexity.*

Table 9.3 Haploid DNA Content (the C *value) of Various Prokaryotes and Eukaryotes*

Organism	C *Value*		
	Daltons	*Base Pairs (bp)*	*Length**
PROKARYOTES			
Bacteria:			
Escherichia coli	2.8×10^9	4.1×10^6	1.4 mm
Salmonella typhimurium	8×10^9	1.1×10^7	3.8 mm
Viruses and phages:			
ΦX174 (double-stranded form)	3.2×10^6	5386	1.8 μm
λ	3.3×10^7	4.65×10^4	16 μm
T2	1.2×10^8	1.75×10^5	60 μm
SV40	3.5×10^6	5226	1.7 μm
EUKARYOTES			
Vertebrates:			
Human	1.9×10^{12}	2.75×10^9	94 cm
Mouse	1.45×10^{12}	2.2×10^9	75 cm
Frog	1.4×10^{13}	2.25×10^{10}	7.65 m
Invertebrates:			
Sea urchin	5×10^{11}	8×10^8	27.2 cm
Drosophila melanogaster	1.2×10^{11}	1.75×10^8	5.95 cm
Plants:			
Lilium longiflorum (lily)	2×10^{14}	3×10^{11}	100 m
Zea mays (maize)	4.4×10^{12}	6.6×10^9	2.24 m
Fungi:			
Neurospora crassa	1.85×10^{10}	2.7×10^7	9.18 mm
Saccharomyces cerevisiae	1.2×10^{10}	1.75×10^7	5.95 mm

* Calculated from the formula 10 bp = 3.4 nm.

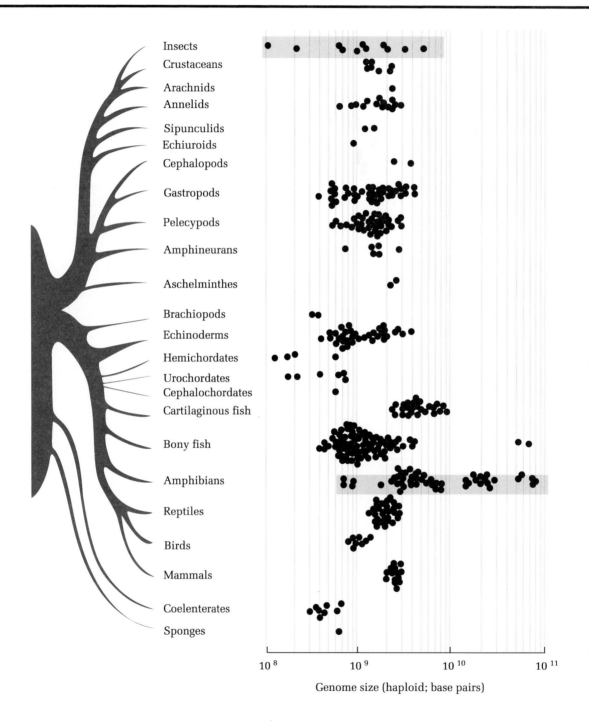

Figure 9.14 The amount of DNA per haploid chromosome set in a variety of animals. Variation within
certain phyla is quite wide (as in insects and amphibians).

The Molecular Structure of the Eukaryotic Chromosome

In comparison with a prokaryotic cell, a eukaryotic cell contains a large amount of DNA in its nucleus. A human cell, for example, has a thousand times as much DNA as does *E. coli.* We saw earlier that the *E. coli* chromosome (about 1 mm long in its fully extended state) is supercoiled and supertwisted so that its 4.1×10^6 base pairs of DNA can be packaged into the nucleoid region. By contrast, the human cell has 5.5×10^9 base pairs of DNA in its diploid nucleus. Without the compacting of this DNA, accomplished by the specific association between DNA and proteins in the chromosomes, the DNA of the chromosomes of a single human cell would be over 6 feet long (about 200 cm) if placed end to end!

Chemical analysis of eukaryotic chromosomes reveals that they consist of DNA and two kinds of proteins. As researchers working in this area know, it is difficult (although not impossible) to isolate intact chromosomes for study. During the processes of releasing the chromosomes from a cell and purifying them, some breakage usually occurs. Thus whole chromosomes are not analyzed. The pieces of DNA-protein complex that are studied are called **chromatin,** and each chromatin fragment reflects the general features of chromosomes but not the specifics of any individual chromosome. Studies of carefully isolated chromosomes have shown, however, that each chromosome consists of one linear, unbroken, double-stranded DNA molecule surrounded by two kinds of protein, the **histones** and **nonhistones,** which play an important role in determining the physical structure of the chromosome. As we will see, the DNA is wrapped around a core of histone molecules, and the nonhistones are somehow associated with that complex. Various studies have shown that the nonhistones have a basic structural role in the chromosomes: If the histones are removed from the chromosome, the DNA unravels and is displaced from the complex, but a skeleton of nonhistone proteins in the shape of the chromosome remains.

The histones. The histones are the most abundant proteins associated with chromosomes. Their role is to bind to the negatively charged DNA in the chromosome. The histones are relatively small basic proteins; that is, at the normal pH of a cell, the histones have a net positive charge, thus facilitating their binding to the DNA. This positive charge is found mainly on the amino groups ($—NH_3^+$) of the basic amino acids lysine and arginine (see Figure 13.2 for the chemical structures of these amino acids). In fact, histone molecules consist of approximately 25 percent arginine and lysine.

Five main types of histones are associated with eukaryotic DNA: H1, H2A, H2B, H3, and H4. Weight for weight there is about an equal amount of histone and of DNA in chromatin. The amount and proportions of these histones are constant from cell to cell in all eukaryotic organisms. The properties of the histones from calf thymus DNA are given in Table 9.4. The amino acid sequences of the four histones H2A, H2B, H3, and H4 have been determined for a wide variety of organisms. In all eukaryotes the basic amino acids are clustered at the amino terminal end of the protein. Comparison of the sequences indicates a high degree of conservation in the sequences among distantly related species. Only two amino acid differences exist in the H4 proteins of cows and peas, for example,

Table 9.4 Characteristics of Histones from Calf Thymus DNA

Histone Type	Properties	Number of Amino Acids	Molecular Weight (daltons)
H1	Very lysine-rich	≈ 215	$\approx 21,500$
H2A	Lysine-rich	129	14,000
H2B	Lysine-rich	125	13,775
H3	Arginine-rich	135	15,320
H4	Arginine-rich	102	11,280

Source: S. C. R. Elgin and H. Weintraub, 1976, *Annu. Rev. Biochem.* 44:725–776.

and there is only one amino acid difference between sea urchin and calf thymus histone H3. In fact these four histones are among the most highly conserved of all known proteins. This remarkable amino acid sequence conservation is a strong indicator that histones perform the same basic pivotal role in organizing the DNA in the chromosomes of all eukaryotes.

The H1 histones represent a different group of histones compared with the other four histones. There are several different but closely related H1 histones in each cell. This class of histones has a central core region of amino acids that has been conserved through evolution, but the rest of the molecule has been much less conserved. Moreover, in some tissues H1 is not present. In avian red blood cells (which, unlike mammalian red blood cells, have nuclei), for example, a histone designated H5 is present that is quite different from H1, instead of H1 in the chromosomes.

Note that we have emphasized the fact that there are five "major" histone types. There are subtypes of the histone classes, the most common of which involve the addition of acetyl and/or phosphate groups.

The nonhistones. The nonhistones are the second type of protein found in chromatin. Like histones, nonhistones are also associated with the DNA. Some nonhistones provide a structural role while others are involved in the regulation of gene expression. As a class, the nonhistones are very different from the histones. They are usually acidic proteins—that is, proteins with a net negative charge—and so they are likely to bind to the positively charged histones in the chromatin. Each eukaryotic cell has many different nonhistones in the nucleus, and weight for weight there may be as much nonhistone protein present as DNA and histone proteins combined.

In contrast to the histones, the nonhistone proteins differ markedly in number and type from cell type to cell type within an organism and from organism to organism. This heterogeneity of nonhistones within and between organisms makes it very likely that at least some of the nonhistones are involved in the regulation of gene expression in eukaryotes. We will discuss this idea in more detail later in the book.

Keynote *The nuclear chromosomes of eukaryotes are complexes of DNA, histone proteins, and nonhistone proteins. There are five main types of histones, which are constant from cell to cell within an organism. Four of the histones, H2A, H2B, H3, and H4, are highly conserved evolutionarily. Nonhistones, of which there are a large number, vary significantly between cell types both within and between organisms.*

Nucleosomes. The DNA helix of each chromosome is coiled within the nucleus in a nonrandom way. It is known that there are several levels of packing that enable chromosomes several millimeters or even centimeters long to fit into a nucleus that is a few micrometers in diameter. The simplest level of packing involves the winding of DNA around a core of histones in a structure called a **nucleosome,** and the most complex level of packing involves higher-order coiling and looping of the DNA-histone complexes, as exemplified by the metaphase chromosomes.

Under the electron microscope the DNA-protein complex is seen as fibers, or **nucleofilaments,** about 10 nm in diameter. Since DNA has a diameter of 2 nm, we must assume that the DNA is complexed with histone and nonhistone proteins in a particular way. In its most unraveled state, the chromatin fiber of a nucleofilament has the appearance of "beads on a string," where the beads are the nucleosomes (Figure 9.15). The thinner "thread" connecting the "beads" is naked DNA.

The nucleosome was first discovered in the 1970s by A. Olins and D. Olins, and by C. L. F. Woodcock. Their data suggested that the nucleosomes were about 10 nm in diameter, flattened spheres in which the DNA was somehow associated with a core of histones. Later, Roger Kornberg and others treated chromatin for a short time with a bacterial endonuclease, micrococcal nuclease. Small particles, the nucleosomes, were released from the chromatin. To this day, the ex-

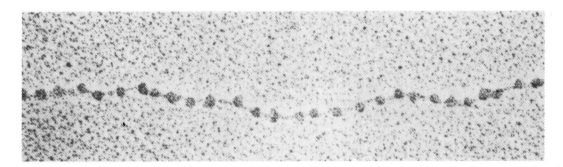

Figure 9.15 *Electron micrograph of unraveled chromatin showing the nucleosomes in a "beads on a string" morphology.*

act association of nonhistone proteins with the nucleosomes is not known in detail.

Association of DNA with histones in nucleosomes. Current information indicates that the nucleosome is found at about 200 base-pair intervals along the DNA molecule. When chromatin is digested for a short time with a bacterial endonuclease, the linker DNA between nucleosomes is degraded, releasing individual nucleosome particles (Figure 9.16). These nucleosomes have been shown by relatively low resolution X-ray diffraction to be flat-ended, cylindrical particles of dimensions 11 × 11 × 5.7 nm. The DNA is wrapped around the histone core with approximately 80 base pairs per turn. There are 146 base pairs of DNA in the nucleosome particle involving one-and-three-quarters turns of the DNA. The way the DNA is wound around the histone core causes negative supercoiling (twisting against the clockwise direction of helix formation) of the DNA strand and altogether the length of the DNA is decreased by a factor of seven by this coiling. The DNA is intimately associated with the histone core of the nucleosome such that the base pairs are protected very well from enzyme attack.

Figure 9.16 *Organization of DNA and histones in a chromatin fiber. Nuclease digestion of the chromatin fiber releases free, 11-nm diameter nucleosome particles that contain 146 base pairs of DNA and a histone core consisting of two copies each of four different histone molecules.*

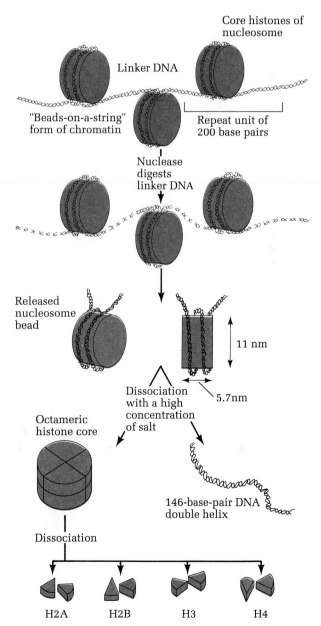

Core histones of nucleosome

Linker DNA

"Beads-on-a-string" form of chromatin

Repeat unit of 200 base pairs

Nuclease digests linker DNA

Released nucleosome bead

11 nm

5.7nm

Dissociation with a high concentration of salt

Octameric histone core

146-base-pair DNA double helix

Dissociation

H2A   H2B   H3   H4

a)

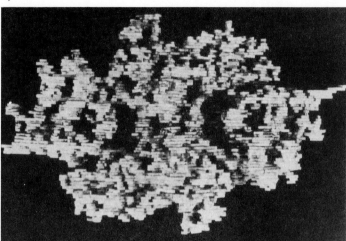

b)

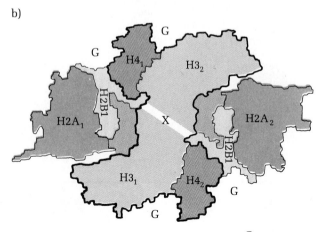

G = grooves

If the free nucleosome particles are treated with a high concentration of salt, the particles dissociate into the histone core and a 146-base pair DNA double helix (Figure 9.16). The histone core has been shown to be an octamer consisting of two subunits each of the four histone types H2A, H2B, H3, and H4. The structure of the histone octamer has been determined using high-resolution X-ray crystallographic techniques. Figure 9.17a shows photographs of a wooden model of the octamer and Figure 9.17b is a schematic drawing of the model showing the different histone domains. The model shows the histone octamer to be a prolate ellipsoid 11 nm long and 6.5 to 7.0 nm in diameter—its general shape is that of a rugby ball or a lemon. At the organizational level there is a central $(H3\text{-}H4)_2$ tetramer (i.e., two molecules each of H3 and H4) flanked by two H2A-H2B dimers. The DNA helix, when it is placed around the histone octamer in a path suggested by the surface features of the model (Figure 9.18), appears like a spring holding the H2A-H2B dimers at either end of the $(H3\text{-}H4)_2$ tetramer.

Higher-order structures in chromatin. In the living cell, chromatin typically will not exist in a "beads on a string" nucleofilament; rather, it is kept in a more highly compacted state. While the first level of DNA packing into nucleosomes is now well established, the details of the higher orders of folding, which are found predominantly in the chromosomes, are still incompletely understood. Moreover, the details of the changes in chromatin folding that take place as a cell goes from interphase to mitosis (or meiosis) and back to interphase are not well understood. There is general agreement, derived from microscopic observations, that the next level of packing above the nucleosome is the *30-nm chromatin fiber* (Figure 9.19). One major

---

Figure 9.17 (a) Photographs of a wooden model showing the molecular organization of the histone octamer. Top photo is an unpainted model and bottom photo is a model painted to show the organization of subunits; (b) interpretative diagram of the top photograph in (a).

a

b

Figure 9.18 Photographs of a wooden model of the histone octamer showing the path of DNA around the structure: (a) view from the front (as in Figure 9.17a); (b) view from the back.

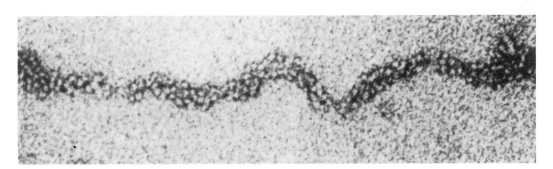

Figure 9.19 Electron micrograph of a 30-nm chromatin fiber.

problem is that as chromatin compaction proceeds, the individual nucleosomes can no longer be distinguished, and hence their arrangement in the 30-nm fiber becomes much more difficult to define. The available biophysical and biochemical data derived from studies of both compact and relaxed (unraveled) chromatin support a model for the 30-nm fiber involving a zigzag ribbon of nucleosomes which is twisted to generate the 30-nm fiber. The stages in folding of a nucleosomal chain, as indicated in this model, are shown in Figure 9.20. The figure shows that the least compact form is the relaxed zigzag ribbon, which undergoes a transition to a compact zigzag ribbon. This transition involves no major changes in basic structure, especially the loss or addition of nucleosome-nucleosome contacts. The 30-nm chromatin fiber is then constructed

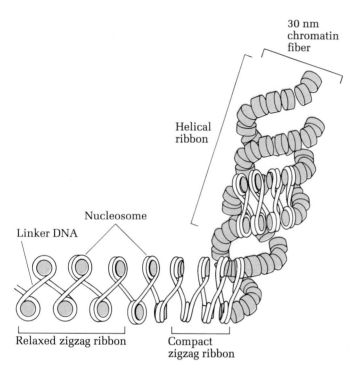

Figure 9.20 *Model for the stages in folding of a nucleosomal chain (the relaxed zigzag ribbon) via the compact zigzag ribbon into the helical ribbon proposed to be the basic structural form of the 30-nm chromatin fiber.*

by folding the compact zigzag ribbon in such a way as to minimize changes in nucleosome-nucleosome contacts (Figure 9.20). This relatively simple coiling of the compact zigzag ribbon creates a double-helical ribbon with a range of possible diameters and with the nucleosomes in face-to-face contact.

Another histone protein that is prevalent in chromosomes is histone H1. There is very good evidence that histone H1 is located at the entry/ exit point of the DNA on the nucleosome. Since histone H1-depleted chromatin forms 10-nm nucleofilaments but not 30-nm fibers, histone H1 proteins must help pack the nucleosomes together into the 30-nm chromatin fibers.

The next levels of packing beyond the 30-nm chromatin fiber are not as clearly understood. An interphase chromosome may have a diameter of 300 nm while a metaphase chromosome may have a diameter of 700 nm.

A major step in modeling the structure of metaphase chromosomes came when data were obtained showing that metaphase chromosomes from which the histones have been removed still retain a residual folded structure with long DNA loops [10 to 90 kbp] extending from a condensed proteinaceous lattice. The central structure is called the *chromosome scaffold* and consists of a very limited set of nonhistone chromosomal proteins. The scaffolding actually exhibits the shape of the metaphase chromosome (Figure 9.21), a shape that remains even when the DNA is digested away by nucleases. The DNA in the loops seems to return to a site very near that from which it emerged. (In regular metaphase chromosomes the DNA loops would be coiled either within the 30-nm fiber or in a supercoiled version of the 30-nm fiber.)

Finally, Figure 9.22 shows a schematic diagram of the different orders of DNA packing that could give rise to the highly condensed metaphase chromosome. Interphase chromosomes would involve less packing than metaphase chromosomes. Very little is known about the mechanisms involved in the transitions between interphase and metaphase chromosome morphologies, or about the more subtle transitions in localized regions of chromosomes when genes are activated for transcription and turned off.

Keynote The large amount of DNA present in the eukaryotic chromosome is compacted by its association with histones in nucleosomes. The functional state of the chromosome is related to the extent of coiling: The more condensed a part of a chromosome is, the less likely it is that the genes in that region will be active.

Centromeres and Telomeres

So far in this chapter we have described eukaryotic chromosomes as linear structures each containing a single linear DNA molecule wrapped around histones and associated with nonhistone

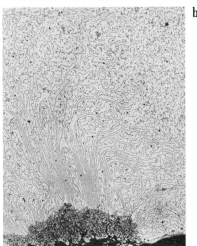

Figure 9.21 Electron micrographs of a metaphase chromosome depleted of histones: (a) the chromosome maintains its general shape by a nonhistone protein scaffolding from which loops of DNA protrude; (b) higher magnification of a section of (a) showing the attachment of the DNA loops to the scaffold.

the sites at which DNA replication is initiated; (3) centromeres, the sites at which chromosomes attach to the mitotic and meiotic spindles; and (4) *telomeres,* the ends of the chromosomes. Genes are, of course, the focus of genetic studies and are the central subjects of this text. Replication origins will be discussed in Chapter 10, the chapter on DNA and chromosome replication. Centromeres and telomeres will be discussed in the following two sections.

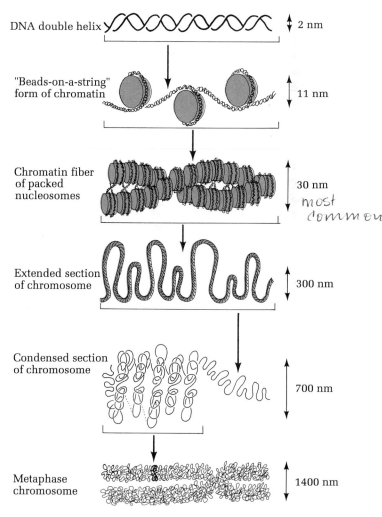

DNA double helix — 2 nm

"Beads-on-a-string" form of chromatin — 11 nm

Chromatin fiber of packed nucleosomes — 30 nm *most common*

Extended section of chromosome — 300 nm

Condensed section of chromosome — 700 nm

Metaphase chromosome — 1400 nm

Figure 9.22 Schematic drawing of the many different orders of chromatin packing which are thought to give rise to the highly condensed metaphase chromosome.

proteins. In terms of size, number, and morphology the chromosome complement of an organism is species-specific. Nonetheless, as described in Chapter 1, all chromosomes behave similarly at the time of cell division. In mitosis, for example, the attached sister chromatids become aligned at the metaphase plate, the chromatids separate at the centromeres, and one chromatid (now daughter chromosome) of each pair is distributed to each daughter cell. How does the structure of the chromosomes explain their behavior in mitosis and meiosis? We can define four functional elements of all eukaryotic chromosomes: (1) genes; (2) replication origins,

a) Specific centromere sequences

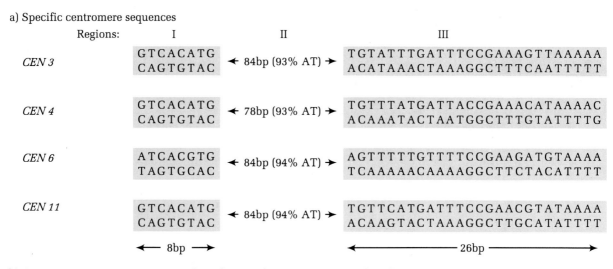

b) Consensus centromere sequences (in at least 7 of 11 centromeres analyzed)

Figure 9.23 (a) Regions of sequence homology between CEN3 *(10),* CEN4, CEN6 *(8), and* CEN11 *(10) in the yeast,* Saccharomyces cerevisiae. *The sequence elements I–III are set up in a spatial arrangement that is nearly identical in the centromeres of four different chromosomes; (b) Consensus sequence of* S. cerevisiae *centromeres.*

Centromeres. The centromere region of each eukaryotic chromosome is responsible for the accurate segregation of the replicated chromosomes to the daughter cells during mitosis and meiosis. The error frequency for this process is very low. In yeast, for example, segregation errors (i.e., nondisjunction—see Chapter 3, pp. 72–75) occur at a frequency of 10^{-5} or less. If the centromere is absent, the chromosome involved is unstable in the cell division process. That is, the chromosome will replicate but the two chromosome copies will not always segregate properly to the daughter cells.

In many higher eukaryotes the centromere is seen as a constricted region at one point along the chromosome that contains the kinetochore (see Chapter 1, p. 17) to which the spindle fibers attach during cell division, and that gives the chromosome its characteristic morphological appearance at metaphase. In most eukaryotes several spindle fibers attach to each centromere, whereas in the yeast *Saccharomyces cerevisiae,* the organism whose centromeres have been best characterized, only one spindle fiber attaches to each centromere.

The DNA sequences (called *CEN* sequences) of at least 10 *S. cerevisiae* centromeres have been determined. While each centromere has the same function, each differs in specific DNA sequence. Nonetheless there is a high degree of organizational homology among the known *CEN* sequences. Figure 9.23a presents the key sequence information for four yeast centromeres and Figure 9.23b gives the consensus sequence (i.e., the sequence that is essentially common) for the known yeast centromeres. The common core centromere region consists of three sequence domains. Element II, a 78–86 bp region of high A + T content (>90 percent A + T bases) is the largest domain. Flanking to one side is element I, a conserved PuTCACPuTG sequence (where Pu is a purine, i.e., A or G), and to the other side is element III, a 25 bp conserved sequence domain.

Some progress has been made in determining the role of the different centromere domains in centromere function. Deletion experiments have shown that the three regions, I, II, and III, are important to the centromere's function. That is, deletions that removed all of regions I, II, and III completely abolished centromere activity. Deletions that went up to region I retained full activity, suggesting that region I is less important to centromere activity than regions II and III.

A centromere from the fission yeast *Schizosaccharomyces pombe* was recently characterized. The centromere region is much more complex than the centromere of *Saccharomyces cerevisiae,* encompassing at least 30 kbp of DNA. Also, *Saccharomyces cerevisiae* centromeres have been shown not to function in *Schizosaccharomyces pombe,* suggesting that the sequences responsible for centromere function are not highly conserved, despite the identical function they must serve in all eukaryotes.

The chromatin organization of *S. cerevisiae* centromeres has been analyzed by mapping sites in the chromatin that are hypersensitive (i.e., more sensitive than average) to digestion by either micrococcal nuclease or DNase I. The data obtained showed that the core centromere regions I, II, and III are contained within a 220–250 bp protected region that is flanked by highly nuclease sensitive sites (Figure 9.24). The protected region is surrounded by an array of nuclease cleavage sites spaced at the usual yeast nucleosomal interval of 160 bp, indicating a normal nucleosomal organization in those areas. As diagramed in Figure 9.24, *CEN* element III is near the center of the 220–250 bp protected region, suggesting that this element might be directly involved with spindle fiber binding. Note in Figure 9.24 that the centromere core diameter closely matches the spindle fiber microtubule diameter.

Finally, it is clear that the complex structure of the centromere and the attachment of the spindle microtubules must be mediated by proteins, some of which presumably interact directly with centromeric DNA. Some progress is

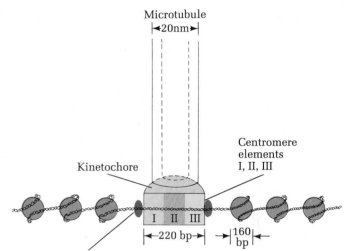

Figure 9.24 *Model for a centromere and adjacent regions in a yeast chromosome.*

being made in identifying and characterizing centromere-binding proteins. One of the major problems to be solved is the mechanism of attachment of the centromere to the spindle.

Telomeres. A telomere is the region of DNA at the end of a linear chromosome that is required for replication and stability of that chromosome. Telomeric regions of chromosomes are characteristically, but not necessarily, heterochromatic in appearance. In most organisms that have been examined, the telomeres are positioned just under the nuclear envelope, and are often found associated with each other as well as with the nuclear envelope.

There is molecular evidence that all telomeres in a given species share a common sequence. Telomeric sequences may be divided into two types:

1. Regions near the ends of chromosomes often contain repeated but still complex DNA sequences extending for many thousands of base pairs from the molecular end of chromosomal DNA. These sequences, called **telomere-associated sequences,** may well mediate many of the telomere-specific interactions, for example, among telomeres and between telomeres and the nuclear envelope.

2. Sequences at, or very close to, the extreme ends of the chromosomal DNA molecules of a number of lower eukaryotes have been determined and have been shown to consist of simple, tandemly repeated DNA sequences. These so-called **simple telomeric sequences** are the essential functional components of telomeric regions in that they are sufficient to supply a chromosomal end with stability. Table 9.5 gives the DNA sequences for the simple telomeric sequences found in a number of lower eukaryotes. In the ciliate *Tetrahymena,* for example, the repeated sequence consists of 5'-CCCCAA-3' elements, and in the flagellate *Trypanosoma,* the repeated sequence is CCCTAA. Much remains to be learned about these sequences in general and the properties of chromosomal ends.

Keynote *The centromere region of each eukaryotic chromosome is responsible for the accurate segregation of the replicated chromosome to the daughter cells during both mitosis and meiosis. The DNA sequences of yeast centromeres (CEN sequences) exhibit individual sequence differences while having a high degree of organizational homology. The common core centromere region consists of three sequence domains, one of which has been postulated to serve as a spindle fiber attachment point.*

Facts about the ends of chromosomes, the telomeres, are also becoming known. Often the telomeres are associated with the nuclear membrane. Telomeres in a species often share a common sequence. Characteristically sequences at or very close to the extreme ends of the chromosomal DNA consist of simple, relatively short, tandemly repeated sequences. Repeated, often complex, DNA sequences, called telomere-associated sequences, are found farther in from the chromosome ends.

Sequence Complexity of Eukaryotic DNA

Now that we know about the basic structure of DNA and its organization in chromosomes, we can examine some of the available information about the distribution of the four bases A, T, G, and C and of certain sequences of these bases in the genomes of prokaryotes and eukaryotes. These data are obtained through denaturation-renaturation analysis of DNA.

Denaturation-Renaturation Analysis of DNA

The two strands of linear, double-helical DNA are held together by the relatively weak hydrogen bonds between the bases. The A–T base pair has two such bonds, while the G–C base pair has three. If a solution of DNA is heated slowly, eventually the hydrogen bonds will break and a population of single-stranded DNA molecules will result; this procedure is called denaturation or melting of the DNA.

If thermally denatured DNA is cooled rapidly, the single-stranded molecules stay single-

Table 9.5 Tandemly Repeated Telomeric DNA Sequences of Lower Eukaryotes

Organism	$5' \to 3'$ Sequence[*]
HOLOTRICHOUS CILIATES	
Tetrahymena	CCCCAA
Glaucoma	
Paramecium	
HYPOTRICHOUS CILIATES	
Stylonchia	CCCCAAAA
Oxytricha	
FLAGELLATES	
Trypanosoma	CCCTAA
Leptomonas	
Leishmania	
Crithidia	
SLIME MOLDS	
Physarum	$CCCTA_n$
Dictyostelium	$C_{1-8}T$
FUNGUS	
S. cerevisiae	$C_{2-3}A(CA)_{1-3}$

[*] The 5' to 3' strand sequence is directed from the end toward the interior of the linear DNA or chromosome.

stranded. However, if the molecules are cooled slowly, the strands start to come together again, and complementary base pairing occurs to produce some double-stranded molecules. This renaturation of denatured DNA depends on two sequential events. The first is the collision of DNA strands carrying complementary sequences and the proper alignment of these sequences. The second is the formation of hydrogen bonds between the base pairs once the complementary sequences have matched. A complete match is not needed; all that is required is a high degree of complementarity between the two sequences. The study of the kinetics of renaturation of denatured DNA, which was pioneered by R. Britten and D. Kohne, gives us useful information about the sequence organization of the DNA in the genome (genomic DNA) and points out yet one more distinct difference between prokaryotes and eukaryotes, as we shall see.

Theory of DNA renaturation. The rate of renaturation of denatured DNA into double-helical DNA is a function of the two reactants, the two populations of dissociated single-stranded DNAs. We can write a simple equation for it:

$$S1 + S2 \xrightarrow{k} D$$

Here $S1$ and $S2$ are the two complementary single strands, D is the renatured double-helical DNA, and k is the rate constant for the reaction.

Experimentally, we start out with a known amount of DNA in single-stranded form, and with time (on the order of seconds to hours) an increasing proportion of it becomes double-stranded by DNA–DNA renaturation. It does so as a function of the rate constant k. The rate equation (called the C_0t equation) for this experiment is usually written as

$$\frac{C}{C_0} = \frac{1}{1 + kC_0t}$$

The symbols are defined as follows:

k is the rate constant, as before.
t is the time of the reaction, in seconds (s).
C_0 is the initial concentration of single-stranded DNA at time zero, in moles per liter (mol/L).

C is the concentration of single-stranded DNA that remains in the reaction, in moles per liter.

The equation shows that the fraction of single-stranded DNA remaining in a renaturation reaction (C/C_0) is a function of C_0t, the product of the initial DNA concentration (C_0) and the elapsed time of the reaction (t). Thus the kinetics of DNA renaturation is usually graphed as a **C_0t plot.** An ideal C_0t plot is shown in Figure 9.25. In this plot the fraction of DNA in single-stranded form is graphed as a logarithmic function of the initial concentration of DNA (C_0), in moles per liter, multiplied by the time (t), in seconds. The assumption is that all single-stranded DNA fragments in the reaction mix had an equal probability of forming a renatured double-stranded DNA. In other words, we assume that all DNA fragments are unique sequences, with none represented more than once. On the Y (vertical) axis the initial state for C/C_0—that is, all single-stranded DNA—is indicated by the value of 1, and the final state—that is, all double-stranded DNA—is indicated by the value 0. A convenient value that can be obtained from the curve is $C_0t_{1/2}$, which is the value of C_0t at which the reaction has been half completed (i.e., $C/C_0 = 1/2$).

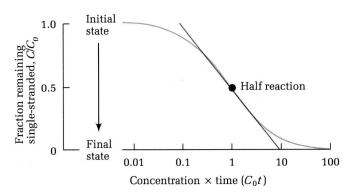

Figure 9.25 Ideal time course for the renaturation of DNA as seen in C_0t (initial DNA concentration × time) plot. In the initial state the DNA is single-stranded, and in the final state it is all double-stranded. Note that 80 percent of the renaturation occurs over a 2 log C_0t interval.

C_0t curves and the sequence organization of DNA. In the production of a C_0t curve, DNA is sheared to a length of about 400 base pairs, the fragments are denatured, and renaturation is allowed to proceed. Samples of the reaction are taken at various times, and the relative proportions of single-stranded and double-stranded DNA are determined. Figure 9.26a shows the C_0t curves for several prokaryotic and viral DNAs. The shape of each curve closely parallels the theoretical curve, indicating that the DNA of these prokaryotic organisms consists of unique sequences.

Note that the curves are not all in the same place on the graph. They occur at different places because the rate of DNA renaturation is related to the amount of DNA in the genome (the genome size) of the organism. Imagine the analogy of a population of nuts and bolts that we are trying to match by randomly trying pairs together. If the population is 100 nuts and 100 bolts in which there are 100 different-sized nut-and-bolt pairs, once a nut is picked, it will take a long time to find the matching bolt since only 1 out of 100 bolts will fit. However, if only 10 different sizes of nuts and bolts are in the population, when one nut is picked, the chance of finding a matching bolt is 1 in 10, and matching the nuts and bolts together would occur more quickly.

This analogy is equivalent to a renaturation experiment in which there is a fixed concentration of DNA, but one genome is a tenth of the size of the other, so the larger genome has one set of sequences and the smaller genome has ten sets of its sequences. As the genome size becomes larger, the $C_0t_{1/2}$ value becomes larger. Further, the $C_0t_{1/2}$ value is related directly to the genome size, so determining the $C_0t_{1/2}$ experimentally allows the genome size to be calculated.

a) C_0t plots of DNAs from organisms with small genomes

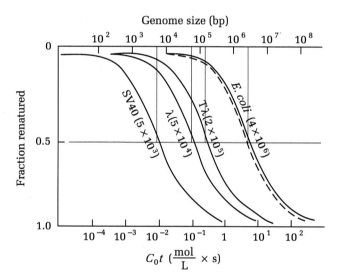

b) C_0t plots of *E. coli* and calf DNA

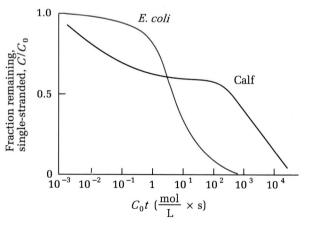

Figure 9.26 C_0t *plots for a variety of DNAs: (a)* C_0t *plots showing the renaturation of DNAs from organisms with small genomes: the bacterium* E. coli, *the bacterial viruses T2 and λ, and the animal virus SV40. The dashed line is a theoretical plot for renaturation of unique-sequence DNA from an organism with a genome size equal to that of* E. coli. *(b) Kinetics of renaturation of DNA from calf thymus and* E. coli *as seen in a* C_0t *plot. The* E. coli *DNA consists almost entirely of unique sequences. However, the shape of the* C_0t *curve for the calf DNA is very different from that of* E. coli *and indicates that there are some sequences (toward the right of the curve) that renature much more slowly and some (toward the left of the curve) that renature much more quickly than the bacterial DNA sequences.*

Effect of repetitive base sequence on DNA renaturation kinetics. If a particular base sequence is present in multiple copies, those sequences will reassociate more rapidly than will unique sequences. Continuing the nuts-and-bolts analogy, if most of the nuts and bolts in the set are unique, then it will take a lot of time to find a matching nut-and-bolt set; whereas if all of the nut-and-bolt pairs are the same (similar to highly repetitive sequences), then matched pairs can be formed very quickly. If some nut-and-bolt types are repeated and some are unique, then pairs of the former will be produced more rapidly than pairs of the latter.

Figure 9.26b presents an illustration of the effects of repetitive sequences on DNA renaturation kinetics. The figure shows a C_0t curve for DNA from a calf alongside that for *E. coli,* which, as we have already seen, has only unique-sequence DNA. The curve for calf DNA is very different. Some of the DNA renatures at very low C_0t values, indicating that the calf DNA contains repetitive sequences, present in many copies, so the probability of their collision in the solution is much higher than it is for the unique-sequence fragments.

At the other end of the scale some of the calf DNA renatures at very high C_0t values, much higher than those for *E. coli.* These higher C_0t values probably represent unique-sequence DNA, which, because of the much larger genome size in the calf, renatures more slowly than the *E. coli* DNA. In between these two extremes there are apparently a variety of sequences that are repeated to various degrees in the genome, resulting in the broadness of the calf C_0t curve.

By following the kinetics of DNA–DNA renaturation and making C_0t curves, geneticists can determine the abundance of particular types of sequences in DNA. For convenience, we speak of three different types of sequences: **unique** (or single-copy), **moderately repetitive,** and **highly repetitive.** Note that because we are studying randomly sheared pieces of DNA, we are not looking directly at genes but rather at DNA sequences that may or may not contain genes. That is, while all DNA sequences can be copied into more DNA during replication, only some

DNA sequences—genes—contain information that is transferred to other molecules in the cell, specifically, RNA and protein. Thus all genes are DNA sequences, but not all DNA sequences are genes. Nonetheless, from extensive investigations much information has been obtained about the relationship between genes and the three types of sequences defined by DNA–DNA renaturation experiments.

As the name suggests, unique sequences (sometimes called single-copy sequences) are present as single copies in the genome. (Thus there are two copies per diploid cell.) In current usage the term usually applies to sequences that have one to a few copies per genome. Moderately repetitive sequences, on the other hand, are reiterated from a few to as many as 10^3 to 10^5 times in the genome. Highly repetitive sequences are repeated between 10^5 and 10^7 times in the genome.

We could discuss the relative proportions of these unique and repetitive sequences. However, since there is a continuum in the number of repeated sequences within and between organisms, such information would only reflect the somewhat arbitrary boundaries set between the types. Suffice it to say that there is a fair degree of variation between the relative proportion of unique and repetitive sequences among eukaryotic organisms. To give but one concrete example, the distribution of unique and repeated sequences in mouse DNA is shown in Figure 9.27. About 70 percent of the DNA is single-copy material. The repeated sequences fall into two main groups, the highly repeated and the moderately repeated types. About 10 percent of the DNA consists of the highly repeated sequences, where the number of copies is approximately a million. The remaining DNA, that is, about 20 percent of the genome, consists of the heterogeneous class of moderately repeated sequences; these sequences are repeated between 1000 and 100,000 times in the genome.

Isolation of Unique-Sequence and Repetitive-Sequence DNA Classes

By taking advantage of the different rates of

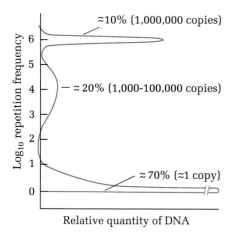

Figure 9.27 Frequency distribution of unique and repeated nucleotide sequences in mouse DNA. The number of copies of each type of sequence is estimated from the C_0t plot for the mouse.

reassociation of unique-sequence and repetitive-sequence DNA, geneticists can isolate DNA that is representative of these types. The strategy is straightforward since it merely involves taking samples from a DNA–DNA reassociation reaction at chosen points along the C_0t curve. For example, if we sample at a relatively early time, then only the highly repetitive DNA will have reassociated to form double-stranded molecules. However, the sample taken will also contain single-stranded molecules of all kinetic classes that have yet to find their partners. Fortunately, it is quite a simple matter to separate double-stranded (duplex) DNA from single-stranded DNA, and thus particular populations of double-stranded DNA molecules can be collected as the reassociation proceeds. And DNA from any cross section of the curve can be isolated in a very similar way. For example, to isolate unique-sequence DNA, we allow a DNA–DNA reassociation experiment to proceed for a time interval long enough so that the DNA we are not interested in (i.e., the highly repetitive and moderately repetitive sequences) form duplex molecules. These molecules are collected, and the DNA that reassociates after that time—that is, at a high C_0t value— will be the unique sequences and can be isolated.

Unique-sequence DNA. Unique sequences, by definition, are present once in the haploid genome. In common application, however, unique sequences are represented from one to a few times in the genome. These sequences take a long time to reassociate after denaturation. In most eukaryotes about 70 percent of the DNA consists of unique-sequence material. Further, most of the genes that we know about—those that code for proteins in the cell—are in the unique-sequence kinetic class of DNA. Conversely, it is probably the case that not all the unique-sequence material contains protein-coding sequences. We make this statement because the amount of DNA present in the unique class would code for more proteins than we believe are produced by the cell.

Moderately repetitive DNA sequences. The moderately repetitive DNA sequence class is wildly heterogeneous in terms of the frequencies of the repeated sequences represented. Some known genes are found in this class, notably the genes for ribosomal RNAs (rRNAs) and transfer RNAs (tRNAs). For example, there are about 200 copies of the major class of rRNA genes in *Neurospora crassa,* 450 copies in the clawed toad, *Xenopus laevis,* 160–200 copies in humans, 260 copies in the sea urchin, 3600 copies in Douglas fir, and 3900 copies in the garden pea.

In most cases these repeated genes are found in one or more clusters in the genome, with each cluster consisting of an uninterrupted tandem array of the genes. Sequences that are repeated one after another in a row are called **clustered repeated sequences.** Clusters also occur for some tRNA genes and, in some organisms, for the multiple copies of the histone genes.

Only part of each repeated unit of a tandemly repeated cluster is transcribed. The general picture, then, is one of long tandem arrays of complex repeated sequences, portions of which are transcriptionally active and portions of which are not. These sequences represent what are called *multigene families* in which the numbers per genome vary from a few hundred to many thousand.

Also found among the moderately repeated sequences are *interspersed repeated sequences,* which are repeated sequences that are scattered, not clustered, in the genome. These sequences are characteristic of most eukaryotic DNAs, although no single unifying description of their arrangement can be applied to all the known examples. The name *interspersed repeated sequences* is applied to those moderately repeated (10^3–10^5, usually) sequence families that are dispersed among unique sequences in the genome.

Two common patterns of interspersion have been found. In the first, the so-called *short-period pattern*—exemplified by *Xenopus* DNA and found, for example, in *Xenopus,* sea urchin, humans, and many other eukaryotes—100–300 base-pair repeated sequences are interspersed with longer unique sequences of about 1000–2000 base pairs in length. Between 50 and 80 percent of the DNA typically shows this pattern. Note, though, that other arrangements can also be found in the same genomes—for example, interspersed repeated sequences not much greater than 100–300 base pairs long, long tandem repeated sequences not interspersed with unique-sequence DNA, and long unique-sequence stretches of DNA not interspersed with short repeated sequences.

In the second, or *long-period pattern,* as exemplified by *Drosophila* DNA, about 5000 base-pair repeated sequences are dispersed among unique sequences that may be up to 35,000 base pairs or more in length. Recent evidence indicates that these long repeated sequences, even though they consist of different sequence families when the genome is considered as a whole, share a common structure in which the 5000 base-pair repeated sequences are flanked on either side by shorter, direct repeated sequences of about 255–400 base pairs. The long-period pattern is also found in yeast and a few other organisms. An intermediate pattern is found in chickens and some other organisms.

A more detailed analysis at the molecular level has revealed additional complexity for the interspersed repeated sequences. On the basis of size and relative abundance two different classes of interspersed and highly repeated sequences have been described. One class consists of dispersed families with unit lengths of under 500 base pairs and with as many as hundreds of thousands of copies. These families are called **SINES,** for **short interspersed repeated sequences.** One example of a SINES family is the *Alu* sequence family in mammals. These involve 150 to 300 bp moderately repetitive sequences. The DNA sequences of about 50 of the human *Alu* family have been determined. No two sequences are exactly alike, but there are enough similarities to show clearly that they are related. There is evidence that at least some of the *Alu* sequences are located in regions of the chromosomes that are transcribed. However, there does not appear to be any specific function for the Alu sequences in controlling transcription.

Other dispersed families of sequences in mammals are several thousand base pairs in length and occur > 20,000 times in the genome. These families are called L1 families (formerly called **LINES,** for **long interspersed repeated sequences**). Only a single L1 family has been identified in each mammalian species investigated. Families in different species have been shown to be homologous to one another over considerable lengths of DNA. At least some of the L1 sequences in a family appear to code for proteins, although the proteins themselves have not been characterized.

Highly repetitive DNA sequences. Among the highly repetitive class of DNA sequences are simple clustered repeated sequences, some with repeated units no more than 6 base pairs long repeated as many as 10^6 to 10^7 times in the genome and others hundreds of base pairs long and repeated millions of times. They reassociate at low C_0t values, thus facilitating their purification. Since these sequences often have a %GC content (base ratio) different from that of the majority of the DNA in the genome, they will usually form a distinct band (or bands) when centrifuged in a solution that separates DNA on the basis of %GC content; such bands are called *satellite bands,* and the DNA contained therein is known as **satellite DNA.**

The amount of satellite DNA found in an organism varies considerably. For example, a small percentage of the human genome consists of satellite DNA, whereas more than 50 percent of the kangaroo rat's genome is satellite DNA. Commonly, the satellite DNA sequences are found associated with heterochromatin and thus are localized around centromeres and at the telomeres (ends) of the chromosomes.

It was once thought that satellite DNAs had simple, redundant repeated units of short base-pair sequences. It is now known that the repeat length can be from a few to several thousands of base pairs, indicating tremendous complexity of these sequences. There is no experimental evidence for the function of satellite DNA sequences. They are not transcribed in the cell; that is, they do not code for a functional product. The universal (although not exclusive) association of satellite DNA sequences with hetero-

chromatin suggests that they have a role in heterochromatin-specific events. Unfortunately, heterochromatin function is itself not understood at the molecular level. Characteristically, these sequences replicate relatively late in the S phase of the cell cycle.

Keynote *Prokaryotic genomes consist mostly of unique-sequence DNA with only a few sequences and genes repeated; eukaryotes have both unique and repetitive sequences in the genome. Particularly in higher eukaryotes, the spectrum of complexity of the repetitive DNA sequences is extensive. The highly repetitive sequences tend to be localized to heterochromatic regions around centromeres and chromosome ends, whereas the unique and moderately repetitive sequences tend to be interspersed.*

*Analytical
Approaches
for Solving
Genetics
Problems*

Q.1 When double-stranded DNA is heated to 100°C, the two strands separate because the hydrogen bonds between the strands break. When cooled again, the two strands typically do not come back together again because they do not collide with one another in the solution in such a way that complementary base sequences can form. In other words, heated DNA tends to remain as single strands. Consider the following DNA double helix:

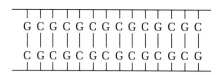

If this DNA is heated to 100°C and then cooled, what might be the structure of the single strands? Assume in formulating your answer that the concentration of DNA is so low that the two strands never find one another.

A.1 This question serves two purposes. First, it reinforces certain information about double-stranded DNA, and, second, it poses a problem that can be solved by simple logic.

In the chapter we examined the structure of DNA and some of the ways to investigate the sequence complexity of the genome, particularly in eukaryotes. One of the experimental approaches discussed was DNA–DNA renaturation in which double-stranded DNA was denatured into single strands and the kinetics of renaturation was measured. Those molecules that renatured rapidly were assumed to be represented in the genome a large number of times, and so on. In the question posed here the double-stranded DNA has been melted to single strands by heat treatment. From the text we know that the renaturation of strands is a function of the product of the concentration at time zero and the time (in minutes)—the C_0t. Conditions are set up here so that the concentration is so low that the two strands do not renature. We are left with an analysis of the base sequences themselves to see whether there is anything special about them in order to avoid an answer of "nothing significant happens." The DNA is a 14-base-pair segment of alternating GC and CG base pairs. By examining just one of the strands, we can see that there is an axis of symmetry at the midpoint such that it is possible for the single strand to form a double-stranded DNA molecule by intrastrand ("within strand") base pairing. The result is a double-stranded hairpin structure, as shown in the following diagram (from the top strand; the other strand will also form a hairpin structure):

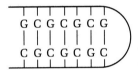

Questions and Problems

*9.1 The nucleotide sequences of two DNA molecules from a population of T2 DNA molecules are as shown:

$$\underset{\longrightarrow}{\underset{1}{\text{T A G C T C C}}} \qquad \underset{\longrightarrow}{\underset{3}{\text{G C T C C T A}}}$$

and

$$\underset{\longleftarrow}{\underset{2}{\text{A T C G A G G}}} \qquad \underset{\longleftarrow}{\underset{4}{\text{C G A G G A T}}}$$

These molecules were heat-denatured and then the separated strands were allowed to renature. Diagram the structures of the renatured molecules most likely to appear when (a) strand 2 renatures with strand 3 and (b) strand 3 renatures with strand 4. Mark the strands and indicate sequences and polarity.

9.2 Chromosomes contain (choose the best answer)

a. protein.
b. DNA and protein.
c. DNA, RNA, histone, and nonhistone protein.
d. DNA, RNA, and histone.
e. DNA and histone.

9.3 List four major features of eukaryotic chromosomes that distinguish them from prokaryotic chromosomes.

9.4 Discuss the structure and role of nucleosomes.

*9.5 What does the term *linkage number* refer to?

9.6 What are topoisomerases?

9.7 What is the relationship between cellular DNA content and phylogeny?

*9.8 What are the main molecular features of the yeast centromeres?

9.9 What are telomeres?

*9.10 Would you expect to find most protein coding genes in unique-sequence DNA, in moderately repetitive DNA, or in highly repetitive DNA?

9.11 Would you expect to find tRNA genes in unique-sequence DNA, in moderately repetitive DNA, or in highly repetitive DNA?

10

DNA Replication

In Chapter 9 we learned that one of the essential properties of genetic material is that it must replicate accurately so that progeny cells have the same genetic information as the parental cell. When Watson and Crick proposed their double-helix model for the structure of DNA, it immediately occurred to them that replication of the molecule would be straightforward if their model was correct. As the DNA double helix unwound, two new strands of DNA could be produced by using the parent strands as templates. From the complementary-base-pairing rules, the result of the process would be two daughter DNA molecules, each of which would consist of one parental strand and one newly synthesized strand bonded together with hydrogen bonds. This DNA replication scheme is called the **semi-conservative model** since each daughter molecule retains one of the parental strands.

At the time another model for DNA replication, the **conservative model,** was also seriously considered. In the conservative model the two parental strands of DNA remain together and serve as a template for the synthesis of a new daughter double helix. Thus one of the two progeny DNA molecules is the parental DNA, and the other consists of totally new material. The two models for replication are presented in Figure 10.1.

In this chapter we will learn about the mechanics of DNA replication in prokaryotes and eukaryotes and about some of the enzymes and other proteins required for replication.

DNA Replication in Prokaryotes

Five years after Watson and Crick developed their model, M. Meselson and F. Stahl obtained experimental evidence that the semiconservative replication model was the correct one. The following section presents their experiments and results.

Experimental Support for the Semiconservative Model of DNA Replication

To examine the process of DNA replication, Meselson and Stahl used the bacterium *Escherichia coli* because it can be easily and quickly grown in a minimal medium. In Meselson and Stahl's experiment (Figure 10.2a) *E. coli* was grown for several generations in a minimal medium in which the only nitrogen source was $^{15}NH_4Cl$ (ammonium chloride); in this compound the normal isotope of nitrogen, ^{14}N, is replaced with ^{15}N, the heavy isotope. As a result, all the bacteria's cellular components, including its DNA, contained ^{15}N instead of ^{14}N. The ^{15}N DNA can be separated from ^{14}N DNA by using equilibrium density gradient centrifugation (described in Box 10.1 on page 317). Briefly, in this technique a solution of cesium chloride (CsCl) is centrifuged at high speed, causing the cesium chloride to form a density gradient. If DNA is present in the solution, it will band to a position where its buoyant density is the same as that of the surrounding cesium chloride.

As the next step in Meselson and Stahl's experiment, the ^{15}N-labeled bacteria were transferred to a medium containing nitrogen in the normal ^{14}N form. The bacteria were allowed to

a) The semiconservative model

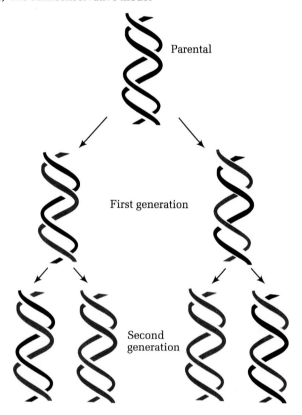

b) The conservative model

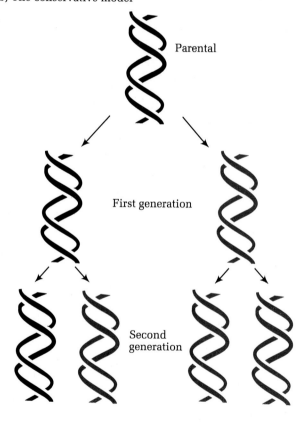

Figure 10.1 Two models for the replication of DNA: (a) the semiconservative model (the correct model); (b) the conservative model. The parental strands are shown in black and the newly synthesized strands are shown in color.

Box 10.1
Equilibrium
Density
Gradient
Centrifugation

In experiments involving equilibrium density gradient centrifugation, a concentrated solution of cesium chloride (CsCl) is centrifuged at high speed. The opposing forces of sedimentation and diffusion produce a stable, linear concentration gradient of the CsCl. The actual densities of CsCl at the extremes of the gradient are related to the CsCl concentration that is centrifuged.

For example, to examine DNA of density 1.70 g/cm^3 (a typical density for DNA), one makes a gradient from 1.60 to 1.80 g/cm^3. If the DNA is mixed with the CsCl when the density gradient is generated in the centrifuge, the DNA will come to equilibrium at the point in the gradient where its buoyant density equals the density of the surrounding CsCl (see Box Figure 10.1). The DNA is said to have banded in the gradient. If DNAs are present that have different densities, as is the case with ^{15}N DNA and ^{14}N DNA, then they will band (i.e., come to equilibrium) in different positions.

In the experiments the gradients contain the stain ethidium bromide, which complexes with DNA. When it does so, the ethidium bromide fluoresces under ultraviolet light, and therefore the bands of DNA can be visualized.

The method can also be used experimentally to purify DNAs on the basis of their different densities. It has been used, for example, to fractionate nuclear and mitochondrial DNA in extracts of eukaryotic cells and to separate bacterial plasmid DNA from the host bacterium's DNA.

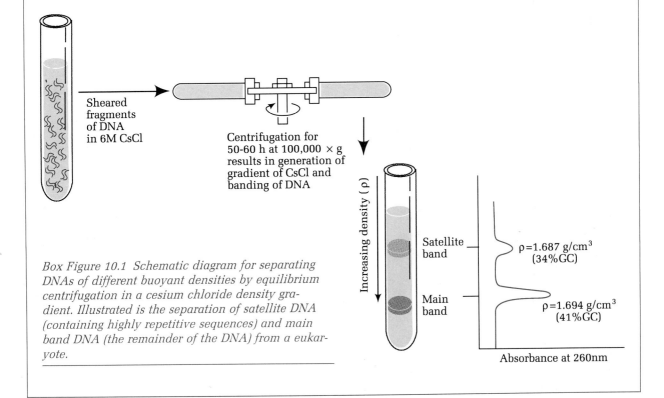

Box Figure 10.1 Schematic diagram for separating DNAs of different buoyant densities by equilibrium centrifugation in a cesium chloride density gradient. Illustrated is the separation of satellite DNA (containing highly repetitive sequences) and main band DNA (the remainder of the DNA) from a eukaryote.

Sheared fragments of DNA in 6M CsCl

Centrifugation for 50-60 h at 100,000 × g results in generation of gradient of CsCl and banding of DNA

Increasing density (ρ)

Satellite band

Main band

ρ=1.687 g/cm^3 (34%GC)

ρ=1.694 g/cm^3 (41%GC)

Absorbance at 260nm

a) Experimental outline of the Meselson and Stahl experiment

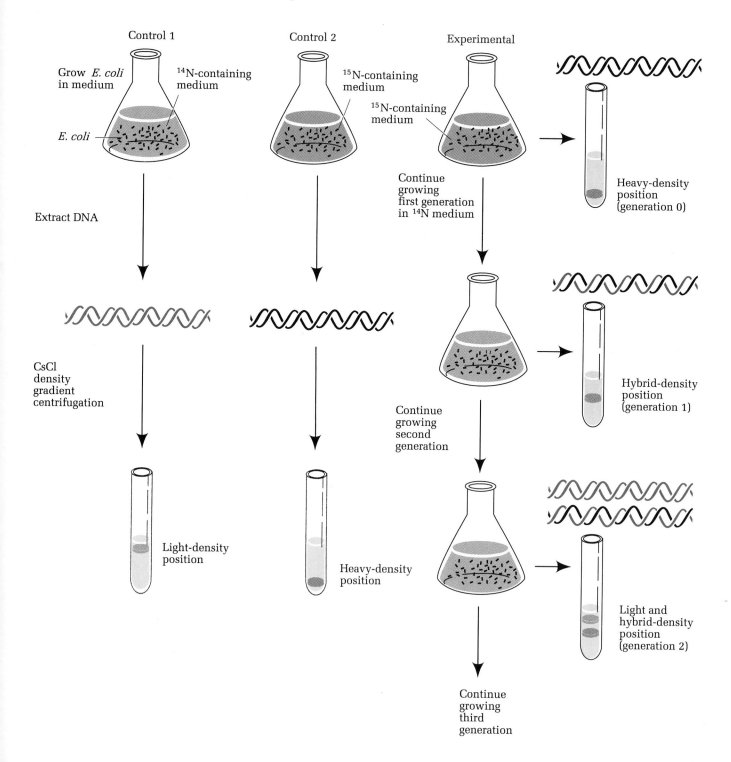

b) Results of the Meselson and Stahl experiment

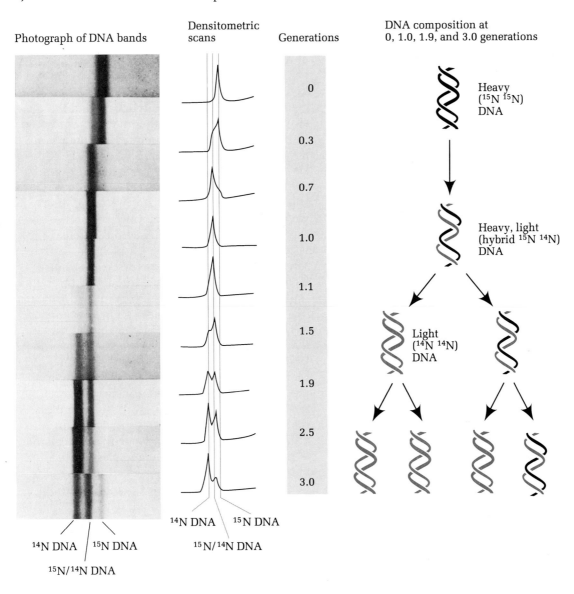

Figure 10.2 Experimental demonstration of semiconservative replication in E. coli: (a) (opposite) experimental outline of the Meselson and Stahl experiment. The controls show the experimental determination of the E. coli *is grown continuously on* ^{14}N-containing (control 1) or ^{15}N-containing (control 2) medium, and the experimental culture slows the experimental determination of the densities of DNA when E. coli *is shifted from a* ^{15}N-containing to a ^{14}N-containing medium. (b) (above) Results of the Meselson and Stahl experiment in which the density of E. coli DNA was analyzed for three generations after a shift from a ^{15}N-containing to a ^{14}N-containing medium. Shown are photographs of the DNA bands (left) and a densitometric scan of the bands (center) and a schematic interpretation of the DNA composition at various generations (right).

replicate in the new conditions for several generations. Throughout the period of incubation in the ^{14}N medium, samples of *E. coli* were lysed to release the cellular contents, and the DNA was analyzed in cesium chloride density gradients. After one generation in ^{14}N medium, all the DNA had a density that was exactly intermediate between that of totally ^{15}N DNA and totally ^{14}N DNA. After two generations, half the DNA was of the intermediate density and half was of the density of DNA containing entirely ^{14}N. These observations, presented in Figure 10.2b, and those for subsequent generations were exactly what the semiconservative model predicted. If the conservative model of DNA replication had been correct, then every generation would have been entirely heavy ^{15}N DNA (the original parental double helix) and newly synthesized ^{14}N DNA. *At no time would any DNA of intermediate density have been found.* The data from the Meselson and Stahl experiment proved the semiconservative model and ruled out the conservative model.

Visualization of DNA Replication in *E. coli*

Meselson and Stahl studied a population of DNA molecules. Thus while the data obtained proved that DNA replication was semiconservative in the population, no information was obtained about replication of a single DNA molecule. An experiment performed in 1963 by J. Cairns provided information of this kind for DNA replication in *E. coli*. Cairns grew *E. coli* cells for less than two generations in a medium containing a radioactive DNA precursor, ^{3}H-thymidine. The DNA that synthesized during the incubation period was therefore radioactive. The cells were then gently lysed, and the cell contents were collected on filter paper. The filter paper was placed against an X-ray film, and the sandwich was kept in the dark for two months, after which the film was developed. The resulting autoradiograph is shown in Figure 10.3. The lines of dark grains were produced on the film by the emission of electrons from the ^{3}H-thymidine incorporated into DNA. Cairns's autoradiograph showed a continuous

line of grains, indicating that the whole *E. coli* chromosome was photographed and that it was circular. The autoradiograph also showed that the chromosome represented was in the process of replication. The stage of replication of each part of the chromosome is shown in the inset of Figure 10.3. This interpretation of the autoradiograph was made by carefully measuring the density of the grains around the intact chromosome. In this way regions in which both strands contained radioactive ^{3}H-thymidine could be discerned from regions in which only one strand contained radioactive ^{3}H-thymidine. Since the cell was labeled for less than two generations, we can conclude that the former region has gone through two rounds of replication, whereas the latter region has only gone through one. Therefore the chromosome we are discussing must have been about two-thirds of the way through a replication cycle, and P_1 and P_2 in the inset are the points at which the replication is occurring.

Thus Cairns's work confirmed the validity of the semiconservative model of DNA replication indicated by Meselson and Stahl's experiments. More recent work has shown that the origin of replication in *E. coli* is at a constant position on the chromosome. This constant origin implies that the enzyme or protein responsible for initiating DNA synthesis must recognize a particular DNA sequence, bind there, and start replication at that site. Furthermore, replication occurs in both directions (i.e., bidirectionally) from the replication origin, so there are two replication forks moving in opposite directions around the chromosome.

Replication of Circular DNA

Cairns's experiments showed that in *E. coli* the parental DNA strands remain in a circular form throughout the replication cycle. This is generally true of circular DNA molecules; they always exhibit a theta-like (θ) shape resulting from a replicating bubble's initiation at the replication origin (Figure 10.4). Replication proceeds bidirectionally from the origin. Since the two DNA strands must untwist for replication to occur, one of the experiment's objectives was to discover how the two intertwined circular DNA strands could unravel during replication. That is, if both strands remain covalently circular, then as a section of DNA double helix is untwisted to make a replication bubble, the formation of positive supercoils will occur elsewhere in the molecule. If you take a circular piece of rope, for example, and try to separate the two strands at one point, at the opposite end of the circle the rope becomes tightly supercoiled. Topoisomerases (see Chapter 9) play an important part in the replication process by preventing excessively knotted DNA from forming and thereby allowing both parental strands to remain intact during the replication cycle (Figure 10.5). That is, periodically the unreplicated part of the theta structure becomes negatively supercoiled by the action of topoisomerase II, and this coun-

terbalances the positive supercoiling that occurs as DNA is untwisted during replication. Thus, topoisomerase II prevents the halting of DNA replication by severely supertwisted DNA.

Keynote DNA replication in E. coli *occurs by a semiconservative mechanism in which the two strands of a DNA double helix separate and two new complementary strands of DNA are synthesized on each of the two parental template strands. Semiconservative replication results in two double-stranded DNA molecules, each of which has one strand from the parent molecule and one newly synthesized strand. This mechanism ensures the faithful copying of the genetic information at each cell division.*

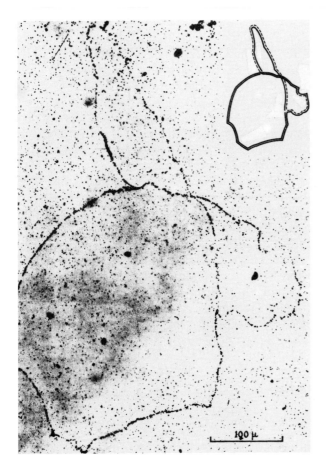

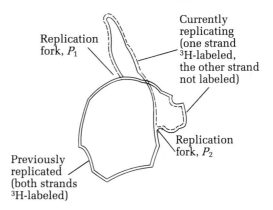

Figure 10.3 Result of Cairns's experiment to visualize the replication of the E. coli *chromosome. The lines of dark grains result from the decay of the ³H-thymidine incorporated into the DNA. The inset, which is also shown enlarged and labeled, presents an interpretation of the autoradiograph. The double solid lines represent DNA in which both strands are radioactive; the solid/dashed lines represent DNA with one radioactive (solid) and one nonradioactive (dashed) strand. P_1 and P_2 are the replication forks.*

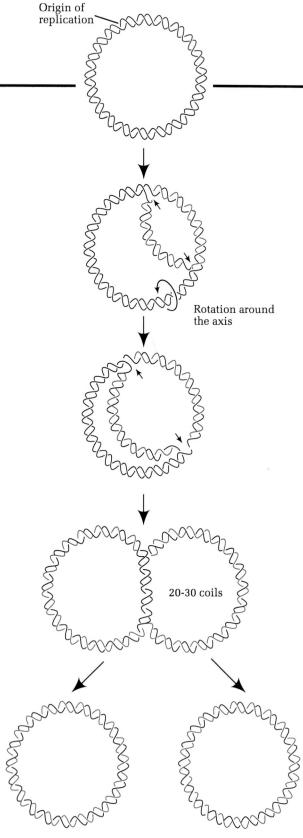

Origin of replication

Rotation around the axis

20-30 coils

Figure 10.4 Bidirectional replication of circular DNA molecules.

Denaturation and Unwinding of DNA

Replication in DNA usually starts at a specific site on the chromosome, the **origin.** At that site the double helix denatures into single strands and continues to unwind as the replication forks migrate. Denaturation is necessary to expose the bases for the synthesis of new strands. The unwound single strands upon which new strands are made (following complementary-base-pairing rules) are called the **template strands.** In the circular *E. coli* chromosome, for example, there is a single origin of replication called *oriC:* from which replication proceeds bidirectionally. Many parts of this relatively long sequence are conserved in closely related bacteria.

Initiation of DNA Synthesis

Enzymes that catalyze the synthesis of DNA are called **DNA polymerases.** No known DNA polymerases can initiate the synthesis of a DNA strand; they can only catalyze the addition of deoxyribonucleotides to a preexisting strand. (To review the processes of enzyme action, see Box 10.2.) The initiation of DNA synthesis involves the synthesis of a short RNA **primer** in a reaction catalyzed by the enzyme **primase.** The primer functions as a pre-existing polynucleotide chain to which new deoxyribonucleotides can be added by reactions catalyzed by DNA polymerase. The *E. coli* primase is a 60,000-dalton protein with 50 to 100 copies per cell. The primase is only partially active by itself, requiring six or seven other polypeptides to complex with it to become functional. The complex is called a **primosome.** In general, only very short RNA primers three to five nucleotides long are made. In a given organism all of the RNA primers synthesized have very similar sequences, indicating that primases synthesize RNA at specific sequences along the template DNA strand. The RNA primers are removed and replaced with DNA by the action of DNA polymerase.

Molecular Details of DNA Synthesis

In 1955, A. Kornberg and his colleagues set out

a)

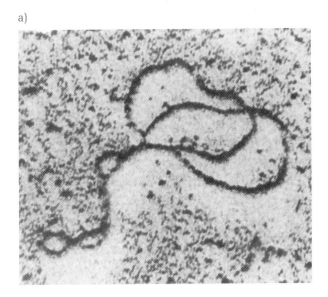

b)

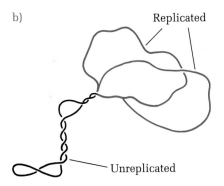

Figure 10.5 Twisted replicating molecules of circular SV40 DNA: (a) Electron micrograph of an early phase of replication; (b) diagrammatic representation showing the unreplicated, supercoiled parent strands and the portions already replicated.

to find the enzymes that were necessary for the replication of DNA so that he could describe the reactions involved in detail. His approach was to identify all the necessary ingredients required for the synthesis of *E. coli* DNA in vitro. The first successful synthesis of DNA was accomplished in a reaction mixture containing DNA fragments, a mixture of four deoxyribonucleotide 5' triphosphates (dATP, dGTP, dTTP, and dCTP) that we will abbreviate as dNTP, and a lysate prepared from *E. coli* cells. So that he could measure the very small amount of DNA expected to be synthesized in the reaction, Kornberg used radioactively labeled dNTPs in the reaction mixture.

Realizing that a crucial component (or components) for DNA synthesis must be present in the *E. coli* lysate, Kornberg next fractionated the lysate in order to find that component. In doing so, he isolated the enzyme that was responsible for DNA synthesis. This enzyme was originally called the *Kornberg enzyme,* but it is now most commonly called **DNA polymerase I.** Once DNA polymerase I was purified, more detailed information could be obtained about DNA synthesis in vitro. The first things studied were the various ingredients necessary for the synthesis of DNA in vitro. Researchers found that four components were needed for the in vitro synthesis

of DNA. If any one of the following four components were omitted, DNA synthesis would not occur:

1. all four dNTPs (If any one dNTP was missing, no synthesis occurred.)
2. magnesium ions (Mg^{2+})
3. a fragment of DNA
4. DNA polymerase I

Subsequent experiments showed that the new DNA that was made in vitro was a faithful base-pair for base-pair copy of the original DNA. That is, the original DNA acted as a template for the new DNA synthesis.

Role of DNA polymerases. All DNA polymerases catalyze the polymerization of nucleotides into a DNA chain. The reaction that they catalyze is

$$\underset{(dNMP)_n}{DNA} + dNTP \underset{\text{DNA polymerase}}{\rightleftharpoons} \underset{(dNMP)_{n+1}}{DNA} + PP_i$$

Here the DNA is represented as a string of a number *(n)* of deoxyribonucleoside 5' monophosphates (dNMP). The next nucleotide to be added is the precursor, dNTP. The DNA polymerase catalyzes a reaction in which one nucleotide is added to the existing DNA chain with the release of inorganic pyrophosphate

Box 10.2 *Enzymatic Activity*

In all organisms a large array of biochemical reactions takes place within the cells: Small molecules are assembled into large molecules, and the large molecules are broken down into small molecules. To proceed, these biochemical reactions, like any chemical reactions, require an initial heat of activation (activation energy)—a spark, so to speak. Because activation energy is needed, most biochemical reactions would take place only infrequently given the temperature, pH, and ionic composition of the cell. Yet the capacity to convert a small number of compounds, rapidly and efficiently, into thousands of other large and small molecules is a unique characteristic of living cells; the same reactions do not occur to any extent in the absence of cells.

So that these reactions are promoted, living cells contain **enzymes,** specialized proteins that, even in low concentrations, assist and accelerate (catalyze) biochemical reactions without being altered by the reactions. Enzymes are essential to all the chemical reactions that together constitute cell metabolism. Every time a chemical bond is made or broken—for example, when a macromolecule like DNA is made or degraded, or when energy is obtained from a compound—a protein with enzymatic activity is required to cause the reaction to proceed at a reasonable rate.

As a catalyst in a chemical reaction, an enzyme acts to lower the activation energy required for the reaction to take place, thereby facilitating and accelerating the reaction itself. A critical part of any enzyme is its *active site,* usually a groove or depression on the surface of the protein molecule. The shape of the active site and the nature of the chemical groups present in and around the active site cause the enzyme to interact only with a certain *substrate* (S). The complex of enzyme and substrate is called the enzyme-substrate complex (ES).

The active sites of most enzymes are not rigid; they can move a bit to aid the enzyme in attaching itself to its specific substrate. When the enzyme and substrate are positioned correctly, one or more molecules of the substrate may react directly (but reversibly) with the chemical groups in the active site. Alternatively, the enzyme may act only to bring two reactive substrate groups close together so that they can react on their own. In either case when the enzyme dissociates from the substrate, the substrate or molecules are changed to a lower total energy state and can be called the *product* (P). The reaction is

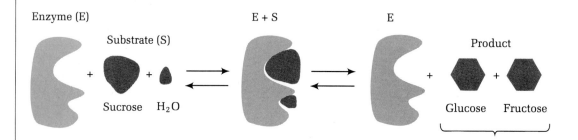

Box Figure 10.2 Schematic diagram of the role of an enzyme in catalyzing the hydrolysis of sucrose.

$$S \quad + \quad E \quad \rightleftharpoons \quad ES$$
substrate   enzyme   enzyme-
                                 substrate

$$\rightleftharpoons \quad P \quad + \quad E$$
              product   enzyme

Since the enzyme is not used up in the reaction, the net reaction is

$$S \rightleftharpoons P$$

Two or more substrates may be involved in an enzyme-catalyzed reaction, and often there are two or more products. That is, enzymes may act not only to join two (or more) molecules (e.g., DNA polymerases), but also to break complex molecules into their component parts (e.g., deoxyribonucleases, which catalyze the breakdown of DNA). For example, in the enzymatic hydrolysis of sucrose the substrates are sucrose and water, and the products are fructose and glucose:

$$\underbrace{\text{sucrose} + H_2O}_{S} + \underset{E}{\text{enzyme}} \rightarrow ES \rightarrow$$

$$\underbrace{\text{glucose} + \text{fructose}}_{P} + \underset{E}{\text{enzyme}}$$

This reaction is presented in a highly simplified way in Box Figure 10.2.

In theory, all enzyme-catalyzed reactions are reversible, as signified by the two-way arrows in the chemical equations. The theory states that if a high enough concentration of the product(s) is provided, the enzyme will change some of the product back into substrate. In fact, under the conditions found in the cell, most enzyme-catalyzed reactions are extremely one-directional.

(PP_i). The enzyme then repeats its action until the new DNA chain is complete.

The action of DNA polymerase in synthesizing a DNA chain is shown at the molecular level in Figure 10.6a. The same reaction in shorthand notation is shown in Figure 10.6b. The reaction has three main features:

1. At the growing end of the DNA chain, DNA polymerase catalyzes the formation of a **phosphodiester bond** between the 3'–OH group of the deoxyribose on the last nucleotide and the 5' phosphate of the deoxyribonucleoside 5'–triphosphate (dNTP) precursor. The formation of the phosphodiester bond results in the release of two of three phosphates from the dNTP. The important concept here is that the lengthening DNA chain acts as a primer in the reaction.

2. The addition of nucleotides to the chain is not random; each deoxyribonucleotide is selected by the DNA polymerase, which is always bound to the DNA and which moves along the template strand as the polynucleotide chain is lengthened. The polymerase finds the precursor (dNTP) that can form a complementary base pair with the nucleotide on the template strand of DNA. Since the DNA polymerase is bound to the template DNA, it ensures that the correct precursor has been chosen. This does not occur with 100 percent accuracy, but the error frequency is extremely low.

3. The direction of synthesis of the new DNA chain is 5' to 3'.

All known DNA polymerases carry out the same reaction and will make new DNA copies from any DNA added to the reaction mixture provided all the necessary ingredients are present. Thus, if given human DNA, *E. coli* DNA polymerase can replicate it faithfully, and vice versa.

One of the best-understood systems of DNA replication is that of *E. coli*. For several years after the discovery of the Kornberg enzyme (DNA polymerase I), scientists believed that this enzyme was the only DNA replication enzyme in *E. coli*. However, evidence was obtained

a) Mechanism of DNA elongation

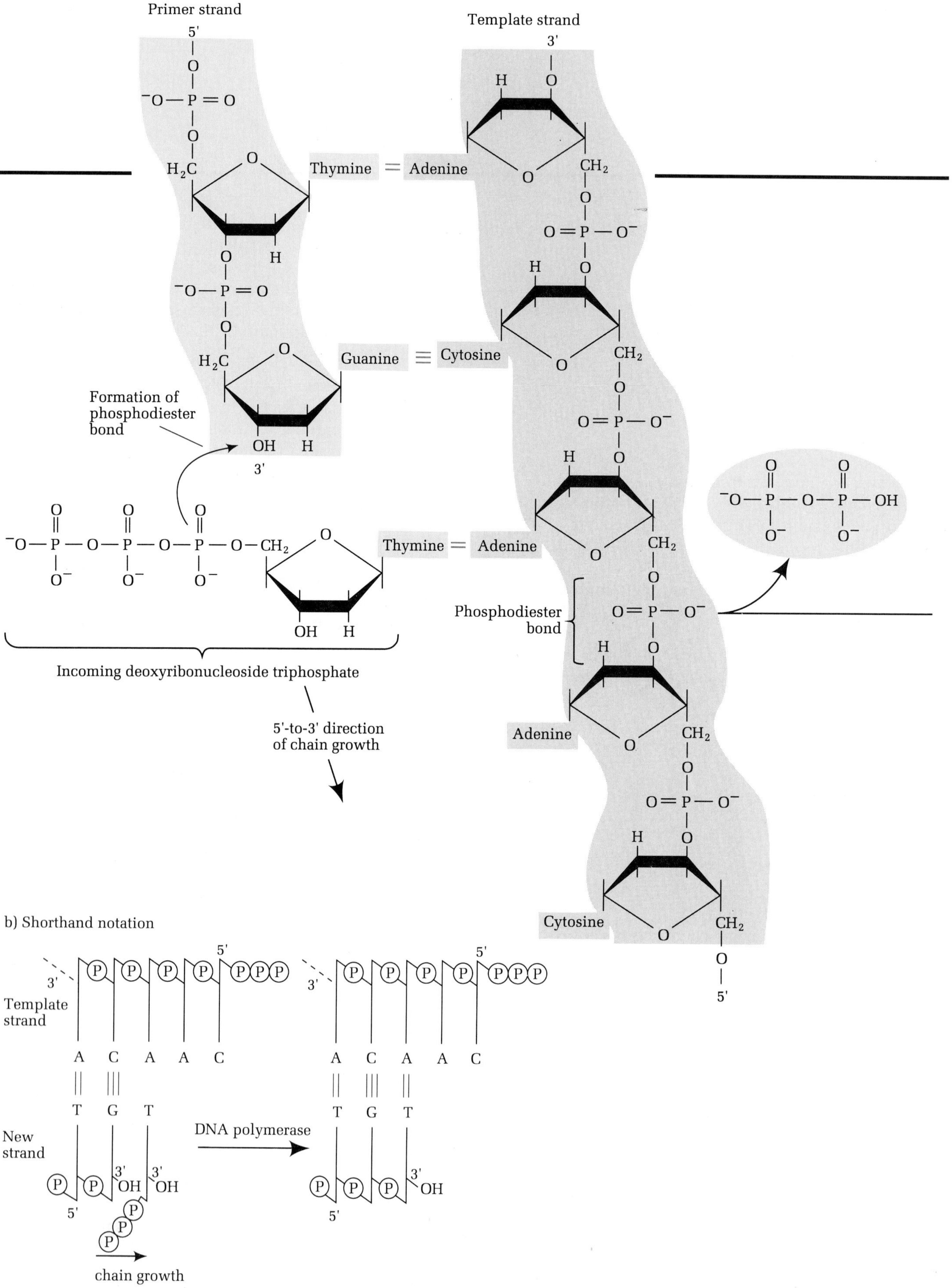

Primer strand

5'

Thymine $=$ Adenine

Guanine $\equiv$ Cytosine

Template strand

3'

Formation of phosphodiester bond

3'

Thymine $=$ Adenine

Phosphodiester bond

Incoming deoxyribonucleoside triphosphate

5'-to-3' direction of chain growth

Adenine

Cytosine

5'

b) Shorthand notation

Template strand

3'

5'

A C A A C

T G T

New strand

P P 3'OH 3'OH

5'

chain growth

DNA polymerase

3'

5'

A C A A C

T G T

P P P 3'OH

5'

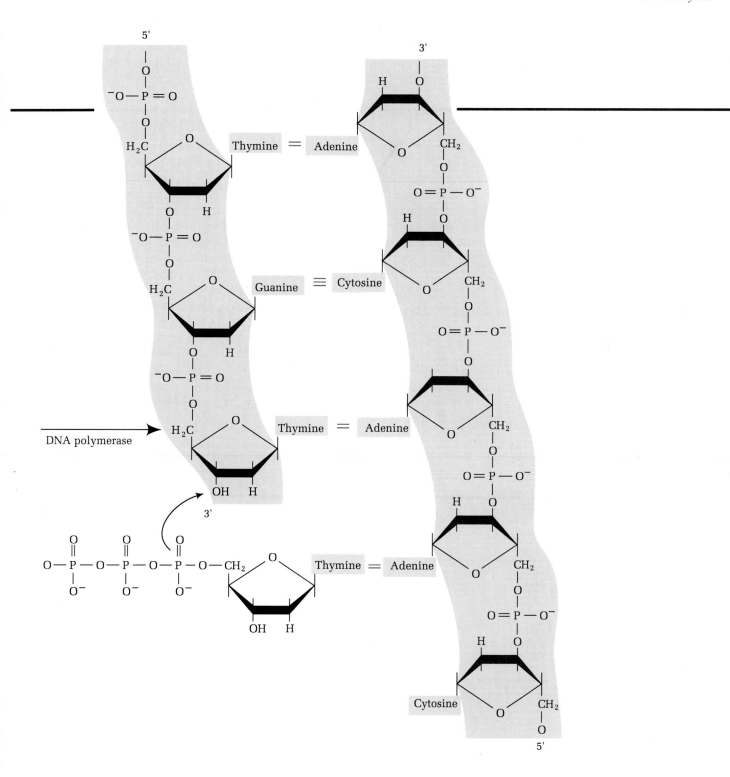

Figure 10.6 DNA chain elongation catalyzed by DNA polymerase: (a) Mechanism at molecular level; (b) mechanism shown using shorthand way of representing DNA.

indicating that the enzyme was not solely responsible for DNA synthesis in vivo.

One such indication came from genetic studies in which strains of *E. coli* with mutations in the gene for DNA polymerase I were isolated. One way to study the action of a particular enzyme in vivo is to induce a mutation in the gene that codes for that particular enzyme. In this

way the phenotypic consequences of the mutation can be compared with the wild-type phenotype. A mutation in the gene coding for an enzyme that is as essential to cell function as DNA polymerase, for instance, would be lethal. The first DNA polymerase I mutant, *polA1*, was isolated in 1969 by P. DeLucia and J. Cairns. The defect in the mutant was a base pair change in the gene that resulted in a shorter-than-normal protein which was non-functional. This type of mutation is called a nonsense mutation (see Chapter 16). Unexpectedly, *E. coli* cells carrying the *polA1* mutation grew and divided normally, showing that DNA polymerase I is not essential to cell function.

Other DNA polymerase I mutants were subsequently isolated which were temperature sensi-

Figure 10.7 The effect of a conditional mutation: (a) A normal gene, when transcribed and the resulting mRNA translated, produces a protein that functions normally at normal and high temperatures. (b) A gene with conditional mutation produces a slightly altered protein that functions normally at normal temperatures but changes its shape at high temperatures so that function is lost or reduced significantly.

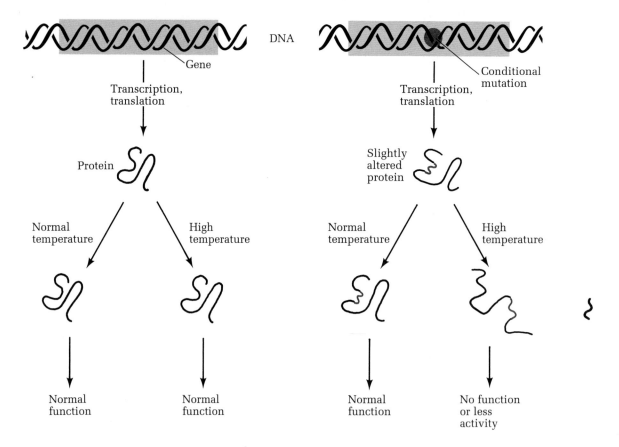

a) Function of a normal gene

DNA

Gene

Transcription, translation

Protein

Normal temperature

High temperature

Normal function

Normal function

b) Function of a gene with a conditional (temperature-sensitive) mutation

Conditional mutation

Transcription, translation

Slightly altered protein

Normal temperature

High temperature

Normal function

No function or less activity

tive. That is, to study the consequences of mutations in genes coding for essential proteins and enzymes, geneticists find it easiest to work with **conditional mutants,** that is, mutant organisms that are normal under one set of conditions but that become seriously impaired or die under other conditions (Figure 10.7). The most common types of conditional mutants are those that are temperature-sensitive—mutant organisms that function normally until the temperature is raised past some threshold level, whereupon some temperature-sensitive defect is manifested.

It was expected that the temperature-sensitive DNA polymerase I mutants would die at elevated temperatures. At *E. coli's* normal growth temperature of 37°C, the temperature-sensitive *polA* mutant strains produce DNA polymerase I with normal catalytic activity. At 42°C, however, the mutant strains produce DNA polymerase I molecules that have about 1 percent of the normal catalytic activity of the wild-type enzyme, an amount so small that we would expect the organisms of these strains to die.

Like *polA1,* the temperature-sensitive *polA* mutant cells multiplied at the same rate as the parental (wild-type) strain, even though the temperature was elevated. These results indicated

that there must be other DNA-polymerizing enzymes in the cell, and a search for these enzymes in extracts of *E. coli* began. In 1970, M. L. Gefter, R. Knippers, and C. C. Richardson, all working independently, discovered DNA polymerase II, and T. Kornberg and M. L. Gefter, working together, discovered DNA polymerase III in 1971. These two new enzymes are coded for by the *polB* and *dnaE, N, Z* loci in *E. coli,* respectively. Since enzymes II and III are present in the cell in much lower amounts than DNA polymerase I, their activities had been masked previously. There are about 400, 50–100, and 10–20 molecules per cell of DNA polymerase I, II, and III, respectively. A comparison of the important properties of the three *E. coli* DNA polymerases is given in Table 10.1.

As Table 10.1 shows, all three *E. coli* DNA polymerases have $3' \rightarrow 5'$ exonuclease activity; that is, they can catalyze the removal of nucleotides from the 3' end of a DNA chain. Thus in addition to selecting the correct nucleotide precursor to add to the primer DNA strand, the DNA polymerases also check the accuracy of the base pair linking the primer strand terminus and the template strand. If the wrong base pair has been added in error, the $3' \rightarrow 5'$ exonuclease

Table 10.1 Comparison of the Structural and Functional Characteristics of the E. coli *DNA Polymerases I, II, and III*

DNA Polymerase	Properties						
	Polymerization: $5' \rightarrow 3'$	Exonuclease: $3' \rightarrow 5'$	Exonuclease: $5' \rightarrow 3'$	Molecular Weight (daltons)	Molecules per Cell (approximately)	Structural Genes	Conditional Lethal Mutants
I	yes	yes	yes	109,000	400	*polA*	yes
II	yes	yes	no	120,000	50–100	*polB*	no
III	yes	yes	yes	140,000[a] + 25,000[a] + 10,000[a]	10–20	*polC* (*dnaE, N, Z*)	yes

[a] Polymerase III consists of at least three subunits with the molecular weights indicated; the total molecular weight is 175,000 daltons. The 140,000-dalton subunit is the predominant one.

activity catalyzes the excision of the erroneous nucleotide on the primer strand. The polymerase activity then catalyzes the formation of the correct base pair. Thus in DNA replication $3' \rightarrow 5'$ exonuclease activity is a **proofreading** mechanism that helps keep the frequency of DNA replication errors very low.

In addition, DNA polymerases I and III have $5' \rightarrow 3'$ exonuclease activities and can remove nucleotides from the 5' end of a DNA strand. This activity, also important in DNA replication, will be examined later in this chapter.

Keynote *The DNA replication in* E. coli *is catalyzed by three known DNA polymerases, I, II, and III. They differ in size, relative enzyme activity, sensitivity to inhibitors, number of molecules present in the cell, and the presence or absence of associated exonuclease and/or endonuclease activities. The semiconservative replication process requires a primer since none of the catalyzing enzymes can initiate a new DNA chain. The primer, a short chain of RNA, is synthesized by an enzyme called primase; then DNA is polymerized in a 5'-to-3' direction. A 3'-to-5' exonuclease activity of DNA polymerase is a proofreading mechanism that minimizes replication errors.*

Semidiscontinuous DNA replication. When a double-stranded DNA molecule unwinds to expose the two single-stranded template strands for DNA replication, a Y-shaped structure is formed, called a **replication fork.** A replication fork moves in one direction. This unidirectional movement poses a problem since DNA polymerases can only catalyze DNA synthesis in the $5' \rightarrow 3'$ direction, yet the two DNA strands are of opposite polarity. As the DNA helix unwinds to provide templates for new synthesis, the new DNA cannot be polymerized continuously on the 3'–5' strand.

The work of R. Okazaki and his colleagues suggested a mechanism to explain this result. They added a radioactive DNA precursor (^{3}H-thymidine) to cultures of *E. coli* for 0.5 percent of a generation time. Next, they added a large amount of nonradioactive thymidine to prevent the incorporation of any more radioactivity into the DNA. They followed what happened to the radioactively labeled thymidine. At various times they extracted the DNA and determined the size of the newly labeled molecules. At intervals soon after the labeling period, most of the radioactive ^{3}H-thymidine was present in relatively low-molecular-weight DNA about 100–1000 nucleotides long. As time increased, a greater and greater proportion of the labeled molecules was found in high-molecular-weight DNA. These results indicated that DNA replication normally involves the synthesis of short DNA segments, called **Okazaki fragments,** which are subsequently linked together by the action of DNA polymerase to remove the RNA primers followed by the action of an enzyme called **DNA ligase,** which catalyzes formation of the final phosphodiester bond between Okazaki fragments to form a long polynucleotide chain. In other words, Okazaki's group had shown that DNA replication is **discontinuous.**

We can relate the discontinuous nature of DNA synthesis to the replication of the circular DNA chromosome in *E. coli.* One location on the chromosome serves as the point of origin for DNA replication. At this point the DNA denatures to expose the two template strands. Since denaturation occurs in the middle of a circular DNA molecule rather than at the end, the result is the two Y-shaped structures linked head-to-head at the points of the Y's. These structures act as two replication forks with DNA synthesis taking place in both directions (bidirectionally) away from the origin point.

A diagram of the early stages of **bidirectional replication** is shown in Figure 10.8. This figure also shows how discontinuous replication occurs. As the replication fork migrates, replication is continuous on one strand (the *leading strand*) since the 3'-to-5' template strand is being copied. Replication on the other strand (the *lagging strand*) must be discontinuous because the helix must unwind to expose a new segment of 3'-to-5' template on which the new strand can be made. Since one new DNA strand is synthesized continuously and the other discontinuously, DNA replication as a whole is

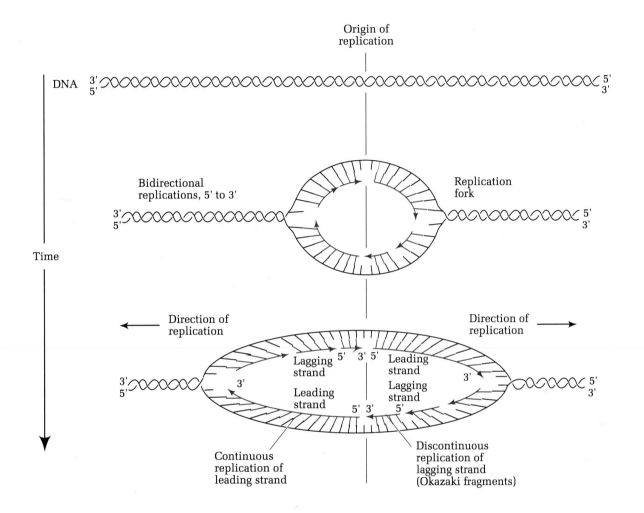

Figure 10.8 Bidirectional DNA replication. Synthesis of DNA is initiated at the origin and proceeds in the 5'-to-3' direction while the two replication forks are migrating in opposite directions. Replication is continuous on the leading strand and semidiscontinuous on the lagging strand. These events occur in the replication of all DNA in prokaryotes and eukaryotes.

considered to occur in a **semidiscontinuous** fashion.

DNA replication model. A model for DNA replication incorporating all the above information is presented in Figure 10.9, which illustrates the formation of a single replication fork. For circular chromosomes with bidirectional replication,

two replication forks are formed on either side of the origin of replication; the events shown in Figure 10.9 occur on both.

The first step in DNA replication is the denaturation and unwinding of the double helix (Figure 10.9a). In *E. coli* the unwinding occurs as a result of the activity of enzymes called **helicases.** The helicase unwinds the DNA in advance of the replicating fork. The *rep* protein (the helicase encoded by the *rep* gene) moves on one strand in the 3'-to-5' direction while another helicase moves on the other strand in the 5'-to-3' direction. Energy released from the hydrolysis of ATP to ADP is used to fuel the migration of the helicases. Since DNA is flexible, it is likely to reform the original hydrogen bonds

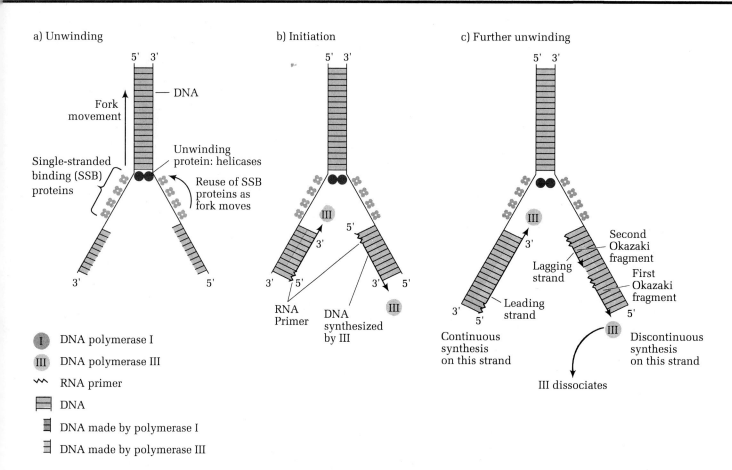

a) Unwinding

Fork movement

Single-stranded binding (SSB) proteins

— DNA

Unwinding protein: helicases

Reuse of SSB proteins as fork moves

b) Initiation

RNA Primer

DNA synthesized by III

c) Further unwinding

Second Okazaki fragment

Lagging strand

First Okazaki fragment

Leading strand

Continuous synthesis on this strand

Discontinuous synthesis on this strand

III dissociates

- Ⓘ DNA polymerase I
- ⒾⒾ DNA polymerase III
- ⌇ RNA primer
- ▥ DNA
- ▤ DNA made by polymerase I
- ▥ DNA made by polymerase III

Figure 10.9 Model for the events occurring around the replication fork of the E. coli *chromosome: (a) unwinding; (b) initiation; (c) further unwinding and continued DNA synthesis; (d) further unwinding with initiation of another DNA fragment; (e) primer removed from first Okazaki fragment by DNA polymerase I; (f) joining of adjacent DNA fragments by the action of DNA ligase.*

once the helicase has moved on. So that the unwound strands are kept in a stable, unbent, single-stranded state, proteins with an affinity for single-stranded DNA, called single-stranded, binding (SSB) proteins, bind to the strands. In *E. coli* the SSB protein is a 177 amino acid polypeptide that exists as a tetramer and that binds to a 32-nucleotide segment of DNA. Over two hundred of the proteins bind to each replication fork. Once the strands have started to unwind,

the internal bases (no longer hydrogen-bonded) are available for the formation of bonds with bases in the new chain. The SSB proteins are recycled as the replication fork migrates (Figure 10.10).

Next, initiation of DNA replication itself takes place (Figure 10.9b). In *E. coli* the primase enzyme complex (primosome) binds to the single-stranded DNA and synthesizes a short RNA primer. The primer chain is lengthened by the action of DNA polymerase III, which synthesizes the complementary DNA chain to the template strand. While the diagram shows the replication enzymes spatially separated, the proteins involved are large relative to DNA. The complex of DNA replication proteins, such as DNA polymerase III, a primosome, helicases, and other proteins, is about the size of a ribosome and has been referred to as a *replisome.*

The DNA helix continues to unwind (Figure

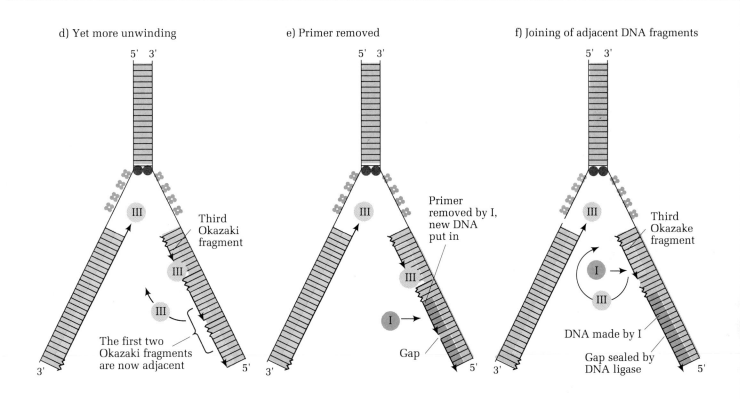

d) Yet more unwinding

e) Primer removed

f) Joining of adjacent DNA fragments

Third Okazaki fragment

The first two Okazaki fragments are now adjacent

Primer removed by I, new DNA put in

Gap

Third Okazake fragment

DNA made by I

Gap sealed by DNA ligase

10.9c), and on the left-hand template strand the leading strand continues to be synthesized continuously. Since DNA synthesis can only proceed in the 5′-to-3′ direction, however, the DNA polymerase synthesis reaction on the right-hand template has gone as far as it can. For DNA replication to continue, a new initiation of DNA synthesis must occur on the newly available single-stranded template. As before, a primer must be lengthened by the action of DNA polymerase III. In Figure 10.9d the process repeats itself: The DNA unwinds, continuous DNA synthesis occurs on the left-hand strand, and discontinuous DNA synthesis occurs on the right-hand strand.

Figure 10.10 Hypothetical scheme demonstrating how single-stranded binding protein (SSB) acts at a replicating fork. The protein binds to single-stranded regions, facilitates replication, and is then recycled.

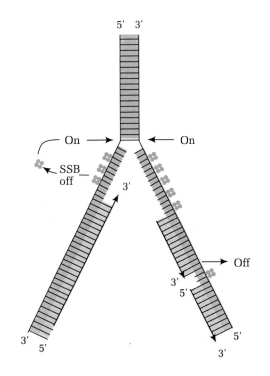

Eventually, the unconnected Okazaki fragments of the right-hand strand are synthesized into a continuous DNA strand, a process that requires the activities of two enzymes, DNA polymerase I and **DNA ligase.** Consider two adjacent Okazaki fragments. The 3' end of the newer DNA fragment is adjacent to, but not joined to, the primer at the 5' end of the previously made fragment. The DNA polymerase III dissociates from the DNA, and DNA polymerase I takes over. This enzyme continues the 5'-to-3' synthesis of the newer DNA fragment made by DNA polymerase III, simultaneously removing the primer section of the older fragment (Figure 10.9e). The removal of the primer takes place nucleotide by nucleotide through the action of the 5' → 3' exonuclease activity of DNA polymerase I. When DNA polymerase I has completed replacement of RNA primer nucleotides with DNA nucleotides, a single-stranded gap exists between adjacent nucleotides on the DNA strand between the two fragments. The two fragments are joined into one continuous DNA strand by the enzyme DNA ligase, which catalyzes the reaction diagramed in Figure 10.11. The result is a longer DNA strand (Figure 10.9f). A larger-scale diagram of how an RNA primer is used in the initiation of lagging strand fragments and how the primer is subsequently removed is shown in Figure 10.12. The whole process is repeated until all the DNA is replicated.

The complicated process of DNA replication in *E. coli* requires many different proteins, some of which function in other cellular processes such as the repair of damaged DNA and genetic recombination. The locations of some of the genes that have been identified as necessary for DNA replication are shown in Figure 10.13 and the functions of the gene products are given in Table 10.2.

Proofreading: Correcting errors in DNA replication. Occasionally an error is made in DNA replication such that an incorrect nucleotide is incorporated into the DNA chain being synthesized. The mismatched base has a very high probability of being excised by the 3' → 5' exonuclease activity of the DNA polymerase before the next base in the chain is added. This process, or proofreading ability, results in an extremely low error frequency in inserting the wrong base during DNA replication.

Some recent computer modeling of the interaction between DNA polymerase I and DNA has given some insight into how the proofreading might operate. Figure 10.14 (p. 337) shows a computer-generated model of DNA binding to what is called the Klenow fragment of DNA polymerase I. The Klenow fragment possesses both the DNA polymerase and the 3' → 5' exonuclease activities of DNA polymerase I. In the Klenow fragment, a 2-nm cleft exists to which DNA binds. Once this occurs, the cleft may then be covered by a flexible subdomain of the protein. The DNA, after binding, probably can only slide forward and backward.

Proofreading might occur next, in the following manner: If an incorrect base is incorporated so that hydrogen bonding between the new base and the base pair to the template strand cannot form properly, the DNA may assume a distorted

Figure 10.11 Action of DNA ligase in sealing the gap between adjacent DNA fragments (e.g., Okazaki fragments) to form a longer, covalently continuous chain. The DNA ligase catalyzes the formation of a phosphodiester bond between the 3'–OH and the 5'-phosphate groups on either side of a gap, thus sealing the gap.

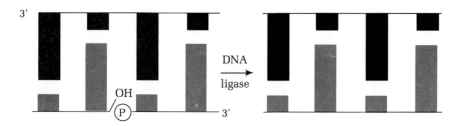

shape and be unable to slip easily through the 2-nm cleft. However, if it were able to move backward and come into contact with the $3' \rightarrow 5'$ exonuclease (proofreading) region, the incorrect base could be removed and polymerization continued (Figure 10.15). As such, the proposed mechanism resembles a correcting typewriter, where a backspace is used to erase the incorrect character and the forward direction is resumed to insert the next correct character.

Rolling circle replication of DNA. The rolling circle model of DNA replication applies to the replication of several viral DNAs and to the replication of the *E. coli F* factor during conjugation and transfer of donor DNA to the recipient (see Chapter 7). The rolling circle model is shown in Figure 10.16 (p. 338). The first step is the generation of a specific cut (nick) in one of the two strands at the origin of replication. The 5′ end of the cut strand is then displaced from the circular molecule. This creates a replication fork structure and leaves a single-stranded stretch of DNA that serves as a template for the addition of deoxyribonucleotides to the free 3′ end. This DNA synthesis is catalyzed by DNA polymerase III. As the 5′ cut end continues to be displaced from the circular molecule, synthesis of new DNA occurs; that is, this is the leading strand of the previous replication fork diagrams. As replication proceeds, the 5′ end of the cut DNA strand is rolled out as a free "tongue" of increasing length. (This is analogous to pulling out the end of a roll of kitchen towels.) This single-stranded DNA tongue becomes covered with SSB proteins.

As the circle continues to roll, the single-stranded tongue converts to a double-stranded form, as was illustrated earlier in our discussion of lagging strand synthesis (see Figure 10.9). That is, primase synthesizes short RNA primers that are extended as DNA (Okazaki fragments) by DNA polymerase III. The RNA primers are ultimately removed and adjacent Okazaki fragments are joined through the action of DNA ligase. As the single-stranded DNA tongue rolls out, DNA synthesis continues on the circular DNA template.

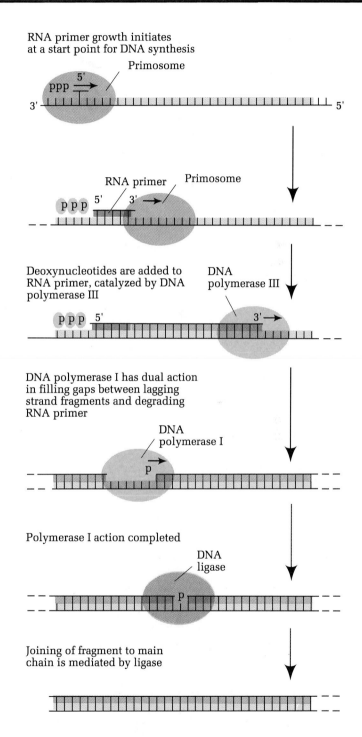

RNA primer growth initiates at a start point for DNA synthesis

Deoxynucleotides are added to RNA primer, catalyzed by DNA polymerase III

DNA polymerase I has dual action in filling gaps between lagging strand fragments and degrading RNA primer

Polymerase I action completed

Joining of fragment to main chain is mediated by ligase

Figure 10.12 Detailed diagram of the initiation of DNA replication with an RNA primer and the subsequent removal of the primer by DNA polymerase I.

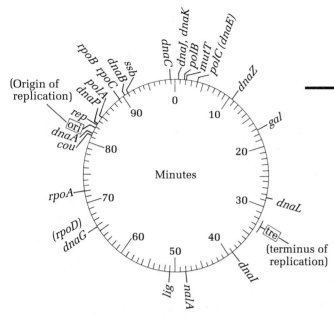

Figure 10.13 Map of the circular E. coli *chromosome showing the locations of some of the genes known to be involved in DNA replication. The units on the map are minutes, where each minute signifies the length of chromosome transferred during 1 min of conjugation at 37°C. The total map is approximately 100 min long.*

Since the parental DNA circle can continue to roll, it is potentially possible to generate a linear double-stranded DNA molecule that is longer than the circumference of the circle. For example, in the later stages of phage lambda DNA replication, which occurs by rolling circle replication, linear tongues are produced that are many times the circumference of the original circle. While it is not known exactly how these are converted into the linear DNA molecules, which are packaged into the lambda, the most likely explanation is that they are cut by the specific endonucleases that convert lambda circles into linear lambda DNA molecules (see Figure 9.9).

The rolling circle model for DNA replication applies directly to the transfer of DNA from donor to recipient cell during conjugation of *E.*

Table 10.2 Functions of Some of the Genes Involved in DNA Replication in E. coli *as Shown in Figure 10.13*

Map Location (min)	Gene Product and/or Function	Gene	Map Location (min)	Gene Product and/or Function	Gene
0.5	DNA replication	*dnaJ*	82	Initiation of chromosomal replication	*dnaA*
0.5	DNA replication	*dnaK*	83	Origin of chromosomal replication	*ori*
2	DNA polymerase II	*polB*	83	Helicase—unwinding activity to generate single-stranded arms of replication fork	*rep*
4	DNA polymerase III	*polC* (*dnaE, N, Z*)	84	Initiation of chromosomal replication	*dnaP*
28	DNA replication	*dnaL*	85	DNA polymerase I	*polA*
27–43	Terminus of chromosomal replication	*ter*	89	RNA polymerase, β sub-unit	*rpoB*
39	DNA replication	*dnaI*	89	RNA polymerase, β′ sub-unit	*rpoC*
51	DNA ligase—seal single-stranded gaps; join Okazaki fragments	*lig*	91	DNA replication	*dnaB*
66	Primase—make primer for extension by DNA polymerase	*dnaG*	91	Single-stranded binding (SSB) protein—stabilize single-stranded arms of replication fork	*ssb*
≈66	RNA polymerase, σ sub-unit	*rpoD*	99	DNA replication	*dnaC*
72	RNA polymerase, α sub-unit	*rpoA*			

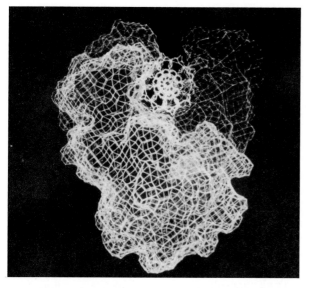
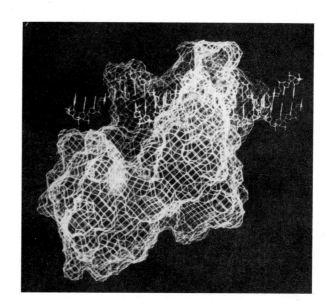

*Figure 10.14 Binding of DNA to the Klenow fragment
(a DNA polymerase I fragment) as illustrated (left)
from above and (right) from the side.*

1. DNA polymerase I binds to single-stranded DNA at growing point

2. Enzyme changes conformation, closing cleft so DNA can only slide forward and backward

3. Enzyme moves forward, catalyzing new DNA synthesis on single-stranded template

4. Formation of mismatched base pair permits only backward movement of the enzyme since the DNA becomes distorted so that it cannot move through the cleft

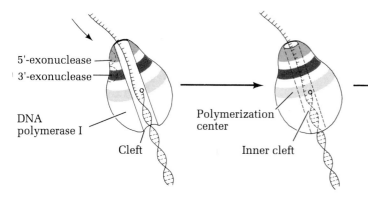

5'-exonuclease

3'-exonuclease

DNA polymerase I

Cleft

Polymerization center

Inner cleft

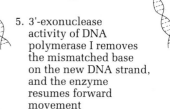

5. 3'-exonuclease activity of DNA polymerase I removes the mismatched base on the new DNA strand, and the enzyme resumes forward movement

*Figure 10.15 Diagrammatic view of proofreading after
incorporation of a mismatched nucleotide base into
DNA.*

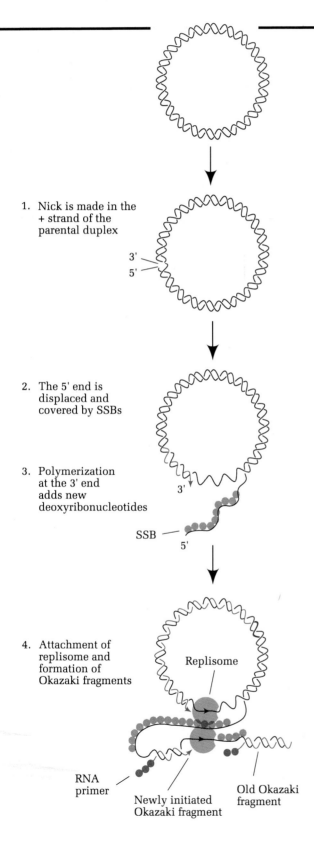

1. Nick is made in the
   + strand of the
   parental duplex

   3'
   5'

2. The 5' end is
   displaced and
   covered by SSBs

3. Polymerization
   at the 3' end
   adds new
   deoxyribonucleotides

   3'

   SSB
   5'

4. Attachment of
   replisome and
   formation of
   Okazaki fragments

   Replisome

   RNA
   primer

   Newly initiated
   Okazaki fragment

   Old Okazaki
   fragment

Figure 10.16 The replication process of double-stranded circular DNA molecules through the rolling circle mechanism. The active force that unwinds the 5' tail is the movement of the replisome propelled by its helicase components.

coli cells (Figure 10.17). In this case leading strand DNA synthesis serves to maintain a complete double-stranded copy of the F factor or Hfr donor chromosome in the donor cell. The displaced 5' tongue passes through the conjugation tube as a single-stranded molecule and enters the recipient cell in that form. It is not known, however, whether it is first converted to double-stranded DNA by lagging strand synthesis and then undergoes recombination with the recipient's chromosome, or whether it recombines with the recipient's chromosome as a single-stranded molecule.

Bacterial DNA Replication and the Cell Cycle

In a bacterium such as E. coli that is growing rapidly, DNA replication occurs throughout the cell division cycle. At the same time, the cell mass is increasing, and the cell is becoming longer. At the beginning of the cell cycle the DNA molecule is already partially replicated (Figure 10.18a). About halfway through the cell cycle the chromosome has completely repli-

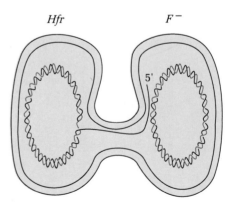

Hfr   F⁻

5'

Figure 10.17 Transfer of single-stranded DNA from a male bacterium into a female via the rolling circle mechanism.

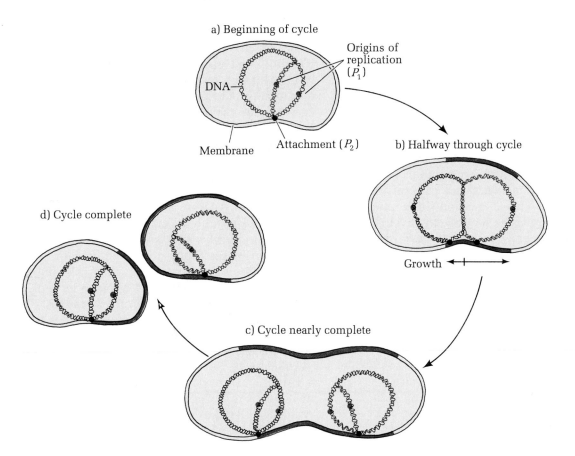

a) Beginning of cycle

Origins of replication (P_1)

DNA

Membrane   Attachment (P_2)

b) Halfway through cycle

Growth

d) Cycle complete

c) Cycle nearly complete

Figure 10.18 Model for chromosome segregation to daughter cells of E. coli, as proposed by Jacob, Brenner, and Cuzin: (a) A newly produced bacterial cell; with the DNA already partially replicated and attached to the membrane at P_2; (b) partway through the division cycle when replication of the chromosome has been completed and the two DNAs are attached to the membrane (the membrane synthesized between the cycles in a and b is colored); (c) near the end of the cycle, when new membrane growth (colored) has moved the chromosome attachment points again; (d) the end of the cycle, with the bacterium divided in two (each daughter cell receives one chromosome).

cated, and the two progeny DNA molecules are attached to the cell membrane with their attachment points quite close to one another (Figure 10.18b). (Note that, while the DNA appears to be attached to the cell membrane during the cell division events, none of the stages of DNA replication requires the presence of any membrane component.) As the cell grows, new membrane material is laid down in the narrow zone between the two attachment points. As that zone grows, the attachment points and the DNA segregate to different ends of the cell (Figure 10.18c). During this stage another round of DNA replication begins. Lastly, a wall is formed across the cell at its midpoint, and two progeny cells are generated by fission. Each progeny cell has a complete replicating chromosome still attached to the cell membrane (Figure 10.18d).

In a nutrient-rich medium *E. coli* can divide every 20 minutes. In such cells, it takes 40 minutes for a replication fork to go from the origin to the terminus; that is, with the two replication forks of bidirectional replication, the chromosome is replicated in 20 minutes. Thus, in the *E. coli* chromosome, each replication fork moves at about 50,000 nucleotides per minute, or 50 kbp/min, where 1 kbp is a kilobase pair (1000 base pairs). Since the helix must unwind to replicate, the DNA is rotating at approximately 50,000 revolutions per minute, or about the same rpm's of an engine in a car cruising at highway speeds! This is an astonishing rate, especially considering the logistics involved and the incredible accuracy of the process. We will see later that the rate of DNA replication in eukaryotes is about four to five times slower.

Keynote *Replication of DNA in* E. coli *requires at least two of the DNA polymerases and several other enzymes and proteins. The DNA helix is unwound to provide templates for the synthesis of new DNA. Since new DNA is made in the 5'-to-3' direction, chain growth is continuous on one strand and discontinuous (i.e., in segments that are later joined) on the other strand. This semidiscontinuous model is applicable to many other prokaryotic replication systems, each of which differs in the number and properties of the enzymes and proteins required.*

DNA Replication in Eukaryotes

Since the chromosomal properties of DNA molecules are identical in prokaryotes and eukaryotes, it is not surprising that the biochemistry of DNA replication is similar in prokaryotes and eukaryotes. However, the added complication in eukaryotes is that DNA is not found in one chromosome but is distributed among many chromosomes, each of which is a complex aggregate of DNA and proteins. In each cell division cycle, each of these chromosomes must be faithfully replicated and a copy of each distributed to each of the two progeny cells.

The cell cycle is qualitatively the same from eukaryote to eukaryote, although there are significant differences both in the relative amount of time spent in each phase of the cycle and in the total time spent in one cycle. Among higher eukaryotes, for example, some cells divide once every 3 hours and some cells divide once every 200 hours; human cells in culture divide once every 24 hours.

Recall from Chapter 1 that the cell cycle in most somatic cells of higher eukaryotes is divided into four stages: gap 1 (G_1), synthesis (S), gap 2 (G_2), and mitosis (M) (see Figure 1.15). The DNA replicates and the chromosomes duplicate during the S phase, and the progeny chromosomes segregate into daughter cells during the M phase. Although DNA is neither replicating nor being segregated during G_1 and G_2, the cell is metabolically active and is growing. Events involved in the subsequent synthesis and mitotic activity also occur during G_1 and G_2.

Collectively, the G_1, S, and G_2 phases of the eukaryotic cell cycle constitute interphase, the time between divisions, while division (mitosis or meiosis) itself occurs within the M phase.

The significant events that occur within the G_1, S, G_2, and M phases are as follows:

1. The G_1 phase is characterized by a change in the morphology of the chromosomes from the condensed state characteristic of mitosis to the extended state of interphase cells; the change is due to a transition to a lower order of coiling of the nucleosomes. In addition, during G_1 the cell becomes prepared biochemically for the S phase. The G_1 phase is a period during which there is very active synthesis of both RNA and protein. In particular, proteins needed for the replication and subsequent segregation of DNA are made in G_1.

   It is during G_1 that normal cells must decide whether or not to proliferate. The point at which the decision is made is called the *restriction point,* and once the biochemical events associated with the restriction point occur, the cell is irreversibly programmed to initiate DNA replication and consequently to divide. The precise molecular signals that operate at the restriction point are not known, although it is clear that the cell must reach a certain mass (i.e., accumulate sufficient cellular protein) before it can commit itself to the rest of the cell cycle.

If a cell is unable to grow and proliferate—for example, as a result of exhaustion of nutrients, of contact inhibition (a phenomenon in which cells in contact somehow inhibit each other), or of the addition of inhibitory drugs—it becomes quiescent. The quiescent state can be reversed by a change in conditions, such as adding nutrients, diluting the cells, or removing the drugs, respectively.

The G_1 phase is absent in some eukaryotes, such as the slime mold *Physarum polycephalum* and the fission yeast *Schizosaccharomyces pombe*. Events required for the initiation of DNA synthesis in these cells probably occur in the G_2 phase preceding mitosis.

2. During the S phase, DNA and chromosome replications take place. Replication involves the coordination of the mechanism for unwinding the DNA of each chromosome and producing two progeny DNA double helices and the mechanism for duplicating the nucleosome organization of the eukaryotic chromosome. These events will be described in more detail a little later in this chapter.

3. The G_2 phase follows the S phase and occupies perhaps 10 to 20 percent of the cell cycle. During G_2 the chromosomes begin condensing in preparation for mitosis. Both RNA and protein are actively synthesized during this phase. One of the important proteins made during G_2 is *tubulin*. Tubulin is a component of the mitotic spindle apparatus, which is used to segregate the chromosomes into daughter cells during mitosis. The end of the G_2 phase is delineated by the beginning of mitosis. It is not clear what events trigger mitosis.

4. In the last stage of the cell cycle, the M phase, the chromosomes are segregated into the progeny cells. Depending on the cell type, chromosome segregation will occur by mitosis or meiosis; both processes were described in Chapter 1.

Comparisons of DNA Replication in Prokaryotes and Eukaryotes

As noted in Chapter 9, the eukaryotic chromosome is a complex of DNA and proteins (histones and nonhistones) organized into nucleosomes. In contrast, the prokaryotic chromosome consists of DNA and only a very few proteins. Thus the replication of the eukaryotic chromosome must involve the replication of the DNA and of the histone core of the nucleosome as well as a doubling of the nonhistones.

The complexity of eukaryotic genetic material makes it difficult to understand the eukaryotic replication process. However, much information has been obtained about the enzymes required for DNA replication in a number of eukaryotic systems and about the propagation of the nucleosome structure from cell generation to cell generation. In addition, studies have shown that many aspects of DNA replication are the same in prokaryotes and eukaryotes, as described next.

DNA replication in eukaryotes is semiconservative. As in prokaryotes, the replication of DNA in eukaryotes takes place by a semiconservative process in which each strand of the double helix serves as a template for the synthesis of a new strand. The two resulting double helices are hybrid molecules in the sense of having one parental and one new strand paired together. At the same time, the protein structure of the chromosome is replicated, so two identical copies (chromatids) of the chromosomes are produced where there was one before. The chromatids are held together at the centromere region and remain together until the M phase, when the centromeres divide and separate and the two chromosomes (now progeny chromosomes) are segregated into the daughter cells. Evidence for semiconservative replication in eukaryotes was provided by J. H. Taylor, who performed the following experiment, which was similar to that of Cairns with *E. coli.*

Cells of a lily plant were labeled with [^{3}H] thymidine during DNA synthesis. Samples of the cell population were taken at intervals and

a) First division of
labeled DNA in
unlabeled medium

b) Second division
in unlabeled medium

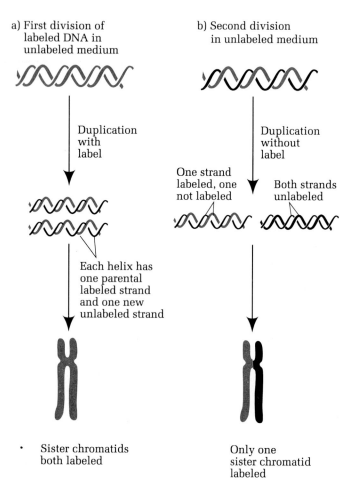

Duplication
with
label

Duplication
without
label

One strand
labeled, one
not labeled

Both strands
unlabeled

Each helix has
one parental
labeled strand
and one new
unlabeled strand

• Sister chromatids
both labeled

Only one
sister chromatid
labeled

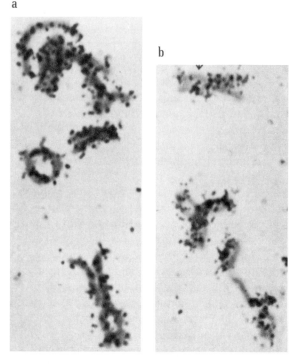

a

b

Figure 10.19 Evidence for semiconservative replication in eukaryotes. Cells of a lily plant were grown in the presence of [³H] thymidine for several generations so that both DNA strands were labeled. The cells were then grown in non-radioactive medium. After the first mitotic division (a), both sister chromatids were labeled since they each had DNA double helices with one labeled and one unlabeled strand. After the second mitotic division (b), one sister chromatid was labeled (one labeled, one unlabeled DNA strand) and the other sister chromatid was unlabeled (both DNA strands unlabeled).

stained. Mitotic chromosomes were located microscopically and subjected to autoradiography. Labeled chromatids may be identified by the presence of the silver grains along them in the resulting autoradiograph. At the first division after labeling, both chromatids of all mitotic chromosomes were labeled (Figure 10.19a). The interpretation of the autoradiograph in this case was that each chromatid represented one double-stranded DNA molecule and one DNA strand in each chromatid was newly synthesized (i.e., labeled) and the other was the parental template strand (i.e., not labeled), thereby suggesting that DNA replication was semiconservative. That conclusion was strengthened by the analysis of autoradiographs made after one more round of cell division in the absence of labeled thymidine. At this point one chromatid in each mitotic pair of chromatids was labeled (again with one strand of the DNA labeled and the other one unlabeled) and the other was unlabeled (Figure 10.19b). It is now accepted that DNA replication

of all eukaryotes occurs by a semiconservative mechanism.

Eukaryotic DNA replication is semidiscontinuous. As in prokaryotes, DNA synthesis in eukaryotes is a semidiscontinuous process. The DNA is synthesized in Okazaki fragments on the lagging strand and is synthesized continuously on the leading strand. This mode of replication is the consequence of the opposite polarity ($5' \rightarrow 3'$ and $3' \rightarrow 5'$) of the two DNA strands of the helix and of the requirement of all known DNA polymerases to synthesize DNA in the $5' \rightarrow 3'$ direction.

Also as in prokaryotes, the double helix of eukaryotic DNA unwinds, and the exposed single strands serve as templates for new DNA synthesis. The best evidence for semidiscontinuous synthesis on these template strands comes from radioactive-labeling experiments like that performed by Okazaki and his colleagues with bacteria and their viruses. In eukaryotes, Okazaki fragments range from 35 to 300 nucleotides long, the average being about 135 nucleotides. These Okazaki fragments, much smaller than those found in prokaryotes, are eventually ligated into a complete new strand.

New DNA chains are initiated by RNA primers
As with prokaryotes, the DNA polymerases of eukaryotes cannot initiate the synthesis of a DNA strand, and there is evidence that DNA synthesis on the discontinuously replicated side of the replication fork in eukaryotes is initiated by RNA primers that are subsequently excised. The RNA primer is synthesized by a primase enzyme, is about 5 to 15 nucleotides long, and is not a unique sequence as it is in prokaryotes. Since eukaryotic DNA polymerases do not have associated exonuclease activities, as do prokaryotic DNA polymerases, other nuclear proteins must excise the RNA primer.

Molecular Details of DNA Synthesis

As we saw earlier, many of the enzymes and proteins involved in DNA replication in prokaryotes have been identified. To date, however, research has not progressed far enough to describe all of the enzymes and proteins involved in eukaryotic DNA replication. Nonetheless, the sequential steps described for DNA synthesis in prokaryotes also occur for DNA synthesis in eukaryotes, namely, denaturation of the DNA double helix and the semiconservative, semidiscontinuous replication of the DNA.

Initiation of the replication process. The DNA replication in eukaryotes takes ten times longer than the same process in prokaryotes. The rate of replication in eukaryotes is 2 to 10 kb/min.

Each eukaryotic chromosome consists of one linear DNA double helix. If there were only one origin of replication per chromosome, the replication of each chromosome would take many, many hours. For example, there are 2.75×10^9 base pairs of DNA in the haploid human genome (23 chromosomes), and the average chromosome is roughly 10^8 base pairs long. With a replication rate of 2 kilobases (2000 bases) per minute in human cells, it would take approximately 830 hours to replicate one chromosome. If each cell cycle were at least that long for a developing human embryo, the gestation period would be many years instead of 9 months.

Actual measurements show that the chromosomes in eukaryotes replicate much faster than would be the case with only one origin or replication per chromosome. The diploid complement of chromosomes in *Drosophila,* for example, replicates in 3 minutes. This is 6 times faster than the replication of the *E. coli* chromosome even though there is about 100 times more DNA in *Drosophila* than there is in *E. coli.*

The rapidity of *Drosophila* replication is possible because DNA replication is initiated at many origins of replication throughout the genome, a state of affairs that is general among eukaryotes. The many replication origins greatly decrease the overall time for duplication of the chromosomal DNA. Mammalian cells, for example, have about 10^4 origins positioned nonrandomly among the chromosomes.

At each origin of replication, the DNA denatures to produce two facing replication forks (as in *E. coli;* see Figure 10.9). Replication proceeds bidirectionally, and the DNA double helix opens to expose single strands that act as templates for new DNA synthesis. Eventually, each replication fork will run into an adjacent replication fork that was initiated at an adjacent origin or replication. In eukaryotes the stretch of DNA from the origin of replication to the two termini of replication (where adjacent replication forks fuse) on each side of the origin is called a **replicon** or a **replication unit.**

Figure 10.20a presents an electron micrograph showing a large number of replicons on a piece of *Drosophila* DNA; Figure 10.20b is an interpretive drawing of that micrograph. The piece of DNA shown is about 500 kb long (about a tenth the size of the *E. coli* chromosome), and many

a)

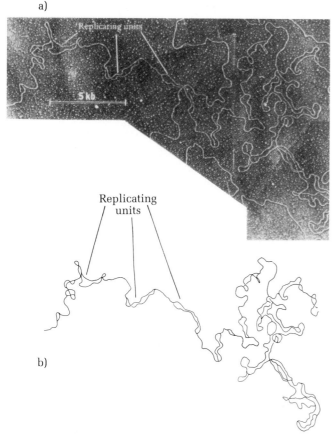

Replicating
units

b)

Figure 10.20 Replicating DNA of Drosophila melano-gaster: *(a) electron micrograph showing replication units (replicons); (b) diagrammatic interpretation of the electron micrograph shown in (a).*

replicons are present. In general, each replicating eukaryotic chromosome has many replicons. From examining replicating chromosomes under the electron microscope geneticists have determined that the length between the origins of replication averages 15 to 500 kbp (5 to 150 μm) in several eukaryotic organisms, such as mammals, plants, and eukaryotic microorganisms. Also, the number of origins and the average size of the replicon can vary with the nature of the cell, for example, its growth rate and tissue of origin.

Replication of DNA does not occur simultaneously in all the replicons in an organism's genome. Instead, a temporal ordering of initiation occurs during the replication phase. Evidence indicates that for many cell types a particular temporal ordering of initiation is characteristic of the cell type; that is, the same pattern occurs cell generation after cell generation. The molecular events that control which replicons replicate early and which replicate late are unknown at this time. Figure 10.21 gives a diagrammatic representation of the temporal ordering of initiation phenomena from autoradiographic studies. The figure shows one segment of one chromosome in which there are three replicons that always begin replicating at distinct times. When the replication forks fuse at the margins of adjacent replicons, the replicated chromosome has replicated into two sister chromatids. In general, replication of a segment of chromosomal DNA occurs following the synchronous activation of a cluster of origins.

It is unclear whether there are any specific sequences that function as origins of replication in eukaryotes. In mammal cells, there is some evidence for the existence of specific origin sequences, but details remain elusive. In baker's yeast, *Saccharomyces cerevisiae,* specific sequences have been identified which, when they are included as part of an extrachromosomal, circular DNA molecule, confer upon that molecule the ability to replicate autonomously. These sequences are called **autonomously replicating sequences,** or **ARS** elements. A variety of sequences from other eukaryotic organisms are able to function as ARS elements in yeast. Unfortunately, while it is provocative to conclude that ARS elements are origins of replication, there is to date no direct substantive evidence that any ARS elements normally situated on one of yeast's 17 chromosomes actually acts as an origin of replication in its native chromosomal site. Hopefully more detailed analysis in the future will shed more light on the origins question.

Denaturation of DNA. As in prokaryotes, topoisomerases are used to facilitate DNA untwisting during the replication process. In *E. coli,* enzymes called helicases catalyze the unwinding of the DNA double helix to expose single strands of DNA to act as templates in DNA replication. Such unwinding enzymes appear to be involved in eukaryotic DNA replication as well. For example, helicases have been identified in a variety of eukaryotic systems including human cells, mouse, rat, calf, and higher plants. The activity of these enzymes increases about twentyfold when a cell begins proliferating, which indicates that they might have a role in DNA replication.

Once the helix unwinds and single-stranded regions are exposed, we expect that proteins similar to the SSB (single-stranded, binding) proteins in prokaryotes might attach and prevent the DNA from renaturing. Since a large number of proteins have the ability to stick to DNA (such as histones and DNA polymerases), it is difficult to identify exactly which proteins are SSBs.

Single-stranded binding proteins *have* been isolated from a number of eukaryotic cells such as yeast, fungi, amphibians, rodents, calf, and human cells. The molecular weight of these proteins is from 20,000 to 36,000 daltons, and there are hundreds of SSBs in a cell. Evidence indicates that some of these proteins promote the denaturation of double-stranded DNA, while some preferentially stimulate the activity of the αDNA polymerase (or its equivalent). Only in the fungus *Ustilago maydis* and in calf thymus has it been shown that these two actions are the result of the same protein.

In mammalian cells the SSB proteins bind to the single-stranded DNA and leave the bases exposed while simultaneously keeping the DNA in an extended configuration and increasing the affinity of the DNA polymerase for the template. Thus these SSB proteins stimulate polymerase activity (at least in mammals) as measured by the amount of DNA synthesized.

Replication enzymes and proteins. At least three DNA polymerases have been identified in higher eukaryotic cells, designated α, β, and γ. The α and β polymerases are located in the nucleus while the γ polymerase is found in the mitochondrion. DNA polymerase α is the enzyme

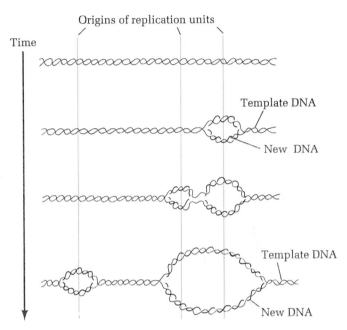

Figure 10.21 Temporal sequencing of DNA replication initiation in replication units of eukaryotic chromosomes.

responsible for nuclear DNA replication, polymerase β serves a DNA repair function, and polymerase γ catalyzes mitochondrial DNA replication. Because the polymerases from only a few eukaryotic organisms have been purified to homogeneity, generalizations at this point should be viewed with caution.

DNA polymerase α is responsible for the replication of nuclear DNA. For example, a correlation exists between α polymerase activity and the rate of DNA synthesis. Thus the onset of DNA replication in mammalian cells sees a three- to tenfold increase in the activity of α polymerase. Growing cells contain perhaps 20,000 to 60,000 molecules of α polymerase per cell, which is more than enough to cope with the rate of replication for the replicons in eukaryotic cells. Whereas *E. coli* DNA polymerases have exonuclease activities associated with them, no DNA polymerase of higher eukaryotes, including α polymerase, has associated exonuclease activities. As a consequence, the polymerases are unable to perform proofreading functions as they catalyze DNA replication. Presumably, separate enzymes carry out the proofreading activities.

DNA polymerase α is characterized by its relatively high molecular weight, complex subunit composition, specific induction during S phase, and ability to use RNA-primed DNA templates. Highly purified DNA polymerase α consists of a 180,000-dalton DNA polymerase catalytic subunit tightly associated with additional polypeptides of 47,000 to 77,000 daltons. The sizes and combinations of subunits have been conserved in evolutionarily diverse organisms. Primase has been shown to be complexed with the DNA polymerase α complex.

The β polymerase is very different from the α polymerase. It is a single polypeptide, 30,000 to 50,000 dalton protein and is the smallest of the known DNA polymerases in either prokaryotes or eukaryotes. In growing cells the activity of β polymerase is about a tenth that of α polymerase; moreover, there is no correlation between the activity of β polymerase and the replicative activity of the cell. Polymerases' role in DNA repair was shown by studies with mammalian tissue culture cells. In these studies the peak levels of β polymerase activity corresponded to periods of maximal DNA repair activity seen in the cells.

The γ polymerase is responsible for the replication of the mitochondrial chromosome. In addition, γ polymerase activity is evident in the nucleus of mammalian cells where it is apparently needed for the replication of adenovirus DNA when cells are infected with that animal virus. Comparatively little γ polymerase is found in the cells of higher eukaryotes; at its highest, less than 2 percent of all DNA polymerase activity is the result of γ polymerase activity.

Of the lower eukaryotes that have been studied, all have a polymerase similar to the γ polymerase found in higher eukaryotes. Here we will consider the polymerases from three extensively studied lower eukaryotes: baker's yeast (*Saccharomyces cerevisiae*), the smut fungus (*Ustilago maydis*), and the simple green alga *Euglena gracilis*. Table 10.3 summarizes the properties of their DNA polymerases.

In general, the lower-eukaryotic DNA polymerases more closely resemble prokaryotic polymerases than higher-eukaryotic polymerases. Also, genetic studies have shown that the poly-

Table 10.3 Properties of DNA Polymerases of Some Lower Eukaryotic Organisms

DNA Polymerases of	Biological Group	Polymerases Found	Molecular Weight[a]	Subunits[a]	Sensitivity to NEM[b]	RNA Primer Needed	Exonuclease Activity	Function
Ustilago maydis	Fungus	Two	100,000	50,000; 55,000	yes	Uncertain	$3' \to 5'$	Replication
Saccharomyces cerevisiae	Yeast	I, II	150,000; 150,000	I = 70,000; 70,000	yes	I, yes; II, uncertain	I, no; II, $3' \to 5'$	Uncertain
Euglena gracilis	Unicellular green alga	A, B	≈200,000		yes	Uncertain	A, no; B, yes	Uncertain

[a] Units in daltons.

[b] NEM is the inhibitor N-ethylmaleimide.

merases that have been identified in lower eukaryotes are essential for DNA replication. For example, mutants that are temperature-sensitive for these enzymes are unable to grow at high temperatures, and this inability to grow is directly correlated with a temperature sensitivity of nuclear DNA replication.

Baker's yeast has two nuclear DNA polymerases, I and II. Polymerase I accounts for approximately 70 to 90 percent of the nuclear enzyme activity and is presumed to be the one responsible for most DNA replication. This enzyme lacks exonuclease activity and therefore is incapable of carrying out proofreading functions as it catalyzes DNA replication. Yeast polymerase II is similar to prokaryotic polymerases in that it appears to have an associated $3' \rightarrow 5'$ exonuclease (proofreading) activity. However, the cellular function of polymerase II is not known.

Ustilago and *Euglena* each contain two polymerases. We know that the *Ustilago* polymerases function in DNA replication in that organism, but the exact roles of the *Euglena* polymerases are not certain. With these simple organisms, as with many of the higher eukaryotes, the studies of the DNA polymerases are still in the descriptive stages.

As we stated previously, DNA replication catalyzed by the DNA polymerases is semidiscontinuous, resulting in the synthesis of Okazaki fragments on the lagging template strand. It appears that the RNA primer that starts an Okazaki fragment is removed by the action of an RNase (possibly RNase H), and the small gap between the adjacent Okazaki fragments is filled, perhaps, through the activity of DNA polymerase β. In prokaryotes the Okazaki fragments are joined into longer chains through the catalytic activity of the enzyme DNA ligase. Similar enzymes must be present in eukaryotes to catalyze the analogous activity. Extracts of mammalian cells have very low ligase activity, about a thousandfold less than that found in *E. coli* cells. Two ligase enzymes, I and II, have been identified in mammalian cells in contrast to the one found in prokaryotes. Both appear to be located in the nucleus, but only ligase I increases when a cell begins proliferating. In that event ligase I becomes the dominant cellular activity. By contrast, ligase II is dominant in resting cells.

Keynote *In eukaryotes DNA replication occurs in the S phase of the cell cycle and is similar to the replication process in prokaryotic cells. Synthesis of DNA is initiated by RNA primers, occurs in the 5'-to-3' direction, is catalyzed by DNA polymerases, requires a large number of other enzymes and proteins, and is a semiconservative and semidiscontinuous process. Replication of DNA is initiated at a large number of sites throughout the chromosomes.*

Assembly of New DNA into Nucleosomes

Eukaryotic DNA is complexed with histones in nucleosome structures, which are the basic units of chromosomes (see Chapter 9). Therefore, when the DNA is replicated, the histone complement must be doubled so that all nucleosomes are replicated. Replication involves two processes: the synthesis of new histone proteins and the assembly of new nucleosomes.

Histone synthesis. For many years geneticists thought that the synthesis of histones was coordinated with DNA synthesis in the cell cycle and that histones were made only during the S phase. Recent evidence indicates that histone synthesis is not coordinated tightly with DNA synthesis.

The first line of evidence comes from experiments that examined at which point in the cell cycle the histone messenger RNAs (mRNAs) are made (mRNAs are RNA molecules that are "read" in the cytoplasm of the cell to produce proteins). The experimental system was the HeLa cell (a human cell). The results of the experiment showed that histone mRNA is made throughout the cell cycle. Another piece of evidence came from experiments with mouse and Chinese hamster ovary (CHO) cells. Cells in G_1 were separated from cells in S by centrifugation techniques. The cells were incubated in the presence of radioactive amino acids, which were, in turn, incorporated into the proteins being manufactured. The proteins were then examined for the presence of radioactivity: Cells from both the G_1 and S phases had synthesized histones. Thus it appears that histones are made at least in G_1 and S and perhaps in G_2 also. The key event that only occurs in the S phase is the association of the new histones with each other and with DNA to form new nucleosomes.

Nucleosomes and newly synthesized DNA.
Newly replicated DNA from *Drosophila* has been examined by electron microscopy to deter-

mine the distribution of nucleosomes. The results of these studies show that unreplicated DNA and newly replicated DNA both have an identical, beaded chromatin appearance in terms of the diameter of the nucleosomes and of the spacing along the DNA fibers. In other words, new DNA is assembled into nucleosomes virtually immediately.

For the systems that have been examined (animal cells), evidence shows that old and new histones do not mix and that the histone octamer (the set of eight histones that comprises the nucleosome core) is conserved from cell generation to cell generation. One model to explain the conservative segregation of the histone octamer (Figure 10.22) proposes that the old histone octamer remains attached to the template DNA on the leading strand of the replication fork, while the new histone octamers become associated with the lagging strand. Other models propose a random distribution of old and new histone octamers between the two arms of the replication

Figure 10.22 Model of the replication fork of a eukaryotic chromosome in which there is conservative segregation of the histone octamer. Mature nucleosomes are shown as darkly shaded spheres and immature nucleosomes as lightly shaded spheres. An RNA-primed Okazaki fragment (the shaded arrow) is on the lagging strand. A helicase is at the origin of the fork itself, and a DNA polymerase and a DNA ligase are on the lagging strand.

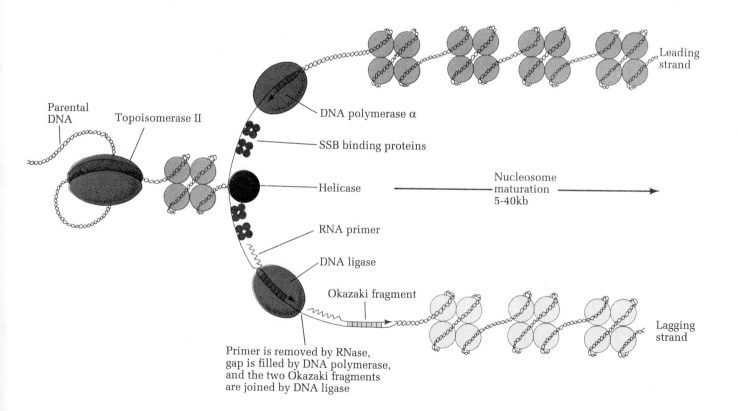

Parental DNA

Topoisomerase II

DNA polymerase α

SSB binding proteins

Helicase

RNA primer

DNA ligase

Okazaki fragment

Primer is removed by RNase, gap is filled by DNA polymerase, and the two Okazaki fragments are joined by DNA ligase

Leading strand

Nucleosome maturation 5-40kb

Lagging strand

fork. Although more work needs to be done to clarify this issue, it is clear that during the replication process the old histone octamers remain associated with the DNA, and furthermore, both new and old histone octamers must be associated with single-stranded DNA at least for a while during the replication events.

Once chromosome replication has been completed, the cell completes the cell cycle by going through the G$_2$ and M phases.

Genetics of the Eukaryotic Cell Cycle

As with any biological activity, useful information can be obtained from studies of mutants that affect the process. With respect to the cell division cycle, relatively few mutants are known, compared with the situation in prokaryotes.

Human cells may carry some naturally occurring genetic diseases in which the defect has been attributed to defects in DNA replication.

Some of these mutants are listed in Table 10.4. Perhaps the most well-known is *xeroderma pigmentosum*. People with this lethal affliction are photosensitive, and portions of their skin that have been exposed to light show intense pigmentation, freckling, and warty growths that may become malignant. The function affected in these people is DNA replication that is associated with the repair of damage caused by ultraviolet light, X rays, or gamma radiation or by chemical treatment. Thus individuals with xeroderma pigmentosum are unable to repair radiation damage to DNA and eventually die, often as a result of malignancies resulting from the damage. Since the disease is inherited, the defect must result from a mutation in a gene coding for a protein involved in the repair of DNA damage.

A number of eukaryotic cell cycle mutants have also been obtained through the use of selection techniques. In general, these mutants were isolated by selection for temperature-

Table 10.4 *Examples of Some Naturally Occurring Human Cell Mutants That Are Defective in DNA Replication*

Disease	Symptoms	Function Affected
Xeroderma pigmento-sum (XP)	Lethal; skin freckling; skin cancerous growths	Repair replication of DNA damaged by irradiation or chemicals
Ataxia telangiectase (AT)	Muscle coordination defect; proneness to respiratory infection; sun-sensitive skin disorder	Repair replication of DNA
Fanconi's anemia (FA)	Aplastic anemia[a]; pigmentary changes in skin; malformation of heart, kidney, and extremities	Repair replication of DNA
Bloom's syndrome (BS)	Dwarfism; sun-sensitive skin disorder	Elongation of DNA chains intermediate in replication

[a] Individuals with aplastic anemia make no, or very few, red blood cells.

Source: After R. Sheinin et al., 1978. *Annu. Rev. Biochem.* 47:277–316.

sensitive growth so that the cells in question do not grow at either a high (heat-sensitive mutants) or a low (cold-sensitive mutants) temperature. The reasoning here is that the functions in the cell cycle are essential functions. So if one of the functions is lost by mutation, the cell dies—or at least it is not able to reproduce, and therefore we cannot obtain high numbers of cells exhibiting the defect for study. However, if a cell is temperature-sensitive for a cell cycle

Table 10.5 Examples of Cell Division Cycle Mutants in Various Eukaryotic Organisms

Cell Type	Mutant Symbol	Cell Cycle Function Affected
Saccharomyces cerevisiae	cdc1	Growth
	cdc2, cdc8, cdc3, cdc10	DNA synthesis; establishment of S phase
	cdc11, cdc12	Mitosis; cytokinesis
	cdc4, cdc6, cdc7	Initiation of DNA synthesis; entry into S phase
	cdc5, cdc14, cdc15	Late nuclear division
	cdc9, cdc13	Medial nuclear division
	cdc25	Start; growth
	cdc28	Start; continuing growth
Ustilago maydis	ts–220	Elongation phase of DNA synthesis
	ts–207	DNA synthesis (S phase)
	pol 1–1	DNA polymerase A
Chinese hamster ovary (CHO) cells	K–12	G_1; entry into S phase blocked
	13B11	DNA synthesis (S phase)
	MS1–1	Mitosis; cytokinesis
Hamster BHK–21 cells	ts AF8	G_1; entry into S blocked
	ts HJ4	Late G_1; possibly initiation of S
	NW1	Mitosis; cytokinesis
Hamster HM–1 cells	ts–655	Mitosis; prophase
	ts–546	Mitosis; metaphase
Mouse leukemia cells	ts 2	Mitosis; cytokinesis
Mouse CAK cells	ts B54	G_1; entry into S blocked
Mouse Ba1B/C cells	ts 2	DNA synthesis (S phase)
Mouse L cells	ts A1S9	Elongation of replication units in DNA synthesis
	ts–C1	DNA synthesis (S phase)
Murine L5178Y cells	ts 2	Mitosis
African green monkey Bsc–1 cells	ts3, ts5, ts9	DNA synthesis (S)

Souce: After R. Sheinin et al., 1978. *Annu. Rev. Biochem.* 47:277–316, and G. Simchen, 1978. *Annu. Rev. Genet.* 12:161–191.

function, then it survives and grows at the normal (permissive) temperature, and only at the nonpermissive temperature is it defective. By switching the temperature experimentally, then, we can get the cells to manifest their defect.

Many temperature-sensitive cell cycle mutants have been isolated in eukaryotes, both in lower eukaryotes and in tissue cultures of cells of higher eukaryotes. Examples of some of the temperature-sensitive mutants that show conditional defects in cell cycle events are listed in Table 10.5. As the data in the table indicate, the temperature-sensitive defect can be in a number of reactions, as we would expect. Some mutants affect the elongation stage of DNA synthesis, some affect the function of the DNA polymerases, and some affect essentially every discrete stage of the cell division cycle.

The yeast mutants are perhaps the best-studied, and they are particularly useful since virtually all stages of the cell division cycle have mutational defects. Studies have been done of the interrelationships of each of the temperature-sensitive events. For example, genetic studies have shown that the initiation and elongation stages of DNA synthesis must be completed before actual nuclear division occurs, and then cytokinesis and cell separation can occur, in that order. While we could have predicted some of these dependencies, it is always useful to have experimental confirmation, and this system shows the power of genetic analysis in providing such basic information.

Keynote Eukaryotic nuclear DNA is complexed with histones and organized into nucleosomes, the basic units of chromosomes. When the DNA replicates, new histones are synthesized and are assembled into new nucleosomes.

Analytical Approaches for Solving Genetics Problems

Because this chapter is primarily descriptive, there are no quantitative problems for this material.

Questions and Problems

10.1 Compare and contrast the conservative and semiconservative models for DNA replication.

10.2 Describe the Meselson and Stahl experiment, and explain how it showed that DNA replication is semiconservative.

*10.3 In the Meselson and Stahl experiment ^{15}N-labeled cells were shifted to ^{14}N medium, at what we can designate as generation 0.

a. For the semiconservative model of replication, what proportion of ^{15}N–^{15}N, ^{15}N–^{14}N, and ^{14}N–^{14}N would you expect to find at generations 1, 2, 3, 4, 6, and 8?

b. Answer the question in part a but this time for the conservative model of DNA replication.

10.4 Describe the semidiscontinuous model for DNA replication. What is the evidence showing that DNA synthesis is discontinuous on at least one template strand?

*10.5 Distinguish between a primer strand and a template strand.

*10.6 The autoradiograph of the *E. coli* chromosome produced by Cairns showed its length to be 1100 μm.

   a. How many base pairs does the *E. coli* chromosome have?
   b. How many complete turns of the helix does this chromosome have?
   c. If this chromosome replicated unidirectionally and if it completed one round of replication in 60 minutes, how many revolutions per minute would the chromosome be turning during the replication process?
   d. The *E. coli* chromosome, like many others, replicates bidirectionally. Draw a simple diagram of a replicating *E. coli* chromosome that is halfway through the round of replication. Be sure to distinguish new and old DNA strands.

10.7 Chromosome replication in *E. coli* commences from a constant point, called the origin of replication. The results of autoradiography experiments suggested to Cairns that chromosome replication was unidirectional. It is now known that DNA replication is bidirectional. Devise a biochemical experiment to prove that the *E. coli* chromosome replicates bidirectionally. (*Hint:* Assume that the amount of gene product is directly proportional to the number of genes.)

10.8 Compare and contrast the three *E. coli* DNA polymerases with respect to their enzymatic activities.

*10.9 Distinguish between the activities of primase, single-stranded, binding protein, helicase, DNA ligase, DNA polymerase I, and DNA polymerase III in DNA replication in *E. coli*.

10.10 Describe the molecular action of the enzyme DNA ligase. What properties would you expect an *E. coli* cell to have if it had a temperature-sensitive mutation in the gene for DNA ligase?

*10.11 Compare and contrast eukaryotic and prokaryotic DNA polymerases.

10.12 When the DNA replicates, the nucleosome structures must duplicate. Discuss the synthesis of histones in the cell cycle, and discuss the model for the assembly of new nucleosomes at the replication forks.

10.13 Suppose *E. coli* cells are grown on an ^{15}N medium for many generations. Then they are quickly shifted to an ^{14}N medium, and DNA is extracted from the samples taken after one, two, and three generations. The extracted DNA is subjected to equilibrium density gradient centrifugation in CsCl. Using the reference positions of pure ^{15}N and pure ^{14}N DNA as a guide, indicate, in the following figure, where the bands of DNA would equilibrate if replication were semiconservative or conservative.

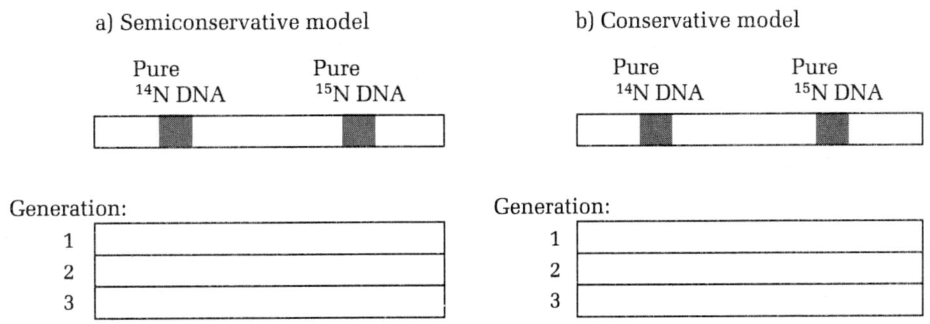

a) Semiconservative model	b) Conservative model

Pure
¹⁴N DNA

Pure
¹⁵N DNA

Pure
¹⁴N DNA

Pure
¹⁵N DNA

Generation:

1

2

3

Generation:

1

2

3

10.14 Assume you have a DNA molecule with the base sequence T–A–T–C–A going from the 5′ to the 3′ end of one of the polynucleotide chains. The building blocks of the DNA are drawn as in the following figure. Use this shorthand system to diagram the completed molecule as proposed by Watson and Crick.

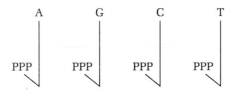

A

G

C

T

PPP   PPP   PPP   PPP

*10.15 List the components necessary to make DNA in vitro by using the enzyme system isolated by Kornberg.

*10.16 Give two lines of evidence that the Kornberg enzyme is not the enzyme involved in the replication of DNA for the duplication chromosomes in growth of *E. coli.*

Transcription

As was stated in Chapter 8, genetic material has three requirements:

1. It must contain all the information for an organism's structure, function, development, and reproduction in a stable form.
2. It must replicate accurately so that progeny cells have the same genetic information as the parental cell.
3. It must be capable of variation.

In the previous three chapters we examined the structure of genetic material and learned how it is replicated so that it can be passed from one cell generation to the next and from a parental to a progeny organism. Now we will examine the processes by which genetic material directs the process of protein synthesis. Since *all* cell and organism functions involve the action of specific proteins, all cell functions are ultimately directed by genetic material.

Each protein consists of one or more chains of amino acids, and the sequence of amino acids in a protein is coded for by a specific base-pair sequence in DNA. When a protein is needed in the cell, the genetic code for that protein's amino acid sequence must be read from the DNA and processed into the finished protein. Two major steps occur in the process of protein synthesis: transcription and translation. **Transcription** (also called RNA synthesis) is the transfer of information from a double-stranded DNA molecule to a single-stranded RNA molecule. **Translation** is the transfer of information from a single-stranded RNA molecule into the amino acid sequence of a protein chain. While replication typically occurs in only part of the cell cycle (at least in eukaryotes), transcription and translation generally occur throughout the cell cycle. In this chapter we will examine the transcription process itself, and in Chapter 12 we will discuss the structures and properties of the different RNA classes and the processing of RNA precursor molecules.

Overview of Transcription

The genome of an organism consists of specific sequences of base pairs distributed among a species-specific number of chromosomes. All base pairs in the genome are replicated during the DNA synthesis phase of the cell cycle, but only *some* of the base pairs are transcribed into RNA. The specific sequences of base pairs that are transcribed are called *genes,* and thus the transcription process is also referred to as *gene expression.* Around the beginning and end of each gene are base-pair sequences called **gene regulatory elements,** which are involved in the regulation of gene expression.

Three different major classes of RNA molecules or transcripts are produced by transcription: **messenger RNA (mRNA), transfer RNA (tRNA),** and **ribosomal RNA (rRNA).** Each of these classes of RNA is found in both prokaryotes and eukaryotes and is involved in protein synthesis, but each has a function different from that of the others. In addition, eukaryotes contain a fourth class of RNA called *small nuclear RNA (snRNA).* The functions of the different RNA molecules are described later in the chapter. For now we can note that only the mRNA molecule is translated to produce a protein molecule. A gene that codes for an mRNA molecule, and hence for a protein, is called a **structural gene.** In this and later chapters we will be discussing primarily structural genes. **Nonstructural genes** are genes whose RNA transcripts (the products of transcription) are the final products of gene expression; they do not function as messenger molecules between DNA and protein. The genes that specify tRNA, rRNA, and snRNA molecules are nonstructural genes because tRNA, rRNA, and snRNA molecules are final products and are never translated into proteins. The four types of RNA transcripts are summarized in Figure 11.1.

Genes are almost always located on the chromosomes, yet they must be able to control functions at various locations throughout the cell. Through transcription the genes produce RNA transcripts that can diffuse away from the chromosomes and be involved in protein synthesis elsewhere in the cell. Also note that while DNA is a very stable structure that endures cell and organism reproduction virtually unchanged, the RNA transcripts of genes typically have a lim-

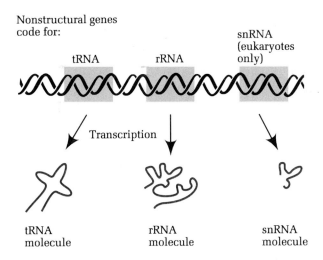

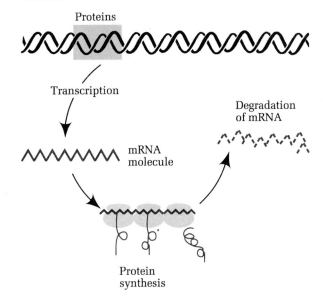

Figure 11.1 Four types of RNA transcripts: messenger RNA (mRNA), transfer RNA (tRNA), ribosomal RNA (rRNA), and small nuclear RNA (snRNA). The latter is found in eukaryotes only, while the other three types are found in both prokaryotes and eukaryotes.

ited existence, being degraded by cellular enzymes called **ribonucleases** when their useful life is over. The transient existence of the RNA transcripts is a key element in the processes that regulate gene expression.

Transcription in Prokaryotes

In both prokaryotes and eukaryotes, transcription occurs by a process that is catalyzed by an enzyme called *RNA polymerase.* In the transcription process only one of the two DNA strands is copied into a single strand of RNA (Figure 11.2).

As in replication, the DNA double helix must unwind before transcription can begin. The DNA strand that acts as a template strand is called the **sense strand.** The strand complementary to the sense strand is called the *antisense strand.* Which of the two strands is the sense strand is gene-specific; that is, one strand may be the sense strand for gene *A,* while the other strand may be the sense strand for gene *B.* If each of the two DNA strands of a gene was transcribed to produce an RNA molecule, each gene would produce two RNA products that were complementary in sequence, and hence two very different proteins. However, genetic evidence indicates that each gene encodes but a single protein, as would be expected if only one RNA molecule were transcribed from a given gene.

Some evidence that only one of the two strands of DNA is copied into RNA in a given region came from the work of J. Marmur and his colleagues in their studies of the RNA species produced when bacteriophage SP8 infects the bacterium *Bacillus subtilis.* The two DNA strands of this double-stranded DNA phage have different densities from each other and from the double-stranded form, and can easily be separated by cesium chloride equilibrium density gradient centrifugation (Figure 11.3).

Marmur and his colleagues allowed SP8 to infect *Bacillus subtilis* growing in a medium containing ^{32}P so that all RNA species transcribed from the phage genome during the life cycle became radioactively labeled. They reasoned that the RNAs produced would complement the DNA strand from which they were transcribed. Next they determined whether the radioactive, single-stranded RNA molecules were able to form stable hybrids with one or both of the separated DNA strands. The results indicated that stable DNA-RNA hybrids were formed only with the heavier of the two strands, from which they concluded that only one of the two DNA strands was used for transcription (Figure 11.4).

In the SP8 case it was fortunate that only one of the two DNA strands of the chromosome is used for the transcription of all genes. In other systems, while it is true that only one of the two strands is transcribed for a given gene, *which* DNA strand functions as the template strand for RNA transcription is gene-specific.

The RNA precursors (i.e., the molecules polymerized into the RNA chain) are the ribonucleoside triphosphates ATP, GTP, CTP, and UTP, col-

---

Figure 11.2 Transcription process. The DNA double helix is denatured by the action of RNA polymerase, which then catalyzes the synthesis of a single-stranded RNA chain beginning at the "start of transcription" point. The RNA chain is made in the 5′ → 3′ direction, using only one strand of the DNA as a template to determine the base sequence.

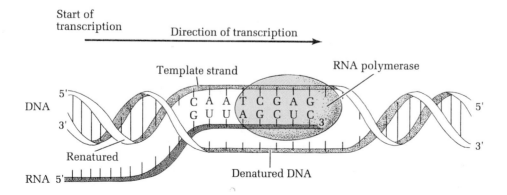

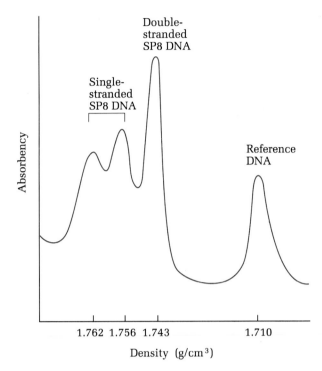

Figure 11.3 Demonstration that the two strands of DNA in phage SP8 have different buoyant densities. The figure shows DNA separated by CsCl equilibrium density gradient centrifugation. Reference DNA, double-stranded SP8 DNA, and the "heavy" and "light" single strands of SP8 are distinguished on the gradient. (D) Reference DNA, (C) Native, double-stranded SP8 DNA, (A,B) Denatured "heavy" and "light" SP8 DNA strands, respectively.

occurs on the DNA template chain, a U nucleotide is placed in the RNA chain.

As in DNA replication, RNA is synthesized in the $5' \rightarrow 3'$ direction, and hence the template DNA strand is copied in the $3' \rightarrow 5'$ direction. Thus, for example, if the DNA template reads

$$3' - A\ T\ A\ C\ T\ G\ G\ A\ C - 5'$$

then the RNA chain produced will read

$$5' - U\ A\ U\ G\ A\ C\ C\ U\ G - 3'$$

The $5' \rightarrow 3'$ direction of RNA synthesis is demonstrated by two observations: (1) The RNA precursors are ribonucleoside 5'-triphosphates.

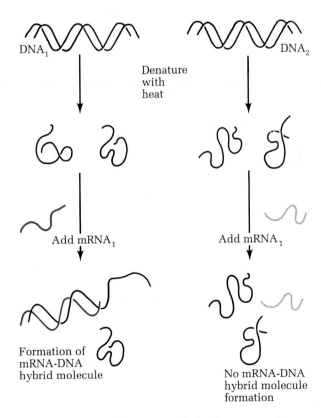

Figure 11.4 Use of DNA-RNA hybridization to show that an mRNA molecule is complementary to only one of the two strands of its DNA template. Left: Formation of DNA-RNA hybrid between an mRNA and one of the two DNA strands. Right: Control experiment showing that no DNA-RNA hybrid forms between mRNA and unrelated DNA.

lectively called NTPs. The enzyme that catalyzes the polymerization reaction is **RNA polymerase;** this enzyme is specific for RNA synthesis and will only use nucleotide triphosphates that contain the ribose sugar rather than the deoxyribose sugar found in DNA precursors.

As Figure 11.5 shows, the polymerization reaction itself is very similar to the DNA polymerization reaction. The next nucleotide to be added to the chain is selected by the RNA polymerase for its ability to pair with the exposed base on the DNA template strand. Recall that RNA chains contain nucleotides with the base uracil instead of thymine. Therefore where an A

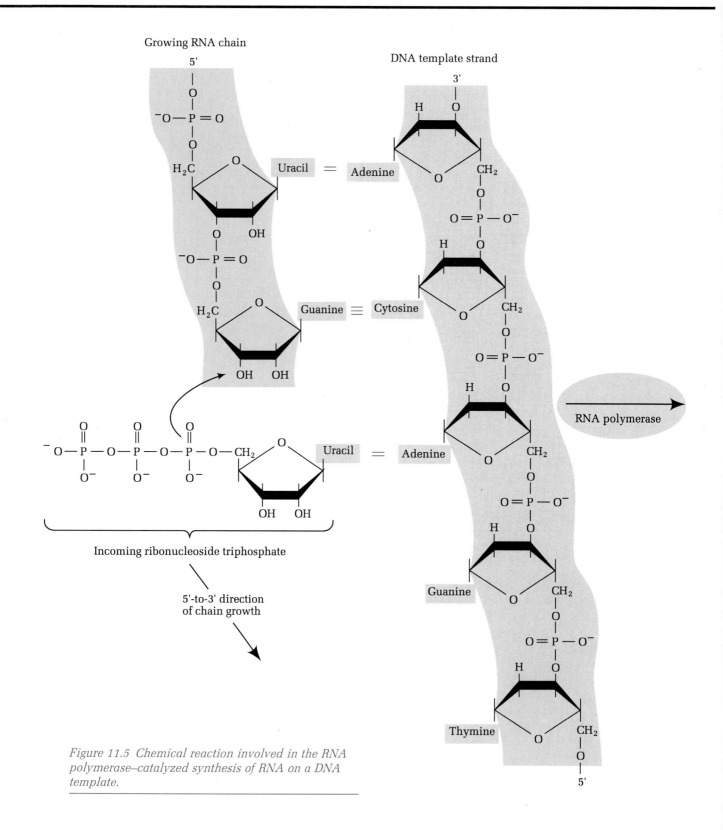

Figure 11.5 Chemical reaction involved in the RNA polymerase–catalyzed synthesis of RNA on a DNA template.

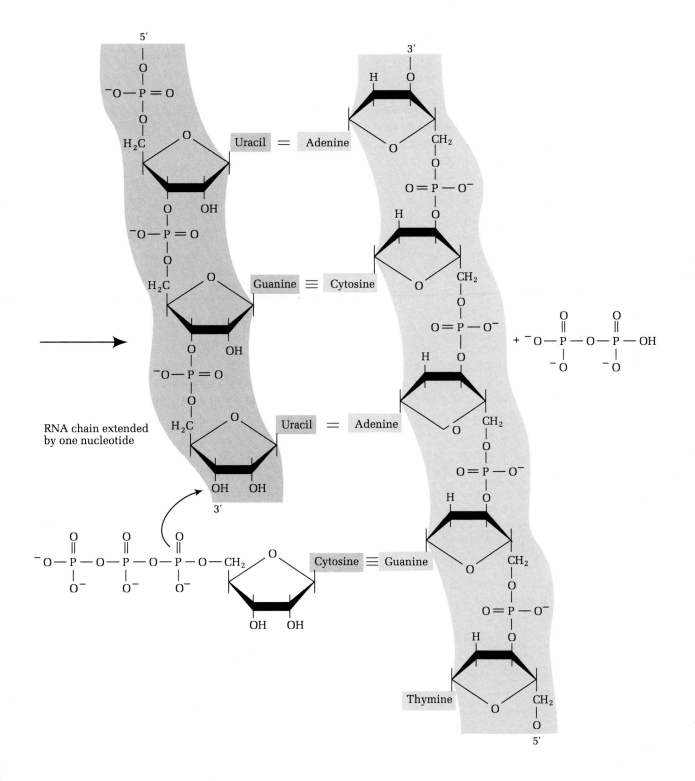

RNA chain extended
by one nucleotide

If the RNA chain grows 5' → 3' then the first nucleotide incorporated will possess a triphosphate group. If, on the other hand, the chain grows 3' → 5' then only the most recently incorporated nucleotide will possess a triphosphate group. All the evidence indicates that the triphosphates are attached to the nucleotides that begin the RNA chain, and newly inserted nucleotides are at the 3' ends. (2) RNA synthesis is inhibited by 3'-deoxyadenosine, an analog of deoxyadenosine that lacks the 3'OH group needed to make the 3' → 5' phosphodiester linkage. When added to cells, 3'-deoxyadenosine is phosphorylated to 3'-deoxyadenosine-5'-triphosphate and is then incorporated into the RNA chain at the 3' growing end. Since it lacks a 3'-OH, RNA chain growth stops. If RNA synthesis were not 5' → 3', the inhibitor could not work because it could not be incorporated.

RNA Polymerases in Prokaryotes

Each living prokaryotic organism has its own specific RNA polymerase. Bacteriophages, however, may use the bacterial host's RNA polymerase or code for their own, depending on the individual phage.

The RNA polymerases are quite similar to DNA polymerases in their action. They require a DNA template, magnesium ions, and the four ribonucleoside triphosphates, ATP, UTP, CTP, and GTP. If any of these components is missing, then transcription cannot proceed. The RNA polymerases differ from DNA polymerases in two important respects: First, the RNA polymerases do not have any proofreading function. Second, RNA polymerases are able to initiate new RNA chains.

The most extensively studied prokaryotic RNA polymerase is that found in the bacterium *Escherichia coli*. Since the *E. coli* RNA polymerase is similar to the RNA polymerases isolated from a large number of other bacteria, it can be considered representative of bacterial RNA polymerases.

The *E. coli* RNA polymerase (Figure 11.6) is a complex enzyme that synthesizes all three classes of RNA: transfer, messenger, and ribo-

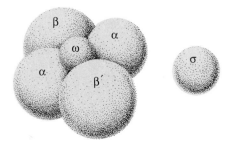

Figure 11.6 Schematic picture of the E. coli *RNA polymerase showing the five subunits of the core enzyme and the dissociable sigma factor.*

somal. The complex enzyme consists of five polypeptide subunits: two alpha (α) subunits, one beta (β) subunit, one beta-prime (β') subunit, and one omega (ω) subunit. The active enzyme is often referred to as the core enzyme or holoenzyme and can be written $\alpha_2\beta\beta'\omega$; its molecular weight is about 450,000 and it has an elongated shape. For the core to begin transcription at the proper place on the chromosome, another polypeptide of weight 98,000, the sigma factor (σ), must become associated with it.

Through mapping techniques, geneticists have found what chromosomal genes control the production of RNA polymerase. They are called the *rpo* genes: *rpoA* (located at 72 min on the *E. coli* map) codes for the α subunits; *rpoB* and *rpoC* (both located at 88.5 min on the map) code for the β and β' subunits, respectively; and *rpoD* (located at 66 min) codes for the sigma factor. The gene for the ω subunit has yet to be defined. Thus the multisubunit RNA polymerase from *E. coli* is coded for by genes that are generally scattered on the chromosome map.

Since the purpose of the transcription process is to make an RNA transcript of a given gene, the RNA polymerase must begin its work at the spot on the DNA strand at which the gene begins. The location on a DNA strand to which RNA polymerase binds to initiate transcription is the transcription-controlling sequence adjacent to the start of the gene; this transcription-controlling sequence is called the **promoter site, promoter sequence,** or, more simply, the **promoter.** In the cell the initiation of transcription is preceded by the binding of a core enzyme–sigma factor complex to the promoter site. Experiments on promoter site recognition performed in vitro have shown that the sigma factor is essential for promoter recognition; if it is absent, the core enzyme initiates transcription randomly. Once transcription is initiated by the core enzyme–sigma factor complex, the sigma factor dissociates from the core enzyme.

The RNA polymerase enzyme must also stop making an RNA molecule once the end of the gene is encountered. Transcription ceases when a controlling element called a **transcription terminator sequence,** or more simply, a **terminator,** is encountered. At some terminators, a special protein factor called the ρ (rho) factor (product of the *rho* gene at 84.5 min on the *E. coli* genetic map) is required for transcription termination. At others, the core enzyme alone carries out the termination. Another protein, the NusA protein (product of the *nusA* gene at 65 min on the map), may also be involved in termination. After termination, the core enzyme is released and is reused on the same or different gene.

Conceptually, then, three types of base-pair sequences are needed for a gene to be transcribed: a sequence of base pairs that codes for the base sequence of the RNA transcript (the gene), a transcription start (promoter) sequence, and a stop (terminator) sequence. These sequences are found in the following arrangement:

promoter . . . gene . . . terminator

For purposes of discussion, the promoter is considered to be "upstream" from the gene, and the terminator is "downstream."

Regarding the subunits of the RNA polymerase core enzyme, biochemical studies with drug-sensitive and drug-resistant RNA polymerases show that the β subunit is involved in the initiation and elongation steps of transcription, while the β′ subunit may function in the initial step of binding the RNA polymerase to the DNA template. The roles of the α and ω subunits are not precisely known.

An RNA polymerase enzyme performs several functions, each of which requires different activities of the enzyme. First, it recognizes a promoter on the double-stranded DNA. Second, it causes the DNA to denature and unwind into single strands at the promoter (similar to the initiation of DNA replication). Third, as a result of reading the promoter sequence, the RNA polymerase orients itself properly and transcribes the entire sense strand of the gene. Lastly, it stops transcribing when it reaches and recognizes the terminator. Figure 11.7 shows the general processes of initiation, elongation, and termination of transcription. These processes will now be discussed in more detail.

Keynote Transcription, the process of transcribing DNA base sequences into RNA sequences (mRNA, tRNA, and rRNA), is similar in prokaryotes and eukaryotes. The DNA denatures and an RNA polymerase catalyzes the synthesis of an RNA molecule in the 5′-to-3′ direction. Only one strand of the double-stranded DNA is transcribed into an RNA molecule. Specific base-pair sequences in the DNA determine where transcription begins (promoter sequence) and ends (terminator sequence).

Initiation and Elongation

As we have just seen, promoter recognition is a function of the sigma factor in association with the core enzyme. Once the core enzyme is bound to the DNA at the correct place, the sigma factor is released and can be reused in other transcription initiation reactions. Soon after the core enzyme is correctly bound to the DNA, it slides along the DNA toward the gene and catalyzes the denaturation of the two DNA strands, thereby exposing the single-stranded sense strand for it to use as a template. Immediately following denaturation, RNA synthesis commences. In *E. coli* the interval from the time the promoter is recognized to the time RNA synthesis is begun is about 0.2 second. Thus RNA synthesis is initiated at a DNA sequence that is different from the sequence to which RNA polymerase first binds.

More precise information about promoter recognition and the initiation of RNA synthesis by RNA polymerase has been obtained from in

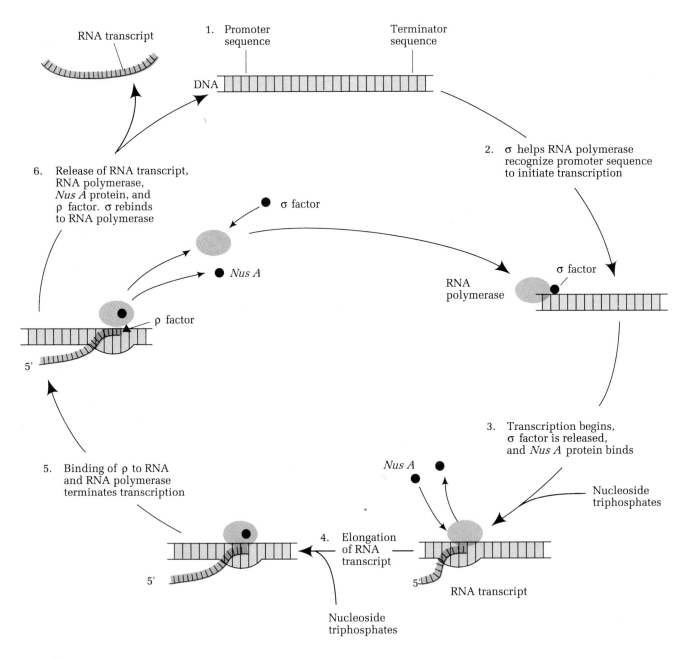

Figure 11.7 Transcription of a gene in E. coli. *The process involves the activities of the RNA polymerase core enzyme, the sigma factor, the NusA factor, and, in this case, the ρ factor. For the gene shown, RNA chain termination requires the ρ factor. This is called ρ-dependent termination. For many other genes, RNA polymerase alone can read the termination sequence—this is called ρ-independent termination.*

vitro experiments. If the NTPs are left out of an in vitro transcription system, the RNA polymerase will bind to the promoter, but it cannot start an RNA chain. The RNA polymerase becomes stalled on the DNA and protects the sequence to which it is bound from digestion by the DNA-degrading enzyme, deoxyribonuclease (DNase). Thus a promoter sequence can be isolated by binding the RNA polymerase to DNA in the absence of NTPs and digesting away other DNA sequences with DNase. The base-pair sequences of the promoters can then be determined.

In *E. coli* two sequences in the promoter are critical for gene expression. These sequences are at two distinct locations upstream from the first base pair to be copied into an RNA chain. Conventionally, this base pair is designated as +1; base pairs upstream from it are given negative numbers, while those downstream from it are given positive numbers. It is also conventional to give the sequences of promoters and other controlling elements for the antisense strand. For the *E. coli* RNA polymerase the two regions of a promoter generally are found at −35 and −10, that is, centered at 35 and 10 base pairs upstream from the base pair at which transcription starts. The −10 region is also called the **Pribnow box** (after the researcher who first discovered it). From examination of the promoters of a large number of genes, the **consensus sequence** (i.e., the most prevalent promoter sequence) for the −35 region is

$$5' — T\ T\ G\ A\ C\ A — 3'$$

and the consensus sequence for the Pribnow box at −10 is

$$5' — T\ A\ T\ A\ A\ T — 3'$$

It is not surprising that RNA polymerase contacts the DNA double helix at two distinct regions, since the RNA polymerase is a large, multisubunit enzyme. Evidence that both promoters are important for gene expression has come from studies of promoter mutations that either diminish or enhance the transcription of a particular gene. The DNA sequence analysis of the promoters in the mutant strains typically reveals base-pair substitutions or deletions in one or the other of these two promoter sequences.

The *E. coli* RNA polymerase binds to a promoter in two distinct steps (Figure 11.8). First, it finds the promoter sequence and binds to it loosely while the DNA is still in double-helical form with all bonds between base pairs intact. This first step involves recognition at the −35 region. The second step involves a shift to a tighter binding between RNA polymerase and DNA that accompanies a local unwinding of about 17 base pairs of the DNA centered around

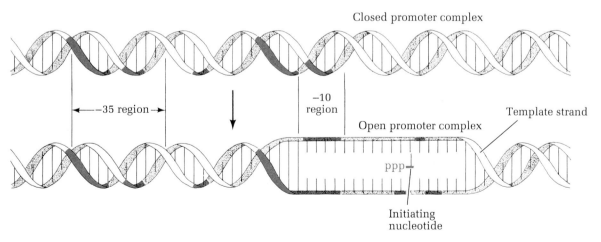

Figure 11.8 Binding of RNA polymerase to the promoter occurs in two steps. First, the enzyme binds loosely at the −35 region and then it binds more tightly at the −10 region, becoming correctly oriented to begin transcription.

the −10 region. It is noteworthy that the −10 region is all AT base pairs; such base pairs are easier to break apart than GC base pairs because they have only two hydrogen bonds rather than three. Once the RNA polymerase is bound to the −35 and −10 regions (Figure 11.9), it is correctly oriented to begin transcription. Conceptually, we can think of the RNA polymerase as holding onto the DNA double helix at specific regions and wedging the two strands apart in preparation for transcription initiation at +1.

Since promoters differ in sequence, the efficiency of RNA polymerase binding varies considerably. As a result, the rate at which transcription is initiated is not constant, a fact that helps to explain why different genes have different rates of expression at the RNA level. There is also evidence that *E. coli* contains other σ factors apart from the major one (σ^{70}) that has been identified. These σ factors generally play a role when significant changes in gene expression are needed, for example, when the environment changes and new genes need to be turned on.

These sequences are usually upstream from the −35 site. Since the rate of initiation of transcription decreases markedly when these sequences are deleted, we can conclude that the sequences play an activation role in transcription. However, the mode of action of the sequences has not been well defined.

As with DNA polymerization, RNA polymerization takes place in the $5' \rightarrow 3'$ direction. It occurs by the formation of complementary base pairs between the sense strand of the DNA and the RNA and the formation of phosphodiester bonds between adjacent ribonucleotides (see Figure 11.2). In this process the RNA polymerase selects a ribonucleoside triphosphate from the pools of the four types of these molecules that are floating in the cell so that the correct complementary base pairing occurs during RNA chain growth. All events take place in a region of the DNA that has denatured to form a transcription bubble that is similar to a DNA replication bubble. Once a few polymerizations have been completed, the sigma factor dissociates from the core enzyme (Figure 11.10; also see Figure 11.7) and can be used again in other transcription initiation reactions.

As the RNA polymerase moves along the DNA, it unwinds the DNA double helix ahead of it, while the helix reforms behind it. About 17 base pairs of the DNA are kept unwound as transcription continues, about the same as are initially unwound. Within the unwound region, about 12 bases of RNA are bonded to the DNA in an RNA-DNA hybrid; the rest is displaced from the DNA. Figure 11.11 presents a model of transcription in progress.

The NusA Protein

The NusA protein plays a role in transcription that is not yet completely understood. The RNA polymerizing activity of RNA polymerase does not require NusA protein, but there is evidence that NusA binds to the core enzyme. After RNA transcription is initiated, σ is released from the RNA polymerase-DNA complex, and it appears that NusA may bind to the core enzyme soon thereafter. Once the core enzyme is released from the template at the termination of transcription, σ probably displaces NusA from the enzyme. At least in vitro, NusA has been shown to have an effect on the rate of RNA synthesis and on the termination of RNA synthesis by RNA polymerase.

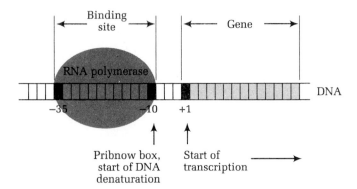

Figure 11.9 RNA polymerase bound at the promoter sequences in E. coli *ready to begin transcription of RNA at + 1.*

1. Initiation
   of transcription

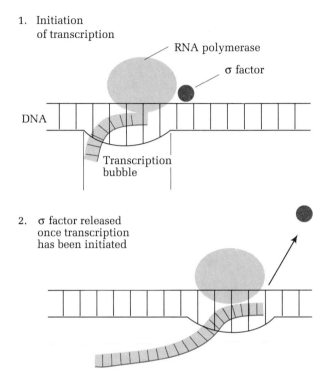

2. σ factor released
   once transcription
   has been initiated

Figure 11.10 Schematic diagram of the catalytic role of the sigma factor in the initiation of transcription in E. coli. At the outset the RNA polymerase core enzyme in a complex with the sigma factor binds to the promoter, the DNA denatures in that region, and transcription begins. After a few nucleotides have been polymerized, the sigma factor dissociates, and the core enzyme continues the RNA synthesis process.

Termination of RNA Synthesis in *E. coli*

The termination of the transcription of a prokaryotic gene is signaled by controlling elements called terminators. Three key events occur at a terminator: (1) RNA synthesis stops, (2) the RNA chain is released from the DNA, and (3) RNA polymerase is released from the DNA. At some terminators a protein called the rho (ρ) factor is required for termination. At other terminators the core RNA polymerase itself can carry out the termination events.

Rho-independent terminators consist of sequences with twofold symmetry that are about 15 to 20 base pairs before the end of the RNA, followed by a string of about six AT base pairs (Figure 11.12). The AT base pairs are transcribed into a string of Us at the end of the RNA. Potentially, the transcript of the region with twofold symmetry can form a hairpin loop. Since mutations that disrupt the twofold symmetry (and therefore affect the transcript's ability to form a hairpin) decrease or prevent termination (see Figure 11.12), it is argued that the hairpin not only forms in vivo but also plays an important role in termination. Similarly, the existence of mutations that disrupt the string of AT's in the terminator sequence of the template cause its inactivation. While it is not clear how the AT string and the hairpin loop facilitate rho-independent termination, it is possible that the rapid formation of the hairpin loop destabilizes the RNA-

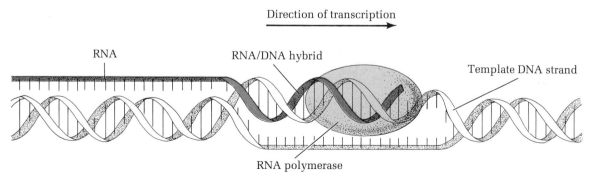

Figure 11.11 Model of RNA polymerase in the process of synthesizing an RNA molecule from a DNA template.

← ——— Dyad symmetry ——— →

Template
(DNA)

5' C C C A G C C C G C C T A A T G A G C G G G C T T T T T T T T G A A C A A A A 3'

3' G G G T C G G G C G G A T T A C T C G C C C G A A A A A A A A C T T G T T T T 5'

Transcript
(RNA)

5' C C C A G C C C G C C U A A U G A G C G G G C U U U U U U U U – OH 3'

Transcript folded to form
termination hairpin

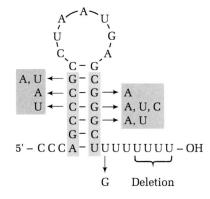

Figure 11.12 Sequence of a ρ-independent terminator and structure of the terminated RNA. The mutations shown in grey partially or completely prevent termination.

DNA hybrid in the region of transcription which in turn leads to the release of the RNA and to transcription termination.

Rho-dependent terminators lack the AT string found in rho-independent terminators and many cannot form hairpin structures. While the mechanism of rho-dependent termination is not clearly understood, it is known that it requires the interaction of rho and the nascent RNA chain upstream of the terminator, at regions that are relatively unstructured, rich in cytosines, and untranslated. Rho-dependent termination requires the hydrolysis of a molecule of ATP. This is accomplished using an ATPase activity that is part of the rho protein. The exact role of ATP hydrolysis in the termination process has yet to be elucidated. Some data suggest that ATP hydrolysis is used to help unwind the RNA-DNA hybrid at the 3' end of the nascent transcript, thereby facilitating the release of the RNA from

the DNA template. If this is true, then the activity of the rho factor that catalyzes this can be considered to be an RNA-DNA helicase.

Summary of RNA Synthesis Events

In sum, the end product of transcription is an RNA molecule that is, in one way or another, involved in protein synthesis. The three general classes of RNA transcripts that are made are messenger RNA, transfer RNA, and ribosomal RNA. Each of these transcripts is made in the 5' → 3' direction, is read from the DNA using similar promoter and terminator sequences, and is synthesized by using the catalytic activity of RNA polymerase core enzyme alone or in association with the sigma factor (for initiation) and the NusA protein (for termination). A third protein factor, the rho factor, is required for the release of the RNA transcript from the DNA.

The Role of RNA Polymerases in Phage Transcription

In general, small bacteriophages use the RNA polymerase of the host cell for their transcription. Larger phages, however, often have a gene that codes for an RNA polymerase that is used to transcribe phage genes. For example, phage T7 has a gene, *gene1,* that codes for an RNA polymerase. The host *E. coli* RNA polymerase transcribes the part of the phage chromosome that includes *gene1.* The *gene1* transcript is translated in the bacterial cell to produce a phage-specific RNA polymerase, which then transcribes the rest of the phage chromosome. At the same time some of the other phage gene transcripts are translated to produce proteins that shut off the activity of the host RNA polymerase. The net effect of all this enzyme activity is that the transcription of the phage chromosome is favored over transcription of the host chromosome. This result occurs because the phage promoters have a special affinity for the phage-coded RNA polymerase while the *E. coli* promoter sequences have a low affinity for the phage-coded enzyme and because the *E. coli* RNA polymerase has been inactivated. Incidentally, the T7 RNA polymerase is a single polypeptide protein with a molecular weight of 98,000. Thus, RNA polymerases need not be as complex as the *E. coli* core enzyme (molecular weight 450,000) to catalyze the transcription process.

The T4 phage exploits the transcription process in a bacterial cell in a different way. Unlike phage T7, phage T4 does not code for its own RNA polymerase. Instead, T4 has a gene that codes for proteins that interact with the *E. coli* RNA polymerase so that its subunits are altered. The altered host polymerase then transcribes phage genes rather than bacterial genes, and the lytic cycle proceeds efficiently.

Transcription in Eukaryotes

In general, transcription in eukaryotes proceeds in a very similar way to that process in prokaryotes. The two main differences are that more than one RNA polymerase enzyme occurs in eukaryotes and that the transcription sequences (promoters and terminators) are different.

RNA Polymerases in Eukaryotes

In contrast to the one RNA polymerase found in bacterial cells, typically three RNA polymerases are found in a eukaryotic cell. Their properties are listed in Table 11.1. As the table shows, each enzyme is involved in the synthesis of a different type of RNA. The **RNA polymerase I,** located exclusively in the nucleolus, catalyzes the synthesis of the two large ribosomal RNA (rRNA) molecules (18S and 28S rRNAs) and one of the two small rRNAs, the 5.8S rRNA. All of these molecules are found in ribosomes, the organelles responsible for protein synthesis. Ribosome assembly, which we will examine in detail later in this chapter, occurs in the nucleolus. The **RNA polymerase II,** found only in the nucleoplasm of the nucleus, is involved in the

Table 11.1 Properties of Eukaryotic RNA Polymerases

RNA Polymerase	Location	Products	α-amanitin Sensitivity
I	Nucleolus	28S, 18S, 5.8S rRNAs	Insensitive
II	Nucleus	hnRNA mRNA	Highly sensitive
III	Nucleus	tRNA 5S rRNA snRNA	Intermediate sensitivity

synthesis of messenger RNAs (mRNAs). Finally, **RNA polymerase III** (also found only in the nucleoplasm) synthesizes the following: (1) the transfer RNAs (tRNAs) which bring amino acids to the ribosome; (2) the 5S rRNAs, one molecule of which is found in each ribosome and (3) the small nuclear RNAs (snRNA), some of which are involved in RNA processing events.

The three RNA polymerases are distinguishable also by their different sensitivities to inhibition by α-amanitin, a product of the poisonous mushroom *Amanita phalloides*. RNA polymerase I is insensitive to inhibition by α-amanitin, polymerase II exhibits the greatest sensitivity to inhibition, and polymerase III shows intermediate sensitivity to α-amanitin inhibition. These properties have made it possible to isolate and purify the three different RNA polymerase classes.

All the eukaryotic RNA polymerases are more complex than prokaryotic polymerases, consisting of several subunits (two large and at least four small). The subunits of any given RNA polymerase type (I, II, or III) have apparently been conserved through evolution, since they are very similar in size and activity (where that is known) in eukaryotes ranging from yeast to humans.

In comparison with the details we know about the structure and function of the RNA polymerases in *E. coli* and a number of other prokaryotes, we know relatively little about the structure and function of eukaryotic RNA polymerases. One reason for this lack of knowledge is that very few RNA polymerase mutants are known in eukaryotes. Another reason is that the amount of RNA polymerase in a eukaryotic cell is relatively low, thus making purification of the polymerase very difficult. In calf thymus, for example, RNA polymerases constitute only 0.05 percent of the total cellular protein, whereas in *E. coli* RNA polymerase accounts for 1 percent of the total cellular protein.

In addition to the nuclear RNA polymerases, all eukaryotes also have RNA polymerases in the cytoplasmic organelles, that is, in mitochondria and in chloroplasts.

Keynote The E. coli *RNA polymerase, which consists of four polypeptide subunits, initiates transcription at the promoter sequence (along with the sigma factor) and ends transcription at the terminator sequence, sometimes in conjunction with the rho factor. The* E. coli *RNA polymerase synthesizes mRNA, tRNA, and rRNA. Eukaryotes have three distinct nuclear RNA polymerases, each of which transcribes different gene types: RNA polymerase I transcribes the genes for the large ribosomal RNAs, II transcribes mRNA genes, and III transcribes genes for the small 5S rRNAs, the tRNAs, and snRNAs.*

Transcription-Controlling Sequences in Eukaryotes

Promoters and terminators. Like prokaryotic genes, eukaryotic genes have promoters and terminators that direct the transcriptional machinery where to start and stop. Much more is known about promoters than about terminators in eukaryotes, Significantly, among the eukaryotes there appear to be differences in promoters and terminators, making it almost impossible to describe the general features of promoters and terminators as we were able to for prokaryotes. Moreover, even within a given organism, the classes of genes transcribed by the three different RNA polymerases have different promoters and terminators. For that reason, we will discuss the features of each type of promoter and terminator when we discuss the synthesis and processing of eukaryotic RNAs in the following chapter.

We can, however, make one generalization about one type of eukaryotic promoter in comparison with the prokaryotic promoter sequences discussed earlier in this chapter. That is, analysis of the base-pair sequences found upstream from protein-coding eukaryotic genes (that is, genes transcribed by RNA polymerase II) that have been sequenced indicates the presence of the consensus sequence TATAAA (reading from 5′ to 3′ on the antisense strand). This sequence is called the **Goldberg-Hogness box** (after its discoverers) or, more simply, the **TATA** box or **TATA element.** In higher eukaryotes, the TATA element is almost always located at position −30 (25 to 30 base pairs upstream from the first base pair transcribed into RNA). In yeast, the distance between the TATA elements and mRNA initiation sites ranges from 40 to 120 bp, depending on the promoter. Interestingly, some highly expressed yeast genes have no recognizable upstream TATA element. Thus, several different types of promoters appear to be used in eukaryotic cells.

In many RNA polymerase II-transcribed genes, another sequence element upstream from the TATA element is necessary for transcription to take place (Figure 11.13). That is, when many different genes are compared, a rough CAAT consensus sequence is found at approximately −80. Among genes and among organisms, there is not as marked conservation of the CAAT sequence in the −80 region as there is for the TATA element. In general, there is an upstream region located from −110 to −40 that contains two sequence elements ("first" and "second" in Figure 11.13) in which GC-rich regions are prevalent. These sequence elements are necessary for transcription to start and are considered part of the promoter.

Although eukaryotic promoter elements such as TATA elements are thought to be necessary, they are not sufficient for initiation of transcription. At least for RNA polymerase II promoters, the promoter elements function to bind the RNA polymerase and aid the enzyme in locating the RNA initiation site.

Enhancers. Another type of DNA sequence element, called an **enhancer sequence** or **enhancer element** (see Figure 11.13) has been identified as having a strong, positive effect on transcription by RNA polymerase II. In some cases, similar elements have been associated with rRNA genes transcribed by RNA polymerase I. These enhancer elements are upstream from the −110 to −40 promoter region. An enhancer sequence is defined as a DNA sequence that somehow, without regard to its position relative to the gene or to its orientation in the DNA, increases the amount of RNA synthesized from the gene it controls. The first enhancer elements were found in the genomes of viruses that infect mammalian cells, and they are now assumed to exist in all eukaryotes and their viruses. Some enhancers function even though they are several thousand base pairs from the gene they control. Moreover, for a given gene the enhancer position relative to the transcription initiation point does

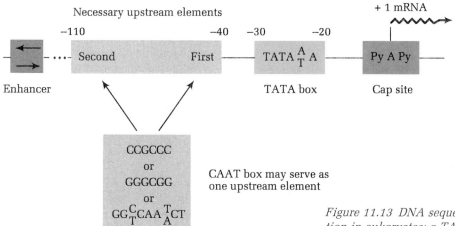

Figure 11.13 DNA sequences necessary for transcription in eukaryotes: a TATA element and two upstream elements near the transcription initiation site. For some genes, an enhancer element may be present at greatly varying distances from the initiation site.

not seem to matter. That is, experiments have been done to place particular enhancers at varying distances from the transcription starting point with no effect on the stimulation of transcription.

There is no consensus sequence for a eukaryotic enhancer element, and the mechanism by which enhancers stimulate transcription is not known. They may bind proteins that serve to regulate transcription (transcription factors) or stimulate the binding of transcription factors to the DNA at other locations upstream of the transcription initiation point. Since the enhancers function independently of their distance from

the transcription start, it is possible that the enhancers, in association with proteins, bring about a conformational change in the DNA, perhaps a looping back that facilitates the binding of stimulatory factors. More research needs to be done to unravel the mystery of how enhancers function.

Lastly, in yeast there is one case of a transcriptional regulatory element that has properties similar to enhancers in terms of location- and orientation-independent function. This element's function, however, is to decrease RNA transcription rather than stimulate it. This new kind of element is called a **silencer element.**

Analytical Approaches for Solving Genetics Problems

Because this chapter is primarily descriptive, there are no quantitative problems for this material.

Questions and Problems

*11.1 Describe the differences between DNA and RNA.

11.2 Compare and contrast DNA polymerases and RNA polymerases.

11.3 Discuss the structure and function of the *E. coli* RNA polymerase. In your answer, be sure to distinguish between RNA core polymerase and RNA core polymerase-sigma factor complex.

*11.4 Discuss the similarities and differences between the *E. coli* RNA polymerase and eukaryotic RNA polymerases.

*11.5 What is the role of each of the *E. coli* RNA polymerase subunits in the transcription process?

11.6 Discuss the molecular events involved in the termination of RNA transcription in prokaryotes.

*11.7 Which classes of RNA do each of the three eukaryotic RNA polymerases synthesize? What are the functions of the different RNA types in the cell?

11.8 What is the Pribnow box? The Goldberg-Hogness box (TATA element)?

*11.9 What is an enhancer element?

RNA Molecules and RNA Processing

In the previous chapter we discussed the process of transcribing genetic material into an RNA molecule in both prokaryotes and eukaryotes. In this chapter we will discuss the structures and properties of the RNA molecules in the cell, namely, messenger RNA, transfer RNA, and ribosomal RNA in prokaryotes and eukaryotes, and small nuclear RNA in eukaryotes. We will see how the transcripts of the genes encoding most of the RNA classes are precursor molecules that must be processed in a highly directed way to produce mature, functional molecules.

Overview of Translation

Once the process of transcription is complete, whether in prokaryotes or eukaryotes, three major end products result: messenger RNA, transfer RNA, and ribosomal RNA. In eukaryotes only, a fourth end product, small nuclear RNA, results. As we shall see, the initial RNA transcripts of most genes must be modified before the biologically active (mature) RNA molecules are produced. These initial transcripts are called **precursor RNA molecules (pre-RNAs),** and their processing may involve the addition and/or removal of bases, the chemical modification of some bases, or the cleavage of sequences from the precursor. Processing of precursor-RNA molecules is similar in both prokaryotes and eukaryotes for tRNA and rRNA precursors. In contrast, prokaryotic mRNA is not processed while many eukaryotic mRNAs are generated by processing of precursor molecules.

Although we will be examining the molecular events that comprise the reading of the genetic information coded in mRNA and the transfer of that information into the amino acid sequence of a protein (a process called **translation**) in Chapter 13, we will better understand the functions of the RNAs if we have a general overview of the translation process here. The overview is as follows:

1. **Messenger RNA,** the RNA molecule that contains the coded information for the amino

acid sequence of a protein, migrates to a **ribo-some,** a cellular organelle that consists of proteins and three (prokaryotes) or four (eukaryotes) rRNA molecules. The ribosome binds to the mRNA near the 5′ end and begins to move along the transcribed message until it recognizes a sequence of bases that indicate where the synthesis of the protein should begin.

2. **Transfer RNA** molecules bring amino acids to the ribosome, where they are matched to the transcribed message on the mRNA. This mRNA message is read in groups of three bases (called **codons**).

3. Two tRNAs can bind to the mRNA at the ribosome at one time. A ribosomal enzyme catalyzes the formation of a bond between the two amino acids of two adjacent tRNAs, thus beginning the synthesis of a protein chain.

4. The ribosome continues to move along toward the 3′ end of the mRNA, and as each new tRNA matches its attached amino acid to the next mRNA sequence, the protein chain lengthens. The process continues until the ribosome recognizes a sequence of bases in the mRNA that indicates the protein is complete. The completed protein is then released.

The Structure and Function of Messenger RNA

Figure 12.1 shows the general structure of the mature, biologically active mRNA as it exists in both prokaryotic and eukaryotic cells. All mRNA molecules are transcribed from structural genes (genes that code for proteins).

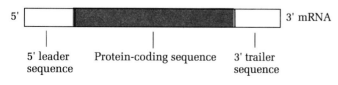

Figure 12.1 *General structure of mature prokaryotic and eukaryotic mRNA molecules.*

The mRNA molecule has three main parts. At the 5′ end is a **leader sequence** whose length varies from mRNA to mRNA. Within this leader sequence is the coded information that the ribosome reads to tell it where to begin the synthesis of the protein; none of the bases of the leader sequence are translated into amino acids of the protein coded for by the mRNA. Following the 5′ leader sequence is the actual **coding sequence** of the mRNA; this sequence determines the amino acid sequence of a protein during translation. Since the number of amino acids varies from protein to protein, it follows that the mRNAs coding for the proteins vary in length. Following the amino acid–coding sequence in the mRNA, and constituting the rest of the mRNA at the 3′ end of the molecule, is a **trailer sequence,** which, like the leader sequence, is not translated and varies in length from mRNA to mRNA. Because of the leader and the trailer sequence, an mRNA molecule contains more nucleotides than are necessary for the protein for which it codes. The trailer sequence indicates that the coding sequence has been completely translated and that all specified amino acids have assembled into the specified protein. The ribosome then detaches from the mRNA, and the completed protein is released.

The production of functioning mRNA is fundamentally different in prokaryotes and eukaryotes. In prokaryotes (Figure 12.2a) the RNA transcript functions directly as the mRNA molecule for translation, while in eukaryotes (Figure 12.2b) the primary RNA transcript must be modified in the nucleus by a series of events known as RNA processing to produce the mature mRNA. In addition, in prokaryotes an mRNA begins to be translated on ribosomes before it has been completely transcribed. An electron micrograph of coupled transcription and translation in *E. coli* is shown in Figure 12.3. This process is called the "coupling" of transcription and translation. In eukaryotes, however, the mRNA must migrate from the nucleus to the cytoplasm (where the ribosomes are located) before it can be translated. Therefore a eukaryotic mRNA is always completely transcribed and processed before it is translated.

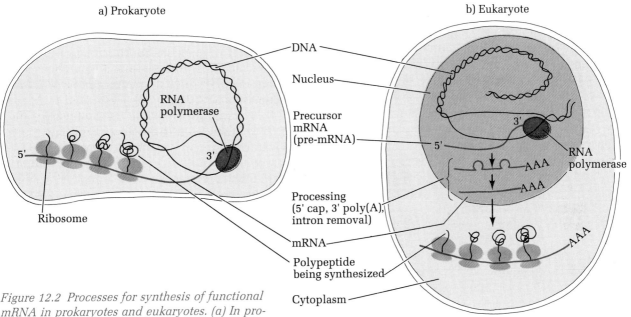

a) Prokaryote

b) Eukaryote

DNA

Nucleus

RNA polymerase

Precursor mRNA (pre-mRNA)

5'

3'

RNA polymerase

Ribosome

Processing (5' cap, 3' poly(A), intron removal)

mRNA

Polypeptide being synthesized

Cytoplasm

AAA

Figure 12.2 Processes for synthesis of functional mRNA in prokaryotes and eukaryotes. (a) In prokaryotes, the mRNA synthesized by RNA polymerase does not have to be processed before it can be translated by ribosomes. Also, since there is no nuclear membrane, translation of the mRNA can begin while transcription continues, resulting in a coupling of the transcriptional and translational processes. (b) In eukaryotes, the primary RNA transcript is a precursor-mRNA (pre-mRNA) molecule which is processed in the nucleus (addition of 5' cap and 3' poly(A) tail, and removal of introns) to produce the mature, functioning mRNA molecule. Only when that mRNA is transported to the cytoplasm can translation occur.

Figure 12.3 Electron micrograph of coupled mRNA transcription and translation in E. coli. From the faint DNA strand, a number of mRNA molecules are emerging. Since the length of the mRNA molecules increases from left to right, we assume that this strand is a single gene with a promoter to the left of the photograph. The globular structures on the mRNAs are ribosomes which are translating the message into a protein before the synthesis of the mRNA is finished.

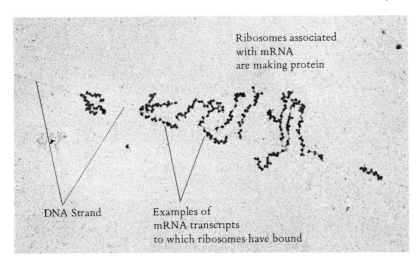

Ribosomes associated with mRNA are making protein

DNA Strand

Examples of mRNA transcripts to which ribosomes have bound

The Relationship between Genes and mRNA in Prokaryotes

In prokaryotes the initial transcript of a structural gene is processed very little, if at all, to produce the mature mRNA molecule. That is, an exact point-by-point relationship exists between the order of base pairs in the structural gene and

a) R-looping procedure

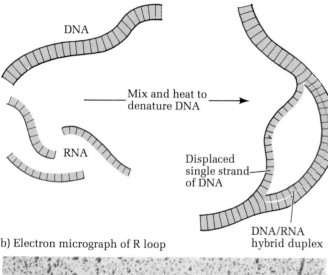

b) Electron micrograph of R loop

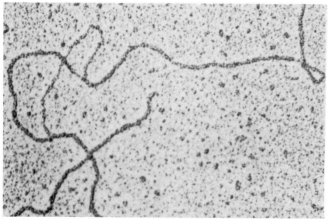

Figure 12.4 The R-looping procedure: (a) Double-stranded DNA is heated in the presence of single-stranded RNA under conditions such that the RNA can form a hybrid with a DNA strand and displace the other strand of DNA. (b) The procedure produces an R loop that can be seen under an electron microscope.

the order of the corresponding bases in both the initial transcript of the gene and the mature mRNA.

One line of evidence for this relationship came from experiments using a technique called **R looping** (Figure 12.4); the procedure was developed by M. Thomas, R. White, and R. Davis. In this procedure molecules of double-stranded DNA are incubated at temperatures below their denaturing temperature to open up (but not completely denature) short stretches of the DNA double helix so that single-stranded RNA molecules in the solution can begin to form RNA/DNA hybrids where the two are complementary (Figure 12.4a). Under these circumstances, RNA/DNA hybrids are more stable than DNA/DNA hybrids. As a consequence, the RNA binds to its complementary sequence in the DNA, thereby displacing a loop of DNA. The DNA outside the region of complementarity between the RNA and the DNA remains double-stranded. The DNA to which the RNA is hybridized is examined under the electron microscope for the presence and characteristics of loops called **R loops** (Figure 12.4b).

When the R-looping experiment uses prokaryotic mRNA, the typical result is as shown in Figure 12.4b. Here one R loop of single-stranded DNA is displaced by the hybridization of the mRNA with the DNA. This displacement indicates that the mature mRNA is an exact base-by-base copy of the base-pair sequence of the gene that codes for it.

The Structure of Eukaryotic mRNA

Unlike prokaryotic mRNAs, which are typically not modified at all, eukaryotic mRNAs are usually modified at both the 5′ and 3′ ends. The modifications that occur after the molecule has been transcribed (posttranscriptional modifications) are catalyzed by specific enzymes.

At the 5′ end the eukaryotic mRNA is modified by the addition of a *cap,* which consists of the addition of a guanine nucleotide (most commonly, 7-methylguanosine) to the terminal 5′ nucleotide by an unusual 5′-to-5′ linkage and the addition of two methyl groups (CH_3) to the

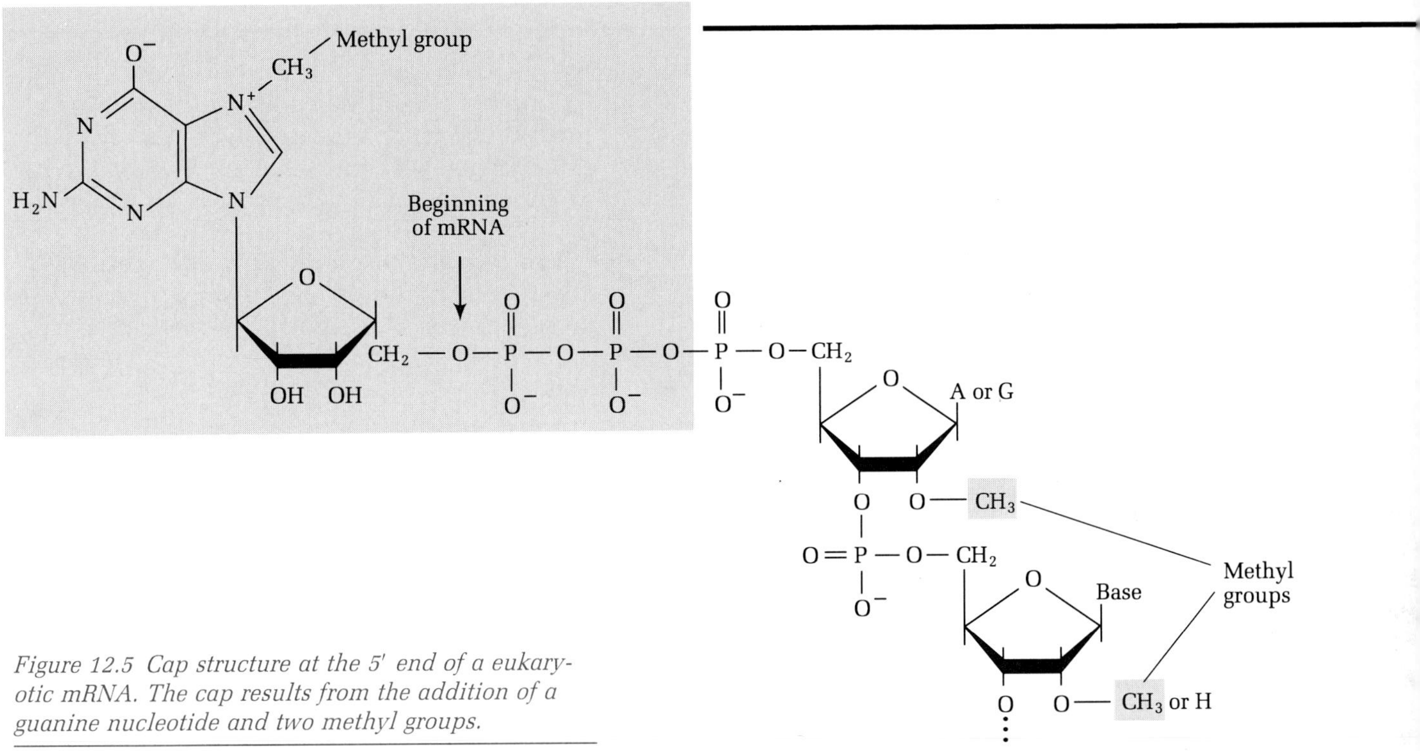

Figure 12.5 Cap structure at the 5′ end of a eukaryotic mRNA. The cap results from the addition of a guanine nucleotide and two methyl groups.

first two nucleotides of the RNA chain (Figure 12.5). This **5′ capping** procedure is found in most eukaryotes, although slight variations are seen in the structure of the cap itself. There is no DNA template for the 5′ cap. Even though its purpose has not been completely elaborated, the cap appears to be essential if the mRNA is to attach to the ribosome. A cap-binding protein makes it possible for the capped end of the mRNA to bind to the small ribosomal subunit. Once bound, the large subunit binds with the complex and translation can begin. Thus capping is necessary in order for the initiation of protein synthesis to take place in eukaryotes. For example, using an in vitro protein synthesizing system consisting of eukaryotic components, prokaryotic mRNAs are not normally translated, but they can be if they are first capped at the 5′ end.

The posttranscriptional modification at the 3′ ends of most eukaryotic mRNAs is an addition of a sequence of about fifty to two hundred and fifty adenine nucleotides. This sequence is called a **poly(A) tail,** and its production is catalyzed by the enzyme polyadenylate, or poly(A), polymerase in the following reaction. (The en-

ergy for the reaction is supplied when the bonds of the ATP molecules are broken.)

$$\text{mRNA} + n\text{ATP} \xrightarrow{\substack{\text{poly(A)}\\\text{polymerase}}} \text{mRNA–(A)}n + n\text{PP}_i$$

where n equals 50–250. Note that there is no DNA template for the production of the poly(A) tail.

Introns in Some Eukaryotic Structural Genes

Eukaryotic mRNA transcripts often contain long insertions of non–amino acid–coding (noncoding) RNA sequences. These noncoding sequences are copied from regions of a gene that are called either **introns** or **intervening sequences.** Such noncoding mRNA sequences must be excised from each mRNA transcript in order to convert the transcript into a mature messenger RNA molecule that can code for the synthesis of a complete protein. Those parts of the gene that correspond to the mature mRNA molecule (and therefore code for amino acids) are called **exons** or **coding sequences.**

The discovery of interrupted genes was totally

unexpected. The first report of interrupted genes was given by A. J. Jeffreys and R. A. Flavell in 1977. They discovered a 600-base-pair (bp) intron in the structural gene for the 146-amino acid β-globin chain in rabbits. (The β-globin chain is part of a hemoglobin molecule.) At about the same time, P. Leder's group discovered a similar intron approximately 550 bp long, that interrupted each of two nonallelic β-globin genes in the mouse. We now will examine the experiments that showed the presence of the intron in the mouse β-globin gene.

Organization of the mouse β-globin gene. P. Leder's group studied the β-globin genes in cultured mouse cells. Because these cells produce

---

Box 12.1
*Sucrose
Density
Gradient
Centrifugation*

Density gradient centrifugation is usually used to separate the components present in a mixture. With this procedure we can measure the rates at which the components sediment in the gradient under centrifugal forces. In sucrose density centrifugation the sedimentation rates are converted to **Svedberg units,** using a formula whose derivation is beyond the scope of this book. Svedberg units, or simply **S values,** are then used as a rough indication of relative sizes of the components being analyzed.

The rate of sedimentation of a component in a density gradient is related both to the molecular weight of the component and to its three-dimensional configuration. Two components of the same molecular weight, for example, will have different S values if one is highly compact and thus sediments rapidly, while the other has a more extended shape and sediments relatively slowly. Density gradient centrifugation is typically used to separate, or estimate, the sizes of RNA molecules, of ribosomal subunits, of ribosomes, of proteins, or of various cellular organelles.

The density gradient centrifugation method involves a supporting column of fluid whose density increases toward the bottom of the tube. The density gradient fluid consists of a suitable low-molecular-weight solute (such as sucrose) in a solvent (water) in which the sample particles can be suspended. For the separation of different-sized RNA molecules, for example, a continuous gradient of sucrose concentration, ranging from 10 percent at the top to 30 percent at the bottom, is prepared in a centrifuge tube (Box Figure 12.1a). Next, a small amount of the sample, in this case a solution containing the RNAs, is very carefully layered on top of the density gradient (Box Figure 12.1b). The sucrose density gradient is then centrifuged at high speeds for several hours, during which time the RNA molecules move through the gradient at different rates depending on their molecular weight and configuration (Box Figure 12.1c). At the completion of centrifugation the RNAs of similar S values are located in discrete zones or bands in the gradient.

So that the different RNA types can be collected, the bottom of the tube is punctured with a needle, and the drops that run out are collected by using a fraction collector (Box Figure 12.1d). Fractions containing the RNAs are identified by measuring the degree to which each fraction absorbs ultraviolet light. Fractions with RNA will absorb ultraviolet light, while those without RNA will not. The relative positions the RNAs occupy in the gradient indicate the S values of the RNAs: Those with higher S values (larger and/or more compact) are found closer to the bottom of the gradient than those with lower S values.

β-globin in large quantities, it is relatively easy to isolate the mRNA for this protein. (Although the techniques for doing this experiment are straightforward, they are beyond the scope of this book.) They analyzed the β-globin mRNA by sucrose density gradient centrifugation (Box 12.1), an experimental procedure that measures the size of the macromolecules or organelles in terms of their sedimentation coefficient (given in S values). The data indicated that the β-globin mRNA has a size of about 10S, which is about the size expected if virtually all the mRNA codes for the β-globin protein.

With the purified, biologically active β-globin mRNA available, a double-stranded complementary DNA (cDNA) copy was made through

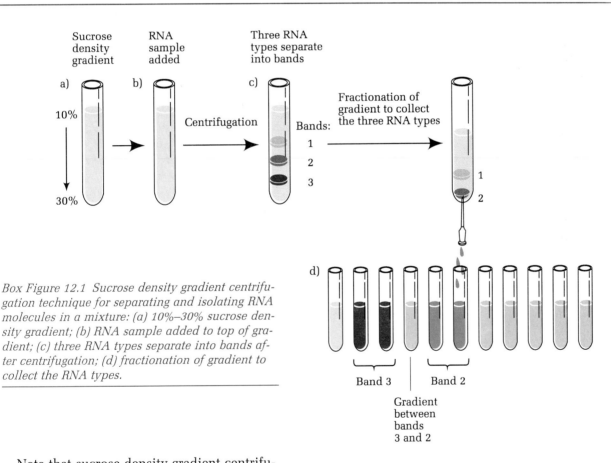

Box Figure 12.1 Sucrose density gradient centrifugation technique for separating and isolating RNA molecules in a mixture: (a) 10%–30% sucrose density gradient; (b) RNA sample added to top of gradient; (c) three RNA types separate into bands after centrifugation; (d) fractionation of gradient to collect the RNA types.

Note that sucrose density gradient centrifugation separates molecules on the basis of their relative rates of sedimentation. Hence it differs from cesium chloride equilibrium density gradient centrifugation, in which molecules are separated on the basis of buoyant density and not size.

reverse transcription, the process of RNA-directed DNA synthesis. In this process a single-stranded RNA molecule (in this case β-globin mRNA) acts as a template, and the enzyme, reverse transcriptase, catalyzes the synthesis (in vitro) of a cDNA strand from the four deoxyribonucleoside triphosphate DNA precursors. The cDNA strand then acts as a template for the synthesis of a DNA strand complementary to it (in a reaction also catalyzed by reverse transcriptase) to produce double-stranded cDNA copies of the original mRNA molecule (Figure 12.6).

The cDNA copies of mature 10S β-globin mRNAs were initially used as probes to find possible precursors to the mature mRNA. The researchers knew that a large population of RNA molecules of various sizes existed in the nucleus. These RNA molecules are called **heterogeneous nuclear RNA, or hnRNA,** and scientists thought at the time of these experiments that they included precursors to mature mRNAs. The cDNA molecules were hybridized with the hnRNA molecules, which resulted in the identification and then purification of a 15S RNA molecule that is a precursor of the 10S mRNA. The precursor mRNA (pre-mRNA, 1500 nucleotides long) is about twice the size of the mature mRNA (700 nucleotides). Further, the pre-mRNA has a 5′ cap and a 3′ poly(A) tail that are indistinguishable from those of the 10S mature mRNA.

Next, the organization of the mouse β-globin gene was investigated in more detail through R-looping experiments. The assumptions were that the 15S pre-mRNA (1500 nucleotides) was processed to produce the 10S mature mRNA (700 nucleotides) and that this processing involved the removal of approximately 800 nucleotides. A question arose: From what segment on the pre-mRNA were these 800 nucleotides removed? They couldn't be from either the 5′ or 3′ end because precursor and mature RNA molecules have the same 5′ and 3′ modifications. Thus they were probably removed from the middle of the pre-mRNA. If the 15S pre-mRNA was R looped to the β-globin gene, the result was as shown in Figure 12.7a: One continuous R loop was seen, indicating that the 15S pre-mRNA is the initial transcript of the β-globin gene. When the mature 10S mRNA was used, however, three small R loops were produced (Figure 12.7b), a result indicating that two sequences occur in the cloned β-globin gene that are complementary to the two ends of the mRNA. Between the two sequences, though, is a sequence that is not complementary to the mRNA. The two outside loops are of single-stranded DNA displaced by the hy-

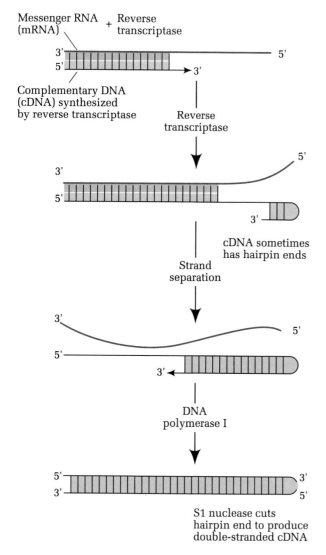

Figure 12.6 Procedure for making a double-stranded complementary DNA (cDNA) copy of a purified mRNA molecule, using reverse transcriptase.

a) 15S β-globin pre-mRNA

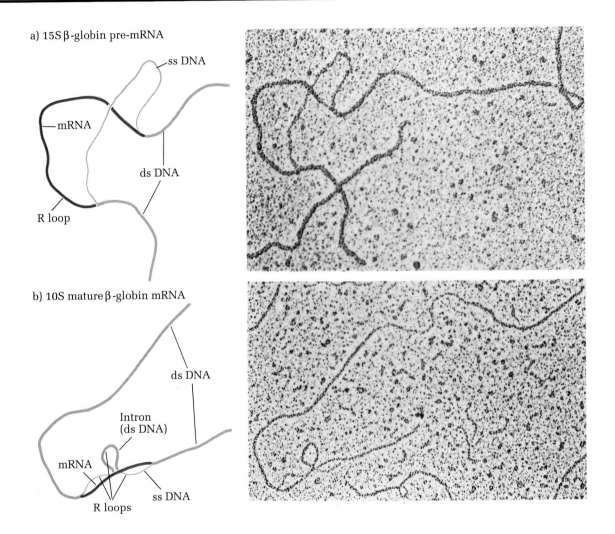

ss DNA

mRNA

ds DNA

R loop

b) 10S mature β-globin mRNA

ds DNA

Intron
(ds DNA)

mRNA

ss DNA

R loops

*Figure 12.7 Electron micrographs of R loops: (a)
formed between 15S β-globin pre-mRNA and the β-
globin gene; (b) formed between 10S mature β-globin
mRNA and the β-globin gene. Interpretative diagrams
are alongside the micrographs.*

bridization between the mRNA and the DNA.
The central loop is double-stranded DNA. Since
the pre-mRNA is a base-by-base copy of the
base-pair sequence of the DNA, whereas the mature mRNA is not, this result was evidence that
the gene contained an intervening sequence of
base pairs that is transcribed as part of the pre-
mRNA, then removed in subsequent processing.

The intron is spliced out to produce the mature
mRNA molecule.

As an aside, we raise a semantic question:
What is a gene? Up until the work just described, geneticists assumed that a gene was a
contiguous stretch of base pairs in the DNA that
was transcribed into a mature mRNA, which,
in turn, was translated into an amino acid sequence. The prokaryotic genes fit this definition
closely. However, we have now seen that in the
case of the mouse β-globin gene the initial 15S
mRNA transcript is not the molecule that is
translated. Before it can be translated, a large
segment of its genetic material must be excised
and the adjacent segments spliced together to

produce the mature 10S mRNA. If we think of only the amino acid–coding regions of the DNA as genes, then the presence of introns clearly means that eukaryotic genes are in pieces. On the other hand, if we define a gene as that region of DNA corresponding to the initial RNA transcript, then the whole stretch of coding sequences plus introns constitute a gene. Both uses may be encountered in your studies, and it is important to keep the distinctions between the two in mind.

Occurrence of genes with introns. Experimenters have now established that a large number of eukaryotic protein-coding genes are interrupted by introns. Moreover, higher eukaryotes tend to have more interrupted genes and longer introns than lower eukaryotes do. We have already seen that early research with the β-globin gene revealed one intron. In subsequent experiments, a second, smaller intron was revealed. In fact, we now know that all mammalian β-globin genes contain two introns. Figure 12.8 shows the structure of the mammalian β-globin gene and a schematic rendering of how it is transcribed and how the introns are removed to generate the mature mRNA.

To give just a few additional examples, the gene for the protein vitellogenin (a precursor to two ovary storage proteins) in *Xenopus* has 33 introns, the gene for chicken collagen has over 50 introns, and the chicken ovalbumin gene has 7 introns. Figure 12.9a shows how genes can be split into many pieces, using the chicken ovalbumin gene as an example; in this case the R loop pattern is complex (Figure 12.9b). Of the 7700 base pairs of the chicken ovalbumin gene, over three-quarters consists of introns; the introns range from 250 to 1600 base pairs in length.

Several eukaryotic genes do not have introns, and prokaryotic genes do not usually contain introns. Recently, however, an intron was found in a prokaryotic system in the bacteriophage T4 thymidylate synthetase gene. In that gene, a single intron split the coding sequence into two parts, one coding for 183 amino acids and the other coding for the other 103 amino acids. The intron is 1017 base pairs long.

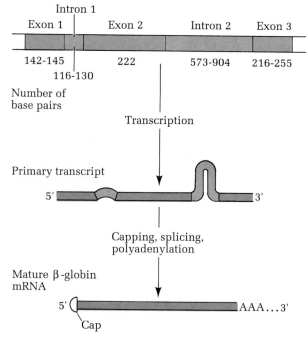

Figure 12.8 *Structure of the mammalian β-globin gene and model for the removal of introns as mature mRNA is produced. Exon 1 is the 5' nontranslated sequence and the sequence for amino acids 1–30; exon 2 codes for amino acids 31–104; and exon 3 contains sequences for the remaining amino acids and the 3' nontranslated sequence.*

Production of Mature Messenger RNA in Eukaryotes

From our understanding of the relationship between gene organization with respect to mRNA structure, we can now describe the details of mRNA production from genes with introns (Figure 12.10). The sequence of steps is the same for genes without introns, except that the step involving the removal of introns is not needed. In brief, the steps are the initiation of transcription by RNA polymerase II, the addition of the methylated 5'-cap, the addition of the poly(A) tail, and the splicing of the primary transcripts, if necessary, in the nucleus to remove the introns and produce the mature mRNAs.

Initiation of transcription by RNA polymerase

II. RNA polymerase II is responsible for transcribing a large number of different RNA molecules including all the protein-coding genes as well as the genes for small nuclear RNAs. Figure 12.11 presents the essential sequence of one of the best-characterized mammalian RNA polymerase II promoters, that for the human β-globin protein. (Some promoter features were described in Chapter 11.) The TATA box that is needed for orienting the RNA polymerase properly for initiating transcription is located at around position −25. Other upstream promoter elements tend to vary. Often there is a CCAAT element that appears to be important for transcription. In addition, there are other sequences farther upstream that are "customized" for the particular gene near where they are located. These sequences are presumed to be important for regulating the expression of the genes they control. That is, some genes in a cell are always being transcribed since the proteins they code for are needed for basic cellular functions—these genes are called **housekeeping genes**—while other genes are turned on and off as needed. As would be expected, these two classes of genes require different signals to control their transcription and this is manifested by different upstream controlling sequences. Presumably the different sequences interact with different protein factors that control RNA polymerase II binding and initiation of transcription.

5′- and 3′-modifications of the primary transcript.

The addition of a methylated cap structure occurs at the 5′ end of each primary transcript when it is about 20 to 30 nucleotides long. All the data support the conclusion that RNA polymerase II starts transcription at the base to which the cap is added. That site in the DNA is called the cap site (see Figure 12.11). There is no DNA template for the 5′ cap.

Poly(A) tails are added to the 3′ ends of most mRNA molecules. Since there is no DNA template for the poly(A) tails of mRNA molecules, the question of how the ends of mRNAs are specified in the genome must be addressed. In contrast to the situation in prokaryotes, where

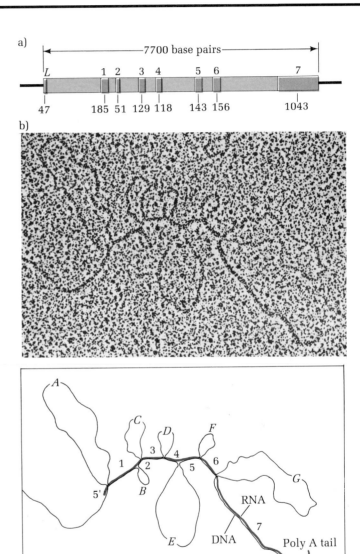

Figure 12.9 The organization of the chicken ovalbumin gene. (a) Arrangement of exons (L, 1→7) and introns (A→G) in the gene. (b) R loop pattern resulting from hybridization of the ovalbumin gene with mature ovalbumin mRNA: the seven introns loop out from the DNA-RNA hybrid.

specific transcription termination sequences exist to specify the end of an mRNA molecule, it appears that in eukaryotes, there are no specific transcription termination sequences near the

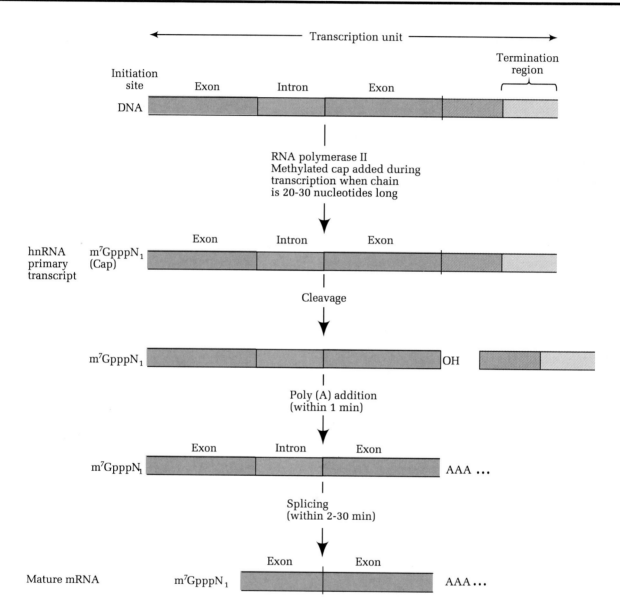

Figure 12.10 General sequence of steps in the formation of eukaryotic mRNA. Not all steps are necessary for all mRNAs.

end of an mRNA molecule. Instead, mRNA transcription catalyzed by RNA polymerase II proceeds for hundreds or thousands of nucleotides transcription past the so-called poly(A) addition site, where cleavage by an RNA-specific endonuclease generates a 3'OH end to which the poly(A) tail is added. The sequences responsible for controlling poly(A) addition have been identified. About 10 to 30 nucleotides upstream from the poly(A) site there is the strongly conserved sequence AAUAAA (Figure 12.12). The available evidence indicates that the AAUAAA sequence and the sequences downstream from the poly(A) site function together to signal the location of the poly(A) site, which leads to cleavage of the primary transcript and addition of the poly(A).

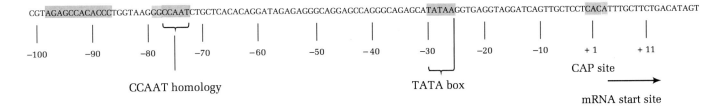

CGT**AGAGCCACACCC**TGGTAAG**GGCCAAT**CTGCTCACACAGGATAGAGAGGGCAGGAGCCAGGGCAGAGCA**TATAA**GGTGAGGTAGGATCAGTTGCTCCT**CACA**TTTGCTTCTGACATAGT

-100   -90   -80   -70   -60   -50   -40   -30   -20   -10   + 1   + 11

CCAAT homology

TATA box

CAP site

mRNA start site

Figure 12.11 *Essential sequences in one of the best-characterized mammalian RNA polymerase II promoters, the human β-globin promoter.*

Experiments have shown that, while cleavage of the primary transcript at the poly(A) site and the addition of the poly(A) tail occur almost simultaneously in the cell, they are actually two separate events. Thus, even for histone mRNAs, to which a poly(A) tail is not added, the primary transcript is cleaved near the 3′ end to produce the mature molecule. Incidentally, the fact that transcription proceeds past the poly(A) addition point before terminating suggests that those two events are separate in the cell. While specific sequences presumably play a role in transcription termination, such sequences have not been well defined to date.

In the nucleus, the length of the poly(A) tail remains essentially constant in a given species. Once the poly(A)-containing mRNA reaches the cytoplasm, however, typically, the poly(A) tail is shortened to produce irregularly sized tails. In mammals, for example, the tail in nuclei is about 250 As long, whereas in the cytoplasm it varies from 30 to 250 As. The purpose or function of the tail is not absolutely clear, although mRNAs with poly(A) tails apparently have longer lives in cells than mRNAs without tails.

Splicing of the primary transcript to produce mature mRNA. The primary transcript is a precursor-mRNA (pre-mRNA) molecule since it contains noncoding as well as coding sequences. The pre-mRNA molecules are processed to remove the introns. That is, the introns are looped out, and the loop is removed by nuclease cleavage. The now-adjacent coding sequences (the exons) are then ligated together to generate a contiguous molecule. These events are called *mRNA splicing.* Little is known about the enzymes involved in splicing. Based on the analysis of the sequences of many genes with introns, it is clear that there are specific nucleotide signals that indicate exon-intron junctions. That is, at the RNA level the introns typically begin with a 5′ GU and end with a 3′ AG, resulting in what is called the *GU-AG rule.* It is clear that more than just those nucleotides are needed to specify a splicing junction and, in fact, the 5′ splice junction probably involves at least seven nucleotides and the 3′ splice junction at least ten nucleotides.

Figure 12.13 diagrams the sequence of events involved in splicing two coding sequences together with the elimination of an intron. Consider a pre-mRNA molecule with two coding sequences, 1 and 2, separated by an intron (Figure 12.13a). The first step in splicing is a cleavage at the 5′ splice junction that results in the separation of coding sequence 1 from an RNA molecule

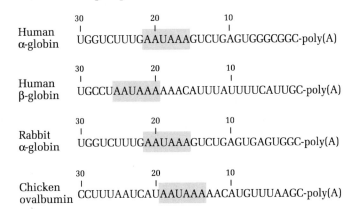

Figure 12.12 *Sequences near poly(A)[A*$_n$*] in four eukaryotic mRNAs. Each contains the sequence AAUAAA about twenty nucleotides upstream of the poly(A) addition site.*

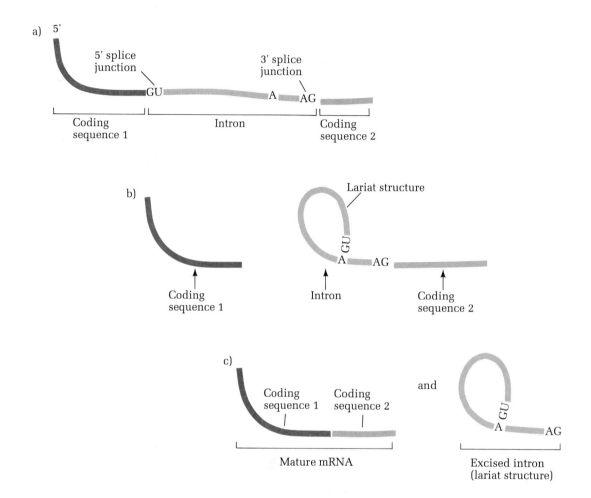

*Figure 12.13 Model for splicing of introns out of
mRNA precursor molecules to produce a mature
mRNA molecule. (a) Pre-mRNA molecule containing
two coding sequences, 1 and 2, separated by an intron. (b) An endonucleolytic cleavage event separates
coding sequence 1 from the rest of the RNA molecule.
The G at the 5′ end of the intron pairs covalently
with an A residue located about 20 to 40 bases before
the 3′ splice junction to produce a lariat structure. (c)
An endonucleolytic cleavage event at the 3′ junction
releases the intron as a lariat structure, and the two
coding sequences are ligated together to produce the
mature mRNA molecule.*

that contains the intron and coding sequence 2.
The free 5′ end of the intron becomes joined to
an A that is part of a sequence about 18 to 40
nucleotides upstream of the 3′ splice junction
where an adenine nucleotide is located (Figure
12.13b). Because of its appearance, the looped
back structure is called an *RNA lariat structure.*
In mammalian systems the *branch-point consensus sequence* is PyXPyPuAPy, where Py is a pyrimidine, Pu is a purine, X is any base, and
within the consensus sequence.

The branch point in the RNA that gives the
lariat structure involves an unusual 2′-5′ phosphodiester bond formation between the 2′ OH of
the adenine nucleotide in the branch-point sequence and the 5′ phosphate of the guanine nucleotide at the end of the intron. The A itself

remains in normal 3'-5' linkage with its adjacent nucleotides of the intron (Figure 12.14).

Next, mRNA and a precisely excised intron, the latter still in a lariat shape, appear simultaneously as a result of a cleavage event at the 3' splice junction and ligation of the two coding sequences together (Figure 12.13c). The lariat RNA is subsequently converted to a linear molecule and then degraded.

Figure 12.14 Details of intron removal from a pre-mRNA molecule. At the 5' end of an intron is the sequence GU and at the 3' end is the sequence AG. Eighteen to 40 nucleotides upstream from the 3' end of the intron is an A nucleotide located within the branch point consensus sequence which, in mammals, is PyXPyPuAPy, where Py = a pyrimidine, X = any base, Pu = purine, and A = adenine. Intron removal begins by a cleavage event at the first exon-intron junction; the G at the released 5' of the intron folds back and forms an unusual 5' → 2' bond with the A of the branch point consensus sequence. This produces a lariat-shaped intermediate. Cleavage at the 3' intron-exon junction and ligation of the two exons completes the removal of the intron.

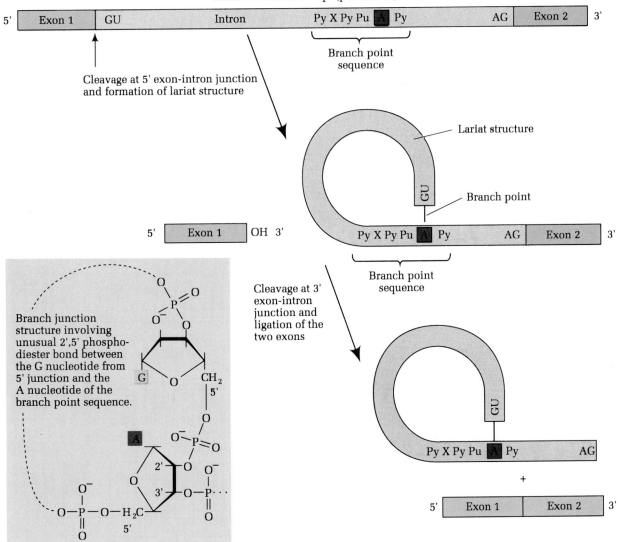

The processing of pre-mRNA molecules occurs exclusively in the eukaryotic nucleus. The processing events do not occur with naked RNA molecules but occur in a large structure called a *spliceosome,* which contains several protein and RNA molecules. Spliceosomes have been described to date in both yeast and mammalian systems; in yeast they are 40S and in mammalian systems they are 60S. Different ribonucleoprotein particles called *small nuclear ribonucleoprotein particles* (snRNPs) have been shown to be essential components of the spliceosome. That is, the nucleoplasm of eukaryotic cells contains a number of different snRNP particles that

a) U1 snRNA

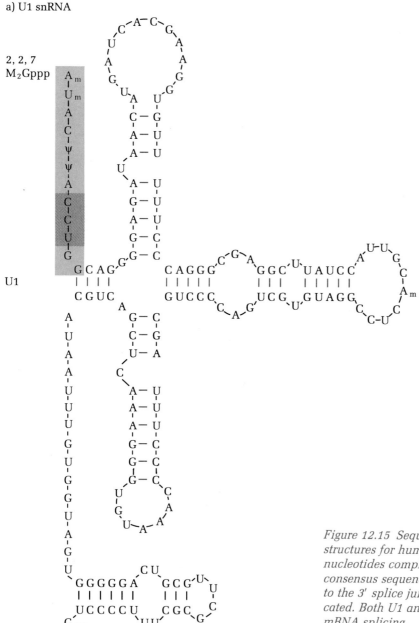

Figure 12.15 Sequences and probable secondary structures for human U1 (a) and U2 (b) snRNAs. U1 nucleotides complementary to the 5' splice junction consensus sequence and those complementary to the 3' splice junction consensus sequence are indicated. Both U1 and U2 appear to be involved in pre-mRNA splicing.

contain small nuclear RNA (snRNA) molecules and between five and nine proteins. In animal cells there are five distinct types of snRNA called U1, U2, U4, U5, and U6 each with an abundance greater than 10^5 per cell, as well as a number of less abundant types. U1, U2, and U5 snRNAs are found individually in their corre-

sponding snRNPs, whereas the U4 and U6 sn-RNAs coexist within a single snRNP particle. Figure 12.15 gives the sequences and possible secondary structures of human U1 and U2 sn-RNAs.

During the splicing reaction, U2 snRNPs have been shown to associate with the intron branch

b) U2 snRNA

point while U1 snRNPs interact with the 5′ splice site. Another component, possibly U5 snRNP, binds the 3′ splice junction in the pre-mRNA. Biochemical experiments have shown that the mammalian spliceosome contains U1, U2, U4, U5, and U6 snRNAs in snRNP particles. Not all of their roles in splicing have been defined yet, but presumably the whole spliceosome structure is required to hold the pre-mRNA molecule in a stable configuration for the splicing events.

Keynote *The transcripts of structural protein-coding genes are messenger RNAs. These molecules are linear and vary in length over a wide range in correspondence to the variation in the size of the polypeptides they specify. Prokaryotic mRNAs are little modified once they are transcribed, whereas eukaryotic mRNAs are modified by the addition of a cap at the 5′ end and a poly(A) tail at the 3′ end. Many eukaryotic mRNAs contain non–amino acid–coding sequences called intervening sequences or introns that must be removed from the mRNA transcript to make a mature, functional mRNA molecule.*

The Structure and Function of Transfer RNA

In a cell, transfer RNAs (tRNAs) bring amino acids to the ribosome-mRNA complex where they are polymerized into protein chains in the translation (protein synthesis) process. The sequence of amino acids in the protein is directed by the sequence of codons in the mRNA molecule.

Molecular Structure of tRNA

Transfer RNA molecules constitute between 10 and 15 percent of the total cellular RNA in both prokaryotes and eukaryotes. They have a size of about 4S, and they consist of a single chain of 75 to 90 nucleotides, whose sequence varies from one tRNA molecule to another. The differences in nucleotide sequences explain the ability of a particular tRNA molecule to bind a particular amino acid.

The nucleotide sequences found in tRNA molecules are different from those seen in mRNAs and rRNAs. Recall that eukaryotic mRNAs are capped at the 5′ end and have a poly(A) tail added to the 3′ end. Otherwise, the nucleotides of the mRNA are not modified. Primary transcripts of tRNA genes in both prokaryotes and eukaryotes, the **precursor tRNAs** (pre-tRNAs), in contrast, are extensively modified as they mature. Two types of modification occur: the addition of a 5′ C–C–A 3′ sequence to the 3′ end of the molecule and the extensive chemical modification of a number of nucleotides at many places within the chain. The type and extent of the modifications vary from tRNA to tRNA, but typically the modifications include the addition of methyl groups to the base, the reduction of certain uridines, say, to dihydrouridine, or the rearrangement of some uridines to produce pseudouridine (Ψ). Figure 12.16 gives examples of the modified bases that are found in tRNAs. These modifications are brought about by specific enzyme action, and the modifications themselves cause the particular two- and three-dimensional configurations of tRNAs, which in turn determine the function of those molecules (e.g., their ability to pick up a specific amino acid).

Cloverleaf model for the structure of tRNA.

Many tRNAs have been sequenced from a number of organisms, and all the sequences can be arranged into what is called a *cloverleaf model for tRNA*. This model is a secondary structure in that only two dimensions are used to display the sequence (although, of course, in reality a tRNA has a well-defined three-dimensional structure). Figure 12.17a shows the general features of the cloverleaf model, and Figure 12.17b shows the complete nucleotide sequence of yeast alanine tRNA. The cloverleaf itself results from complementary-base pairing between different sections of the molecule. Three base-paired "stems" appear in the molecule, with the number of base pairs in each stem varying from tRNA to tRNA.

All tRNAs have three unpaired loops, I, II, and IV. Loop I usually contains 9 bases. Loop II has 7 bases and contains within it the three-nucleotide sequence called the **anticodon,** which pairs with a codon (three-nucleotide sequence) in mRNA by complementary-base pairing. This codon-anticodon pairing is crucial for adding the correct amino acid (as specified by the mRNA) to the growing polypeptide chain. Some, but not all, tRNAs have a small loop of variable length, called loop III, or, sometimes, the lump. Loop IV has 7 bases, and within this loop the sequence 5′ T–ψ–C 3′ appears to be universal among tRNAs. Another universal feature of tRNAs is the 5′ C–C–A 3′ sequence at the 3′ end of the molecule.

In sum, the cloverleaf model shows four stems and three (sometimes four) loops containing unpaired nucleotides. The particular nucleotides present in the stems vary greatly from tRNA to tRNA, but the number of nucleotides in a particular stem is fairly constant.

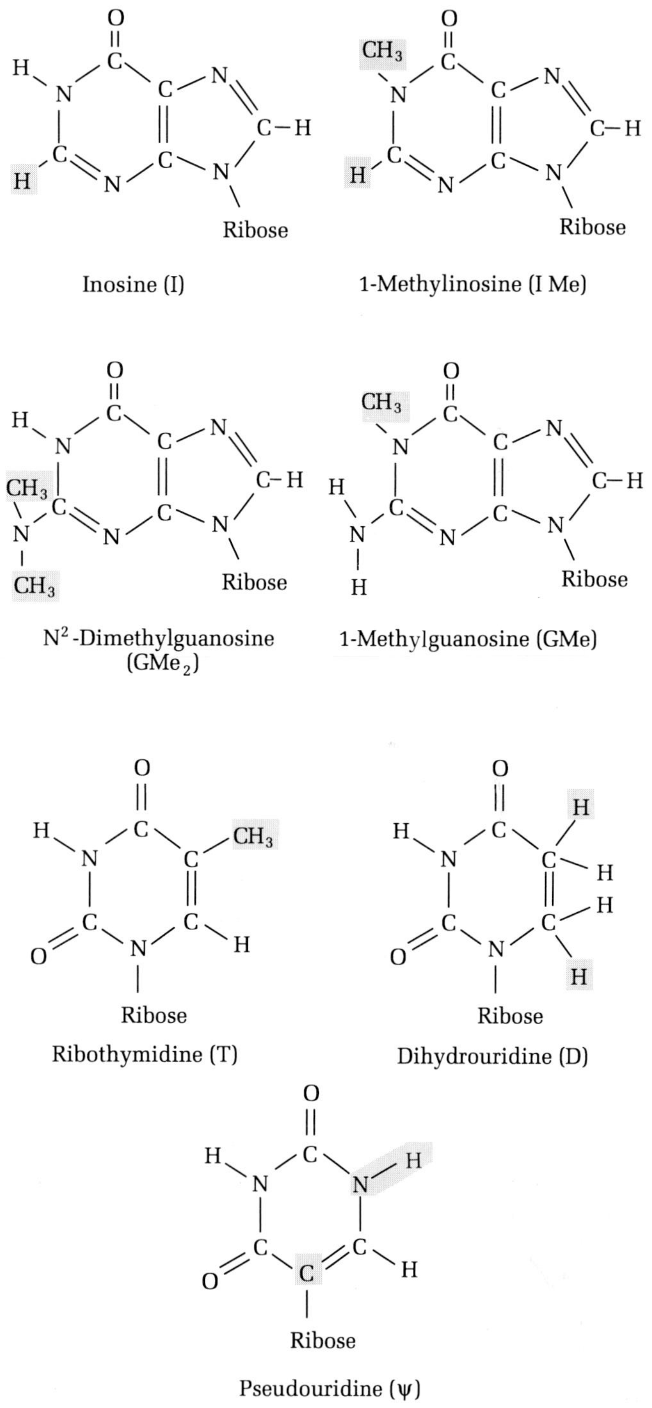

Three-dimensional structure of tRNA. The two-dimensional cloverleaf model is a schematic drawing based on an analysis of the primary nucleotide sequences of tRNAs. However, like all molecules, the tRNA has a three-dimensional structure. Because tRNA molecules are so small, they can be crystallized, and X-ray crystallography can be used to develop a three-dimensional model. Figure 12.18 shows the tertiary-structure model for yeast tRNA.Phe (this terminology indicates the amino acid specified by the anticodon of the tRNA, in this case phenylalanine) which was the first tRNA to be analyzed in this way: Figure 12.18a presents a schematic drawing, Figure 12.18b is a photograph of a backbone model, and Figure 12.18c is a photograph of a space-filling molecular model. All other tRNAs that have been examined in this way show similar three-dimensional structures.

From the data they obtained, the crystallographers concluded that all hydrogen-bonded stem structures proposed in the cloverleaf model do exist in the tRNA molecule. But in addition, the results showed that other hydrogen bonding occurs in the molecule. These additional hydrogen

Figure 12.16 Some modified bases found in tRNA molecules.

a) Schematic of tRNA molecule

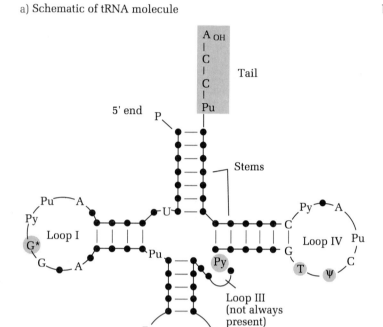

b) Yeast alanine tRNA

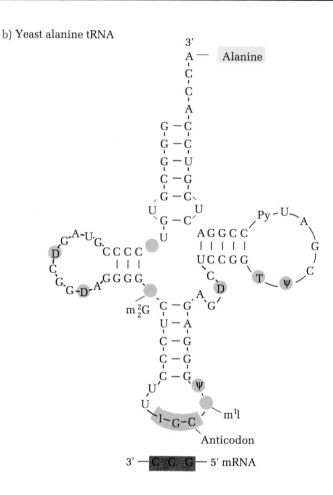

Figure 12.17 (a) Schematic diagram of a tRNA molecule shown in the two-dimensional cloverleaf configuration; (b) The complete nucleotide sequence of yeast alanine tRNA showing the unusual bases and anticodon position.

bonds fold the cloverleaf into a more compact shape, like an upside-down L. In this L-shaped structure the 3′ end of the tRNA (the end to which the amino acid attaches) is at the opposite end of the L from the anticodon loop.

Transfer RNA Genes

A three-base sequence (codon) in an mRNA specifies each amino acid to be added to a polypeptide chain. While only 20 different amino acids can be used to make a protein, actually 61 different codons can be used in an mRNA to specify amino acids, and 3 additional codons do not specify amino acids but instead are used as termination signals for protein synthesis. (These concepts will be further elaborated in the next chapter.) Since each amino acid–specifying codon must be matched by an appropriate anticodon on a tRNA molecule, theoretically any cell must have at least 61 different tRNA types. And eukaryotes commonly have a number of different tRNA molecules with the same anticodon. So that the instructions for the manufacture of such large numbers of tRNAs can be coded, the genome has many tRNA genes. The number and the distribution of tRNA genes in prokaryotes and eukaryotes have been extensively studied in a number of organisms.

a)

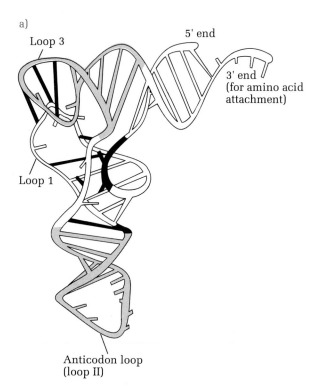

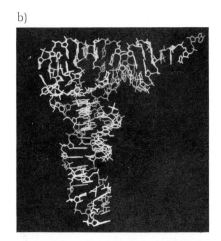

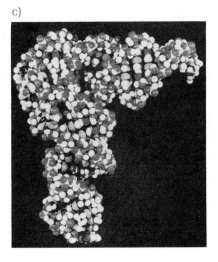

Figure 12.18 (a) Schematic drawing of yeast phenyl-alanine tRNA as determined by X-ray diffraction of tRNA crystals. Note the characteristic L-shaped structure. (b) Photograph of a backbone model and (c) photograph of a space-filling molecular model of yeast phenylalanine tRNA. The CCA end of the molecule is at the upper right, and the anticodon loop at the bottom.

The tRNA genes of *E. coli*. Examples of virtually all possible sorts of arrangements of tRNA genes occur in *E. coli*. Some tRNA genes occur only once on the *E. coli* chromosome, while others are present two or more times, a situation called **gene redundancy.** In some cases redundant genes are not linked on the chromosome, but in other cases they are linked, side by side, in what is called a *redundant gene cluster.* Unique tRNA genes can also be linked side by side on the chromosome in a *nonredundant tandem gene cluster.*

Eukaryotic tRNA genes. In general, many more tRNA genes occur in eukaryotes than in prokaryotes. In the lower eukaryote yeast, for example, about 400 tRNA genes occur in the genome. Since 61 different tRNAs are identified, the genome has approximately six copies of each gene. Higher eukaryotes have a tendency toward even greater redundancy of these genes. *Xenopus laevis,* for example, has over 200 copies of each tRNA gene per genome. As a class, then, tRNA genes are found in the middle-repetitive kinetic class of DNA.

Although we have not located every tRNA gene for any given eukaryote, experiments have shown that, as in *E. coli,* all possible arrangements of the genes occur. Thus some genes are found singly, whereas others are found in clusters, which can be redundant or nonredundant tandem clusters. There is also evidence of tandem clusters, in which some genes are a single copy while others are redundant.

At least some (nuclear) tRNA genes from many eukaryotic organisms, like many structural genes, contain introns. (Recall that an intron is a segment within a gene that is transcribed as part of the gene but that is removed from the RNA transcript as the mature mRNA is produced.) Depending on the tRNA, the intron is 14 to 60 base pairs long and is almost always located between the first and second nucleotides 3' to the anticodon. The anticodon itself often, but not always, pairs with intron sequences in the tRNA precursor. For reasons that are not understood, introns are found only in tRNAs for tyrosine, phenylalanine, tryptophan, lysine, proline, serine, leucine, and isoleucine. The introns present in any given family of tRNA genes (e.g., the eight yeast tyrosine tRNAs) are identical or nearly so.

The first tRNA gene in which an intron was found was the yeast gene for tRNA.Tyr, the gene for the tRNA that carries the amino acid tyrosine to the ribosome. Figure 12.19 shows the nucleotide sequence for the initial transcript of the tRNA gene, the pre-tRNA (Figure 12.19a), and the mature tRNA (Figure 12.19b), as well as the organization of the two sequences into cloverleaf models. The 14-base intron (i.e., the tRNA copy of the intron in the gene) is located in the pre-tRNA just to the 3' side of the anticodon in this tRNA and in all similar pre-tRNAs analyzed to date. The presence of the transcript of the intron sequence in the pre-tRNA results in a significant change in the anticodon loop, as Figure 12.19 shows. J. Abelson's group has identified an enzyme, called **RNA ligase,** that splices together the RNA pieces once the intron is removed from the pre-tRNA. Removal of the intron involves cleavage with a splicing endonuclease that is associated with the nuclear membrane. About 10 percent of the 400 tRNA genes in yeast have intervening sequences, and all are spliced in the same way.

Biosynthesis of Transfer RNAs

In prokaryotes the tRNA genes are transcribed by the same RNA polymerase enzyme that transcribes the structural genes and the rRNA genes. In eukaryotes the tRNA genes are transcribed by RNA polymerase III.

Like mRNA genes and rRNA genes, prokaryotic tRNA genes have promoter sequences located upstream from the starting point of transcription. Recent evidence has shown that the promoter sequences for eukaryotic tRNA genes are actually *within* the gene sequences rather than upstream from the starting point of transcription, as is typical of structural genes. The model shows that RNA polymerase III is a large molecule, one part of which recognizes the promoter sequence (called an *internal control region*) within the gene sequence while the other part initiates transcription at the appropriate point upstream from the promoter. Internal control regions are also found in the eukaryotic 5S rRNA genes, the other class of genes transcribed by RNA polymerase III. Two regions of conserved sequences (called A and B blocks) are important for RNA polymerase III to transcribe the tRNA and 5S rRNA genes (Figure 12.20). The distance between A and B blocks can vary considerably, especially in tRNA genes with introns. The A and B blocks are found in sections of tRNAs that are highly conserved in both eukaryotes and prokaryotes.

While internal control regions are essential sequences for RNA polymerase III to initiate transcription, there are other specific upstream and downstream sequences that are needed for the expression of some RNA polymerase III-transcribed genes. These sequences are presumed to be sites for specific DNA binding proteins whose functions are not yet clearly defined.

The termination signals for RNA polymerase III are also located within the transcribed region of the gene. The termination sequence is a group of three or more Ts on the nontemplate strand, surrounded by GC-rich regions.

As in the case with mRNAs, precursor tRNAs are longer than mature tRNAs and have a 5' leader sequence and a 3' trailer sequence. Un-

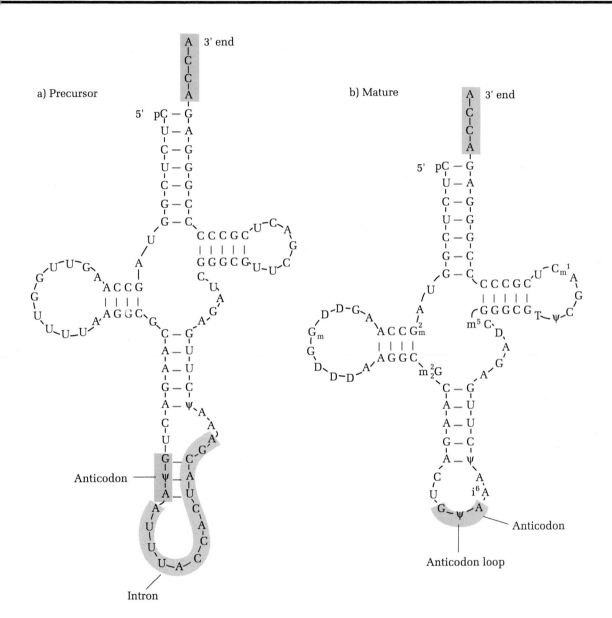

Figure 12.19 Cloverleaf models for precursor yeast tRNA.Tyr and mature tRNA.Tyr: (a) the 14-base intervening sequence in the precursor, located adjacent to the anticodon-coding region; (b) the intervening sequence removed from the precursor to produce a mature tRNA.Tyr molecule.

like mRNA, however, these sequences are removed during the processing of the pre-tRNA into the mature tRNA. In eukaryotes the pre-tRNA processing occurs in the nucleus, so only mature tRNAs are found in the cytoplasm. Figure 12.21 shows the nucleotide sequence of a prokaryotic pre-tRNA and the leader and trailer sequences that are removed during the processing steps. A similar arrangement exists in eukaryotic systems.

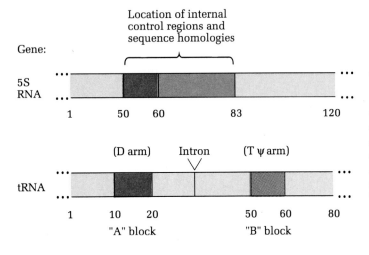

Gene:

Location of internal
control regions and
sequence homologies

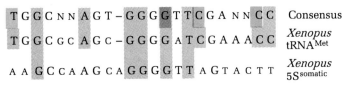

Figure 12.20 *Summary of the internal control regions
required for genes transcribed by RNA polymerase III.
In the consensus sequences for A block and the B
block, the nucleotides in color are the same in all
non-5S genes, whereas those in grey are the same in
all tRNA genes.*

T G G C N N A G T – G G G G T T C G A N N C C   Consensus

T G G C G C A G C – G G G G A T C G A A A C C   *Xenopus*
tRNA^Met

A A G C C A A G C A G G G G T T A G T A C T T   *Xenopus*
5S^somatic

Invariant in all non-5S genes transcribed by
RNA polymerase III

Invariant in all tRNA

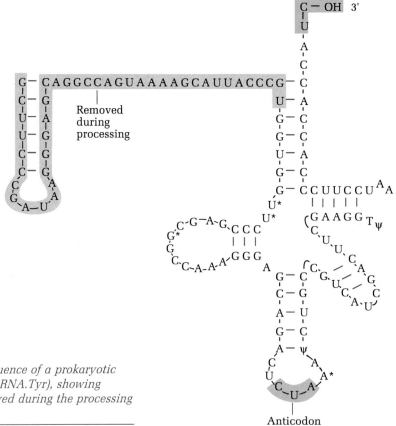

Figure 12.21 *Nucleotide sequence of a prokaryotic
precursor tRNA (*E. coli *pre-tRNA.Tyr), showing
(color) the nucleotides removed during the processing
steps.*

As we mentioned earlier, tRNA genes may exist in clusters. In prokaryotes, but not in eukaryotes, there is evidence that a cluster of tRNA genes may be transcribed to produce a single RNA transcript containing a number of tRNA sequences. The general organization of the multi-tRNA pre-tRNA molecules is

5′–leader–(tRNA–spacer)n–tRNA–trailer–3′

where n is a number of tRNA-spacers characteristic of a cluster. For example, a multi-tRNA pre-tRNA molecule with three tRNA sequences would look like this:

5′–leader–tRNA$_1$–spacer$_1$–tRNA$_2$–spacer$_2$–tRNA$_3$–trailer–3′

Figure 12.22 gives an example of a pre-tRNA molecule from *E. coli* that contains two tRNA sequences. In such cases the leader, the trailer, and the spacer sequences must be removed during the processing of the pre-tRNA to produce two discrete, mature tRNAs.

The non-tRNA sequences are removed from the pre-tRNA molecules by the action of specific enzymes. In *E. coli* at least two enzymes are needed. Enzyme RNase P catalyzes the removal of the 5′ leader sequences, and RNase Q catalyzes the removal of the 3′ trailer sequence. Evidence for this result has come from studies of temperature-sensitive mutants that have defects in these two enzyme activities. At high temperatures partially processed pre-tRNAs accumulate

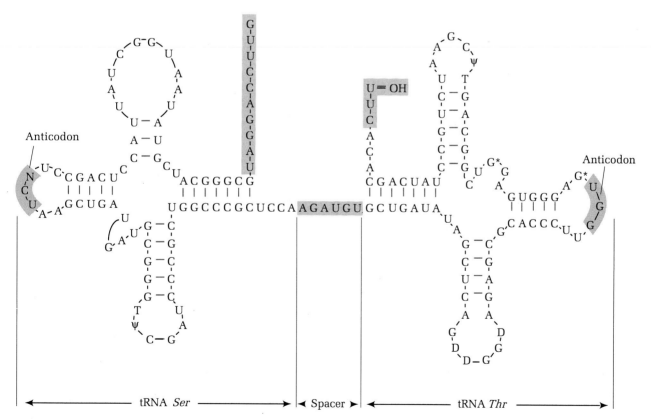

Figure 12.22 A tRNA precursor from E. coli *that contains two tRNA sequences, one for tRNA.Ser and the other for tRNA.Thr. The nucleotides removed during processing are shown in color.*

in the mutant strains. Another enzyme (or enzymes) is involved in removing the spacer region between clustered tRNAs. Little is known about the enzymes needed for the removal of the terminal noncoding sequences of eukaryotic pre-tRNAs, although they must act in the nucleus, since pre-tRNAs are not found in the cytoplasm.

Keynote *Molecules of transfer RNA bring amino acids to the ribosomes where the amino acids are polymerized into a protein chain. All the tRNA molecules are between 75 and 90 nucleotides long, contain a number of modified bases, and have similar three-dimensional shapes. A CCA sequence is found at the 3' end of all tRNAs.*

Ribosomal RNA

Ribosomes are the organelles within the cell on which protein synthesis takes place. Ribosomes bind to mRNA and facilitate the binding of the anticodon on the tRNA to the codon on the

mRNA so that the polypeptide chain can be synthesized. Ribosomes are complex structures, and we have yet to understand fully how they function. In both prokaryotes and eukaryotes the ribosomes consist of two unequally sized subunits, the large and small ribosomal subunits, each of which consists of a complex between RNA molecules and proteins. Each subunit contains at least one **ribosomal RNA (rRNA)** molecule and a large number of **ribosomal proteins.** Ribosomal subunits plus mRNA plus tRNAs are required for protein synthesis.

Structure of the Prokaryotic Ribosome in *E. coli*

The *E. coli* ribosome has been the subject of intensive study, and because the features of *E. coli* are generally found in other bacterial ribosomes, it is used as a model of a prokaryotic ribosome. Figure 12.23 shows the general structure and composition of the *E. coli* ribosome.

The *E. coli* ribosome has a size of 70S, and the sizes of the two subunits are 50S (large sub-

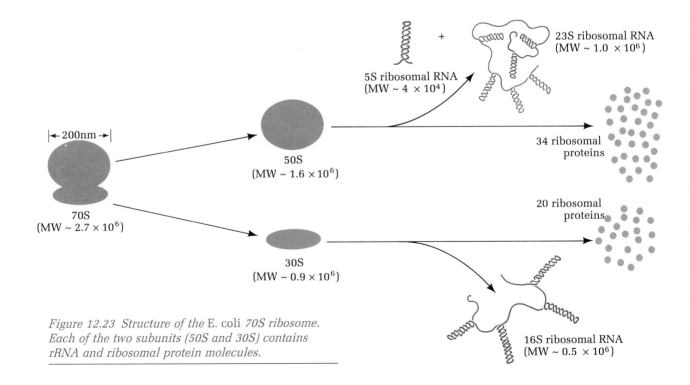

Figure 12.23 Structure of the E. coli *70S ribosome. Each of the two subunits (50S and 30S) contains rRNA and ribosomal protein molecules.*

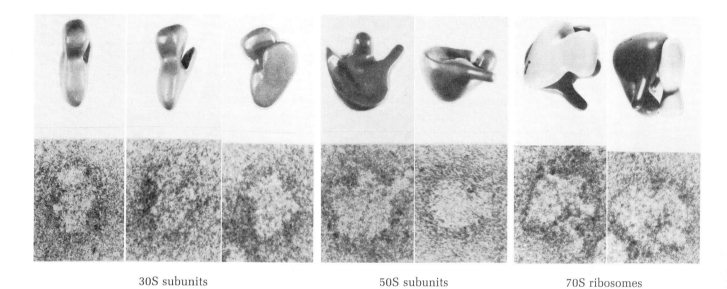

30S subunits   50S subunits   70S ribosomes

Figure 12.24 Electron micrographs of E. coli *small
(30S) ribosomal subunits, large (50S) ribosomal subunits, and complete (70S) ribosomes, shown with
photographs of interpretive models of the subunits.*

unit) and 30S (small subunit). The shape of the
E. coli ribosomes and its two subunits has been
discovered through electron microscopic examination of purified components. Figure 12.24 presents electron micrographs of *E. coli* ribosomes
and ribosomal subunits and interpretive models
that have been constructed from the electron mi-
crography studies. In interpreting these models,
however, we must exercise caution. Material
studied under the electron microscope must be
placed in a vacuum, which causes dehydration.
Therefore the organelle in the living cell may be
more plumped up. Nonetheless, we see that the
two subunits have distinct three-dimensional
shapes, unlike the two-dimensional circles or el-
lipses of textbook drawings.

Approximately two-thirds of the *E. coli* ribosome consists of ribosomal RNA (rRNA), the rest
of the mass consisting of ribosomal proteins.
The large 50S subunit has 34 different proteins
and two rRNA molecules, one with a size of 23S
(2904 nucleotides) and the other with a size of
5S (120 nucleotides). The small 30S ribosomal
subunit has 20 different proteins and one rRNA
molecule that has a size of 16S (1542 nucleotides).

The rRNA Genes of *E. coli*

In prokaryotes and eukaryotes the regions of
DNA that contain the genes for the three types
of rRNA are called **ribosomal DNA (rDNA)**. In *E.
coli* production of equal amounts of the three
rRNAs is ensured by the transcription of the
three adjacent genes for 16S, 23S, and 5S in
rDNA into a single **precursor rRNA (pre-rRNA)**
molecule. One transcription unit consists of one
gene each for the three rRNAs, and *E. coli* has
seven such transcription units (*rrn* regions). Figure 12.25a gives the general organization of a
transcription unit, and Figure 12.25b shows a
schematic drawing of the organization of rRNAs,
tRNAs, and spacers in the transcribed pre-rRNA
molecule.

In each transcription unit, the three rRNA
genes are arranged in the order 16S-23S-5S; so
the arrangement of the rRNAs on the pre-rRNA
transcript is 5′ 16S-23S-5S 3′. Transfer RNA
genes are associated with the rRNA transcription
units. In all seven transcription units, one or
two tRNA genes are found in the spacer between
the 16S and 23S rRNA coding sequences and
another one or two tRNA genes are found

tween the 5S rRNA gene and the 3' end of the transcription unit in three of the seven transcription units. During RNA synthesis from the transcription units, the tRNA genes are transcribed as part of the pre-rRNA molecule. The tRNAs are then removed from the precursor by the action of specific processing enzymes.

The Ribosomal Protein Genes of *E. coli*

The *E. coli* ribosome has one gene for each of the 54 ribosomal proteins. The genes are located at many places on the *E. coli* genetic map. Some of the ribosomal genes are found clustered on the map; for example, 27 ribosomal protein

a) *E. coli* rRNA transcription unit

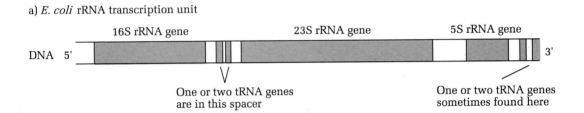

b) Ribosomal RNA (rrn) operons of *E. coli* K-12

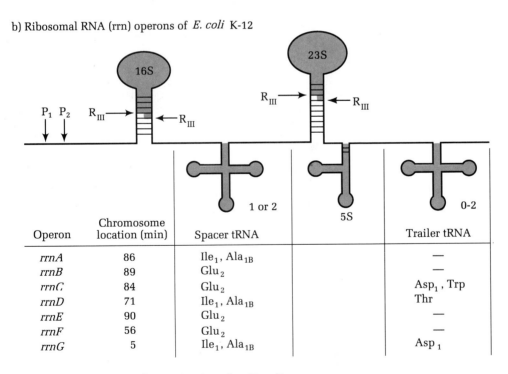

Operon	Chromosome location (min)	Spacer tRNA		Trailer tRNA
rrnA	86	Ile_1, Ala_{1B}		—
rrnB	89	Glu_2		—
rrnC	84	Glu_2		Asp_1, Trp
rrnD	71	Ile_1, Ala_{1B}		Thr
rrnE	90	Glu_2		—
rrnF	56	Glu_2		—
rrnG	5	Ile_1, Ala_{1B}		Asp_1

Figure 12.25 (a) General organization of an E. coli *rRNA transcription unit (an* rrn *region); (b) The seven ribosomal RNA* (rrn) *operons of* E. coli *K-12 showing the relationship of rRNA and tRNA molecules. P_1 and P_2 are the two transcription initiation sites, and R_{III} indicates the sites of cleavage by RNaseIII.*

genes are in the *strA* (streptomycin) region (at 72 min on the map), and 4 ribosomal protein genes are in the *rif* (rifampicin) region (at 88 min). Each of these two regions also contains other genes, such as genes for subunits of RNA polymerase.

Biosynthesis of the *E. coli* Ribosome

In outline, prokaryotic ribosomes are made as follows: The rRNA genes are transcribed onto a pre-rRNA, which is then modified and cleaved at specific sites to yield the mature 16S, 23S, and 5S rRNAs. Assembly with ribosomal proteins occurs while the pre-rRNA is being transcribed, and the two resultant ribosomal subunits are immediately able to synthesize proteins.

In 1969, M. Nomura and his colleagues investigated the question of how the *E. coli* ribosome is assembled. They dissociated the 30S subunit into its component rRNA and ribosomal proteins and then attempted to reconstitute the subunit in vitro. They found that at 37°C (the normal growth temperature for *E. coli*) and in the presence of an ideal ionic environment, the parts could be reconstituted into a completely functional 30S subunit. This phenomenon was called **self-assembly.** The result was significant because it indicated that all the information needed to form a functional subunit was somehow embedded into the structure of individual components (mRNA and proteins). A few years later, K. Nierhaus and F. Dohme were able to reconstitute the 50S *E. coli* subunit from its component parts.

The reconstitution of a ribosomal subunit occurs in a step-by-step series of reactions, each of which adds specific ribosomal proteins to the rRNA(s). Many of the steps have been elucidated by omission experiments in which one ribosomal protein is left out of the reactions and the resulting particle is analyzed. In this way geneticists have shown definite pathways for the assembly of the subunits. In addition, even where the omission of a protein results in an essentially complete subunit, it is not functional in protein synthesis. Thus all proteins must be

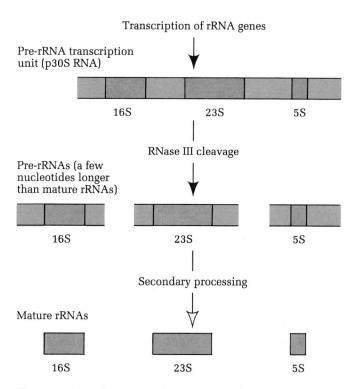

Figure 12.26 Scheme for the processing of a precursor rRNA (p30S) to the mature 16S, 23S, and 5S rRNAs of E. coli. *(The tRNAs have been omitted from the diagram.)*

available in order for the ribosomal subunit to function.

Information is also available about the way the *E. coli* ribosome is assembled in vivo. Figure 12.26 shows the scheme for the production of the mature rRNAs in *E. coli*. Each transcription unit is transcribed into a 30S precursor (p30S) RNA, which contains a 5′ leader sequence, the 16S, 23S, and 5S rRNA sequences (each separated by spacer sequences), and a 3′ trailer sequence.

The p30S molecule is cleaved by RNase III to produce three precursor molecules, one each of the p16S, p23S, and p5S molecules. In normal cells, RNase III cleaves the transcript of the rRNA transcription unit to produce the three rRNA precursors while transcription is still occurring. As a result of this rapid processing, the p30S is not seen in normal cells. However, in mutants that have a temperature-sensitive RNase III activity, the p30S molecule accumulates at the high temperature. Through studies of these mutants the pathway of p30S synthesis and processing has been discovered. The mature 16S, 23S, and 5S rRNAs are produced by the action of RNases.

In order for ribosome assembly to take place correctly, the 16S and 23S rRNAs must be methylated at a few places. The methylation is catalyzed by specific methylases, and most of the methyl groups are added to the bases (recall the tRNA modifications), with only a few added to the 2′–OH of the ribose group. Methylation of the 23S sequence occurs on the p30S molecule, whereas methylation of the 16S rRNA occurs when it is in its mature state.

As the rRNA genes are being transcribed, the transcript rapidly becomes associated with ribosomal proteins. The cleavage of the transcript takes place, then, within a complex formed between the rRNA transcript (as it is being transcribed) and ribosomal proteins. It is most probably directed by the conformational state of the rRNA-ribosomal protein particle at the time of the cleavage.

Structure of the Eukaryotic Ribosome

In general, eukaryotic ribosomes are larger and more complex than their prokaryotic counterparts. The size of the ribosome and the molecular weights of the rRNA molecules differ from organism to organism. Lower eukaryotes have the smallest ribosomes (although they are larger than *E. coli* ribosomes), while mammalian ribosomes are the largest. Nonetheless, all eukaryotic ribosomes have many common structural and chemical features. We will use the mammalian ribosome as a model for discussion, and we will occasionally compare it with lower-eukaryotic ribosomes.

Figure 12.27 shows the general composition of mammalian ribosomes. They have a size of 80S (versus 70S for the bacterial ribosomes) and have two subunits, a large one with a size of 60S and a small one with a size of 40S. The 40S subunit is quite similar in size and mass throughout the eukaryotes, but the 60S subunit varies quite a lot.

The 80S mammalian ribosome consists of about equal weights of rRNA and ribosomal proteins. There are four rRNA types, with sizes of 18S (~2300 nucleotides), 28S (~4200 nucleo-

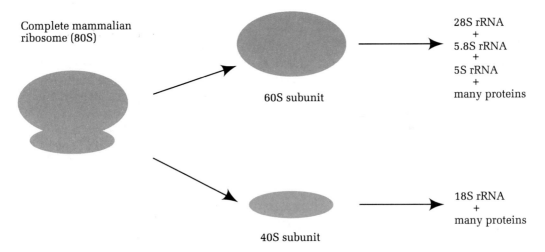

Figure 12.27 Composition of whole ribosomes and of ribosomal subunits in mammalian cells.

tides) (the large rRNAs), 5.8S (156 nucleotides), and 5S (120 nucleotides) (the small rRNAs). The ribosomal proteins in eukaryotes have not been identified as clearly as those for prokaryotes, but in general, the ribosomes of higher eukaryotes have 70–80 different ribosomal proteins (about 30 in the small subunit and about 50 in the large subunit in mammalian cells), and the ribosomes of lower eukaryotes have 60–70.

In all eukaryotes the 40S ribosomal subunit contains only one rRNA type, 18S rRNA. The large 60S ribosomal subunit contains the other three types of rRNA, the 28S, 5.8S, and 5S rRNAs. The 5.8S rRNA is hydrogen-bonded to the 28S rRNA in the functional ribosome.

Ribosomal RNA Genes in Eukaryotes

Location and organization of rRNA genes in the genome. Within the nucleus of all eukaryotes, the rRNA genes are located in regions of DNA called ribosomal DNA (rDNA), and around these rDNA regions organelles, called nucleoli, form within the nucleus. The nucleus of every eukaryote has one or more nucleoli within which rRNA is transcribed and within which the resulting rRNAs associate with ribosomal proteins to form the two ribosomal subunits. Because the ribosomal proteins are made in the cytoplasm, the production of ribosomes requires the ribosomal proteins to migrate back into the nucleus and into the nucleolus. Very little information is available on such directional movement of proteins. Once the ribosomal subunits are made, they must be transported to the cytoplasm before they can function in protein synthesis. The ribosomal subunits do not carry out protein synthesis in the nucleus.

Most eukaryotes that have been examined have a large number of copies of the genes for each of the four rRNA species. The genes for 18S, 5.8S, and 28S rRNAs are found in one or more clusters in the genome, and around each cluster a nucleolus is formed. The extent of gene repetition varies from eukaryote to eukaryote and is in the range of 100 to 1000, with a few exceptions. Thus the rRNA genes are representative of the middle-repetitive DNA in the ge-

nome. For example, yeast has 140 rRNA gene sets, *Neurospora* has 200, *Drosophila* has 260, and HeLa (human) cells have 1250.

In most organisms the 5S rRNA genes are located in the genome at a site or sites distinct from the sites for the other rRNA genes. In humans, for example, the 5S genes are clustered, whereas in *Xenopus* they are scattered throughout the genome. The number of 5S genes relative to the other genes also sets no pattern. The organism may have more of them, the same number, or less of them than the other rRNA genes. In some organisms, such as baker's yeast *(Saccharomyces cerevisiae)* and the cellular slime mold *(Dictyostelium discoideum),* the 5S rRNA genes are interspersed with the other rRNA gene set.

At the chromosome level the 18S, 5.8S, and 28S rRNA genes are usually found adjacent to one another in the order 18S–5.8S–28S, with each set of three genes repeated many times to form repeated tandem arrays of rRNA genes called rDNA repeat units. Just as in *E. coli,* each set of three rRNA genes is transcribed onto a single pre-rRNA molecule (Figure 12.28). The rDNA repeat units of four eukaryotes are diagramed in Figure 12.29. In general, variation in transcribed spacer lengths is largely responsible for variation in length of the transcribed region (the rRNA transcription unit) of the rDNA repeat unit.

At the gene level several types of noncoding sequences have been discovered. Because the pre-rRNA is longer than the three mature rRNAs together, two types of transcribed spacer sequences in the DNA have been identified. The external transcribed spacers (ETS) are DNA sequences that are transcribed into the pre-rRNA and are located immediately adjacent to the 5' end of the 18S sequence and to the 3' end of the 28S sequence in the pre-rRNA molecule. The internal transcribed spacer (ITS) is transcribed into the pre-rRNA on either side of the 5.8S sequence. Finally, another class of rDNA spacer sequences, nontranscribed spacer (NTS) sequences, are found at the extreme 5' and 3' ends of each tandem rRNA repeating unit in the rDNA. As the name indicates, these regions are

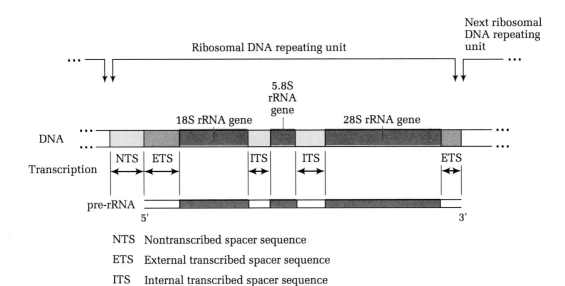

Figure 12.28 *Generalized diagram of a eukaryotic, ribosomal DNA-repeating unit. The coding sequences for 18S, 5.8S, and 28S rRNAs are indicated in color.*

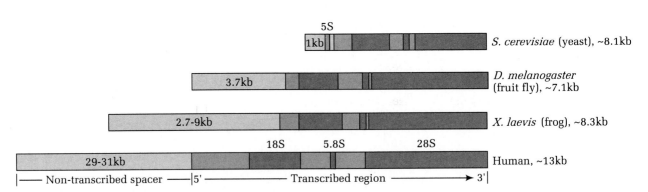

Figure 12.29 *The ribosomal transcription units of four eukaryotes. The 18S, 5.8S, and 28S regions code for the rRNAs found in all ribosomes. The transcribed spacer regions vary in length and result in large differences in the lengths of the transcription units. Variation in the length of the nontranscribed spacer regions is also considerable.*

not transcribed into the pre-rRNA molecule. NTS sequences vary considerably in length among eukaryotes, from 2 kb in frogs to 30 kb in humans.

Some evidence also exists for the presence of introns in the rRNA genes of eukaryotes. However, they have been found in only a few systems (e.g., *Drosophila* and *Tetrahymena*), so they are certainly not as widespread as the introns in mRNA genes. The introns are spliced out of the pre-rRNA transcript as it is processed to produce the mature rRNAs. The splicing reactions involved in this case are distinct from those involved in removing introns from pre-mRNA and pre-tRNA.

Keynote *Ribosomes, the organelles within the cell in which protein synthesis takes place, consist of two unequally sized subunits in both prokaryotes and eukaryotes. Each subunit contains ribosomal RNA and ribosomal proteins. Prokaryotic ribosomes contain three distinct rRNA molecules, whereas eukaryotic cytoplasmic ribosomes (larger and more complex) contain four.*

Biosynthesis of eukaryotic ribosomes. Eukaryotic ribosomes are assembled in a process similar to that found in prokaryotes. Ribosomal proteins are made in the cytoplasm from mRNAs that are transcribed by RNA polymerase II. The 5S rRNA is transcribed by RNA polymerase III in the nucleus directly from individual rRNA genes. Ribosomal proteins and the 5S rRNA molecule both migrate to the nucleolus in the cell's nucleus by unknown means.

In the nucleolus the 18S, 5.8S, and 28S rRNA genes are transcribed by RNA polymerase I into a large pre-rRNA molecule [such as the 45S pre-rRNA (~13,700 nucleotides) found in human HeLa cells or the 40S pre-rRNA of the frog *Xenopus laevis*]. Figure 12.30 shows an electron micrograph of pre-rRNA from an rDNA repeat unit of the newt *Triturus viridescens.* Even though rRNA sequences are highly conserved in the evolutionary sense and RNA polymerase I only synthesizes one type of product, it is striking that different organisms have very different sequences around their initiation sites for pre-rRNA synthesis (Figure 12.31). Strong homologies are seen only for closely related species such as mouse, rat, and human.

As with prokaryotic pre-rRNA, the bases and ribose sugars of eukaryotic pre-rRNA are methylated after transcription. In eukaryotes the methylation occurs primarily on the 2′–OH group of

the ribose. In prokaryotes it is primarily the bases that are methylated.

After the pre-rRNA is modified, it is cleaved at specific sites by specific RNases to remove ITS and ETS. Unlike processing for prokaryotes, processing the pre-rRNA does not occur until the molecule is completely synthesized. These discrete, intermediate rRNAs are finally processed into mature 18S, 5.8S, and 28S rRNAs. As an example, Figure 12.32 shows the pre-rRNA-processing pathway in HeLa cells. The pathway is similar in all eukaryotes that have been examined. The first cleavage removes the 5′ ETS sequence and produces a precursor molecule that still contains all three rRNA sequences. In the next step an internal cleavage occurs to produce the precursor to 18S rRNA and the precursor to 28S and 5.8S rRNAs. For the former the generation of 18S rRNA involves the removal of the ITS sequence to the 3′ side of the sequence. For the latter the molecule folds so that the 5.8S sequence becomes hydrogen-bonded to the 28S sequence. After that the ITS sequence is removed from between those two rRNAs, and the mature rRNAs are produced.

All the pre-rRNA-processing events take place in complexes formed between the pre-rRNA, 5S rRNA, and the ribosomal proteins. As the pre-

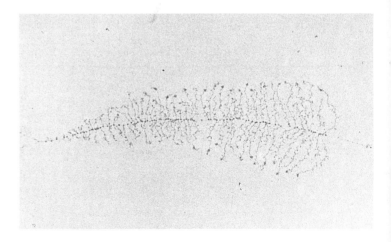

Figure 12.30 *Electron micrograph of transcription of pre-rRNA from an rRNA gene in an oocyte cell from the spotted newt,* Triturus viridescens. *Transcription of the rRNA gene is from left to right.*

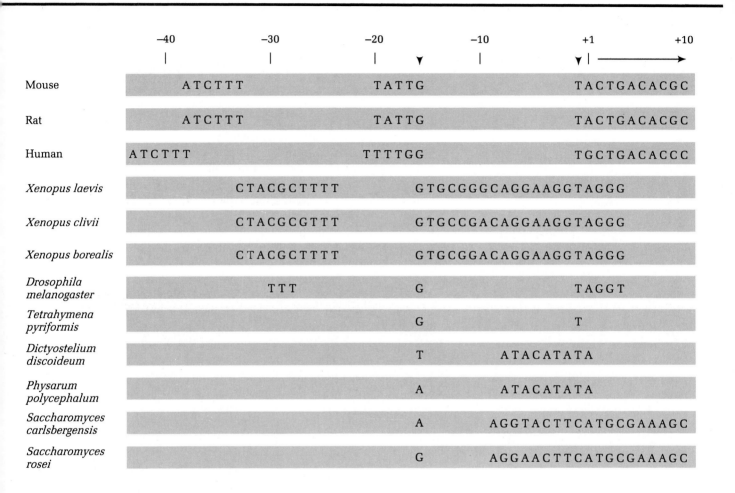

	−40	−30	−20	−10	+1	+10

Mouse	ATCTTT		TATTG		TACTGACACGC
Rat	ATCTTT		TATTG		TACTGACACGC
Human	ATCTTT		TTTTGG		TGCTGACACCC
Xenopus laevis	CTACGCTTTT		GTGCGGGCAGGAAGGTAGGG
Xenopus clivii	CTACGCGTTT		GTGCCGACAGGAAGGTAGGG
Xenopus borealis	CTACGCTTTT		GTGCGGACAGGAAGGTAGGG
Drosophila melanogaster	TTT		G		TAGGT
Tetrahymena pyriformis			G		T
Dictyostelium discoideum			T	ATACATATA
Physarum polycephalum			A	ATACATATA
Saccharomyces carlsbergensis			A	AGGTACTTCATGCGAAAGC
Saccharomyces rosei			G	AGGAACTTCATGCGAAAGC

rRNA is cleaved, conformational changes take place in the complexes so that when the processing is finished, the 60S and 40S ribosomal subunits are produced, each with the appropriate mature rRNAs and ribosomal proteins. Once assembled, the two ribosomal subunits then migrate out of the nucleolus and into the cytoplasm, where they associate with mRNAs and tRNAs and begin the process of protein synthesis.

Figure 12.31 Sequence homologies around the initiation sites for RNA polymerase I in rDNA of various eukaryotic organisms: Little homology is evident except in closely related species. No recognizable homologies appear where the sequence is not given.

405
*Analytical
Approaches
for Solving
Genetics
Problems*

Figure 12.32 Processing of 45S pre-rRNA in HeLa cells to produce the mature 18S, 5.8S, and 28S rRNAs.

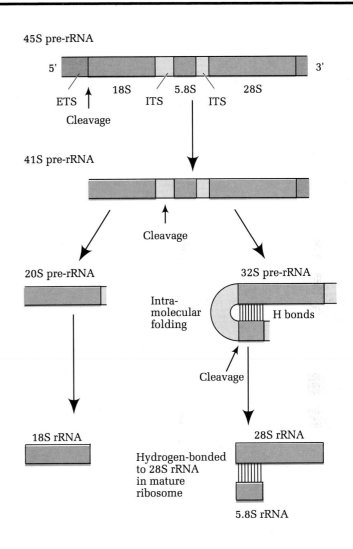

*Analytical
Approaches
for Solving
Genetics
Problems*

Because this chapter is primarily descriptive, we have only a few types of quantitative problems for this material. The following question is an example of an experimental approach question in which concepts from the chapter are applied to a typical molecular situation.

Q.1 You are given four different RNA samples. Sample I has a short lifetime; sample II has a homogeneous molecular weight; sample III is produced by processing of a larger precursor RNA; and sample IV has an additional sequence added onto the original transcript. For each sample, state whether the RNA could be rRNA, mRNA, or tRNA. If it is not one of those, state what it might be. Note that for each sample more than one RNA could apply. Give reasons for your choices.

A.1 Sample I: A short lifetime is characteristic of mRNAs in prokaryotes, and many mRNAs in eukaryotes. Heterogeneous nuclear RNA of eukaryotes also has a short lifetime. Sample II: rRNA and tRNA species have homogeneous molecular weights since they carry out specific functions within the cell for which their length and three-dimensional configuration are important. Messenger RNA and heterogeneous nuclear RNA are heterogeneous in length. Sample III: rRNA, tRNA, and mRNA are all produced by the processing of a larger precursor RNA molecule. Sample IV: Both tRNA and eukaryotic mRNA have additional sequences added after they are transcribed. For tRNA this sequence is the CCA at the 3′ end, and for mRNA this sequence is the poly(A) tail at the 3′ end.

Questions and Problems

12.1 The RNA polymerases bind to promoter sequences in the DNA.

a. Compare and contrast promoter sequences from prokaryotes and eukaryotes.

b. Apart from the sequence information available for promoters for a number of genes, what evidence is there that promoters do not all have the same base sequence?

12.2 Compare and contrast the structures of prokaryotic and eukaryotic mRNAs.

*12.3 Compare the structures of the three classes of RNA found in the cell.

12.4 Many eukaryotic mRNAs, but not prokaryotic mRNAs, contain introns. What is the evidence for the presence of introns in genes? Describe how these sequences are removed during the production of mature mRNA.

*12.5 Discuss the posttranscriptional modifications that take place on the primary transcripts of tRNA, rRNA, and protein-coding genes.

12.6 Distinguish between leader sequence, trailer sequence, coding sequence, intron, spacer sequence, nontranscribed spacer sequence, external transcribed spacer sequence, and internal transcribed sequence. Give examples of actual molecules in your answer.

12.7 Describe the organization of the ribosomal DNA repeating unit of a higher-eukaryotic cell.

13

The Genetic Code and the Translation of the Genetic Message

As we learned earlier, one of the three requirements for genetic material is that it must contain all the information for an organism's structure, function, development, and reproduction in a stable form. All these properties of an organism involve the action of specific proteins, and the information for the proteins is coded in the structural genes of the cell's genetic material. The expression of a structural gene—that is, the production of the protein coded for by the gene—occurs in two major steps: transcription and translation. In transcription or RNA synthesis (discussed in Chapter 11), the base-pair sequence of the DNA segment constituting the gene is read, and that information is transferred to the base sequence of a single-stranded mRNA molecule. The base sequence of the mRNA carries the specific information for the amino acid sequence of a polypeptide. The conversion in the cell of the mRNA base sequence information into an amino acid sequence of a polypeptide is called **translation** or **protein synthesis.** The base-pair information in the DNA that specifies the amino acid sequence of a polypeptide is called the genetic code.

In this chapter we will study the translation of the genetic message and how the information for the amino acid sequence of proteins is encoded in the nucleotide sequence of messenger RNA. We will see that the three classes of RNA—messenger RNA, transfer RNA, and ribosomal RNA—are involved in the translation process. Before examining translation, though, we will analyze the structure of proteins so that we will have a good understanding of their chemical composition and how their structure relates to their functional activities.

Protein Structure

Chemical Structure of Proteins

A **protein** is one of a group of high-molecular-weight, nitrogen-containing organic compounds of complex shape and composition. Each cell type has a characteristic set of proteins that give that cell type its functional properties. A protein consists of one or more molecular subunits called **polypeptides,** which are themselves composed of smaller building blocks, the **amino acids,** linked together to form long chains. The composition of a given protein is uniform in that each molecule consists of the same number and kind of polypeptide chains and each of these in turn is composed of the same number, kind, and sequence of amino acid. As we shall see, it is the sequence of amino acids in a polypeptide that gives the polypeptide its three-dimensional shape and its properties in the cell.

The basic chemical structural units of proteins are amino acids. With the exception of proline, the amino acids have a common structure, which is shown in Figure 13.1. Specifically, the structure consists of a central carbon atom (α-carbon) to which is bonded an amino group (α–NH$_2$), a carboxyl group (α–COOH), and a hydrogen atom. At the pH commonly found within cells, the α–NH$_2$ and α–COOH groups of the free amino acids are in a charged state; that is, –NH$_3{}^+$ and –COO$^-$, respectively.

Each amino acid has an additional chemical group bound to the α-carbon, called the *radical* or *R group.* It is the R group that varies from one amino acid to another and gives each amino acid its distinctive properties. Since different proteins have different sequences and proportions of amino acids, the organization of the R groups gives a protein its structural and functional properties.

There are 20 naturally occurring amino acids; their names and chemical structures are shown in Figure 13.2. The 20 amino acids are divided into subgroups based on whether the R group is acidic (e.g., aspartic acid), basic (e.g., lysine), neutral-polar (e.g., leucine), or neutral-nonpolar (e.g., serine).

The amino acids of a polypeptide are held together by a **peptide bond,** a covalent bond that joins the α-carboxyl group of one amino acid to the α-amino group of another amino acid. The formation of the peptide bond is the result of the reaction depicted in Figure 13.3. A polypeptide, then, is a linear, unbranched molecule that consists of many amino acids (usually 100 or more) joined by peptide bonds. (The presence of peptide bonds along the molecule, in fact, gives

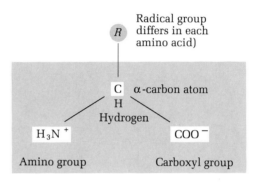

Structures common to all amino acids

Figure 13.1 General structural formula for an amino acid.

a polypeptide its name.) Regardless of its specific chemical composition, every polypeptide has a free α-amino group at one end (called the N terminus or the N terminal end) and a free α-carboxyl group at the other end (called the C terminus or the C terminal end). Like the nucleic acids, polypeptides have polarity. By convention and because the polypeptide is made that way, the N terminal end is defined as the beginning of a polypeptide chain.

Molecular Structure of Proteins

The molecular structure of a protein is relatively complex; it has four levels of structural organization, as shown in Figure 13.4.

1. The primary structure of the polypeptide chain(s) that constitute a protein is the amino acid sequence (Figure 13.4a).
2. The secondary structure of a protein refers to the folding and twisting of a single polypeptide chain into a variety of shapes. A polypeptide's secondary structure is the result of weak bonds (e.g., electrostatic or hydrogen) that form between R groups of amino acids that are relatively nearby on the chain. One

Figure 13.2 Structures of the 20 naturally occurring amino acids.

Aspartic acid (asp)

$$H_3N^+ - \overset{\overset{\displaystyle H}{|}}{C} - CH_2 - C\overset{\displaystyle O^-}{\underset{\displaystyle O}{\diagdown\diagup}}$$
^-OOC

Glutamic acid (glu)

$$H_3N^+ - \overset{\overset{\displaystyle H}{|}}{C} - CH_2 - CH_2 - C\overset{\displaystyle O^-}{\underset{\displaystyle O}{\diagdown\diagup}}$$
^-OOC

Basic

Lysine (lys)

$H_3N^+ - \overset{H}{\underset{}{C}} - (CH_2)_3 - CH_2 - {}^+NH_3$
^-OOC

Arginine (arg)

$H_3N^+ - \overset{H}{\underset{}{C}} - (CH_2)_2 - CH_2 - N - C - {}^+NH_3$
^-OOC , with H and N (double bond)

Histidine (his)

$H_3N^+ - \overset{H}{\underset{}{C}} - CH_2 - C - N$
^-OOC
ring: C=CH, HC—N—H

Neutral, nonpolar

Tryptophan (trp)

$H_3N^+ - \overset{H}{\underset{}{C}} - CH_2 - C$
^-OOC
HC, N, H (indole ring with benzene)

Phenylalanine (phe)

$H_3N^+ - \overset{H}{\underset{}{C}} - CH_2 -$ (benzene ring)
^-OOC

Glycine (gly)

$H_3N^+ - \overset{H}{\underset{}{C}} - H$
^-OOC

Alanine (ala)

$H_3N^+ - \overset{H}{\underset{}{C}} - CH_3$
^-OOC

Valine (val)

$H_3N^+ - \overset{H}{\underset{}{C}} - CH \overset{CH_3}{\underset{CH_3}{\diagup\diagdown}}$
^-OOC

Isoleucine (ile)

$H_3N^+ - \overset{H}{\underset{}{C}} - CH \overset{CH_2 - CH_3}{\underset{CH_3}{\diagup\diagdown}}$
^-OOC

Leucine (leu)

$H_3N^+ - \overset{H}{\underset{}{C}} - CH_2 - CH \overset{CH_3}{\underset{CH_3}{\diagup\diagdown}}$
^-OOC

Methionine (met)

$H_3N^+ - \overset{H}{\underset{}{C}} - CH_2 - CH_2 - S - CH_3$
^-OOC

Proline (pro)

ring: H_3N^+, H—C—COO$^-$, CH$_2$, H$_2$C, C H$_2$

Neutral, polar

Tyrosine (tyr)

$H_3N^+ - \overset{H}{\underset{}{C}} - CH_2 -$ (benzene ring) $- OH$
^-OOC

Serine (ser)

$H_3N^+ - \overset{H}{\underset{}{C}} - CH_2 - OH$
^-OOC

Threonine (thr)

$H_3N^+ - \overset{H}{\underset{}{C}} - \overset{H}{\underset{CH_3}{C}} - OH$
^-OOC

Asparagine (asn)

$H_3N^+ - \overset{H}{\underset{}{C}} - CH_2 - C - NH_2$
^-OOC , C=O

Glutamine (gln)

$H_3N^+ - \overset{H}{\underset{}{C}} - (CH_2)_2 - C - NH_2$
^-OOC , C=O

Cysteine (cys)

$H_3N^+ - \overset{H}{\underset{}{C}} - CH_2 - SH$
^-OOC

410
The Genetic
Code and the
Translation
of the
Genetic
Message

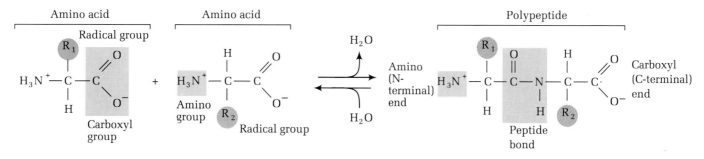

Figure 13.3 Mechanism for the formation of the peptide bond between the carboxyl group of one amino acid and the amino group of another amino acid.

a) Primary structure

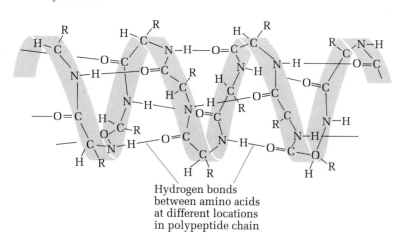

b) Secondary structure

Hydrogen bonds between amino acids at different locations in polypeptide chain

c) Tertiary structure

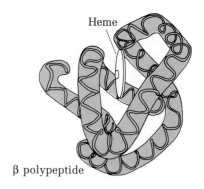

Heme

β polypeptide

d) Quarternary structure

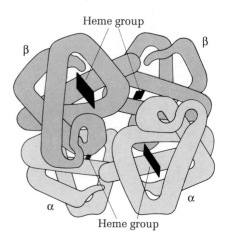

Heme group

β   β

α   α

Heme group

type of secondary structure found in regions of many polypeptides is the α-*helix*, discovered by Linus Pauling around 1960. Figure 13.4b shows the α-helix and diagrams the hydrogen bonding between the NH group of one amino acid (i.e., an NH group that is part of a peptide bond) and the CO group (also part of a peptide bond) of an amino acid four amino acids away in the chain. The repeated formation of this bonding results in the helical coiling of the chain. The α-helix content of proteins varies.

3. A protein's tertiary structure (Figure 13.4c) is the three-dimensional structure into which the helices and other parts of a polypeptide chain are folded. The three-dimensional shape of a polypeptide is often called its conformation. Tertiary folding is a direct property of the amino acid sequence of the chain and, hence, is related to the relative frequency and distribution of the R groups along the chain. The folding also places hydrophobic (water-hating) parts of the molecule on the inside and hydrophilic (water-loving) parts on the outside.

4. Quaternary structure of a protein is shown in Figure 13.4d. Note that primary, secondary, and tertiary structures all refer to single polypeptide chains. As we said earlier, however, proteins may consist of more than one polypeptide chain; such proteins are called multimeric ("many subunits") proteins. Quaternary structure refers to how polypeptides are packaged into the whole protein molecule for those proteins that consist of more than one polypeptide chain. Figure 13.5 diagrams the quaternary structure of a well-known example of a multimeric protein, the oxygen-

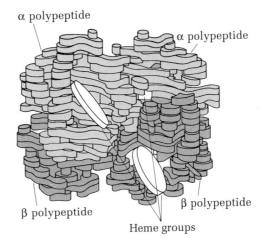

Figure 13.5 *An example of quaternary structure in proteins. This model of the hemoglobin molecule shows the two α chains, the two β chains, and the four heme groups.*

carrying protein hemoglobin, which consists of four polypeptide chains (two α polypeptides and two β polypeptides), each of which is associated with a heme group (involved in the binding of oxygen). The two types of polypeptides (α and β) have different amino acid sequences, each sequence produced by the translation of different mRNA transcripts specified by different genes. The α polypeptide contains 141 amino acids, and the β polypeptide contains 146 amino acids. In the quaternary structure of hemoglobin, each α chain is in contact with each β chain; however, little interaction occurs between the two α chains or between the two β chains.

The three-dimensional structures of the α and β polypeptides of the hemoglobin protein are similar to the three-dimensional structure of the single-polypeptide, oxygen-carrying protein myoglobin (Figure 13.6), which is involved in the transport of oxygen in muscle tissue. Like hemoglobin, myoglobin contains a heme group. The great similarity between the tertiary structures of the α hemoglobin polypeptide, the β hemoglobin polypeptide, and the myoglobin polypeptide is interesting because the amino acid sequences

Figure 13.4 *Four levels of protein structure: (a) primary, the sequence of amino acids in a polypeptide chain; (b) secondary, the folding of the amino acid chain into an α-helix stabilized by hydrogen bonds; (c) tertiary, the specific three-dimensional folding of the polypeptide chain; (d) quaternary, the specific aggregate of polypeptide chains (here four chains of hemoglobin are shown).*

412
*The Genetic
Code and the
Translation
of the
Genetic
Message*

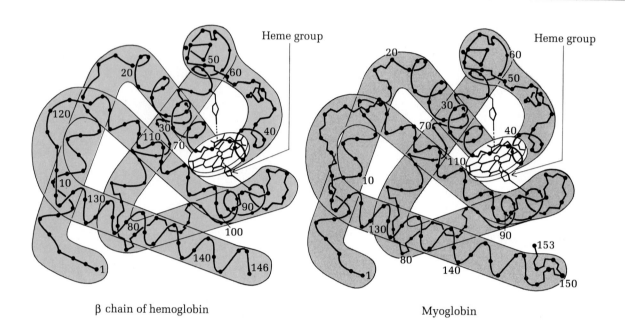

β chain of hemoglobin   Myoglobin

Figure 13.6 Comparison of the tertiary structures of the main polypeptide chain of myoglobin (153 amino acids long) and of the β polypeptide chain of hemoglobin (146 amino acids long). The numbers refer to the amino acid positions in the polypeptide chains. Despite differences in their primary structures, these two oxygen-carrying proteins have very similar three-dimensional conformations.

are quite different. Indeed, only 24 identities of amino acid placements occur among the three chains. Thus widely different amino acid sequences can give rise to similar three-dimensional conformations. In fact, the structure shown in Figures 13.5 and 13.6 is common among all vertebrate myoglobins and hemoglobins, and it clearly represents the fundamental protein structure required for oxygen-carrying capability.

Figure 13.7 Experimental details of a cell-free, protein-synthesizing system.

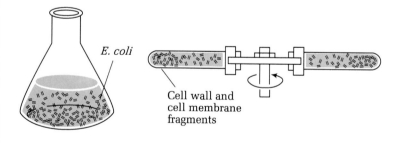

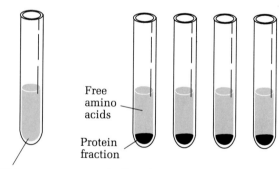

1. *E. coli* cells are collected by centrifugation and lysed. DNase is added to degrade DNA.

2. Cell lysate is centrifuged to pellet sediment cell wall and cell membrane fragments. Supernatant, containing polysomes, ribosomes, mRNA, tRNA and enzymes, is transferred to test tubes. ATP, GTP, and radioactive amino acids are added to the tubes, which are incubated at 37°C for various times.

tRNA
ATP and GTP
Enzymes
Free ribosome subunits
  and polyribosomes
  containing mRNA
Radioactive amino
  acids

3. Proteins are precipitated by acid, leaving amino acids in solution. Radioactivity in the precipitate indicates the amount of amino acids incorporated into newly-synthesized protein.

Keynote *A protein consists of one or more molecular subunits called polypeptides, which are themselves composed of smaller building blocks, the amino acids, linked together by peptide bonds to form long chains. The primary amino acid sequence of a protein determines its secondary, tertiary, and quaternary structure and hence its functional state.*

The Nature of the Genetic Code

In Chapter 12 we learned that nucleotides in the mRNA molecule are read in groups of three (called **codons**) to specify the amino acid sequence in proteins. Since four different nucleotides (A, C, G, U) occur in the message and 20 different amino acids occur in proteins, we deduce that the genetic code for each amino acid must be at least three letters. If it were a one-letter code, only four amino acids could be encoded. If it were a two-letter code, then only $4 \times 4 = 16$ amino acids could be encoded. A three-letter code, however, generates $4 \times 4 \times 4 = 64$ possible codes, more than enough to code for the 20 amino acids. The assumption of a three-letter code also suggests that some amino acids may be specified by more than one codon, which is, in fact, the case. Experimental evidence for the three-letter genetic code was obtained by F. Crick, L. Barnett, S. Brenner, and R. Watts-Tobin in the early 1960s.

Deciphering the Genetic Code

The exact relationship of the 64 codons to the 20 amino acids was determined by experiments done mostly in the laboratories of M. Nirenberg and G. Khorana. Essential to these experiments was the use of **cell-free, protein-synthesizing systems,** which contained ribosomes, tRNAs with amino acids attached, and all the necessary protein factors for polypeptide synthesis. These cell-free, synthesizing systems were assembled from components isolated and purified from *Escherichia coli* (Figure 13.7). Protein synthesis in these systems is inefficient in that very little protein is made. Therefore, radioactively labeled

amino acids have to be used to measure the incorporation of amino acids into new proteins.

Figure 13.8 shows the typical time course for the incorporation of radioactive amino acids into proteins in an in vitro system. As can be seen, protein synthesis takes place rapidly for about 15 minutes and then stops gradually because the mRNA that was in the *E. coli* extract degrades. If fresh mRNA is added, protein synthesis resumes and proceeds until the new mRNA is degraded. Thus, once the *E. coli* extract has been made and incubated to allow the system's original mRNA to degrade, it becomes an excellent system for testing the coding capacity of synthetic mRNAs. The system's usefulness lies in the fact that a single mRNA molecule can be added to the system, enabling the investigators to analyze the result of this addition on the composition of the polypeptide produced. As

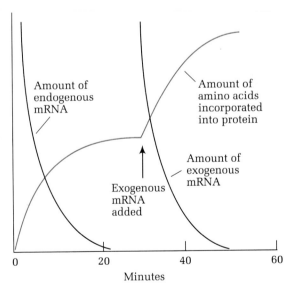

Figure 13.8 Typical time course for the incorporation of radioactive amino acids into proteins in an in vitro system. Initially radioactivity is incorporated into proteins until the endogenous mRNA breaks down. Incorporation of radioactive amino acids into proteins resumes when exogenous mRNA is added and continues until that mRNA breaks down. (From Watson, et al., Molecular biology of the gene, *vols. I and II, 4th ed. Copyright © 1987 Benjamin/Cummings Publishing Company, Inc. Reprinted by permission.)*

414
The Genetic
Code and the
Translation
of the
Genetic
Message

we shall see, the mRNAs used were synthetically made in the laboratory; their relative simplicity generated valuable information about the relationships between codons and amino acids that would not have been possible with the very complex, natural mRNAs.

In one approach to establish in vitro which codons specify which amino acids, synthetic polyribonucleotides (i.e., synthetic mRNAs) containing one, two, or three different types of bases were made and added to cell-free, protein-synthesizing systems. The polypeptides made in these systems were then analyzed. When the synthetic mRNA contained only one type of ribonucleotide, the results were unambiguous. The synthetic poly(U) mRNA, for example, directed the synthesis of a polypeptide consisting of phenylalanines (a polyphenylalanine chain). Since the genetic code is a triplet code, this result indicated that UUU is a codon for phenylalanine. Similarly, a synthetic poly(A) mRNA directed the synthesis of a polylysine, and poly(C) directed the synthesis of polyproline, indicating that AAA is a codon for lysine and CCC is a codon for proline. The results from poly(G) were inconclusive since the poly(G) folds up upon itself and hence cannot easily be translated in vitro.

Synthetic mRNAs made by the random incorporation of two different bases (called random copolymers) were also analyzed in the cell-free, protein-synthesizing systems. When mixed copolymers are made, the bases are assumed to be incorporated into the synthetic molecule in a random way. Thus poly(AC) molecules can contain eight different codons (CCC, CCA, CAC, ACC, CAA, ACA, AAC, and AAA), and the proportions of each will depend on the A-to-C ratio used to make the polymer. In the cell-free, protein-synthesizing system, poly(AC) synthetic mRNAs cause the incorporation of asparagine, glutamine, histidine, and threonine into polypeptides, in addition to the lysine expected from AAA codons and the proline expected from CCC codons. The proportions of asparagine, glutamine, histidine, and threonine incorporated into the polypeptides produced depend on the A/C ratio used to make the mRNA, and these observations can be used to deduce information about the codons that specify the amino acids. For example, since an AC random copolymer containing much more A than C results in the incorporation of many more asparagines than histidines, researchers concluded that asparagine is coded by two A's and one C and histidine by two C's and one A. With experiments of this kind, then, the base composition *(but not the base sequence)* of the codons for a number of amino acids was determined.

A second experimental approach also used copolymers, but these copolymers had been synthesized so that they had a known sequence, not a random one. For example, a repeating copolymer of U and C gives a synthetic mRNA of UCUCUCUCUC . . . , which, when it is tested in a cell-free, protein-synthesizing system, directs the synthesis of a polypeptide with a repeating amino acid pattern of leucine-serine-leucine-serine. . . . From this result, researchers could conclude that UCU and CUC specify leucine and serine, although they could not determine from this information alone which coded for which.

A third approach was more direct. It utilized the fact that in the absence of protein synthesis, specific tRNA molecules will bind to complexes formed between ribosomes and mRNAs. For example, when the synthetic mRNA poly(U) is mixed with ribosomes, it forms a poly(U)-ribosome complex, and only tRNA.Phe (i.e., the tRNA that will bring phenylalanine to an mRNA and that has the appropriate anticodon, AAA, for the UUU codon) will bind to the UUU codon. Importantly, the specific binding of the appropriate tRNA to the mRNA-ribosome complex does not require the presence of long mRNA molecules, since the binding of a trinucleotide (i.e., a codon-length piece of RNA) will suffice. The discovery of the trinucleotide-binding property made it possible to determine relatively easily the specific relationships between many codons and the amino acids for which they code. Note that in this particular approach, the specific nucleotide sequence of the codon is determined, not merely the nucleotide composition. With this approach, for example, GUU was found to promote the binding of a tRNA.Val;

that is, a tRNA that brings the amino acid valine to the message. Thus GUU must code for valine. Similarly, UUG causes tRNA.Leu binding, so UUG codes for leucine. All in all, about 50 codons were clearly identified by using this approach.

In sum, none of the approaches by themselves produced an unambiguous set of codon assignments, but information obtained through all three approaches enabled all codon assignments to be deduced with high degrees of certainty.

Nature and Characteristics of the Genetic Code

From the types of experiments just described, all 64 codons were assigned with the result shown in Figure 13.9. Note that each codon is written as it appears in mRNA and reads in a 5′-to-3′ direction. What are the characteristics of this code?

1. *The code is a triplet code.* Each codon in the mRNA that specifies an amino acid in a polypeptide chain consists of three nucleotides.
2. *The code is comma-free.* In other words, the mRNA is read continuously, three nucleotides (one codon) at a time, without skipping any sections of the message.
3. *The code is nonoverlapping.* That is, the mRNA is read in successive groups of three nucleotides. A message of AAGAAGAAG . . . in the cell would be read as lysine-lysine-lysine . . . , which is what the AAG specifies. Theoretically, though, three readings are possible from this message, depending on where the reading is begun (i.e., the reading frame in which message is read); namely, the repeating AAG, the repeating AGA, and the repeating GAA. Later in this chapter we will examine the mechanisms in the cell that ensure that the translation of the genetic code in an mRNA is begun at the correct point.
4. *The code is universal.* All organisms share the same genetic language. Thus, for example, lysine is coded for by AAA or AAG in the mRNA of all organisms, arginine by CGU, CGC, CGA, CGG, AGA, and AGG, and so on.

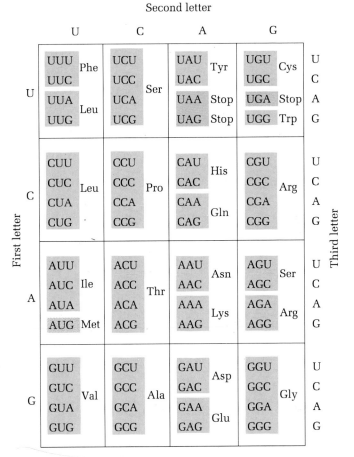

Second letter

Figure 13.9 *The genetic code. Of the 64 codons 61 specify 20 amino acids (one codon, AUG, also initiates protein synthesis). The other 3 codons are chain-terminating codons and do not specify any amino acid.*

Hence we can isolate an mRNA from one organism, translate it by using the machinery isolated from another organism, and produce the protein as if it had been translated in the original organism. The code, however, is not as universal as was once thought. For example, the mitochondria of some organisms, such as mammals, have minor changes in the code, as does the nuclear genome of the protozoan, *Tetrahymena*.

416
*The Genetic
Code and the
Translation
of the
Genetic
Message*

5. With two exceptions [AUG (methionine) and UGG (tryptophan)], more than one codon occurs for each amino acid. This multiple coding is called the **degeneracy** of the code. A close examination of Figure 13.9 reveals particular patterns in this degeneracy. Thus when the first two nucleotides in a codon are identical and the third letter is U or C, the codon often codes for the same amino acid. For example, UUU and UUC specify phenylalanine; similarly, CAU and CAC specify histidine. Also, when the first two nucleotides in a codon are identical and the third letter is A or G, the same amino acid is often speci-

fied. For example, UUA and UUG specify leucine, and AAA and AAG specify lysine. In some cases, with identity in the first two positions the base in the third position may be U, C, A, or G and the same amino acid will be specified. Examples are CUU, CUC, CUA, and CUG, all for leucine, and GCU, GCC, GCA, and GCG, all for alanine.

Even though there is degeneracy of the code, that does not mean that all codons are used equally. Studies have shown that codon usage is not random. Rather, some codons are used repeatedly, while others almost never are used. Figure 13.10 illustrates the results

Second letter

First letter	U	C	A	G	Third letter
U	13 UUU / 28 UUC — Phe 2 UUA / 9 UUG — Leu	16 UCU / 18 UCC / 9 UCA / 2 UCG — Ser	10 UAU / 23 UAC — Tyr UAA Stop / UAG Stop	10 UGU / 13 UGC — Cys UGA Stop / 12 UGG Trp	U C A G
C	9 CUU / 27 CUC / 7 CUA / 47 CUG — Leu	14 CCU / 17 CCC / 10 CCA / 5 CCG — Pro	10 CAU / 21 CAC — His 10 CAA / 28 CAG — Gln	8 CGU / 11 CGC / 4 CGA / 5 CGG — Arg	U C A G
A	11 AUU / 24 AUC — Ile 4 AUA / 16 AUG Met	15 ACU / 28 ACC / 11 ACA / 6 ACG — Thr	8 AAU / 28 AAC — Asn 19 AAA / 49 AAG — Lys	12 AGU / 21 AGC — Ser 8 AGA / 10 AGG — Arg	U C A G
G	9 GUU / 21 GUC / 5 GUA / 33 GUG — Val	28 GCU / 38 GCC / 14 GCA / 6 GCG — Ala	16 GAU / 24 GAC — Asp 21 GAA / 34 GAG — Glu	22 GGU / 32 GGC / 16 GGA / 11 GGG — Gly	U C A G

Figure 13.10 Codon usage in the genes of animals. The numbers next to the codons are the instances of the codons in 2244 codons tabulated.

of a study of a number of genes in animals. This study showed, for example, a significant bias toward the use of certain codons for particular amino acids. For example, UUC is the more frequently used codon for phenylalanine, followed by UUU.

6. Specific start and stop signals for protein synthesis are contained in the code. In both eukaryotes and prokaryotes AUG (methionine) is most commonly used as a start codon for protein synthesis, although in rare cases GUG may also be used. Thus in examining the sequence of a particular mRNA for the location of the amino acid–coding sequence, we look at the 5′ end for the occurrence of the AUG start codon and begin reading the amino acid sequence from there.

As Figure 13.9 shows, only 61 of the 64 codons specify amino acids; these codons are called the **sense codons.** The other three codons, UAG, UAA, and UGA, do not specify an amino acid, and no tRNAs in normal cells carry the appropriate anticodons. These three codons are the **stop, nonsense, or chain-terminating codons.** They are used singly or in tandem groups (UAG UAA, for example) to specify the end of the translation process of a polypeptide chain. Again, when we read a particular mRNA sequence we look for the presence of a stop codon in the same reading frame as the AUG start codon to determine where the amino acid–coding sequence for the polypeptide ends. We will examine starting and stopping protein synthesis when we study the details of the translation process.

7. *Wobble occurs in the anticodon.* Since 61 sense codons specify amino acids in mRNA, we may assume that 61 corresponding tRNA molecules have the appropriate anticodons. In most organisms they do. Theoretically, though, the complete set of 61 sense codons can be read by fewer than 61 distinct tRNAs because of the wobble in the anticodon. The wobble hypothesis, proposed by F. Crick, is shown in Table 13.1. Sequence analysis had shown that the base at the 5′ end of the anticodon (complementary to the base at the 3′ end of the codon; i.e., the third letter) is not

Table 13.1 Base-Pairing Wobble in the Genetic Code

Base at 5′ End of Anticodon	Can Pair With	Base at 3′ End of Codon
G		U or C
C		G
A		U
U		A or G
I (inosine)		A, U, or C

as constrained as the other two bases. This feature allows for less exact base pairing so that the base at the 5′ end of the anticodon can potentially pair with one of three different bases at the 3′ end of the codon; it can wobble. As Table 13.1 shows, no single tRNA molecule can recognize four different codons. But if the tRNA molecule contains the modified base inosine (see Figure 12.16) at the 5′ end of the anticodon, then three different codons can be read by a single tRNA. Figure 13.11a gives an example of how a single leucine tRNA can read two different leucine codons by base-pairing wobble, and Figure 13.11b shows how a single glycine tRNA can read three different alanine codons by wobble pairing.

Keynote The genetic code is a triplet code in which each codon (a contiguous set of three bases) in an mRNA specifies one amino acid. Since 64 code words are possible and since 20 amino acids exist, some amino acids are specified by more than one codon. The genetic code in mRNA is read without gaps and in successive, nonoverlapping codons. The code is universal: The same codons specify the same amino acids in most systems. Protein synthesis is typically initiated by codon AUG (methionine), and it is terminated by three codons singly or in combination: UAG, UAA, UGA (these codons do not code for any amino acid).

Translation of the Genetic Message

Since we now know the types of RNA made in the cell, the structure of ribosomes (from

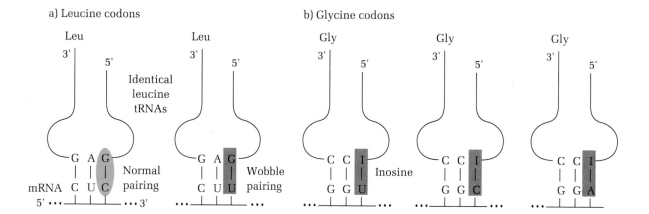

a) Leucine codons

b) Glycine codons

Figure 13.11 Example of base-pairing wobble. (a) Two different leucine codons (CUC, CUU) can be read by the same leucine tRNA molecule, contrary to regular base-pairing rules. (b) Three different glycine codons (GGU, GGC, GGA) can be read by the same glycine tRNA molecule by base-pairing wobble involving inosine in the anticodon.

Chapter 12), and the genetic code, we are ready to examine the details of the translation of the genetic message (also called protein synthesis). In brief, protein synthesis takes place on ribosomes where the genetic message encoded in mRNA is translated. The mRNA molecule is translated in the 5′-to-3′ direction (the same direction in which it is made), and the polypeptide is made in the N-terminal to C-terminal direction. Amino acids are brought to the ribosome bound to tRNA molecules. The correct amino acid sequence is achieved as a result of (1) the specific binding between the codon of the mRNA and the complementary anticodon in the tRNA and (2) the specific binding of each amino acid to its own specific tRNA. The three basic stages of protein synthesis—*initiation, elongation,* and *termination*—are similar in prokaryotes and eukaryotes. In the following sections we will discuss each of these stages in turn, concentrating, as before, on the processes in *E. coli,* with which most work has been done. In the discussions we will note where significant differences in translation occur in prokaryotes and eukaryotes.

Aminoacyl-tRNA Molecules

The important function of tRNA molecules in protein synthesis is that they bring specific amino acids to the mRNA-ribosomal complex so that the correct polypeptide chain can be assembled. In this section we will see how an amino acid becomes attached to its appropriate tRNA molecule to produce an aminoacyl-tRNA molecule.

The mRNA codon recognizes tRNA and not the amino acid. That the mRNA codon recognizes the tRNA anticodon and not the amino acid carried by the tRNA was proved by G. von Ehrenstein, B. Weisblum, and S. Benzer. They attached cysteine to tRNA.Cys in vitro, using cysteinyl-tRNA synthetase, the enzyme that, in vivo, is responsible for adding cysteine to tRNA.Cys. Then they chemically converted the attached cysteine to alanine. The resulting compound, alanyl-tRNA.Cys, was used in the in vitro synthesis of hemoglobin. In vivo the α and β chains of hemoglobin each contain one cysteine. When the hemoglobin made in vitro was examined, however, the amino acid alanine was found in both chains at the positions normally occupied by cysteine. This result could only mean that the alanyl-tRNA.Cys had read the cysteine codon and had inserted the amino acid it carried, in this case alanine. Therefore, the researchers concluded that the specificity of codon recognition lies in the tRNA molecule and not in the amino acid it carries.

Attachment of amino acid to tRNA. In another part of this process the researchers wanted to find out how the amino acid attaches to the correct tRNA molecule. We know that although all tRNAs have similar secondary and tertiary structures, their primary sequences are very different. The correct amino acid is attached to the tRNA by an enzyme called an **aminoacyl-tRNA synthetase.** The aminoacyl-tRNA synthetases have been studied in a number of organisms. They differ in size, subunit structure, and amino acid composition even within a single organism. Since 20 different amino acids exist, 20 species of aminoacyl-tRNA synthetases also exist. Further, since degeneracy is common in the genetic code (i.e., a single amino acid may be specified by more than one codon), all the tRNAs that are specific for a particular amino acid must have a common recognition site for the aminoacyl-tRNA synthetase for that amino acid. The aminoacyl-tRNA synthetases themselves are highly specific in their function, so few errors are made in the bonding of an amino acid to its tRNA. As with so many other processes in genetics, we do not know at this point exactly how the aminoacyl-tRNA synthetase recognizes the appropriate tRNA as its substrate.

Figure 13.12 shows how an amino acid is attached to a tRNA molecule to produce an **aminoacyl-tRNA,** in this case seryl-tRNA, and presents the generalized structure of an aminoacyl-tRNA molecule. As the figure shows, the amino acid attaches at the 3' end of the tRNA by a linkage between the carboxyl group of the amino acid and the 3'–OH group of the ribose of the adenine nucleotide found at the end of every tRNA. The act of adding the amino acid to the tRNA is called *charging,* and the product is commonly referred to as a charged tRNA.

Keynote *Protein synthesis occurs on ribosomes where the genetic message encoded in mRNA is translated. Amino acids are brought to the ribosome on charged tRNA molecules. The correct amino acid sequence is achieved as a result of (1) the specific binding between the codon of the mRNA and the complementary anticodon in the tRNA and (2) the specific binding of each amino acid to its own specific tRNA.*

Initiation of Translation

Initiator codon and initiator tRNA. In both prokaryotes and eukaryotes, protein synthesis begins at the initiator codon in the mRNA (the AUG codon, which specifies methionine). As a result, newly made proteins in both types of organisms begin with methionine; in some cases this methionine is subsequently removed.

In prokaryotes the initiator methionine is actually a specially modified form of methionine, called **formylmethionine** (abbreviated fMet), in which a formyl group has been added to the amino group of the amino acid. The fMet is brought to the ribosome attached to a special tRNA, called tRNA.fMet, which has the anticodon 5'–CAU–3'. This tRNA is special since it is involved specifically with the initiation process of protein synthesis. The aminoacylation of tRNA.fMet occurs as follows: First, methionyl-tRNA synthetase catalyzes the addition of methionine to the tRNA. Then an enzyme called *transformylase* catalyzes the addition of the formyl group to the methionine. The resulting molecule is designated fMet-tRNA.fMet (this nomenclature indicates that the tRNA is specific for the attachment of fMet and that, in fact, fMet is attached to it).

The tRNA.fMet molecule is only used in the initiation of translation. When an AUG codon is encountered in an mRNA molecule at a position other than at the start of the amino acid–coding sequence, another species of tRNA is used to insert methionine in the polypeptide chain at that point. This tRNA is called tRNA.Met, and it is aminoacylated by the same aminoacyl-tRNA synthetases as is tRNA.fMet to produce Met.tRNA.Met. However, tRNA.Met and tRNA.fMet are coded for by different genes and have different primary nucleotide sequences.

In eukaryotes, protein synthesis initiation occurs in much the same way. That is, AUG is the usual initiation codon, and hence the N-terminal amino acid in polypeptides is methionine. This amino acid may or may not be removed after the

420
*The Genetic
Code and the
Translation
of the
Genetic
Message*

a) First step

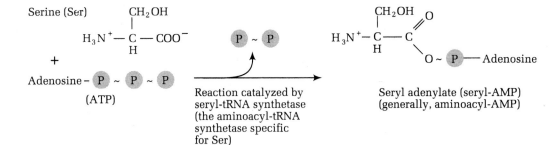

Serine (Ser)   CH₂OH

$H_3N^+ - C - COO^-$

H

+

Adenosine – P ~ P ~ P

(ATP)

P ~ P

Reaction catalyzed by
seryl-tRNA synthetase
(the aminoacyl-tRNA
synthetase specific
for Ser)

CH₂OH   O

$H_3N^+ - C - C$

H

O ~ P —— Adenosine

Seryl adenylate (seryl-AMP)
(generally, aminoacyl-AMP)

b) Second step

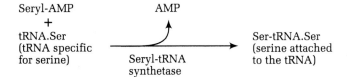

Seryl-AMP

+

tRNA.Ser
(tRNA specific
for serine)

AMP

Seryl-tRNA
synthetase

Ser-tRNA.Ser
(serine attached
to the tRNA)

c) General structure

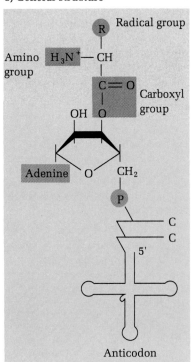

R   Radical group

Amino
group   H₃N⁺ — CH

C = O

Carboxyl
group

OH   O

Adenine   O   CH₂

P

C

C

5'

Anticodon

Amino acid
attached by
carboxyl group
to ribose of
last ribonucleotide
of tRNA chain

Last 3 nucleotides
of all tRNAs
are -C-C-A-3"

*Figure 13.12 Molecular details of the attachment of
an amino acid to a tRNA molecule: (a) In the first
step the amino acid (serine here) reacts with ATP to
produce an aminoacyl-AMP complex. This reaction is
catalyzed by an aminoacyl-tRNA synthetase (seryl-
tRNA synthetase). (b) In the second step (also catalyzed by aminoacyl-tRNA synthetase) the aminoacyl-
AMP complex then reacts with the appropriate tRNA
molecule (tRNA.Ser) to produce an aminoacyl-tRNA
(Ser-tRNA.Ser). (c) In the general molecular structure
of an aminoacyl-tRNA molecule (charged tRNA), the
carboxyl group of the amino acid is attached to the
3'−OH group of the 3' terminal adenine nucleotide
of the tRNA.*

		Binding site sequences					Initiation codon			
Phage R17 A protein	UCC	UAG	GAG	GUU	UGA	CCU	AUG	CGA	GCU	UUU
Phage Qβ replicase	UAA	CUA	AGG	AUG	AAA	UGC	AUG	UCU	AAG	ACA
Phage λ Cro	AUG	UAC	UAA	GGA	GGU	UGU	AUG	GAA	CAA	CGC
Phage φX174 A	AAU	CUU	GGA	GGC	UUU	UUU	AUG	GUU	CGU	UCU
Phage fd III		UUG	GAG	AUU	UUC	AAC	GUG	AAA	AAA	UUA
E. coli trpB	AUA	UUA	AGG	AAA	GGA	ACA	AUG	ACA	ACA	UUA
E. coli lacZ	UUC	ACA	CAG	GAA	ACA	GCU	AUG	ACC	AUG	AUU
E. coli RNA polymerase β	AGC	GAG	CUG	AGG	AAC	CCU	AUG	GUU	UAC	UCC

Figure 13.13 Sequences of some prokaryotic ribosome-binding sites. The initiation codon, AUG, is lightly shaded. The colored regions indicate the regions of contiguous complementarity between mRNA and the 3' end of 16S rRNA.

completion of the polypeptide. The N-terminal methionine is not formylated since the enzymes needed to carry out this function do not exist in the eukaryotic cytoplasm. A special methionine tRNA is used for initiation of translation in eukaryotes, and a distinct species of tRNA.Met is used to read AUG codons elsewhere in an mRNA molecule.

Initiation sequence. In addition to the AUG initiation codon, the initiation of protein synthesis in prokaryotes requires other information coded in the base sequence of an mRNA molecule upstream (to the 5' side) of the initiation codon. These sequences serve to align the ribosome on the message in the proper reading frame so that polypeptide synthesis can proceed correctly. To identify these aligning sequences, researchers sequenced a number of mRNAs in the region upstream from the AUG initiation codon. To do so, they allowed ribosomal subunits to bind to the mRNA under conditions whereby protein synthesis could not commence. They found that the 30S ribosomal subunits freeze on the mRNA message at the **ribosome-binding site,** the site at

which the ribosome becomes oriented in the correct reading frame for the initiation of protein synthesis. When the subunits are bound to the mRNAs, the message is protected from RNase attack, and so the protected regions can be isolated and studied.

Figure 13.13 shows the sequences of some prokaryotic ribosome-binding sites. The AUG codon can be clearly identified. Most of the binding sites have a purine-rich sequence about 8 to 12 nucleotides upstream from the initiation codon. Evidence from the work of J. Shine and L. Dalgarno indicates that this purine-rich sequence and other nucleotides in this region are complementary to a pyrimidine-rich region (which always contains the sequence CCUCC) at the 3' end of 16S rRNA (Figure 13.14). The mRNA region that binds in this way has become known as the Shine-Dalgarno sequence. Apparently, then, the formation of complementary base pairs between the mRNA and 16S rRNA allows the ribosome to locate the true initiator regions of the message.

None of the eukaryotic rRNAs have the CCUCC sequence found in prokaryotic 16S rRNA that binds with the mRNA to facilitate the initiation of translation. Instead, the eukaryotic ribosome uses another way to initiate protein synthesis on mRNA. That is, the 40S eukaryotic ribosomal subunit binds to the 5' cap of the mRNA (see Chapter 12) and then migrates to the

422
The Genetic
Code and the
Translation
of the
Genetic
Message

a) Sequence at 3' end

16S rRNA 3' end   3' ... AUUCCUCCAUAG ... 5'

b) Example of mRNA leader and 16S rRNA pairing

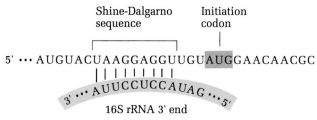

Shine-Dalgarno   Initiation
sequence   codon

5' ... AUGUACUAAGGAGGUUGUAUGGAACAACGC

16S rRNA 3' end

Figure 13.14 Sequences involved in the binding of ribosomes to the mRNA in the initiation of protein synthesis in prokaryotes: (a) nucleotide sequence at the 3' end of E. coli 16S rRNA; (b) example of how the 3' end of 16S rRNA can base-pair with the nucleotide sequence 5' upstream from the AUG initiation codon.

nearest AUG codon to the 5' end of the molecule. After the 60s subunit binds, protein synthesis is initiated at that codon.

Formation of the initiation complex. As Figure 13.15 shows, protein synthesis in *E. coli* commences with the formation of a 30S initiation complex. At the beginning of this process, three protein **initiation factors,** IF1, IF2, and IF3, are bound to the 30S ribosomal subunit along with a molecule of GTP (guanosine triphosphate). The presence of IF3 is known to alter the charge of the 30S subunit, thereby preventing the 50S ribosomal subunit from binding to it. The fMet-tRNA.fMet and the mRNA then attach in either order to the 30S-IF-GTP complex to form the completed *30S initiation complex.* IF3 is released as a result of this process. Next, the 50S subunit binds, leading to GTP hydrolysis and

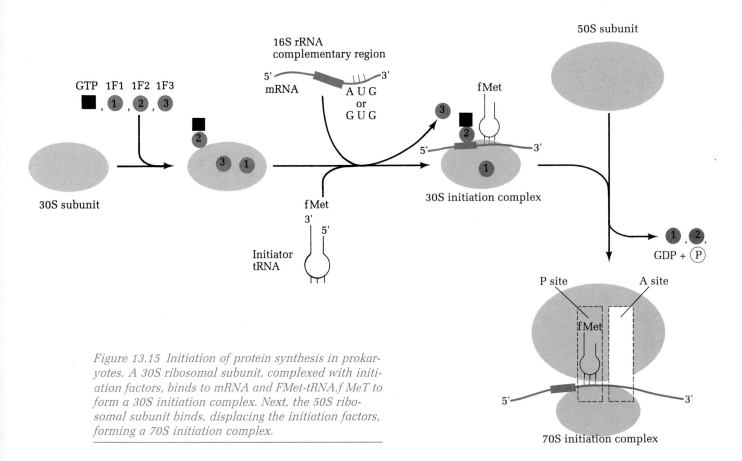

Figure 13.15 Initiation of protein synthesis in prokaryotes. A 30S ribosomal subunit, complexed with initiation factors, binds to mRNA and FMet-tRNA.f MeT to form a 30S initiation complex. Next, the 50S ribosomal subunit binds, displacing the initiation factors, forming a 70S initiation complex.

the release of IF1 and IF2. The final complex is called the *70S initiation complex.* All three IF molecules are recycled for use in other initiation reactions. The 70S ribosome has two binding sites for aminoacyl-tRNA, the peptidyl (P), and aminoacyl (A) sites; the fMet-tRNA.fMet is bound to the mRNA in the P site.

The formation of the eukaryotic 80S initiation complex is similar to that of prokaryotes, except that the methionine on the initiator tRNA is not formulated, and more initiation factors (*eukaryotic initiation factors,* or eIFs) are involved. At least five eIFs are known in mammalian cells. The initiator AUG codon is recognized first by the 40S ribosomal subunit scanning from the 5′ cap to the first AUG. The 60S subunit then binds to form the *80S initiation complex.*

The initiation process of prokaryotes and eukaryotes differs in two important respects:

1. While prokaryotic ribosomes can initiate translation at internal sites on mRNA molecules, initiation in eukaryotes can only occur at a single site located within the 5′ part of each mRNA molecule, that is, at the AUG codon nearest the 5′ end.
2. While mRNAs in most prokaryotes have similar nucleotide sequences in the initiation region, the nucleotide sequences surrounding the initiation region in eukaryotic mRNAs vary greatly.

In addition, great variation is shown between mRNAs in the location of the initiation AUG codon relative to the 5′ terminus of the mRNA molecule.

Elongation of the Polypeptide Chain

After the initiation events have been completed, the elongation phase of protein synthesis begins. This phase has three steps: the binding of aminoacyl-tRNA to the ribosome, the formation of peptide bonds, and the movement (translocation) of the ribosome along the mRNA, one codon at a time. Figure 13.16 diagrams these events. As before, we shall start with a discussion of the events in *E. coli* and then compare them with the processes in eukaryotes.

Binding of aminoacyl-tRNA. At the outset of the elongation phase, in the 70S initiation complex, the fMet-tRNA.fMet is hydrogen-bonded to the AUG initiation codon in the peptidyl site of the ribosome. There is insufficient evidence to show whether the initiator tRNA binds directly in the P site or whether it goes first to the A site and then moves to the P site. The orientation of this tRNA-codon complex exposes the next codon in the mRNA in the aminoacyl site. In Figure 13.16 this codon (UCC) specifies serine (Ser).

The next step in the translation process is the binding of the appropriate aminoacyl-tRNA (in this case Ser-tRNA.Ser) to the newly exposed codon [Figure 13.16(1)]. The aminoacyl-tRNA is brought to the ribosome complexed with the protein **elongation factor** EF-Tu and a molecule of GTP. The association of the aminoacyl-tRNA with the codon in the aminoacyl site is accompanied by the hydrolysis of GTP to GDP and Ⓟ, and the release of EF-Tu to which the GDP is bound.

The EF-Tu is recycled in the following steps. First, a second elongation factor, EF-Ts, binds to EF-Tu and displaces the GDP. Next, GTP binds to the EF-Tu–EF-Ts complex to generate an EF-Tu-GTP complex simultaneously with the release of EF-Ts. The aminoacyl-tRNA binds to the EF-Tu–GTP, and that complex can then bind to the A site in the ribosome if the correct codon-anticodon pairing forms.

Peptide bond formation. In Figure 13.16(2) fMet-tRNA.fMet is bound in the P site, and the second specified aminoacyl-tRNA is bound in the A site. The ribosome maintains the two aminoacyl-tRNAs in the correct configuration so that a peptide bond can form between the two amino acids.

The two steps involved in the formation of a peptide bond are shown in Figure 13.17. The first step is the breakage of the bond between the carboxyl group of the amino acid and the tRNA in the P site. In this case the breakage is between fMet and its tRNA. The second step is the formation of the peptide bond between the now-freed fMet and the Ser attached to the

424
The Genetic
Code and the
Translation
of the
Genetic
Message

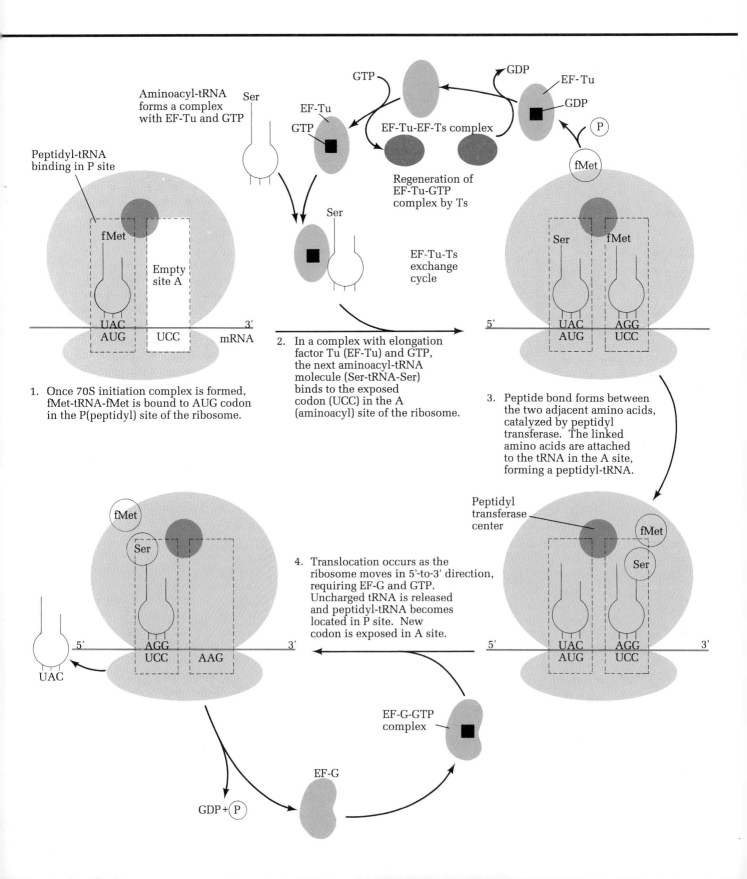

Aminoacyl-tRNA forms a complex with EF-Tu and GTP

EF-Tu-EF-Ts complex

Regeneration of EF-Tu-GTP complex by Ts

EF-Tu-Ts exchange cycle

Peptidyl-tRNA binding in P site

Empty site A

1. Once 70S initiation complex is formed, fMet-tRNA-fMet is bound to AUG codon in the P(peptidyl) site of the ribosome.

2. In a complex with elongation factor Tu (EF-Tu) and GTP, the next aminoacyl-tRNA molecule (Ser-tRNA-Ser) binds to the exposed codon (UCC) in the A (aminoacyl) site of the ribosome.

3. Peptide bond forms between the two adjacent amino acids, catalyzed by peptidyl transferase. The linked amino acids are attached to the tRNA in the A site, forming a peptidyl-tRNA.

Peptidyl transferase center

4. Translocation occurs as the ribosome moves in 5'-to-3' direction, requiring EF-G and GTP. Uncharged tRNA is released and peptidyl-tRNA becomes located in P site. New codon is exposed in A site.

EF-G-GTP complex

EF-G

GDP + P

Figure 13.16 Elongation stage of polypeptide synthesis in prokaryotes.

tRNA in the A site. This reaction is catalyzed by the enzyme *peptidyl transferase.* The activity of this enzyme results from interaction among a few ribosomal proteins of the 50S ribosomal subunit.

Once the peptide bond has formed [Figure 13.16(3)], a tRNA without an attached amino acid (an uncharged tRNA) is left in the P site.

Figure 13.17 The formation of a peptide bond between the first two amino acids (fMet and Ser) of a polypeptide chain is catalyzed on the ribosome by peptidyl transferase: (a) adjacent aminoacyl-tRNAs bound to the mRNA at the ribosome; (b) following peptide bond formation, an uncharged tRNA is in the P site and a dipeptidyl-tRNA is in the A site.

The tRNA (now called peptidyl-tRNA) in the A site has the first two amino acids of the polypeptide chain attached to it.

Translocation. The last step in the elongation cycle is **translocation** [Figure 13.16(4)]. Once the peptide bond is formed and the growing polypeptide chain is on the tRNA in the A site, the ribosome moves one codon along the mRNA toward the 3' end. Translocation requires the activity of another protein elongation factor, EF-G. An EF-G–GTP complex binds to the ribosome, and translocation then takes place along with ejection of the uncharged tRNA from the P site. The EF-G is then released in a reaction requiring GTP hydrolysis; EF-G can then be reused. During the translocation step the peptidyl-tRNA remains attached to its codon. And since the ribosome has moved, the peptidyl-tRNA is now located in the P site (hence the name, the *peptidyl* site). The exact mechanism for the physical

a) Adjacent aminoacyl-tRNAs

b) Following peptide bond formation

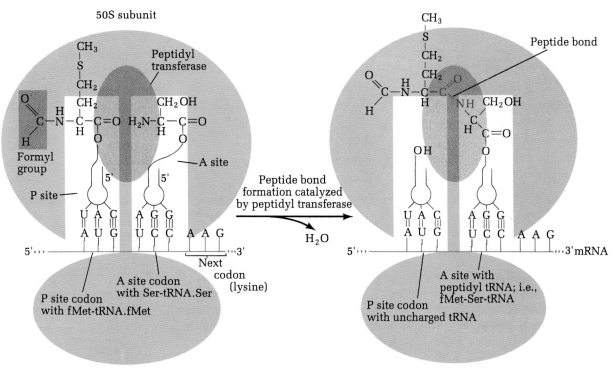

426
*The Genetic
Code and the
Translation
of the
Genetic
Message*

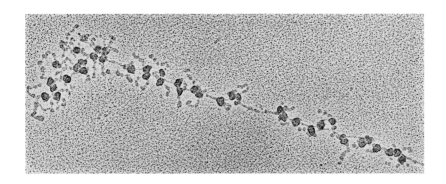

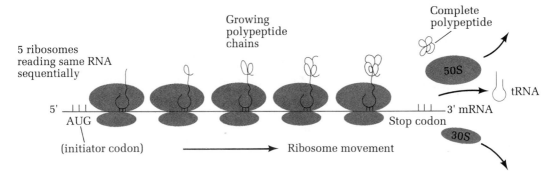

Figure 13.18 Electron micrograph and diagrammatic representation of a polysome, a number of ribosomes each translating the same mRNA sequentially.

translocation of the ribosome is not known.

After translocation is completed, the A site is vacant. An aminoacyl-tRNA with the correct anticodon binds to the newly exposed codon in the A site using the process already described. The whole process is repeated until translation is terminated.

The elongation steps in eukaryotes are similar to those in prokaryotes, although there are differences in the number and properties of elongation factors and in the exact sequences of events. Once the ribosome moves away from the initiation site, the initiation site is open for another initiation event to occur. Thus many ribosomes may simultaneously be translating each mRNA. The complex between an mRNA molecule and all the ribosomes that are translating it simultaneously is called a *polyribosome,* or *polysome* (Figure 13.18). An average length

mRNA may have eight to ten ribosomes synthesizing protein from it, thus enabling a large amount of protein to be produced from each mRNA molecule.

Keynote *The AUG (methionine) initiator codon signals the start of translation in prokaryotes and eukaryotes. Elongation proceeds when a peptide bond forms between the amino acid attached to the tRNA in the A site of the ribosome and the growing polypeptide attached to the tRNA in the P site. Translocation occurs when the now-uncharged tRNA in the P site is released from the ribosome and the ribosome moves one codon down the mRNA.*

Termination of Protein Synthesis

Elongation continues until the polypeptide coded for in the mRNA is completed. The completion is signaled by one of three stop codons, UAG, UAA, and UGA, which are the same in prokaryotes and eukaryotes. The stop codons do not code for any amino acid, and so no tRNAs in the cell have anticodons for them. The ribosome recognizes a chain termination codon only with the help of proteins called **termination** or **release factors (RF),** which read the chain termination codons and then initiate a series of specific termination events (Figure 13.19).

The *E. coli* has three RFs, RF1, RF2, and RF3, and each is a single polypeptide. Factor RF1 recognizes UAA and UAG, while RF2 recognizes UAA and UGA. Thus these release factors have overlapping specificity in codon recognition. Factor RF3, which does not recognize any of the stop codons, plays a stimulatory role in the termination events. The release factors in eukaryotes have not been clearly characterized. One RF (found in rabbit reticulocytes) recognizes all three stop codons. No stimulatory factor analogous to RF3 has been detected in eukaryotes.

Since no amino acid exists in the A site for the polypeptide (attached to the tRNA in the P site) to be transferred to, the polypeptide is released from the tRNA in the P site of the ribosome in a reaction catalyzed by peptidyl transferase. The polypeptide and the tRNA to which it was attached are then released from the ribosome. The ribosome dissociates into the two subunits, which are recycled and used in further protein synthesis events.

Keynote *Protein synthesis continues until a chain-terminating codon is located in the A site of the ribosome. These codons are read by one or more release factor proteins. Then the polypeptide and its tRNA are released from the ribosome, and the ribosome disengages from the reading frame of the mRNA.*

Protein Localization in Eukaryotes

In eukaryotes, some proteins may be secreted, depending on cell type, while other proteins need to be located in different cell compartments in order to function. That is, eukaryotic cells are compartmentalized, and the examination of the contents of these compartments (such as the nucleus, mitochondria, and vesicles) shows that each contains a specific set of proteins. Therefore proteins must end up in the compartments in which they are to function, since errors in protein transport can cause cell malfunction or death, which may, in turn, affect the whole organism. The rare human genetic disease I-cell disease, for example, is caused by misdirected protein transport. In this disease digestive enzymes that are supposed to remain inside the cell are instead shipped out of the cell. As a consequence, the cells become filled with inclusion bodies filled with debris that would normally be broken down by the missing enzymes (hence the *I* or I-cell disease). The I-cell disease is usually fatal in childhood.

Proteins Distributed by the Endoplasmic Reticulum

The electron microscope was instrumental in studying the transport of newly made secretory proteins. Electron microscopic studies showed that secretory cells, such as those found in the liver and pancreas, have an extensive membrane structure called the endoplasmic reticulum (ER). Some of the ER has a rough appearance due to the presence of ribosomes on the membrane. This observation has resulted in the terms *membrane-bound* and *free* ribosomes. The Nobel Prize–winning research of G. Palade and research of other investigators showed that proteins synthesized on the rough ER are extruded into the space between the two membranes of the ER (the cisternal space) and then transferred via the Golgi complex to secretory vesicles. From these vesicles the proteins are secreted to the outside of the cell by the fusion of the vesicles with the cell membrane. Figure 13.20 diagrams all these events on a simplistic level. For our purposes, it is necessary to understand that proteins destined to be secreted from the cell, proteins that will become embedded in the plasma membrane, and proteins that are packaged into lysosomes all initially are translocated into the ER cisternal space by a common mechanism. Thereafter, sorting to their final destinations occurs primarily at the Golgi complex.

In the mid 1970s, G. Blobel and D. Sabatini proposed that the signal for attachment might be

428

The Genetic
Code and the
Translation
of the
Genetic
Message

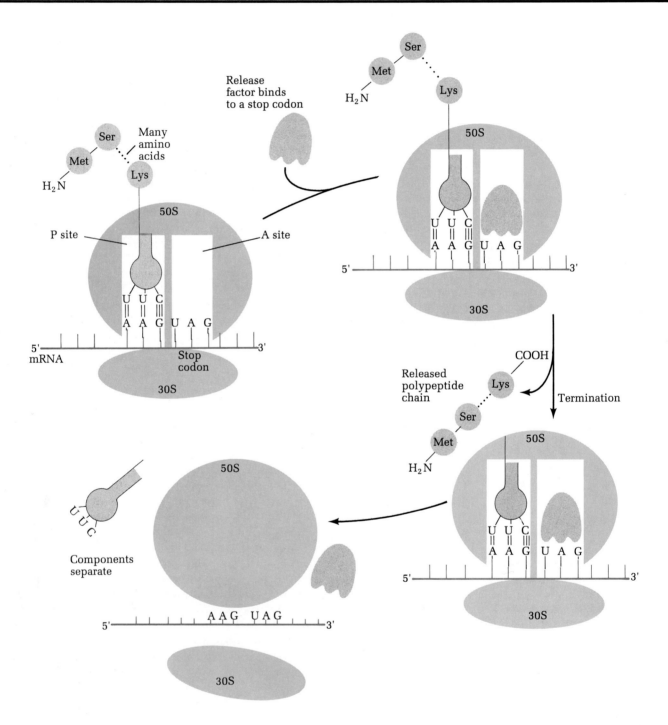

Figure 13.19 Termination of protein synthesis. The ribosome recognizes a chain termination codon (UAG), and after a release factor binds to the codon, the completed polypeptide is released.

in the structure of the growing polypeptide itself. The most logical candidate for the signal was the amino terminal end of the chain, since it is the first to be synthesized.

One piece of evidence for the nature of the signal came from the work of C. Milstein and G. Brownlee in 1972. They found that the smaller of the two polypeptide chains that make up antibody molecules is produced by processing of a larger precursor molecule. The extra material was shown to be about 15 amino acids at the amino terminal end of the polypeptide. Later, in 1975, Blobel and his colleagues found that all the major proteins secreted by the pancreas initially contain extra amino acids at the amino terminal end. They established that the extra amino acids gave hydrophobic properties to the amino terminal extension, just the properties a protein would need to pass through a membrane. As a result of their studies, G. Blobel and B. Dobberstein proposed the **signal hypothesis,** which states that the secretion of proteins out of a cell occurs through the binding of a hydrophobic, amino terminal extension (the *signal sequence*) to the membrane and the subsequent removal and degradation of the extension in the cisternal space of the ER.

Our current understanding of the signal hypothesis is as follows (Figure 13.21): The growing proteins (i.e., those being synthesized on ribosomes) that are to be secreted from the cell or inserted into membranes all have an N terminal extension of about 15 to 30 amino acids, called the signal sequence. Only proteins that have signal sequences can be transferred across or inserted into the membrane of the rough ER while translation is taking place.

When a protein destined for the ER exposes its signal sequence, a cytoplasmic receptor particle called the **signal recognition particle** (SRP: a complex of a small RNA molecule with six proteins) recognizes the signal sequence, binds to it, and blocks further translation of the mRNA. The temporary halt in protein synthesis brought about by SRP occurs when the polypeptide chain is long enough so that the signal sequence has completely emerged from the ribosome and can be recognized by SRP. Translation stops un-

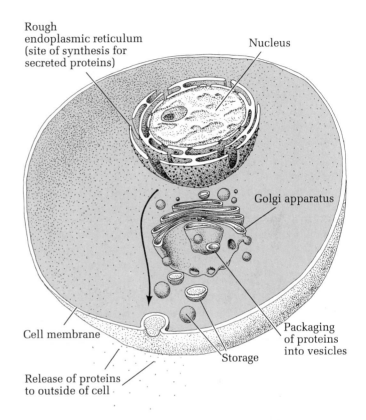

Figure 13.20 Movement of secretory proteins. These proteins move from their site of synthesis on the rough endoplasmic reticulum (RER), through the Golgi apparatus, into vesicles, and then into the extracellular space. The general path of the proteins to be secreted is shown by an arrow.

til the nascent polypeptide-SRP-ribosome complex reaches and binds to the ER. The SRP recognizes an integral membrane protein of the ER called the **docking protein,** and the association of the SRP and docking protein facilitates the polypeptide's signal sequence and associated ribosome's binding to the ER. Translation now resumes, and the SRP is released. Once this occurs, the growing polypeptide (with its signal sequence) is translocated through the membrane into the cisternal space of the ER. This translocation step is accompanied by the attachment of polysomes to the ER's membrane.

Once the signal sequence is fully into the cisternal space of the ER, it is removed from the

430
*The Genetic
Code and the
Translation
of the
Genetic
Message*

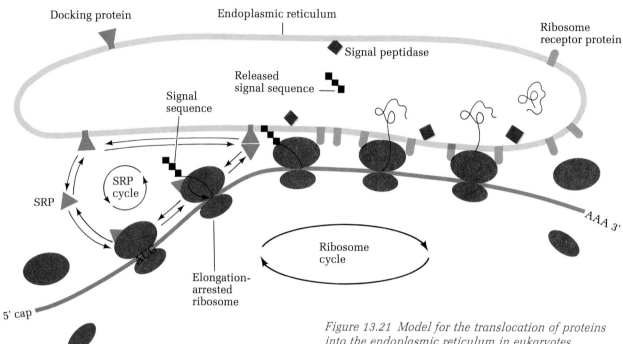

Docking protein

Endoplasmic reticulum

Ribosome receptor protein

Signal peptidase

Released signal sequence —

Signal sequence

SRP cycle

SRP

Ribosome cycle

AAA 3'

AUG

Elongation-arrested ribosome

5' cap

Figure 13.21 Model for the translocation of proteins into the endoplasmic reticulum in eukaryotes.

polypeptide by the action of an enzyme called **signal peptidase.** When the complete polypeptide is entirely within the ER cisternal space, it is typically modified further by the addition of specific carbohydrate groups to produce *glycoproteins.* The glycoproteins must now be sorted for their final destinations. The Golgi complex is the main distribution center where most of the sorting decisions are made. Proteins targeted for the plasma membrane are carried from the Golgi complex to the plasma membrane in transport vesicles. Proteins destined to be secreted are packaged into secretory storage vesicles, which form by budding from the Golgi complex. The secretory vesicles migrate to the cell surface where they fuse with the plasma membrane and release their contents (the proteins to be secreted) to the outside of the cell. Lysosomal proteins are collected into a specific class of Golgi export vesicles, which eventually become functional lysosomes.

Although there is no conclusive information about how sorting decisions are made, there is speculation that a protein destined for a particular location via the ER must contain a sorting

signal. Such signals are thought to be composed of regions on the surface of the protein called *signal patches.* Unlike signal sequences, which are discrete amino acid sequences near the amino end of a growing polypeptide, signal patches are likely to form from noncontiguous regions of the polypeptide that are brought together during protein folding. Signal sequences, then, direct the translocation of polypeptides across membranes because the polypeptides are unfolded during translocation. However, the polypeptides fold into their tertiary configurations after translocation, and it is in that form they are sorted in the ER-Golgi system. Thus, surface features of the proteins, that is, signal patches, could be used to direct their traffic.

Currently, significant progress is being made in defining the precise steps involved in protein translocation across the ER membrane and the subsequent sorting events in yeast through the study of defined genetic mutants, which affect different stages of those processes. This kind of combined genetic–cell biology approach will likely contribute significant new knowledge to this area in the future.

Mitochondrial Proteins

A large number of proteins, including the proteins of the mitochondrial ribosomes and some enzymes of the electron transport chain, find their way into mitochondria. Most proteins destined for mitochondria are synthesized as precursor proteins with specific N-terminal signal sequences that are recognized by the mitochondria and that allow the proteins to be taken up. Like the signal sequences of proteins that enter the ER, these signal sequences contain significant numbers of hydrophobic amino acids. However, unlike secreted proteins and proteins destined for the plasma membrane or lysosomes, mitochondrial proteins are synthesized on non-membrane-bound ribosomes and no ribosomes associate with the mitochondrial outer membrane. The signal sequences of proteins destined for the mitochondria are not recognized by the SRP. Once inside the mitochondria, the mitochondrial signal sequences are cleaved from the proteins.

Thirty to 40 inherited metabolic disorders are characterized by specific deficiencies of imported mitochondrial enzymes. It is possible that some of these disorders result from mutations affecting the signal sequences, making them analogous to the defect responsible for I-cell disease (see p. 427).

Nuclear Proteins

Many proteins find their way into the nucleus, including the ribosomal proteins that are needed for ribosome assembly in the nucleus, the histone and nonhistone chromosomal proteins, and gene regulatory proteins, as well as the many enzymes needed for DNA replication, DNA repair, recombination, transcription, and RNA processing. Nonetheless, the nucleus is extremely selective about which proteins it allows in. Like the other proteins we have discussed, nuclear proteins have simple signal sequences at their N-terminal ends that specify their translocation into the nucleus. The recognition of the signal sequence may be by a component of the nuclear membrane. Some evidence indicates that nuclear proteins enter the nucleus via the nuclear pore complex.

Unlike the other compartmentalized proteins we have discussed, the signal sequences of many nuclear proteins are not removed once they enter the nucleus. The reason for this is that each time the cell divides, the nuclear envelope is degraded, then reforms prior to cytokinesis. Thus, during cell division, the nuclear proteins are free in the cytoplasm, and they must retain their ability to reenter the nucleus selectively once the nuclear envelope reforms.

Keynote *Eukaryotic proteins that enter the endoplasmic reticulum, the mitochondrion, or the nucleus are distinguished from proteins that remain free in the cytoplasm by the presence of specific signal sequences at their N-terminal ends. Characteristically the signal sequences contain a significant number of hydrophobic amino acids. Proteins destined for the ER are translocated into the cisternal space of the ER where the signal sequence is removed by signal peptidase. They are then sorted to their final destinations via the Golgi complex.*

Proteins destined for the mitochondria and for the nucleus also have specific signal sequences to direct them into their respective organelles. Once mitochondrial proteins are inside the mitochondria, the signal sequence is removed. For nuclear proteins, however, the signal sequence remains to allow those proteins to become localized in the nucleus again after cell division.

Protein Secretion in Bacteria

Most bacterial proteins function within the cell. However, a number of proteins are destined either to function in the inner membrane of the cell (membrane proteins) or to be exported from the cell (secreted proteins). The membrane proteins and secreted proteins are synthesized as precursor molecules that have signal sequences that are very similar in principle to those found in eukaryotic membrane proteins and secreted proteins. These proteins typically have signal sequences about 15 to 30 amino acids long at the N-terminal end of the protein. The signal se-

432
*The Genetic
Code and the
Translation
of the
Genetic
Message*

quences usually include a high proportion of hydrophobic amino acids (Table 13.2), which facilitate the association of this region of the protein with the membrane.

The model for insertion of a protein into the membrane or the secretion of a protein in bacteria is similar to that for alike proteins in eukaryotes. That is, as soon as the signal sequence emerges from the ribosome translating the mRNA that codes for the protein, the hydrophobic nature of the signal sequence causes an association of the nascent polypeptide with the membrane (Figure 13.22). This event results in the ribosome's becoming bound to the membrane. As soon as the N-terminus of the protein is embedded in the membrane, protein synthesis continues and, in the case of exported proteins, will push the rest of the polypeptide through to the other side of the membrane. For the membrane proteins that function while embedded in the membrane, there is a second hydrophobic region of the polypeptide that keeps the protein firmly within the membrane. For both types of proteins, the mature protein is completed when the N-terminal signal sequence is removed in a reaction catalyzed by the signal peptidase enzyme.

Keynote *Bacterial proteins that are destined to function in the bacterial membrane or to be secreted from the cell are synthesized as precursor molecules that have an extra 15 to 30 amino acids, called a signal sequence, at their N-terminal end. A number of these extra amino acids are hydrophobic and serve to attach the growing polypeptide and the associated ribosome to the membrane. Subsequent protein synthesis pushes the polypeptide into or through the membrane. The signal sequence is removed by signal peptidase to generate the mature polypeptide molecule.*

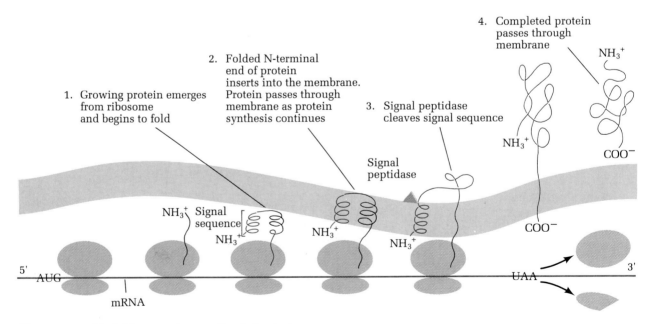

1. Growing protein emerges from ribosome and begins to fold

2. Folded N-terminal end of protein inserts into the membrane. Protein passes through membrane as protein synthesis continues

3. Signal peptidase cleaves signal sequence

4. Completed protein passes through membrane

Signal peptidase

NH_3^+ Signal sequence

NH_3^+

NH_3^+

NH_3^+

NH_3^+

COO^-

COO^-

5' AUG

mRNA

UAA

3'

Figure 13.22 *Model for protein secretion in bacteria.*

433
*Analytic
Approaches
for Solving
Genetics
Problems*

Table 13.2 Signal Sequences of Some Exported Bacterial Proteins*

Protein	Charged Segment	Hydrophobic Segment
INNER MEMBRANE PROTEINS:		
Phage fd, major coat protein	Met Lys Lys Ser Leu Val Leu Lys	Ala Ser Val Ala Val Ala Thr Leu Val Pro Met Leu Ser Phe Ala ↓Ala Glu Gly
Phage fd, minor coat protein	Met Lys Lys	Leu Leu Phe Ala Ile Pro Leu Val Val Pro Phe Tyr Ser His Ser ↓Ala Glu Thr
PERIPLASMIC PROTEINS:		
Alkaline phosphatase	Met Lys	Gln Ser Thr Ile Ala Leu Ala Leu Leu Pro Leu Leu Phe Thr Pro Val Thr Lys Ala ↓Arg Thr Pro
Maltose binding protein	Met Lys Ile Lys Thr Gly Ala Arg	Ile Leu Ala Leu Ser Ala Leu Thr Thr Met Met Phe Ser Ala Ser Ala Leu Ala ↓Lys Ile Glu
Leucine-specific binding protein	Met Lys Ala Asn Ala Lys	Thr Ile Ile Ala Gly Met Ile Ala Leu Ala Ile Ser His Thr Ala Met Ala ↓Asp Asp Ile
β-lactamase of pBR322	Met Ser Ile Gln His Phe Arg	Val Ala Leu Ile Pro Phe Phe Ala Ala Phe Cys Leu Pro Val Phe Ala ↓His Pro Glu
OUTER MEMBRANE PROTEINS:		
Lipoprotein	Met Lys Ala Thr Lys	Leu Val Leu Gly Ala Val Ile Leu Gly Ser Thr Leu Leu Ala Gly ↓Cys Ser Ser
LamB	Met Met Ile Thr Leu Arg Lys	Leu Pro Leu Ala Val Ala Val Ala Ala Gly Val Met Ser Ala Gln Ala Met Ala ↓Val Asp Phe
OmpA	Met Lys Lys	Thr Ala Ile Ala Ile Ala Val Ala Leu Ala Gly Phe Ala Thr Val Ala Gln Ala ↓Ala Pro Lys

* The hydrophobic portion (color) of the signal sequence is invariably preceded by a short charged segment. The arrows indicate the sites of cleavage by signal peptidase.

Source: S. Michaelis and J. Beckwith, *Annu. Rev. Microbiol.* 36 (1982):435.

Analytical Approaches for Solving Genetics Problems

This chapter is primarily descriptive, and hence only a limited number of quantitative problems are suitable for this material. The following question and answer are therefore largely illustrative.

Q.1 a. How many of the 64 codon permutations can be made from the three nucleotides A, U, and G?

b. How many of the 64 codon permutations can be made from the four nucleotides A, U, G, and C, with one or more C's in each codon?

A.1 a. This question involves probability. There are four bases, so the probability of a cytosine at the first position in a codon is 1/4. Conversely, the probability of a base other than cytosine in the first position is $(1 - 1/4) = 3/4$. These same probabilities apply to the other two positions in the codon. Therefore the probability of a codon without a cytosine is $(3/4)^3 = 27/64$.

b. This question involves the relative frequency of codons that have one or more cytosine. We have already calculated the probability of a codon *not* having a cytosine, so all the remaining codons have one or more cytosines. The answer to this question, therefore, is $(1 - 27/64) = 37/64$.

434
The Genetic
Code and the
Translation
of the
Genetic
Message

Questions and Problems

*13.1 Proteins are (choose the correct answer):

a. Branched chains of nucleotides
b. Linear, folded chains of nucleotides
c. Linear, folded chains of amino acids
d. Invariably enzymes

*13.2 The form of genetic information used directly in protein synthesis is (choose the correct answer):

a. DNA
b. mRNA

c. rRNA
d. Ribosomes

13.3 The process in which ribosomes engage is (choose the correct answer):

a. Replication
b. Transcription
c. Translation

d. Disjunction
e. Cell division

13.4 What are the characteristics of the genetic code?

13.5 Base-pairing wobble occurs in the interaction between the anticodon of the tRNAs and the codons. On the theoretical level, determine the minimum number of tRNAs needed to read the 61 sense codons.

13.6 Antibiotics have been very useful in elucidating the steps of protein synthesis. If you have an artificial messenger of the sequence of AUGUUUUUUUUUUUUU . . . , it will produce the following polypeptide in a cell-free, protein-synthesizing system: fMet-Phe-Phe-Phe . . . . In your search for new antibiotics you find one called putyermycin, which blocks protein synthesis. When you try it with your artificial mRNA in a cell-free system, the product is fMet-Phe. What step in protein synthesis does putyermycin affect? Why?

13.7 Describe the reactions involved in the aminoacylation (charging) of a tRNA molecule.

13.8 Compare and contrast the following in prokaryotes and eukaryotes: (a) protein synthesis initiation; (b) protein synthesis elongation; (c) protein synthesis termination.

13.9 Discuss the two species of methionine tRNA, and describe how they differ in structure and function. In your answer, include a discussion of how each of these tRNAs binds to the ribosome.

*13.10 Random copolymers were used in some of the experiments that revealed the characteristics of the genetic code. For each of the following ribonucleotide mixtures, give the expected codons and their frequencies, and give the expected proportions of the amino acids that would be found in a polypeptide directed by the copolymer in a cell-free, protein-synthesizing system:

a. 4 A:6 C   c. 1 A:3 U:1 C
b. 4 G:1 C   d. 1 A:1 U:1 G:1 C

*13.11 Other features of the reading of mRNA into proteins being the same as they are now (i.e., codons must exist for 20 different amino acids), what would the minimum WORD (CODON) SIZE be if the number of different bases in the mRNA were, instead of four:

a. Two
b. Three
c. Five

13.12 Suppose that at stage A in the evolution of the genetic code only the first two nucleotides in the coding triplets led to unique differences and that *any* nucleotide could occupy the third position. Then, suppose there was a stage B in which differences in meaning arose depending upon whether a purine (A or G) or pyrimidine (C or T) was present at the third position. Without reference to the number of amino acids or multiplicity of tRNA molecules, how many triplets of different meaning can be constructed out of the code at stage A? At stage B?

*13.13 A gene makes a polypeptide 30 amino acids long containing an alternating sequence of phenylalanine and tyrosine. What are the sequences of nucleotides corresponding to this sequence in:

a. The DNA strand which is read to produce the mRNA, assuming Phe = UUU and Tyr = UAU in mRNA
b. The DNA strand which is not read
c. tRNA

*13.14 Antibiotics have been useful in determining whether cellular events depend on transcription or translation. For example, actinomycin D is used to block transcription, and cycloheximide (in eukaryotes) is used to block translation. In some cases, though, surprising results are obtained after antibiotics are administered. The addition of actinomycin D, for example, may result in an increase and not a decrease in the activity of a particular enzyme. Discuss how this result might come about.

Genetic
Fine Structure
and Gene
Function

In the past few chapters we have learned about the structure of genetic material and how it replicates and is transcribed and translated to produce proteins, which, as we have discovered, play a crucial role in the structure and functions of cells and organisms. We have also viewed genes from a molecular perspective without much regard to their hereditary properties. Until the mid 1950s and early 1960s the results of genetic analysis had suggested that the gene was the *unit of function, mutation,* and *recombination.* Indeed, it was thought that genes were rather like beads on a string, with recombination occurring between the beads and mutation causing a change in the bead, thereby resulting in an alteration in gene expression.

Following the development of the molecular model for DNA in 1953, geneticists were able to develop new experimental approaches to investigate the nature of the gene. Thus the mid 1950s and early 1960s saw a series of classical genetic experiments that investigated the fine structure of the gene, that is, the detailed molecular organization of the gene as it relates to the mutational, recombinational, and functional events in which the gene is involved. These experiments indicated that the gene was not like a bead on a string but, rather, was a linear sequence of nucleotides in DNA, a concept that had not been proved by any other set of genetic experiments. In the past few years, of course, the nucleotide sequences of many genes have been obtained from DNA-sequencing procedures.

In this chapter we examine genetic fine structure and also some aspects of gene function. We will encounter the classical experiments that demonstrated that genes code for enzymes and for other proteins. In particular, we examine the involvement of particular sets of genes in directing and controlling a particular biochemical pathway; that is, the series of enzyme-catalyzed steps required for the breakdown or synthesis of a particular chemical compound. We will see that the gene, rather than working in isolation as we have regarded it in the past few chapters, must often work in cooperation with other genes in order for cells to function properly.

Fine-Structure Analysis of a Gene

The genetic-mapping experiments described in Chapter 5, 6, and 7 all involved the use of mutant alleles of different genes: The recombinational mapping of the distance between genes, called *intergenic mapping* (inter = between), can be used to construct chromosome maps for organisms. The same general principles of recombinational mapping can be applied to mapping the distance between mutational sites within the same gene, a type of mapping called *intragenic mapping* (intra = within).

Intragenic mapping is possible because each gene consists of many nucleotide pairs of DNA linearly arranged along the chromosome. This fact was not always known. In the 1950s and early 1960s, Seymour Benzer performed a series of experiments to define mutational and recombinational sites within a gene. In essence, his genetic experiments gave new insights into what alleles were and laid to rest the concept of a gene being a bead on a string.

In his experiments Benzer used phage T4, principally because bacteriophages produce large numbers of progeny, thereby facilitating the potential determination of very low recombination frequencies. Benzer's initial experiments involved the detailed genetic mapping of sites within a gene, an approach called **fine-structure mapping.**

When cells of *Escherichia coli* strain *B* or strain *K12*(λ), growing on solid medium in a petri dish, are infected with wild-type T4, small turbid plaques with fuzzy edges are produced as a result of the phage life cycle (see Figure 14.1). [Strain *E. coli K12*(λ) contains the lambda phage chromosome incorporated into the bacterial chromosome.] On the other hand, when cells of *E. coli* strain *B* are infected with *r* (rapid-lysis) mutants of phage T4, large, clear plaques with distinct edges are produced (Figure 14.1). From intergenic-mapping experiments, the *r* mutations were found to map at several locations in the phage's genome, defining several *r* genes (see Figure 7.22).

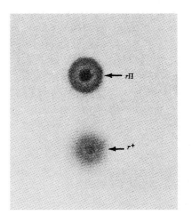

Figure 14.1 The r⁺ *and mutant* rII *plaques on a lawn of* E. coli B. *The* r⁺ *plaque is turbid with a fuzzy edge, and the* rII *plaque is larger, clear, with a distinct boundary.*

In his experiments Benzer used particular *r* mutants from two adjacent genes in the *rII* region of the T4 map. These two genes, *rIIA* and *rIIB,* affect the same phenotypic trait. Of special importance to the experiments that we are about to describe was the finding by Benzer in 1953 that *rII* mutants, in addition to their *plaque morphology* phenotypic distinction from wild-type (*r⁺*) T4, also have distinct *host range properties;* that is, they can only grow on certain host strains of *E. coli.* Specifically, while wild-type T4 is able to grow in and lyse cells of either *E. coli B* or *K12*(λ), *rII* mutants can grow in and lyse cells of *B* (producing large, clear plaques on solid medium: Figure 14.1) but are unable to grow in cells of *K12*(λ). Strain *B* is defined as the permissive host for *rII* mutants, while strain *K12*(λ) is the nonpermissive host. Thus *rII* mutants are conditional mutants because they can grow under one set of conditions but not others. The reasons for the inability of *rII* phages to multiply in the *K12*(λ) strain are not known, although it is known that the presence of the incorporated lambda phage genome is necessary for blocking the phage's propagation.

Recombination Analysis of *rII* Mutants

Benzer realized that the growth defect of *rII* mutants on *E. coli K12*(λ) could serve as a powerful selective tool for detecting the presence of a very small proportion of r^+ phages within a large population of rII mutants. Between 1953 and 1963 Benzer collected thousands of rII mutants, some of which had arisen spontaneously and some of which had been induced by mutagen treatment.

Initially, Benzer set out to construct a fine-

structure genetic map of the rII region. To do so, he crossed 60 independently isolated rII mutants in all possible pairwise combinations in $E.\ coli$ B (the permissive host), and he collected the progeny phages once the cells had lysed (Figure 14.2). (For a phage such as T4, about 10^{10}–10^{11} phages per milliliter of phage lysate would be typical.) For each cross he plated a sample of the phage progeny on $E.\ coli\ B,$ the permissive host, in order to count the total number of progeny phage per milliliter. He plated another sample of the phage progeny on $E.\ coli\ K12(\lambda)$, the nonpermissive host, to find the total number of

Figure 14.2 General procedure used by Benzer for determining the number of r^+ recombinants from a cross involving two rII *mutants of T4.*

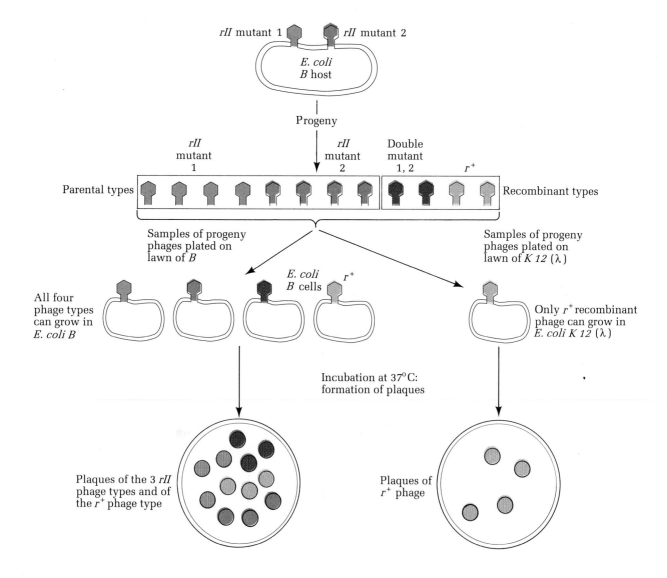

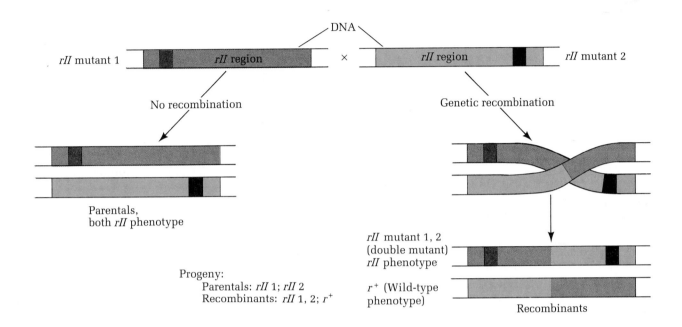

Figure 14.3 Production of parental and recombinant progeny from a cross of two rII *mutants with mutation in different sites within the* rII *region. Progeny phages of parental genotype are produced if no crossing-over occurs, whereas progeny phages with recombinant genotypes are produced if a single crossover occurs between the sites of the two mutations.*

the very rare r^+ recombinants per milliliter resulting from genetic recombination between the mutations carried by the two *rII* mutants used in the cross. In this way Benzer was able to calculate the percentage of very rare r^+ recombinants between closely linked genetic sites.

For each cross of two *rII* mutants, such as *rII*1 and *rII*2 (Figure 14.3), four genetic classes of progeny were usually found: the two parental *rII* types (*rII*1 and *rII*2) and two recombinant types resulting from a single crossover. One of the recombinant types was the wild type (r^+), and the other was a double mutant carrying both *rII* mutations (*rII*1, 2), which had an *r* phenotype and was indistinguishable phenotypically from one or the other of the parental *rII* mutants. Recall from Chapters 5, 6, and 7 that the map distance between two mutations is given by the percentage of recombinant progeny among all progeny

from crosses of the two mutants. For the crosses of *rII* mutants a single crossover event between the two mutations will give the r^+ and double *rII* mutant recombinants (see Figure 14.3). Therefore the frequencies of the two recombinant classes of phages are expected to be the same. Consequently, the total number of recombinants from each *rII* × *rII* cross is approximated by twice the number of r^+ plaques counted on plates of strain *K12*(λ). Thus the general formula for the map distance between two *rII* mutations is

$$\frac{2 \times \text{number of } r^+ \text{ recombinants}}{\text{total number of progeny}} \times 100\%$$

$$= \text{map distance (in map units)}$$

From the recombination data obtained from all possible pairwise crosses of the initial set of 60 *rII* mutants, Benzer was able to construct a linear genetic map (Figure 14.4). Among the mutants some pairs produced no r^+ recombinants when they were crossed. This result was interpreted to mean that those pairs carried mutations at exactly the same site; that is, the same nucleotide pair in the DNA had been changed; there was no possibility of recombination between the mutations. Mutations that change the

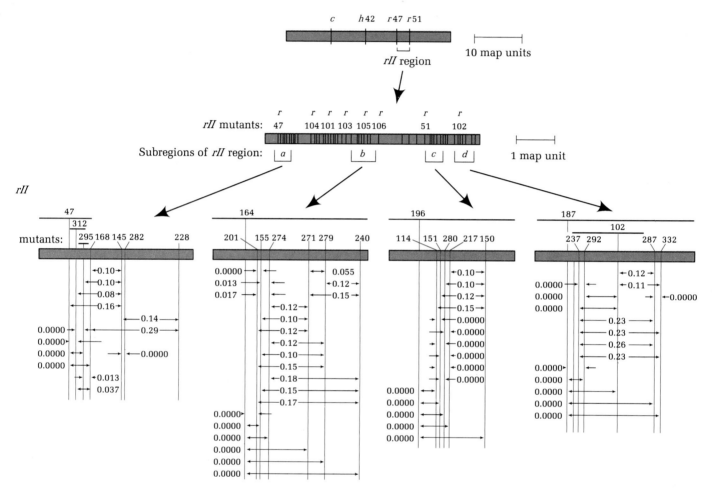

Figure 14.4 Preliminary fine-structure genetic map of the rII *region of phage T4 derived by Benzer from crosses of an initial set of 60* rII *mutants. Lower levels in the figure show finer detail of the map. In the lowest level the numbered vertical lines indicate individual* rII *mutants, and the decimals indicate the percentage of* r+ *recombinants found in crosses between the two* rII *mutants connected by an arrow.*

same nucleotide pair within a gene are termed *homoallelic.* Most pairs of *rII* mutants did produce r^+ recombinants when crossed, indicating that the mutations they carried had altered different nucleotide pairs in the DNA. Mutations that change different nucleotide pairs within a gene are termed *heteroallelic.* The map showed that the lowest frequency with which r^+ recombinants were formed in any pairwise crosses of

rII mutants carrying heteroallelic mutations was 0.01 percent.

The minimum map distance figure of 0.01 percent can be used to make a rough calculation of the molecular distance—the distance in base pairs—involved in the recombination event. From extensive mapping experiments with T4, its circular genetic map is known to be about 1500 map units. If two *rII* mutants produce 0.01 percent r^+ recombinants, the mutations are separated by 0.02 map units, or by about 0.02/1500 = 1.3×10^{-5} of the total T4 genome. Since the total T4 genome contains about 2×10^5 base pairs, the minimum recombination distance is $(1.3 \times 10^{-5}) \times (2 \times 10^5)$, or about 3 base pairs; genetic recombination can occur in distances of 3 base pairs or less. Since genes were known to consist, typically, of a few hundred base pairs,

Benzer's mapping data clearly indicated that recombination could occur *within* a gene. We now know that the nucleotide pair is the *unit of recombination.* Equally important, Benzer's work and its extensions established that the nucleotide pair is also the *unit of mutation.* These definitions clearly replace the classical ones that the whole gene is the unit of recombination and the unit of mutation.

Deletion Mapping

Following his initial series of crossing experiments, Benzer continued to map the over 3000 *rII* mutants he had so that he could complete his fine-structure map. To map these 3000 mutants would have required approximately 5 million crosses—an overwhelming task even in phages, with which up to 50 crosses can be done per day. Therefore Benzer developed some genetic tricks to simplify his mapping studies. These tricks involved the use of *deletion mapping* to localize unknown mutations.

Most of the *rII* mutants isolated by Benzer were point mutants; their phenotype resulted from an alteration of a single nucleotide pair. A point mutant can revert to the wild-type state spontaneously or following treatment with an appropriate mutagen. However, some of Benzer's *rII* mutants did not behave as point mutants in that they did not revert and they did not produce r^+ recombinants in crosses with a number of *rII* point mutants that were known to be located at different places on the *rII* map. These mutants were *deletion mutants* that involved the loss of a section of DNA. A wide range in the extent and location of deleted genetic material was found among the *rII* deletion mutants studied by Benzer. Some deletion mutants are shown in Figure 14.5.

In actual practice an unknown *rII* point mutant was first crossed with each of seven standard deletion mutants that defined seven main segments of the *rII* region (segments *A1–A6* and *B* in Figure 14.5). For example, if an *rII* point mutant gave r^+ recombinants when crossed with deletions defining *A6* and *B* (*rA105* and *r638,* respectively) but not with the other five

deletions, the point mutant must be in segment *A5.* That is to say, r^+ recombinants can only be produced in crosses with deletions if the deleted segment does not include the region of DNA containing the point mutation. If a point mutant does not produce r^+ recombinants with the deletions *r1272, r1241, rJ3, rPT1,* and *rPB242,* then the point mutation must be in the segment of DNA that these mutants lack. The fact that r^+ recombinants are produced with *rA105* and *r638* indicates that the point mutation must be in the *A5* region that is not missing in either deletion mutant.

Once the main segment in which the mutation occurred was known, the point mutant was crossed with each of the relevant secondary set of reference deletions, *r1605, r1589,* and *rPB230* (Figure 14.5). With segment *A5,* for example, there are three deletions that divide *A5* into the four subsegments *A5a* through *A5d.* The presence or absence of r^+ recombinants in the progeny of the crosses of the *A5 rII* mutant with the secondary set of deletions enabled Benzer to more precisely localize the mutation to a smaller region of the DNA. For example, if the mutation was in segment *A5c,* then r^+ recombinants will be produced with deletion *rPB230* but not with either of the other two deletions. Yet other deletion mutants defined smaller regions of each of the four subsegments *A5a* through *A5d;* for example, *A5c* was divided into *A5c1, A5c2a1, A5C2a2,* and *A5c2b* by deletions *r1993, r1695,* and *r1168* (Figure 14.5).

In three sequential sets of crosses of point mutants with deletion mutants, it was possible to localize any given *rII* point mutant to one of 47 regions defined by the deletions, as shown in Figure 14.6. Then all those point mutants within a given region could be crossed in all possible pairwise crosses to construct a detailed genetic map. In this way Benzer used the greater than 3000 *rII* mutants to prove that the *rII* region is subdivisible into more than 300 mutable sites (nucleotide pairs) that were separable by recombination (Figure 14.7, p. 444.). The distribution of mutants is not random; certain sites (called hot spots) are represented by a large number of independently isolated point mutants.

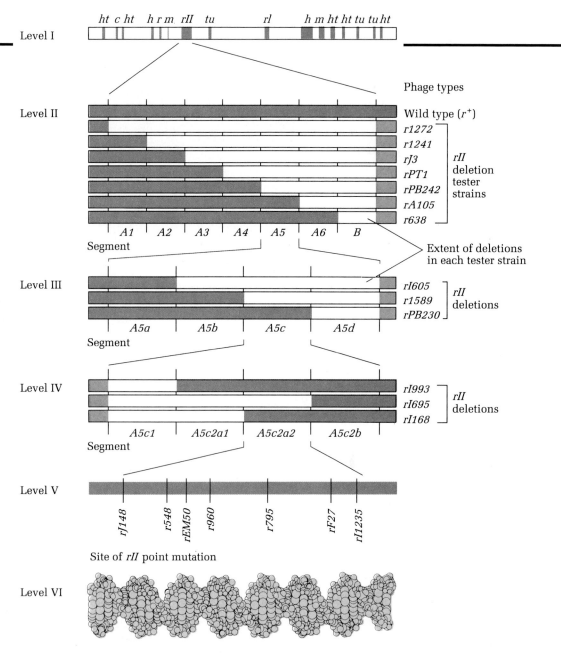

Figure 14.5 Segmental subdivision of the rII *region of
phage T4 by means of deletion. Level I shows the
whole T4 genetic map. In Level II seven deletions define seven segments of the* rII *region. In Level III three
deletions define four subsegments of the* A5 *segments.
In Level IV three deletions define four subsegments of
the* A5c *subsegment. Level V shows the order and
spacing of the sites of the* rII *mutations in the* A5c2a2
*subsegment, as established by pairwise crosses of
seven point mutants. Level VI is a model of the DNA
double helix to indicate the approximate scale of the
level V map.*

Keynote *Benzer's detailed recombination analysis of the rII region of bacteriophage T4 indicated that the unit of mutation and of recombination is the base pair in DNA. This definition replaced the classical view that genes were indivisible by mutation and recombination.*

Defining Units of Function by Complementation Tests

From the classical point of view the gene is a *unit of function;* it is the sequence of nucleotide pairs in DNA that specifies the sequence of amino acids in the functional protein coded for by the gene. Benzer designed genetic experiments to determine whether this classical view was indeed true of the *rII* region. To find out whether two different *rII* mutants belonged to the same gene (unit of function), Benzer adapted the **cis-trans** or **complementation test** developed by E. Lewis to study the nature of the functional unit of the gene in *Drosophila.* For clarification of the following discussion, it will help to know that the complementation tests indicated that the *rII* region consists of two genes (units of function), *rIIA* and *rIIB.* A mutation at any point in either gene will produce the *rII* phenotypes [the production of large, clear plaques on *E. coli* B and the inability to grow in *E. coli* *K12*(λ)]. In

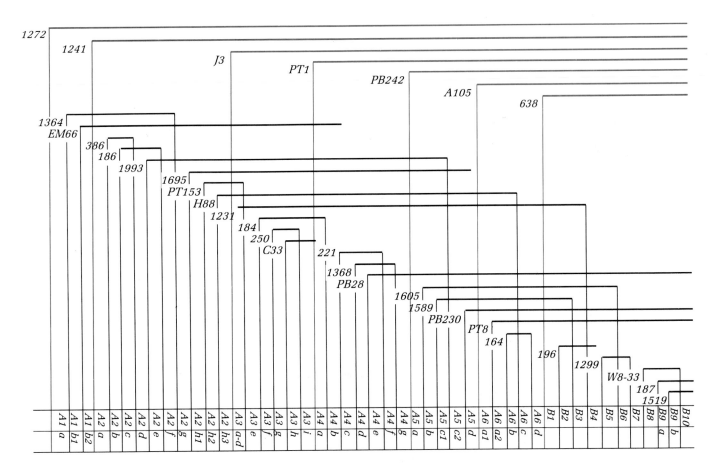

Figure 14.6 *Map of deletions used to divide the* rII *region into 47 small segments (shown as small boxes at the bottom of the figure).*

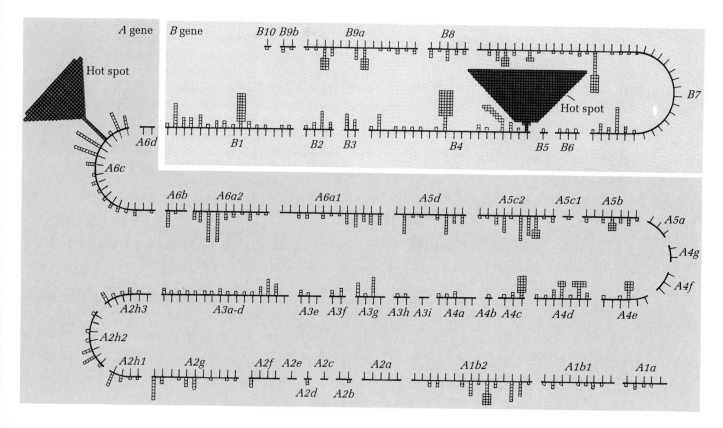

Figure 14.7 Fine-structure map of the rII *region derived from the extensive experiments of Benzer. The number of independently isolated mutations that mapped to a given site is indicated by the number of blocks at the site. Hot spots are represented by a large number of blocks.*

other words, there are two distinct genes in the *rII* region, *rIIA* and *rIIB,* each of which specifies a functional product needed for growth in *E. coli K12*(λ).

The complementation test, then, is used to establish how many units of functions (genes) define a given set of mutations expressing the same mutant phenotypes. In Benzer's work with the *rII* mutants the nonpermissive strain *K12*(λ) was infected with a pair of *rII* mutants to see whether the two mutants, each alone unable to grow in strain *K12*(λ), are able to work together to produce progeny phages. If the phages multiply and produce progeny, the two mutants are

said to complement each other, meaning that the two mutations must be in different genes (units of function) that specify different functional products. If no progeny phages are produced, the mutants have not complemented, indicating that the mutations are in the same functional unit.

These two situations are diagrammed in Figure 14.8. In the first case (Figure 14.8a), complementation occurs because the *rIIA* mutant still makes a functional *B* product and the *rIIB* mutant makes a functional *A* product. Thus between the two mutants both products necessary for phage propagation in *E. coli K12*(λ) are generated, and progeny phages are assembled and released. In the second case (Figure 14.8b), no complementation occurs because, while both produce a functional *rIIB* product, the two mutants have in common the lack of *A* function, thus preventing phage reproduction in *E. coli K12*(λ).

On the basis of the results of complementation

tests, Benzer found that *rII* mutants fall into two units of functions, *rIIA* and *rIIB* (also called complementation groups, which in this case directly correspond to genes). That is, all *rIIA* mutants complement all *rIIB* mutants. In contrast, *rIIA* mutants fail to complement other *rIIA* mutants, and *rIIB* mutants fail to complement other *rIIB* mutants. The dividing line between the *rIIA* and *rIIB* units of function is indicated in the fine-structure map of Figure 14.7. Point mutants and deletion mutants in the *rII* region obey the same rules in the complementation tests. The only exceptions are deletions that span parts of

both the *A* and the *B* functional units. These deletion mutants do not complement either *A* or *B* mutants.

For the complementation test examples shown in Figure 14.8, each of the two phages that co-infect the nonpermissive *E. coli* strain *K12*(λ) carries an *rII* mutation, a configuration of mutations called the *trans* configuration. As a control, it is usual to co-infect *E. coli K12*(λ) with an r^+ (wild-type) phage and an *rII* mutant phage carrying *both* mutations to see whether the expected wild-type function results. When both mutations under investigation are carried by the same phage, the configuration is called the *cis* configuration of mutations. (Because of the *cis* and *trans* configurations of mutations used, the complementation test is also called the *cis-trans* test.) The r^+ is expected to be dominant over the two mutations carried by the *rII* mutant phage and that progeny phages will be pro-

---

Figure 14.8 Complementation tests for determining the units of function in the rII *region of phage T4; the nonpermissive host* E. coli K12(λ) *is infected with two different* rII *mutants: (a) Complementation occurs; (b) complementation does not occur.*

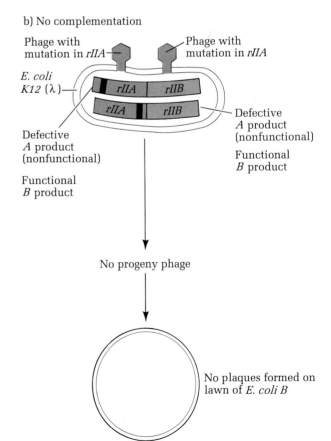

duced. Thus the mutants which are non-complementary in *trans* are in the same functional unit, i.e. in the same cistron.

Benzer referred to the genetic unit of function revealed by the cis-trans test as the *cistron.* When it was coined in 1955, the term *cistron* became widely used and replaced the ambiguous term *gene* in many writings. At the present time, *gene* is commonly used and *cistron* is being used less. Nonetheless, *gene* and *cistron* are equivalent in referring to the genetic unit of function. It is appropriate to refer to the *A* and *B* functional units as the *rIIA* and *rIIB* cistrons or genes. Presumably, their two products act in common processes necessary for T4 propagation in strain *K12*(λ). Genetically, the *rIIA* cistron is about 6 map units and 800 base pairs, and the *rIIB* cistron is about 4 map units and 500 base pairs.

The complementation test is commonly used to define the functional units (complementation groups or genes) for mutants with the same phenotype. The principles for performing a complementation test are always the same; only the practical details of performing the test are organism-specific. For example, in yeast two haploid cells of differing mating types (*a* and α) and carrying different mutations conferring the same mutant phenotype would be mated to produce a diploid, which would then be analyzed for complementation of the two mutations. In animal cells two cells, each exhibiting the same mutant phenotype, can be fused together and analyzed for wild-type or mutant phenotype; a wild-type phenotype would indicate that complementation had occurred.

Keynote *The complementation, or cis-trans, test is used to determine how many units of functions (genes) define a given set of mutations expressing the same mutant phenotypes. If two mutants, each carrying a mutation in a different gene, are combined, the mutations will complement and a wild-type function will result. If two mutants, each carrying a mutation in the same gene, are combined, the mutations will not complement and the mutant phenotype will be exhibited.*

Gene Control of Enzyme Structure

Our studies so far have given us a firm notion of what a gene is at the DNA level, how it is expressed, and how genetic analysis can provide insights about the relationships between mutations and phenotypic change. Now we can analyze some of the genetic data that showed that genes code for enzymes and for nonenzymatic proteins.

Garrod's Hypothesis of Inborn Errors of Metabolism

In 1902 Archibald Garrod, an English physician, provided the first evidence of a specific relationship between genes and enzymes. (As we now know, genes specify the amino acid sequence of all proteins, including enzymes.) Garrod studied *alkaptonuria,* a human disease characterized by urine that turns black upon exposure to the air and by the tendency to develop arthritis later in life.

In studying the occurrence of alkaptonuria in families of individuals with the disease, Garrod and his colleague W. Bateson discovered two interesting facts that indicated to them that alkaptonuria is a genetically controlled trait: (1) Several members of the families had alkaptonuria, and (2) the disease was much more common among children of marriages involving first cousins than among children of marriages between unrelated partners. This finding was significant because first cousins have many genes in common and, therefore, the chance is greater for recessive genes to become homozygous in children of first-cousin marriages.

Garrod then found that people with alkaptonuria excrete all ingested homogentisic acid (HA) in their urine whereas normal people excrete none. Moreover, he showed that it is the HA in the urine that turns black in the air. This result indicated to Garrod that normal people are able to metabolize HA to its breakdown products but that people with alkaptonuria cannot, indicating that alkaptonurics lack the enzyme that metabolizes HA. Figure 14.9 shows part of the biochemical pathway in which HA is involved and the step blocked in people with alkaptonuria. From his results and the genetic evidence Garrod concluded that alkaptonuria is a genetic disease that results from the absence of a particular enzyme in the steps in the metabolism of HA to its breakdown products. In Garrod's terms this result is an example of an *inborn error of metabolism.* We now know that the mutation responsible for alkaptonuria is recessive, so only people homozygous for the mutant gene express the defect. The gene has been localized to an autosome, but which autosome is unknown.

Garrod also studied three other human genetic diseases that affected biochemical processes, and in each case he was able to conclude correctly that a metabolic pathway was blocked. An important aspect of Garrod's analysis of these human diseases was his understanding that the position of a block in a metabolic pathway can be determined by the accumulation of the chemical compound (HA in the case of alkaptonuria) that precedes the blocked step.

Genetic Control of *Drosophila* Eye Pigments

After Garrod's work the next significant piece of evidence linking genes and enzymes was obtained in 1935. In that year George Beadle and Boris Ephrussi published the results of their experiments on gene-enzyme relationships in a biochemical pathway responsible for the synthesis of eye pigments in the fruit fly *Drosophila melanogaster.* Wild-type *Drosophila* have brick red eyes, the result of the blending of two distinct pigments, bright red and brown.

At the time of these experiments three genes were known to be involved in the production of brown eye pigment. Mutations in any one of these genes resulted in the absence of brown pigment and, as a consequence, flies with bright orange or scarlet eyes. Mapping experiments had shown that the three genes occupy three distinct loci. In other words, the genes are not clustered even though they are involved in the same bio-

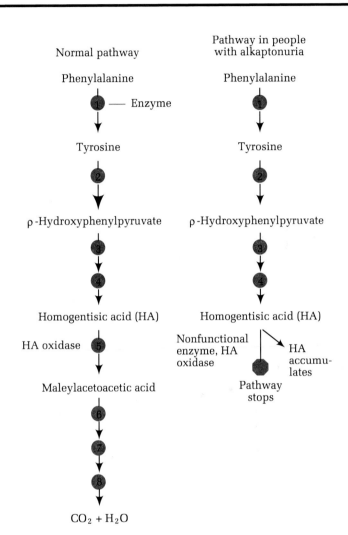

Figure 14.9 Part of the phenylalanine-tyrosine metabolic pathways, showing the biochemical steps that operate in normal individuals (left) and the metabolic block found in individuals with alkaptonuria (right).

chemical pathways. Each mutant allele is recessive to its wild-type allele, so the mutant phenotype is only exhibited in flies homozygous for the mutation. The three mutant, brown-pigment genes are designated *st* (scarlet), *cn* (cinnabar), and *v* (vermilion). The wild-type forms of these genes are designated *st*$^+$, *cn*$^+$, and *v*$^+$, respectively.

First, Beadle and Ephrussi isolated two groups of cells, called *imaginal disks,* from *Drosophila* larvae, cells that develop into the two eyes in an adult. When an eye disk is transplanted from a larva into the abdomen of a second larva, the disk will develop into a recognizable eye structure that can be isolated from the abdomen of the adult that develops from the second larva. Beadle and Ephrussi transplanted imaginal eye disks taken from each of the three types of mutant larvae (*st, cn,* and *v*) separately into a wild-type larval host and analyzed the eye color of the eye structure produced by the transplanted disk once the wild-type host had become an adult. They also investigated the fate of a wild-type disk transplanted into a mutant larval host. The results of their studies are summarized in Table 14.1.

In the case of the *st* mutant, the *st* disk in the wild-type host developed into a scarlet eye (Figure 14.10). In the reciprocal transplant, the wild-type implant developed into a wild-type eye structure with a wild-type, brick red eye color. These results meant that the wild-type host was not able to provide substances to the mutant *st* disk to enable the brown pigment to be produced. The transplanted *st* disk developed according to the genetic information in its cells and was not affected by the environment. When a cell develops in this way, it is said to undergo **autonomous development.**

The results of similar transplants of mutant *v*

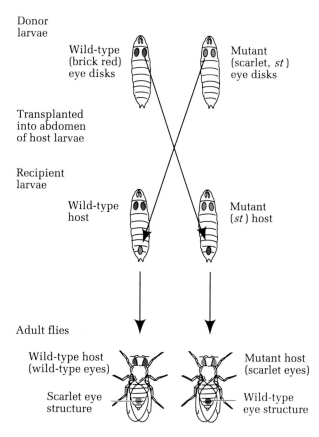

Figure 14.10 *Imaginal eye disk transplant experiment in* Drosophila, *showing autonomous development.*

Table 14.1 Results of Eye Disk Transplantation Experiments Involving Eye Color Mutants of Drosophila

Source of Eye Disk	Host Fly	Color of Transplanted Eye after Metamorphosis
+	*v*	+
v	+	+
+	*cn*	+
cn	+	+
cn	*v*	Cinnabar
v	*cn*	+
+	*st*	+
st	+	Scarlet

and *cn* disks (Figure 14.11a and b) illustrate **nonautonomous development:** The development of a cell's phenotype is affected by the environment in which it develops. When eye disks are transplanted from either *v* or *cn* larvae into a wild-type host, they develop into eye structures with wild-type eye color. In contrast, wild-type eye disks transplanted into *cn* or *v* hosts exhibit autonomous development, producing wild-type, colored eye structures. The explanation is that the wild-type host is able to provide substances that the *v* and *cn* disks can use to bypass the genetic block they have and can, therefore, make brown pigment. The needed substances must diffuse into the implanted disk from the host circulation.

Next, Beadle and Ephrussi did reciprocal

a) Vermilion mutant experiment

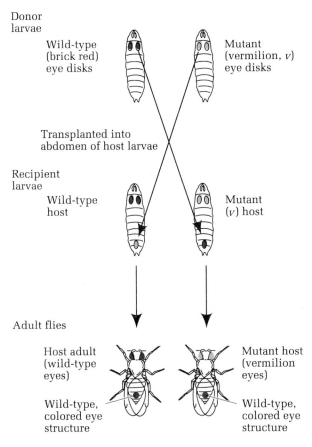

b) Cinnabar mutant experiment

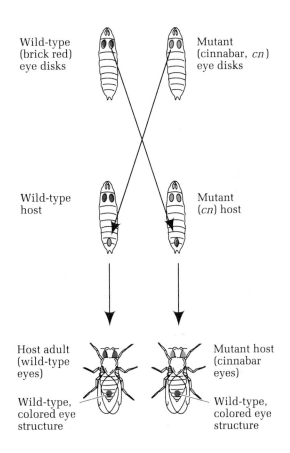

Figure 14.11 Imaginal eye disk transplant experiments in Drosophila, *showing nonautonomous development: (a) reciprocal transplant between the wild type and the vermilion mutant* v; *(b) reciprocal transplant between the wild type and the cinnabar mutant* cn.

transplants of eye disks between *v* and *cn* larvae (see Figure 14.12). These experiments showed that *v* disks transplanted into *cn* hosts develop nonautonomously into wild-type, colored eye structures; this result shows that *cn* larvae can provide the *v* disks with some substance needed for wild-type pigment production. In the reciprocal experiment, however, *cn* disks transplanted into *v* hosts develop autonomously into *cn* eyes; this result shows that *v* larvae cannot provide *cn* disks with the material necessary to

produce wild-type eye pigment.

As a result of these experiments, Beadle and Ephrussi concluded that the production of the brown pigment necessary for wild-type brick red eyes involves a biochemical pathway with at least two precursor pigment substances. Wild-type flies possess both precursors, *cn* flies have only one precursor, and *v* flies have neither. Figure 14.13 shows the postulated relationships in the gene reaction steps.

As the figure shows, the v^+ gene codes for a product that makes v^+ substance, and the cn^+ gene makes a product that converts the v^+ substance to the cn^+ substance, which is then converted to the brown pigment. This hypothesis explains how a wild-type, colored eye structure develops from a mutant *v* disk transplanted into a *cn* host. The implanted disk comes from a fly

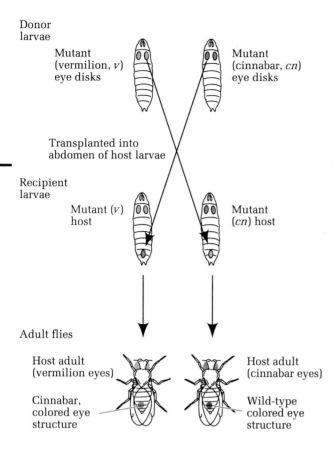

with a mutation in the *v* gene so that the *v*⁺ substance cannot be made, but it has the wild-type *cn*⁺ gene. The *cn* host has a mutant *cn* gene, but because it has a normal *v*⁺ gene, it is able to make the *v*⁺ substance. This substance diffuses into the implanted *v cn*⁺ disk, where it is converted to the *cn*⁺ substance. Since the genes (including the *st*⁺ gene) for the rest of the pathway are also normal in the implanted *v* disk the *cn*⁺ substance is converted to the brown pigment, and the eye derived from the implanted disk is wild type. Stated in another way, the *cn* host is able to make up for the deficiency of the *v* disk by supplying it with a diffusible substance so that it can develop into a wild-type colored eye.

In the reciprocal transplant, a *cn* disk implanted into a *v* host cannot convert its own *v*⁺ substance to the *cn*⁺ substance because it possesses the mutant *cn* gene. Because the *v* host is deficient in the production of *v*⁺ substance, the

Figure 14.12 Reciprocal imaginal eye disk transplant experiments involving v *and* cn *mutants of* Drosophila, *showing nonautonomous development of* v *disks in a* cn *host and autonomous development of* cn *disks in a* v *host.*

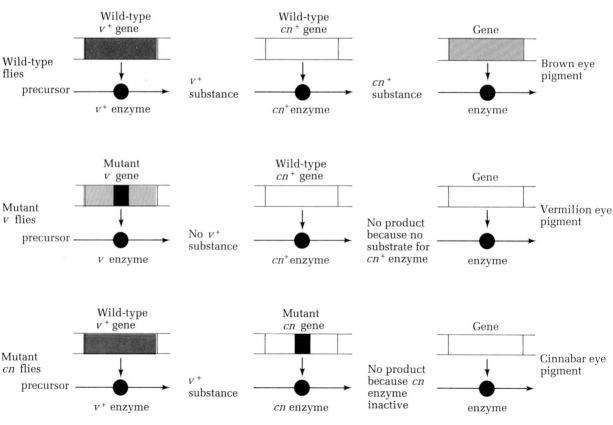

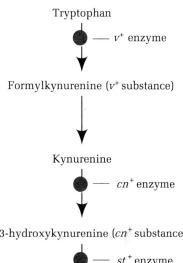

Tryptophan

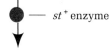 — v^+ enzyme

Formylkynurenine (v^+ substance)

Kynurenine

 — cn^+ enzyme

3-hydroxykynurenine (cn^+ substance)

— st^+ enzyme

Hydroxyxanthommatin

Xanthommatin
(brown pigment that
mixes with red pigment
to produce wild-type eyes)

Figure 14.14 Actual biochemical pathway for the production of the brown eye pigment in Drosophila *from the amino acid tryptophan. The steps catalyzed by the enzymes encoded by the* v+, cn+, *and* st+ *genes are indicated.*

host also cannot produce the cn^+ substance that the implanted cn disk would need to give rise to a wild-type, colored eye structure, and thus the resulting eye structure is cinnabar (bright orange), showing autonomous development.

From subsequent biochemical analysis researchers were able to identify the chemical nature of the v^+ and cn^+ substances, to work out the biochemical pathway, and to localize the steps catalyzed by the v^+ and cn^+ gene products (see Figure 14.14). The v mutation results in a nonfunctional enzyme, with the consequence that tryptophan cannot be converted to formylkynurenine. Similarly, the cn mutation results in a nonfunctional enzyme that catalyzes a subsequent step in the pathway so that kynurenine cannot be converted to 3-hydroxykynurenine. Lastly, the st (scarlet) mutation affects the enzyme that catalyzes the step after the cn^+ step; in st mutants, 3-hydroxykynurenine cannot be converted to hydroxyxanthommatin.

In all three mutants no brown pigment is produced because the pathway cannot be completed. The eye colors produced in each mutant are slightly different because at each step a slight coloration is added to the bright orange pigment that is produced by a separate biochemical pathway controlled by a distinct set of genes. In other words, the closer the pathway functions toward the brown pigment itself, the more colored the compound is and the more the eye color resembles the wild-type color.

One Gene–One Enzyme Hypothesis

The work on *Drosophila* eye pigments led to more refined studies on the relationship between genes and enzymes that are generally regarded as heralding the beginnings of biochemical genetics, a branch of genetics that combines genetics and biochemistry to elucidate the nature of metabolic pathways. These studies, car-

Figure 14.13 Proposed biochemical sequence and relationships in the gene reaction steps for wild-type, v, *and* cn *flies, as deduced from the results of the Beadle and Ephrussi transplantation experiments diagramed in Figures 14.10–14.12.*

ried out by George Beadle and Edward Tatum with the fungus *Neurospora crassa,* showed that there was a direct relationship between genes and enzymes and resulted in the *one gene–one enzyme hypothesis,* an important landmark in the history of genetics. In Chapter 5 (pp. 147–149) we discussed the life cycle of *N. crassa* in detail. Briefly, *Neurospora* is a haploid organism that propagates asexually or sexually.

Isolation of nutritional mutants of *Neurospora.* One important attribute of *Neurospora* is that it has simple growth requirements. Wild-type *Neurospora,* by definition, is prototrophic, meaning that it can grow and propagate on a *minimal medium* that contains a basic set of ingredients (inorganic salts, a carbon source such as glucose or sucrose, and the vitamin biotin). Beadle and Tatum reasoned that *Neurospora* synthesized the materials it needed for growth (amino acids, nucleotides, vitamins, nucleic acids, proteins, etc.) from the chemicals present in the minimal medium. They also realized that it should be possible to isolate *nutritional mutants* (i.e., auxotrophs) of *Neurospora:* mutant strains that required nutritional supplements in the culture medium in order to grow. These auxotrophic mutants could be isolated because they would not grow on the minimal medium.

Figure 14.15 shows how Beadle and Tatum isolated and characterized nutritional mutants. They treated asexual spores (conidia) with X rays to induce genetic mutants; then they crossed the cultures derived from the surviving spores with a wild-type (prototrophic) strain of the opposite mating type. This sexual cross was done because they wanted to study the genetics of biochemical events, and they had to identify those nutritional mutants that were heritable. By crossing the mutagenized spores with the wild type, they ensured that any nutritional mutant they isolated had gone through a cross and therefore had a genetic basis and did not exhibit the requirement for a nutrient for nongenetic reasons.

Each progeny spore from the crosses was allowed to germinate in a nutritionally complete medium that contained all necessary amino acids, purines, pyrimidines, and vitamins in addition to sucrose, salts, and biotin found in minimal medium. Thus any strain that could not make one or more of these compounds from the basic ingredients found in minimal medium could still grow by using the compounds supplied in the growth medium. Each culture grown from a progeny spore was then tested for growth on minimal medium. Those strains that did not grow were assumed to be auxotrophic mutants, and these mutants were, in turn, individually tested for their abilities to grow on various supplemented minimal media. In this screening the media used were minimal medium plus amino acids and minimal medium plus vitamins. Theoretically, an amino acid auxotroph—a mutant strain that requires a particular amino acid in order to grow—would grow on minimal medium plus amino acids but not on either of the two other media. Vitamin auxotrophs would only grow on minimal medium plus vitamins, and so on.

Having categorized the strains into nonauxotrophs, amino acid auxotrophs, and vitamin auxotrophs, Beadle and Tatum next conducted a second round of screening to specify which chemical the strains needed to be supplemented with to grow. Let us imagine an amino acid auxotroph. To find out which of the 20 amino acids is required for a particular amino acid auxotroph to grow, we would test the strain in 20 tubes, each containing minimal medium plus one of the 20 amino acids. For example, a tryptophan auxotroph would be identified if it grows only in the tube containing minimal medium plus tryptophan.

Lastly, Figure 14.16 shows the method used to confirm that an auxotroph identified by the above procedures has a genetic rather than a nongenetic basis. For a tryptophan auxotroph, the auxotrophic strain is crossed with the wild-type strain of opposite mating type. As a result of meiosis and ascus development, the eight spores from a single meiosis (a meiotic tetrad) can be isolated and analyzed as described in Chapter 6 (pp. 184–194). If the auxotrophic property (the requirement to have tryptophan in the medium in order to grow) is caused by a gene mutation, then half the spores should be wild types and half should be auxotrophs. This hypothesis is tested by first germinating the spores individually on minimal medium plus tryptophan so all will grow and then checking for the ability of each progeny strain to grow on unsupplemented minimal medium. On this latter medium only the wild-type strains will grow; the auxotrophic strains will not. If, however, the tryptophan auxotroph is not the result of a gene mutation, then all progeny spores will likely be wild types—or at least there will not be a 4:4 segregation of wild type:auxotroph when tetrads are tested for growth on minimal medium.

Genetic dissection of a biochemical pathway.
Once Beadle and Tatum had isolated and identified nutritional mutants, they set out to investigate the biochemical pathways affected by the mutations. They assumed that *Neurospora* cells,

Figure 14.15 Method devised by Beadle and Tatum to isolate auxotrophic mutations in Neurospora. *Here the mutant strain isolated is a tryptophan auxotroph.*

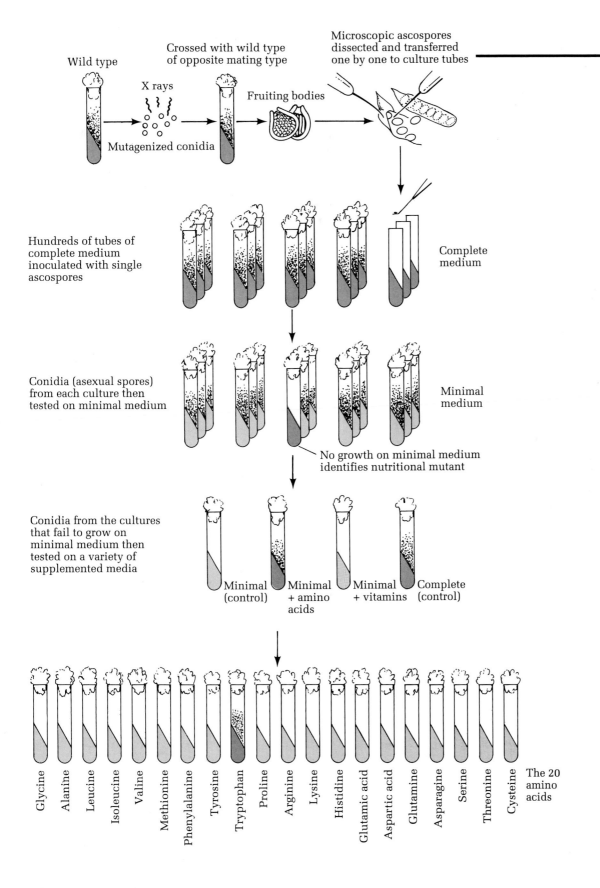

Wild type

X rays

Mutagenized conidia

Crossed with wild type
of opposite mating type

Fruiting bodies

Microscopic ascospores
dissected and transferred
one by one to culture tubes

Hundreds of tubes of
complete medium
inoculated with single
ascospores

Complete
medium

Conidia (asexual spores)
from each culture then
tested on minimal medium

Minimal
medium

No growth on minimal medium
identifies nutritional mutant

Conidia from the cultures
that fail to grow on
minimal medium then
tested on a variety of
supplemented media

Minimal
(control)

Minimal
+ amino
acids

Minimal
+ vitamins

Complete
(control)

Glycine

Alanine

Leucine

Isoleucine

Valine

Methionine

Phenylalanine

Tyrosine

Tryptophan

Proline

Arginine

Lysine

Histidine

Glutamic acid

Aspartic acid

Glutamine

Asparagine

Serine

Threonine

Cysteine

The 20
amino
acids

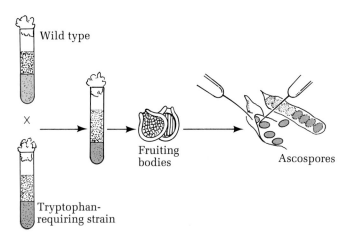

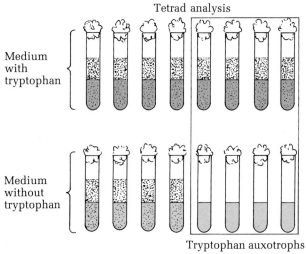

Figure 14.16 Procedure used for confirming the genetic basis of a nutritional defect in Neurospora.

the enzymes necessary for carrying out the steps in a biochemical pathway the **one gene–one enzyme hypothesis.** That is, a gene controls the production and/or activity of one enzyme. Consequently, mutations that result in the loss of enzyme activity lead to the accumulation of precursors in the pathway (and to possible side reactions) as well as to the absence of the end product of the pathway.

Using the one gene–one enzyme hypothesis, we can make some predictions about the consequences of mutations affecting enzymes in the hypothetical pathway shown. If, for example, a mutation in gene C results in a nonfunctional enzyme C, then product B in the pathway cannot be converted to the end product C. The consequences to the organism are that the organism now requires substance C to be provided in its growth medium in order to grow, and that product B accumulates in the cells since it cannot be converted to C. Mutations in gene A or gene B will also result in auxotrophy for chemical C.

However, the three mutants A, B, and C are distinguishable on other grounds, namely, their ability to grow or not to grow on various intermediates in the pathway. Gene C mutants can grow only if supplemented with C; gene B mutants can grow if supplemented with either B or C; and gene A mutants can grow on minimal medium plus A, B, or C. A reciprocal pattern of pathway precursor accumulation occurs in these strains. Gene C mutants accumulate product B; gene B mutants accumulate product A; and gene A mutants accumulate the initial substrate for the pathway. In actual experiments where the pathway is unknown, the logic is carried out in the opposite direction. As a result, from the pattern of growth supplementation and precursor accumulation, the sequence of steps in a pathway can be deduced.

Let us take an actual example in the biochemical pathway for the synthesis of the amino acid arginine in *Neurospora crassa*. Figure 14.18 shows this pathway along with the genes that code for the enzymes that catalyze each step. The starting point is a set of arginine auxotrophs. Genetic crosses and complementation tests determine that four distinct genes are in-

like all cells, function by the interaction of the products of a very large number of genes. Furthermore, they imagined that wild-type *Neurospora* converted the constituents of minimal medium into amino acids and so forth by a series of reactions organized into biochemical pathways. In this way the synthesis of cellular components occurs by a series of small steps, each catalyzed by its own enzyme. As the hypothetical pathway in Figure 14.17 shows, the product of each step is used as the substrate for the next enzyme. Beadle and Tatum called the proposed relationship between an organism's genes and

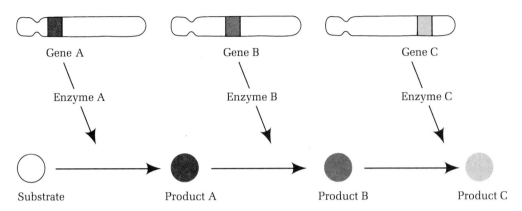

Figure 14.17 Hypothetical biochemical pathway for the conversion of a precursor substrate to an end product C in three enzyme-catalyzed steps. Each enzyme is coded for by one gene.

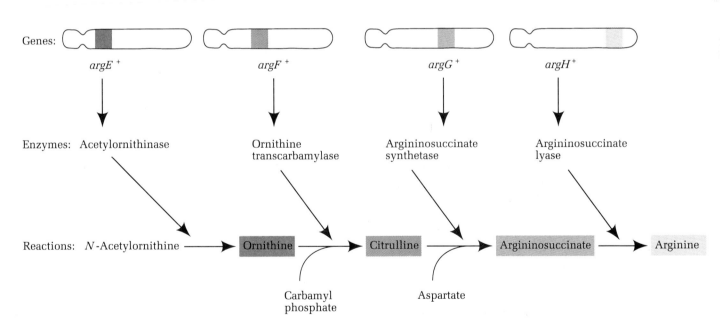

Figure 14.18 Arginine biosynthetic pathway, with the four genes in Neurospora crassa *that code for the enzymes that catalyze the reactions shown.*

volved; a mutation in any one of them gives rise to auxotrophy for arginine. These four genes in a wild-type cell are designated $argE^+$, $argF^+$, $argG^+$, and $argH^+$.

Next, the sequence of biochemical steps in the pathway can be deduced by the growth pattern of the mutant strains on media supplemented with presumed arginine precursors. Table 14.2 shows the results obtained for this set of four mutants. By definition, all four mutant strains can grow on arginine, and none can grow on unsupplemented minimal medium.

As shown in the table, the *argH* mutant strain can grow when supplemented with arginine but not when supplemented with any of the intermediates in the pathway. This result indicates that the *argH* gene codes for the enzyme that controls the last step in the pathway, which leads to the formation of arginine. The *argG* mutant strain grows on media supplemented with arginine or argininosuccinate, the *argF* mutant strain grows on media supplemented with arginine, argininosuccinate, or citrulline, and the *argE* strain grows on arginine, argininosuccinate, citrulline, or ornithine. The later in a pathway the mutant strain is blocked, the fewer are the number of intermediate chemicals in the pathway that can be added to the growth medium to enable the strain to grow. The earlier in the pathway the mutant strain is blocked, the larger is the number of intermediates in the pathway that can be added to enable the mutant strain to grow. With these results in mind we conclude that argininosuccinate must be the immediate precursor to arginine since it permits all but the *argH* mutant to grow.

By similar reasoning, the pathway can be hypothesized to be as shown in Figure 14.18. Gene *argF*$^+$ codes for the enzyme that converts ornithine to citrulline. An *argF* mutant strain can, therefore, grow on minimal medium plus citrul-line, argininosuccinate, or arginine. As was discussed before, the intermediate chemical prior to the step blocked by the genetic mutation generally accumulates, and that information can be used to confirm the conclusions about the sequence of steps in a pathway. The *argF* mutant strains, for example, accumulate ornithine, and hence ornithine would be placed prior to citrulline, argininosuccinate, and arginine in the pathway.

With this sort of approach we can dissect a biochemical pathway genetically—that is, determine the sequence of steps in the pathway and relate each step to a specific gene or genes. We should not conclude, though, that there is necessarily one gene for each step in a pathway, although that is usually true. In some cases an enzyme may consist of more than one polypeptide chain, each of which is coded for by a specific gene. In that event more than one gene would specify that enzyme. Therefore Beadle and Tatum's original name for their hypothesis is modified to the *one gene–one polypeptide hypothesis.*

Keynote A number of classical studies indicated the specific relationship between genes and enzymes, eventually embodied in Beadle and Tatum's one gene–one enzyme hypothesis, which states that each gene controls the synthesis or activity of a single enzyme. Since enzymes may consist of more than one polypeptide, a more modern title for this hypothesis is the one gene–one polypeptide hypothesis.

Table 14.2 Growth Responses of Arginine Auxotrophs

Mutant Strain	Growth Response on Minimal Medium and . . .				
	Nothing	Ornithine	Citrulline	Arginino-succinate	Arginine
Wild-type	+	+	+	+	+
argE	–	+	+	+	+
argF	–	–	+	+	+
argG	–	–	–	+	+
argH	–	–	–	–	+

Genetically Based Enzyme Deficiencies in Humans

Many genetic diseases in humans are caused by a single gene mutation that alters the function of an enzyme. In general, an enzyme deficiency caused by a mutation may have simple, or pleiotropic, consequences. Table 14.3 presents several of these diseases (which Garrod would have called inborn errors of metabolism). The studies of these diseases have offered further evidence that some genes code for enzymes.

Phenylketonuria

Like alkaptonuria, *phenylketonuria (PKU)* is a genetic disease caused most commonly by a mutation in the gene for phenylalanine hydroxylase that results in a block in a metabolic pathway for the metabolism of the amino acids phenylalanine and tyrosine. As a result, phenylalanine cannot be converted to tyrosine (see Figure 14.19a and b).

An individual with PKU cannot make tyrosine, an amino acid required for protein synthesis and for the production of the hormones thyroxine and adrenaline and the skin pigment melanin. This aspect of the phenotype is not very serious because tyrosine can be obtained from food. Yet food does not normally have a lot of tyrosine; people with PKU make relatively little melanin, since only tyrosine rather than phenylalanine and tyrosine in food can be used in melanin synthesis. Hence people with PKU tend to have very fair skin and blue eyes (even if they have brown-eye genes). In addition, they have relatively low adrenaline levels.

Figure 14.19 Phenylalanine-tyrosine metabolic pathways: (a) biochemical steps that operate in normal individuals; (b) metabolic block in the pathway exhibited by individuals with phenylketonuria—phenylalanine cannot be metabolized to tyrosine and unusual metabolites of phenylalanine accumulate; (c) metabolic block in the pathway exhibited by individuals with albinism—no or very little melanin pigment is produced.

a) Normal pathway

Phenylalanine

Enzyme, phenylalanine hydroxylase

Tyrosine

Pathway of Figure 14.9

Melanin (brown pigment)

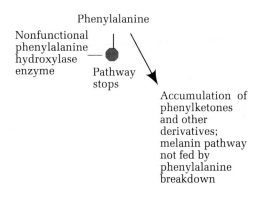

b) Pathway in people with phenylketonuria (PKU)

Phenylalanine

Nonfunctional phenylalanine hydroxylase enzyme

Pathway stops

Accumulation of phenylketones and other derivatives; melanin pathway not fed by phenylalanine breakdown

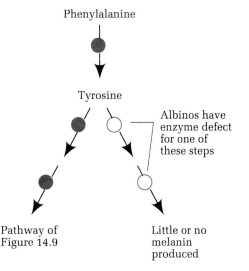

c) Pathway in people with albinism

Phenylalanine

Tyrosine

Albinos have enzyme defect for one of these steps

Pathway of Figure 14.9

Little or no melanin produced

Table 14.3 Selected Human Genetic Disorders with Demonstrated Enzyme Deficiencies

Condition	Enzyme Deficiency	Condition	Enzyme Deficiency
Acid phosphatase deficiency	Acid phosphatase	Leigh's necrotizing encephalomelopathy[a]	Pyruvate carboxylase
Albinism	Tyrosinase	Lesch-Nyhan syndrome[a]	Hypoxanthine phosphoribosyl transferase
Alkaptonuria	Homogentisic acid oxidase	Lysine intolerance	Lysine: NAD-oxidoreductase
Ataxia, intermittent[a]	Pyruvate decarboxylase	Male pseudohermaphroditism	Testicular 17,20-desmolase
Disaccharide intolerance	Invertase	Maple sugar urine disease[a]	Keto acid decarboxylase
Fructose intolerance	Fructose-1-phosphate aldolase	Orotic aciduria[a]	Orotidylic decarboxylase
Fructosuria	Liver fructokinase	Phenylketonuria	Phenylalanine hydroxylase
G6PD deficiency (favism)[a]	Glucose-6-phosphate dehydrogenase	Porphyria, acute[a]	Uroporphyrinogen I synthetase
Glycogen storage disease[a]	Glucose-6-phosphatase	Porphyria, congenital erythropoietic[a]	Uroporphyrinogen III cosynthetase
Gout, primary	Hypoxanthine phosphoribosyl transferase	Pulmonary emphysema	α-1-Antitrypsin
Hemolytic anemia	Glutathione peroxidase	Pyridoxine-dependent infantile convulsions	Glutamic acid decarboxylase
Hemolytic anemia	Hexokinase	Pyridoxine-responsive anemia	λ-Aminolevulinic synthetase
Hemolytic anemia	Pyruvate kinase	Kidney tubular acidosis with deafness	Carbonic anhydrase B
Hypoglycemia and acidosis	Fructose-1,6-diphosphatase	Ricketts, vitamin D–dependent	25-Hydroxycholecalciferol 1-hydroxylase
Immunodeficiency[a]	Adenosine deaminase	Tay-Sachs disease[a]	Hexosaminidase A
Immunodeficiency	Purine nucleoside phosphorylase	Thyroid hormone synthesis, defect in	Iodide peroxidase
Immunodeficiency	Uridine monophosphate kinase	Thyroid hormone synthesis, defect in	Deiodinase
Intestinal lactase deficiency (adult)	Lactase	Tyrosinemia	para-Hydroxyphenylpyruvate oxidase
Ketoacidosis[a]	Succinyl CoA:3-ketoacid CoA-transferase	Xeroderma pigmentosum[a]	DNA-specific endonuclease

[a] Prenatal diagnosis possible or potentially possible. From A. Milunsky, 1976. *N Eng. J. Med.* 295:377. Note that some similar conditions result from various enzyme deficiencies, and single enzyme deficiencies can produce multiple defects.

Source: Modified from J. B. Standbury, J. W. Wyngaarden, and D. S. Fredrickson, eds., 1978. *The Metabolic Basis of Inherited Disease,* 4th ed., Table 1–4. New York: McGraw-Hill.

The absence of phenylalanine hydroxylase also results in the accumulation of phenylalanine. However, unlike the accumulated precursor HA for alkaptonuria, which is excreted in the urine, the accumulated phenylalanine in phenylketonuriacs is converted to a number of phenylalanine derivatives. The accumulation of one of these derivatives, phenylpyruvic acid, drastically affects the cells of the central nervous system and produces the serious symptoms associated with PKU: severe mental retardation, a slow growth rate, and early death.

As with alkaptonuria, PKU is caused by a relatively rare recessive mutation; in Caucasians, only about one in 10,000 newborns are homozygous for the mutant gene and hence have PKU. Heterozygous children born to PKU mothers stand a high risk of having secondary (nongenetic) PKU due to the higher levels of phenylalanine in the mother's blood, which can cause brain damage in fetuses.

In the United States, by law, all newborns must be screened for PKU. This screening assays the infant's blood for phenylalanine hydroxylase or treats their wet diapers with ferric oxide. (In the latter test, diapers of phenylketonuriacs turn green in the presence of the excreted phenylpyruvic acid and other ketones.) A better test, devised by Guthrie, is the *inhibition assay test,* which bases its efficacy on the fact that the growth of a bacterial strain is inhibited in the presence of phenylalanine in the growth medium. Once PKU is identified, it is treated by modifying the infant's diet to limit phenylalanine intake. In this way enough phenylalanine is available for protein synthesis, but it and its derivatives do not accumulate. This dietary correction must be implemented rigidly over a prolonged period of time to be effective. If the modified diet is dropped after about 5 years of age, for example, there is subsequently a small but significant decline in IQ. In sum, PKU is one of the very few genetic diseases that can be treated in a relatively easy way.

Albinism

Albinism is caused by a recessive mutation, and individuals must be homozygous for the mutation in order to exhibit albinism. About one in 17,000 individuals is albino. The mutation is in a gene for an enzyme used in the pathway from tyrosine to the brown pigment melanin (see Figure 14.19c). Since melanin is not produced, albinos have white skin, white hair, and red eyes (owing to the lack of pigment in the iris). Melanin absorbs light in the UV range and is important in protecting the skin against harmful UV irradiation from the sun. Albinos are therefore very light-sensitive. No apparent problems result from the accumulation of precursors in the pathway prior to the block.

There are at least two kinds of albinism since at least two biochemical steps can be blocked to prevent melanin formation. Therefore, if two albinos of different types mate, normal children can be produced as a result of complementation of the two nonallelic mutations.

Lesch-Nyhan Syndrome

Lesch-Nyhan syndrome is a fatal human trait caused by a recessive, single gene mutation on the X chromosome. Only males (who are XY) exhibit Lesch-Nyhan syndrome because females who are homozygous for the mutation die as embryos.

Lesch-Nyhan syndrome is the result of a deficiency in the enzyme hypoxanthine guanine phosphoribosyl transferase (HGPRT), an enzyme essential to the synthesis of purines (see Figure 14.20). When the biosynthetic pathway is highly impaired, as in this case, excess purines accumulate, which are converted to uric acid.

At birth, Lesch-Nyhan individuals are healthy and develop normally for several months. The uric acid excreted in the urine leads to the deposition of orange uric acid crystals in the diapers, an indication that an infant has this genetic disease. Between three and eight months, delays in motor development occur that lead to weak muscles. Later, the muscle tone changes radically, producing uncontrollable movements and involuntary spasms, and, as a result, feeding activities are seriously affected.

After two or three years, Lesch-Nyhan chil-

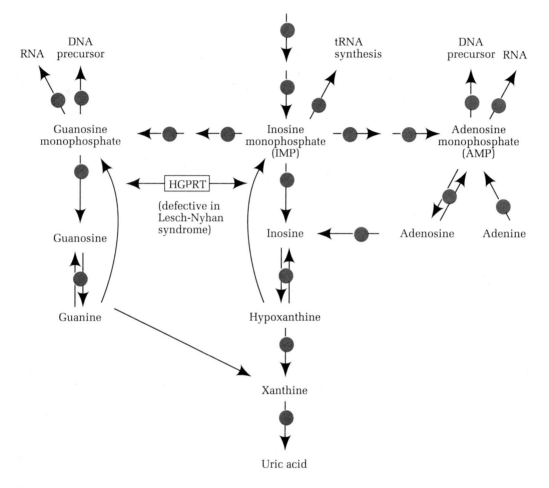

Figure 14.20 Some steps in the purine biosynthetic pathways, showing the reactions catalyzed by hypoxanthine guanine phosphoribosyl transferase (HGPRT), an enzyme that is defective in individuals with Lesch-Nyhan syndrome.

dren begin to show extremely bizarre activity, such as compulsive biting of fingers, lips, and the inside of the mouth. This self-mutilation is difficult to control and is not without severe pain to those afflicted. Typically, aggressive behavior is shown toward others. In intelligence tests the patients score in the severely retarded region, although the difficulties they show in communicating with others may be a major contributing factor to the low scores. Most affected individuals die before they reach their twenties, usually from pneumonia, kidney failure, or uremia (uric acid in the blood).

Understandably, the elevated uric acid levels that result from an HGPRT deficiency might give rise to uremia, kidney failure, and mental deficiency. No clear explanation has been advanced, however, of how such a deficiency can lead to the self-mutilating behavior.

Tay-Sachs Disease

A number of human diseases are caused by mutations in genes that code for lysosomal enzymes, enzymes that function in intracellular digestion. These diseases, collectively called *lysosomal-storage diseases,* are generally caused by a recessive mutation.

The best-known genetic disease of this type is

Tay-Sachs disease, which is caused by homozygosity for a rare recessive mutation that maps to chromosome 15 (an autosome). While Tay-Sachs disease is rare in the population as a whole, it has a high incidence in Jews of Central European origin because of the tendency for these individuals to pair with other individuals of the same origin.

The mutant gene codes for the enzyme *N*-acetylhexosaminidase A (hex A), which catalyzes the reaction shown in Figure 14.21. Enzyme hex A cleaves a terminal *N*-acetylgalactosamine

Figure 14.21 Schematic diagram of the biochemical step for the conversion of the brain ganglioside G_{M2} to the ganglioside G_{M3} catalyzed by the enzyme N-acetylhexosaminidase A (hex A): (a) normal pathway; (b) pathway in individuals with Tay-Sachs disease, where the activity of hex A is deficient, resulting in the abnormal buildup of ganglioside G_{M2} in the brain cells.

group from a brain chemical called a ganglioside. Tay-Sachs infants are deficient in that enzyme activity, so the uncleaved ganglioside accumulates in the brain cells, causing cerebral degeneration and death usually by three years of age.

Individuals with Tay-Sachs disease exhibit a number of clinical symptoms, illustrating the pleiotropic effects of the gene mutation. Typically, the symptom first recognized is an unusually enhanced reaction to sharp sounds. Early diagnosis is also made possible by the presence of a cherry-colored spot on the retina surrounded by a white halo. About a year after birth, a rapid degeneration occurs as the uncleaved ganglioside accumulates and the brain begins to lose control over normal function and activities. This degeneration involves generalized paralysis, blindness, a progressive loss of hearing, and serious feeding problems. By two years of age the infants are essentially immobile,

a) Normal pathway

b) Pathway in individuals with Tay-Sachs disease

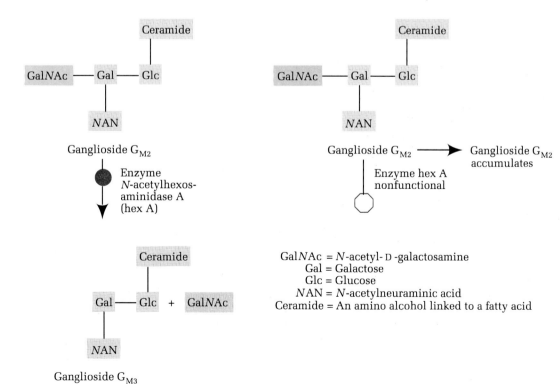

GalNAc = *N*-acetyl-D-galactosamine
Gal = Galactose
Glc = Glucose
NAN = *N*-acetylneuraminic acid
Ceramide = An amino alcohol linked to a fatty acid

and death ensues at about three to four years of age, often from respiratory infections.

No cure is known for Tay-Sachs disease, but as we will see later in this chapter, there are ways to determine whether a fetus has the disease while it is still developing in the uterus.

Genetic Counseling

Genetic counseling refers to the procedures whereby the risks of prospective parents having a child who expresses a genetic disease are evaluated and explained to them. In this chapter we will confine our discussion to analyses involving biochemical assays. In the next chapter we will return to this topic as we discuss analyses involving molecular approaches.

Genetic counseling generally starts with pedigree analysis of both families to ascertain the likelihood that a genetic disease is segregating in one or both families. (Pedigree analysis was described in Chapter 3, pp. 82–90.) Once evidence has been obtained that there is a genetic disease in a family or families, prospective parents need to be informed of the probability that they will produce a child with that disease. Early detection of a genetic disease that has a known biochemical basis occurs at one or both of two levels. One is the detection of heterozygotes (*carriers*) of recessive mutations, and the other is the determination of whether or not the developing fetus shows the biochemical defect. In genetic diseases with known biochemical defects, the two types of procedures are limited to those diseases in which the biochemical defect is expressed in the parents or the developing fetus.

Carrier detection. *Carrier detection* is the detection of individuals who are heterozygous for a recessive-gene mutation. The heterozygous carrier of a mutant gene is normal in phenotype but carries a recessive mutation that does not contribute to the phenotype yet can be passed on to the progeny. In the case of a recessive mutation that has serious deleterious consequences to an individual who is homozygous for that mutation, there is great value in determining whether

people who are contemplating having a child are both carriers, because in that situation a fourth of the children would get the trait. Carrier detection is limited to those cases in which a gene product can be assayed and in which a direct correlation exists between the amount of gene product and the number of copies present of the wild-type gene that codes for the product.

As an example of how carrier detection might proceed in humans, let us suppose that the a^+ gene produces a certain amount of a functional enzyme that can be measured easily and that the mutant gene a is a null allele—it produces no functional enzyme. Homozygous a^+/a^+ individuals would produce twice the amount of enzyme as the heterozygous a^+/a people, and the a/a types would produce none of the enzyme. Therefore, one possible way to identify heterozygotes (the carriers) would be to measure the amount of enzyme they have.

Such measurements, however, are difficult to make. As with many other human characteristics, when one assays for an enzyme among a number of individuals with the same genotype, a range of enzyme activity typically is found. The range of enzyme activity in a^+/a^+ individuals, for example, might be measured by a bell-shaped curve with a high average enzyme activity, while the range of enzyme activity of a^+/a individuals might be measured by a second bell-shaped curve with an average enzyme activity about a half that of a^+/a^+ individuals (Figure 14.22). In most cases heterozygotes can be clearly identified, but at the region of overlap it is impossible to make the crucial distinction between a^+/a^+ and a^+/a.

Carriers of Tay-Sachs disease can be tested in this way because carriers have 50 percent of the normal activity for the enzyme hex A. Similarly, Lesch-Nyhan syndrome carriers may be detected by a lower HGPRT enzyme level, and PKU carriers tend to have elevated levels of phenylalanine in their blood compared with normal individuals.

Identifying carriers is useful since, if both parents are carriers for a recessive genetic disease, the risk is very significant that they will have a child with the genetic disease. From the princi-

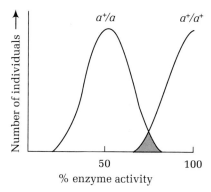

Figure 14.22 Theoretical distribution of enzyme activity in populations of a⁺/a⁺ (homozygous) and a⁺/a (heterozygous) individuals.

ples of genetics the chance is one in four that their progeny will be homozygous for the mutant allele and will have the disease.

Amniocentesis. A second important aspect of genetic counseling is finding out whether a fetus is normal. This test can be done in many instances by a procedure called *amniocentesis* (Figure 14.23). As a fetus develops in the amniotic sac, it is surrounded by amniotic fluid that serves to cushion it against shock. In amniocentesis a sample of the amniotic fluid is taken by carefully inserting a syringe needle through the mother's uterine wall and into the amniotic sac. Ultrasound imaging monitors the position of the fetus so that the needle does not damage it. The amniotic fluid contains cells that have sloughed off the fetus's skin; these cells can be cultured in the laboratory. Once cultured, they are examined for chromosomal abnormalities and for the presence or absence of enzymes. Because amniocentesis is complicated and costly, it is primarily used in high-risk cases.

Amniocentesis analysis can reveal the chromosome composition, or karyotype, of the cells and can thus show any obvious chromosomal abnormalities, such as the extra chromosome 21 found in humans with trisomy-21 (Down's syndrome). In general, chromosome abnormalities have serious consequences on the development and function of an individual, and karyotype

analysis offers early detection of such defects. Additionally, karyotyping gives advance information about the sex of the fetus.

Cells can also be examined for enzyme defects. If, for example, both parents are carriers for a known enzyme deficiency disease and the enzyme is normally detectable in the fetus, then an enzyme assay on an extract of the cells will answer directly whether the fetus will have the disease when it is born. This type of analysis is

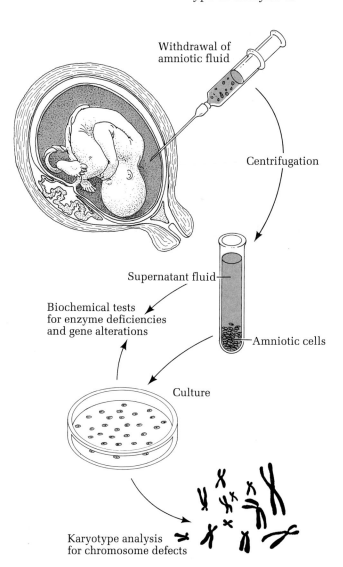

Figure 14.23 Steps in amniocentesis, a procedure used for prenatal diagnosis of genetic defects.

applicable to Tay-Sachs disease, Lesch-Nyhan syndrome, and any other enzyme deficiency disease in which the enzyme involved is expressed in the developing fetus.

Although we can identify carriers of many genetic diseases and can determine if fetuses express genetically based enzyme deficiencies, in most cases there is no cure for the diseases. Carrier detection and amniocentesis serve mainly to inform parents of the risks and probabilities of having a child with the defect. It is then up to the parents to act on the information in the way they see fit. If both are carriers for a serious (perhaps fatal) genetic disease, then they could choose not to have children. If they do conceive a child, they might then have amniocentesis to see if the fetus has the disease. If it does, then they can prepare themselves to rear the child, or they can opt for an abortion.

Keynote *Genetic counseling refers to the procedures whereby prospective parents are informed about the risk that their children might be born with a genetically based disease. Early detection of a genetic disease is done by carrier detection and by amniocentesis, which is also used to determine whether a fetus exhibits any chromosomal abnormalities.*

Gene Control of Protein Structure

The preceding section of this chapter provided evidence that genes code for enzymes; this evidence was given because of the historical significance of those studies. While all enzymes are proteins, however, not all proteins are enzymes. For a complete understanding of how genes function, we will look at the experimental evidence that genes are responsible for the structure or nonenzymatic proteins such as hemoglobin, actin, and myosin (the latter two are muscle proteins). Nonenzymatic proteins are often easier to study than enzymes. Enzymes are usually present in small amounts whereas nonenzymatic proteins occur in relatively large quantities in

the cell, which makes them easier to isolate and purify.

Sickle-Cell Anemia

Sickle-cell anemia is a genetically based human disease that results from an amino acid change in the hemoglobin protein that transports oxygen to the blood.

Symptoms and basis of the disease. Sickle-cell anemia was first described in 1910 by J. Herrick. He found that red blood cells from individuals with the disease lose their characteristic platelet shape in the absence of oxygen and assume the shape of a sickle (Figure 14.24). The sickled red blood cells are more fragile than normal red blood cells and tend to break sooner. In addition, sickled cells are not as flexible as normal cells and therefore tend to jam up in the capillaries rather than squeeze through them. As a consequence, blood circulation is impaired, and tissues served by the capillaries become deprived of oxygen. Although oxygen deprivation occurs particularly at the extremities, the heart, lungs, brain, kidneys, gastrointestinal tract, muscles, and joints can also suffer from oxygen deprivation and its subsequent damage. The afflicted individual may therefore suffer from a variety of health problems including heart failure, pneumonia, paralysis, kidney failure, abdominal pain, and rheumatism.

Sickle-cell anemia is caused by homozygosity for a mutation in the gene for the β polypeptide of hemoglobin. The sickle-cell mutation β^S is codominant with the wild-type allele β^A. Thus people who are heterozygous $\beta^A\beta^S$ make two types of hemoglobins. One, called Hb–A (hemoglobin A), is completely normal, with two normal α chains and two normal β chains specified by two wild-type α genes and one wild-type β gene (β^A). The other, called Hb–S, is the defective hemoglobin with two normal α chains specified by wild-type α genes and two abnormal β chains specified by the mutant β gene β^S. In the heterozygotes $\beta^A\beta^S$ the red blood cells appear normal, although enough defective hemoglobin is present so that the individuals suffer from a

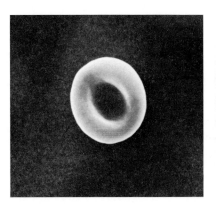
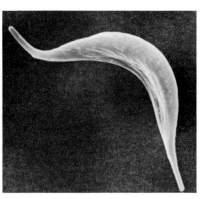

Figure 14.24 *Scanning electron micrograph of normal* (left) *and sickled* (right) *red blood cells.*

mild form of sickle-cell anemia called *sickle-cell* trait; that is, these individuals experience some red blood cell sickling under conditions of very low oxygen, such as when climbing mountains.

The evidence that an abnormal hemoglobin molecule was present in individuals with sickle-cell anemia was obtained by Linus Pauling and his co-workers in the 1950s, using the procedure of electrophoresis to separate the electrically charged protein molecules in an electric field: in this technique, proteins of the same molecular weight but different charge will migrate at different rates. Pauling isolated hemoglobin from the red blood cells of three groups: normal individuals, individuals with sickle-cell trait, and individuals with sickle-cell anemia. He then subjected each sample to electrophoresis. Figure 14.25 shows the results he obtained. Under the conditions he used, the hemoglobin from normal people (Hb–A) migrated more slowly than the hemoglobin from people who had sickle-cell anemia (Hb–S). The hemoglobin from sickle-cell trait individuals behaved in electrophoresis like a 1:1 mixture of Hb–A and Hb–S. Pauling concluded that sickle-cell anemia results from a mutation that alters the chemical structure of the hemoglobin molecule. This experiment was one of the first rigorous proofs that protein structure is controlled by a unique gene.

The precise molecular change in the hemoglobin molecule of sickle-cell anemia individuals

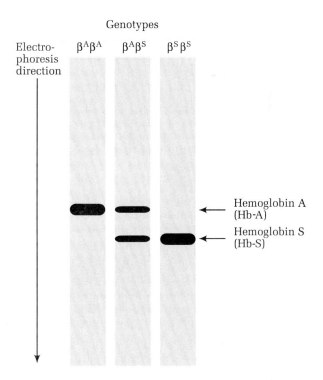

Figure 14.25 *Electrophoresis of hemoglobin found in normal* $\beta^A\beta^A$ *individuals* (left), *in* $\beta^A\beta^S$ *individuals with sickle-cell trait* (center), *and in* $\beta^A\beta^S$ *individuals who exhibit sickle-cell anemia* (right). *The two hemoglobins migrate to different positions in an electric field and hence must differ in electric charge.*

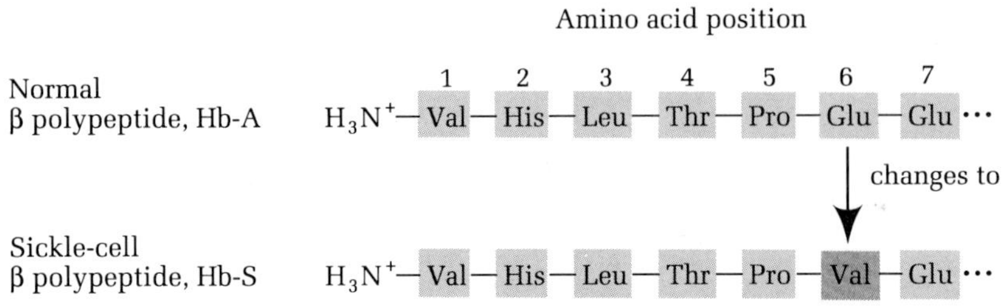

Amino acid position

Normal
β polypeptide, Hb-A

H_3N^+ — Val — His — Leu — Thr — Pro — Glu — Glu ···

changes to

Sickle-cell
β polypeptide, Hb-S

H_3N^+ — Val — His — Leu — Thr — Pro — Val — Glu ···

Figure 14.26 The first seven N terminal amino acids in normal and sickled hemoglobin β polypeptides, showing the single amino acid change from glutamic acid to valine at the amino acid in the sixth position.

was determined by J. Ingram in 1957. He sequenced the α and β polypeptide chains of the two types of hemoglobin, Hb–A and Hb–S, and found that the α chain was identical in the two. In the β chain, however, Ingram found a single amino acid substitution at the sixth position from the N terminal end. Figure 14.26 shows this substitution, which is the replacement of the acidic amino acid glutamic acid with the neutral amino acid valine. The simplest explanation for this change is that the $β^S$ mutation results in the change of a GAG codon (for glutamic acid) to a GUG codon (for valine). This result was undeniable proof that genes specify the amino acid sequence of proteins. Furthermore, since the hemoglobin protein consists of two types of polypeptides, this experiment gave additional support for the one-gene–one-polypeptide hypothesis.

The substitution of valine for glutamic acid results in the pleiotropic consequences characteristic of sickle-cell anemia in the following way. The amino acid at position 6 of the β chain is on the outside of the molecule. In Hb–A (the normal hemoglobin) the glutamic acid at that position is a negatively charged, hydrophilic ("water-loving") molecule. In the folding process that occurs (usually in an aqueous solution), hydrophilic regions of a polypeptide (in this case the β polypeptide) find themselves on the outside of the molecule and hydrophobic ("water-hating") regions generally are folded into the interior of the protein, away from water. Valine is a hydrophobic amino acid, so this

amino acid would prefer to be in the interior of the molecule rather than on the surface. As a consequence, the Hb–S molecules in individuals with sickle-cell anemia tend to align side by side to form long tubules. It is the tubular organization of the hemoglobin molecules that is responsible for the sickling of the red blood cells.

Occurrence of sickle-cell anemia in human populations. The mutant gene responsible for sickle-cell anemia does not have a random distribution among human populations. Instead, it occurs with relatively high frequency in individuals whose ancestors lived for many generations in parts of the world where malaria was present, such as Africa, India, and southern Europe.

Populations in which the mutant gene is common have significant numbers of people with sickle-cell anemia. In the United States these individuals tend to die at a relatively young age, usually about 40 years; in less developed countries, afflicted individuals die much earlier. In addition, females who have sickle-cell anemia and who reach childbearing age tend to abort because insufficient oxygen is available to the developing fetus.

The connection between sickle-cell anemia (an inherited blood disorder) and malaria (a disease that results from a parasite injected by a mosquito) is not as obscure as it might seem. People who are homozygous normals are much more likely to die from malaria than are heterozygous carriers for sickle-cell anemia because the malarial parasite cannot thrive in abnormal red blood cells. As a result, heterozygous individuals in areas where malaria is endemic will survive in higher numbers than those with normal red blood cells. Consequently, the mutant gene continues to survive in the population. For this reason the sickle-cell anemia gene in the American black population, whose origins lie in Africa, has a much higher frequency than in Caucasian or Oriental populations in the United States.

Other Hemoglobin Mutants

Many other mutant hemoglobins have been detected in general screening programs in which hemoglobin is isolated from red blood cells and subjected to electrophoresis. Altered hemoglobins were detected by a difference in the electrophoretic mobility compared with the Hb–A molecule. Over two hundred hemoglobin mutants have been detected through electrophoresis; Figure 14.27 lists some of these mutants along with the amino acid substitutions that have been identified. Some mutations affect both the α and the β chain, and there is a wide variety in the types of amino acid substitutions that occur. From the codon changes that are assumed to be responsible for the substitutions, it is apparent that a single base change is involved in each case.

Not all the hemoglobin mutants that have been identified have as drastic an effect as the sickle-cell anemia mutant because the consequences of the amino acid substitution depend both on the amino acids involved and on the location of the amino acid in the chain. For example, the Hb–C hemoglobin molecule has a change in the β chain from a glutamic acid to a lysine. Like the Hb–S change, this substitution involves the sixth amino acid from the N terminal end of the polypeptide. But unlike the Hb–S change, this change is not a serious defect because both amino acids involved are hydrophilic. Therefore people homozygous for the β^C mutation that results in the abnormal β chain of the Hb–C hemoglobin molecule experience only a mild form of anemia.

Biochemical Genetics of the Human ABO Blood Groups

The mechanism of inheritance of the human ABO blood groups was mentioned in Chapter 4,

pp. 104–106. To summarize, the blood group depends on the presence of chemical substances called antigens on the red blood cell surface. When antigens are injected into a host organism, they may be recognized as foreign and removed from the circulation by the host's antibodies. The molecular basis of the ABO blood groups further supports the direct relationship between genes and amino acid sequences in proteins.

An individual generally has antibodies circulating in blood serum that are directed toward those antigens that the individual is not carrying on his or her red blood cells. People of blood group A (genotypes $I^A I^A$ or $I^A i$) carry the A antigen on their blood cells, and in their serum they have antibodies against the B antigen (the β antibody). Similarly, people of blood group B (genotypes $I^B I^B$ or $I^B i$) have circulating antibodies against the A antigen (the α antibody). An individual with AB blood type (genotype $I^A I^B$), however, has no circulating antibodies. Finally, an individual of blood group O (genotype ii) has neither the A nor the B antigens on their blood cells and so has both the α and β antibodies in circulation. As we learned in Chapter 4, I^A, I^B, and i are all alleles of one gene (the ABO locus) and hence constitute a multiple allelic series.

The ABO locus is involved in the production of enzymes that are used in the biosynthesis of polysaccharides. The enzymes, called *glycosyltransferases,* act specifically to add sugar groups to a preexisting polysaccharide. The polysaccharides important here are those that attach to lipids to produce compounds called glycolipids, which then associate with red blood cell membranes to form the blood group antigens.

Figure 14.28 shows the structures of the terminal sugar organization of the polysaccharide components of the glycolipids involved in the ABO blood type system. Most individuals produce a glycolipid called the H antigen that has the terminal sugar sequence shown in the figure. The I^A allele produces a glycosyltransferase enzyme called α-*N*-acetylgalactosamyl transferase, which recognizes the H antigen and adds the sugar α-*N*-acetylgalactosamine to the end of the polysaccharide to produce the A antigen. The I^B allele, on the other hand, produces an α-D-galac-

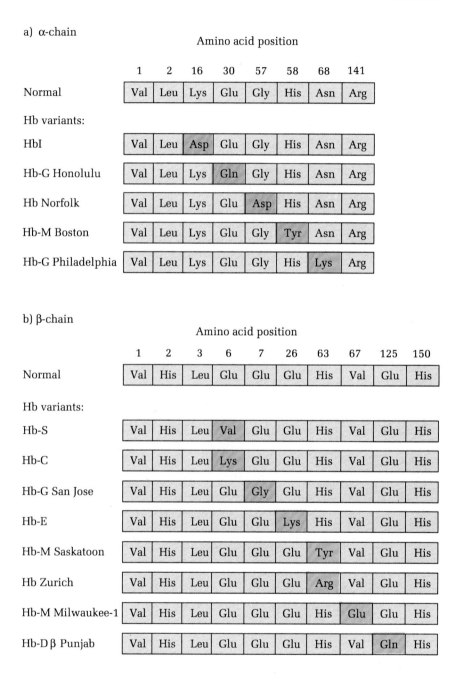

Figure 14.27 Examples of amino acid substitutions found in (a) α *and (b)* β *polypeptides of various human hemoglobin variants.*

tosyltransferase, which also recognizes the H antigen but adds galactose to its polysaccharide to produce the B antigen. (Note that this small difference in the structure of the A and the B antigens is sufficient to be recognized by an individual and to induce an antibody response if the blood types are incompatible.) In both cases

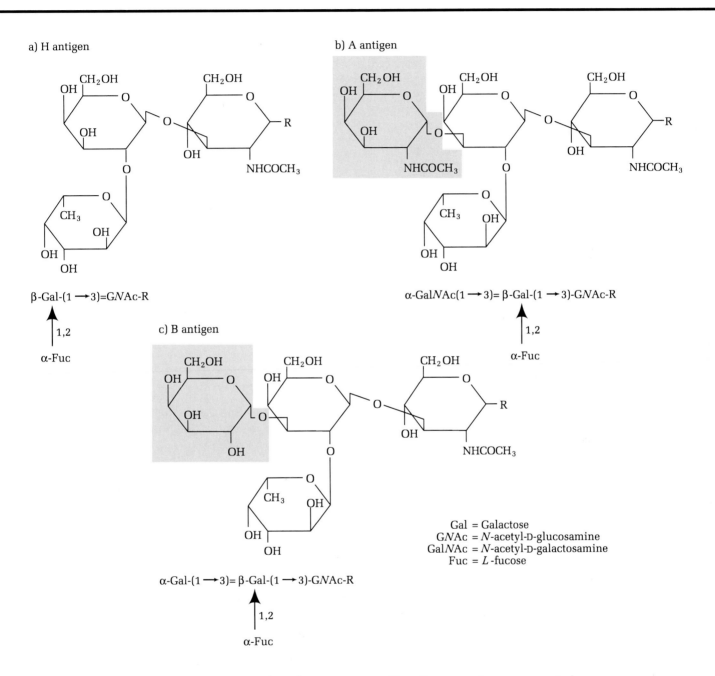

a) H antigen

β-Gal-(1 $\longrightarrow$ 3)=G*N*Ac-R

$\uparrow$ 1,2

α-Fuc

b) A antigen

α-Gal*N*Ac(1 $\longrightarrow$ 3)= β-Gal-(1 $\longrightarrow$ 3)-G*N*Ac-R

$\uparrow$ 1,2

α-Fuc

c) B antigen

α-Gal-(1 $\longrightarrow$ 3)= β-Gal-(1 $\longrightarrow$ 3)-G*N*Ac-R

$\uparrow$ 1,2

α-Fuc

Gal = Galactose
G*N*Ac = *N*-acetyl-D-glucosamine
Gal*N*Ac = *N*-acetyl-D-galactosamine
Fuc = *L*-fucose

Figure 14.28 Terminal sugar sequences in the poly-saccharide chains of glycolipids involved in the human ABO blood types: (a) H antigen; (b) A antigen; (c) B antigen.

some H antigen remains unconverted.

For the $I^A I^B$ heterozygote both enzymes are produced, so some H antigen is converted to the A antigen and some to the B antigen. The blood cell has both antigens on the surface, and the individual is of blood type AB. This situation provides the molecular basis of the codominance of the I^A and I^B alleles.

People who are homozygous for the i allele produce no enzymes to convert the polysaccharide component of the H antigen glycolipid. As a consequence, their red blood cells carry neither the A nor the B antigen, although they do carry the H antigen. The H antigen does not elicit an antibody response in people of other blood groups because its polysaccharide component is the basic component of the A and the B antigens as well, so it is not exactly a foreign substance. Lastly, people who are heterozygous for the i allele will have the blood type of the other allele. For example, in $I^B i$ individuals the I^B allele will result in the conversion of some of the H antigen to the B antigen, which will determine the blood type of the individual.

Lastly, the H antigen is produced by the action of the dominant H allele at a locus that is distinct from the ABO locus. Individuals homozygous for the recessive mutant allele, h, do not make the H antigen and, therefore, regardless of the presence of I^A or I^B alleles at the ABO locus, no A or B antigens can be produced. Thus, hh individuals are blood type O; they are said to have the Bombay blood type. As a consequence of this phenomenon, it is possible for two blood type O individuals to produce a child who has blood type A or B. That is, if one parent is hh I^B- (i.e., hh $I^B I^B$ or hh $I^B i$) and the other is HH ii, the child could be Hh $I^B i$ and, therefore, be blood type B.

Keynote *From the study of alterations in proteins other than enzymes, such as the mutations responsible for sickle-cell anemia, convincing evidence was obtained that genes control the structure of all proteins.*

Analytical Approaches for Solving Genetics Problems

Q.1 Five different *rII* deletion strains of phage T4 were tested for recombination by pairwise crossing in *E. coli* B. The following results were obtained, where $+$ = r^+ recombinants produced, and 0 = no r^+ recombinants produced:

	A	B	C	D	E
E	0	+	0	+	0
D	0	0	0	0	
C	0	0	0		
B	+	0			
A	0				

Draw a deletion map compatible with these data.

A.1 The principle here is that if two deletion mutations overlap, then no r^+ recombinants can be produced. Conversely, if two deletion mutations do not overlap, then r^+ recombinants can be produced. To approach a question of this kind, we must draw overlapping and nonoverlapping lines from the given data.

Starting with A and B, these two deletions do not overlap since r^+ recombinants are produced. Therefore these two mutations can be represented as follows:

A B

_____   _____

The next deletion, C, does not give r^+ recombinants with any of the other four deletions. We must conclude, therefore, that C is an extensive deletion that overlaps all of the other four, with endpoints that cannot be determined from the data given. One possibility is as follows:

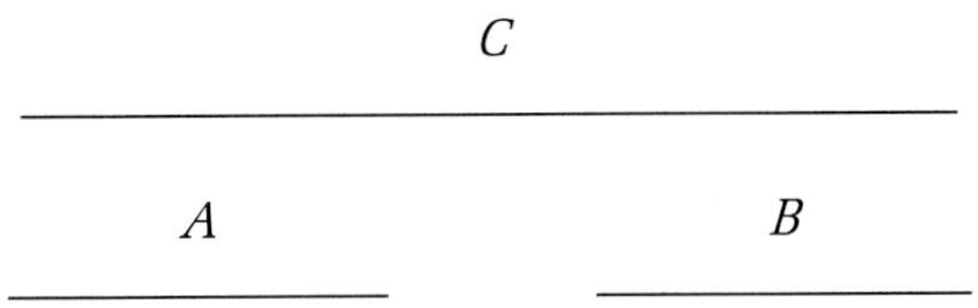

C
_____

A   B
_____   _____

Deletion D does not give r^+ recombinants with A, B, or C, but it does with E. In turn, E gives r^+ recombinants with B and D but not with A or C. Thus D must overlap both A and B but not E, and E must overlap A and C but not B. A compatible map for this situation follows. Other maps can be drawn in terms of the endpoints of the deletions.

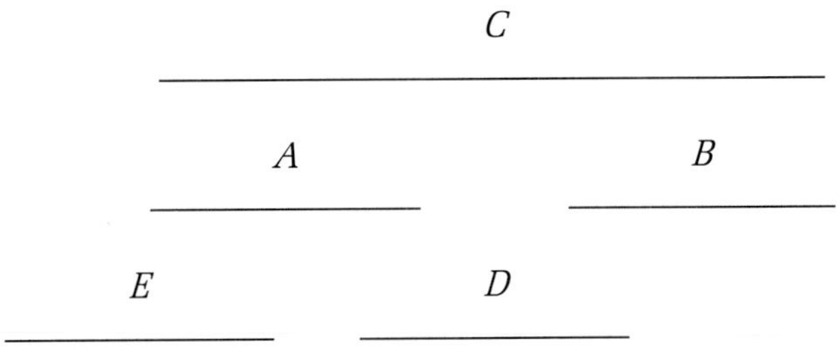

C
_____

A   B
_____   _____

E   D
_____   _____

Q.2 Seven different *rII* point mutants (*1* to *7*) of phage T4 were tested for recombination crosses in *E. coli B* with the five deletion strains described in Question 1. The following results were obtained, where $+ = r^+$ recombinants produced and $0 =$ no r^+ recombinants produced:

	A	B	C	D	E
1	0	+	0	+	+
2	+	0	0	+	+
3	0	+	0	+	0
4	+	+	0	+	0
5	+	0	0	0	+
6	0	+	0	0	+
7	+	+	0	0	+

In which regions of the map can you place the seven point mutations?

A.2 If an r^+ recombinant is produced, the *rII* point mutation must be in the region covered by the deletion mutation with which it was crossed. Thus the matrix of results localizes the point mutations to the regions defined by the deletion mutants. Potentially, the results define the relative extents of deletion overlap. For example, point mutation *7* gives r^+ recombinants with A, B, and E but not with D. Logically, then, *7* is located in the region defined by the part of deletion D that is not involved in the overlap with A and B. Similarly, point mutation *4* gives r^+ recombinants with A, D, and B but not with E. Thus *4* must be in a region defined by a segment of deletion E that does not overlap deletion A. Furthermore, since *4* does not give r^+ recombinants with C either, deletion C must overlap the site defined by point mutation *4*. (This result, then, refines the deletion map with regard to the *E,* *C,* and *A* endpoints.) The map we can draw from the matrix of results is as follows:

Q.3 The ABO blood group locus determines the production of blood group antigens. People known as secretors have the same AB blood group antigens in their saliva as are found on the surfaces of the red blood cells. Nonsecretors do not have these antigens in their saliva. The secretion phenotype is controlled by a single pair of alleles: *Se* (secretor) and *se* (nonsecretor). The *Se* allele is completely dominant to the *se* allele. The secretor locus is unlinked to the ABO locus.

A child has the blood group O, M and is a secretor. Five matings are indicated in the following list. Explain for each case whether or not the child could have resulted from the mating.

a. O, M secretor × O, MN, secretor
b. A, MN, nonsecretor × B, MN, secretor
c. A, M, secretor × A, M, secretor
d. AB, MN, secretor × O, M, nonsecretor
e. A, M, nonsecretor × A, M, nonsecretor

A.3 The child must be *ii MM Se–,* where the secretor locus can be either homozygous or heterozygous for the dominant *Se* allele.

a. The genotypes of the parents are *ii MM Se–* and *ii MN Se–,* so it is possible to produce the child described. There are no major assumptions that we have to make since it does not matter whether the parents are homozygous or heterozygous for the secretor locus.
b. The genotypes of the parents are *I^A– MN se se* and *I^B– MN Se–,* so it is possible to produce the child. The assumptions here are that both parents carry the *i* allele and that the secretor parent is heterozygous at that locus.
c. The parents are *I^A– MM Se–* and *I^A– MM Se–.* On the assumption that both parents are heterozygous *I^Ai,* it is possible to produce the child. Since both parents are secretors, the genotype is no problem with respect to producing a secretor child.
d. The parents are *I^AI^B MN Se–* and *ii MM se se.* The child could not be produced from this mating since one parent is AB in blood group, meaning that the child must be either A or B blood group. There is no problem

with either of the other two loci.

e. The parents are $I^A- MM$ *se se* and $I^A- MM$ *se se*. The child could not be produced from this mating since both parents are nonsecretors and hence are homozygous *se se*. Therefore, all children must be *se se* and be nonsecretors. There is no problem with either of the other two loci.

14.1 Choose the correct answer in each case.

a. If one wants to know if two different *rII* point mutants lie at exactly the same site (nucleotide pair), one should:

   i. Coinfect *E. coli K12(λ)* with both mutants. If phage are produced, they lie at the same site.
   ii. Coinfect *E. coli K12(λ)* with both mutants. If phage are not produced, they lie at the same site.
   iii. Coinfect *E. coli B* with both mutants and plate the progeny phage on both *E. coli B* and *E. coli K12(λ)*. If plaques appear on *B* but not *K12(λ)*, they lie at the same site.
   iv. Coinfect *E. coli K12(λ)* with both mutants and plate the progeny phage on both *E. coli B* and *E. coli K12(λ)*. If plaques appear on *K12(λ)* but not *B*, they lie at the same site.

b. If one wants to know if two different *rII* point mutants lie in the same cistron, one should:

   i. Coinfect *E. coli K12(λ)* with both mutants. If phage are produced, they lie at the same cistron.
   ii. Coinfect *E. coli K12(λ)* with both mutants. If phage are not produced, they lie in the same cistron.
   iii. Coinfect *E. coli B* with both mutants and plate the progeny phage on both *E. coli B* and *E. coli K12(λ)*. If plaques appear on *B* but not *K12(λ)*, they lie in the same cistron.
   iv. Coinfect *E. coli K12(λ)* with both mutants and plate the progeny phage on both *E. coli B* and *E. coli K12(λ)*. If plaques appear on *K12(λ)* but not *B*, they lie in the same cistron.

*14.2 Construct a map from the following two-factor phage cross data (show map distance):

	% recombination
$r_1 \times r_2$	0.10
$r_1 \times r_3$	0.05
$r_1 \times r_4$	0.17
$r_2 \times r_3$	0.15
$r_2 \times r_4$	0.10
$r_3 \times r_4$	0.32

*14.3 Given the following map with point mutants, and given the data in the table below, draw a topological representation of deletion mutants *r21, r22, r23, r24,* and *r25*. (Be sure to indicate clearly the endpoints of the deletions.)

$(+ = r^+$ recombinants are obtained. $0 = r^+$ recombinants are not obtained.)

Map:

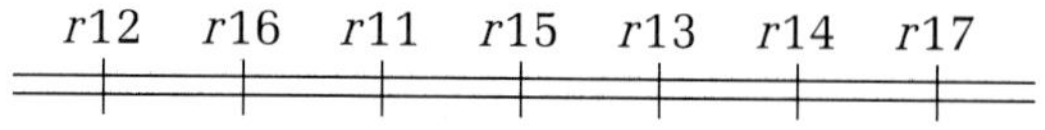

	r12	r16	r11	r15	r13	r14	r17

| Deletion | Point Mutants | | | | | | |
Mutants	r11	r12	r13	r14	r15	r16	r17
r21	0	+	0	+	0	+	+
r22	+	+	0	0	+	+	0
r23	0	0	0	+	0	0	+
r24	+	+	0	0	+	+	+
r25	+	+	0	0	0	+	+

14.4 A set of seven different *rII* deletion mutants of bacteriophage T4, *1* through *7*, were mapped, with the following result:

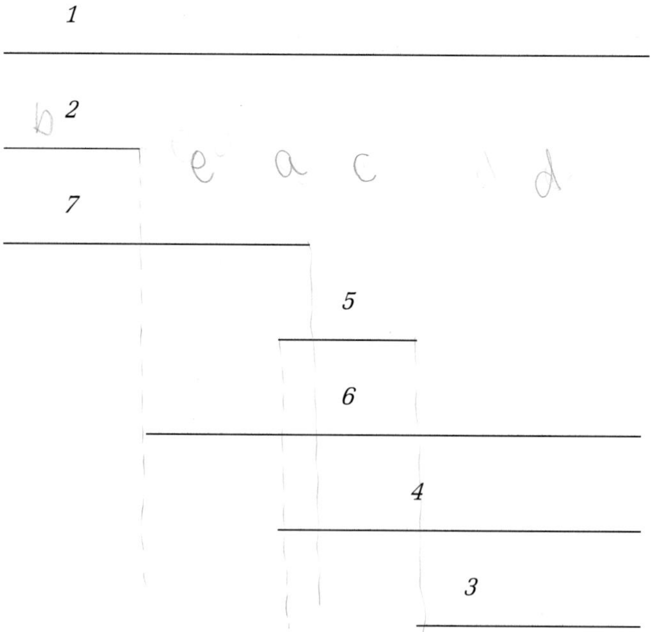

Five *rII* point mutants were crossed with each of the deletions, with the following results, where $+ = r^+$ recombinants were obtained, $0 = $ no r^+ recombinants were obtained:

| Point | Deletion Mutants | | | | | | |
Mutants	1	2	3	4	5	6	7
a	0	+	+	+	0	0	0
b	0	0	+	+	+	+	0
c	0	+	+	0	0	0	+
d	0	+	0	0	+	0	+
e	0	+	+	+	+	0	0

Map the locations of the point mutants.

*14.5 Given the following deletion map with deletions r31, r32, r33, r34, r35, and r36, place the point mutants r41, r42, and so on, on the map. Be sure you show where they lie with respect to endpoints of the deletions.

r31

r32

r33

r34

r35   r36

	Deletion Mutants (+ = r⁺ recombinants produced, 0 = no r⁺ recombinants produced)					
Point Mutants	r31	r32	r33	r34	r35	r36
r41	0	0	0	0	+	0
r42	0	0	0	+	0	+
r43	0	0	+	+	+	0
r44	0	0	0	0	+	+
r45	0	+	0	+	-|-	+
r46	0	0	+	0	+	0

Show the dividing line between the A cistron and the B cistron on your map above from the following data [+ = growth on strain $K12(\lambda)$, 0 = no growth on strain $K12(\lambda)$]:

Mutant	Complementation with rIIA	rIIB
r31	0	0
r32	0	0
r33	0	+
r34	0	0
r35	0	+
r36	0	0
r41	0	+
r42	0	+
r43	+	0
r44	0	+
r45	0	+
r46	0	+

*14.6 Factors *A, B, C,* and *D* are independently assorting Mendelian factors controlling the production of a black pigment. The alternate alleles that give abnormal functioning of these genes are *a, b, c,* and *d*. A black *AA BB CC DD* is crossed with a colorless *aa bb cc dd* to give a black F$_1$. The F$_1$ is then selfed. Assume that A, B, C, and D act in a pathway as follows:

$$\text{colorless} \xrightarrow{A} \text{colorless} \xrightarrow{B} \text{colorless} \xrightarrow{C} \text{brown} \xrightarrow{D} \text{black}$$

a. What proportion of the F_2 are colorless?
b. What proportion of the F_2 are brown?

14.7 Using the genetic information given in Problem 14.6, now assume that A, B, and C act in a pathway as follows:

$$\begin{array}{l} \text{colorless} \xrightarrow{A} \text{red} \\ \qquad\qquad\qquad\qquad \searrow \xrightarrow{C} \text{black} \\ \text{colorless} \xrightarrow{B} \text{red} \nearrow \end{array}$$

Black can be produced only if both red pigments are present, i.e. c converts the two red pigments together into a black pigment.

a. What proportion of the F_2 are colorless?
b. What proportion of the F_2 are red?
c. What proportion of the F_2 are black?

*14.8 a. Three genes on different chromosomes are responsible for three enzymes that catalyze the same reaction in corn:

$$\text{colorless compound} \xrightarrow{A, B, C} \text{red compound}$$

The normal functioning of any one of these genes is sufficient to convert the colorless compound to the red compound. The abnormal functioning of these genes is designated by a, b, and c respectively. A red AA BB CC is crossed by a colorless aa bb cc to give a red F_1, Aa Bb Cc. The F_1 is selfed. What proportion of the F_2 are colorless?

b. It turns out that another step is involved in the pathway. It is controlled by gene D, which assorts independently of A, B, and C:

$$\begin{array}{ccc} & D & A, B, C \\ \text{colorless} & \longrightarrow \text{colorless} & \longrightarrow \text{red} \\ \text{compound} & \text{compound 2} & \text{compound} \end{array}$$

The inability to convert colorless 1 to colorless 2 is designated d. A red AA BB CC DD is crossed by a colorless aa bb cc dd. The F_2 are all red. The red F_1's are now selfed. What proportion of the F_2 are colorless?

14.9 In hypothetical diploid organisms called mongs, the recessive bw causes a brown eye, and the (unlinked) recessive st causes a scarlet eye. Organisms homozygous for both recessives have white eyes. The genotypes and corresponding phenotypes, then, are as follows:

bw^+ —	st^+ —	red eye
bw bw	st^+ —	brown
bw^+ —	st st	scarlet
bw bw	st st	white

Outline a hypothetical biochemical pathway that would give this type of gene interaction. Demonstrate why each genotype shows its specific phenotype.

*14.10 The Black Riders of Mordor in *Lord of the Rings* ride steeds with eyes of fire. As a geneticist, you are very interested in the inheritance of the fire red eye color. You discover that the eyes contain two types of pigments, brown and red, that are usually bound to core granules in the eye. In wild-type steeds precursors are converted by these granules to the above pigments, but in steeds homozygous for the recessive X-linked gene *w* (white eye), the granules remain unconverted and a white eye results. The metabolic pathways for the synthesis of the two pigments are shown in the figure below. Each step of the pathway is controlled by a gene: A mutation *v* gives vermilion eyes; *cn* gives cinnabar eyes; *st* gives scarlet eyes; *bw* gives brown eyes; and *se* gives black eyes. All the mutations are recessive to their wild-type alleles and all are unlinked. For the following genotypes, show the phenotypes and proportions of steeds that would be obtained in the F_1 of the given matings.

a. $ww\ bw^+\ bw^+\ st st \times w^+\ Y\ bw bw\ st^+\ st^+$
b. $w^+\ w^+\ se se\ bw bw \times wY\ se^+\ se^+\ bw^+\ bw^+$
c. $w^+\ w^+\ v^+\ v^+\ bw bw \times wY\ vv\ bw bw$
d. $w^+\ w^+\ bw^+\ bw\ st^+\ st \times wY\ bw bw\ st st$

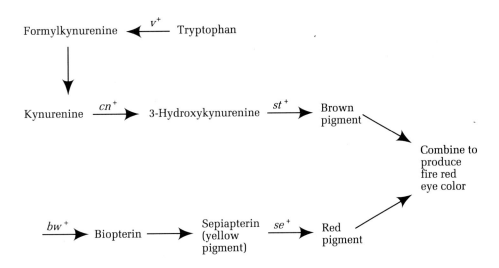

*14.11 Upon infection of *E. coli* with bacteriophage T4, a series of biochemical pathways result in the formation of mature progeny phages. The phages are released following lysis of the bacterial host cells. Let us suppose that the following pathway exists:

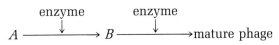

Let us also suppose that we have two temperature-sensitive mutants that involve the two enzymes catalyzing these sequential steps. One of the mutations is cold-sensitive (*cs*) in that no mature phages are produced at 17°C. The other is heat-sensitive (*hs*) in that no mature phages are produced at 42°C. Normal progeny phages are produced when phages carrying either of the mutations infect bacteria at 30°C. However, let us assume that we do not know the sequence of the two mutations. Two models are therefore apparent:

$$(1)\ A \xrightarrow{hs} B \xrightarrow{cs} \text{phage}$$
$$(2)\ A \xrightarrow{cs} B \xrightarrow{hs} \text{phage}$$

Outline how you would experimentally determine which model is the correct model without artificially breaking phage-infected bacteria.

14.12 In *Drosophila*, mutants *A, B, C, D, E, F,* and *G* all have the same phenotype: the absence of red pigment in the eyes. In pairwise combinations in complementation tests the following results were produced, where + = complementation and − = no complementation:

	A	B	C	D	E	F	G
G	+	−	+	+	+	+	−
F	−	+	+	−	+	−	
E	+	+	−	+	−		
D	−	+	+	−			
C	+	+	−				
B	+	−					
A	−						

a. How many genes are present?
b. In which genes are the mutations located?

*14.13 Two mutant strains of *Neurospora* lack the ability to make compound Z. When crossed, the strains usually yield asci of two types: (1) those with spores that are all mutant and (2) those with four wild-type and four mutant spores. The two types occur in a 1:1 ratio.

a. Let *c* represent one mutant, and let *d* represent the other. What are the genotypes of the two mutant strains?
b. Are *c* and *d* linked?
c. Wild-type strains can make compound Z from the constituents of the minimal medium. Mutant *c* can make Z if supplied with X but not if supplied with Y, while mutant *d* can make Z from either X or Y. Construct the simplest linear pathway of the synthesis of Z from the precursors X and Y, and show where the pathway is blocked by mutations *c* and *d.*

14.14 The following growth responses (where + = growth and 0 = no growth) of *Neurospora* mutants *1–4* were seen on the related biosynthetic intermediates A, B, C, D, and E. Assume all intermediates are able to enter the cell, that each mutant carries only one mutation, and that all mutants affect steps after B in the pathway.

Mutant	Growth on				
	A	B	C	D	E
1	+	0	0	0	0
2	0	0	0	+	0
3	0	0	+	0	0
4	0	0	0	+	+

Which of the schemes in the figure fits best with the data with regard to the biosynthetic pathway?

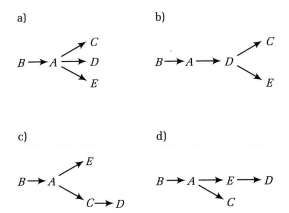

a)

$B \rightarrow A \rightarrow \begin{matrix} C \\ D \\ E \end{matrix}$

b)

$B \rightarrow A \rightarrow D \rightarrow \begin{matrix} C \\ E \end{matrix}$

c)

$B \rightarrow A \begin{matrix} \nearrow E \\ \searrow C \rightarrow D \end{matrix}$

d)

$B \rightarrow A \begin{matrix} \rightarrow E \rightarrow D \\ \searrow C \end{matrix}$

14.15 Four strains of *Neurospora,* all of which require arginine but have unknown genetic constitution, have the following characteristics. The nutrition and accumulation characteristics are as follows:

		Growth on			
Strain	Minimal Medium	*A*→ Ornithine	*B*→ Citrulline	*C*→ Arginine	Accumulates
1	−	−	+	+	Ornithine *enzyme B defective*
2	−	−	−	+	Citrulline *c defective*
3	−	−	−	+	Citrulline *c defective*
4	−	−	−	+	Ornithine *b defective +c*

The pairwise complementation tests of the four strains gave the following results (+ = growth on minimal medium and 0 = no growth on minimal medium):

	4	3	2	1
1	0	+	+	0
2	0	0	0	
3	0	0		
4	0			

2×3 same gene defective

Crosses among mutants yielded prototrophs in the following percentages:

> 1 × 2: 25 percent
> 1 × 3: 25 percent
> 1 × 4: none detected among 1 million ascospores
> 2 × 3: 0.002 percent
> 2 × 4: 0.001 percent
> 3 × 4: none detected among 1 million ascospores

Analyze the data and answer the following questions.

a. How many distinct mutational sites are represented among these four strains?
b. In this collection of strains, how many types of polypeptide chains (normally found in the wild type) are affected by mutations?
c. Give the genotypes of the four strains, using a consistent and informative set of symbols.
d. Give the map distances between all pairs of linked mutations.
e. Give the percentage of prototrophs that would be expected among ascospores of the following types: (1) strain 1 × wild type; (2) strain 2 × wild type; (3) strain 3 × wild type; (4) strain 4 × wild type.

*14.16 A breeder of Irish setters has a particularly valuable show dog that he knows is descended from the famous bitch Rheona Didona, who carried a recessive gene for atrophy of the retina. Before he puts the dog to stud, he must ensure that it is not a carrier for this allele. How should he proceed?

*14.17 Suppose you were on a jury to decide the following case. What would you conclude?

The Jones family claims that Baby Jane, given to them at the hospital, does not belong to them but to the Smith family, and that the Smith's baby Joan really belongs to the Jones's family. It is alleged that the two babies were accidentally exchanged soon after birth. The Smiths deny that such an exchange has been made. Blood group determinations show the following results:

> Jones mother, AB
> Jones father, O
> Smith mother, A
> Smith father, O
> Baby Jane, A
> Baby Joan, O

14.18 Phenylketonuria (PKU) is a heritable metabolic disease of humans; its symptoms include mental deficiency. The gross phenotypic effect is due to:

a. Accumulation of phenylketones in the blood
b. Accumulation of maple sugar in the blood
c. Deficiency of phenylketones in the blood
d. Deficiency of phenylketones in the diet

14.19 In evaluating my teacher, my sincere opinion is that:

  a. He (she) is a swell person whom I would be glad to have as a brother-in-law/sister-in-law.
  b. He (she) is an excellent example of how tough it is when you don't have either genetics or environment going for you.
  c. He (she) may have okay DNA to start with, but somehow all the important genes got turned off.
  d. He (she) ought to be preserved in tissue culture for the benefit of other generations.

15

Recombinant DNA Technology and the Manipulation of DNA

In recent years, experimental procedures have been developed that have allowed researchers to construct **recombinant DNA molecules** in test tubes. This process entails combining genetic material from two different sources into a single DNA molecule. The technology has opened the way for new and exciting research possibilities and affirms the plausibility of *genetic engineering.* In this chapter we discuss the techniques that can be used to make recombinant DNA and present some examples of how **recombinant DNA technology** is furthering our knowledge of the structure and function of prokaryotic and eukaryotic genomes, how cloned DNA sequences are being used for genetic diagnosis, and how new routes for producing commercial products have been opened because of the ability to manipulate DNA sequences.

Recombinant DNA is made by taking a piece of DNA from an organism and splicing it into a **cloning vector.** A cloning vector is a DNA molecule capable of replication in a host organism, such as a bacterium, and into which a gene or a piece of DNA can be specifically inserted at known positions. The resulting molecule is introduced (transformed) into a host cell. The host is often a special strain of *Escherichia coli* that is unable to reproduce without special culture conditions. Reproduction of the *E. coli* results in the replication **(molecular cloning)** of the recombinant DNA molecule, thus producing many copies for analysis and/or manipulation.

Restriction Endonucleases

The construction of recombinant DNA molecules could not occur without the use of **restriction endonucleases** (or **restriction enzymes**), which are able to cleave double-stranded DNA molecules at specific nucleotide pair sequences. These enzymes are used not only in the production of a pool of DNA molecules to be cloned, but also in the assembly of the recombinant DNA molecules from the DNA fragments and cloning vectors.

The Restriction-Modification Phenomenon

The rapid development of recombinant DNA technology was made possible by the discovery of restriction endonucleases, or restriction enzymes, which catalyze the cleavage of DNA at defined, short base-pair sequences. (Recall from previous discussions that an endonuclease results in a cut within a nucleic acid chain.)

In the 1950s, S. Luria and his colleagues found that the progeny phages produced after infection of the B/4 strain of *E. coli* by phage T2 could no longer reproduce in the normal host strain of *E. coli*. They reasoned that the phages released from strain B/4 had become modified in some way such that the normal bacterial host could no longer support growth of T2. This phenomenon of variation of phages that takes place under the direct influence of the host cell was called **host-controlled modification and restriction.**

In the 1960s, W. Arber and D. Dussoix shed some light on the restriction phenomenon. They grew phage λ on *E. coli K* and used the progeny phages to infect *E. coli B*. While most of the λ phage DNA was degraded in the *E. coli B* host (i.e., the DNA was *restricted*), a few complete phage chromosomes remained intact and these produced some progeny phages. All these progeny phages could now infect *E. coli B* normally and about 2 percent of them could also infect *E. coli K*.

To investigate this outcome in more detail, Arber's group used phage λ to infect *E. coli K* growing in a culture medium containing the heavy isotopes ^{15}N and ^2H(deuterium), so that the DNA of the resulting progeny phage was denser than normal phage DNA. The "heavy" phages were then used to infect *E. coli B* growing in normal "light" medium and the densities of the progeny phage's DNA were determined by cesium chloride density gradient centrifugation. Most of the progeny phages contained normal "light" DNA, indicating that they had been entirely synthesized from new material and could not grow in *E. coli K*. A few of the progeny phages, however, were "heavy" and retained the ability to grow in *E. coli K*. Arber's group rea-

soned, therefore, that phages grown in particular bacterial strains become modified in a way specific for that particular strain. Further, the modification occurred on the DNA itself and was not lost by passage through the bacterial host. It now appears that in many bacteria the modification involves the addition of methyl groups on certain bases, in particular DNA sequences in the genome.

In 1970, a major breakthrough occurred when H. Smith identified a **restriction enzyme,** an enzyme that plays a role in the restriction phenomenon. Restriction enzymes recognize a specific nucleotide sequence in DNA and then the enzymes cleave the DNA. Restriction enzymes are integral parts of bacterial host-controlled modification and restriction systems. That is, in a bacterial cell that has a modification/restriction system, there are two enzymes (or sets of two enzymes depending on the number of modifications of which it is capable) both of which have a recognition site for the same, specific nucleotide pair sequence. The *modification enzyme* catalyzes the addition of a modifying group (often a methyl group) to one or more nucleotides in the sequence. Thus, the bacterial DNA becomes modified by the action of the modification enzyme. The *restriction enzyme* can only recognize the unmodified nucleotide pair sequence and hence is unable to restrict the modified bacterial DNA. The restriction enzyme can digest invading DNA, however, unless it, too, is modified at the same specific nucleotide pair sequence in the same way as the host DNA. The restriction enzymes function, therefore, as site-specific endonucleases. To date, a number of restriction-modification systems have been identified in bacteria. However, not all of the enzymes used in recombinant DNA experiments have a role in a restriction-modification system in the bacterial strain from which they have been isolated, even though they are called restriction enzymes.

Properties of Restriction Enzymes

Over 100 restriction enzymes have been discovered, all of which have been found in prokaryotes.

484
*Recombinant
DNA Technology
and the
Manipulation
of DNA*

No similar enzymes have been identified in the few eukaryotic organisms that have been examined.

Restriction enzymes fall into two classes. Type I restriction enzymes recognize a specific nucleotide pair sequence in DNA and then cleave the DNA at a nonspecific site away from that sequence. The enzymes involved in the *E. coli K* and *B* restriction system described earlier belong to this class. Since the cut site is not nucleotide pair specific, Type I restriction enzymes are not useful for the construction or analysis of recombinant DNA molecules.

Type II restriction enzymes also recognize a specific nucleotide pair sequence in DNA and, in this case, they cleave the DNA within that sequence. Thus, since all DNA fragments generated by cleavage with a particular Type II restriction enzyme have been cut at the same sequence, these enzymes are very valuable for constructing recombinant DNA molecules. The recognition sequences for Type II restriction enzymes have an axis of symmetry through the midpoint of the recognition sequence so that the base sequence from 5′ to 3′ on one DNA strand is the same as the base sequence from 5′ to 3′ on the complementary DNA strand (Figure 15.1); that is, the sequences are said to have twofold rotational symmetry. Type II restriction enzymes have been isolated from a large number of differ-

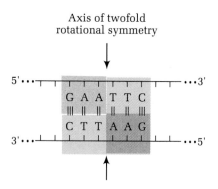

Figure 15.1 *Restriction enzyme recognition sequence in DNA showing twofold rotational symmetry of the sequence. The sequence shown is the recognition sequence for the restriction enzyme EcoRI.*

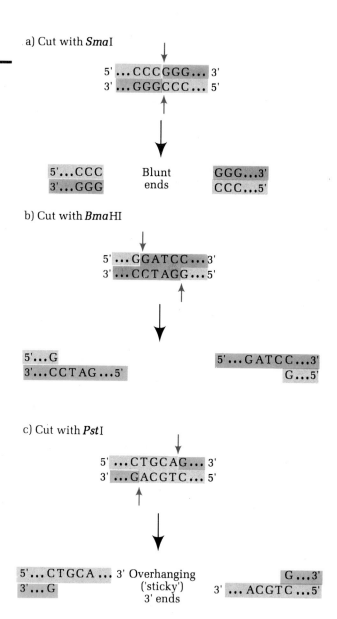

Figure 15.2 *Examples of how restriction enzymes cleave DNA. (a)* SmaI *results in blunt ends; (b)* BamHI *results in overhanging ("sticky") 5′ ends; (c)* PstI *results in overhanging ("sticky") 3′ ends.*

ent bacterial strains. Since each of the enzymes cuts DNA at an enzyme-specific nucleotide pair sequence, the number of cuts the enzyme makes in a particular DNA molecule or molecules depends on the number of times the particular recognition/cleavage sequence is present in the DNA. In some cases, restriction enzymes (called *isoschizomers*) isolated from different bacteria recognize and cleave the same nucleotide pair sequences.

The cleavage sites for a number of restriction enzymes are shown in Table 15.1 along with the number of cuts each enzyme makes in phage λ DNA and in the plasmid pBR322. This information can be used as a guideline for choosing an enzyme for a particular application. As Table 15.1 indicates, some enzymes, such as *Hae*III or *Sma*I, cut both strands of DNA between the same two nucleotide pairs to produce *blunt ends* (Figure 15.2a), while others, such as *Eco*RI, *Bam*HI, and *Hind*III, make staggered cuts in the symmetrical nucleotide pair sequence to produce *sticky* or *staggered ends.* The staggered ends may either have an overhanging 5′ end, as

Table 15.1 Characteristics of Some Restriction Endonucleases

Enzyme Name	Organism from which Enzyme is Located	Recognition Sequence and Position of Cut	Number of Cleavage Sites in DNA from:	
			λ	pBR322
*Bam*HI	*Bacillus amyloliquefaciens H*	5′ G↓G A T C C 3′* 3′ C C T A G↑G 5′	5	1
*Bgl*II	*Bacillus globigi*	A↓G A T C T T C T A G↑A	5	0
*Eco*RI	*E. coli RY13*	G↓A A T T C C T T A A↑G	5	1
*Hae*III	*Haemophilus egyptius*	G G↓C C C C↑G G	> 50	22
*Hha*I	*Haemophilus hemolyticus*	G C G↓C C↑G C G	> 50	31
*Hind*III	*Haemophilus influenzae* R_d	A↓A G C T T T T C G A↑A	6	1
*Pst*I	*Providencia stuartii*	C T G C A↓G G↑A C G T C	18	1
*Sma*I	*Serratia marcescens*	C C C↓G G G G G G↑C C C	3	0
*Hpa*II	*Haemophilus parainfluenzae*	C↓C G G G G C↑C	> 50	26
*Sal*I	*Streptomyces albus* G	G↓T C G A C C A G C T↑G	2	1
*Hae*II	*Haemophilus aegyptius*	Pu G C G C↓Py Py↑C G C G Pu	> 30	11

* In this column the two strands of DNA are shown with the sites of cleavage indicated by arrows. Since there is an axis of twofold rotational symmetry in each recognition sequence, the DNA molecules resulting from the cleavage are symmetrical. In some cases, the two cuts are staggered and hence the DNA molecules produced have complementary, single-stranded ends (e.g., *Bam*HI), while in other cases the point of cleavage is at the axis of symmetry and the DNA molecules produced are double-stranded with no single-stranded ends; these are called "blunt" ends (e.g., *Hae*III). Key: Pu-purine; Py-pyrimidine.

After R. J. Roberts. 1976. *Crit. Rev. Biochem.* 4:123.

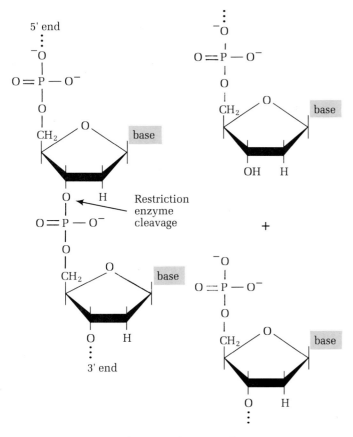

Figure 15.3 Cleavage of DNA with a restriction enzyme results in DNA fragments with a phosphate on their 5' ends and a 3' hydroxyl on their 3' ends.

in the case of cleavage with *Bam*HI or *Eco*RI (Figure 15.2b), or an overhanging 3' end, as in the case of cleavage with *Hae*II or *Pst*I (Figure 15.2c). In each case, the restriction enzyme makes its cut by cleaving the DNA backbone between the 3' carbon of the deoxyribose sugar and the phosphate on the subsequent deoxyribose in the chain (Figure 15.3). Thus, all restriction enzyme-cut DNA fragments have a phosphate on their 5' ends and a hydroxyl group on their 3' ends.

Restriction enzymes that produce staggered ends are of particular value in cloning DNA fragments because each and every DNA fragment generated by cutting a piece of DNA has the same base sequence at the two staggered ends. Therefore, if the ends of two pieces of DNA produced by the action of the same restriction enzyme (such as *Eco*RI) come together in solution, perfect base pairing occurs (Figure 15.4). In the presence of DNA ligase, the sugar-phosphate backbones can then be sealed to produce a

whole DNA molecule with a reconstituted restriction enzyme site. As we will see in the following section on cloning vectors, this annealing of sticky ends followed by DNA ligase-catalyzed ligation is the principle behind the formation of recombinant DNA molecules.

Keynote *Restriction enzymes, or restriction endonuclease, are enzymes that recognize specific nucleotide pair sequences in DNA and cleave it either nonspecifically at a site away from a twofold symmetrical recognition site (Type I restriction enzymes) or at a specific site within the recognition site (Type II restriction enzymes). Cleavage of the DNA with a Type II restriction enzyme can be staggered (producing DNA fragments with single-stranded, "sticky," ends) or blunt. Because they cut DNA at specific sites, the Type II restriction enzymes are generally used in recombinant DNA cloning experiments.*

Cloning Vectors

To clone a piece of DNA, two things are needed: a disabled host that has no significant chance of surviving outside the laboratory, and a **cloning vector.** A cloning vector is a DNA molecule capable of replication in a host organism, such as a bacterium, and into which a gene or a piece of DNA can be specifically inserted at known positions.

While this chapter concentrates mostly on bacterial hosts, recombinant DNA molecules are also routinely introduced into yeast and other fungi, into other lower eukaryotes, and into mammalian cells, among other systems. In all these cases the host organism is a derivative of the natural form found in the environment. The host organism might, for example, contain mutations that minimize the possibility of its survival outside the culturing environment of the laboratory. A host cell with such mutations can only grow in a specially supplemented medium in the laboratory.

Three types of vectors are commonly used for cloning DNA sequences: plasmids, bacterio-

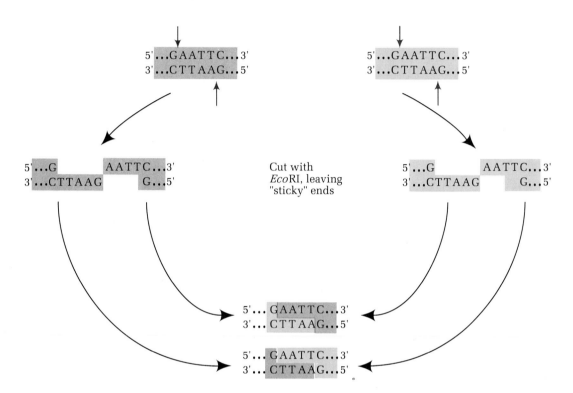

Figure 15.4 Cleavage of DNA by the restriction enzyme EcoRI. EcoRI makes staggered, symmetrical cuts in DNA leaving "sticky" ends. A DNA fragment with a sticky end produced by EcoRI digestion can anneal to any other DNA fragment with a sticky end produced by EcoRI cleavage. The gaps may then be sealed by DNA ligase.

phages (e.g., λ), and cosmids. Each differs in the way it must be manipulated to clone pieces of DNA and in the maximum amount of DNA that can be cloned using it. Each type of vector has been specially constructed in the laboratory to possess the features necessary to make it efficient cloning vectors, and to prevent its spread throughout the environment if it were to escape from the laboratory.

pBR322: A Plasmid Cloning Vector

Plasmids (see Chapter 7, page 226, and the chapter on "Transposable Genetic Elements," Chapter 18.) are extrachromosomal genetic elements that replicate autonomously within bacterial cells.

Their DNA is circular and double-stranded and it carries the sequences (called origins or *ori*) required for the replication of the plasmid, as well as for the plasmid's other functions. The particular plasmids used for cloning experiments are all derivatives of naturally occurring plasmids "engineered" to have features that facilitate gene cloning. Because they are most commonly used, our discussion focuses on features of *E. coli* plasmid cloning vectors.

A plasmid cloning vector must have three features:

1. An *ori* sequence, which directs the replication of the plasmid within the *E. coli* host.
2. A dominant selectable marker, which enables *E. coli* cells that carry the plasmid to be easily distinguished from cells that lack the plasmid. The usual selectable marker is a gene that confers an antibiotic resistance phenotype upon the *E. coli* host, for example, the amp^R gene for ampicillin resistance, the tet^R gene for tetracycline resistance, or the cam^R gene for chloramphenicol resistance. When plasmids carrying antibiotic resistance genes

488
Recombinant
DNA Technology
and the
Manipulation
of DNA

such as these are introduced by transformation (see Chapter 7) into a plasmid-free and therefore antibiotic-sensitive *E. coli* host cell, the uptake of an antibiotic-resistance plasmid can be detected easily.

3. Unique restriction enzyme cleavage sites for the insertion of the DNA sequences that are to be cloned.

Figure 15.5 shows a stylized plasmid cloning vector to summarize the important features of such a vector.

Figure 15.6 diagrams one of the plasmids that has been used extensively for molecular cloning, *pBR322*. This plasmid is 4363 nucleotide pairs [4.363 kbp (kbp = kilobase pair = 1000 base pairs)] long. By comparison, the *E. coli* chromosome is 4720 kbp long. The plasmid pBR322 contains two antibiotic-resistance genes, amp^R and tet^R, and has unique restriction enzyme cleavage sites for a number of enzymes, such as *Eco*RI, *Hind*III, *Bam*HI, *Sal*I, *Pst*I, *Pvu*I, *Pvu*II, *Ava*I, and *Cla*I. These sites are ideal for insert-

ing pieces of DNA for cloning since cutting the plasmid with any of the enzymes simply converts the circular molecule into a linear molecule. After a piece of DNA is added to the linear molecule, the two ends of the molecule can be sealed to the linear plasmid, thus creating a larger circular DNA molecule.

The sites for *Hind*III, *Bam*HI, and *Sal*I lie within the tet^R gene, and the sites for *Pst*I and *Pvu*I are within the amp^R gene. Therefore, DNA fragments inserted into the *Hind*III, *Bam*HI, or *Sal*I site inactivate the tetracycline-resistance gene, and fragments inserted into the *Pst*I or *Pvu*I site inactivate the ampicillin-resistance gene.

A number of other restriction enzymes cut the pBR322 plasmid. Since these other enzymes cleave pBR322 at two or more sites, however, they are not used in cloning experiments. If the plasmid is cleaved into two or more pieces by their action, it is not possible simply to insert a piece of DNA into the plasmid and reseal the pieces into a circle containing the pBR322 parts in the correct order.

DNA fragments of up to 5 to 10 kbp are efficiently cloned in plasmid vectors like pBR322. Plasmids carrying larger DNA fragments, however, are often unstable and tend to lose most of the inserted DNA. Moreover, the replication rate of plasmids tends to decrease as their size increases. As far as replication and stability are concerned, then, there is an advantage for smaller plasmids.

Figure 15.7 illustrates how a piece of DNA can be inserted into a plasmid cloning vector such as pBR322. In the first step, pBR322 is cut with a restriction enzyme that has only one cleavage site in the plasmid. Next, the piece of DNA to be cloned is generated by cutting high-molecular-weight DNA with the same restriction enzyme. This DNA fragment is then inserted between the two cut ends of the plasmid. If the DNA is inserted into the amp^R gene (as shown in Figure 15.7), the resulting recombinant DNA plasmid would, when inserted into an *E. coli* host by transformation, confer tetracycline resistance (since the tet^R gene is still intact), but not ampicillin resistance (since the amp^R gene is disrupted). The presence of *E. coli* cells that were

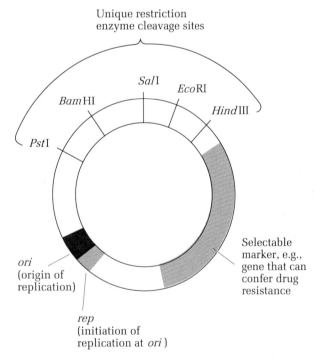

Figure 15.5 *Important features of a plasmid cloning vector.*

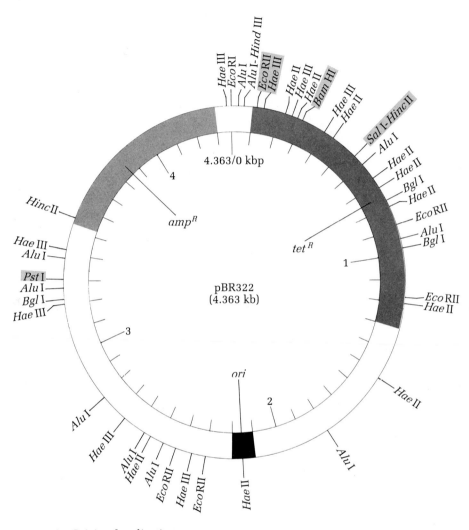

Figure 15.6 Restriction map of the plasmid cloning vector pBR322.

ori = Origin of replication sequence
tet ^{R} = Tetracycline resistance gene
   kbp scale is approximate
amp^R = Ampicillin resistance gene
   kbp scale is approximate

tetracycline-resistant but ampicillin-sensitive would therefore indicate that they had been transformed by a recombinant DNA plasmid and not by pBR322 that had no DNA inserted into it.

Bacteriophage Lambda

As discussed in Chapter 7, wild-type λ phages have a choice between the lysogenic and lytic cycles when they infect the *E. coli* host cell. Re-

call that for the λ phage to follow the lysogenic cycle, it first integrates itself into the *E. coli* chromosome. In the lytic cycle, on the other hand, the λ chromosome replicates, and genes are expressed for the production of progeny phages and the subsequent lysis of the bacterial cell. Since present governmental regulations forbid genetic researchers to use cloning vectors that can integrate into the *E. coli* chromosome, specially engineered derivatives of λ have been

490
Recombinant
DNA Technology
and the
Manipulation
of DNA

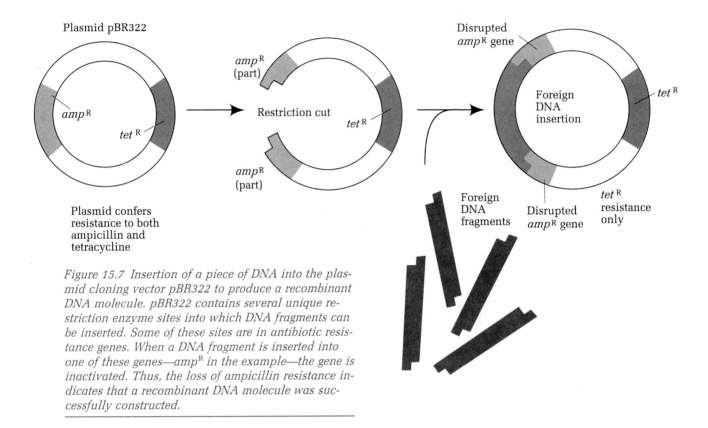

Figure 15.7 Insertion of a piece of DNA into the plasmid cloning vector pBR322 to produce a recombinant DNA molecule. pBR322 contains several unique restriction enzyme sites into which DNA fragments can be inserted. Some of these sites are in antibiotic resistance genes. When a DNA fragment is inserted into one of these genes—amp^R *in the example—the gene is inactivated. Thus, the loss of ampicillin resistance indicates that a recombinant DNA molecule was successfully constructed.*

constructed that cannot follow the lysogenic pathway but that can follow the lytic pathway. These λ derivatives also possess restriction enzyme sites so that DNA fragments can be cloned using them. One such derivative, the lambda replacement vector (Figure 15.8), has a chromosome in which there is a "left" arm and a "right" arm that collectively contain all the essential genes for the lytic cycle. Between the two arms is a disposable segment of DNA ("disposable" since it does not contain any genes needed for phage propagation). The junctions between the disposable central segment and the two arms each have a cleavage site for one restriction enzyme, *Eco*RI; no other *Eco*RI sites are present in the λ replacement vector's DNA.

Cloning a DNA fragment using a λ replacement vector is achieved by the following process. First the λ vector is cut with the enzyme *Eco*RI to separate the two arms from the replace-

able segment. Next, the DNA fragments to be cloned are generated by cutting high-molecular-weight DNA from a given organism with *Eco*RI. The foreign DNA fragments are then mixed with the λ DNA fragments and the pieces ligated together in the presence of DNA ligase.

While a variety of DNA splicing events can occur, the only ones that produce viable, functional λ chromosomes are those in which the foreign DNA is inserted between the left arm and right arm and in which the total length of the recombined DNA fragments is approximately 45 kbp. Since only DNA fragments of about 45 kbp can be packaged into lambda DNA particles, DNA molecules that are either too long or too short are unable to be packaged. Given the number of essential genes that must be present on the left and right arms for phage propagation, a λ replacement vector can only clone a DNA fragment of about 15 kbp.

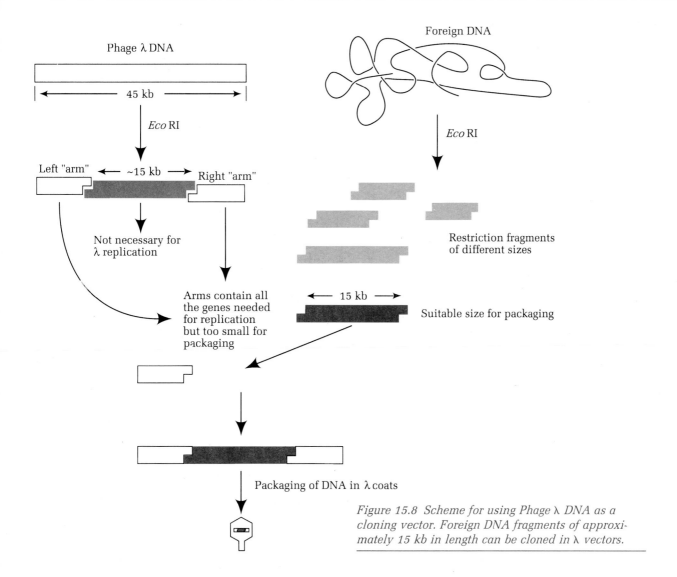

Figure 15.8 Scheme for using Phage λ DNA as a cloning vector. Foreign DNA fragments of approximately 15 kb in length can be cloned in λ vectors.

Cosmids

Plasmid and λ cloning vectors can only be used to clone small DNA fragments (5 to 10 kbp in plasmids and about 15 kbp in λ). Since some genes from higher eukaryotes may be as long as 30 kbp, they could not be cloned in one piece using these cloning vectors.

Currently, DNA fragments of approximately 35 to 40 kbp can be cloned in vectors called **cosmids.** Unlike plasmid and phage vectors that are

derived from naturally occurring genetic elements, cosmids do not occur naturally—they are constructed in the laboratory by combining features of both plasmids (e.g., pBR322) and phage λ. Thus, a cosmid has an *ori* sequence to permit its replication in *E. coli,* a dominant selectable marker such as *tet*[R], and unique restriction enzyme cleavage sites for the insertion and cloning of DNA fragments. In addition, cosmids have a *cos* site that is derived from phage λ. Recall from Chapter 9 that the *cos* site is the site at

492
Recombinant
DNA Technology
and the
Manipulation
of DNA

which multiple copies of the λ genome, attached in one long piece called a concatamer, are cleaved into 34- to 45-kbp pieces, which are packaged into phage heads. Recall also that the cut at the *cos* site is staggered and results in a linear DNA molecule with sticky (single-stranded), complementary ends. All that is needed for packaging DNA into a phage head are *cos* sites that are a specified distance apart. Thus the *cos* sites on a cosmid permit packaging of DNA into a λ phage particle that facilitates the introduction of large DNA molecules into a bacterial cell. This property is absent from plasmid cloning vectors.

λ *cos* sites have been cloned into plasmid cloning vectors to produce cosmids as small as 5 kbp. When a DNA fragment of about 40 kbp is cloned into such a cosmid, the recombinant DNA molecule is then the right size to be packaged into a phage head. The phage is then used to introduce the recombinant cosmid into the *E. coli* host cell where it replicates as a plasmid does. Recombinant cosmids that are either too small (<35 kbp) or too large (>~50 kbp) cannot be packaged.

Shuttle Vectors

So far all the cloning vectors we've discussed (plasmids, λ phage, and cosmids) must be used to clone DNA within *E. coli* cells. Other vectors have been developed to introduce recombinant DNA molecules into a variety of prokaryotic and eukaryotic organisms. Although there are vectors that can be used to transform mammalian cells in culture, the most developed vectors are used to transform yeast cells. A *shuttle vector* is a cloning vector that carries dominant selectable markers that allow it to be detected in two or more host organisms. Shuttle vectors are used for experiments in which recombinant DNA is to be introduced into organisms other than *E. coli*. Figure 15.9 shows the yeast–*E. coli* shuttle vector YEp24, which can be introduced into yeast or *E. coli* cells.

Like *E. coli* cloning vectors, YEp24 has an *ori* sequence that allows it to replicate in *E. coli* and has a dominant selectable marker that con-

fers tetracycline resistance upon *E. coli* cells that contain this vector. YEp24 also contains the selectable marker *URA3* (a wild-type yeast gene for an enzyme required for uracil biosynthesis—in yeast, wild-type genes are symbolized by uppercase letters and the corresponding mutant genes by lowercase letters) that enables yeast *ura3* mutant host cells containing YEp24 to be identified. That is, if YEp24 were used to transform a yeast cell carrying a *ura3* mutation, the yeast cell's phenotype would be changed from uracil-requiring to uracil-independent by the presence of the *URA3* gene in the YEp24 vector. This particular shuttle vector also carries a yeast-specific sequence, 2μ, that allows YEp24 to replicate autonomously when it is in a yeast cell. Thus, YEp24 is able to replicate in yeast and in *E. coli*. Not all shuttle vectors have this ability to replicate in the nonbacterial host. Those that do not will typically integrate into the host cell's chromosomes and be replicated as the host cell's chromosomes replicate.

Keynote *Different kinds of vectors have been developed to construct and clone recombinant DNA molecules. To be useful, all vectors must have three characteristics: they must be able to replicate within their host organism, they must have one or more restriction enzyme cleavage sites at which foreign DNA fragments can be inserted, and they must have one or more dominant selectable markers (such as those that confer antibiotic resistance). Three primary cloning vectors are used to clone recombinant DNA in* E. coli: *plasmids, which are introduced by transformation into the* E. coli *host where they replicate, and bacteriophage λ and cosmids, both of which are packaged into λ phage particles which, in turn, inject the DNA into the* E. coli *host. In λ clones, the injected λ DNA replicates and progeny phages are produced, while in cosmid clones, the cosmids replicate like plasmids.*

Shuttle vectors have the three important features of all cloning vectors, but they can also be used to clone recombinant DNA in more than one organism, for example, in E. coli *and yeast, or in* E. coli *and mammalian cells.*

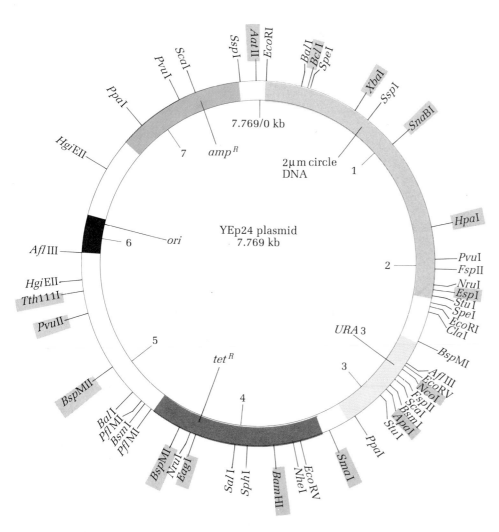

*Figure 15.9 The yeast–*E. coli *shuttle vector YEp24. Unique cestriction enzyme cleavage sites are shown in shaded areas.*

Construction of Genomic Libraries and cDNA Libraries

We have already discussed cloning vectors and restriction enzymes and have outlined in principle how they can be used to form recombinant DNA molecules. In general, researchers are interested in cloning individual genes so that those genes can be studied in detail. It is relatively easy to isolate high-molecular-weight DNA from cells of an organism and to cleave

that DNA into fragments using restriction enzymes. The restriction fragments collectively represent the entire genome of the organism, and they may be inserted into an appropriate cloning vector and cloned. The collection of clones that contains at least one copy of every DNA sequence in the genome is called a **genomic library.** A genomic library can be searched to identify and single out a recombinant DNA molecule that contains a particular gene of interest.

Alternatively, mRNA molecules can be iso-

494
Recombinant
DNA Technology
and the
Manipulation
of DNA

lated from cells and DNA copies, called **complementary DNA (cDNA),** can be made. These cDNA molecules can be inserted into a cloning vector and cloned. If a specific mRNA molecule can be isolated, the corresponding cDNA can be made and cloned. The analysis of that cloned cDNA molecule can then provide information about the gene that encoded the mRNA. More typically, the entire mRNA population of a cell is isolated and a corresponding set of cDNA molecules is generated and inserted into a cloning vector to produce what is called a **cDNA library.** The following sections describe the construction of genomic and cDNA libraries.

Genomic Libraries

A genomic library is a collection of clones that contains at least one copy of every DNA sequence in the genome. Researchers typically want to study only a small segment of an organism's genome—that is, they want to study a specific gene rather than a whole chromosome. One approach to obtaining a clone of this small segment is to isolate it from a genomic library through the use of a specific probe. In this section we focus on the construction of genomic libraries of eukaryotic DNA.

Production of genomic libraries. There are two ways to produce genomic libraries:

1. Genomic DNA is completely digested by a restriction enzyme and the resulting DNA fragments are inserted into a particular cloning vector, either a plasmid, λ, or a cosmid, then put into a host cell and cloned. If the specific gene the researcher wants to study contains restriction sites for the enzyme, the gene will be split into two or more fragments when the DNA is digested by the restriction enzyme. In this case, the gene would then be cloned in two or more pieces. Another drawback is that the average size of the fragment produced by digestion of eukaryotic DNA with restriction enzymes is relatively small (about 4 kbp for restriction enzymes that have six-base-pair recognition sequences). Thus, an entire li-

brary would need to contain a very large number of recombinant DNA molecules, and screening for the specific gene would be very laborious.

2. The problems of genes split into fragments and the large number of recombinant DNA molecules can be avoided by cloning longer DNA fragments (e.g., 40-kbp fragments in cosmids). Longer DNA fragments can be generated by shearing high-molecular-weight (usually 100 to 150 kbp) DNA by mechanical means. If we start with genomic DNA isolated from a large population of cells, the shearing process (e.g., passage through a syringe needle) will produce a population of overlapping DNA fragments. Random fragmentation of DNA molecules can only be achieved by mechanical shearing. However, the DNA fragments produced in this way cannot be inserted directly into a cloning vector since their ends were not generated by cutting with restriction enzymes and, hence, they will not pair with the ends produced by restriction enzyme digestion of the vector. Thus, to clone these DNA fragments, they first must be subjected to several additional enzymatic manipulations to add appropriate ends to the molecules so that the fragments can be cloned.

Since enzymatic manipulations of randomly sheared DNA fragments are relatively complex, a more common approach for producing DNA fragments of appropriate size for constructing a genomic library is to perform a partial digestion of the DNA with restriction enzymes that recognize frequently occurring four-base-pair recognition sequences (Figure 15.10a). The result is a population of overlapping fragments representing the entire genome. Sucrose density gradient centrifugation is then used to collect fragments of the desired size. Those fragments can be cloned directly since the ends of the fragments were produced by restriction enzyme digestion. For example, if the DNA is digested with the enzyme Sau3A, which has the recognition sequence 5'-GATC-3', the ends are complementary to the ends produced by digestion of an appropriate cloning vector with *Bam*HI, which has the recognition sequence 5′-GGATCC-3′ (Figure 15.10b). Recombinant DNA molecules are produced by ligating the *Sau*3A-cut fragments and the *Bam*HI-cut vectors together. These melecules are used to transform *E. coli* cells which, in the case of plasmid and cosmid libraries, are then plated on selective medium to clone the sequences. Each colony that is produced represents a different cloned DNA sequence since each bacterium that gave rise to a colony contained a different recombinant DNA molecule. The library is preserved by picking each colony off the plate using a toothpick and placing it into the medium in a well of a microtiter dish. Each microtiter dish contains 96 wells in an 8-by-12 array, so a large number of dishes are needed for an entire genomic library which, for mammalian cells, may require many thousands of individual clones, even for cosmid clones.

In the case of λ genomic libraries, the recombinant DNA is packaged into λ particles. Those particles are used to infect a culture of *E. coli* and each DNA fragment is cloned by the repeated rounds of infection and lysis that each original λ phage goes through in the culture. Eventually, the culture becomes transparent as all the bacteria have been lysed and a population of progeny λ phages, with a concentration of 10^{10} to 10^{11} phages/ml, is produced with many, many representatives of each of the recombinant DNA molecules that were originally constructed in the test tube. Such phage suspensions may be stored in the refrigerator for years.

The aim of the methods just described is to produce a library of recombinant molecules that is as complete as possible. However, not all sequences of the eukaryotic genome are equally represented in such a library. For example, if the restriction sites are very far apart or extremely close together in a particular region, the chances of obtaining a fragment of clonable size are small. Additionally, some regions of eukaryotic chromosomes may contain sequences that affect

a) Partial digestion of DNA by a restriction enzyme, e.g., *Sau*3A, generates a series of overlapping fragments, each with identical 5′ GATC sticky ends

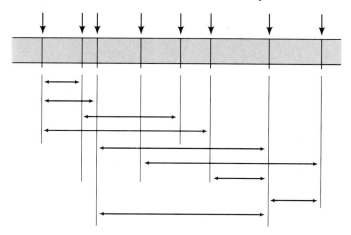

b) Resulting fragments may be inserted into *Bam*HI sites of plasmid cloning vector

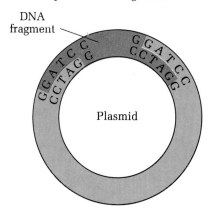

Figure 15.10 Use of restriction enzyme to produce DNA fragments of appropriate size for constructing a genomic library.

the ability of vectors containing them to replicate in *E. coli:* these sequences would then be lost from the library. There is also evidence that the presence of tandemly repeated sequences in the cloned eukaryotic DNA segment can lead to deletion of some of the sequence by recombination during vector replication in *E. coli*.

Lastly, the probability of having any DNA sequence represented in the genomic library can

496

*Recombinant
DNA Technology
and the
Manipulation
of DNA*

be calculated from the formula

$$N = \frac{\ln(1 - P)}{\ln(1 - f)}$$

where N is the necessary number of recombinant DNA molecules, P is the probability desired, and f is the fractional proportion of the genome in a single recombinant DNA molecule (i.e., f is the average size, in kilobase pairs, of the fragments used to make the library divided by the size of the genome, in kilobase pairs). For example, for a 99 percent chance that a particular yeast DNA fragment is represented in a library of 40-kbp fragments in a cosmid genomic library, where the yeast genome size is 14,000 kbp,

$$N = \frac{\ln(1 - 0.99)}{\ln[1 - (40/14{,}000)]}$$
$$= 1610$$

That is, 1610 recombinant DNA molecules of 40-kbp size would be needed to be 99 percent sure that a given DNA sequence was represented in the library.

Keynote   *A genomic library is a collection of clones that contains at least one copy of every DNA sequence in an organism's genome. Like regular book libraries, genomic libraries are great resources of information; in this case, the information is about the genome. They are used for isolating specific genes and for studying the organization of the genome, among many other things.*

cDNA Libraries

Another approach to constructing fragments to be cloned is a process called **cDNA cloning.** In this process, mRNA molecules are isolated from cells and DNA copies (called **complementary DNA** or **cDNA**) of those mRNAs are made and inserted into a cloning vector. When the entire poly(A)-mRNA population of a cell is isolated, and cDNA copies are made and inserted into cloning vectors, the result is a **cDNA library.** Since a cDNA library reflects the gene activity of the cell type at the time the mRNAs are isolated,

the construction and analysis of cDNA libraries is useful for comparing gene activities in different cell types of the same organism, since there would be similarities and differences in the clones represented in the cDNA libraries of each cell type.

Synthesis of cDNA molecules. RNA extracted from cells contains ribosomal RNA, transfer RNA, and messenger RNA molecules (plus small nuclear RNA molecules in the case of eukaryotes). For eukaryotic organisms (the focus of our discussion), the mRNA in this mixture is unique among the RNAs in that it contains a poly(A) tail. The mRNAs can be purified from the mixture, then, by passing the RNA molecules over a column to which short chains of deoxythymidylic acid, called oligo(dT) chains, have been attached. As the RNA molecules pass through an oligo(dT) column, the poly(A) tails on the mRNA molecules form complementary base pairs with the oligo(dT) chains, with the result that the mRNAs are captured on the column while the tRNAs and rRNAs pass through. The mRNAs are subsequently released and collected, for example, by increasing the ionic strength of the buffer passing over the column so that the hydrogen bonds will be disrupted.

Once a pure population of mRNA molecules has been isolated, double-stranded complementary DNA (cDNA) copies are made in vitro using the enzyme reverse transcriptase (Figure 15.11). This process was described generally in Chapter 12, pp. 377–378. First, a short oligo(dT) chain (a short, single-stranded DNA chain containing only Ts) is hybridized to the poly(A) tail at the 3′ end of each mRNA strand. The oligo(dT) acts as a primer for reverse transcriptase, which makes a DNA strand that is complementary in sequence to the mRNA strand. In this process, a small hairpin loop is produced at the 3′ end of the single-stranded DNA. The mRNA strands of the RNA-DNA hybrids are degraded by NaOH treatment and the 3′ end of the hairpin loop serves as a primer for DNA polymerase I, which catalyzes the synthesis of a DNA strand complementary to the first DNA strand. The result is a double-stranded DNA molecule, the sequence of

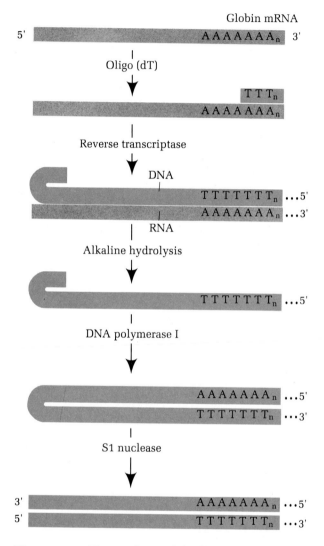

Globin mRNA

Oligo (dT)

Reverse transcriptase

DNA

RNA

Alkaline hydrolysis

DNA polymerase I

S1 nuclease

Figure 15.11 The synthesis of double-stranded cDNA from a polyadenylated globin mRNA using reverse transcriptase and DNA polymerase I.

molecules and produce a cDNA library: One method involves using *homopolymeric tails* (i.e., adding a chain of one type of nucleotide to the DNA), and the other involves using specific, short DNA sequences, called *linkers,* that contain cleavage sites for a particular restriction enzyme. The latter method is easier and is used more frequently.

Insertion of cDNA into a vector using homopolymeric tails. Figure 15.12 shows an example of cDNA cloning by using homopolymeric tails. In this case, the plasmid cloning vector pBR322 is cleaved with *Pst*I, thereby disrupting the amp^R gene. Then, in a reaction mixture containing dGTP and the enzyme *terminal transferase,* short oligo(dG) tails (i.e., short, single strands of DNA containing the base guanine) are added to the two 3' ends. Similarly, using dCTP and terminal transferase, short oligo(dC) tails are added to the two 3' ends of the cDNA molecules. Because the two types of tails added to the vector and to the cDNA are complementary, a recombinant DNA molecule is produced with the cDNA inserted into the plasmid when the vector and the cDNA are mixed. The recombinant DNA molecule has gaps in the sugar-phosphate backbone at the junctions between the vector and the cDNA and is held together only by hydrogen bonds between the oligo(dG) and oligo(dC) tails. The gaps are repaired by *E. coli* enzymes once the recombinant DNA molecule is introduced into *E. coli* cells by transformation. In this particular case, transformant *E. coli* cells will be tetracycline-resistant but not ampicillin-resistant (since the cDNA was inserted into the amp^R gene).

Insertion of cDNA into a vector using linkers. Figure 15.13 (see p. 499) illustrates the cloning of cDNA using a **restriction enzyme cleavage site linker,** or **linker,** which is a relatively short, double-stranded oligodeoxyribonucleotide about 8 to 12 nucleotide pairs long that is synthesized by chemical means and contains the cleavage site for a particular restriction enzyme within its sequence. The linker shown in Figure 15.13 is the *Bam*HI linker (abbreviated in the figure to just the six-base-pair recognition sequence for the enzyme).

which is derived from the original poly(A)-mRNA molecule. By the use of the enzyme S1 nuclease, which recognizes and cleaves any single-stranded regions of DNA, the hairpin loop that remains is cleaved to produce a regular, double-stranded cDNA molecule.

Production of cDNA libraries. Once cDNA molecules have been synthesized, there are two principal ways to construct recombinant DNA

498
*Recombinant
DNA Technology
and the
Manipulation
of DNA*

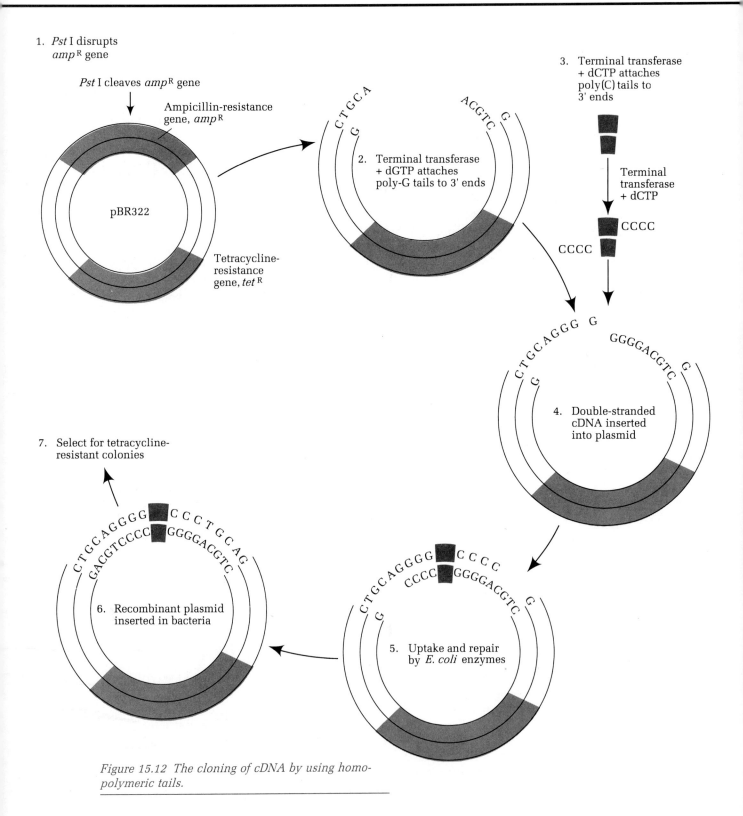

1. *Pst* I disrupts
   *amp*R gene

   Pst I cleaves *amp*R gene

   Ampicillin-resistance
   gene, *amp*R

   pBR322

   Tetracycline-
   resistance
   gene, *tet*R

2. Terminal transferase
   + dGTP attaches
   poly-G tails to 3' ends

3. Terminal transferase
   + dCTP attaches
   poly(C) tails to
   3' ends

   Terminal
   transferase
   + dCTP

   CCCC

   CCCC

4. Double-stranded
   cDNA inserted
   into plasmid

5. Uptake and repair
   by *E. coli* enzymes

6. Recombinant plasmid
   inserted in bacteria

7. Select for tetracycline-
   resistant colonies

Figure 15.12 *The cloning of cDNA by using homopolymeric tails.*

499
Identifying
Specific Cloned
Sequences in
cDNA Libraries
and Genomic
Libraries

Both the cDNA molecules and the linkers have blunt ends. At high concentrations of T4 DNA ligase, *Bam*HI linkers can be added to each end of a cDNA molecule. Sticky ends are produced in the cDNA molecule by cleaving the cDNA + linker with *Bam*HI restriction enzyme. The resulting DNA fragment can then be ligated into a suitable cloning vector, such as pBR322, that has also been cleaved with *Bam*HI. The resulting recombinant DNA molecule can then be transformed into an *E. coli* host cell for cloning.

Keynote   *If a specific mRNA or a population of mRNAs can be purified from a cell, it is possible to make DNA copies of those mRNA molecules. First, the enzyme reverse transcriptase makes a single-stranded DNA copy of the mRNA and then DNA polymerase I makes a double-stranded DNA copy called complementary DNA (cDNA). This cDNA can be spliced into cloning vectors using homopolymeric tails or restriction enzyme cleavage site linkers.*

Identifying Specific Cloned Sequences in cDNA Libraries and Genomic Libraries

Identifying Specific Clone Sequences in a cDNA Library

A cDNA clone can be identified in a cDNA library in a number of ways. One way is to select the cDNA clone that codes for a specific protein (Figure 15.14). This approach requires that a protein can be purified in sufficient quantities so that antibodies can be made against the protein.

Each transformed *E. coli* cell theoretically contains a different cDNA clone and each transformed cell is grown separately; the DNA is extracted from the resulting population of cells (including the many clonal cDNA copies) and denatured to single strands. The single-stranded DNA molecules are bound to nitrocellulose filters (or nylon-based membranes containing nitrocellulose) and a population of mRNA mole-

cules isolated from a cell is passed through the filters. The mRNA molecules that are complementary in sequence (i.e., that can base pair) to the pure cDNA on the filter hybridize to the cDNA and become bound to the filter; the rest of the RNA passes through. Next, the bound mRNA is washed off the filter by denaturing the RNA-DNA hybrids and the mRNA is added to a cell-free system that can translate the mRNA into protein. Since the cDNA on the filter is of only

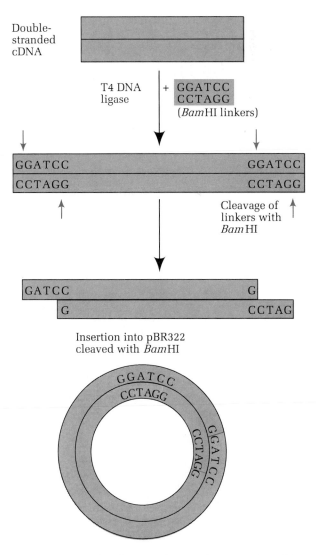

Figure 15.13 The cloning of cDNA by using Bam*HI linkers.*

500
*Recombinant
DNA Technology
and the
Manipulation
of DNA*

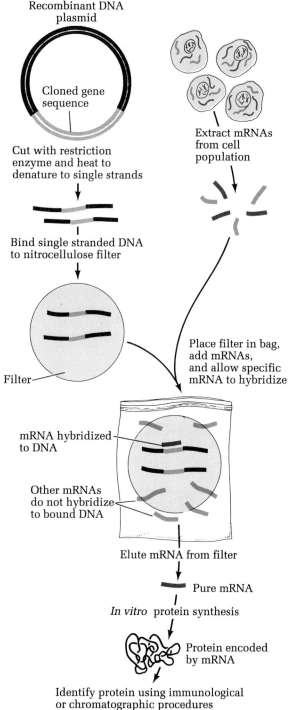

Figure 15.14 *Identifying specific cDNA plasmids in a cDNA library.*

one type, only one mRNA sequence can hybridize to it, and hence only one type of protein is synthesized in the cell-free system. If we have an antibody to a specific protein, the cDNA clone coding for that protein can be identified. That cDNA clone can be used as a probe in other experiments, for example, to analyze the genome of the same or other organisms for homologous sequences, or for quantifying mRNA production.

The more abundant an mRNA species is in a cell extract, the easier it is to identify the corresponding cDNA sequence in a cDNA library. For that reason, the first cDNA probes that were made were those for abundant proteins such as hemoglobin.

Identifying Specific Cloned Sequences in a Genomic Library

Ideally, a genomic library contains every DNA sequence of the organism represented as an overlapping array of cloned DNA fragments. Given the existence of a probe, such as a cloned cDNA probe, it is possible to identify in the library the cloned gene that codes for the mRNA molecule from which the cDNA was made, and then to isolate it for characterization.

Screening plasmid and cosmid libraries. The process of screening for specific DNA sequences in a plasmid or cosmid library is shown in Figure 15.15. The starting point is the set of microtiter dishes that comprise the library. Replicas of the set of cultures in each dish are printed onto a nitrocellulose filter that has been placed on a petri plate of selective medium appropriate for the recombinant plasmids or cosmids. Incubation of the plate gives rise to a matrix of colonies growing on the surface of the filter as a result of nutrients diffusing through the filter from the medium. Each filter is peeled from the growth medium and placed sequentially on filter papers that have been soaked in different solutions to lyse the bacteria, denature the DNA to single strands, and bind the single-stranded DNA to the filter.

Next, the filter is placed in a plastic bag (e.g.,

501
Identifying
Specific Cloned
Sequences in
cDNA Libraries
and Genomic
Libraries

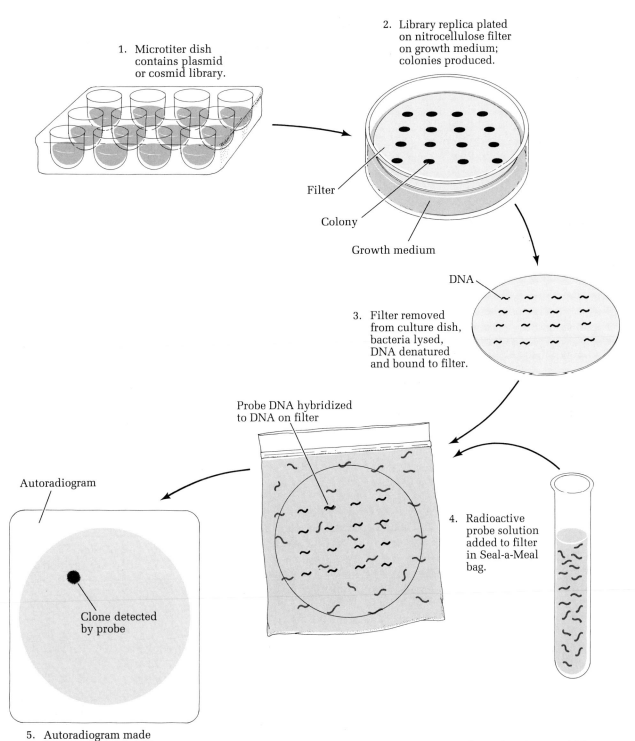

1. Microtiter dish contains plasmid or cosmid library.

2. Library replica plated on nitrocellulose filter on growth medium; colonies produced.

Filter

Colony

Growth medium

DNA

3. Filter removed from culture dish, bacteria lysed, DNA denatured and bound to filter.

Probe DNA hybridized to DNA on filter

Autoradiogram

Clone detected by probe

4. Radioactive probe solution added to filter in Seal-a-Meal bag.

5. Autoradiogram made from washed, dried filter. Dark spots indicate clones detected by probe.

Figure 15.15 Screening plasmid and cosmid libraries for specific DNA sequences.

502
*Recombinant
DNA Technology
and the
Manipulation
of DNA*

Seal-a-Meal bags, which are the type used for boiling vegetables), and incubated with the cDNA probe, which has been made radioactive by incorporating ^{32}P. The radioactive labeling of the cDNA probe may be done by the process of **nick translation** (Figure 15.16). In nick translation, the DNA to be labeled is incubated in a buffer containing the enzymes DNase I and DNA polymerase I, and the four deoxyribonucleoside triphosphate DNA precursors (dATP, dGTP, dCTP, and dTTP). In the reaction mixture, one or more of the deoxyribonucleoside triphosphate precursors has ^{32}P in the phosphate group, which is attached to the 5′ carbon of the deoxyribose sugar. (In Figure 15.16, the dCTP is ^{32}P-labeled.) This phosphate group is called the α-phosphate because it is the first in the chain of three; the α-phosphate is used in forming the phosphodiester bonds of the sugar-phosphate backbone. In the reaction mixture, the DNase I enzyme makes "nicks," that is, single-stranded gaps in the backbones of the DNA molecules. DNA polymerase I recognizes the nicks and proceeds to attempt to repair the DNA by excising a number of nucleotides in the 3′-to-5′ direction

starting at the nick by using its 3′-to-5′ exonuclease activity (see Chapter 10, pp. 329–330). The resulting single-stranded regions are filled in by the 5′-to-3′ polymerizing activity of DNA polymerase I. Wherever a G nucleotide is on the intact template strand, the new DNA strand will have a ^{32}P-labeled C nucleotide incorporated. In this way, the DNA becomes radioactively labeled with ^{32}P.

To prepare the radioactive DNA for use as a probe, the DNA is boiled and then quickly cooled on ice to produce single-stranded DNA molecules. When a suspension of these molecules is added to the filters to which the denatured (single-stranded) DNA from each colony has been bound, the potential exists for the formation of DNA-DNA hybrids between the probe and the colony DNA. This will occur if the two sequences are complementary. If the cDNA probe derived from the mRNA for β-globin, for example, that probe would hybridize with DNA bound to the filter that encodes the β-globin mRNA, that is, the β-globin gene. After sufficient time for hybridization has elapsed (approximately 24 hours), the filters are washed to re-

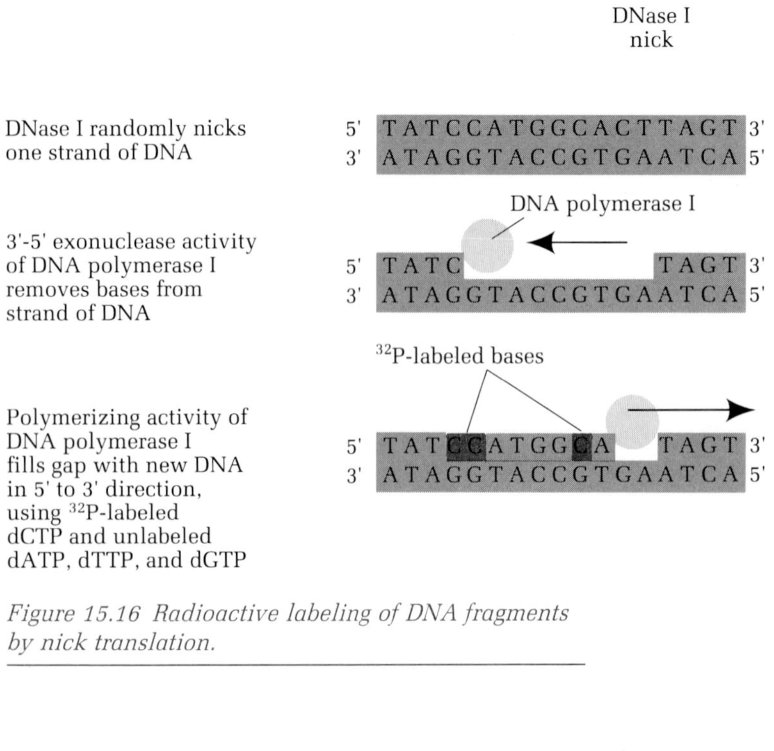

Figure 15.16 Radioactive labeling of DNA fragments by nick translation.

503
Identifying
Specific Cloned
Sequences in
cDNA Libraries
and Genomic
Libraries

move unbound probe, dried, placed against X-ray film, and left in the dark for a period of time (1 hour to overnight) to produce what is called an *autoradiogram.* When the film is developed, dark spots are evident on the autoradiogram wherever the radioactive probe bound to the colony-derived DNA as a result of the decay of the ^{32}P atoms. This decay results in changes in the silver grains of the film and produces the dark spots. From the position(s) of the spot(s) on the film, the locations of the original bacterial culture(s) in the microtiter dish(es) can be determined and the clone(s) of interest isolated for further characterization.

Screening λ genomic libraries. In the case of screening a genomic library made in λ for a clone of interest, samples of the phage suspension are used to infect susceptible bacteria, which are then spread onto nutrient plates so that plaques are produced (Figure 15.17). Each plaque consists of a pool of progeny phages produced by successive rounds of infection and lysis of bacteria initiated originally by a single phage infection. Each time a bacterium is lysed by a phage infection, not only are progeny phages released, but also many copies of replicated λ DNA that were not packaged into particles are also released. A piece of nitrocellulose filter is laid carefully on top of the plate to sit for a few minutes, permitting the λ DNAs to bind to the filter; this is called making a *plaque lift.* The filter is then removed and placed on blotting paper soaked in an alkaline solution, which causes the DNA to denature into single strands. After neutralization, the filters are incubated with probe as described for the screening of the plasmid and cosmid libraries.

A number of plaque lifts can be made from each plate of λ plaques, allowing the investigator to screen the same library simultaneously with a number of different probes. Evidence of hybridization of the probe with DNA from a plaque is seen as a spot on the autoradiogram. The position of the spot on the film is used to find the plaque on the plate from which the plaque lift was made. The plaque can be punched out with the end of a Pasteur pipet,

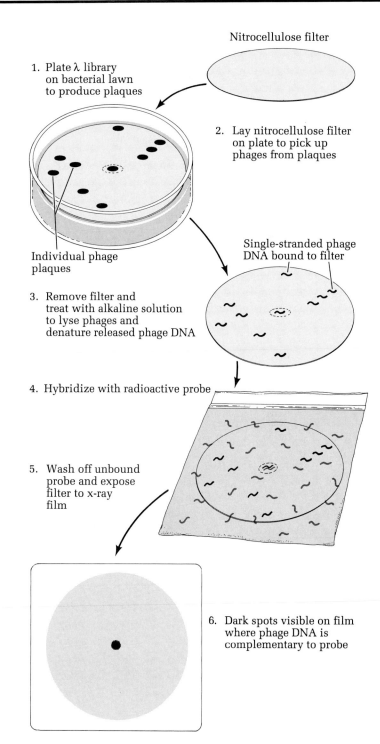

1. Plate λ library on bacterial lawn to produce plaques

Nitrocellulose filter

2. Lay nitrocellulose filter on plate to pick up phages from plaques

Individual phage plaques

Single-stranded phage DNA bound to filter

3. Remove filter and treat with alkaline solution to lyse phages and denature released phage DNA

4. Hybridize with radioactive probe

5. Wash off unbound probe and expose filter to x-ray film

6. Dark spots visible on film where phage DNA is complementary to probe

Figure 15.17 *Screening a bacteriophage λ library for a specific gene clone.*

504
*Recombinant
DNA Technology
and the
Manipulation
of DNA*

and the phages with the cloned DNA fragment of interest can be propagated in a bacterial culture to produce large quantities of phage DNA for analysis.

Identifying Specific DNA Sequences in Libraries Using Heterologous Probes

Originally cDNA probes were used to identify and isolate genes. To date, a very large number of genes have been cloned from both prokaryotes and eukaryotes. Thus, it is more common these days to identify genes of interest in a genomic library by using clones of equivalent genes from other organisms as probes. Such probes are called *heterologous probes* and their effectiveness depends upon a good degree of homology between the probes and the genes. For that reason, the greatest success with this approach has come either with highly conserved genes or with probes from a species closely related to the organism from which a particular gene is to be isolated.

Identifying Genes in Libraries by Complementation of Mutations

For those organisms in which genetic systems of analysis have been well developed and for which there are well-defined mutations, it is possible to clone genes by complementation of those mutations. Consider as an example the yeast *Saccharomyces cerevisiae*. This organism has been extremely well exploited genetically, a large number of mutations have been generated and characterized, and integrative and replicative transformation systems using *E. coli*-yeast shuttle vectors (see earlier) have been developed.

To clone a yeast gene by complementation, first a genomic library is made of DNA fragments from the wild-type yeast strain in a replicative shuttle vector such as YEp24 (see pp. 493: Figure 15.9). The library is then used to transform a host yeast strain carrying a mutation to enable transformants to be selected—*ura3* in the case of YEp24—and a mutation in the gene for which the wild-type gene is to be cloned. As an

example, let us consider an *arg1* mutation, which would result in an inactive enzyme for arginine biosynthesis and, hence, a growth requirement for arginine. When a population of *ura3 arg1* yeast cells is transformed with the YEp24 genomic library, some cells will receive plasmids containing the normal gene for the arginine biosynthesis enzyme, which is called the *ARG1* gene in the yeast nomenclature. The plasmid's *ARG1* gene will be transcribed and the resultant mRNA will be translated to produce a normal, functional enzyme for arginine biosynthesis, thereby allowing the cell to grow in the absence of arginine despite the presence of a defective *arg1* gene on the cell's chromosomes. The *ARG1* gene is said to overcome the functional defect of the *arg1* mutation by *complementation* of that mutation. The cells transformed by the *ARG1*-containing plasmid vector are isolated in this experiment selection on a growth medium lacking arginine (and lacking uracil to select for transformation). The plasmid is then isolated from the cells and the cloned gene is characterized.

Keynote Specific sequences in cDNA libraries and genomic libraries can be identified using a number of approaches, including the use of specific cDNA probes, heterologous probes, and complementation of mutations.

Techniques for the Analysis of Genes and Gene Transcripts

Restriction Enzyme Analysis of Genes

As part of the analysis of genes, it is often useful to determine the arrangement and specific locations of restriction enzyme cleavage sites. This information is useful, for example, for comparing homologous genes in different species, for analysing intron organization, or for planning experiments to clone parts of a gene, such as its promoter or controlling sequences, into a vector. The arrangement of restriction sites in a gene can be analyzed without actually cloning the gene by using a cDNA probe or a closely related heterologous gene probe. The process of analysis proceeds as follows:

1. Samples of high-molecular-weight DNA are cut with different restriction enzymes (Figure 15.18), each of which will produce DNA fragments of different lengths (depending on the locations of the recognition sequences on the DNA molecules). For eukaryotic DNA, for example, a restriction enzyme will produce a large number of fragments of various sizes because of the distribution of cleavage sites in the genome.

2. The DNA fragments are then separated according to their molecular size by *agarose gel electrophoresis.* The gel is a rectangular slab of a firm, gelatinous material (agarose), with a matrix of pores through which the DNA passes in an electric field. Each gel is divided into a number of lanes so that a number of different samples can be analyzed simultaneously. Since DNA is negatively charged because of its phosphates, the DNA migrates toward the positive pole. Because they can "squirm" more readily through the small pores in the gel, small DNA fragments move more rapidly through the gel than large DNA fragments. Thus, the smallest fragments migrate the farthest distance while the largest fragments migrate the least.

   After electrophoresis, the DNA is stained with ethidium bromide so that it can be seen under ultraviolet light. When total cellular DNA is digested with a restriction enzyme, the result is usually a smear of fluorescence down the length of the gel lane. This smear is continuous because the enzyme produces fragments of all sizes.

3. The DNA fragments are then transferred to a nitrocellulose filter so that they are in exactly the same relative position on the filter as they were on the gel (Figure 15.18). The transfer to the nitrocellulose filter is done by the **Southern blot technique** (named after its inventor, E. M. Southern), diagramed in Figure 15.19. In brief, the gel is first soaked in alkali to denature the double-stranded DNA into

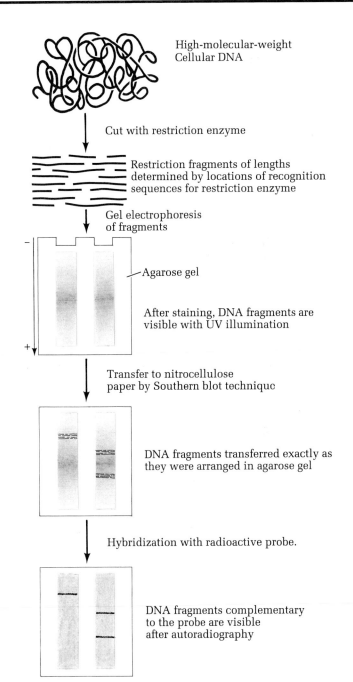

Figure 15.18 Procedure for the analysis of cellular DNA for the presence of sequences complementary to a radioactive probe, such as a cDNA molecule made from an isolated mRNA molecule. The hybrids, shown as three bands in this theoretical example, are visualized by autoradiography.

506
*Recombinant
DNA Technology
and the
Manipulation
of DNA*

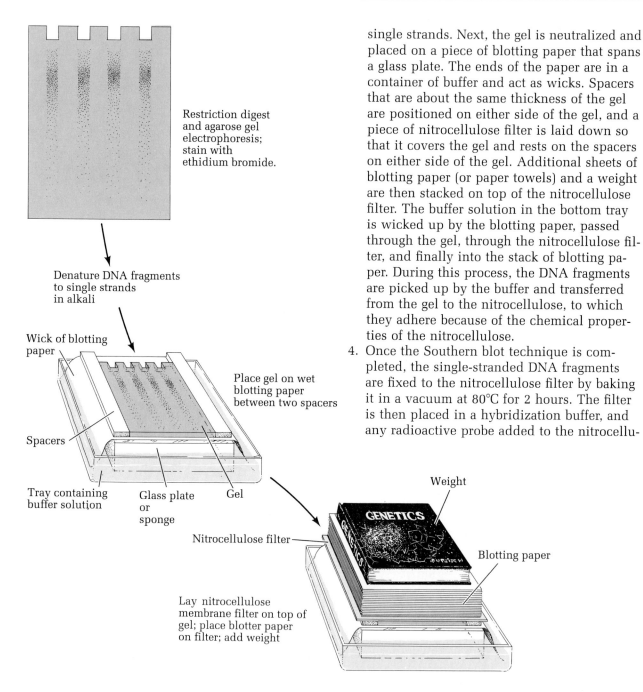

Restriction digest and agarose gel electrophoresis; stain with ethidium bromide.

Denature DNA fragments to single strands in alkali

Wick of blotting paper

Spacers

Tray containing buffer solution

Glass plate or sponge

Gel

Place gel on wet blotting paper between two spacers

Nitrocellulose filter

Lay nitrocellulose membrane filter on top of gel; place blotter paper on filter; add weight

GENETICS

Weight

Blotting paper

Figure 15.19 Southern blot technique for the transfer of DNA fragments from an electrophoretic gel to nitrocellulose paper.

single strands. Next, the gel is neutralized and placed on a piece of blotting paper that spans a glass plate. The ends of the paper are in a container of buffer and act as wicks. Spacers that are about the same thickness of the gel are positioned on either side of the gel, and a piece of nitrocellulose filter is laid down so that it covers the gel and rests on the spacers on either side of the gel. Additional sheets of blotting paper (or paper towels) and a weight are then stacked on top of the nitrocellulose filter. The buffer solution in the bottom tray is wicked up by the blotting paper, passed through the gel, through the nitrocellulose filter, and finally into the stack of blotting paper. During this process, the DNA fragments are picked up by the buffer and transferred from the gel to the nitrocellulose, to which they adhere because of the chemical properties of the nitrocellulose.

4. Once the Southern blot technique is completed, the single-stranded DNA fragments are fixed to the nitrocellulose filter by baking it in a vacuum at 80°C for 2 hours. The filter is then placed in a hybridization buffer, and any radioactive probe added to the nitrocellulose filter at this point will bind to any complementary DNA fragment (Figure 15.18).

5. After the hybridization of the probe and the DNA fragments, the filter is washed extensively to remove radioactive probes that have not hybridized. The filter is then dried, and

an autoradiograph is prepared to determine the position(s) of the hybrids. If DNA fragment size markers (e.g., λ DNA cut with *Hind*III or with *Hind*III and *Eco*RI) are separated in a different lane in the agarose gel electrophoresis process, the sizes of the genomic restriction fragments that hybridized with the probe can be calculated. From the fragment sizes obtained, a **restriction map** can be generated to show the relative positions of the restriction sites. Suppose, for example, that using only BamHI produces a DNA fragment of 3 kbp that hybridizes with the radioactive probe. If a combination of *Bam*HI and *Pst*I is then used and produces two DNA fragments, one of 1 kbp and the other of 2 kbp, we would deduce that the 3-kbp *Bam*HI fragment contains a *Pst*I restriction site 1 kbp from one end and 2 kbp from the other end. Further analysis with other enzymes, individually and combined, enables the researcher to construct a map of all the enzyme sites relative to all other sites.

Restriction Enzyme Analysis of Cloned DNA Sequences

Cloned DNA sequences are also analyzed to determine the arrangement and specific locations of restriction sites for essentially the same reasons as were stated in the previous section. Because cloned DNA sequences represent a homogeneous population of relatively low-molecular-weight DNA molecules, restriction enzyme cleavage of cloned DNA sequences produces a relatively small number of discretely sized DNA fragments rather than a continuous array of different sized DNA fragments as is the case with cleavage of genomic DNA. These DNA fragments are readily visualized following agarose gel electrophoresis and ethidium bromide staining, permitting restriction maps to be constructed without the need for hybridization with a radioactive probe and autoradiography.

Analysis of Gene Transcripts

A related blotting technique to the Southern blot

technique (called **northern blotting**) has been developed to analyze RNA rather than DNA. (In this case, the name is derived, not from a person, but from the opposite of Southern.) In northern blotting, RNA extracted from a cell is separated by size using gel electrophoresis, and the RNA molecules are transferred and bound to a filter using an essentially identical procedure as for Southern blotting. After hybridization with a radioactive probe, an autoradiograph is taken of the bands corresponding to the RNA species that were complementary to the probe. If appropriate RNA size markers are present in one of the gel lanes, the sizes of the RNA species identified with the probe can be determined. Northern blotting is useful in several kinds of experiments. For example, northern blotting can reveal the size of the specific mRNA encoded by a gene. In some cases, a number of different mRNA species encoded by the same gene have been identified in this way, suggesting that either different promoter sites can be used, different terminator sites can be used, or alternative mRNA processing can occur. Northern blotting can also be used to investigate whether or not an mRNA species is present in a cell, or how much of it is present. This type of experiment is useful for determining levels of gene activity, for instance, during development, in different cell types of an organism, or in cells before and after they are subjected to specific physiological stimuli.

Keynote *Genes and cloned DNA sequences are often analyzed to determine the arrangement and specific locations of restriction sites. The analytical process involves cleavage of the DNA with restriction enzymes, followed by separation of the resulting DNA fragments by agarose gel electrophoresis, and staining of the DNA fragments with ethidium bromide so that they may be visualized with ultraviolet light. The DNA fragments produced by cleavage of cloned DNA sequences can be seen as discrete bands, enabling restriction maps to be constructed based on the calculated molecular lengths of the DNA in the bands. DNA fragments produced by cleavage of genomic DNA*

508
Recombinant
DNA Technology
and the
Manipulation
of DNA

show a wide range of sizes, resulting in a continuous smear of DNA fragments in the gel. In
this case, specific gene fragments can only be
visualized by transferring the DNA fragments to
a nitrocellulose filter in a procedure called the
Southern blot technique, hybridizing a specific
radioactive probe with the DNA fragments, and
detecting the hybrids by autoradiography. At
that point, a restriction map can be made.

Gene transcripts can be analyzed by separating mRNA species by gel electrophoresis, transferring the mRNAs to a filter by the northern
blot technique, and hybridizing with a specific
radioactive probe.

Sequencing of DNA

Cloned DNA fragments may be analyzed to determine the base pair sequence of the DNA. This
information is useful, for example, for identifying gene sequences and controlling sequences
within the fragment, and for comparing the sequences of homologous genes from different organisms.

Two techniques for the *rapid sequencing of
DNA molecules* were developed independently
by F. Sanger and A. R. Coulson in 1975 and by
A. M. Maxam and W. Gilbert in 1977. The Sanger and Coulson method uses enzymes to analyze
DNA sequences, while the Maxam and Gilbert
method uses chemicals. Coupled with the technology for cloning specific DNA fragments, these
methods (both of which are used currently) have
allowed us to rapidly advance our knowledge of
gene structure and function, as we have seen in
previous chapters where various "consensus sequences" were determined from DNA sequencing experiments. The Maxam and Gilbert
method is described below.

The Maxam and Gilbert technique uses specific chemical reactions to break the DNA at specific nucleotides. The starting point is a homogeneous population of DNA molecules of about
two hundred to a thousand base pairs in length.
In our hypothetical example, however, we will
sequence a DNA fragment of only ten base pairs.

In brief, the steps for DNA sequencing by the
Maxam and Gilbert technique are as follows:

1. *Labeling the DNA:* Each strand of the double-
stranded DNA molecule is radiolabeled with
^{32}P (a radioactive isotope of phosphorus) at
the 5′ or 3′ end of a chain. The DNA is then
either denatured into single strands, each of
which is labeled at only one end, or cut with
an enzyme somewhere along its length to
generate two double-stranded fragments, each
of which has only one strand with a terminal
^{32}P molecule.

2. *Chemical modification and cleavage of the
DNA:* In our hypothetical sequencing experiment we will begin with a sample homogeneous population of single-stranded DNA
molecules ten base pairs long, radiolabeled at
the 5′ end:

$$^{32}\text{P 5}'\ \frac{?}{1}\ \frac{?}{2}\ \frac{?}{3}\ \frac{?}{4}\ \frac{?}{5}\ \frac{?}{6}\ \frac{?}{7}\ \frac{?}{8}\ \frac{?}{9}\ \frac{?}{10}\ 3'$$

After the sample is divided into four fractions, each fraction is chemically treated in a
different way to modify and remove one of
the four bases, thus breaking the DNA backbone at that point.

The first fraction is treated to indicate the
position of every G in the strand by cleaving
the DNA at each G. Because this reaction
works to a lesser extent on A's in the chain,
it is designated G > A; that is, this reaction
works to identify G more than it does A. The
second fraction is treated to cleave the DNA
at A's. Since this reaction cleaves G's to a
lesser extent, it is designated A > G. The
third fraction is treated to cleave the DNA at
T's and C's, and the fourth fraction is treated
to cleave only at C's in the chain.

3. *Electrophoretic analysis of DNA fragments:*
The cleaved chains from each reaction are
then separated according to their length by
polyacrylamide gel electrophoresis. In this
technique, a jellylike slab of a substance
called polyacrylamide in the presence of an
appropriate buffer is used to separate electrically charged molecules in an electric field.

The single-stranded DNA molecules migrate at a rate related to their length (the smallest fragment migrates fastest, the largest migrates slowest). After electrophoresis this ladder gel is subjected to autoradiography (see pp. 502–503); that is, it is placed against an X-ray film to determine the distribution of fragments. Figure 15.20 diagrams the autoradiograph expected of the four groups. Figure 15.21 is a photograph of an actual autoradiograph from a Maxam-Gilbert DNA sequencing experiment.

The first lane of the autoradiograph shows the result of the G > A reaction. Since the first reaction shows preference for G over A, the fragments generated by cleavage at G's appear as thicker bands than those generated by cleavage at A's. The actual results may thus be interpreted to mean that the second and seventh bases were G's and that the first, sixth, and eighth were A's.

The second lane shows the A > G results. In this case, as we would expect, the band intensities are exactly the opposite of those in the first lane, confirming the location of A's and G's.

The last two lanes locate the C and T bases in the chain. Lane 3 shows the C + T positions, and lane 4 shows the C positions. Comparison of the two lanes therefore enables the C and T positions to be deduced: A band appearing in both lanes indicates a C, while a band appearing only in lane 3 indicates a T. The DNA sequence is read from the band positions, starting at position 1 (the smallest fragment) and deducing the base from the banding patterns at each step of the ladder. In the example position 1 has two bands, the one in the A > G lane being denser than the one in the G > A lane, thereby indicating an A. Position 2 has a dense band in G > A and a faint band in A > G, indicating a G. Positions 3 and 4 have bands in both the C + T and C lanes, indicating C's. Position 5 has a band in the C + T lane but no band in the C lane, indicating a T. Following similar logic, positions 6 through 10 are A, G, A, C, and T, respectively.

Sequence of strand being analyzed:

5' ^{32}P–A–G–C–C–T–A–G–A–C–T   3'
Nucleotide   1 2 3 4 5 6 7 8 9 10

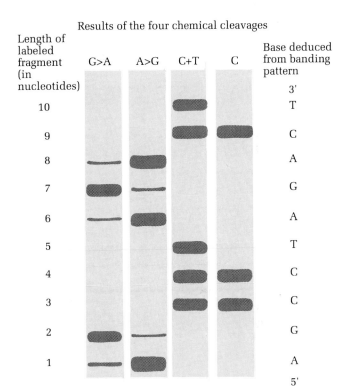

Results of the four chemical cleavages

Figure 15.20 Autoradiogram expected for the four Maxam and Gilbert sequencing reactions with the DNA fragment 5'–A–G–C–C–T–A–G–A–C–T–3'.

Sequences determined by these kinds of experiments are usually entered into computer data bases so that other researchers can compare a variety of sequences for homologous regions, controlling site similarities, and so on. Appropriate computer programs can search DNA sequences for protein-coding regions by looking for an initiator codon in frame with a chain terminating codon (called an **open reading frame**). Other programs can be used to translate a cloned DNA sequence theoretically into an amino acid sequence and to predict secondary and tertiary structures of the polypeptide.

510
Recombinant
DNA Technology
and the
Manipulation
of DNA

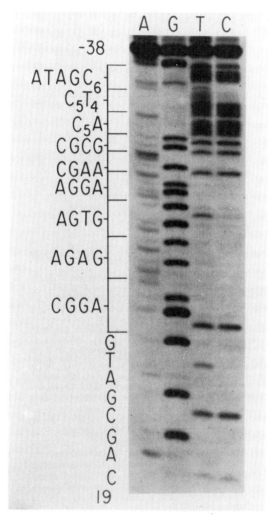

Figure 15.21 Photograph of an autoradiograph from a Maxam-Gilbert DNA sequencing experiment.

Applications of Recombinant DNA Technology

Recombinant DNA technology has many applications, including the diagnosis of human genetic diseases such as sickle-cell anemia; the synthesis of commercially important products such as human insulin, human growth hormone, and interferon; in vitro modifications of genes; and genetic engineering of plants. Some of these applications will be discussed in this section.

Recombinant DNA Approaches to Diagnoses of Genetic Disease

In Chapter 14, we discussed biochemical approaches to diagnosing genetic diseases (pp. 462–464). Recently, recombinant DNA techniques have been used to investigate a number of structural genes whose mutations result in human diseases, such as sickle-cell anemia, Huntington's chorea, and diabetes. This research has led, in a number of instances, to simple diagnostic procedures that can be used for carrier detection and for the analysis of the genetic condition of the fetus. Most of the basic techniques used in these procedures were described in Chapter 14.

Recombinant DNA approaches to detecting genetically based human disease (or genetically based disease in other organisms, for that matter) require cellular DNA as the starting point. Cellular DNA can be isolated from white blood cells in parental blood (in the case of carrier detection) or from fetal cells isolated by amniocentesis and cultured (in the case of fetal diagnosis) (see Figure 14.23). The DNA is digested with a restriction enzyme, which will cut the DNA at the specific recognition sequence characteristic of the enzyme, thereby producing restriction fragments of lengths determined by the locations of the recognition sequences along the DNA molecules. The restriction fragments are then analyzed as described earlier in the section on Techniques for the Analysis of Genes and Gene Transcripts. That is, the fragments are separated according to size by agarose gel electrophoresis, and then Southern blotted to a nitrocellulose filter for hybridization with a ^{32}P-labeled DNA probe.

These agarose gel electrophoresis and Southern blot procedures are now routine in many molecular biology laboratories and are being used in a number of instances in genetic counseling. These procedures are most useful when the genetic mutation that causes a disease is associated with a change in the recognition site number or distribution for the restriction endonuclease, either within the structural gene involved or within the adjacent region. Indeed, more and more cases give evidence of *polymorphism in restriction sites*—that is, of the existence of many patterns of distribution of restriction sites in regions of DNA containing a gene of interest—with a clear distinction between the pattern of restriction sites in the wild-type DNA and the pattern in the mutant DNA.

An example of the use of a recombinant DNA approach to diagnose a genetic disease involves *sickle-cell anemia* (discussed in more detail in Chapter 14 (pp. 464-7). In brief, sickle-cell anemia results from a single base-pair change in the gene for the β-globin polypeptide of hemoglobin, resulting in an abnormal form of hemoglobin, Hb–S, instead of the normal Hb–A form. This base-pair change is from AT to TA, resulting in the substitution of a valine for a glutamic acid in the sixth amino acid of the polypeptide. Valine and glutamic acid have very different properties, and these differences lead to abnormal associations of hemoglobin molecules, sickling of the red blood cells, tissue damage, and sometimes death.

Using a cDNA probe for human β-globin, researchers have shown that there is a polymorphism in the arrangement of sites for the restriction enzyme *Hpa*I such that the presence of the mutant β-globin gene characteristic of people with sickle-cell anemia can be detected. Specifically, following digestion with *Hpa*I, Southern blotting, and probing with cDNA probe (as described in Figure 15.20), normal individuals who produce Hb–A will display fragments on the autoradiogram of either 7600 or 7000 base pairs in length. In contrast, people with the sickle-cell anemia mutation, who produce Hb–S, will display an *Hpa*I fragment of 13,000 base pairs in length (see Figure 15.22). Interestingly, this *restriction fragment length polymorphism* (RFLP) is not a direct result of the base-pair change since it involves differences that are outside the coding region of the gene itself.

An RFLP that is related to the genetic disease PKU (see Chapter 14 (pp. 457-9) has also been discovered. Recall that PKU results from a deficiency in the activity of the enzyme phenylalanine hydroxylase. S. Woo and his colleagues found that following digestion of genomic DNA

Normal genes; product found in Hb-A molecule

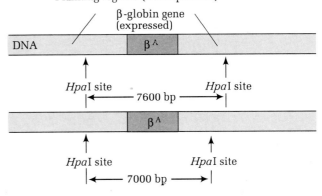

Sickle-cell mutant gene; product found in Hb-S molecule

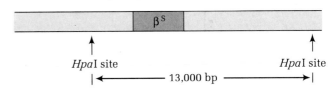

Figure 15.22 Restriction fragment length polymorphism (RFLP) associated with the DNA flanking the β-globin gene.

with *Hpa*I, Southern blotting, and probing with a cDNA probe derived from phenylalanine hydroxylase mRNA, different-sized restriction fragments were produced from DNA isolated from PKU individuals and from DNA isolated from homozygous normal individuals. Like the sickle-cell anemia example, this RFLP results from a difference outside the coding region of the gene, in this case to the 3' side of the gene. The RFLP will be used for diagnosing PKU in fetuses following amniocentesis.

The beauty of the recombinant DNA approach is that it directly assays for the presence or absence of a particular mutation in the DNA and therefore does not depend on the expression of the gene. Thus this approach is not limited to those genes that are active in the parent or the fetus at the time of analysis. For example, phenylalanine hydroxylase, the enzyme defective in individuals with PKU, is found in the liver but is not found either in blood serum or in

512
*Recombinant
DNA Technology
and the
Manipulation
of DNA*

fibroblast cells, the cells usually cultured following amniocentesis. The DNA of fibroblast cells can be analyzed, however, for restriction site polymorphism. Clearly, this approach is promising for the detection of a much larger number of genetic disorders, including those that are expressed later in life, such as Huntington's chorea. It is also able to detect carriers; for instance, a heterozygote for the sickle-cell anemia mutation would display *Hpa*I fragments of both 13,000 base pairs (the mutant form) and either 7600 or 7000 base pairs (the normal form).

Other examples of human genetic diseases for which recombinant DNA technology can provide early diagnosis include four types of thalassemia (hemoglobin diseases resulting in anemia): α-antitrypsin deficiency (which is a deficiency of a serum protein), hemophilia A, hemophilia B, and Duchenne muscular dystrophy (which is a progressive disease resulting in muscle atrophy and muscle dysfunction). Many more human genetic diseases will be subject to diagnosis using recombinant DNA approaches in the future.

Synthesis of Commercial Products

Many new biotechnology companies have emerged since the mid-1970s, when basic recombinant DNA techniques were developed. The development of these techniques and related procedures has spawned a surge in the biotechnology industry and has involved many long-standing companies, such as drug and chemical companies, which have made large investments in the industry. Most biotechnology companies are focusing on making commercial products that are useful in agriculture and in the health industry.

Of the many agricultural products resulting from recombinant DNA technology, one is a genetically engineered bacterium which, when sprayed on fields (e.g., strawberries or potatoes) lowers the temperature at which the leaves will freeze by a degree or two, thus providing some protection against frost damage. Because the freezing process kills plant cells, the ability to decrease frost damage is of great value. The re-

lease of genetically engineered microorganisms into the environment, however, has resulted in extensive public and legal debate.

Another example of a genetically engineered product for agricultural use is the use of genes to confer resistance to the harmful effects of common herbicides. When these genes are introduced into crop plants (e.g., corn and wheat), this resistance will protect the crop plants against the deleterious effects of herbicides sprayed on the fields to kill weeds. Although there appears to be an economic benefit to this approach, in that crop yields should increase, this product still raises questions about possible adverse effects to the environment.

Clinical applications for recombinant DNA technology include a number of pharmaceuticals and other products currently being produced commercially. These products are synthesized using specially engineered plasmids containing an appropriate DNA sequence inserted into the cloning vector between transcription and translation initiation and termination sequences. In some cases, the protein is isolated from disrupted, harvested cells, but in other cases, special systems are used in which the protein is secreted from the cell and is harvested from the culture medium. The latter approach is economically more efficient. Microorganisms, such as *E. coli,* yeast, and *Aspergillus,* are used most often as the reaction vessels for product synthesis. Among the products that can be synthesized in commercial quantities using recombinant DNA technology are human blood clotting factor (for use by hemophiliacs), human growth hormone (to help correct pituitary dwarfism), bovine growth hormone (to increase cattle yields), human insulin (avoids possible side effects that can be caused by the use of the porcine pancreatic insulin), human interferon, and tissue plasminogen activator (a protein that potentially can dissolve blood clots).

Genetic Engineering of Plants

For many centuries the traditional genetic engineering of plants involved selective breeding experiments in which plants with desirable traits were used as parents for the next generation in order to reproduce offspring with those traits. As a result, humans have produced a hardy variety of plants (e.g., corn, wheat, oats) and have been successful in breeding varieties with increased yields, all by using standard plant breeding techniques. (Similar experiments have also been done with animals, e.g., dogs and horses, to produce desired breeds.) In this section, we will discuss the application of recombinant DNA technology to plant breeding.

Tissue culture systems for plants. Within the past 40 to 50 years, progress has been made in developing tissue culture systems for plants. One such system involves excising a piece of plant tissue and placing it on a sterile, solid nutrient medium (containing salts, sugars, amino acids, vitamins, and appropriate growth hormones). The cells in the tissue divide and produce a mass of undifferentiated cells called a **callus culture,** somewhat analogous to a tumor cell mass in an animal system. In some plant species, if the cells of a tissue are teased apart and placed on a medium containing the appropriate nutrients and growth hormones, some of the cells will be induced to differentiate into shoots, roots, or whole plants, depending on the type and concentration of the growth hormones present. Since the single cells are able to regenerate a complete new plant, they still must contain all the genetic material for the development and differentiation of that plant; those cells are referred to as **totipotent.**

For some kinds of plants such as petunia, tobacco, tomato, and potato, it is also possible to generate complete new plants from protoplasts, that is, cells from which the rigid cellulose cell wall has been removed. Protoplasts are useful for two kinds of experiments. In one, protoplasts of two different plants can be fused to produce a hybrid cell from which a hybrid plant can be produced. This hybrid plant can then be subjected to conventional plant breeding techniques, if desired. Potato-tomato hybrids can be produced using this method, for example. In another kind of experiment, the protoplasts can be used as recipient cells for the introduction of

specific genes being carried on vectors; that is, they can be transformed just as *E. coli,* yeast, and other eukaryotic cells can be transformed.

In the long run, the latter approach is expected to bring forth a plethora of genetically engineered plants, including ones with increased yield, insect resistance, and herbicide resistance. With more sophisticated approaches, some of which are already available, it will be possible to introduce genes into plants and control their expression in different tissues. One example of this process is controlling the point at which fruit ripens. If all fruits in a field ripen at the same time, harvesting will become easier and the rate of spoilage will be reduced.

Transformation of plant cells. Historically, introducing genes into plant cells has been a problem that has placed plant genetic engineering's rate of progress behind bacterial, fungal, and animal genetic engineering. Work is being done, however, to resolve this problem. Currently, one solution that has been found is to exploit features of a soil bacterium, *Agrobacterium tumefaciens,* that infects many kinds of plants so that techniques are now being developed for introducing genes into those plants.

Agrobacterium tumefaciens causes crown gall disease, characterized by tumors (the gall) at wounding sites. Most dicotyledonous plants (called *dicots*), but not monocotyledonous plants (called *monocots*) are susceptible to crown gall disease. Dicots are classified on the basis that the embryo of the seed has two cotyledons; examples are potatoes, petunias, and apples. Monocots are classified on the basis that the embryo of the seed has a single cotyledon; examples are corn, wheat, and grass. *Agrobacterium tumefaciens* transforms plant cells at the wound site, causing them to grow and divide autonomously and, therefore, to produce the tumor. Depending upon the particular type of infecting bacterium, the transformed cells in the tumor begin to synthesize an unusual derivative of the amino acid arginine, namely, octopine or nopaline. Both of these are examples of chemicals called *opines,* which are the food and energy source for the infecting bacteria.

514
Recombinant
DNA Technology
and the
Manipulation
of DNA

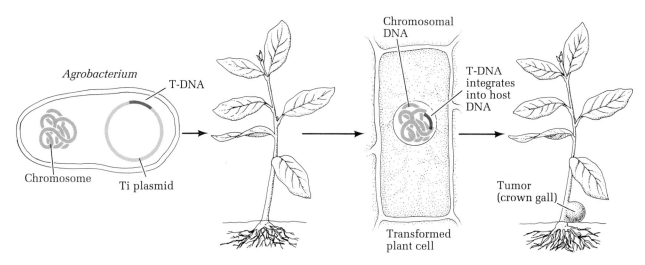

Figure 15.23 Formation of tumors (crown galls) in plants by infection with certain species of Agrobacterium. *Tumors are induced by the Ti plasmid, which is carried by the bacterium and which integrates some of its DNA (the T, or transforming DNA) into the plant cell's chromosome.*

The transformation of plant cells is mediated by a plasmid in the *Agrobacterium* called the *Ti plasmid* (the Ti stands for tumor-inducing) (Figure 15.23). Ti plasmids are circular DNA plasmids analogous to pBR322, but enormous in size (about 200 kbp vs. 4.36 kbp for pBR322).

The interaction between the infecting bacterium and the plant cell of the host stimulates the bacterium to excise a 30-kbp region of the Ti plasmid called *T-DNA* (so-called because it is transforming DNA), which is flanked by two repeated 25-bp sequences. Excising T-DNA from the Ti plasmid involves a specific recombination event, involving these two homologous repeated sequences, which produces a T-DNA circle with just one copy of the 25-bp repeat sequence. The circular T-DNA molecules, which are held together by proteins, are transferred from the bacterium to the nucleus of the plant cell, where the T-DNA integrates into the nuclear genome. As a result, the plant cell acquires the genes found on the T-DNA, including the genes for plant cell transformation, as well as the gene for the appropriate opine synthetase (either octo-

pine synthetase or nopaline synthetase). However, the genes needed for the excision, transfer, and integration of the T-DNA into the host plant cell are not part of the T-DNA but are found elsewhere on the Ti plasmid in a region called the *vir* (for virulence) region.

Using recombinant DNA approaches, researchers have found that excision, transfer, and integration of the T-DNA require only the 25-bpp terminal repeat sequences, because the alteration or removal of the rest of the T-DNA does not affect those processes. This makes the Ti plasmid and the T-DNA it contains a potentially useful vector for introducing new DNA sequences into the nuclear genome of somatic cells from susceptible plant species.

Recall from earlier in the chapter, one of the essential features of a cloning vector is the presence of unique restriction enzyme cleavage sites into which DNA fragments can be inserted. Because the Ti plasmid is so large, most restriction sites are represented more than once so it is not a particularly useful cloning vector. More useful cloning vectors (called *binary vectors:* Figure 15.24) have been made by subcloning some important part of the Ti plasmid into pBR322 to make a shuttle vector that enables DNA sequences to be amplified in *E. coli* for use in the plant transforming experiments and to replicate autonomously in *Agrobacterium*.

In the example shown in Figure 15.24, a segment of the Ti plasmid containing the T-DNA

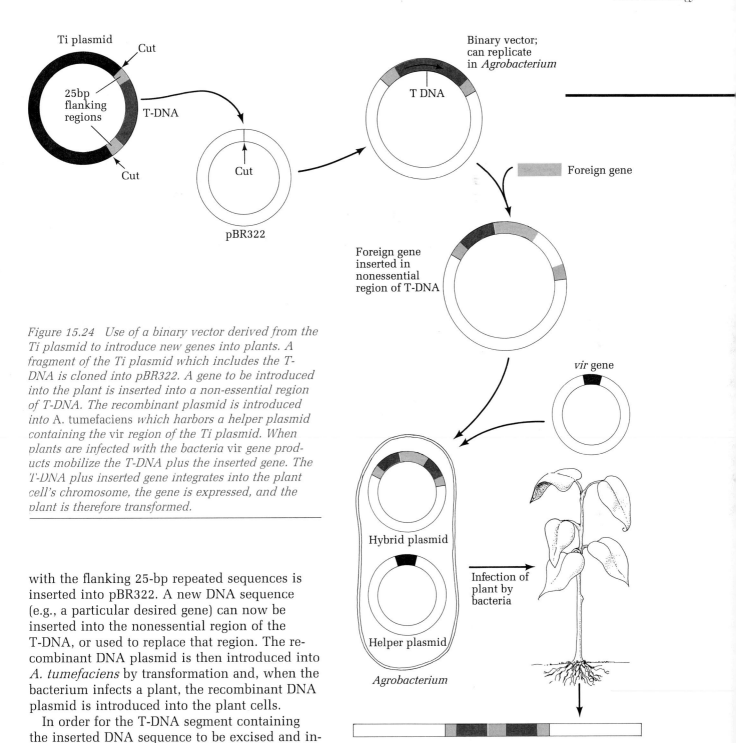

Figure 15.24 Use of a binary vector derived from the Ti plasmid to introduce new genes into plants. A fragment of the Ti plasmid which includes the T-DNA is cloned into pBR322. A gene to be introduced into the plant is inserted into a non-essential region of T-DNA. The recombinant plasmid is introduced into A. tumefaciens *which harbors a helper plasmid containing the* vir *region of the Ti plasmid. When plants are infected with the bacteria* vir *gene products mobilize the T-DNA plus the inserted gene. The T-DNA plus inserted gene integrates into the plant cell's chromosome, the gene is expressed, and the plant is therefore transformed.*

with the flanking 25-bp repeated sequences is inserted into pBR322. A new DNA sequence (e.g., a particular desired gene) can now be inserted into the nonessential region of the T-DNA, or used to replace that region. The recombinant DNA plasmid is then introduced into *A. tumefaciens* by transformation and, when the bacterium infects a plant, the recombinant DNA plasmid is introduced into the plant cells.

In order for the T-DNA segment containing the inserted DNA sequence to be excised and integrated into the plant genome when the bacteria infect a plant, the bacteria must also contain the *vir* region. The *vir* region may be introduced into the bacteria by using a second plasmid (called a *helper plasmid*) into which the *vir* region has been inserted. When plants are infected with these bacteria, the gene products of the *vir* region on the helper plasmid mobilize the T-DNA with the inserted gene. As a result, the

T-DNA plus the inserted gene is integrated into the plant genome where the foreign gene can be expressed. At the moment, only dicots can be transformed using the T-DNA transformation system.

516
*Recombinant
DNA Technology
and the
Manipulation
of DNA*

Keynote *Recombinant DNA technology is finding ever-increasing applications in the world. With appropriate probes, a number of genetic diseases can now be diagnosed using recombinant DNA methods, and many products in the clinical, veterinary, and agricultural areas can be synthesized in commercial quanti-ties using recombinant DNA procedures. Transformation of plants with vectors based on the Ti plasmid has opened the way for genetic engineering of plants using recombinant DNA technology. It is expected that many types of improved crops will result from applications of this new technology.*

Analytical Approaches for Solving Genetics Problems

While this is a rather descriptive area, it is often necessary to interpret data derived from restriction enzyme analysis of DNA fragments in order to generate a restriction map, that is, a map of the locations of restriction enzymes. The logic used for this type of analysis is very similar to that used in generating a genetic map of loci from two-point mapping crosses.

Q.1 A piece of DNA 900 bp long is cloned and then cut out of the vector for analysis. Digestion of this linear piece of DNA with three different restriction enzymes singly and in all possible pairwise arrangements gave the following restriction fragment size data:

Enzyme(s)	Restriction Fragment Sizes
*Eco*RI	200 bp, 700 bp
*Hind*III	300 bp, 600 bp
*Bam*HI	50 bp, 350 bp, 500 bp
*Eco*RI + *Hind*III	100 bp, 200 bp, 600 bp
*Eco*RI + *Bam*HI	50 bp, 150 bp, 200 bp, 500 bp
*Hind*III + *Bam*HI	50 bp, 100 bp, 250 bp, 500 bp

Construct a restriction map from these data.

A.1 The approach to this kind of problem is to consider a pair of enzymes and to analyze the data from the single and double digestions. First let us consider the *Eco*RI and *Hind*III data. Cutting with *Eco*RI produces two fragments, one of 200 bp and the other of 700 bp, while cutting with *Hind*III also produces two fragments, one of 300 bp and the other of 600 bp. Thus, we know that both restriction sites are asymmetrically located along the linear DNA fragment with the *Eco*RI site 200 bp from an end and the *Hind*III site 300 bp from an end. When we consider the *Eco*RI + *Hind*III data we can determine the positions of these two restriction sites relative to one another. If, for example, the *Eco*RI site is 200 bp from the fragment end, and the *Hind*III site is 300 bp from that same end, then we would predict that cutting with both enzymes would produce three fragments of sizes 200 bp (end to *Eco*RI site), 100 bp (*Eco*RI site to *Hind*III site), and 600 bp (*Hind*III site to other end). On the other hand, if the *Eco*RI site is 200 bp from one fragment end and the *Hind*III site is 300 bp from the other fragment end, cutting with both enzymes would produce three fragments of sizes 200 bp (end to *Eco*RI site), 400 bp (*Eco*RI site to *Hind*III site), and 300 bp (*Hind*III site to end). The actual data support the first model.

Now we pick another pair of enzymes, for example, *Hind*III and *Bam*HI.

Cutting with *Hind*III produces fragments of 300 bp and 600 bp as we have seen, and cutting with *Bam*HI produces three fragments of sizes 50 bp, 350 bp, and 500 bp, indicating that there are two *Bam*HI sites in the DNA fragment. Again the double digestion products are useful in locating the sites. Double digestion with *Hind*III and *Bam*HI produces four fragments of 50 bp, 100 bp, 250 bp, and 500 bp. The simplest interpretation of the data is that the 300-bp *Hind*III fragment is cut into the 50-bp and 250-bp fragments by *Bam*HI, and that the 600-bp *Hind*III fragment is cut into the 100-bp and 500-bp fragments by *Bam*HI. Thus, the restriction map shown in the accompanying figure can be drawn:

*Bam*HI		*Eco*RI		*Hind*III		*Bam*HI	
50	150	100	100			500	

The *Bam*HI + *Eco*RI data are compatible with this model.

*15.1 A new restriction endonuclease is isolated from a bacterium. This enzyme cuts DNA into fragments that average 4096 base pairs long. Like all other known restriction enzymes, the new one recognizes a sequence in DNA that has twofold rotational symmetry. From the information given, how many base pairs of DNA constitute the recognition sequence for the new enzyme?

15.2 Restriction endonucleases are used to construct restriction maps of linear or circular pieces of DNA. The DNA is usually produced in large amounts by recombinant DNA techniques. The generation of restriction maps is similar to the process of putting the pieces of jigsaw together. Suppose we have a circular piece of double-stranded DNA that is 5000 base pairs long. If this DNA is digested completely with restriction enzyme I, four DNA fragments are generated: fragment *a* is 2000 base pairs long; fragment *b* is 1400 base pairs long; *c* is 900 base pairs long; and *d* is 700 base pairs long. If, instead, the DNA is incubated with the enzyme for a short time, the result is incomplete digestion of the DNA; not every restriction enzyme site in every DNA molecule will be cut by the enzyme, and all possible combinations of adjacent fragments can be produced. From an incomplete digestion experiment of this type, fragments of DNA were produced from the circular piece of DNA, which contained the following combinations of the above fragments: *a-d-b, d-a-c, c-b-d, a-c, d-a, d-b,* and *b-c.* Lastly, after digesting the original circular DNA to completion with restriction enzyme I, the DNA fragments were treated with restriction enzyme II under conditions conducive to complete digestion. The resulting fragments were: 1400, 1200, 900, 800, 400, and 300. Analyze all the data to locate the restriction enzyme sites as accurately as possible.

*15.3 Draw the banding pattern you would expect to see on a DNA-sequencing gel if you applied the Maxam and Gilbert DNA-sequencing method to the following single-stranded DNA fragment (which is labeled at the 5′ end with ^{32}P): 5′^{32}P-A-A-G-T-C-T-A-C-G-T-A-T-A-G-G-C-C-3′.

16

Gene Mutations

Throughout the book so far, especially the first seven chapters, we have used the terms *normal, wild type, variants,* and *mutants* rather loosely. And in Chapters 8–14 we have learned about the *normal* functions of genes at the molecular level; specifically, what genes are, how genes are replicated, and how genes are expressed through the processes of transcription and translation. Our picture of gene functions shows these processes occurring with great accuracy and without variation. The existence of mutants, though, indicates that changes *do* occur in genes and that such changes often have such significant consequences to the organism that normal function is no longer possible (see Figure 16.1). Furthermore, were it not for the production of variant organisms with the potential to adapt to new environments and new situations, evolution could not occur.

In this chapter and the next we will examine the various mechanisms by which genetic variations can arise; they can arise through changes at the base-pair level or at the chromosomal level. (A mutation is defined as any detectable and heritable change in the genetic material.) Along with genetic recombination, which we studied in Chapters 5–7, these mechanisms are the ways by which genetic variants are produced in all organisms.

In this chapter we focus on some of the mechanisms that result in genetic changes at the gene level—the changes in the base-pair sequences of genes and the phenotypic consequences of these changes. We also examine some of the methods used to select for genetic mutants, methods that are important in the geneticist's approach to understanding a process. As we have seen throughout our studies thus far, geneticists often study a number of organisms having mutations that affect a particular process in order to determine the differences between the normal and mutant organisms. These comparative investigations may take place at a number of levels, but most frequently they involve physiological comparisons and investigations of biochemical and/or molecular events, because such defects in the mutants may be defined in some detail. From the information gained from studying the mu-

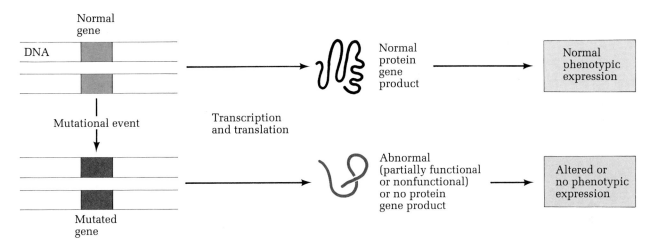

Normal
gene

DNA

Mutational event

Transcription
and translation

Normal
protein
gene
product

Normal
phenotypic
expression

Mutated
gene

Abnormal
(partially functional
or nonfunctional)
or no protein
gene product

Altered or
no phenotypic
expression

Figure 16.1 Concept of a mutation.

tant process, geneticists can extrapolate the function of the normal gene.

Types of Gene Mutation

A **mutation** is any detectable and heritable change in the genetic material not caused by genetic recombination. A mutation can be transmitted to daughter cells and even to succeeding generations, thereby giving rise to mutant cells or mutant individuals. If a mutant cell gives rise only to somatic cells (in multicellular organisms), a mutant spot or area is produced, but the mutant characteristic is not passed on to the succeeding generation. However, mutations in the germ line of sexually reproducing organisms may be transmitted by the gametes to the next generation, giving rise to an individual with the mutant state in both its somatic and germ line cells. Since the genetic material is usually DNA, a mutation may be the result of any detectable, unnatural change that affects DNA's chemical or physical constitution, its replication, its phenotypic function, or the sequence of one or more DNA base pairs (base pairs may, for example, be added, deleted, substituted, reversed in order or inverted, or transposed to new positions).

Mutations can occur at the level of the chromosome or at the level of the gene. A change in

the organization of a chromosome or chromosomes is called a **chromosomal aberration,** also called a **chromosomal mutation** (see Chapter 17). When a chromosomal mutation involves a change in the number of sets of chromosomes in the genome, the mutation is called a **genome mutation.** A mutation that occurs at the level of a gene is called a **gene mutation,** and it can involve any one of a number of alterations of the DNA sequence of the gene, including base-pair substitutions and additions or deletions of one of more base pairs. Those gene mutations that affect a single base pair of DNA are called **point mutations.**

Mutations can occur spontaneously, but they can also be induced experimentally by the application of a **mutagen,** any physical or chemical agent that significantly increases the frequency of mutational events above a spontaneous mutation rate. Mutations that result from treatment with mutagens are called **induced mutations;** naturally occurring mutations are **spontaneous mutations.** There are no qualitative differences between spontaneous and induced mutations.

The manifestation of a mutant phenotype is typically the result of a change in DNA that results in the altered function of a protein. To summarize some relevant aspects of the genetic code presented in an earlier chapter, the base-pair sequences of a protein-coding gene are transcribed into an mRNA molecule. The ribosome

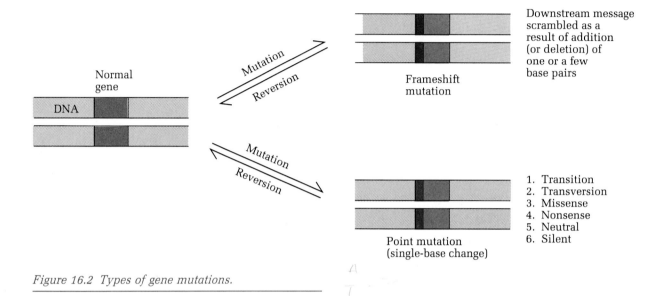

Figure 16.2 Types of gene mutations.

translates the base sequence in sequential groups of three bases (codons) to generate the amino acid sequence of the protein. Thus, each amino acid is specified by a three-base sequence of the mRNA molecule (and, hence, by a three-base-pair sequence of the DNA). For many of the 20 amino acids, more than one codon specifies the same amino acid in the translation process.

With the above background we can now move to the definitions of several terms related to gene mutation. These terms are listed in Figure 16.2. Keep in mind that since the tertiary structure of a polypeptide is a function of its primary amino acid sequence which is coded for by a gene, a polypeptide synthesized by a mutant strain may be structurally different from the wild-type polypeptide and, if so, will be partially functional, nonfunctional, or not produced.

Base-pair substitution mutation (point mutation) is a change in a gene such that one base pair is replaced by another base pair. Thus an AT may be replaced by a GC pair.

A **transition mutation** is a specific type of base-pair substitution mutation that involves a change in the chemical structure of the DNA from one purine-pyrimidine base pair to the

other purine-pyrimidine base pair at a particular site. The four types of transition mutations are AT to GC, GC to AT, TA to CG, and CG to TA.

A **transversion mutation** is another specific type of base-pair substitution mutation that involves a change in the chemical structure of the DNA from a purine-pyrimidine base pair to a pyrimidine-purine base pair at the same site. Examples of transversion mutations are AT to TA, GC to CG, AT to CG, and GC to TA.

A **missense mutation** is a gene mutation in which a base-pair change in the DNA causes a change in an mRNA codon with the result that a different amino acid is inserted into the polypeptide in place of the one specified by the wild-type codon. A base-pair substitution mutation can result in a missense mutation. Suppose, for example, that a mutation occurs that changes a nucleotide pair, perhaps from GC to AT (a transition mutation). If this mutation occurs in the amino acid–coding region of a gene, then the mutation may have serious consequences for the cell. That is, when the mutated gene is transcribed, the mRNA will have a different codon from the one specified in the normal or wild-type gene. If the mutated codon causes the wrong amino acid to be incorporated into the protein, the function of the protein would probably be impaired, and the mutant organism would probably function differently from a wild-type organism.

Whether a mutant phenotype is readily detectable depends on the particular amino acid substitution. In humans, for example, a particular single nucleotide pair change in the β-globin gene leads to an amino acid substitution in the beta hemoglobin chain. If the individual is homozygous for that mutation, he or she will have sickle-cell anemia. A mutation may escape detection (and unless a phenotypic change occurs, it will not be detected by a geneticist) in those cases where the insertion of the incorrect amino acid in the polypeptide chain does not lead to an appreciable change in the function of the protein.

A **nonsense mutation** is a base-pair change in the DNA that results in the change of an mRNA codon from one that specifies an amino acid to a chain-terminating (nonsense) codon (UAG, UAA, or UGA). Because a nonsense mutation gives rise to chain termination at the wrong place in a polypeptide, it prematurely ends the polypeptide. Instead of complete polypeptides, polypeptide fragments (usually nonfunctional) are released from the ribosomes (Figure 16.3). Nonsense mutations are known for many structural genes.

A **neutral mutation** is a base-pair change in the gene that changes a codon in the mRNA such that the resulting amino acid substitution produces no change in the function of the protein translated from that message. A neutral mutation occurs when the new codon codes for a different amino acid that is chemically equivalent to the original and hence does not affect the function of the protein. An example would be a change from the codon AGG to AAG, which would substitute the amino acid lysine for the amino acid arginine. Both amino acids are basic amino acids and are sufficiently similar in properties so that the protein's function may well not be significantly diminished.

A **silent mutation** is a special case of a neutral mutation. That is, a silent mutation is a base-pair change in the gene that alters a codon in the mRNA such that the *same* amino acid is inserted in the protein. The resulting protein has wild-type function because its amino acid sequence is unchanged. An example of a silent mutation would be a change from mRNA codon AGG to AGA, both of which specify arginine.

A **frameshift mutation** results from the addition or deletion of a base pair in a gene. Such additions or deletions shift the mRNA's reading frame by one base so that incorrect amino acids are added to the polypeptide chain after the mutation site and a nonfunctional polypeptide results. Box 16.1 on page 524 shows how Crick and his colleagues used frameshift mutations to determine that the genetic code was a triplet code.

Point mutations generally fall into two classes with respect to the effects on the phenotype, with reference to the wild type. **Forward mutations** are mutations that occur in the direction wild-type to mutant, and **reverse mutations** (or **reversions**) occur in the direction mutant to wild-type.

A reversion can restore the partially functional or nonfunctional polypeptide produced by a mutant gene to its full (or partial) function. Reversion of a nonsense mutation, for instance, occurs by a base-pair change in DNA corresponding to the incorrect nonsense codon in the mRNA so that the codon again specifies an amino acid, either the original amino acid of the wild type (in which case the mutation is a true reversion) or some other amino acid (in which case the mutation is a partial reversion, since complete function is unlikely to be restored). Reversion of missense mutations can occur in the same way.

A **suppressor mutation** is a mutation at a second site (also called a secondary or **second-site mutation**) that totally or partially restores a function lost because of a primary mutation at another site. A suppressor mutation does not result in a reversal of the original mutation; instead, it masks or compensates for the effects of the primary mutation (Figure 16.4). Suppressor mutations can occur within the gene in which the primary mutation occurs. If they occur

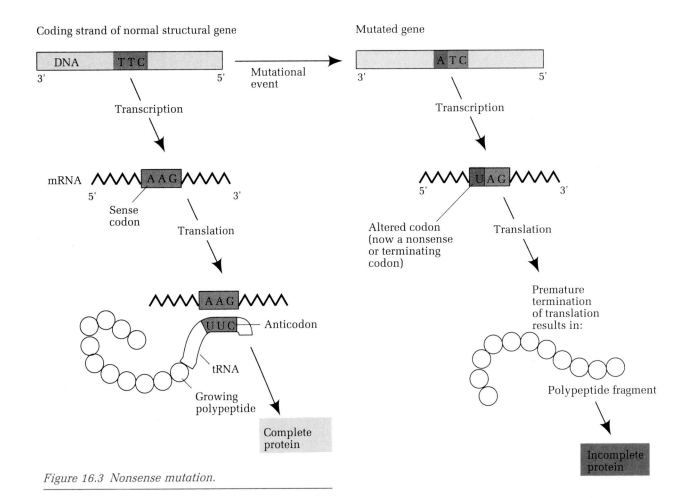

Coding strand of normal structural gene

Mutated gene

Figure 16.3 Nonsense mutation.

within genes separate from the genes in which the primary mutation occurred, they are called **intergenic suppressors.**

Suppressors of nonsense mutations are typically intergenic suppressors that are produced when particular tRNA genes are mutated so that (in contrast to what occurs with wild-type tRNAs) their anticodons recognize a chain-terminating codon and put an amino acid into the chain. Thus instead of polypeptide chain synthesis being stopped prematurely, an amino acid is inserted at that position by the altered tRNA, and that amino acid may restore full or partial function to the polypeptide. Since there are three types of nonsense codons, there are three classes of nonsense suppressors: one for UAG, one for UAA, and one for UGA. If, for ex-

ample, a gene for tRNA.Tyr (which has the anticodon 5′–GUA–3′) is mutated so that the tRNA has the anticodon 5′–CUA–3′, the mutated tRNA (which still has Tyr attached) will read the nonsense codon 5′–UAG–3′. But instead of terminating the chain, it will insert its tyrosine at that point in the chain.

If we have changed this particular class of tRNA.Tyr so that its anticodon can now read a nonsense codon, it cannot read the original codon that specifies the amino acid it carries. Thus nonsense suppressor tRNAs are typically produced by mutation of tRNA genes that are redundant in the genome; that is, for which several different genes all specifying the same tRNA base sequence exist. Therefore if one of the redundant genes mutates so that the tRNA will

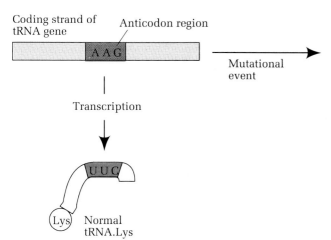

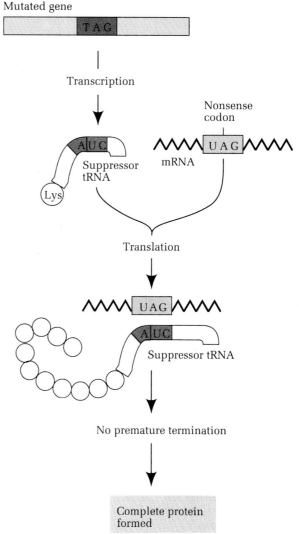

Figure 16.4 Mechanism for action of a nonsense suppressor mutation.

read, say, UAG, the other genes specifying the same tRNA produce a tRNA molecule that will read the normal Tyr codon.

Keynote *A mutation is any detectable and heritable change in the genetic material that is not caused by genetic recombination. Mutations of genes may occur spontaneously or may be induced experimentally by the application of mutagens. Mutations that affect a single base pair of DNA are called base pair substitution mutations or point mutations.*

Causes of Mutation

Now that we have studied the various types of gene mutations, we are ready to discuss how mutations occur. The two general types of mutation we will consider are spontaneous mutations and induced mutations.

Spontaneous Mutations

Spontaneous mutations are mutations that occur in nature without the apparent intervention of humans. All types of point mutations described in the previous section occur spontaneously. For a long while geneticists thought that sponta-

neous mutations were produced by mutagens indigenous to the environment, such as radiation and chemicals. However, recent evidence indicates that the rate at which spontaneous mutations appear, while it is extremely low, is nonetheless too high to be accounted for by indigenous mutagens alone. In *Drosophila,* for example, the spontaneous mutation rates for individual genes is about 10^{-4} to 10^{-5} per gene per gamete. In humans the spontaneous mutation rate varies between 10^{-4} and 4×10^{-6}.

Box 16.1 Use of Frameshift Mutations to Prove the Triplet Nature of the Genetic Code

The evidence that three contiguous nucleotides in mRNA code for one amino acid in a polypeptide chain (i.e., that the genetic code is a triplet code) came from experiments involving frameshift mutations of phage T4 that were performed by F. Crick, L. Barnett, S. Brenner, and R. Watts-Tobin in the early 1960s. The experiments used *rII* mutant strains and wild-type (r^+) strains. This system was chosen for two reasons. First, the two strains can be distinguished by their plaque morphologies. (When phages are produced on a lawn of bacteria growing on solid medium in a petri dish, the successive rounds of infection and bacterial lysis yield cleared regions in the lawn called phage plaques.) The *rII* mutant strains produce clear plaques, whereas the r^+ strain produces slightly smaller, cloudy plaques. Second, the two strains differ in their host range properties; that is, they differ in their ability to grow in different strains of *E. coli*. The *rII* strains are able to grow in *E. coli B* but not in *E. coli K12(λ)*, while the wild-type r^+ strain is able to grow in both strains.

Crick and his colleagues began with an *rII* mutant strain that had been produced by mutagenic treatment of a wild-type (r^+) strain. The mutagen used was the chemical proflavin, which induces mutation by causing the addition or deletion of a base pair in the DNA. They reasoned correctly that if the *rII* mutant phenotypes resulted from either an addition or a deletion, a second treatment of proflavin could reverse the mutation to the wild-type (r^+) state. Thus they were able to isolate a number of r^+ revertants by plating the population of *rII* cells that had been treated with proflavin onto a lawn of *E. coli K12(λ)* in which only r^+ phages can grow,

making it very easy to isolate the low numbers of r^+ revertants in the population of phages.

Genetic analysis showed that three types of revertants resulted from a second proflavin treatment of proflavin-induced *rII* mutants. One type was the true revertant in which the second mutational event exactly corrected the original mutational change. In this case the revertant is a true r^+ phage. In the second revertant type the mutation after the second application of proflavin occurred in a gene other than the *rII* gene. The third type of revertant was the most useful for proving the triplet code hypothesis. This revertant type resulted from a second mutation *within* the *rII* gene at a site nearby, but distinct from, the original mutation site. Here, then, the combination of two mutations gave an almost wild-type (pseudowild) phenotype; that is, the plaques did not look totally normal. Strains with just the second mutation alone are *rII* mutants.

If, for example, the first mutation was a deletion of a single base pair, the reversion of

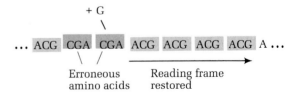

a) Normal DNA

... ACG ACG ACG ACG ACG ACG ACG A ...

b) Frameshift mutation by deletion

... ACG CGA CGA CGA CGA CGA CGA ...

– A

c) Reversion by addition

+ G

... ACG CGA CGA ACG ACG ACG ACG A ...

Erroneous amino acids Reading frame restored

Box Figure 16.1

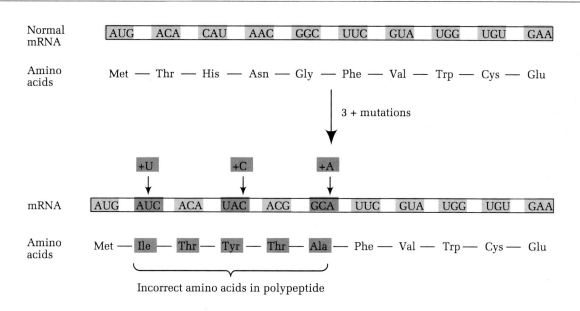

Normal mRNA: AUG ACA CAU AAC GGC UUC GUA UGG UGU GAA

Amino acids: Met — Thr — His — Asn — Gly — Phe — Val — Trp — Cys — Glu

3 + mutations

+U   +C   +A

mRNA: AUG AUC ACA UAC ACG GCA UUC GUA UGG UGU GAA

Amino acids: Met — Ile — Thr — Tyr — Thr — Ala — Phe — Val — Trp — Cys — Glu

Incorrect amino acids in polypeptide

Box Figure 16.2

this mutation by a second-site event within the same gene would involve an addition of a nearby base pair in order to restore the reading frame of the message. A pictorial representation of this reversion phenomenon is given in Box Figure 16.1. The figure shows a hypothetical segment of DNA (Box Figure 16.1a). On the assumption that the genetic code is, in fact, a triplet code, the result of a deletion of a base pair is that the genetic message is thrown out of frame by one (Box Figure 16.1b). Reversion is accomplished by inserting a base nearby so that the reading frame is restored to the way it was (Box Figure 16.1c). The short segment between the two mutational changes contains incorrect coding information that leads to the insertion of incorrect amino acids in the polypeptide chain. Providing that the incorrect amino acids do not significantly affect the function of the polypeptide coded for by this segment of DNA, the organism will function normally (or essentially normally). In fact, the slightly

less-than-normal functioning of a polypeptide that results from a double-mutation event like this is responsible for the pseudowild phenotype of the revertants we are discussing.

To generalize, we can revert a deletion mutation by a nearby addition mutation, and we can revert an addition mutation by a nearby deletion mutation. We symbolize addition mutations as + mutations and deletion mutations by − mutations. In actuality, Crick and his colleagues did not know whether an individual *rII* mutant resulted from an addition or a deletion. But they did know which of their single-mutant *rII* strains were of one sign and which were of the other sign.

Their next step was to combine genetically distinct *rII* mutations of the same type (either all additions or all deletions) in various numbers and see whether this combination reverted the *rII* phenotypes. They found that the combination of either three nearby addition mutations or three nearby deletion mutations gave revertants. No other multiple combinations, except multiples of three, worked. Box Figure 16.2 gives a hypothetical interpre-

(Continued on next page)

Spontaneous mutations can result from any one of a number of events, including errors in DNA replication and spontaneous chemical changes in DNA. Spontaneous mutations can also result from the movement of transposable genetic elements, which will be discussed in Chapter 18.

DNA replication errors. A mismatched base pair such as A–C might form during DNA replication. When the DNA with the A–C mismatch replicates again, one progeny helix will result with a GC base pair in place of the original AT base pair; that is, a transition mutation is produced. The other progeny helix will still have an AT base pair in that position. Because the bases themselves are able to exist in alternate chemical forms, called **tautomers,** erroneous base pairs form. In DNA, the normal form of each base is the *keto* form. When each base is in its rare form (*imino* form for cytosine and adenine, *enol* form for thymine and guanine), different hydrogen bondings become possible and lead to mismatched base pairs (Figure 16.5). As an example, the rare imino form of cytosine can pair with adenine, and the rare enol form of guanine can pair with thymine. The change in the chemical form of a base is called a **tautomeric shift.** Figure 16.6 illustrates how a transition mutation can be produced as a result of a tautomeric shift. Note that many more mutations would be seen as the result of tautomeric shifts than are actually observed if it were not for the proofreading activity of DNA polymerase III,

which recognizes many of the mismatches, excises them, and replaces them with the correct base pairs.

Spontaneous chemical changes. Two of the most common chemical events that occur to produce spontaneous mutations are depurination and deamination of particular bases. Depurination, as the name suggests, involves the removal of a purine, either adenine or guanine, from the DNA as a result of breakage of the bond between the purine and the deoxyribose (Figure 16.7, page 529). Thousands of purines are lost by depurination in a typical generation time of a mammalian cell in tissue culture. If such lesions are not repaired, there is no base to specify a complementary base during DNA replication. Instead, a randomly chosen base is inserted, and this can produce a mismatched base pair. Genetic mutations may then result as described in the previous section.

Deamination is the removal of an amino group from a base. As an example, a susceptible base for this event is cytosine, the deamination of which produces uracil (Figure 16.8, page 529). If the uracil is not repaired, it will direct that an adenine be incorporated in the new DNA strand during replication, ultimately resulting in the conversion of a CG base pair to a TA base pair, that is, a transition mutation. (See later discussion of the deaminating effects of nitrous acid for other examples of mutations resulting from deamination.)

Since the rate of spontaneous mutation is so

(Box 16.1   Continued from preceding page)

tation of this result. The figure shows a 30-nucleotide segment of mRNA that codes for 10 distinct amino acids. If we add three base pairs at nearby locations in the DNA coding for this mRNA segment, the result will be a 33-nucleotide segment that codes for 11 amino acids, that is, one more than the origi-

nal. Note, though, that the amino acids between the first and third insertions are not the same as in the wild-type mRNA. In essence, the reading frame is correct before the first insertion and again after the third insertion. The incorrect amino acids between those points give the pseudowild phenotype.

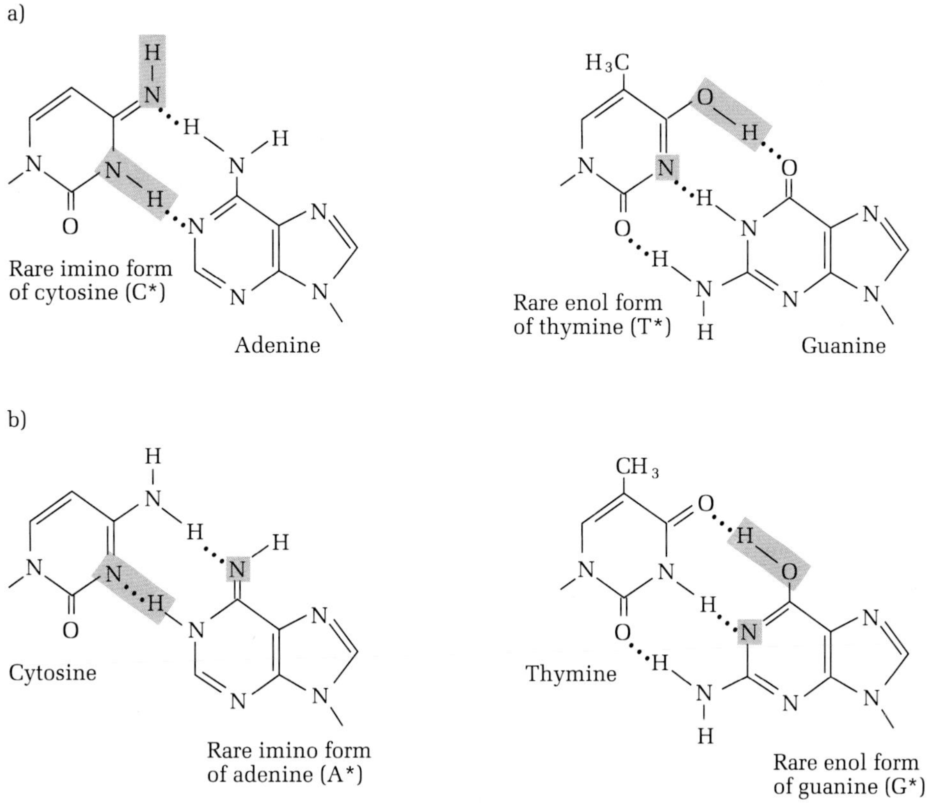

a)

Rare imino form
of cytosine (C*)

Adenine

Rare enol form
of thymine (T*)

Guanine

b)

Cytosine

Rare imino form
of adenine (A*)

Thymine

Rare enol form
of guanine (G*)

*Figure 16.5 Examples of mismatched bases. in DNA:
(a) mismatched based resulting from rare forms of pyrimidines; (b) mismatched bases resulting from rare
forms of purines.*

low, genetic researchers generally use mutagens to enhance the frequency of mutation so that a significant number of the organisms have mutations in the gene being studied. Two classes of mutagens are used—radiation and chemical— and the mechanisms of action for some examples of these mutagens will now be described.

Induced Mutations—Radiation

Ultraviolet (UV) light is a useful mutagen, and at high enough doses it can kill cells. For geneticists, other scientists, and medical personnel,

this property is a useful one: UV light is used as a sterilizing agent in some applications. Ultraviolet light causes mutations because DNA absorbs light very strongly if it has a wavelength of 254–260 nm; that is, in the ultraviolet range. At this wavelength UV light induces gene mutations primarily by causing photochemical (light-induced chemical) changes in the DNA. This mutagenic activity is attenuated to some extent because UV radiation has lower-energy wavelengths than X rays and, hence, has very limited penetrating power.

Our sun is a very powerful source of UV radiation, but much of the UV light is screened out by the atmospheric layers. Nonetheless, significant amounts of UV radiation are present in sunlight, as evidenced by the tanning (or burning) of humans who sunbathe. Tanning is more rapid at higher altitudes, where the thinner layer

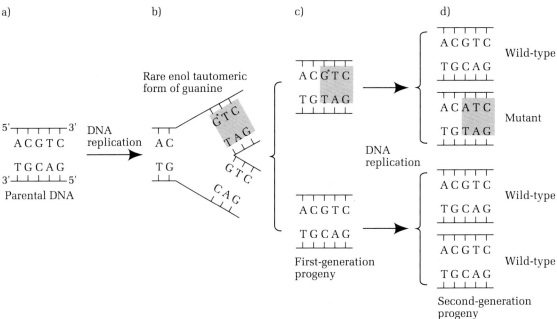

a)

5' ‖‖‖‖‖ 3'
A C G T C

T G C A G
3' ‖‖‖‖‖ 5'
Parental DNA

DNA
replication →

b)

Rare enol tautomeric
form of guanine

A C
‖‖
T G

G̈T C
T A G

G T C
C A G

c)

A C G̈T C
‖‖‖‖‖
T G T A G

DNA
replication

A C G T C
‖‖‖‖‖
T G C A G

First-generation
progeny

d)

A C G T C
‖‖‖‖‖
T G C A G
Wild-type

A C A T C
‖‖‖‖‖
T G T A G
Mutant

A C G T C
‖‖‖‖‖
T G C A G
Wild-type

A C G T C
‖‖‖‖‖
T G C A G
Wild-type

Second-generation
progeny

Figure 16.6 Production of a mutation as a result of a tautomeric shift in a DNA base. (a)A guanine on the top strand shifts to the rare enol form (G*) during DNA replicaton; (b)In its enol form, thymine is paired with it on the new DNA strand; (c) and (d) The guanine shifts back to its more stable normal form during the next DNA replication. The mismatched thymine specifies an adenine on the new strand and the overall result is a GC-to-AT transition mutation for the DNA strand lineage in which the tautomeric shift occurred. The other DNA strand lineages produce no mutations.

of the atmosphere above has screened out less of the UV radiation. In fact, we can think of tanning as our way of mobilizing a defense against UV-induced damage to our skin cells. Individuals who are particularly prone to skin cancer typically stay covered in the sun in order to avoid the induction of cancerous lesions.

One of the effects of UV radiation on DNA is the formation of abnormal chemical bonds between adjacent pyrimidine molecules in the same strand or between pyrimidines on the opposite strands of the double helix. This bonding is induced mostly between adjacent thymines on the same or opposite strands of the DNA, forming what are called *thymine dimers* (Figure

16.9), usually designated T̂T̂. This unusual pairing disrupts the normal pairing of the T's with the corresponding A's on the opposite strand, causes a bulge in the DNA strand, and weakens the hydrogen bonds between T and A on opposite strands. Mutations are produced by UV radiation usually because of the persistence of the thymine dimers or their imperfect repair. The damage to DNA caused by ultraviolet light is severe enough to block DNA synthesis completely, thereby resulting in cell death. The thymine dimers, however, can be repaired completely by specific cellular systems. Generating viable mutants requires a special host enzyme system that has been characterized in bacteria, and is presumed to exist in eukaryotic cells. This system, called the *SOS system*, is induced as an emergency response to prevent cell death when significant DNA damage has been incurred. The action of the SOS system is not confined to UV-induced damage; it is also activated after different kinds of DNA damage.

The following describes how the SOS system works. When a replication fork encounters significant DNA damage, for example, a thymine dimer induced by UV light, it stops and reinitiates DNA synthesis some distance away. This leaves a gap in the new DNA strand to which a specific

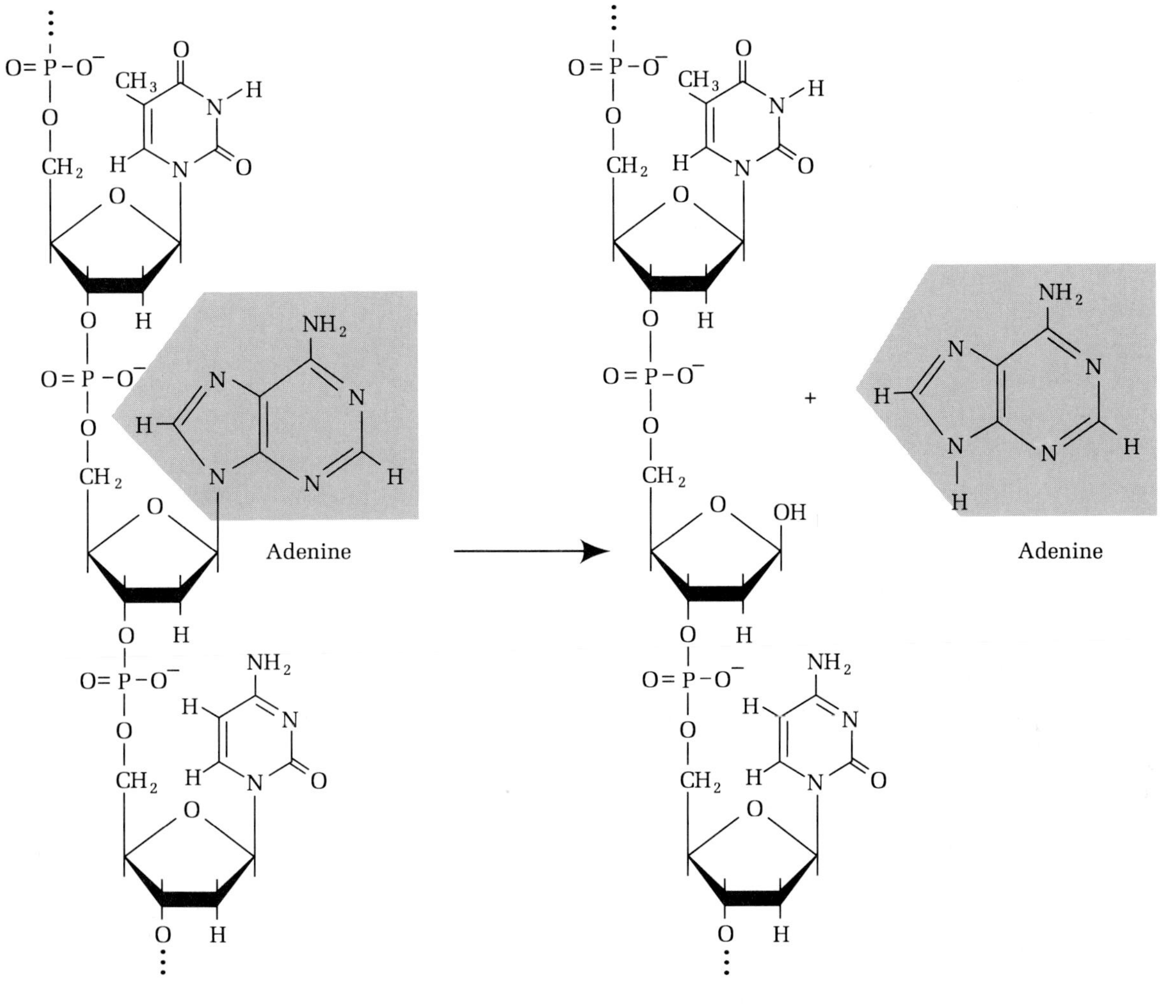

Figure 16.7 Depurination (loss of a purine) from a single strand of DNA. The sugar-phosphate backbone is unbroken.

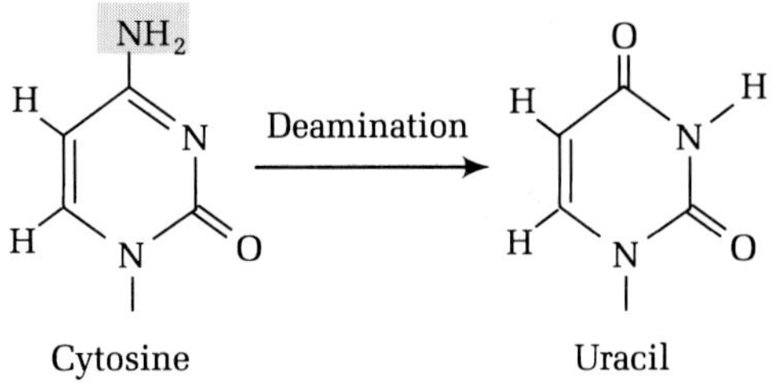

protein called the RecA protein (encoded by the *recA* gene) binds. In other circumstances the RecA protein functions in recombinational repair; but when it binds to single-stranded DNA (i.e., in the gap), a new function of the RecA protein is activated. In its new capacity, the

Figure 16.8 Deamination of cytosine.

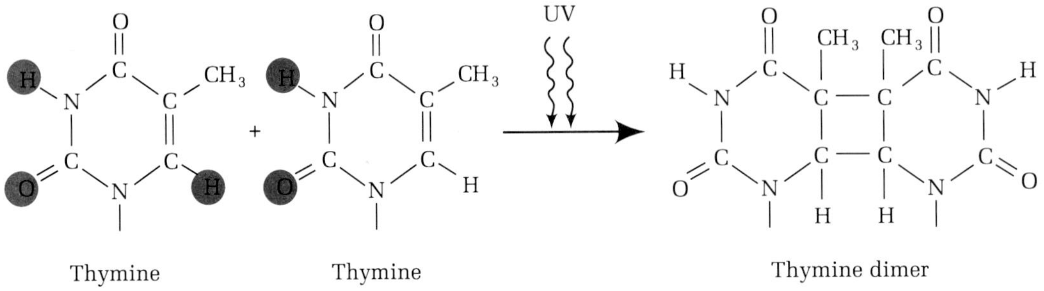

Figure 16.9 Production of thymine dimers by ultraviolet light irradiation. The two components of the dimer are covalently linked in such a way that the DNA double helix is distorted at that position.

RecA protein destroys a repressor, which prevents the expression of the group of about 15 genes, the SOS genes, which are responsible for the SOS system. As a result, the SOS genes are activated and, through a variety of mechanisms, the gap is repaired. The key to producing mutations is that the SOS repair system facilitates the insertion of a base in the new DNA strand even though no template base exists. This process is normally prevented in DNA synthesis and, therefore, has been named *bypass synthesis.* Since a random base will be inserted, both transition and transversion mutagens may result from SOS-induced bypass synthesis.

Keynote Mutations may be induced by the use of radiation or chemical mutagens. Radiation may cause genetic damage by breaking chromosomes, by producing chemicals that affect the DNA (as in the case of X rays), or by causing the formation of unusual bonds between DNA bases, such as thymine dimers (as in the case of ultraviolet light). If DNA is damaged significantly (e.g., by UV light), mutations are produced as a result of the activation of SOS genes, the function of which is to permit the insertion of bases in new DNA where no base exists on the template strand. This process is called bypass synthesis or error-prone repair. If the damage is not repaired, it may be lethal to the cell.

Chemical Mutagens

Base analogs. **Base analogs** are chemicals that have molecular structures that are extremely similar to the bases normally found in DNA. The base analog 5-bromouracil (5BU), for example, has a bromine residue instead of the methyl group of thymine. These base analogs require DNA replication for their incorporation, and because they often can form base pairs with more than one of the usual bases, mutations can result. We shall consider two commonly used base-analog mutagens, 5-bromouracil and 2-aminopurine.

The mutagen 5-bromouracil exists in two states. In the normal state it resembles thymine and will pair only with adenine in DNA (Figure 16.10a). In its alternate, rarer state it pairs only with guanine (Figure 16.10b). Mutagen 5BU induces mutations by switching between the two forms once the base analog has been incorporated into the DNA (Figure 16.10c). This is essentially similar to the way spontaneous mutations occur by DNA replication errors. The difference here is that base analogs are much more frequently in the rare state than are normal bases. Hence, the frequency of mutations induced by base analogs is much higher than the frequency of spontaneous mutations. If, for example, 5BU is incorporated in its normal state, it pairs with adenine. If it changes into its rare state during replication, then guanine instead of adenine will be put in as the complementary base pair. In the next round of replication the G–5BU base pair will be resolved into a G–C base pair instead of the A–T base pair. By this process a transition mutation is produced, in

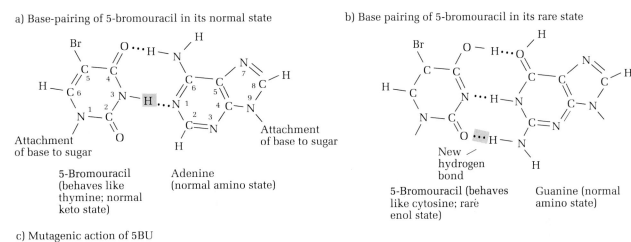

a) Base-pairing of 5-bromouracil in its normal state

b) Base pairing of 5-bromouracil in its rare state

5-Bromouracil
(behaves like
thymine; normal
keto state)

Adenine
(normal amino state)

5-Bromouracil (behaves
like cytosine; raré
enol state)

Guanine (normal
amino state)

c) Mutagenic action of 5BU

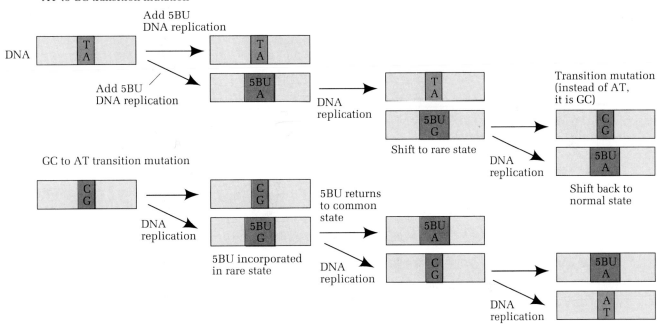

this case from AT to GC. The 5BU will induce a mutation from GC to AT if it is first incorporated into DNA in its rare state and then switches to the normal state during replication. Thus 5BU-induced mutations can be reverted by a second treatment of 5BU.

The base analog 2-aminopurine (2AP) acts as a mutagen in essentially the same way as 5BU does, and like 5BU, 2AP exists in normal and rare states (Figure 16.11a and b). In its normal

Figure 16.10 Mutagenic effects of the base analog 5-bromouracil (5BU): (a) In its normal state 5BU pairs with adenine. (b) In its rare, enol state 5BU pairs with guanine. (c) 5BU induces transition mutations when it incorporates into DNA in one state and then shifts to its alternate state during the next round of DNA replication. The two possible ways this mutation can happen are shown in diagrams (1) and (2).

transitions

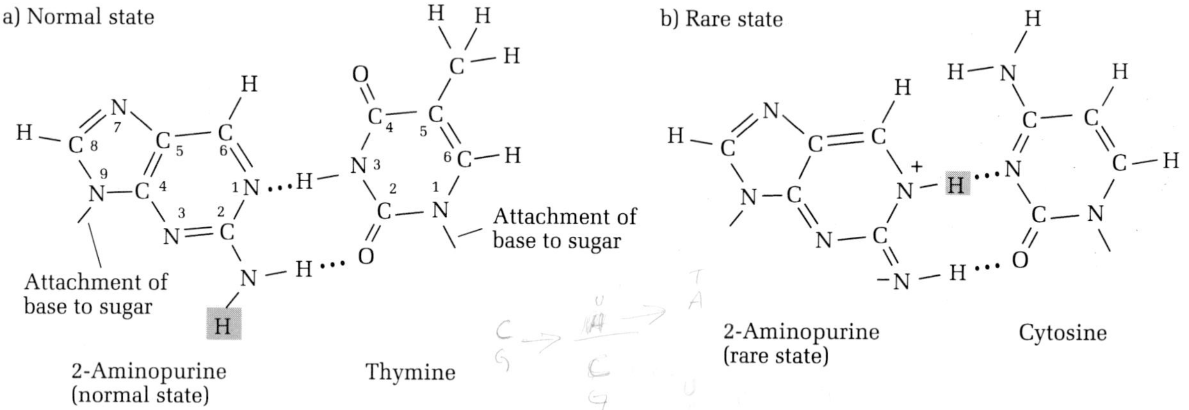

a) Normal state

2-Aminopurine
(normal state)

Thymine

b) Rare state

2-Aminopurine
(rare state)

Cytosine

Figure 16.11 Base-pairing properties of the normal and rare states of the base-analog mutagen 2-aminopurine (2AP): (a) In its normal state 2AP pairs with thymine. (b) In its rare state 2AP pairs with cytosine.

state 2AP resembles adenine but has an amino group at a different position on the purine ring than does adenine; in this state it base-pairs with thymine. In its rare state 2AP resembles guanine and base-pairs with cytosine. As with 5BU, 2AP induces transition mutations, which can be reverted by a second application of 2AP. The mutation will be either AT to GC or GC to AT, depending on its state during its initial incorporation into the DNA and its state during replication. Note that since the mutations involved are transitions in both cases, 5BU can revert mutations induced by 2AP and vice versa.

Base-modifying agents. Unlike base analogs, which actually replace bases in the DNA and require DNA replication for their incorporation, a number of chemicals act as mutagens by directly modifying the chemical structure and properties of the bases. Figure 16.12 shows the action of three types of mutagens that work in this way that is, a deaminating agent, a hydroxylating agent, and an alkylating agent.

Nitrous acid, HNO_2 (Figure 16.12a), is a deaminating agent that removes amino groups (—NH_2) from the bases guanine, cytosine, and adenine. Treatment of guanine with nitrous acid produces xanthine, but this purine base has the

same pairing properties as guanine, and so only a neutral mutation results (Figure 16.12a, part 1). When cytosine is treated with nitrous acid, however, uracil (which pairs with adenine) is produced (Figure 16.12a, part 2). The deamination of cytosine by nitrous acid, then, produces a CG-to-TA transition mutation during replication. Likewise, nitrous acid modifies adenine to produce hypoxanthine, a base that pairs with cytosine rather than thymine, thus resulting in an AT-to-GC transition mutation (Figure 16.12a, part 3). A nitrous acid–induced mutation can be reverted by a second treatment with nitrous acid.

Another base-modifying mutagen is hydroxylamine, NH_2OH (Figure 16.12b), which reacts specifically with cytosine, modifying it by adding a hydroxyl group (OH) so that it can pair only with adenine instead of with guanine. The mutations induced by hydroxylamine are, therefore, CG-to-TA transitions. Since only a one-way transition can occur with this mutagen, hydroxylamine-induced mutations cannot be reverted by a second treatment with this chemical. They can be reverted, however, by treatment with

Figure 16.12 Action of three base-modifying agents: (a) Nitrous acid (HNO_2) modifies guanine, cytosine, and adenine. The cytosine and adenine modifications result in mutations while the guanine modification does not. (b) Hydroxylamine (NH_2OH) reacts only with cytosine. (c) Ethylmethane sulfonate (EMS), an alkylating agent, alkylates guanine and thymine.

Original base	Mutagen	Modified base	Pairing partner	Predicted transition

Guanine — Nitrous acid (HNO₂) → Xanthine ··· Cytosine — None

Cytosine — Nitrous acid (HNO₂) → Uracil ··· Adenine — CG → TA

Adenine — Nitrous acid (HNO₂) → Hypoxanthine ··· Cytosine — AT → GC

Cytosine — Hydroxylamine (NH₂OH) → Hydroxylaminocytosine ··· Adenine — CG → TA

Guanine — Ethylmethane sulfonate (EMS) (alkylating agent) → O⁶-Ethylguanine ··· Thymine — GC → AT

Thymine — Ethylmethane sulfonate (EMS) (alkylating agent) → O⁴-Ethylthymine ··· Guanine — TA → CG

other mutagens (such as 5BU, 2AP, and nitrous acid) that cause TA-to-CG transition mutations.

The last base-modifying agent we will consider is ethylmethane sulfonate, EMS (Figure 16.12c). This agent is one of a diverse group of alkylating agents that introduce alkyl groups (e.g., —CH_3, —CH_2CH_3) onto the bases at a number of places. An alkylating agent alters a base so that it pairs with the wrong base. As shown in Figure 16.12c, after treatment with EMS, the now-ethylated guanine pairs with thymine, giving GC-to-AT transitions (Figure 16.12c, part 1), and the ethylated thymine pairs with guanine, giving TA-to-CG transitions (Figure 16.12c, part 2).

Intercalating agents. The intercalating mutagens (including proflavin, acridine, ethidium bromide, and ICR–170) act by inserting themselves (*intercalating*) between adjacent bases in one or both strands of the DNA double helix (Figure 16.13). The chemical structures of proflavin and acridine orange are shown in Figure 16.13a. If the intercalating agent inserts between adjacent base pairs of the DNA strand that is the template for new DNA synthesis (Figure 16.13b), then an extra base (x′) must be inserted in the new DNA strand opposite the intercalating agent. After one more round of replication accompanied by loss of the intercalating agent, the overall result is a frameshift mutation due to the insertion of one base pair (x-x′).

If the intercalating agent inserts into the new DNA strand instead of a base (Figure 16.13c), then when that DNA double helix replicates after the intercalating agent is lost, the result is a frameshift mutation due to the deletion of one base pair (3-3′). Following the insertion of an intercalating agent into a strand of DNA, the consequence is that when the DNA replicates, a base is either added or deleted, resulting in a frameshift mutation. As a consequence of the frameshift mutation, all the amino acids past the mutation point are likely to be incorrect, and the resulting protein will most likely be nonfunctional. Frameshift mutations induced by intercalating agents can be reverted by treatment with these agents.

Keynote *Chemical mutagens act in a variety of ways. Base analogs physically replace the proper bases during DNA replication and then shift form so that their base-pairing capabilities change. Base-modifying agents cause chemical changes in existing bases, thereby altering their base-pairing properties. Base analogs and base-modifying agents result in base-pair substitution mutations. Intercalating agents are inserted between existing base pairs during replication, causing single base-pair additions or deletions, which are frameshift mutations.*

The Ames Test: A Screen for Potential Mutagens and Carcinogens

All the mutagens (radiation and chemicals) we have examined so far are those commonly used to induce mutations during genetic research. Presently, scientists are aware that many chemicals in our environment might have mutagenic effects. Some of these chemicals have been shown to be carcinogenic; that is, they cause cancer in laboratory animals. Indeed, research indicates that about 90 percent of all known carcinogens are also mutagens. This finding has led to the proposal of theories that cancer is caused by the induction of particular mutations or by the accumulation of mutational damage in the somatic cells. However, we should not think of cancer as one disease but as many; some cancers may be caused by mutational damage, others by viruses. Nevertheless, the possible link between cancer and mutagens means that we should be aware of the chemicals we breathe and ingest in our modern environment.

Several methods are used to test new or old environmental chemicals for carcinogenic effects. One method is to test a chemical directly for carcinogenicity by setting up carefully controlled experiments with inbred lines of mice or rats. The inbred lines are genetically identical so that a statistical analysis of treated versus sham-treated animals (i.e., animals treated in exactly the same way as the experimental animals but without use of the potential carcinogen) for the incidence of cancers provides information about the probability that a chemical is carcinogenic. Although these studies are expensive and time-consuming, they are done routinely by research laboratories, often under contract to federal agencies or to private companies.

In the early 1970s a more rapid test was developed by Bruce Ames. The **Ames test** uses the bacterium *Salmonella typhimurium* as the test organism. Two strains are used, both of which are auxotrophic for histidine. An **auxotrophic mutation** affects an organism's ability to make a particular molecule which is essential for growth. Histidine (*his*) auxotrophs, then, require the presence of histidine in the growth medium in order to grow. In one *his* strain the *his* mutation can be reverted to *his*⁺ by a base-pair substitution, and in the other strain the *his* mutation can be reverted by a frame-shift mutation. (These strains also carry other mutations that make them particularly suitable for the experimental manipulations used, but a discussion of these features is beyond the scope of this text.)

The procedure for the Ames test is summarized in Figure 16.14. Rat livers are obtained, homogenized, and centrifuged to sediment the cellular debris. The enzymes of the rat liver are collected from the supernatant and are added to a liquid culture of the auxotrophic *Salmonella typhimurium* along with the chemical being tested. At the same time a control experiment is run in which no potential carcinogen is present.

Enzymes from the rat liver are used in these experiments because in the living organism, as in humans, the liver enzymes perform the tasks of detoxifying and toxifying various chemicals,

a) Representative agents

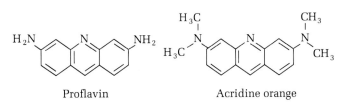

Proflavin   Acridine orange

b) Mutation by addition

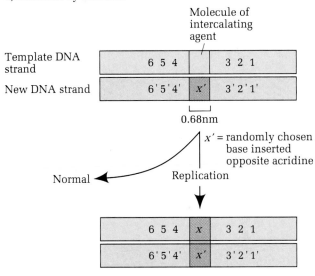

Result: frameshift mutation due to insertion of one base pair (*x, x′*)

c) Mutation by deletion

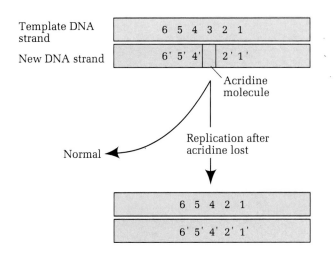

Figure 16.13 Intercalating mutations: (a) structures of representative intercalating agents, proflavin and ac-ridine orange; (b) frameshift mutation by addition, when agent inserts into template strand; (c) frame-shift mutation by deletion, when agent inserts into newly synthesizing strand.

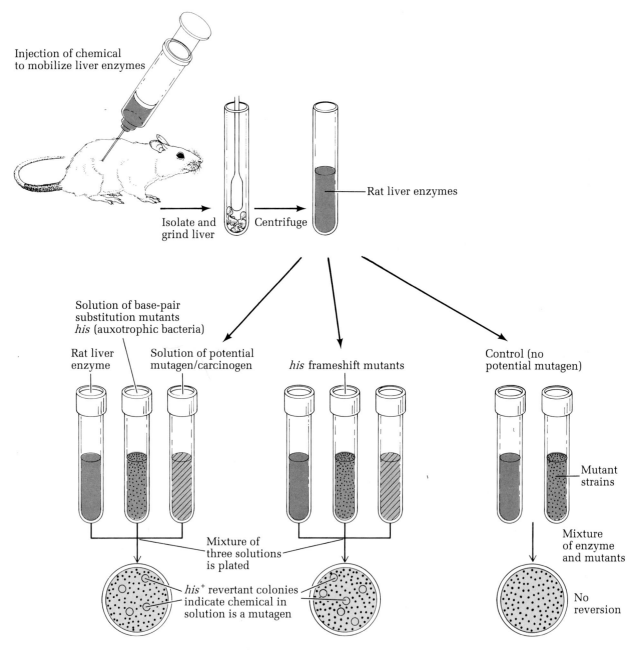

Figure 16.14 *Ames test for potential mutagens.*

including many potential mutagens. Their presence allows the investigators to determine whether a chemical that by itself is not mutagenic can become mutagenic by being processed in the liver.

The mixture is then plated onto a medium containing no histidine. Revertants of a mutant *Salmonella typhimurium* strain containing the base-pair substitution mutation or a strain containing a frameshift mutation are sought. The control mixture is plated in the same way.

The *his*⁺ revertants are detected since they will produce colonies on the histidine-lacking plates. The presence of *his*⁺ revertants on the plates spread with the chemical-treated mixtures indicates that the chemical is a mutagen. And since about 90 percent of all known carcinogens are also mutagens, the chemical may be a carcinogen as well.

The control plate is important in this test because it indicates the spontaneous reversion rate in the test bacteria. For the investigator to conclude that the chemical being tested is a mutagen, the rate of reversion must be significantly higher than the spontaneous rate. To date, the Ames test has identified a large number of mutagens among environmental chemicals, such as hair dye additives, vinyl chloride, and particular food colorings.

DNA Repair Mechanisms

Both prokaryotic and eukaryotic cells have a number of repair systems to deal with damage to DNA (e.g., the SOS system). All of the systems use enzymes to make the correction. Some of the systems directly correct the mutational lesion while others first excise creating a single-stranded gap and then synthesize new DNA for the resulting gap.

Direct Correction of Mutational Lesions

An example of direct correction of mutational lesions involves the repair of UV-light-induced thymine (or other pyrimidine) dimers. This system is called **photoreactivation** or **light repair** (Figure 16.15). By this process the dimers are reverted directly to the original form by exposure to visible light in the wavelength range of 320–370 nm (blue light). Photoreactivation is catalyzed by an enzyme called photolyase (encoded by the *phr* gene), which, when activated by a photon of light, tears or splits the dimers apart. Since photolyase has been found in all organisms that have been studied, it may be a ubiquitous enzyme. Presumably, it functions by scav-

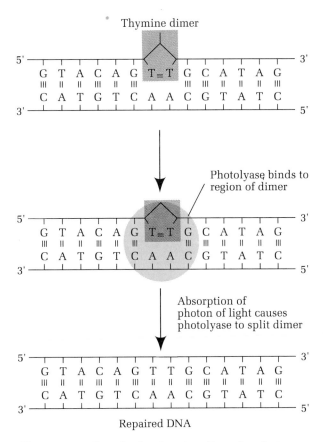

Figure 16.15 *Repair of a thymine dimer by photoreactivation.*

enging along the double helix, seeking the bulges that result when thymine dimers are present. Photolyases are apparently very effective since few thymine dimers (and, hence, mutations) are left after photoreactivation.

Another system of this kind corrects bases that have become alkylated, for example, by the action of mutagens such as EMS. In this system an enzyme called *alkyltransferase* catalyzes the removal of the alkyl group from the base and its temporary attachment to one of the amino acids of the enzyme itself. In this repair system, then, the modified base is not removed from the DNA.

Repair Involving Excision of Base Pairs

Excision repair. A second repair process for

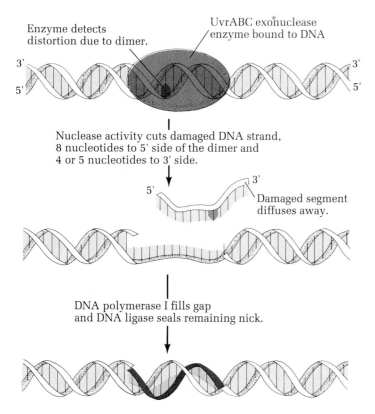

Enzyme detects distortion due to dimer.

UvrABC exonuclease enzyme bound to DNA

Nuclease activity cuts damaged DNA strand, 8 nucleotides to 5' side of the dimer and 4 or 5 nucleotides to 3' side.

Damaged segment diffuses away.

DNA polymerase I fills gap and DNA ligase seals remaining nick.

Figure 16.16 Excision repair of pyrimidine dimer and other damage-induced distortions of DNA initiated by the UvrABC endonuclease.

UV-light-induced pyrimidine dimers is called **excision repair** or **dark repair** since it does not depend on the presence of light. This repair mechanism was discovered in 1964, both by R. P. Boyce and P. Howard-Flanders and by R. Set-low and W. Carrier. They isolated some UV-sensitive mutants of *E. coli* that exhibited a higher-than-normal rate of induced mutation in the dark after UV irradiation. These mutants were *uvrA* mutants, where uvr means "UV repair." They can repair dimers only with the input of light; that is, they lack the dark-repair system.

Our present understanding of the excision repair system in *E. coli* is as follows (Figure 16.16). This system has been shown to correct not only pyrimidine dimers but also other serious damage-induced distortions of the DNA helix. These helix distortions are recognized by the UvrABC endonuclease, a multi-subunit enzyme encoded by the three genes *uvrA, uvrB,* and *uvrC.* This enzyme makes one cut in the damaged DNA strand eight nucleotides to the 5' side of the damage (e.g., the dimer) and four nucleotides to the 3' side of the damage, thereby releasing a 12-nucleotide stretch of single-stranded DNA containing the damaged base(s). The 12-base gap is f lled by DNA polymerase I and sealed by DNA ligase.

Although it is not clear exactly how the UvrABC endonuclease recognizes damaged DNA, it seems to function by recognizing the absence of a normal helix. Somehow the enzyme must wrap around a 12-base-pair segment (a little over one helical turn) and "feel" for unnatural shapes.

The excision repair system is found in most organisms that have been studied, and its mode of action is thought to be essentially like the one found in *E. coli.* Occasionally, errors are introduced in the repair synthesis of the DNA, and such errors are another source of mutations resulting from UV radiation. The most common cause of the errors is incorrect pairing of new nucleotides with those in the template strand.

Genetic mutations affecting the excision repair process are also known in a number of organisms. In general, such mutations lead to an increased rate of mutation after UV treatment. For example, the human disease xeroderma pigmentosum is the result of a recessive mutation that causes a def ciency in one of the excision repair enzymes. People with this disease are particularly UV-sensitive and tend to get skin cancers after exposure to sunlight. The simplest interpretation is that unrepaired UV damage disrupts the normal control functions of the skin cells so that malignant growths are generated.

Repair by glycosylases. Damaged bases can also be excised in another way involving the action of a glycosylase enzyme that detects an individual unnatural base and catalyzes its removal from the deoxyribose sugar to which it is attached (Figure 16.17). This catalytic activity leaves a hole in the DNA without a base. This is called an *AP site* (for apurinic, where there is no

539
Screening
Procedures
for the
Isolation
of Mutants

A or G, or apyriminic, where there is no C or T). Such holes can also occur through natural, spontaneous losses of bases, which we discussed earlier in the chapter. The hole is recognized by a specific enzyme called *AP endonuclease*, which cuts the backbone besides the missing base (Figure 16.17). This leaves a primer end from which DNA polymerase I (in *E. coli*) initiates repair synthesis, using its $5' \rightarrow 3'$ exonuclease activity to remove a few nucleotides ahead of the missing base and its polymerizing activity to fill in the gap. DNA ligase seals the remaining nick. DNA polymerase I activity is displayed in this repair system in which the enzyme simultaneously polymerases DNA and removes nucleotides ahead of the growing chain. This process is called *nick translation.*

Keynote Both prokaryotes and eukaryotes have a number of repair systems that deal with different kinds of DNA damage. Without such repair systems mutational lesions would accumulate and be lethal to the cell or organism. Not all mutational lesions are repaired, and hence mutations are produced.

Screening Procedures for the Isolation of Mutants

Geneticists have made great progress over the years in understanding how normal processes take place by studying mutants that have defects in those processes. Researchers use mutagens to induce mutations at a greater rate than the rate at which spontaneous mutations occur. Mutagens, however, are not directed in their action toward particular genes. Instead, they indiscriminately change base pairs without regard to the

Figure 16.17 Excision repair of damaged bases involving the action of glycosylase. Glycosylase detects the damaged base and catalyzes its removal. A few adjacent bases are removed by AP endonuclease and the resulting single-stranded gap is repaired by DNA polymerase I and DNA ligase.

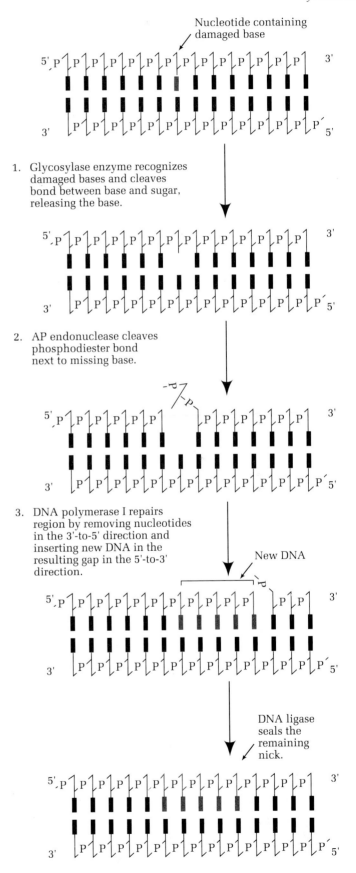

Nucleotide containing damaged base

1. Glycosylase enzyme recognizes damaged bases and cleaves bond between base and sugar, releasing the base.

2. AP endonuclease cleaves phosphodiester bond next to missing base.

3. DNA polymerase I repairs region by removing nucleotides in the 3'-to-5' direction and inserting new DNA in the resulting gap in the 5'-to-3' direction.

New DNA

DNA ligase seals the remaining nick.

positions of the base pairs in the genetic material. In other words, the mutation process is random. Thus an observable phenotypic change will only result if the mutation affects the function of a gene that governs a phenotype researchers can observe. Mutations in regulatory or protein-coding genes usually result in a phenotypic change unless the mutation is a neutral one.

Particularly with microorganisms, a number of screening procedures have been developed to help geneticists obtain particular mutants of interest from among a heterogeneous mixture in a mutagenized population. A few screening procedures for the isolation of particular types of mutations are described in the following subsections.

Visible Mutations

Visible mutations affect the morphology or physical appearance of an organism. Examples of visible mutations are eye color mutants or wing shape mutants of *Drosophila,* coat color mutants of animals (e.g., albino organisms), colony-size mutants of yeast, and plaque morphology mutants of bacteriophages. Since visible mutations, by definition, are readily apparent, screening for them is done by inspection.

Nutritional Mutations

An **auxotrophic mutation** (also called a **nutritional** or **biochemical mutation**) affects an organism's ability to make a particular molecule essential for growth. Auxotrophic mutations are most readily detected in microorganisms such as *E. coli,* yeast, *Neurospora,* or some unicellular algae that grow on simple and defined growth media from which they synthesize the enzymes used to make all the molecules essential to their growth. A number of screening procedures can isolate auxotrophic mutants, and some of them will be described now.

In 1952 Esther and Joshua Lederberg developed a procedure to screen for auxotrophic mutants of any microorganism that grows in discrete colonies on solid medium. This procedure

is called **replica plating;** Figure 16.18 diagrams how replica plating can be used to isolate arginine auxotrophs of a microorganism.

In the replica-plating technique, samples from a culture of a colony-forming organism or cell type that has or has not been mutagenized are plated onto a complete medium that contains all possible nutrients. Upon incubation, colonies grow wherever a cell landed on the medium. These colonies consist of clones of the original cell; that is, they are genetically identical copies of the original cell. The population of colonies will be a mixture of prototrophic (wild-type) colonies and auxotrophic mutant colonies. The pattern of the colonies is transferred onto sterile velveteen cloth. Replicas of the original colony pattern on the cloth are then made by gently pressing new plates onto the velveteen. If the new plate contains minimal medium, only prototrophic colonies can grow. So by comparing the patterns on the original master plate with the minimal replica plate, researchers can readily identify the potential auxotrophic colonies. If another plate contains minimal medium *plus* arginine, then the colonies that grow on this medium but not on the minimal medium are the arginine auxotrophs, that is, the mutants that require arginine in order to grow. They can then be selected from the original master plate and cultured for further study.

A useful replica-plating technique to screen for a range of mutant types employs an antibiotic to kill many nonmutant cells but not many or any mutant cells. For example, the antibiotic nystatin (named after New York state) may be used to enrich for mutants of yeast; that is, to increase the proportion of mutant to nonmutant cells in the population. Nystatin is an effective killer of growing yeast cells.

Suppose we wish to enrich selectively for adenine auxotrophic (*ade*) mutants of yeast. We incubate the mutagenized population of cells in a culture medium that lacks adenine but contains nystatin. Under these conditions, wild-type cells grow and are killed by the nystatin, while the *ade* (adenine-requiring) cells that either do not grow or grow very slowly are selectively spared. If the exposure to the antibiotic is care-

541
Screening
Procedures
for the
Isolation
of Mutants

fully limited, the population of surviving cells has a much higher proportion of *ade* mutant cells than is found in the original, unselected population of cells. This method can be modified to select for a number of other classes of mutants. An essentially identical method with penicillin as the selective antibiotic may be used with *E. coli.*

As we have frequently seen, the mycelial fungus *Neurospora crassa* has played a significant role in biochemical genetics. This organism grows as a multinucleate mycelium rather than as individual cells, and it is a model organism for a large number of mycelial fungi of agricultural and medical significance. (The genetics of this organism was discussed in Chapter 6.) Since individual cells are not formed when colonies are grown on solid medium, the use of replica plating to screen for mutants is not possible. Antibiotic selection can be used, however, under certain circumstances. For example, a population of conidia (the asexual spores) can be incubated in a medium with a fungicidal antibiotic under the appropriate conditions for isolating a particular class of mutants. When the nonmutant conidia germinate, they are killed by the antibiotic, which leads to a significant enrichment for the mutant type.

Another routinely used method achieves the same purpose of enriching for mutants by taking advantage of the mycelial growth habit of *Neurospora* cultures. This method is called the *filtration enrichment procedure* (Figure 16.19). Conidia are added to a culture medium appropriate for the enrichment desired.

For example, if we wish to isolate mutants that are auxotrophic for the vitamin thiamine, we use a medium that lacks thiamine. So that other types of auxotrophs are not picked up, the medium must contain all the other vitamins, all amino acids, all purines and pyrimidines, and so on. When the culture is incubated, the mutant conidia do not germinate, or they germinate only very slowly. The conidia that are not auxotrophic for thiamine, though, do germinate and begin to form the weblike growth characteristic of mycelial fungi. So by periodically filtering the culture through a cheesecloth filter that removes

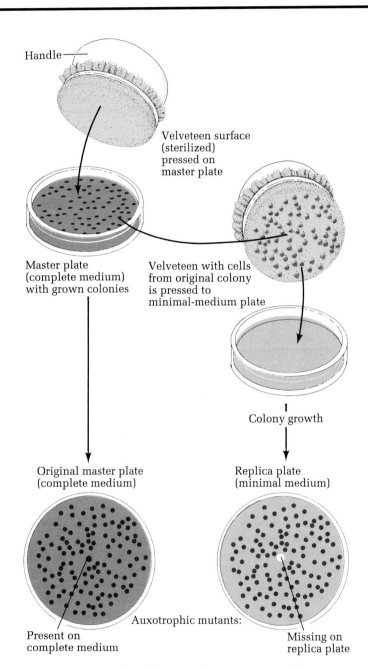

Figure 16.18 Replica-plating technique to screen for mutant strands of a colony-forming microorganism.

the relatively large germlings but permits the much smaller conidia to pass through, we can achieve a significant enrichment for thiamine auxotrophs. As with any of the selection procedures, the enrichment is not total, but at least

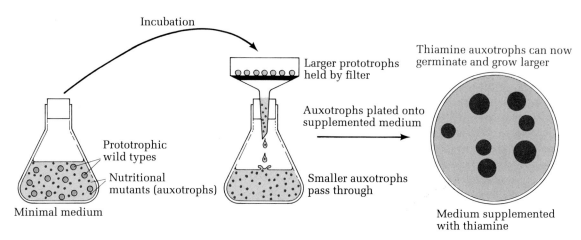

Figure 16.19 Filtration enrichment procedure for selecting mutants of mycelial fungi such as Neurospora
crassa.

the task of identifying the mutants of interest is
made a bit easier.

Conditional Mutations

The products of many genes—DNA polymerase
and RNA polymerase, for example—are important for the growth and division of cells, and
most mutations in these genes result in a lethal
phenotype. The structure and function of this
class of genes may be studied by inducing conditional mutations in the genes. A common type
of conditional mutation to study is a heat-sensitive mutation that is characterized by normal
function at the normal growth temperature and
no or severely impaired function at a higher
temperature. In yeast, for instance, heat-sensitive
mutations typically are isolated at 36°C, while
the normal growth temperature is 23°C.

Essentially the same procedures to screen for
heat-sensitive mutations of microorganisms can
be used to screen for auxotrophic mutations.
Replica plating can select for temperature-sensi-

tive mutants when the replica plate is incubated
at a higher temperature than the master plate.
Nystatin selection could also be used to enrich
for heat-sensitive mutations of yeast that do not
grow at 36°C.

Keynote Geneticists have made great progress in understanding how cellular
processes take place by studying mutants that
have defects in those processes. With microorganisms, a number of screening procedures enrich selectively for mutants of interest from a
heterogeneous mixture of cells in a mutagenized
population of cells.

In summary, in this chapter we saw how
gene mutations occur or may be induced, and
we learned how populations of cells can be
screened for mutants of interest. This has given
us the conceptual tools necessary for studying
the processes of mutation itself, for examining a
variety of chemicals for possible mutagenic effects, and for using mutations for the genetic
analysis of biological functions.

543
*Analytical
Approaches
for Solving
Genetics
Problems*

*Analytical
Approaches
for Solving
Genetics
Problems*

Q.1 Five strains of *E. coli* containing mutations that affect the tryptophan synthetase A polypeptide have been isolated. The following diagram shows the changes produced in the protein itself in the indicated mutant strains:

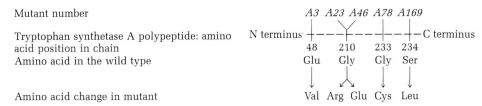

Mutant number

Tryptophan synthetase A polypeptide: amino acid position in chain
Amino acid in the wild type

Amino acid change in mutant

In addition, *A23* can be further mutated to insert Ile, Thr, Ser, or the wild-type Gly into position 210.

a. Using the genetic code (Figure 13.9), explain how the two mutations *A23* and *A46* can result in two different amino acids being inserted at position 210. Give the nucleotide sequence of the wild-type gene at that position and the two mutants.

b. Can mutants *A23* and *A46* recombine? Why or why not?

c. From what you can infer of the nucleotide sequence in the wild-type gene, indicate, for the codons specifying amino acids 48, 210, 233, and 234, whether or not a nonsense mutant could be generated by a single nucleotide substitution in the gene.

A.1 a. There are no simple ways to answer questions like this one. The best approach is to scrutinize the genetic code dictionary and use a pencil and paper to try to define the codon changes that are compatible with all the data. The number of amino acid changes in position 210 of the polypeptide is helpful in this case. The answer is that the original codon for glycine in the wild type was GGA; also, *A23* is AGA, and *A46* is GAA. In other words, the two different amino acid substitutions in the two mutants result from different, single base-pair changes in the part of the gene that corresponds to this codon.

b. The answer to this question follows from the answer deduced in part a. Mutants *A23* and *A46* can recombine since the mutations in the two mutant strains are in different base pairs. The results of a single recombination event at position 210 are a wild-type GGA codon (Gly) and a double mutant AAA codon (Lys).

c. Amino acid 48 had a Glu-to-Val change. This change must have involved GAA to GUA or GAG to GUG. In either case the Glu codon can mutate with a single base-pair change to a nonsense codon, that is, UAA or UAG, respectively.

   Amino acid 210 in the wild type has a GGA codon, as we have already discussed. This gene could mutate to the UGA nonsense codon with a single base-pair change.

   Amino acid 233 had a Gly-to-Cys change. This change must have involved either GGU to UGU or GGC to UGC. In either case the Gly codon cannot mutate to a nonsense codon with one base-pair change.

   Amino acid 234 had a Ser-to-Leu change. This change was either UCA to UUA or UCG to UUG. If the Ser codon was UCA, it changed to UGA in one step. But if the Ser codon was UCG, it cannot change to a nonsense codon in one step.

Q.2 The chemically induced mutations, *a, b, c,* and *d* show specific reversion patterns when subjected to treatment by the following mutagens: 2-aminopurine (AP), 5-bromouracil (BU), proflavin (pro), hydroxylamine (HA), and ethylmethane sulfonate (EMS). The reversion patterns are shown in the following table:

Mutation	Mutagens tested in reversion studies				
	AP	BU	pro	HA	EMS
a	−	−	+	−	−
b	−	+	−	+	+
c	+	−	−	−	−
d	−	−	−	−	+

(*Note:* + indicates many reversions to wild type were found; − indicates no reversions, or very few, to wild type were found.)

Indicate, for each original mutation (i.e., a^+ to *a,* b^+ to *b,* etc.), the probable base-pair change (i.e., AT to GC, deletion of GC, etc.) and the mutagen that was most probably used to induce the original change.

A.2 This question tests knowledge of the base-pair changes that can be induced by the various mutagens used. Mutagen AP induces mainly AT-to-GC changes and can cause GC-to-AT changes also. Thus AP-induced mutations can be reverted by AP. Mutagen BU induces mainly GC-to-AT changes and can cause AT-to-GC changes, so BU-induced mutations can be reverted by BU. Proflavin causes single base-pair deletions or additions with no specificity. Proflavin-induced changes can be reverted by a second treatment with proflavin, as we discovered in the discussion of the experiments used to prove the three-letter basis of the genetic code. Mutagen HA causes one-way transitions from GC to AT, and so HA-induced mutations cannot be reverted by HA. Mutagen EMS alkylates G, which results in its loss from the DNA; it is replaced at random by one of the four bases. If the replacing base is G, there is no mutation, and the effect of EMS is not detected. Substitution with any one of the other three bases, though, results in a mutation. So a third of the time there will be a GC-to-AT change, a third of the time there will be a GC-to-TA change, and a third of the time there will be a GC-to-CG change. With these mutagen specificities in mind, we can answer the questions for each mutation in turn.

Mutation a^+ to *a:* The *a* mutation was reverted only by proflavin, indicating that it was a deletion or an addition. Therefore the original mutation was induced by proflavin, since that is the only mutagen in the list that can cause an addition or a deletion.

Mutation b^+ to *b:* The *b* mutation was reverted by BU, HA, or EMS. A key here is that HA only causes GC-to-AT changes. Therefore *b* must be GC, and the original b^+ must have been an AT. Thus the mutational change of b^+ *to* b must have been caused by treatment with AP or BU, since these are the two mutagens in the list that are capable of inducing that change.

Mutation c^+ to c: The c mutation was reverted only by AP. Since it could not be reverted by HA, c must be an AT and c^+ a GC. The mutational change from c^+ to c therefore involved a GC-to-AT transition and could have resulted from treatment with BU, HA, or EMS.

Mutation d^+ to d: The d mutation was reverted only by EMS, indicating that something other than a transition mutation was involved in the d^+-to-d mutational event. Mutagen EMS causes GC-to-CG transversions, so the most likely event in d^+ to d was a GC-to-CG change (or CG to GC). Hence the reversion was a CG-to-GC change (or GC to CG). Thus the original mutagen could only have been EMS.

*Questions
and
Problems*

*16.1   Mutations are (choose the correct answer):

a. Caused by genetic recombination.
b. Heritable changes in genetic information.
c. Caused by faulty transcription of the genetic code.
d. Usually but not always beneficial to the development of the individuals in which they occur.

16.2   Answer true or false: Mutations occur more frequently if there is a need for them.

*16.3   The following is not a class of mutation (choose the correct answer):

a. Frameshift.
b. Missense.
c. Transition.
d. Transversion.
e. None of the above (i.e., all are classes of mutation).

*16.4   Ultraviolet light usually causes mutations by a mechanism involving (choose the correct answer):

a. One-strand breakage in DNA.
b. Light-induced change of thymine to alkylated guanine.
c. Induction of thymine dimers and their persistence or imperfect repair.
d. Inversion of DNA segments.
e. Deletion of DNA segments.
f. All of the above.

*16.5   a. The sequence of nucleotides in an mRNA is:

5′-UUUACCCAUUGGUCUCGU-3′

How many amino acids long would you expect the polypeptide chain made with this messenger to be?

b. Hydroxylamine is a mutagen that results in the replacement of an AT base pair for a CG base pair in the DNA; that is, it induces a transversion mutation. When applied to the organism that made the mRNA molecule shown in part a, a strain was isolated in which a mutation occurred at the 11th position of the DNA that coded for the mRNA. How many amino acids long would you expect the polypeptide made by this mutant to be? Why?

*16.6 Three of the codons in the genetic code are chain-terminating codons for which no naturally occurring tRNAs exist. Just like any other codons in the DNA, though, these codons can change as a result of base-pair changes in the DNA. Confining yourself to single base-pair changes at a time, determine which amino acids could be inserted in a polypeptide by mutation of these chain-terminating codons: (a) UAG; (b) UAA; (c) UGA. (The genetic code is listed in Figure 13.9.)

16.7 The amino acid substitutions in the figure occur in the α and β chains of human hemoglobin. Those amino acids connected by lines are related by single nucleotide changes. Propose the most likely codon or codons for each of the numbered amino acids. (Refer to the genetic code listed in Figure 13.9.)

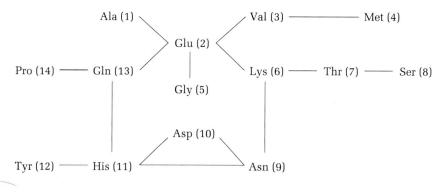

*16.8 Yanofsky studied the tryptophan synthetase of *E. coli* in an attempt to identify the base sequence specifying this protein. The wild type gave a protein with a glycine in position 38. Yanofsky isolated two *trp* mutants, *A23* and *A46*. Mutant *A23* had Arg instead of Gly at position 38, and mutant *A46* had Glu at position 38. Mutant *A23* was plated on minimal medium, and four spontaneous revertants to prototrophy were obtained. The tryptophan synthetase from each of four revertants was isolated, and the amino acids at position 38 were identified. Revertant 1 had Ile, revertant 2 had Thr, revertant 3 had Ser, and revertant 4 had Gly. In a similar fashion, three revertants from *A46* were recovered, and the tryptophan synthetase from each was isolated and studied. At position 38 revertant 1 had Gly, revertant 2 had Ala, and revertant 3 had Val. A summary of these data is given in the figure. Using the genetic code shown in Figure 13.9, deduce the codons for the wild type, for the mutants *A23* and *A46,* and for the revertants, and place each designation in the space provided in the figure.

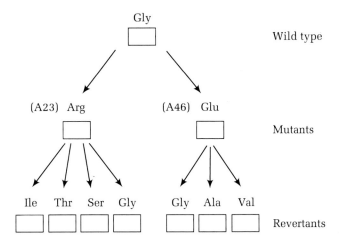

16.9 Consider an enzyme chewase from a theoretical microorganism. In the wild-type cell the chewase has the following sequence of amino acids at positions 39 to 47 (reading from the amino end) in the polypeptide chain:

$$-Met-Phe-Ala-Asn-His-Lys-Ser-Val-Gly-$$
$$39\quad 40\quad 41\quad 42\quad 43\quad 44\quad 45\quad 46\quad 47$$

A mutant of the organism was obtained; it lacks chewase activity. The mutant was induced by a mutagen known to cause single base-pair insertions or deletions. Instead of making the complete chewase chain, the mutant makes a short polypeptide chain only 45 amino acids long. The first 38 amino acids are in the same sequence as the first 38 of the normal chewase, but the last 7 amino acids are as follows:

$$-Met-Leu-Leu-Thr-Ile-Arg-Val-$$
$$39\quad 40\quad 41\quad 42\quad 43\quad 44\quad 45$$

A partial revertant of the mutant was induced by treating it with the same mutagen. The revertant makes a partly active chewase, which differs from the wild-type enzyme only in the following region:

$$-Met-Leu-Leu-Thr-Ile-Arg-Gly-Val-Gly-$$
$$39\quad 40\quad 41\quad 42\quad 43\quad 44\quad 45\quad 46\quad 47$$

Using the genetic code given in Figure 13.9, deduce the nucleotide sequences for the mRNA molecules that specify this region of the protein in each of the three strains.

16.10 Two mechanisms in *E. coli* were described for the repair of DNA damage (thymine dimer formation) after exposure to ultraviolet light: photoreactivation and excision (dark) repair. Compare and contrast these mechanisms, indicating how each achieves repair.

*16.11 After a culture of *E. coli* cells was treated with 5-bromouracil, it was noted that the frequency of mutants was much higher than normal. Mutant colonies were then isolated and grown and treated with nitrous acid; some of the mutant strains reverted to wild type.

a. In terms of the Watson-Crick model, diagram a series of steps by which 5BU may have produced the mutants.
b. Assuming the revertants were not caused by suppressor mutations, indicate the steps by which nitrous acid may have produced the back mutations.

16.12 A single, very hypothetical strand of DNA is composed of the base sequence indicated in the figure. In the figure, A indicates adenine, T indicates thymine, G indicates guanine, C denotes cytosine, U denotes uracil, BU is 5-bromouracil, 2AP is 2-aminopurine, BU-enol is an enol tautomer of 5BU, 2AP-imino is an imino tautomer of 2AP, HX is hypoxanthine, and X is xanthine; 5′ and 3′ are the numbers of the free, OH-containing carbons on the deoxyribose part of the terminal nucleotides.

a. Opposite the bases of the hypothetical strand, and using the shorthand of the figure, indicate the sequence of bases on a complementary strand of DNA.
b. Indicate the direction of replication of the new strand by drawing an arrow next to the new strand of DNA from part a.
c. When postmeiotic germ cells of a higher organism are exposed to a chemical mutagen before fertilization, the resulting offspring expressing an induced mutation are almost always mosaics for wild-type and mutant tissue. Give at least one reason that in the progenies of treated individuals these mosaics are found and not the so-called complete or whole-body mutants.

The following information applies to Problems 16.13–16.17: A solution of single-stranded DNA is used as the template in a series of reaction mixtures. The structure of the DNA is presented in the figure where A = adenine, G = guanine, C = cytosine, T = thymine, H = hypoxanthine, and HNO_2 = nitrous acid. For Problems 16.13 through 16.17, use this shorthand system and draw the products expected from the reaction mixtures. Assume that a primer is available in each case.

$$5' - T - HX - U - A - G - BU\text{-enol} - 2AP - C - BU - X - 2AP\text{-imino} - 3'$$

*16.13 The DNA template + purified Kornberg enzyme + dATP + dGTP + dCTP + dTTP + Mg^{2+}.

16.14 The DNA template + purified Kornberg enzyme + dATP + dGMP + dCTP + dTTP + Mg^{2+}.

16.15 The DNA template + purified Kornberg enzyme + dATP + dHTP + dGMP + dTTP + Mg^{2+}.

*16.16 The DNA template is pretreated with HNO_2 + purified Kornberg enzyme + dATP + dGTP + dCTP + dTTP + Mg^{2+}.

16.17 The DNA template + purified Kornberg enzyme + dATP + dGMP + dHTP + dCTP + dTTP + Mg^{2+}.

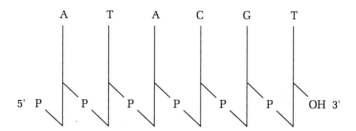

16.18 A strong experimental approach to determining the mode of action of mutagens is to examine the revertibility of the products of one mutagen by other mutagens. The following table represents collected data on revertibility of various mutagens on *rII* mutations in phage T2; + indicates majority of mutants reverted, − indicates virtually no reversion; BU = 5-bromouracil, AP = 2-aminopurine, NA = nitrous acid, and HA = hydroxylamine. Fill in the empty spaces.

Mutation induced by	Proportion of mutations reverted by				Base-pair substitution inferred
	BU	AP	NA	HA	
BU	−	__	__	−	_____
AP	__	−	__	+	_____
NA	+	+	__	+	_____
HA	__	__	+	−	GC → AT

*16.19 Three *ara* mutants of *E. coli* were induced by mutagen X. The ability of other mutagens to cause the reverse change (*ara* to *ara*$^+$) was tested, with the results shown in Table 16.1.

Assume all *ara*$^+$ cells are true revertants. What base changes were probably involved in forming the three original mutations? What kind(s) of mutations are caused by mutagen X?

Table 16.1 Frequency of ara$^+$ *Cells Among Total Cells After Treatment*

Mutant	Mutant				
	None	*BU*	*AP*	*HA*	*Frameshift*
ara-1	1.5 × 10^{-8}	5 × 10^{-5}	1.3 × 10^{-4}	1.3 × 10^{-8}	1.6 × 10^{-8}
ara-2	2 × 10^{-7}	2 × 10^{-4}	6 × 10^{-5}	3 × 10^{-5}	1.6 × 10^{-7}
ara-3	6 × 10^{-7}	10^{-5}	9 × 10^{-6}	5 × 10^{-6}	6.5 × 10^{-7}

17

Chromosome Aberrations

Chapter 16 discussed gene mutations, that is, heritable changes that occur within the limits of a single gene. This chapter discusses another kind of genetic change, namely chromosome aberrations, which are changes in normal chromosome structure or chromosome number. Discrimination between gene mutations and chromosomal structural changes, especially minute ones, is, in practice, frequently impossible. However, many chromosome aberrations are significant enough to be visible under the microscope. Thus the study of chromosome behavior in normal cells and of how this behavior is altered in cells with chromosome aberrations involves both genetic and cytological approaches and hence falls into the field of cytogenetics.

Types of Chromosome Changes

Cells of the same organism characteristically have the same number of chromosomes, with the exception of gametes, which (in diploid organisms) have half as many chromosomes as somatic cells. Moreover, the organization and the number of genes on the chromosomes of an organism are the same from cell to cell. These characteristics of chromosome number and gene organization typically hold for all members of the same species. Exceptions to these tenets do occur and are known as **chromosome aberrations.** Chromosome aberrations will be defined here as variations from the wild-type condition in either chromosome number or chromosome structure. Chromosome aberrations can arise spontaneously or can be induced experimentally by chemical or radiation mutagens, and in many instances they can be detected cytologically during mitosis and meiosis. The ability to detect different kinds of chromosome aberrations depends on the size, structure, and number of chromosomes concerned and the ease with which they can be handled.

Variations in Chromosome Number

Most eukaryotic organisms are diploid, meaning that they have two sets of chromosomes in their somatic cells; the gametes of these organisms are haploid since they have only one set of chromosomes. The following discussions are related to a diploid organism.

Variations may occur from the wild-type state in the total amount of genetic material in a cell. Such changes may occur spontaneously or may be experimentally induced. Variations in chromosome number, as opposed to variations in parts of chromosomes, give rise to **aneuploidy, monoploidy,** and **polyploidy.**

Changes in One or a Few Chromosomes

Aneuploidy. Aneuploidy describes the abnormal case in which one to a few whole chromosomes are lost from or added to the normal set of chromosomes (Figure 17.1). In most cases, aneuploidy is lethal in animals, and is detected mainly in aborted fetuses. Aneuploidy can occur, for example, from the loss of individual chromosomes in mitosis or meiosis; in this case nuclei are formed with less than the normal chromosome numbers. Aneuploidy may also result from nondisjunction, the irregular distribution to the cell poles of sister chromatids in mitosis or of homologous chromosomes in meiosis (see Chapter 3). In nondisjunction, nuclei with more and nuclei with less than the normal number of chromosomes are produced.

In the case of diploid organisms, aneuploid variations take four main forms (see Figure 17.1):

1. **Nullisomy** (a nullisomic cell) involves a loss of one homologous chromosome pair; that is, the cell is 2N − 2.
2. **Monosomy** (a monosomic cell) involves a loss of a single chromosome; that is, the cell is 2N − 1.
3. **Trisomy** (a trisomic cell) involves a single extra chromosome; that is, the cell has three

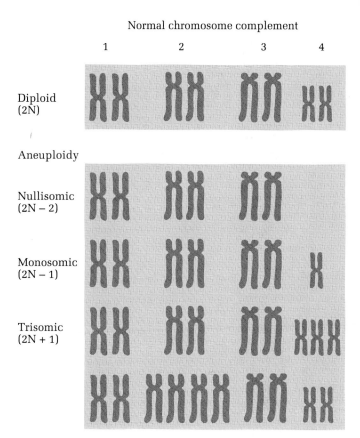

Normal chromosome complement

1   2   3   4

Diploid (2N)

Aneuploidy

Nullisomic (2N − 2)

Monosomic (2N − 1)

Trisomic (2N + 1)

*Figure 17.1 Normal (theoretical) set of metaphase chromosomes in a diploid (2N) organism (*top*) and examples of aneuploidy (*bottom*).*

copies of one chromosome type and two copies of every other chromosome type. A trisomic cell is 2N + 1.
4. **Tetrasomy** (a tetrasomic cell) involves an extra chromosome pair, resulting in the presence of four copies of one chromosome type and two copies of every other chromosome type. A tetrasomic cell is 2N + 2.

Aneuploidy may involve the loss or the addition of more than one specific chromosome or a chromosome pair. Such situations are characterized, for example, by the term *doubly monosomic* (2N − 1 − 1), *doubly tetrasomic* (2N + 2 + 2), and so forth.

All forms of aneuploidy have serious consequences in meiosis. Monosomics, for example,

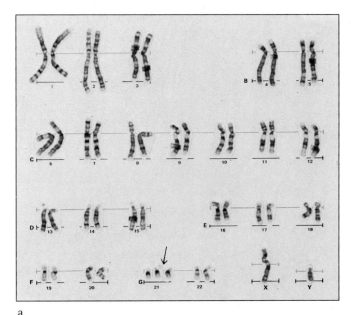

a

b

Figure 17.2 Trisomy-21 (Down syndrome): (a) karyotype; (b) individual.

will produce two kinds of haploid gametes, N and N − 1. Alternatively, the odd, unpaired chromosome in the 2N − 1 cell may be lost during anaphase and not be included in either daughter nucleus, thereby producing two N − 1 gametes. In the following sections we examine some examples of aneuploidy as they are found in the human population. In general, aneuploidy is more easily tolerated by plants than by animals.

Trisomy-21. **Trisomy-21** occurs when there are three copies of the chromosome 21, as shown in the karyotype in Figure 17.2a. Individuals with trisomy-21 have **Down syndrome** and exhibit such abnormalities as low IQ, epicanthal folds over the eyes, short and broad hands, and below-average height. Figure 17.2b is a photograph of a Down syndrome individual. Trisomy of other chromosomes in humans is rarely found with the exception of the sex chromosomes, which we shall study in the following subsection. Trisomy-21 is probably more common in living humans than other trisomics are

because chromosome 21 is a very small chromosome with fewer genes than most of the other chromosomes. The disruptions caused by the extra dose of chromosome 21 genes are less severe than would be the case with, say, the largest autosome (nonsex chromosome). Trisomics are produced for all autosomes since they are detected among fetuses that are aborted spontaneously, and it is from such fetuses that we have obtained knowledge about this sort of chromosome aberrancy. Such studies have also revealed that all monosomics in humans are lethal.

The relationship between the age of the mother and the probability of her having a trisomy-21 individual is indicated in Table 17.1. (The correlation with age of the father is much more tenuous.) In females the eggs that are released monthly until menopause go through meiosis and arrest at diplotene at the age the individual becomes fertile. Apparently, the probability of nondisjunction occurring as an egg matures increases with the storage time of the egg in the ovary. It is important, then, that older mothers-to-be consider genetic counseling to determine whether the fetus has a normal complement of chromosomes. This test involves taking a sample of the amniotic fluid surrounding the

Table 17.1 *Relationship between Age of Mother and Risk of a Trisomy-21
Child*

Age of Mother	Risk of Trisomy-21 in Child
<29	1/3000
30–34	1/600
35–39	1/280
40–44	1/70
45–49	1/40
All mothers combined	1/665

fetus by inserting a needle through the abdominal wall (a procedure called amniocentesis), culturing the fetal cells contained therein, and performing a karyotype analysis (see Chapter 14, p. 463).

Aneuploidy in sex chromosomes. Aberrancies in the number of the X and Y chromosomes are much more readily found among living humans than are aberrancies in the number of autosomes (see Chapter 3, pp. 77–78). The reason they are more readily found is that a mechanism (called *dosage compensation*) exists in mammals to compensate for X chromosomes

in females which are in excess of the number in males. Dosage compensation does not occur for autosomes. As we know, normal human males are XY and normal females are XX. The nuclei of normal females have a highly condensed mass of chromatin not found in the nuclei of normal male cells. This mass, which represents a cytologically condensed and inactivated X chromosome, was first discovered by Maurice Barr, and it has been named the **Barr body.** Therefore, XX individuals normally have two X chromosomes and one Barr body in their somatic cells, and XY individuals have one X chromosome, one Y chromosome, and no Barr bodies (Figure 17.3).

In the somatic cells of individuals with a greater-than-normal number of X chromosomes, all but one of the X chromosomes become

Figure 17.3 *Nucleus of (a) normal human female cells (XX) showing Barr bodies and of (b) normal human male cells (XY) showing no Barr bodies.*

a

b

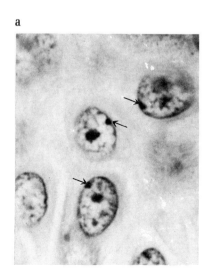
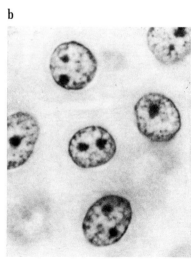

inactivated and produce Barr bodies. (A general formula for the number of Barr bodies is the number of X chromosomes minus one.) This concept was expanded in 1961 by Mary Lyon in what is called the *Lyon hypothesis.* Lyon proposed that the Barr body is an inactive (or mostly inactive) X chromosome, and her hypothesis provided an explanation for the survival of individuals with X chromosome aberrations. The components of the Lyon hypothesis are that the Barr body is a genetically inactive X chromosome (i.e., it has become lyonized; the process is called **lyonization**), that the inactivation occurs at about the sixteenth day following fertilization, and that the X chromosome to be inactivated is randomly chosen from the maternal and paternal X chromosomes in a process that is independent from cell to cell. Lyonization, then, is a mechanism that, when it operates in cells with extra X chromosomes, serves to minimize the effects of multiple copies of X-linked genes.

Individuals with **Turner syndrome** have two sets of autosomes but only one sex chromosome, an X, and hence are designated XO (Figure 17.4a

and b). The XO condition exemplifies monosomy of the X chromosome in humans. These people are female (because of the absence of the Y chromosome) and have a genomic complement of 45, X (indicating that they have a total of 45 chromosomes—in contrast to the normal 46—and that the sex chromosome complement consists of one X chromosome). The XO individuals have no Barr bodies and exhibit few abnormalities until puberty, when they do not produce normal secondary sexual characteristics. Phenotypically, XO individuals tend to be short and have a weblike neck, poorly developed breasts, and immature internal sexual organs.

A number of human syndromes result from extra copies of the X and Y chromosomes. **Klinefelter syndrome,** for example, results from the presence of an extra X chromosome in a male, giving a 47, XXY male (Figure 17.5). The XXY individuals also have a single Barr body, which is not found in normal males.

All XXY males have poorly developed sexual characteristics and are often taller than the average male. Some haver lower-than-average IQs. Individuals with similar phenotypes are also found with higher numbers of X chromosomes, such as XXXY, XXXXY, and so on. The number

Figure 17.4 Turner syndrome (XO) (a) karyotype; (b) individual.

a

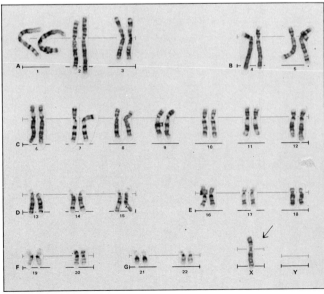

b

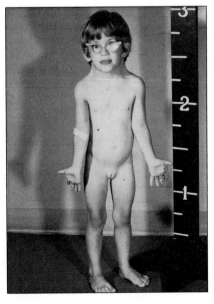

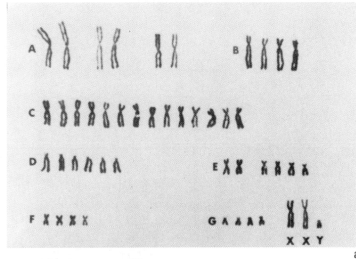

of Barr bodies in these cases follows the rule proposed earlier. In all cases nondisjunction is the likely cause for the chromosome aberrancy, and survival of these individuals is ensured by the X inactivation mechanism.

Some people may have multiple Y chromosomes, sometimes with multiple X chromosomes also. Such individuals are males because of the presence of the Y chromosome, and tend to be taller than average. Occasionally, there are effects on fertility.

Finally, some females are trisomic for the X chromosome. Their chromosome complement is 47, XXX, and they have two Barr bodies per cell. These XXX individuals are sometimes mentally deficient and infertile.

The effects manifested in individuals with the X chromosome aberrancies are probably the result of the activities of the X chromosome genes during embryonic development up until the time all but one of the X chromosomes are deactivated.

Table 17.2 presents a summary of various aneuploid abnormalities for autosomes and for sex chromosomes.

Changes in Complete Sets of Chromosomes

Monoploidy. **Monoploidy** and polyploidy involve variations from the normal state in the number of complete sets of chromosomes (Figure 17.6). Again considering an organism that is usually diploid, a monoploid individual has only one set of chromosomes (Figure 17.6a). Monoploidy is sometimes called haploidy, although haploidy is typically used to describe the chromosome complement of gametes.

Monoploidy occurs only rarely. Owing to the presence of recessive lethal mutations in the chromosomes of many diploid eukaryotic organisms, which are usually counteracted by dominant alleles in heterozygous individuals, many monoploids probably do not survive and thus are undetected. Certain organisms have monoploid organisms as a normal part of their life cycle. Some male wasps, ants, and bees, for example, are normally monoploid because they develop from unfertilized eggs (see p. 79).

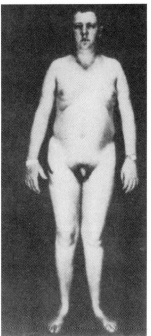

Figure 17.5 Klinefelter syndrome (XXY) (a) karyotype; (b) individual.

Because monoploid organisms are unable to go through meiosis (which must start with a diploid cell), they are sterile. Currently, monoploids are used extensively in plant-breeding experiments in which it is possible to isolate monoploid cells from the normal haploid products of meiosis in the anthers of the plants. These monoploids are then induced to grow and are studied for genetic traits of interest.

Cells of a monoploid individual can also be mutagenized. The mutants can then be isolated for study without researchers having to take into account the masking of recessive mutations, as they must do when they are studying diploid cells.

Polyploidy. **Polyploidy** describes the condition

of a cell or organism that has more than its normal number of sets of chromosomes (Figure 17.6b). With reference to a normal diploid cell, a cell with three sets of chromosomes is called a triploid, and one with four sets of chromosomes is called a tetraploid. Polyploids, which may arise spontaneously or may be induced experimentally, often occur as a result of a breakdown in the spindle apparatus in one or more meiotic divisions or in mitotic divisions. In **autopolyploidy** all the sets of chromosomes are of the same type, a condition that probably results from a defect in meiosis, which in turn leads to diploid or triploid gametes. In **allopolyploidy** the sets of chromosomes differ, a situation that usually arises when different species interbreed after which the hybrid becomes polyploid when meiosis is defective. For example, fusion of haploid gametes of two plants that can intercross may produce an N1 + N2 hybrid plant. These plants are usually sterile owing to the differences between the two chromosome sets, resulting in no tetrads being produced at meiosis. Rarely, tissue of 2N1 + 2N2 genotype is produced from which N1 + N2 gametes can be generated. Fusion of two gametes like this can produce fully fertile, allotetraploid, 2N1 + 2N2 plants.

There are two general classes of polyploids: those that have an even number of chromosome sets and those that have an odd number of sets. Polyploids with an even number of chromosome sets have a better chance of being at least partially fertile since there is the potential for chromosomes to come together in pairs during meiosis. However, polyploids with an odd number of chromosome sets always have an unpaired chromosome for each chromosome type, so the probability of a normal gamete is extremely low, and such organisms are usually sterile. An example of an important polyploid with an even number of chromosome sets is the cultivated bread wheat plant *Triticum aestivum*, which is hexaploid (6N). This plant is fertile and produces gametes with 21 chromosomes. The cultivated banana is an example of a polyploid organism with an odd number of chromosome sets; it is triploid. As a result, the gametes

Table 17.2 Aneuploid Abnormalities in the Human Population

Chromosomes	Syndrome	Frequency at Birth
AUTOSOMES Trisomic 21	Down	1/700
Trisomic 13	Patau	1/5000
Trisomic 18	Edwards	1/10,000
SEX CHROMOSOMES, FEMALES XO, monosomic	Turner	1/5000
XXX, trisomic XXXX, tetrasomic XXXXX, pentasomic		1/700
SEX CHROMOSOMES, MALES XYY, trisomic	Normal	1/1000
XXY, trisomic XXYY, tetrasomic XXXY, tetrasomic XXXXY, pentasomic XXXXXY, hexasomic	Klinefelter	1/500

Normal chromosome complement

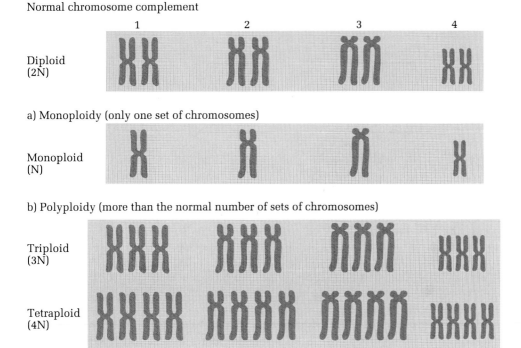

Figure 17.6 Variations in number of complete chromosomes sets: (a) monoploidy (only one set of chromosomes instead of two); (b) polyploidy (more than the normal number of chromosomes).

have a variable number of chromosomes and few fertile seeds are set, thereby making the banana a highly palatable fruit to eat.

Animals appear to have evolved to the point that the consequences of more than two sets of chromosomes in a diploid organism are often severe or lethal. The molecular basis for the severe or lethal effects is not well understood. Thus the evidence for polyploidy in animals often comes from studies of spontaneously aborted fetuses, so we know that polyploidy is one of the major bases for early, spontaneous abortions of human or other animal fetuses. In contrast, plants are much more likely to tolerate polyploidy although, as we have seen, they may be sterile.

Keynote *Variations in the chromosome number of a cell or an organism give*
rise to aneuploidy, monoploidy, and polyploidy. In aneuploidy a cell or organism has one, two, or a few whole chromosomes more or less than the basic number of the species under study. In monoploidy an organism that is usually diploid has only one set of chromosomes. And in polyploidy an organism has more than its normal number of sets of chromosomes. Any or all of these abnormal conditions may have serious consequences to the organism.

Variations in Chromosome Structure

The second type of chromosome aberration involves changes in parts of individual chromosomes rather than whole chromosomes or sets of chromosomes in a genome. There are four types of structural alterations: deletions and duplications (which involve a change in the amount of genetic material on a chromosome), inversions (which involve a change in the arrangement of a chromosomal segment), and translocations (which involve a change in the location of a chromosomal segment).

Much information about changes in chromosome structure has been gained from the study of **polytene chromosomes,** a special type of chromosome found in certain insects that consists of a bundle of chromatids (see Figure 17.15a). These chromatid bundles result from repeated cycles of chromosome duplication without nuclear or cell division, and they are easily detectable through microscopic examination. Polytene chromosomes are characteristic of various tissues of the Order Diptera and may be a thousand times the size of corresponding chromosomes at meiosis or in the nuclei of ordinary somatic cells. In Diptera, homologous chromosomes are somatically paired, and therefore the number of polytene chromosomes per cell corresponds to half the diploid chromosome of somatic cells. The number of chromatids per polytene chromosome is species-specific.

As a result of the intimate pairing of the multiple copies of chromatids, characteristic banding patterns are produced, which enable cytologists to identify unambiguously any segment of a chromosome and therefore to see clearly where chromosome changes have occurred. The banding patterns have been particularly useful for defining chromosomal mutations in Diptera. In *Drosophila melanogaster,* for example, over 5000 bands and interbands can be counted in the four polytene chromosomes. For a long time each band was thought to represent a single protein-coding gene and the region between bands to represent intergenic DNA. It is now known that each band contains on the average 30,000 base pairs (30 kbp) of DNA, which is enough to encode several proteins. Careful analysis involving cloning and sequencing has shown that many bands contain a number of genes (up to seven) that are transcribed independently. Sometimes the genes are related in function.

Deletion

A **deletion** is a chromosome mutation resulting in the loss of a segment of the genetic material

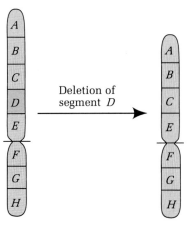

Figure 17.7 Deletion with a chromosome segment (here, D) missing.

and the genetic information contained therein from a chromosome (Figure 17.7). The deleted segment may be located in the middle of a chromosome or at an end. A deletion is initiated by breaks in chromosomes induced by agents such as heat, radiation, viruses, and chemicals or by errors in recombination enzymes.

The consequences of the deletion depend on the genes or parts of genes that have been removed. In diploid organisms the effects may be lessened by the presence of another set of the deleted genes in the homologous chromosome. However, if the homolog contains recessive genes with deleterious results, then the consequences can be severe. We are most knowledgeable, therefore, about the deletions that cause a detectable phenotypic change. If the deletion involves the loss of the centromere, for example, the result is an acentric chromosome, which is usually lost during meiosis, thus leading to a deletion of an entire chromosome from the genome producing a monosomic cell.

In organisms in which karyotyping is practical, deletions can be detected by that procedure if the deficiencies are large enough so that one can see an apparently mismatched pair of homologous chromosomes, with one shorter than the other. Deletions may also be recognized cytologically by the appearance, in heterozygous individuals, of characteristic unpaired loops when the two homologous chromosomes pair at meiosis (Figure 17.8a). Such loops are also readily observable in the polytene chromosomes of certain insects that are heterozygous for a deletion (i.e., one homolog is normal and one homolog has a deletion) (Figure 17.8b).

A number of human disorders are caused by deletions. In all cases the disorders are found in heterozygous individuals, from which we can conclude that homozygotes are lethal if the deletion is large enough. One human disorder caused by a heterozygous deletion is cri-du-chat syndrome, which results from an observable deletion of part of the short arm of chromosome 5, one of the larger human chromosomes (see Figure 17.9). Children with cri-du-chat syndrome are severely mentally retarded, have a number of physical abnormalities, and cry with a sound like the mew of a cat (hence the name, which is French for "cry of the cat").

Duplication

A **duplication** is a chromosome aberration that results in the doubling of a segment of a chromosome (Figure 17.10). Duplications may or may not be lethal to an organism. The size of the duplicated segment may vary considerably, and duplicated segments may occur at different locations in the genome or in a tandem configuration (i.e., adjacent to each other). When the order of genes in the duplicated segment is the opposite of the order of the original, it is a reverse tandem duplication; when the duplicated segments are tandemly arranged at the end of a chromosome, it is a terminal tandem duplication (Figure 17.11). Cytologically, heterozygous duplications result in unpaired loops similar to those described for chromosome deletions.

Evidence suggests that duplications of particular genetic regions can have unique phenotypic effects, as in the *bar* mutant on the X chromosome of *Drosophila melanogaster* that we discussed in Chapter 3 (pp. 70–71). In strains carrying the *bar* mutation (not to be confused with the Barr body discussed earlier in the chapter) the number of facets of the compound eye is fewer than for the normal eye (shown in

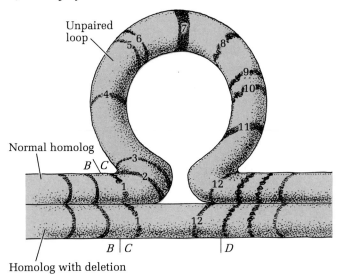

Figure 17.8 Cytological effects at meiosis of heterozygosity for a deletion: (a) formation of an unpaired loop when the homologous chromosomes pair; (b) paired Drosophila *salivary gland polytene chromosomes in a strain with a heterozygous deletion, showing the unpaired region; numbers refer to bands, which are presumed to indicate genes.*

Figure 17.12a), giving the eye a bar-shaped (slitlike) rather than an oval appearance. *Bar* acts like an incomplete dominant mutation since females heterozygous for *bar* have more facets and hence a somewhat larger bar-shaped eye than do females homozygous for *bar*. Males hemizygous for *bar* have very small eyes like those of homozygous *bar* females. Cytological examination of polytene chromosomes has shown that the *bar* trait is the result of a duplication of a small

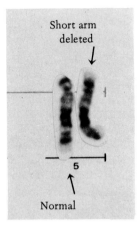

**Short arm
deleted**

5

Normal

Figure 17.9 Cri-du-chat syndrome results from the deletion of part of one of the copies of human chromosome 5.

segment of the X chromosome (Figure 17.12b). Some flies have been found that have three copies of that segment instead of the normal one copy, a condition called *double bar* (Figure 17.12c). These flies have very small, slitlike eyes.

A duplication such as *bar* probably arose as a

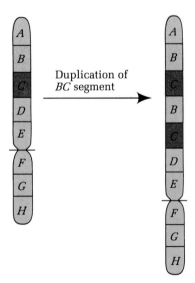

Duplication of
BC segment

Figure 17.10 Duplication with a chromosome segment (here, BC) repeated.

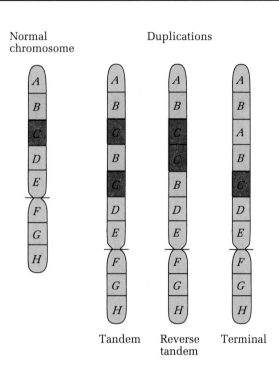

Normal
chromosome

Duplications

Tandem

Reverse
tandem

Terminal

Figure 17.11 Forms of chromosome duplications are tandem, reverse tandem, and terminal duplications.

result of a process called *unequal crossing-over* (see Figure 17.13). Unequal crossing-over may result when homologous chromosomes pair inaccurately, perhaps because similar DNA sequences occur in neighboring regions of chromosomes. Crossing-over in the mispaired region will result in gametes with a duplication or a deletion. Figure 17.13 shows how the duplicated 16*A* segment of the X chromosome of *Drosophila* that is associated with *bar* (Figure 17.12b) could have arisen.

Inversions

An **inversion** is a chromosome mutation that results when a segment of a chromosome is excised and then reintegrated in an orientation 180° from the original orientation (see Figure 17.14). When the inverted segment occurs on one chromosome arm and does not include the centromere, the inversion is called a **paracentric inversion.** When the inverted segment includes

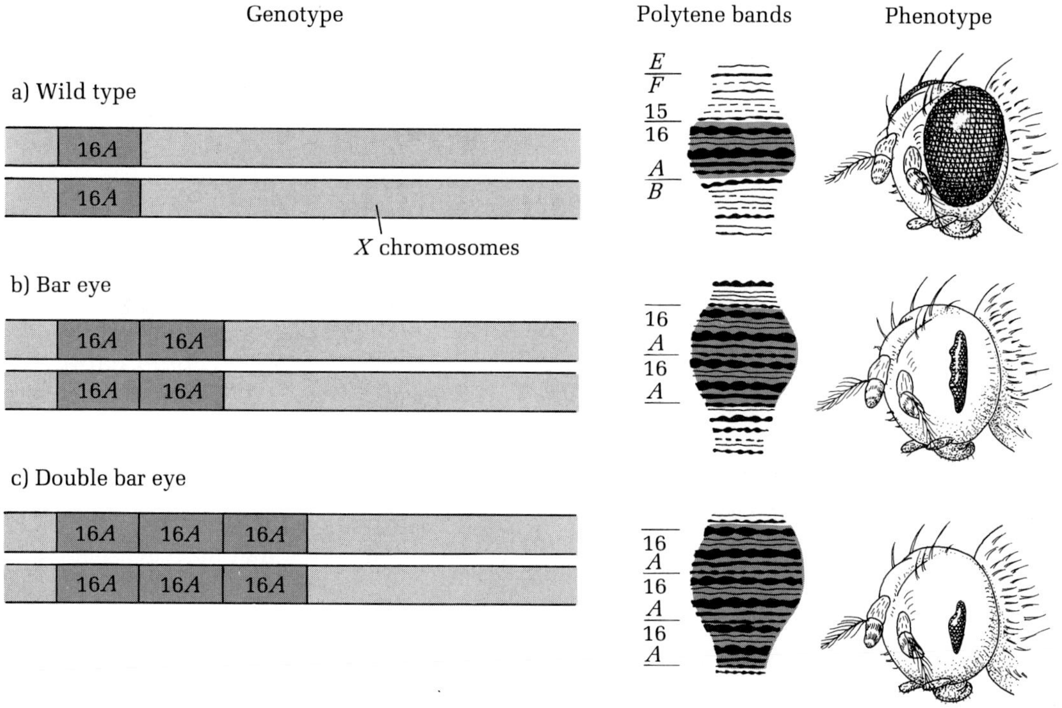

Genotype
Polytene bands
Phenotype
a) Wild type
16A
16A
X chromosomes
E
F
15
16
A
B
b) Bar eye
16A 16A
16A 16A
16
A
16
A
c) Double bar eye
16A 16A 16A
16A 16A 16A
16
A
16
A
16
A

Figure 17.12 Chromosome constitutions of Drosophila *strains showing the relationship between duplications of region 16*A *of the X chromosome and the production of reduced–eye size phenotypes: (a) wild type; (b)* bar *mutant; (c)* double-bar *mutant.*

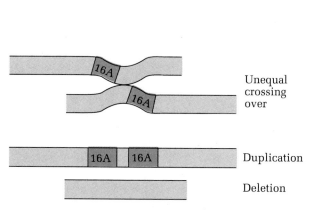

16A
16A
Unequal crossing over
16A
16A
Duplication
Deletion

Figure 17.13 Duplications and deletions of chromosome segments from unequal crossing-over.

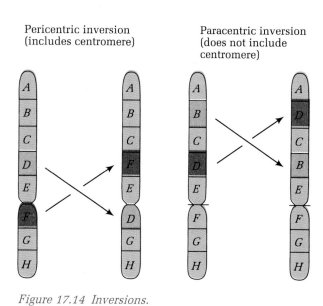

Pericentric inversion (includes centromere)
Paracentric inversion (does not include centromere)
A
B
C
D
E
F
G
H
A
B
C
F
E
D
G
H
A
B
C
D
E
F
G
H
A
D
C
B
E
F
G
H

Figure 17.14 Inversions.

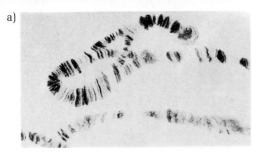

a)

b) Products of meiotic crossover

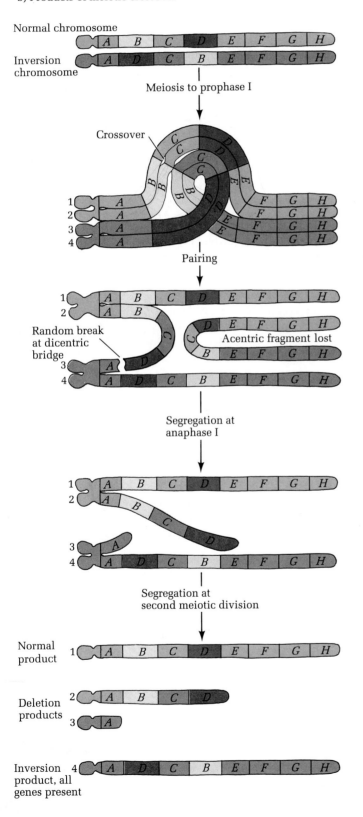

Normal chromosome

Inversion chromosome

Meiosis to prophase I

Crossover

Pairing

Random break at dicentric bridge

Acentric fragment lost

Segregation at anaphase I

Segregation at second meiotic division

Normal product 1

Deletion products 2 3

Inversion product, all genes present 4

the parts of both chromosome arms and, hence, includes the centromere, the inversion is called a **pericentric inversion.**

In general, genetic material is not lost when an inversion takes place, although there can be phenotypic consequences when the break points (inversion points) occur within genes or within regions that control gene expression. Homozygous inversions can be detected because of the unusual linkage relationships that result for the genes within the inverted segment and the genes that flank the inverted segment. For example, if the normal chromosome is *ABCDEFGH* and the *BCD* segment is inverted, the gene order will now be *ADCBEFGH.*

The meiotic consequences of a chromosome inversion depend on whether the inversion occurs in a homozygote or a heterozygote. If the inversion is homozygous (e.g., *ADCBEFGH/ ADCBEFGH,* where the *BCD* segment is the inverted segment), then meiosis will take place normally and there are no problems related to gene duplications or deletions. The situation differs if the individual is heterozygous with a normal chromosome (e.g., *ABCDEFGH/ADCBEFGH,* where one chromosome has a normal gene order and the other has an inverted *BCD* segment). In inversion heterozygotes, as these individuals are called, the homologous chromosomes attempt to pair so that the best possible match of the base-pair sequences is made. Because of the inverted segment on one homolog, pairing of homologous chromosomes necessitates the formation of loops containing the inverted segments, called inversion loops (Figure 17.15a; also see Figure 17.16). Inversion heterozyotes, then, may be identified

Figure 17.15 Consequences of a paracentric inversion: (a) photomicrograph of an inversion loop in polytene chromosomes of a strain of Cryptochironomus (a Dipteran insect) that is heterozygous for a paracentric inversion; (b) meiotic products resulting from a single crossover within a heterozygous, paracentric inversion loop. Crossing-over occurs at the four-strand stage involving two ninsister homologous chromatids.

by looking for loops in cells undergoing meiosis or, where appropriate, in polytene chromosomes.

Heterozygous inversions can also be identified in genetic experiments since recombination is reduced significantly or suppressed. Actually, the frequency of crossing-over is not diminished in inversion heterozygotes in comparison with normal cells, but gametes derived from recombined chromatids are inviable, as illustrated in Figure 17.15b, where the effects of a single crossover in the inversion loop of an individual heterozygous for a paracentric inversion are shown. During the first meiotic anaphase, the two centromeres migrate to opposite poles of the cell. As a result of the crossover between genes *B* and *C* in the inversion loop, one recombinant chromatid becomes stretched across the cell as the two centromeres begin to migrate, forming a **dicentric bridge.** As the two centromeres continue to migrate to opposite poles in the cell, the dicentric bridge, under physical tension, breaks randomly. The other recombinant product of the crossover event, a chromosome without a centromere (an acentric fragment), is unable to continue through meiosis and is lost (i.e., it is not found in the resultant gametes).

In the second meiotic division the centromeres divide, and the chromosomes are segregated to the four gametes. Two of the gametes have complete sets of genes and are viable: the gamete with the normal order of genes (*ABCDEFGH*) and the gamete with the inverted segment (*ADCBEFGH*). The other two gametes are inviable since they both are deficient for many of the genes of the chromosome. In general, the only gametes produced that can give rise to viable progeny are those containing the chromosomes that were not involved in the crossover event.

Figure 17.16 shows the consequences of a single crossover in the inversion loop of an individual heterozygous for a pericentric inversion. The results of the crossover event and the ensuing meiotic divisions are two viable gametes, with the nonrecombinant chromosomes *ABCDEFGH* (normal) and *ADCBEFGH* (inversion), and two recombinant gametes that are in-

viable, each as a result of the deletion of some genes and the duplication of other genes.

It should be noted that not all crossover events within an inversion loop lead to inviable recombinants. One exception occurs if a double crossover takes place in the inversion loop and the same two chromatids are involved in each crossover (i.e., it is a two-strand double crossover; see Chapter 6, p. 191). A second exception occurs when the duplicated and deleted segments of the recombinant chromatids do not affect gene expression and hence viability to a significant degree, as when the chromosome segments involved are very small.

Translocation

A **translocation** (also called a **transposition**) is a chromosome mutation in which there is a change in position of chromosome segments and the gene sequences they contain (see Figure 17.17). Simple translocations may be classified as follows:

1. Intrachromosomal (within a chromosome) translocations (Figure 17.17a) involve a change in position of a chromosome segment within the same chromosome, either from one chromosome arm to the other or within the same chromosome arm.
2. Interchromosomal (between chromosomes) translocations involve the transfer of a chromosome segment from one chromosome into another chromosome (called a nonreciprocal translocation; see Figure 17.17b) or the reciprocal exchange of segments between nonhomologous chromosomes (called a reciprocal translocation; see Figure 17.17c). A variety of nonreciprocal and reciprocal translocation types occur, depending on the locations of the chromosome segment or segments involved.

The genetic consequence of translocations is that, in homozygotes for the translocations, the linkage relationships of genes are altered in comparison with the relationships in strains with normal chromosomes. For example, in the nonreciprocal translocation shown in Figure

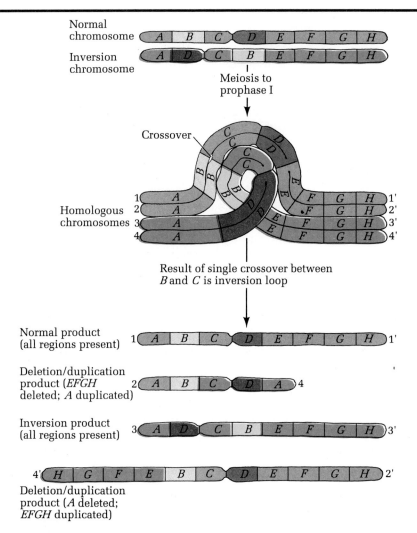

Normal
chromosome

Inversion
chromosome

Meiosis to
prophase I

Crossover

Homologous
chromosomes

Result of single crossover between
B and *C* is inversion loop

Normal product
(all regions present)

Deletion/duplication
product (*EFGH*
deleted; *A* duplicated)

Inversion product
(all regions present)

Deletion/duplication
product (*A* deleted;
EFGH duplicated)

*Figure 17.16 Meiotic products resulting from a single
crossover within a heterozygous, pericentric inversion
loop. Crossing-over occurs at the four-strand stage involving two nonsister homologous chromatids.*

17.17a, the *BC* segment has moved to the other
chromosome arm and has become inserted between the *F* and *G* segments. As a result, genes
in the *F* and *G* segments are now farther apart
than they are in the normal strain, and genes in
the *A* and *D* segments are now more closely
linked. Similarly, in reciprocal translocations
new linkage relationships are produced.

The consequences of translocations to the production of meiotic products depend on the type
of translocation involved. In many cases, some
of the gametes produced have duplications and/
or deletions and, hence, are inviable. We concentrate here on reciprocal translocations since
they are the most frequent and the most important to genetics.

In strains homozygous for a reciprocal translocation, meiosis proceeds normally since all
chromosome pairs can synapse to produce bivalents. In strains heterozygous for a reciprocal
translocation, however, as a result of the requirements of pairing between homologous chromosomes, crosslike configurations are seen in
meiotic prophase I (see Figure 17.18). These
crosslike figures consist of four associated chromosomes, each chromosome being partially homologous to two other chromosomes in the
group. Depending on how the chromosomes seg-

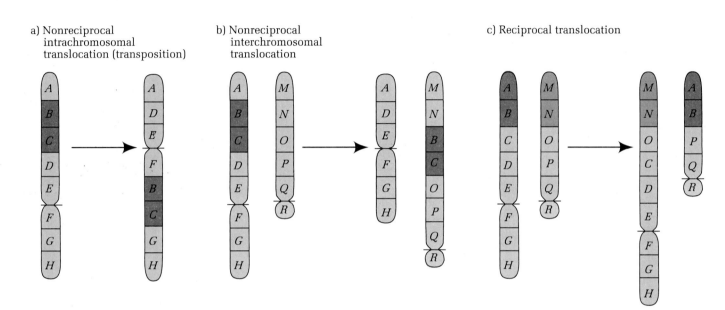

a) Nonreciprocal
intrachromosomal
translocation (transposition)

b) Nonreciprocal
interchromosomal
translocation

c) Reciprocal translocation

*Figure 17.17 Translocations: (a) nonreciprocal in-
trachromosomal; (b) nonreciprocal interchromosomal;
(c) reciprocal.*

regate, three possible paths produce the ring and
figure eight configurations seen at metaphase I.
Segregation at anaphase I may occur in three
different ways, producing six types of gametes.
Of these six gametes, two are functional because
alternate disjunction produces germ cells which
contain a complete set of genes, no more, no
less (one gamete contains a pair of normal chromosomes and the other contains the pair of re-
ciprocally translocated chromosomes). The other
four types of gametes are often nonfunctional
since each type contains duplications and deletions of chromosome segments.

Therefore, if we assume that the chromosomes
in a reciprocal-translocation heterozygote migrate randomly in pairs to opposite poles in
meiosis, we expect that two-thirds of the gametes generated will be nonfunctional because
of duplications and deletions. Our assumption is
not quite valid, though, in the sense that not all
possible segregation patterns occur with the
same frequency: The middle pair of gametes diagramed in Figure 17.18 actually occur infrequently, and when they do occur, they are non-

functional, as previously stated. In actuality,
then, approximately half the gametes produced
from a heterozygous reciprocal translocation are
nonfunctional—hence the term *semisterility* is
applied to this event.

In practice, gametes in animals that have duplicated or deleted chromosome segments may
function, but the zygotes formed by such gametes typically die. The gametes may function
normally and viable offspring may result if the
duplicated and deleted chromosome segments
are small. In plants, pollen grains with duplicated or deleted chromosome segments typically
do not develop completely and hence are nonfunctional.

Keynote *Chromosome aberrations may involve parts of individual chromosomes rather than whole chromosomes or sets of
chromosomes. The four major types of structural
alterations are deletions and duplications (both
of which involve a change in the amount of
DNA on a chromosome), inversions (which involve no change in the amount of DNA on a
chromosome but rather a change in the arrangement of a chromosomal segment), and translocations (which also involve no change in DNA
amount but involve a change in location of one
or more DNA segments).*

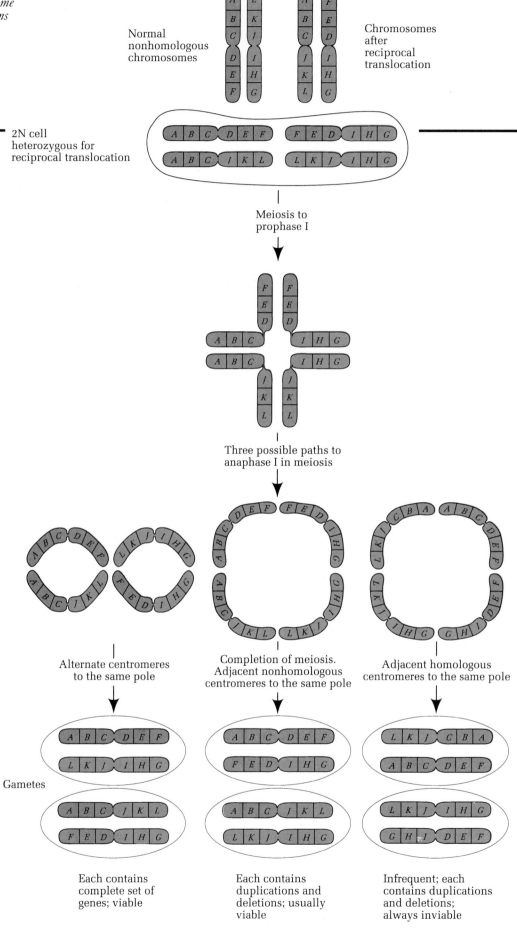

Normal
nonhomologous
chromosomes

Chromosomes
after
reciprocal
translocation

2N cell
heterozygous for
reciprocal translocation

Meiosis to
prophase I

Three possible paths to
anaphase I in meiosis

Alternate centromeres
to the same pole

Completion of meiosis.
Adjacent nonhomologous
centromeres to the same pole

Adjacent homologous
centromeres to the same pole

Gametes

Each contains
complete set of
genes; viable

Each contains
duplications and
deletions; usually
viable

Infrequent; each
contains duplications
and deletions;
always inviable

Chromosome Aberrations and Human Tumors

A number of human tumors have been shown to be associated with chromosome aberrations, either through a change in the number of chromosomes resulting from nondisjunction, or through a change in chromosome structure(s) involving deletions, duplications, inversions, and translocations. It is not clear whether the tumor state is caused by chromosomal aberrations, or whether the chromosomal aberration results from the growth activities of the tumor cell. There are a number of examples, however, that support the first hypothesis. In some cases, for example, a tumor that exhibits a chromosomal abnormality early in its inception will develop other chromosomal aberrations in time, and this often correlates with a progression to an uncontrolled growth state. Examples of tumors associated with a consistent chromosome aberration are chronic myelogenous leukemia (reciprocal translocation involving chromosomes 9 and 12), Burkitt's lymphoma (reciprocal translocation involving chromosomes 8 and 14), constitutional retinoblastoma (deletion of part of chromosome 13), and meningioma (monosomy, chromosome 22).

Chronic Myelogenous Leukemia and the Philadelphia Chromosome

Chronic myelogenous leukemia is an invariably fatal cancer involving uncontrolled replication of myeloid stem cells. Ninety percent of chronic myelogenous leukemia patients have a chromosome aberration in the leukemic cells called the *Philadelphia chromosome* (Ph^1). The Philadelphia chromosome results from a reciprocal translocation event in which a part of the long arm of chromosome 22 (the smallest human chromosome) is translocated to chromosome 9, and a small part from the tip of chromosome 9

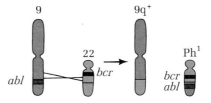

Figure 17.19 Origin of the Philadelphia chromosome in chronic myelogenous leukemia by a reciprocal translocation involving chromosome 9 and 22.

is translocated to produce a small, abnormal chromosome (Figure 17.19). Thus, chronic myelogenous leukemia actually results from two chromosomal aberrations, one involving chromosome 22 and the other involving chromosome 9. This reciprocal translocation event apparently activates genes (called *oncogenes:* see Chapter 21) that initiate the transition from a differentiated cell to a tumor cell with an uncontrolled pattern of growth.

Chromosomal Aberration and Burkitt's Lymphoma

Burkitt's lymphoma, a particularly common disease in Africa, is a viral induced tumor that affects cells of the immune system called B cells. Characteristically, the tumorous B cells secrete antibodies. Ninety percent of tumors in Burkitt's lymphoma patients are associated with a reciprocal translocation involving chromosomes 8 and 14 (Figure 17.20). As in the case for chronic myelogenous leukemia and the Philadelphia chromosome, there is evidence that an oncogene (in this case c-*myc*) becomes activated as a result of the reciprocal translocation event, in this case on chromosome 8. In translocations involving chromosome 14, the oncogene plus a distal part of the chromosome moves to chromosome 14 near DNA encoding parts of immunoglobulin chains. This translocation event then leads to altered transcription of the gene and is associated with the transition to the tumor state as well as to the transcription of the immunoglobulin genes, the latter being responsible for the observed secretion of antibodies associated with this disease.

Figure 17.18 Meiosis in a translocation heterozygote in which no crossover occurs.

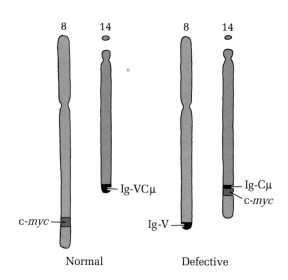

8   14   8   14

Ig-VCμ

Ig-Cμ
c-*myc*

c-*myc*

Ig-V

Normal   Defective

*Figure 17.20 The common translocation involving chromosomes 8 and 14 seen in Burkitt's lymphoma. Diagrams of normal and aberrant chromosomes are shown. The approximate locations of the c-*myc* and immunoglobulin genes are shown.*

Analytical Approaches for Solving Genetics Problems

Q.1 Eyeless is a recessive gene (*ey*) on chromosome 4 of *Drosophila melanogaster*. Flies homozygous for *ey* have tiny eyes or no eyes at all. A male fly trisomic for chromosome 4 with the genotype + + *ey* is crossed with a normal, eyeless female of genotype *ey ey*. What are the expected genotypic and phenotypic ratios that would result from random assortment of the chromosomes to the gametes?

A.1 To answer this question, we must apply our understanding of meiosis to the unusual situation of a trisomic cell. Regarding the *ey ey* female, only one gamete class can be produced, namely, eggs of genotype *ey*. Gamete production with respect to the trisomy for chromosome 4 occurs by a random segre-

a) Segregation

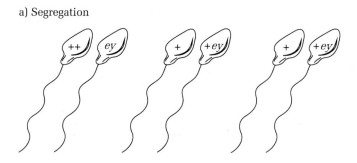

$1/6$ ++, $1/3$ + *ey*, $1/3$ + *ey*, $1/6$ *ey*

b) Union

		Eggs	Phenotype
		ey	
	++	++ *ey*	+
	ey	*ey ey*	*ey*
Sperm	+	+ *ey*	+
	+ *ey*	+ *ey ey*	+
	+	+ *ey*	+
	+ *ey*	+ *ey ey*	+

c) Summary of genotypes and phenotypes

Ratios:	Genotypes	Phenotypes
	$1/6$ ++ *ey*	$5/6$ wild type
	$1/3$ + *ey ey*	$1/6$ eyeless
	$1/3$ + *ey*	
	$1/6$ *ey ey*	

569
Analytical
Approaches
for Solving
Genetics
Problems

gation pattern in which two chromosomes migrate to one pole and the other chromosome migrates to the other pole in meiosis I. (This pattern is similar to the meiotic segregation pattern shown in secondary nondisjunction of XXY cells; see Chapter 3, p. 74.) To illustrate the types of segregation possible, we will distinguish one of the +'s as (+). Three types of segregation are possible in the formation of gametes in the trisomy, as shown in part a of the figure. The union of these sperm at random with eggs of genotype *ey* occurs as shown in part b.

In sum, the resulting genotypic and phenotypic ratios are illustrated in part c of the figure on p. 568.

Q.2 Diagram the meiotic-pairing behavior of the four chromatids in an inversion heterozygote *abcdefg/a'b'f'e'd'c'g'*. Assume that the centromere (symbolized ○) is to the left of gene *a*. Next, diagram the early anaphase configuration if a crossover occurred between genes *d* and *e*.

A.2 This question requires a knowledge of meiosis and the ability to draw and manipulate an appropriate inversion loop. Part a of the figure on the next page gives the diagram for the meiotic pairing. Note that the lower pair of chromatids (*a'*, *b'*, etc.) must loop over in order for all the genes to align; this looping is characteristic of the pairing behavior expected for an inversion heterozygote.

Once the first diagram has been constructed, answering the second part of the question is straightforward. We diagram the crossover and then trace each chromatid from the centromere end until the other end is reached. It is convenient to distinguish maternal and paternal genes, perhaps by *a'* versus *a*, and so on, as we did in part a in the figure. The result of the crossover between *d* and *e* is shown in part b.

In anaphase I of meiosis the two centromeres, each with two chromatids attached, migrate toward the opposite poles of the cell. The noncrossover chromatids (top and bottom chromatids in the figure) segregate to the poles normally at anaphase. As a result of the single crossover between the other two chromatids, however, unusual chromatid configurations are produced, and these configurations are found by tracing the chromatids from left to right. If we begin by tracing the second chromatid from the top, we get ○ *a b c d e' f' b' a'* ○; in other words, we have a single chromatid attached to two centromeres. This chromatid also has duplications and deletions for some of the genes. Thus during anaphase this so-called dicentric chromosome becomes stretched between the two poles of the cells as the centromeres separate, and the chromosome will eventually break at a random location. The other product of the single crossover event is an acentric fragment that can be traced starting from the *right* with the second chromatid from the top. This chromatid is *g f e d'c'g'*, which contains neither a complete set of genes nor a centromere—it is an acentric fragment that will be lost as meiosis continues.

Thus the consequence of a crossover event within the inversion in an inversion heterozygote is the production of gametes with duplicated or deleted genes. Hence these gametes often will be inviable. Viable gametes are produced, however, from the noncrossover chromatids: One of these chromatids

(1 in part b of the figure) has the normal gene sequence, and the other (4 in part b of the figure) has the inverted gene sequence.

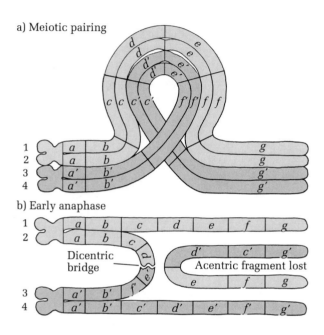

a) Meiotic pairing

b) Early anaphase

*Questions
and
Problems*

17.1 Define the terms *aneuploidy, monoploidy, and polyploidy.*

17.2 If a normal diploid cell is 2N, what is the chromosome content of the following: (a) a nullisomic; (b) a monosomic; (c) a double monosomic; (d) a tetrasomic; (e) a double trisomic; (f) a tetraploid; (g) a hexaploid?

*17.3 In humans, how many chromosomes would be typical of nuclei of cells that are (a) monosomic; (b) trisomic; (c) monoploid; (d) triploid; (e) tetrasomic?

*17.4 An individual with 47 chromosomes, including an additional chromosome 15, is said to be:

a. A triplet
b. Trisomic
c. Triploid
d. Tricycle

*17.5 A Turner syndrome individual would be expected to have the following number of Barr bodies in the majority of cells:

a. 0
b. 1
c. 2
d. 3

*17.6 A color-blind man marries a homozygous normal woman and after four joyful years of marriage they have two children. Unfortunately, both children have Turner syndrome, although one has normal vision and one is colorblind. The type of color blindness involved is a sex-linked recessive trait.

    a. For the color-blind child with Turner syndrome, did nondisjunction occur in the mother or the father? Explain your answer.

    b. For the Turner child with normal vision, in which parent did nondisjunction occur? Explain your answer.

17.7 In general, polyploids with even multiples of the chromosome set are more fertile than polyploids with odd multiples of the chromosome set. Why?

*17.8 From Mendel's first law genes A and a segregate from each other and appear in equal numbers among the gametes. However obvious that phenomenon may seem now, it was not obvious to Mendel. Mendel did not know that his plants were diploid. In fact, since plants are frequently tetraploid, he could have been unlucky enough to have started with peas that were 4N rather than 2N. Let us assume that Mendel's peas were tetraploid, that every gamete contains two alleles, and that the distribution of alleles to the gamete is random. Suppose we have a cross of $AA\ AA \times aa\ aa$, where A is dominant regardless of the number of a alleles present in an individual.

    a. What will be the genotype of the F_1?

    b. If the F_1 is selfed, what will be the phenotypic ratios in the F_2?

17.9 What phenotypic ratio of A to a is expected if $AA\ aa$ plants are testcrossed against $aa\ aa$ individuals? (Assume that the dominant phenotype is expressed whenever at least one A is present, no crossing-over occurs, and each gamete receives two chromosomes.)

17.10 The root tip cells of an autotetraploid plant contain 48 chromosomes. How many chromosomes did the gametes of the diploid from which this plant was derived contain?

*17.11 How many chromosomes would be found in somatic cells of an allotetraploid derived from two plants, one with N = 7 and the other with N = 10?

17.12 Plant species A has a haploid complement of 4 chromosomes. A related species B has N = 5. In a geographical region where A and B are both present, C plants are found that have some characters of both species and somatic cells with 18 chromosomes. What is the chromosome constitution of the C plants likely to be? With what plants would they have to be crossed in order to produce fertile seed?

*17.13 A normal chromosome has the following gene sequence:

$$A\quad B\quad C\quad D\quad E\quad F\quad G\quad H$$

Determine the chromosome mutation in each of the following chromosomes:

a. A B C F E D G H

b. A B D E F B C G H

c. A B C D E F E F G H

d. A B C D E F F E G H

e. A B D E F G H

17.14 Define pericentric and paracentric inversions.

17.15 Very small deletions behave in some instances like recessive mutations. Why are some recessive mutations known not to be deletions?

*17.16 Inversions are said to affect crossing-over. The following homologs with the indicated gene order are given:

A B C D E

A D C B E

a. Diagram the way these chromosomes would align in meiosis.
b. Diagram what a single crossover between homologous genes B and C in the inversion would result in.
c. Considering the position of the centromere, what is this sort of inversion called?

17.17 Single crossovers within the inversion loop of inversion heterozygotes give rise to chromatids with duplications and deletions. What happens when, within the inversion loop, there is a two-strand double crossover in such an inversion heterozygote when the centromere is outside the inversion loop?

17.18 An inversion heterozygote possesses one chromosome with genes in the normal order:

a b c d e f g h

It also contains one chromosome with genes in the inverted order:

a b f e d c g h

A four-strand double crossover occurs in the areas e–f and c–d. Diagram and label the four strands at synapsis (showing the crossovers) and at the first meiotic anaphase.

*17.19 The following gene arrangements in a particular chromosome are found in *Drosophila* populations in different geographical regions. Assuming the arrangement in part a is the original arrangement, in what sequence did the various inversion types most likely arise?

a. *ABCDEFGHI*
b. *HEFBAGCDI*
c. *ABFEDCGHI*
d. *ABFCGHEDI*
e. *ABFEHGCDI*

*17.20 Chromosome I in maize has the gene sequence *ABCDEF*, whereas chromosome II has the sequence *MNOPQR*. A reciprocal translocation resulted in *ABCPQR* and *MNODEF*. Diagram the expected pachytene configuration of the F_1 of a cross of homozygotes of these two arrangements.

17.21 Diagram the pairing behavior at prophase of meiosis I of a translocation heterozygote with normal chromosomes of gene order *abcdefg* and *tuvwxyz* and the translocated chromosomes *abcdvwxyz* and *tuefg*. Assume that the centromere is at the left end of all chromosomes.

18

Transposable Genetic Elements

Overview

Since the rediscovery of Mendel's laws at the beginning of the twentieth century, the classical picture of genes has been one in which the genes are at fixed loci on a chromosome. There is now evidence showing that certain genetic elements of chromosomes of both prokaryotes and eukaryotes have the capacity to mobilize themselves and move from one location to another in the genome. These mobile genetic elements have been given a number of names in the literature, including controlling elements, jumping genes, insertion sequences, and transposons. We shall use the term that has become fairly widely accepted, **transposable genetic elements (TGE),** since this generic term reflects the transposition (translocation) events associated with these elements.

Transposable genetic elements are defined as unique DNA segments that have the capacity to insert themselves at one or more of several sites in a genome. Also, TGEs are similar both in prokaryotes (in which they are found in bacteria, phages, and plasmids) and in eukaryotes and their viruses. Indeed, the existence of TGEs in both prokaryotes and eukaryotes suggests that they are a general feature of genomes. In eukaryotes, TGEs have the property to move to new positions within the same chromosome or to a different chromosome. In both prokaryotes and eukaryotes TGEs are responsible for genetic change; for example, they can produce various kinds of chromosome aberrations, and by inserting into a gene or the region adjacent to a gene that regulates its expression, they can affect gene expression. These effects of TGEs have been established through genetic, cytological, molecular, and recombinant DNA procedures.

The molecular nature of transposable genetic elements is best understood in bacteria and their viruses, but increasingly we are acquiring new information about eukaryotic TGEs. It is notable that the eukaryotic elements that have been studied are very similar to the prokaryotic ones. We shall consider the elements in the two types of organisms separately.

Prokaryotic Transposable Genetic Elements

There are four types of transposable genetic elements in prokaryotes: insertion sequence (IS) elements, transposons (Tn), plasmids, and certain temperate bacteriophages.

Insertion Sequence Elements

An **IS element,** the simplest transposable genetic element found in prokaryotes, is a mobile segment of DNA. An IS element contains genes required for inserting the DNA segment into a chromosome and for mobilizing the element to different locations. The IS elements have been detected at many locations in prokaryotic genomes.

Discovery of IS Elements

The IS elements were first identified in *E. coli* as a result of their effects on the expression of a set of three genes whose products are needed to metabolize the sugar galactose when that sugar is supplied as a carbon source. A certain set of mutations affecting the expression of these genes did not have properties typical of the classes of gene mutations discussed in Chapter 16. Careful analysis of this unusual set of mutations revealed that the mutant phenotypes resulted from the insertion of an approximately 800-base-pair (bp) DNA segment into a gene. This particular DNA segment is now called insertion sequence 1, or IS1 (Figure 18.1).

At present geneticists think that IS1 is one of a family of genetic elements capable of integrating into a chromosome at locations with which it has *no* homology. This event is an example of a *transposition event.* Integration can occur independently of normal DNA recombination activity. If the integration of an IS element places it within a gene, then that gene is usually inactivated. The exact effect depends on the IS element involved. For example, the 1327-bp IS2 element decreases gene expression if it integrates into the chromosome in one orientation, but it increases gene expression if it integrates in the opposite orientation. By contrast, the 768-bp IS1 element decreases the expression of the gene into which it integrates no matter what its orientation is.

An IS element also has the ability to excise itself from the chromosome. If the excision event is perfect, then a wild-type revertant results. Occasionally, however, excision is imprecise, so some or all parts of the genes surrounding the IS element may be deleted and transposed with the element. Inversions can also result at the site of the insertion, or the IS element can transpose to another site in the same or a different chromosome while leaving a copy of itself at the original site. In other words, the IS elements can cause a number of types of chromosome aberrations in addition to affecting gene activity.

Sequence organization of IS elements. There are a number of IS elements that have been identified in *E. coli,* including IS1, IS2, IS3, and IS4, each present in 5–30 copies per genome. Altogether, they constitute approximately 0.3 percent of the cell's genome. Along prokaryotes as a whole, the IS's range in size from 768 bp to over 5000 bp and are normal cell constituents (i.e., they are found in all *E. coli*). Table 18.1 lists the properties of some of the IS elements. All IS elements that have been sequenced (using rapid DNA-sequencing methods; see Chapter 15, pp. 508–509) end in a perfect or nearly perfect inverted repeating segment of between 20 and 40 bp. Thus if the stretch of DNA containing an IS is denatured and allowed to reanneal, complementary base pairing can occur within the same strand of DNA (see Figure 18.2a). Between

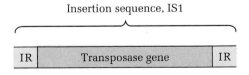

Figure 18.1 An insertion sequence (IS) transposable genetic element. The IS element has inverted repeat (IR) sequences at the ends.

Table 18.1 Prokaryotic Insertion Sequences

Insertion Sequence	Normal Occurrence in E. coli	Length (bp)	Inverted Repeating Segment (bp)	Comments
IS1	5–8 copies on chromosome	768	18/23	Full sequence known for two different examples
IS2	5 on chromosome, 1 on *F* factor	1327	32/41	Full sequence known
IS3	5 on chromosome, 2 on *F* factor	1400	32/38	Sequence of 700 bp known
IS4	1 or 2 copies on chromosome	1400	16/18	Sequence of ends known
IS5	Unknown	1250	Short	No sequence information
Gamma-delta	1 on *F* factor, one or more on chromosome	5700	35	Sequence of ends known

the complementary ends, though, is a segment of unpaired DNA, which gives the reannealed molecule a lollipop appearance (called a stem-and-loop structure) under the electron microscope. The base pair sequence of the stem produced when IS1 is denatured and the single strands allowed to anneal is shown in Figure 18.2b.

Transposons

The next most complex type of prokaryotic transposable genetic elements is the **transposon (Tn)**. It is a mobile DNA segment that, like IS elements, contains genes for the insertion of the DNA segment into the chromosome and for the mobilization of the element to other locations on the chromosome; but unlike IS elements, it also contains genes of identifiable function (e.g., for drug resistance). Despite this difference, there is no question that IS's and Tn's are related. For example, Tn's often end in long (800–1500 bp) direct or inverted repeats, which are IS elements themselves. In fact, many Tn's are mobile in the genome because they have IS elements (called modules) at their ends. Note that IS elements themselves end in short, inverted repeats, so a

Tn flanked by two IS elements will also be flanked by inverted repeats. In other words, essentially all IS elements and Tn's are characterized by terminal inverted repeats. The properties of a number of transposons are summarized in Table 18.2.

To illustrate the concepts just presented, Figure 18.3 (page 579) shows the genetic organization of Tn10. This transposon is 9300 bp long and consists of 6500 bp of nonrepeated material flanked at each end with a 1400-bp IS-like module. The IS modules of this transposon are designated IS10L (left) and IS10R (right) and are arranged in an inverted orientation. Since IS10L and IS10R are inverted repeats, then like IS elements, when Tn10 is denatured to single strands and allowed to renature, a stem-and-loop structure forms as in (Figure 18.2a).

Within the 6500-bp segment are genes for tetracycline resistance. Thus cells containing Tn10 are resistant to the antibiotic tetracycline. In vivo, Tn10 promotes its own transposition, a process in which a copy of the Tn element is placed in a new location on the chromosome while the original insert is retained. These events are catalyzed by enzymes encoded by genes in the transposon. Deletion and insertion events also occur as a result of the gene activities of Tn10.

Figure 18.4 shows the structure of another transposon, Tn3. Tn3 has inverted repeat (IR) sequences 38 bp long at each end, and three known genes. One gene encodes β-lactamase, an enzyme that can degrade ampicillin. Therefore,

bacteria carrying the transposon are ampicillin resistant. The two other genes are involved in transposition events: one encodes *transposase,* an enzyme that catalyzes insertion of the Tn into new sites, and the other encodes *resolvase,* an enzyme found in some transposons. Resolvase is involved in particular recombinational events associated with transposition.

a) Formation of stem-and-loop (lollipop) structure

b) Base-pair sequence of IS1

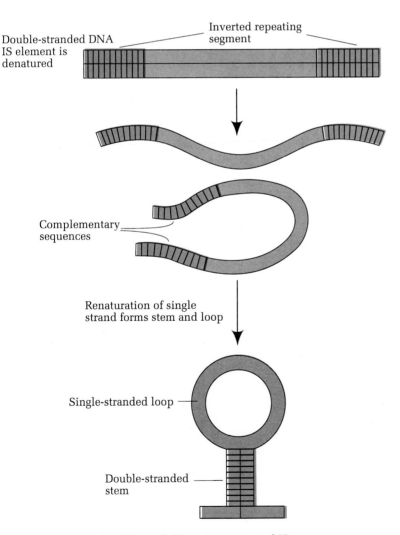

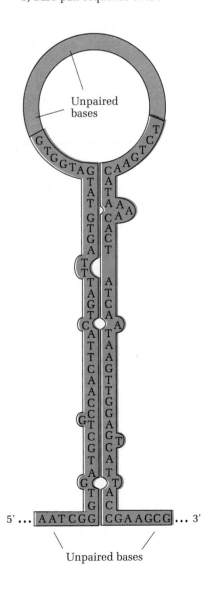

Figure 18.2 Stem-and-loop (lollipop) structure of IS1: (a) formation after it is denatured and renatured; (b) actual base-pair sequence of IS1 from E. coli.

Table 18.2 Prokaryotic Transposons

Transposon	Gene Marker(s)	Length (bp)	Inverted Repeat (bp)	Comments
Tn3	Resistance to ampicillin	4,957	38	Full sequence of Tn3 known
Tn4	Resistance to ampicillin, streptomycin, sulfona-mide	20,500	Short	Contains Tn3
Tn5	Resistance to kanamycin	5,700	1500 (IS50 at end)	Sequence of ends known
Tn6	Resistance to kanamycin	4,200	Not detectable with electron microscopy	
Tn7	Resistance to trimetho-prim, streptomycin	14,000	Not detectable with electron microscopy	
Tn9	Resistance to chloram-phenicol	2,638	18/23	Full sequence known, flanked by direct repeats of IS1
Tn10	Resistance to tetracycline	9,300	1400 (IS10 at end)	Sequence of ends known
Tn1681	Heat-stable enterotoxin	2,088	768 (IS1 at end)	

Transposition of IS Elements and Transposons

The most frequent event of a transposon's life is transposition; that is, the integration of the Tn into new sites in the genome. The site for the integration of the transposon is called the *target site.*

The transposition of IS1 into the *E. coli* chromosome was the first to be well characterized at the molecular level. It was discovered that the transposition event resulted in the repeat of a 9-bp segment of the target DNA on either side of the IS element, where only one copy was present prior to the integration of IS1 (see Figure 18.5). This phenomenon appears to be general since repeated sequences have been found for every IS element and Tn studied: Tn5, Tn9, and Tn10 produce a 9-bp repeat, while IS2 and Tn3 give rise to a 5-bp repeat. Figure 18.6 shows the duplication of a 5-bp sequence that accompanies the transposition of Tn3. Evidence suggests that

the repeated sequence is generated during the integration process since it is not present in the target DNA before transposition. Thus any model for transposition of IS elements and Tn's must account for this production of a short repeated sequence of the target DNA on each side of the integrated element.

All transposons move precisely, which strongly indicates that an enzyme or enzymes recognize the ends of the elements, excise the elements, and facilitate their insertion to new locations. It is likely that the enzyme, transposase, recognizes the transposon ends, which results in the mobilization of the transposon. Given that one enzyme is responsible for transposition, it is not surprising that both ends are repeated sequences. The transposition event in prokaryotes is rare, occurring once in 10^7 cell generations for Tn10. This is the case because less than one transposase molecule is made by Tn10 per cell generation. The other major feature of transposition is the duplication of the target site sequence.

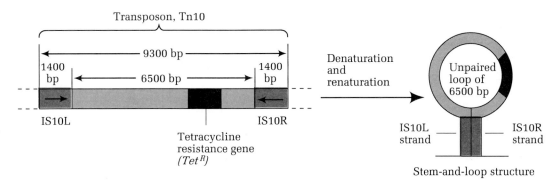

Figure 18.3 Structure of bacterial transposon Tn10.

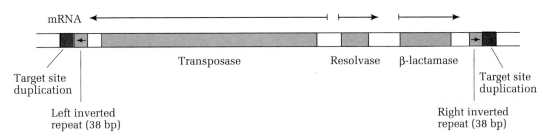

Figure 18.4 Structure of bacterial transposon Tn3.
Tn3 encodes three enzymes: β-lactamase (destroys
β-lactam antibiotics like penicillin and ampicillin),
transposase and resolvase.

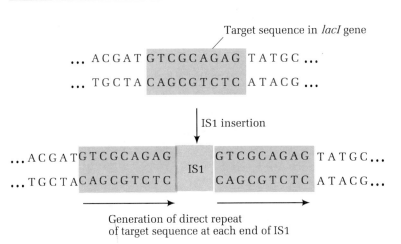

Figure 18.5 Molecular consequences of IS1 integra-
tion into the lacI *gene of E. coli. The integration re-*
sults in the generation of a direct repeat of a 9-bp
DNA sequence at either end of the inserted sequence,
where only one copy of the sequence existed prior to
integration.

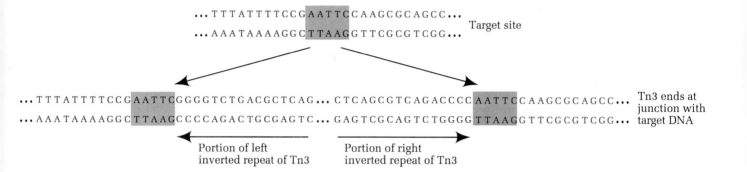

Figure 18.6 DNA sequence of a target site of Tn3; the important five-nucleotide segment is highlighted. After Tn3 inserts, the same 5-nucleotide sequence is found at each end.

Thus, it may be concluded that some DNA synthesis must take place during transposition to produce this duplication.

A number of models have been generated for transposition, one of which is shown in Figure 18.7. The first step in this model (proposed in 1979) is a staggered cut producing a single-stranded nick at each end of the transposable genetic element (either an IS or a Tn). At the same time a staggered cut of opposite polarity is made in the target DNA (Figure 18.7, part 1). Next, one end of the TGE is attached by a single strand to each single-stranded end of the staggered cut on the DNA, thereby generating two replication forks (Figure 18.7, part 2). By semiconservative replication involving both forks, two transposons are generated, each with one old and one new DNA strand and with short, direct repeating strands representing a duplication of the target site (Figure 18.7, part 3). Thus if *ab* and *cd* were originally part of circular molecules, as is often the case, then *a, b, c,* and *d* are now covalently connected in one large, circular structure called a *cointegrate*. A transposon is present at both junction points between the two "parental" types of DNA. Lastly, crossing-over between the two transposons, catalyzed by resolvase, will resolve the cointegrate into the starting replicon *ab* and the target replicon *cd* (Figure 18.7, part 4). The target DNA replicon

now has a copy of the transposable genetic element flanked by a short, direct repeat that was generated by DNA replication, filling in the staggered cut that started the process.

Plasmids

When we studied bacterial genetics in Chapter 7, we learned that the transfer of genetic material between conjugating bacteria is the result of the function of the fertility factor *F*. The *F* factor, a circular, double-stranded DNA molecule, is one example of a bacterial **plasmid,** an extrachromosomal genetic element capable of self-replication. Plasmids such as *F* that are also capable of integrating into the bacterial chromosome are called **episomes.**

Figure 18.8a shows the genetic organization of the *E. coli F* factor. It consists of 94,500 bp of DNA that code for a variety of proteins:

1. A number of genes, called *tra* (for "transfer") genes, are required for the conjugal transfer of the DNA from a donor bacterium to a recipient bacterium (see Chapter 6).

Figure 18.7 Transposition of a transposable genetic element (TGE): (1) Single-stranded cuts in the two starting circular DNA molecules, one of which has a TGE and the other does not. (2) The two cut DNAs come together, forming two replication forks. (3) A cointegrate molecule is produced by semiconservative replication. (4) Crossing-over between the two TGEs catalyzed by resolvase resolves the cointegrate into two circular DNA molecules, each with a TGE.

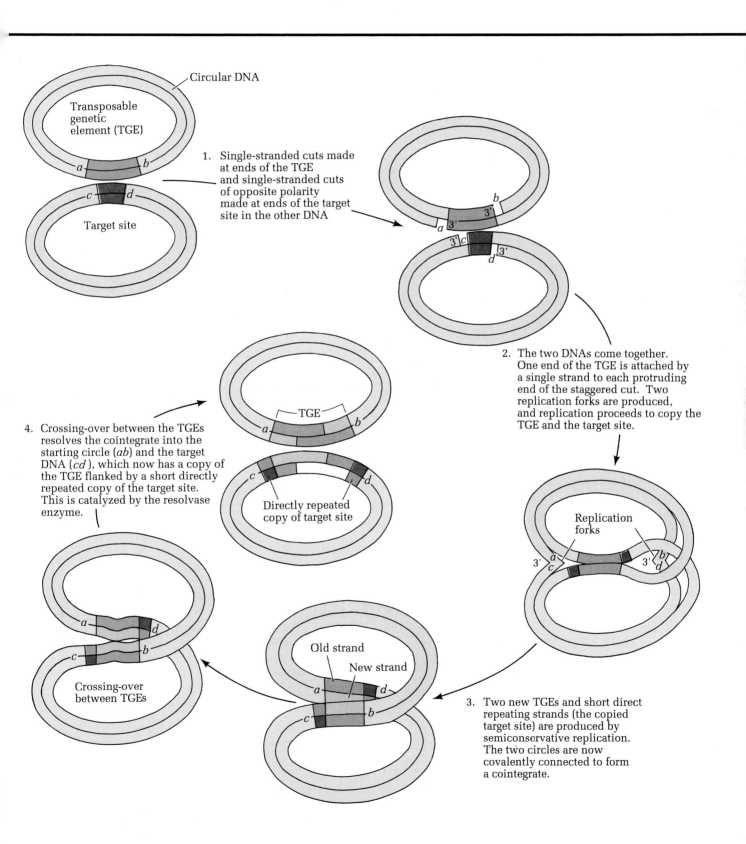

Circular DNA

Transposable genetic element (TGE)

a *b*

c *d*

Target site

1. Single-stranded cuts made at ends of the TGE and single-stranded cuts of opposite polarity made at ends of the target site in the other DNA

b
3'
a 3'
3' *c*
d 3'

2. The two DNAs come together. One end of the TGE is attached by a single strand to each protruding end of the staggered cut. Two replication forks are produced, and replication proceeds to copy the TGE and the target site.

Replication forks

3' *a*
c
3' *b*
d

TGE

a *b*

c *d*

Directly repeated copy of target site

4. Crossing-over between the TGEs resolves the cointegrate into the starting circle (*ab*) and the target DNA (*cd*), which now has a copy of the TGE flanked by a short directly repeated copy of the target site. This is catalyzed by the resolvase enzyme.

a *d*
c *b*

Crossing-over between TGEs

Old strand
New strand

a *d*
c *b*

3. Two new TGEs and short direct repeating strands (the copied target site) are produced by semiconservative replication. The two circles are now covalently connected to form a cointegrate.

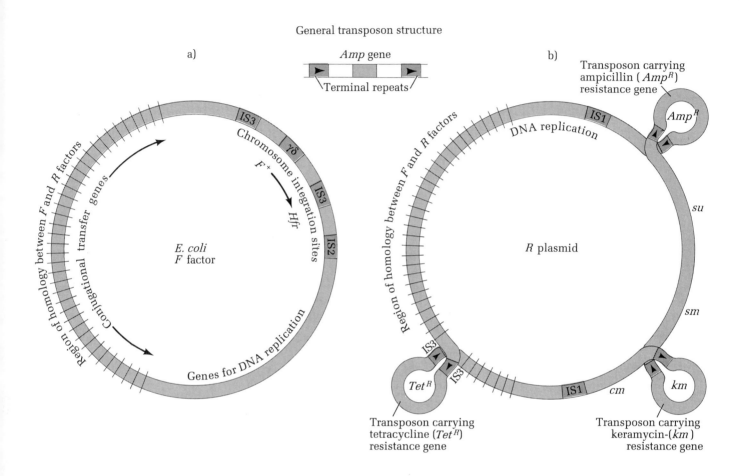

Figure 18.8 Organizational maps: (a) the E. coli F factor. The map shows the locations of genes required for transfer of the F factor during conjugation and for the replication of F; also shown are the insertion elements responsible for the integration into the bacterial chromosome; (b) a typical R plasmid. The map shows the region of homology with the F factor (the hatch marks), the area needed for DNA replication, and the sites of three antibiotic-resistance genes (amp, km, tet). These genes are flanked by insertion sequences and hence are true transposons.

2. Genes are involved in the plasmid's replication.
3. There are four insertion sequences: two copies of IS3, one of IS2, and one of another type of insertion sequence called gamma-delta.

It is because the *E. coli* chromosome has copies

of these same insertion sequences at various positions that the *F* factor can integrate into the *E. coli* chromosome at different sites and in different orientations. Thus *F* factor integration occurs by conventional recombination between the homologous sequences of the insertion elements. Once integrated into the *E. coli* chromosome, the *tra* genes in the *F* factor direct the conjugal transfer functions.

A second class of plasmids that has become medically significant is the *R* plasmid group, which was discovered in Japan in the 1950s. This discovery came about as a result of research into a cure for dysentery, an intestinal disease that is the result of infection by the pathogenic bacterium *Shigella*. The usual treatment of bacterial infections is the administration of antibiotics such as penicillin or ampicillin. However, the *Shigella* strain that occurred in dysentery patients in the Japanese hospitals was found to be resistant not only to penicillin but also to tetracycline, sulfanilamide, chloramphenicol, and streptomycin—all antibiotics that are usually effective in killing bacteria or stopping their growth.

Scientists discovered that the multiple-resistance phenotype was transmissible to other nonresistant strains of *Shigella* as well as to other bacteria commonly found in the human intestine. Subsequently, they found that the genes responsible for the drug resistances were carried on *R* plasmids, which can promote the transfer of genes between bacteria by conjugation, just as the *F* factor can.

Figure 18.8b shows the genetic organization of one type of *R* plasmid. One segment of an *R* plasmid that is homologous to a segment in the *F* factor is the part needed for the conjugal transfer of genes. That segment and the plasmid-specific genes for DNA replication constitute what is called the *RTF* (resistance transfer factor) region. The rest of the *R* plasmid differs from type to type and includes the antibiotic-resistance genes or other types of genes of medical significance, such as resistance to heavy-metal ions.

The resistance genes in *R* plasmids are, in fact, transposons; that is, each resistance gene is located between flanking, directly repeated segments such as one of the IS modules. These transposons can be transposed from plasmid to plasmid within a cell and from plasmid to chromosome within a cell, using the basic mechanism outlined in the previous section.

Thus we can think of each transposon with its resistance gene in the *R* plasmid as a "cassette" that can be inserted into new locations on other plasmids or on the bacterial chromosome while at the same time leaving a copy of itself behind in the original position. Different *R* plasmids contain various combinations of these transposons (cassettes). In an individual to whom drugs are being administered, any bacterium containing an *R* plasmid that is resistant to the drugs will be able to survive and propagate, whereas bacteria without resistance genes will be killed. Indeed, it is because of the widespread occurrence of *R* plasmids among bacterial species and of their ability to be transferred rapidly by conjugation between cells that drug-resistant cells predominate when drugs are present. Therefore the long-term administration of antibiotics should always be carefully evaluated.

Temperate Bacteriophages

The final class of prokaryotic transposable genetic elements that we shall examine are the **temperate bacteriophages** that have the capacity to integrate their chromosome into the bacterial chromosome. (Note that not all temperate bacteriophages can integrate.) The lambda bacteriophage, for example, normally inserts and excises at one particular attachment site in the chromosome (see Chapter 7, p. 242; Figure 7.19). Unlike the other transposable genetic elements, which can integrate at any site on the chromosome by the use of recombination processes that do not require the DNA sequences to be homologous, the lambda phage integrates by using the conventional recombination process between homologous regions. In the lambda phage, the homologous region of attachment is 15 bp long. If the normal attachment site of the lambda phage is deleted, lambda can integrate at one of a number of imperfectly homologous secondary sites.

Another temperate bacteriophage, Mu, functions much like typical transposons. Mu is similar to bacteriophage λ in that it is found as a prophage integrated into the host's chromosomes. Like other transposons, it has genes for regulating its transposition, and inverted repeats at or near the ends of its DNA. When it transposes, Mu makes a 5-bp duplication of the bacterial target site.

As is the case for most transposons, Mu transposes into random DNA sites. When the site is a gene, the gene is inactivated and, in effect, a mutation is produced by the transposition event. Thus, Mu is considered to be a mutator phage; in fact, it gets its name from the first two letters of mutator. Compared with other transposons, Mu transposes much more frequently. Mu also carries the genes required to make infectious phage particles.

Keynote *Transposable genetic elements are unique DNA segments that can insert themselves at one or more sites in a genome. The presence of TGEs in a cell is usually detected by the changes they bring about in the expression and activities of the genes at or near the chromosomal sites into which they integrate. In prokaryotes four types of transposable genetic elements have been identified: insertion sequence (IS) elements, transposons (Tn), plasmids, and some temperate bacteriophages.*

Eukaryotic Transposable Genetic Elements

Transposable genetic elements have been identified in many eukaryotes, and they have been studied mostly in yeast, *Drosophila,* corn, and humans. Their structure and function are very similar to those of prokaryotic transposable genetic elements. Eukaryotic transposable genetic elements have been shown to be capable of integrating into chromosomes at a number of sites and thus may be able to affect the function of virtually any gene, turning it on or off, depending on the element involved and how it integrates into the gene. The integration events themselves, like those of prokaryotic TGEs, involve nonhomologous recombination. Many of the eukaryotic TGEs carry genes, and there is good evidence that these genes are transcribed from the integrated elements and that at least some of the resulting mRNA is translated. However, the role for the products of such mRNA is unknown in most cases.

Transposons in Corn

One of the earliest observations of a transposable genetic element came from Barbara Mc-Clintock's work with corn (*Zea mays*) in the 1950s. She described a number of what she called, on the basis of her genetic work, "controlling elements" that modify or suppress gene activity in corn. One of the characteristics she studied was the extent of pigmentation in the

corn kernels. It was known at the time that a number of different genes must function together for the synthesis of red anthocyanin pigment in the kernel. Classical genetic experiments had shown that mutation of any one of these genes causes an unpigmented kernel. McClintock studied kernels that, rather than being red or white, exhibited spots of red pigment on an otherwise white kernel (Figure 18.9). From her careful genetic and cytological studies she came to the conclusion that the spotted phenotype was not the result of any conventional kind of mutation but, rather, was due to a controlling element, which we now know is a transposon.

The explanation for the spotted kernels was as follows: A mobile controlling element had transposed into one of the pigment-producing genes, causing that gene to become nonfunctional. As the kernel developed, the cells were unpigmented. However, the controlling element is unstable in the sense of having the capacity to excise from a pigment gene and transpose to another location in the genome. When this event occurs, the pigment gene is again functional and all cells derived from that cell produce red pigment; so a red spot on the kernel results. Therefore, depending on the stage in the development of the kernel, the red spot may be large (indicating an early excision of the controlling element), or it may be very small (indicating a late excision of the controlling element).

The remarkable fact of McClintock's conclusion was that at the time there was no precedent for the existence of transposable genetic elements—indeed, the genome was thought to be very static with regard to gene locations. Only recently have transposable genetic elements been widely identified and studied, and only in 1983 was direct evidence obtained for the movable genetic element proposed by McClintock, over thirty years after she proposed the hypothesis. For her pioneering studies and model building she was awarded the Nobel Prize for medicine in 1983.

In recent years, a number of transposons have been identified in corn and other plants. Some of these transposable elements are presented in Table 18.3. Most corn transposons are found in

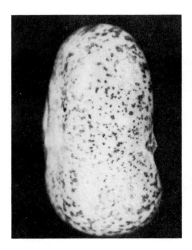

Figure 18.9 Corn kernels showing spots of pigment produced by cells in which a transposable genetic element had transposed out of a pigment-producing gene, thereby allowing function of that gene to be restored. The cells in the white areas of the kernel lack pigment because a pigment-producing gene continues to be inactivated by the presence of a TGE within that gene.

results in a null mutation, that is, a mutation where the function of a gene is reduced to zero. Transposition into controlling regions of genes can result in either decreased or increased gene expression. If a transposon moves into a gene's promoter, the efficiency of that promoter can be decreased or obliterated. The transposon may provide promoter activity itself, however, and

one of two forms: *autonomous elements*, which can transpose by themselves, and *nonautonomous elements*, which can only transpose if an autonomous element is also present.

One of the best-analyzed nonautonomous element systems is the *Ac-Ds system* (Figure 18.10). In this system, there are *Ds* elements with sizes ranging from 0.4 to 4 kbp. These different *Ds* elements have been produced by deletion of various internal segments of the 4.5-kbp *Ac* element. Because the *Ac* element contains a transposase gene, these elements can move autonomously. However, *Ds* elements cannot transpose on their own because part or all of the transposase gene has been deleted. Only when an *Ac* element is present in the cell to provide "helper" transposase activity can a *Ds* element transpose. Like bacterial transposons, maize (and other plant) transposons have terminal repeated sequences. Both the *Ds* and *Ac* elements have inverted terminal repeats of 11 bp.

Transposition of transposons into genes causes mutations. Disruption of a gene typically

Table 18.3 Transposable Genetic Elements in Plants

Transposable Element	Properties
Ds	In corn, *Ds* is an autonomous element whose transposition is dependent on the presence of an active *Ac* element. *Ds* (*Dissociation*) refers to the ability of particular *Ds* elements to induce chromosome breakage at their insertion site
Ac	In corn, *Ac* (*Activator*) is the autonomous element of the *Ac/Ds* family
Mu1	*Mu1* (*Mutator 1*) is a transposable element in corn originally identified in strains with a very high frequency of spontaneous mutations
Tam	Transposable element in *Antirrhinum majus*
Tgm	Transposable element in *Glycine max*

From H.P. Doring and P. Starlinger, 1986. *Annu. Rev. Genet.* 20:175–200.

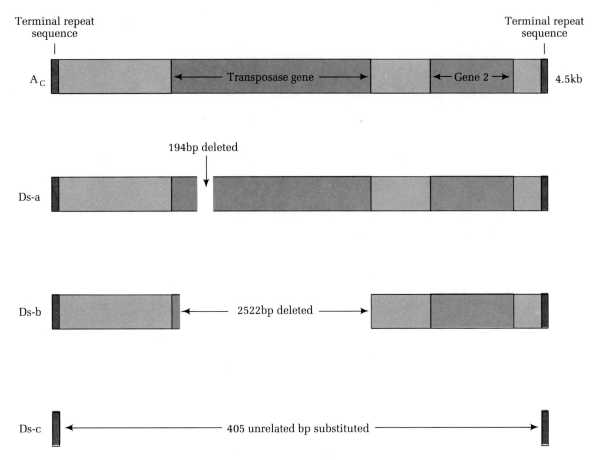

*Figure 18.10 The structure of the Ac element of corn
and several Ds elements. (From N. V. Federoff, Transposable genetic elements in maize, illustration by
Bunji Tagawa. Copyright © 1984 by Scientific American, Inc. All rights reserved. Reprinted by permission.)*

lead to an increase in gene expression. Most mutations caused by transposons are unstable, with
reversion to the wild-type phenotype occurring
at a frequency of up to a few percent. This reversion is the result of perfect excision of the
transposable element.

Our picture of plant transposition, then, is
quite similar to transposition in prokaryotes.
Transposons integrate at a target site by a precise mechanism so that the integrated elements
are flanked at the insertion site by a short duplication of target site DNA of characteristic length.
For *Ac*, *Ds* element transposition, for example,
integration results in 6- to 8-bp direct repeats of

the target site flanking the elements. Such target
site duplication is also found in many other
kinds of eukaryotic transposons. For plant transposons, unlike most other eukaryotic transposons, the duplication is retained (usually in an
altered form), in most cases, upon transposon
excision. The duplication left behind is called a
transposon *footprint* and is a useful record of
the sites in the genome where transposons have
integrated.

Ty Elements in Yeast

The *Ty* elements are segments of DNA found in
yeast that are capable of transposition around
the yeast genome. Figure 18.11 diagrams the *Ty1*
transposable genetic element of yeast, which is
about 6000 bp long and includes two 334-bp-
long, directly repeated sequences called long terminal repeats (LTR) or delta (δ), one at each end

of the element. The deltas consist of about 70 percent AT, and each contains a promoter and sequences recognized by transposing enzymes. There are about 35 copies of the *Ty1* element in some yeast strains, although the number of copies of this element varies between strains. At least two other types of *Ty* elements occur in yeast; they differ not only in the unrepeated DNA region but also in the composition of the delta segments (whose compositions differ by 10–15 percent from the *Ty1* deltas).

The *Ty* elements have a number of structural properties in common with prokaryotic transposons:

1. They are found inserted in many different target sequences that share no obvious homology to each other or to the ends of the transposable genetic elements.
2. Upon insertion the elements generate a duplication of a few base pairs of the target sequence located immediately adjacent to the ends of the element.
3. The elements contain terminal repeated sequences bounding the main body of the element.

Because the *Ty1* element has no genetic marker associated with it, it is much more difficult to follow its movements around the genome than it is to follow the drug-resistance markers of prokaryotic transposons. In at least twelve cases in which *Ty* element movements have been followed, a 5-bp terminal repeat of the yeast target DNA has been generated upon integration of the element. The element inserted into the regulatory site of the gene and not into the structural gene itself in all cases where there were detectable changes in expression of the genes.

Since there are multiple *Ty* elements in a cell, a frequent event is recombination between two deltas by homologous recombination. The consequence of such recombination is that one delta is left behind in the host chromosome. Studies have shown that there are at least a hundred copies of delta per yeast cell. These delta "droppings" are incapable of transposition, but because they contain transcription initiation signals, they can affect the expression of the genes near the location where they are left. The *Ty* elements encode a single, 5700 nucleotide mRNA that begins at the promoter in the delta at the 5′ end of the element (see Figure 18.11). The mRNA transcript contains two open reading frames (ORFs), that is, regions with a start codon in frame with a chain-terminating codon, indicating that two proteins could be produced from the mRNA. Transcription of *Ty* elements accounts for between 5 and 10 percent of all the mRNAs in yeast.

Ty Elements and Retroviruses

Recent evidence indicates that *Ty* elements transpose via an RNA intermediate in a way very similar to viruses known as *retroviruses.* The retrovirus genome and *Ty* elements have a common evolutionary origin. Because of their similarity with retroviruses, *Ty* elements have been referred to as *retrotransposons.* Retroviruses are viruses surrounded by an envelope that infects animal cells. Resulting from these infections are illnesses such as Rous sarcoma virus, feline leukemia virus, mouse mammary tumor virus, and human immunodeficiency virus (HIV, or the AIDS virus) (Figure 18.12).

Retroviruses have single-stranded RNA genomes that could theoretically function directly as mRNAs (i.e., the RNA is known as " + strand RNA" because of this potential). Two copies of

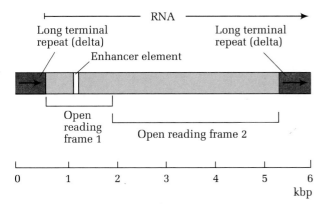

Figure 18.11 The Ty1-*transposable element of yeast. ORF = open reading frame. LTR = long terminal repeat, called "delta." (Courtesy of Dr. Gerry Fink.)*

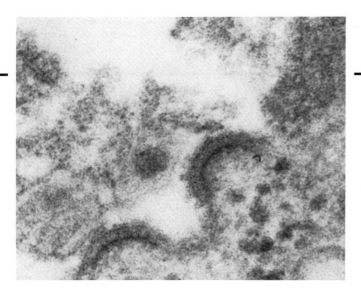

Figure 18.12 Human immunodeficiency virus (HIV), the causative agent of AIDS.

the 7- to 10-kb RNA genome are present in each virus particle; that is, retroviruses are diploid. When a retrovirus infects a cell, the RNA genome does not act as mRNA but serves as a template for the synthesis of double-stranded DNA (the *provirus*) catalyzed by *reverse transcriptase* (see pp. 377–378; 496), an enzyme brought into the cell as part of the virus particle. The double-stranded DNA then integrates into the host chromosome (Figure 18.13) where it can serve as a template for transcription of viral mRNA (i.e., transcription of parts of the inserted DNA) or of progeny genomes (i.e., transcription of the entire, inserted DNA).

Typical retroviruses have three protein-coding genes: *gag* encodes a precursor protein that is cleaved to produce the proteins of the virus particle; *pol* encodes a precursor protein that is cleaved to produce reverse transcriptase, and an enzyme involved in the integration of the provirus into the host chromosome; and *env* encodes the precursor to an envelope glycoprotein. Figure 18.14 shows the genome organization of a generalized retrovirus. Figure 18.15 shows the LTRs as having the structure U_3-R-U_5 and containing signals for initiation of transcription by RNA polymerase II and for 3' cleavage and polyadenylation of the mRNA transcripts.

Analysis of yeast *Ty* elements revealed a number of similarities in the organization of those elements and of retroviruses. This analysis led to the hypothesis that the *Ty* elements move to new chromosomal locations using the same mechanisms as those involved in retrovirus provirus production and integration. That is, rather than transposing DNA to DNA as is the case with bacterial transposons, it was hypothesized that *Ty* elements transpose by making an RNA copy of the integrated DNA sequence and then by creating a new *Ty* element by reverse transcription. The new element would then integrate at a new chromosome location.

Evidence substantiating the above hypothesis was obtained through experiments using *Ty* elements that had been constructed using recombinant DNA techniques; these *Ty* elements had special features that enabled their transposition to be monitored easily. One compelling piece of evidence came from experiments in which an intron was placed into the *Ty* element (there is none in normal *Ty* elements) and the element was monitored from the beginning through the transposition event. At the new location, the *Ty* element no longer had the intron sequence, which was consistent with the notion that transposition occurred via an RNA intermediate. The intron was removed by cellular splicing. Subsequently it has been demonstrated that *Ty* elements encode a reverse transcriptase. Moreover, *Ty* viruslike particles (*Ty*-VLPs) containing *Ty* RNA and reverse transcriptase activity have recently been identified (Figure 18.16, p. 591). These particles are about 60 nm in diameter. It is not yet known which *Ty* genes encode the structural components of the *Ty*-VLPs or how the particles are formed.

Copia and Other Transposable Elements in *Drosophila*

Figure 18.17 (p. 591) diagrams the structure of a *copia* transposable genetic element of *Drosophila*. There are several families of *copia*-like elements. The members of each family are highly conserved and are located at about 5 to 100 widely scattered sites in the genome. *Copia*

Figure 18.13 A model for integration of retroviral DNA into a chromosome.

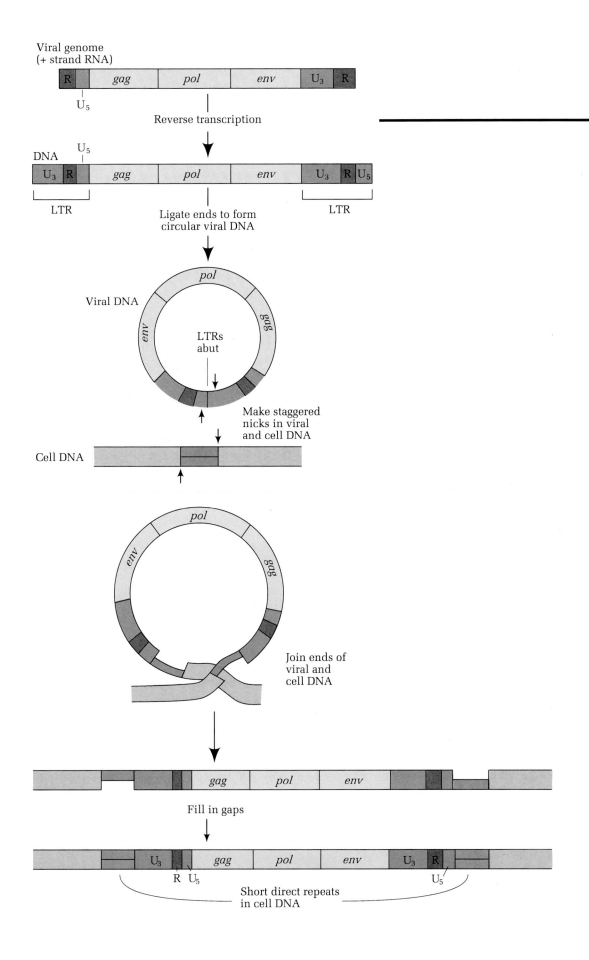

Viral genome
(+ strand RNA)

Reverse transcription

DNA

LTR   LTR

Ligate ends to form
circular viral DNA

Viral DNA

LTRs
abut

Make staggered
nicks in viral
and cell DNA

Cell DNA

Join ends of
viral and
cell DNA

Fill in gaps

Short direct repeats
in cell DNA

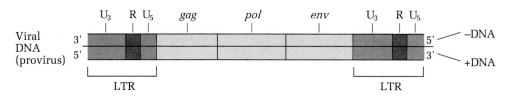

Figure 18.14 Genome organization of a generalized retrovirus.

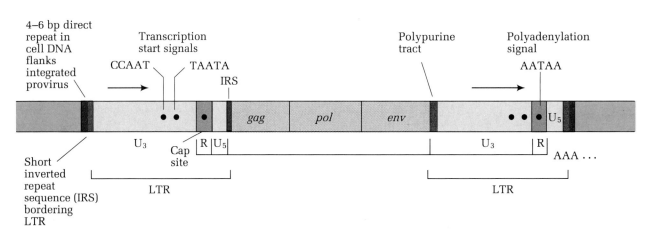

Figure 18.15 Features of LTRs. Most retroviruses (and many "transposable elements" of Drosophila and yeast) have LTRs with these features.

elements code for abundant mRNAs found in *Drosophila*. All *copia* elements found in *Drosophila* are capable of transposition, although there are differences among *Drosophila* strains in the number and distribution of these elements.

The structure of the *copia* element is similar to that of the *Ty* elements of yeast. Directly repeated termini (LTRs) of 300 bp are found at either end of a 4400-bp segment of DNA. The ends have significant sequence homology with the yeast *Ty* elements, which suggests some common evolutionary origin of these two elements. Evidence suggests that *copia* elements, like yeast *Ty* elements, transpose via an RNA intermediate, using a reverse transcriptase catalyzed process. *Copia* viruslike particles (VLPs) have also recently been identified. Integration of

a *copia* element into the genome, like other transposons, results in a duplication of a short target site sequence, in this case 3 to 6 bp long.

At least two other classes of *Drosophila* transposons, called *P elements* and *fold-back* (*FB*) *elements*, have been well characterized. Each class has distinctive properties, and each is different from *copia* and *copia*-like elements.

Keynote Transposable genetic elements in eukaryotes are typically transposons. Insertion sequences and transposons can transpose to new sites while leaving a copy behind in the original site, or they may excise themselves from the chromosome. When the excision is imperfect, deletions can occur; and by various recombination events other chromosomal rearrangements such as inversions and *duplications may occur. Some eukaryotic transposons such as yeast* Ty *elements and* Drosophila copia *elements transpose via an RNA intermediate (using a transposon encoded reverse transcriptase), thereby resembling retroviruses.*

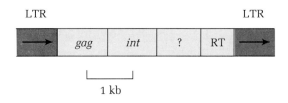

Figure 18.17 Structure of the transposable element copia, *a transposon found in* Drosophila melanogaster.

Perspectives on Transposable Genetic Elements

This chapter has outlined the essential features of what is known about prokaryotic and eukaryotic transposable genetic elements:

1. They are capable of transposition from place to place in the genome.
2. Integration of a transposon into a target site typically results in a duplication of a short sequence at the target sequence at each end of the integrated transposon DNA.

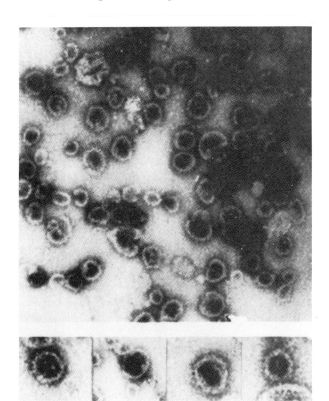

Figure 18.16 Electron micrographs of Ty-VLPs

3. When they transpose, they may leave a copy at their original site.
4. All types of chromosomal rearrangements can be caused by the transposition of transposable genetic elements: They can excise perfectly or imperfectly (which can lead to deletions), and through other types of events during transposition, duplications and inversions can be generated.
5. They can affect gene expression depending on the transcriptional and translational signals they carry and where they integrate relative to the gene itself.
6. All bacterial transposons involve a DNA to DNA information transfer, while some eukaryotic transposons (e.g., yeast *Ty* elements) move via an RNA intermediate.

Since such a large number of transposable elements with homologous regions seem to occur in prokaryotes and eukaryotes, it is remarkable that the genome is not more scrambled than it is by homologous recombination. The reason that this scrambling does not take place is a mystery at the moment, as is the gene composition of all these elements. The picture we are left with is a genome that is dynamic rather than static in that DNA segments are moving around the genome altering its basic organization and affecting the activities of genes.

Analytical Approaches for Solving Genetics Problems

Because this chapter is primarily descriptive, there are no quantitative problems for this material.

Questions and Problems

*18.1 Compare and contrast the four types of transposable genetic elements in prokaryotes.

18.2 What are the properties in common between prokaryotic and eukaryotic transposons?

19

Genetic Recombination

In this last of four chapters on the mechanisms of genetic change, we focus on genetic recombination. One of the fundamental activities of geneticists has been to construct maps of genes for both prokaryotic and eukaryotic organisms. The maps were made by calculating distance based on the frequency of genetic recombination between genetic markers. In the discussions of gene mapping, we assumed that crossovers occur randomly along the chromosomes and that the probability of occurrence of crossovers is proportional to distance. In both prokaryotes and eukaryotes, crossing-over occurs by breakage and reunion of adjacent DNA double helices. There is great precision in the way homologous sequences align (synapsis) and the accuracy of the breakage and reunion events between the same two base pair positions of the two homologous DNA sequences is very high. Clearly, genetic recombination has been highly significant in generating genetic diversity, which is an essential ingredient of the evolutionary process. In this chapter, we will discuss the mechanisms and enzymology of genetic recombination at the molecular level, with emphasis on eukaryotes.

Crossing-Over Involves Breakage and Reunion of DNA

Crossing-over in eukaryotes occurs in meiosis at the point when the chromosomes have duplicated; there are four chromatids for each pair of synapsed chromosomes (see Chapter 5). Figure 19.1 diagrams how crossing-over can occur by *breakage and reunion* of chromatids at the four-chromatid stage.

Given that crossing-over involves the breakage and reunion of DNA, a very early step in recombination must be the generation of a break in one double helix. One possible sequence of events is that cellular enzymes introduce breaks along a pair of synapsed chromosomes in meiosis (and mitosis), thereby stimulating the recombination process. Some evidence for this

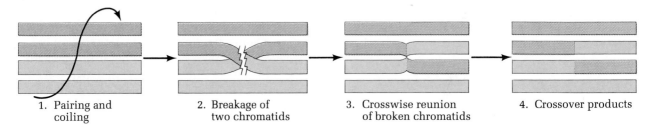

1. Pairing and coiling   2. Breakage of two chromatids   3. Crosswise reunion of broken chromatids   4. Crossover products

Figure 19.1 Representation of a mechanism for cross-ing-over between chromosomes. Each line represents a separate chromatid.

is that the artificial introduction of double-stranded DNA breaks and single-stranded nicks, such as by UV- or X-irradiation, results in a significant stimulation of crossing-over. Similarly, crossing-over is stimulated by mutations in *E. coli* genes for DNA ligase and DNA polymerase I. These mutations result in double-stranded breaks, single-stranded nicks, and gaps because DNA fragments do not become covalently joined during replication or repair. However, molecular details of the initiation of recombination are not known at this time and, thus, it is conceivable that an entirely different sequence of events occurs than that just described.

The breakage and reunion model for crossing-over was experimentally investigated by M. Meselson and J. Weigle in 1961 using bacteriophage λ. Meselson and Weigle infected *E. coli* simultaneously with two λ strains. One λ strain had DNA, which had a higher than normal density as a result of the incorporation of the heavy isotopes ^{13}C and ^{15}N. The chromosome of this strain carried the mutant alleles *c* (clear plaques) and *mi* (minute plaques): these mutant genes are located near one end of the chromosome and give characteristic plaque morphologies either singly or in combination. The other λ strain had normal density DNA (i.e., with ^{12}C and ^{14}N), and had the wild-type alleles c^+ and mi^+ (Figure 19.2a). The infected *E. coli* cells were grown in a culture medium in which only normal density carbon and nitrogen isotopes (^{12}C, ^{14}N) were present.

The progeny phages released from the cells were analyzed both for density (by cesium chlo-

ride density gradient centrifugation: see Chapter 10, Box 10.1, p. 317) and for genotype. In the CsCl gradient, the density of progeny phage DNAs ranged from the heavy parental density to the light parental density (Figure 19.2b). Some recombinant phages of the genotype c mi^+ were identified as having values not quite as dense as the heavy parental DNAs. These recombinant phages indicated that recombination could occur without DNA replication. The recombinant phages were interpreted to have resulted from a breakage and reunion event involving the two parental DNAs.

Based on these experiments, breakage and reunion of double helices was found to be the mechanism for crossing-over in bacteriophages. The conclusion that recombination results from the physical exchange of DNA is generally accepted for both prokaryotic and eukaryotic organisms. Figure 19.3 shows the physical exchange of DNA between sister chromatids of mammalian cells growing in tissue culture. The exchanges were visualized by a special staining technique that enabled the chromatids to be distinguished from one another, resulting in what are called *harlequin chromosomes.*

The breakage and reunion process occurs between the same two base pairs of each of the two participating DNA double helices. No base pairs are added or deleted to either DNA as a result of the breakage and reunion events. Since homologous DNA sequences are involved in genetic recombination, crossing-over is considered to generate homologous recombination. For most organisms, any two homologous DNA sequences can undergo recombination in the cell. For example, if a recombinant DNA plasmid containing a piece of yeast DNA is introduced into a yeast cell by transformation, that yeast DNA may pair with the homologous nuclear DNA sequence and, by crossing-over, become integrated into the chromosome at that location (Figure 19.4).

Keynote In both prokaryotes and eukaryotes, recombination involves the breakage and reunion of homologous DNA double helices.

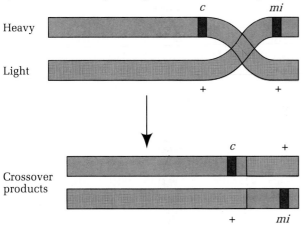

a) Crossover of heavy and light chromosomes in λ phage

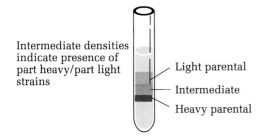

b) Centrifugation in CsCl density gradient of progeny phages

Figure 19.2 (a) The chromosomes of the two parental λ strains and one class of recombinant progeny. (b) Gradient appearance of progeny phages produced after multiple injection.

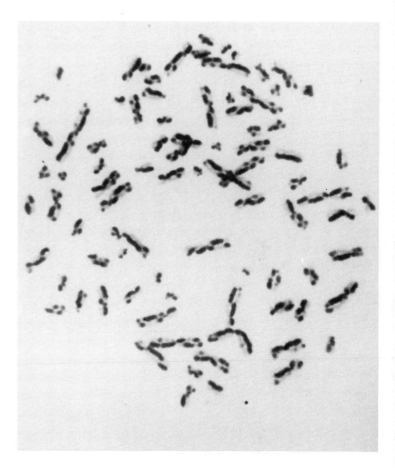

Figure 19.3 Reciprocal crossing-over visualized in hamster cells in tissue culture. Chromosomes were stained so that chromatids could be distinguished from one another, resulting in harlequin chromosomes in which the results of recombination can readily be seen.

The Holliday Model for Recombination

In the mid-1960s R. Holliday proposed a model for reciprocal recombination. The *Holliday model* has been refined by other geneticists, particularly M. Meselson and C. M. Radding.

The Holliday model is diagramed in Figure 19.5 for genetically distinguishable homologous chromosomes, one with alleles a^+ and b^+ near the two ends, and the other with alleles a and b.

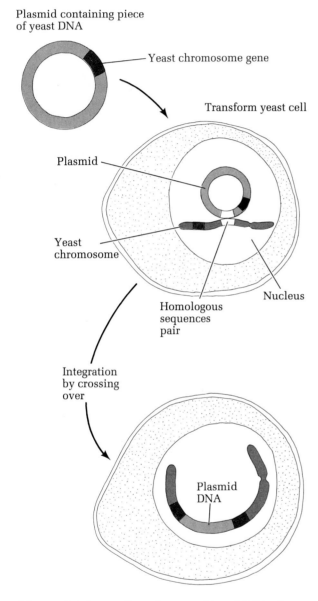

Plasmid containing piece
of yeast DNA

Yeast chromosome gene

Transform yeast cell

Plasmid

Yeast
chromosome

Homologous
sequences
pair

Nucleus

Integration
by crossing
over

Plasmid
DNA

*Figure 19.4 Integration of a recombinant DNA plas-
mid into a yeast chromosome by homologous recombination.*

The first stage of the recombination process is
recognition and alignment (Figure 19.5, part 1),
in which two homologous DNA double helices
become aligned precisely so that no genetic material is added or deleted to either DNA in the
subsequent exchange events. (In eukaryotes, this

synapsis is accompanied by the formation of a
synaptinemal complex, which mediates the
meiotic exchange of genetic information. See
Chapter 1, pp. 21, 24). In the second stage,
one strand of each double helix breaks; each
broken strand then invades the opposite double
helix and base pairs with the complementary
nucleotides of the invaded helix (Figure 19.5,
part 2). (Alternatively, it is possible that a break
occurs in one strand of one double helix and the
broken end then invades the homologous double
helix resulting in a break in one of its strands.)
Each of these steps is catalyzed by specific cellular enzymes. This leaves gaps that are closed
by DNA polymerase and sealed by DNA ligase
as observed during DNA replication (see Chapter
10, pp. 331–334) and DNA repair (see Chapter
16, pp. 537–539) producing what is called a
Holliday intermediate with an internal branch
point (Figure 19.5, part 3). Potentially, the two
DNA double helices in the Holliday intermediate can rotate, causing the branch point to move
to the right or to the left. Figure 19.5, part 4
shows a branch migration event that has occurred to the right; also the four-armed structure
for the DNA strand organization produced simply by pulling the four chromosome ends apart.
Holliday intermediate structures have been demonstrated to be very stable, with no unpaired
bases even in the region of the branch. Figure
19.6 shows an electron micrograph of a Holliday
intermediate with some single-stranded DNA in
the branch point region.

The result of branch migration is the generation of complementary regions of hybrid DNA in
both double helices (diagramed as stretches of
DNA helices with two different colors in Figure
19.5, parts 5 through 8). Careful measurements
have indicated that branch migrations can occur
very rapidly, perhaps moving as much as 50
base pairs per second in eukaryotes. Thus, long
regions of hybrid DNA could be formed in the
few seconds that might elapse between the formation of a Holliday intermediate and the

*Figure 19.5 Holliday model for reciprocal genetic re-
combination.*

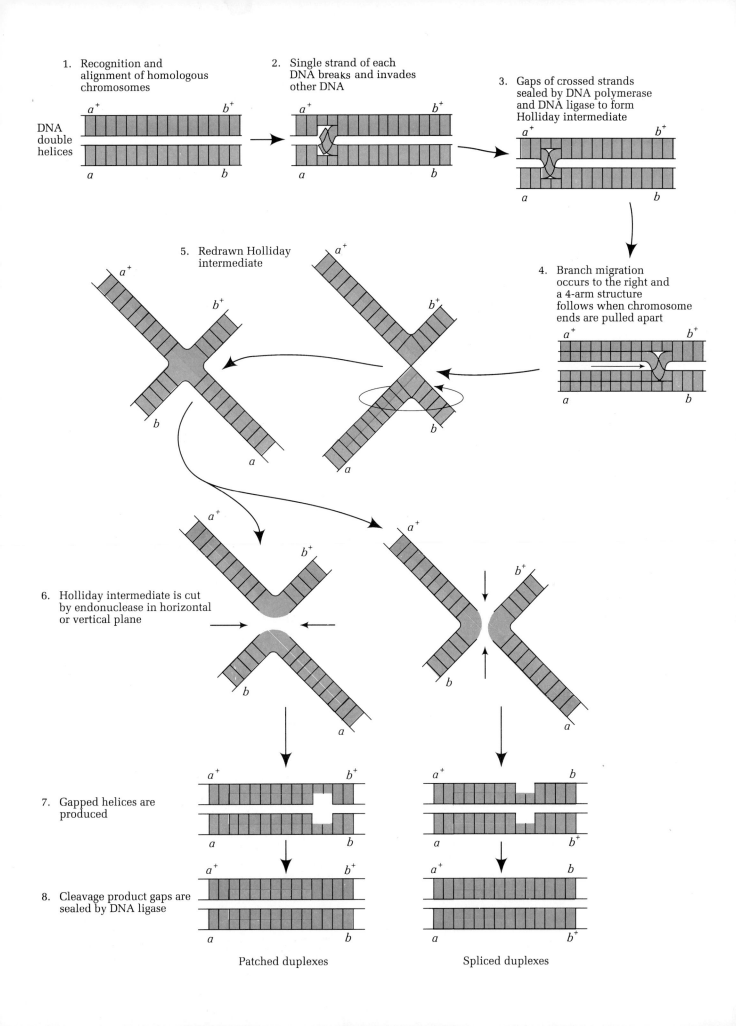

1. Recognition and alignment of homologous chromosomes

DNA double helices

a^+ b^+

a b

2. Single strand of each DNA breaks and invades other DNA

a^+ b^+

a b

3. Gaps of crossed strands sealed by DNA polymerase and DNA ligase to form Holliday intermediate

a^+ b^+

a b

4. Branch migration occurs to the right and a 4-arm structure follows when chromosome ends are pulled apart

a^+ b^+

a b

5. Redrawn Holliday intermediate

a^+ b^+

b a

a^+ b^+

b a

6. Holliday intermediate is cut by endonuclease in horizontal or vertical plane

a^+ b^+

b a

a^+ b^+

b a

7. Gapped helices are produced

a^+ b^+

a b

a^+ b

a b^+

8. Cleavage product gaps are sealed by DNA ligase

a^+ b^+

a b

a^+ b

a b^+

Patched duplexes   Spliced duplexes

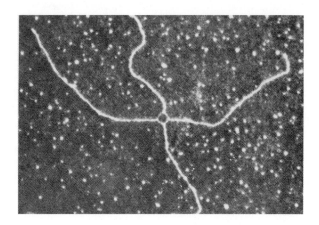

Figure 19.6 Electron micrograph of a Holliday intermediate with some single-stranded DNA in the branch point region.

cleavage and ligation phase of the model that occurs next.

The *cleavage and ligation* phase of recombination is best visualized if the Holliday intermediate is redrawn so that no DNA strand passes over or under another DNA strand. Thus, if the four-armed Holliday intermediate following Figure 19.5, part 4 is taken as a starting point and the lower two arms are rotated 180° relative to the upper arms, the structure shown in Figure 19.5, part 5 is produced.

Next, the Holliday intermediate is cut by endonucleases at two points in the single-stranded DNA region of the branch point. The cuts can be in either the horizontal or vertical planes, with both kinds of cuts occurring with equal probability. Endonuclease cleavage of the intermediates in Figure 19.5, part 5 in the horizontal plane (Figure 19.5, part 6, left) produces the two double helices shown in Figure 19.5, part 7, left, each of which has a single-stranded gap. The gaps are sealed by DNA ligase to produce the double helices shown in Figure 19.5, part 8, left. Since each of the resulting helices contains a segment of single-stranded DNA from the other helix flanked by nonrecombinant DNA, these double helices are called *patched duplexes.*

If the endonuclease cleavage of the intermediates in Figure 19.5, part 5 is in the vertical plane, as shown in Figure 19.5, part 6, right, the gapped double helices of Figure 19.5, part 7, right are produced. In this case, there are segments of hybrid DNA in each duplex, but they are formed by what looks like a splicing together of two helices each with a very long sticky end of one-to-several thousand base pairs. The result of this activity is the double helices in Figure 19.5, part 8, right, which are called *spliced duplexes.*

Keynote The Holliday model for general recombination involves a precise alignment of homologous DNA sequences of two parental double helices, endonucleolytic cleavage, invasion of the other helix by each broken end, ligation to produce the Holliday intermediate, branch migration, and finally, cleavage and ligation to resolve the Holliday intermediate into the recombinant double helices. Depending on the orientation of the cleavage events, the resulting double helices will be patched duplexes or spliced duplexes.

Gene Markers and the Holliday Model for General Recombination

Fate of outside genetic markers. The parental duplexes in Figure 19.5 contained different genetic markers at the ends of the molecules. One parent was $a^+ b^+$ and the other was $a\ b$, as would be the case for a doubly heterozygous parent. However, the reciprocal recombination events shown in Figure 19.5 result in different products. In the patched duplexes (Figure 19.5, part 8, left) the markers are $a^+ b^+$ and $a\ b$, which is the parental configuration. On the other hand, the spliced duplexes (Figure 19.5, part 8, right) have the markers in the recombinant configuration, that is, $a^+ b$ and $a\ b^+$. Since the endonucleases that cut in the branch region during the final cleavage and ligation phase (Figure 19.5, part 6, left and right) most likely act randomly with respect to the pair of strands they cut, the Holliday model predicts that a physical exchange between homologous chromosomes should result in the genetic exchange of

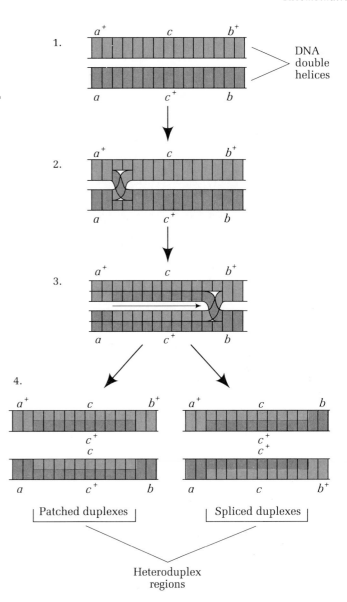

the outside chromosomal markers about half of the time.

Fate of inside genetic markers. An important element of the Holliday model is the generation of hybrid segments of DNA as a result of branch migration. As we have just seen, these hybrid segments may or may not be accompanied by recombination of the outside genetic markers.

Two possibilities exist for the hybrid segments. One possibility is that both parental DNAs have identical DNA sequences so that the hybrid segments are exactly like the parental DNAs in that area. The second possibility is that the two parental DNAs are genetically dissimilar. Consider a case where the inside genetic marker involves one allelic difference in the hybrid region so that one parental duplex is c and the other parental duplex is c^+ (Figure 19.7, part 1). At the DNA sequence level this could be the result of simple base pair differences between the two DNAs as shown in Figure 19.8a. In this theoretical sequence, the c allele resulted from a CG $\rightarrow$ TA transition mutation. If we consider strand cleavage and strand invasion events to the left of the c allele (Figure 19.7, part 2) then after branch migration to the right, the Holliday intermediate shown in Figure 19.7, part 3 will result. This Holliday intermediate is resolved to either patched duplexes or spliced duplexes depending on endonuclease cut orientation (Figure 19.7, part 4). As already discussed above, the patched duplexes have a parental configuration of the outside genetic markers while the spliced duplexes are recombinant for the outside markers. Inspection of these two types of duplexes for the c inside genetic marker in the hybrid segments indicates that the inside marker in both cases always remains in its parental configuration for a single DNA strand. The hybrid segments, however, are hybrid for DNA base pair sequence and, therefore, for the genetic markers, as shown in Figure 19.8b. As can be seen, the hybrid DNA contains mismatched base pair, T-G, at the c/c^+ site. DNA containing a mismatched base pair is called *heteroduplex DNA* and the two strands are said to contain heterologous nucleotide sequences.

Figure 19.7 Formation of heteroduplex DNA during reciprocal recombination. (1) pairing; (2) strand cleavage and strand cleavage events to left of the 5 allele; (3) Holliday intermediate produced by branch migration to the right resolution of Holliday intermediates to produce; (4) resolution of Holliday intermediates to produce patched and spliced duplexes.

Heteroduplex DNA can be resolved in two ways (Figure 19.9). One way is by DNA replication (Figure 19.9a). For example, if we consider the mismatched T-G base pair, replication will produce progeny homoduplexes, one of which will now have a T-A base pair at the site, making it a c allele, and the other of which will now have a C-G base pair, making it a c^+ allele (see Figure 19.8). The second way is by *mismatch repair* (Figure 19.9b) in which the mismatched

a)

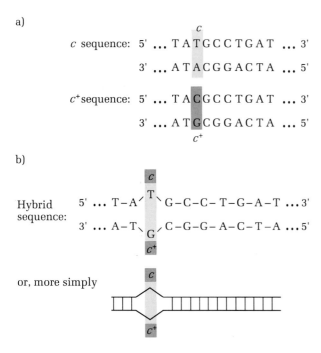

b)

Figure 19.8 (a) Parental DNA nucleotide sequences of an inside genetic marker. (b) Heteroduplex DNA resulting from genetic recombination containing heterologous nucleotide sequences.

base pair is recognized by special DNA repair machinery, a short segment of one DNA strand is removed, and the gap is filled in. This results in a correct base pair at the original mismatch site. Depending upon which DNA strand is excised and repaired, the T-G mismatch can either be corrected to T-A (c allele, i.e., nonrecombinant) or to C-G (c^+ allele, i.e., recombinant). Note that mismatch repair does not recognize "right" (i.e., c^+) or "wrong" (i.e., c) DNA.

Keynote *With regard to recombination of outside markers, the Holliday model for genetic recombination predicts that a physical exchange between homologous chromosomes should result in the genetic exchange of the outside genetic markers about half the time. By contrast, for both patched and spliced duplexes, the inside genetic markers always remain in their parental configuration for a given single DNA strand. Heteroduplex DNA is produced by the recombination process, however,*

Figure 19.9 Heteroduplex DNA resolution by (1) DNA replication and segregation or (2) mismatch repair.

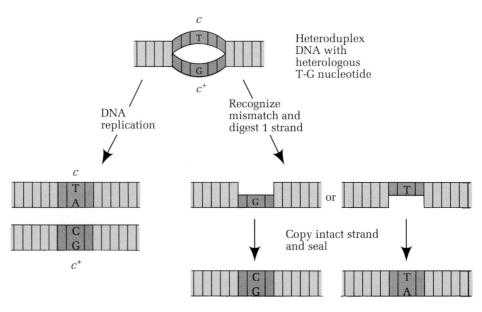

1. Parental homologous chromatids heterozygous for 2 genes

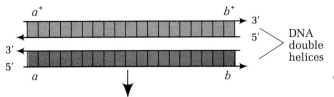

and if mismatched base pairs exist in this DNA (as is the case when there are allelic differences in the region), they are resolved either by replication or by mismatch repair.

Double-Strand Break Repair Model for Recombination

The double-strand break repair model for recombination is shown in Figure 19.10 for the cross of $a^+ b^+ \times a b$ (Figure 19.10, part 1). This model differs from the Holliday model for recombination in that recombination is initiated by a double-strand break. That is, both strands of one of the two parental DNA double helices are broken at one point and the break is enlarged by the action of exonucleases to produce a gap with 3′ single-stranded ends (Figure 19.10, part 2). One of the free 3′ ends invades a homologous region on the intact (unbroken) DNA double helix, generating a small D loop (Figure 19.10, part 3). The D loop is enlarged by repair DNA synthesis until the other 3′ end can form base pairs with complementary single-stranded sequences (Figure 19.10, part 4). The process of gap repair is completed by repair synthesis from the second 3′ end, and branch migration results in the formation of two Holliday intermediates (Figure 19.10, part 5). Resolution of the two intermediates by cutting either inner or outer strands generates two possible noncrossover ($a^+ b^+$ and $a b$) and two possible crossover ($a^+ b$ and $a b^+$) progeny double helices (shown in Figure 19.10, part 6). (Note that in Figure 19.10, part 6, the right-hand Holliday intermedi-

2. Double strand cut in one DNA and formation of gap

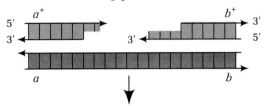

3. Invasion of homologous DNA by one 3′ end, generating a D loop

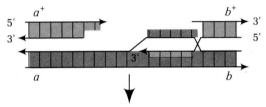

4. Enlargement of D loop by repair synthesis and completion by second 3′ end

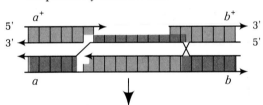

5. Gap repair completed by repair synthesis from second 3′ end; branch migration produces two Holliday intermediates

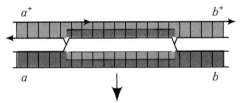

6. Resolution of Holliday intermediates by cleavage and ligation

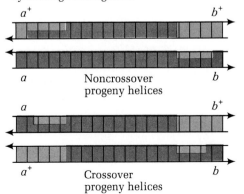

Figure 19.10 Double-strand break repair model for recombination. (1) Parental, homologous chromatids heterozygous for two genes; (2) Double-strand cut in one DNA and formation of gap; (3) Invasion of homologous DNA by one 3′ end, generating a D loop; (4) Enlargement of D loop by repair synthesis; (5) Gap repair completed by repair synthesis from second 3′ end. Branch migration produces two Holliday intermediates; (6) Resolution of Holliday intermediates by cleavage and ligation to produce noncrossover or crossover progeny helices.

1. Parent homologues in meiotic prophase

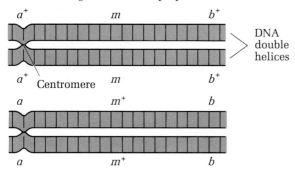

2. Recombination event generates two mismatches

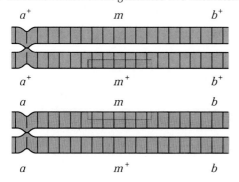

3. Excision and repair by DNA synthesis

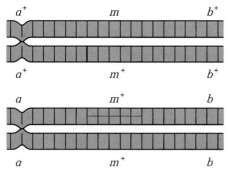

4. Tetrad produced showing
   3:1 gene conversion for m^+

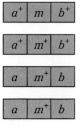

ate was resolved by cutting the inner, crossed strand.)

Mismatch Repair and Gene Conversion

In Chapter 6 we discussed tetrad analysis in which all products of each meiotic event can be isolated and analyzed. Tetrad analysis is limited to relatively few organisms; among them are fungi such as *Saccharomyces cerevisiae* (yeast) and *Neurospora crassa.* Recall that from a cross of $a\ b \times a^+\ b^+$ in which a and b are linked on the same chromosome, but still reasonably distant, three types of tetrads can result: parental ditype (PD), $a\ b,\ a\ b,\ a^+\ b^+,\ a^+\ b^+$; tetratype (T), $a\ b^+,\ a\ b,\ a^+\ b^+,\ a^+\ b;$ and the rare nonparental ditype (NPD), $a\ b^+,\ a\ b^+,\ a^+\ b,\ a^+\ b.$ For each tetrad type, there is a 2:2 segregation of the alleles.

Occasionally, the Mendelian law of 2:2 segregation of alleles is not obeyed in that 3:1 and 1:3 ratios are seen. Tetrads showing these unusual ratios are proposed to derive from a process called *gene conversion.* Any pair of alleles is subject to gene conversion. Consider, for example, the cross $a^+\ m\ b^+ \times a\ m^+\ b$, where gene m is located between genes a and b. The generation by mismatch repair of a tetrad showing gene conversion is diagramed in Figure 19.11. The starting point is parental homologous chromosomes synapsed in meiotic prophase (Figure 19.11, part 1). If a recombination event occurs between the two inner chromatids (Figure 19.11, part 2), a patched duplex can be produced with two mismatches (heteroduplexes involving m

Figure 19.11 Gene conversion by mismatch repair at two sites. (1) Parent homologues; (2) Recombination between inner two chromatids produces a patched duplex with two mismatches; (3) both mismatches are repaired by excision and DNA synthesis; (4) two meiotic divisions produce a tetrad with a 3:1 conversion for the m^+ allele.

and m^+ sequences). Both mismatches can be excised and repaired as shown in Figure 19.11, part 3. Following the subsequent two meiotic divisions, a tetrad is produced showing a 3:1 gene conversion for the + allele of m, while the outside genetic markers a^+/a and b^+/b retain their parental configuration (Figure 19.11, part 4).

Lastly, if two markers are very close together, they can undergo *co-conversion* so that both will exhibit a 1:3 or 3:1 segregation in a tetrad, while outside markers will segregate in a normal 2:2 ratio.

Keynote In organisms in which all four products of meiosis can be analyzed, mismatch repair of heteroduplex DNA can result in a non-Mendelian segregation of alleles, typically a 3:1 or 1:3 ratio rather than a 2:2 segregation. This phenomenon is known as gene conversion. Co-conversion of alleles can occur if the two alleles are very close together.

Questions and Problems

19.1 Explain the chain of events resulting in recombination of two linked genetic markers.

*19.2 How is mismatch repair related to recombination and repair of DNA?

*19.3 Crosses were made between strains each of which carried one of three different alleles of the same gene, a, in yeast. For each cross, some unusual tetrads resulted at low frequencies. Explain the origin of each of these tetrads:

Cross: $a1\ a2^+$ $a1\ a3^+$ $a2\ a3^+$
          $\times$ $\times$ $\times$
         $a1^+a2$ $a1^+a3$ $a2^+a3$

Tetrads: $a1^+\ a2$ $a1^+a3$ $a2^+a3$
          $a1^+\ a2^+$ $a1^+a3$ $a2^+a3^+$
          $a1\ \ a2^+$ $a1^+a3^+$ $a2\ a3^+$
          $a1\ \ a2^+$ $a1\ a3^+$ $a2\ a3^+$

19.4 From a cross of $y1\ y2^+ \times y1^+\ y2$, where $y1$ and $y2$ are both alleles of the same gene in yeast, the following tetrad type occurs at very low frequencies:

$y1^+\ y2$
$y1\ \ y2$
$y1\ \ y2$
$y1\ \ y2^+$

Explain the origin of this tetrad at the molecular level.

20

Regulation of Gene Expression in Bacteria and Bacteriophages

Organisms do not live and reproduce in a constant environment. Through evolutionary processes they have developed ways to compensate for environmental changes and hence to function in a variety of environments. One way that an organism can adjust to a new environment is to alter its gene activity so that gene products appropriate to the new conditions are synthesized, with the result that the organism is optimally adjusted to grow and reproduce in that environment.

A change in the array of gene products synthesized in a cell involves regulatory mechanisms that control gene expression—that is, the synthesis of the product of a gene by transcription and, in the case of protein-coding genes, translation. In general, genes whose activity is controlled in response to the needs of a cell or organism are called **regulated genes.** An organism also possesses a large number of genes whose products are essential to the normal functioning of the cell, no matter what the life-supporting environmental conditions are. These genes are always active in growing cells and are known as **constitutive genes;** examples include genes that code for the synthesis of proteins, and enzymes needed for protein synthesis and glucose metabolism.

In this chapter we examine the mechanisms by which gene expression is regulated in prokaryotic organisms, particularly in bacteria and bacteriophages. In the wild a bacterium lives in a potentially unstable environment that does not ensure a constant food supply or access to the same types of nutrients at all times. To survive, then, a bacterium must be able to adapt physiologically and adjust its cellular biochemical activities to changing conditions. To adapt to alterations in their environments, bacteria have evolved several regulatory mechanisms for turning off genes not needed in new environmental conditions and for turning on different genes needed for survival in the new conditions. In this chapter we learn in detail about some of the basic gene regulation mechanisms in bacteria and how this regulation relates to the organization of genes in the bacterial genome.

Since bacteriophages are parasites, relying on

the activities of a host bacterium to reproduce, they must direct their own reproductive cycle in a carefully programmed way; this cycle is directed through the regulation of gene expression. In this chapter we examine specifically the gene regulation events required for the production of progeny phages and, in the case of phage lambda, for the establishment of the lysogenic state.

Regulated and Constitutive Genes

When gene expression is turned on by the addition of a substance (such as lactose) to the medium, the genes are said to be *inducible.* The regulatory substance that brings about this gene induction is called an **inducer,** and it is a member of a class of small molecules, called **effectors** or effector molecules, that are involved in the control of expression of many regulated genes.

To make the distinction clear, Figure 20.1 diagrams the difference between an inducible gene and a constitutive gene. For an inducible gene the transcription of the gene occurs only in re-

sponse to a particular molecular event occurring at a specific sequence of nucleotide pairs adjacent to the gene, a location called a **controlling site.** The molecular event typically involves an inducer and a regulatory protein, and when the molecular event occurs, RNA polymerase binds to the promoter (usually adjacent to the controlling site) upstream from the gene(s) and initiates transcription of the gene. The gene is "turned on," mRNA is made, and the synthesis of the enzyme coded for by the gene(s) is induced under the regulatory effect of the controlling site. The controlling site itself does not code for any product. The phenomenon of producing a gene product only in response to an inducer is called **induction.**

Constitutive genes, on the other hand, are always expressed in a living cell. They have a promoter that is always accessible to RNA polymerase binding so that transcription of the gene(s) is continuous and there is a constant supply of the essential gene products. Thus constitutive genes are said to be under *promoter control.* Constitutive genes may also have controlling sites associated with them that are involved in regulatory systems for controlling the rates of transcription of these genes.

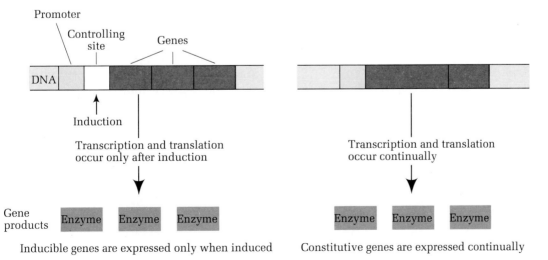

Figure 20.1 Inducible and constitutive gene systems.

606
*Regulation
of Gene
Expression in
Bacteria and
Bacteriophages*

Gene Regulation of Lactose Utilization in E. coli

Lactose as a Carbon Source for *E. coli*

Escherichia coli is able to grow in a simple medium containing salts and a carbon source such as glucose. These chemicals provide molecules that can be manipulated by the enzymatic machinery of the cell to produce everything the cell needs to grow and reproduce, such as nucleic acids, proteins, lipids, and fats. The energy for these biochemical reactions comes from the metabolism of glucose, a process that is of central importance to the function of a bacterial cell and to the function of cells of all organisms. The enzymes required for the metabolism of glucose are coded for by constitutive genes.

If lactose, or one of several other sugars, is provided to *E. coli* as a carbon source instead of glucose, a number of enzymes are rapidly synthesized; these enzymes are needed for the metabolism of the particular sugar added. The enzymes are synthesized because the genes that code for them become actively transcribed in the presence of the sugar; these same genes are inactive if the sugar is absent. In addition, the mRNAs for the enzymes have a relatively short lifetime, so the transcripts must continually be made in order for the enzymes to be produced. In other words, the genes are regulated genes whose products are needed only at certain times.

When lactose is the sole carbon source present in the growth medium, three proteins are synthesized, at least two of which are needed for lactose utilization in *E. coli:*

1. β-*galactosidase* has two functions that are shown in Figure 20.2. It catalyzes the isomerization ("conversion to a different form") of lactose to allolactose, a compound important in the regulation of expression of the lactose utilization genes; and it catalyzes the breakdown of lactose into its two component monosaccharides, glucose and galactose. In the growing cell the galactose is then converted to glucose through the action of enzymes encoded by a gene system specific for galactose catabolism. The glucose is then utilized by constitutively produced enzymes.

2. *Lactose permease* (also called *M protein*) is found in the *E. coli* membrane and is needed for the active transport of lactose from the growth medium into the cell.

3. *Transacetylase*, a protein whose function is poorly understood, may not be related to lactose utilization.

In a wild-type *E. coli* that is growing in a medium containing glucose (or another carbon source) but no lactose, only a few molecules of each of the three proteins are produced, indicating a low level of expression of the three genes that code for the proteins. If lactose but no glucose is present in the growth medium, however, the number of molecules of each of the three proteins increases about a thousandfold, indicating that the three previously (essentially) inactive genes are now being actively transcribed and translated. The inducer molecule directly responsible for the increased production of the three proteins is actually allolactose, which is produced from lactose as a result of one of the activities of the β-galactosidase enzyme.

Organization of the Lactose Utilization Genes

Through mapping experiments (Chapter 7) and complementation tests (Chapter 14) done with mutants affecting the function of the three proteins induced by lactose, geneticists have shown that three structural genes are involved in lactose utilization and that they are adjacent to one another (i.e., clustered) in the genome.

In the presence of lactose the three structural genes are *coordinately induced;* all three genes are simultaneously (and rapidly) transcribed, and their protein products are rapidly produced. The regulation of the three adjacent genes as a unit is under the control of a regulator protein, which interacts with a controlling site, called an operator, adjacent to the gene cluster. The operator does not code for any product; it is simply a unique DNA sequence. The promoter is located next to the operator. A cluster of genes whose

607
Gene
Regulation
of Lactose
Utilization
in E. coli

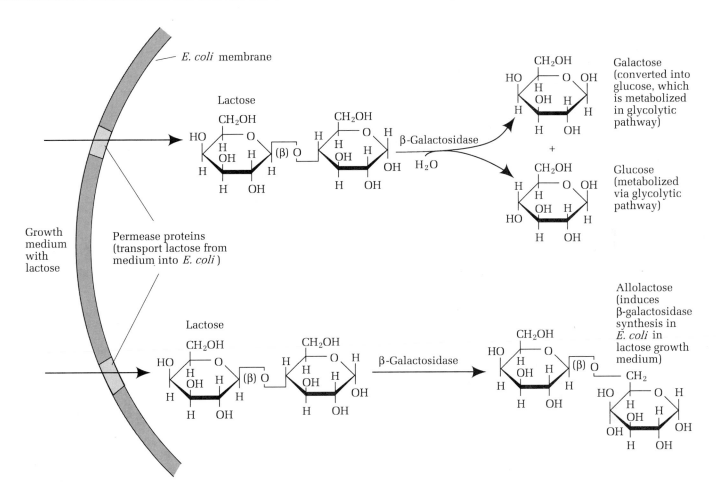

Figure 20.2 Reactions catalyzed by the enzyme β-galactosidase. Lactose brought into the cell by the permease is either converted to glucose and galactose (top) *or to allolactose* (bottom), *the true inducer for the lactose operon of* E. coli.

expressions are regulated together by operator-regulator protein interactions, plus the operator region itself and the promoter, is called an **operon.** Thus it is appropriate to speak of the lactose operon, or simply the *lac* operon, which consists of three genes (for β-galactosidase, permease, and transacetylase) and the adjacent operator and promoter.

Figure 20.3 shows the arrangement of the five components of the *lac* operon along with the approximate number of base pairs for each structural gene. The order of the genes is β-galactosi-

dase, permease, and transacetylase, and they are symbolized respectively by *lacZ, lacY,* and *lacA.* (In a wild-type cell the genes are designated as *lacZ*[+], *lacY*[+], and *lacA*[+].) The regulatory protein involved in controlling the expression of the *lac* operon is a repressor molecule, which is the product of the *lacI* gene (*lacI*[+] in the wild type) that maps a short distance away from the promoter for the *lac* operon (Figure 20.3). The promoter and terminator for the *lacI* gene are distinct from those elements for the *lac* operon itself. Expression of the *lacI* gene is under promoter control, and so this gene is another example of a constitutive gene.

Experimental Evidence for the Regulation of the *lac* Operon

Our basic understanding of the organization of

608
*Regulation
of Gene
Expression in
Bacteria and
Bacteriophages*

the genes and the controlling sites involved in lactose utilization and of the control of the *lac* operon came largely from the genetic experiments of F. Jacob and J. Monod, for which they received the Nobel Prize.

Mutants of the structural genes. Mutants of the structural genes were obtained following treatment with mutagens; they were identified on the basis of enzyme assays since in every case the activity of a particular enzyme was reduced greatly as a consequence of a mutation in the structural gene coding for that enzyme. The *lacZ⁻, lacY⁻*, and *lacA⁻* mutations obtained were used to map the locations of the three genes, and from these experiments it was established that the order of the structural genes in the genome was *lacZ-lacY-lacA* and that the three genes were tightly linked in a cluster.

In Chapter 13 we studied the types of mutations that affect the reading of the genetic message. In the *lac* operon, missense mutations, which result in the substitution of one amino acid for another in a polypeptide, only affect the expression of the gene in which the mutation maps. A *lacZ⁻* missense mutation results in a nonfunctional or partially functional β-galactosidase, but permease and transacetylase are totally normal. In contrast, the effect of a chain-terminating (nonsense) mutation will depend on the location of the mutation in the gene.

For example, for a nonsense mutation near the normal chain termination codon of the *lacZ* gene, active or partially active β-galactosidase may still be produced, with little or no affect on permease and transacetylase production. However, the closer the nonsense mutation is to the 5′ end of the *lacZ* gene, the more serious are the effects. First, a short β-galactosidase polypeptide is produced, which is most likely to be nonfunctional. Second, no functional products from the permease and transacetylase genes are produced, even though there are no mutations in those genes. Nonsense mutations in the *lacY* gene that lead to the loss of permease function also lead to the loss of transacetylase function but have no effect on β-galactosidase activity. Nonsense mutations in the *lacA* gene that lead to the loss of transacetylase function have no effect on either β-galactosidase or permease activities. In sum, nonsense mutations in the cluster of three genes involved in lactose utilization have different effects, depending on where they are located within the cluster. In other words, the nonsense mutations exhibit *polar effects,* and the phenomenon is called **polarity.** (Nonsense mutations that show polar effects are often called *polar mutations.*)

The interpretation of the polar effects of nonsense mutations in the *lac* operon structural genes is that all three genes are transcribed onto a single mRNA molecule—called a **polygenic mRNA**—rather than onto three separate mRNAs. Unequivocal evidence now indicates that RNA

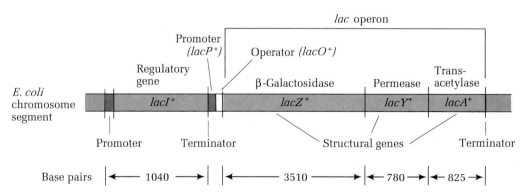

Figure 20.3 *Organization of the lactose operon of* E. coli *and its regulatory gene. The number of base pairs for each gene is indicated.*

609
*Gene
Regulation
of Lactose
Utilization
in* E. coli

polymerase initiates transcription at the single promoter that is adjacent to the operator and that a polygenic mRNA is synthesized with the structural gene transcripts in the order $5'-lacZ^+-lacY^+-lacA^+-3'$. Translation of this mRNA begins near the 5' end and proceeds toward the 3' end. So a ribosome will (1) synthesize β-galactosidase; (2) slide along the mRNA until it recognizes the initiation sequence for permease; (3) synthesize permease; (4) slide along the mRNA until it recognizes the initiation sequence for transacetylase; (5) synthesize transacetylase; and (6) dissociate from the mRNA. Because polygenic mRNAs are characteristic of bacterial operons, nonsense mutations in structural genes in an operon can potentially prevent expression of normal genes that are further along the polygenic mRNA.

Promoter mutants. The promoter for the structural genes (located at the *lacZ* gene end of the cluster of *lac* genes; see Figure 20.3) is also susceptible to mutation. Most of the known promoter mutants ($lacP^-$) affect all three structural genes. Even in the presence of lactose, the lactose utilization enzymes are not made—or are made only at very low rates. Since the promoter is the recognition sequence for RNA polymerase and does not code for any product, the effect of a $lacP^-$ mutation is confined to the genes that it controls on the same chromosomal strand. A mutation whose effects are confined only to the DNA to which it is linked is said to be a *cis-dominant mutation*.

Operator mutants. Of special interest to Jacob and Monod were mutants that affected the regulation of the lactose operon. Recall that in wild-type *E. coli* the three structural gene products are not made in the absence of lactose, while in the presence of lactose all three genes are induced. Jacob and Monod isolated a number of mutants in which all gene products of the lactose operon were synthesized constitutively; that is, all three proteins were synthesized in the presence or absence of lactose. These mutants were hypothesized to be regulatory mutants that did not affect the functions of the enzymes

themselves but, instead, affected the cellular mechanisms responsible for regulating the expression of the genes coding for the enzymes. As a result of their mapping experiments, Jacob and Monod recognized two classes of constitutive mutants: One class mapped to a relatively small DNA region adjacent to the *lacZ* gene that they called the operator (*lacO*), and the other mapped to a gene-sized DNA region a short distance away that they called the *lacI* gene (see Figure 20.3).

The mutations of the operator were called operator-constitutive, or $lacO^c$, mutations. All $lacO^c$ mutants exhibit the phenotype of synthesis of the lactose utilization enzymes in the presence *or* absence of lactose. Through the use of partial-diploid strains (i.e., *F'* strains; see Figure 7.10), Jacob and Monod were able to define better the role of the operator in regulating the expression of the *lac* operon. One such partial diploid was $lacO^+ lacZ^- lacY^+/lacO^c lacZ^+ lacY^-$ (both *lac* operons have a normal promoter and the *lacA* gene is omitted since it is not germane to our discussions). One of the chromosomes in the partial diploid has a normal operator ($lacO^+$), a mutant β-galactosidase gene ($lacZ^-$), and a normal permease gene ($lacY^+$). The other chromosome has a constitutive operator mutant ($lacO^c$), a normal β-galactosidase gene ($lacZ^+$), and a mutant permease gene ($lacY^-$). This partial diploid was tested for the production of β-galactosidase (from the $lacZ^+$ gene) and of permease (from the $lacY^+$ gene), both in the presence and the absence of the inducer lactose.

In the absence of inducer, the β-galactosidase is synthesized, but permease is not. Only when lactose is added to the culture does M protein synthesis occur. In other words, in this partial diploid the $lacZ^+$ gene is constitutively expressed (i.e., the gene is active in the presence or absence of lactose) whereas the $lacY^+$ gene is under normal inducible control (i.e., the gene is inactive in the absence of lactose and active in the presence of lactose). The interpretation of these results was that $lacO^c$ mutations affect only those genes that are adjacent to it on the same chromosome strand. Furthermore, the $lacO^+$ region only controls those genes adjacent to it and

610
Regulation
of Gene
Expression in
Bacteria and
Bacteriophages

has no effect on the genes on the other chromosome strand. Thus the $lacO^c$ mutation is cis-dominant since the defect affects the adjacent genes only and cannot be dominated by a normal $lacO^+$ region elsewhere in the genome. Jacob and Monod also concluded from these results that the operator region does not produce a diffusible product that functions in the cell, because if it did, then in the $lacO^+/lacO^c$ diploid state one or the other of the regions would have controlled all the lactose utilization genes wherever they were.

The *lacI* gene regulatory mutants. The second class of constitutive mutants of the *lac* operon defined the *lacI* gene (see the previous section). The study of these $lacI^-$ mutants in partial-diploid strains was instrumental in obtaining an understanding of the normal regulation of the *lac* operon.

In the partial diploid $lacI^+$ $lacO^+$ $lacZ^-$ $lacY^+/lacI^-$ $lacO^+$ $lacZ^+$ $lacY^-$ (in which both *lac* operons have normal operators and normal promoters), one operon has a normal *lacI* gene ($lacI^+$), a mutant β-galactosidase gene ($lacZ^-$), and a wild-type permease gene ($lacY^+$): the other operon has a constitutive mutant *lacI* gene ($lacI^-$), a normal β-galactosidase gene ($lacZ^+$), and a mutant permease gene ($lacY^-$). This partial diploid was tested for structural gene expression in the absence and the presence of the inducer lactose. The result was that no β-galactosidase or permease was produced in the absence of lactose, but both were synthesized in the presence of lactose. In other words, the expression of *both* genes was inducible. This means that the $lacI^+$ gene in the cell can overcome the defect of the $lacI^-$. Hence $lacI^+$ is dominant to $lacI^-$, and the $lacI^-$ mutants are therefore recessive.

Since the $lacI^+$ gene did control the genes on the other chromosome strand (that is, it is trans-dominant), the *lacI* gene must produce a diffusible product. Jacob and Monod proposed that the $lacI^+$ gene produces a functional **repressor molecule** (the *lacI* gene is also called a **repressor gene**) and that no functional repressor molecules are produced in $lacI^-$ mutants. Thus in a haploid bacterial strain that has a $lacI^-$ mutation,

the *lac* operon is constitutive. In a partial diploid with both a $lacI^+$ and a $lacZ^-$, however, the functional repressor molecules produced by the $lacI^+$ gene control the expression of both *lac* operons present in the cell, making both operons inducible.

Jacob and Monod's Model for the Regulation of the *lac* Operon

Based on their studies of genetic mutants affecting the regulation of the synthesis of the lactose utilization enzymes, Jacob and Monod proposed their now-classical operon model. The following description of the Jacob-Monod model for the regulation of the *lac* operon has been embellished with more up-to-date molecular information.

Figure 20.4 diagrams the state of the *lac* operon in wild-type *E. coli* growing in the absence of lactose. In this case RNA polymerase molecules bind to the repressor gene ($lacI^+$) promoter and transcribe the *lacI* gene. Translation of the resultant mRNA produces a polypeptide consisting of 360 amino acids. Four of these polypeptides associate together to form the repressor protein. Because synthesis of the repressor is under promoter control, the gene product is synthesized constitutively. The repressor has affinity for the base-pair sequence of the operator to which it binds. When the repressor is bound to the operator, RNA polymerase cannot bind to the operon's promoter, and hence transcription of the three structural genes in the operon cannot occur. The *lac* operon is said to be under *negative control* since the binding of the repressor at the operator site blocks transcription of the structural genes. (The low level of transcription of the genes that results in the presence of a few molecules of each protein, even in the absence of lactose, occurs because repressors do not just bind and stay; they bind and unbind. In the split second when one repressor unbinds and before another binds, an RNA polymerase can initiate transcription of the operon even in the absence of lactose.)

When wild-type *E. coli* is growing in the presence of lactose as the sole carbon source (Figure

611
Gene
Regulation
of Lactose
Utilization
in E. coli

20.5), some of the lactose transported into the cell is converted by existing molecules of β-galactosidase into allolactose, which, in turn, induces the production of the *lac* operon enzymes. (Thus, allolactose, not lactose, is the actual inducer of the *lac* operon structural genes.) In addition to having a recognition site for the *lac* operator, the *lac* repressor protein also has a recognition site for allolactose. When allolactose binds to the repressor, it changes the shape of the repressor and hence its recognition site for the operator. As a result, the repressor loses its affinity for the *lac* operator, and it dissociates from the site. Free repressor proteins are also altered so that they cannot bind to the operator.

In the absence of repressor, RNA polymerase is now able to bind to the operon's promoter and initiate the synthesis of a single polygenic mRNA molecule that contains the transcripts for the *lacZ*⁺, *lacY*⁺, and *lacA*⁺ genes. The polygenic mRNA for the *lac* operon is then translated by a string of ribosomes to produce the three

proteins specified by the operon. This efficient mechanism ensures the coordinate (simultaneous) production of proteins of related function.

Effect of *lacO*ᶜ mutations. The *lacO*ᶜ mutations lead to constitutive production of the *lac* operon genes and are cis-dominant to *lacO*⁺. The explanation of the phenotype of *lacO*ᶜ mutations is that base-pair alterations of the operator DNA sequence make it unrecognizable by the repressor protein. Since the repressor cannot bind, the structural genes become constitutively expressed. The cis dominance of *lacO*ᶜ is due to the fact that repressor cannot bind to a *lacO*ᶜ mutant, but can bind to another wild-type operator. The cis dominance of a *lacO*ᶜ mutant is illustrated for the partial diploid described earlier, *lacI*⁺ *lacO*⁺ *lacZ*⁻ *lacY*⁺/*lacI*⁺ *lacO*ᶜ *lacZ*⁺ *lacY*⁻ growing in the absence (Figure 20.6a, p. 614) and presence (Figure 20.6b, p. 615) of inducer.

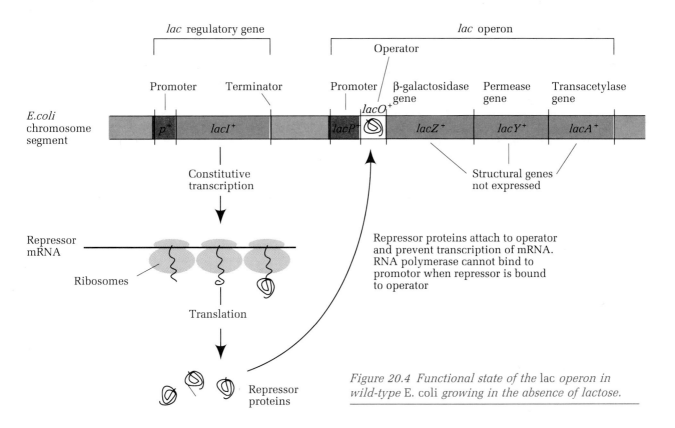

Figure 20.4 Functional state of the lac *operon in wild-type* E. coli *growing in the absence of lactose.*

612
Regulation
of Gene
Expression in
Bacteria and
Bacteriophages

Effects of *lacI* gene mutations. The effect of *lacI*⁻ mutations on *lac* operon expression is shown in Figure 20.7a (p. 616). The *lacI*⁻ mutations map within the repressor structural gene and result in amino acid changes in the repressor such that the repressor's shape is changed and it cannot now recognize and bind to the operator. As a consequence, transcription cannot be prevented even in the absence of lactose, and there is constitutive expression of the *lac* operon. The dominance of the *lacI*⁺ (wild-type) gene over *lacI*⁻ mutants is illustrated for the partial diploid described earlier, *lacI*⁺ *lacO*⁺ *lacZ*⁻ *lacY*⁺/*lacI*⁺ *lacO*ᶜ *lacZ*⁺ *lacY*⁻, in Figure 20.7b and c. In the absence of the inducer (Figure 20.7b), the defective *lacI*⁻ repressor is unable to bind to either normal operator (*lacO*⁺) in the cell. But sufficient normal repressors, produced from the *lacI*⁺ gene, are diffusing through the cell; they will bind to the two operators and hence block transcription of both operons. In the presence of inducer (Figure 20.7c) the wild-type repressors become inactivated and both operons are transcribed. One produces a defective β-galactosidase and a normal permease, while the other produces a normal β-galactosidase and a defective permease; between them, active β-galactosidase and permease are produced. Thus, in *lacI*⁺/*lacI*⁻ partial diploids, both operons present in the cell are under inducible control.

Other classes of *lacI* gene mutants have been identified since the time Jacob and Monod studied the *lacI*⁻ class of mutants. One of these classes, the *lacI*ˢ (*superrepressor*) mutants, shows no production of *lac* enzymes in the presence or absence of lactose. In partial diploids with a *lacI*⁺/*lacI*ˢ genotype, the *lacI*ˢ allele is trans-dominant, affecting both chromosomal strands. Figure 20.8 (p. 618) diagrams the effect of *lacI*ˢ mutations in the partial diploid. Our interpretation here is that the mutant repressor gene does not produce a diffusible polypeptide product since, if there were no product, the *lacI*ˢ mutants would not be dominant to the *lacI*⁺ gene. The superrepressor protein is not altered in its ability to bind to the operator region but, instead, cannot recognize the inducer allolactose. Therefore the mutant superrepressors

get stuck on the operators, and transcription of the operons can never occur. The presence of normal repressors in the cell has no effect on this situation, since once a *lacI*ˢ repressor is on the operator, nothing can remove it. As a consequence, cells with a *lacI*ˢ mutation cannot use lactose as a carbon source.

A third type of repressor gene mutation is the *lacI*⁻ᵈ (*dominance*) class. These mutations are clustered toward the 5′ end of the *lacI* gene. In haploid cells the *lacI*⁻ᵈ mutants have a constitutive phenotype like the *lacI*⁻ mutants; the *lac* enzymes are made in the presence or absence of lactose. Unlike the *lacI*⁻ mutations, the *lacI*⁻ᵈ mutants are trans-dominant to *lacI*⁺ in *lacI*⁻ᵈ/*lacI*⁺ partial diploids, so *lac* enzymes are produced constitutively even in the presence of the *lacI*⁺ gene.

The dominance of these *lacI*⁻ᵈ mutants seems to relate to the structure of the *lac* repressor. As we know, the repressor protein has four identical polypeptides. In the cell are relatively few, perhaps a dozen, repressor molecules. In the *lacI*⁻ᵈ mutants it appears that the repressor subunits do not combine normally, so no operator-specific binding is possible. The *lacI*⁻ᵈ/*lacI*⁺ diploids have a mixture of normal and mutant polypeptides, which combine randomly to form repressor tetramers. The presence of one or more defective polypeptides in the repressor is apparently enough to block normal binding to the operator. Under these conditions there is a good chance that no normal repressor proteins will be produced since there are so few molecules per cell, and therefore a constitutive enzyme phenotype results.

Lastly, some mutations in the repressor gene promoter affect the expression of the repressor gene itself. We mentioned earlier that the extent of transcription of a gene is a function of the affinity of that gene's promoter for RNA polymerase molecules. Clearly, since relatively few repressor molecules are synthesized in wild-type *E. coli* cells, the repressor gene promoter must be of low affinity. As in any other region of the DNA, the repressor gene promoter is subject to base-pair changes by mutation. Base-pair mutations have been found that decrease and that in-

613
Gene
Regulation
of Lactose
Utilization
in E. coli

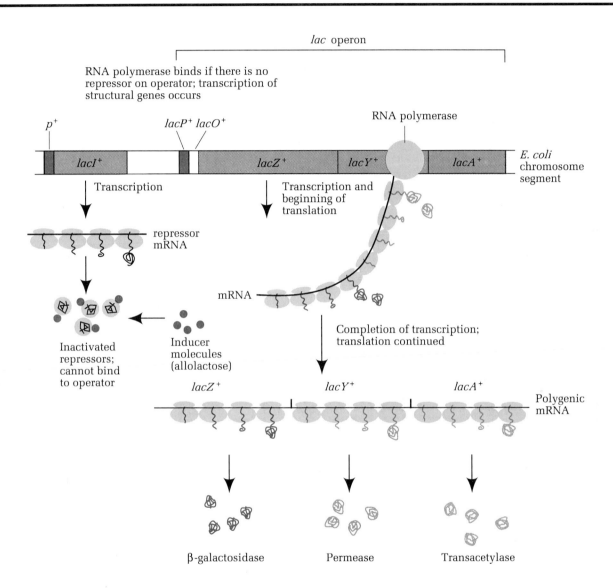

Figure 20.5 Functional state of the lac *operon in wild-type* E. coli *growing in the presence of lactose as the sole carbon source.*

crease transcription rates. (Remember that the repressor gene is a constitutively expressed gene under promoter control.) The most useful mutants in this regard are *lacI*[Q] and *lacI*[SQ] mutants (where Q stands for "quantity" and SQ for "superquantity"). Both mutations result in an increase in the rate of transcription of the repressor gene, with the *lacI*[SQ] mutants giving the greater increase. These mutants were useful historically because they produce large numbers of repressor molecules, which facilitated their isolation and purification and, consequently, the determination of the amino acid sequence of the constituent polypeptide subunit. Since *lacI*[Q] and *lacI*[SQ] mutants produce more repressor molecules than the wild type, these mutants reduce the efficiency of induction of the *lac* operon. They can be induced, however, at very high lactose concentrations.

The mutants of the *lacI* gene point out the functions of the product of that gene, the repres-

614
*Regulation
of Gene
Expression in
Bacteria and
Bacteriophages*

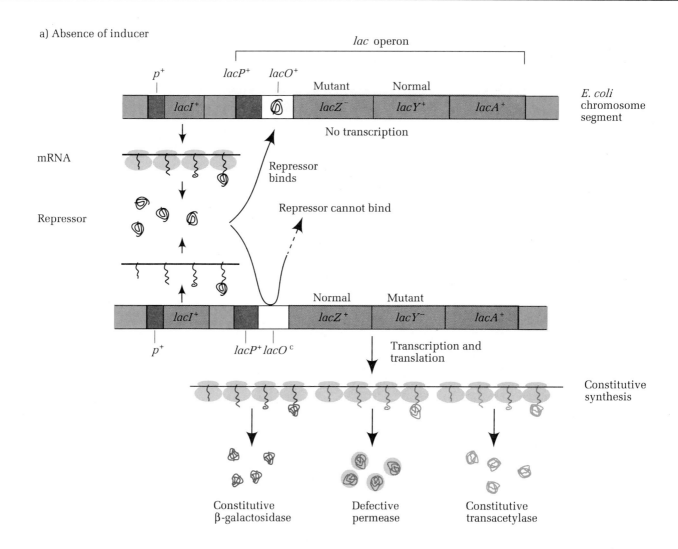

a) Absence of inducer

Figure 20.6 Cis-dominant effect of a lacO^c *mutation in a partial-diploid strain of* E. coli: *(a) In the absence of the inducer the* lacO$^+$ *operon is turned off while the* lacO^c *operon produces functional β-galactosidase and nonfunctional permease molecules from the* lacY$^-$ *gene. (b) In the presence of the inducer the functional β-galactosidase and defective permease are produced from the* lacO^c *operon, while the* lacO$^+$ *operon produces nonfunctional β-galactosidase from the* lacZ$^-$ *gene and functional permease from the* lacY$^+$ *gene. Between the two operons present in the cell, functional β-galactosidase and permease are produced.*

sor. Specifically, the repressor is involved in three recognition interactions, any one of which can be affected by mutation: (1) binding of the repressor to the operator region; (2) binding of the inducer to the repressor; and (3) binding of individual repressor polypeptides to each other to form the active repressor tetramer.

Positive Control of the *lac* Operon

In the previous sections we have learned that in the *lac* operon the repressor protein functions to prevent the transcription of the operon's structural genes by binding to the operator, providing

615
Gene
Regulation
of Lactose
Utilization
in E. coli

b) Presence of inducer

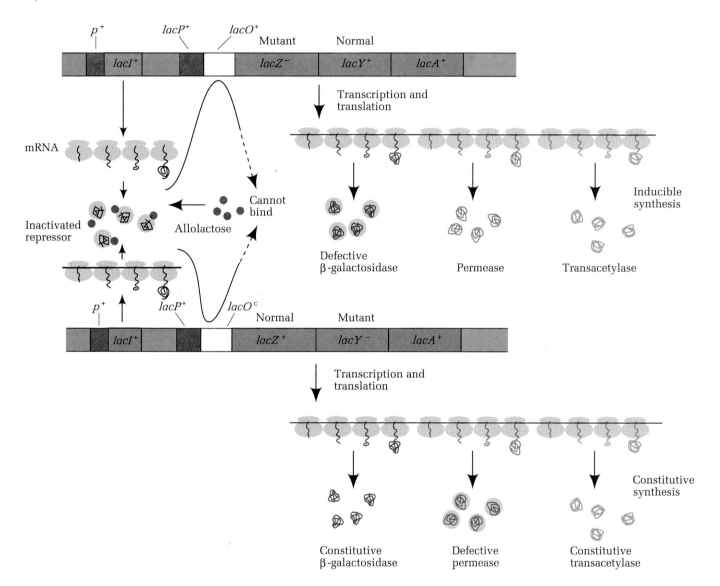

the inducer is not present. The repressor is exerting a negative effect on the expression of the *lac* operon, and this operon is said to be under *negative control.* Several years after Jacob and Monod proposed their operon model, researchers also found a positive control system that regulates the *lac* operon, a system that functions to turn on the expression of the operon. This system is used to ensure that the *lac* operon will be

expressed if lactose is the sole carbon source but not if glucose is present as well.

We have carefully noted all along in our discussion of the *lac* operon that expression of the structural genes occurs when lactose is added as the sole carbon source. Recall also that glucose is preferred as a carbon source to all other sugars. Lactose is converted to glucose and galactose (which is, in turn, converted to glucose).

a) Haploid strain (in presence or absence of inducer)

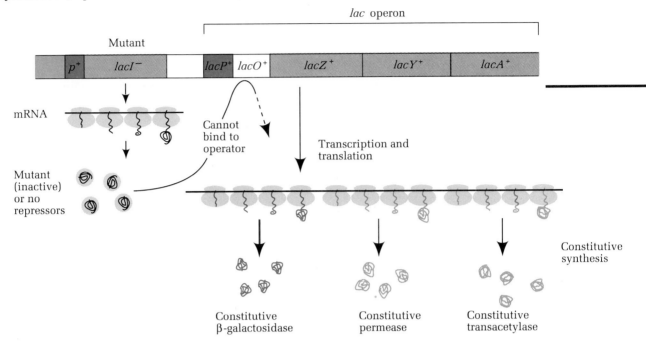

lac operon

Mutant

| p^+ | $lacI^-$ | | $lacP^+$ | $lacO^+$ | $lacZ^+$ | $lacY^+$ | $lacA^+$ |

mRNA

Mutant (inactive) or no repressors

Cannot bind to operator

Transcription and translation

Constitutive synthesis

Constitutive β-galactosidase

Constitutive permease

Constitutive transacetylase

b) Partial diploid in the absence of inducer

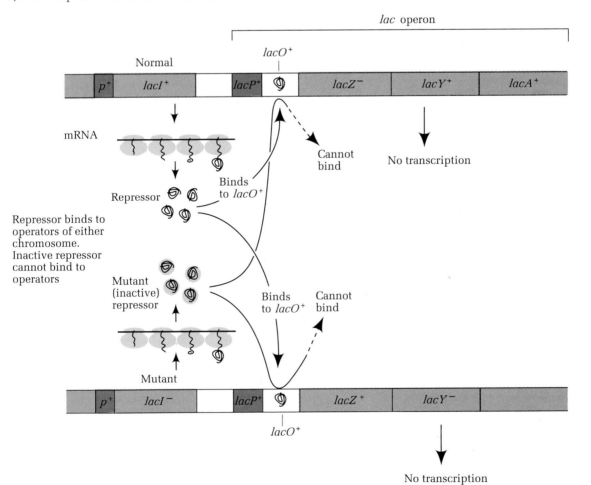

lac operon

$lacO^+$

Normal

| p^+ | $lacI^+$ | | $lacP^+$ | 🧬 | $lacZ^-$ | $lacY^+$ | $lacA^+$ |

mRNA

Repressor

Binds to $lacO^+$

Cannot bind

No transcription

Repressor binds to operators of either chromosome. Inactive repressor cannot bind to operators

Mutant (inactive) repressor

Mutant

Binds to $lacO^+$

Cannot bind

| p^+ | $lacI^-$ | | $lacP^+$ | 🧬 | $lacZ^+$ | $lacY^-$ | |

$lacO^+$

No transcription

617
*Gene
Regulation
of Lactose
Utilization
in* E. coli

c) Partial diploid in the presence of inducer

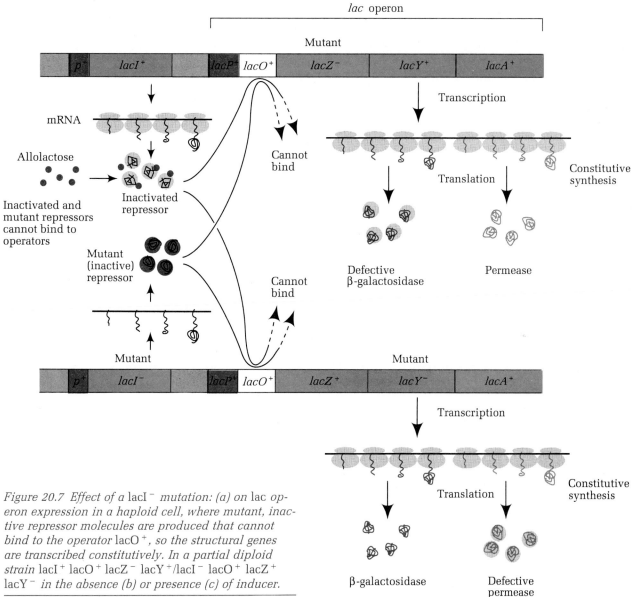

Figure 20.7 Effect of a lacI $^-$ *mutation: (a) on* lac *operon expression in a haploid cell, where mutant, inactive repressor molecules are produced that cannot bind to the operator* lacO $^+$, *so the structural genes are transcribed constitutively. In a partial diploid strain* lacI $^+$ lacO $^+$ lacZ $^-$ lacY $^+$/lacI $^-$ lacO $^+$ lacZ $^+$ lacY $^-$ *in the absence (b) or presence (c) of inducer.*

If both glucose and lactose are present in the medium, the glucose is used preferentially and the *lac* operon is not expressed. The *lac* operon is repressed under these conditions because a *positive regulator* must bind to the *lac* operon before transcription can occur. Figure 20.9 shows the mechanism involved here. First, a protein called *CAP* (catabolite activator protein) binds with *cAMP* (cyclic AMP, or cyclic adenosine 3',5'-monophosphate; see Figure 20.10, p. 620) to form a CAP-cAMP complex. This complex is the positive-regulator molecule. The CAP protein itself is a dimer that consists of two identical polypeptides. Next, the CAP–cAMP

618
*Regulation
of Gene
Expression in
Bacteria and
Bacteriophages*

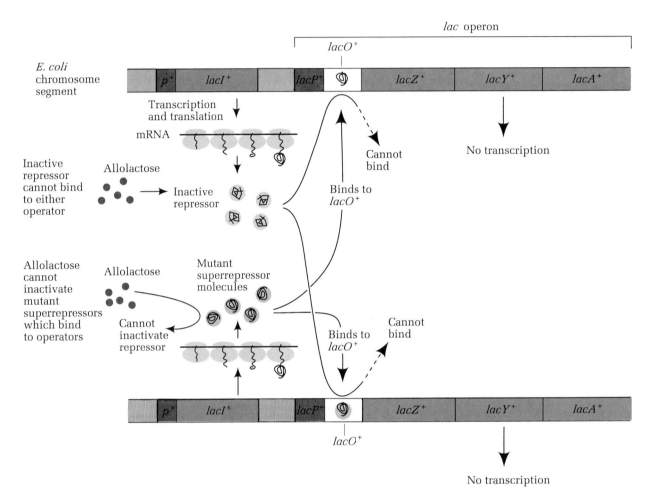

Figure 20.8 Dominant effect of lacI^s *mutation over wild-type* lacI$^+$ *in a partial-diploid cell growing in the presence of lactose.*

complex binds to the CAP site in the DNA, and this facilitates RNA polymerase binding to the promoter sequence in the presence of an inducer. The polymerase then moves toward the structural genes and begins transcription near the start of the β-galactosidase gene.

In the presence of glucose, **catabolite repression (glucose effect)** occurs. A breakdown product, or catabolite, of glucose (as yet unknown) causes a rapid decrease in the cellular levels of cAMP. (See Figure 20.10 for the steps involved in the synthesis and degradation of cAMP.) In-

sufficient CAP–cAMP complex is then available to facilitate polymerase binding to the *lac* operon promoter, and transcription is blocked even though repressors are removed from the operator by the presence of allolactase.

Catabolite repression occurs in a number of other bacterial operons related to catabolism of sugars other than glucose. In all instances, if glucose is present, catabolite repression occurs and, through a similar mechanism, blocks the expression of those operons. From studies of a number of glucose-sensitive operons, CAP bind-

619
*Gene
Regulation
of Lactose
Utilization
in* E. coli

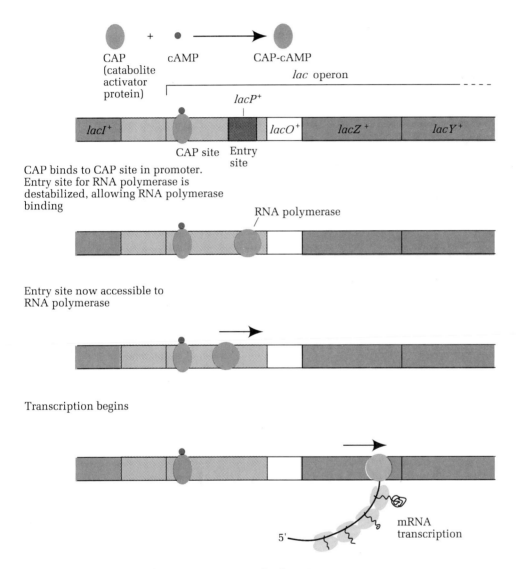

Figure 20.9 Role of cyclic AMP (cAMP) in the functioning of glucose-sensitive operons such as the lactose operon of E. coli.

ing site consensus sequences have been determined: they are shown in Figure 20.11. It has been shown that the CAP binding site is not located at a constant distance from the transcription start site among the operons, indicating that direct protein-protein interaction between CAP and RNA polymerase is unlikely.

Molecular Details of *lac* Operon Regulation

From DNA- and RNA-sequencing experiments much information is now available concerning the nucleotide sequences of the significant *lac* operon regulatory sequences. One general approach to obtain this information has been to purify the protein known to bind to a regulatory

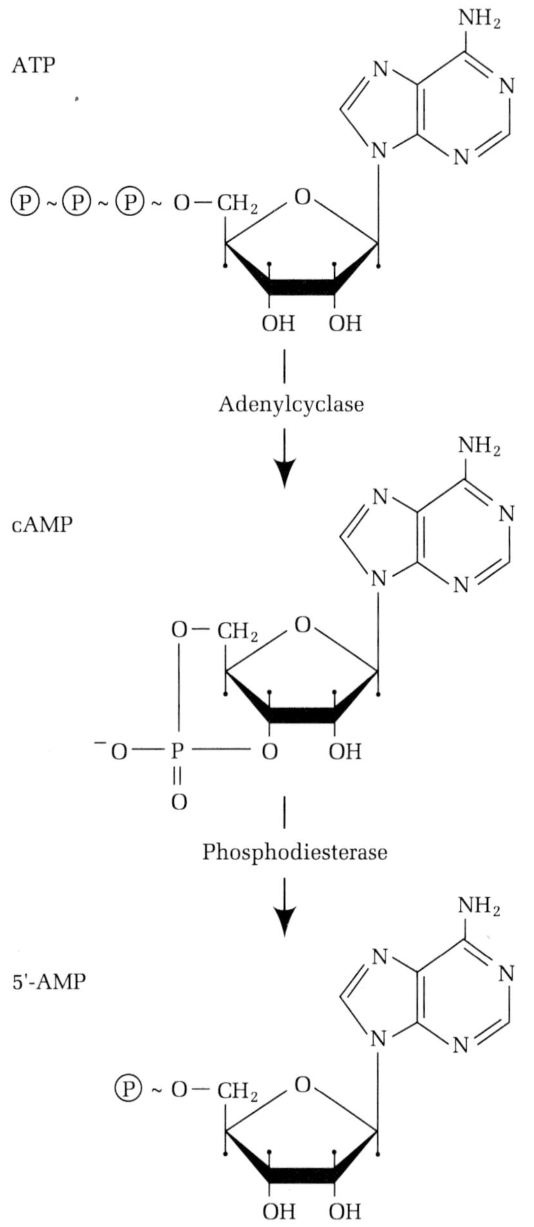

620
*Regulation
of Gene
Expression in
Bacteria and
Bacteriophages*

ATP

cAMP

Adenylcyclase

Phosphodiesterase

5'-AMP

Figure 20.10 *Structure of cyclic AMP (cAMP, or cyclic adenosine 3', 5'-monophosphate). cAMP is synthesized from ATP in a reaction catalyzed by adenylcyclase, and it is broken down in a reaction catalyzed by phosphodiesterase.*

site and to let it bind to isolated *lac* operon DNA in the test tube. For example, if the repressor is bound to the *lac* operator, it will protect that region of the operon from deoxyribonuclease digestion. If DNase is allowed to digest the rest of the DNA, the operator sequence can be isolated, cloned by using recombinant DNA technology, and sequenced.

Promoter region of the *lac* repressor gene (*lacI*). Figure 20.11 shows the nucleotide pair sequence for the *lacI* gene promoter region, the sequence for the 5' end of the repressor mRNA, and the first few amino acids of the repressor protein itself. The nucleotide sequence of the repressor mRNA can be aligned with this promoter sequence, with its start approximately in the middle. As is usually the case with all gene transcripts, translation does not start right at the end of the mRNA molecule: The translation start sequence (ribosome binding site) is a Shine-Dalgarno sequence (AGGA) at 11 to 8 bases upstream from the AUG start codon. Here the first AUG codon (the actual start codon) is nucleotides 27–29 from the 5' end of the messenger. Figure 20.11 also shows the single base-pair change found for a particular *lacI*^{Q} mutant; this change; from CG to TA, brings about a tenfold increase in repressor production.

***Lac* operon controlling sites.** Figure 20.12 shows the nucleotide pair sequence of the *lac* operon controlling sites. The orientation of this sequence was put together from several different pieces of information. First, the amino acid sequences of the repressor protein and of β-galactosidase are completely known, and this information enables us to identify the coding regions of the *lacI* gene and of the *lacZ*$^+$ gene. Then the other regions were identified on the basis of "protection" experiments of the kind described a little earlier. Here CAP–cAMP complex, RNA polymerase, and repressor protein were used separately to bind to the DNA, and the DNase-resistant regions were then sequenced. (Although studies indicate the results that follow, we should perhaps be cautious before we definitely assume that the discrete boundaries be-

621
*Gene
Regulation
of Lactose
Utilization
in* E. coli

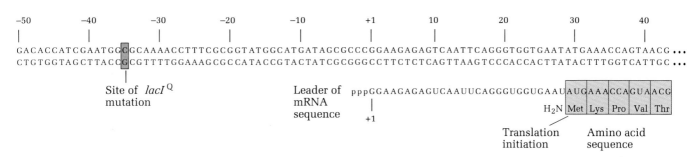

Figure 20.11 *Base-pair sequence of the* lac *operon*
lacI⁻ *gene promoter and of the 5' end of the repressor mRNA. Also shown is the amino acid sequence of the first part of the repressor protein itself.*

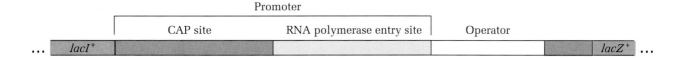

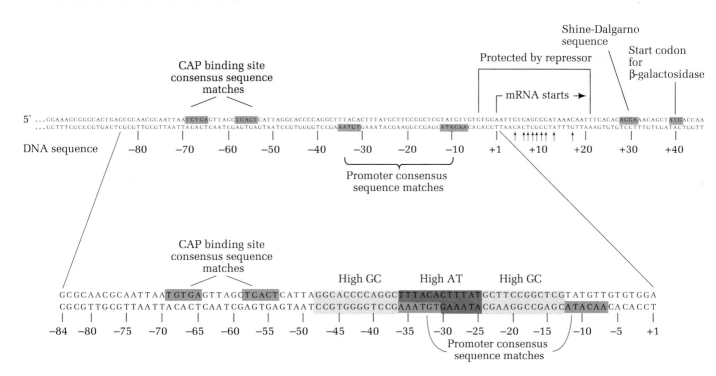

Figure 20.12 *Base-pair sequence of the controlling sites, promoter and operator, for the lactose operon of* E. coli. *Also shown are locations of some known* lacO^c *mutations (indicated by arrows).*

622
*Regulation
of Gene
Expression in
Bacteria and
Bacteriophages*

tween the various parts of the controlling region are real.)

The beginning of the promoter region is defined as position -84 in the figure (i.e., 84 base pairs upstream from the mRNA initiation site), immediately next to the stop codon for the *lacI* gene. Nucleotide pairs -54 to -58, and -65 to -69 are the consensus sequence matches for the CAP complex binding site. That this location is upstream from the polymerase binding site (the region spanned by nucleotide pairs -47 to -8 including -10 and -35 consensus sequence matches) fits in well with the model that CAP–cAMP binding destabilizes the DNA to facilitate RNA polymerase binding (see Figure 20.9). Together the region from -8 to -84 that includes the CAP protein and the RNA polymerase interaction sites (including a Pribnow box) essentially define the *lac* operon promoter region.

Immediately next to the promoter region is the operator. The region protected by the repressor protein is the area containing nucleotide pairs -3 to $+21$. When the repressor is bound to the operator, it is impossible for polymerase to bind to the DNA, and thus it cannot transcribe the genes.

The β-galactosidase mRNA has a *leader region* before the start codon is encountered. The actual start of the mRNA here is nucleotide pair $+1$ in Figure 20.12, which is very close to the beginning of the repressor binding site. Transcription of the *lac* operon includes a large proportion of the operator region in addition to the structural genes themselves. The start codon for β-galactosidase, which defines the beginning of the *lacZ* gene, is at nucleotide pairs $+39$ to $+41$. Thus the first 38 bases of the *lac* mRNA are not translated.

Figure 20.12 also shows the base-pair substitutions that have been identified for some of the *lacO^c* mutations that have been studied. In each case a single base-pair change is responsible for the altered control of the *lac* operon.

In conclusion, the lactose operon has proved to be a model system for understanding gene regulation in prokaryotic organisms. Jacob and Monod's original work on this system had a great impact on further studies. As the first mo-lecular model for the regulation of gene expression in any organism, it sparked numerous studies in both prokaryotes and eukaryotes to see whether operons were generally the case. We now know that operons are prevalent in bacteria and bacteriophages but that they do not exist in the eukaryotes. Each operon is regulated through the interaction of a regulatory protein with an operator, and this regulation takes place in several different ways. In the *lac* operon, for example, the repressor exerts a negative effect on structural gene expression in the absence of lactose; the genes are turned on when the inducer removes this negative effect by binding to the repressor. Other inducible operons use a similar mechanism; yet others are turned on by an inducer binding to a regulatory protein, altering it so that it binds to a controlling site, thereby facilitating operon transcriptions.

Keynote Studies of the synthesis of the lactose-utilizing enzymes of E. coli generated a model that is the basis for the regulation of gene expression in a large number of bacterial and bacteriophage systems. In the lactose system the addition of lactose to the cells brings about a rapid synthesis of three enzymes required for lactose utilization. The genes for these enzymes are contiguous on the E. coli chromosome and are adjacent to a controlling site (an operator) and a single promoter. The genes, the operator, and the promoter constitute an operon, which is transcribed as a single unit. In the absence of lactose the operon is turned off.

Tryptophan Operon of E. coli

Just as glucose may not always be available in a bacterial growth medium for use as a carbon source, all necessary amino acids may not be present in a growth medium to enable bacteria to assemble proteins and to produce compounds whose synthesis requires amino acids as precursors or as donators of specific chemical groups.

If an amino acid is missing, a bacterium has certain operons and other gene systems that enable it to manufacture that amino acid so that it may grow and reproduce. Each step in the biosynthetic pathway through which amino acids are assembled is catalyzed by a specific enzyme coded by a specific gene.

When all 20 amino acids are present, the genes encoding the enzymes for the amino acid biosynthetic pathways are turned off. If an amino acid is not present in the medium, however, then the genes must be turned on in order for the biosynthetic enzymes to be made. Unlike the case in the *lac* operon where gene activity is induced when a chemical (lactose) is added to the medium, in this case there is a repression of gene activity when a chemical (an amino acid) is added. Customarily, we refer to amino acid biosynthesis operons controlled in this way as *repressible operons.* One repressible operon in *E. coli* that has been extensively studied is the operon for the biosynthesis of the amino acid tryptophan (Trp). In *E. coli,* tryptophan is used sparingly in proteins, perhaps occurring once in every hundred amino acids or so. Although the regulation of the *trp* operon shows some basic similarities to the regulation of the classical *lac* operon, some intriguing differences also appear to be common among similar, repressible, amino acid biosynthesis operons in bacteria.

Gene Organization of the Tryptophan Biosynthesis Genes

Figure 20.13 shows the organization of the controlling sites and of the structural genes that code for the tryptophan biosynthetic enzymes and how they relate to the biosynthetic steps. Much of the work that we will discuss is that of C. Yanofsky and his collaborators.

Five structural genes (*A–E*) occur in the tryptophan operon. The promoter and operator regions are closely integrated in the DNA and are upstream from the *trpE* gene. Between the promoter-operator region and *trpE* is a 162-base-pair region called *trpL,* or the leader region. Within *trpL,* relatively close to *trpE,* is an *attenuator site* (*a*) that plays an important role in the

regulation of the tryptophan operon, as we will see later.

The entire tryptophan operon is approximately 7000 base pairs long. Transcription of the operon results in the production of a polygenic mRNA containing the transcripts for the five structural genes. Each of these transcripts is translated to an equal extent.

Regulation of the *trp* Operon

An expanded schematic rendering of the regulatory elements of the tryptophan operon is shown in Figure 20.14. The initiation of transcription of the *trp* operon is regulated at the operator. A regulatory gene *trpR* for the operon is unlinked to the structural genes (and therefore does not appear in Figure 20.13). The product of *trpR* is an aporepressor protein, which, alone, has no affinity for the operator. However, the aporepressor has two binding sites for tryptophan and is converted to an active repressor by binding with tryptophan. The active complex has affinity for the operator. When the repressor-tryptophan complex is bound to the operator, the promoter is not accessible to RNA polymerase, and initiation of transcription cannot occur. (Tryptophan is an example of an effector molecule, just as allolactose is the effector molecule for the *lac* operon.) As a result of repression, transcription of the *trp* operon can be reduced about seventy-fold.

A second regulatory mechanism is involved in the expression of the *trp* operon, usually under conditions of relatively severe tryptophan starvation. In brief, not all RNA transcripts initiated at the promoter site are continued through the structural genes, and the proportion that are is related to the amount of tryptophan that is present. Within the *trpL* region is the attenuator, a transcription termination site (Figures 20.13 and 20.14). Those RNA transcripts that do not extend into the structural genes are terminated at the attenuator in a process called **attenuation.** Attenuation can reduce transcription of the *trp* operon by 8-fold to 10-fold. Thus repression and attenuation together can regulate the transcription of the *trp* operon over a range of about 560-

624
*Regulation
of Gene
Expression in
Bacteria and
Bacteriophages*

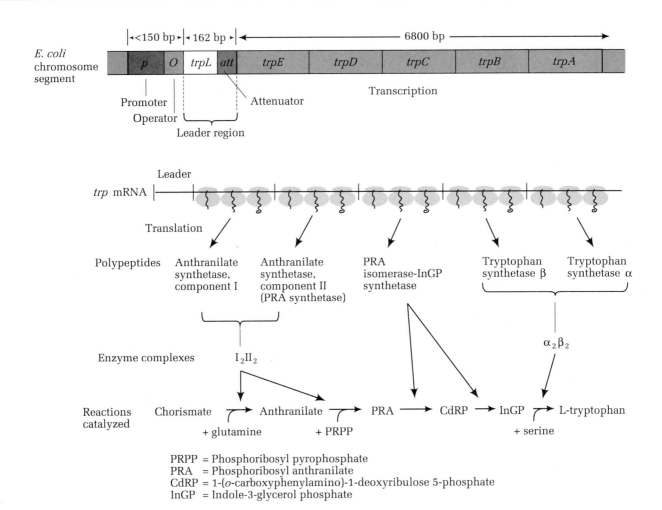

Figure 20.13 *Organization of controlling sites and the structural genes of the* E. coli *tryptophan operon. Also shown are the steps catalyzed by the products of the structural genes* trpA, trpB, trpC, trpD, *and* trpE.

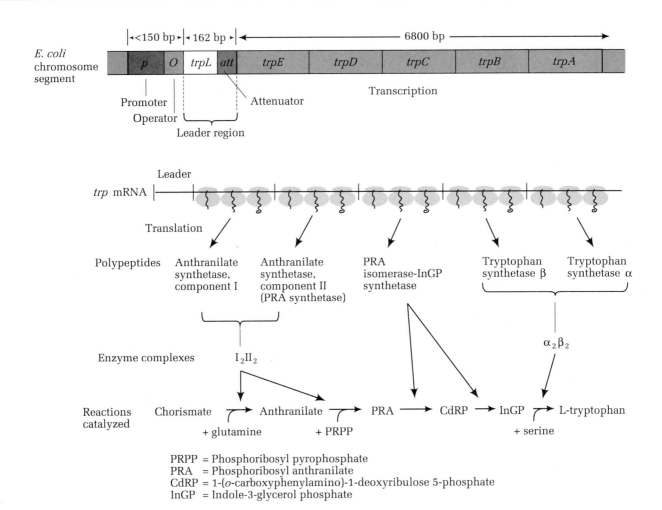

Figure 20.14 *Regulatory elements of the* E. coli *tryptophan operon.*

fold to 700-fold. The details of these regulatory mechanisms are more fully elaborated next.

Expression of the *trp* operon in the presence of tryptophan. When tryptophan is abundant in the growth medium, the repressor protein is activated by tryptophan and binds to the operator, thereby preventing the initiation of transcription of the *trp* operon structural genes by RNA polymerase. As a result, none of the tryptophan biosynthesis enzymes are produced.

Expression of the *trp* operon in the absence of or in the presence of limiting amounts of tryptophan. In the absence of tryptophan (or if tryptophan is present in very low levels) in the growth medium, a tryptophan-repressor complex is not bound to the operator, and transcription of the operon is initiated at the promoter. Following the binding of RNA polymerase, transcription can proceed toward the structural genes. The first region to be transcribed is the leader region (*trpL*), a region of 162 base pairs that does not code for any of the tryptophan biosynthesis enzymes. After the *trpL* region is transcribed, the structural genes are transcribed, in order, onto a polygenic mRNA.

When no tryptophan is present, all the RNAs produced contain transcripts of the structural genes. However, when some, but not a great deal of, tryptophan is present, *E. coli* produces two types of mRNA transcripts of the tryptophan operon. One transcript is of the entire operon—the leader region plus the structural genes. The second transcript is an mRNA transcript of the first 140 base pairs of the operon, which includes most of the leader region. The relative proportions of the two transcripts depend on the amount of tryptophan present—the more tryptophan, the greater the number of short transcripts.

The existence of a number of partial transcripts suggests that a transcription termination site exists within the *trpL* region at the 140-base-pair position. The regulation that occurs at this site is attenuation, and the site itself is the attenuator (*att*).

The DNA sequence of the whole tryptophan operon is known. Inspection of the sequence for the leader region revealed that the transcript of this region has a ribosome binding site with an AUG start codon in frame with a UGA chain termination codon such that translation of the transcript would produce a 14–amino acid–long polypeptide. This polypeptide has never been isolated from cells, but there are solid indications that it is made and that it is involved in the attenuation process. Of particular interest is the existence of two adjacent codons for tryptophan near the stop codon of the leader transcript. If cells are starved for tryptophan, then the amount of Trp-tRNA.Trp molecules drops dramatically, because no tryptophan molecules are available for the aminoacylation of the

tRNA.Trp molecules. A ribosome translating the leader transcript would stall at the Trp codons, because the next specified amino acid in the peptide could not be added to the growing chain. As a result, the leader peptide could only be completed if sufficient tryptophan were present.

The sequence data therefore suggested that the position of the ribosome on the leader transcript has an important role in the regulation of transcription termination at the attenuator. In Chapter 12 we learned that transcription and translation are closely coupled in prokaryotes, so this hypothesis was by no means out of line. In fact, evidence from other systems suggests that the tandem arrangement of the Trp codons in this case is not a chance event. Figure 20.15 shows the predicted leader peptides of a number of other amino acid biosynthetic operons of *E. coli* and of *Salmonella typhimurium,* which also have an attenuation mechanism for regulation of gene expression. In every case a significant number of the codons specify the amino acid for which the operon specifies the biosynthetic enzymes. For example, the histidine operon of *E. coli* has a string of 7 histidines in the leader peptide, and 7 of the 15 amino acids in the leader for the phenylalanine A operon of *E. coli* are phenylalanine.

Molecular model for attenuation. The main question now is, How does amino acid starvation and the possible stalling of the ribosome at the Trp codons control transcription by RNA polymerase? C. Yanofsky proposed a model to answer this question; the model describes different secondary structures of the transcript of the leader region.

Analysis of the nucleotide sequence of the leader peptide mRNA indicated that there were four regions that could form secondary structures by complementary base pairing (Figure 20.16). As we will see, pairing of regions 1 and 2 results in a pause signal, pairing of 3 and 4 is a termination of transcription signal, and pairing of 2 and 3 is an *antitermination signal* for transcription.

Figure 20.17 shows the possible alternative

626
*Regulation
of Gene
Expression in
Bacteria and
Bacteriophages*

pheA: Met – Lys – His – Ile – Pro – Phe – Phe – Phe – Ala – Phe – Phe – Phe – Thr – Phe – Pro

his: Met – Thr – Arg – Val – Gln – Phe – Lys – His – His – His – His – His –His – His – Pro – Asp

leu: Met – Ser – His – Ile – Val – Arg – Phe – Thr – Gly – Leu – Leu – Leu – Leu – Asn – Ala – Phe –
         Ile – Val – Arg – Gly – Arg – Pro – Val – Gly – Ile – Gln – His

thr: Met – Lys – Arg – Ile – Ser – Thr – Thr – Ile – Thr – Thr – Thr – Ile – Thr – Ile – Thr – Thr –
         Gly – Asn – Gly – Ala – Gly

ilv: Met – Thr – Ala – Leu – Leu – Arg – Val – Ile – Ser – Leu – Val – Val – Ile – Ser – Val – Val –
         Val – Ile – Ile – Ile – Pro – Pro – Cys – Gly – Ala – Ala – Leu – Gly – Arg – Gly – Lys – Ala

Figure 20.15 Predicted amino acid sequences of the leader peptides of a number of attenuator-controlled bacterial operons. Shown are the peptides for the pheA, his, leu, thr, *and* ilv *operons of* E. coli *or* Salmonella typhimurium. *The amino acids that regulate the respective operons are capitalized.*

Leader peptide coding region:

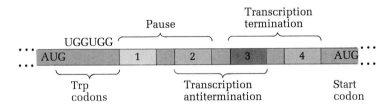

Alternate RNA structures:

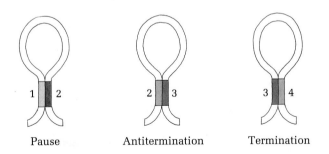

Figure 20.16 Four regions of the tryptophan operon leader mRNA that can form alternate secondary structures by complementary base pairing.

secondary structure of nucleotides 50–141 of the *trp* leader transcript. Figure 20.17a shows the two hairpin loops that are formed by hydrogen bonding between complementary base pairs. The four strands (actually regions of the same single strand) that are involved in the pairing are designated 1–4 as in Figure 20.16 (see inset in the figure). Thus in this form strands 1 and 2 are paired, as are strands 3 and 4. The two Trp codons are located in strand 1, starting at position 54. Figure 20.17b shows the alternative secondary structure that could form. In this case strand 2 is hydrogen-bonded to strand 3.

The model proposed by Yanofsky is shown in Figure 20.18. In the following discussion a ribosome is considered to be translating the RNA transcript a little way behind the point at which RNA polymerase is at work extending the transcript—transcription and translation are very tightly coupled in this regulatory system. This coupling is accomplished by the RNA polymerase pausing just after the RNA of the 1:2 structure has been synthesized. The pausing is dependent upon the NusA protein, which stabilizes the transcription-pause complex of RNA polymerase + DNA + RNA. The pausing gives enough time for the ribosome to load and to begin translating the leader peptide, resulting in the tight coupling of transcription and translation.

The position of the ribosome on the leader transcript determines which of the two RNA secondary structures forms. If the cells are starved for tryptophan (Figure 20.18a), the ribosome stalls on the Trp codons, and this stalling physically prevents pairing of strands 1 and 2.

a) Regions 1 + 2 and 3 + 4 paired

b) Regions 2 and 3 paired

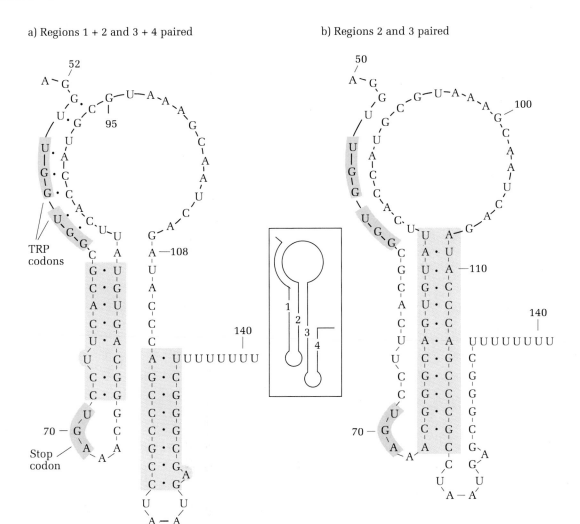

Figure 20.17 Secondary structure alternatives for the trp *operon leader transcript; the inset* (center) *shows the numbering of the four stems that can participate in hydrogen bonding: (a) Regions (1 + 2) and (3 + 4) are hydrogen-bonded together in a structure assumed to terminate transcription at the attenuator. (b) Regions 2 and 3 are hydrogen-bonded together, while regions 1 and 4 are unbonded; the 2:3 binding prevents transcription termination at the attenuator.*

Strand 2 then pairs with strand 3 once the latter is synthesized. As a result, when strand 4 is synthesized, it does not pair with strand 3 since strand 3 is already paired with strand 2. With

this secondary structure of the leader transcript, RNA synthesis is able to continue past the attenuator, and the structural genes are then transcribed. The polygenic mRNA is then translated into the biosynthetic enzymes necessary to begin tryptophan synthesis.

If, instead, sufficient tryptophan is present (Figure 20.18b) so that the ribosome can read the Trp codons, then the ribosome stalls at the stop codon for the leader peptide at positions 69–71 in the leader transcript. As a result, region 2 is unable to pair with region 3, and the latter is then able to pair with region 4 when that region is synthesized. This secondary structure then somehow stimulates RNA polymerase

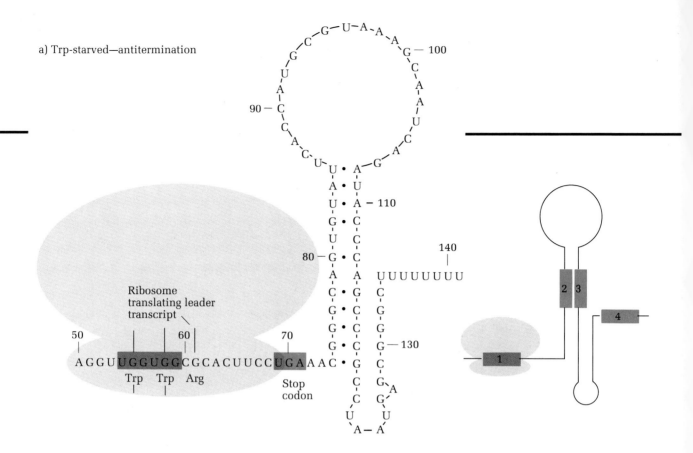

a) Trp-starved—antitermination

b) Nonstarved—termination

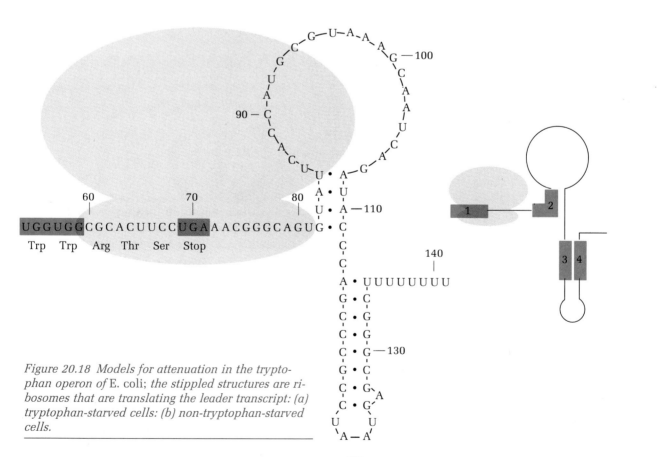

Figure 20.18 Models for attenuation in the tryptophan operon of E. coli; the stippled structures are ribosomes that are translating the leader transcript: (a) tryptophan-starved cells: (b) non-tryptophan-starved cells.

to terminate transcription. This structure is referred to as the attenuator. That is, the RNA polymerase is sensing the secondary structure of the RNA it is making. It is also important to realize that the RNA polymerase and ribosome are moving together and that the secondary structure loops are forming as they are synthesized.

Genetic evidence for the attenuation model has been obtained through the study of mutants. One type of mutant shows less efficient transcription termination at the attenuator, and structural gene expression is increased. The mutations involved are single base-pair changes that lead to the base changes in the leader transcript shown in Figure 20.19. In each case the change is in the regions of 3:4 pairing, and each causes a disruption of the pairing so that the structure is less stable. In the less stable state the structure is less able to prevent transcription from proceeding into the structural genes.

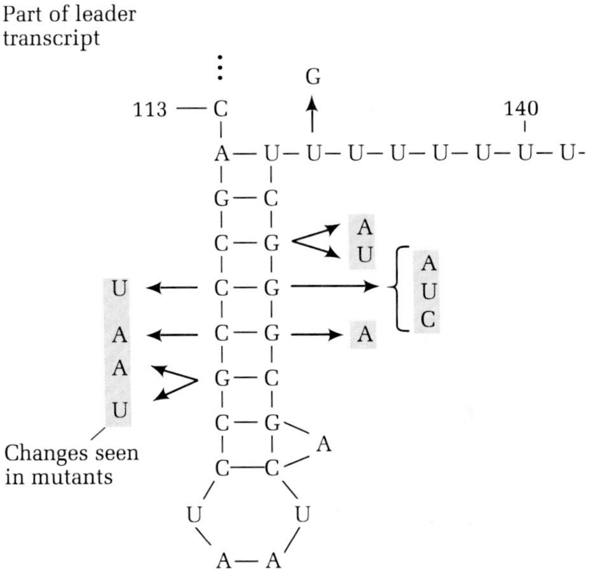

Fine-scale regulation of the *trp* operon. Clearly, everything we have said so far relates to regulation at the gene expression level in the *trp* operon. Two other mechanisms operate in addition to this regulation to give an overall regulatory system. One mechanism is the degradation of the polygenic mRNA transcript of the structural genes. As is the case with the lactose operon and with most other prokaryotic genes, the transcripts have a relatively short lifetime, so mRNA must continually be made in order for the enzymes to be produced. The short lifetime allows the cell to change its physiological state when the environment changes. When tryptophan is added to the medium, transcription termination at the attenuator and repression at the operator block further transcription of the structural genes, and the structural gene mRNAs already made are rapidly degraded ("turned over"). In this way the cell does not waste energy making unnecessary materials.

Another part of the overall regulatory system is a fine-tuning system. Gene regulation (turning genes on and off) involves control processes that function in response to long-term changes in the environment. Cells must also respond to short-term changes that occur. If, for example, trypto-

Part of leader transcript

Figure 20.19 Sites of mutations in the trpL *region that show less efficient transcription termination at the attenuator site. The mutations map to the regions in the DNA that correspond to regions 3 and 4 in the RNA.*

phan being made by the tryptophan biosynthetic pathway accumulates at a rate faster than it is being used, a mechanism called *feedback inhibition* exists to regulate this problem. We will discuss feedback inhibition (also called end-product inhibition) only at a general level since it has nothing to do with gene regulation.

In feedback inhibition the end product of a biosynthetic pathway can often be recognized by the first enzyme in the biosynthetic pathway. When too much tryptophan is produced, for example, the tryptophan binds to the first enzyme in the tryptophan biosynthetic pathway, thereby altering its three-dimensional conformation. (The technical term for this configurational change is *allosteric shift,* and the phenomenon is called *allostery.*) When the shape of the enzyme is changed, the function of the enzyme is impaired, because the substrate for the enzyme can no longer bind to the enzyme. As a result, the tryptophan pathway is turned off temporarily at the first step in the biosynthetic pathway.

630
*Regulation
of Gene
Expression in
Bacteria and
Bacteriophages*

The enzyme remains nonfunctional until the tryptophan level in the cell drops and the tryptophan dissociates from the enzyme. This step reverses the feedback inhibition of the enzyme's function, and the pathway is restarted. This feedback mechanism is a common attribute of many biosynthetic pathways in cells of all kinds, serving to halt the flow of intermediates in a pathway.

Regulation of Other Amino Acid Biosynthesis Operons

Other amino acid biosynthetic operons also have attenuator mechanisms. The molecular information that has been gathered for those operons indicates that the attenuation mechanism is very similar in each case. Interestingly, some of the operons use only attenuation and not repression. For example, no histidine repressor protein, or any mutation that gives a phenotype that we would expect of an altered repressor protein, has been found for the histidine operon of *Salmonella typhimurium*.

Attenuation has also been shown to regulate a number of genes not involved with amino acid biosynthesis, for example, the ribosomal protein gene S10, rRNA operons (*rrn*), and the *ampC* gene of *E. coli*, as well as some animal virus and animal cell genes. Among all these examples, a number of attenuation mechanisms have been found to be involved.

Keynote *Expression of the tryptophan* (trp) *operon of* E. coli *is accomplished primarily through a repressor-operator system and through attenuation at a second controlling site called an attenuator, which is located downstream from the operator in the leader region. The attenuator is a partially effective termination site that allows only a fraction of RNA polymerases to transcribe the rest of the operon. In the presence of tryptophan enough Trp-tRNA.Trp is present so that the ribosome can move past the attenuator and allow the leader transcript to form a secondary structure that causes transcription to be blocked. In the absence of tryptophan the ribosomes stall at*

the attenuator, and the leader transcript forms a secondary structure that permits continued transcription activity.

Summary of Operon Function

There are numerous examples of operons in bacteria and their bacteriophages. The following generalizations can be made about the regulation of operons:

1. A regulator protein (e.g., the *lac* repressor) plays a key role in the regulation process since it is able to bind to a controlling site in the operon, the operator.
2. The transcription of a set of clustered structural genes is controlled by an adjacent operator through the interaction with a regulator protein.
3. The trigger for changing the state of an operon from off to on and vice versa is an effector molecule (e.g., allolactose), which controls the conditions under which the regulator protein will bind or not bind to the operator.

It should be pointed out, however, that not in all cases are genes for a related function clustered in the prokaryotic genome.

Global Control of Transcription

As we have seen, specific mechanisms allow the control of gene expression for individual catabolic and biosynthetic pathways. In addition, the bacteria are able to respond at a more general level to major environmental changes. Clearly, it is advantageous in the evolutionary sense for a bacterium to be able to withstand difficult conditions. Such conditions can be mimicked in the laboratory. For example, an amino acid auxotroph starved for that amino acid suffers severe hardship. A trauma may be induced by shifting cells from a nutrient-rich medium to a nutrient-poor medium in what is called a nutritional shift-down experiment. The cell responds to this stress with a **stringent re-**

sponse (or **stringent control**), that is, a rapid shutdown of essential cellular activities.

The study of the stringent response has shown that regulatory mechanisms in bacteria coordinate protein synthesis with a number of other basic cellular activities. Starvation conditions, for example, result in a rapid inhibition of glycolysis and transport mechanisms and of the syntheses of proteins, rRNAs, tRNAs, mRNAs, lipids, carbohydrates, and nucleotides. Were these events not to occur, the cell would grow in an uncontrolled fashion, with all macromolecular components (except proteins) accumulating in large amounts. The stringent response is very complex, and we shall only deal with some of the information known.

Mutants have been isolated that do not show the stringent response under starvation conditions. These mutants are called *relaxed* (*relA*) mutants, and their study has helped us understand the stringent response. When starved, wild-type cells rapidly accumulate an unusual nucleotide, ppGpp (guanosine 5′-diphosphate-3′-diphosphate) (Figure 20.20), which *relA* cells do not accumulate. The ppGpp is synthesized by a ribosomal protein enzymatic activity that is activated on ribosomes that are starved of aminoacylated tRNA molecules. The ppGpp is an example of an effector molecule, a regulatory molecule that responds to the state of the cell. Apparently, the ppGpp brings about a change in the preference the RNA polymerase has for promoters so that different genes are transcribed. This reaction occurs because of its direct interaction with RNA polymerase, resulting in a lowered affinity of RNA polymerase for DNA. The RNA polymerase sigma factor plays some role in this change in promoter affinity, but its exact involvement is not understood.

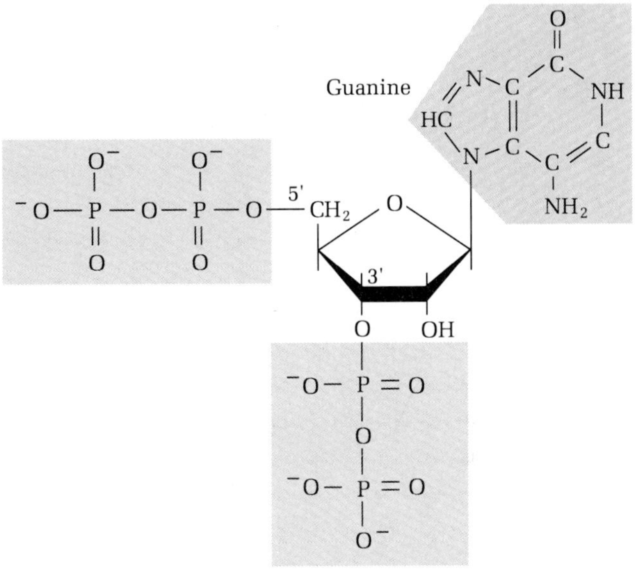

Gene Regulation in Bacteriophages

Because bacteriophages exist by parasitizing bacteria, they do not themselves need to code for enzymes required for all the biosynthetic pathways needed to make progeny bacteriophages.

Figure 20.20 Structure of the unusual nucleotide ppGpp (guanosine tetraphosphate, or guanosine 5′-diphosphate-3′-diphosphate).

Instead, the essential components for phage reproduction are provided by the host cell, and those components are directed by the products of phage genes. Most genes of a phage, then, code for products that control the production of progeny phage particles. Since there is a temporal sequence of events for phage particle assembly, there must be specific regulatory pathways to turn genes on and off at particular times. In this section we will learn about the genomic organization and the regulation of gene expression in two phages that we have discussed before: T4 and lambda (λ).

Regulation of Gene Expression in Phage T4

In Chapter 1 we saw that the T4 phage is a complex array of different proteins: head proteins, core proteins, baseplate proteins, and tail fiber proteins. In the mature phage the genetic material (double-stranded DNA) is packaged within the head.

632
*Regulation
of Gene
Expression in
Bacteria and
Bacteriophages*

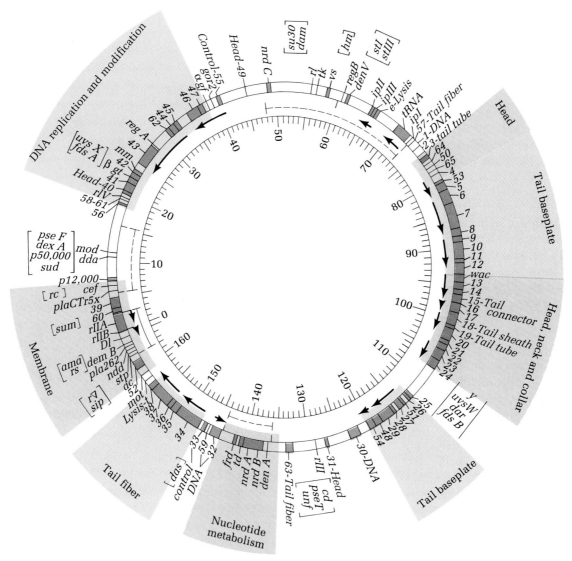

Figure 20.21 T4 genetic map, showing the clustering of genes with related functions. In most cases a number of genes are transcribed from a single promoter onto a single polygenic mRNA, indicated by arrows on the inside of the circle. (Shaded blocks indicate the relative sizes of the genes.)

Functional organization of the T4 genome.

Figure 20.21 shows the gene organization of the T4 chromosome. (Recall that although the chromosome in the phage particle is linear, the genetic map is circular, owing to terminal redundancy and circular (permutation.) As indicated on the map, genes with related functions are often found in clusters, many of which are operons. The genes required for DNA replication, for example, are clustered in two principal areas: one between 5 and 6 o'clock on the map and the other between 8 and 11 o'clock. These two clusters are separated by a gene cluster (from about 6 to 7 o'clock) that controls tail fiber production. The gene clusters found throughout the rest of the genome direct the synthesis of the various components of the phage progeny and direct the synthesis of enzymes required for the lysis of the host cell when mature progeny phages have accumulated within. It has been well established that the arrangement of related genes in clusters is important in regulating gene expression.

Control of T4 gene transcription during the life cycle. One of the fundamental roles of gene regulation mechanisms in any system is to cause different gene sets to be transcribed at different times in the organism's life cycle. In T4, for instance, the genes that specify DNA replication functions are transcribed within the first few minutes after the T4 has infected a host. Genes like these that are the first to be transcribed after infecting a host are called *early genes.* The genes for the phage particle components and for bacterial lysis, on the other hand, are transcribed after the early genes and are called the *middle* and *late genes,* respectively. This timing of gene transcription is logical given the phage life cycle. The enzyme lysozyme, for example, is needed for cell lysis after progeny phages have been made and is therefore specified by a late gene. If it were transcribed early, the host cell would lyse prematurely and the phage's life cycle could not be completed. Clearly, then, control mechanisms exist to bring about the transcription of the T4 genes in a particular temporal order.

The sequential transcription of phage genes is regulated by phage-coded proteins that direct the host RNA polymerase to recognize different phage promoters as the life cycle proceeds. In this way the series of events leading to the assembly and release of mature progeny phages proceeds in a sequential and orderly fashion. When the phage DNA is first injected into the cell, transcription of a specific set of phage genes occurs under the control of the bacterial sigma factor, which facilitates the binding of the bacterial RNA polymerase to about twenty-five early-gene T4 promoters (see Figure 20.22). About 5 minutes after infection at 37°C, initiation of transcription at the early-gene promoters stops. The end of initiation is brought about by an early-gene product that prevents the host cell's sigma factor from further interaction with RNA polymerase core molecules. From that

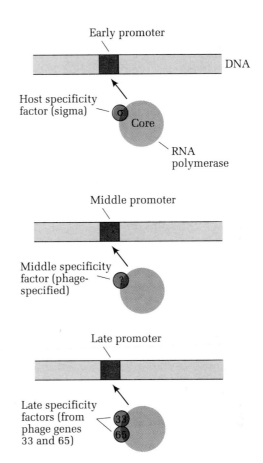

Figure 20.22 Diagrammatic representation showing how T4 genes are transcribed in a sequential manner.

point on, RNA polymerase cannot recognize host promoters. The core polymerase molecules are then activated by T4-specified factors, which then bind to middle promoters. Products of the "middle" genes include late specificity factors. These factors bind to core RNA polymerases and act as sigma-like factors to recognize late gene promoters. Thus, the sequential production of σ factors ensures the proper order of gene expression; this approach is utilized by many bacteriophages.

Genetic control of T4 phage assembly. Applying the genetic approach described in Chapter 14 for dissecting the steps in a biochemical pathway, R. Edgar and W. Wood delineated the

634
*Regulation
of Gene
Expression in
Bacteria and
Bacteriophages*

morphogenesis (genetic control of morphology) of phage T4. They isolated conditional lethal mutants of T4 with mutations in genes essential to the assembly of mature phages. Two types of mutants were identified: nonsense mutants and temperature-sensitive mutants.

Edgar and Wood infected *E. coli* cells with these phage mutants under nonpermissive conditions so that the genetic defects would be manifested. For each mutant the phage assembly proceeded until the step at which an essential protein (coded for by the mutated gene) was absent or inactive. They found that depending on the protein that was missing, different precursor phage particles accumulated in the bacterial cell. These particles were isolated by artificially lysing the cells and centrifuging the contents, and they were then examined under the electron microscope. By correlating the precursor particle with the genetic mutant, they constructed a morphogenetic pathway that involved the functions of about fifty genes, as shown in Figure 20.23. The figure indicates that the pathway has three different parts. One of the branches is concerned with the assembly of the head and the packaging of the DNA within the head. The second is involved with tail assembly, and the third with tail fiber production. In the final step of the pathway the head attaches to the tail, after which the tail fibers are added.

Regulation of Gene Expression in Phage Lambda

As we learned in Chapter 7, phage lambda is a temperate phage. When it infects *E. coli*, it can go through a lytic cycle (like that of T4) in which DNA replication and phage assembly occur, or it can establish a lysogenic state in which the phage chromosome integrates into the bacterial chromosome and is then called the prophage. When the phage is in the lytic cycle, phage genes for reproduction are active. Those same genes are repressed when the phage is in the lysogenic (prophage) state. In this section we will examine the proposed regulatory mechanisms that determine whether a λ phage enters the lytic or lysogenic cycle.

Functional organization of the λ genome. Figure 20.24 shows the genetic map of the λ phage. Like the T4 chromosome, the mature λ chromosome is linear, in this case with complementary "sticky" ends. Once it is free in the host cell, the λ chromosome circularizes, and thus it is conventional to show the genetic map in a circular form. The point of joining of the two ends is at 12 o'clock in the figure.

As we have come to expect in prokaryotes, functionally related genes are clustered in the genome. The genes for DNA replication that are active during the lytic cycle are clustered in an operon at about 1 o'clock (the early right operon), whereas the genes for lysogeny (e.g., those controlling integration and excision of the λ chromosome) are clustered in an operon at 10 to 12 o'clock (the early left operon). The genes for the various structural components of the phage particles (active only in the lytic cycle)—namely, the heads and the tails—are clustered together in the late operon (from 2 to 7 o'clock).

Early transcription events. When λ infects *E. coli,* it chooses between the lytic and lysogenic pathways soon after the λ chromosome is injected into the host cell. This decision involves a sophisticated *genetic switch.* If the lytic pathway is followed, progeny phages are assembled, the bacterial cells are lysed, and the phages are released. If the lysogenic pathway is followed, the λ chromosome becomes integrated into the *E. coli* chromosome at a specific site and no progeny phages are produced. In this integrated, prophage state, the lytic pathway genes are repressed and the λ genome replicates as part of the *E. coli* chromosome replication.

The two pathways are followed as the result of the expression of different genes. When the λ chromosome first infects a cell, however, some stages of phage growth are the same, regardless of whether the lytic or lysogenic pathway is

Figure 20.23 Morphogenetic pathway of T4 phage production, which has three branches and involves over fifty genes (indicated by numbers in the diagram).

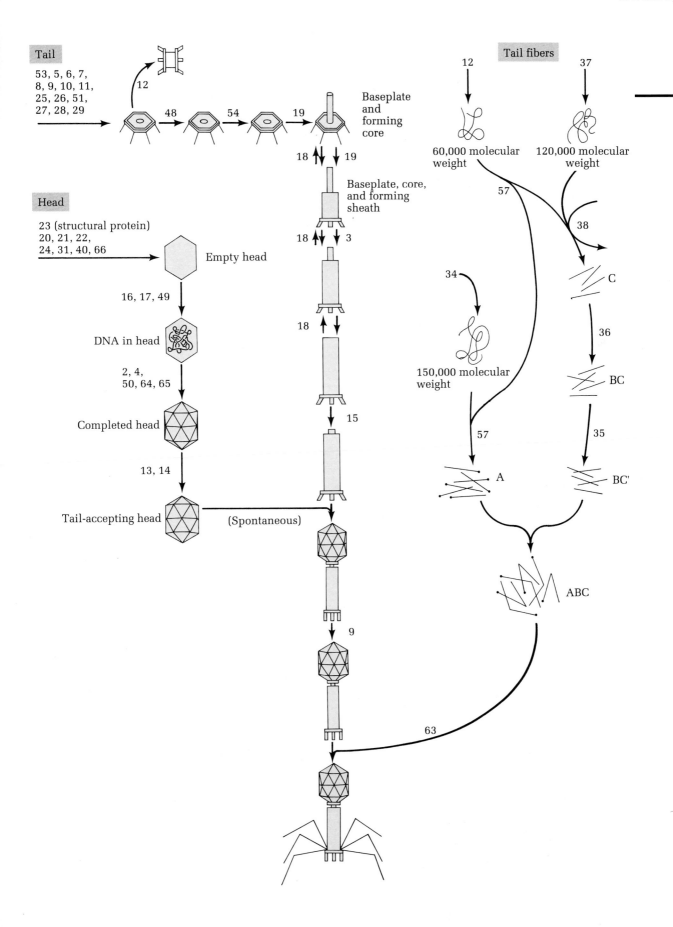

Tail

53, 5, 6, 7,
8, 9, 10, 11,
25, 26, 51,
27, 28, 29

Baseplate
and
forming
core

Baseplate, core,
and forming
sheath

Head

23 (structural protein)
20, 21, 22,
24, 31, 40, 66

Empty head

16, 17, 49

DNA in head

2, 4,
50, 64, 65

Completed head

13, 14

Tail-accepting head

(Spontaneous)

Tail fibers

60,000 molecular
weight

120,000 molecular
weight

150,000 molecular
weight

636
*Regulation
of Gene
Expression in
Bacteria and
Bacteriophages*

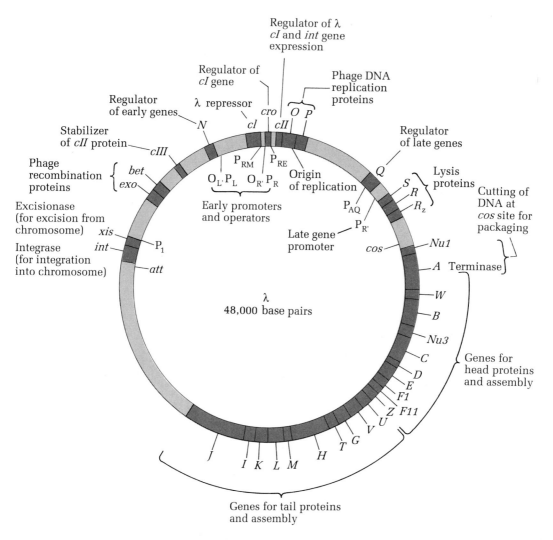

Regulator of λ
cI and *int* gene
expression

Regulator of
cI gene

λ repressor

cro

Phage DNA
replication
proteins

Regulator
of early genes

N

cI

cII

O P

Regulator
of late genes

Stabilizer
of *cII* protein

cIII

P$_{RM}$

P$_{RE}$

Q

Lysis
proteins

Phage
recombination
proteins

bet

exo

O$_L$ P$_L$

O$_{R'}$ P$_R$

Origin
of replication

P$_{AQ}$

S

R

R$_z$

Cutting of
DNA at
cos site for
packaging

Excisionase
(for excision from
chromosome)

xis

Early promoters
and operators

P$_{R'}$

cos

Nu1

Integrase
(for integration
into chromosome)

int

P$_1$

att

Late gene
promoter

A Terminase

λ
48,000 base pairs

W

B

Nu3

C

D

E

F1

Z *F11*

U

V

G

T

H

J

I *K* *L* *M*

Genes for
head proteins
and assembly

Genes for tail proteins
and assembly

*Figure 20.24 A map of phage λ showing the major
genes.*

chosen. First, the linear chromosome becomes
circular through the pairing of the 12-nucleo-
tide-long complementary ends and the subsequent action of DNA ligase to form a covalent
circle. Since no repressor molecules have been
made, at this time, to prevent expression of the
genes for the lytic pathway, transcription begins
at promoters P_L and P_R (Figure 20.25, part 1).
(P_L is the promoter for leftward transcription of
the left early operon, and P_R is the promoter for
rightward transcription of the right early op-

eron.) Promoter P_L is on a different DNA strand
from the P_R promoter, hence transcription occurs in opposite directions, counterclockwise
and clockwise, respectively.

From P_R, the first gene to be described is *cro*
(*c*ontrol of *r*epressor and *o*ther), the product of
which is the Cro protein. This protein plays an
important role in setting the genetic switch to
the lytic pathway, as will be discussed later.
From P_L, the first gene to be transcribed is *N*.
The resulting N protein is a transcription *anti-*

Figure 20.25 Expression of λ genes after infecting E.
coli *and the transcriptional events that occur when
either the lysogenic or lytic pathways are followed.*

1. Phage growth begins when RNA polymerase binds early promoters P_L and P_R, making mRNA for *N* and *cro* genes   (Throughout figure, ☐ are inactive promoters and ▪ are active promoters)

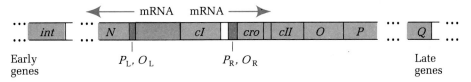

2. N protein, acting at 3 sites, extends RNA synthesis to other genes; O and P proteins permit DNA synthesis to begin; cII protein is made and acts as shown below

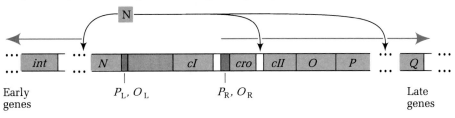

3. Two competing pathways occur:

Toward lysogenic development: dominance of repressor

Toward lytic development: dominance of Cro

4a. cII protein stimulates synthesis of mRNA:
    from P_{RE} for repressor
    from P_I for integrase

4b. Cro protein occupies operators O_R and O_L, prevents synthesis of repressor mRNA but allows enough rightward transcription for Q protein to accumulate

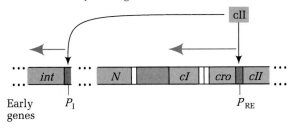 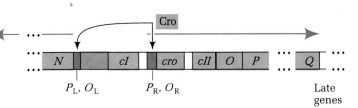

5a. Sufficient repressor binds O_R and O_L, blocking RNA synthesis; repressor bound to O_R stimulates synthesis of more repressor. Integrase promotes integration of phage DNA. Lysogeny is established

5b. Q protein stimulates late gene transcription; these genes encode structural proteins that package DNA into new phage particles

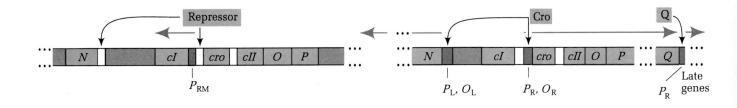

638
*Regulation
of Gene
Expression in
Bacteria and
Bacteriophages*

terminator that allows RNA synthesis to proceed past certain transcription terminators. This process is called *antitermination* or *readthrough transcription.* The action of N protein requires the presence of sites in the DNA called *nut* (*N utilization) sites. The *nut* sites are about 15 base pairs long. N protein binds to these sites, then alters RNA polymerases that are transcribing the region so that almost all transcription terminators that follow are ignored by the enzyme.

N protein acts to allow RNA polymerase to proceed leftward of gene *N,* and rightward of gene *cro,* thereby including all of the early genes (Figure 20.25, part 2). In the right early operon, genes transcribed as a result of the action of N protein include *cII,* the product of which stimulates the synthesis of λ repressor (product of *cI* gene—see later), *O* and *P,* which are genes for DNA replication, and *Q,* which is needed to turn on late genes for lysis and phage proteins. The Q protein is a positive regulatory protein that functions as an antiterminator protein, similar to the function of N protein. That is, Q protein binds to a site called *qut* (*Q ut*ilization), alters RNA polymerase, transcribing the right operon. This permits readthrough transcription into the late genes, which are involved in the lytic pathway. However, only when the lytic pathway is chosen and transcription continues from P_R for a sufficient time does enough Q protein accumulate to function effectively.

The antiterminator activity of N protein in the left early operon leads to the expression of genes *cIII* through *int* (see Figure 20.24). The cIII protein stabilizes the cII protein which, as already mentioned, stimulates the synthesis of λ repressor, and *int* encodes an integrase enzyme that is needed for integration of the λ genome at the bacterial λ attachment site.

The lysogenic pathway. After the early transcription events, either the lysogenic or lytic pathway is followed (Figure 20.25, part 3). The choice for the lysogenic pathway is made in the following way.

As we have learned, the integration of the prophage into the *E. coli* chromosome depends on protein products coded by genes in the left early operon. These proteins are synthesized after transcription termination is overcome by the action of the N protein. The establishment of the lysogenic state requires the protein products of the *cII* (right early operon) and *cIII* (left early operon) genes. The cII protein (stabilized by cIII protein) activates transcription of the *cI* gene (located between the P_L and P_R promoters) (Figure 20.25, part 4a). The product of the *cI* gene, the λ repressor protein, acts to block the expression of particular λ genes (In fact, the λ repressor was identified before the lactose repressor.)

The precise mechanism involved in this activation is not known, but as a result, transcription of *cI* is initiated from a promoter called P_{RE} (promoter for repressor establishment), located to the right of the *cI* gene. Transcription of the strand opposite the strand controlled by the P_R promoter, and of the same strand as that controlled by the P_L promoter, proceeds in a counterclockwise direction. Thus this transcription event covers part of the antisense strand of the *cro* gene before the *cI* gene.

The λ repressor is a diffusible molecule that maintains the lysogenic state of the integrated λ prophage by preventing the expression of the genes necessary for chromosome replication, progeny phage assembly, and cell lysis.

Based on the work of M. Ptashne, the λ repressor is a protein of 236 amino acids that folds into two nearly equal-sized domains, connected by a string of 40 amino acids (Figure 20.26a). The domains represent the amino carboxyl parts of the polypeptide. The functional λ repressor is a dimer of the polypeptide, formed largely by contacts between the carboxyl domains (Figure 20.26b). Almost all of the repressor in the cell is in the dimeric form. The λ repressor action is not limited to the integrated prophage but is capable of binding to and repressing any phage chromosomes that are subsequently injected into the cell. Thus a bacterium carrying an integrated prophage has superinfection immunity: it is protected against lysis by another infecting λ.

As soon as enough λ repressor is produced within the cell, it binds to two operator regions O_L and O_R (Figure 20.24), whose sequences

a) λ repressor

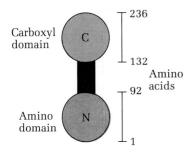

b) Dimerization of λ repressor

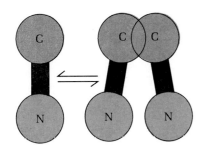

Figure 20.26 (a) The λ repressor showing the amino (N) and carboxyl (C) domains and the short connecting segment. (b) Repressor monomers from dimers which can dissociate to monomers.

overlap the P_L and P_R promoters, respectively. The binding of the λ repressor thus prevents the further initiation of transcription by RNA polymerase at the early operons controlled by P_L and P_R. As a result, transcription of the N and *cro* genes is blocked, and since the two proteins specified by these two genes are unstable, the levels of these two proteins in the cell drop dramatically. The binding of the λ repressor also blocks the N protein–controlled readthrough transcription of the O, P, and Q genes in the right early operon. Further, repressor bound to O_R stimulates the synthesis of more repressor mRNA from a different promoter, P_{RM}, thereby maintaining its concentrations in the cell (Figure 20.25, part 5a). Thus in the presence of enough λ repressors, the lysogenic state is established by the binding of the repressor to the O_L and O_R operator regions and integration of λ DNA catalyzed by integrase. With recombinant DNA tech-

niques the O_L and O_R operators have been cloned and sequenced (see Figure 20.27). What is most interesting about the sequences is that there are three 17-nucleotide-long repressor binding sites for both the O_L and the O_R regions: O_{L1}, O_{L2}, and O_{L3}, and O_{R1}, O_{R2}, and O_{R3}, respectively.

Binding studies using purified repressor show that the relative binding strength (binding affinity) of the repressor for each of the operator binding sites in O_L is $O_{L1} > O_{L2} > O_{L3}$; and in O_R it is $O_{R1} > O_{R2} > O_{R3}$. Let us consider in detail the events that occur at the right operator (O_R) sites, which are major sites where the genetic switch controlling the choice between the lysogenic and lytic pathways operates: O_L sites are not part of the switch. Figure 20.28 shows the arrangement of the O_R repressor binding sites relative to the *cI* (λ repressor) and *cro* (*c*ontrol of *r*epressor and *o*ther) genes. Note that the O_{R1} site overlaps the P_R promoter and that the O_{R3} site overlaps the P_{RM} (RM = repressor maintenance) promoter. Because of the different binding affinities of the three binding sites for repressor, if we started with a repressor-free operator, a repressor dimer would bind first to O_{R1} (Figure 20.29a). Once a repressor becomes bound to O_{R1}, the affinity of O_{R2} for the second repressor increases because the second repressor dimer not only binds to O_{R2}, but it also contacts the repressor that is bound at O_{R1} (Figure 20.29b). This enhancement of binding a second repressor by the presence of the initial repressor is called *cooperativity of repressor binding*. At higher repressor concentrations, repressor will also bind to O_{R3} (Figure 20.29c). The affinity of the repressor for this operator site is lower than for O_{R2} because in this case the repressor must bind independently without also contacting a previously bound repressor.

For determining the lysogenic state, then, repressor is typically present at concentrations such that repressor dimers are bound to both O_{R1} and O_{R2}, thereby preventing RNA polymerase from binding to P_R (see Figure 20.24 and Figure 20.25, part 5a) and transcribing the *cro* gene. However, since no repressor is bound to O_{R3}, RNA polymerase can bind to P_{RM} and the

640
*Regulation
of Gene
Expression in
Bacteria and
Bacteriophages*

a) Right operator and adjacent genes

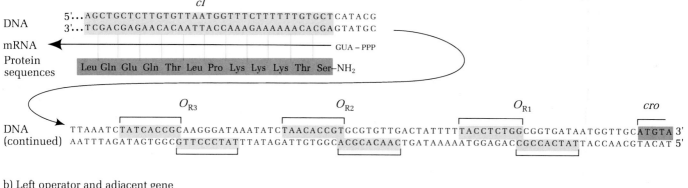

b) Left operator and adjacent gene

Figure 20.27 The DNA, RNA, and protein sequences for (a) right operator (O_R) and parts of the adjacent genes (cI and cro) and (b) left operator (O_L) and adjacent gene (N) regions of phage λ. The brackets indicate the six binding sites for the λ repressor. The starting points for transcription of genes N, cro, and cI are indicated.

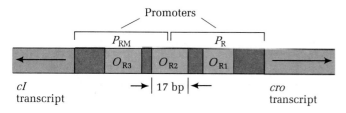

Figure 20.28 Part of the λ chromosome showing the two promoters P_{RM} and P_R and the right operator (O_R) which overlaps them. Transcription from P_{RM} occurs leftward to transcribe the repressor gene (cI). Transcription from P_R proceeds rightwards to transcribe the cro gene.

repressor gene *cI* can be transcribed (Figure 20.30). As repressor concentration increases as a result of this transcriptional activity, repressor dimers bind also to the weak O_{R3} site. This blocks repressor synthesis by preventing RNA polymerase from binding and transcribing the *cI* gene. The repressor concentration drops because no new repressor transcripts are being made and existing repressor molecules are being degraded. Once the repressor concentration drops sufficiently, the O_{R3} site is freed again, and RNA polymerase can then bind to P_{RM}, and *cI* transcription can resume. In other words, the λ repressor plays a negative regulatory role in blocking transcription of the *cro* gene, and a positive regulatory role in maintaining optimum levels of repressor in the cell by modulating RNA polymerase binding at P_{RM}.

Repressor binding to O_R is not permanent, but it is a dynamic event in which repressor dimers continuously fall off the operator binding sites to be replaced by other repressor molecules. The concentration of repressor molecules in the cell is sufficiently high to ensure that O_{R1} and O_{R2} are likely to have repressor bound to them most of the time, enabling the lysogenic state to be maintained. As we will see, this state changes dramatically when the genetic switch is thrown and the lytic pathway is induced.

The lytic pathway. The lysogenic pathway is favored when enough λ repressor is made so that early promoters are turned off, thereby repress-

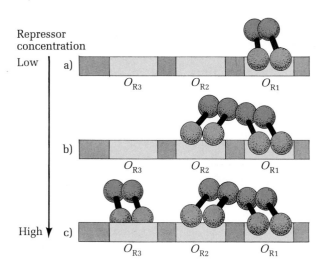

Figure 20.29 λ repressor binds to three sites in the right operator (O_R). (a) The affinity of repressor for O_{R1} is about 10 times higher than for O_{R2} or O_{R3} so repressor first binds to O_{R1}. (b) Once repressor is bound to O_{R1}, a second repressor rapidly binds to O_{R2}. (c) Repressor binds to O_{R3} only at high repressor concentrations.

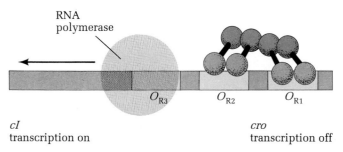

Figure 20.30 In the lysogenic state, repressors are bound to O_{R1} and O_{R2}, but not to O_{R3}, thereby allowing RNA polymerase to bind to P_{RM} and transcribe the cI (repressor) gene. Polymerase cannot bind to P_R since that promoter is covered by repressors: hence, the cro gene is not transcribed.

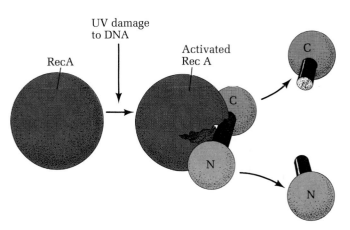

Figure 20.31 Cleavage of repressor by RecA protein. Repressor is cleaved in the linker between the C and N domains of the repressor.

ing all the genes needed for the lytic pathway. An important gene in this regard is Q, since the Q protein is a positive regulatory protein required for the production of lysis proteins and phage coat proteins.

Let us consider the induction of the lytic pathway caused by ultraviolet light irradiation. Inducers such as ultraviolet light typically damage the DNA, and this somehow causes a change in the function of the bacterial protein RecA (product of the *recA* gene). Normally RecA functions enzymatically in DNA recombination events, but when DNA is damaged, it changes its function to that of a protease, which cleaves λ repressor monomers within the linker segment, thereby separating the amino and carboxyl domains (Figure 20.31). This event causes the chain reaction shown in Figure 20.32: the cleaved monomers are unable to form dimers and, as a result, when repressors fall off the operator (see earlier), they are not replaced. The absence of repressor at O_{R1} that results leads to RNA polymerase binding at P_R and the *cro* gene is then transcribed. The resulting Cro protein

then controls the subsequent lytic growth. Cro protein action results in a decrease in RNA synthesis from P_L and P_R, thereby reducing synthesis of cII protein, the regulator of λ repressor synthesis, and blocking synthesis of λ repressor mRNA from P_{RM} (Figure 20.25, part 4b). At the same time, transcription of the right early operon genes from P_R is decreased, but enough Q proteins are accumulated to stimulate transcription (by antitermination) of the late genes for lytic growth (Figure 20.25, part 5b).

642
*Regulation
of Gene
Expression in
Bacteria and
Bacteriophages*

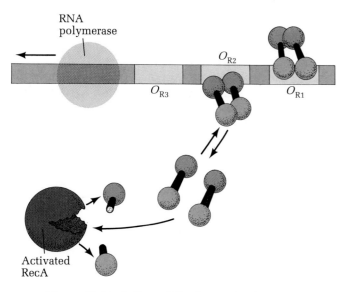

Figure 20.32 Activated RecA protein cleaves repressors which, then, cannot dimerize. As a result, when repressors dissociate from operators they are not replaced, cro *is transcribed, and induction of lytic growth follows.*

The mechanism of action of Cro protein is well understood. Cro is a 66-amino acid protein that is folded into one globular domain; virtually all of the monomeric Cro polypeptides exist in the cell in dimers (Figure 20.33a) and each dimer can bind to the 17-nucleotide pair operator binding sites (Figure 20.33b). Cro dimers bind independently to the three O_R sites, although the order of affinity for the three sites is *exactly opposite* that of the λ repressor dimer; that is, the strongest affinity of Cro protein is for O_{R3} (Figure 20.34a) and after that the affinity is approximately equal for O_{R2} and O_{R1} (Figure 20.34b). Unlike the λ repressor, which has both a positive and negative regulatory action, Cro acts only as a negative regulator. The first Cro synthesized binds to O_{R3}, blocking the ability of RNA polymerase to bind to P_{RM}, thereby stopping further repressor synthesis. At this point, the switch has been reset and the lytic pathway is followed. Through continued P_R function, *cro* transcription continues, along with transcription of the right early operon genes that are needed

for the early stages of the lytic pathway. As with the repressor protein, the Cro protein controls its own transcription. At high Cro protein concentrations, all three O_R binding sites become occupied by Cro protein (Figure 20.34c), blocking transcription initiation at the P_R promoter. This results in decreased transcription of the *cro* gene, and of the other early lytic pathway genes.

In summary, lambda uses complex regulatory systems to direct either the lytic or the lysogenic pathway. The choice is made between the two pathways, using an elaborate genetic switch, most of the details of which are now in hand. This switch involves two regulatory proteins with opposite affinities for three binding sites within each of the two major operators that control the λ life cycle.

Keynote Bacteriophages are especially adapted for undergoing their reproduction within a bacterial host. Many of the genes related to the production of progeny phages, or to the establishment or reversal of lysogeny in temperate phages, are organized into operons. These operons (like bacterial operons) are controlled through the interaction of regulatory proteins with operators that are adjacent to clusters of structural genes.

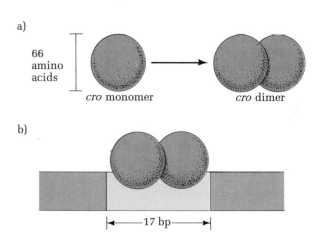

Figure 20.33 (a) λ *Cro protein. This globular polypeptide is found in the cell as dimers. (b) Cro dimer bound to a 17-bp* O_R *site.*

643
*Analytical
Approaches
for Solving
Genetics
Problems*

cro concentration

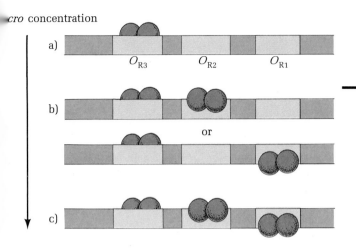

Figure 20.34 *Cro protein binds to the same three sites in the right operator* (O_R) *as λ repressor, but with different affinities for the three sites. (a) The affinity of Cro protein for* O_{R3} *is about 10 times higher than for* O_{R2} *or* O_{R1}. *(b) Once a Cro protein is bound to* O_{R3}, *the second dimer binds with equal chance to* O_{R2} *or* O_{R1}. *(c) At high Cro dimer concentration, all three sites are occupied.*

*Analytical
Approaches
for Solving
Genetics
Problems*

Q.1 In the laboratory you are given ten strains of *E. coli* with the following lactose operon gentoypes:

(1) $lacI^+$ $lacP^+$ $lacO^+$ $lacZ^+$;
(2) $lacI^-$ $lacP^+$ $lacO^+$ $lacZ^+$;
(3) $lacI^+$ $lacP^+$ $lacO^c$ $lacZ^+$;
(4) $lacI^-$ $lacP^+$ $lacO^c$ $lacZ^+$;
(5) $lacI^+$ $lacP^+$ $lacO^c$ $lacZ^-$;
(6) F $lacI^+$ $lacP^+$ $lacO^c$ $lacZ^-$/$lacI^+$ $lacP^+$ $lacO^+$ $lacZ^+$;
(7) F $lacI^+$ $lacO^+$ $lacZ^-$/$lacI^+$ $lacP^+$ $lacO^c$ $lacZ^+$;
(8) F $lacI^+$ $lacP^+$ $lacO^+$ $lacZ^+$/$lacI^-$ $lacP^+$ $lacO^+$ $lacZ^-$;
(9) F $lacI^+$ $lacP^+$ $lacO^c$ $lacZ^-$/$lacI^-$ $lacP^+$ $lacO^+$ $lacZ^+$;
(10) F $lacI^-$ $lacP^+$ $lacO^+$ $lacZ^-$ /$lacI^-$ $lacP^+$ $lacO^c$ $lacZ^+$.

(Note: In the partial-diploid strains (6–10) one copy of the *lac* operon is in the host chromosome and the other copy is in the *F* episome.)

For each strain, predict whether β-galactosidase will be produced (a) if lactose is absent from the growth medium and (b) if lactose is present in the growth medium.

A.1 The answers are as follows, where $+$ = β-galactosidase produced and $-$ = it is not produced:

Genotype	Lactose absent	Lactose present
(1)	−	+
(2)	+	+
(3)	+	+
(4)	+	+
(5)	−	−
(6)	−	+
(7)	+	+
(8)	−	+
(9)	−	+
(10)	+	+

To answer this question completely, we need a good understanding of how the lactose operon is regulated in the wild type and of the consequences of particular mutations on the regulation of the operon.

644
*Regulation
of Gene
Expression in
Bacteria and
Bacteriophages*

Strain (1) is the standard wild-type operon. No enzyme is produced in the absence of lactose since the repressor produced by the *lacI* gene binds to the operator (*lacO*) and blocks the initiation of transcription. When lactose is added, it binds to the repressor, changing its conformation so that it no longer can bind to the *lacO* region, thereby facilitating transcription of the structural genes by RNA polymerase.

Strain (2) is a haploid strain with a mutation in the *lacI* gene (*lacI$^-$*). The consequence is that the repressor protein cannot bind to the (normal) operator region, so there is no inhibition of transcription even in the absence of lactose. This strain, then, is constitutive, meaning that the functional enzyme is produced by the *lacZ$^+$* gene in the presence or absence of lactose.

Strain (3) is the other possible constitutive mutant. In this case the repressor gene is a wild type and the β-galactosidase gene *lacZ$^+$* is a wild type, but ther is a mutation in the operator region (*lacO^c*). Therefore repressor protein cannot bind to the operator, and the transcription occurs in the presence or absence of lactose.

Strain (4) carries both of the regulatory mutations of the previous two strains. Functional repressor is not produced, but even if it were, the operator is changed so that it cannot bind. The consequence is the same: constitutive enzyme production.

Strain (5) produces functional repressor, but the operator (*lacO^c*) is mutated. Therefore transcription cannot be blocked, and lactose polygenic mRNA is produced in the presence or absence of lactose. However, because there is also a mutation in the β-galactosidase gene (*lacZ$^-$*), no functional enzyme is generated.

In the partial-diploid strain (6) one lactose operon is completely wild type, and the other carries a constitutive operator mutation and a mutant β-galactosidase gene. In the absence of lactose no functional enzyme is produced. For the wild-type operon the functional repressor binds to the operator and blocks transcription. For the operon with the two mutations the operator region is mutated and cannot bind repressor, so the mRNA for the mutated operon is produced since transcription cannot be inhibited, and the *lacZ* gene is mutated so that functional enzyme cannot be produced. In the presence of lactose, functional enzyme is produced, since repression of the wild-type operon is relieved so that the *lacZ$^+$* gene can be transcribed. This type of strain provided one of the pieces of evidence that the operator region does not produce a diffusible substance.

In diploid (7) functional enzyme will be produced in the presence or absence of lactose, because one of the operons has an *lacO^c* mutation that does not respond to a repressor and that is linked to a wild-type *lacZ$^+$* gene. That operon is transcribed constitutively. The other operon is inducible, but because there is a *lacZ$^-$* mutation, no functional enzyme is produced.

Diploid (8) has a wild-type operon and an operon with an *lacI$^-$* regulatory mutation and a *lacZ$^-$* mutation. The *lacI$^+$* gene product is diffusible so that it can bind to the *lacO$^+$* region of both operons, thereby putting both operons under inducer control. This strain gives a classic demonstration that the *lacI$^+$* gene is trans-dominant to an *lacI$^-$* mutation. In this case the particular location of the one *lacZ$^-$* mutation is irrelevant: The same result would have been obtained had the *lacZ*⁺ and *lacZ*⁻ been switched between the two operons. In this diploid β-galactosidase is not produced unless lactose is present.

In strain (9) β-galactosidase is produced only when the lactose is present, because the *lacO*ᶜ region controls only those genes that are adjacent to it on the same chromosome (cis dominance), and in this case one of the adjacent genes is *lacZ*⁻, which codes for a nonfunctional enzyme. The diploid is heterozygous *lacI*⁺/*lacI*⁻, but *lacI*⁺ is trans-dominant, as discussed for strain (8). Thus the only normal *lacZ*⁺ gene is under inducer control.

Diploid (10) has defective repressor protein as well as an *lacO*ᶜ mutation adjacent to a *lacZ*⁺ gene. On the latter ground alone this diploid is constitutive. The other operon is also constitutively transcribed, but because there is a *lacZ*⁻ mutation, no functional enzyme is generated from it.

Questions
and
Problems

20.1 How does lactose bring about the induction of synthesis of β-galactosidase, permease, and transacetylase? Why does this event not occur when glucose is also in the medium?

20.2 Operons produce polygenic mRNA when they are active. What is a polygenic mRNA? What advantages, if any, do they confer on a cell in terms of its function?

*20.3 If an *E. coli* mutant strain synthesizes β-galactosidase whether or not the inducer is present, what genetic defect(s) might be responsible for this phenotype?

*20.4 Distinguish the effects you would expect from (a) a missense mutation and (b) a nonsense mutation in the *lacZ* (β-galactosidase) gene of the *lac* operon.

20.5 The elucidation of the regulatory mechanisms associated with the enzymes of lactose utilization in *E. coli* was a landmark in our understanding of regulatory processes in microorganisms. In formulating the operon hypothesis as applied to the lactose system, Jacob and Monod found that results from particular partial-diploid strains were invaluable. Specifically, in terms of the operon hypothesis, what information did the partial diploids provide that the haploids could not?

*20.6 For the *E. coli lac* operon, write the partial-diploid genotype for a strain that will produce β-galactosidase constitutively and permease by induction.

20.7 Mutants were instrumental in the elaboration of the model for the regulation of the lactose operon.

a. Discuss why *lacO*ᶜ mutants are cis-dominant but not trans-dominant.
b. Explain why *lacI*ˢ mutants are trans-dominant to the wild-type *lacI*⁺ allele but *lacI*⁻ mutants are recessive.
c. Discuss the consequences of mutations in the repressor gene promoter as compared with mutations in the structural gene promoter.

646
*Regulation
of Gene
Expression in
Bacteria and
Bacteriophages*

*20.8 This question involves the lactose operon of *E. coli.* Complete the following table, using + to indicate if the enzyme in question will be synthesized and − to indicate if the enzyme will *not* be synthesized.

Genotype	Inducer absent		Inducer present	
	β-galactosidase	Permease	β-galactosidase	Permease
a. $lacI^+ lacP^+ lacO^+ lacZ^+ lacY^+$				
b. $lacI^+ lacP^+ lacO^+ lacZ^- lacY^+$				
c. $lacI^+ lacP^+ lacO^+ lacZ^+ lacY^-$				
d. $lacI^- lacP^+ lacO^+ lacZ^+ lacY^+$				
e. $lacI^s lacP^+ lacO^+ lacZ^+ lacY^+$				
f. $lacI^+ lacP^+ lacO^c lacZ^+ lacY^+$				
g. $lacI^s lacP^+ lacO^c lacZ^+ lacY^+$				
h. $lacI^+ lacP^+ lacO^c lacZ^+ lacY^-$				
i. $lacI^{-d} lacP^+ lacO^+ lacZ^+ lacY^+$				
j. $lacI^- lacP^+ lacO^+ lacZ^+ lacY^+/$ $lacI^+ lacP^+ lacO^+ lacZ^- lacY^-$				
k. $lacI^s lacP^+ lacO^+ lacZ^+ lacY^-/$ $lacI^+ lacP^+ lacO^+ lacZ^- lacY^+$				
l. $lacI^s lacP^+ lacO^+ lacZ^+ lacY^-/$ $lacI^+ lacP^+ lacO^+ lacZ^- lacY^+$				
m. $lacI^+ lacP^+ lacO^c lacZ^- lacY^+/$ $lacI^+ lacP^+ lacO^+ lacZ^+ lacY^-$				
n. $lacI^- lacP^+ lacO^c lacZ^+ lacY^-/$ $lacI^+ lacP^+ lacO^+ lacZ^- lacY^+$				
o. $lacI^s lacP^+ lacO^+ lacZ^+ lacY^+/$ $lacI^+ lacP^+ lacO^c lacZ^+ lacY^+$				
p. $lacI^{-d} lacP^+ lacO^+ lacZ^+ lacY^-/$ $lacI^+ lacP^+ lacO^+ lacZ^- lacY^+$				
q. $lacI^+ lacP^- lacO^c lacZ^+ lacY^-/$ $lacI^+ lacP^+ lacO^+ lacZ^- lacY^+$				
r. $lacI^+ lacP^- lacO^+ lacZ^+ lacY^-/$ $lacI^+ lacP^+ lacO^c lacZ^- lacY^+$				
s. $lacI^- lacP^- lacO^+ lacZ^+ lacY^+/$ $lacI^+ lacP^+ lacO^+ lacZ^- lacY^-$				
t. $lacI^- lacP^+ lacO^+ lacZ^+ lacY^-/$ $lacI^+ lacP^- lacO^+ lacZ^- lacY^+$				

20.9 A new sugar, sugarose, induces the synthesis of two enzymes from the *sug* operon of *E. coli.* Some properties of deletion mutations affecting the appearance of these enzymes are as follows (+ = enzyme induced normally, i.e., synthesized only in the presence of the inducer; C = enzyme synthesized constitutively; 0 = enzyme cannot be detected):

Mutation of	Enzyme 1	Enzyme 2
Gene *A*	+	0
Gene *B*	0	+
Gene *C*	0	0
Gene *D*	C	C

a. The genes are adjacent in the order *ABCD*. Which gene is most likely to be the structural gene for enzyme 1?
b. Complementation studies using partial-diploid (*F'*) strains were made. The episome (*F'*) and chromosome each carried one set of *sug* genes. The results were as follows (symbols are the same as in previous table):

Genotype of *F'*	Chromosome	Enzyme 1	Enzyme 2
$A^+ B^- C^+ D^+$	$A^- B^+ C^+ D^+$	+	+
$A^+ B^- C^- D^+$	$A^- B^+ C^+ D^+$	+	0
$A^- B^+ C^- D^+$	$A^+ B^- C^+ D^+$	0	+
$A^- B^+ C^+ D^+$	$A^+ B^- C^+ D^-$	+	+

From all the evidence given, determine whether the following statements are true or false:

1. It is possible that gene *D* is a structural gene for one of the two enzymes.
2. It is possible that gene *D* produces a repressor.
3. It is possible that gene *D* produces a cytoplasmic product required to induce genes *A* and *B.*
4. It is possible that gene *D* is an operator locus for the *sug* operon.
5. The evidence is also consistent with the possibility that gene *C* could be a gene that produces a cytoplasmic product *required* to induce genes *A* and *B.*
6. The evidence is also consistent with the possibility that gene *C* could be the controlling end of the *sug* operon (end from which mRNA synthesis presumably commences).

20.10 Four different polar mutations, *1, 2, 3,* and *4,* in the *lacZ* gene of the lactose operon were isolated following mutagenesis of *E. coli.* Each caused total loss of β-galactosidase activity. Two revertant mutants, due to suppressor mutation in genes unlinked to the *lac* operon, were isolated from each of the four strains: Suppressor mutations of polar mutation *1* or *1A* and *1B;* those of polar mutation *2* or *2A* and *2B;* and so on. Each of the eight suppressor

648
*Regulation
of Gene
Expression in
Bacteria and
Bacteriophages*

mutations was then tested, by appropriate crosses, for its ability to suppress each of the four polar mutations; the test involved examining the ability of a strain carrying the polar mutation and the suppressor mutation to grow with lactose as the sole carbon source. The results follow (+ = growth on lactose and − = no growth):

Polar Mutation	Suppressor Mutation							
	1A	1B	2A	2B	3A	3B	4A	4B
1	+	+	+	+	+	+	+	+
2	+	−	+	+	+	+	−	−
3	+	−	+	−	+	+	−	−
4	+	+	+	+	+	+	+	+

A mutation to a UAG codon is called an amber nonsense mutation, and a mutation to a UAA codon is called an ochre nonsense mutation. Suppressor mutations allowing reading of UAG and UAA are called amber and ochre suppressors, respectively.

a. Which of the polar mutations are probably amber? Which are probably ochre?
b. Which of the suppressor mutations are probably amber suppressors? Which are probably ochre suppressors?
c. How would you explain the anomalous failure of suppressor *2B* to permit growth with polar mutation *3*? How could you test your explanation most easily?
d. Explain precisely why ochre suppressors suppress amber mutants but amber suppressors do not suppress ochre mutants.

*20.11 What consequences would a mutation in the catabolite activator protein (CAP) gene of *E. coli* have for the expression of a wild-type *lac* operon?

20.12 The lactose operon is an inducible operon, whereas the tryptophan operon is a repressible operon. Discuss the differences between these two types of operons.

20.13 In the bacterium *Salmonella typhimurium* seven of the genes coding for histidine biosynthetic enzymes are located adjacent to one another in the chromosome. If excess histidine is present in the medium, the synthesis in all seven enzymes is coordinately repressed, whereas in the absence of histidine all seven genes are coordinately expressed. Most mutations in this region of the chromosome result in the loss of activity of only one of the enzymes. However, mutations mapping to one end of the gene cluster result in the loss of all seven enzymes, even though none of the structural genes have been lost. What is the counterpart of these mutations in the *lac* operon system?

*20.14 What is stringent control, and how does this regulatory system work?

20.15 Bacteriophage λ, upon infecting an *E. coli* cell, has a choice between the lytic and lysogenic pathways. Discuss the molecular events that determine which pathway is taken.

20.16 How do the lambda repressor protein and the cro protein regulate their own synthesis?

*20.17 If a mutation was induced in the cI gene of phage lambda such that the resulting cI gene product was nonfunctional, what phenotype would you expect the phage to exhibit?

20.18 Bacteriophage λ can form a stable association with the bacterial chromosome because the virus manufactures a repressor. This repressor prevents the virus from replicating its DNA, making lysozyme and all the other tools used to destroy the bacterium. When you induce the virus with UV light, you destroy the repressor, and the virus goes through its normal lytic cycle. This repressor is the product of a gene called the cI gene and is a part of the wild-type viral genome. A bacterium that is lysogenic for λ^+ is full of repressor substance, which confirms immunity against any λ virus added to these bacteria. These added viruses can inject their DNA, but the repressor from the resident virus prevents replication, presumably by binding to an operator on the incoming virus. Thus this system has many analogous elements to the lactose operon. We could diagram a virus as shown in the figure. Several mutations of the cI gene are known. The c_i mutation results in an inactive repressor.

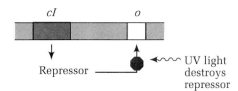

a. If you mix λ containing a c_i mutation, can it lysogenize (form a stable association with the bacterial chromosome)? Why?
b. If you infect a bacterium simultaneously with a wild-type c^+ and a c_i mutant of λ, can you obtain stable lysogeny? Why?
c. Another class of mutants called c^{IN} makes a repressor that is insensitive to UV destruction. Will you be able to induce a bacterium lysogenic for c^{IN} with UV light? Why?

21

Regulation of Gene Expression and Development in Eukaryotes

In prokaryotes, gene expression is commonly regulated in a unit of structural genes and adjacent controlling sites (promoter and operator) collectively called an operon. The molecular mechanisms in operon function provide a relatively simple means of controlling the coordinate synthesis of proteins with related functions. The discovery of operons in prokaryotes stimulated searches for similar regulatory systems in eukaryotes. To date, no operons have been found in eukaryotes, although some gene regulation systems in lower eukaryotes do have similar features. Gene regulation in eukaryotes must occur by means other than the use of operons. In this chapter we look at some of the well-studied regulatory systems in eukaryotes.

Eukaryotic gene regulation falls into two categories. Short-term regulation involves regulatory events in which gene sets are quickly turned on or off in response to changes in environmental or physiological conditions in the cell's or organism's environment. Short-term regulatory systems show parallels with regulatory events in bacteria and their viruses as exemplified by operon regulation. Long-term gene regulation involves regulatory events other than those required for rapid adjustment to local environmental or physiological changes; that is, those events that are required for an organism to develop and differentiate.

Levels of Control of Gene Expression in Eukaryotes

Prokaryotic organisms are unicellular and free-living. As we discussed in Chapter 20, the control of gene expression in prokaryotes occurs primarily at the transcriptional level. This is accomplished through the interaction of regulatory proteins with upstream regulatory DNA sequences. A rapid transition between protein synthesis and no protein synthesis is achieved by both switching off gene transcription and rapid degradation of the mRNA molecules involved. It has also been shown that some control of gene expression occurs at the translational level.

Eukaryotes fall into two major groups: unicellular organisms and multicellular organisms. In both of these eukaryote groups the control of gene expression is often more complicated than is the case in prokaryotes. Figure 21.1 diagrams six different levels at which gene expression can be controlled in eukaryotes. The first level of control is **transcriptional control,** which involves the regulation of whether or not a gene is to be transcribed and of the rate at which transcripts are produced. The former involves regulatory systems that control the accessibility of a promoter for RNA polymerase. The rate of transcript production is a function of the rate at which RNA polymerase molecules can be loaded onto the promoter to initiate transcription. This process is typically controlled by the interaction of regulatory proteins with upstream regulatory sequences such as enhancers (see Chapter 11, pp. 369–370).

A second level of control is **processing control** in which the production of mature RNA molecules from precursor-RNA molecules is regulated. As we discussed in Chapter 12, all three major classes of RNA molecules, which include mRNA, tRNA, and rRNA, can be synthesized as precursor molecules. Perhaps the most significant range of control is exhibited by the mRNA class and involves control of processing of mRNA precursor molecules. One possible mechanism for this involves differential processing of the precursor-mRNA molecule by the choice of alternate poly(A) sites (Figure 21.2a) and/or alternate splice sites (Figure 21.2b). Both of these examples of differential processing have been shown to occur in cells. Small nuclear ribonucleoprotein particles (snRNPs) may well play a role in the differential processing process. The other mechanism involves a choice between processing of the primary transcript and discarding the transcript. This second mechanism has yet to be demonstrated in cells and therefore must be considered to be a hypothetical model.

The next level of control is **transport control,** which is the regulation of the number of transcripts that exit the nucleus to the cytoplasm. A profound difference between prokaryotes and eukaryotes is the presence of a nucleus with a surrounding membrane in the latter. This nuclear membrane can be considered to be a signif-

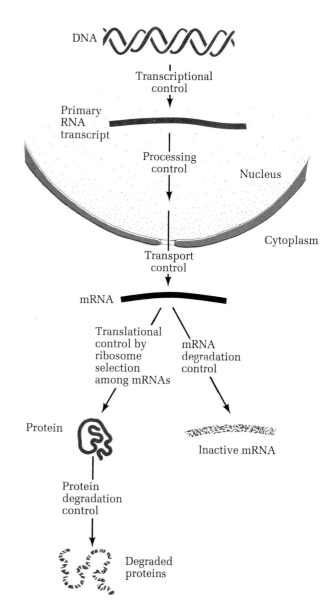

Figure 21.1 Levels at which gene expression can be controlled in eukaryotes.

icant control point in the cell, although the regulatory factors responsible for modulating the flow of RNA molecules across the membrane are not defined.

Once in the cytoplasm, all RNA species are subjected to **degradation control** in which the rate of RNA breakdown (also called RNA turn-

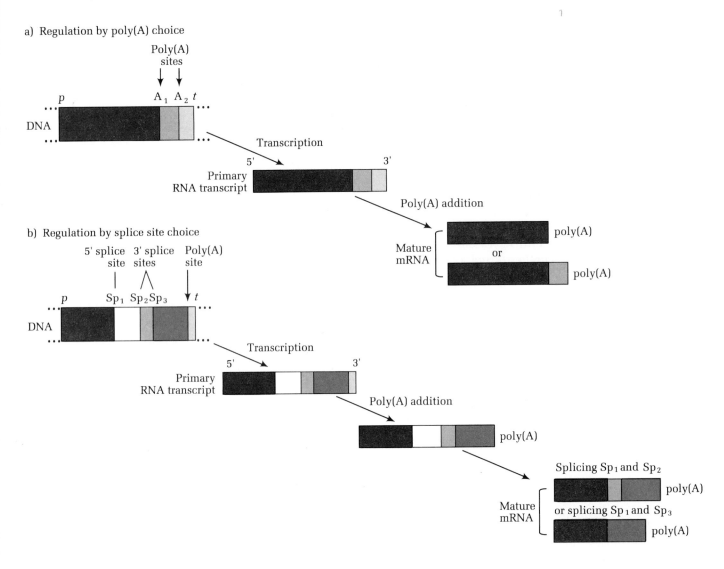

a) Regulation by poly(A) choice

b) Regulation by splice site choice

Figure 21.2 Models for control of mRNA precursor processing in eukaryotes by choice of poly(A) sites (a) and/or splice sites (b). p = transcription initiation site (promoter); t = transcription termination site.

over) is regulated. Usually, both rRNA (in ribosomes) and tRNA are very stable species. By contrast, mRNA molecules exhibit a diverse range of stability, with some mRNA types known to be stable for many months while others degrade within minutes. Examples of changes in the stability of mRNA molecules have been described. For example, the addition of a regulatory molecule to a cell type can lead to an increase in synthesis of a particular protein or proteins. This is accomplished by both an increase in the rate of transcription of the gene(s) involved and/or an increase in the stability of the mRNAs produced. Table 21.1 presents examples of systems in which changes in mRNA sta-

bility for a number of cell types occur in the presence and absence of specific effector molecules.

Messenger RNA molecules are subject to **translational control** by ribosome selection among mRNAs. Thus, differential translation can greatly affect gene expression. For example, mRNAs are stored in many vertebrate and invertebrate unfertilized eggs. In the unfertilized state, the rate of protein synthesis is very slow; however, protein synthesis increases significantly after fertilization. Since this increase occurs without new mRNA synthesis, it can be concluded that translational control is responsible. Lastly, the resulting proteins are subject to **protein degradation control** in which the rate of protein degradation is regulated.

Keynote In prokaryotes control of gene expression occurs mainly at the transcriptional level in association with rapid turnover of mRNA molecules. In eukaryotes, gene *expression is regulated at a number of distinct levels. That is, regulatory systems appear to exist for the control of transcription, of precursor-RNA processing, transport out of the nucleus, degradation of the mature RNA species, translation of the mRNA, and degradation of the protein product.*

Molecular Aspects of Gene Regulation in Eukaryotes

Within the past ten years or so, as a result of recombinant DNA techniques, DNA sequencing, and other modern molecular biology approaches, we have begun to learn about some of the molecular events that occur when genes are turned on and off. In this section we examine the chromosomal changes that are associated with the onset of gene expression and the possible rela-

tionship between methylation of the DNA and gene expression. The molecular processes that are discussed are presumably fundamental to gene expression in eukaryotes and, hence, may be involved in both short-term and long-term regulation of gene expression.

Chromosome Changes and Transcriptional Control

As we learned in Chapter 9, the eukaryotic chromosome consists of DNA complexed with histones and nonhistone proteins. Ultimately, the regulation of gene transcription must involve particular nucleotide sequences of the DNA that are recognized by the appropriate regulatory molecules.

The histones are extremely conserved in the evolutionary sense; their amino acid sequence, structure, and function are all very similar among all eukaryotic organisms. Furthermore,

Table 21.1 Examples of Tissues or Cells in Which Regulation of mRNA Stability Occurs in Response to Specific Effector Molecules[a]

mRNA	Tissue or Cell	Regulatory Signal (= Effector Molecule)	Half-life of mRNA	
			With Effector	Without Effector
Vitellogenin	Liver (frog)	Estrogen	500 h	16 h
Vitellogenin	Liver (rooster)	Estrogen	~24 h	<3 h
Apo-very low density lipoprotein (apoVLDL)	Liver (rooster)	Estrogen	~20–24 h	<3 h
Ovalbumin, conalbumin	Oviduct (hen)	Estrogen, progesterone	>24 h	2–5 h
Casein	Mammary gland (rat)	Prolactin	92 h	5 h
Prostatic steroid-binding protein	Prostate (rat)	Androgen	Increases 30 ×	

[a] Note that the effector molecule in each case results in an increase in transcription as well as stabilization of the mRNA.

the five different types of histones are arranged in an essentially constant way along the DNA from cell to cell in an organism. Clearly, this uniform distribution of histones along the DNA gives nowhere near the specificity a system needs for controlling the expression of thousands of genes. Nonetheless, evidence suggests that histones do become modified, primarily through acetylation and phosphorylation, thereby altering their ability to bind to the DNA. If the gene that they cover is to be expressed, the histones must be modified to loosen their grip on the DNA so that the DNA strands can interact with RNA polymerase or regulatory proteins. In essence, then, geneticists believe that histones act primarily as negative controlling elements in eukaryotic gene regulation—that is, they are gene repressors.

If histones are not the molecules responsible for turning specific genes on, then this process must be controlled by the nonhistones, which not only are more numerous (there are hundreds of different types) but also vary significantly in number and distribution from one cell to another. Since it is to be expected that cells with different functions will have different arrays of genes being expressed, the variations in nonhistones from cell to cell strongly suggest that nonhistones play an important role in controlling which genes are expressed and therefore in controlling cell differentiation and organism development. Evidence for a specific regulatory role of nonhistones has come from chromatin reconstitution experiments. Chromatin from a cell may be dissociated into its component DNA, histones, and nonhistones, and it may be reconstituted, a process that makes it possible to "mix and match" chromatin components from cells of different types. If the DNA and histones of transcriptionally inactive cells are mixed with nonhistones of transcriptionally active cells, transcriptional activity will occur.

Evidence indicates that transcriptionally active chromatin is in a much looser configuration than transcriptionally inactive chromatin. Thus the highly compact secondary and tertiary structures of DNA maintained in the nucleosomes unwind to allow the DNA to be accessible to regulatory proteins and to RNA polymerase. Over the past few years a number of studies have revealed significant physical differences between active and inactive genes, involving either the packing of protein along the DNA or the folding and supercoiling of the DNA-protein structure.

DNase sensitivity and gene expression. The work of several researchers has shown unequivocally that when a gene becomes transcriptionally active, the region of the chromosome containing the gene becomes much more sensitive to digestion by DNase I (an endonuclease) than transcriptionally inactive regions. The interpretation for this general result is that the DNA-protein structure of the chromosome becomes loosened in regions where genes are active so that DNase I can access the DNA and degrade it.

One well-studied system involves globin and ovalbumin synthesis in chickens. In the nuclei of **erythroblasts** (red blood cell precursors) from chick embryos, the genes for globin are susceptible to digestion by DNase I whereas the sequences encoding ovalbumin, which is not produced in erythroblasts, are resistant to DNase digestion. Conversely, in hen oviduct nuclei in which ovalbumin mRNA is actively transcribed but globin mRNA is not produced, the ovalbumin gene sequences are much more sensitive to DNase digestion than are the globin gene sequences.

The experimental approach used to demonstrate changes in DNase I sensitivity of DNA sequences involves the use of specific cloned genes as probes and the Southern blotting technique (see Chapter 15). Figure 21.3 shows how this approach was used for the chicken globin gene and ovalbumin gene example just described. Chromatin was extracted from erythroblast nuclei and, as a control, from cells growing in culture that were not making globin, and samples were treated with different concentrations of DNase I. DNA was then extracted from each chromatin sample and digested with the restriction enzyme *Bam*HI (see Chapter 15), which cuts the DNA at sites on either side of the globin gene sequence. The two *Bam*HI sites are 4.5 kbp

a) DNA from 14-day erythroblasts making globin and probed with the β-globin gene.

DNase (μg/ml)

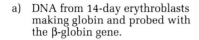

0 0.01 0.05 0.1 0.5 1.0 1.5

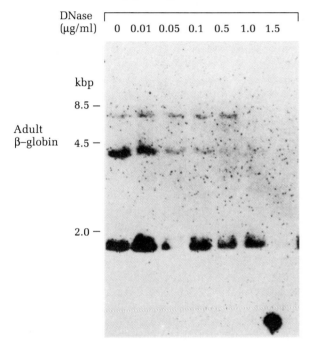

kbp

8.5 —

Adult
β–globin

4.5 —

2.0 —

b) DNA from 14-day erythroblasts making globin and probed with the ovalbumin gene.

DNase (μg/ml)

0 0.01 0.05 0.1 0.5 1.0 1.5

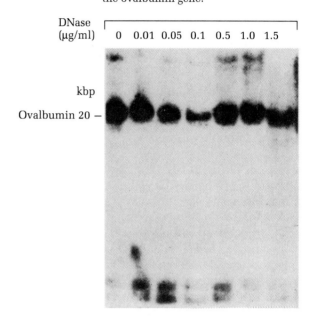

kbp

Ovalbumin 20 —

Figure 21.3 Relationship of gene expression and sensitivity to DNase. (a) In erythroblasts, cells making globin, globin DNA is sensitive to DNase I. (b) In erythroblasts, which do not make ovalbumin, ovalbumin DNA is resistant to DNase.

apart on the chromosomal DNA. The resulting DNA fragments were separated by agarose gel electrophoresis and transferred to nitrocellulose filters using the Southern blotting technique. The filters were hybridized with a radioactive globin DNA probe to determine the sizes of the globin gene DNA fragments that were present following DNase digestion.

Analysis of the results showed that for the chromatin extracted from erythroblasts, the globin DNA sequence was sensitive to DNase digestion as evidenced by the gradual disappearance of the 4.5-kbp *Bam*HI-*Bam*HI DNA fragments containing the globin gene with increasing DNase concentration (Figure 21.3a). That is, the chromatin organization was loosened suffi-

ciently in the globin gene region such that the globin DNA sequence was accessible to DNase I. By contrast, when chromatin isolated from control cells not making globin was analyzed in the same way, the globin gene was not sensitive to DNase-catalyzed degradation, as evidenced by the presence of the 4.5-kbp DNA fragment at all DNase concentrations (data not shown). Moreover, analysis of the ovalbumin gene (which is also flanked on either side by a *Bam*HI site so that cutting with *Bam*HI produces a 20-kbp DNA fragment) in chromatin from erythroblasts using a cloned ovalbumin gene probe showed that the ovalbumin gene was not sensitive to digestion by DNase I, as evidenced by the presence of the 20-kbp *Bam*HI-*Bam*HI DNA fragment at all DNase I concentrations (Figure 21.3b). The latter result is explained by the fact that ovalbumin is not made in erythroblasts, so that region of the genome has not become loosened.

Many experiments of this sort lead us to conclude that almost all transcriptionally active genes have an increased DNase I sensitivity. The extent of the DNase sensitive region varies with the gene in question, ranging from a few kilobase pairs around the gene to as much as 20 kbp of flanking DNA.

DNase hypersensitive sites. Careful analysis of the regions of DNA around transcriptionally active genes has shown that certain sites, called **hypersensitive sites** or **hypersensitive regions** are even more highly sensitive to digestion by DNase I. These sites or regions are typically the first to be cut with DNase I. Most, but not all, DNase-hypersensitive sites are in the regions upstream from the start of transcription, probably corresponding to the DNA sequences where RNA polymerase and other gene regulatory proteins bind. In fact, it was predicted that such sequences would be hypersensitive to DNase I digestion on the grounds that they would have the most loosened form of chromatin so that they would be readily accessible to polymerase and to regulatory proteins.

Keynote *Transcriptionally active chromatin is characterized by a sensitivity to digestion with DNase I as a result of the loosened DNA-protein structure in those areas. By contrast, transcriptionally inactive chromatin is resistant to DNase I digestion. For transcriptionally active genes, certain sites, probably corresponding to the binding sites for RNA polymerase and regulatory proteins, are especially sensitive to DNase I digestion. These sites are called hypersensitive sites.*

DNA Methylation and Transcriptional Control

Once DNA has been synthesized within the eukaryotic cell, a small proportion of bases become chemically modified through the action of enzymes. Potentially, modified bases in DNA could act as signals in the processes in which DNA is involved, such as DNA replication, DNA repair, and DNA expression, although in most cases we have no information about this activity. Recently, a particular modified base has received significant attention since it has been correlated in a number of cases with gene activity. This base, *5-methylcytosine* (5mC), is generated by the modification of DNA shortly after replication by the action of the enzyme DNA methylase (Figure 21.4). (The DNA of some prokaryotes also contains 5mC, though there is no direct role known in the regulation of gene expression.)

The 5mC content of DNA can be measured by a variety of techniques; their discussion is beyond the scope of this book. The percentage of 5mC in eukaryotic DNA varies over a wide range. In mammalian DNA, for example, about 3 percent of cytosines are present as 5mC. The DNAs of lower eukaryotes, such as *Drosophila*,

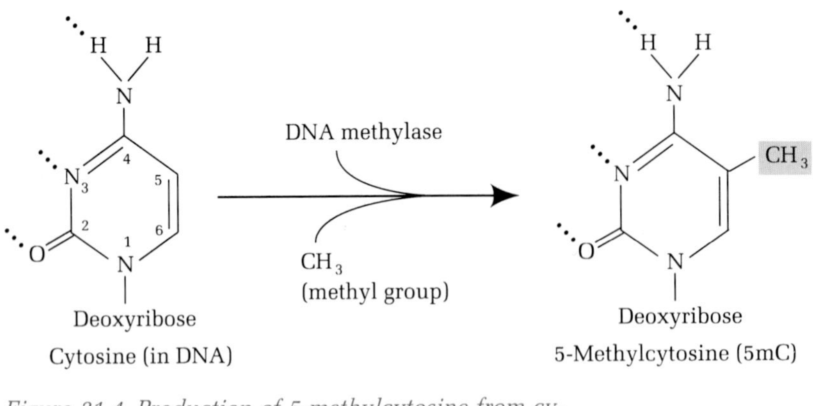

Figure 21.4 Production of 5-methylcytosine from cytosine in DNA by the action of the enzyme DNA methylase.

Tetrahymena, and yeast, appear to contain very little, if any, 5mC. There is no clear relationship, however, between the amount of 5mC and the complexity of an organism, evolutionarily speaking, since the fungus *Neurospora crassa* has significant amounts of 5mC in its genome.

The distribution of 5mC is nonrandom, with most of the 5mC (greater than 90 percent in mammalian DNA) being found in the sequence CG. Since this sequence has twofold rotational symmetry in DNA, the cytosines are symmetrically located on opposite strands of the DNA molecule. These CG sequences also form part of restriction enzyme recognition sites, which allow the use of restriction enzymes to investigate the extent of methylation of a segment of DNA.

Certain restriction enzymes (enzymes that cut within a defined DNA base sequence that exhibits twofold rotational symmetry; see Chapter 15, pp. 482–486) are used as tools for studies of 5mC methylation since many enzymes with cytosine in their recognition sequence fail to cleave the DNA when this base is methylated. The enzyme *Hpa*II, for example, will cleave DNA at the sequence 5′–CCGG–3′ (the opposite strand is complementary to this sequence, of course) but not if the internal cytosine is methylated, that is, if it is 5′–C^{m}CGG–3′. The enzyme *Msp*I also will cleave the CCGG sequence recognized and cleaved by *Hpa*II, but unlike *Hpa*II, it will also cleave the methylated sequence C^{m}CGG. (Conversely, *Hpa*II will cleave DNA in which the external C of its recognition sequence is methylated—that is, ^{m}CCGG —whereas *Msp*I will not. However, this positioning of 5mC in DNA sequences is rare, at least in vertebrate DNA, and hence need not concern us in the following discussion.)

The use of *Hpa*II and *Msp*I in analyzing the extent of methylation of a segment of DNA is illustrated by the following theoretical example. Consider a fragment of genomic DNA that contains three CCGG sequences (Figure 21.5). If the sequences are not methylated (Figure 21.5a), then both *Hpa*II and *Msp*I will cleave the DNA at these sequences to produce four DNA fragments of discrete sizes. These DNA fragments can be identified following electrophoresis of the digested DNA on agarose gels, transferring the fragments to nitrocellulose (Southern blotting), and probing them with a radioactively labeled DNA probe that will hybridize with any and all of the fragments. (These techniques were described in Chapter 15.) In this case DNA fragments of identical sizes will be produced from both digests. If one of the CCGG sequences is methylated—for example, the first one from the left (Figure 21.5b)—then all three sequences will be cleaved by *Msp*I, whereas only the two unmethylated sequences will be cleaved by *Hpa*II. As a result, the array of DNA fragments produced by digestion of the same DNA piece with the two enzymes will be different, as indicated in the stylized blot result. In actual practice, a piece of DNA is digested by the two enzymes *Hpa*II and *Msp*I, and if the results show that the *Msp*I has digested the DNA into more pieces than *Hpa*II, the interpretation is that one of the CCGG sequences is methylated; that is, C^{m}CGG.

One of the first applications of this approach to study DNA methylation involved the rabbit β-globin gene. Restriction enzyme-mapping experiments had located a single *Msp*I site within an intron of the rabbit β-globin gene. Genomic DNA prepared from a number of rabbit tissues was digested by *Hpa*II and by *Msp*I; the fragments were separated by agarose gel electrophoresis and then transferred to nitrocellulose for blotting with a radioactive probe, in this case the cloned β-globin gene. The prediction was that if the CCGG site was unmethylated, then identical DNA fragment patterns would result from *Hpa*II and *Msp*I digestion. Alternatively, if the CCGG was methylated, then one less DNA fragment would be produced following *Hpa*II digestion than would be produced following *Msp*I digestion since *Hpa*II cannot cleave C^{m}CGG. The results of many experiments indicated that there are tissue-specific differences in the degree of methylation of this particular *Hpa*II site in the rabbit β-globin gene. In tissues producing β-globin the *Hpa*II site was not methylated, while in tissues not producing β-globin the HpaII site was methylated.

With the *Hpa*II/*Msp*I approach a large number of systems have been examined in an attempt to

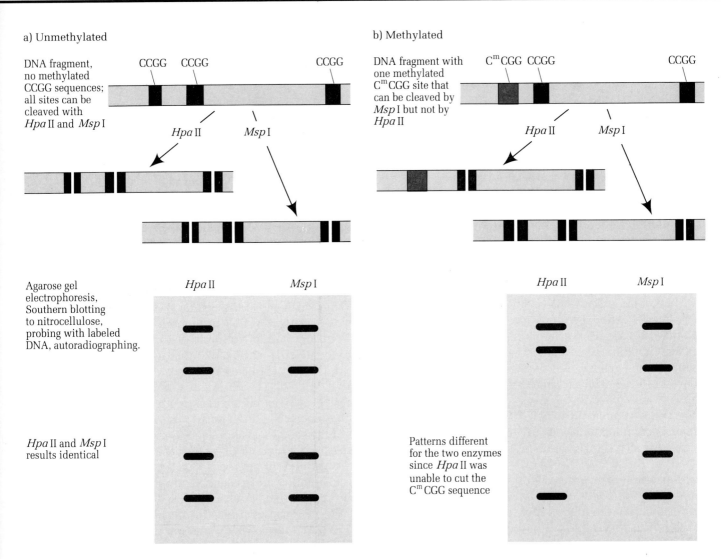

Figure 21.5 Effect of 5-methylcytosine on cleavage of DNA with Hpa II Msp I: (a) unmethylated (CCGG) sequence; (b) methylated (C^mCGG) sequence.

determine whether there is a relationship between gene methylation and transcriptional activity of the gene. For at least thirty genes a negative correlation exists between DNA methylation and transcription; there is a low level of methylated DNA (hypomethylated DNA) in transcriptionally active genes in comparison with transcriptionally inactive genes. Caution should be exercised in taking this broad generalization too far because not all methylated CCGG se-

quences in a gene region become demethylated when the gene is expressed and because, as was stated earlier, not all organisms have been shown to have significant amounts of methylated C in their DNA. Moreover, the fact that a correlation exists between the level of DNA methylation and the transcriptional state of a gene provides no information about cause and effect. In other words, at this time we do not know whether methylation changes are prerequisites for the onset of transcriptional activity (in those organisms in which there is significant DNA methylation) or whether the changes are secondary consequences of the transcriptional activity initiated by other means. Some recent experiments suggest that the level of methylation of a gene does not by itself determine whether the gene will be transcribed.

Given that, at least in some organisms, there is a correlation between hypomethylation and transcriptional activity, we might predict that changes in cellular activities will result if the amount of DNA methylation in a cell can be altered through experimental techniques. Researchers have been able to test this prediction through the use of the base analog, 5-azacytidine (5AzaC) (compared with cytosine and 5mC in Figure 21.6), which has a nitrogen atom instead of a carbon atom at position 5 of the pyrimidine ring. Base 5AzaC can be incorporated into DNA during replication, but unlike cytosine, it cannot be methylated. In at least thirty systems 5AzaC treatment has resulted in the activation of previously inactive genes. In Chapter 17 we discussed the inactivation of one X chromosome in normal human females. Those genes inactivated on the X chromosome have been well established to be very stable in their inexpression. Nevertheless, at least three genes that have been studied in this system have been shown to be activated by 5AzaC treatment. Therefore, it has been concluded that inactivation was the result of methylation of X chromosome sequences.

The ability of 5AzaC treatment to cause hypomethylation of DNA has led to its therapeutic use. Recall that people with sickle-cell anemia have a life-threatening condition. In some cases treatment with 5AzaC in patients who have sickle-cell anemia causes an activation of the γ-globin gene, which codes for a β-globin–related polypeptide found normally in fetal hemoglobin (Hb–F). In these patients the level of Hb–F, which is normally very low in adults (about 2 percent), was increased to about 8 percent by 5AzaC. This population of hemoglobin molecules was sufficient to reduce some of the symptoms of the patients.

In summary, the evidence presented indicates that in at least some systems (notably mammals and many other animals) 5-methylcytosine functions as part of a gene-silencing system. A number of cases have demonstrated that removal of methyl groups is correlated with increases of transcriptional activity. The 5-methylcytosine changes cannot explain all changes in transcriptional activity, however, since a number of organisms that have been examined (notably many invertebrates and some fungi) do not appear to have significant amounts of 5mC in their DNA.

Keynote The DNA of most eukaryotes has been shown to be methylated, with most methylations occurring on cytosine residues. Transcriptionally active genes exhibit lower levels of DNA methylation than transcriptionally inactive genes.

Short-Term Gene Regulation in Eukaryotes

In this section we shall look at some examples of the short-term regulation of gene expression in eukaryotes. This short-term regulation involves rapid responses to changes in environmental or physiological conditions. The responses occur primarily at the transcriptional level. As we examine these systems, keep in mind how short-term regulation of gene expression can occur in bacteria through regulatory units called operons so that you can compare the molecular events involved.

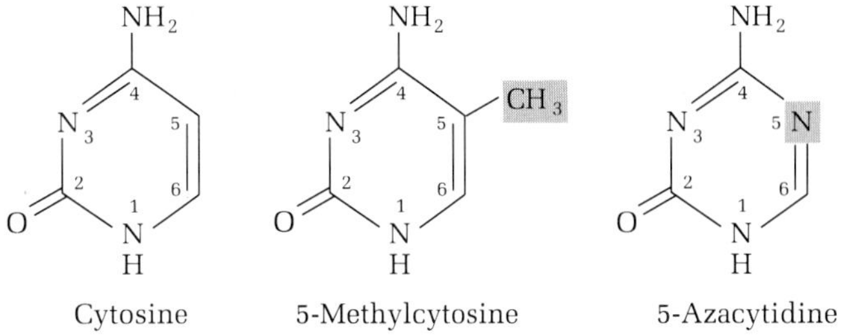

Figure 21.6 Structures of cytosine, 5-methylcytosine, and 5-azacytidine.

Quinic Acid Metabolism Genes in *Neurospora Crassa*

An interesting gene regulation system in *Neurospora crassa* involves the metabolism of quinic acid (QA) as a carbon source. The genes involved in the metabolism of quinic acid to protocatechuic acid (PCA), the *qa* genes, have been shown by N. H. Giles, M. E. Case, and their collaborators to constitute a relatively simple and genetically well-characterized system for studying gene regulation in eukaryotes. The *qa* genes are arranged as a cluster in the *Neurospora* chromosome and consist of five structural genes and two regulatory genes, arranged as shown in Figure 21.7. Transcription of the structural genes can be induced 300- to 1000-fold by the addition of QA to the medium. Thus QA acts as a regulatory effector molecule. Three of the structural genes, *qa–3, qa–2,* and *qa–4,* code for the enzymes that catalyze the first three steps of QA catabolism (breakdown); namely, quinate dehydrogenase, catabolic dehydroquinase, and dehydroshikimate dehydrase, respectively. Note that these three structural genes are not arranged on the chromosome in the same order as the biochemical steps in which their products participate. The structural genes *qa–3, qa–2,* and *qa–4* were identified and mapped through studies of mutations affecting each of the three enzyme activities specified by the genes.

More recently, the region of the *Neurospora* chromosome containing the *qa* genes has been cloned by using recombinant DNA techniques, and it has been sequenced by using rapid DNA-sequencing techniques. With the DNA sequence in hand researchers were able to use a computer to search the sequence or sequences that might be significant for gene expression, including possible promoter and terminator sequences. In

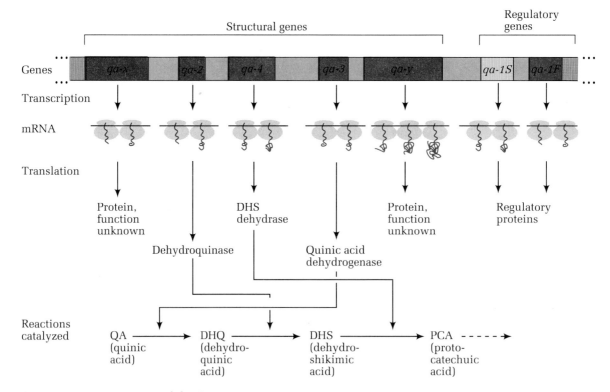

Figure 21.7 Organization of the Neurospora crassa *quinic acid* (qa) *catabolism genes.*

addition, they could look for what are called *open reading frames* (ORFs), segments of the sequence that might code for a protein product. The ORFs are located by searching for an initiation codon (ATG at the DNA level) that is in frame with a chain termination codon. It must be stressed, of course, that the existence of an ORF in a DNA sequence does not necessarily mean that the sequence is expressed in the cell. Two ORFs have been identified in the *qa* region in addition to the known genes; these ORFs have likely promoter sequences upstream. By isolating all the RNAs from cells induced with QA, separating the RNAs by electrophoresis, transferring these single-stranded molecules to a special type of filter paper to which the RNA will bind (a technique similar to Southern blotting but called northern blotting Chapter 15, p. 507), and using a radioactive probe consisting of cloned pieces of DNA corresponding to the ORFs, researchers have shown that both ORFs are transcribed and hence do, in fact, correspond to structural genes. While we know that these two genes (called *qa–x* and *qa–y*) are transcribed, we have no idea of what role the proteins specified by the transcripts have in the catabolism of QA.

Near the structural genes, at the *qa–y* end of the *qa–y, qa–3, qa–4, qa–2, qa–x* cluster, are two regulatory genes *qa–1S* and *qa–1F* (Figure 21.7). The presence of regulatory genes close to a cluster of structural genes is reminiscent of prokaryotic operons, and for a time this system was studied intensely to determine whether it was a eukaryotic operon. We now know, however, that the *qa* system is not an operon, and evidence also indicates that each structural gene specifies a distinct mRNA transcript. Mutations of the regulatory genes have given us insights into how the *qa* gene cluster is regulated. The regulatory mutants have been examined in the haploid condition (recall that *Neurospora* is a haploid organism throughout most of its life cycle) and in heterokaryons (systems in which two genetically different strains are fused so that their nuclei coexist, but do not fuse, in a common cytoplasm). The heterokaryons allowed Giles and his colleagues to investigate the functions of the

qa–1S and *qa–1F* genes in much the same way as partial diploids allowed Jacob and Monod to determine the nature of the regulatory mutants affecting the *lac* operon of *E. coli.*

Giles found that *qa–1S* mutants were either constitutive for quinic acid metabolism and recessive (*qa–1Sc*) or noninducible and semidominant (*qa–1S⁻*), and *qa–1F* mutants (*qa–1F⁻*) were noninducible for quinic acid metabolism and recessive even in the presence of a *qa–1Sc* mutation. Both *qa–1S* and *qa–1F* are known to produce an mRNA transcript, and *qa–1F* mRNA is induced over a 50-fold range by QA, this induction requiring the presence of wild-type *qa–1F* and *qa–1S* genes. The current model for the role of *qa–1* regulatory genes in the regulation of the *qa* genes is as follows: *qa–1F* codes for a protein that has an activator (positive) function needed both for transcription of its own mRNA and for transcription of all the structural genes except *qa–x*. Gene *qa–1S*, on the other hand, codes for a protein that has a repressor (negative) function. If QA is absent, the repressor protein apparently interacts with QA, the effector molecule, and blocks transcription of *qa–1F*. Transcription of the *qa–x* structural gene seems to be controlled mainly by *qa–1S* and to a much lesser extent by *qa–1F*. Current work is focusing on precisely defining the mechanism by which the regulatory gene products function. Some evidence has been accumulated, for example, that each of the four structural genes under *qa–1F* control is transcribed from two to four *qa–1F*-dependent promoters.

In conclusion, the *qa* gene cluster is a true gene cluster. It contains five structural genes that are inducible and whose products are synthesized coordinately, thus showing parallels with some bacterial operons. The regulatory genes for the cluster are immediately adjacent to the structural genes and code for an activator (*qa–1F* product) and a repressor protein (*qa–1S* product). Unlike bacterial operons, this cluster gives no evidence for operator-like controlling regions, and so it is not an operon. It is, however, another example of a well-characterized regulatory system in a lower eukaryote that functions to maintain cellular levels of particular chemicals below a certain level. It is an example of a short-term gene regulation system analogous to those used by bacteria to adjust to a rapidly changing physiological environment.

Keynote *No operons have been found in eukaryotes although the genes for related functions in some biochemical pathways are sometimes closely linked or contiguous. More often, genes that are coordinately regulated are dispersed throughout the eukaryotic genome.*

Steroid Hormone Regulation of Gene Expression

Short-term regulation enables an organism or cell to adapt rapidly to changes in its physiological environment so that it can function as optimally as possible. While lower eukaryotes exhibit some cell specialization, higher eukaryotes are differentiated into a number of cell types, each of which carries out a specialized function or functions. The cells of higher eukaryotes are not exposed to rapid changes in environment as are cells of bacteria and of microbial eukaryotes. The environment to which most cells of higher eukaryotes are exposed, blood, is relatively constant in the nutrients, ions, and other important molecules it supplies. Blood, in a sense, is analogous to a bacterium's growth medium. The constancy of the blood environment is, in turn, maintained through the action of chemicals called hormones, which are secreted by various cells in response to signals and which circulate in the blood until they stimulate their target cells. Elaborate feedback loops control the amount of hormone secreted and hence modulate the response so that appropriate levels of chemicals in the blood and tissues are maintained.

A hormone, then, is an effector molecule that is produced in low concentrations by one cell and that causes a physiological response in another cell. As shown in Figure 21.8, some hormones exert their action by binding directly to the cell's genome and causing changes in gene expression. Other hormones act at the cell sur-

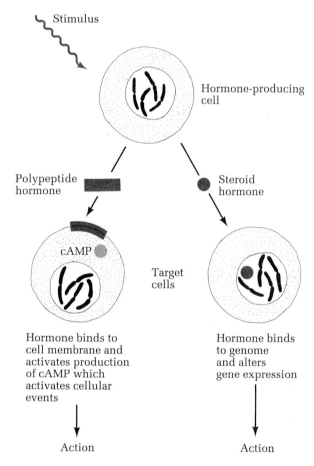

Figure 21.8 Mechanisms of action of polypeptide hormones and steroid hormones.

face to activate a membrane-bound enzyme, adenyl cyclase, which produces cyclic AMP (cAMP) from ATP (see Chapter 20, p. 620). The cAMP acts as a second messenger molecule to activate the cellular events normally associated with the release of the hormone involved. The exact molecular mechanisms of these reactions are not known.

A variety of types of molecules can act as hormones, such as polypeptides, fatty acid derivatives, and steroids. Since the action of steroid hormones on gene expression has been well studied, we shall turn to steroid hormones here. Steroid hormones have been shown to be important to the development and physiological regu-

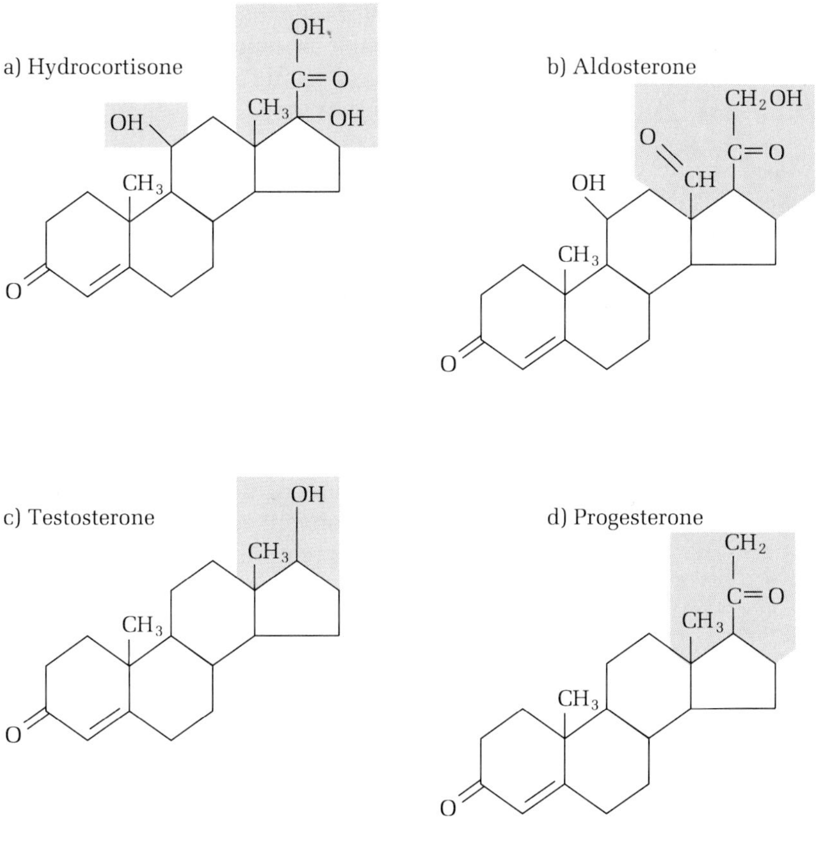

Figure 21.9 Structures of some mammalian steroid hormones: (a) hydrocortisone, which helps regulate carbohydrate and protein metabolism; (b) aldosterone, which regulates salt and water balance; (c) testosterone, which is used for the production and maintenance of male sexual characteristics; (d) progesterone, which, with estrogen, prepares and maintains the uterine lining for embryo implantation.

lation of organisms ranging from fungi to humans. Figure 21.9 gives the structures of four of the most common mammalian steroid hormones. All have a common four-ring structure; the differences in the side groups are responsible for their different physiological effects.

The key to hormone action is that hormones act on specific target cells that have receptors capable of recognizing and binding that particular hormone. For most of the polypeptide hormones, the receptors are on the cell surface, whereas the receptors for steroid hormones are inside the cell.

Mammalian cells contain between 10,000 and 100,000 steroid receptor molecules. The receptor molecules have a very high affinity for their respective steroid hormones. It is not certain whether free receptor molecules are in the cytoplasm or in the nucleus, although recent data support the latter location. Thus, a steroid hormone enters a cell and, by diffusion, encounters its specific receptor molecule to which it binds to form a steroid-receptor complex (Figure 21.10). This complex binds to specific DNA sequences in the nucleus, resulting in the synthesis of specific mRNA molecules (Figure 21.10). These new mRNAs can appear within minutes after a steroid hormone encounters its target cell, enabling new proteins to be produced rapidly.

These DNA sequences are hormone-dependent transcriptional enhancers that are upstream and/

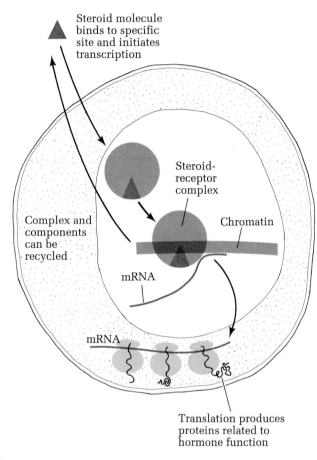

Steroid molecule binds to specific site and initiates transcription

Steroid-receptor complex

Chromatin

Complex and components can be recycled

mRNA

mRNA

Translation produces proteins related to hormone function

Figure 21.10 Model for the action of steroid hormones in vertebrate cells.

exception of receptors for the steroid hormone glucocorticoid, which are widely distributed among tissue types, steroid receptors are found in a limited number of target tissues. Table 21.2 presents examples of the induction of specific proteins by selected steroid hormones. As can be seen, there are tissue-specific effects. Moreover, some hormones, such as estrogen, are known to regulate gene expression in different tissues. While steroid hormones have well-substantiated effects on transcription, steroids also can affect the stability of mRNAs, and possibly the processing of mRNA precursors.

In comparison with the bacterial operons, which have a better-characterized regulatory system, the steroids act as positive effector molecules, and the receptors act as regulatory molecules. When the two combine, the resulting complex activates gene transcription and results in a large (and specific) increase in cellular mRNA levels.

Keynote *In higher eukaryotes one of the well-studied systems of short-term gene regulation is the control of enzyme synthesis by steroid hormones. Some hormones exert their action by binding directly to the cell's genome, while others act at the cell surface, thereby activating a system to produce cAMP, which acts as a second messenger to control gene activity. The specificity of hormone action is caused by the specific array of receptors in the genome for each hormone.*

Hormone Control of Gene Expression in Plants

Many chemicals have been identified that play important roles in controlling growth and development in plants; these chemicals are called plant hormones. Plant hormones fall into five main classes, namely, ethylene gas, abscisic acid, auxins, cytokinins, and gibberellins (Figure 21.11). These extracellular hormones are responsible for many activities. Ethylene gas, for example, induces fruit ripening, flower maturation, and plant senescence, among other things. The molecular actions of plant hormones are generally ill-defined.

or downstream from the genes' promoters. A number of such enhancers may be associated with each transcription unit. Once a gene is activated by a steroid hormone, significant changes in chromatin structure occur. That is, the region of the gene becomes sensitive to DNase I digestion. Discrete DNase I-hypersensitive region persists after the hormone is removed, at which time the DNase-hypersensitive sites disappear.

The specificity of the response to steroid hormone release is controlled by the hormone receptors. Thus, only target tissues that contain receptors for a particular steroid hormone can respond to that hormone, and this directly controls the sites of the hormone's action. With the

Among the best-characterized plant hormones are *gibberellins.* These hormones have actions similar to those of steroid hormones in mammals. That is, gibberellins stimulate transcription and thereby result in an increase in the production of specific proteins. Those proteins are responsible for profound changes in plant cell form and cell differentiation.

Gibberellins have been shown to have many effects when applied to plants, such as making dwarf plants grow tall or making normal plants grow taller. Gibberellins are also important in stimulating germination of some seeds; for example, those of wheat (Figure 21.12). In this process the embryo of the seed produces gibber-

ellins that diffuse to the aleurone layer, the outermost layer of the endosperm. As a result of the effect of the gibberellins on gene expression, and particularly on the gene for the enzyme α-amylase, the aleurone layer synthesizes and secretes α-amylase, which breaks down the endosperm, releasing nutrients that enable the embryo to grow. Once the endosperm has been exhausted, the young, developing plant can rely on photosynthesis to obtain nutrients for continued growth.

Even though the action of some gibberellins has been well described, no gibberellin receptor has been detected to date, and the molecular aspects of their action are poorly understood.

Table 21.2 Examples of Proteins Induced by Steroid Hormones

Hormone	*Induced Protein*	*Tissue*
Estrogen	Conalbumin[a] Lysozyme[a] Ovalbumin[a] Ovomucoid[a]	Oviduct (hen)
	Vitellogenin Apo-very low density lipoprotein (apoVLDL)	Liver (rooster)
	Vitellogenin	Liver (frog)
	Prolactin	Pituitary gland (rat)
Glucocorticoids	Tyrosine aminotransferase Tryptophan oxygenase	Liver (rat)
	Phosphoenolypyruvate carboxykinase	Kidney (rat)
	Growth hormone	Pituitary gland (rat)
Progesterone	Avidin[a] Conalbumin[a] Lysozyme[a] Ovomucoid[a]	Oviduct (hen)
	Uteroglobin	Uterus (rat)
Testosterone	α-2u Globulin	Liver (rat)
	Aldolase	Prostate gland (rat)

[a] Egg white proteins.

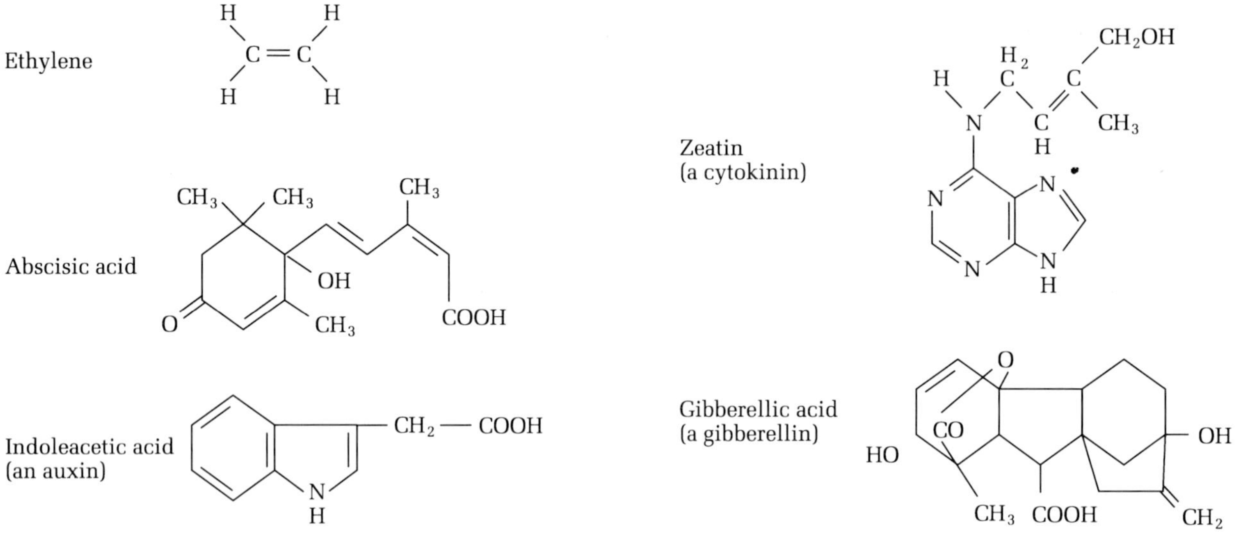

Ethylene

Abscisic acid

Indoleacetic acid
(an auxin)

Zeatin
(a cytokinin)

Gibberellic acid
(a gibberellin)

Figure 21.11 Chemical structures of five plant hormones.

Gene Regulation in Development and Differentiation

Most higher eukaryotes have many different cells, tissues, and organs with specialized functions (e.g., in animals, liver cells and nerve cells perform different functions; in plants, root cells and leaf cells perform different functions), and yet all cells of the same organism have the same genotype. During an organism's growth throughout its life cycle, cells which were genetically identical become physiologically and phenotypically differentiated in terms of their structure and function. Two terms are used in the discussion of long-term gene regulation. **Development** refers to the process of regulated growth that results from the interaction of the genome with cytoplasm and the environment and that involves a programmed sequence of phenotypic events that are typically irreversible; the total of the phenotypic changes constitutes the life cycle of an organism. **Differentiation,** the most spectacular aspect of development, involves the formation of different types of cells, tissues, and organs from a zygote through the processes of specific regulation of gene expression; differentiated cells have characteristic structural and functional properties.

At a very general level we can safely assume that the processes in differentiation and development are the result of a highly programmed pattern of gene activation and gene repression. But while we know a great deal about the molecular events that turn the bacterial *lac* operon on and off, we are a long way from understanding the molecular bases for the differentiation and development events in higher eukaryotes. We can describe these events well at the morphological level, and we can describe them to some extent at the biochemical level. We are only in the early stages of knowing the details of how the complex activation repression patterns are programmed, because of the greater complexity of the processes in comparison with the processes of bacterial operons.

Structural Genes in Higher Eukaryotes

Compared with the prokaryotic genome, the genome of higher eukaryotes is much larger, and as discussed in Chapter 7, it has a large amount of highly repetitive and moderately repetitive DNA. The highly repetitive DNA is localized mainly in centromeric heterochromatin and does not appear to encode structural information. This DNA constitutes perhaps 20 to 40 percent of the genomic DNA of higher eukaryotes. The remaining 60 to 80 percent of the DNA is distributed between the moderately repetitive and unique-sequence DNA.

A number of researchers have done experiments to determine the function of the moderately repetitive and unique-sequence DNA. For instance, in his studies of the sea urchin, E. H. Davidson has isolated total mRNA from sea urchins, and by determining the total percentage of unique-sequence DNA to which this RNA hybridized, he was able to quantify the amount of DNA that encodes structural information. Instead of confining his observations to a specific stage in sea urchin development, where he expected only a subset of the total structural information to be expressed, Davidson extended his hybridizations to mRNAs obtained from populations of sea urchins in all stages of development and from mature sea urchin tissues. The results indicated the percentage of unique-sequence DNA to which RNA hybridized at various stages of development. Maximal transcription was found to occur in the oocyte, with about 6 per-

cent of the unique-sequence DNA hybridized. Transcriptional activity remains high during the blastula, gastrula, and larval stages of development (>2 percent unique-sequence DNA hybridized) and then drops to low levels (<0.8 percent unique-sequence DNA hybridized) in mature tissue such as the intestine and tubefoot. These results mean that at any one time a maximum of about 6 percent of the unique-sequence DNA is transcriptionally active and producing proteins.

Because higher eukaryotes are more complex than prokaryotes, it was expected that a larger proportion of their DNA would encode structural information. It was shown that the majority of the sea urchin's DNA must have other functions. We now know that the structural genes are located within long stretches of DNA that are not transcribed. The function of the large amount of nontranscribed DNA in higher eukaryotes (less is found in lower eukaryotes) has long been controversial. It could be "junk" DNA left over from evolutionary changes, or it could

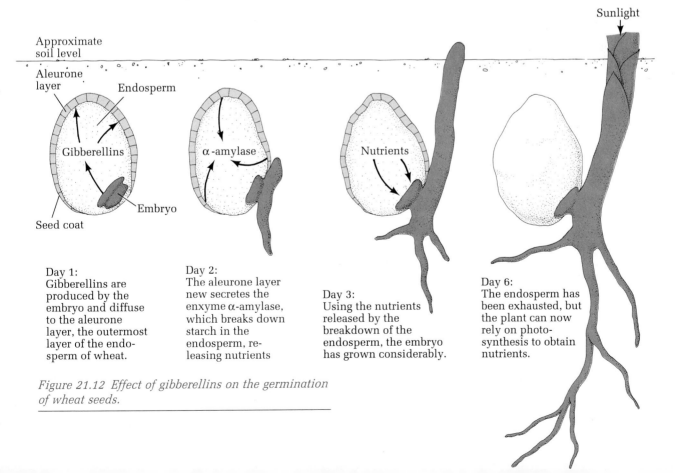

Day 1:
Gibberellins are produced by the embryo and diffuse to the aleurone layer, the outermost layer of the endosperm of wheat.

Day 2:
The aleurone layer new secretes the enxyme α-amylase, which breaks down starch in the endosperm, releasing nutrients

Day 3:
Using the nutrients released by the breakdown of the endosperm, the embryo has grown considerably.

Day 6:
The endosperm has been exhausted, but the plant can now rely on photosynthesis to obtain nutrients.

Figure 21.12 Effect of gibberellins on the germination of wheat seeds.

currently be undergoing undetectable changes as it is molded by evolution. Or it might have important regulatory roles as yet undetermined, or just a structural role. As more and more information becomes available, many scientists are coming to believe that there is a purpose for this DNA, although much research is needed before this purpose can be identified.

Constancy of DNA in the Genome During Development

In early studies of the genetic processes associated with differentiation and development, an important area of research focused on whether differentiation and development involve a *loss* of genetic information (i.e., of DNA) or whether these processes involve a programmed sequence of gene activation and repression events in a genome that is unchanged from the sequence found in the nucleus of the zygote from which the mature organism develops. The most direct way to decide is to determine whether the genome of a differentiated cell contains the same genetic information found in the zygote. Two important studies were especially instrumental in answering this question: One involved the study of carrots; the other focused on frogs.

Regeneration of carrot plants from mature single cells. The first study was done in the 1950s

by F. C. Steward, who worked with carrot plants. He took a carrot, dissociated the tissue so that the cells were separated, and then attempted to culture new carrot plants from those cells by using plant tissue culture techniques (see Figure 21.13). He was successful, and mature plants with edible carrots were produced (cloned) from phloem cells. That these cells had the potential to act as zygotes and develop into mature plants indicated that mature cells had all the DNA found in zygotes, thereby supporting the notion that the DNA content of a cell remains constant during development and differentiation and providing evidence against the notion that development and differentiation do not involve losses of genetic information.

Experiments similar to Steward's provided another important piece of information. If instead of carefully dissociating the carrot into individual cells, researchers cut and culture a small piece of tissue, a new plant does not develop. Instead, the carrot tissue grows larger and larger to form what is called a *callus.* This result indicates that cell-to-cell contact plays an important role in determining and maintaining the differentiated state of a cell. Apparently, once that contact is broken, the cell is often free to divide and, perhaps, act as a zygote. In a sense cancer cells provide a similar example. Such cells no longer respond to the differentiated-state-maintaining effects of the adjacent cells in the tissue

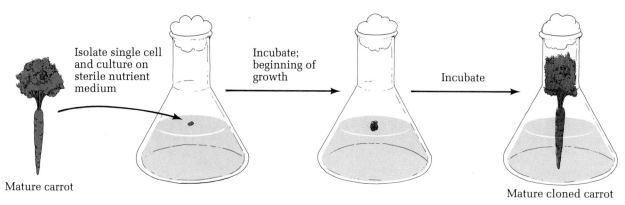

Isolate single cell and culture on sterile nutrient medium

Incubate; beginning of growth

Incubate

Mature carrot

Mature cloned carrot

Figure 21.13 Cloning of a mature carrot plant from a cell of a mature carrot.

and, instead, begin to proliferate uncontrollably, much like a rapidly dividing zygote cell.

Nuclear transplantation experiments with *Xenopus.* An objection to the general applicability of the results of the carrot experiments is that plants are much more able to propagate themselves vegetatively than are animals. For example, horticulturists have long been able to regenerate plants from cuttings. Researchers also had to determine whether the DNA content of animal cells remained constant during development. This result was indicated by a study involving the South African clawed toad, *Xenopus laevis.* This work, done in 1964 by J. Gurdon, was based on earlier work done in the early 1950s by R. Briggs and T. J. King with another amphibian, the leopard frog, *Rana pipiens.*

Gurdon's experiments (Figure 21.14) tested whether or not a nuclcus taken from a tadpole (one differentiated stage of the *Xenopus* life cycle) could direct the development of an egg into a new tadpole—or even into an adult toad. The nucleus that he used was tagged with a genetic marker so that he could be sure that it was responsible for any of the developmental changes that occurred. *Xenopus* has a mutation called *O–nu,* which is a deletion for the nucleolus organizer region, and the nuclei of homozygous *O–nu/O–nu* cells have no nucleoli. Since the nucleolus organizer region normally contains the ribosomal RNA genes, homozygous *O–nu/O–nu* organisms cannot make their own ribosomes, and hence this genotype is ultimately lethal. However, a *O–nu/O–nu* egg can develop to the swimming larval stage if the experiment is set up so ribosomes are deposited in the egg from a heterozygous *+/O–nu* mother; after that point the tadpoles die, because no new ribosomes are made.

Heterozygous *+/O–nu* individuals develop in a normal fashion, and their nuclei have one nucleolus per nucleus instead of the two found in *+/+* cells. Since it is easy to determine whether a nucleus has one or two nucleoli by the use of a microscope, Gurdon used *+/O–nu* (uninucleolate) tadpoles as the source for donor nuclei and *+/+* (binucleolate) eggs as recipients.

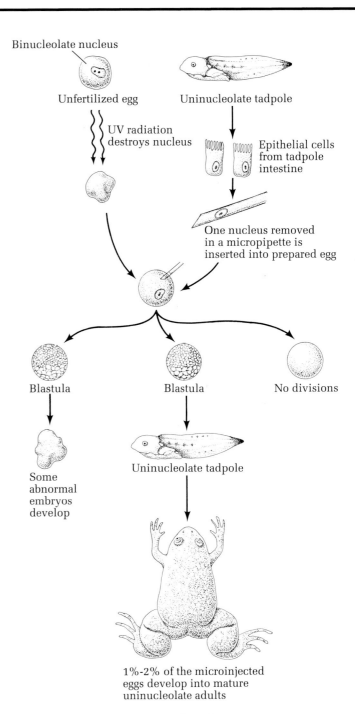

Figure 21.14 Representation of Gurdon's experiments, which showed the totipotency of the nucleus of a differentiated cell of Xenopus laevis.

The experiment required some very skillful manipulations of cells and nuclei under the microscope. First, intestinal epithelial cells (i.e., cells of the gut lining) were isolated from uninucleolate tadpoles, and their nuclei were removed by using a micropipette larger than the diameter of the nucleus but smaller than the diameter of a whole cell. Gentle sucking of a cell caused the cell to break, and the nucleus could be sucked into the pipette. Second, an unfertilized egg from a binucleolate strain was isolated and then irradiated with ultraviolet light to destroy the genetic information it contained. Next, the intestinal $+/O–nu$ nucleus was introduced into the recipient egg, which was then incubated to see whether development would occur.

In some cases no divisions occurred; in others abnormal embryos developed; and in 1 to 2 percent of the cases a fertile, adult $+/O–nu$ toad developed. Most frequently, however, swimming tadpoles were generated that later died. Because the nuclei of the tadpoles and adult toads produced were always uninucleolate, Gurdon concluded that the donor tadpole nucleus must have contained all the genetic information needed to specify an adult toad. The nucleus from the *Xenopus* intestine is said to be totipotent, where **totipotency** is the capacity of a nucleus to direct a cell through all the stages of development and therefore produce a normal adult. In this particular study, only a few cells (1 to 2 percent) were totipotent.

The results of these and many other studies of various plants and animals indicate that differentiation and development generally do not involve a loss of genetic information from the genome. Thus because there is constancy of DNA in the genome over an organism's life cycle, differentiation and development must result from regulatory processes affecting gene expression. (As always, there are some rare exceptions. Certain organisms, for example, have somatic cells that do not contain all of the genetic information that is found in germ line cells, while other organisms display mature differentiated cells that have genetic material arranged differently from the way it is found in its embryonic precursor cells.)

Keynote Long-term regulation is involved in controlling the events that occur to activate and repress genes during development and differentiation. Development and differentiation result from differential gene activity of a genome that contains a constant amount of DNA, from the zygote stage to the mature organism stage, and thus these events do not result from a loss of genetic information.

Differential Gene Activity in Tissues and During Development

We can see that specialized cell types have different cell morphologies; for example, nerve cells are clearly different from intestinal epithelial cells. We can also distinguish different organs and tissues. Since researchers have shown that the amount of DNA remains constant during development, the simplest hypothesis is that the observable phenotypic differences reflect differential gene activity, and the following examples discuss some of the evidence supporting this hypothesis.

Tissue-specific protein differences. Based on the assumption that different tissues carry out different functions that require specific enzymes and proteins, a large body of evidence has been compiled indicating that different tissues exhibit a spectrum of proteins. One particular experimental procedure that is used to analyze the array of proteins present in a tissue is *two-dimensional polyacrylamide gel electrophoresis* (2D-PAGE). In this procedure, proteins are extracted from cells or from a tissue and are separated, in the first dimension on the basis of their charge in a process called isoelectric focusing. The gel strip is then laid alongside a rectangular slab of polyacrylamide gel, which contains sodium dodecyl sulfate (SDS) in the buffer. Electrophoresis is carried out in the second dimension, then, in the direction perpendicular to the direction of electrophoresis in the first dimension. Because SDS is present in the gel, the proteins are further separated in the second dimension on the basis of their size. This combination of two different parameters for protein separation permits hundreds or even thousands of proteins to be re-

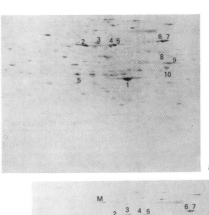

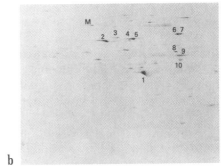

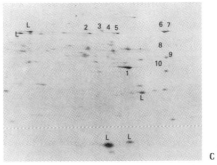

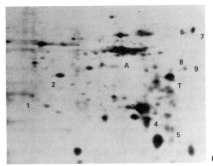

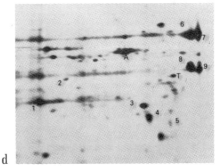

solved in a single slab gel. The proteins in the gel are detected by staining or, if they are radioactively labeled, by autoradiography. Both detection methods only resolve proteins that are relatively abundant; proteins that are present in low numbers, such as many regulatory proteins, are likely not to be detected in the 2D gels.

When 2D-PAGE analysis is done for proteins extracted from different cell types or tissues, different protein patterns are typically seen (Figure 21.15). Often many different protein spots can be discerned. The cell-specific proteins that are detected are presumably a subset of the proteins responsible for giving each cell type its particular structure and function. Note, however, that while differences may be readily detected on the gels, it is extremely difficult to determine the functions of the proteins involved.

The presence of different proteins in different cells means that different mRNAs are being translated. In most cases the different cell types and tissues exhibit transcriptional control so that the mRNAs are cell-type-specific or tissue-specific.

Hemoglobin types and human development. In this book so far, human adult hemoglobin Hb–A has been examined in many contexts. For simplicity, it is usually described as a tetrameric protein made up of two α and two β polypeptides, where each type of polypeptide is coded by a separate gene, α and β. The two genes appear to have arisen by duplication of a single ancestral gene followed by alteration of the base sequences in each gene during evolution. The organization of the genes (Figure 21.16) shows that each contains two introns (intron 1 and intron 2), which are transcribed with the coding sequences but are removed to produce a mature mRNA transcript. While the introns are of different sizes in the two genes, they are placed in very similar positions in the two genes. (The demonstration that the globin genes actually contain two introns came later than the pioneering studies described in Chapter 9, in which the β-globin gene was shown to have only one intron.)

Hemoglobin Hb–A is only one type of hemo-

Figure 21.15 Two-dimensional gel analysis of proteins from different mouse tissues, showing tissue-specific protein patterns: (a) kidney; (b) muscle; (c) liver; (d) embryonic heart; and (e) adult heart.

globin found in humans. Genetic studies have shown that several distinct genes code for α- and β-like globin polypeptides that are assembled in specific combinations to form different types of hemoglobin, which are synthesized and function at different times during human development. Figure 21.17 shows the globin chains synthesized at different stages of human development. In the human embryo the hemoglobin initially made in the yolk sac is a tetramer of two ζ polypeptides and two ε polypeptides. From comparisons of the amino acid sequences, ζ is an α-like polypeptide, and ε is a β-like polypeptide. After about three months of development, synthesis of embryonic hemoglobin ceases (i.e., the ζ and ε genes are no longer transcriptionally active), and the site of hemoglobin synthesis shifts to the liver and the spleen. Here true α polypeptides are made, and the two other polypeptide chains in the tetramer are either β-like γA chains or β-like γG chains, producing two forms of fetal hemoglobin. Each type of γ chain is coded for by distinct genes. The hemoglobin formed here is called *fetal hemoglobin* (Hb–F).

Fetal hemoglobin is made until just before birth, at which time synthesis of the two types

of γ chains stops and the site of hemoglobin switches to the bone marrow, where α and true β polypeptides are made along with some β-like δ polypeptides. In the newborn through adult human, most of the hemoglobin is our familiar $\alpha_2\beta_2$ tetramer (Hb–A), with about one in forty molecules having the constitution $\alpha_2\delta_2$. Thus, there is clear evidence for switching of globin gene types during human development, and this switching must involve a sophisticated gene regulatory system that turns appropriate globin genes on and off over a long time period.

With the use of recombinant DNA techniques, some further interesting facts have been elucidated about this developmental phenomenon. A number of research groups have succeeded in cloning and sequencing all seven (α, β, δ, γA, γG, ζ, and ε) human globin gene types. It has also been possible to identify their locations in the genome. The α-like genes (α, ζ) are located on chromosome 16, while all the β-like genes (ε, γA, γG, and δ) are on chromosome 11 (see Figure 21.18). Note that there are two ζ genes and two α genes—all of these genes are transcribed at the appropriate times. The α-like genes are arranged in the same orientation on the chromosome with regard to their transcription; that is,

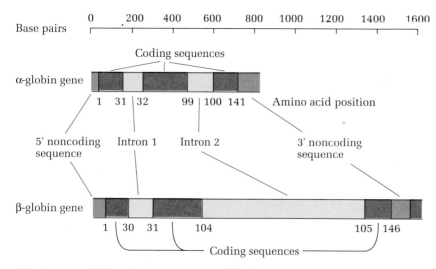

Figure 21.16 *Molecular organization of the human α-globin and β-globin genes.*

the promoter for each is to the left of the gene in the figure, and transcription is from left to right. Between the ζ and α genes is a sequence that closely resembles the base-pair sequence of the α-globin gene. Careful scrutiny of this sequence reveals that it does not have all the features required for producing a functional polypeptide product. Sequences such as this one, which are highly related to known functional genes but which themselves are defective so that they cannot produce a functional product, are called *pseudogenes.* Pseudogenes are thought to be nonfunctional relics of gene duplication and gene modification that occurred some time ago in evolutionary time.

The β-like genes are also each arranged in the same orientation with respect to transcription. This cluster has two pseudogenes, one at the extreme left and the other between the γ genes and the δ gene. Most intriguing is the fact that for both clusters the order of the genes, from left to right, exactly parallels the order in which the genes become transcriptionally active during human development. Recall that embryonic hemoglobin consists of ζ and ε polypeptides, and these genes are the first functional genes at the left of the clusters. Next, the α and γ genes are transcribed to produce fetal hemoglobin (Hb–F), and these genes are the next functional genes

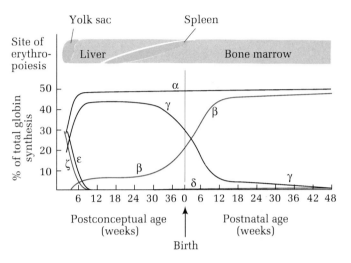

Figure 21.17 *A representation of the synthesis of different globin chains at various stages of embryonic and fetal development.*

that can be transcribed from the clusters if one moves from left to right. Lastly, the δ and β polypeptides are produced, and these genes are last in line in the β-like globin gene cluster. Although these arrangements surely must occur by more than coincidence, there is no insight as yet about how the gene order relates to the regula-

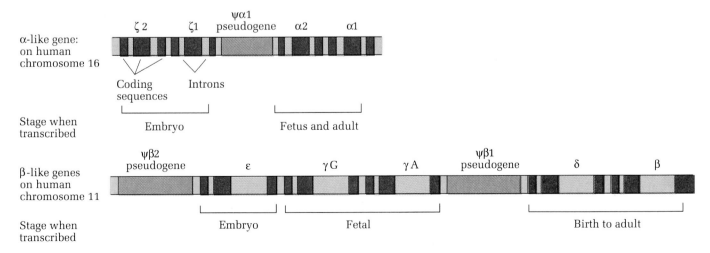

Figure 21.18 *Linkage maps of human globin gene clusters. (Map is not to scale.)*

tion of expression of these genes during development.

Finally, a developmental mutation in humans disrupts the normal switching of globin gene activity. This mutation, called *hereditary persistence of fetal hemoglobin* (HPFH), prevents the switch from the synthesis of the β-like γ polypeptide to the β and δ chains. There is some molecular evidence indicating that changes in the DNA and chromatin are associated with this developmental regulation phenomenon. For example, there is strong evidence that DNA sequences within the promoter region influence the developmental expression of individual genes. Also, there is evidence for the existence of trans-acting regulatory proteins that influence gene expression by interacting with specific DNA sequences. When a globin gene becomes activated, its chromatin becomes more sensitive to DNase I, and there is a decreased methylation of the DNA sequence. The latter has resulted in a possible therapeutic approach for treating sickle cell anemia patients. That is, administration of 5-azacytidine, a drug which demethylates DNA, activates the fetal hemoglobin (Hb–F) gene in sickle cell anemia patients. The resulting γ-globin polypeptide can substitute for the defective β-globin chain in these patients' hemoglobin molecules.

Polytene chromosome puffs during Dipteran (two-winged fly) development. In this subsection we will see further evidence that developmental processes are the result of differential gene activity.

Polytene chromosomes (see Chapter 17, p. 559) are the result of multiple rounds of DNA replication without nuclear division or chromosome segregation (a process called *endoreduplication*). Polytene chromosomes are characteristic of certain tissues of Diptera; for example, the salivary glands in the larval stages, and they may be 1000 times the size of corresponding chromosomes at meiosis or in the nuclei of ordinary somatic cells. After staining, distinct and characteristic bands, called chromomeres, can be seen along each chromosome. We do not know how many genes are contained in each chromomere.

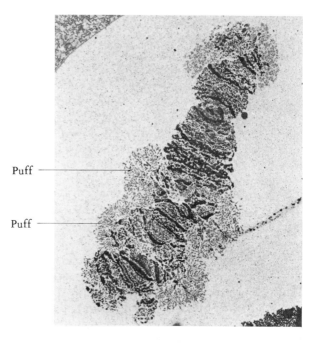

Puff

Puff

Figure 21.19 Light micrograph of a polytene chromosome from Chironomus, *showing two puffs that result from localized uncoiling of the chromosome structure and indicate transcription of those regions.*

Examination of polytene chromosomes during the developmental stages of various dipteran insects, including *Drosophila melanogaster* and *Chironomus*, have shown that specific chromomeres exhibit *puffs* at characteristic and reproducible times during development (see Figure 21.19). In other words, the puffs appear and disappear at certain chromosomal loci as development proceeds, so it is fair to say that they are developmentally controlled. Figure 21.20 diagrams the appearance and disappearance of puffs in a particular segment of a *Drosophila* chromosome during the period before the molt from larval to prepupal and from prepupal and pupal stages. Since some of these puffs can also be induced by injecting the molting hormone ecdysone into nonmolting larva, we can conclude that hormones play a role in inducing at least some puffs during development. As the figure shows, a characteristic puffing pattern is seen at each stage, indicating a specific progression of puffing activity during development. For example, puffs at 74EF and 75B appear early in the period before the molt from larval to prepupal stages, and again in about the middle of the period between the prepupal and pupal stages, but they disappear at other stages. Other puffs, at 66B, for example, appear characteristically late in both periods examined.

Electron microscopic studies have shown that puffing involves a loosening of the tightly coiled DNA into long, looped structures that are more accessible to RNA polymerase. At the molecular level this event must involve an uncoiling of the supercoiled DNA-protein structure of the chromosome.

Evidence that the puffs are the visual manifestation of transcriptionally active genes has come from radioactive tracer experiments. If radioactive uridine is added to developing flies, it is incorporated into the RNA being synthesized. If salivary glands are dissected from larvae and their polytene chromosome is prepared and autoradiographed, we find that radioactivity is localized at the puff sites, indicating that RNA,

presumably transcripts, is associated with the puffs. Very simply, when a polytene chromosome gene needs to be transcribed to produce a protein required for salivary gland activities during development, the chromosome structure loosens in order to permit efficient transcription of that region of the DNA. When transcription is completed, the chromosome reassumes its compact configuration and the puff disappears.

While we do not know what molecular event is responsible for the induction of a puff, it appears that hormonal control is involved in many cases. Dipteran larvae go through different stages, each separated by a molt. The molting process is controlled by the molting hormone ecdysone, which is made in the prothoracic gland of the larvae. The prothoracic gland is, in turn, stimulated to produce ecdysone by the action of a brain hormone synthesized by neurosecretory cells. If we inject ecdysone into larvae that are between molts, it causes the formation of the same sequence and types of puffs that are characteristically seen in larvae undergoing molting under endogenous ecdysone control.

Polytene puffs, the sites of RNA synthesis in dipteran insects, are directly related to the developmental processes in the organism, and at least some of the puffs are regulated by hormone action. We can conclude, therefore, that at least some gene regulation of developmental pro-

---

Figure 21.20 Developmentally programmed formation (induction) and disappearance (regression) of seven puffs in the third chromosome of Drosophila melanogaster, *during the period before the molt from larval to prepupal and from prepupal to pupal stages.*

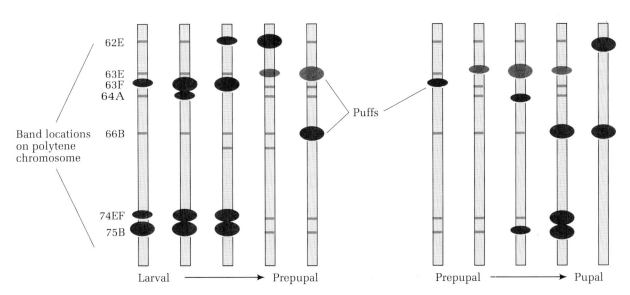

cesses in dipteran flies is done by hormones, and that this regulation occurs at the transcriptional level.

Determination of Cell Type During Development

The previous section presented evidence that developmental processes occur as a result of differential gene activity. In this section we will explore some of the details of what is known about how a particular cell type is determined during differentiation and development by discussing the role of imaginal disks in *Drosophila* development, and mutants that affect the developmental process.

***Drosophila* development.** The production of an adult *Drosophila* from a fertilized egg involves a well-ordered sequence of developmentally programmed events under strict genetic control, although, as with other eukaryotes, the genetic basis of these events is not yet understood. About 24 hours after fertilization, a *Drosophila* egg hatches into a larva, which then undergoes two molts, after which it is called a pupa. The pupa metamorphoses into an adult fly. The whole process from egg to adult fly takes about nine days at 25°C (Figure 21.21).

Drosophila adult structures develop from imaginal disks, which are determined early in larval development. The determined state so programmed is very stable, although rare changes do occur. That is, there is a clear distinction between the stage when a phenotype is determined and the stage when the differentiation processes that give rise to the phenotype occur.

Thus larvae contain two groups of cells: Cells of one group are involved exclusively with larval development and function, while cells of the other group are grouped into the clusters of cells called imaginal disks. The imaginal disk cells remain in an embryonic state throughout larval development, even though larval cells differentiate around them. Each imaginal disk eventually differentiates into a particular adult structure. Figure 21.22 shows the positions of some imaginal disks in a mature larva and the adult structures that develop from them. Figure 21.23

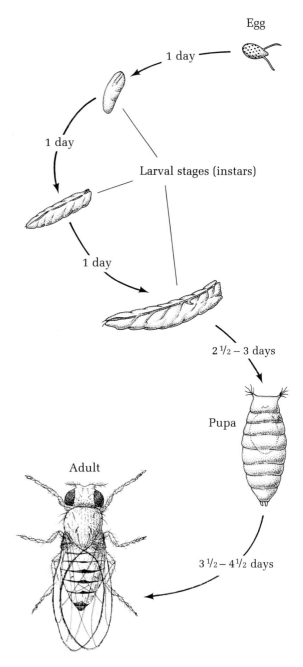

Figure 21.21 Development of an adult Drosophila *from a fertilized egg.*

Imaginal disks in larva   Adult structures

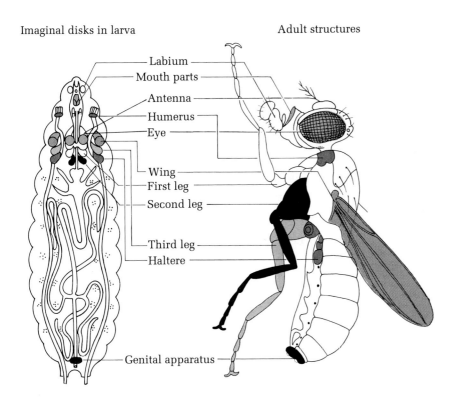

- Labium
- Mouth parts
- Antenna
- Humerus
- Eye
- Wing
- First leg
- Second leg
- Third leg
- Haltere
- Genital apparatus

Figure 21.22 Locations of imaginal disks in a mature Drosophila *larva and the adult structures derived from each disk.*

presents an exploded diagram of an adult fly. In this figure each distinct part shown develops during pupation from a separate imaginal disk.

From the genetic point of view imaginal disks are excellent subjects for study because each disk develops during the first larval stage, and when it consists of about 20–50 cells, it is already programmed to specify its given adult structure. From then on the number of cells in each disk increases by mitotic division, until by the end of the larval stages, there are many thousand cells per disk. At this point a disk is visible under the microscope and can be isolated and used in transplantation studies (see Chapter 14 pp. 447–451). If a disk is removed from a larva and injected into the abdomen of an adult fly, it never differentiates into its adult structure. Instead, the disk merely increases in size by continued cell divisions. In other words,

the cells proliferate but do not differentiate. If, however, a disk is transplanted into the abdomen of a larva in a late stage of development, when that larva molts to produce a pupa, the disk differentiates into the adult structure. Studies have indicated that the stimulus for a disk to differentiate into its adult structure is the molting hormone ecdysone. Ecdysone, then, not only causes specific puffing patterns in polytene chromosomes but also stimulates the differentiation of imaginal disks.

The nature of the determination process is unknown, although at a simple level it probably involves a programming of those genes that are accessible to hormone-stimulated activation during the pupal stage. Evidence suggests, though, that the determination process is a remarkably stable event; some mechanism apparently keeps all the cells in a particular disk programmed in the same way. This evidence comes from transplantation experiments in which disks or parts of disks are transplanted from a larva into the abdomen of an adult fly. In the adult abdomen the cells grow and divide so that the disk gets

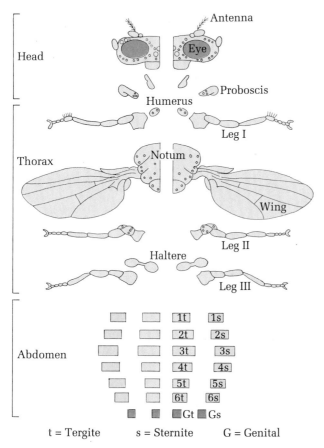

Antenna

Head

Eye

Proboscis

Humerus

Leg I

Thorax

Notum

Wing

Leg II

Haltere

Leg III

Abdomen

		1t	1s
		2t	2s
		3t	3s
		4t	4s
		5t	5s
		6t	6s
		Gt	Gs

t = Tergite   s = Sternite   G = Genital

Figure 21.23 Exploded diagram of the external body parts of an adult Drosophila *fly.*

larger, but the disk does not differentiate into its adult structure because the adult abdomen lacks the necessary hormones to induce this differentiation. In essence, the disk cells are being cultured in the adult abdomen environment. After the disk has grown, it can be reisolated from the abdomen and part of it transplanted into the abdomen of another adult. Continued growth, isolation, and transplantation to a new host (called serial transplantation) at least 150 times does not change the programming of the disk since, when it is now transplanted into a larval host (in which the necessary hormones for differentiation exist), it will still, with very rare exceptions, develop into the same adult structure as the ancestral disk.

Transdetermination. Rarely, the determined state of an imaginal disk does change, as the transplantation studies of E. Hadorn have shown. The disk does not totally dedifferentiate but switches to another determined path in a process called *transdetermination.* For example, an eye disk may become altered so that it specifies a wing structure, or an antenna disk may become altered so that it specifies a leg structure. Since the position of the disks in the larvae relates to the positions of adult structures, the consequence of such transdetermination events is that the adult fly has familiar structures at unusual places, such as a leg where an antenna should be. Significantly, the entire disk changes its determined state; no studies have reported, for instance, that half a disk becomes transdetermined while the other half does not. The results of such transdetermination would be easy to detect by visual scrutiny of adult flies since they would be hybrid structures. Figure 21.24, which shows the pathways of transdetermination in *Drosophila* that have been detected by transplantation experiments such as those of Hadorn, indicates that certain disks transdetermine only into certain other disk types. Leg disks, for example, transdetermine readily into proboscis disks, but the opposite transdetermination is not seen. Leg disks transdetermine into wing disks, but the opposite transdetermination occurs far less frequently.

Selector genes and homeotic mutants. For many decades scientists have addressed the question of how the fertilized egg gives rise to a complex organism, which consists of many different cell types, each with a different function. Geneticists have focused, in particular, on defining the genes responsible for developmental events and the time course of regulation of those genes. In this section we will concentrate on special regulatory genes called **selector genes,** which play a major role in laying down the basic body plan of an organism, starting early in embryonic development. Our discussion will focus on selector genes in the fruit fly, *Drosophila melanogaster,* and their role in development. These selector genes are expressed only in certain regions of the embryo. They are believed to act as master switches turning on or off other regulatory genes that control the developmental pathways cells will follow.

Two categories of selector genes have been defined, **segmentation genes** and **homeotic genes.** *Drosophila,* like all insects, is a segmented organism with four to six segments of the embryo fusing and developing into a head; three segments, each with a pair of legs and the middle of which also has a pair of wings, develop into the thorax; and eight segments develop into the abdomen (Figure 21.25). Over 20 segmentation genes have been shown to be involved in controlling the number and correct orientation of the segments.

Homeotic mutations. Given the basic segmentation plan, the homeotic genes give a specific developmental identity to each of the segments. The homeotic genes have been defined by mutations that affect the development of the fly. The principal pioneer of genetic studies of homeotic mutants is E. Lewis, and the more recent molecular analysis has been done in many laboratories, including T. Kaufman's, W. Gehring's, W. McGinnis's, M. Scott's, and W. Bender's.

Homeotic mutations affect the developmental program of the cells, in particular, imaginal disks resulting in strange, developmental aberrations of the fly in which one part of the adult body has been transformed into another, much like Hadorn's transdetermination phenomenon. Lewis's pioneering studies involved mutations in a large cluster of homeotic genes called the *bithorax* complex (BX-C). BX-C contains three complementation groups called *Ultrabithorax* (*Ubx*), *abdominal-A* (*abd-A*), and *Abdominal-B* (*Abd-B*), each of which contains more than one gene. Mutations in these homeotic genes are often lethal so that the fly typically does not survive past embryogenesis. Some nonlethal mutations have been characterized, however, which permit development into an adult fly to occur. Figure 21.26 shows the abnormal adult structures that can result from *bithorax* mutations. A schematic diagram of a normal adult fly showing the segments is shown in Figure 21.26a: note that the wings are located on segment Thorax 2

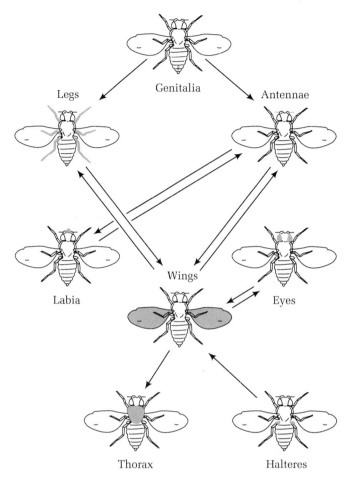

Figure 21.24 *Pathways of transdetermination in* Drosophila *disk cells as derived from transplantation experiments. The colored areas indicate the parts of the adult that develop from serially transplanted disk material. The arrows indicate the observed transdetermination changes. (From E. Hadorn, Transdetermination in cells, illustration by Bunji Tagawa. Copyright © 1968 by Scientific American, Inc. All rights reserved. Reprinted by permission.)*

(T2) while the pair of halteres (rudimentary wings used as "balancers" in flight) are on segment T3. A photograph of a normal adult fly clearly showing the wings and halteres is presented in Figure 21.26b. Figure 21.26c shows one type of developmental abnormality that can result from nonlethal homeotic mutations in BX-C: shown is a fly that is homozygous for three separate mutations in the *Ubx* complementation group, *abx, bx*[3], and *pbx.* Collectively these mutations transform segment T3 into an adult

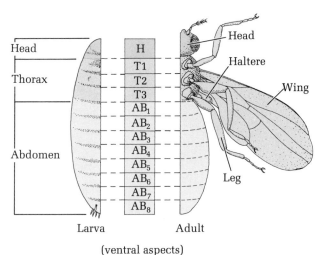

a)

b)

c)

Figure 21.25 Body segmentation patterns of a Drosophila *larva (left) and adult (right); H = head; T = thorax; AB = abdomen.*

structure similar to T2. The transformed segment exhibits a fully developed set of wings. The fly lacks halteres, however, because no normal T3 segment is present.

Another well-studied group of mutations defines another large cluster of homeotic genes called the *Antennapedia* complex (ANT-C). ANT-C contains at least four genes called *Deformed* (*Dfd*), *fushi tarazu* (*ftz*), *Sex combs reduced* (*Scr*), and *Antennapedia* (*Antp*). As is the case for BX-C, most mutations in ANT-C are lethal. Among the nonlethal mutations is a group of mutations in *Antp* that result in leg parts instead of an antenna growing out of the cells near the eye during the development of the eye disk (Figure 21.27). Note that the leg has a normal structure, but is obviously in an abnormal location. A different mutation in *Antp*, called *Aristapedia*, has a different effect in that only the distal part of the antenna, the arista, is transformed into a leg (Figure 21.28).

The homeotic mutants of ANT-C and BX-C, therefore, are deficient in one or more gene products that are involved in controlling the normal development of the relevant adult fly structures. The accepted conclusion is that the normal products of the two complexes function to prevent certain abnormal developmental

Figure 21.26 *(a) Drawing of a normal fly. T = thoracic segment. A = abdominal segment. The haltere (rudimentary wing) is on T3. (b) Photograph of a normal fly with a single set of wings. (c) Photograph of a fly homozygous for three mutant alleles (*bx, abx, *and* pbx*) that results in the transformation of segment T3 into a structure like T2; namely, a segment with a pair of wings. These flies therefore have two sets of wings.*

a)

b)

c)

Eye

Antenna

Mouth parts

d)

e)

Coxa

Claw

Tarsal
structures

Trochanter

Tibia

Femur

Figure 21.27 The homeotic mutant in Drosophila Antennapedia, *in which the antenna is converted into a
leg. (a) scanning electron micrograph of the antennal
area of a wild-type fly; (b) scanning electron micrograph of the antennal area of the mutuant* Antennapedia, *in which the antenna is transformed into a leg;
(c) drawing of the head of a wild-type fly with the
antenna highlighted in color; (d) drawing of the head
of an* Antennapedia *fly: (e) correspondence between
antennal and leg structures.*

events from occurring. Thus, the mutations lack
those inhibitory functions.

Molecular characterization of homeotic genes.
Most of the *Antennapedia* and *bithorax* homeotic genes have now been cloned using recombinant DNA techniques. Some unexpected findings have derived from analysis of the DNA
sequences cloned from these two complexes. For
example, a number of the homeotic genes are
extremely large compared with most *Drosophila*
genes and even compared with genes from
higher eukaryotes (where genes are generally
larger than genes in *Drosophila*). In particular,

a)

b)

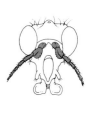

Figure 21.28 The homeotic mutant of Drosophila, Aristapedia, *in which the arista is transformed into a
leg. (a) scanning electron micrograph of the antenna
region of an* Aristapedia *fly where the arista has been
replaced by part of a leg. (b) drawing of the head of
an* Aristapedia *fly.*

these *Drosophila* homeotic genes have numerous and very long introns, while most *Drosophila* genes generally have very few, small introns. The *Antp* gene, for example, consists mostly of introns, with mature mRNAs of a few kilobase pairs produced by splicing a 103-kbp primary transcript. For both ANT-C and BX-C it may be the case that the mRNAs are spliced out of very long primary transcripts.

Interestingly, evidence reveals that the exceptionally long primary transcripts from ANT-C and BX-C are regulated at the level of splicing. For *Antennapedia,* for example, an embryonic

mRNA species is spliced from four exons, while a pupal mRNA species includes two of those exons plus one pupal-specific exon (Figure 21.29). Presumably, different functional proteins are being produced by the modular assembly of protein domains. However, to date, the functions of the proteins produced from these mRNAs, or any protein encoded by ANT-C and BX-C, remain unknown.

Detailed comparisons of the DNA sequences of *Drosophila* homeotic and segmentation genes have revealed a consensus sequence in many of the genes. This sequence is a short, 180-bp sequence called the *homeobox,* which is located within the most 3′ exon and therefore is a translated sequence. The homeobox sequence is very highly conserved, with many of the differences at the DNA level resulting in amino acid differences in the protein that would probably not affect protein function significantly. Homeobox se-

Figure 21.29 Developmental regulation of Antennapedia *mRNAs by alternative splicing of gene transcripts. (top) Section of the polytene chromosome 3 showing the location of the* Antennapedia *complex. (bottom) Alternative splicing patterns in pupae and embryos.*

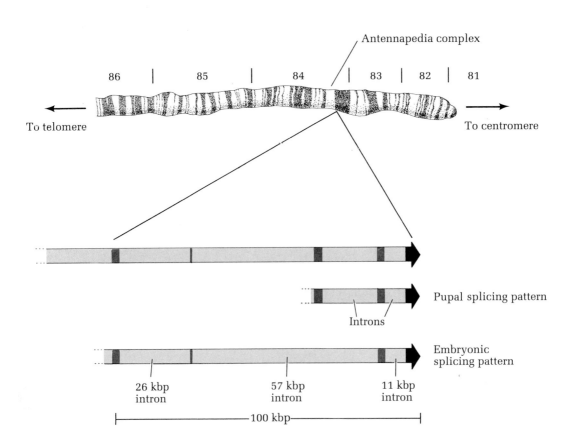

quences have been discovered in the genomes of many other organisms of greater complexity than *Drosophila,* including frogs, mice, and humans. Further, there is sequence similarity between the product of the homeobox sequence and a yeast cell protein encoded by the *MAT* locus, which controls mating type in that organism. The MAT protein is thought to be a DNA-binding regulatory protein. Thus one can speculate that the *Drosophila* homeotic gene product and the possible homologous proteins from other organisms are also regulatory proteins. However, we have no firm clues as yet as to the molecular function of these genes. What we can say is that these genes appear to act as master switches to control certain developmental events.

Keynote *In many instances the structure and the function of a cell are determined early, even though the manifestations of this determination process are not seen until later in development. Such early determination events may involve some preprogramming of the genes that will be turned on later. Neither the nature of these determination events nor the mechanism of timing in developmental processes is well understood.*

Genetic Defects in Human Development

Much of the information about the regulation of gene expression during development has come from studies of nonhuman organisms. The reason for the relative dearth of information about human development is that studies of early human embryos is both technically and legally difficult. Moreover, in most instances a mutation that affects development generates a pleiotropy (variety) of phenotypic changes, making it difficult to pinpoint the primary source of the defect. Nevertheless, several human developmental mutants have been identified. These mutants presumably result from defects in the regulation of expression of genes that are crucial for the developmental processes affected or from defects in structural genes whose products are needed for the developmental defects.

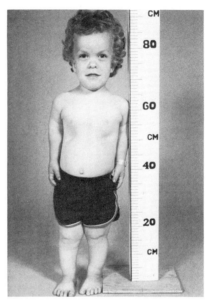

Figure 21.30 Human showing phenotypes typical of achondroplasia.

Achondroplasia. Phenotypically, people with achondroplasia (Figure 21.30) have a normal-sized trunk, enlarged head, and unusually short long-bones in the arms and legs. As a result, adults with achondroplasia average about 4 ft in height. Often other skeletal abnormalities can be seen upon X-ray examination.

Approximately twenty children in a million births exhibit the disorder. Pedigree analysis shows that this disorder is inherited as an autosomal dominant mutation. The best guess is that the mutation involved is in a gene that functions in bone formation, although we do not know the exact primary defect involved.

Nail-patella syndrome. The phenotypes of an individual with nail-patella syndrome (Figure 21.31) include deformed or absent fingernails and toenails and a patella (kneecap) that is abnormally small or absent. These defects are the result of some lesion in embryonic mesodermal and ectodermal development, although the exact cause(s) is unknown. This syndrome is also the result of an autosomal dominant mutation.

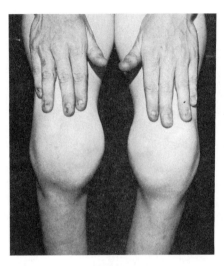

Figure 21.31 Hands and knees of a person afflicted with nail-patella syndrome. The nails are almost nonexistent, and the patellae (kneecaps) are absent.

Cleft lip and cleft palate. Cleft lip and cleft palate (Figure 21.32) are also the result of gene mutations. In embryonic development the palate (roof of the mouth) halves do not fuse since they are not in the correct position at the appropriate time. The seriousness of the cleft lip and cleft palate phenotypes depends on the degree to which the palate halves are abnormally oriented at the critical time during development. Unlike the achondroplasia and nail-patella syndrome, most forms of cleft lip and palate are not the result of a single gene mutation. Rather, there appear to be a number of rare mutations that, as part of the phenotypic abnormalities they produce, include cleft lip, cleft palate, or both. It is thought that at least two gene loci are involved, and in some cases an unknown environmental component may affect the expression of the mutation or mutations. One form of cleft lip and palate has been shown to result from a single gene mutation.

Thalidomide: The effect of an environmental agent on human development. The development of the human embryo (and of embryos of other organisms, for that matter) is a highly complex and precisely controlled sequence of events. The timing of events during development is crucial if a normal organism is to result. Clearly, mutations that affect the presence or absence of an important regulatory molecule or that decrease the amount of a chemical below a critical threshold level would be expected to result in developmental problems that would be detected as developmental mutants or, in the extreme, as naturally aborted fetuses. All the available evidence shows that the embryo stage is the most sensitive stage of human development, and hence a mother must be particularly careful about the drugs and other chemicals she introduces into her bloodstream in this period.

One example of historical significance concerns the drug thalidomide, a sedative prescribed to pregnant women in Germany, the Netherlands, and the United Kingdom from 1959 to 1961. Years went by before it was realized that thalidomide was responsible for children being born without arms or legs, a condition known as *phocomelia.* Apparently, the drug acted at a crucial time in development when the cells that would develop into arms and legs were being determined—were genetically programmed for development. Unfortunately, about 5400 children with the disorder were born before the causal connection with the use of thalidomide was made. (Agents like thalidomide that cause gross developmental abnormalities are called *teratogens.*) Thalidomide is, in fact, one of the very few examples of a teratogenic agent that has been directly related to a congenital malformation in humans.

Two important points can be made from the thalidomide story. First, thalidomide had been thoroughly tested in laboratory mammals (mice in this case) to determine whether there were any undesirable side effects. There was none. Thus while laboratory animals are necessary in drug-testing programs, biochemical differences between the chosen animal and humans may result in unexpected side effects in humans but not in the animal, as happened in the thalidomide case. Second, a recessive mutation in humans in the homozygous state gives rise to phocomelia. The thalidomide-induced condition is very similar in phenotype to the inherited con-

685
*Oncogenes
and the
Molecular
Biology
of Cancer*

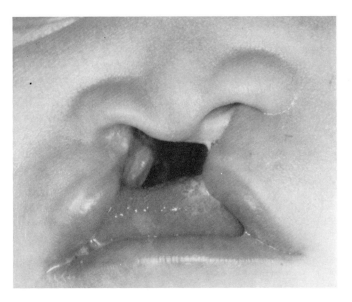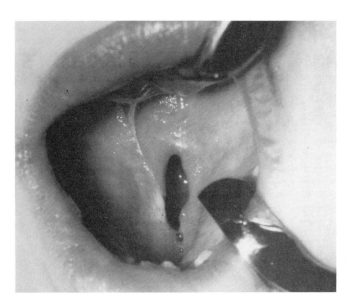

Figure 21.32 Individuals with a cleft lip (left) *and a cleft palate* (right).

dition. Thus thalidomide had induced a phenocopy.

Many examples of developmental mutants are found in humans and other eukaryotes. For example, a large number of mutants of *Drosophila melanogaster* could be classed as developmental mutants since one or more adult structures are abnormal. These mutants include changes in eye shape, wing shape, body size, body segmentation, bristle number, and bristle shape. It is hoped that studies of developmental mutants in laboratory-bred organisms such as *Drosophila* will provide valuable information about how genes are regulated during development.

Oncogenes and the Molecular Biology of Cancer

We have seen in the preceding sections that the normal development of a complex multicellular organism involves processes that lead to the differentiation of various cell types. These cell types are organized into tissues and organs that have characteristic structures and functions. Oc-

casionally, differentiated cells revert to an undifferentiated state (a process called *dedifferentiation*), and instead of remaining in a nondividing mode, they begin to divide mitotically and give rise to tissue masses called *tumors*. Tumors may be either benign, in which case their cells divide and they increase in size without threat to the life of the organism, or malignant, in which case the cells divide and the tumor increases in size, and the life of the organism becomes threatened. In some cases, cells break off from the tumor and migrate to adjacent tissues and throughout the organism, leading to tumor formation at many sites. The spreading of a tumor around the body is called *metastasis,* and metastasizing tumors are called **cancers.** Tumor initiation in an organism is called **oncogenesis** (*onkos,* "mass" or "bulk"; *genesis,* "birth"). Clearly, cancer is an important topic for humans. However, cancer is not one disease but many, and there can be many causes of cancer, such as spontaneous genetic changes (e.g., spontaneous gene mutations or chromosome aberrations), exposure to mutagens, radiation, or cancer-inducing viruses (tumor viruses). There is also evidence for hereditary predisposition to cancer.

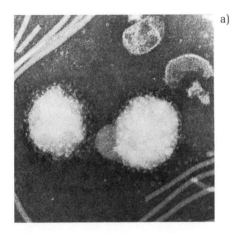

a)

b)

Outer coat (envelope)

v-*src* (oncogene)

Capsid (gag)

RNA

Reverse transcriptase gene

| LTR | gag | pol | env | v-src | LTR |

Viral genome

Figure 21.33 Rous sarcoma virus, an RNA virus: (a) electron micrograph; (b) diagram of the virus and gene organization of the viral genome. env = *enve- lope;* gag = *group-specific antigen;* pol = *RNA-de- pendent DNA polymerase (reverse transcriptase);* v- src = *sarcoma virus oncogene; LTR = long terminal repeats.*

We must assume that cancer cells, however they are induced, do not respond to the normal gene regulatory processes responsible for normal cell growth and division. Every cell in a cancer mass is believed to be descended from a single cell that for some reason became genetically al- tered so that it ceased to obey the regulatory rules for cell proliferation. Tumors have been studied for at least seven decades in attempts to understand the genetic changes responsible, but it is only since the development of recombinant DNA technology that tools have become avail- able to probe the molecular events associated with oncogenesis in any detail. In the following subsections we will discuss some of the molecu- lar information that relates to oncogenesis.

Tumor Viruses and Viral Oncogenes

In 1910, F. P. Rous showed that when pieces of a sarcoma (a particular kind of tumor) of a chicken were transplanted into other chickens from the same stock, those chickens developed sarcomas. More significantly, Rous showed that cell-free filtrates of the sarcoma inoculated into

chickens also resulted in tumor development. These results were interpreted to mean that some kinds of cancer might result from infection with a "filter-passing microbe" or a "virus." We now know that Rous's results (published in 1911) are explained by the existence of a tumor- causing virus, called Rous sarcoma virus (RSV) (Figure 21.33), which is an RNA virus of the re- trovirus type (see Chapter 18, pp. 587–588). Re- call that when a cell is infected by a retrovirus, a DNA copy of the single-stranded RNA genome is made by reverse transcriptase. The DNA copy (now called a provirus) becomes integrated into the nuclear DNA. Moreover, evidence shows that a number of viruses containing either DNA or RNA as the genetic material are tumor-induc- ing. We will limit our discussions to the RNA tumor viruses, all of which are retroviruses.

It is now understood that tumor induction by RSV results from the activity of a particular gene present in the retroviral genome. The discovery that the product of a single gene was both neces- sary and sufficient for tumor induction and for- mation was a significant breakthrough. RSV's transforming gene is called *src* after the sarcoma

687
*Oncogenes
and the
Molecular
Biology
of Cancer*

tumor it induces (Figure 21.34). The *src* gene is an example of an *oncogene,* that is, a gene that induces tumor formation. *Viral oncogenes* (which will be called v-*onc*'s) are responsible for inducing tumors of many animals including mouse mammary tumor virus (MMTV); leukemia viruses of, for example, mice, cats, and humans; sarcoma viruses of many species; and a few carcinoma-inducing viruses. In each case, the v-*onc* is carried by a retrovirus and all the retroviruses involved are morphologically alike and are probably descended from one ancestral virus. Since retroviruses all replicate via a DNA intermediate that is integrated into the host's genome, we can conclude that, like many of the other cancer-causing agents, oncogenic retroviruses cause cancer by altering the DNA. We must note here, however, that not all retroviruses induce tumors: only retroviruses that contain a v-*onc* gene are tumor viruses. The latter retroviruses are called **transducing retroviruses** because they have

picked up DNA other than *gag, pol,* and *env,* that is, they have picked up an oncogene from the cellular genome. (We will learn how this happens later in this section.) When introduced into a normal cell, a retrovirus carrying an oncogene causes that cell to transform into a cancer cell. The v-*onc* genes on the transducing retroviruses are named for the tumor from which the virus was obtained, with the prefix "v" to indicate that it is of viral origin. Thus, the v-*onc* gene of RSV is v-*src,* defined in 1976 by J. M. Bishop. Table 21.3 lists some transducing retroviruses and their oncogenes.

Cells infected by RSV rapidly transform into the tumorous state because of the presence of the v-*src* gene. Since RSV contains all the genes necessary for viral replication, a RSV-transformed cell produces progeny RSV particles (Figure 21.34). All other transducing retroviruses are defective in some of their genes (Figure 21.35) so that they can transform cells but are

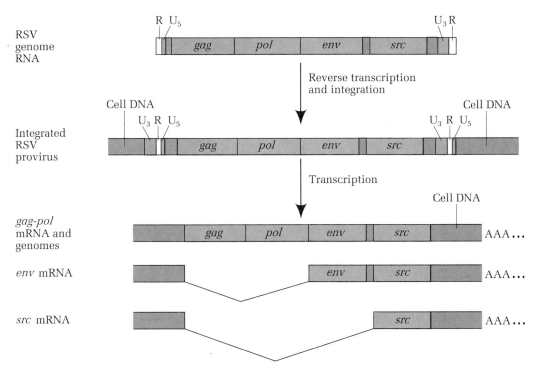

Figure 21.34 The Rous sarcoma virus genome and its mRNAs.

Table 21.3 Some Acute Transforming Retroviruses and Their Oncogenes

onc	Retrovirus Isolates	v-onc Origin	v-onc Protein	Virus Disease
src	Rous sarcoma virus (RSV)	Chicken	pp60^{src}	Sarcoma
*fps**	Fujinami sarcoma virus (FuSV)	Chicken	P130$^{gag\text{-}fps}$	Sarcoma
	PRCII-avian sarcoma virus (ASV)	Chicken	P105$^{gag\text{-}fps}$	Sarcoma
*fes**	Snyder-Theilen feline sarcoma virus (FeSV)	Cat	P85$^{gag\text{-}fes}$	Sarcoma
	Gardner-Arnstein FeSV	Cat	P110$^{gag\text{-}fes}$	Sarcoma
yes	Y73-ASV	Chicken	P90$^{gag\text{-}yes}$	Sarcoma
fgr	Gardner-Rasheed FeSV	Cat	P70$^{gag\text{-}actin\text{-}fgr}$	Sarcoma
ros	UR2-ASV	Chicken	P68$^{gag\text{-}ros}$	Sarcoma
abl	Abelson murine leukemia virus (MLV)	Mouse	P90-P160$^{gag\text{-}abl}$	Pre-B cell leukemia
	Hardy-Zuckerman (HZ2)-FeSV	Cat	P98$^{gag\text{-}abl}$	Sarcoma
ski	SKV	Chicken	P110$^{gag\text{-}ski\text{-}pol}$	Squamous carcinoma
erbA	Avian erythroblastosis virus (AEV)	Chicken	P75$^{gag\text{-}erb\text{-}A}$	Erythroblastosis and sarcoma
erbB			gp65^{erbB}	
fms	McDonough (SM)-FeSV	Cat	gp180$^{gag\text{-}fms}$	Sarcoma
fos	FBJ (Finkel-Biskis-Jinkins)-MSV	Mouse	pp55^{fos}	Osteosarcoma
mos	Moloney MSV	Mouse	P37$^{env\text{-}mos}$	Sarcoma
sis	Simian sarcoma virus (SSV)	Monkey	P28$^{env\text{-}sis}$	Sarcoma
	Parodi-Irgens FeSV	Cat	P76$^{gag\text{-}sis}$	Sarcoma
myc	MC29	Chicken	P100$^{gag\text{-}myc}$	Sarcoma, carcinoma, and myelocytoma
myb	Avian myeloblastosis virus (AMV)	Chicken	p45^{myb}	Myeloblastosis
	AMV-E26	Chicken	P135$^{gag\text{-}myb\text{-}ets}$	Myeloblastosis and erythroblastosis
rel	Reticuloendotheliosis virus (REV)	Turkey	p64^{rel}	Reticuloendotheliosis
kit	HZ4-FeSV	Cat	P80$^{gag\text{-}kit}$	Sarcoma
*raf**	3611-MSV	Mouse	P75$^{gag\text{-}raf}$	Sarcoma
H-*ras*	Harvey MSV (Ha-MSV)	Rat	pp21^{ras}	Sarcoma and erythroleukemia
	RaSV	Rat	P29$^{gag\text{-}ras}$	Sarcoma?
K-*ras*	Kirsten MSV (Ki-MSV)	Rat	pp21^{ras}	Sarcoma and erythroleukemia
ets	AMV-E26	Chicken	P135$^{gag\text{-}myb\text{-}ets}$	See above for *myb*

fps, the oncogene of several chicken sarcoma viruses, is derived from a chicken gene, c-*fps*, that is the chicken equivalent of c-*fes*, a gene of cats first discovered as the oncogene *fes* in several feline sarcoma viruses. It was not realized that *fes* and *fps* were homologous when they were assigned different names. Note that AGV and AMV-E26 each have two oncogenes.

SOURCE: R. Weiss, N. Teich, H. Varmus, and J. Coffin (eds.), *RNA Tumor Viruses,* 2d ed. Cold Spring Harbor Laboratory, Cold Spring Harbor, N.Y., 1985.

689
*Oncogenes
and the
Moleculuar
Biology
of Cancer*

unable to produce progeny viruses. These retroviruses can produce progeny viral particles,
however, if cells containing them are also infected with another virus that can supply the
gene products that are missing. These viruses
are called *helper viruses* (Figure 21.36).

Cellular Proto-oncogenes

In the mid-1970s, J. M. Bishop and H. Varmus,
and a number of other researchers demonstrated
that normal animal cells contain genes that are
very closely related to the viral oncogenes in
DNA sequence. In the early 1980s, R. A. Weinberg and M. Wigler showed independently that
a variety of human tumor cells contain oncogenes. These genes, when introduced into other
cells growing in culture, transformed those cells
into tumor cells (Figure 21.37). These human
oncogenes were very similar to viral oncogenes
that had been characterized previously, even
though there was no evidence for viral induction of the human cancers involved. Moreover,
the human oncogenes were shown also to be
closely related to genes in normally growing
cells. In general, both for the animal and human
oncogenes, the related genes in normal cells
were not identical in sequence but there were
long sequences showing high degrees of similarity. The genes in normal cells that are related to
oncogenes are called **proto-oncogenes** and they
are referred to as c-*onc* where the *onc* is the
same three-letter sequence as for the related
viral oncogenes. Thus, for RSV a *src* gene is
present in all chickens, both in normal cells and
in tumor cells. The normal cellular gene is
termed c-*src*. The research data obtained allow
us to conclude that transducing retroviruses
pick up cellular proto-oncogenes in a modified
form and then use them to transform other animal cells into tumor cells.

There are many significant differences between a cellular proto-oncogene and its viral oncogene counterpart:

1. Most c-*onc*'s contain introns that are not
   present in the corresponding v-*onc*. The v-*src*
   oncogene of RSV, for example, is 1700

a) Avian myeloblastosis virus (AMV) genomic RNA

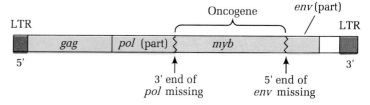

b) Avian defective leukemia virus (DLV) genomic RNA

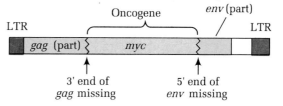

c) Feline sarcoma virus (FeSV) genomic RNA

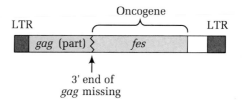

d) Abelson murine leukemia virus (AbMLV) genomic RNA

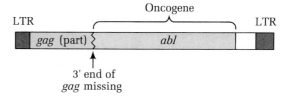

*Figure 21.35 Structure of four defective transducing
viruses (not drawn to scale). (a) Avian myeloblastosis
virus (AMV) contains the v-myb oncogene which replaces some of the 3' end of pol and most of the env.
(b) Avian defective leukemia virus (DLV) contains the
v-myc oncogene which replaces the 3' end of gag, all
of pol, and the 5' end of env. (c) Feline sarcoma virus
(FeSV) contains the v-fes oncogene which replaces
the 3' end of gag and all of pol and env. (d) Abelson
murine leukemia virus (AbMLV) contains the v-abl
oncogene which replaces the 3' end of gag and all of
pol and env.*

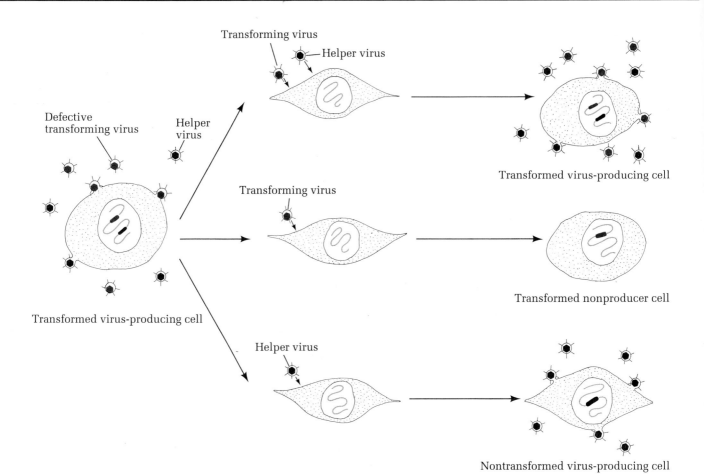

Transforming virus

Helper virus

Defective
transforming virus

Helper
virus

Transformed virus-producing cell

Transformed virus-producing cell

Transforming virus

Transformed nonproducer cell

Helper virus

Nontransformed virus-producing cell

*Figure 21.36 Top: When a cell is infected by both a
defective transforming virus and a nondefective
helper virus, a transformed cell results which produces both types of virus. Middle: When a cell is infected by a defective transforming virus alone, trans
formed virus-nonproducer cells are generated.
Transforming viruses can be rescued from these cells
by infection with a helper virus. Bottom: When a cell
is infected by a helper virus alone, a nontransformed,
virus-producing cell results. (From Watson, et al.,*
Molecular biology of the gene, *vols. I and II, 4th ed.
Copyright © 1987 Benjamin/Cummings Publishing
Company, Inc. Reprinted by permission.)*

nucleotides long and is transcribed to produce a spliced mRNA containing viral sequences and the *src* sequence (Figure 21.34).
The chicken c-*src* gene is over 8 kbp long, consisting of 13 exons (Figure 21.38). The mRNA

transcribed from this gene is about 4 kbp
long with no associated retroviral sequences.

The v-*onc*'s do not contain introns. This is
not surprising since viral propagation involves the production of RNA from the integrated DNA prophage, and any introns present would be removed from that RNA molecule.
The RNA is then packaged into progeny virus
particles.

2. When a normal cellular gene (proto-oncogene) becomes a retroviral oncogene, it is
altered significantly. That is, transducing
retroviruses typically arise by complex rearrangements of the DNA after integration of a
retrovirus near a cellular proto-oncogene (Figure 21.39). Usually, the newly acquired genetic information replaces some or all of the
gag, pol, and *env* genes making the new retrovirus defective (and incapable, therefore, of
replication).

691
*Oncogenes
and the
Molecular
Biology
of Cancer*

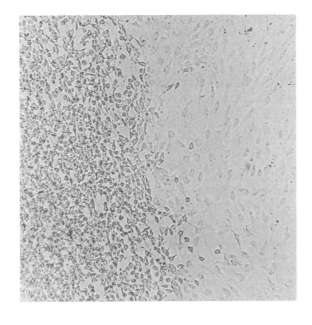

Figure 21.37 Mouse fibroblast cells transformed by DNA from a line of tumor cells. Untransformed cells (right) *form a monolayer, whereas transformed cells* (left) *grow and divide and form a dense, multilayered mass of cells.*

3. The v-*onc* is transcribed in a different range of cells, and results in larger amounts of mRNA than the corresponding proto-oncogene. These result from the fact that the gene is now under viral control, using retroviral promoter, enhancer, and poly(A) addition signals. Thus, there is a significant quantitative change in the protein encoded by the oncogene. In addition, since the proto-oncogene may also have changed when picked up by the retroviruses (e.g., point mutations, additions, deletions, rearrangements), there is typically a qualitative change in that protein, ex-

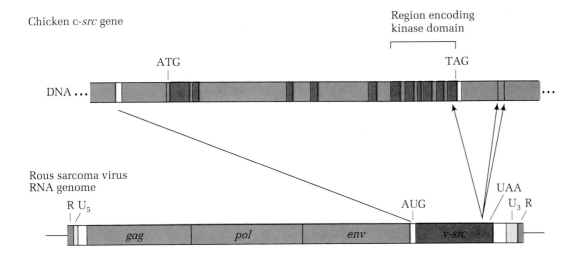

Figure 21.38 Molecular organization of the chicken c-src *gene. The gene contains thirteen exons (shown as colored boxes). Untranslated parts of exons are shown as white boxes. Below is the molecular organization of the rous sarcoma virus RNA genome to indicate the relationship of nucleotide sequences in* c-src *and* v-src; *the arrows signify regions of the* c-src *gene and flanking sequences from which* v-src *was generated, mostly by intron removal.*

hibited as functional differences from the proto-oncogene product.

Interestingly, the c-*onc*'s are highly conserved in the evolutionary sense. For example, similar sequences to the RSV v-*src* are present in the DNA of all vertebrates, as well as in DNA of *Drosophila*. Similar results have been obtained for other v-*onc*'s.

In at least one case, sequence similarities have

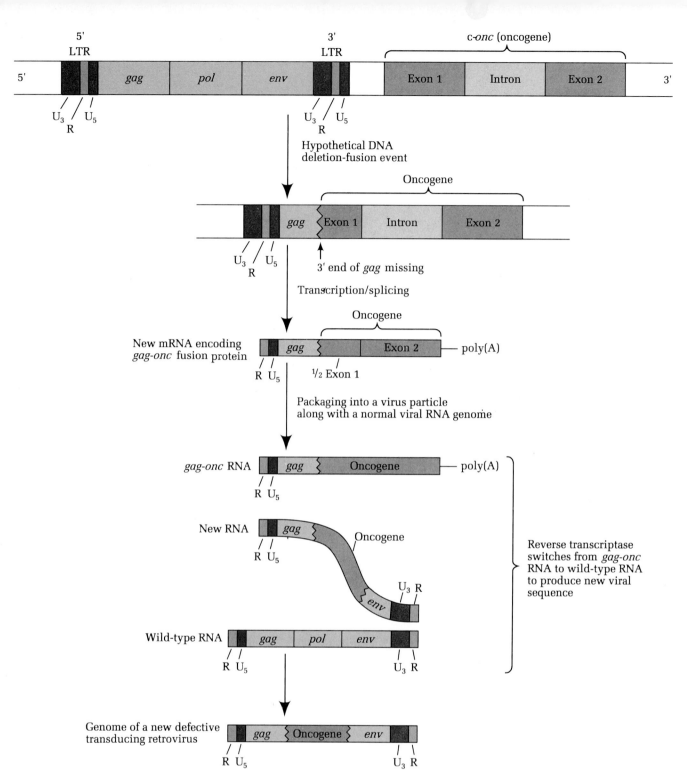

Figure 21.39 Model for the formation of a transducing retrovirus. First, a wild-type provirus integrates near a cellular proto-oncogene (c-onc). Next, a hypothetical deletion-fusion event fuses a c-onc exon into the gag region of the retrovirus. The LTR in this fusion directs the synthesis of a transcript that undergoes splicing to produce an mRNA for the gag-onc fusion protein. If the cell is also infected with a wild-type (helper) retrovirus, as part of the virus life-cycle, a wild-type RNA and a gag-onc RNA can be packaged into one virus particle. In a virus particle containing the two RNAs, reverse transcriptase can start copying the gag-onc RNA and then switch to copying the wild-type RNA to produce a new defective transducing retrovirus. (From James D. Watson, et al., Molecular biology of the gene, 4th ed. Copyright © 1965, 1970, 1976, 1987 by The Benjamin/Cummings Publishing Company, Inc. Reprinted by permission.)

693
Oncogenes
and the
Molecular
Biology
of Cancer

been found between v-*onc*'s and DNA sequences from the lower eukaryotic microorganism *Saccharomyces cerevisiae* (yeast). Here the v-*ras* oncogene of Harvey sarcoma virus (H-*ras:* derived from the rat proto-oncogene c-*ras*) was shown to hybridize weakly with two DNA sequences of yeast, defining the *RAS1* and *RAS2* genes. Mutation studies have indicated that the function of either *RAS* gene alone may be knocked out without affecting cell growth, but that double mutants (*ras1 ras2*) do not grow. Thus, these *RAS* homologs appear to play an important role in cellular metabolism. In fact, it has been generally concluded that proto-oncogenes (i.e., the *normal* gene counterpart of oncogenes) must have essential, basic functions involved with cellular metabolism and its regulation.

Protein Products of Proto-oncogenes

There are approximately 30 known proto-oncogenes. However, there are similarities between them at the DNA level and, interestingly, at the amino acid sequence level, indicating that the proto-oncogenes fall into several distinct classes, each with a characteristic type of protein product (Table 21.4). The major classes of protein products are growth factors (e.g., *sis* product), protein kinases (e.g., *src* product), GTP-binding proteins (e.g., H-*ras* product), nuclear proteins (e.g., *myc* product), and hormone receptors (e.g., *erbA* product). In the following, then, we will consider just two examples of protein products, growth factors and protein kinases.

Growth factors. Based on the effect of oncogenes on cell growth and division, it was hypothesized early that proto-oncogenes might be "switching" genes involved with the control of cell multiplication during differentiation. Evidence supporting this hypothesis came in the early 1980s when the product of a viral oncogene v-*sis* was shown to be identical to part of platelet-derived growth factor (PDGF), a factor found in blood platelets in mammals, which is released after tissue damage. PDGF affects only one type of cell, fibroblasts, causing them to grow and divide. The fibroblasts are part of the

wound-healing system. PDGF itself consists of two polypeptides, one of which has been shown to be encoded by c-*sis.* The causal link between PDGF and tumor induction was demonstrated in an experiment in which the cloned PDGF gene was introduced into a cell that normally does not make PDGF (i.e., a fibroblast); that cell was transformed into a tumor cell.

We can generalize and say that tumor cells can result from the synthesis of growth factors. These growth factors are not normally produced in those cells, but the introduction of a certain class of transducing retrovirus can cause the growth factor genes to come under viral and not cellular control.

Protein kinases. A large number of proto-oncogenes appear to encode protein kinases. These enzymes catalyze the addition of phosphate groups to proteins, thereby modifying their functions. The *src* gene product, for example, has been shown to be a protein kinase and has been termed pp60^{src}. The viral protein, pp60^{v-src}, and the proto-oncogene, pp60^{c-src}, differ in only a few amino acids, and both proteins are found bound to the inner surface of the plasma membrane. What is particularly interesting about the *src* protein kinases is that both versions add a phosphate group to the amino acid tyrosine; that is, they are tyrosine protein kinases. Before this discovery, the protein kinases that had been characterized had all been shown to add phosphates only to the amino acids serine or threonine. Since protein phosphorylation was known to be an important event in effecting a multitude of metabolic changes in cells, the *src* discovery was an exciting one since it suggested a possibility of how the *src* and other tyrosine protein kinase-coding oncogenes might transform a normal cell into a cancer cell in which many metabolic differences are evident compared with normal cells. For example, a large class of proteins, including the receptors for growth factors, use protein phosphorylation to transmit their signals through the membrane. In the specific case of the *src* product, a cascade of regulatory events occurs. First, the protein kinase catalyzes the phosphorylation of phosphatidyl inositol, which

Table 21.4 Oncogene Families and Their Protein Products

Oncogene	Subcellular Location of Protein	Properties or Normal Function of Protein	Oncogene Found in Animal Retrovirus
Class I: Protein kinases			
src	Plasma membrane	Tyrosine-specific protein kinase	Rous avian sarcoma
yes	Plasma membrane	Tyrosine-specific protein kinase	Yamaguchi avian sarcoma
fgr	?	Tyrosine-specific protein kinase	Gardner-Rasheed feline sarcoma
abl	Plasma membrane	Tyrosine-specific protein kinase	Abelson murine leukemia
fps (*fes*)	Cytoplasm	Tyrosine-specific protein kinase	Fujinami avian sarcoma (and feline sarcoma)
*erb*B	Plasma membrane (transmembrane)	EGF receptor/tyrosine-specific protein kinase	Avian erythroblastosis
fms	Plasma membrane (transmembrane)	CSF-1 receptor/tyrosine-specific protein kinase	McDonough feline sarcoma
ros	Plasma membrane (transmembrane)	Tyrosine-specific protein kinase	UR II avian sarcoma
kit	Plasma membrane		Feline sarcoma
mos	Cytoplasm	Serine/Threonine protein kinase	Moloney murine sarcoma
raf (*mil*)	?	Serine/Threonine protein kinase	3611 Murine sarcoma
Class II: GTP binding proteins			
H-*ras*	Plasma membrane	Guanine nucleotide binding protein with GTPase activity	Harvey murine sarcoma
K-*ras*	Plasma membrane	Guanine nucleotide binding protein with GTPase activity	Kirsten murine sarcoma
Class III: Growth factors			
sis	Secreted	Derived from a gene that encodes platelet derived growth factor (PDGF)	Simian sarcomca
Class IV: Nuclear proteins			
myc	Nucleus		Avian MC29 myelocytomatosis
myb	Nucleus		Avian myeloblastosis
fos	Nucleus		FBJ osteosarcoma
ski	Nucleus		Avian SKV770
Class V: Hormone receptor			
*erb*A	Cytoplasm	Thyroid hormone receptor	Avian erythroblastosis

Adapted from James D. Watson, et al., *Molecular biology of the gene,* 4th ed. Copyright © 1965, 1970, 1976, 1987 by The Benjamin/Cummings Publishing Company, Inc. Reprinted by permission.

695
*Analytical
Approaches
for Solving
Genetics
Problems*

results in an increase in the concentration of diacylglycerol. This compound, in turn, activates the enzyme protein kinase C, which is thought to play a key role in the sequence by which growth factor receptors transmit their message to the nucleus. Thus, the action of protein kinases with regard to oncogene functions appears also to be linked to growth factors and their activities.

Keynote *Some forms of cancer are caused by tumor viruses. The RNA tumor viruses are retroviruses which contain tumor-inducing genes called oncogenes. Both normal cells and non-viral-induced cancer cells contain sequences that are homologous to the viral oncogenes. Apparently, retroviruses that induce tumors have picked up certain normal cellular genes, called proto-oncogenes, while simultaneously, they lose part of their genetic information. The proto-oncogenes function in normal cells in various ways to regulate cell differentiation. In the retrovirus, these genes have become modified so that the protein product, now produced under viral control, is both quantitatively and qualitatively changed, as well as being expressed in cells in which their products are not usually found. These gene products, which include growth factors, are directly responsible for the transformation of cells to the cancerous state.*

Analytical Approaches for Solving Genetics Problems

Q.1 A region of the yeast chromosome specifies three enzyme activities in the histidine biosynthesis pathway; these activities are synthesized coordinately. How would you distinguish among the following three models?

a. Three genes are not organized into an operon. They code for three discrete mRNAs that are translated into three distinct enzymes.
b. Three genes are arranged in an operon. The operon is transcribed to produce a single polygenic mRNA, whose translation produces three distinct enzymes.
c. One gene (a supergene) is transcribed to produce a single mRNA, whose translation produces a single polypeptide with three different enzyme activities.

A.1 A key feature of an operon (model b) is that a contiguously arranged set of genes is transcribed onto a single polygenic mRNA. Thus a nonsense mutation in a structural gene will result in the loss of not only the enzyme activity coded for by that gene but also the enzyme activities coded for by the structural genes that are more distant from the promoter (see Chapter 20, p. 608). If there is a supergene coding for a single polypeptide with three different enzyme activities (model c), polar effects of nonsense mutations will also be seen. However, if there are three genes that are closely linked but each with its promoter (model a), transcription will produce three distinct mRNAs, and a nonsense mutation will affect only the gene in which it is located and no other gene. Therefore, if nonsense mutations are shown to have polar effects, then either model b or model c is correct, but model a cannot be correct. If nonsense mutations do not have polar effects, no matter in which gene they are located, then model a must be correct.

Characterization of the enzyme activities coded for by the three genes would enable us to distinguish between models b and c. That is, in the operon model b three distinct polypeptides would be produced. These polypeptides could be isolated and purified individually by using standard techniques.

Note that it would be particularly important to make sure that inhibitors of protein-degrading enzymes, proteases, are present during cell disruption so that if there is a trifunctional polypeptide present, it is not cleared by the proteases to produce three separable enzyme activities.

Thus if the operon model b is correct, we could show that there are three distinct polypeptides, each exhibiting only one of the enzyme activities— that is, three polypeptides and three enzyme activities. On the other hand, if the supergene model c is correct, it should *only* be possible to isolate a large polypeptide with all three enzyme activities (again assuming careful isolation and purification procedures); no polypeptides with only one of the enzyme activities should exist.

Questions and Problems

21.1 Are there operons in eukaryotes? Discuss the features of the *qa* gene cluster in *Neurospora crassa* that suggested that it might be a eukaryotic operon. What characteristic feature or features of bacterial operons does this eukaryotic system lack?

21.2 Discuss the regulation of gene expression of the *qa* gene cluster of *Neurospora crassa*.

*21.3 In what ways do the functions of eukaryotic multifunctional proteins (which have several enzyme activities for a biosynthetic pathway) resemble and differ from the functions of prokaryotic operons?

*21.4 Eukaryotic organisms have a large number of copies (usually more than a hundred) of the genes that code for ribosomal RNA, yet they have only one copy of each gene that codes for each ribosomal protein. Explain why.

21.5 What is a hormone?

21.6 How do hormones participate in the regulation of gene expression in eukaryotes?

21.7 Distinguish between the terms *development* and *differentiation*.

21.8 What role does cell-to-cell contact play in maintaining the differentiated state of a tissue?

21.9 What is totipotency? Give an example of the evidence for the existence of this phenomenon.

21.10 Discuss some of the evidence for differential gene activity during development.

*21.11 The enzyme lactate dehydrogenase (LDH) consists of four polypeptides (a tetramer). Two genes are known to specify two polypeptides, A and B, which combine in all possible ways (A_4, A_3B, A_2B_2, AB_3, and B_4) to produce five LDH isozymes. If, instead, LDH consisted of three polypeptides (i.e., it was a trimer), how many possible isozymes would be produced by various combinations of polypeptides A and B?

21.12 Discuss the expression of human hemoglobin genes during development.

21.13 Discuss the organization of the hemoglobin genes in the human genome. Is there any correlation with the temporal expression of the genes during development?

21.14 What are the polytene chromosomes? Discuss the molecular nature of the puffs that occur in polytene chromosomes during development.

21.15 Puffs of regions of the polytene chromosomes in salivary glands of *Drosophila* are surrounded by RNA molecules. How would you show that this RNA is single-stranded and not double-stranded?

*21.16 In experiment A, ³H-thymidine (a radioactive precursor of DNA) is injected into larvae of *Chironomus,* and the polytene chromosomes of the salivary glands are later examined by autoradiography. The radioactivity is seen to be distributed evenly throughout the polytene chromosomes. In experiment B, ³H-uridine (a radioactive precursor of RNA) is injected into the larvae, and the polytene chromosomes are examined. The radioactivity is first found only around puffs; later, radioactivity is also found in the cytoplasm. In experiment C, actinomycin D (an inhibitor of transcription) is injected into larvae and then ³H-uridine is injected. No radioactivity is found associated with the polytene chromosomes, and few puffs are seen. Those puffs that are present are much smaller than the puffs found in experiments A and B. Interpret these results.

21.17 Define *imaginal disk, homeotic mutant,* and *transdetermination.*

*21.18 If actinomycin D, an antibiotic that inhibits RNA synthesis, is added to newly fertilized frog eggs, there is no significant effect on protein synthesis in the eggs. Similar experiments have shown that actinomycin D has little effect on protein synthesis in embryos up until the gastrula stage. After the gastrula stage, however, protein synthesis is significantly inhibited by actinomycin D, and the embryo does not develop any further. Interpret these results.

*21.19 It is possible to excise small pieces of early embryos of the frog, transplant them to older embryos, and follow the course of development of the transplanted material as the older embryo develops. A piece of tissue is excised from a region of the late blastula or early gastrula that would later develop into an eye and is transplanted to three different regions of an older embryo host (see a in the figure). If the tissue is transplanted to the head region of the host, it will form eye, brain, and other material characteristic of the head region. If the tissue is transplanted to other regions of the host, it will form organs and tissues characteristic of those regions in normal development (e.g., ear, kidney, etc.). In contrast, if tissue destined to be an eye is excised from a neurula and transplanted into an older embryo host to exactly the same places as used for the blastula/gastrula transplants, in every case the transplanted tissue differentiates into an eye (see b in the figure). Explain these results.

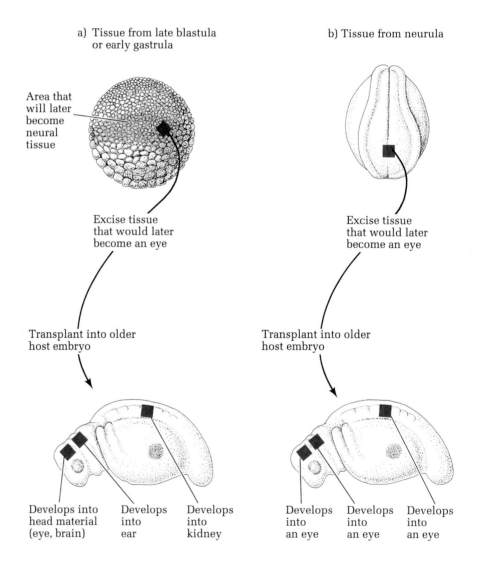

a) Tissue from late blastula or early gastrula

b) Tissue from neurula

Area that will later become neural tissue

Excise tissue that would later become an eye

Excise tissue that would later become an eye

Transplant into older host embryo

Transplant into older host embryo

Develops into head material (eye, brain)

Develops into ear

Develops into kidney

Develops into an eye

Develops into an eye

Develops into an eye

22

Organization and Genetics of Extranuclear Genomes

To this point in the text, we have analyzed the structure and expression of the genes located on chromosomes in the nucleus of eukaryotic organisms and have defined rules for the segregation of nuclear genes. The nucleus is not the only place in the cell in which DNA is located, however. Outside the nucleus, DNA is found in two principal organelles, the mitochondria (found in both animals and plants) and chloroplasts (found only in green plants). The genes in these mitochondrial and chloroplast genomes have become known as extrachromosomal genes, cytoplasmic genes, non-Mendelian genes, organellar genes, or extranuclear genes.

Although *extranuclear* is used in the discussions that follow, the term *non-Mendelian* is also informative because extranuclear genes do not follow the rules of Mendelian inheritance we expect of nuclear genes in segregation and recombination. Recently, the application of modern molecular biology techniques has led to rapid advances in knowledge about the organization of extranuclear genomes. In this chapter we examine some of these advances and discuss the inheritance patterns that extranuclear genes follow from generation to generation. First we examine the genetic structure of mitochondria and chloroplasts, and then we discuss how their genomes segregate. There are nonchromosomal, nuclear genetic elements that segregate in a non-Mendelian fashion, but they will not be discussed in this text.

Organization of Extranuclear Genomes

Mitochondrial Genome

Mitochondria, organelles found in the cytoplasm of all aerobic animal and plant cells, are the principal sources of energy in the cell. They contain the enzymes of the Krebs cycle, carry out oxidative phosphorylation, and are involved in fatty acid biosynthesis. The presence of DNA in mitochrondria has been shown by a variety of techniques, including electron microscopy,

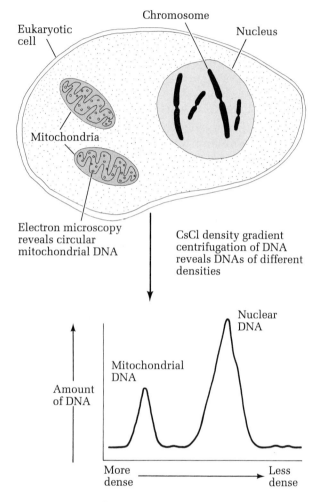

Figure 22.1 *Nuclear and mitochondrial DNA. Mito-chondrial DNA can be detected under the electron microscope as small circles and can often be separated from nuclear DNA by CsCl density gradient centrifugation since the two DNAs differ in buoyant density.*

autoradiography, and the isolation from mito-chondria of a DNA species with a buoyant density different from that of nuclear DNA (see Figure 22.1). Some of the properties of mitochondrial DNA, or mtDNA, are presented in Table 22.1 along with properties of nuclear and chloroplast DNAs.

Structure. Mitochondrial DNA is a double-stranded, supercoiled, circular chromosome de-void of histones and, therefore, not organized into nucleosomes (Figure 22.2). (As we shall see shortly, a mitochondrion may contain a number of identical copies of the mitochondrial DNA molecules.) The size of the chromosome varies from organism to organism but is constant among mitochondria within any particular spe-cies. Interestingly, the chromosomes of higher-animal mitochondria are significantly smaller than those of fungi and plants. The mitochon-drial chromosome in humans, for example, is about 14,000 base pairs, while that from *Droso-phila* is 18,000 base pairs, that from *Neurospora crassa* is 60,000 base pairs, and that from yeast is 75,000 base pairs. The mitochondrial genomes of plants are even larger, ranging from 250,000 to 2 million base pairs (80 to 800 μm), depend-ing on the species. Maize, for example, has a mi-tochondrial genome of 600,000 base pairs (200 μm).

Despite the differences in the amount of DNA, evidence indicates that mitochondria from all organisms contain about the same amount of single-copy DNA; that is, DNA that codes for functional and mitochondrial product. The dif-ference is that essentially all the mitochondrial genomes of animals code for products whereas the mitochondrial genomes of fungi and plants include a lot of DNA that does not code for products, in addition to coding sequences.

While the informational content of the mito-chondrial genome is very small relative to that of the nuclear genome, the relative amount of mitochondrial DNA is actually quite large. Within mitochondria are nucleoid regions (simi-lar to those of bacterial cells), each of which contains several copies of the mitochondrial chromosome. Yeast, for example, has 4–5 mtDNA molecules per nucleoid, and each mito-chondrion has 10–30 nucleoids. Since each yeast cell has between 1 and 45 mitochondria per cell, there are many mtDNA molecules per cell. This multiplicity of mtDNA molecules within mitochondria has been found in other studied organisms, and so it can be assumed that the mitochondrial genome is *polyploid* in a given mitochondrion. As a consequence of the

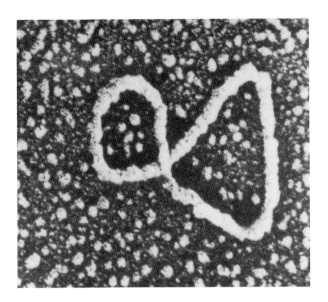

Figure 22.2 EM of mitochondrial DNA.

multiplicity of mitochondrial chromosomes, the contribution of mitochondrial activities to overall cellular activities can be relatively significant when compared with nuclear contributions.

Replication. Replication of mtDNA uses mitochondrial DNA polymerases that function totally independently of nuclear DNA polymerases. Moreover, the replication process occurs throughout the cell cycle without any preference for the S phase of the cell cycle when nuclear DNA replicates. Some differences appear in the details of mtDNA replication among eukaryotes, and the displacement loop (D loop) model (deduced from observing animal mitochondria in vivo) is useful as a general scheme. Figure 22.3 shows an electron micrograph of an early stage of D loop replication; Figure 22.4 diagrams the model in more detail.

Table 22.1 Properties of Nuclear, Mitochondrial, and Chloroplast DNAs

	Nuclear DNA		*Mitochondrial DNA*		*Chloroplast DNA*	
	%GC	*Genome Size (bp for all chromosomes)*	*%GC*	*Genome Size (bp per chromosome)*[a]	*%GC*	*Genome Size (bp per chromosome)*[b]
ANIMALS						
Human	41	2.75×10^9	41	14×10^3	—	—
Chick	43	1.2×10^9	49	15×10^3	—	—
Sea urchin	39	7.2×10^8	45	13×10^3	—	—
D. melanogaster	42	1.75×10^8	22	18×10^3	—	—
FUNGI						
N. crassa	53	4.3×10^7	42	60×10^3	—	—
S. cerevisiae	44	1.75×10^7	18	75×10^3	—	—
PLANTS						
C. reinhardi	64	1×10^8	71	2×10^5	36	30×10^3
Corn	49	6.6×10^9	48	2.2×10^5	39	13×10^3
Tobacco	38	1.1×10^9	51	2×10^5	47	15×10^3
Sweet pea	35	9.4×10^9	38	9×10^4	46	12×10^3

[a] Each mitochondrion is believed to contain about five copies of the mitochondrial chromosome.

[b] Each chloroplast is believed to contain about twenty copies of the chloroplast chromosome.

Source: Data mainly compiled from "Chloroplast DNA: Physical and genetic studies" by Ruth Sager and Gladys Schlanger and "Mitochondrial DNA" by Margit M. K. Nass, 1976. In *Handbook of genetics,* Robert C. King, ed., v. 5, *Molecular genetics,* p. 371 and p. 477. New York: Plenum Press.

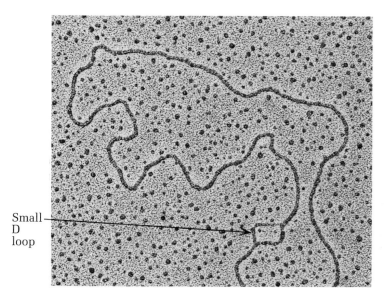

Figure 22.3 Electron micrograph of the D loop structure found in replicating mitochondrial DNA; a small D loop is shown.

Small
D
loop

In most animals the two strands of mtDNA have different densities, and they are called the H (heavy) and L (light) strands. The D loop model of replication shows relative asynchrony in DNA replication for the two complementary H and L strands. In the D loop model the synthesis of the new H strand is started in one origin (the L strand origin) and forms a D loop structure (which can be seen by electron microscopy). As the new H strand extends, initiation of synthesis of the new L strand at a second origin (the H strand origin) takes place. Both strands are completed by continuous replication. Lastly, the circular DNAs are each converted to a supercoiled form with approximately a hundred superhelical twists.

As in replication of nuclear DNA, RNA primers are synthesized for initiation. Apparently, the excision of these primers is not precise since interspersed ribonucleotides are found in the mtDNA. The evidence is that mtDNA can be cleaved by treatment with alkali or RNase, both of which are known to cleave at phospho-

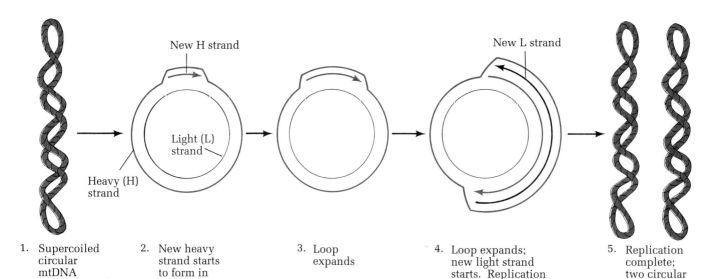

1. Supercoiled circular mtDNA (approximately 100 coils) uncoils

New H strand

Light (L) strand

Heavy (H) strand

2. New heavy strand starts to form in displacement loop

3. Loop expands

New L strand

4. Loop expands; new light strand starts. Replication structure resembles a letter D

5. Replication complete; two circular mtDNAs supercoil

Figure 22.4 Model for mitochondrial DNA replication that involves the formation of a D loop structure.

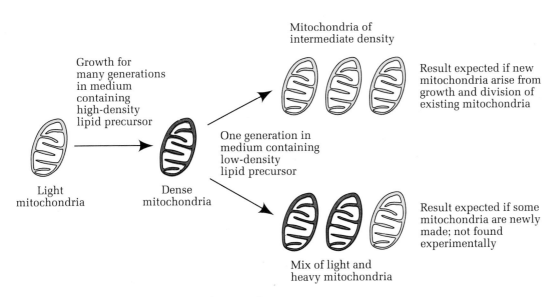

Figure 22.5 *Luck's experiment for showing that mitochondria arise by division of preexisting mitochondria and not by assembly from simple components.*

diester linkages in which a ribose sugar is involved. About ten of these ribonucleotides are
found in each mtDNA, with most of them clustered around the replication origins.

Another facet of the replication process was
examined to discover how mitochondria themselves reproduce so that in cell division each
cell gets mitochondria. Evidence suggests that
mitochondria (and chloroplasts) grow and divide rather than assemble from simple components. The classical experiment for replication
of complete mitochondria was done in 1963 by
David Luck (see Figure 22.5), who used density-
labeling procedures to study mitochondrial division in *Neurospora crassa.* In essence, this experiment is similar to the one done by Meselson
and Stahl that showed that DNA replication is
semiconservative (see Chapter 10). Luck grew *N.
crassa* on a medium that was very low in protein but high in a lipid precursor, choline. Under these conditions the mitochondria produced
were less dense than those found in *Neurospora*
cultures grown in normal medium.

Next, Luck shifted the *Neurospora* cells to a
medium with a high protein content and a low

lipid precursor content. After one generation he
examined the mitochondria for their density.
There are two possible outcomes: If mitochondria grow and divide, then all progeny mitochondria would contain half of the material
made in the low-density medium and half of the
material made in the high-density medium.
These mitochondria would have an intermediate
density between light and heavy mitochondria.
If, however, mitochondria are made de novo,
then there would be two populations of mitochondria: those of low density made early in the
experiment and those of high density made after
the shift to high-density medium. Luck's results
paralleled the f rst prediction and indicated that
mitochondria grow and divide.

Gene organization of mtDNA. Mitochondrial
DNA contains information for a number of mitochondrial components such as tRNAs, rRNAs,
and proteins. Mitochondria contain ribosomes
that are responsible for all protein-synthetic activity that occurs within mitochondria. Messenger RNAs synthesized within the mitochondria
remain in the organelle and are translated by

mitochondrial ribosomes. The two rRNA molecules found in mitochondrial ribosomes are coded for in the mtDNA, while most, if not all, of the ribosomal proteins of mitochondrial ribosomes are coded for in the nuclear genome. In general, mtDNA codes for 2 rRNAs, 22–23 tRNAs, and 10–12 proteins that are part of the inner mitochondrial membrane and are hydrophobic to some degree. The mtDNA molecule has been studied in a variety of organisms, and since it is relatively small, a great deal of progress has been made in determining the exact organization of its genes.

One approach to mapping mtDNA physically has been through denaturation. The DNA is placed in a medium that causes stretches of AT base pairs to denature while GC-rich regions remain hydrogen-bonded. This selective denaturation leads to the formation of denaturation bubbles that can be visualized by electron microscopy. This physical map shows AT-rich regions of the DNA that can be used as landmarks in other studies.

A more modern approach is to sequence the DNA by using rapid DNA-sequencing procedures (Chapter 15, pp. 508–509). Today the sequence of human (and several other organisms') mtDNA is completely known. Some of its features were unexpected and may be general for mtDNAs. Figure 22.6 presents the map of the genes and transcripts of human mtDNA. The genes coding for the 16S and 12S rRNAs of the large and small mitochondrial ribosomal subunits are located adjacent to one another on the H strand. The genes for tRNAs are present at various positions around the H strand and the L strand, some in clusters, some individually, and one between the two rRNA genes.

The structural genes are also found in both strands. Their positions were identif ed in two ways. One method involved searching the DNA sequence for possible start and stop signals in the same reading frame (an open reading frame, or ORF; see Chapter 21, p. 661) that were separated by significant lengths and that could code for amino acids. This search was accomplished by using computer programs designed for this purpose. The second method (used pri-

marily by G. Attardi and his group) was to align the 5′ and 3′ end proximal sequences of mitochondrial mRNAs with the DNA sequence. The mitochondrial mRNAs were isolated and were found to have a poly(A) tail at the 3′ end but no 5′ cap. As in the case of cytoplasmic mRNAs (i.e., the mRNAs coded for by nuclear genes but found in the cytoplasm), the poly(A) tail is a great aid in isolating these molecules.

At this time the products of all ORFs of the human mitochondrial genome have been functionally def ned. Thus the genetic content of the genome is now completely known. Among the approaches taken in elucidating the functional content of the human mitochondrial genome, one exploited the expected evolutionary similarity between human and animal mitochondrial proteins. For example, the amino acid sequence of subunit II of the enzyme cytochrome oxidase (COII) from humans, an enzyme found in all mitochondria, was known for the enzyme found in bovine mitochondria from direct-sequencing experiments. (Cytochromes are proteins located in mitochondria that are used in the electron transport chain and are needed for the oxidation of metabolic products to generate ATP as an energy source.) Since the chemical structure of the enzyme was expected to be the same for the human and bovine enzymes, on the basis of evolutionary arguments, it was assumed that there is a high degree of homology between the amino acid sequence of human and bovine COII.

Thus researchers were able to identify a region of homology in the human mtDNA sequence that was probably the gene for human COII. By inspecting the presumed COII gene sequence of human mtDNA, using the bovine enzyme amino acid sequence as a guide, scientists found a TGA nucleotide sequence in the reading frame that would produce the UGA codon in a mitochondrial mRNA. In the translation of mRNAs coded for by nuclear genes, the UGA codon is a chain-terminating codon (see Chapter 13, p. 417). Inspection of the bovine protein revealed, however, that the translation of the COII mRNA is not terminated at the UGA; instead, the amino acid tryptophan is found at that position. This result suggested that the genetic code

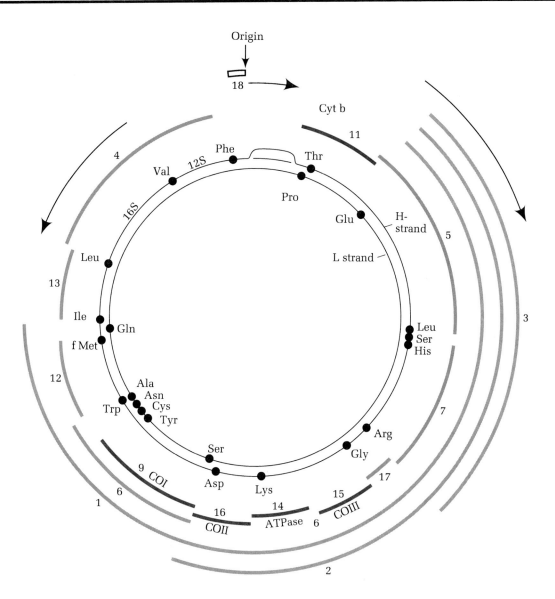

Origin

18

Cyt b

11

Phe

12S

Val

Thr

16S

4

Pro

Glu

H-strand

L strand

5

Leu

13

Leu
Ser
His

Ile

Gln

f Met

3

12

7

Ala
Asn
Cys
Tyr

Trp

Arg

Gly

Ser

17

Asp

Lys

9 COI

6

1

16

14

15

COII

ATPase 6

COIII

2

Figure 22.6 Map of the genes and transcripts of human mitochondrial DNA. The two inner circles show the positions of the 12S and 16S rRNA genes at 10 to 12 o'clock and of the tRNA genes. The curved solid bars and the hatched bars show the mapping positions of the H strand and L strand poly(A)+ RNA transcripts, respectively. Five structural genes and the corresponding mRNAs have been identified as specific for cytochrome c oxidase subunits I, II, and III (COI, COII, and COIII), ATPase, and cytochrome b.

may not be read in quite the same way in the mitochondria as in the cytoplasm. Other differences were found; for example, ATA in a mitochondrial gene codes for methionine, whereas in the cytoplasm it codes for isoleucine.

Figure 22.7 shows the human mitochondrial genetic code, the coding properties of the tRNAs, and the total number of codons used in the whole genome. All the codons are used except the chain-terminating codons UAA and UAG, which have the same function in the classical genetic code. No tRNAs have been found

Second letter

	U	C	A	G	
U	77 UUU Phe 140 UUC 73 UUA Leu 17 UUG	32 UCU 99 UCC 83 UCA Ser 7 UCG	46 UAU Tyr 89 UAC UAA Stop UAG Stop	5 UGU Cys 17 UGC 93 UGA Trp 10 UGG	U C A G
C	65 CUU 167 CUC Leu 276 CUA 45 CUG	41 CCU 119 CCC Pro 52 CCA 7 CCG	18 CAU His 79 CAC 81 CAA Gln 9 CAG	7 CGU 25 CGC Arg 29 CGA 2 CGG	U C A G
A	125 AUU Ile 196 AUC 166 AUA Met 40 AUG	51 ACU 155 ACC Thr 133 ACA 10 ACG	33 AAU Asn 130 AAC 85 AAA Lys 10 AAG	14 AGU Ser 39 AGC AGA Stop AGG Stop	U C A G
G	30 GUU 49 GUC Val 71 GUA 18 GUG	43 GCU 124 GCC Ala 80 GCA 8 GCG	15 GAU Asp 55 GAC 64 GAA Glu 24 GAG	24 GGU 88 GGC Gly 67 CGA 34 GGG	U C A G

First letter / Third letter

Figure 22.7 The genetic code of the human mito-chondrial genome. Each box indicates that one tRNA is used to read those codons within. The numbers within each box indicate the total number of those codons found in the whole mitochondrial DNA molecule. Highlighted are the codons that have different coding properties in the human mitochondrion than in the nuclear genome of all eukaryotic organisms or in the genomes of prokaryotic organisms.

with anticodons for AGA and AGG, so perhaps these codons also have a chain-terminating function. In fact, some unidentified reading frames found by computer searches of the mtDNA sequence may end with these codons. Interestingly, the genetic code of yeast mitochondria is not identical with that of human mitochondria, and Table 22.2 shows the points of difference. Differences in the genetic code have also been detected in ciliated protozoans. These differences suggest that the mitochondrial genetic code may be read differently in many eukaryotic organisms, that is, the genetic code is not as universal as was originally thought.

The tRNA genes in human mtDNA have been identified by using a computer program designed to search the DNA sequence for polynucleotide stretches that could form the classical cloverleaf, two-dimensional structure. However, a number of features thought to be invariable in cytoplasmic tRNAs, in terms of the biological function, vary in mitochondrial tRNAs. For example, one serine tRNA has one complete loop missing, yet it is still functional in mitochondria.

Table 22.2 Differences between Human and Yeast Mitochondrial Genetic Codes

Codon[a]	Amino Acid		
	Nuclear Code	Mitochondrial Code	
		Mammal	Yeast
UGA	Termination	Tryptophan	Tryptophan
AUA	Isoleucine	Methionine	Isoleucine
CUN[b]	Leucine	Leucine	Threonine
AGA, AGG	Arginine	Termination?	Arginine
CGN[b]	Arginine	Arginine	Termination?

[a] All sequences read 5′ to 3′.

[b] N = any one of the four bases A, G, U, and C.

With base-pairing wobble a minimum of 32 tRNAs are needed to read all the 61 sense codons in the cytoplasm. As we saw in Chapter 13, there are many instances in the genetic code where the first two bases of a codon are the same, the same amino acid is specified whether the third base is U or C, or A or G, or U, C, A, or G. (Each group of codons that specifies the same amino acid with the same first two bases is called a *family box*.) No cytoplasmic tRNA can read more than three of these related codons even with wobble. So even when the same amino acid is specified by the first two bases no matter what the third base is (a four-codon family box), the four codons need two tRNAs in order for them to be read. Many of the mitochon- drial tRNAs, however, apprently have the ability to read all four members in a four-codon family box. Table 22.3 contrasts the two systems. As a result of this extended wobble situation in mito- chondria, only 22 tRNA genes are needed in mammalian mitochondria. For all four-codon family boxes only one tRNA is used, and this tRNA has U (uracil) in the anticodon at the posi- tion corresponding to the third letter of the co- don. Similar conclusions have been reached in studies of *Neurospora crassa* and yeast mito- chondria.

From quantitative DNA hybridization we know that each mitochondrial genome contains one copy of each of the two rRNA genes. Figure 22.8 diagrams the organization of the two rRNA

Table 22.3 Example of Four-Position, Base-Pair Wobble in Mitochondrial tRNAs[a]

Codon	Anticodon	
	Nuclear Wobble	Mitochondrial Wobble
GCU	GGC	UGC
GCC		
GCA	UGC	
GCG		

[a] All sequences read 5′ to 3′.

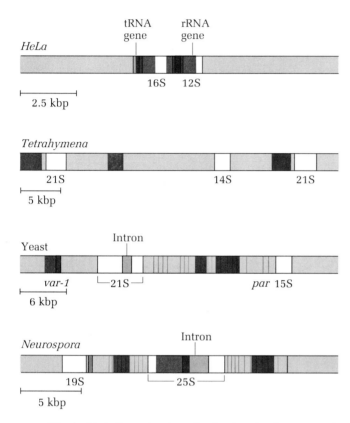

Figure 22.8 Organization of mitochondrial rRNA and mitochondrial tRNA genes in a number of organisms.

has presumably arisen by a duplication event and is not commonly found in mitochondrial genomes.

One of the striking features of human mtDNA is that it is efficiently organized into a compact structure. The reading frames for the protein-coding and rRNA genes are mostly adjacent to tRNA genes, with no, or very few, nucleotides in between. Apparently, very large transcripts are produced during the expression of mitochondrial genes. By processing, the tRNA transcript sequences are spliced out, since they are easily recognized by their cloverleaf structure. With essentially no spacers in between the mRNA transcripts and tRNA transcripts in the large RNA precursor molecules, this process generates mature tRNAs and monogenic mRNAs.

Another aspect of this compact organization of genes in the mitochondrial genome is that the AUG initiation codon of a mitochondrial mRNA is right at the 5′ end of the molecule, in contrast to being located internally in a cytoplasmic mRNA after a leader sequence that includes a ribosome binding site (see Chapter 13). Mitochondrial mRNAs also lack the 5′ cap structure found in cytoplasmic mRNAs. Since cytoplasmic mRNAs require the 5′ cap and an internally located AUG codon for translation, the different features of mitochondrial mRNAs indicate that the mitochondrial protein synthesis machinery must be quite different—the ribosome and the rest of the initiation mechanism may simply recognize the end of the mRNA without any requirement for a specific ribosome binding site or a 5′ cap structure. This mechanism of translation initiation is also different from that for prokaryotic systems in which an internal ribosome-binding site located upstream of the AUG initiation codon is needed.

Another way in which human mtDNA is efficiently organized is that the DNA sequence for most of the mitochondrial mRNAs does not encode a complete chain-terminating codon. Instead, the transcripts end with either U or UA in the same reading frame as the initiation codon, and apparently the polyadenylation (addition of a poly(A) tail at the 3′ end) of the mRNA supplies the missing part(s) of the UAA stop codon.

genes and local tRNA genes in a number of organisms and illustrates the variation in the organization of rRNA genes in mitochondrial genomes.

In animals, as exemplified by HeLa (human) cells, the genes for the small (12S) and large (16S) rRNAs are very closely linked. In the spacer separating them, one tRNA gene is found. By contrast, in fungi the two genes are separated by a large amount of DNA: 6 kbp in *Neurospora* and about 30 kbp in yeast. A number of tRNA genes are located in these spacer regions, and in most, if not all, cases these genes are in the same strand as the rRNA genes. The rRNA genes of *Neurospora* and yeast both contain introns (Figure 22.8). In the protozoan *Tetrahymena* there is an extra copy of the large rRNA gene, but not of the small rRNA gene. This extra copy

While this efficient organization is prevalent for animal mtDNA, the situation is completely different in those organisms with larger mitochondrial genomes. In yeast, whose genome is five times larger while coding for about the same number of components, the genes are separated by long AT-rich sequences that have no obvious biological function. Intervening sequences are also found in some genes in these organisms, while they are never found in animal mitochondrial genomes.

Mitochondrial ribosomes. Mitochondrial ribosomes are structurally diverse among organisms, with all of them having a higher protein/RNA ratio than *E. coli* ribosomes. Table 22.4 shows examples of structural characteristics of mitochondrial ribosomes in selected organisms. In animals mitochondrial ribosomes are about the same size or perhaps slightly larger than *E. coli* ribosomes in terms of molecular weight and volume. However, owing to the high protein/RNA content, they have a relatively low buoyant density of 55S. In such fungi as yeast and *Neurospora,* the mt ribosomes have a sedimentation coefficient of 70S–75S.

We can make some rough generalizations about the rRNA content of mt ribosomes. In most cases the ribosomes lack the 5S and 5.8S rRNA components characteristic of cytoplasmic ribosomes. One exception is higher plants, which have a 5S rRNA molecule in mt ribosomes that is distinct from that molecule found in cytoplasmic ribosomes. In other words, most mt ribosomes have two unequal-sized subunits with a single rRNA molecule in each subunit. At least in animal mitochondria, these rRNA molecules, which have sedimentation coefficients of 12S and 16S, are significantly smaller than the large rRNAs of *E. coli* ribosomes, which have sedimentation coefficients of 16S and 23S.

The number of ribosomal proteins found in mt ribosomes is not as well defined, and estimated numbers vary greatly depending on the method of electrophoresis used to separate the isolated proteins. Yeast and *Neurospora* appear to have 60–70 proteins in the mt ribosomes (significantly more than the 54 found in *E. coli* ribosomes), while animal mitochondrial ribosomes have between 70 and 100 ribosomal proteins.

The proteins of mitochondrial ribosomes appear to be totally distinct from the proteins found in cytoplasmic ribosomes, and with one possible exception each in *Neurospora* and yeast, they are encoded by nuclear genes. Hence they are synthesized on cytoplasmic ribosomes and subsequently transported into the mitochondria. This process points again to the remarkable abilities of cells to direct the products of protein synthesis to the specific sites where they are needed. We know that cytoplasmic ribosomes in eukaryotes contain about 65–70 ribosomal proteins, and lower eukaryotes appear to have 60–70 ribosomal proteins in mitochondrial ribosomes. Thus 125–140 ribosomal proteins are made on cytoplasmic ribosomes. About half of

Table 22.4 Relative Size (in S Values) of Mitochondrial (mt) Ribosomes and Cytoplasmic (cyto) Ribosomes in Three Organisms

Component	Human (HeLa)		Yeast		Neurospora	
	mt	*cyto*	*mt*	*cyto*	*mt*	*cyto*
RIBOSOMES	60S	74S	75S	80S	73S	77S
Large subunit	45S	60S	53S	60S	50S	60S
Small subunit	35S	40S	35S	40S	37S	37S
RIBOSOMAL RNA Large ribosomal subunit	16S	28S	21S	26S	25S	26S
Small ribosomal subunit	12S	18S	14S	18S	19S	17S

them subsequently migrate into the nucleus to be used in the assembly of cytoplasmic ribosomes, while the remainder enter the mitochondria, where they are used in the assembly of mitochondrial ribosomes along with the transcripts of the two rRNA genes located on the mitochondrial chromosome. But we have a lot to learn about this migration between cellular compartments.

Mitochondrial protein synthesis. Mitochondria have their own protein-synthetic machinery, and the process of protein synthesis is, in some ways, analogous to the process of protein synthesis in bacteria. A species of tRNA.fMet is present in mitochondria, and it appears to function in the initiation of protein synthesis. The initiation factors (IFs) are also very similar to those in bacteria. In a number of cases mitochondrial and bacterial components are interchangeable in an in vitro, protein-synthesizing system. For example, *Neurospora crassa* ribosomes are able to recognize, bind, and translocate *E. coli* fMet-tRNA.fMet in response to AUG, and *E. coli* IFs can substitute for mt IFs in catalyzing the initiation of protein synthesis. Moreover, the tRNA.fMet species from *Neurospora* mitochondria must have a high degree of structural similarity with its *E. coli* counterpart, since once charged with methionine, that amino acid can be formulated by using enzymes isolated from *E. coli*.

Another piece of evidence for the similarity between *E. coli* and mt ribosomes is that the latter are sensitive to most inhibitors of bacterial ribosome function, including streptomycin, spectinomycin, neomycin, and chloramphenicol. In some cases the degree of sensitivity to these agents, which are mostly antibiotics, is quite different from the situation in *E. coli*. The inhibitors may not work on mt ribosomes by the classical mechanisms worked out in studies of bacterial ribosomes. Also, mt ribosomes are generally insensitive to antibiotics or other agents to which cytoplasmic ribosomes are sensitive, such as cycloheximide. By the selective use of antibiotics we can investigate the site of synthesis of proteins found in mitochondria. For example,

those mt proteins that are synthesized in the presence of cycloheximide, an inhibitor of cytoplasmic ribosomes, must be made on mt ribosomes. Conversely, those mt proteins made in the presence of chloramphenicol, an inhibitor of mt ribosomes, must be made on cytoplasmic ribosomes. Using this approach, scientists have discovered that of the seven subunits of the mitochondrial enzyme cytochrome oxidase in yeast mitochondria, four are made in the cytoplasm and three are made in the mitochondria (see Figure 22.9).

Chloroplast Genome

Chloroplasts are cellular organelles found only in green plants and are the site of photosynthesis in the cells containing them. Chloroplasts are characterized by a double membrane surrounding an internal, chlorophyll-containing lamellar structure embedded in a protein-rich stroma. Like mitochondria, chloroplasts contain their own genomes, although we simply do not know as much about the chloroplast (cp) genome as we do about the mitochondrial genome.

Structure. In many respects the structure of the chloroplast genome is similar to the structure of mitochondrial genomes. Table 22.1 gives some properties of cpDNA. In all cases the DNA is double-stranded, circular, devoid of structural proteins, and supercoiled. An electron micrograph of cpDNA is shown in Figure 22.10a. In many cases the GC content of cpDNA differs greatly from that of the nuclear and mitochondrial DNA, which allows us to isolate cpDNA by using CsCl equilibrium density gradient centrifugation. Figure 22.10b shows the result of such an experiment using total cellular DNA from the unicellular alga *Chlamydomonas* (Figure 22.11). Two discrete bands of DNA are seen (apart from the two marker bands) and indicate the presence of molecular species with different base-pair compositions. Since the cpDNA has a GC content of 36 percent compared with 64 percent for nuclear DNA and 71 percent for mtDNA, the cpDNA produces a *satellite band* that can be isolated and studied further. The relatively large

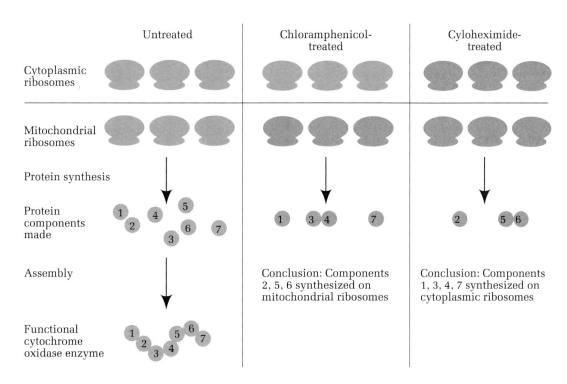

	Untreated	Chloramphenicol-treated	Cyloheximide-treated

Cytoplasmic ribosomes

Mitochondrial ribosomes

Protein synthesis

Protein components made

Assembly

Conclusion: Components 2, 5, 6 synthesized on mitochondrial ribosomes

Conclusion: Components 1, 3, 4, 7 synthesized on cytoplasmic ribosomes

Functional cytochrome oxidase enzyme

Figure 22.9 Synthesis of the multi-subunit protein cytochrome oxidase on cytoplasmic and mitochondrial ribosomes.

amount of nuclear DNA results in a fairly broad peak that does not permit resolution of the mtDNA peak.

The number of copies of cpDNA per chloroplast varies from species to species. In all cases there are multiple copies per chloroplast, and these copies are found in nucleoid regions that are also present in multiple copies. For example, leaf cells of the garden beet have 4–8 cpDNA molecules per nucleoid, 4–18 nucleoids per chloroplast, and about 40 chloroplasts per cell; giving almost 6000 cpDNA molecules per cell. In *Chlamydomonas* the one chloroplast in a cell generally contains between 500 and 1500 cpDNA molecules.

With a length of about 40–45 μm in many plants (62 μm in *Chlamydomonas*), cpDNA is approximately 8–9 times longer than animal mtDNA. Although its great length indicates that a potentially large amount of genetic information is coded for in cpDNA, how much of the

genome codes for products has yet to be determined.

Replication. The replication of cpDNA is a semiconservative process that probably is similar to the process for mitochondria. It is clear that chloroplast-specific enzymes are used in this replication process. These enzymes are distinct from nuclear replication enzymes and mitochondrial replication enzymes, and most of them are coded for by nuclear genes and transported into the chloroplast. Chloroplasts themselves grow and divide in essentially the same way as mitochondria do.

Gene organization of cpDNA. The genes encoded by cpDNA have been analyzed by using antibiotics selectively to define chloroplast protein products and by using hybridization of chloroplast rRNAs and tRNAs. From this analysis we know that the chloroplast genome

a)

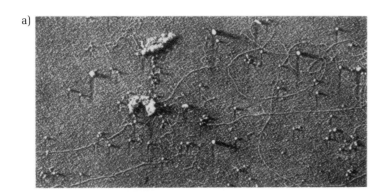

Figure 22.10 Chloroplast (cp) DNA: (a) electron micrograph of a chloroplast DNA molecule; (b) example of the results of a CsCl equilibrium density gradient centrifugation experiment done with total DNA extracted from the unicellular alga Chlamydomonas. *The large amount of nuclear DNA results in a broad peak that obscures the mtDNA peak.*

b) Satellite cpDNA band in CsCl gradient

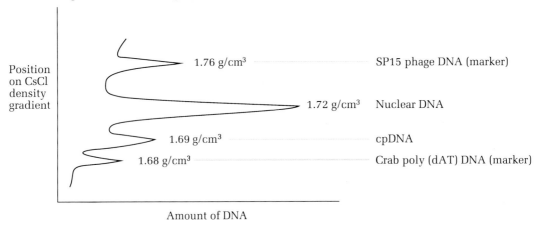

Position
on CsCl
density
gradient

1.76 g/cm³ .... SP15 phage DNA (marker)

1.72 g/cm³ Nuclear DNA

1.69 g/cm³ .... cpDNA

1.68 g/cm³ .... Crab poly (dAT) DNA (marker)

Amount of DNA

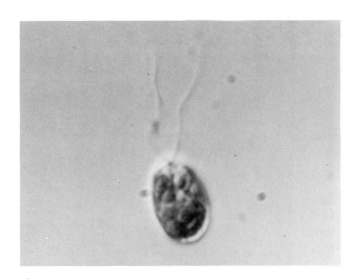

Figure 22.11 Chlamydomonas

contains genes for the cp rRNAs and perhaps for some ribosomal proteins, for a number of tRNAs, and for some cp enzymes and membrane proteins that are needed for electron transport in photosynthesis. However, relatively few cp structural genes have been identified, and only a few of these genes have been correlated directly with any chloroplast phenotype. This situation is in stark contrast to our state of knowledge of mitochondrial genomes for a number of organisms. All of the mRNAs from chloroplast genes are translated by chloroplast ribosomes.

One of the chloroplast proteins that *has* been well characterized is part of ribulose bisphosphate decarboxylase, the first enzyme used in the pathway for the fixation of carbon dioxide in the photosynthetic process. This enzyme is a major protein and is found in the chloroplasts of all plant tissues. Since it constitutes about 50 percent of the protein found in green-plant tissue, it is the most prevalent protein in the world. Ribulose bis-phosphate decarboxylase contains eight polypeptides, four identical small ones and four identical large ones. The small polypeptide is coded by a nuclear gene, and the large polypeptide is coded by a chloroplast gene. (This result was determined by using selective antibiotics in essentially the same way as they are used in mitochondrial protein synthesis studies.)

The organization of the rRNA genes in chloroplast genomes has been studied in detail in a few organisms, including *Chlamydomonas, Euglena* (Figure 22.12), corn, and spinach. Figure 22.13 shows the location and orientation of the rRNA genes in the cp genome of these plants, as well as the location of some of the known tRNA genes. With the exception of *Euglena,* these plants have two sets of the rRNA genes in each genome, and these sets are generally localized in two regions widely separated from one another on the DNA. *Euglena* has three sets of genes.

Figure 22.14 shows the detailed organization of cp rRNA genes in the four plants, again with the nearby tRNA genes. The details vary, but there are some generalizations. Each plant has a transcriptional order of 16S–23S–5S, which is extremely similar to the situation in *E. coli.* In addition, all the rRNA genes are in the same

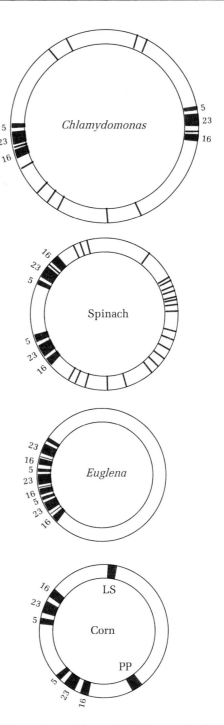

Figure 22.13 *Maps of rRNA genes in the chloroplast genome of* Chlamydomonas, Euglena, *maize, and* spinach. *The color bars show the positions of tRNA genes. In maize LS and PP are two structural genes.*

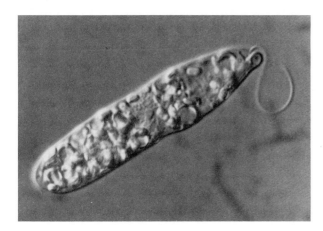

Figure 22.12 Euglena

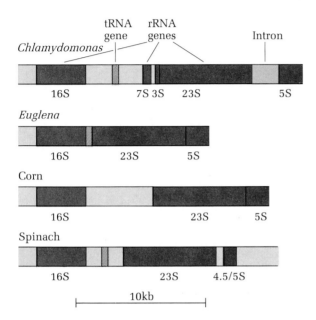

*Figure 22.14 Organization of chloroplast rRNA genes
in* Chlamydomonas, Euglena, *corn, and spinach.*

strand of cpDNA, and evidence for some systems suggests that a large precursor is processed into the three species. In three of the four plants in the figure, a tRNA gene is found between the 16S and 23S rRNA genes. Intervening sequences are present in some species (as seen in the 23S sequence in *Chlamydomonas*).

Chloroplast ribosomes. Chloroplast ribosomes are completely distinct from mitochondrial ribosomes and from cytoplasmic ribosomes. They have a sedimentation coefficient of 70S and consist of two unequal-sized subunits, 50S and 30S—thus they are similar in sedimentation coefficients to prokaryotic ribosomes. The contribution of chloroplast ribosomes to total protein synthesis in the cell can be quite considerable. In *Chlamydomonas,* for example, 30–35 percent of the ribosomes of the cell are in chloroplasts; of the remainder, the majority of the cell's ribosomes are found in the cytoplasm, with negligible numbers in the mitochondria. In higher plants between a third and a quarter of the cell's ribosomes are in the chloroplasts.

The large subunit of the cp ribosome contains at least two species of rRNA present in one copy each: 23S and 5S. In addition, some plants have other species of small rRNA molecules in the large subunit; for example, 4.5S rRNA is in spinach, and 7S and 3S rRNAs are in *Chlamydomonas.* The small subunit of the cp ribosome contains one copy of a 16S rRNA. The two large rRNAs, 23S and 16S, are similar in size to the 23S and 16S rRNAs found in *E. coli.* The presence of two small rRNAs in the large subunit is akin to the situation in cytoplasmic ribosomes, which contain one copy each of a 5S and a 5.8S rRNA, but is different from the situation in prokaryotic ribosomes, which have only a single small 5S rRNA, and from the situation in mitochondrial ribosomes, which have no small rRNA species.

As we saw for mitochondria, the assigned number of ribosomal proteins varies in chloroplast ribosomes, depending on the methods of analysis used. In general, 20–25 proteins are in the small subunit, and 30–38 proteins are in the large subunit. Many of these proteins are made on cytoplasmic ribosomes and are, therefore, encoded in the nuclear genome. The most detailed study done so far, which used *Euglena,* has shown that 9 ribosomal proteins are made in the chloroplast and 12 are made in the cytoplasm.

Chloroplast protein synthesis. Protein synthesis in chloroplasts is similar to the process in prokaryotes. Formylmethionyl tRNA is used to initiate all proteins, and the formylation reaction is catalyzed by a transformylase localized in the chloroplast. The chloroplast uses IFs and EFs (elongation factors) that are distinct from those of the cytoplasmic protein synthesis system. In the alga *Chlorella* the cp elongation factors, EF–G and EF–Tu, are not made in the presence of the antibiotic chloramphenicol and are therefore apparently synthesized in the chloroplast itself.

As we saw for mitochondrial ribosomes, cp ribosomes are resistant to cycloheximide, an inhibitor of cytoplasmic ribosomes, but are sensitive to virtually all inhibitors known to block prokaryotic protein synthesis. The ribosomes themselves are found both free and membrane-bound in the chloroplast. The free ribosomes appear to make the large subunit of the ribulose bis-phosphate decarboxylase enzyme, while the membrane-bound ribosomes are presumed to make the hydrophobic proteins that are important in photosynthesis. These hydrophobic proteins are located in the membranes of the chloroplast.

Keynote   *The genomes of both mitochondria and chloroplasts are naked, circular, double-stranded DNAs that contain genes for the rRNA components of the ribosomes of these organelles, for many (if not all) of the tRNAs used in organellar protein synthesis, and for a few proteins that remain in the organelles and perform functions specific to the organelles. At least in mitochondria, the genetic code is different from that found in nuclear protein-coding genes.*

Rules of Extranuclear Inheritance

Since the pattern of inheritance shown by genes located in organelles differs strikingly from the pattern shown by nuclear genes, the term **non-Mendelian inheritance** is appropriate to use when we are discussing extranuclear genes. In fact, the criteria for establishing that a phenotype is the result of an extranuclear (or extrachromosomal) gene rather than a nuclear gene are essentially results that do not conform to results expected from crosses involving nuclear genes.

The four main characteristics of *extranuclear inheritance* are illustrated next.

1. In higher eukaryotes the results of reciprocal crosses differ. Reciprocal crosses are a pair of crosses that involve the same genetic differences between the parents but with the sexes of the parents (the sources of male and female gametes) interchanged. For example, A ♀ × B ♂ and B ♀ × A ♂, where A and B represent different genotypes, are a pair of reciprocal crosses (see also Chapters 2 and 3).

   Extranuclear genes usually exhibit a phe-

nomenon of **uniparental inheritance** from generation to generation; that is, all progeny (both males and females) have the phenotype of only one parent. (The fact that both males and females resemble one of the parents distinguishes uniparental inheritance of an extranuclear gene from that of a sex-linked nuclear gene.) Typically, for higher eukaryotes it is the mother's phenotype that is expressed exclusively, a phenomenon called **maternal inheritance.** Maternal inheritance occurs because the zygote receives most of its cytoplasm (containing the extranuclear genes in the organelles; i.e., the mitochondria and, where applicable, the chloroplasts) from the female parent and a negligible amount from the male parent. In other words, the amount of cytoplasm in the female gamete, the egg, usually greatly exceeds that in the male gamete, the sperm.

By contrast, the results of reciprocal crosses between a wild-type and a mutant strain are ordinarily identical if the genes being transmitted are located on nuclear chromosomes. The only exception for nuclear genes occurs when sex-linked genes are involved, but even then the results (see Chapter 3, pp. 66–72) are distinct from those for extranuclear inheritance. That is, in the case of a sex-linked gene (recessive in this example), a cross between a wild-type female and a mutant male will produce F_1's that are all wild type. The F_1's of the reciprocal cross between a mutant female and a wild-type male are different, however; the females will be wild type and the males will be mutant.

2. Extranuclear genes cannot be mapped to known nuclear chromosomes. If the nuclear chromosomes of an organism are well mapped, any new nuclear mutations can be located on the genetic linkage map by mapping experiments that follow Mendelian principles of segregation and recombination. If a new mutation does not show linkage to any of the nuclear genes, it is probably an allele of an extranuclear gene.

3. Ratios typical of Mendelian segregation are not found. For nuclear gene mutations the

rules of Mendelian segregation predict that all F₁ progeny of a cross between a homozygous mutant (recessive) and a wild type will be wild type in phenotype and that a 3:1 ratio of wild-type:mutant phenotypes will be exhibited among the F₂ progeny. Such a segregation ratio is not characteristic of extranuclear genes—the uniparental inheritance pattern characteristic of extranuclear genes is clearly a deviation from Mendelian segregation.

4. Extranuclear inheritance is indifferent to nuclear substitution. When a particular phenotypic characteristic persists after the nucleus is removed and replaced with a nucleus having a different genotypic constitution associated with alternative characteristics, this indicates that the characteristic is likely to be controlled by a genome other than the nuclear genome, such as that of the mitochondrion or the chloroplast.

Keynote *Numerous examples of mutations in mitochondrial and chloroplast genomes are known. The inheritance of these extranuclear genes follows rules different from those for nuclear genes. In particular, no meiotic segregation is involved, generally uniparental (and often maternal) inheritance is exhibited, the trait persists even after nuclear substitution, and extranuclear genes are non-mappable to the known nuclear-linkage groups.*

Maternal Effect

The maternal inheritance pattern of extranuclear genes is distinct from the phenomenon of *maternal effect*, which is defined as the predetermination of gene-controlled traits by the maternal nuclear genotype prior to the fertilization of the egg. Maternal effect does not involve any extranuclear genes.

Maternal effect is seen, for instance, in the inheritance of the direction of coiling of the shell of the snail *Limneae peregra* (Figure 22.15). In this snail the direction of coiling of the shell is

Figure 22.15 The snail, Limneae peregra.

determined by a single pair of nuclear alleles, *D* for coiling to the right (dextral coiling) and *d* for coiling to the left (sinistral coiling). Breeding experiments have shown that the *D* allele is completely dominant to the *d* allele and that the shell-coiling phenotype is always determined by the genotype of the mother. Figure 22.16 shows the results of reciprocal crosses between a true-breeding, dextral-coiling, and a sinistral-coiling snail. All the F₁'s have the same genotype since a nuclear gene is involved, yet the phenotype is different for the reciprocal crosses. Phenotypically different results from a reciprocal cross are characteristic of extranuclear inheritance. In addition, because the F₁ phenotypes are those of the mothers in the cross in each case, maternal inheritance would *seem* to be indicated.

When we self the dextral F₁'s from the cross involving a dextral mother (Figure 22.16a), all the F₂'s are phenotypically dextral coiling. From Mendelian principles we predict a 1:2:1 ratio of *DD, Dd,* and *dd* genotypes in these snails. As a result, three-fourths of these F₂ snails give rise to dextral-coiling progeny upon selfing, whereas the *dd* snails, despite being dextral-coiling themselves, give rise to sinistral-coiling progeny upon selfing; that is, the progeny snails in this case do *not* resemble the mother in snail-coiling

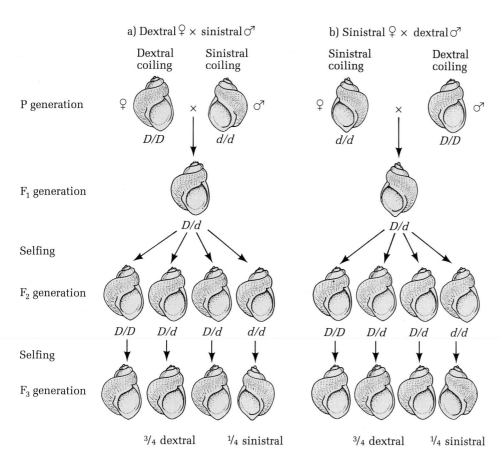

Figure 22.16 Inheritance of the direction of shell coiling in the snail Limneae peregra *is an example of maternal effect: (a) cross between true-breeding dextral-coiling female* (DD) *and sinistral-coiling male* (dd); *(b) cross between sinistral-coiling female* (dd) *and dextral-coiling male* (DD).

phenotype. This result does not fit our criteria for extranuclear inheritance, since if the coil direction phenotype were controlled by an extranuclear gene, the progeny would always exhibit the phenotype of the mother, owing to maternal inheritance. In the reciprocal cross (Figure 22.16b) the sinistral-coiling F_1's are genotypically identical to the dextral-coiling F_1's of Figure 22.16a, and the F_2 and F_3 genotypes and phenotypes are similarly identical. Thus the coiling phenotype is governed directly by the nuclear genotype of the mother and is an exam-

ple of maternal effect. The exact molecular basis of the phenotype is not known, although it is known that the coiling results from the particular spiraling cleavage patterns of the cells in the first few divisions after eggs are fertilized.

Keynote *Maternal effect is different from extranuclear inheritance. The maternal inheritance pattern of extranuclear genes occurs because the zygote contains most of its organelles (containing the extranuclear genes) from the female parent, whereas in the maternal effect the trait inherited is controlled by the maternal* nuclear *genotype before the fertilization of the egg and does not involve extranuclear genes.*

Examples of Extranuclear Inheritance

In this section we shall discuss the properties of a selected number of mutations in extranuclear chromosomes in order to illustrate the principles of extranuclear inheritance.

Leaf Variegation in the Higher Plant *Mirabilis jalapa*

One of the first exceptions to Mendelian inheritance was demonstrated in 1909 through the work of C. Correns, a plant geneticist. Correns had been studying the inheritance of a number of variegated-leaf forms in a number of flowering plants. Many of these phenotypes were con-

trolled by genes that showed typical Mendelian inheritance. One phenotype that did not follow the expected pattern was the yellowish white patches of a variegated strain of *Mirabilis jalapa* (also called the four o'clock or the marvel of Peru) (Figure 22.17). This strain, called *albomaculata,* also has occasional shoots that are wholly green or wholly yellow-white (see Figure 22.18).

All types of shoots (variegated, green, white) give rise to flowers, so it is possible to make crosses by taking pollen from one type of flower and using it to fertilize a flower on the same or different type of shoot. Table 22.5 summarizes the results of these intercrosses. The flowers on green shoots give only green progeny, regardless of whether the pollen is from green, white, or variegated shoots. Flowers on white shoots give only white progeny, regardless of the source of the pollen. However, because the whiteness indicates the absence of chlorophyll and hence an inability to carry out photosynthesis, the white progeny die soon after seed germination. Last, flowers on variegated shoots all give rise to three types of progeny—completely green, completely white, and variegated—regardless of the nature of the pollen used to fertilize them. No pattern is seen for the relative proportions of these three types. In subsequent generations maternal inheritance is always seen in the same patterns just described.

In sum, these breeding experiments with *M. jalapa* show results that are different from the results expected for genes following Mendelian inheritance patterns in that they showed maternal inheritance. That is, the progeny phenotype in each case was the same as that of the maternal parent (i.e., the color of the progeny shoots resembled the color of the parental flower), thus demonstrating the phenomenon of maternal inheritance. Moreover, the results of reciprocal crosses differed, and there was a lack of constant proportions of the different phenotypic classes in the segregating progeny. These last two properties are characteristic of traits showing extranuclear inheritance and are not expected for traits showing Mendelian inheritance.

The basis for the green color of higher plants

Figure 22.17 The four o'clock, Mirabilis jalapa.

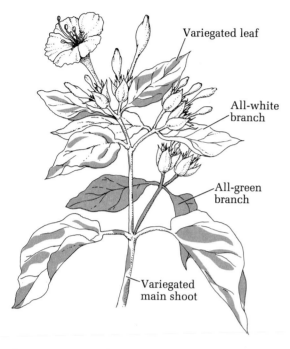

Figure 22.18 Leaf variegation in Mirabilis jalapa.
Shoots that are all green, all white, and variegated
are found on the same plant, and flowers may form
on any of these shoots.

Table 22.5 Results of Crosses of Variegated Plants of Mirabilis jalapa

Phenotype of Branch Bearing ♀ (Egg) Parent	Phenotype of Branch Bearing ♂ (Pollen) Parent	Phenotype of Progeny
White	White	White
White	Green	White
White	Variegated	White
Green	White	Green
Green	Green	Green
Green	Variegated	Green
Variegated	White	Variegated, green, or white
Variegated	Green	Variegated, green, or white
Variegated	Variegated	Variegated, green, or white

is the presence of the green pigment chlorophyll in large numbers of chloroplasts (the organelles responsible for photosynthesis). Green shoots in *Mirabilis* have a normal complement of chloroplasts. White shoots have abnormal, colorless chloroplasts called leukoplasts, which lack chlorophyll and hence are incapable of carrying out photosynthesis.

The simplest explanation for the inheritance of leaf color in the *albomaculata* strain of *Mirabilis jalapa* is that the abnormal chloroplasts are defective as a result of a mutant gene in the cpDNA. During plant growth the two types of organelles, chloroplasts and leukoplasts, may segregate so that a particular cell and its progeny cells have *only* chloroplasts (leading to green tissues), *only* leukoplasts (leading to white tissues), or a mixture of chloroplasts and leukoplasts (leading to variegation). This model is shown in Figure 22.19.

In a variegated plant, then, the white shoots derive from cells in which, through segregation, only leukoplasts are present. Green shoots are similarly derived from cells containing only chloroplasts. Variegated shoots are generated from cells that contain chloroplasts and leukoplasts, so by later segregation, patches of white tissue are produced on the shoots and leaves. Let us take this model one step further. The flowers on a green shoot have only chloroplasts, and so through maternal inheritance these chloroplasts form the basis of the phenotype of the next generation. Similar, logical arguments can be made for the flowers on white or variegated shoots.

Implicit in the simple model we have just described are three assumptions. One is that the pollen contributes essentially no cytoplasmic information (i.e., no chloroplasts or leukoplasts) to the egg (a reasonable assumption in view of the far greater size of the egg compared with the pollen). Thus in the zygote the extranuclear genetic determinants in the egg overwhelm those in the pollen. The second assumption is that the chloroplast genome replicates autonomously and that, by growth and division of plastids (the general term for photosynthetic organelles), the wild-type and mutant cpDNA molecules have

the potential to segregate randomly to the new plastids so that pure lines can be generated from a mixed line. The third assumption is that segregation of plastids to daughter cells is random so that some daughters receive chloroplasts, some receive leukoplasts, and some receive mixtures.

Keynote The leaf color phenotypes in a variegated strain of the four o'clock, Mirabilis jalapa, *show maternal inheritance. The abnormal chloroplasts in white tissue are the result of a mutant gene in the cpDNA, and the observed inheritance patterns follow the segregation of cpDNA.*

The *poky* Mutant of *Neurospora*

The *poky* mutant, which involves a change in mtDNA, illustrates a number of the classical expectations of extranuclear inheritance. The *poky* mutant is also called [*mi–1*], where the square brackets indicate an extranuclear gene and *mi* represents maternal inheritance.

In 1952 Mary and Herschel Mitchell first described the *poky* mutant of *Neurospora*. The phenotype that first indicated that it was different from the wild type was its relatively slow growth either on solid medium or in liquid medium. Thus slow-growth mutants can be monitored because they produce smaller colonies than the wild type. The name *poky* was given to reflect the poor growth characteristics of this mutant strain.

Neurospora crassa is an obligate aerobe; that is, it requires oxygen in order to grow and survive, so mitochondrial functions are essential for its growth. Biochemical analysis showed that the *poky* mutant is defective in aerobic respiration as a result of changes in the cytochrome complement of the mitochondria. The three principal cytochromes are a + a_3, b, and c. Compared with the wild type, *poky* lacks cytochromes a + a_3 and b and has a greater amount of cytochrome c. This change in the cytochrome spectrum clearly affects the ability of the mitochondria to generate sufficient ATP to support rapid growth, and slow growth results.

In *Neurospora* the sexual phase of the life

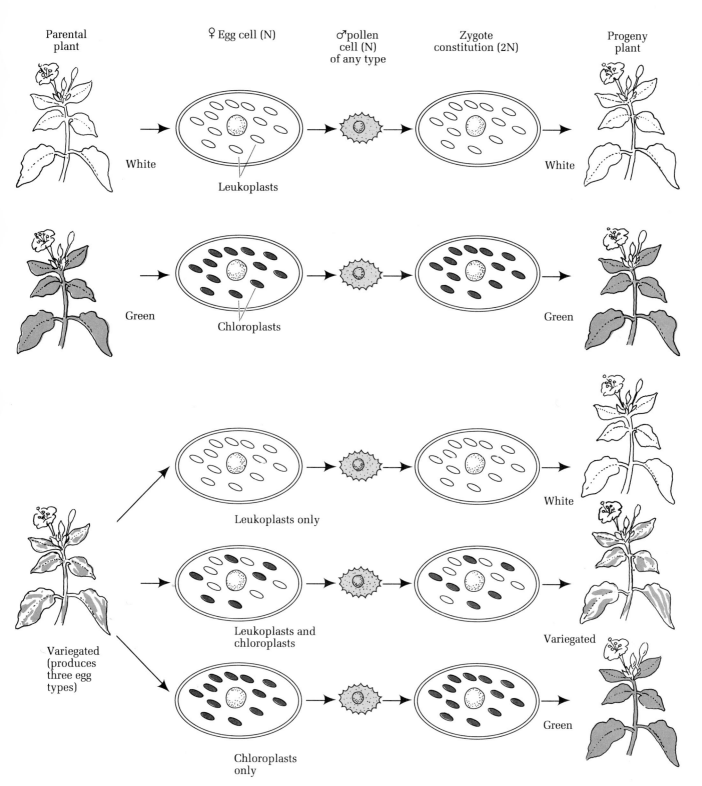

Figure 22.19 *Model for the inheritance of leaf color in the four o'clock,* Mirabilis jalapa.

cycle is initiated under nitrogen starvation conditions when there is a fusion of nuclei from *A* and *a* parents. A sexual cross can be made in one of two ways: either by putting both parents on the crossing medium simultaneously or by inoculating the medium with one strain, and after three or four days at 25°C, adding the other parent. In the latter case the first parent on the medium produces all the protoperithecia, the bodies that will give rise to the true fruiting bodies containing the asci with the sexual spores in them.

Compared with the conidia, the asexual spores, the protoperithecia have a tremendous amount of cytoplasm and hence can be considered the female parent in much the same way as an egg of a plant or an animal is the female parent. Now by adding conidia (which have relatively little cytoplasm in comparison with protoperithecia) of a strain of the opposite mating type to the crossing medium, we have what is called a *controlled cross* in which one strain acts as the female and the other as the male parent. Using a strain to produce the protoperithecia as the female parent and conidia of another strain as a male parent, geneticists can make reciprocal crosses to determine whether any trait shows extranuclear inheritance. This experiment is now illustrated for the *poky* mutant.

The Mitchells did reciprocal crosses between *poky* and the wild type, with the following results:

$$poky \; ♀ \times \text{wild type} \; ♂ \rightarrow \text{all } poky$$
$$\text{wild type} \; ♀ \times poky \; ♂ \rightarrow \text{all wild type}$$

In other words, all the progeny show the same phenotype as the maternal parent, indicating maternal inheritance as a characteristic for the *poky* mutation.

This analysis can be made more refined by using tetrad analysis (Chapter 6) to follow the phenotype more closely. Recall that in *Neurospora* the eight products of a meiosis and subsequent mitosis are retained in linear order within the ascus. The eight ascospores can be removed from the asci, and strains germinated from the spores can be analyzed for the particular phenotype being followed. By doing tetrad analysis for the *poky* ♀ × wild type ♂ cross, we find an 8:0 ratio of *poky*:wild-type progeny, and for the wild type ♀ × *poky* ♂ cross, we get a 0:8 ratio of *poky*:wild-type progeny with regard to the growth phenotype. At the same time nuclear genes show the 4:4 segregation expected of Mendelian inheritance. Again, the simplest explana-

tion for these results is that *poky* is determined by an extranuclear gene (located in this case in the mitochondria) that exhibits a pattern of maternal inheritance (see Figure 22.20).

What we know about the molecular basis for the *poky* mitochondrial mutation is based partially on the work of D. Luck, A. Lambowitz, and their colleagues. Their findings indicate that in *poky* a severe deficiency of small ribosomal

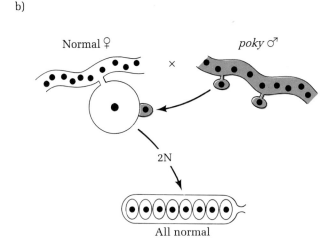

Figure 22.20 Results of reciprocal crosses of poky *and normal* Neurospora. *(a)* poky ♀ × normal ♂ *; (b)* normal ♀ × poky ♂.

subunits relative to the large subunits results in defective mitochondrial ribosomes. Because the protein-synthetic capacity of these defective mitochondrial ribosomes is impaired, cytochrome deficiencies develop, which in turn cause problems with aerobic respiration and ultimately result in slow growth. Additional molecular evidence confirms this hypothesis and indicates a deficiency in the accumulation of the 19S rRNA component of the small ribosomal subunit, while the accumulation of 25S rRNA component of the large subunit appears to be normal. The deficiency in the amount of 19S rRNA has been shown to be the result of a defect in the processing of a higher-molecular-weight precursor RNA to 19S rRNA that leads to significant degradation of the precursor molecules such that relatively few 19S rRNA molecules are produced. Lambowitz's group has also shown that the small ribosomal subunits in *poky* mitochondria are deficient in the ribosomal protein S5, which is known to be the only mitochondrial ribosomal protein actually coded for by mtDNA and synthesized within the mitochondria of *Neurospora*. The researchers hypothesized that ribosomal protein S5 plays a role in assembling the small ribosomal subunits in the mitochondria, in facilitating the processing of 19S rRNA from a higher-molecular-weight precursor RNA, and in stabilizing the binding of other mitochondrial, small, ribosomal subunit proteins during the assembly process. The absence of S5 in *poky*, then, has serious consequences to the assembly of small, mitochondrial, ribosomal subunits, with the secondary consequences of diminished mitochondrial protein synthesis and slow growth.

Keynote The slow-growing *poky* mutant of *Neurospora crassa* shows maternal *inheritance and deficiencies for some mitochondrial cytochromes.*

Yeast *petite* Mutants

Yeast grows as single cells rather than as a mycelium like *Neurospora*. Thus on solid medium yeast forms discrete colonies consisting of many thousands of individual cells clustered together. Yeast can grow with or without oxygen. In the absence of oxygen yeast can grow anaerobically; that is, it switches gears and obtains energy for cell growth and cellular metabolism through fermentation metabolism in which the mitochondria are not used.

Characteristics of *petite* mutants. In the late 1940s, Boris Ephrussi and his colleagues studied the growth of yeast cells on a solid medium on which growth could occur either by aerobic respiration or by fermentation. Occasionally, they noticed a colony that was much smaller than wild-type colonies. Since Ephrussi was French, the small colonies were called *petites* (French for "small"), and the wild-type colonies were called *grandes* (French for "big"). Ephrussi found that *petite* colonies were small not because the cells were small but because the growth rate of the mutant *petite* strain is significantly slower than that of the wild-type. Thus there are fewer cells in the *petite* colonies.

Biochemical analysis shows that the *petites* are essentially incapable of carrying out aerobic respiration and so must obtain their energy primarily from fermentation, which is an inefficient process relative to aerobic respiration.

What surprised Ephrussi and his colleagues, and what is still puzzling, is the frequency with which *petites* occur. In an unmutagenized population of cells between 0.1 and 1 percent of the cells spontaneously become *petite*. In the presence of an intercalating agent (a chemical that can wedge itself between adjacent base pairs in DNA) such as ethidium bromide, 100 percent of the cells become *petite*! The yeast *petite* system is particularly useful in studies of extranuclear inheritance because there is an abundance of *petite* mutants and because yeast cells that lack mitochondrial functions can still survive and grow. In other words, mutations to respiratory deficiencies in yeast are automatically conditional mutants. On a medium that can support fermentation, the *petites* grow more slowly than the *grandes*, whereas on a medium that supports only aerobic respiration, *petites* are unable to grow.

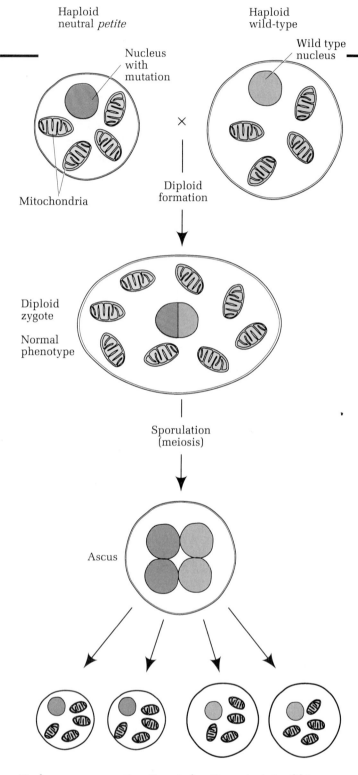

Nuclear gene segregation: 2 neutral *petite* mutant: 2 wild-type
Extranuclear segregation: 0 neutral *petite* mutants; 4 wild type

Figure 22.21 Extrachromosomal inheritance of neutral petite.

Nuclear, neutral, and suppressive *petite* mutants. Let us now turn to the genetics of *petite* mutants. Some *petites* have as their basis a mutation in the nuclear genome—a characteristic that is not surprising since at least some parts of every mitochondrion are coded by nuclear genes. The nuclear gene mutations are called *pet⁻*. When a *pet⁻* mutant is crossed with the wild type (*pet⁺*), the diploid is a *pet⁺/pet⁻ grande*. When this *pet⁺/pet⁻* cell goes through meiosis, each tetrad shows a 2:2 segregation of *grande* (*pet⁺*):*petite* (*pet⁻*) phenotype. This result is typical of Mendelian inheritance, so we will consider these *petites* no further, except to say that nuclear *petites* occur much less frequently than extranuclear *petites*.

One class of *petites* that exhibits the traits of extranuclear inheritance is the *neutral petite* class (symbolized [*rho⁻N*]). Figure 22.21 shows the inheritance pattern of [*rho⁻N*] *petites*. When a neutral *petite* is crossed with wild type ([*rho⁺N*]), the resulting diploid strain, [*rho⁺N*]/[*rho⁻N*], has a normal (*grande*) phenotype. When these diploids go through meiosis, all resulting meiotic tetrads show a 0:4 ratio of *petite*:*grande* strains, while at the same time nuclear markers segregate 2:2. This result is a classical example of uniparental inheritance, in which all progeny have the phenotype of only one parent. We cannot ascribe this phenomenon to maternal inheritance, however, since the two haploid cells that fuse to produce the diploid are the same size and contribute equally to the cytoplasm.

What is the nature of the neutral *petite* mutation? As we have said, *petites* can be induced readily by treatment with intercalating agents like ethidium bromide. With prolonged treatment the majority of *petites* are of the neutral type. The mitochondrial genome is implicated because cytochromes are altered, because there is evidence for extranuclear inheritance, and because the mitochondria are the only other site in the yeast cell in which genetic material is found.

An examination of the mitochondrial genetic material in the neutral *petites* reveals a remarkable result: Essentially 99–100 percent of the mtDNA is missing. Not surprisingly, then, the neutral *petites* are unable to perform mitochondrial functions. In genetic crosses with wild-type yeasts normal (wild-type) mitochondria form a population from which new, normal (wild-type) mitochondria are produced in all progeny, and hence the *petite* trait is lost after one generation. The *petite* is a neutral mutation in the sense that it does not affect the normal phenotype.

The suppressive *petites* are different from the neutrals. Most [*rho⁻*] mutants are of the suppressive type. Like the neutrals and the nuclear *petites*, the [*rho⁻ S*] *petites* are deficient in mitochondrial protein synthesis, but they differ from the other two classes in that they exert an effect in the diploid zygote when crossed with a normal [*rho⁺*] strain (see Figure 22.22).

When a [*rho⁺*]/[*rho⁻ S*] diploid is formed, it has respiratory properties intermediate between those of normal and *petite* strains. If this 2N is sporulated immediately after it is produced, the resulting ascospores show a 4:0 segregation of *petite*:normal. If, instead, the [*rho⁺*]/[*rho⁻ S*] diploid is allowed to divide mitotically a number of times, the diploid progeny population will have between 1 and 99 percent *petites*, depending on the particular suppressive *petite* strain. Sporulation of any of the normals present in the population gives 0 *petites*:4 normals.

The mechanism by which the suppressive phenotype takes place is not very well understood. One tenet is that yeast mitochondria can fuse, permitting the intermingling and recombination of different mtDNA molecules.

Some of the properties of suppressive *petites* are known. Suppressive *petites* can be generated by treatment with ethidium bromide, a chemical known to cause losses of parts or all of mtDNA, as mentioned earlier; so the problem apparently originates with changes in the mtDNA.

The suppressive mutations start out as deletions of part of the mtDNA, and then, by some correction mechanism whose reason for being is unclear, sequences that are not deleted become reiterated until the normal amount of mtDNA is restored. During these events rearrangements of the mtDNA sometimes occur (see Figure 22.23). Since the protein-coding genes in the mitochon-

drial genome are widely scattered, any significant deletion and/or rearrangement of mtDNA is likely to give rise to deficiencies in the assembly of the mitochondrial protein synthesis machinery and hence to defects in mitochondrial functions. In other words, no matter how the mtDNA is disrupted, we can expect a change in the organism's abilities to synthesize proteins within the mitochondria. As exemplified by the neutral and suppressive *petites*, this change leads to a slow-growth phenotype since the cell can no longer carry out aerobic respiration.

Keynote Yeast petite *mutants grow slowly and have various deficiencies in mitochondrial functions as a result of alterations in mitochondrial DNA. The* petite *mutants show extranuclear inheritance, with particular patterns of inheritance varying with the type of* petite *involved.*

Extranuclear Genetics of *Chlamydomonas*

One of the most thorough analyses of chloroplast genetics was begun by Ruth Sager and her colleagues in their work with the unicellular alga *Chlamydomonas reinhardii*. This motile, haploid organism has two flagella and a single chloroplast that contains many copies of cpDNA. The life cycle of this organism was given in Figure 6.2. There are two mating types, *mt⁺* and *mt⁻*. A zygote is formed by fusion of two equal-sized cells (which, therefore, contribute an equal amount of cytoplasm), one of each mating type. A thick-walled cyst develops around the zygote. After meiosis, four haploid progeny cells are produced, and since mating type is determined by a nuclear gene, a 2:2 segregation of the mating-type phenotype results.

From a systematic study of traits affecting this organism, researchers identified a number of traits that showed the expected patterns for extranuclear inheritance. Some of these extranuclear traits were shown to involve resistance or dependence on antibiotics that affect chloroplast protein synthesis by changing ribosome structure and/or function.

One trait that is inherited in an extranuclear

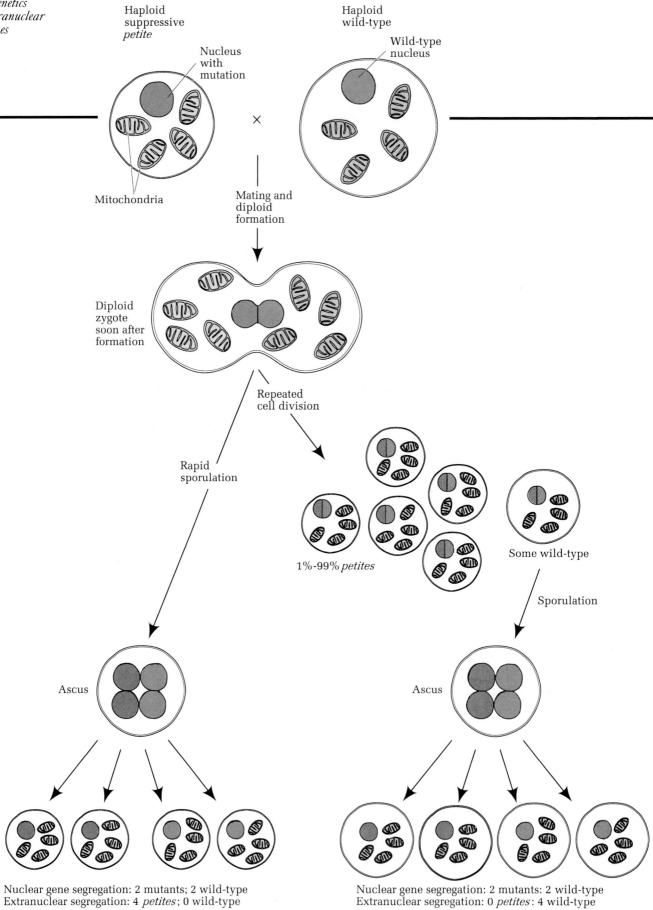

Haploid
suppressive
petite

Nucleus
with
mutation

Haploid
wild-type

Wild-type
nucleus

Mitochondria

×

Mating and
diploid
formation

Diploid
zygote
soon after
formation

Repeated
cell division

Rapid
sporulation

1%-99% *petites*

Some wild-type

Sporulation

Ascus

Ascus

Nuclear gene segregation: 2 mutants; 2 wild-type
Extranuclear segregation: 4 *petites*; 0 wild-type

Nuclear gene segregation: 2 mutants: 2 wild-type
Extranuclear segregation: 0 *petites*: 4 wild-type

manner confers erythromycin resistance ($[ery^{-r}]$) on the organism. Wild-type *Chlamydomonas* cells are erythromycin-sensitive ($[ery^{-s}]$). If we cross mt^+, $[ery^{-r}]$ × mt^-, $[ery^{-s}]$, about 95 percent of the offspring are erythromycin-resistant phenotypes. This result is a classic instance of uniparental inheritance, an attribute we have come to expect of extranuclear traits. The reciprocal cross, mt^-, $[ery^{-r}]$ × mt^+, $[ery^{-s}]$, also shows uniparental inheritance about 95 percent of the time, although in this case it is the erythromycin-sensitive phenotype that is inherited. Thus even though both parents contribute equal amounts of cytoplasm to the zygote, the progeny always resemble the mt^+ parent with regard to chloroplast-controlled phenotypes. Somehow, preferential segregation of one chloroplast type occurs in this organism.

As we indicated, only 95 percent of the zygotes in the above crosses show uniparental inheritance. The other 5 percent of the zygotes show extranuclear traits from both parents (biparental inheritance), indicating that both types of chloroplast chromosomes are present in these zygotes. In many instances the extranuclear traits of biparental zygotes segregate into pure types (i.e., either erythromycin-sensitive or erythromycin-resistant) on successive mitotic divisions (see Figure 22.24). This phenomenon presumably involves the segregation of the different chloroplast chromosomes, and hence the different chloroplasts, into pure types. This situation parallels the situation we discussed for *Mirabilis*.

Geneticists are able to make crosses between strains carrying different extranuclear traits that are chloroplast-controlled. The same principles of inheritance described for the erythromycin resistance trait apply to these traits. That is, in most cases there is uniparental inheritance with progeny resembling the mt^+ parent in extranu-

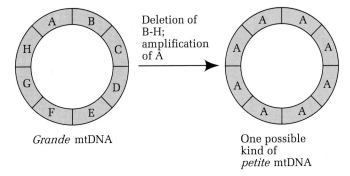

Figure 22.23 Origin of different types of suppressive petite *yeast mutants. Characteristically, a large segment of DNA is deleted from the* grande *mtDNA, and then whatever region remains is amplified by tandem duplication to generate an mtDNA molecule of approximately normal length.*

clear phenotype, and in the remaining cases there is biparental inheritance. Occasionally, a biparental zygote does not segregate the two traits; instead, both traits continue to be expressed in subsequent generations. The explanation is that a recombination event takes place between the two types of cpDNA so that a recombinant chromosome carrying both alleles originally carried by the two parents is produced. Sager and her colleagues have performed extensive analyses of these recombinants and, on the basis of the relative frequencies of recombination, have produced a genetic map (Figure 22.25a) and a restriction map (Figure 22.25b) of the cpDNA of *Chlamydomonas*.

Keynote *A number of genes have been identified and mapped in the* Chlamydomonas *chloroplast genome. These genes are inherited in an extranuclear manner.*

Figure 22.22 Extrachromosomal inheritance of suppressive petite. *Nuclear gene segregation: 2 mutants; 2 wild type extranuclear segregation, 0 petites: 4 wild types.*

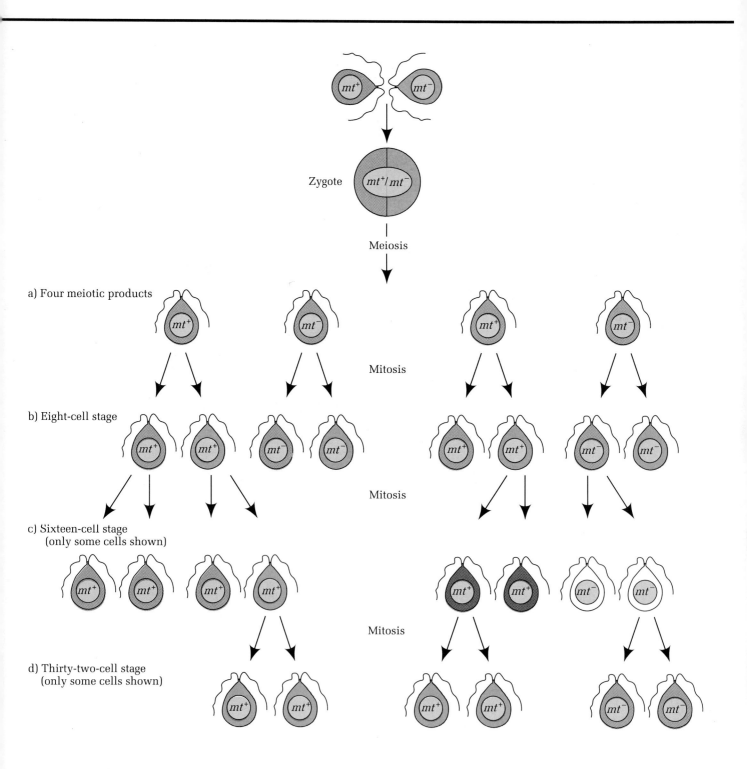

a) Four meiotic products

Zygote

Meiosis

Mitosis

b) Eight-cell stage

Mitosis

c) Sixteen-cell stage
   (only some cells shown)

Mitosis

d) Thirty-two-cell stage
   (only some cells shown)

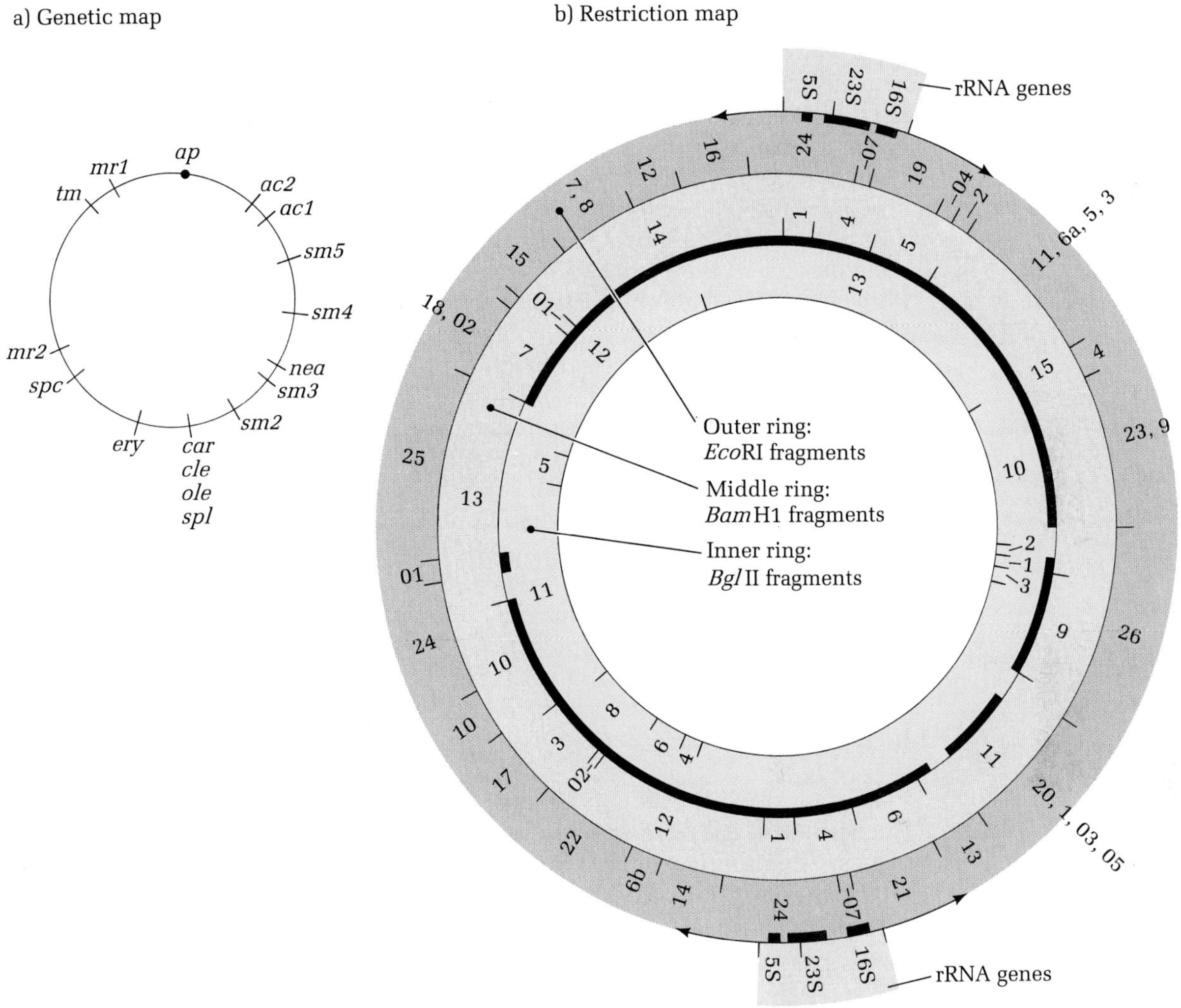

a) Genetic map

b) Restriction map

Figure 22.24 Segregation of extranuclear factors in
Chlamydomonas *in postmeiotic, mitotic divisions: (a)
The four products of meiosis undergo mitosis. (b) The
results are cell pairs that segregate for extranuclear
(shading differences) but not for nuclear* (mt⁺/mt⁻)
*traits. (c) Some cells continue to segregate the two
types of extranuclear traits, while others generate
equal numbers of two true-breeding extranuclear
types. (d) Segregation of the extranuclear traits so
that they breed true may be completed as early as the
32-cell stage.*

Figure 22.25 Maps of Chlamydomonas *cpDNA: (a)
circular genetic map; (b) restriction map.*

*Analytical
Approaches
for Solving
Genetics
Problems*

Q.1 In *Neurospora* strains of *poky* have been isolated that have reverted to nearly wild-type growth, although they still retain the abnormal metabolism and cytochrome pattern characteristic of *poky*. These strains are called *fast-poky*. A cross of a *fast-poky* female with a wild-type male gives a 1:1 segregation of *poky:fast-poky* ascospores in all asci. Interpret these results, and predict the results you would expect from the reciprocal of the stated cross.

A.1 The *poky* mutation shows maternal inheritance, so when it is used as a female parent, all the progeny ascospores will carry the *poky* determinant. Conventionally, extranuclear genes are designated within square brackets so in this case the cross with respect to the *poky* [*po*] determinant was [*po*] × [*N*], where *N* signifies normal cytoplasm. All progeny are [*po*]. The asci show a 1:1 segregation for the *poky* and *fast-poky* phenotypes. This ratio is characteristic of a nuclear gene segregating in the cross. Therefore the simplest explanation is that the factor that causes *poky* strains to be *fast-poky* is a nuclear gene mutation; we can call it *F* (its actual designated name). The *F* gene segregates in meiosis, as do all nuclear genes. The cross can be rewritten as [*po*]*F* ♀ × [*N*]$^+$ ♂, which gives a 1:1 segregation of [*po*]$^+$ and [*po*]*F*. The former is *poky* and the latter is *fast-poky*.

 With these results behind us, the reciprocal cross may be diagramed as [*N*]$^+$ ♀ × [*po*]*F* ♂. All the progeny spores from this cross are [*N*], with half of them being *F* and half of them being +. If *F* has no effect on normal cytoplasm, then these two classes of spores would be phenotypically indistinguishable, which is the case.

Q.2 Four slow-growing mutant strains of *Neurospora crassa*, coded *a, b, c,* and *d*, were isolated. All have an abnormal system of respiratory mitochondrial enzymes. The inheritance patterns of these mutants were tested in controlled crosses (see p. 722) with the wild type, with the following results:

Protoperithecial (Female) Parent		Conidial (Male) Parent	Progeny (Ascospores)	
			Wild Type	Slow Growing
Wild type	×	*a*	847	0
a	×	Wild type	0	659
Wild type	×	*b*	1113	0
b	×	Wild type	0	2071
Wild type	×	*c*	596	590
Wild type	×	*d*	1050	1035

Give a genetic interpretation of these results.

A.2 This question asks us to consider the expected transmission patterns for nuclear genes and for extranuclear genes. The nuclear genes will have a 1:1 segregation in the offspring, since this organism is a haploid organism and hence should exhibit no differences in the segregation patterns whichever strain is the maternal parent. On the other hand, a distinguishing characteristic of extranuclear genes is a difference in the results of reciprocal crosses. In *Neurospora* this characteristic is usually manifested by all progeny having the phenotype of the maternal parent. With these ideas in mind we can analyze each mutant in turn.

Mutant *a* shows a clear difference in its segregation in reciprocal crosses and is, in fact, a classic case of maternal inheritance. The interpretation here is that the gene is extranuclear, and since *Neurospora* is not green, the gene must be in the mitochondrion. The *poky* mutant described in this chapter shows this type of inheritance pattern.

By the same reasoning used to analyze mutant *a*, the mutation in strain *b* must also be extranuclear.

Mutants *c* and *d* segregate 1:1, indicating that the mutations involved are in the nuclear genome. In these cases we need not consider the reciprocal cross since there is no evidence for maternal inheritance. In fact, the actual mutations that are the basis for this question are female and sterile, so the reciprocal cross cannot be done. We can confirm that the mutations are in the nuclear genome by doing mapping experiments, using known nuclear markers. Evidence of linkage to such markers would confirm that the mutations are not extranuclear.

*Questions
and
Problems*

22.1 Compare and contrast the structure of the nuclear genome, the mitochondrial genome, and the chloroplast genome.

22.2 How do mitochondria reproduce? What is the evidence for the method you describe?

22.3 What genes are present in the human mitochondrial genome?

22.4 What conclusions can you draw from the fact that most nuclear-encoded mRNAs and all mitochondrial mRNAs have a poly(A) tail at the 3′ end?

*22.5 Discuss the differences between the universal genetic code of the nuclear genes and the code found in mammalian and fungal mitochondria. Is there any advantage to the mitochondrial code?

22.6 When the DNA sequences for most of the mRNAs in human mitochondria are examined, no nonsense codons are found at their termini. Instead, either U or UA is found. Explain this result.

22.7 Compare and contrast the cytoplasmic and mitochondrial protein-synthesizing systems.

22.8 Compare and contrast the organization of the ribosomal RNA genes in mitochondria and in chloroplasts.

*22.9 What features of extranuclear inheritance distinguish it from the inheritance of nuclear genes?

22.10 Distinguish between maternal effect and extranuclear inheritance.

22.11 Distinguish between nuclear (segregational), neutral, and suppressive *petite* mutants of yeast.

*22.12 The snail *Limneae peregra* has been studied extensively with respect to the inheritance of the direction of shell coiling. A snail produced by a cross between two individuals has a shell with a right-hand twist (dextral coiling). This snail produces only left-hand (sinistral) progeny on selfing. What are the geneotypes of the F_1 snail and its parents?

22.13 *Drosophila melanogaster* has a sex-linked, recessive, mutant gene called *maroon-like* (*ma–l*). Homozygous *ma–l* females or hemizygous *ma–l* males have light-colored eyes owing to the absence of the active enzyme xanthine dehydrogenase, which is involved in the synthesis of eye pigments. When heterozygous *ma–l*$^+$/*ma–l* females are crossed with *ma–l* males, all the offspring are phenotypically wild-type. However, half the female offspring from this cross, when crossed back to *ma–l* males, give all *ma–l* progeny. The other half of the females, when crossed to *ma–l* males, give all phenotypically wild-type progeny. What is the explanation for these results?

*22.14 When females of a particular mutant strain of *Drosophila melanogaster* are crossed to wild-type males, all the viable progeny flies are females. Hypothetically, this result could be the consequence of either a sex-linked, lethal mutation or a maternally inherited factor that is lethal to males. What crosses would you perform in order to distinguish between these alternatives?

22.15 Reciprocal crosses between two *Drosophila* species, *D. melanogaster* and *D. simulans,* produce the following results:

> *melanogaster* female × *simulans* male → females only
> *simulans* female × *melanogaster* male → males, with few or no females

Explain these results.

22.16 Some *Drosophila* flies are very sensitive to carbon dioxide; they become anesthetized when it is administered to them. The sensitive flies have a cytoplasmic particle called *sigma* that has many properties of a virus. Resistant flies lack *sigma.* The sensitivity to carbon dioxide shows strictly maternal inheritance. What would be the outcome of the following two crosses: (a) sensitive female × resistant male and (b) sensitive male × resistant female?

*22.17 In yeast a haploid nuclear (segregational) *petite* is crossed with a neutral *petite.* Assuming that both strains have no other abnormal phenotypes, what proportion of the progeny ascospores are expected to be *petite* in phenotype if the diploid zygote undergoes meiosis?

22.18 When grown on a medium containing acriflavin, a yeast culture produces a large number of very small (*tiny*) cells that grow very slowly. How would you determine whether the slow-growth phenotype was the result of a cytoplasmic factor or a nuclear gene?

*22.19 In *Neurospora* a chromosomal gene *F* suppresses the slow-growth characteristic of the *poky* phenotype and makes a *poky* culture into a *fast-poky* culture, which still has abnormal cytochromes. Gene *F* in combination with normal cytoplasm has no detectable effect. (Hint: Since both nuclear and extranuclear genes have to be considered, it will be convenient to use symbols to distinguish the two. Thus cytoplasmic genes will be designated in square brackets; e.g., [*N*] for normal cytoplasm, [*po*] for *poky*.)

   a. A cross in which *fast-poky* is used as the female (protoperithecial) parent and a normal wild-type strain is used as male parent gives half *poky* and half *fast-poky* progeny ascospores. What is the genetic interpretation of these results?
   b. What would be the result of the reciprocal cross of the cross described in part a, that is, normal female × *fast-poky* male?

22.20 Reciprocal crosses between two types of the evening primrose, *Oenothera hookeri* and *Oenothera muricata,* produce the following effects on the plastids:

   O. hookeri female × *O. muricata* male → yellow plastids
   O. muricata female × *O. hookeri* male → green plastids

Explain the difference between these results, noting that the chromosome constitution is the same in both types.

*22.21 A form of male sterility in corn is maternally inherited. Plants of a male-sterile line crossed with normal pollen give male-sterile plants. Some lines of corn carry a dominant, so-called restorer (*Rf*) gene, which restores pollen fertility in male-sterile lines.

   a. If a male-sterile plant is crossed with pollen from a plant homozygous for gene *Rf*, what will be the genotype and phenotype of the F_1?
   b. If the F_1 plants of part a are used as females in a testcross with pollen from a normal plant (*rf/rf*), what would be the result? Give genotypes and phenotypes, and designate the type of cytoplasm.

23

Quantitative Genetics

In the previous chapters, much of our attention focused on the molecular aspects of gene structure, function, and expression. By isolating mutants that affect a particular process and by comparing the mutants with the wild-type, geneticists can often describe the molecular basis for the mutant phenotype and can piece together an understanding of how the events for normal cells and organisms proceed. The mutations used in these molecular studies, and in fact all the traits we have studied up to this point, have been characterized by the presence of a few distinct phenotypes. The seed coats of pea plants, for example, were either grey or white, the seed pods were green or yellow, and the plants were tall or short. In each trait the different phenotypes were distinct, and each phenotype was easily separated from all other phenotypes. A trait such as this, with only a few distinct phenotypes, is called a **discontinuous trait.** The phenotypes of a discontinuous trait can be described in qualitative terms, and all individuals can be placed into a few phenotypic categories, as shown in Figure 23.1. Some additional examples of discontinuous traits are the ABO blood types, coat colors in mice, the prototrophic versus auxotrophic mutants of bacteria, and the presence or absence of extra digits on the hand.

For discontinous traits, a simple relationship usually exists between the genotype and the phenotype. In most cases, each genotype produces only a single phenotype, and frequently each phenotype results from a single genotype. When dominance or epistasis occurs, the same phenotype may be produced by several different genotypes, but the relationship between the genes and the trait remains simple. Because of this simple relationship, genotypes can be inferred by studying the phenotypes of parents and offspring, and in this way Mendel was able to work out the basic principles of heredity.

Not all traits, however, exhibit phenotypes that fall into a few distinct categories. Many traits, such as birth weight and adult height in humans, protein content in corn, and number of eggs laid by *Drosophila,* exhibit a wide range of possible phenotypes. Traits such as these, with a

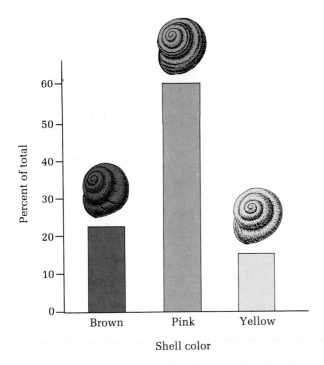

Figure 23.1 *Discontinuous distribution of shell color in the snail* Cepaea nemoralus *from a population in England.*

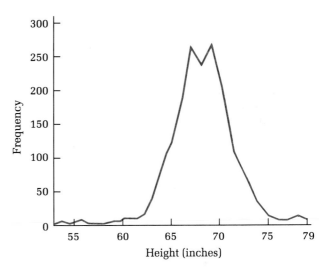

gerprint pattern, and birth weight (see Figure 23.3). The traditional techniques of transmission genetics provide little information about the inheritance of these complex traits, and different, more quantitative methods are required.

Figure 23.2 *Continuous distribution of height for male students at the University of Minnesota in 1924–25.*

continuous distribution of phenotypes, are called **continuous traits.** The distribution of a continuous trait is illustrated in Figure 23.2. Since the phenotypes of continuous traits must be described in quantitative terms, such traits are also known as **quantitative traits,** and the study of the inheritance of quantitative traits comprises the field of **quantitative genetics.**

Numerous traits have a continuous distribution, and quantitative genetics plays an important role in many areas of basic and applied biology. In agriculture, for example, crop yield, weight gain, milk production, and fat content are all continuous traits which are studied with the techniques of quantitative genetics. In psychology, methods of quantitative genetics are frequently employed for the study of complex behavioral traits, such as IQ, learning ability, and personality. Geneticists also use these methods to study other continuous traits found in humans, such as blood pressure, antibody titer, fin-

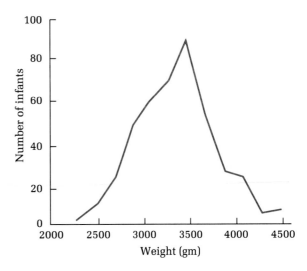

Figure 23.3 *Distribution of birth weight for first-born males born in New York City around 1930.*

The Nature of Continuous Traits

Biologists began developing statistical techniques for the study of continuous traits during the latter part of the 19th century, even before they were aware of Mendel's principles of heredity. Francis Galton and his associate Karl Pearson studied a number of continuous traits in humans, such as height, weight, and mental traits; by demonstrating that the traits of parents and their offspring are statistically associated, they were able to show that these traits are inherited, but they were not successful in determining how genetic transmission occurs. After the rediscovery of Mendel's work, considerable controversy arose over whether continuous traits also follow Mendel's principles, or whether they are inherited in some different fashion. Around 1903, Wilhelm Johannsen conducted a series of experiments on the inheritance of seed weight of beans, and he demonstrated that continuous variation is partly genetic and partly environmental. Several years later, Herman Nilsson-Ehle, working with wheat, proposed that continuous traits are determined by multiple genes, each of which segregates according to Mendelian principles.

Why Some Traits Have Continuous Phenotypes

Traits are continuous because they possess many phenotypes. To understand the inheritance of continuous traits, we must first determine why some traits have many phenotypes.

Multiple phenotypes of a trait arise in several ways. Frequently many phenotypes occur because numerous genotypes exist among the individuals of a group—this happens when the trait is influenced by a large number of loci. For example, when a single locus with two alleles determines a trait, three genotypes are present: *AA, Aa,* and *aa.* With two loci, each with two alleles, the number of genotypes is $3^2 = 9$ (*AABB, AaBB, AABb, AaBb, AAbb, aaBB, aaBb, Aabb,* and *aabb*). In general, the number of genotypes is 3^n, where *n* equals the number of loci with two alleles; if more than two alleles are present at a locus, the number of genotypes is even greater. As the number of loci influencing a trait increases, the number of genotypes quickly becomes large. Traits that are encoded by many loci are referred to as **polygenic traits.** If each genotype in a polygenic trait encodes a separate phenotype, many phenotypes will be present. And because many phenotypes are present, and the differences between phenotypes are slight, the trait appears to be continuous.

More frequently, several genotypes of a polygenic trait produce the same phenotype. For example, with dominance, only two phenotypes are produced by the three genotypes at a locus (*AA* and *Aa* produce one phenotype and *aa* produces a different phenotype). If dominance occurs among the alleles at each of three loci, the number of genotypes is $3^3 = 27$, but the number of phenotypes is only $2^3 = 8$. When epistatic interactions occur among the alleles at different loci, as discussed in Chapter 4, a number of genotypes may code for the same phenotype. In many polygenic traits, multiple genotypes code for the same phenotype and the relationship between genotype and phenotype is obscured.

A second reason that a trait may have many phenotypes is that environmental factors also affect the trait. When environmental factors exert an important influence on the phenotype, each genotype is capable of producing a range of phenotypes. Which phenotype is expressed depends both on the genotype and on the specific environment in which the genotype is found. For most continuous traits, both multiple genotypes and environmental factors influence the phenotype; such a trait is **multifactorial.**

When multiple genes and environmental factors influence a trait, no simple relationship exists between genotype and phenotype, as is present in discontinuous traits. Therefore, the simple rules of inheritance that we learned in classical genetics provide us with little information about the genes involved in continuous traits. In order to understand the genetic basis of these traits and their inheritance, we must employ special concepts and analytical procedures.

Keynote *Discontinuous traits exhibit only a few distinct phenotypes and can be described in qualitative terms. Continuous traits, on the other hand, display a spectrum of phenotypes and must be described in quantitative terms. Many phenotypes are present in a continuous trait because the trait is encoded by many loci, producing many genotypes, and because environmental factors cause each genotype to produce a range of phenotypes.*

Questions Studied in Quantitative Genetics

Not only are the methods used in quantitative genetics different from those we have previously studied, but the fundamental nature of the questions asked is also different. In transmission genetics, we frequently determined the probability of inheriting a discontinuous trait. With most continuous traits, however, numerous genes are involved and no simple relationship exists between the genotype and the phenotype; thus, we cannot make precise predictions about the probability of inheriting a continuous trait, as we did for simple discontinuous traits. For the same reason, we can rarely identify and study the DNA sequences that code for the trait, so molecular studies of continuous traits are usually impossible. In spite of these limitations, some important questions about the genetic nature of continuous traits can be addressed. Questions frequently studied by quantitative geneticists include the following.

1. To what degree is the observed variation in phenotype the result of differences in genotype and to what degree does this variation reflect the influence of different environments? In our study of discontinuous traits, this question assumed little importance, because the differences in phenotype were usually entirely genetic.
2. How many genes influence the phenotype of the trait? When only a few loci are involved and the trait is discontinuous, the number of loci involved can often be determined by examining the phenotypic ratios in genetic

crosses. With complex, continuous traits, however, determining the number of loci involved is a more difficult undertaking.
3. Are the contributions of the genes equal? Or do a few genes have major effects on the trait and other genes only modify the phenotype slightly?
4. To what degree do alleles at the different loci interact with one another? Are the effects of alleles additive, or do epistatic interactions occur?
5. When selection occurs for a particular phenotype, how rapidly does the trait change? Do other traits change at the same time?
6. What is the best method for selecting and mating individuals so as to produce desired phenotypes in the progeny?

Statistics

With discontinuous traits, we described the phenotypes of a group of individuals by listing the numbers or frequencies of each phenotype; we might state that three-fourths of the progeny were yellow and one-fourth were green. With continuous traits, such a simple description of the phenotypes is often impossible, because a spectrum of phenotypes is present, and usually the phenotype can only be described in terms of a measurement. Furthermore, the relationship between genotype and phenotype is complex, so we do not observe the simple ratios predicted by Mendelian principles. Consequently, quantitative genetics requires an understanding of statistical procedures for describing continuous traits and for answering many of the questions we have listed above.

Samples and Populations

Suppose we want to describe some aspect of a trait in a large group of individuals. For example, we might be interested in the average birth weight of infants born in New York City during the year 1987. Since thousands of babies are born in New York City every year, collecting information on each baby's weight might not be practical. An alternative method would be to collect information on a subset of the group, say birth weights on 100 infants born in New York City during 1987, and then use the average obtained on this subset as an estimate of the average for the entire city. Biologists and other scientists commonly employ this sampling procedure in data collection, and statistics are necessary for analyzing such data. The group of ultimate interest (in our example, all infants born in New York City during 1987) is called the **population,** and the subset used to give us information about the population (our set of 100 babies) is called a **sample.** For a sample to give us reliable information about the population, it must be large enough so that chance differences between the sample and the population are not misleading. If our sample consisted of only a single baby, and that infant was unusually large, then our estimate of the average birth weight of all babies would not be very accurate. The sample must also be a random subset of the population. If all the babies in our sample came from Hope Hospital for Premature Infants, then we would grossly underestimate the true birth weight of the population. Figure 23.3 shows a distribution of birth weight in humans.

Keynote *To describe and study a large group of individuals, scientists frequently examine a subset of the group. This subset is called a sample and the sample provides information about the larger group, which is termed the population. The sample must be of reasonable size and it must be a random subset of the larger group in order to provide accurate information about the population.*

Distributions

When we studied discontinuous traits, we were able to describe the phenotypes found among a group of individuals by stating the proportion of individuals falling into each phenotypic class. As we discussed earlier, continuous traits exhibit a range of phenotypes, and describing the phenotypes found within a group of individuals is more complicated. One means of summarizing the phenotypes of a continuous trait is with a **frequency distribution,** which is a description of the population in terms of the proportion of individuals that have each phenotype.

To make a frequency distribution, classes are constructed consisting of individuals falling within a specified range of the phenotype, and the number of individuals in each class is counted. Table 23.1 presents a frequency distribution constructed from the data in Johannsen's study of the inheritance of weight in the dwarf bean, *Phaseolus vulgaris.* Johannsen weighed 5494 beans from the F_2 progeny of a cross and classified them into nine groups or classes, each of which covered a 10-centigram range of weight, as is illustrated in Table 23.1. A frequency distribution such as this can be displayed graphically by plotting the phenotypes in a frequency histogram, as shown in Figure 23.4 for Johannsen's beans. In the histogram, the phenotypic classes are indicated along the horizontal axis and the number present in each class is plotted on the vertical axis. If a curve is drawn tracing the outline of the histogram, the curve assumes a shape that is characteristic of the frequency distribution. Several different types of distributions are illustrated in Figure 23.5.

Many continuous phenotypes exhibit a sym-

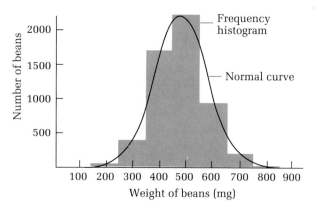

Figure 23.4 *Frequency histogram for bean weight in* Phaseolus vulgaris *plotted from the data in Table 23.1. A normal curve has been fitted to the data and is superimposed on the frequency histogram.*

Table 23.1 Weight of 5494 F₂ Beans (Seeds of Phaseolus vulgaris) Observed by Johannsen in 1909

Weight (cg) (midpoint of range)	5–15 (10)	15–25 (20)	25–35 (30)	35–45 (40)	45–55 (50)	55–65 (60)	65–75 (70)	75–85 (80)	85–95 (90)
Number of beans	5	38	370	1676	2255	928	187	33	2

metrical, bell-shaped distribution similar to the one shown in Figure 23.5a. This type of distribution is called a **normal distribution.** The normal distribution is produced when a large number of independent factors influence the measurement. Since many continuous traits are multifactorial (influenced by multiple genes and multiple environmental factors), observing a normal distribution for these traits is not surprising. Two other types of distribution are illustrated in 23.5b and 23.5c.

Binomial Theorem

The **binomial distribution** is the theoretical frequency distribution of events that have two possible outcomes. For example, among humans two sexes are present, male and female. In a

family of three children, several different combinations of the sexes are possible: three boys, two boys and one girl, one boy and two girls, and three girls. Suppose that we want to know what the probability is that in a family of three children, two will be boys and one will be a girl. The probability is not ½ × ½ × ½ = ⅛, but rather ⅜. This is because there are three different ways, or three permutations, in which a family of three children can have two boys and one girl: The first two children might be boys and the third child a girl (BBG). Alternatively, the first child might be a boy, followed by a girl and then another boy (BGB), or the first child might be a girl followed by two boys (GBB). In each case, the probability is ½ × ½ × ½ = ⅛ (using the product rule; see Chapter 2), and so the total probability is the sum of these individual probabilities (using the sum rule; see Chapter 2), or ⅛ + ⅛ + ⅛ = ⅜.

In a family with three children, four different combinations of the sexes are possible: three boys, two boys and a girl, one boy and two girls, and three girls. Three boys can occur in only one way: BBB. Therefore, the probability of having all boys in a family of three children is ½ ×

Figure 23.5 Three different types of distributions. (a) a normal distribution of percent sucrose in sugar beets; (b) a skewed distribution representing coat color in guinea pigs; (c) a bimodal distribution of the size of female Crotalus viridis.

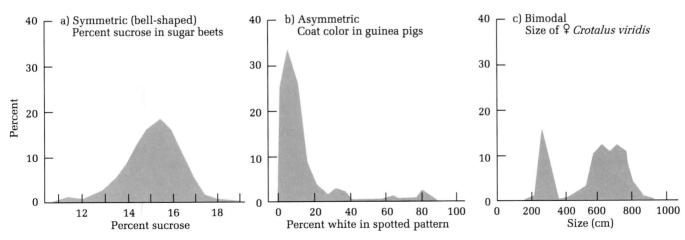

$\frac{1}{2} \times \frac{1}{2} = \frac{1}{8}$. As we have seen, the probability of two boys and one girl is $\frac{3}{8}$. Using the same reasoning, the probability of two girls and one boy is $\frac{3}{8}$, because three permutations of this combination are possible (GGB, GBG, and BGG). Since three girls can be present in only one way (GGG), the probability of this combination is $\frac{1}{2} \times \frac{1}{2} \times \frac{1}{2} = \frac{1}{8}$. In summary, we have probabilities of $\frac{1}{8}:\frac{3}{8}:\frac{3}{8}:\frac{1}{8}$ for three boys: two boys and one girl: two girls and one boy: three girls.

As you might imagine, such computations get rather laborious when the number of combinations is large. A convenient and efficient way around the tedium is to use the *binomial theorem.* According to this theorem, the frequencies (or probabilities of occurrence) of the various combinations correspond to the terms of the *binomial expansion.* The first three binominal expansions are as follows:

$$(a + b)^2 = a^2 + 2ab + b^2$$
$$(a + b)^3 = a^3 + 3a^2b + 3ab^2 + b^3$$
$$(a + b)^4 = a^4 + 4a^3b + 6a^2b^2 + 4ab^3 + b^4$$

Using these general formulas, we could consider *a*, for example, to be the probability of having a boy and *b* the probability of having a girl. The exponents of *a* and *b* in the expansions correspond, then, to the number of children of each type in the family to which the term applies.

Let us consider the first expansion. The exponent indicates that we are dealing with a two-child family. The term *a* is the probability of having a boy, namely, $\frac{1}{2}$, and *b* is the probability of having a girl, also $\frac{1}{2}$. To the right of the equals sign are the terms for the various two-child families that can result, and from these terms the probability of each combination can be determined. The a^2 term, for example, is the probability of having two boys, which is $(\frac{1}{2})^2 = \frac{1}{4}$. The probability of having a boy and a girl is $2ab = 2(\frac{1}{2})(\frac{1}{2}) = \frac{2}{4}$; and so on.

The general expression for the binomial expansion is $(a + b)^n$, where *a* is the probability of one event occurring and *b* is the probability of the alternative event occurring. In our example of the sex of children, $a = b$, but this does

not always have to be the case in a binomial expansion; however, $a + b$ must equal 1. The term *n* is the number of trials. For our purposes we must be able to compute the coefficients for terms in the binomial expansion. As Figure 23.6 shows, there is a symmetry in the coefficients of successive powers of the binomial such that they can be arranged in what is called Pascal's triangle. In this triangle each number is the sum of the numbers in the row immediately above and one place to the left and right. There is also a symmetry in the exponents of the expansion. Consider, for example, the expansion of $(a + b)^5$:

$$(a + b)^5 = a^5 + 5a^4b + 10a^3b^2 + 10a^2b^3 + 5ab^4 + b^5$$

Note that the exponents of *a* begin at the power to which the binomial is raised and then decrease incrementally to zero; at the same time the reverse happens to the exponents of *b*. Also, note that the number of terms in the expansion is always $n + 1$.

Last, we need a reasonably quick method of determining the value of a particular coefficient in an expansion without searching for the nearest copy of Pascal's triangle. The principle here is as follows: If the probability of occurrence of event *A* is *a,* and the probability of occurrence of the alternative event *B* is *b* (i.e., $1 - a$), then the probability that in *n* attempts event *A* will take place *s* times and event *B* will take place *t* $(n - s)$ times is calculated as follows:

$$\frac{n!}{s!t!} a^s b^t$$

The symbol ! in this expression means "factorial" and is the product of all the integers from 1 to the term *n* involved. For example, 5! is $1 \times 2 \times 3 \times 4 \times 5$. By definition, 0! is 1, not zero.

The Mean

A distribution of phenotypes can be summarized in the form of two convenient statistics, the mean and the variance. The **mean,** which is also known as the average, gives us information

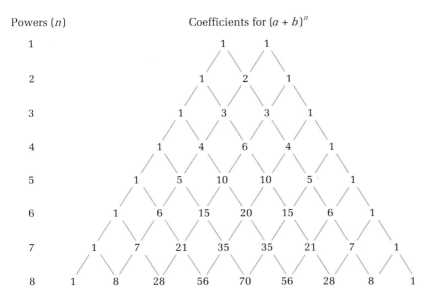

Powers (*n*)   Coefficients for $(a + b)^n$

Figure 23.6 Coefficients for terms in the binomial expansion (a + b)ⁿ *for values of* n *from 1 to 8.*

about the center of the phenotypes in a sample. The mean of a sample is calculated by simply adding up all the individual measurements (x_1, x_2, x_3, etc.) and dividing by the number of measurements we added (*n*). We can represent this calculation with the following formula:

$$\text{Mean} = \bar{x} = \frac{x_1 + x_2 + x_3 + \cdots x_n}{n}$$

This equation can be abbreviated by using the symbol Σ, which means the summation, and x_i, which means the *i*th value of *x*:

$$\text{Mean} = \bar{x} = \frac{\Sigma x_i}{n}$$

When data are taken from a frequency distribution, such as the distribution of bean weights seen in Table 23.1, several individuals have the same phenotype. Here, the mean can be estimated by first multiplying the midpoint of each class, m_i (in this case the middle weight in each class) by the number of individuals in that class f_i. These products are then added up and divided by the total number of measurements:

$$\bar{x} = \frac{\Sigma f_i m_i}{n}$$

Table 23.2 presents a sample calculation of the mean.

The mean is frequently used in quantitative genetics to summarize the phenotypes of a group of individuals. For example, in an early study of continuous variation, E. M. East examined the inheritance of flower length in several strains of the tobacco plant. He crossed a strain of tobacco with short flowers to a strain with long flowers. Within each strain, however, flower length varied some, so East reported that the mean phenotype of the short strain was 40.4 mm and the mean phenotype of the other strain was 93.1 mm. The F₁ progeny, which consisted of 173 plants, had a mean flower length of 63.5 mm. In this type of situation, the mean provides a convenient way of quickly summarizing the phenotypes of parents and offspring.

The Variance and the Standard Deviation

A second statistic that provides key information about a distribution is the **variance.** The variance is a measure of how much the individual measurements spread out around the mean— how variable the measurements are. Two distri-

Table 23.2 Sample Calculations of the Mean, Variance, and Standard Deviation for Body Length of 10 Spotted Salamanders from Penobscot County Maine

Body Length x_i *(mm)*	$(x_i - \bar{x})$	$(x_i - \bar{x})^2$
65	$(65 - 57.1) = 7.9$	$7.9^2 = 62.41$
54	$(54 - 57.1) = -3.1$	$-3.1^2 = 9.61$
56	$(56 - 57.1) = -1.1$	$-1.1^2 = 1.21$
60	$(60 - 57.1) = 2.9$	$2.9^2 = 8.41$
56	$(56 - 57.1) = -1.1$	$-1.1^2 = 1.21$
55	$(55 - 57.1) = -2.1$	$-2.1^2 = 4.41$
53	$(53 - 57.1) = -4.1$	$-4.1^2 = 16.81$
55	$(55 - 57.1) = -2.1$	$-2.1^2 = 4.41$
58	$(58 - 57.1) = 0.9$	$0.9^2 = 0.81$
59	$(59 - 57.1) = 1.9$	$1.9^2 = 3.61$
$\Sigma x_i = 571$		$\Sigma(x_i - \bar{x})^2 = 112.9$

Mean $= \bar{x} = \Sigma x_i/n = 571/10 = 57.1$

Variance $= s_x^2 = \Sigma(x_i - \bar{x})^2/n - 1 = 112.9/9 = 12.54$

Standard deviation $= s_x = \sqrt{12.54} = 3.54$

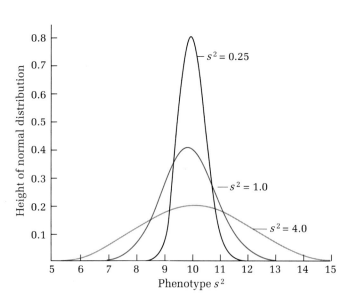

Figure 23.7 Graphs showing three distributions with the same mean but different variances.

butions may have the same mean but very different variances, as shown in Figure 23.7. The variance, symbolized as s^2, is defined as the average squared deviation from the mean. It can be calculated by first subtracting each individual measurement from the mean. Each value obtained from this subtraction is squared and all the squared values are added up. The sum of this calculation is then divided by the number of original measurements minus one. (For mathematical reasons, which we will not discuss here, an unbiased estimate of the variance is obtained by dividing by $n - 1$ instead of n.) The formula for calculating the variance is:

$$\text{Variance} = s^2 = \frac{\Sigma(x_i - \bar{x})^2}{n - 1}$$

Another statistic, closely related to the variance, is the **standard deviation,** which is simply the square root of the variance:

$$\text{Standard deviation} = s = \sqrt{s^2}$$

The standard deviation is often preferred to the variance, because the standard deviation is in the same units as the original measurements, while the variance is in the units squared. Sample calculations for the variance and the standard deviation are presented in Table 23.2.

In a normal distribution about two-thirds of the individual observations have values within one standard deviation above or below ($\pm 1s$) the mean of the distribution (Figure 23.8). About 95 percent of the values fall within two standard deviations ($\pm 2s$) of the mean, and over 99 percent fall within three standard deviations ($\pm 3s$). Therefore, a broad curve implies a considerable variability in the quantity measured and a correspondingly large standard deviation. A narrow curve, in contrast, indicates relatively little variability in the quantity measured and a correspondingly small standard deviation.

The variance and the standard deviation can provide us with valuable information about the phenotypes of a group of individuals. In our discussion of the mean, we saw how East used the mean to describe flower length of parents and offspring in crosses of the tobacco plant. When East crossed a strain of tobacco with short flowers to a strain with long flowers, the F_1 offspring had a mean flower length of 63.5 mm, which was intermediate to the phenotypes of the parents. When he intercrossed the F_1, the mean flower length of the F_2 offspring was 68.8 mm, approximately the same as the mean phenotype of the F_1. However, the F_2 progeny differed from the F_1 in an important attribute which is not apparent if we only examine the means of the phenotype—the F_2 were more variable in phenotype than the F_1. The variance in the flower length of the F_2 was 42.4, whereas the variance in the F_1 was only 8.6. This finding indicated that more genotypes were present among the F_2 progeny than in the F_1. Thus, the mean and the variance are both necessary for fully describing the distribution of phenotypes among a group of individuals.

Correlation

Often two variables are associated or correlated.

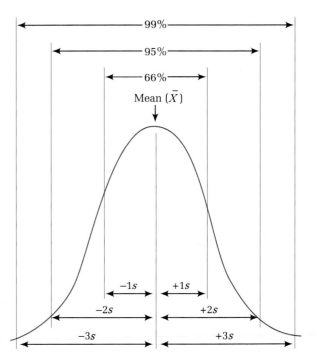

Figure 23.8 Normal distribution curve showing the proportions of the data in the distribution that are included within certain multiples of the standard deviation.

This means that if one variable changes, the other is also likely to change. For example, head size and body size are correlated in most animals—individuals with large bodies have correspondingly large heads, and those with small bodies tend to have small heads. The **correlation coefficient** is a statistic that measures the strength of the association between two variables. Suppose we have two variables, x and y (x might equal head length and y might equal body length), and we wish to calculate the correlation between them. We begin by obtaining the covariance of x and y; the covariance is computed by taking the first x value and subtracting it from the mean of x and then multiplying the result by the deviation of the first y value from the mean of y. This is done for each pair of x and y values and the products are added together. The sum is then divided by

$n - 1$ to give the covariance of x and y, where n equals the number of xy pairs:

$$\text{cov}_{xy} = \frac{\Sigma(x_i - \bar{x})(y_i - \bar{y})}{n - 1}$$

An algebraically equivalent equation, which is easier to compute, is

$$\text{cov}_{xy} = \frac{\Sigma x_i y_i - 1/n(\Sigma x_i \Sigma y_i)}{n - 1}$$

where $\Sigma x_i y_i$ is the sum of each value of x multiplied by each value of y, Σx_i is the sum of all x values, and Σy_i is the sum of all y values. The correlation coefficient r, can then be obtained by dividing the covariance by the product of the standard deviations of x and y.

$$\text{Correlation coefficient} = r = \frac{\text{cov}_{xy}}{s_x s_y}$$

where s_x equals the standard deviation of x and s_y equals the standard deviation of y. Table 23.3 gives a sample calculation of the correlation coefficient between two variables.

The correlation coefficient can range from -1 to $+1$. The sign of the correlation coefficient, whether it is positive or negative, indicates the direction of the correlation. If the correlation coefficient is positive, then an increase in one variable tends to be associated with an increase in the other variable. If seed size and seed number are positively correlated in sunflowers, for example, plants with larger seeds also tend to produce more seeds. A negative correlation coefficient indicates that an increase in one variable is associated with a decrease in the other. If seed size and seed number are negatively correlated, plants with large seeds tend to produce fewer seeds on the average than plants with smaller seeds. The absolute value of the correlation coefficient (its magnitude if the sign is ignored) provides information about the strength of the association. When the correlation coefficient is close to -1 or to $+1$, the correlation is strong, meaning that a change in one variable is almost always associated with a corresponding change in the other variable. On the other hand, a correlation coefficient near 0 indicates a weak relationship between the variables. Figure 23.9 shows several different correlations between two variables, illustrating the meaning of negative and positive correlations and strong and weak correlations.

Several important points about correlation coefficients warrant emphasis at this point. First, a correlation between variables means only that the variables are associated; correlation does not imply that a cause-effect relationship exists. The classic example of a noncausal correlation between two variables is the positive correlation that exists between number of Baptist ministers and liquor consumption in cities with population size over 10,000. One should not conclude from this correlation that Baptist ministers are consuming all the alcohol. Alcohol consumption and number of Baptist ministers are associated because both are positively correlated with a third factor, population size; larger cities contain more Baptist ministers and also have higher alcohol consumption. Assuming that two factors are causally related because they are correlated may lead to erroneous conclusions.

Another important point is that correlation is not the same thing as identity. Correlation only means that a change in one variable is associated with a corresponding change in the other variable. Two variables can be highly correlated, and yet have very different values. For example, height of today's college-age males is correlated with the height of their fathers; tall fathers tend to produce tall sons and short fathers tend to produce short sons. This correlation results from the fact that genes influence human height. However, most college-age males are taller than their fathers, probably because better diet and health care have increased the average height of all individuals in recent years. Thus, father and son exhibit a correlation in height, but they are not the same height.

Keynote *The correlation coefficient is a measure of how strongly two variables are associated. A positive correlation coefficient indicates that the two variables change in the same direction; an increase in one variable is usually associated with a corresponding increase in the other variable. When the correlation coefficient is negative, the variables are inversely related; an increase in one variable is most often associated with a decrease in the other. The absolute value of the correlation coefficient provides information about the strength of the association.*

Regression

The correlation coefficient tells us about the strength of association between variables and indicates whether the relationship is positive or negative, but it provides little information about the precise relationship between the variables. Often we are interested in knowing how much

Table 23.3 Sample Calculation of the Correlation Coefficient for Body Length and Head Width of Tiger Salamanders

Body Length (mm) x_i	$x_i - \bar{x}$	$(x_i - \bar{x})^2$	Head Width (mm) y_i	$y_i - \bar{y}$	$(y_i - \bar{y})^2$	$x_i y_i$
72.00	-7.92	62.67	17.00	-0.75	0.56	1224
62.00	-17.92	321.01	14.00	-3.75	14.06	868
86.00	6.08	37.01	20.00	2.25	5.06	1720
76.00	-3.92	15.34	14.00	-3.75	14.06	1064
64.00	-15.92	253.34	15.00	-2.75	7.56	960
82.00	2.08	4.34	20.00	2.25	5.06	1640
71.00	-8.92	79.51	15.00	-2.75	7.56	1065
96.00	16.08	258.67	21.00	3.25	10.56	2016
87.00	7.08	50.17	19.00	1.25	1.56	1653
103.00	23.08	532.84	23.00	5.25	27.56	2369
86.00	6.08	37.01	18.00	0.25	0.06	1548
74.00	-5.92	35.01	17.00	-0.75	0.56	1258
$\Sigma x_i =$ 959.00		$\Sigma(x_i - \bar{x})^2 =$ 1686.92	$\Sigma y_i =$ 213.00		$\Sigma(y_i - \bar{y})^2 =$ 94.25	$\Sigma x_i y_i =$ 17,385

$\bar{x} = \Sigma x_i / n = 959/12 = 79.92$
$\bar{y} = \Sigma y_i / n = 213/12 = 17.75$

Variance of $x = s_x^2 = \Sigma(x_i - \bar{x})^2 / n - 1 = \underline{1686.92}/11 = 153.35$
Standard deviation of $x = s_x = \sqrt{s_x^2} = \sqrt{153.35} = 12.38$

Variance of $y = s_y^2 = \Sigma(y_i - \bar{y})^2 / n - 1 = 94.25/11 = 8.57$
Standard deviation of $y = s_y = \sqrt{s^2} = \sqrt{8.57} = 2.93$

Covariance $= cov_{xy} = (\Sigma x_i y_i - 1/n(\Sigma x_i \Sigma y_i))/n - 1$
$cov_{xy} = (17385 - 1/12(959 \times 213))/1 - 12$
$cov_{xy} = 32.97$

Correlation coeff cient $= r = cov_{xy}/(s_x s_y) = 32.97/(12.38 \times 2.93)$
$r = 0.91$

of a change in one variable is associated with a given change in another variable. Returning to our example of the correlation between heights of father and son, we might ask: If a father is 6 feet tall, what is the most likely height of his son? To answer this question, **regression** analysis is used.

The relationship between two variables can be expressed in the form of a **regression line,** as shown in Figure 23.10 for the relationship between heights of father and son. Each point on the graph represents the actual height of a father (value on the horizontal or x axis) and the height of his son (value on the vertical or y axis). The regression line is a mathematically computed line that represents the best fit of a line to the points; what this means is that the squared vertical distance from the points to the

regression line is minimized. The regression line can be represented with the equation

$$y = a + bx$$

where x and y represent the values of the two variables (in Figure 23.10 the heights of father and son, respectively), b represents the **slope** (also called the **regression coefficient**), and a is the **y intercept.** The slope can be calculated from the covariance of x and y and the variance of x in the following manner:

$$\text{Slope} = b = \frac{\text{cov}_{xy}}{s^2_x}$$

The slope indicates how much of an increase in the variable on the y axis is associated with a unit increase in the variable on the x axis. For

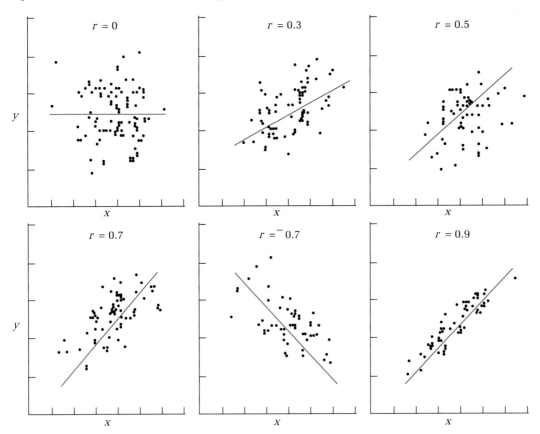

Figure 23.9 Scatter diagrams of x *and* y *variables having various degrees of correlation* (r).

example, a slope of 2 for the regression of father and son height would mean that for each 1-inch increase in height of a father, the expected height of the son would increase 2 inches. The y intercept is the expected value of y when x is zero (the point at which the regression line crosses the y axis). Examples of several regression lines are presented in Figure 23.11. Regression analysis is one method that is commonly used for measuring the extent to which variation in a trait is genetically determined.

Analysis of Variance

One last statistical technique that we will mention briefly is **analysis of variance.** Analysis of variance, often called simply **ANOVA,** is a powerful series of statistical procedures for examining differences in means and for partitioning variance. We might be interested in knowing if males with the XYY karyotype (see Chapter 17) differ in IQ from males with a normal XY karyotype. We could proceed by first calculating the mean IQ of a sample of XYY males and the mean IQ of a sample of XY males. Suppose we found the mean IQ of the XYY males was 85, while the mean IQ of the XY males was 102. The IQs appear to be different, but considerable variation in IQ exists among all individuals, and our difference could be due to chance deviations between the samples. Analysis of variance can be used to determine the probability that the difference in mean IQ of the XY males and the mean IQ of XYY males results from chance differences in our samples. The analysis of variance can also be used to partition the variation in IQ into components associated with different factors, such as number of siblings, socioeconomic status, sex, and karyotype. All these factors may influence IQ; the analysis of variance might indicate that the mean IQ of XYY males is significantly lower than the mean IQ of XY males, but that differences in the karyotypes explain only 10 percent of the overall variation in IQ. Other factors, such as number of siblings, may be more important in determining the differences in IQ.

The calculations involved in analysis of vari-

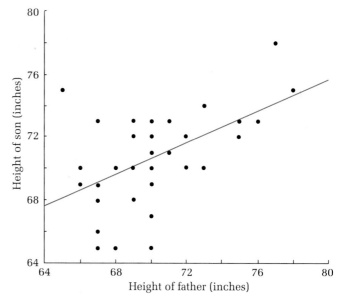

Figure 23.10 *Regression of son's height on father's height. Each point represents a pair of data for the height of a father and his son. The regression equation is* y = 36.05 + 0.49x.

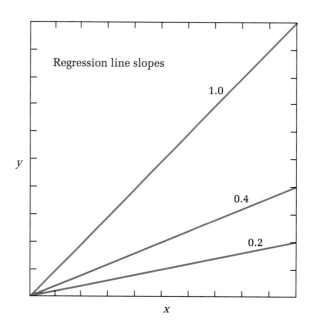

Figure 23.11 *Slopes of straight lines passing through the origin.*

ance are beyond the scope of this book, but the concept of partitioning variance is important in quantitative genetics. Frequently, we are interested in how much of the variation in a trait is associated with genetic differences among individuals and how much of the variation is associated with environmental factors. Suppose, for example, that we wanted to increase milk production in a breed of dairy cattle. We might use analysis of variance to determine how much of the variation in milk production results from environmental differences among cows and how much arises because of genetic differences. If much of the variation is genetic, we might increase milk production by selective breeding. On the other hand, if most of the variation is environmental, selective breeding will do little to increase milk production, and our efforts might be better directed toward providing the optimum environment for high production. Analysis of variance is often used in this type of problem.

Quantitative Inheritance

In the examples of quantitative inheritance described in the following sections, geneticists did not know at f rst how these traits were inherited, although it was apparent that their pattern of inheritance differed from that of discontinuous traits.

Inheritance of Ear Length in Corn

An organism that has been the subject of genetic and cytological studies for many years is corn, *Zea mays.* Ear length is one of the traits of the corn plant that came under scrutiny in a classic study that demonstrated quantitative inheritance for that trait. In this study, reported in 1913, R. A. Emerson and E. M. East started their experiments with two pure-breeding strains of corn, each of which displayed little variation in ear length. The two varieties were Black Mexican sweet corn (which had short ears of mean length 6.63 cm) and Tom Thumb popcorn (which had long ears of mean length 16.80 cm).

Emerson and East crossed the two strains and

then interbred the F_1 plants. The results are shown pictorially in Figure 23.12a and in histogram form in Figure 23.12b. Note that the F_1's have a mean ear length of 12.12 cm, which is approximately intermediate between the mean ear lengths of the two parental lines. The parental plants are pure breeding, and therefore we can assume that each is homozygous for whatever genes control the lengths of their ears. Since the two parental plants differ in ear length, though, each must be genetically different. When two pure-breeding strains are crossed, the F_1 plants will be heterozygous for all genes and all plants should have the same genotype. Therefore the range of ear length phenotypes seen in the F_1 plants must be due to factors other than genetic differences; these other factors are probably environmental factors, since it is impossible to grow plants in exactly identical conditions.

In the F_2 the mean ear length of 12.89 cm is about the same as the mean for the F_1 population, but the F_2 population has a much larger variation around the mean than the F_1 population has. This variation is easy to see in Figure 23.12b; it can also be shown by calculating the standard deviation s. The standard deviation of the long-eared parent is 0.816, and that of the short-eared parent is 1.887. In the F_1, $s = 1.519$, and in the F_2, $s = 2.252$. These numbers confirm that the F_2 has greater variability, something we could conclude by looking at the data.

Is this variation the result of the effects of environmental factors? Certainly, if the environment was responsible for variation in the parental and the F_1 generations, we have every reason to believe that it would have a similar effect on the F_2. However, we have no reason to suppose that the environment would have a greater influence on the F_2 than on the other two generations, and thus there must be another explanation for the greater variation in ear length in the F_2 generation. A more reasonable hypothesis is that the increased variability of the F_2 results from the presence of greater genetic variation in the F_2.

Setting aside the environmental influence for the moment, the data reveal four observations

a)

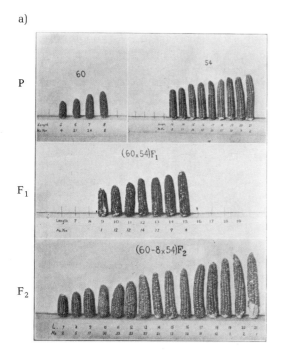

b)

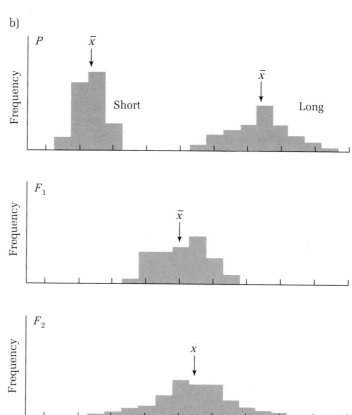

Figure 23.12 Inheritance of ear length in corn: (a) representative corn ears from the parental, F₁, and F₂ generations from an experiment in which two pure-breeding corn strains that differ in ear length were crossed and then the F₁'s interbred. (b) Histograms of the distributions of ear length (in centimeters) of ears of corn from the experiment pictorially represented in part (a); the vertical axes represent the percentages of the different populations found at each ear length.

that apply generally to quantitative-inheritance studies similar to this one:

1. The mean value of the quantitative trait in the F_1 is approximately intermediate between the means of the two true-breeding parental lines.
2. The mean value for the trait in the F_2 is approximately equal to the mean for the F_1 population.
3. The F_2 shows more variability around the mean than the F_1 does.
4. The extreme values for the quantitative trait in the F_2 extend further into the distribution of the two parental values than do the extreme values of the F_1.

Can the data presented be explained in terms of standard Mendelian genetic principles that govern the inheritance of discontinuous traits? No, the data cannot support such an interpretation. That is, if a single gene was responsible for the two phenotypes of the original parents (AA = homozygous, long; aa = homozygous, short), then the F_1 data could be explained if we assume incomplete dominance. However, by crossing the F_1 heterozygote (Aa), we would expect a 1:2:1 ratio of AA, Aa, and aa, or long, intermediate, and short phenotypes. Clearly, the data do not fall into such discrete classes.

Keynote *For a quantitative trait, the F_1 progeny of a cross between two phenotypically distinct, pure-breeding parents has a phenotype intermediate between the parental phenotypes. The F_2 shows more variability than the F_1, with a mean phenotype close to that of the F_1. The extreme phenotypes of the F_2 extend well beyond the range of the F_1 and into the ranges of the two parental values.*

Polygene Hypothesis for Quantitative Inheritance

The simplest explanation for the data obtained from Emerson and East's experiments on corn ear length and from other experiments with quantitative traits is that quantitative traits are controlled by not one but by many genes. This explanation, called the **polygene,** or **multiple-gene, hypothesis for quantitative inheritance,** is regarded as one of the landmarks of genetic thought.

The polygene hypothesis can be traced back to 1909 and the classic work of Hermann Nilsson-Ehle, who studied the color of the wheat kernel. Like Mendel and like Emerson and East, Nilsson-Ehle started by crossing true-breeding lines of plants with red kernels and plants with white kernels. The F_1 had grains that were all the same shade of an intermediate color between red and white. At this point he could not rule out incomplete dominance as the basis for the F_1 results. However, when he intercrossed the F_1's, a number of the F_2 progeny showed a ratio of approximately 15 red (all shades):1 white kernels, clearly a deviation from a 3:1 ratio expected for a monohybrid cross. He recognized four discrete shades of red, in addition to white, among the progeny. He counted the relative number of each class and found a 1:4:6:4:1 phenotypic ratio of wheat with dark red, medium red, intermediate red, light red, and white kernels. Note that $\frac{1}{16}$ of the F_2 has a kernel phenotype as extreme as the original red parent, and $\frac{1}{16}$ has a kernel phenotype as extreme as the original white parent.

How can the data be interpreted in genetic terms? Recall from Chapter 4 (Analytical Approaches for Solving Genetics Problems, Q3c, pp. 133–134) that a 15:1 ratio of two alternative characteristics resulted from the interaction of the products of two genes each affecting the same trait; the two genes are known as duplicate genes. The explanation for the 15:1 ratio was that two allelic pairs are involved in determining the phenotypes segregating in the cross. Since several of the F_2 populations from the wheat crosses exhibited a 15 red:1 white ratio, we can apply that explanation to the kernel trait.

Let us hypothesize that there are two pairs of independently segregating alleles that control the production of red pigment: alleles R (red) and C (crimson) result in red pigment and alleles r and c result in the lack of pigment. Nilsson-Ehle's parental cross and the F_1 genotypes can then be shown as follows:

$$\begin{array}{ccc} \text{P} & RR\ CC & \times & rr\ cc \\ & \text{(dark red)} & & \text{(white)} \\ F_1 & & Rr\ Cc \\ & & \text{(intermediate red)} \end{array}$$

When the F_1 is interbred, the distribution of genotypes in the F_2 is that typical of dihybrid inheritance, that is, $\frac{1}{16}$ $RR\ CC$ + $\frac{2}{16}$ $Rr\ CC$ + $\frac{1}{16}$ $rr\ CC$ + $\frac{2}{16}$ $RR\ Cc$ + $\frac{4}{16}$ $Rr\ Cc$ + $\frac{2}{16}$ $rr\ Cc$ + $\frac{1}{16}$ $RR\ cc$ + $\frac{2}{16}$ $Rr\ cc$ + $\frac{1}{16}$ $rr\ cc$. If R and C are dominant to r and c, the 9:3:3:1 phenotypic ratio characteristic of dihybrid inheritance will result. For the wheat kernel color phenotype, then, dominance is not the simple answer, since the observed phenotypic ratio approximates 1:4:6:4:1.

These numbers in the phenotypic ratio are the same as the coefficients in the binomial expansion of $(a + b)^4$. A simple explanation is that each dose of a gene controlling pigment production allows the synthesis of a certain amount of pigment. Therefore the intensity of red coloration is a function of the number of dominant R or C alleles in the genotype; $RR\ CC$ (term a^4 in the expansion) would be dark red and $rr\ cc$ (b^4 in the expansion) would be white. Table 23.4 summarizes this situation with regard to the five phenotypic classes observed by Nilsson-Ehle. In other words, the genes with dominant alleles

Table 23.4 Genetic Explanation for the Number and Proportions of F₂ Phenotypes for the Quantitative Trait Red Kernel Color in Wheat

Genotype	Number of Genes for Red	Phenotype	Fraction of F_2
RR CC	4	Dark red	$\frac{1}{16}$
RR Cc or Rr CC	3	Medium red	$\frac{4}{16}$
RR cc or rr CC or Rr Cc	2	Intermediate red	$\frac{6}{16}$
rr Cc or Rr cc	1	Light red	$\frac{4}{16}$
rr cc	0	White	$\frac{1}{16}$

code for products that add to the phenotypic characteristic; for example, each dominant allele, *R* or *C,* causes more red pigment to be added to the wheat kernel color phenotype. Alleles that contribute to the phenotype are called **contributing alleles.** The alleles that do not have any effect on the phenotype of the quantitative trait, such as the *r* and *c* alleles in the wheat kernel color trait, are called **noncontributing alleles.** Thus the inheritance of red kernel color in wheat is an example of a multiple gene or polygene series of as many as four contributing alleles.

We must be cautious in interpreting the genetic basis of this particular quantitative trait, though. Some F₂ populations show only three phenotypic classes with a 3:1 ratio of red to white, while other F₂ populations show a 63:1 ratio of red to white, with discrete classes of color between the dark red and the white. These results indicate that the genetic basis for the quantitative trait can vary with the strain of wheat involved. The 3:1 case could be explained by a single-gene system with two contributing alleles, while the 63:1 case could indicate a polygene series with six contributing alleles. The number of discrete classes in the latter case would be seven, with the proportion of each class following the coefficients in the binomial expansion of $(a + b)^6$, that is, 1:6:15:20:15:6:1.

The multiple-gene hypothesis that fits the wheat kernel color example so well has been applied to other examples of quantitative inheri-

tance, including ear length in corn. In its basic form the multiple-gene hypothesis proposes that a number of the attributes of quantitative inheritance can be explained on the basis of the action and segregation of a number of allelic pairs that have identical (or nearly identical) additive effects on the phenotype with no complete dominance involved. Simplified, the assumptions of the hypothesis are as follows:

1. No allelic pairs exhibit dominance of one allele over another. Instead, a series of contributing and noncontributing alleles is involved.
2. Each contributing allele in a series has an equal effect.
3. The effect of each contributing allele is additive.
4. No genetic interaction (epistasis) occurs among the alleles at different loci in a polygenic series.
5. No genetic linkage is exhibited between the genes in a polygenic series, and therefore they assort independently.
6. There are no environmental effects.

It would be naive to suppose that each of these assumptions is uniformly, or even widely, valid. Since each assumption is open to criticism from theoretical principles, it is surprising that so many examples conform so well to the assumptions. In fact, many polygenic effects do conform to the first four assumptions.

When a large number of polygenes are involved, however, it is hard to imagine that linkage or environmental effects can be ignored, and we have no reason to suppose that evolution has made genes in a small polygenic series unlinked. Also, environmental effects cannot be ignored. In the ear length data, for example, the variability of length in the true-breeding strains is probably the result of environmental effects. With plants, variations in environmental agents such as soil composition, water availability, and local pest concentrations can affect the quantitative-trait phenotypes. Thus we should always remember that an organism's genotype merely specifies the organism's *potential* phenotype and that the genotypic blueprint is interpreted during the organism's development and differentiation in a way that is influenced by internal and external environments.

For the most part the multiple-gene hypothesis is satisfactory as a working hypothesis for interpreting many quantitative traits. The whole picture of quantitative traits is very complicated, though, and there are still many gaps in our understanding of quantitative inheritance. In addition, we have a lot to learn about the molecular aspects of quantitative traits. For example, while the proposal that a number of alleles each function to produce a particular amount of pigment is an attractive hypothesis, what does that hypothesis mean at the molecular level? Is some regulation of product output exerted at the translational level or in a biochemical pathway? In a large polygenic series, how many biochemical pathways are controlled? Thus quantitative inheritance provides an explanation for the inheritance of continuous traits that is compatible with Mendel's laws, but many aspects of quantitative inheritance are still unknown.

Keynote *Quantitative traits are based on genes in a multiple-gene series, or polygenes. The multiple-gene hypothesis assumes that contributing and noncontributing alleles in the series operate so that as the number of contributing alleles increases, there is an additive (or occasionally multiplicative) effect on the phenotype, that no genetic interaction occurs between the gene loci, that the loci are unlinked, and that there are no environmental effects. Since the last three assumptions are not always true, the analysis of multiple-gene series that control quantitative traits is difficult.*

Determining the Number of Polygenes for a Quantitative Trait

We must be able to estimate the number of genes in a multiple-gene system so that we can apply our knowledge in a predictive way to genetic crosses involving quantitative traits. This predictive ability is important, for example, to plant and animal breeders who want to breed features into the organisms with which they deal. Such features are often phenotypes con-trolled by a polygene system or systems, and the breeders need to be able to predict the probabilities of obtaining progeny with particular combinations of phenotypes from breeding programs. Examples here usually relate to yields, such as number and size of tobacco leaves from a tobacco plant, quantity of milk from a dairy cow, beef from beef cattle, or ears of corn from a corn plant. Related to these factors, of course, is the quality of the yield, which can also be affected by polygene systems.

As you might imagine, it is not an easy task to estimate the number of genes controlling a quantitative trait. If a large number of genes are involved, many different phenotypes are present, and the progeny are not easily placed into distinct phenotypic classes.

Where discrete phenotypic classes can be identified, however, we can estimate the number of genes involved. In Nilsson-Ehle's studies of the inheritance of red kernel color in wheat, for example, different F_2 populations were seen. While crosses of some red-kerneled varieties gave F_2 progeny that fell into five phenotypic classes, others gave seven phenotypic classes. The latter crosses showed an approximately 1:6:15:20:15:6:1 ratio of different shades, ranging from dark red to white. Since the relative proportions of the phenotypic classes correspond to the coefficients in the binomial expansion of $(a + b)^6$, it is likely that three pairs of alleles controlled kernel color in this particular cross.

In general, the relationship between the number of independently segregating allelic pairs in a polygene series and the number of quantitative-trait phenotypes in the F_2 is as shown in Table 23.5. When one pair of alleles controls a trait, then $\frac{1}{4}$ of the F_2 should show one phenotypic extreme (e.g., the darkest red) and $\frac{1}{4}$ should show the other phenotypic extreme (e.g., white). The general formula, $(\frac{1}{2})^n$, describes the predicted probability for an extreme phenotype when n is the number of segregating alleles involved. For two genes, and therefore four alleles, the fraction of the F_2 population with an extreme character is $(\frac{1}{2})^4 = \frac{1}{16}$. For three genes (six alleles) the fraction is $\frac{1}{64}$, as in the wheat kernel example. As the number of genes increases, the fraction of the F_2 with an extreme phenotype decreases, and it does so very rapidly. For five genes the fraction is 1/1024, and for ten genes it is 1/1,048,576.

The number of genotypic classes also increases rapidly as the number of pairs of alleles increases. The beginnings of this increase are shown in Table 23.5. Three pairs of alleles yield 27 possible genotypic classes in the F_2; five pairs of alleles yield 243 genotypes; and ten pairs of alleles yield 59,049 genotypes. In general, the number of genotypic classes in the F_2 is $(3)^n$, where n is the number of pairs of alleles involved.

As the number of pairs of alleles in a multiple-gene series increases, the F_2 population rapidly assumes a continuum of phenotypic variation for which distinctions between classes are impossible. In short, when the number of pairs of alleles is five or more, it is difficult for us to be sure of the number of pairs of alleles involved in controlling a quantitative effect. Other methods are available for determining the number of pairs of alleles; they have as their basis more sophisticated mathematical treatments. Even with these methods, however, we still encounter significant complications caused by the environment, different dominance relationships of the genes in the series, and linkage.

Heritability

Heritability is the proportion of phenotypic variation in a population attributable to genetic factors. The concept of heritability is used to examine the relative contributions of genes and environment to variation in a specific trait. As we have seen, continuous traits are frequently influenced by multiple genes and by environmental factors. One of the important questions addressed in quantitative genetics is: How much of the variation in phenotype results from genetic differences among individuals and how much is a product of environmental variation? For example, multifactorial traits such as weight of cattle, number of eggs laid by chickens, and the amount of fleece produced by sheep are important for breeding programs and agricultural management. Many ecologically important traits, such as variation in body size, fecundity, and developmental rate are also multifactorial, and the genetic contribution to this variation is important for understanding how natural populations evolve. The extent to which genetic and environmental factors contribute to human variation in traits like blood pressure and birth weight is important for health care. Also, the extent to which genes affect human behaviors, such as alcoholism and criminality, may be

Table 23.5 Probability Information for a Quantitative Character Controlled by a Polygenic Series in Which Independently Segregating Allelic Pairs Have Duplicate, Cumulative Effects

Number of Allelic Pairs	Number of Segregating Alleles	Fraction of Population Showing an Extreme Expression of the Character	Number of Genotypes in F_2	Number of Phenotypes in F_2	F_2 Phenotypic Ratios are the Coefficients in the Binomial Expansion of
1	2	$(\frac{1}{2})^2 = \frac{1}{4}$	$(3)^1 = 3$	3	$(a + b)^2$
2	4	$(\frac{1}{2})^4 = \frac{1}{16}$	$(3)^2 = 9$	5	$(a + b)^4$
3	6	$(\frac{1}{2})^6 = \frac{1}{64}$	$(3)^3 = 27$	7	$(a + b)^6$
4	8	$(\frac{1}{2})^8 = \frac{1}{256}$	$(3)^4 = 81$	9	$(a + b)^8$
n	$2n$	$(\frac{1}{2})^{2n}$	$(3)^n$	$2n + 1$	$(a + b)^{2n}$

useful in establishing social policy. Hence, the study of heritability is crucial in a variety of scientific disciplines.

The following section deals with two types of heritability: **broad-sense heritability** and **narrow-sense heritability.** To assess heritability, we must first measure the variation in the trait and then we must partition that variance into components attributable to different causes.

Components of the Phenotypic Variance

The **phenotypic variance,** represented by V_P, is a measure of the variability of a trait. It is calculated by computing the variance of the trait for a group of individuals, as was outlined in the statistics section on pp. 741–742. Differences among individuals arise from several factors, and therefore we can partition the phenotypic variance into several components attributable to different sources. First, some of the phenotypic variation arises because of genetic differences among individuals (different genotypes within the group). This contribution to the phenotypic variation is called **genetic variance** and is represented by the symbol V_G. As noted, additional variation often results from environmental differences among the individuals; in other words, different environments experienced by individuals may contribute to the differences in their phenotypes. The **environmental variance** is symbolized by V_E, and by definition, it includes any nongenetic source of variation. Temperature, nutrition, and parental care are examples of obvious environmental factors that may cause differences among individuals; environmental variance also includes random factors that occur during development, which are sometimes referred to as "developmental noise." A third source of phenotypic variance is genetic-environmental interaction, represented by V_{GE}. Genetic-environmental interaction occurs when the relative effects of the genotypes differ among environments. For example, in a cold environment, genotype AA of a plant may be 40 cm in height and genotype Aa may be 35 cm in height. However, when the genotypes are moved to a warm climate, genotype Aa is now 60 cm, as

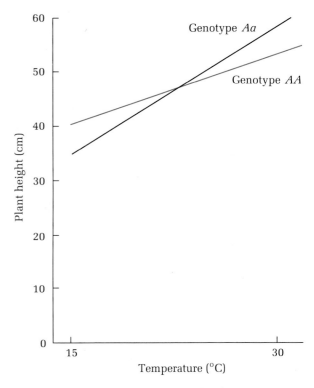

Figure 23.13 *Hypothetical example of genetic-environmental interaction. At 15°C, the genotype* AA *is tallest with a height of 40 cm. Genotype* Aa *is only 35 cm tall. However, at 30°C,* Aa *is tallest (60 cm) while* AA *is 55 cm tall. Both genotype and environment influence plant height, but the effects of the genotypes differ between environments.*

compared with genotype AA, which is only 50 cm in height. This relationship is shown in Figure 23.13. In this example, both genotypes grow taller in the warm environment, but the relative performance of the genotypes switches in the two environments. Therefore, both environmental differences (temperature) and genetic differences (genotypes) contribute to the phenotypic variance, but the effects of genotype and environment cannot simply be added together. An additional component that accounts for how genotype and environment interact must be considered, and this is V_{GE}.

The phenotypic variance, composed of differences arising from genetic variation, environmental variation, and genetic-environmental variation, can be represented by the following equation:

$$V_P = V_G + V_E + V_{GE}$$

The relative contributions of these three factors to the phenotypic variance depend upon the genetic composition of the population, the specific environment, and the manner in which the genes interact with the environment.

The genetic variance V_G can be further subdivided into components arising from different types of interactions between genes. Some of the genetic variance occurs as a result of the average effects of the different alleles on the phenotype. For example, an allele g may, on the average, contribute 2 centimeters in height to a plant, and the allele G may contribute 4 centimeters. In this case, the gg homozygote would contribute $2 + 2 = 4$ centimeters in height, the Gg heterozygote would contribute $2 + 4 = 6$ centimeters in height, and the GG homozygote would contribute $4 + 4 = 8$ centimeters in height. To determine the genetic contribution to height, we would then add the effects of alleles at this locus to the effects of alleles at other loci which might influence the phenotype. Genes such as these are said to have additive effects, and this type of genetic variation is called **additive genetic variance.** You may recall that the genes studied by Nilsson-Ehle, which determine kernel color in wheat, are strictly additive in this way. Some alleles contribute to the pigment of the kernel and others do not; the added effects of all the individual contributing alleles determine the phenotype of the kernel. Thus, the genotypes *AAbb, aaBB,* and *AaBb* all produce the same phenotype, since each genotype has two contributing alleles. The phenotypic variance arising from the additive effects of genes is the additive genetic variance and is symbolized by V_A.

Some genes may exhibit dominance, and this comprises another source of genetic variance, the **dominance variance** (V_D). Dominance occurs when one allele masks the expression of the other allele at the locus. If dominance is present, the individual effects of the alleles are not strictly additive; we must also consider how alleles at a locus interact. In the presence of dominance, the heterozygote Gg would contribute 8 centimeters in height to the phenotype, the same amount as the GG homozygote. Finally, epistatic interactions may occur among alleles. Recall that in epistasis, alleles at different loci interact to determine the phenotype. The presence of epistasis adds another source of genetic variation, called epistatic or **interaction variance** (V_I). So we can partition the genetic variance as follows:

$$V_G = V_A + V_D + V_I$$

The total phenotypic variance can then be summarized as

$$V_P = V_A + V_D + V_I + V_E + V_{GE}$$

Partitioning the phenotypic variance into these components is useful for thinking about the contribution of different factors to the variation in phenotype. When determining the narrow-sense heritability of a trait (which is discussed in the following section), for example, we are primarily concerned with the relative contribution of the additive genetic component V_A. However, precisely estimating all, or even some, of these components is frequently a formidable task.

Broad-Sense and Narrow-Sense Heritability

Geneticists frequently partition the phenotypic variance of a trait to determine the extent to which variation among individuals results from genetic differences. Thus, they are interested in how much of the phenotypic variance V_P can be attributed to genetic variance V_G. This quantity, the proportion of the phenotypic variance that consists of genetic variance, is called the **broad-sense heritability** and is expressed as follows:

$$\text{Broad-sense heritability} = H^2 = \frac{V_G}{V_P}$$

The heritability of a trait can range from 0 to 1. A broad-sense heritability of 0 indicates that none of the variation in phenotype among individuals results from genetic differences. A heritability of 0.5 means that 50 percent of the phenotypic variation arises from genetic differences among individuals, and a heritability of 1 would suggest that all the phenotypic variance is genetically based.

The broad-sense heritability includes genetic variation from all types of genes. Frequently, we are more interested in the proportion of the phenotypic variation that results only from additive genetic effects. This is because the additive genes are those that allow us to predict the phenotype of the offspring from the phenotypes of the parents. To understand the reason for this, consider a cross involving a trait that results from the effect of alleles at a single locus. One parent is A^1A^1 and 10 cm tall; the other parent is A^2A^2 and is 20 cm tall. All the offspring from this cross (the F_1) will be A^1A^2. If the alleles are additive and contribute equally to height, the offspring should be 15 cm tall, exactly intermediate between the parents. However, if A^2 is dominant, all the offspring will be 20 cm, resembling only one of the parents. Thus, if dominant genes are important, the offspring will not be intermediate to the parental phenotypes. In a similar fashion, epistatic genes will not always contribute to the resemblance between parents and offspring.

Because the additive genetic variance allows one to make accurate predictions about the resemblance between offspring and parents, quantitative geneticists frequently determine the proportion of the phenotypic variance that results from additive genetic variance, a quantity referred to as the **narrow-sense heritability.** The additive genetic variance is also that variation that responds to selection in a predictable way, and thus the narrow-sense heritability provides information about how a trait will evolve. The narrow-sense heritability is represented mathematically as

$$\text{Narrow-sense heritability} = h^2 = \frac{V_A}{V_P}$$

Understanding Heritability

Despite their utility, heritability estimates have a number of significant limitations. Unfortunately, these limitations are often ignored, and as a result, heritability is one of the most misunderstood and widely abused concepts in all of genetics. Before we discuss how heritability is determined, it is important that we list some of the important qualifications and limitations of heritability estimates.

1. Heritability does not indicate the extent to which a trait is genetic; heritability measures the *proportion of the phenotypic variance* among individuals in a population that results from genetic differences. Although saying that genes determine a trait and saying that genes affect the variation in a trait sound like similar statements, they actually mean two very different things. Genes are almost always important in the development of a trait. Thus our ability to learn about the game of football is dependent upon proper development of our nervous system, which is clearly controlled by genes. However, the differences we see among individuals in their knowledge of football is not genetic but results from different environments and personal interests. Heritability measures the degree to which variation among individuals results from different genotypes. Genes may be critical in determining a trait, but if all individuals in the population are homozygous for the same genes, then the genetic variance will be zero. Consequently, the heritability will be zero. Although the heritability in this case is zero, it would be incorrect to assume that genes play no role in the development of the trait. Similarly, a high heritability does not negate the importance of environmental influences on a trait; a high heritability simply means that environmental factors influencing the trait are relatively uniform among the individuals studied.

2. Heritability does not indicate what proportion of an individual's phenotype is genetic. Since it is based on the variance, which is calculated on a group of individuals, herita-

bility is a population parameter. An individual does not have heritability—a population does.

3. Heritability is not a fixed characteristic of a trait. Rather, it is a population parameter that depends upon the genetic makeup and the specific environment of the population. Heritability calculated for one population and a specific environment cannot be applied to another group, or even to the same group in a different environment.

To illustrate this point, suppose that we calculated heritability for adult height on individuals living in a small New England town and obtained a value of 0.7, suggesting that 70 percent of the variation in adult height among these individuals is genetic. This value of heritability cannot be extended to cover other populations. Residents of San Francisco, for example, might be more racially heterogeneous than the inhabitants of the small New England town; therefore the San Francisco population would have more genetic variation for height. Let us assume that the environmental variance of the two populations is about the same. If the environmental variance of the two populations is similar, but the genetic variance is greater in San Francisco, then heritability calculated for the height of San Francisco residents also would be greater.

Genes are not the only factor influencing height in humans; diet, an environmental effect, is also a major determinant of height. Since most individuals in our small New England town probably receive an adequate diet, at least in terms of calories, the environmental variance for height would not be large there. In a third-world nation, however, some individuals may receive adequate nutrition, while the diet of others might be severely deficient. Because greater differences in diet exist in such a nation, the environmental variance for height would be larger, and as a result, the heritability of height would be less.

These examples illustrate that heritability can be applied only to a specific group of individuals and to a specific environment. If the genetic composition of the group changes or the environment changes, the original heritability value obtained is no longer valid. Changing the genetic composition of the group or the environment does not alter the way in which genes affect the trait, but it does change the amount of genetic and environmental variance in the trait.

4. Even if heritability is high in each of two populations and the populations differ markedly in a particular trait, one cannot assume that the populations are genetically different. For example, suppose that you obtain a number of genetically variable mice and divide them into two groups. You feed one group a nutritionally rich diet and you are careful to provide each mouse with exactly the same amount of food, space, water, and so on. The mice grow to a large size because of the rich diet. When you measure heritability for adult body weight, you obtain a high value of 0.93. The high heritability is not surprising since the mice were genetically variable and environmental differences were kept at a minimum. The second group of mice comes from the same genetic stock, but you feed them an impoverished diet, lacking in calories and essential nutrients; again, each mouse gets exactly the same amount of food, space, water, and so on. Because of the poor diet, the mice of this second group are all smaller than those in the first group. When you calculate heritability for adult weight in the small mice, you again obtain a high value of 0.93, because the mice were genetically variable and the environmental differences were kept to a minimum. On the basis of the observation that heritability of body weight is high in both groups and the mice of the two groups differ in adult weight, some people might suggest that the mice are genetically different in body size. Yet, any claim that the mice of the two groups differ genetically is clearly wrong: both groups came from the same stock. The important point is that heritability cannot be used to draw conclusions about the nature of population differences in a trait.

5. Traits shared by members of the same family do not necessarily have high heritability. When members of the same family share a trait, the trait is said to be **familial.** Familial traits may arise because family members share genes or because they are exposed to the same environmental factors. Thus, familiality is not the same as heritability.

Keynote *Broad-sense heritability of a trait represents the proportion of the phenotypic variance that results from genetic differences among individuals. The narrow-sense heritability is more restricted—it measures the proportion of the phenotypic variance that results from additive genetic variance. The narrow-sense heritability allows quantitative geneticists to make predictions about the resemblance between parents and offspring, and it represents that part of the phenotypic variance that responds to selection in a predictable manner.*

How Heritability Is Calculated

A number of different methods are available for calculating heritability. Many of the methods involve comparing related and unrelated individuals, or comparing individuals with different degrees of relatedness. If genes are important in determining the phenotypic variance, then closely related individuals should be more similar in phenotype, since they have more genes in common. Alternatively, if environmental factors are responsible for determining differences in the trait, then related individuals should be no more similar in phenotype than unrelated individuals. An important point to remember is that the related individuals studied must not share a more common environment. We assume that if related individuals are more similar in phenotype, it is because they share similar genes. If related individuals also share a more common environment than unrelated individuals, separating the effects of genes and environment is much more difficult and frequently impossible. The absence of common environmental factors among related individuals can often be achieved in domestic plants and animals, and common

environments may not exist among family members in the wild; however, this is very difficult to obtain in humans, where family structure and extended parental care create common environments for many related individuals.

Heritability from parent-offspring regression. Some of the methods used to calculate heritability include comparison of parents and offspring, comparison of full and half sibs, comparison of identical and nonidentical twins, and response-to-selection experiments. Here, we will discuss the parent-offspring method; later we will examine the calculation of heritability through response-to-selection.

If additive genes are important in determining the differences among individuals, then we expect that offspring should resemble their parents. To quantify the degree to which genes influence a trait, we must first measure the phenotypes of parents and offspring in a series of families, and then statistically analyze the relationship between their phenotypes. Correlation and regression are appropriate for this type of problem.

We can represent the relationship between offspring phenotype and parental phenotype by plotting the mean phenotype of the parents against the mean phenotype of the offspring, as shown in Figure 23.14. Each point on the graph represents one family. If the points are randomly scattered across the plot, as in Figure 23.14c, then no relationship exists between the traits of parents and offspring. We would conclude that additive genetic differences are not important in determining the phenotypic variability and that the heritability is low. On the other hand, if a definite relationship exists between the phenotypes of parents and offspring, as shown in Figure 23.14a and 23.14b, then additive genetic variance is more important and heritability is high (assuming that no common environmental effects between parents and offspring influence the trait).

In a plot of the parent and offspring phenotypes, the slope of the regression line can provide us with information about the magnitude of the heritability (see section of this chapter on

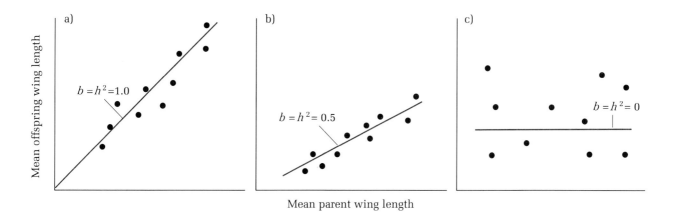

Figure 23.14 Three hypothetical regressions of mean parental wing length on mean offspring wing length in Drosophila. *In each case, the slope of the regression line* (b) *equals the narrow-sense heritability* (h²).

statistics). If the slope is 0, as shown in Figure 23.14c, then the narrow-sense heritability h^2 is zero. When the slope of the parent-offspring regression is 1, as in Figure 23.14a, the mean offspring phenotype is exactly intermediate to the phenotype of the two parents, and genes with additive effects determine all the phenotypic differences. If the slope is less than 1 but greater than zero, as in Figure 23.14b, both additive genes and nonadditive factors (genes with dominance, genes with epistasis, and environmental factors) affect the phenotypic variation. It is possible to show mathematically that when the mean phenotype of the offspring is regressed against the mean phenotype of the parents, the narrow-sense heritability (h^2) equals the slope of the regression line (b):

$h^2 = b$ (for the regression of mean offspring phenotype and mean parental phenotype)

When the mean phenotype of the offspring is regressed against the phenotype of only one parent, the narrow-sense heritability is twice the slope:

$h^2 = 2b$ (for the regression of mean offspring phenotype and one parental phenotype)

Heritability values for a number of traits in different species are given in Table 23.6. These heritability values are based upon various populations and have been determined using a variety of methods, including the parent-offspring method. Estimates of heritability are rarely precise, and most heritability values have large standard errors. This lack of precision is reflected in the fact that heritabilities calculated for the same trait in the same organism often vary widely. Heritability values calculated for human traits must be viewed with special caution, given the difficulties of separating genetic and environmental influences in humans.

Response to Selection

Two fields of study in which quantitative genetics has played a particularly important role are plant and animal breeding and evolutionary biology. Both fields are concerned with genetic change within groups of organisms: in the case of plant and animal breeding, genetic change can lead to improvement in yield, hardiness, size, and other agriculturally important qualities; in the case of evolutionary biology, genetic change occurs in natural populations as a result of the forces of nature. **Evolution** can be defined as genetic change within a group of organisms that takes place over time. Therefore, both evolutionary biologists and plant and animal breeders are interested in the process of evolution and both use the methods of quantitative genetics to predict the rate and magnitude of genetic change that will occur.

Evolution results from a variety of forces, many of which will be explored more thoroughly in the next chapter. Here we concentrate on genetic change that results from selection. **Natural selection** is undoubtedly one of the most powerful forces of evolution in natural populations. First fully described by Charles Darwin, the concept of natural selection can be summarized in four statements:

1. More individuals are produced every generation than can survive and reproduce.
2. Much phenotypic variation exists among these individuals, and some of this variation is heritable.
3. Individuals with certain traits survive and reproduce better than others; these individuals leave more offspring in the next generation.
4. Since individuals with certain traits leave more offspring and the traits are heritable, the traits that allowed increased survival and reproduction will be more common in the next generation.

Table 23.6 Heritability Values for Some Traits in Humans, Domesticated Animals, and Natural Populations

Organism	Trait	Heritability	Ref*
Humans	Stature	0.65	1
	Serum immunoglobulin (IgG) level	0.45	1
Cattle	Milk yield	0.35	1
	Butterfat content	0.40	1
	Body weight	0.65	1
Pigs	Back-fat thickness	0.70	1
	Litter size	0.05	1
Poultry	Egg weight	0.50	1
	Egg production (to 72 weeks)	0.10	1
	Body weight (at 32 weeks)	0.55	1
Mice	Body weight	0.35	1
Drosophila	Abdominal bristle number	0.50	1
Jewelweed	Germination time	0.29	2
Milkweed bugs	Wing length (females)	0.87	3
	Fecundity (females)	0.50	3
Spring peeper (frog)	Size at metamorphosis	0.69	4
Wood frog	Development rate (mountain population)	0.31	5
	Size at metamorphosis (mountain population)	0.62	5

Note: The estimates given in this table apply to particular populations in particular environments; heritability values for other individuals may differ.

* References: (1) Falconer, D.S. 1981. *Introduction to Quantitative Genetics,* 2d ed. Longman, New York; (2) Mitchell-Olds, T. 1986. *Evolution* 40:107–116; (3) Palmer, J.O., and H. Dingle. 1986. *Evolution* 40:767–777; (4) Travis, J., et al. 1987. *Evolution* 41:145–156; (5) Berven, K.A. 1987. *Evolution* 41:1088–1097.

The essential element of natural selection is that some genotypes leave more offspring than others. In this way, groups of individuals change or evolve over time and become better adapted to their particular environment.

Humans bring about evolution in domestic plants and animals through the similar process of **artificial selection.** In artificial selection, humans select the individuals that survive and reproduce. If the selected traits have a genetic basis, they too will change over time and evolve, just as traits in natural populations evolve as a result of natural selection. Artificial selection can be a powerful force in bringing about rapid evolutionary change, as evidenced by the extensive variation observed among domesticated plants and animals. For example, all breeds of domestic dogs are derived from one species that was domesticated some 10,000 years ago. The large number of breeds that exists today, encompassing a tremendous variety of sizes, shapes, colors, and even behaviors, has been produced by artificial selection and selective breeding (Figure 23.15).

Both the process of natural selection, as described by Charles Darwin, and artificial selection practiced by plant and animal breeders depend upon the presence of genetic variation. Only if genetic variation is present within a population of individuals can that population change genetically and evolve. Furthermore, the amount and the type of genetic variation present is extremely important in determining how fast evolution will occur. Both evolutionary biologists and breeders are interested in the question of how much genetic variation for a particular trait exists within a population. As we have seen, quantitative genetics is often employed to answer this question.

Estimating the Response to Selection

When natural or artificial selection is imposed upon a phenotype, the phenotype will change from one generation to the next, provided that genetic variation underlying the trait is present in the population. The amount that the pheno-type changes in one generation is termed the **selection response.** To illustrate the concept of selection response, suppose a geneticist wishes to produce a strain of *Drosophila melanogaster* with large body size. In order to increase body size in fruit flies, the geneticist would first examine flies from a genetically diverse culture and would measure the body size of these *unselected flies.* Suppose that our geneticist found the mean body weight of the unselected flies to be 1.3 mg. After determining the mean body weight of this population, the geneticist would select flies that were endowed with large bodies (assume that the mean body weight of these selected flies was 3.0 mg). He would then place the large, selected flies in a separate culture vial and allow them to interbreed. After the F_1 offspring of these selected parents emerged, the geneticist would measure the body weights of the F_1 flies and compare them with the body weights of the original, unselected population. What our geneticist has done in this procedure is to apply selection for large body size to the population of fruit flies. If genetic variation underlies the variation in body size of the original population, the offspring of the selected flies will resemble their parents and the mean body size of the F_1 generation will be greater than the mean body size of the original population. If the F_1 flies have a mean body weight of 2.0 mg, which is considerably larger than the mean body weight of 1.3 mg observed in the original, unselected population, a response to selection has occurred.

The amount of change that occurs in one generation, or the selection response, is dependent upon two things: the narrow-sense heritability and the **selection differential.** The selection differential is defined as the difference between the mean phenotype of the selected parents and the mean phenotype of the unselected population. In our example of body size in fruit flies, the original population had a mean weight of 1.3 mg, and the mean weight of the selected parents was 3.0 mg, so the selection differential is 3.0 mg − 1.3 mg = 1.7 mg. The selection response is related to the selection differential and the heritability by the following formula:

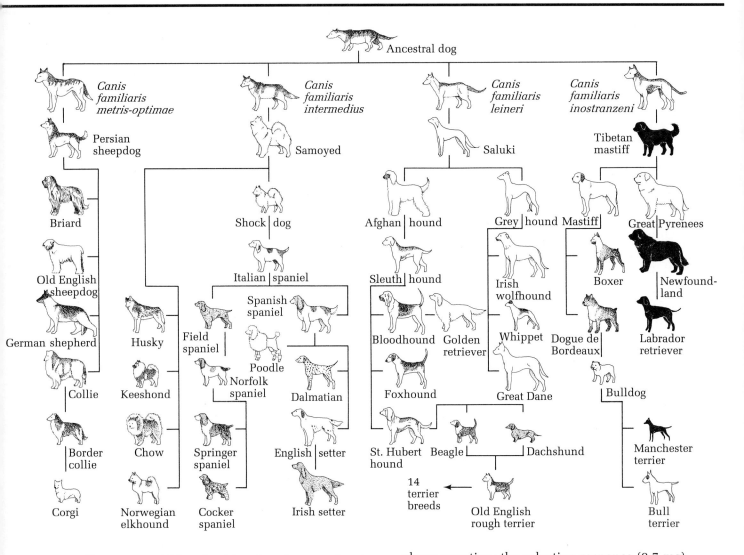

Figure 23.15 All breeds of dogs have been produced by artificial selection, a demonstration of the tremendous power of artifical selection for bringing about change.

Selection response = narrow-sense heritability × selection differential

When the geneticist applied artificial selection to body size in fruit flies, the difference in the mean body weight of the F_1 flies and the original population was 2.0 mg − 1.3 mg = 0.7 mg, which is the response to selection. We now have values for two of the three parameters in the above equation: the selection response (0.7 mg) and the selection differential (1.3 mg). By rearranging the formula for the selection response, we can solve for the narrow-sense heritability.

Narrow-sense heritability = h^2 = selection response/selection differential

$$h^2 = 0.7 \text{ mg}/1.3 \text{ mg} = 0.54$$

Measuring the response to selection provides another means for determining the narrow-sense heritability, and heritabilities for many traits are determined in this way.

A trait will continue to respond to selection, generation after generation, as long as genetic variation for the trait remains within the population. The results from an actual, long-term selection experiment on phototaxis in *Drosophila pseudoobscura* are presented in Figure 23.16. Phototaxis is a behavioral response to light. In this study, flies were scored for the number of times the fly moved toward light in a total of 15 light-dark choices. Two different response-to-selection experiments were carried out. In one, attraction to light was selected, and in the other, avoidance of light was selected. As can be seen in Figure 23.16, the fruit flies responded to selection for positive and negative phototactic behavior for a number of generations. Eventually, however, the response to selection tapered off, and finally no further directional change in phototactic behavior occurred. One possible reason for this lack of response in later generations is that no more genetic variation for phototactic behavior exists within the population. In other words, all flies at this point are homozygous for all the alleles affecting the behavior. If this were the case, phototactic behavior could not undergo further evolution in this population unless input of additional genetic variation occurred. More often, some variation still exists for the trait even after the selection response levels off, but the population fails to respond to selection because the genes for the selected trait have detrimental effects on other traits. These detrimental effects occur because of genetic correlations, which are discussed in the next section.

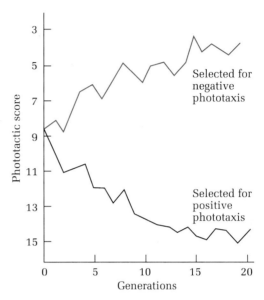

Figure 23.16 Drosophila pseudoobscura, *phototaxis. The upper graph is the line selected for avoidance of light. The phototactic score is the number of times the fly moved toward the light out of a total of 15 light–dark choices. (After Dobzhansky and Spassky, 1969 Proc. Natl. Acad. Sci. 62: 75–80).*

Keynote *The amount that a trait changes in one generation as a result of selection for the trait is called the selection response or the response to selection. The magnitude of selection response depends on both the intensity of selection, what is termed the selection differential, and the narrow-sense heritability.*

Genetic Correlations

The phenotypes of two or more traits may be associated or correlated; this means that the traits do not vary independently. For example, fair skin, blond hair, and blue eyes are often found together in the same individual. The association is not perfect—we sometimes see individuals with dark hair, fair skin, and blue eyes—but the traits are found together with enough regularity for us to say that they are correlated. The **phenotypic correlation** between two traits can be computed by measuring two phenotypes on a number of individuals and then calculating a correlation coefficient for the two traits. (See the section on statistics, pp. 743–744, for a discussion of correlation coefficients.) One reason for a phenotypic correlation among traits is that the traits are influenced by a common set of genes. Indeed, this is the most likely reason for the association among hair color, eye color, and skin color in humans. Rarely do genes affect only a single trait. More commonly, each gene influences a number of traits, and this is particularly true for the polygenes that influence continuous traits. When genes affect multiple phenotypes, we say they are *pleiotropic,* a concept we introduced in Chapter 4. If a gene is pleiotropic, it simultaneously affects two or more traits, and thus the phenotypes of these two traits will be correlated. For example, the genes that affect growth rates in humans influence both weight and height, so these two phenotypes tend to be correlated. When pleiotropy is present and some of the genes influencing two traits are the same, we say that a **genetic correlation** exists for the traits.

Pleiotropy is not the only cause of phenotypic correlations among traits. Environmental factors may also influence several traits simultaneously and may cause nonrandom associations between phenotypes. For example, adding fertilizer to the soil causes plants both to grow taller and to produce more flowers. If we measured plant height and counted the number of flowers on a group of plants, some of which received fertilizer and some of which did not, we would find that the two traits are correlated; those plants receiving fertilizer would be tall and would have many flowers, while those without fertilizer would be short and have few flowers. However, this phenotypic correlation does not result from any genetic correlation (pleiotropy), but from the common effect of the environmental factor, the fertilizer, on both traits.

Genetic correlations may be positive or negative. A positive correlation means that genes causing an increase in one trait bring about a simultaneous increase in the other trait. In chickens, body weight and egg weight have a positive genetic correlation. If breeders select for larger chickens, thereby favoring the genes for large body size, the size of the chickens will increase and the mean weight of the eggs produced by these chickens will also increase. This increase in egg weight occurs because the genes that produce larger chickens have a similar effect on egg weight.

Other traits exhibit negative genetic correlations. In this case, genes that cause an increase in one trait tend to produce a corresponding decrease in the other trait. Egg weight and egg number in chickens have a negative genetic correlation. When breeders select for chickens that produce larger eggs, the average egg size increases, but the number of eggs laid by each chicken decreases.

Genetic correlations are important for predicting the changes in phenotype that result from selection. If two traits are genetically correlated, they evolve together. Any genetic change in one trait occurring as a result of selection will simultaneously alter the other trait, since both are influenced by the same genes. This often presents problems for plant and animal breeders. As an example, milk yield and butterfat content have a negative genetic correlation in cattle. The same genes that cause an increase in milk production bring about a decrease in butterfat content of the milk. Thus, when breeders select for increased milk yield, the amount of milk produced by the cows may go up, but the butterfat content will decrease at the same time. Negative correlations among traits often place practical constraints on the ability of breeders to select for desirable traits. Knowing about the presence of genetic correlations before undertaking an expensive breeding program is essential in order to avoid the production of associated undesirable traits in the selected stock.

An organism's ability to adapt to a particular environment is also strongly influenced by genetic correlations among traits, and thus genetic correlations are of considerable interest to evolutionary biologists. As an illustration of the significance of genetic correlations for understanding the evolutionary process, consider two traits in tadpoles: developmental rate and size at metamorphosis. Most tadpoles are found in small ponds and pools, where fish (potential predators) are absent and food is abundant. A major liability in using this type of aquatic habitat is that ponds often dry up, frequently before the tadpoles have developed sufficiently to metamorphose into frogs and leave the water. One might expect, then, that natural selection would favor a maximum rate of development in tadpoles, so that the tadpoles could quickly complete development, metamorphose into frogs, and leave the pond before drying occurs. However, many species of tadpoles fail to develop at maximum rates, contrary to what we might expect to evolve under natural selection. One reason that tadpoles develop slowly, in spite of the danger of dying when the pond dries, is that a negative genetic correlation may exist between developmental rate and body size at metamorphosis. Genes that accelerate development also tend to cause metamorphosis at a smaller size, at least in some populations. Thus, selection for fast metamorphosis will also produce smaller frogs, and size is extremely important in determining the survival of young frogs. Small frogs tend to lose water more rapidly in the terrestrial environment, are more likely to be eaten by predators, and have more difficulty finding sufficient food. The negative genetic correlation between developmental rate and body size at metamorphosis places constraints on the frogs' ability to evolve rapid development and on their ability to evolve large body size at metamorphosis. Knowing about such genetic correlations is important for understanding how animals adapt or fail to adapt to a particular environment. Table 23.7 presents some genetic correlations that have been detected in studies of quantitative genetics.

Keynote Genetic correlations arise when two traits are influenced by the same genes. When a trait is selected, any genetically correlated traits will also exhibit a selection response.

Inbreeding and Hybrid Vigor

Now that we have an understanding of quantitative inheritance, heritability, and selection, we may consider the consequences of **inbreeding,** or repeated self-fertilization in individuals. In addition, we will consider **heterosis,** the phenomenon in which hybrids made between true-breeding (inbred) lines of animals or plants often exhibit greater fitness than the parentals.

Inbreeding. Human populations have been aware of the possible serious and unfavorable effects of inbreeding for centuries. In fact, laws in many countries prevent marriages between close

Table 23.7 Genetic Correlations among Traits in Humans, Domesticated Animals, and Natural Populations

Organism	Traits	Genetic Correlation	Ref*
Humans	IgG, IgM	0.07	1
Cattle	Butterfat content, milk yield	−0.38	1
Pigs	Weight gain, back-fat thickness	0.13	1
	Weight gain, efficiency	0.69	1
Chickens	Egg weight, egg production	−0.31	1
	Body weight, egg weight	0.42	1
	Body weight, egg production	−0.17	1
Mice	Body weight, tail length	0.29	1
Jewelweed	Seed weight, germination time	−0.81	2
Milkweed bugs	Wing length, fecundity	−0.57	3
Wood frogs	Developmental rate, size at metamorphosis	−0.86	5
Drosophila	Early life fecundity, resistance to starvation	−0.91	6

Note: The estimates given in this table apply to particular populations in particular environments; genetic correlations for other individuals may differ.

* References: (1) Falconer, D.S. 1981. *Introduction to Quantitative Genetics.* 2d ed. Longman, New York; (2) Mitchell-Olds, T. 1986. *Evolution* 40:107–116; (3) Palmer, J.O., and H. Dingle. 1986. *Evolution* 40:767–777; (4) Travis, J., et al. 1987. *Evolution* 41:145–156; (5) Berven, K.A. 1987. *Evolution* 41:1088–1097; (6) Service, P.M., and M.R. Rose. 1985. *Evolution* 39:943–944.

relatives. The logic behind such thinking is that close relatives (e.g., first cousins) have many genes in common because of common ancestry. As a result, there is a greater chance of bringing deleterious recessive mutations into the homozygous state by a mating of close relatives than by a mating of unrelated individuals.

A substantial amount of data from animal and plant breeders indicates the deleterious effects of inbreeding. However, a number of plants and some animals reproduce predominantly by self-fertilization, and many of these species are quite successful. For example, the land snail, *Rumina decollata,* was introduced into the United States from its native habitat in Europe sometime before 1822. By 1915 it had spread to Florida, Texas, Oklahoma, and Mexico, and it is common today across the southern United States. Genetic studies indicate that *Rumina decollata* is completely homozygous in North America and reproduces only through self-fertilization. This species illustrates that inbreeding is not always harmful.

In other plants and animals, however, inbreeding is deleterious, and mechanisms have evolved to minimize self-fertilization. In hermaphroditic organisms, for example, the two types of gametes often mature at different times, thus forcing cross-fertilization between individuals, or *outbreeding.* In other organisms genetic systems cause self-incompatibility. Figure 23.17 illustrates the genetic basis of incompatibility for a number of groups of plants, such as evening primroses and tobacco. In this example a polygenic series controls compatibility in sexual reproduction. When pollen of one incompatibility type falls onto a style of the same genetic type, the pollen tube does not grow properly. Thus fertilization depends on the pollen and style having different incompatibility alleles, thereby virtually assuring that only cross-fertilization occurs. These incompatibility systems can be very complicated, with at least forty different alleles being involved in some plants. Incompatibility alleles are generally designated S_1, S_2, S_3 and so on, where the *S* stands for sterility.

In plants where self-fertilization *is* possible, some mechanisms may still operate to minimize

that process and facilitate cross-fertilization. For example, orchids have complex flower structures designed so that pollen must be carried from one plant to another by insects.

The consequences of inbreeding can be examined by undertaking an inbreeding program with an organism that does not have an incompatibility system. At the outset the population would show a range of phenotypic variation reflecting the heterogeneity of the alleles the organisms carry. After repeated self-fertilization distinct phenotypic traits would become fixed, or pure, in the lines. In plants, for example, lines would sort out that would be relatively uniform for height and flower size. Some lines would sort out to show deleterious effects, such as lethal white seedlings. Thus a number of lines produced by an inbreeding program will not be able to survive. At the same time the surviving lines would almost certainly show less vigor than the original parental line because of the deleterious effects of some of the homozygous pairs of alleles produced.

As an example, Figure 23.18 shows the consequences of 30 generations of inbreeding on the grain yield of three different lines of corn, A, B, and C, derived from the same parental line. Note that yield gradually declines as the number of generations of self-fertilization increases. Eventually, each line becomes fixed at a particular yield level, probably because all the alleles that control grain yield have become homozygous. Since this trait is under polygenic control, the different fixed yields reflect different patterns of homozygosity of the genes involved.

Homozygosity and Inbreeding

The homozygosity brought about by inbreeding is unquestionably responsible for many of the deleterious effects associated with inbreeding, since as inbreeding continues, more and more loci become homozygous. In this situation there is no preference for either dominant or recessive alleles becoming homozygous. However, the deleterious effects of inbreeding are mostly caused by homozygosity for certain recessive alleles. In a genetically heterogeneous individual the ef-

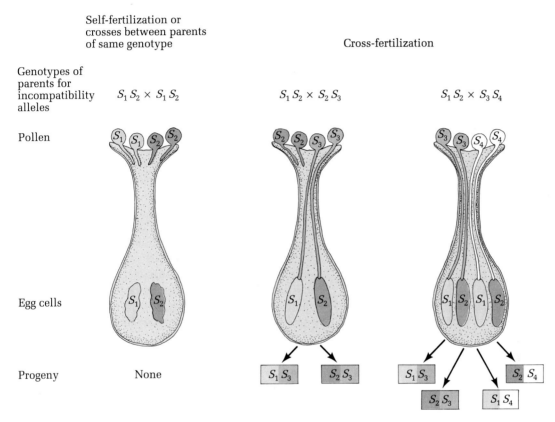

Self-fertilization or
crosses between parents
of same genotype

Cross-fertilization

Genotypes of
parents for
incompatibility
alleles

Pollen

Egg cells

Progeny

Figure 23.17 Multiple allelic series determining compatibility in sexual reproduction.

fects of deleterious recessive alleles are masked by wild-type alleles. In contrast, deleterious dominant alleles always exhibit their effects, whether they are heterozygous or homozygous. Thus deleterious dominant alleles (which are also rare) are seen directly in the heterogeneous population without resorting to inbreeding. If the deleterious dominant alleles affect vigor significantly, their frequencies are soon reduced in the population by natural selection. Deleterious recessive alleles concern geneticists because the frequencies of such alleles are much higher in populations than the frequencies of dominant deleterious alleles, since the recessive alleles can be maintained in a population in the heterozygous state.

To see how self-fertilization leads to a reduction of heterozygosity, let us consider a cross be-

tween a population of two heterozygous individuals, $Aa \times Aa$, the result of which is ¼ AA, ½ Aa, and ¼ aa offspring. In other words, while 100 percent of the population is heterozygous initially, after one generation only 50 percent of the population is heterozygous, and the rest are homozygous. When the progeny are self-

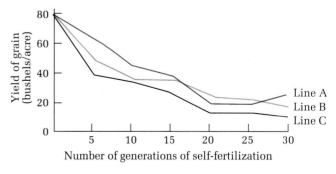

Figure 23.18 Consequences of continued inbreeding of three different lines of corn, A, B, and C.

fertilized, the *AA* and *aa* types breed true, while the *Aa* individuals give progeny of the same type and in the same proportions as in the original cross. So after a second round of self-fertilization the proportion of heterozygotes is further reduced by a factor of 2; that is, now the proportions of the three genotypes are ⅜ *AA*, ¼ *Aa*, and ⅜ *aa*. Generally speaking, the relative number of heterozygotes in a population is reduced 50 percent for each self-fertilized generation.

Figure 23.19 illustrates how heterozygosity in a theoretical population is reduced by inbreeding over four generations. After the four generations the original number of *Aa* individuals is reduced by a factor of $(2)^4 = 16$, that is, from 1600 to 100. Note, though, that inbreeding does not favor an increase in the number of either the recessive or the dominant alleles. Instead, the proportion of homozygotes increases and leads

specifically to an increase in the relative proportion of the population expressing the recessive trait. Thus as shown in Figure 23.19, the same number of *A* and *a* alleles occurs in each generation, but their distribution among the three genotypic classes has changed. (The effects of inbreeding on genotype frequencies will be discussed in more detail in the next chapter.)

The example just presented is greatly simplified since it considers only one gene locus. In real situations, unless the population is already inbred, a large number of loci are heterozygous. Nonetheless, the basic effects operate on all gene loci so that both continuous (quantitative) and discontinuous traits are affected.

Many studies of outcrossing organisms provide examples of deleterious expression of recessive traits that result from homozygosity brought about by inbreeding. In Chapter 14, for example, we discussed Tay-Sachs disease, which causes early death in infants homozygous for a particular recessive allele. Tay-Sachs disease is very rare in the general human population, but it occurs with a fairly high frequency

---

Figure 23.19 Theoretical cascade diagram showing the reduction in heterozygosity over four generations when only self-fertilization is permitted.

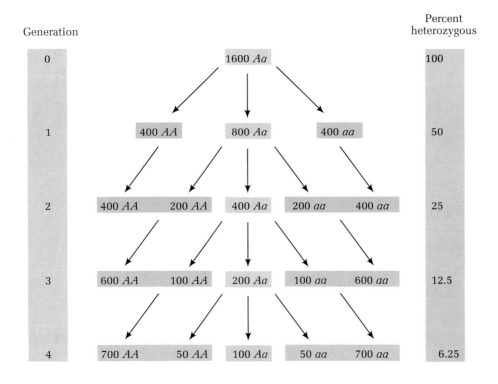

in particular Ashkenazi Jewish populations, such as in the New York City area. This high frequency occurs because religious beliefs lead to marriages between individuals in a relatively small population, and since the population has remained discrete for many generations, there are many genes in common among the individuals. As a consequence, the probability of heterozygotes for Tay-Sachs disease is much higher in the Ashkenazi population than in the population at large, and thus marriages between heterozygotes are much more likely in the Ashkenazi population. A fourth of the children of such marriages would be expected to have Tay-Sachs disease.

Similar situations are found in relatively small human populations in which marriages between group members are more common than marriages to new blood (and new genes) from outside the population. The Amish population of Pennsylvania, for example, exhibits a number of genetic abnormalities attributable to homozygosity for one or more deleterious recessive alleles. It is quite common, for instance, to find Amish people with supernumerary digits (extra fingers or toes).

Inbreeding does not always result in the expression of deleterious recessive alleles. It can bring together particular combinations of dominant alleles with deleterious or beneficial effects or can bring out recessive alleles with nondeleterious effects. The primary result of extensive inbreeding, however, is to fix phenotypes by making the genes controlling them homozygous. The fixation process is a random one (unless the genes involved are linked closely), so a number of pure-breeding lines can be generated from one starting point. Once established, the pure-breeding lines change genetically only in response to mutation. Thus the inbreeding of a heterozygous organism of genotype *Aa Bb* will produce four distinct, pure lines by producing all possible homozygous types, each with a distinct phenotype: *AA BB, AA bb, aa BB,* and *aa bb.*

The fixation consequence of inbreeding is important in animal and plant breeding (and in evolution, since it reduces the genetic variation in

a population, as we will see in Chapter 24). Professional animal and plant breeders make their living by generating stable breeding lines that give rise to homogeneous and predictable offspring. Their breeding programs balance inbreeding and outbreeding to fix desirable characters in the absence of deleterious characters. In crop plants, for example, it may be important for all plants to have the same height to facilitate machine harvesting or for all plants to bear fruit at the same time to accommodate shipping schedules. Similarly, various breeds of dogs represent the result of extensive breeding programs to accentuate some features (e.g., the flattened face of the English bulldog or the short legs of a dachshund) while minimizing other aspects of the general phenotype. If you are familiar with purebred animals of any type, you may be aware of the physiological problems that arise with fair frequency in purebreds (e.g., hip dysplasia in certain dog breeds). Many of these problems result from breeding programs that must inevitably involve some inbreeding as breeders try to coax along a desirable phenotype that has arisen.

Keynote *Inbreeding involves the mating of closely related individuals of a species. At the beginning of an inbreeding program the population shows much phenotypic variation, since the individuals carry a heterogeneous array of alleles. With continued inbreeding loci become homozygous, resulting in the formation of a variety of pure lines, each with a characteristic set of phenotypes that remain constant with further inbreeding. Inbreeding leads to the fixation of either beneficial, neutral, or deleterious traits, depending on the alleles that become homozygous.*

Hybrid Vigor

In an earlier example we saw that repeated inbreeding of corn led to a decrease in vigor (grain yield). After many generations the yield stabilized at a plateau and did not increase or decrease on further self-fertilization. We interpreted this result to mean that the genes controlling grain yield had become fixed in the homo-

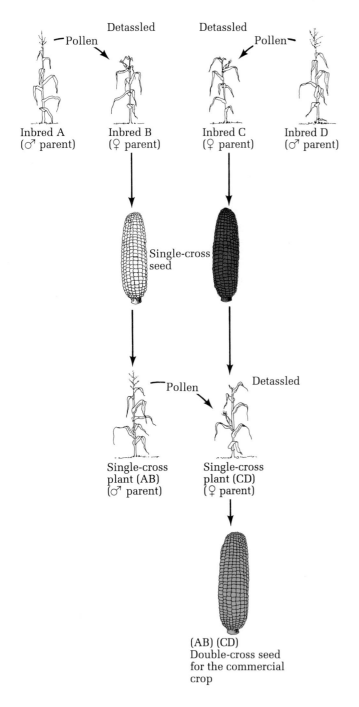

Inbred A
($\male$ parent)

Inbred B
($\female$ parent)

Inbred C
($\female$ parent)

Inbred D
($\male$ parent)

Detassled

Pollen

Detassled

Pollen

Single-cross
seed

Pollen

Detassled

Single-cross
plant (AB)
($\male$ parent)

Single-cross
plant (CD)
($\female$ parent)

(AB) (CD)
Double-cross seed
for the commercial
crop

Figure 23.20 Representation of the production of vigorous hybrid seed for commercial corn by double crosses to ensure hybrid vigor.

zygous state. (The yield does vary in actuality owing to environmental effects, but we will continue to ignore this feature in our theoretical arguments.)

If self-fertilization leads to pure-breeding lines of corn, each of which has significantly less vigor than the original strain, what happens in a cross of two inbred lines? Often the progeny show more vigor than the two inbred parents. This phenomenon, called **hybrid vigor,** or **heterosis,** is common in both animals and plants. Yet while the yield of hybrid corn is much better than the yield of inbred lines, it is not as high as the yield of some outbred lines. Farmers have a good reason, however, for choosing the lower-vigor hybrid corn over the high-yield outbreds: The F_1 hybrids have a uniform phenotype, while outbreds do not—and as we noted before, uniformity of a plant or an animal type is strongly favored in agriculture. The reason outbreds do not have a uniform phenotype is that an outbred line is heterozygous at many loci, and any breeding done with it will give rise to a plethora of genotypes in the progeny that translate into heterogeneity of yield among all the plants produced. Also, as we saw in the discussion of quantitative traits, if the F_1 hybrid plants are allowed to interbreed, the F_2 will show a much greater phenotypic variability than the F_1, since more gene combinations result from genetic segregation and recombination.

No consensus can be found on the genetic basis for hybrid vigor. One model suggests that the interaction of alleles in a heterozygous situation is responsible. Simply stated, a heterozygous combination of alleles at a locus is superior to any homozygous combination of alleles. This assumption implies that the different alleles at the relevant loci do different things and that the sum of the different products, or the interactions between them, gives superior vigor. Another term used for the superiority of the heterozygote at an individual locus in this model is *overdominance.*

While we do not have an exact understanding of the basis for hybrid vigor, the formation of hybrids is nonetheless of great practical importance in agriculture. As an example, hybrid

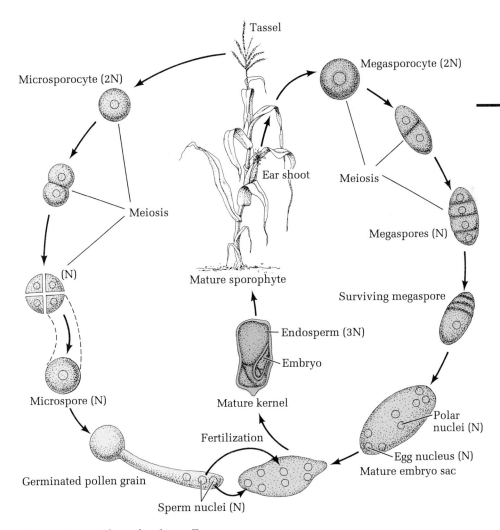

Figure 23.21 Life cycle of corn Zea mays.

vigor might be used to devleop a commercial corn crop. Four inbred lines (A, B, C, and D), each of which produces small plants with limited vigor, are used in double crosses (a valid genetic term), as shown in Figure 23.20. Lines A and B are crossed to produce an AB hybrid, and lines C and D are crossed to give a CD hybrid. Because of the life cycle of corn (Figure 23.21), these crosses are not suitable for providing seeds for the farmer to plant, as we will see in the following discussion.

Corn (*Zea mays*) is a monoecious plant; that is, both male and female flowers develop on the same plant. The female flowers are borne in the ears of corn, and the male flowers are in the tassel (silk) at the top of the stalk. When a mature, physiologically reactive pollen grain contacts the silk of a corn ear, the fertilization process begins as a pollen tube is initiated, which grows down through the silk to the embryo sac in the

flower. In the embryo sac one pollen nucleus fuses with the egg to form a zygote; at the same time other events take place to give rise to the endosperm (a tissue that has an important role in kernel development). By cell division and differentiation the zygote develops into an embryo, which is within the corn kernel. When the kernel germinates, the embryo enlarges rapidly and develops into the mature plant.

Thus the seeds (kernels) for the next generation are carried on the *original* maternal plant. If the maternal plant is from an inbred line (e.g., A, B, C, or D) and has little vigor, then the ears will be relatively small and the number of seeds produced will also be small. More seeds of a high quality are obtained by collecting and planting the AB and CD F_1 hybrid seeds. The F_1 plants that grow will show *hybrid vigor*—they will be larger and much more vigorous than the parental plants.

The next step is to cross the single-cross F_1 AB and CD plants. To prevent self-fertilization, geneticists control the cross, with one plant acting as the female and the other as the male (CD and AB in Figure 23.20, respectively). For this cross, they physically remove the tassel (the male flower parts) from the plant that is to act as the female (CD) and use pollen from the tassel of the AB plant as the male parent to fertilize the CD female flowers. From this AB male × CD female cross, the female, F_1 single-cross, CD hybrid plant produces large and uniform ears when fertilized by pollen from the other, male,

AB, F_1 single-cross hybrid. The double-cross hybrid seeds are harvested from the ears of the maternal parent plants and are sold to farmers.

When the double-cross seeds are planted, they develop into vigorous, high-yielding, and uniform double-cross plants, which is just what the farmer wants. As we would expect from the basic principles of quantitative inheritance, though, the farmer cannot use those plants to propagate the next year's crop, since the progeny (F_2) would show a lot of variation rather than uniformity. Therefore each year the farmer must buy a new batch of double-cross seed.

Analytical Approaches for Solving Genetics Problems

Q.1 Assume that genes *A, B, C,* and *D* are members of a multiple-gene series that control a quantitative trait. Each of these genes has a duplicate, cumulative effect in that each contributes 3 cm of height to the organism when it is present. Each gene assorts independently. In addition, gene *L* is always present in the homozygous state, and the *LL* genotype contributes a constant 40 cm of height. The alleles *a, b, c,* and *d* do not contribute anything to the height of the organism. If we ignore height variation caused by environmental factors, an organism with genotype *AA BB CC DD LL* would be 64 cm high, and one with genotype *aa bb cc dd LL* would be 40 cm. A cross is made of *AA bb CC DD LL* × *aa BB cc DD LL* and is carried into the F_2 by selfing of the F_1.

a. How does the size of the F_1 individuals compare with the size of each of the parents?

b. Compare the mean of the F_1 with the mean of the F_2, and comment on your findings.

c. What proportion of the F_2 population would show the same height as the *AA bb CC DD LL* parent?

d. What proportion of the F_2 population would show the same height as the *aa BB cc DD LL* parent?

e. What proportion of the F_2 population would breed true for the height shown by the *aa BB cc DD LL* parent?

f. What proportion of the F_2 population would breed true for the height characteristic of F_1 individuals?

A.1 This question explores our understanding of the basic genetics involved in a multiple-gene series that in this case controls a quantitative trait. The approach we will take is essentially the same as the approach used with a series of independently assorting genes that control distinctly different traits. That is, we make predictions on the basis of genotypes and relate the results to phenotypes, or we make predictions on the basis of phenotypes and relate the results to genotypes.

a. Each allele represented by a capital letter contributes 3 cm of height to the

773
*Analytical
Approaches
for Solving
Genetics
Problems*

base height of 40 cm, which is controlled by the ever-present *LL* homozygosity. Therefore the *AA bb CC DD LL* parent, which has six capital-letter alleles from the *A*-through-*D*, multiple-gene series, is $40 + (6 \times 3) = 58$ cm high. Similarly, the *aa BB cc DD LL* parent has four capital-letter alleles and therefore is $40 + 12 = 52$ cm high. The F_1 from a cross between these two individuals would be heterozygous for the *A, B,* and *C* loci and homozygous *D* and *L*, that is, *Aa Bb Cc DD LL*. This progeny has five capital-letter alleles apart from *LL* and hence is $40 + 15 = 55$ cm high.

b. The F_2 is derived from a self of the *Aa Bb Cc DD LL* F_1. All the F_2 individuals will be *DD LL*, making them at least $40 + 6 = 46$ cm high. Now we must deal with the heterozygosity at the other three loci. What we need to calculate is the relative proportions of individuals with all the various possible numbers of capital-letter alleles. This calculation is equivalent to determining the relative distribution of three independently assorting traits, each showing incomplete dominance. In other words, we must calculate directly the relative frequencies of all possible genotypes for the three loci and collect those with no, one, two, three, four, five, and six capital-letter alleles. Thus the probability of getting an individual with two capital-letter alleles for each locus is ¼, the probability of getting an individual with one capital-letter allele for each locus is ½, and the probability of getting an individual with no capital-letter alleles for each locus is ¼. So the probability of getting an F_2 individual with six capital-letter alleles for the *A, B,* and *C* loci is $(¼)^3 = ^1/_{64}$, and the same probability is obtained for an individual with no capital-letter alleles. This analysis gives us a clue about how we should consider all the possible combinations of genotypes that have the other numbers of capital-letter alleles. That is, the simplest approach is to compute the coefficients in the binomial expansion of $(a + b)^6$. The expansion gives a 1:6:15:20:15:6:1 distribution of zero, one, two, three, four, five, and six capital-letter alleles, respectively. Now since each capital-letter allele in the *A, B,* and *C* set contributes 3 cm of height over the 46-cm height given by the *DD LL* genotype common to all, then the F_2 individuals would fall into the following distribution:

Number of Capital-Letter Alleles	Height Added to Basic Height of 46 cm for Common *DD LL* Genotype (cm)	Height of Individuals (cm)	Frequency
6	18	64	1
5	15	61	6
4	12	58	15
3	9	55	20
2	6	52	15
1	3	49	6
0	0	46	1

The distribution is clearly symmetrical, giving an average of 55 cm, the same height shown in F_1 individuals.

c. The *AA bb CC DD LL* parent was 58 cm, so we can read the proportion of F_2 individuals that show this same height directly from the table in part b. The answer is $^{15}/_{64}$.

d. The *aa BB cc DD LL* parent was 52 cm, and from the table in part b the proportion of F_2 individuals that show this same height is $^{15}/_{64}$.

e. We are asked to determine the proportion of the F_2 population that would breed true for the height shown by the *aa BB cc DD LL* parent, which was 52 cm. To breed true, the organism must be homozygous. We have also established that *DD LL* is a constant genotype for the F_2 individuals, giving a basic height of 46 cm. Therefore for a height of 52 cm, two additional, active, capital-letter alleles must be present, apart from those at the *D* and *L* loci. With the requirement for homozygosity there are only three genotypes that give a 52-cm height; they are *AA bb cc DD LL, aa BB cc DD LL,* and *aa bb CC DD LL.* The probability of each combination occurring in the F_2 is $^1/_{64}$, so the answer to the problem is $^1/_{64} + ^1/_{64} + ^1/_{64} = ^3/_{64}$. (Note that the individual probability for each genotype may be calculated. That is, probability of *AA* = $^1/_4$, probability of *bb* = $^1/_4$, probability of *cc* = $^1/_4$, probability of *DD LL* = 1, giving an overall probability for *AA bb cc DD LL* of $^1/_{64}$.)

f. We are asked to determine the proportion of the F_2 population that would breed true for the height characteristic of F_1 individuals. Again, the basic height given by *DD LL* is 46 cm. The F_1 height is 55 cm, so three capital-letter alleles must be present in addition to *DD LL* to give that height, since (3×3) cm = 9 cm, and 9 cm + 46 cm = 55 cm. However, since an individual must be homozygous to be true breeding, the answer to this question is none, since 3 is an odd number, meaning that at least one locus must be heterozygous in order to get the 55-cm height.

Q.2 Five field mice collected in Texas had weights of 15.5 g, 10.3 g, 11.7 g, 17.9 g, and 14.1 g. Five mice collected in Michigan had weights of 20.2 g, 21.1 g, 20.4 g, 22.0 g, and 19.7 g. Calculate the mean weight and the variance in weight for mice from Texas and for mice from Michigan.

A.2 To answer this question, we use the formulas given in the chapter section on statistics. The formula for the mean is

$$\bar{x} = \frac{\Sigma x_i}{n}$$

The symbol Σ means to add, and the x_i represents all the individual values. We begin by summing up all the weights of the mice from Texas:

$$\Sigma x_i = 15.5 + 10.3 + 11.7 + 17.9 + 14.1 = 69.5$$

Next, we divide this summation by n, which represents the number of values added together. In this case, we added together five weights, so $n = 5$.

775
Analytical
Approaches
for Solving
Genetics
Problems

The mean for the Texas mice is therefore

$$\frac{\Sigma x_i}{n} = \frac{69.5}{5} = 13.9$$

To calculate the variance in weight among the Texas mice, we utilize the formula

$$s2 = \frac{\Sigma(x_i - \overline{x})^2}{n - 1}$$

We must take each individual weight and subtract it from the mean weight of the group. Each value obtained from this subtraction is then squared, and all the squared values are added up, as shown below.

$$
\begin{array}{llll}
15.5 - 13.9 = & 1.6 & (1.6)^2 = & 2.56 \\
10.3 - 13.9 = & -3.6 & (-3.6)^2 = & 12.96 \\
11.7 - 13.9 = & -2.2 & (-2.2)^2 = & 4.84 \\
17.9 - 13.9 = & 4.0 & (4.0)^2 = & 16.0 \\
14.1 - 13.9 = & 0.2 & (0.2)^2 = & \underline{0.04} \\
 & & & 36.4
\end{array}
$$

The sum of all the squared values is 36.4. All that remains for us to do is to divide this sum by $n - 1$, which is $5 - 1 = 4$.

$$s^2 = \frac{\Sigma(x_i - \overline{x})^2}{n - 1} = \frac{36.4}{4} = 9.1$$

The mean and the variance for the Texas mice are therefore 13.9 and 9.1.
  We now repeat these steps for the mice from Michigan.

$$\Sigma x_i = 20.2 + 21.2 + 20.4 + 22.0 + 19.7 = 103.5:$$

$$\frac{\Sigma x_i}{n} = \frac{103.5}{5} = 20.7$$

$$s^2 = \frac{\Sigma(x_i - \overline{x})^2}{n - 1}$$

$$
\begin{array}{llll}
20.2 - 20.7 = & -0.5 & (-0.5)^2 = & 0.25 \\
21.2 - 20.7 = & 0.5 & (0.5)^2 = & 0.25 \\
20.4 - 20.7 = & -0.3 & (-0.3)^2 = & 0.9 \\
22.0 - 20.7 = & 1.3 & (1.3)^2 = & 1.69 \\
19.7 - 20.7 = & -1.0 & (-1.0)^2 = & \underline{1.0} \\
 & & & 4.09
\end{array}
$$

$$s^2 = \frac{\Sigma(x_i - \overline{x})^2}{n - 1} = \frac{4.09}{4} = 1.23$$

The mean and the variance for the Michigan mice are 20.79 and 1.23.
  We conclude that the Michigan mice are considerably heavier than the Texas mice and the Michigan mice also exhibit less variance in weight.

Questions and Problems

*23.1 The following measurements of head width and wing length were made on a series of steamer-ducks:

Specimen	Head Width (cm)	Wing Length (cm)
1	2.75	30.3
2	3.20	36.2
3	2.86	31.4
4	3.24	35.7
5	3.16	33.4
6	3.32	34.8
7	2.52	27.2
8	4.16	52.7

a. Calculate the mean and standard deviation of head width and of wing length for these eight birds.
b. Calculate the correlation coefficient for the relationship between head width and wing length in this series of ducks.
c. What conclusions can you make about the association between head width and wing length in steamer-ducks?

23.2 a. In a family of six children, what is the probability that 3 will be girls and 3 will be boys?
b. In a family of five children, what is the probability that 1 will be a boy and 4 will be girls?
c. What is the probability that in a family of six children, all will be boys?

*23.3 In flipping a coin, there is a 50 percent chance of obtaining heads and a 50 percent chance of obtaining tails on each flip. If you flip a coin 10 times, what is the probability of obtaining exactly 5 heads and 5 tails?

*23.4 The F_1 generation from a cross of two pure-breeding parents that differ in a size character is usually no more variable than the parents. Explain.

23.5 If two pure-breeding strains, differing in a size trait, are crossed, is it possible for F_2 individuals to have phenotypes that are more extreme than either grandparent (i.e., be larger than the largest or smaller than the smallest in the parental generation)? Explain.

23.6 Two pairs of genes with two alleles each, A/a and B/b, determine plant height additively in a population. The homozygote AA BB is 50 cm tall, the homozygote aa bb is 30 cm tall.

a. What is the F_1 height in a cross between the two homozygous stocks?
b. What genotypes in the F_2 will show a height of 40 cm after an $F_1 \times F_1$ cross?
c. What will be the F_2 frequency of the 40-cm plants?

*23.7 Three independently segregating genes (A, B, C), each with two alleles, determine height in a plant. Each capital-letter allele adds 2 cm to a base height of 2 cm.

   a. What are the heights expected in the F_1 progeny of a cross between homozygous strains *AA BB CC* (14 cm) $\times$ *aa bb cc* (2 cm)?

   b. What is the distribution of heights (frequency and phenotype) expected in an $F_1 \times F_1$ cross?

   c. What proportion of F_2 plants will have heights equal to the heights of the original two parental strains, *AA BB CC* and *aa bb cc*?

   d. What proportion of the F_2 will breed true for the height shown by the F_1?

23.8 Repeat Problem 23.7, but assume that each capital-letter allele acts to double the existing height; for example, *As bb cc* = 4 cm, *AA bb cc* = 8 cm, *AA Bb cc* = 16 cm, and so on.

*23.9 The mean internode length in spikes of the barley variety *asplund* in a particular experiment was found to be 2.12 mm. In the variety *abed binder* the mean internode length was found to be 3.17 mm. The mean of the F_1 of a cross between the two varieties was approximately 2.7 mm. The F_2 gave a continuous range of variation from one parental extreme to the other. Analysis of the F_3 generation showed that in the F_2 8 out of the total 125 individuals were of the *asplund* type, giving a mean of 2.19 mm. Eight other individuals were similar to the parent *abed binder,* giving a mean internode length of 3.24 mm. Is the internode length in spikes of barley a discontinuous or a quantitative trait? Why?

23.10 From the information given in Problem 23.9, determine how many gene pairs involved in the determination of internode length are segregating in the F_2.

23.11 Assume that the difference between a type of oats yielding about 4 g per plant and a type yielding 10 g is the result of three equal and cumulative multiple-gene pairs *AA BB CC.* If you cross the type yielding 4 g with the type yielding 10 g, what will be the phenotypes of the F_1 and the F_2? What will be their distribution?

*23.12 Assume that in squashes the difference in fruit weight between a 3-lb type and a 6-lb type is due to three allelic pairs, *A/a, B/b,* and *C/c.* Each capital-letter allele contributes a half pound to the weight of the squash. From a cross of a 3-lb plant (*aa bb cc*) with a 6-lb plant (*AA BB CC*), what will be the phenotypes of the F_1 and the F_2? What will be their distribution?

23.13 Refer to the assumptions stated in Problem 23.12. Determine the range in fruit weight of the offspring in the following squash crosses: (a) *Aa Bb CC* $\times$ *aa Bb Cc*; (b) *AA bb Cc* $\times$ *Aa BB cc*; (c) *aa BB cc* $\times$ *AA BB cc*.

*23.14 Assume that the difference between a corn plant 10 dm (decimeters) high and one 26 dm high is due to four pairs of equal and cumulative multiple alleles, with the 26-dm plants being *AA BB CC DD* and the 10-dm plants being *aa bb cc dd*.

a. What will be the size and genotype of an F_1 from a cross between these two true-breeding types?
b. Determine the limits of height variation in the offspring from the following crosses: (1) *Aa BB cc dd* × *Aa bb Cc dd*; (2)*aa BB cc dd* × *Aa Bb Cc dd*; (3) *AA BB Cc DD* × *aa BB cc Dd*; (4)*Aa Bb Cc Dd* × *Aa bb Cc Dd*.

23.15 Refer to the assumptions given in Problem 23.14. But for this problem two 14-dm corn plants, when crossed, give nothing but 14-dm offspring (case A). Two other 14-dm plants give one 18-dm, four 16-dm, six 14-dm, four 12-dm, and one 10-dm offspring (case B). Two other 14-dm plants, when crossed, give one 16-dm, two 14-dm, and one 12-dm offspring (case C). What genotypes for each of these 14-dm parents (cases A, B, and C) would explain these results? Would it be possible to get a plant taller than 18 dm by selection in any of these families?

23.16 A quantitative geneticist determines the following variance components for leaf width in a population of wild flowers growing along a roadside in Kentucky:

Additive genetic variance (V_A)	= 4.2
Dominance genetic variance (V_D)	= 1.6
Interaction genetic variance (V_I)	= 0.3
Environmental variance (V_E)	= 2.7
Genetic-environmental variance (V_{GE})	= 0.0

a. Calculate the broad-sense heritability and the narrow-sense heritability for leaf width in this population of wildflowers.
b. What do the heritabilities obtained in part a indicate about the genetic nature of leaf width variation in this plant?

*23.17 A Kansas farmer is using a variety of wheat called TK138. On his farm in Kansas, the farmer grows TK138 and calculates the narrow-sense heritability for yield (the amount of wheat produced per acre). He finds that the heritability of yield for TK138 is 0.95. The next year he visits a farm in Poland and observes that a Russian variety of wheat, UG334, growing there has only about 40 percent as much yield as TK138 grown on his farm in Kansas. Since he found the heritability of yield in his wheat to be very high, he concludes that the American variety of wheat (TK138) is genetically superior to the Russian variety (UG334), and he tells the Polish farmers that they can increase their yield by using TK138. What is wrong with his conclusion?

23.18 Dermatoglyphics are the patterns of the ridged skin found on the fingertips, toes, palms, and soles. (Fingerprints are dermatoglyphics.) Classification of dermatoglyphics is frequently based on the number of triradii; a triradius is a point from which three ridge systems separate at angles of 120°. The number of triradii on all ten fingers was counted for each member of several families, and the results are tabulated below.

Family	Mean Number of Triradii in the Parents	Mean Number of Triradii in the Offspring
I	14.5	12.5
II	8.5	10.0
III	13.5	12.5
IV	9.0	7.0
V	10.0	9.0
VI	9.5	9.5
VII	11.5	11.0
VIII	9.5	9.5
IX	15.0	17.5
X	10.0	10.0

a. Calculate the narrow-sense heritability for the number of triradii by the regression of the mean phenotype of the parents against the mean phenotype of the offspring.

b. What does your calculated heritability value indicate about the relative contributions of genetic variation and environmental variation to the differences observed in number of triradii?

*23.19 A scientist wishes to determine the narrow-sense heritability of tail length in mice. He measures tail length among the mice of a population and finds a mean tail length of 9.7 cm. He then selects the ten mice in the population with the longest tails; mean tail length in these selected mice is 14.3 cm. He interbreeds the mice with the long tails and examines tail length in their progeny. The mean tail length in the F_1 progeny of the selected mice is 13 cm.

Calculate the selection differential, the response to selection, and the narrow-sense heritability for tail length in these mice.

23.20 Suppose that the narrow-sense heritability of wool length in a breed of sheep is 0.92, and the narrow-sense heritability of body size is 0.87. The genetic correlation between wool length and body size is -0.84. If a breeder selects for sheep with longer wool, what will be the most likely effects on wool length and on body size?

*23.21 The heights of 10 college-age males and the heights of their fathers are presented below.

Height of Son (inches)	Height of Father (inches)
70	70
72	76
71	72
64	70
66	70
70	68
74	78
70	74
73	69

a. Calculate the mean and the variance of height for the sons and do the same for the fathers.
b. Calculate the correlation coefficient for the relationship between the height of father and height of son.
c. Determine the narrow-sense heritability of height in this group by regression of the son's height on the height of father.

*23.22 The narrow-sense heritability of egg weight in a particular flock of chickens is 0.60. A farmer selects for increased egg weight in this flock. The difference in the mean egg weight of the unselected chickens and the selected chickens is 10 g. How much should egg weight increase in the offspring of the selected chickens?

23.23 Would a law banning marriages between individuals and their stepparents be founded on genetic principles?

23.24 Is there any circumstance for which inbreeding does not have deleterious consequences?

*23.25 What would a breeder's purpose be in inbreeding normally cross-pollinated crop plants that show reduction in vigor when they are inbred?

24

Genetics of Populations

The science of genetics can be broadly divided into three major subdisciplines: transmission genetics (also called classical genetics), molecular genetics, and population genetics. Each of these three areas focuses on a different aspect of heredity. **Transmission genetics** is primarily concerned with genetic processes that occur within individuals, and how genes are passed from one individual to another. Thus, our unit of study for transmission genetics is the individual. In **molecular genetics,** we are largely interested in the molecular nature of heredity—how genetic information is encoded within the DNA and how biochemical processes of the cell translate the genetic information into the phenotype. Consequently, in molecular genetics we focus on the cell.

Population genetics, which we introduce in this chapter, is the field of genetics that studies heredity in groups of individuals. Population geneticists investigate the patterns of genetic variation found within groups and how these patterns change, or evolve over time. Because population genetics is concerned with the genetic basis of evolution, the field is also called **evolutionary genetics.** In this discipline, our perspective shifts away from the individual and the cell, and focuses instead upon a **Mendelian population.** A Mendelian population is not just any group of individuals, but a group of *interbreeding* individuals, who share a common set of genes. The genes shared by the individuals of a Mendelian population are called the **gene pool.** To understand the genetics of the evolutionary process, we study the gene pool of a Mendelian population, rather than the genotypes of its individual members. Except for rare mutations, individuals are born and die with the same set of genes; what changes genetically over time (evolves) is the hereditary makeup of a group of individuals, reproductively connected in a Mendelian population.

Questions frequently studied by population geneticists include:

1. How much genetic variation is found in natural populations, and what forces of nature control the amount of variation observed?

2. What evolutionary forces shape the genetic structures of populations?

3. What forces are responsible for producing genetic divergence among populations?

4. How do biological characteristics of a population, such as breeding system, fecundity, and age structure, influence the gene pool of the population?

To answer these questions, population geneticists frequently develop mathematical models and equations to describe what happens to the gene pool of a population under various conditions. An example is the set of equations that describes the influence of random mating on the gene and genotypic frequencies of an infinitely large population, a model called the **Hardy-Weinberg law,** which we discuss later in the chapter. At first, students often have difficulty grasping the significance of models such as these, because the models are frequently simple and require numerous assumptions that are unlikely to be met by organisms in the real world. Beginning with simple models is useful, however, because with such models we can examine the effects of individual evolutionary factors on the genetic structure of a population. Once we understand the results of the simple models, we can incorporate more realistic conditions into the equations. In the case of the Hardy-Weinberg law, the assumptions of infinitely large size and random mating may seem unrealistic, but these assumptions are necessary for simplifying the mathematical analysis. We cannot begin to understand the impact of nonrandom mating and limited population size on gene frequencies until we first know what happens under the simpler conditions of random mating and large population size.

Genotypic and Gene Frequencies

Genotypic Frequencies

To study the genetic composition of a Mendelian population, population geneticists must first quantitatively describe the gene pool of the group. This is done by calculating genotypic fre-

quencies and allelic frequencies for the individuals within the population. A frequency is a percent, or a proportion, and always ranges between 0 and 1. If 43 percent of the people in a group have red hair, the frequency of red hair in the group is 0.43. The **genotypic frequencies** of a population consist of the frequencies of all the genotypes in the population. To calculate the genotypic frequencies at a specific locus, we count the number of individuals with one particular genotype and divide this number by the total number of individuals in the population. We do this for each of the genotypes at the locus, and the sum of all the genotypic frequencies should add up to 1. Consider a locus that determines the pattern of spots in the scarlet tiger moth, *Panaxia dominula* (Figure 24.1). Three genotypes are possible, and each genotype produces a different phenotype. E. B. Ford collected moths at one locality in England and found the following numbers of genotypes: 452 *BB,* 43 *Bb,* and 2 *bb,* out of a total of 497 moths. The genotypic frequencies are therefore:

$$
\begin{aligned}
P &= f(BB) = 452/497 = 0.909 \\
H &= f(Bb) = 43/497 = 0.087 \\
Q &= f(bb) = 2/497 = \underline{0.004} \\
&\qquad\qquad\qquad\text{Total} \quad 1.000
\end{aligned}
$$

Population geneticists often use the capital letters *P, H,* and *Q* to represent the frequencies (*f*) of the three genotypes at a locus. Be careful that you do not confuse these symbols with the small letters *p* and *q* which are used to represent the allelic frequencies, as discussed in the next section. Box 24.1 presents a sample calculation of genotypic frequencies for some hemoglobin variants in a human population.

Gene Frequencies

While genotypic frequencies are useful for examining the effects of certain evolutionary forces on a population, in most cases population geneticists use frequencies of alleles to describe the gene pool. The frequencies of alleles at a locus are called the **allelic frequencies** or, more commonly, the **gene frequencies.** The use of gene frequencies offers several advantages over the genotypic frequencies. First, there are always fewer alleles than genotypes, and so the gene pool can be described with fewer parameters when gene frequencies are used. If a locus has three alleles, six genotypic frequencies must be calculated to describe the gene pool, whereas use of the allelic frequencies requires the calculation of only three allelic frequencies. Furthermore, in sexually reproducing organisms genotypes break down to alleles when gametes are formed, and alleles, not genotypes, are passed from one generation to the next. Consequently, only alleles have continuity over time, and the gene pool evolves through changes in the frequencies of alleles.

Gene frequencies may be calculated in either of two ways. First, we can calculate the gene frequencies directly from the *numbers* of genotypes. In this method, we count the number of alleles of one type and divide it by the total number of alleles in the population:

$$\text{Gene frequency} = \frac{\text{number of copies of a given allele in the population}}{\text{sum of all alleles in the population}}$$

As an example, imagine a population of 1000 diploid individuals with 353 *AA,* 494 *Aa,* and 153 *aa* individuals. Each *AA* individual has two *A* alleles, while each *Aa* heterozygote possesses only a single *A* allele. Therefore, the number of *A* alleles in the population is 2 × (the number of *AA* homozygotes) + (the number of *Aa* heterozygotes), or (2 × 353) + 494 = 1200. Because every diploid individual has two alleles, the total number of alleles in the population will be twice the number of individuals, or 2 × 1000. Using the equation given above, the gene frequency is 1200/2000 = 0.60.

When two alleles are present at a locus, we can use the following formulas for calculating gene frequencies:

$$p = f(A) = \frac{2 \times (\text{number of } AA \text{ homozygotes}) + (\text{number of } Aa \text{ heterozygotes})}{2 \times (\text{total number of individuals})}$$

Figure 24.1 Panaxia Dominula, *the scarlet tiger moth. The top two moths are normal homozygotes (BB), those in rows 2 and 3 are heterozygotes (Bb) and the bottom moth is the rare homozogote (bb).*

$$q = f(a) = \frac{2 \times (\text{number of } aa \text{ homozyotes}) + (\text{number of } Aa \text{ heterozygotes})}{2 \times (\text{total number of individuals})}$$

The frequencies (f) of two alleles are commonly symbolized as p and q. The gene frequencies for a locus should always add up to 1; therefore, once p is calculated, q can be easily obtained by subtraction: $1 - p = q$. A sample calculation of gene frequencies is presented in Box 24.1.

Gene frequencies with multiple alleles. Suppose we have three alleles—A^1, A^2, and A^3—

Box 24.1
Sample Calculation of Gene and Genotypic Frequencies for Hemoglobin Variants among Nigerians

Data are from Livingstone, F. B. 1973. Data on the Abnormal Hemoglobins and Glucose-6-Phosphate Dehydrogenase Deficiency in Human Populations of 1967–1973. Contributions In Human Biology No. 1. Museum of Anthropology, University of Michigan.

Hemoglobin Genotypes

AA	AS	SS	AC	SC	CC	Total
2017	783	4	173	14	11	3002

Calculation of Genotypic Frequencies

$$\text{Genotypic frequency} = \frac{\text{number of individuals with the genotype}}{\text{total number of individuals}}$$

$$f(SS) = \frac{4}{3002} = 0.0013$$

$$f(AS) = \frac{783}{3002} = 0.261$$

$$f(AA) = \frac{2017}{3002} = 0.672$$

$$f(SC) = \frac{14}{3002} = 0.0047$$

$$f(AC) = \frac{173}{3002} = 0.058$$

$$f(CC) = \frac{11}{3002} = 0.0037$$

Calculation of Gene Frequencies

$$\text{Gene frequency} = \frac{\text{number of copies of a given allele in the population}}{\text{sum of all alleles in the population}}$$

$$f(S) = \frac{2(\text{number of } SS \text{ individuals}) + (\text{number of } AS \text{ individuals}) + (\text{number of } SC \text{ individuals})}{2(\text{total number of individuals})}$$

$$f(S) = \frac{2(4) + 783 + 14}{2(3002)} = \frac{805}{6004} = 0.134$$

$$f(A) = \frac{2(\text{number of } AA \text{ individuals}) + (\text{number of } AS \text{ individuals}) + (\text{number of } AC \text{ individuals})}{2(\text{total number of individuals})}$$

$$f(A) = \frac{2(2017) + 783 + 173}{2(3002)} = \frac{4990}{6004} = 0.831$$

$$f(C) = \frac{2(\text{number of } CC \text{ individuals}) + (\text{number of } AC \text{ individuals}) + (\text{number of } SC \text{ individuals})}{2(\text{total number of individuals})}$$

$$f(C) = \frac{2(11) + 173 + 14}{2(3002)} = \frac{209}{6004} = 0.035$$

at a locus, and we want to determine the gene frequencies. Here, we employ the same rule that we used with two alleles: we add up the number of alleles of each type and divide by the total number of alleles in the population:

$$p = f(A^1) = \frac{2 \times (A^1A^1) + (A^1A^2) + (A^1A^3)}{2 \times (\text{total number of individuals})}$$

$$q = f(A^2) = \frac{2 \times (A^2A^2) + (A^1A^2) + (A^2A^3)}{2 \times (\text{total number of individuals})}$$

$$r = f(A^3) = \frac{2 \times (A^3A^3) + (A^1A^3) \times (A^2A^3)}{2 \times (\text{total number of individuals})}$$

We will illustrate the calculation of gene frequencies when more than two alleles are present with data from a study on genetic variation in milkweed beetles. Walter Eanes and his coworkers examined gene frequencies at a locus that codes for the enzyme phosphoglucomutase (PGM). Three alleles were found at this locus; each allele codes for a different molecular variant of the enzyme. In one population, the following numbers of genotypes were collected:

AA	=	4
AB	=	41
BB	=	84
AC	=	25
BC	=	88
CC	=	32
Total	=	274

The frequencies of the alleles are

$$f(A) = p = \frac{(2 \times 4) + 41 + 25}{(2 \times 274)} = 0.135$$

$$f(B) = q = \frac{(2 \times 84) + 41 + 88}{(2 \times 274)} = 0.542$$

$$f(C) = r = \frac{(2 \times 32) + 88 + 25}{(2 \times 274)} = 0.323$$

As seen in these calculations, we add twice the number of homozygotes that possess the allele and one times each of the heterozygotes that have the allele. We then divide by twice the number of individuals in the population, which represents the total number of alleles present. Notice that for each allelic frequency, we do not add all the heterozygotes, because some of the heterozygotes do not have the allele; for example, in calculating the gene frequency of A, we do not add the number of *BC* heterozygotes in the top part of the equation, because we are adding the number of *A* alleles, and *BC* individuals do not have an *A* allele. We can use the same procedure for calculating gene frequencies when four or more alleles are present.

Gene frequencies from genotypic frequencies. A second method for calculating gene frequencies is from the genotypic *frequencies.* This calculation may be quicker if we have already determined the frequencies of the genotypes. For two alleles, the gene frequencies are

$p = f(A) =$ (frequency of the *AA* homozygote) $+ \frac{1}{2} \times$ (frequency of the *Aa* heterozygote)

$q = f(a) =$ (frequency of the *aa* homozygote) $+ \frac{1}{2} \times$ (frequency of the *Aa* heterozygote)

The frequency of the homozygote is added to half of the heterozygote frequency because half of the heterozygote's alleles are *A* and half are *a*. If three alleles (A^1, A^2, and A^3) are present in the population, the gene frequencies are

$p = f(A^1) = f(A^1A^1) + \frac{1}{2} \times f(A^1A^2) + \frac{1}{2} \times f(A^1A^3)$

$q = f(A^2) = f(A^2A^2) + \frac{1}{2} \times f(A^1A^2) + \frac{1}{2} \times f(A^2A^3)$

$r = f(A^3) = f(A^3A^3) + \frac{1}{2} \times f(A^1A^3) + \frac{1}{2} \times f(A^2A^3)$

Although calculating the gene frequencies from genotypic frequencies may be quicker than calculating them directly from the numbers of genotypes, more rounding error will occur. As a result, calculations from direct counts are usually preferred.

Gene frequencies at an X-linked locus. Calculation of gene frequencies at an X-linked locus is slightly more complicated, because males have only a single X-linked allele. However, we can use the same rules we used for autosomal loci. Remember that each homozygous female carries two X-linked alleles; heterozygous fe-

males and all males have only a single X-linked allele. To determine the number of alleles at an X-linked locus, we multiply the number of homozygous females by 2 and add the number of heterozygous females and the number of hemizygous males. We then divide by the total number of alleles in the population. When determining the total number of alleles, we add twice the number of females (because each female has two X-linked alleles) to the number of males (who have a single allele at X-linked loci). Using this reasoning, the frequencies of two alleles at an X-linked locus (X^A and X^a) are determined with the following equations:

$$p = f(X^A) = \frac{2 \times (X^AX^A \text{ females}) + (X^AX^a \text{ females}) + (X^AY \text{ males})}{2 \times (\text{number of females}) + (\text{number of males})}$$

$$q = f(X^a) = \frac{2 \times (X^aX^a \text{ females}) + (X^AX^a \text{ females}) + (X^aY \text{ males})}{2 \times (\text{number of females}) + (\text{number of males})}$$

Allelic frequencies at an X-linked locus can be determined from the genotypic frequencies by

$p = f(X^A) = f(X^AX^A) + \frac{1}{2} \times f(X^AX^a) + f(X^AY)$

$q = f(X^a) = f(X^aX^a) + \frac{1}{2} \times f(X^AX^a) + f(X^aY)$

Students should strive to understand the logic behind these calculations, and not just memorize the formulas. If the basis of the calculations is fully understood, the exact equations need not be remembered and gene frequencies can be determined for any situation.

The Hardy-Weinberg Law

The Hardy-Weinberg law is the most important principle in population genetics, for it provides an explanation of how reproduction influences gene and genotypic frequencies of a population. The Hardy-Weinberg law is named after the two individuals who independently discovered it in the early 1900s (Box 24.2). We begin our discussion of the Hardy-Weinberg law by simply stating what it tells us about the gene pool of a population. We then explore the implications of this principle, and briefly discuss how the Hardy-Weinberg law is derived. Finally, we present some applications of the Hardy-Weinberg law and test a population to determine if the genotypes are in Hardy-Weinberg proportions.

The Hardy-Weinberg law isdivided into three parts—a set of assumptions and two major results. A simple statement of the law is:

Part 1: In an infinitely large, randomly mating population, free from outside evolutionary forces (mutation, migration, natural selection);

Part 2: The frequencies of the alleles do not change over time; and

Part 3: After one generation of random mating, the genotypic frequencies will remain in the proportions p^2 (frequency of AA), $2pq$ (frequency of Aa), and q^2 (frequency of aa), where p is the allelic frequency of A and q is the allelic frequency of a.

In short, the Hardy-Weinberg law explains what happens to gene and genotypic frequencies of a population as the genes are passed from generation to generation in the absence of evolutionary forces.

Assumptions of the Hardy-Weinberg Law

Part 1 of the Hardy-Weinberg law presents certain conditions, or assumptions, which must be present for the law to apply. First, the law indi-

Box 24.2
Hardy,
Weinberg,
and the
History
of Their
Contribution
to Population
Genetics

Godfrey H. Hardy (1877–1947) a mathematician at Cambridge University, often met R. C. Punnett, the Mendelian geneticist, at the faculty club. One day in 1908 Punnett told Hardy of a problem in genetics that he attributed to a strong critic of Mendelism, G. U. Yule (Yule later denied having raised the problem). Supposedly, Yule said that if the gene for short fingers (brachydactyly) was dominant (which it is) and its allele for normal-length fingers was recessive, then short fingers ought to become more common with each generation. In time virtually everyone in Britain should have short fingers. Punnett believed the argument was incorrect, but he could not prove it.

Hardy was able to write a few equations showing that, given any particular frequency of alleles for short fingers and alleles for normal fingers in a population, the relative numbers of people with short fingers and people with normal fingers will stay the same generation after generation providing no natural selection is involved that favors one phenotype or the other in producing offspring. Hardy published a short paper describing the relationship between genotypes and phenotypes in populations, and within a few weeks a paper was published by Wilhelm Weinberg (1862–1937), a German physician of Stuttgart, that clearly stated the same relationship. The Hardy-Weinberg law signaled the beginning of modern population genetics.

To be complete, we should note that in 1903 the American geneticist W. E. Castle of Harvard University was the first to recognize the relationship between allele and genotypic frequencies, but it was Hardy and Weinberg who clearly described the relationship in mathematical terms. Thus the law is sometimes referred to as the Castle-Hardy-Weinberg law.

cates that the population must be infinitely large. If a population is limited in size, chance deviations from expected ratios can cause changes in gene frequency. Changes in gene frequency such as these are termed **genetic drift.** The assumption of infinite size in part 1, however, is unrealistic—no population has an infinite number of individuals. Yet, population size determines the magnitude of genetic drift, and sampling error has a significant effect on gene frequencies only in "fairly small populations," which will be discussed later, when we examine genetic drift in more detail. At this point, we merely want to stress that populations need not be infinitely large for the Hardy-Weinberg law to hold true.

A second condition of the Hardy-Weinberg law is that mating must be random. **Random mating** refers to matings between genotypes occurring in proportion to the frequencies of the genotypes in the population. More specifically, the probability of a mating between two genotypes is equal to the product of the two genotypic frequencies.

To illustrate random mating, consider the MN blood types in humans. The MN blood type is an antigen on the surface of a red blood cell, similar to the ABO antigens except that incompatibility at the MN system does not cause problems during blood transfusion. The MN blood type is determined by one locus with two codominant alleles, *M* and *N*. In a population of Eskimos, the frequencies of the three *MN* genotypes are *MM* = 0.835, *MN* = 0.156, and *NN* = 0.009. If Eskimos interbreed randomly, the probability of a mating between an *MM* male and an *MM* female is equal to the frequency of *MM* times the frequency of *MM* = 0.835 × 0.835 = 0.697. Similarly, the probabilities of other possible matings are equal to the products of the genotypic frequencies when mating is random.

The requirement of random mating for the Hardy-Weinberg law is often misinterpreted. Many students assume, incorrectly, that the population must be interbreeding randomly for all traits. Humans, as well as many other organisms, do not mate randomly. Humans mate preferentially for height, IQ, skin color, socioeco-

nomic status, and other traits. While mating may be nonrandom for some traits, most humans still mate randomly for the *MN* blood types; few of us even know what our *MN* blood type is. The principles of the Hardy-Weinberg law apply to any locus for which random mating occurs, even if mating is nonrandom for other loci.

A third condition of the Hardy-Weinberg law is that the population must be free from outside evolutionary forces. Remember, in the Hardy-Weinberg law we are interested in whether heredity alone changes gene frequencies and how reproduction influences genotypic frequencies. Therefore, the influence of other evolutionary factors must be excluded. Later we will discuss these outside evolutionary forces and their effect on the gene pool of a population. This condition (no evolutionary forces acting on the population) applies only to the locus in question—a population may be subject to evolutionary forces acting on some genes, while still meeting the Hardy-Weinberg assumptions at other loci.

Predictions of the Hardy-Weinberg Law

If the conditions in part 1 of the Hardy-Weinberg law are met, the population is in genetic equilibrium and two results are expected. First, the frequencies of the alleles do not change from one generation to the next, and therefore the gene pool is not evolving at this locus. Second, the genotypic frequencies will be in the proportions p^2, $2pq$, and q^2 after one generation of random mating. The genotypic frequencies also remain in these proportions as long as all the conditions required by the Hardy-Weinberg law are met. When the genotypes are in these proportions, the population is said to be in **Hardy-Weinberg equilibrium.** An important use of the Hardy-Weinberg law is that the genotypic frequencies can be determined from the gene frequencies when the population is in equilibrium.

To summarize, the Hardy-Weinberg law makes several predictions about the gene frequencies and the genotypic frequencies of a population when certain conditions are satisfied. The necessary conditions are that the population is large, randomly mating, and free from outside evolutionary forces. When these conditions are met, the Hardy-Weinberg law indicates that gene frequencies will not change, genotypic frequencies will change only in the first generation, and thereafter the genotypic frequencies will be determined by the gene frequencies, occurring in the proportions p^2, $2pq$, and q^2.

Derivation of the Hardy-Weinberg Law

The Hardy-Weinberg law states that when a population is in equilibrium, the genotypic frequencies will be in the proportions p^2, $2pq$, and q^2. To understand the basis of these frequencies at equilibrium, consider a hypothetical population in which the frequency of allele A is p and the frequency of allele a is q. In producing gametes, each genotype passes on both alleles that it possesses with equal frequency; therefore, the frequencies of A and a in the gametes are also p and q. Table 24.1 shows the combinations of gametes when mating is random. This table illustrates the relationship between the allelic frequencies and the genotypic frequencies, which forms the basis of the Hardy-Weinberg law. We see that when gametes pair randomly, the genotypes will occur in the proportions p^2 (AA), $2pq$ (Aa), and q^2 (aa). These genotypic proportions result from the expansion of the square of the allelic frequencies $(p + q)^2 = p^2 + 2pq + q^2$, and the genotypes reach these proportions after one generation of random mating.

The Hardy-Weinberg law also states that gene and genotypic frequencies remain constant generation after generation if the population remains large, randomly mating, and free from outside evolutionary forces. This result can be demonstrated by considering a hypothetical, randomly mating population, as illustrated in Table 24.2. In Table 24.2, all possible matings are given. By definition, random mating means that the frequency of mating between two genotypes will be equal to the product of the genotypic frequencies. For example, the frequency of an $AA \times AA$ mating will be equal to p^2 (the frequency of AA) $\times p^2$ (the frequency of AA) $= p^4$. The frequencies of the offspring produced from each mating are also presented in Table 24.2.

Table 24.1 Possible Combinations of A *and* a *Gametes from Gametic Pools for a Population*

Gametes		Sperm	
		$p\ A$	$q\ a$
Eggs	$p\ A$	$p^2\ AA$	$pq\ Aa$
	$q\ a$	$pq\ Aa$	$q^2\ aa$

In sum, $p^2\ AA + 2pq\ Aa + q^2\ aa = 1.00$

We see that the probability of $AA \times Aa$ ($2p^3q$) and $Aa \times AA$ ($2p^3q$) matings is $4p^3q$, and we know from Mendelian principles that these crosses produce ½ AA and ½ Aa offspring. Therefore, the probability of obtaining AA offspring from these matings is $4p^3q \times ½ = 2p^3q$. The frequencies of offspring produced by each type of mating are presented in the body of the table. At the bottom of the table, the total frequency for each genotype is obtained by addition. As we can see, after random mating the genotypic frequencies are still p^2, $2pq$, and q^2 and the gene frequencies remain at p and q. The population can thus be represented in the zygotic and gametic stages as follows:

Zygotes	Gametes
$p^2\ AA + 2pq\ Aa + q^2\ aa$	$p\ A + q\ a$

Each generation of zygotes produces A and a gametes in proportions p and q. The gametes unite to form AA, Aa, and aa zygotes in the proportions $p^2 + 2pq + q^2$, and the cycle is repeated indefinitely as long as the assumptions of the Hardy-Weinberg law hold. This short proof gives the theoretical basis for the Hardy-Weinberg law.

The Hardy-Weinberg law indicates that at equilibrium, the genotypic frequencies depend upon the frequencies of the alleles. This relationship between allelic frequencies and genotypic frequencies for a locus with two alleles is represented in Figure 24.2. Several aspects of this relationship should be noted: (1) The maximum frequency of the heterozygote is 0.5, and this maximum value occurs only when the frequencies of A and a are both 0.5; (2) if gene frequencies are between 0.33 and 0.66, the heterozygote is the most numerous genotype; (3) when the frequency of one allele is low, the homozygote for that allele is the rarest of the genotypes. When $f(a) = q = 0.1$, the frequencies of the genotypes are $f(AA) = 0.81$, $f(Aa) = 0.18$, and $f(aa) = 0.01$. This point is also illustrated by the distribution of genetic diseases in humans, which are frequently rare and recessive. For a rare recessive trait, the frequency of the gene causing the trait will be much higher than the frequency of the trait itself, because most of the rare alleles are in nonaffected heterozygotes. Albinism, for example, is a rare recessive condition in humans. One form of albinism is *tyrosinase-negative albinism*. In this type of albinism, affected individuals have no tyrosinase activity, which is required for normal production of pigment. Among North American whites, the frequency of tyrosinase-negative albinism is roughly 1 in 40,000, or 0.000025. Since albinism is a recessive disease, the genotype of affected

individuals is *aa*. According to the Hardy-Weinberg law, the frequency of the *aa* genotypes equals q^2. If $q^2 = 0.000025$, then $q = 0.005$ and $p = 1 - q = 0.995$. The heterozygote frequency is therefore $2pq = 2 \times 0.995 \times 0.005 = 0.00995$ (almost 1 percent). Thus, while the frequency of albinos is low (1 in 40,000), individuals heterozygous for albinism are much more common (almost 1 in 100). Heterozygotes for recessive traits can be common, even when the trait is rare.

Keynote *The Hardy-Weinberg law describes what happens to gene and genotypic frequencies of a large population as a result of random mating, assuming that no outside evolutionary forces act on the population. If these conditions are met, gene frequencies do not change from generation to generation, and the genotypic frequencies stabilize after one generation in the proportions p², 2pq, and q², where p and q equal the frequencies of the alleles in the population.*

Table 24.2 Algebraic Proof of Genetic Equilibrium in a Randomly Mating Population for Two Alleles in a Population in which $(p + q)^2 = p^2 + 2pq + q^2 = 1.00$

Type of Mating ♀ ♂	Mating Frequency	Offspring Frequencies		
		AA	Aa	aa
$p^2\ AA \times p^2\ AA$	p^4	p^4	—	—
$\left.\begin{array}{l} p^2\ AA \times 2pq\ Aa \\ 2pq\ Aa \times p^2\ AA \end{array}\right\}$*	$4p^3q$	$2p^3q$	$2p^3q$	—
$\left.\begin{array}{l} p^2\ AA \times q^2\ aa \\ q^2\ aa \times p^2\ AA \end{array}\right\}$	$2p^2q^2$	—	$2p^2q^2$	—
$2pq\ Aa \times 2pq\ Aa$	$4p^2q^2$	p^2q^2	$2p^2q^2$	p^2q^2
$\left.\begin{array}{l} 2pq\ Aa \times q^2\ aa \\ q^2\ aa \times 2pq\ Aa \end{array}\right\}$	$4pq^3$	—	$2pq^3$	$2pq^3$
$q^2\ aa \times q^2\ aa$	q^4	—	—	q^4
Totals	$(p^2 + 2pq + q^2)^2$ $= 1$	$p^2(p^2 + 2pq + q^2)$ $= p^2$	$2pq(p^2 + 2pq + q^2)$ $= 2pq$	$q^2(p^2 + 2pq + q^2)$ $= q^2$

Genotype frequencies $= (p + q)^2 = p^2 + 2pq + q^2 = 1$ in each generation afterward
Gene (allele) frequencies $= p(A) + q(a) = 1$ in each generation afterward

* For example, matings between AA and Aa will occur at $p^2 \times 2pq = 2p^3q$ for $AA \times Aa$ and at $p^2 \times 2pq = 2p^3q$ for $Aa \times AA$, for a total of $4p^3q$. Two progeny types, AA and Aa, result in equal proportions from these matings. Therefore offspring frequencies are $2p^3q$ (i.e., ½ × $4p^3q$) for AA and for Aa.

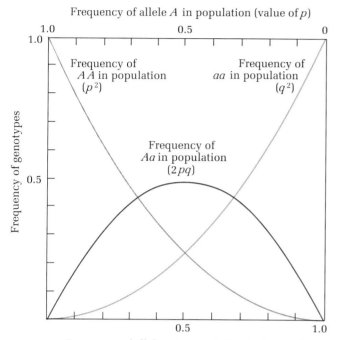

Frequency of allele *A* in population (value of *p*)

Frequency of
A A in population
(p^2)

Frequency of
aa in population
(q^2)

Frequency of
Aa in population
($2pq$)

Frequency of genotypes

Frequency of allele *a* in population (value of *q*)

*Figure 24.2 Relation of the frequencies of the geno-
types* AA, Aa, *and* aa *and the frequencies of alleles* A
and a *(in values of* p *and* q, *respectively) in a randomly mating population, as predicted by the Hardy-
Weinberg formula. It is assumed that all three genotypes have equal reproductive success.*

Extensions of the Hardy-Weinberg Law

When two alleles are present at a locus, the
Hardy-Weinberg law tells us that at equilibrium
the frequencies of the genotypes will be p^2, $2pq$,
and q^2, which is the square of the allelic frequencies $(p + q)^2$. If three alleles are present—
for example, alleles *A, B,* and *C*—with frequencies equal to *p, q,* and *r,* the frequencies of the
genotypes at equilibrium are also given by the
square of the allelic frequencies:

$$(p + q + r)^2 = p^2 (AA) + 2pq (AB) + q^2 (BB) + 2pr (AC) + 2qr (BC) + r^2(CC)$$

In the blue mussel found along the Atlantic
coast of North America, three alleles are common at a locus coding for the enzyme leucine
amino peptidase (LAP). Geographic variation in

allele frequencies of this locus is shown in Figure 24.3. For a population of mussels inhabiting
Long Island Sound, R. K. Koehn and colleagues
determined that the frequencies of the three alleles were as follows:

Allele	Frequency
LAP⁹⁸	$p = 0.52$
LAP⁹⁶	$q = 0.31$
LAP⁹⁴	$r = 0.17$

If the population were in Hardy-Weinberg equilibrium, the expected genotypic frequencies
would be

Genotype		Expected Frequency		
LAP⁹⁸/LAP⁹⁸	p^2	$= (0.52)^2$	=	0.27
LAP⁹⁸/LAP⁹⁶	$2pq$	$= 2(0.52)(0.31)$	=	0.32
LAP⁹⁶/LAP⁹⁶	q^2	$= (0.31)^2$	=	0.10
LAP⁹⁶/LAP⁹⁴	$2qr$	$= 2(0.31)(0.17)$	=	0.10
LAP⁹⁴/LAP⁹⁸	$2pr$	$= 2(0.52)(0.17)$	=	0.18
LAP⁹⁴/LAP⁹⁴	r^2	$= (0.17)^2$	=	0.03

The square of the gene frequencies can be used
in the same way to estimate the expected frequencies of the genotypes when four or more alleles are present at a locus.

If alleles are X-linked, females may be homozygous or heterozygous, but males carry only a
single allele for each X-linked locus. For X-
linked alleles in females, the Hardy-Weinberg
frequencies are the same as those for autosomal
loci: p^2 ($X^A X^A$), $2pq$ ($X^A X^B$), and q^2 ($X^B X^B$). In
males, however, the frequencies of the genotypes
will be p ($X^A Y$) and q ($X^B Y$), the same as the
frequencies of the alleles in the population. For
this reason, recessive X-linked traits are more
frequent among males than among females. Red-
green color blindness is an X-linked recessive
trait, and the frequency of the color-blind allele
among American blacks is 0.039. At equilibrium, the expected frequency of color-blind
males is $q = 0.039$, but the frequency of color-
blind females is only $q^2 = (0.039)^2 = 0.0015$.

When random mating occurs within a population, the equilibrium genotypic frequencies are
reached in one generation. However, if the alleles are X-linked and the sexes differ in allelic
frequency, the equilibrium frequencies are ap-
proached over several generations. This is

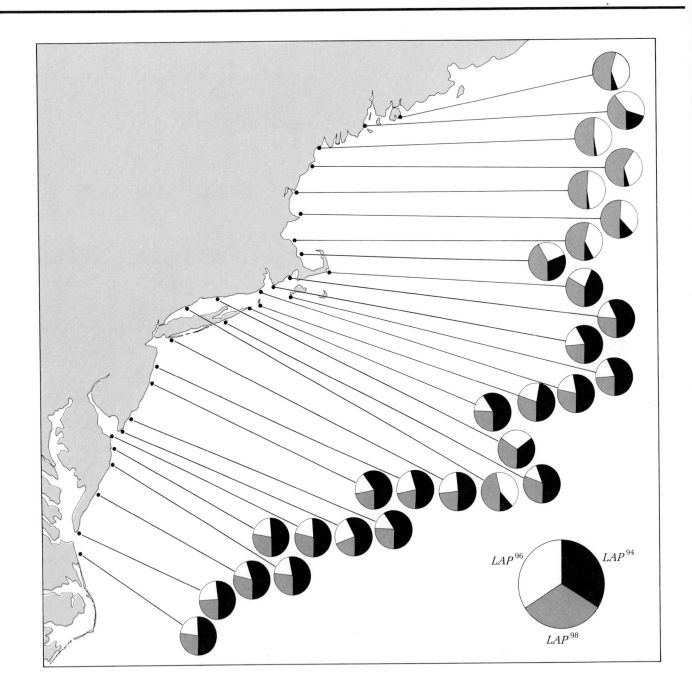

Figure 24.3 Geographic variation in allelic frequencies of the locus coding for leucine amino peptidase (LAP) in the blue mussel.

because males receive their X chromosome from their mother only, while females receive an X from both the mother and the father. Consequently, the frequency of an X-linked allele in males will be the same as the frequency of that allele in their mothers, whereas the frequency in females will be the average of that in mothers and fathers. With random mating, the allelic frequencies in the two sexes oscillate back and forth each generation, and the difference in gene frequency between the sexes is reduced by half each generation, as illustrated in Figure 24.4. Once the allelic frequencies of the males and females are equal, the frequencies of the genotypes will be in Hardy-Weinberg proportions after one more generation of random mating.

Testing for Hardy-Weinberg Proportions

To determine whether the genotypes of a population are in Hardy-Weinberg proportions, we first compute p and q from the observed frequencies of the genotypes. Once we have obtained these gene frequencies, we can calculate the expected genotypic frequencies (p^2, $2pq$, and q^2) and compare these frequencies with the actual observed frequencies of the genotypes using a chi-square test.

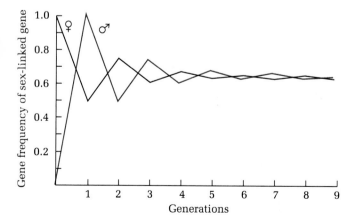

Figure 24.4 Representation of the gradual approach to equilibrium of an X-linked gene with an initial frequency of 1.0 in females and 0 in males.

To illustrate this procedure, consider a locus that codes for transferrin (a blood protein) in the red-backed vole, *Clethrionomys gapperi*. Three genotypes are found at the transferrin locus: *MM, MJ,* and *JJ.* In a population of voles trapped in the Northwest Territories of Canada in 1976, 12 *MM* individuals, 53 *MJ* individuals, and 12 *JJ* individuals were found. To determine if the genotypes are in Hardy-Weinberg proportions, we first calculate the gene frequencies for the population using our familiar formula:

$$p = 2 \times \frac{\text{(number of homozygotes)} + \text{(number of heterozygotes)}}{2 \times \text{(total number of individuals)}}$$

Therefore:

$$p = f(M) = \frac{(2 \times 12) + (53)}{(2 \times 77)} = 0.50$$

$$q = f(J) = 1 - p = 0.50$$

Using p and q calculated from the observed genotypes, we can now compute the expected Hardy-Weinberg proportions for the genotypes: $f(MM) = p^2 = (0.50)^2 = 0.25$; $f(MJ) = 2pq = 2(0.50)(0.50) = 0.50$; and $f(JJ) = q^2 = (0.50)^2 = 0.25$. However, for the chi-square test, actual numbers of individuals are needed, not the proportions. To obtain the expected numbers, we simply multiply each expected proportion times the total number of individuals counted (N), as shown below.

	Expected	Observed
$f(MM) = p^2 \times N$		
$\quad = 0.25 \times 77 =$	19.3	12
$f(MJ) = 2pq \times N$		
$\quad = 0.50 \times 77 =$	38.5	53
$f(JJ) = q^2 \times N$		
$\quad = 0.25 \times 77 =$	19.3	12

With observed and expected numbers, we can compute a chi-square value to determine the probability that the differences between observed and expected numbers could be the result of chance. The chi-square is computed using the same formula that we employed for analyzing genetic crosses; that is, d, the deviation, is calculated for each class as (observed − expected); d^2, the deviation squared, is divided by the expected number e for each class; and chi-square (χ^2) is computed as the sum of all d^2/e values. For this example, $\chi^2 = 10.98$. We now need to find this value in the chi-square table (Table 5.2 p. 153) under the appropriate degrees of freedom. This step is not as straightforward as in our previous χ^2 analyses. In those examples the number of degrees of freedom was the number of classes in the sample minus 1. Here, however, while there are three classes, there is only one degree of freedom, because the frequencies of alleles in a population have no theoretically expected values. Thus p must be estimated from the observations themselves. So one degree of freedom is lost for every parameter (p in this case) that must be calculated from the data. Another degree of freedom is lost because, for the fixed number of individuals, once all but one of the classes have been determined, the last class has no degree of freedom and is set automatically. Therefore with three classes (*MM*, *MJ*, and *JJ*), two degrees of freedom are lost, leaving one degree of freedom.

In the chi-square table under the column for one degree of freedom, the chi-square value of 10.98 indicates a P value less than 0.05. Thus the probability that the differences between the observed and expected values is due to chance is very low. That is, the observed numbers of genotypes do not fit the expected numbers under Hardy-Weinberg law.

Using the Hardy-Weinberg Law to Estimate Gene Frequencies

An important application of the Hardy-Weinberg law is the calculation of gene frequencies when one or more alleles is recessive. For example, we have seen that albinism in humans results from an autosomal recessive gene. Normally, this trait is rare, but among the Hopi Indians of Arizona, albinism is remarkably common. Charles M. Woolf and Frank C. Dukepoo visited the Hopi villages in 1969 and observed 26 cases of albinism in a total population of about 6000 Hopis (Figure 24.5). This gave a frequency for

Figure 24.5 Three Hopi girls, taken about 1900. The middle child is an albino. Albinism, an autosomal recessive disorder, occurs in high frequency among the Hopi Indians of Arizona.

the trait of 26/6000, or 0.0043, which is much higher than the frequency of albinism in most populations. Although we have calculated the frequency of the trait, we cannot directly determine the frequency of the albino gene, because we cannot distinguish between heterozygous individuals and those individuals homozygous for the normal allele. Recall that our computation of the gene frequency involves counting the number of alleles:

$$p = \frac{2 \times \text{(number of homozygotes)} + \text{(number of heterozygotes)}}{2N}$$

But, since heterozygotes for a recessive trait such as albinism cannot be identified, this is impossible. Nevertheless, we can determine the gene frequency from the Hardy-Weinberg law if the population is in equilibrium. At equilibrium, the frequency of the homozygous recessive genotype is q^2. For albinism among the Hopis, $q^2 = 0.0043$, and q can be obtained by taking the square root of the frequency of the trait. Therefore, $q = \sqrt{0.0043} = 0.065$, and $p = 1 - q = 0.935$. Following the Hardy-Weinberg law, the frequency of heterozygotes in the population will be $2pq = 2 \times 0.935 \times 0.065 = 0.122$. Thus, one out of eight Hopis carries an allele for albinism!

We should not forget that this method of calculating gene frequency rests upon the assumptions of the Hardy-Weinberg law. If the conditions of the Hardy-Weinberg law do not apply, then our estimate of gene frequency will be inaccurate. Also, once we calculate gene frequencies with these assumptions, we cannot then test the population to determine if the genotypic frequencies are in the Hardy-Weinberg expected proportions. To do so would involve circular reasoning, for we assumed Hardy-Weinberg proportions in the first place to calculate the gene frequencies.

Genetic Variation in Natural Populations

One of the most significant questions addressed in population genetics is how much genetic variation exists within natural populations. The amount of heritable variation within populations is important for several reasons. First, it determines the potential for evolutionary change and adaptation. The amount of variation also provides us with important clues about the relative importance of various evolutionary forces, since some forces increase variation and others decrease it. Furthermore, the manner in which new species arise may depend upon the amount of genetic variation harbored within populations. For all these reasons, population geneticists are interested in measuring genetic variation and in attempting to understand the evolutionary forces that control it.

During the 1940s and 1950s, population geneticists developed two opposing hypotheses about the amount of genetic variation within natural populations. These hypotheses, or models of genetic variation, have become known as the **balance model** and the **classical model.** The classical model emerged primarily from the work of laboratory geneticists, who proposed that most natural populations possess little genetic variation. According to this model, within each population one allele functions best, and this allele is strongly favored by natural selection. As a consequence, almost all individuals in the population are homozygous for this "best" or "wild-type" allele. New alleles arise from time to time through mutation, but almost all are deleterious and are kept in low frequency by selection. Once in a long while, a new mutation arises that is better than the wild-type allele. This new allele increases the survival and reproduction of the individuals that carry it, and the frequency of the allele increases over time because of this selective advantage. Eventually the new allele reaches high frequency and becomes the new wild-type. In this way a population evolves, but little genetic variation is found within the population at any one time.

In contrast, the balance model predicts that no single allele is best and widespread. The gene pool of a population consists of many alleles at each locus, and therefore individuals in the populations are heterozygous at numerous loci. Evolution occurs through gradual shifts in the frequencies and kinds of alleles. To account for the presence of this variation, the proponents of the balance model suggest that natural selection actively maintains genetic variation within a population through balancing selection. Balancing selection is selection that maintains a balance between alleles, preventing any single allele from reaching high frequency. One form of balancing selection is overdominance, where the heterozygote has higher fitness than either homozygote. The balance view of genetic variation arose among those population geneticists who came from backgrounds in natural history and who studied wild populations.

For many years, population geneticists were unable to establish how much variation existed within natural populations. The problem in determining which model of genetic variation was correct stemmed from the fact that no general procedure was available for assessing the amount of genetic variation in nature. To distinguish between the classical and balance models, population geneticists required data on genotypes at many loci of many individuals from multiple species. Until 1966, no technique existed for obtaining this information. Naturalists recognized that plants and animals in nature frequently differ in phenotype (Figure 24.6), but the genetic basis of most traits is too complex to assign specific genotypes to individuals. A few species do possess simple genetic traits, such as spot patterns in butterflies and shell color in snails, but these isolated cases were too few to provide any general estimate of genetic variation. Indirect methods suggested that many populations possess large amounts of genetic variation, but no direct measure was available.

In 1966, population geneticists began to apply the principle of protein electrophoresis to the study of natural populations. Electrophoresis is a biochemical technique that allows proteins with different molecular structures to be separated. (See Box 24.3 for a description of protein electrophoresis.) For population geneticists, electrophoresis provides a technique for quickly genotyping many individuals at many loci. This procedure is now used to examine genetic variation in hundreds of plant and animal species. The amount of genetic variation within a population is commonly measured with two parameters, the **proportion of polymorphic loci** and **heterozygosity.** A polymorphic locus is any locus that has more than one allele present within a population. The proportion of polymorphic loci (P) is calculated by determining the number of polymorphic loci and dividing by the total number of loci examined. For example, suppose we examined 33 loci in a population of green frogs and found that 18 were polymorphic. The proportion of polymorphic loci would be $18/33 = 0.55$. Heterozygosity (H) is the proportion of individuals heterozygous at a locus. In our green frogs, suppose we genotyped individuals from one population at a locus coding for esterase and found that the frequency of heterozygotes was 0.09. Heterozygosity for this locus would be

0.09. We would average this heterozygosity with those for other loci and obtain an estimate of heterozygosity for the population.

Table 24.3 (p. 801) presents estimates of the proportion of polymorphic loci and heterozygosity for many species that have been surveyed with electrophoresis. The results of these studies are unambiguous—most species possess large amounts of genetic variation in their proteins, and the classical model is clearly wrong. Actually, the technique of standard electrophoresis misses a large proportion of the genetic variation that is present, and so the true amount of genetic variation is even greater than that revealed by this technique.

Studies of electrophoretic variation in natural populations showed that the classical model was incorrect, because populations possess large amounts of genetic variation. However, this did not prove that the balance model was correct. The balance model predicts that much genetic variation is found within natural populations, and this turns out to be the case. But the balance model also proposes that large amounts of genetic variation are maintained by balancing selection, and this fact was not conclusively demonstrated by the electrophoretic studies.

At this point, the nature of the controversy changed. Previously, the central question was how much genetic variation exists. Now, emphasis shifted to the question of what maintains the extensive variation observed. A new hypothesis arose to replace the classical model. This idea, called the **neutral-mutation hypothesis,** acknowledges the presence of extensive genetic variation in proteins but proposes that this variation is neutral with regard to natural selection. This does not mean that the proteins detected by electrophoresis have no function, but rather that the different genotypes are physiologically equivalent. Therefore, natural selection does not act on the neutral alleles, and random processes such as mutation and genetic drift shape the patterns of genetic variation that we see in natural populations. The neutral-mutation hypothesis proposes that variation at some loci does affect fitness, and natural selection eliminates variation at these loci.

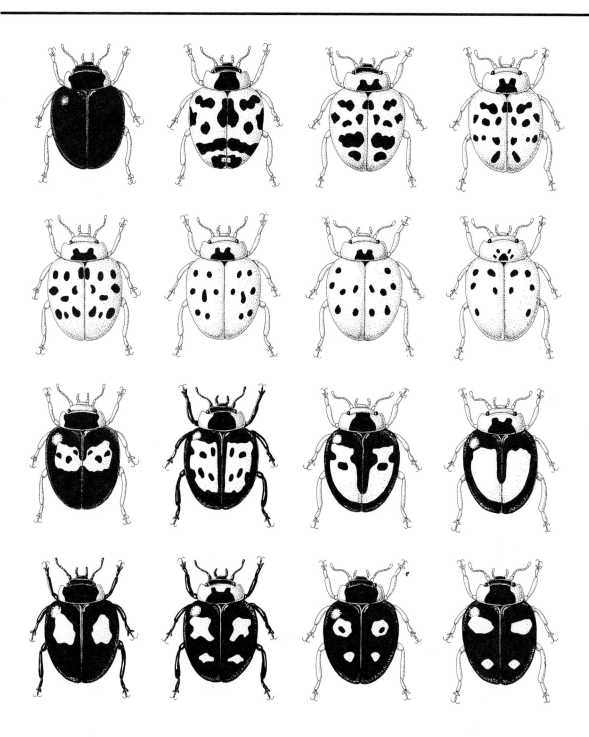

*Figure 24.6 Extensive phenotypic variation exists in
most natural populations, as is apparent in the color
patterns of the lady bettle,* Harmonia axyridis. *This
species occurs in Siberia, China, Korea, and Japan.*

Distinguishing between the balance model and the neutral-mutation hypothesis has been difficult. Population geneticists still argue over the relative merits of the two, and no clear consensus has emerged as to which model of genetic variation is correct. In reality, both models may be partly correct; at some loci genetic variation may be essentially neutral, while at others, it may be acted upon by natural selection. Techniques in molecular genetics now provide the means to directly examine genetic variation in the DNA and to unambiguously determine the amount of genetic variation present. Population geneticists are already exploiting these new tools and are developing new insights into the forces that maintain genetic variation at specific loci.

Keynote *During the 1940s and 1950s, two hypotheses were developed to explain the amount of variation in natural populations. The classical model predicted that little variation would occur, and the balance model proposed that much genetic variation would be maintained in natural populations. Through the use of electrophoresis, population geneticists*

Box 24.3
Analysis
of Genetic
Variation
with Protein
Electrophoresis

Gel electrophoresis has been widely used to study genetic variation in natural populations. This process is a biochemical technique that separates large molecules on the basis of size, shape, and charge. To examine genetic variation in proteins with electrophoresis, tissue samples from a number of individuals are ground up separately; this releases the proteins into an aqueous solution. The solutions from different individuals are then inserted into a gel made of starch, agar, polyacrylamide, or some other porous substance. Electrodes are placed at the ends of the gel as shown in Box Figure 24.1a and a direct electrical current is supplied, which sets up an electrical field across the gel.

Because proteins are charged molecules, they will migrate within the electrical field. Molecules with different shape, size, and/or charge will migrate at different rates and will separate from one another within the gel over time. If two individuals have different genotypes at a locus coding for a protein, they may produce slightly different molecular forms of the protein, which can be separated and identified with electrophoresis.

After a current has been supplied to the gel for several hours and the proteins allowed to separate, the current is turned off and the proteins visualized. The gel now contains a large number of different enzymes and proteins; to identify the product of a single locus, one must stain for a specific enzyme or protein. This is frequently accomplished by having the enzyme in the gel carry out a specific biochemical reaction. The substrate for the reaction is added to the gel, along with a dye that changes color when the reaction takes place. A colored band will appear on the gel wherever the enzyme is located. If two individuals have different molecular forms of the enzyme, indicating differences in genotypes, bands for those individuals will appear at different locations on the gel. By examining the pattern of bands produced, it is also possible to determine whether an individual is homozygous or heterozygous for a particular allele. A picture of a stained electrophoretic gel is shown in Box Figure 24.1b, and a diagram of the banding pattern is shown in Box Figure 24.1a. Histochemical stains are available for dozens of enzymes, so many individuals may be quickly genotyped for variation at a number of loci.

a)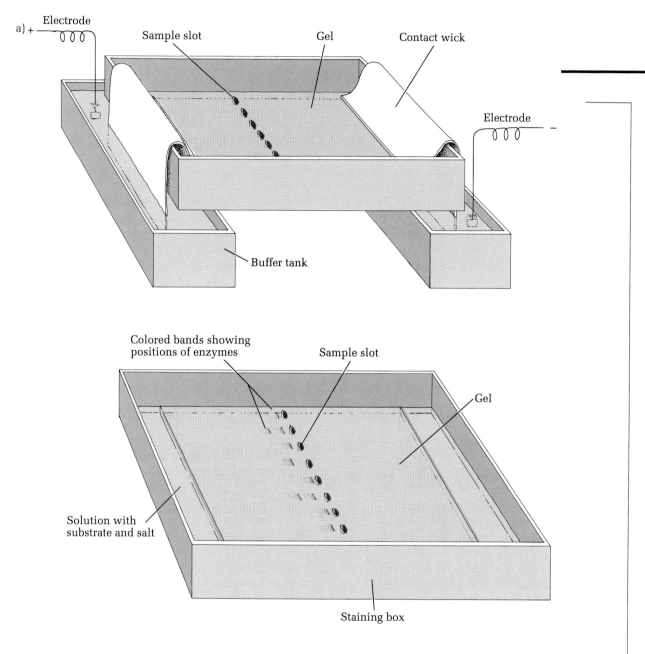

Electrode +

Sample slot

Gel

Contact wick

Electrode −

Buffer tank

Colored bands showing positions of enzymes

Sample slot

Gel

Solution with substrate and salt

Staining box

b)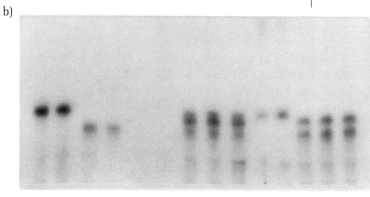

Box Figure 24.1 The technique of protein electorphores is used to measure genetic variation in natural populations. (a) Solutions of proteins from different individuals are inserted into the slots in the gel, and a direct current is supplied to separate the proteins. (b) After electrophoresis, enzyme-specific staining reactions are used to reveal the proteins. The pattern of bands on the gel indicates the genotype of each individual.

eventually demonstrated that natural populations harbor much genetic variation, thus disproving the classical model. However, the forces responsible for keeping this variation within populations are still controversial. The neutral-mutation hypothesis proposes that the genetic variation detected by electrophoresis is neutral with regard to natural selection, whereas the balance model proposes that this variation is favored by balancing selection.

Changes in Gene Frequencies of Populations

We have seen that heredity does not, by itself, generate changes in the gene frequencies of a population. When the population is large, randomly mating, and free from outside forces, no evolution occurs. For many populations, however, the conditions required by the Hardy-Weinberg law do not hold. Populations are frequently small, mating may be nonrandom, and other evolutionary forces may occur. In these situations, gene frequencies do change, and the gene pool of the population evolves in response to the interplay of different evolutionary factors. In the following sections we discuss the role of four evolutionary forces—mutation, genetic drift, migration, and natural selection—in changing the gene frequencies of a population.

Mutation

One force that can potentially alter the frequencies of alleles within a population is mutation. As we discussed in Chapter 16, gene mutations consist of heritable changes in the DNA that occur within a locus. Usually a mutation converts one allelic form of the gene to another. The rate at which mutations arise is generally low, but varies between loci and among species (Table 24.4). Certain genes affect overall mutation rates, and many environmental factors, such as chemicals, radiation, and infectious agents, may increase the number of mutations.

Evolution is a two-step process: first, genetic

variation arises, and then different alleles increase or decrease in frequency in response to evolutionary forces. Mutation is potentially important in both steps.

Mutation is ultimately the source of all new genetic variation; new combinations of alleles may arise through recombination, but new alleles only occur as a result of mutation. Thus, mutation provides the raw genetic material upon which evolution acts. Most mutations will be detrimental and will be eliminated from the population. A few mutations, however, will convey some advantage to the individuals that possess them and will spread through the population. Whether a mutation is detrimental or advantageous depends upon the specific environment, and if the environment changes, previously harmful mutations may become beneficial. For example, after the widespread use of the insecticide DDT, insects with mutations that conferred resistance to DDT were capable of surviving and reproducing; because of this advantage, the mutations spread and many insect populations quickly evolved resistance to DDT. Mutations are of fundamental importance to the process of evolution, since they provide the genetic variation upon which other evolutionary forces act.

Mutations can potentially affect evolution also by contributing to the second step in the evolutionary process—by changing the frequency of alleles within a population. Consider a population that consists of 50 individuals, all with the genotype AA. The frequency of A, p, is 1.00 [$(2 \times 50)/100$]. If one A allele mutates to a, the population now consists of 49 AA individuals and 1 Aa individual. The frequency of p is now [$(2 \times 49) + 1]/100 = 0.99$. When another mutation occurs, the frequency of A drops to 0.98. If these mutations continue to occur at a low but steady rate over long periods of time, the frequency of A will eventually decline to zero, and the frequency of a will reach a value of 1.00.

This effect of the mutational process on allelic frequencies is analogous to the following situation: If we have a large jar of black marbles and every year or so we take out one black marble

Table 24.3 *Genic Variation in Some Major Groups of Animals and in Plants*

Group	Number of Species or Forms	Mean Number of Loci per Species	Mean Proportion of Loci	
			Polymorphic per Population	Heterozygous per Individual
INSECTS *Drosophila*	28	24	0.529 ± 0.030	0.150 ± 0.010
Others	4	18	0.531	0.151
Haplodiploid wasps	6	15	0.243 ± 0.039	0.062 ± 0.007
MARINE INVERTEBRATES	9	26	0.587 ± 0.084	0.147 ± 0.019
SNAILS Land	5	18	0.437	0.150
Marine	5	17	0.175	0.083
FISH	14	21	0.306 ± 0.047	0.078 ± 0.012
AMPHIBIANS	11	22	0.336 ± 0.034	0.082 ± 0.008
REPTILES	9	21	0.231 ± 0.032	0.047 ± 0.008
BIRDS	4	19	0.145	0.042
RODENTS	26	26	0.202 ± 0.015	0.054 ± 0.005
LARGE MAMMALS[a]	4	40	0.233	0.037
PLANTS[b]	8	8	0.464 ± 0.064	0.170 ± 0.031

[a] Man, chimpanzee, pigtailed macaque, and southern elephant seal.

[b] Predominantly outcrossing species; mean gene diversity is 0.233 ± 0.029.

From Selander, R. K. 1976. Genetic variation in natural populations, in F. J. Ayala. *Molecular Evolution.* Sinauer, Sunderland, Mass.

and replace it with a white marble, the jar eventually will contain only white marbles. Similarly, in populations the slow, persistent pressure of recurrent mutation may lead to significant changes in gene frequency, assuming other evolutionary forces are not present.

The mutation of *A* to *a* is referred to as a *forward mutation.* For most genes, mutations also occur in the reverse direction; *a* may mutate to *A.* These mutations are called *reverse mutations* and they typically take place at a somewhat lower rate than the forward mutations. To include reverse mutation in our marble jar analogy, suppose we start with a jar of black marbles, but instead of taking out a black marble at

each dip and replacing it with a white one, we take out any marble at random and replace it with one of the opposite color. With this sampling system there is no selection against either marble type. At the start we are more likely to pull out black marbles and replace them with white ones, but as white marbles become more common in the jar, we are more and more likely to pull out a white one and replace it with black. At equilibrium, black and white marbles will occur in equal frequency. At this point no further changes in black and white marble frequencies will occur as long as the same method of marble replacement is applied.

In our example with marbles, the forward and

Table 24.4 Spontaneous Mutation Rates at Specific Loci for Various Organisms[a]

Organism	Trait	Mutation per 100,000 Gametes[b]
VIRUSES: T2 BACTERIOPHAGE	To rapid lysis ($r^+ \rightarrow r$)	7
	To new host range ($h^+ \rightarrow h$)	0.001
BACTERIA *E. coli* (K12)	To streptomycin resistance	0.00004
	To phage T1 resistance	0.003
	To leucine independence ($leu^- \rightarrow leu^+$)	0.00007
	To arginine independence	0.0004
	To tryptophan independence	0.006
	To arabinose dependence ($ara^+ \rightarrow ara^-$)	0.2
Salmonella typhimurium	To threonine resistance	0.41
	To histidine dependence	0.2
	To tryptophan independence	0.005
Diplococcus pneumoniae	To penicillin resistance	0.01
NEUROSPORA CRASSA	To adenine independence	0.0008–0.029
	To inositol independence (one *inos* allele, JH5202)	0.001–0.010 1.5
DROSOPHILA MELANOGASTER MALES	y^+ to *yellow*	12
	bw^+ to *brown*	3
	e^+ to *ebony*	2
	ey^+ to *eyeless*	6
CORN	*Wx* to *waxy*	0.00
	Sh to *shrunken*	0.12
	C to *colorless*	0.23
	Su to *sugary*	0.24
	Pr to *purple*	1.10
	I to *i*	10.60
	R^r to r^r	49.20
MOUSE	a^+ to *nonagouti*	2.97
	b^+ to *brown*	0.39
	c^+ to *albino*	1.02
	d^+ to *dilute*	1.25
	ln^+ to *leaden*	0.80
	Reverse mutations for above genes	0.27

Table 24.4 Spontaneous Mutation Rates at Specific Loci for Various Organisms[a] cont.

Organism	Trait	Mutation per 100,000 Gametes[b]
CHINESE HAMSTER SOMATIC CELL TISSUE CULTURE	To azaguanine resistance	0.0015
	To glutamine independence	0.014
HUMANS	Achondroplasia	0.6–1.3
	Aniridia	0.3–0.5
	Dystrophia myotonica	0.8–1.1
	Epiloia	0.4–1
	Huntington's chorea	0.5
	Intestinal polyposis	1.3
	Neurofibromatosis	5–10
	Osteogenesis imperfecta	0.7–1.3
	Pelger's anomaly	1.7–2.7
	Retinoblastoma	0.5–1.2

[a] Mutations to independence for nutritional substances are from the auxotrophic condition (e.g., *leu*$^-$) to the prototrophic condition (e.g., *leu*$^+$).

[b] Mutation rate estimates of viruses, bacteria, *Neurospora,* and Chinese hamster somatic cells are based on particle or cell counts rather than gametes.

From Strickberger, M. W. 1985. *Genetics,* 3d ed. Macmillan, New York.

reverse mutation rates were equal, since every marble drawn from the jar, whether black or white, was replaced with one of the opposite color. In practice, forward and reverse mutation rates usually differ. The rate at which A mutates to a ($A \rightarrow a$), the forward mutation rate, is symbolized with u, and the rate at which a mutates to A ($A \leftarrow a$), the reverse mutation rate, is symbolized with v. Consider a hypothetical population in which the frequency of A is p and the frequency of a is q. We assume that the population is large and no selection occurs on the alleles. Each generation, a proportion u of all A alleles mutates to a. The actual number mutating depends upon both u and the frequency of A alleles. For example, suppose the population consists of 100,000 alleles. If u equals 10^{-4}, one out of every 10,000 A alleles mutates to a. When $p = 1.00$, all 100,000 alleles in the population are A, and free to mutate to a, so 10^{-4}

$\times$ 100,000 = 10 A alleles should mutate to a. However, if $p = 0.10$, only 10,000 alleles are A and free to mutate to a. Therefore, with a mutation rate of 10^{-4} only 1 of the A alleles will undergo mutation. The decrease in the frequency of A resulting from mutation of $A \rightarrow a$ is equal to up, and the increase in frequency resulting from $A \leftarrow a$ is equal to vq. The amount that A decreases in one generation as a result of mutation is equal to the increase in A alleles due to reverse mutations minus the decrease in A alleles due to forward mutations. That is, the change in the frequency of A (Δp) is

$$\Delta p = vq - up$$

As we have seen, when p is high the number of A alleles mutating to a is relatively large, but as more and more A alleles mutate to a, the change in p decreases. At the same time, when q is small, few alleles will be available to undergo reverse mutation, but as q increases, the number of alleles undergoing reverse mutation increases. Eventually, an equilibrium point is reached where the number of alleles undergoing forward mutation is exactly equal to the number of alleles undergoing reverse mutation. At this point, no further change in gene frequency occurs, in spite of the fact that forward and reverse mutations continue to take place.

When equilibrium is reached, the change in gene frequency is zero:

$$\Delta p = vq - up = 0$$

Therefore,

$$vq = up$$

and

$$vq = u(1 - q)$$

since $p = (1 - q)$. Solving this equation, we obtain the equilibrium value of q (symbolized by $\hat{q}$):

$$vq = u - uq$$

$$vq + uq = u$$

$$q(v + u) = u$$

$$\hat{q} = \frac{u}{u + v}$$

The equilibrium value for p, $\hat{p}$, is $1 - \hat{q}$. If we solve the above equations for p instead of q, we obtain

$$\hat{p} = \frac{v}{u + v}$$

Consider a population in which the initial gene frequencies are $p = 0.9$ and $q = 0.1$ and the forward and reverse mutation rates are $u = 5 \times 10^{-5}$ and $v = 2 \times 10^{-5}$, respectively (these values are similar to forward and reverse mutation rates observed for many genes). The change in gene frequency in the first generation is

$$\Delta p = vq - up$$
$$= (2 \times 10^{-5} \times 0.1) - (5 \times 10^{-5} \times 0.9)$$

$$\Delta p = -0.000043$$

The frequency of A decreases by only four-thousandths of 1 percent. At equilibrium, the frequency of the a allele, $\hat{q}$, equals

$$\hat{q} = \frac{u}{u + v}$$

$$\hat{q} = \frac{5 \times 10^{-5}}{(5 \times 10^{-5}) + (2 \times 10^{-5})} = 0.714$$

If no other forces act on a population, after many generations the alleles will reach equilibrium. Mutation rates therefore determine the allelic frequencies of the population in the absence of other evolutionary forces. However, because mutation rates are so low, the change in gene frequency due to mutation pressure is exceedingly slow. Furthermore, as gene frequencies approach their equilibrium values, the change in frequency becomes smaller and smaller. Suppose that the mutation rate of $A \rightarrow a$ is $u = 10^{-5}$ and no reverse mutation occurs. With no reverse mutation, the equilibrium frequency of A is 0.0. If the frequency of A is initially 1.00, 1000 generations are required to change the frequency of A from 1.00 to 0.99. To change the frequency from 0.50 to 0.49, 2000 generations are required, and to change it from 0.1 to 0.09, 10,000 generations are necessary. If some reverse mutation occurs, the rate of change is even slower. In practice, mutation by itself changes the gene frequencies at such a slow rate that populations are rarely in mutational equilibrium. Other forces have more profound effects on gene frequencies, and mutation alone rarely determines the gene frequencies of a population. For example, achondroplastic dwarfism is an autosomal dominant trait in humans that arises through recurrent mutation. However, the frequency of this disorder in human populations is determined by an interaction of mutation pressure and natural selection.

Keynote *Recurrent mutation can alter the gene frequencies of a population over time, provided that other evolutionary forces are not active. Eventually, a mutational equilibrium is reached in which the gene frequencies of the population remain constant in spite of continuing mutation; the gene frequencies at this equilibrium are a function of the forward and reverse mutation rates.*

Genetic Drift

A major assumption of the Hardy-Weinberg law is that the population is infinitely large. Real populations are not infinite in size, but frequently they are large enough that expected ratios are realized and chance factors have insignificant effects on gene frequencies. Some populations are small, however, and in these groups chance factors may produce random changes in gene frequencies. Random change in gene frequency due to chance is called **genetic drift,** or simply *drift* for short. Sewall Wright (Figure 24.7), a brilliant population geneticist who laid much of the theoretical foundation of the discipline, championed the importance of genetic drift in the 1930s, and sometimes genetic drift is called the *Sewall Wright effect* in his honor.

Figure 24.7 Sewall Wright, an important figure in the field of population genetics, made a major contribution to our understanding of the role of genetic drift to the evolutionary process.

Chance changes in gene frequency. Random changes in gene frequency resulting from chance can be an important evolutionary force in small populations. Imagine a small group of humans inhabiting a South Pacific island. Suppose that this population consists of only 10 individuals, five of whom have green eyes and five of whom have brown eyes. For this example, we assume that eye color is determined by a single locus (actually, eye color is polygenic) and that the allele for green eyes is recessive to brown (*BB* and *Bb* codes for brown eyes and *bb* codes for green). The frequency of the allele for green eyes is 0.6 in the island population. A typhoon strikes the island, killing 50 percent of the population; five of the inhabitants perish in the storm. Just by chance, those five individuals who die all have brown eyes. Eye color in no way affects the probability of surviving; strictly as a result of chance only those with green eyes survive. After the typhoon, the gene frequency for green eyes is 1.0. Evolution has occurred in this population—the frequency of the green-eye allele has changed from 0.6 to 1.0, simply as a result of chance.

Now, imagine the same scenario, but this time with a population of 1000 individuals. As before, 50 percent of the population has green eyes and 50 percent has brown eyes. A typhoon strikes the island and kills half of the population. How likely is it that all 500 people who perish this time will have brown eyes, just by chance? In a population of 1000 individuals, the probability of this occurring by chance is extremely remote. The example illustrates an important characteristic of genetic drift—chance factors are likely to produce significant changes in gene frequencies only in small populations.

Random factors producing mortality in natural populations, such as the typhoon in the above example, is only one of several ways in which genetic drift arises. Chance deviations from expected ratios of gametes and zygotes also produce genetic drift. We have seen the importance of chance deviations from expected ratios in the genetic crosses we studied earlier in the book. For example, when we cross a heterozygote with a homozygote (*Aa* × *aa*), we expect 50 percent of the progeny to be heterozygous and 50 percent to be homozygous. We do not expect to get exactly 50 percent every time, however, and if the number of progeny is small, the observed ratio may differ considerably from the expected. Recall that the Hardy-Weinberg law is based upon random mating and expected ratios of progeny resulting from each type of mating (Table 24.2). If the actual number of progeny differs from the expected ratio due to chance, genotypes may not be in Hardy-Weinberg proportions and changes in gene frequencies may occur.

Chance deviations from expected proportions arise from a general phenomenon called **sampling error.** Imagine that a population produces an infinitely large pool of gametes, with alleles in the proportions p and q. If random mating occurs and all the gametes unite to form zygotes, the proportions of the genotypes will be equal to p^2, $2pq$ and q^2, and the frequencies of the alleles in these zygotes will remain p and q. If the number of progeny is limited, however, the gametes that unite to form the progeny constitute a sample from the infinite pool of potential gametes. Just by chance, or by "error," this sample may deviate from the larger pool, and the smaller the sample, the larger the potential deviation.

Flipping a coin is an analogous situation in which sampling error occurs. When we flip a coin, we expect 50 percent heads and 50 percent tails. If we flip the coin 1000 times, we will get very close to that expected fifty-fifty ratio. But, if we flip the coin only four times, we would not be surprised if by chance we obtain 3 heads and 1 tail, or even all tails. When the sample—in this case the number of flips—is small, the sampling error can be large. All genetic drift arises from such sampling error.

Measuring genetic drift. Genetic drift is random, and thus we cannot predict what the gene frequencies will be after drift has occurred. However, since sampling error is related to the size of the population, we can make predictions about the magnitude of genetic drift. Ecologists often measure population size by counting the number of individuals present, but not all individuals contribute gametes to the next generation. To determine the magnitude of genetic drift, we must know the **effective population size,** which equals the equivalent number of adults contributing gametes to the next generation. If the sexes are equal in number and all individuals have an equal probability of producing offspring, the effective population size equals the number of breeding adults in the population. However, when males and females are not present in equal numbers, the effective population size is

$$N_e = \frac{4 \times N_f \times N_m}{N_f + N_m}$$

where N_f equals the number of breeding females and N_m equals the number of breeding males.

Students often have difficulty understanding why this equation must be used—why the effective population size is not simply the number of breeding adults. The reason is that males, as a group, contribute half of all genes to the next generation and females, as a group, contribute the other half. Therefore, in a population of 70 females and 2 males, the two males are not genetically equivalent to two females; each male contributes $\frac{1}{2} \times \frac{1}{2} = 0.25$ of the genes to the next generation, whereas each female contributed $\frac{1}{2} \times \frac{1}{70} = 0.007$ of all genes. The small number of males disproportionately influences what alleles are present in the next generation. Using the above equation, the effective population size is $N_e = (4 \times 70 \times 2)/(70 + 2) = 7.8$, or approximately 8 breeding adults. What this means is that in a population of 70 females and 2 males, genetic drift will occur as if the population had only four breeding males and four breeding females. Therefore, genetic drift will have a much greater effect in this population than in one with 72 breeding adults equally divided between males and females.

Other factors, such as differential production of offspring, fluctuating population size, and overlapping generations can further reduce the effective population size. Considering these complications, it is quite difficult to accurately measure effective population size.

The amount of variation among populations

resulting from genetic drift is measured by the **variance of gene frequency,** which equals

$$s_p^2 = \frac{pq}{2N_e}$$

where N_e equals the effective population size and p and q equal the gene frequencies. A more useful measure is the **standard error of gene frequency,** which is the square root of the variance of gene frequency:

$$s_p = \sqrt{\frac{pq}{2N_e}}$$

The standard error can be used to calculate the 95 percent confidence limits of gene frequency, which indicate the expected range of p in 95 percent of such populations. The 95 percent confidence limits equal approximately $p \pm 2s_p$. Suppose, for example, that $p = 0.8$ in a population with N_e equal to 50. The standard error in gene frequency is $s_p = \sqrt{pq/2N_e} = 0.04$. The 95 percent confidence limits for p are therefore $p \pm 2s_p = 0.72 \le p \le 0.88$. To interpret the 95 percent confidence limits, imagine that 100 populations with N_e of 50 have p initially equal to 0.8. Genetic drift may cause gene frequencies in some populations to change; the 95 percent confidence limits tell us that in the next generation, 95 of the original 100 populations should have p within the range of 0.72 to 0.88. Therefore, if we observe a change in p greater than this, say from 0.8 to 0.68, we know that the probability that this change will occur by genetic drift is less than 0.05. Most likely we would conclude that some force other than genetic drift contributed to the observed change in gene frequency.

Causes of genetic drift. All genetic drift arises from sampling error, but there are several ways in which sampling error occurs in natural populations. First, genetic drift arises when population size remains continuously small over many generations. Undoubtedly, this situation is frequent, particularly where populations occupy marginal habitats, or when competition limits population growth. In such populations, genetic drift plays an important role in the evolution of

gene frequencies. The effect of genetic drift arising from small population size is seen in an experiment conducted by Buri with *Drosophila melanogaster.* Buri examined the frequency of two alleles, bw^{75} and bw, at a locus that determines eye color in the fruit flies. He set up 107 experimental populations, and the initial frequency of bw^{75} was 0.5 in each. The flies in each population interbred randomly, and in each generation Buri randomly selected 8 males and 8 females to be the parents for the next generation. Thus, effective population size was always 16 individuals. The distribution of allelic frequencies in these 107 populations is presented in Figure 24.8. Notice that the gene frequencies in the early generations were clumped around 0.5, but genetic drift caused the frequencies in the populations to spread out or diverge over time. By generation 19, most of the populations were fixed for one of the two alleles.

Another way in which genetic drift arises is through **founder effect.** Founder effect occurs when a population is initially established by a small number of individuals. Although the population may subsequently grow in size and later consist of a large number of individuals, the gene pool of the population is derived from the genes present in the original founders. Chance may play a significant role in determining which genes were present among the founders, and this has a profound effect upon the gene pool of subsequent generations.

Many excellent examples of founder effect come from the study of human populations. Consider the inhabitants of Tristan da Cunha, a small, isolated island in the South Atlantic. This island was first permanently settled by William Glass, a Scotsman, and his family in 1817. (Several earlier attempts at settlement failed.) They were joined by a few additional settlers, some shipwrecked sailors, and a few women from the distant island of St. Helena, but for the most part the island remained a genetic isolate. In 1961, a volcano on Tristan da Cunha erupted, and the population of almost 300 inhabitants was evacuated to England. During their two-year stay in England, geneticists studied the islanders and reconstructed the genetic history of the

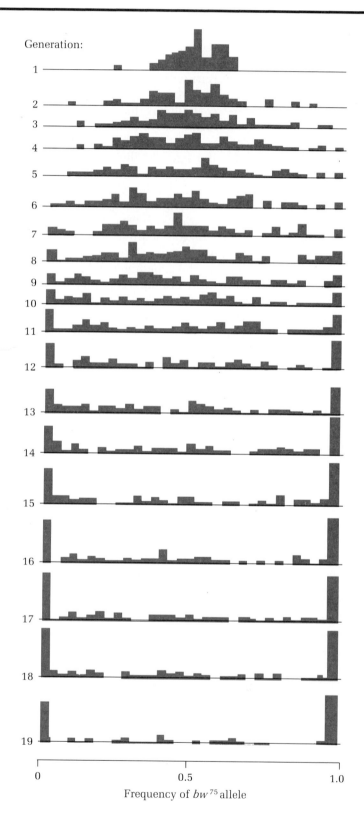

Generation:

Frequency of *bw*⁷⁵ allele

population. These studies revealed that the current gene pool of Tristan da Cunha has been strongly influenced by genetic drift.

Three forms of genetic drift occurred in the evolution of the Tristan da Cunha population. First, founder effect took place at the initial settlement. By 1855, the population of Tristan da Cunha consisted of about 100 individuals, but 26 percent of the genes of the population in 1855 were contributed by William Glass and his wife. Even in 1961, these original two settlers contributed 14 percent of all the genes in the 300 individuals of the population. The particular genes that Glass and other original founders carried heavily influenced the subsequent gene pool of the population. Second, population size remained small throughout the history of the settlement, and sampling error continually occurred.

A third form of sampling error, called **bottleneck effect,** also played an important role in the population of Tristan da Cunha. Bottleneck effect occurs when a population is drastically reduced in size. During such a population reduction, some genes may be lost from the gene pool as a result of chance. Recall our earlier example of the population consisting of 10 individuals inhabiting a South Pacific island. When a typhoon struck the island, the population size was reduced to five, and by chance, all individuals with brown eyes perished in the storm, changing the frequency of green eyes from 0.6 to 1.0. This is an example of bottleneck effect. Bottleneck effect can be viewed as a type of founder effect, since the population is refounded by those few individuals that survive the reduction.

Two severe bottlenecks occurred in the history of Tristan da Cunha. The first took place around 1856 and was precipitated by two events: the death of William Glass and the arrival of a missionary who encouraged the inhab-

Figure 24.8 Results of Buri's study of genetic drift in 107 populations of Drosophila melanogaster. *Shown are the distributions of allelic frequency among the populations in 19 consecutive generations. Each population consisted of 16 individuals.*

Table 24.5 Frequencies of Alleles Controlling the ABO Blood Group System in Three Human Populations

Population	Allele Frequencies			Phenotype (Blood Group) Frequencies			
	I^A	I^B	i	A	B	AB	O
Dunker	0.38	0.03	0.59	0.593	0.036	0.023	0.348
United States	0.26	0.04	0.70	0.431	0.058	0.021	0.490
West Germany	0.29	0.07	0.64	0.455	0.095	0.041	0.410

itants to leave the island. At this time many islanders emigrated to America and to South Africa, and the population dropped from 103 individuals at the end of 1855 to 33 in 1857. A second bottleneck occurred in 1885. The island of Tristan da Cunha has no natural harbor, and the islanders intercepted passing ships for trade by rowing out in small boats. On November 28, 1885, fifteen of the adult males on the island put out in a small boat to make contact with a passing ship. In full view of the entire island community, the boat capsized and all fifteen men drowned. Following this disaster, only four adult males were left on the island, one of whom was insane and two of whom were old. Many of the widows and their families left the island during the next few years, and the population size dropped from 106 to 59. Both bottlenecks had a major effect on the gene pool of the population. All the genes contributed by several settlers were lost and the relative contributions of others were altered by these events. Thus, the gene pool of Tristan da Cunha has been influenced by genetic drift in the form of founder effect, small population size, and bottleneck effect.

As we shall see later when we discuss migration, gene flow among populations increases the effective population size and reduces the effects of genetic drift. Small breeding units that lack gene flow are genetically isolated from other groups and often experience considerable genetic drift, even though surrounded by much larger populations. A good example is a religious sect found in eastern Pennsylvania known as the Dunkers. Between 1719 and 1729, fifty Dunker families emigrated from Germany and settled in the United States. Since that time, the Dunkers have remained an isolated group, rarely marrying outside of the sect, and the number of individuals in their communities has always been relatively small.

Geneticists studied one of the original Dunker communities in Franklin County, Pennsylvania during the 1950s. At the time of the study, this population had about 300 members, and the population size had remained relatively constant for many generations. The investigators found that some of the gene frequencies in the Dunkers were very different from the frequencies found among the general population of the United States and were different from the frequencies of the West German population from which the Dunkers descended. Table 24.5 presents some of the gene frequencies at the ABO blood group locus. The frequencies of the ABO alleles among the Dunkers are not the same as those in either the United States population or those of the West Germans. Nor are the Dunker frequencies intermediate between the United States and German frequencies, which might be expected if intermixing of Dunkers and Americans had occurred. The most likely explanation for the unique gene frequencies observed among the Dunkers is that genetic drift has been responsible for producing random change in the gene pool. Founder effect probably occurred when the original 50 families emigrated from Germany, and genetic drift has most likely continued to influence gene frequencies each generation since 1729, because the population size has remained small.

Effects of genetic drift. Genetic drift produces changes in gene frequencies and these changes have several effects on the genetic structure of populations. First, genetic drift causes the gene frequencies of a population to change over time. This is illustrated in Figure 24.9. The different lines represent gene frequencies in several populations over a number of generations. Although all populations begin with an allelic frequency equal to 0.50, the frequencies in each population change over time as a result of sampling error. In each generation, the gene frequency has an equal probability of increasing or decreasing, and over time the frequencies wander randomly or drift (hence the name "genetic drift"). Sometimes, just by chance, the gene frequency reaches a value of 0.0 or 1.0. At this point, one allele is lost from the population and the population is said to be *fixed* for the remaining allele. Once an allele has reached fixation, no further change in gene frequency can occur, unless the other allele is reintroduced through mutation or

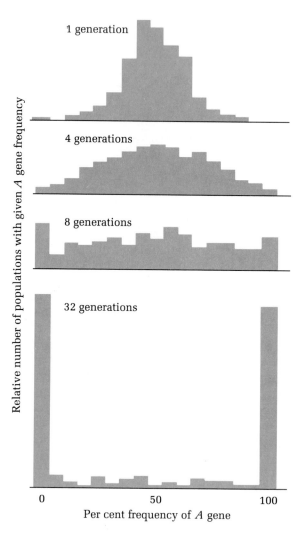

Figure 24.10 *Fixation of alleles through genetic drift. The graphs illustrate the results of 400 experiments; in each experiment a population has 8 diploid individuals with an initial frequency of 0.5 for allele A. Gene frequencies in each population are allowed to drift randomly over time. The upper graph shows the distribution of A in the 400 populations after a single generation. The distributions of A after 4, 8, and 32 generations are shown in the lower graphs. (From Mettler, Gregg, and Shaffer,* Population Genetics and Evolution, *2nd ed. Copyright © 1988 by Prentice-Hall, Inc. Reprinted by permission of Prentice-Hall, Inc., Englewood Cliffs, NJ.)*

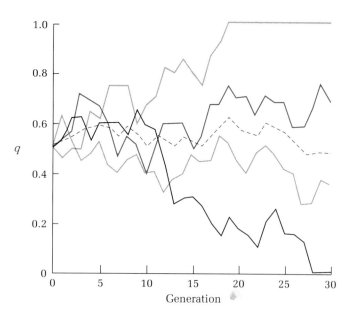

Figure 24.9 *The effect of genetic drift on the frequency (q) of allele A² in four populations. Each population begins with q equal to 0.5 and the effective population size for each is 20. The mean frequency of allele A² for the four replicates is indicated by the broken line. These results were obtained by a computer simulation.*

Table 24.6 Variance in Gene Frequency as a Function of Population Size

Type of Population	Number of Populations	Mean Allelic Frequency		Variance in Allelic Frequency	
		Est-3[b]	Hbb[8]	Est-3[b]	Hbb[8]
Small ($N < 50$)	29	0.418	0.849	0.051	0.188
Large ($N > 50$)	13	0.372	0.843	0.013	0.008

Data from Selander, R. K. 1970. Behavior and genetic variation in natural populations *(Mus musculus). Am Zoologist* 10:53–66.

mirgation. The probability of fixation in a population increases with time, as shown in Figure 24.10. If the initial gene frequencies are equal, which allele becomes fixed is strictly random. If, on the other hand, initial gene frequencies are different, the rare allele is more likely to be lost. During this process of genetic drift and fixation, the number of heterozygotes in the population also decreases, and after fixation, the population heterozygosity is zero. Populations lose genetic variation through such decreases in heterozygosity and random fixation of alleles, and thus, the second effect of genetic drift is a reduction in genetic variation within populations.

Since genetic drift causes random change in gene frequency, the allelic frequencies in different populations will not change in the same direction. Therefore populations diverge in their gene frequencies through genetic drift. This is illustrated in Figures 24.9 and 24.10; in both figures all the populations begin with p and q equal to 0.5. After a few generations, the gene frequencies of the populations diverge, and this divergence increases over generations. The maximum divergence in gene frequencies is reached when all populations are fixed for one or the other allele. If gene frequencies are initially equal to 0.5, approximately half of the populations will be fixed for one allele and half will be fixed for the other allele.

Because genetic drift is greater in small populations and leads to genetic divergence, we expect more variance in gene frequency among small populations than among large populations. Such a relationship has been observed in studies of natural populations. R. K. Selander, for exam-

ple, studied genetic variation in populations of the house mouse inhabiting barns in Texas. Through systematic trapping, he was able to estimate population size, and using electrophoresis he examined the variance in gene frequency at two loci, a locus coding for the enzyme esterase (Est-3) and a locus coding for hemoglobin (Hbb). Selander found that the variance in gene frequency among small populations was several times larger than that among large populations (Table 24.6), an observation consistent with our understanding of how genetic drift leads to population divergence.

Keynote *Genetic drift, chance changes in gene frequency due to sampling error, can be an important evolutionary force in small populations. Genetic drift leads to loss of genetic variation within populations, genetic divergence among populations, and random fluctuation in the gene frequencies of a population over time.*

Migration

One of the assumptions of the Hardy-Weinberg law is that the population is closed and not influenced by outside evolutionary forces. Many populations are not completely isolated, however, and exchange genes with other populations of the same species. Individuals migrating into a population may introduce new alleles to the gene pool and alter the frequencies of existing alleles. Thus **migration** has the potential to disrupt Hardy-Weinberg equilibrium and may influence the evolution of gene frequencies within populations.

The term *migration* usually implies movement of organisms. In population genetics, however, we are interested in the movement of genes, which may or may not occur when organisms move. Movement of genes takes place only when organisms migrate and then reproduce, contributing their genes to the gene pool of the recipient population. This process is also referred to as **gene flow.**

Gene flow has two major effects on a population. First, it introduces new alleles to the population. Because mutation is generally a rare event, a specific mutant allele may arise in one population and not in another. Gene flow spreads unique alleles to other populations and, like mutation, is a source of genetic variation for the population. When the gene frequencies of migrants and the recipient population differ, gene flow also changes the allelic frequencies within the recipient population. Through exchange of genes, populations remain similar, and thus, migration is a homogenizing force that tends to prevent populations from evolving genetic differences.

To illustrate the effect of migration on gene frequencies, we will consider a simple model in which gene flow occurs in only one direction, from population I to population II. Suppose that the frequency of allele A in population I (p_I) is 0.8 and the frequency of A in population II (p_{II}) is 0.5. Each generation some individuals migrate from population I to population II, and these migrants are a random sample of the genotypes in population I. After migration has occurred, population II actually consist of two groups of individuals: the migrants with $p_I = 0.8$ and the residents with $p_{II} = 0.5$. The migrants now make up a proportion of population II, which we will designate m. The frequency of A in population II after migration (p'_{II}) is

$$p'_{II} = mp_I + (1 - m)p_{II}$$

We see that the frequency of A after migration is determined by the proportion of A alleles in the two groups that now comprise population II. The first component, mp_I, represents the A alleles in the migrants—we multiply the propor-

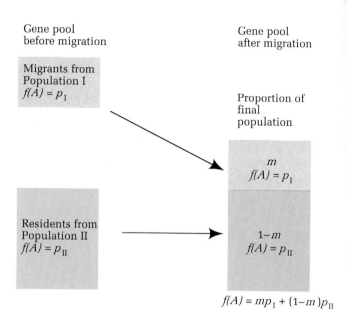

$$f(A) = mp_I + (1-m)p_{II}$$

Figure 24.11 Theoretical model illustrating the effect of migration on the gene pool of a population. After migration, population II consists of two groups, the migrants with gene frequency of p_I and the original residents with gene frequency p_{II}.

tion of the population that consists of migrants (m) by the gene frequency of the migrants (p_I). The second component represents the A alleles in the residents, and equals the proportion of the population consisting of residents ($1 - m$) multiplied by the gene frequency in the residents (p_{II}). Adding these two components together gives us the gene frequency of A in population II after migration. This model of gene flow is diagramed in Figure 24.11.

The change in gene frequency in population II as a result of migration (Δp) equals the original frequency of A subtracted from the frequency of A after migration:

$$\Delta p = p'_{II} - p_{II}$$

In the previous equation, we found that p'_{II} equaled $mp_I + (1 - m)p_{II}$, so the change in gene frequency can be written as

$$\Delta p = mp_I + (1 - m)p_{II} - p_{II}$$

Multiplying $(1 - m)$ by p_{II} in the above equation, we obtain

$$\Delta p = mp_{\mathrm{I}} + p_{\mathrm{II}} - mp_{\mathrm{II}} - p_{\mathrm{II}}$$

$$\Delta p = mp_{\mathrm{I}} - mp_{\mathrm{II}}$$

$$\Delta p = m\,(p_{\mathrm{I}} - p_{\mathrm{II}})$$

This final equation indicates that the change in gene frequency from migration depends upon two factors: the proportion of the migrants in the final population and the difference in gene frequency between the migrants and the residents. If no differences exist in the gene frequency of migrants and residents ($p_{\mathrm{I}} - p_{\mathrm{II}} = 0$), then we can see that the change in gene frequency is zero. Populations must differ in their gene frequencies in order for migration to affect the makeup of the gene pool. With continued migration, p_{I} and p_{II} become increasingly similar, and the change in gene frequency as a result of migration decreases. Eventually, gene frequencies in the two populations will be equal, and no further change will occur. This is only true, however, when other factors besides migration do not influence gene frequencies.

An additional point to be noted is that migration among populations tends to increase the effective size of the populations. As we have seen, small population size leads to genetic drift, and genetic drift causes populations to diverge. Migration, on the other hand, reduces divergence among populations, effectively increasing the size of the individual populations. Extensive gene flow has been shown to occur, for example, among populations of the monarch butterfly (Figure 24.12). Even a small amount of gene flow can reduce the effect of genetic drift. Calculations have shown that a single migrant moving between two populations each generation will prevent the two populations from becoming fixed for different alleles.

Keynote *Migration of individuals into a population may alter the makeup of the population gene pool, if the genes carried by the migrants differ from the residents of the population. Migration, also termed gene flow, tends to reduce genetic divergence among populations and increases the effective size of the population.*

a

b

Figure 24.12 Extensive gene flow occurs among populations of the monarch butterfly (a). The butterflies overwinter in Mexico (b) and then migrate north during spring and summer to breeding grounds as far away as northern Canada. Extensive gene flow occurs during the migration period and monarch populations display relatively little genetic divergence.

Natural Selection

We have now examined three major evolutionary forces capable of changing gene frequencies and producing evolution—mutation, genetic drift, and migration. These forces alter the gene pool of a population and certainly influence the evolution of a species. However, mutation, migration, and genetic drift do not result in adaptation. Adaptation is the process by which traits evolve that make organisms more suited to their immediate environment; these traits increase the organism's chances of surviving and reproducing. Adaptation is responsible for the many extraordinary traits seen in nature—wings that enable a hummingbird to fly backward, leaves of the pitcher plant that capture and devour insects, brains that allow humans to speak, read, and love. These biological features and countless other exquisite traits are the product of adaptation (Figure 24.13). Genetic drift, mutation, and migration all influence the pattern and process of adaptation, but adaptation arises chiefly from natural selection. **Natural selection** is the dominant force in the evolution of many traits and has shaped much of the phenotypic variation observed in nature.

Charles Darwin (Figure 24.14) and Alfred Russel Wallace independently developed the concept of natural selection in the middle of the nineteenth century, although some earlier naturalists had similar ideas. In 1858, Darwin's and Wallace's theory was presented to the Linnaean Society of London and was enthusiastically received by other scientists. Darwin pursued the theory of evolution further than Wallace, amassing hundreds of observations to support it, and publishing his ideas in a book entitled *On the Origin of Species.* For his innumerable contributions to our understanding of natural selection, Darwin is frequently regarded as the "father" of evolutionary theory.

Natural selection can be defined as differential reproduction of genotypes. It simply means that individuals with certain genes survive and reproduce better than others, and therefore those genes increase in frequency in the next generation, as discussed in Chapter 23. Through natural selection, traits that contribute to survival and reproduction increase over time, and in this way organisms adapt to their environment.

Selection in natural populations. A classic example of selection in natural populations is the evolution of melanic (dark) forms of moths in association with industrial pollution, a phenomenon known as "industrial melanism." Melanic phenotypes have appeared in a number of different species of moths found in the industrial regions of Europe, North America, and England. One of the best studied cases involves the peppered moth, *Biston betularia.* The common phenotype of this species, termed the *typical* form, is a greyish white color with black mottling over the body and wings. Prior to 1848, all peppered moths collected in England possessed this *typical* phenotype, but in 1848, a single black moth was collected near Manchester, England. This new phenotype, called *carbonaria,* presumably arose by mutation, and rapidly increased in frequency around Manchester and in other industrial regions. By 1900, the *carbonaria* phenotype had reached a frequency of more than 90 percent in several populations. High frequencies of *carbonaria* appeared to be associated with industrial regions, whereas the *typical* phenotype remained common in more rural districts. Labo-

Figure 24.13 Natural selection produces organisms that are finely adapted to their environment, such as this lizard with cryptic coloration that allows it to blend in with its natural surroundings.

Figure 24.14 Charles Darwin first fully developed the theory of evolution through natural selection.

and covered the tree trunks with black soot. Against this black background, the *typical* phenotype was quite conspicuous and was readily consumed by birds. In contrast, the *carbonaria* form was well camouflaged against the blackened trees and had a higher rate of survival than the *typical* phenotype in polluted areas (Figure 24.15b). Because *carbonaria* survived better in polluted woods, more *carbonaria* genes were transmitted to the next generation, and in this way the *carbonaria* phenotype increased in frequency in industrial areas. In rural areas, where pollution was absent, the *carbonaria* phenotype was conspicuous and the *typical* form was camouflaged; in these regions the frequency of the *typical* form remained high.

Kettlewell demonstrated that selection affected the frequencies of the two phenotypes by conducting a series of release-and-recapture experiments involving dark and light moths in smoky, industrial Birmingham, England, and in nonindustrialized Dorset. As predicted, the *typical* phenotype was favored in Dorset, and *carbonia* was favored in Birmingham.

ratory studies by a number of investigators, including E. B. Ford and R. Goldschmidt, demonstrated that the *carbonaria* phenotype was dominant to the *typical* phenotype. A third phenotype was also discovered, which was somewhat intermediate to *typical* and *carbonaria;* this phenotype, *insularia,* was produced by a dominant allele at a different locus.

H. B. D. Kettlewell investigated this color polymorphism in the peppered moth, and demonstrated that the increase in the *carbonaria* phenotype occurred as a result of strong selection against the *typical* form in polluted woods. Peppered moths are nocturnal; during the day they rest on the trunks of lichen-covered trees. Birds frequently prey upon the moths during the day, but because the lichens that cover the trees are naturally grey in color, the *typical* form of the peppered moth is well camouflaged against this background (Figure 24.15a). In industrial areas, however, extensive pollution beginning with the industrial revolution in the mid-nineteenth century had killed most of the lichens

Fitness and coefficient of selection. Darwin described natural selection primarily in terms of survival, and even today, many nonbiologists think of natural selection in terms of a struggle for survival. However, what is most important in the process of natural selection is the relative number of genes that are contributed to future generations. Certainly the ability to survive is important, but survival alone will not ensure that genes are passed on—reproduction must also occur. Therefore, we measure natural selection by assessing reproduction. Natural selection is measured with **fitness,** and fitness is defined as the relative reproductive ability of a genotype.

Fitness is usually symbolized as W. Since fitness is a measure of the relative reproductive ability, population geneticists usually assign a fitness of 1 to the genotype that produces the most offspring. The fitnesses of the other genotypes are assigned relative to this. For example, suppose that the genotype G^1G^1 on the average produces 8 offspring, G^1G^2 produces an average

Figure 24.15 Biston betularia, *the peppered moth, and its dark form* carbonaria *on the trunk of a lichened tree in the unpolluted countryside (a), and (b) on the trunk of a tree blackened by industrial pollution from Birmingham, England. On the lichened tree, the dark form is readily seen, whereas the light form is well camouflaged. On the polluted tree, the dark form is well camouflaged.*

of 4 offspring, and G^2G^2 produces an average of 2 offspring. The G^1G^1 genotype has the highest reproductive efficiency, so its fitness is 1 (W_{11} = 1.0). Genotype G^1G^2 produces on the average 4 offspring for the 8 produced by the most fit genotype, so the fitness of G^1G^2 (W_{12}) is 4/8 = 0.5. Similarly, G^2G^2 produces 2 offspring for the 8 produced by G^1G^1, so the fitness of G^2G^2 (W_{22}) is 2/8 = 0.25. Table 24.7 illustrates the calculation of fitness values.

Fitness tells us how well a genotype is doing in terms of natural selection. A related measure is the **selection coefficient,** which is a measure of the relative intensity of selection against a genotype. The selection coefficient is symbolized

by s and equals $1 - W$. In our example, the selection coefficients are for G^1G^1, $s = 0$; for G^1G^2, $s = 0.5$; for G^2G^2, $s = 0.75$.

Effect of selection on gene frequencies. Natural selection produces a number of different effects. At times, natural selection eliminates genetic variation and at other times it maintains variation; it can change gene frequencies or prevent gene frequencies from changing; it can produce genetic divergence among populations or maintain genetic uniformity. Which of these effects occurs depends primarily on the relative fitness of the genotypes and on the frequencies of the alleles in the population.

The change in gene frequency that results from natural selection can be calculated by constructing a table such as Table 24.8. This "table method" can be used for any type of single-locus trait, whether the trait is dominant, codominant, recessive, or overdominant. To use the table method we begin by listing the genotypes (A^1A^1, A^1A^2, and A^2A^2) and their initial frequencies. If random mating has just taken place,

Table 24.7 *Computation of Fitness Values and Selection Coefficients of Three Genotypes*

	Genotypes		
	G^1G^1	G^1G^2	G^2G^2
Number of breeding adults in one generation	16	10	20
Number of offspring produced by all adults of the genotype in the next generation	128	40	40
Average number of offspring produced per breeding adult	128/16 = 8	40/10 = 4	40/20 = 2
Fitness W (relative number of offspring produced)	8/8 = 1	4/8 = 0.5	2/8 = 0.25
Selection coefficient $(1 - W)$	1 − 1 = 0	1 − 0.5 = 0.5	1 − 0.25 = 0.75

Table 24.8 *General Method of Determining Change in Gene Frequency Due to Natural Selection*

	Genotypes			
	A^1A^1	A^1A^2	A^2A^2	
Initial genotypic frequencies	p^2	$2pq$	q^2	
Fitness*	W_{11}	W_{12}	W_{22}	
Frequency after selection	p^2W_{11}	$2pqW_{12}$	q^2W_{22}	$\overline{w} = p^2W_{11} + 2pqW_{12} + q^2W_{22}$
Relative genotypic frequency after selection	$P' = \dfrac{p^2W_{11}}{\overline{w}}$	$H' = \dfrac{2pqW_{12}}{\overline{w}}$	$Q' = \dfrac{q^2W_{22}}{\overline{w}}$	

Gene frequency after selection = $p' = P' + 1/2(H')$
$q' = 1 - p'$

Change in gene frequency due to selection = $\Delta p = p' - p$

* Note: For simplicity, fitness in this example is considered to be the probability of survival. Change in gene frequency due to differences in the number of offspring produced by the genotypes is calculated in the same manner.

the genotypes are in Hardy-Weinberg proportions and the initial frequencies are p^2, $2pq$, and q^2. We then list the fitnesses for each of the genotypes, W_{11}, W_{12}, and W_{22}. Now, suppose that selection occurs and only some of the genotypes survive. The contribution of each genotype to the next generation will be equal to the initial frequency of the genotype times its fitness. For

A^1A^1 this will be $p^2 \times W_{11}$. Notice that the contributions of the three genotypes do not add up to one. The sum of the contributions equals $p^2W_{11} + 2pqW_{12} + q^2W_{22} = \overline{W}$, which is called the *mean fitness of the population.* Next, we normalize the relative contributions by dividing each by the mean fitness of the population, and this gives us the relative frequencies of the genotypes after selection. We then calculate the new gene frequencies (p') from the genotypes after selection using our familiar formula, $p' =$ (frequency of A^1A^1) + 1/2 × (frequency of A^1A^2). Finally, the change in gene frequency resulting from selection equals $p' - p$. A sample calculation using some actual gene frequencies and fitness values is presented in Table 24.9.

Selection against a recessive trait. Many important traits and most new mutations are recessive and have reduced fitness. When a trait is completely recessive, both the heterozygote and the dominant homozygote have a fitness of 1, while the recessive homozygote has reduced fitness, as shown below.

Genotype	Fitness
AA	1
Aa	1
aa	$1 - s$

If the genotypes are initially in Hardy-Weinberg proportions, the contribution of each genotype to the next generation will be the frequency times the fitness.

AA	$p^2 \times 1 = p^2$
Aa	$2pq \times 1 = 2pq$
aa	$q^2 \times (1 - s) = q^2 - sq^2$

The mean fitness of the population is $p^2 + 2pq + q^2 - sq^2$. Since $p^2 + 2pq + q^2 = 1$, the mean fitness becomes $1 - sq^2$, and the

Table 24.9 *General Method of Determining Change in Gene Frequency Due to Natural Selection when initial gene frequencies are p-0.6 and q-0.4*

	Genotypes		
	A^1A^1	A^1A^2	A^2A^2
Initial genotypic frequencies	p^2 $(0.6)^2 = 0.36$	$2pq$ $2(0.6)(0.4) =$ 0.48	q^2 $(0.4)^2 = 0.16$
Fitness	$W_{11} = 0$	$W_{12} = 0.4$	$W_{22} = 1$
Frequency after selection	$p^2W_{11} =$ $(0.36)(0) = 0$	$2pqW_{12} =$ $(0.48)(0.4) =$ 0.19	$q^2W_{22} =$ $(0.16)(1) = \overline{w} = p^2W_{11} +$ $0.16 \qquad 2pqW_{12} +$ q^2W_{22} $\overline{w} = 0 + 0.19$ $+ 0.16$ $\overline{w} = 0.35$
Relative genotypic frequency after selection	$P' = \dfrac{p^2W_{11}}{\overline{w}}$ $P' = 0/0.35 = 0$	$H' = \dfrac{2pqW_{12}}{\overline{w}}$ $H' = 0.19/0.35$ $= 0.54$	$Q' = \dfrac{q^2W_{22}}{\overline{w}}$ $Q' = 0.16/0.35$ $= 0.46$

Gene frequency after selection $p' = P' + 1/2(H')$
$\qquad\qquad p' = 0 + 1/2(0.54) = 0.27$
$\qquad\qquad q' = 1 - p' = 1 - 0.27 = 0.73$

Change in gene frequency due to selection $= \Delta p = p' - p$
$\qquad\qquad\qquad\qquad\qquad \Delta p = 0.6 - 0.27 = 0.33$

normalized genotypic frequencies after selection are

AA $\qquad \dfrac{p^2}{1 - sq^2}$

Aa $\qquad \dfrac{2pq}{1 - sq^2}$

aa $\qquad \dfrac{q^2 - sq^2}{1 - sq^2}$

To obtain q', the frequency after selection, we add the frequency of the aa homozygote and half the frequency of the heterozygote:

$$q' = \frac{q^2 - sq^2}{1 - sq^2} + \frac{1\,(2pq)}{2\,(1 - sq^2)}$$

$$q' = \frac{q^2 - sq^2}{1 - sq^2} + \frac{pq}{1 - sq^2}$$

$$q' = \frac{q^2 - sq^2 + pq}{1 - sq^2}$$

$$q' = \frac{q^2 + pq - sq^2}{1 - sq^2}$$

$$q' = \frac{q(q + p) - sq^2}{1 - sq^2}$$

Since $(q + p) = 1$,

$$q' = \frac{q - sq^2}{1 - sq^2}$$

Therefore, the change in the frequency of a after one generation of selection is

$$\Delta q = q' - q$$

$$= \frac{q - sq^2}{1 - sq^2} - q$$

$$= \frac{q - sq^2}{1 - sq^2} - \frac{q(1 - sq^2)}{(1 - sq^2)}$$

$$= \frac{q - sq^2 - q(1 - sq^2)}{1 - sq^2}$$

$$= \frac{q - sq^2 - q + sq^3}{1 - sq^2}$$

$$= \frac{-sq^2 + sq^3}{1 - sq^2}$$

$$= \frac{-sq^2\,(1 - q)}{1 - sq^2}$$

So $\Delta q = -spq^2/(1 - sq^2)$ because $1 - q = p$. When $\Delta q = 0$, no further change occurs in allelic frequencies. Notice that there is a negative sign in the equation to the left of spq^2; because the values of $s, p,$ and q are always positive or zero, Δq is negative or zero. Thus, the value of q will decrease with selection.

Selection also depends on the actual frequencies of genes in the population. This is because the relative proportions of Aa and aa individuals at various frequencies of allele a influence how effectively selection can reduce a detrimental recessive trait. When the frequency of a recessive gene is relatively high, many homozygous recessive individuals are present in the population and will have low f tness, causing a large change in the gene frequency. When the gene frequency is low, however, the homozygous recessive genotype is rare, and little change in gene frequency occurs. Table 24.10 shows the magnitude of change in gene frequency each generation in three populations with different initial gene frequencies. Population 1 begins with gene frequency q equal to 0.9, population 2 begins with q equal to 0.5, and population 3 begins with q equal to 0.1. In this example the homozygous recessive genotype (aa) has a f tness of 0 and the other two genotypes (AA and Aa) have a f tness of 1. When the frequency of q is high, as in population 1, the change in gene frequency is large; in the f rst generation, q drops from 0.9 to 0.47. However, when q is small, as in population 3, the change in q is much less; here q drops from 0.1 to 0.091 in the f rst generation. So, as q becomes smaller, the change in q becomes less. Because of this diminishing change in frequency, it is virtually impossible to eliminate entirely a recessive trait from the population. This result only applies, however, to completely recessive traits; if the f tness of the heterozygote is also reduced, the change in gene frequency will be more rapid, because now selection also acts against the heterozygote in addition to the homozygote. The effect of dominance on changes in gene frequency as a result of selection is illustrated in Figure 24.16.

We have gone through a lengthy discussion of the effects of selection on a recessive trait to

Table 24.10 Effectiveness of Complete Selection against a Recessive Phenotype at Different Initial Frequencies

| Generation | Genotype (q^2 aa) and Allele (q a) Frequencies | | | | | |
| | Population 1 | | Population 2 | | Population 3 | |
	q^2	q	q^2	q	q^2	q
0	0.810	0.900	0.250	0.500	0.010	0.100
1	0.224	0.474	0.112	0.333	0.008	0.091
2	0.103	0.321	0.063	0.250	0.007	0.083
3	0.059	0.243	0.040	0.200	0.006	0.077
4	0.038	0.195	0.028	0.167	0.005	0.071
5 . . .	0.027	0.163	0.020	0.143	0.004	0.066
10			0.007	0.08		

Table 24.11 Formulas for Calculating Change in Gene Frequency After One Generation of Selection

| Type of Selection | Fitnesses of Genotypes | | | Calculation of Change in Gene Frequency |
	A^1A^1	A^1A^2	A^2A^2	
Selection against recessive homozygote	1	1	$1 - s$	$\Delta q = \dfrac{-spq^2}{1 - sq^2}$
Selection against a dominant allele	$1 - s$	$1 - s$	1	$\Delta q = \dfrac{-spq^2}{1 - s + sq^2}$
Selection with no dominance	1	$(1 - a/2)$	$1 - s$	$\Delta q = \dfrac{-spq/2}{1 - sq}$
Selection which favors the heterozygote (overdominance)	$1 - s$	1	$1 - t$	$\Delta q = \dfrac{pq(sp - tq)}{1 - sp^2 - tq^2}$
Selection against the heterozygote	1	$1 - s$	1	$\Delta q = \dfrac{spq(q - p)}{1 - 2spq}$
General	W_{11}	W_{12}	W_{22}	$\Delta q = \dfrac{pq[p(W_{12} - W_{11}) + q(W_{22} - W_{12})]}{\overline{W}^*}$

*Note: For calculation of w, see Table 24.8.

illustrate how the formula for change in gene frequency can be derived from our general table method of gene frequency change under selection. Similar derivations can be carried out for dominant traits and codominant traits. We will

not present those derivations here, but the appropriate formulas for calculating changes in gene frequency under different types of dominance are presented in Table 24.11. Remember, however, it is possible to calculate changes in gene frequency for any type of trait using the table method.

Keynote *Natural selection involves differential reproduction of genotypes and is measured in terms of fitness, the relative reproductive contribution of a genotype. The effects of selection depend not only on the fitness of the different genotypes, but also on the dominance, or lack of dominance, of the alleles and on the frequencies of the alleles in the population.*

Mutation and Selection

As we have seen, natural selection reduces the frequency of a deleterious recessive allele, but as the frequency becomes low, the change in frequency each generation diminishes. When the allele is rare, the change in frequency is very slight. Opposing this reduction in the frequency of the allele due to selection is mutation pressure, which will continually produce new alleles and tend to increase the frequency. Eventually a balance, or equilibrium, is reached, in which the input of new alleles by recurrent mutation is exactly counterbalanced by the loss of alleles through natural selection. When this equilibrium is obtained, the frequency of the allele remains stable, in spite of the fact that selection and mutation continue, unless perturbed by some other evolutionary force.

Consider a population in which selection occurs against a deleterious recessive allele, *a*. As we saw on p. 819, the amount *a* will change in one generation (Δq) as a result of selection is

$$\Delta q = \frac{-spq^2}{1 - sq^2}$$

For a rare recessive allele, q^2 will be near 0 and the denominator in this equation, $1 - sq^2$, will be approximately 1, so that the decrease in frequency caused by selection is given by

$$\Delta q = -spq^2$$

At the same time, the frequency of the *a* allele increases as a result of mutation from *A* to *a* (pp. 803–804). Provided the frequency of *a* is

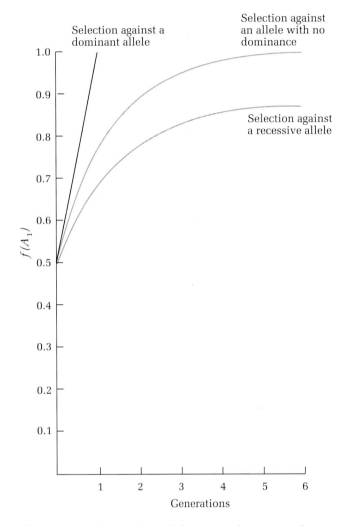

Figure 24.16 *Comparison of changes in frequency of an allele with different types of dominance, holding selection coefficient constant. A_1 is the favored allele.*

low, the reverse mutation of *a* to *A* essentially can be ignored. Equilibrium between selection and mutation occurs when the decrease in gene frequency produced by selection is the same as the increase produced by mutation:

$$spq^2 = up$$

We can predict what the frequency of *a* at equilibrium ($\hat{q}$) will be by rearranging this equation:

$$sq^2 = u$$
$$q^2 = u/s$$

and

$$\hat{q} = \sqrt{u/s}$$

If the recessive homozygote is lethal ($s = 1$), the equation becomes

$$\hat{q} = \sqrt{u}$$

As an example of the balance between mutation and selection, consider a recessive gene for which the mutation rate is 10^{-6} and s is 0.1. At equilibrium, the frequency of the gene will be $\hat{q} = \sqrt{10^{-6}/0.1} = 0.0032$. Most recessive deleterious traits remain within a population at low frequency because of this equilibrium between mutation and selection.

For a dominant allele A, the frequency at equilibrium ($\hat{q}$) is

$$\hat{q} = u/s$$

If the mutation rate is 10^{-6} and s is 0.1, as in the above example, the frequency of the dominant gene at equilibrium will be $10^{-6}/0.1 = 0.00001$, which is considerably less than the equilibrium frequency for a recessive allele with the same fitness and mutation rate. This is because selection cannot act on a recessive allele in the heterozygote state, whereas both the homozygote and the heterozygote for a dominant allele have reduced fitness. For this reason, detrimental dominant alleles generally are less common than recessive ones.

Overdominance

An equilibrium of gene frequencies also arises when the heterozygote has higher fitness than either of the homozygotes. This situation is called **overdominance** or **heterozygote advantage**. In overdominance, both alleles are maintained in the population, because both are favored in the heterozygote genotype. Gene frequencies will change as a result of selection until the equilibrium point is reached and then will remain stable. The gene frequencies at

which the population reaches equilibrium depend upon the relative fitnesses of the two homozygotes. If the selection coefficient of AA is s and the selection coefficient of aa is t, it can be shown algebraically that at equilibrium

$$\hat{p} = f(A) = t/(s + t) \quad \text{and} \quad \hat{q} = f(a) = s/(s + t)$$

An example of overdominance in humans is sickle-cell anemia. Sickle-cell anemia results from a mutation in the gene coding for beta hemoglobin. In some populations, three hemoglobin genotypes occur: $Hb\text{-}A/Hb\text{-}A$, $Hb\text{-}A/Hb\text{-}S$, and $Hb\text{-}S/Hb\text{-}S$. Individuals with the $Hb\text{-}A/Hb\text{-}A$ genotype have completely normal red blood cells; $Hb\text{-}S/Hb\text{-}S$ individuals have sickle-cell anemia; and $Hb\text{-}A/Hb\text{-}S$ individuals have sickle-cell trait, a mild form of sickle-cell anemia (see Chapter 14). In an environment in which malaria is common, the heterozygotes are at a selective advantage over the two homozygotes. Apparently, the abnormal hemoglobin mixture in the heterozygotes provides an unfavorable environment for the growth or maintenance of the malarial parasite in the red cell. The heterozygotes therefore have greater resistance to malaria and thus higher fitness than the $Hb\text{-}A/Hb\text{-}A$ people. The $Hb\text{-}S/Hb\text{-}S$ individuals are at a serious selective disadvantage because they have sickle-cell anemia. As a result, in malaria-infested areas in which the $Hb\text{-}S$ gene is also found, an equilibrium state is established in which a significant number of $Hb\text{-}S$ alleles are found in the heterozygotes because of the selective advantage of this genotype. The distributions of malaria and the $Hb\text{-}S$ allele are illustrated in Figure 24.17.

Nonrandom Mating

A fundamental assumption of the Hardy-Weinberg law is that members of the population mate randomly. But, many populations do not mate randomly for some traits, and when nonrandom mating occurs, the genotypes will not be in the proportions predicted by the Hardy-Weinberg law.

One form of nonrandom mating is **positive assortative mating**, which occurs when individu-

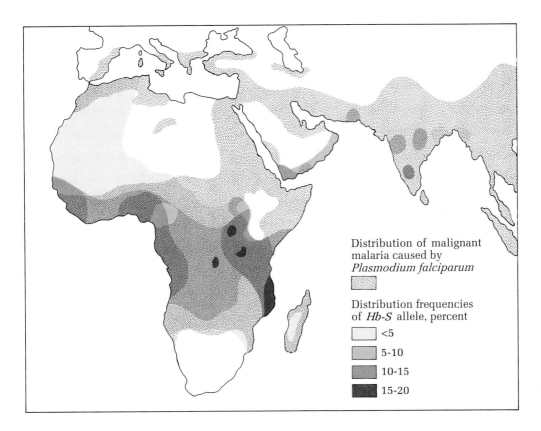

Figure 24.17 The distribution of malaria caused by the parasite Plasmodium falciparum *coincides with distribution of the* Hb-S *allele for sickle-cell anemia. The frequency of* Hb-S *is high in areas where malaria is common, because* Hb-A Hb-S *heterozygotes are resistant to malarial infection. (From Francisco J. Ayala and John A. Kiger, Jr.,* Modern genetics, *2nd ed. Copyright © 1984 and 1980 by The Benjamin/Cummings Publishing Company, Inc. Reprinted by permission.)*

als with similar phenotypes mate preferentially. Positive assortative mating is common in natural populations. For example, humans mate assortatively for height; tall men marry tall women and short men marry short women more frequently than would be expected on a random basis. **Negative assortative mating** occurs when phenotypically dissimilar individuals mate more often than randomly chosen individuals. If humans exhibited negative assortative mating for height, tall men would preferentially marry short women and short men would marry tall women. Neither positive nor negative assortative mating af-

fects the gene frequencies of a population, but they may influence the genotypic frequencies if the phenotypes for which assortative mating occurs are genetically determined.

Two other departures from random mating are **inbreeding** and **outbreeding.** As we learned in Chapter 23, inbreeding involves preferential mating between close relatives, and thus, inbreeding is really positive assortative mating for relatedness. Outbreeding is preferential mating between nonrelated individuals and is a form of negative assortative mating. The most extreme case of inbreeding is self-fertilization, which occurs in many plants and a few animals, such as some snails. The effects of self-fertilization are illustrated in Table 24.12. Assume that we begin with a population consisting entirely of *Aa* heterozygotes, and that all individuals in this population reproduce by self-fertilization. After one generation of self-fertilization, the progeny will consist of ¼ *AA*, ½ *Aa,* and ¼ *aa.* Now only half of the population consists of heterozygotes.

When this generation undergoes self-fertilization, the *AA* homozygotes will produce only *AA* progeny, and the *aa* homozygotes will produce only *aa* progeny. When the heterozygotes reproduce, however, only half of their progeny will be heterozygous like the parents, and the other half will be homozygous (¼ *AA* and ¼ *aa*). So, in each generation of self-fertilization, the percentage of heterozygotes decreases by 50 percent. After an infinite number of generations, there will be no heterozygotes and the population will be divided equally between the two homozygous genotypes. Note that the population was in Hardy-Weinberg proportions in the first generation after self-fertilization, but after further rounds the proportion of homozygotes is greater than that predicted by the Hardy-Weinberg law.

The result of continued self-fertilization is to increase homozygosity at the expense of heterozygosity. The frequencies of alleles *A* and *a* remain constant, while the frequencies of the three genotypes change significantly. When less intensive inbreeding occurs, similar, but less pronounced, effects occur. On the other hand, outbreeding increases heterozygosity.

Keynote *Inbreeding involves preferential mating between close relatives, and outbreeding is preferential mating between unrelated individuals. Continued inbreeding increases homozygosity within a population, whereas outbreeding tends to increase heterozygosity.*

Summary of the Effects of Evolutionary Forces on the Gene Pool of a Population

Let us now review the major effects of the different evolutionary forces on (1) changes in gene frequency within a population, (2) genetic divergence among populations, and (3) increases and decreases in genetic variation within populations.

Mutation, migration, genetic drift, and selection all potentially change the gene frequencies of a population over time. Mutation, however, usually occurs at such a low rate that the change resulting from mutation pressure alone is frequently negligible. Genetic drift will produce substantial changes in gene frequency only when the population size is small. Furthermore, mutation, migration, and selection may lead to an equilibrium situation, where the evolutionary forces continue to act, but the gene frequencies no longer change. Nonrandom mating does not change gene frequencies, but it does affect the genotypic frequencies of a population: inbreeding leads to increases in homozygosity and outbreeding produces an excess of heterozygotes.

Several evolutionary forces lead to genetic divergence among populations. Because genetic drift is a random process, gene frequencies in different populations may drift in different directions, and genetic drift produces genetic divergence among populations. Migration among populations has just the opposite effect, increasing effective population size and equalizing gene frequency differences among populations. If the population size is small, different mutations may arise in different populations, and, therefore, mutation may contribute to population differentiation. Natural selection can increase genetic differences among populations by favoring different alleles in different populations, or it can prevent divergence by keeping gene frequencies among populations uniform. Nonrandom mating will not, by itself, generate genetic differences among populations, although it may contribute to the effects of other forces by increasing or decreasing effective population size.

Gene flow and mutation tend to increase genetic variation within populations by introducing new alleles to the gene pool. Genetic drift produces the opposite effect, decreasing genetic variation within small populations through loss of alleles. Because inbreeding leads to increases in homozygosity, it also diminishes genetic variation within populations; outbreeding, on the other hand, increases genetic variation by increasing heterozygosity. Natural selection may increase or decrease genetic variation. If one particular allele is favored, other alleles decrease in frequency and may be eliminated from the population by selection. Alternatively, natural selection may increase genetic variation within populations through overdominance and other forms of balancing selection.

In practice, these evolutionary forces never act in isolation but combine and interact in complex ways. In most natural populations, the combined effects of these forces and their interaction determine the pattern of genetic variation observed in the gene pool over time.

Molecular Evolution

In recent years population geneticists have begun to apply recombinant DNA techniques to studies of genetic variation within populations and to questions about the molecular basis of evolution. By using restriction mapping and DNA sequencing methods (see Chapter 15), biologists can now examine evolution at the most basic genetic level, the level of the DNA. These studies have not altered our basic principles of population genetics, but they have provided a more complete and more detailed picture of how the forces of nature produce genetic change. Here we will mention just a few of the important findings of this new research.

DNA Sequence Evolution

Studies of nucleotide sequences in numerous genes have revealed that different parts of a gene evolve at different rates. Recall from our discussion of molecular genetics that a typical eukaryotic gene is made up of some nucleotides that specify the amino acid sequence of a protein (coding sequences) and other nucleotides that do not code for amino acids in a protein (noncoding sequences). The noncoding sequences include introns, the leader regions, trailer regions, and 5' and 3' flanking sequences that are not transcribed. In addition, we find pseudogenes, which are nucleotide sequences that no longer produce a functional gene product because of mutations. Even within the coding regions of a functional gene, not all nucleotide substitutions produce a corresponding change in the amino acid sequence of a protein. In particular, many mutations occurring at the third position of the codon have no effect on the amino acid sequence of the protein, because such mutations produce a synonymous codon (see Chapter 16).

Different regions of the gene are apparently subject to different evolutionary forces, and evolve at different rates. For example, nucleotide sequences within the introns and within pseudogenes accumulate genetic differences at a

Table 24.12 Relative Genotype Distributions Resulting from Self-Fertilization over Several Generations Starting with an Aa Individual

Generation	Genotypes			Inbreeding Coefficient F
	AA	Aa	aa	
0		1		0
1	$\frac{1}{4}$	$\frac{1}{2}$	$\frac{1}{4}$	$\frac{1}{2}$
2	$\frac{1}{4} + \frac{1}{8} = \frac{3}{8}$	$\frac{1}{4}$	$\frac{1}{4} + \frac{1}{8} = \frac{3}{8}$	$\frac{3}{4}$
3	$\frac{3}{8} + \frac{1}{16} = \frac{7}{16}$	$\frac{1}{8}$	$\frac{3}{8} + \frac{1}{16} = \frac{7}{16}$	$\frac{7}{8}$
4	$\frac{7}{16} + \frac{1}{32} = \frac{15}{32}$	$\frac{1}{16}$	$\frac{7}{16} + \frac{1}{32} = \frac{15}{32}$	$\frac{15}{16}$
5	$\frac{15}{32} + \frac{1}{64} = \frac{31}{64}$	$\frac{1}{32}$	$\frac{15}{32} + \frac{1}{64} = \frac{31}{64}$	$\frac{31}{32}$
n	$[1 - (\frac{1}{2})^n]/2$	$(\frac{1}{2})^n$	$[1 - (\frac{1}{2})^n]/2$	$1 - (\frac{1}{2})^n$
∞	$\frac{1}{2}$	0	$\frac{1}{2}$	1

faster rate than the coding sequences of functional genes. Among pseudogenes found in the human globin gene family, the rate of nucleotide change is approximately 10 times that observed in the coding sequences of functional globin genes. The faster rate of evolution in the noncoding sequences most likely occurs because mutations that arise within introns and pseudogenes have no effect on the phenotype and do not change an individual's fitness. On the other hand, mutations that occur in coding sequences may change the amino acid sequence of the protein and influence an individual's physiology. Most mutations that arise within coding sequences are detrimental to fitness and are eliminated by natural selection. So selection pressure appears to be more intense within the coding regions, and selection limits the amount of variation observed there. Most mutations in introns and pseudogenes appear to be neutral mutations.

Another observation from the study of nucleotide sequences is that change in the coding region of a gene is more frequent at the third position of the codon than at the first and second positions. This finding is also consistent with the idea that natural selection limits variation at nucleotides that determine the amino acid sequence of a protein. In our discussion of the genetic code (Chapter 13), we learned that most synonymous codons differ at the third nucleotide position. All nucleotide changes in the second position of a codon produce a corresponding change in the amino acid determined by that codon. About 95 percent of the changes in the first position of the codon alter the amino acid, but only 28 percent of changes in the third position change the amino acid. If selection limits variation in sequences that affect the protein structure, we might expect to find more frequent genetic change in the third position of the codons. This is exactly what molecular geneticists have observed in a number of genes. However, the rate of evolution for synonymous nucleotide changes at the third position is not quite as high as that observed in introns and pseudogenes. This observation suggests that synonymous mu-

tations at the third position are not totally neutral; natural selection may favor some synonymous codons over others. This hypothesis is reinforced by the finding that synonymous codons are not used equally throughout the coding sequences. For example, six different codons specify the amino acid leucine (UUA, UUG, CUU, CUC, CUA, and CUG), but 60 percent of the leucine codons in bacteria are CUG, and in yeast, 80 percent are UUG. Since the synonymous codons specify the same amino acid, selection must be favoring some synonymous codons over others through the availability of tRNAs. Remember that some synonymous codons pair with different tRNAs that carry the same amino acid. Therefore, a mutation to a synonymous codon does not change the amino acid, but it may change the tRNA. Studies of the different tRNAs reveal that within a cell, the amounts of the isoacceptor tRNAs (different tRNAs that accept the same amino acid) differ, and the most abundant tRNAs are those that pair with the most frequently used codons. Selection may favor one synonymous codon over another because the tRNA for the codon is more abundant, and translation of that codon is more efficient.

DNA Length Polymorphisms

In addition to evolution of nucleotide sequences through nucleotide substitution, variation frequently occurs in the number of nucleotides found within a gene. These variations are called *DNA length polymorphisms.* DNA length polymorphisms arise through deletions and insertions of relatively short stretches of nucleotides. Most of these occur in noncoding regions of the DNA, but some are also found in coding regions. Another class of DNA length polymorphisms involves variation in the number of copies of a particular gene. For example, among individual fruit flies, the number of copies of ribosomal genes varies extensively. The number of copies of transposons also varies extensively among individuals and is responsible for some DNA length polymorphisms.

Evolution in Mitochondrial DNA Sequences

Nucleotide sequences within animal mitochondrial DNA (mtDNA) evolve at a faster rate than coding sequences in nuclear genes. The reason that mtDNA undergoes more genetic divergence is not entirely clear, but because of its rapid evolution, mtDNA sequences are now being used widely to assess the evolutionary relationships among groups of closely related organisms. Mitochondrial DNA also differs from nuclear DNA in that all mtDNA is inherited clonally from the mother; mitochondria are located in the cytoplasm and only the egg cell contributes cytoplasm to a zygote. As a consequence, mtDNA does not undergo meiosis, and all offspring should be identical to the maternal genotype for mtDNA sequences (the offspring are clones for mtDNA genes). This pattern of inheritance allows matriarchal lineages (descendants from one female) to be traced, and provides a means for examining family structure in some populations. An example of geographic variation in mtDNA sequences is shown in Figure 24.18.

Concerted Evolution

One of the surprising findings from studies of DNA sequence variation is the observation that some molecular force, or forces, maintain uniformity of sequence in multiple copies of a gene. This phenomenon has been termed **concerted evolution** or **molecular drive.** As we discussed in previous chapters, some genes in eukaryotes exist in multiple copies. Ribosomal RNA genes in complex organisms, for example, typically exist in hundreds or thousands of copies. Undoubtedly, these multiple copies arose through duplication. Following duplication, individual copies of a gene might be expected to acquire mutations and diverge. Selection might limit mutations in coding regions, but if many copies exist, we would expect some divergence to occur, especially in the noncoding sequences. Contrary to this expectation, numerous studies have revealed that nucleotide sequences in the different copies of a gene are frequently quite homo-

geneous. Furthermore, the noncoding sequences are also homogeneous, which suggests that purifying selection is not responsible. When the same genes are examined in closely related species, the sequences are also homogeneous but are frequently different from the homogeneous sequence found in the first species.

These observations have led to the conclusion that some molecular process continually enforces uniformity among multiple copies of the same sequence within a species. At the same time, the process allows for rapid differentiation among species. The mechanism of concerted evolution is not fully understood, but concerted evolution has important consequences for how genes evolve, and it represents an evolutionary force that was unknown before the application of modern molecular techniques to population genetics.

These recent findings represent only a few of

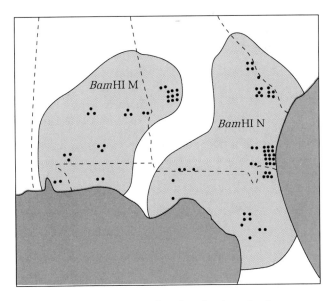

Figure 24.18 The geographic distribution of mitochondrial DNA phenotypes in the pocket gopher, Geomys pinetis. *Analysis of mitochondrial DNA with restriction enzymes is one of the modern molecular tools used by population geneticists to study the genetics of natural populations.*

the exciting discoveries that have resulted from the application of recombinant DNA technology to questions in evolutionary genetics. As these techniques become more widely used in popula-

tion genetics research, a more detailed understanding of how evolution shapes the patterns of genetic variation within and among populations will surely emerge.

*Analytical
Approaches
for Solving
Genetics
Problems*

Q.1 In a population of 2000 gaboon vipers a genetic difference with respect to venom exists at a single locus. The alleles are incompletely dominant. The population shows 100 individuals homozygous for the *t* allele (genotype *tt*, nonpoisonous), 800 heterozygous (genotype *Tt*, mildly poisonous), and 1100 homozygous for the *T* allele (genotype *TT*, deadly poisonous).

a. What is the frequency of the *t* allele in the population?
b. Are the genotypes in Hardy-Weinberg equilibrium?

A.1 This question addresses the basics of calculating gene frequencies and relating them to the genotype frequencies expected of a population in Hardy-Weinberg equilibrium.

a. The *t* frequency can be calculated from the information given since the trait is an incompletely dominant one. There are 2000 individuals in the population under study, meaning a total of 4000 alleles at the *T/t* locus. The number of *t* alleles is given by

$(2 \times tt \text{ homozygotes}) + (1 \times Tt \text{ heterozygotes})$
$$= (2 \times 100) + (1 \times 800) = 1000$$

This calculation is straightforward since both alleles in the nonpoisonous snakes are *t*, while only one of the two alleles in the mildly poisonous snakes is *t*. Since the total number of alleles under study is 4000, the frequency of *t* alleles is $1000/4000 = 0.25$. This system is a two-allele system, so the frequency of *T* must be 0.75.

b. For the genotypes to be in Hardy-Weinberg equilibrium, the distribution must be p^2 *TT* + $2pq$ *Tt* + q^2 *tt* genotypes, where *p* is the frequency of the *T* allele and *q* is the frequency of the *t* allele. This expression is simply the algebraic statement of the Hardy-Weinberg law customized to this particular example. In part a we established that the frequency of *T* is 0.75 and the frequency of *t* is 0.25. Therefore $p = 0.75$ and $q = 0.25$. Using these values, we can determine the genotype frequencies expected if this population is in Hardy-Weinberg equilibrium:

$$(0.75)^2 \text{ } TT + 2(0.75)(0.25) \text{ } Tt + (0.25)^2 \text{ } tt$$

This expression gives 0.5625 *TT* + 0.3750 *Tt* + 0.0625 *tt*. Thus with 2000 individuals in the population we would expect 1125 *TT*, 750 *Tt*, and 125 *tt*. These values are close to the values given in the question, suggesting that the population is indeed in genetic equilibrium.

To check this result, we should perform a chi-square analysis (see Chapter 5, pp. 151–153), using the given numbers (not frequencies of the three genotypes as the observed numbers and the calculated numbers

829
Analytical
Approaches
for Solving
Genetics
Problems

as the expected numbers. The chi-square analysis is as follows, where $d = $ (observed $-$ expected):

Genotype	Observed	Expected	d	d^2	d^2/e
TT	1100	1125	-25	625	0.556
Tt	800	750	$+50$	2500	3.334
tt	100	125	-25	625	5.000
Totals	2000	2000	0		8.890

Thus the chi-square value (i.e., the sum of all the d^2/e values) is 8.89. For the reasons discussed in the text for a similar example, there is only one degree of freedom. Looking up the chi-square value in the chi-square table (Table 5.2, p. 153), we find a P value of approximately 0.0025. So about 25 times out of 10,000 we would expect chance deviations of the magnitude observed. In other words, our hypothesis that the population is in Hardy-Weinberg equilibrium is not substantiated. In this case our guess that it was in equilibrium was inaccurate. Nonetheless, the population is not greatly removed from an equilibrium state.

Q.2 About one normal allele in 30,000 mutates to the X-linked recessive allele for hemophilia in each human generation. Assume for the purposes of this problem that one h allele in 300,000 mutates to the normal alternative in each generation. (Note that in reality it is difficult to measure the reverse mutation of a human recessive allele that is essentially lethal, like the allele for hemophilia.) The mutation frequencies are indicated in the following equation:

$$h^+ \underset{v}{\overset{u}{\rightleftarrows}} h$$

where $u = 10v$. What gene frequencies would prevail at equilibrium under mutation pressures alone in these circumstances?

A.2 This question seeks to test our understanding of the effects of mutation on gene frequencies. In the chapter we discussed the consequences of mutation pressure. The conclusion was that if A mutates to a at n times the frequency that a mutates back to A, then at equilibrium the value of q will be $\hat{q} = u/(u + v)$ or $\hat{q} = nv/(n + 1)v$. Applying this general derivation to this particular problem, we simply use the values given. We are told that the forward mutation rate is 10 times the reverse mutation rate, or $u = 10v$. At equilibrium the value of q will be $\hat{q} = u/(u + v)$. Since $u = 10v$, this equation becomes $\hat{q} = 10v/11v$, so $\hat{q} = 10/11$, or 0.909. Therefore at equilibrium brought about by mutation pressures the frequency of h (the hemophilia allele) will be 0.909, and the frequency of h^+ (the normal allele) will be $\hat{p}$, that is, $(1 - q) = (1 - 0.909) = 0.091$.

Questions and Problems

*24.1 In the European land snail, *Cepaea nemoralis,* multiple alleles at a single locus determine shell color. The allele for brown (C^B) is dominant to the allele for pink (C^P) and to the allele for yellow (C^Y). Pink is recessive to brown, but is dominant to yellow, and yellow is recessive to pink and brown. Thus, the dominance hierarchy among these alleles is $C^B > C^P > C^Y$. In one population of *Cepaea,* the following color phenotypes were recorded:

Brown	236
Pink	231
Yellow	33
Total	500

Assuming that this population is in Hardy-Weinberg equilibrium (large, randomly mating, and free from outside evolutionary forces), calculate the frequencies of the C^B, C^Y, and C^P alleles.

24.2 Three alleles are found at a locus coding for malate dehydrogenase (MDH) in the spotted chorus frog. Chorus frogs were collected from a breeding pond and each frog's genotype at the MDH locus was determined with electrophoresis. The following numbers of genotypes were found:

M^1M^1	8
M^1M^2	35
M^2M^2	20
M^1M^3	53
M^2M^3	76
M^3M^3	62
Total	254

a. Calculate the frequencies of the M^1, M^2, and M^3 alleles in this population.
b. Using a chi-square test, determine whether the MDH genotypes in this population are in Hardy-Weinberg proportions.

24.3 In a large interbreeding population 81 percent of the individuals are homozygous for a recessive character. In the absence of mutation or selection, what percentage of the next generation would be homozygous recessives? Homozygous dominants? Heterozygotes?

*24.4 Let A and a represent dominant and recessive alleles whose respective frequencies are p and q in a given interbreeding population at equilibrium (with $p + q = 1$).

a. If 16 percent of the individuals in the population have recessive phenotypes, what percentage of the total number of recessive genes exist in the heterozygous condition?
b. If 1.0 percent of the individuals were homozygous recessive, what percentage of the recessive genes would occur in heterozygotes?

*24.5 A population has eight times as many heterozygotes as homozygous recessives. What is the frequency of the recessive gene?

24.6 In a large population of range cattle the following ratios are observed: 49 percent red (*RR*), 42 percent roan (*Rr*), and 9 percent white (*rr*).

    a. What percentage of the gametes that give rise to the next generation of cattle in this population will contain allele *R*?

    b. In another cattle population only 1 percent of the animals are white and 99 percent are either red or roan. What is the percentage of *r* alleles in this case?

24.7 In a population gene pool the alleles *A* and *a* have initial frequencies of *p* and *q*, respectively. Prove that the gene frequencies and zygotic frequencies do not change from generation to generation as long as there is no selection, mutation, or migration, the population is large, and the individuals mate at random.

*24.8 The *S–s* antigen system in humans is controlled by two codominant alleles, *S* and *s*. In a group of 3146 individuals the following genotypic frequencies were found: 188 *SS*, 717 *Ss*, and 2241 *ss*.

    a. Calculate the frequency of the *S* and *s* alleles.

    b. Determine whether the genotypic frequencies conform to the Hardy-Weinberg equilibrium by using the chi-square test.

24.9 Refer to Problem 24.8. A third allele is sometimes found at the *S* locus. This allele S^u is recessive to both the *S* and the *s* alleles and can only be detected in the homozygous state. If the frequencies of the alleles *S*, *s*, and S^u are *p*, *q*, and *r*, respectively, what would be the expected frequencies of the phenotypes *S-*, *Ss*, *s-*, and $S^u S^u$?

24.10 In a large interbreeding human population 60 percent of individuals belong to blood group O (genotype *ii*). Assuming negligible mutation and no selective advantage of one blood type over another, what percentage of the grandchildren of the present population will be type O?

*24.11 A selectively neutral, recessive character appears in 0.40 of the males and in 0.16 of the females in a randomly interbreeding population. What is the gene's frequency? How many females are heterozygous for it? How many males are heterozygous for it?

24.12 Suppose you found two distinguishable types of individuals in wild populations of some organism in the following frequencies:

	Type 1	Type 2
Females	99%	1%
Males	90%	10%

The difference is known to be inherited. What is its genetic basis?

*24.13 Red-green color blindness is due to a sex-linked recessive gene. About 64 women out of 10,000 are color-blind. What proportion of men would be expected to show the trait if mating is random?

24.14 About 8 percent of the men in a population are red-green color-blind (owing to a sex-linked recessive gene). Answer the following questions, assuming random mating in the population, with respect to color blindness.

    a. What percentage of women would be expected to be color-blind?
    b. What percentage of women would be expected to be heterozygous?
    c. What percentage of men would be expected to have normal vision two generations later?

24.15 List some of the basic differences in the classical, balance, and neutral-mutation models of genetic variation.

*24.16 Two alleles of a locus, A and a, can be interconverted by mutation:

$$A \; \overset{u}{\underset{v}{\rightleftharpoons}} \; a$$

and u is a mutation rate of 6.0×10^{-7}, and v is a mutation rate of 6.0×10^{-8}. What will be the frequencies of A and a at mutational equilibrium, assuming no selective difference, no migration, and no random fluctuation caused by genetic drift?

*24.17 a. Calculate the effective population size (N_e) for a breeding population of 50 adult males and 50 adult females.
    b. Calculate the effective population size (N_e) for a breeding population of 60 adult males and 40 adult females.
    c. Calculate the effective population size (N_e) for a breeding population of 10 adult males and 90 adult females.
    d. Calculate the effective population size (N_e) for a breeding population of 2 adult males and 98 adult females.

24.18 In a population of 40 adult males and 40 adult females, the frequency of allele A is 0.6 and the frequency of allele a is 0.4.

    a. Calculate the 95 percent confidence limits of the gene frequency for A.
    b. Another population with the same gene frequencies consists of only 4 adult males and 4 adult females. Calculate the 95 percent confidence limits of the gene frequency for A in this population.
    c. What are the 95 percent confidence limits of A if the population consists of 76 females and 4 males?

24.19 The land snail *Cepaea nemoralis* is native to Europe but has been accidentally introduced into North America at several localities. These introductions occurred when a few snails were inadvertently transported on plants, building supplies, soil, or other cargo. The snails subsequently multiplied and established large, viable populations in North America.

Assume that today the average size of *Cepaea* populations found in North America is equal to the average size of *Cepaea* populations in Europe. What predictions can you make about the amounts of genetic variation present in European and North American populations of *Cepaea*? Explain your reasoning.

*24.20 A population of 80 adult squirrels reside on campus, and the frequency of the Est^1 allele among these squirrels is 0.70. Another population of squirrels is found in a nearby woods, and there, the frequency of Est^1 allele is 0.5. During a severe winter, 20 of the squirrels from the woods population migrate to campus in search of food and join the campus population. What will be the gene frequency of Est^1 in the campus population after migration?

24.21 Upon sampling three populations and determining genotypes, you find the following three genotype distributions. What would each of these distributions imply with regard to selective advantages of population structure?

Population	*AA*	*Aa*	*aa*
1	0.04	0.32	0.64
2	0.12	0.87	0.01
3	0.45	0.10	0.45

24.22 The frequency of two adaptively neutral alleles in a large population is 70 percent *A*:30 percent *a*. The population is wiped out by an epidemic, leaving only four individuals, who produce many offspring. What is the probability that the population several years later will be 100 percent *AA*? (Assume no mutations.)

*24.23 A completely recessive gene, owing to changed environmental circumstances, becomes lethal in a certain population. It was previously neutral, and its frequency was 0.5.

a. What was the genotype distribution when the recessive genotype was not selected against?
b. What will be the gene frequency after one generation in the altered environment?
c. What will be the gene frequency after two generations?

24.24 Human individuals homozygous for a certain recessive autosomal gene die before reaching reproductive age. In spite of this removal of all affected individuals, there is no indication that homozygotes occur less frequently in succeeding generations. To what might you attribute the constant rate of appearance of recessives?

*24.25 A completely recessive gene (Q^1) has a frequency of 0.7 in a large population, and the Q^1Q^1 homozygote has a relative fitness of 0.6.

a. What will be the frequency of Q^1 after one generation of selection?
b. If there is no dominance at this locus (the fitness of the heterozygote is intermediate to the fitnesses of the homozygotes), what will the gene fre-

quency be after one generation of selection?

c. If Q^1 is dominant, what will the gene frequency be after one generation of selection?

24.26 As discussed earlier in this chapter, the gene for sickle-cell anemia exhibits overdominance. The *Hb-A/Hb-S* heterozygote has increased resistance to malaria and therefore has greater fitness than the *Hb-A/Hb-A* homozygote, who is susceptible to malaria, and the *Hb-S/Hb-S* homozygote, who has sickle-cell anemia. Suppose that the fitness values of the genotypes in Africa are as presented below:

$$Hb\text{-}A/Hb\text{-}A = 0.88$$
$$Hb\text{-}A/Hb\text{-}S = 1.00$$
$$Hb\text{-}S/Hb\text{-}S = 0.14$$

Give the expected equilibrium frequencies of the sickle-cell gene *(Hb-S)*.

*24.27 Achondroplasia, a type of dwarfism in humans, is caused by an autosomal dominant gene. The mutation rate for achondroplasia is about 5×10^{-5} and the fitness of achondroplastic dwarfs has been estimated to be about 0.2, compared with unaffected individuals. What is the equilibrium frequency of the achondroplasia gene based on this mutation rate and fitness value?

Glossary

acrocentric chromosome A chromosome whose centromere is somewhat off-center. (13)

additive genetic variance Genetic variance that arises from the additive effects of genes on the phenotype. (755)

adenine (A) A purine base found in RNA and DNA; in double-stranded DNA adenine pairs with the pyrimidine thymine. (263)

allele One of two or more alternative forms of a single gene locus. Different alleles of a gene each have a unique nucleotide sequence, and their activities are all concerned with the same biochemical and developmental process, although their individual phenotypes may differ. (43)

allelic frequencies (gene frequencies) The frequencies of alleles at a locus occurring among individuals in a population. (782)

allelomorph (allele) A term coined by William Bateson; literally means "alternative form"; later shortened by others to *allele.* (57)

allopolyploidy Polyploidy involving two or more genetically distinct sets of chromosomes. (556)

alternation of generations The two distinct reproductive phases of green plants in which stages alternate between haploid cells and diploid cells (gametophyte cells and sporophyte cells). (30)

Ames test A test developed by Bruce Ames in the early 1970s that investigates new or old environmental chemicals for carcinogenic effects. It uses the bacterium *Salmonella typhimurium* as a test organism for mutagenicity of compounds. (535)

amino acids The building blocks of polypeptides; 20 amino acids are normally found in polypeptides. (408)

aminoacyl-tRNA A tRNA molecule covalently bound to an amino acid. This complex brings the amino acid to the ribosome so that it can be used in polypeptide synthesis. (419)

aminoacyl-tRNA synthetase An enzyme that catalyzes the addition of a specific amino acid to a tRNA molecule. Since there are 20 amino acids, there are also 20 synthetases. (419)

analysis of variance (ANOVA) A series of statistical procedures for examining differences in means and for partitioning variance. (747)

anaphase The stage in mitosis or meiosis during which the sister chromatids (mitosis) or homologous chromosomes (meiosis) separate and migrate toward the opposite poles of the cell. (19)

Numbers in parentheses indicate page on which term is defined.

aneuploidy The abnormal condition in which one or more whole chromosomes of a normal set of chromosomes either are missing or are present in more than the usual number of copies. (551)

antibody A protein molecule that recognizes and binds to a foreign substance introduced into the organism. (78)

anticodon A three-nucleotide sequence that pairs with a codon in mRNA by complementary base pairing. (389)

antigen Any large molecule which stimulates the production of specific antibodies or which binds specifically to an antibody. (78)

antiparallel A term that refers to the opposite orientations, in the DNA double helix, of the two strands in which the 5' end of one strand aligns with the 3' end of the other strand. (269)

artificial selection Human determination as to which individuals will survive and reproduce. If the selected traits have a genetic basis, they will change and evolve. (761)

asexual reproduction Reproduction in which a new individual develops from either a single cell or from a group of cells in the absence of any sexual process. (13)

attenuation A regulatory mechanism in certain bacterial biosynthetic operons that controls gene expression by causing RNA polymerase to terminate transcription. (623)

autonomous development The development of a cell solely according to the genetic information it contains, with no influence by the environment. (448)

autonomously replicating sequences (ARS elements) Specific sequences (e.g., in baker's yeast, *Saccharomyces cerevisiae*) which, when included as part of an extrachromosomal, circular DNA molecule, confer upon that molecule the ability to replicate autonomously. (344)

autopolyploidy Polyploidy involving more than two chromosome sets of the same species. (556)

autosomal dominant inheritance A mechanism of inheritance of a trait that is due to a dominant mutant gene carried in an autosome. (84)

autosomal dominant trait A trait due to a dominant mutant gene carried on an autosome. (84)

autosomal recessive inheritance A mechanism of inheritance of a trait that is due to a homozygous recessive mutant gene carried in an autosome. (86)

autosomal recessive trait A trait due to a homozygous recessive mutant gene carried on an autosome. (86)

autosome A chromosome other than a sex chromosome. (12)

auxotroph A mutant strain of a given organism that is unable to synthesize a molecule required for growth and therefore must have the molecule supplied in the growth medium in order for it to grow. (225)

auxotrophic mutation (nutritional, biochemical) A mutation that affects an organism's ability to make a particular molecule essential for growth. (535)

bacteria Spherical, rod-shaped, or spiral-shaped, single-cell prokaryotic organisms that vary in size. (4)

balance model A hypothesis of genetic variation proposing that balancing selection maintains large amounts of genetic variation within populations. (796)

balanced polymorphism The situation in which a genetic polymorphism is maintained by a selective advantage of the heterozygote over either homozygote type and in which the genotypes are maintained in stable proportions.

Barr body A highly condensed mass of chromatin found in the nuclei of normal females, but not in the nuclei of normal male cells. It represents a cytologically condensed and inactivated X chromosome. (553)

base analog A chemical whose molecular structure is extremely similar to the bases normally found in DNA. (530)

base-pair substitution mutation A change in a gene such that one base pair is replaced by another base pair, e.g., an AT is replaced by a GC pair. (520)

bidirectional replication The DNA synthesis that takes place in both directions away from the origin of the replication point. (330)

binary fission The process in cell reproduction in which a parental cell divides into two daughter cells of approximately equal size. (4)

binomial distribution The theoretical frequency distribution of events that have two possible outcomes. (739)

biochemical mutation *See* auxotrophic mutation. (540)

bivalent A pair of homologous, synapsed chromosomes during the first meiotic division. (21)

bottleneck effect A form of genetic drift that occurs when a population is drastically reduced in size. Some genes may be lost from the gene pool as a result of chance. (808)

broad-sense heritability A quantity representing the proportion of the phenotypic variance that consists of genetic variance. (755)

C_0t plot A graphical representation of the kinetics of renaturation of a sample of single-stranded DNA molecules. The graph is a plot of the fraction of remaining single-stranded DNA molecules versus the product of DNA concentration at time zero (C_0) and time (t)—hence the term C_0t. (307)

C value The amount of DNA found in the haploid set of chromosomes. (294)

callus culture A mass of undifferentiated cells in plant tissue, somewhat analogous to a tumor cell mass in an animal system. (513)

cancer Diseases characterized by the uncontrolled and abnormal division of eukaryotic cells and by the spread of the disease (metastasis) to disparate sites in the organism. (685)

capping The addition of a methylated guanine nucleotide (a "cap") to the 5′ end of a premessenger RNA molecule; the cap is retained on the mature mRNA molecule. (375)

carpel (also called pistil) The female reproductive organ in a flowering plant. A carpel is a leaflike floral structure which encloses the ovule or ovules and which is typically divided into ovary, style, and stigma. A flower may have one or more carpels, either single or fused. (28)

catabolite repression (glucose effect) The inactivation of an inducible bacterial operon in the presence of glucose even though the operon's inducer is present. (618)

cDNA DNA copies made from an RNA template catalyzed by the enzyme reverse transcriptase. (494)

cDNA cloning A process of constructing fragments to be cloned whereby the mRNA molecules are isolated from cells, and DNA copies of these RNAs are made and inserted into a cloning vector. The end result is a cDNA library. (*See* cDNA; cDNA library.) (496)

cDNA library The collection of molecular clones that contains cDNA copies of the entire mRNA population of a cell. (*See* cDNA.) (494)

cell The smallest building unit of a multicellular organism and, as a unit, in itself an elementary organism. (2)

cell cycle The cyclical process of growth and cellular reproduction in unicellular and multicellular eukaryotes. The cycle includes nuclear division, or mitosis, and cell division, or cytokinesis. (14)

cell division A process whereby one cell divides to produce two cells. (14)

cell-free, protein-synthesizing system A system, isolated from cells, which contains ribosomes, mRNA, tRNAs with amino acids attached, and all the necessary protein factors for the in vitro synthesis of polypeptides. (413)

cell theory The basic principles defining cells and their functions. It has three basic tenets: (1) The cell is the smallest building unit of a multicellular organism; (2) each cell in a multicellular organism has a specific role; (3) a cell can be produced from another cell only by cell division. (2)

cellular oncogenes The genes, present in a functional state in cancerous cells, which are responsible for the cancerous state. (689)

centi-Morgan (cM) The map unit. It is sometimes called a centi-Morgan in honor of T. H. Morgan. (154)

centromere (kinetochore) A specialized region of a chromosome seen as a constriction under the microscope. This region is important in the activities of the chromosomes during cellular division. (12)

chain-terminating codon One of three codons for which no normal tRNA molecule exists with an appropriate anticodon. A nonsense codon in an mRNA specifies the termination of polypeptide synthesis. (417)

character An observable phenotypic feature of the developing or fully developed organism that is the result of gene action. (36)

chiasma A cross-shaped structure formed during crossing-over and visible during the diplonema stage of meiosis. (21)

chiasma interference (chromosomal interference) The physical interference caused by the breaking and rejoining chromatids that reduces the probability of more than one crossing-over event's occurring near another one in one part of the meiotic tetrad. (165)

chi-square test A statistical procedure that determines what constitutes a significant difference between observed results and results expected on the basis of a particular hypothesis; a goodness-of-fit test. (140)

chloroplast The cellular organelle found only in green plants that is the site of photosynthesis in the cells containing it. (710)

chromatid One of the two visibly distinct longitudinal subunits of all replicated chromosomes that becomes visible between early prophase and metaphase of mitosis. (15)

chromatin The piece of DNA-protein complex that is studied and analyzed. Each chromatin fragment reflects the general features of chromosomes but not the specifics of any individual chromosome. (297)

chromosomal interference *See* chiasma interference. (165)

chromosomal mutation A change in the structure or number of chromosomes. (519)

chromosome The genetic material of the cell, complexed with protein and organized into a number of linear structures. It literally means "colored body," because the threadlike structures are visible under the microscope only after they are stained with dyes. (8)

chromosome aberration The variation from the wild-type condition in either chromosome number or chromosome structure. (550)

chromosome theory of heredity The theory that the chromosomes are the carriers of the genes. The first clear formulation of the theory was made by both Sutton and Boveri, who independently recognized that the transmission of chromosomes from one generation to the next closely paralleled the pattern of transmission of genes from one generation to the next. (64)

cis-trans (complementation) test A test, developed by E. Lewis, used to determine whether two different mutations are within the same cistron (gene). (443)

classical model A hypothesis of genetic variation proposing that natural populations contain little genetic variation as a result of strong selection for one allele. (796)

cloning The generation of many copies of a DNA molecule (e.g., a recombinant DNA molecule) by replication in a suitable host. (482)

cloning vehicle (cloning vector) A double-stranded DNA molecule that is able to replicate autonomously in a host cell and with which a DNA fragment (or fragments) can be bonded to form a recombinant DNA molecule for cloning. (482)

clustered repeated sequences The DNA sequences that are repeated one after another in a row. (310)

coding sequence The part of an mRNA molecule that specifies the amino acid sequence of a polypeptide during translation. (372)

codominance The situation in which the heterozygote exhibits the phenotypes of both homozygotes. (108)

codon A group of three adjacent nucleotides in an mRNA molecule that specifies either one amino acid in a polypeptide chain or the termination of polypeptide synthesis. (372)

coefficient of coincidence A number that expresses the extent of chiasma interference throughout a genetic map; the ratio of observed double-crossover frequency to expected double-crossover frequency. Interference is equal to 1 minus the coefficient of coincidence. (165)

complementary-base pairing The hydrogen bonding between a particular purine and a particular pyrimidine in double-stranded nucleic acid molecules (DNA–DNA, DNA–RNA, or RNA–RNA). The major specific pairings are guanine with cytosine and adenine with thymine or uracil. (268)

complementation test *See* cis-trans test. (443)

complementary DNA *See* cDNA. (494)

complete dominance The case in which one allele is dominant to the other so that at the phenotypic level the heterozygote is essentially indistinguishable from the homozygous dominant. (106)

complete recessiveness The situation in which an allele is phenotypically expressed only when it is homozygous. (106)

concerted evolution (molecular drive) A poorly understood evolutionary process that produces uniformity of sequence in multiple copies of a gene. (827)

concordant A condition in twins indicating that both possess or do not possess a trait. (130)

conditional mutant A mutant organism that is normal under one set of conditions but becomes seriously impaired or dies under other conditions. (329)

conjugation A process having a unidirectional transfer of genetic information through direct cellular contact between a donor ("male") and a recipient ("female") bacterial cell. (222)

consensus sequence The most commonly encountered nucleotide sequence believed to provide the same function in different genes, such as promoter sequences for the initiation of RNA transcription or splicing sites for the excision of introns from pre-messenger RNA. (363)

conservative model A DNA replication scheme in which the two parental strands of DNA remain together and serve as a template for the synthesis of a new daughter double helix. (315)

constitutive gene A gene whose products are essential to the normal functioning of the cell, no matter what the life-supporting environmental conditions are. (604)

continuous trait *See* quantitative (continuous) traits. (735)

contributing allele An allele that contributes to the phenotype. (751)

controlling site A specific sequence of nucleotide pairs adjacent to the gene where the transcription of a gene occurs in response to a particular molecular event. (605)

correlation coefficient A statistic that measures the strength of the association between two variables. (743)

cosmids Cloning vectors derived from plasmids that are capable of cloning large fragments of DNA. In addition to the features of plasmid cloning vectors (i.e., origin of replication and selectable marker(s) for growth in bacteria), cosmids contain phage lambda *cos* sites which permit recombinant DNA molecules that are constructed to be packaged into the lambda phage head in vitro. (491)

cotransduction The simultaneous transduction of two or more bacterial genes; a good indication that the bacterial genes are closely linked. (241)

coupling An arrangement in which the two wild-type alleles are on one homologous chromosome and the two recessive mutant alleles are on the other. (154)

crisscross inheritance A type of gene transmission passed from a male parent to a female child to a male grandchild. (68)

cross *See* cross-fertilization. (40)

cross-fertilization (cross) A term used for the fusion of male and female gametes from different individuals; the bringing together of genetic material from different individuals for the purpose of genetic recombination. (40)

crossing-over A term introduced by Morgan and E. Cattell, in 1912, to describe the process of chromosomal interchange by which recombinants arise. (21)

cytokinesis A term that refers to the division of the cytoplasm. The two new nuclei compartmentalize into separate daughter cells, and the mitotic cell division process is completed. (20)

cytological marker A cytologically distinguishable feature of chromosomes. (144)

cytoplasm The semisolid matrix of cell material in the surrounding cell body. The term was coined by E. Strasburger. (9)

cytosine (C) A pyrimidine base found in RNA and DNA. In double-stranded DNA, cytosine pairs with the purine guanine. (263)

dark repair   *See* excision repair. (538)

daughter chromosome   The chromatid after metaphase of mitosis or metaphase II of meiosis. (15)

degeneracy   A multiple coding; more than one codon per amino acid. (416)

degradation control   Regulation of the RNA breakdown rate in the cytoplasm. (651)

deletion (deficiency)   A chromosome mutation resulting in the loss of a segment of the genetic material and the genetic information contained therein from a chromosome. (207)

deoxyribonuclease (DNase)   An enzyme that catalyzes the degradation of DNA to nucleotides. (258)

deoxyribonucleic acid (DNA)   A polymeric molecule consisting of deoxyribonucleotide building blocks that in a double-stranded, double-helical form is the genetic material of most organisms. (2)

deoxyribonucleotide   The basic building block of DNA, consisting of a sugar (deoxyribose), a base, and a phosphate. (263)

deoxyribose   The pentose (five-carbon) sugar found in DNA. (263)

development   The process of regulated growth that results from the interaction of the genome with cytoplasm and the environment. It involves a programmed sequence of phenotypic events that are typically irreversible. (666)

diakinesis   The stage that follows diplonema and during which the four chromatids of each tetrad are most condensed and the chiasmata often terminalize. (24)

dicentric bridge   *See* dicentric chromosome

dicentric chromosome   A chromosome with two centromeres. For example, as a result of the crossover between genes *B* and *C* in the inversion loop, one recombinant chromatid becomes stretched across the cell as the two centromeres begin to migrate, forming a dicentric bridge. (563)

differentiation   An aspect of development that involves the formation of different types of cells, tissues, and organs from a zygote through the processes of specific regulation of gene expression. (666)

dihybrid cross   A cross between two dihybrids of the same type. Individuals that are heterozygous for two pairs of alleles at two different loci are called dihybrid. (50)

dioecious   A term referring to plant species that have both male and female sex organs on different individuals. (81)

diploid   A eukaryotic cell with two sets of chromosomes. (12)

diplonema   The second stage of prophase I in which the chromosomes begin to repel one another and tend to move apart. (21)

discontinuous DNA replication   A DNA replication involving the synthesis of short DNA segments, which are subsequently linked to form a long polynucleotide chain. (330)

discontinuous trait   A heritable trait in which the mutant phenotype is sharply distinct from the alternative, wild-type phenotype. (734)

discordant   A condition in twins, indicating that one possesses a trait and the other twin does not. (130)

disjunction   The process in anaphase during which sister chromatid pairs undergo separation. (19)

dizygotic twins   *See* fraternal twins. (129)

DNA   *See* deoxyribonucleic acid.

DNA gyrase   The only topoisomerase that catalyzes the formation of negatively supercoiled DNA. It is found only in bacteria. (280)

DNA ligase (polynucleotide ligase)   An enzyme that catalyzes the formation of a covalent bond between free single-stranded ends of DNA molecules during DNA replication and DNA repair. (330)

DNA polymerase   An enzyme that catalyzes the synthesis of DNA. (322)

DNA polymerase I   An *E. coli* enzyme that catalyzes DNA synthesis, originally called the Kornberg enzyme. (323)

docking protein   An integral protein membrane of the endoplasmic reticulum (ER) to which the nascent polypeptide-signal recognition particle (SRP)-ribosome complex binds to facilitate the binding of the polypeptide's signal sequence and associated ribosome to the ER. (429)

dominance variance   Genetic variance that arises from the dominance effects of genes. (755)

dominant   An allele or phenotype that is expressed in either the homozygous or the heterozygous state. (43)

double crossover   Two crossovers occurring in a particular region of a chromosome in a meiosis. (158)

double fertilization   An event found only in the life cycle of flowering plants. It is the fusing of the sperm cell with the two nuclei of the gametophyte's central cell to form the cell that will become the endosperm of the seed. (30)

Down syndrome (trisomy-21)   A human clinical condition characterized by various abnormalities. It is caused by the presence of an extra copy of chromosome 21. (552)

duplicate genes   Two identical allele pairs showing

the same phenotypic action and occupying distinct locations in the genome.

duplication   A chromosome mutation that results in the doubling of a segment of a chromosome. (559)

dyad   The two sister chromatids, connected at the centromere, which make up one chromosome in meiosis I. (24)

effective population size   The effective number of adults contributing gametes to the next generation. (806)

effector   A small molecule involved in the control of expression of many regulated genes. (605)

elongation factor   The protein involved in bringing aminoacyl-tRNA molecules to the ribosome so that the amino acid can be added to the growing poly-peptide chain. (423)

endosperm   A triploid (3N) tissue in flowering plants that surrounds and nourishes the embryo. It devel-ops from the union of a sperm nucleus and the po-lar nuclei of the egg. (28)

enhancer sequence (or enhancer element)   In eukar-yotes, a type of DNA sequence element having a strong, positive effect on transcription by RNA polymerase II. (369)

environmental variance   Any nongenetic source of phenotypic variation among individuals. (754)

enzymes   Specialized proteins that assist and accel-erate (catalyze) biochemical reactions without being altered by the reactions. (324)

episome   An autonomously replicating plasmid (a circular, double-stranded DNA molecule) that is capable of integrating into the host cell's chromo-some. (580)

epistasis   A form of gene interaction in which one gene interferes with the phenotypic expression of another nonallelic gene so that the phenotype is governed by the former gene and not by the latter gene when both genes are present in the genotype. (113)

erythroblasts   Red blood cell precursors. (654)

eukaryote   A term that literally means "true nu-cleus." Eukaryotes are organisms that have cells in which the genetic material is located in a mem-brane-bound nucleus. Eukaryotes can be unicellular or multicellular. (4)

evolution   Genetic change within a group of organ-isms that takes place over time. (759)

evolutionary genetics   *See* population genetics. (781)

excision repair (dark repair)   An enzyme-catalyzed, light-independent process of repair of ultraviolet light–induced thymine dimers in DNA that involves removal of the dimers and synthesis of a new piece of DNA complementary to the undamaged strand. (538)

ex-conjugant   *See* trans-conjugant. A pair of bacterial cells, one male and one female, which has sepa-rated during or following conjugation. The female ex-conjugant contains a fragment of male cell DNA. (231)

exon   *See* coding sequence. (375)

expressivity   The degree to which a particular geno-type is expressed in the phenotype. (85)

F duction (sexduction)   The incorporation of bacte-rial genes into an F factor to produce an F' factor and the subsequent transfer of the F' factor to a re-cipient female, F^-, cell. (229)

F_1 generation   The first filial generation produced by crossing two parental strains. (41)

F_2 generation   The second filial generation produced by selfing the F_1. (41)

F-pili (or sex pili)   Hairlike cell surface components produced by cells containing the F factor, which al-low the physical union of F^+ and F^- cells, or Hfr and F^- cells, to take place. (226)

familial trait   A trait shared by members of a family. (758)

fine-structure mapping   A high-resolution mapping of allelic sites within a gene. (437)

first filial generation   *See* F_1 generation; the offspring that result from the first experimental crossing of animals or plants. (41)

first law   *See* Mendel's first law. (44)

fitness   The relative reproductive ability of a geno-type. (815)

formylmethionine (fMet)   A specially modified amino acid involving the addition of a formyl group to the amino group of methionine. It is the first amino acid incorporated into a polypeptide chain in prokaryotes and in eukaryotic cellular organelles. (419)

forward mutation   A mutational change from a wild-type allele to a mutant allele. (521)

founder effect   A phenomenon that occurs when the isolate effect is exhibited by a small breeding unit that has formed by migration of a small number of individuals from a large population. (807)

frameshift mutation   A mutational addition or dele-tion of a base pair in a gene that disrupts the nor-mal reading frame of an mRNA, which is read in groups of three bases. (521)

fraternal (dizygotic) twins   A result of the fertilization of two eggs released by the ovaries at the same time. The two eggs are likely to be very different genetically since most humans are heterozygous for many of their genes. (129)

frequency distribution A means of summarizing the phenotypes of a continuous trait whereby the population is described in terms of the proportion of individuals that have each phenotype. (738)

gametogenesis The formation of male and female gametes by meiosis. (20)

gametophyte The haploid sexual generation in the life cycle of plants that produces the gametes. (27)

gene (Mendelian factor) The determinant of a characteristic of an organism. Genetic information is coded in the DNA, which is responsible for species and individual variation. A gene's nucleotide sequence specifies a polypeptide or RNA and is subject to mutational alteration. (36)

gene flow The movement of genes that takes place when organisms migrate and then reproduce, contributing their genes to the gene pool of the recipient population. (812)

gene frequency (allele frequency) The proportion of one particular type of allele to the total of all alleles at this genetic locus in a Mendelian breeding population. (783)

gene locus *See* locus.

gene mutation A heritable alteration of the genetic material, usually from one allelic form to another. (519)

gene pool The total genetic information encoded in the total genes in a breeding population existing at a given time. (781)

gene redundancy A situation in which tRNA genes occur two or more times in the *E. coli* chromosome. (391)

gene regulatory elements *See* transcription-controlling sequence. (355)

gene segregation *See* Mendel's first law, the principle of segregation. (44)

generalized transduction A type of transduction in which any gene may be transferred between bacteria. (238)

genetic correlation An association between the genes that determine two traits. (764)

genetic counseling The procedures whereby the risks of prospective parents having a child who expresses a genetic disease are evaluated and explained to them. The genetic counselor typically makes predictions about the probabilities of partic-

ular traits (deleterious or not) occurring among children of a couple. (83)

genetic drift Any change in gene frequency in a population. (788)

genetic load The proportion, for any genetic locus of a genotype or gene pool, by which the fitness of the optimum genotype at the locus is decreased as a result of the presence of deleterious genes.

genetic map (linkage map) A representation of the genetic distance separating nonallelic gene loci in a linkage structure. (151)

genetic mapping The using of genetic crosses to locate genes on chromosomes relative to one another. (151)

genetic marker Any genetically controlled phenotypic difference used in genetic analysis, particularly in the detection of genetic recombination events. (144)

genetic polymorphism The regular occurrence in the same population of two or more genotypes that are not the result of recurrent mutation.

genetic recombination A process by which parents with different genetic characters give rise to progeny so that genes in which the parents differed are associated in new combinations. For example, from *A B* and *a b* the recombinants *A b* and *a B* are produced. (21)

genetic variance (represented by V_G) Genetic sources of phenotypic variation among individuals of a population. Includes dominance genetic variance, additive genetic variance, and epistatic genetic variance. (754)

genetics A term coined by W. Bateson in 1905 for the science of heredity and variation. (1)

genome The total amount of genetic material in a cell; in eukaryotes the haploid set of chromosomes of an organism. (12)

genome mutation A mutation characterized by a change in the number of chromosomes. (519)

genomic library The collection of molecular clones that contains at least one copy of every DNA sequence in the genome. (493)

genotype The genetic constitution of an organism. (36)

genotypic frequencies The frequencies or percentages of different genotypes found within a population. (782)

glucose effect *See* catabolite repression. (618)

Goldberg-Hogness box (TATA box, or TATA element) Found approximately at position -30 from the transcription initiation site. The Goldberg-Hogness sequence is considered to be the likely eukaryotic promoter sequence. The consensus sequence for the Goldberg-Hogness box is TATAAAAA. (368)

guanine (G) Purine base found in RNA and DNA. In double-stranded DNA guanine pairs with the pyrimidine cytosine. (263)

haploid A cell or an individual with one copy of each nuclear chromosome. (12)

haploidization The formation of a haploid nucleus from a diploid heterokaryon. (198)

Hardy-Weinberg equilibrium (Hardy-Weinberg law, Hardy-Weinberg law of genetic equilibrium) An extension of Mendel's laws of inheritance that describes the expected relationship between gene frequencies in natural populations and the frequencies of individuals of various genotypes in the same populations. (782)

helicase An enzyme that catalyzes the unwinding of the DNA double helix during replication in *E. coli;* product of the *rep* gene. (331)

hemizygous The condition of X-linked genes in males. Males that have an X chromosome with an allele for a particular gene but do not have another allele of that gene in the gene complement are hemizygous. (68)

hereditary trait A characteristic under control of the genes that is transmitted from one generation to another. (36)

heritability The proportion of phenotypic variation in a population attributable to genetic factors. (753)

hermaphroditic The plant species that have both male and female sex organs on the same flower. (83)

heterogametic sex The sex that has sex chromosomes of different types (e.g., XY) and therefore produces two kinds of gametes with respect to the sex chromosomes. (66)

heterogeneous nuclear RNA (hnRNA) The RNA molecules of various sizes that exist in a large population in the nucleus. Some of the RNA molecules are precursors to mature mRNAs. (378)

heterokaryon A cell or collection of cells (as in a mycelium) possessing genetically different nuclei (regardless of their number) in a common cytoplasm. (198)

heterosis The phenomenon in which the heterozygous genotypes with respect to one or more characters are superior in comparison with the corresponding homozygous genotypes in terms of growth, survival, phenotypic expression, and fertility. (765)

heterozygosity The proportion of individuals heterozygous at a locus. (797)

heterozygote advantage *See* overdominance. (822)

heterozygous A term describing a diploid organism having different alleles of one or more genes and therefore producing gametes of different genotypes. (44)

***Hfr* (high-frequency recombination) cell** A male cell in *E. coli,* with the *F* factor integrated into the bacterial chromosome. When the *F* factor promotes conjugation with a female (*F⁻*) cell, bacterial genes are transferred to the female cell with high frequency. (227)

highly repetitive sequence A DNA sequence that is repeated between 10^5 and 10^7 times in the genome. (309)

histocompatibility The acceptance by one organism of another organism's tissue graft. (79)

histone One of a class of basic proteins which are complexed with DNA in chromosomes and which play a major role in determining the structure of eukaryotic nuclear chromosomes. (297)

holandric The type of inheritance controlled entirely by genes located on the Y chromosome. (89)

homeotic genes One of the two categories of selector genes. (*See* selector genes.) (679)

homogametic sex The gender in the species, most often the female, which produces only the X sex chromosome. (66)

homolog Each individual member of a pair of homologous chromosomes. (12)

homologous chromosomes The members of a chromosome pair that are identical in the arrangement of genes they contain and in their visible structure. (12)

homozygous A term describing a diploid organism having the same alleles at one or more genes and therefore producing gametes of identical genotypes. (44)

homozygous dominant A diploid organism that has the same dominant allele for a given gene locus on both members of a homologous pair of chromosomes. (46)

homozygous recessive A diploid organism that has the same recessive allele for a given gene locus on both members of a homologous pair of chromosomes. (46)

host-controlled modification and restriction The variation of phages that takes place under direct influence of the host bacterial cell. (483)

housekeeping genes Genes that are always being transcribed because the proteins for which they code are needed for basic cellular functions. (380)

H-Y antigen The product of the Y-linked H-Y locus (the histocompatibility locus). (79)

hybrid vigor *See* heterosis. (770)

hypersensitive sites (hypersensitive regions) Sites in the regions of DNA around transcriptionally active genes that are highly sensitive to digestion by DNase I. (654)

identical (monozygotic) twins The result of the fertilization of a single egg. These two individuals are genetically identical because they come from the same fertilized egg. (129)

inbreeding Repeated self-fertilization in individuals. (765)

incomplete (partial) dominance The condition resulting when one allele is not completely dominant to another allele so that the heterozygote has a phenotype between that shown in individuals homozygous for either individual allele involved. An example of partial dominance is the frizzle chicken. (106)

independent assortment, principle of *See* Mendel's second law. (50)

induced mutation A mutation that results from treatment with mutagens. (519)

inducer A chemical or environmental agent for bacterial operons that brings about the transcription of an operon. (605)

induction The synthesis of a gene product (or products) in response to the action of an inducer, i.e., a chemical or environmental agent. (605)

initiation complex The formation of a complex between mRNA, ribosomal subunits, initiator tRNA, and initiation factors in preparation for the commencement of polypeptide synthesis. (422)

initiation factor The protein factor required for initiating ribosome-directed polypeptide synthesis. (422)

insertion sequence (IS) element The simplest transposable genetic element found in prokaryotes. It is a mobile segment of DNA that contains genes required for the process of insertion of the DNA segment into a chromosome and for the mobilization of the element to different locations. (575)

interaction variance Genetic variance that arises from epistatic interactions among genes. (755)

intergenic suppressor A mutation whose effect is to suppress the phenotypic consequences of another mutation in a gene distinct from the gene in which the suppressor mutation is located. (522)

internal control region Promoter sequence, recognized by RNA polymerase III, that is located within the gene sequence, e.g., in tRNA genes and 5SrRNA genes of eukaryotes. (392)

intervening sequence (ivs) *See* intron.

intron A nucleotide sequence in eukaryotes that must be excised from a structural gene transcript in order to convert the transcript into a mature messenger RNA molecule containing only coding sequences that can be translated into the amino acid sequence of a polypeptide. (375)

inversion A chromosome mutation that results when a segment of a chromosome is excised and then reintegrated in an orientation 180° from the original orientation. (560)

IS element *See* insertion sequence (IS) element. (575)

isogenic strain A group of individuals (e.g., produced by inbreeding) that have the same genotype. (78)

isolate effect The effect on gene frequencies of breeding in small, self-contained breeding units within a larger population.

isozyme Each form of an enzyme (literally, "same enzyme") when there are a number of different forms of the enzyme.

karyokinesis The division of the nucleus in eukaryotic cell division. (14)

karyotype A complete set of all the metaphase chromatid pairs in a cell (literally, "nucleus type"). (290)

kinetochore *See* centromere.

Klinefelter's syndrome A human clinical syndrome that results from disomy for the X chromosome in a male, which results in a 47,XXY male. Many of the affected males are mentally deficient, have underdeveloped testes, and are taller than average. (78)

leader sequence One of three main parts of the mRNA molecule. The leader sequence is located at the 5' end of the mRNA molecule and contains the coded information that the ribosome and special proteins read to tell it where to begin the synthesis of the polypeptide. (372)

leptonema The stage during meiosis in prophase I at which the chromosomes have begun to coil and are visible. (21)

lethal allele An allele that results in the death of an organism. (120)

light repair *See* photo-reactivation. (537)

LINES (long interspersed repeated sequences) The dispersed families of repeated sequences in mammals that are several thousand base pairs in length and occur on the order of 10^4 times or fewer in the genome. (311)

linkage A term describing genes located on the same chromosome. (140)

linkage map *See* genetic map. (151)

linkage number In linear DNA, the number of times the two strands of the double helix twist around each other. (280)

linked genes Genes that are located on the same chromosome. (140)

linker *See* restriction enzyme cleavage site linker. (497)

locus (*plural,* **loci**) The position of a gene on a genetic map; the specific place on a chromosome where a gene is located. (154)

lyonization A mechanism in mammals that allows them to compensate for X chromosomes in excess of the normal complement. The excess X chromosomes are cytologically condensed and inactivated, and they do not play a role in much of the development of the individual. (78)

lysogenic A term describing a bacterium that contains a temperate phage in the prophage state. The bacterium is said to be lysogenic for that phage. Upon induction phage reproduction is initiated, progeny phages are produced, and the bacterial cell lyses. (235)

lysogenic pathway A path, besides the lytic cycle, that a phage can follow. The chromosome does not replicate; instead, it inserts itself physically into a specific region of the host cell's chromosome in a way that is essentially the same as *F* factor integration. (235)

lysogeny The phenomenon of the insertion of a temperate phage chromosome into a bacterial chromosome, where it replicates when the bacterial chromosome replicates. In this state the phage genome is repressed and is said to be in the prophage state. (235)

lytic cycle A type of phage life cycle in which the phage takes over the bacterium and directs its growth and reproductive activities to express the phage's genes and to produce progeny phages. (234)

macromolecule A large molecule (such as DNA, RNA, and proteins) that has a molecular weight of at least 1000 daltons. (263)

map unit (mu) A unit of measurement used for the distance between two gene pairs on a genetic map. A crossover frequency of 1 percent between two genes equals 1 map unit. *See also* centi-Morgan. (154)

maternal inheritance A phenomenon in which the mother's phenotype is expressed exclusively. (715)

mating types A genic system in which two sexes are morphologically indistinguishable but carry different alleles and will mate. (81)

mean The average of a set of numbers, calculated by adding all the values represented and dividing by the number of values. (740)

megasporogenesis The formation in flowering plants of megaspores and the production of the embryo sac (the female gametophyte). (28)

meiosis Two successive nuclear divisions of a diploid nucleus that result in the formation of haploid gametes or of meiospores having one-half the genetic material of the original cell. (20)

meiosis I The first meiotic division that results in the reduction of the number of chromosomes. This division consists of four stages: prophase I, metaphase I, anaphase I, and telophase I. (21)

meiosis II The second meiotic division, resulting in the separation of the chromatids. (21)

Mendel, Gregor Johann (1822–1884) An Austrian priest whose breeding experiments with garden peas established the foundation of modern genetics. (37)

Mendelian factor *See* gene.

Mendelian population An interbreeding group of individuals sharing a common gene pool; the basic unit of study in population genetics. (781)

Mendel's first law, principle of segregation The law that two members of a gene pair (alleles) segregate (separate) from each other during the formation of gametes. As a result, one-half the gametes carry one allele and the other half carry the other allele. (44)

Mendel's second law, principle of independent assortment The law that the factors (genes) for different traits assort independently of one another. In other words, genes on different chromosomes behave independently in the production of gametes. (50)

messenger RNA (mRNA) One of the three classes of RNA molecules involved in protein synthesis; the RNA molecule that contains the coded information for the amino acid sequence of a protein. (355)

metacentric chromosome A chromosome that has the centromere approximately in the center of the chromosome. (12)

metaphase A stage in mitosis or meiosis in which

chromosomes become aligned along the equatorial plane of the spindle. (17)

metaphase plate The plane where the chromosomes become aligned during metaphase. (17)

microsporogenesis The formation in flowering plants of microspores in the anthers and the production of the male gametophyte (pollen), normally from diploid microsporocytes. (29)

migration Movement of organisms from one location to another. (811)

missense mutation A gene mutation in which a base-pair change in the DNA causes a change in an mRNA codon, with the result that a different amino acid is inserted into the polypeptide in place of the one specified by the wild-type codon. (520)

mitochondria The organelles found in the cytoplasm of all aerobic animal and plant cells; the principal sources of energy in the cell. (699)

mitosis The process of nuclear division in haploid or diploid cells producing daughter nuclei which contain identical chromosome complements and which are genetically identical to one another and to the parent nucleus from which they arose. (14)

mitotic crossing-over (mitotic recombination) A genetic recombination that occurs following the rare pairing of homologs during mitosis of a diploid cell. (195)

moderately repetitive sequence A DNA sequence that is reiterated from a few to as many as 10^3 to 10^5 times in the genome. (309)

molecular cloning *See* cloning. (482)

molecular genetics A subdivision of the science of genetics involving how genetic information is encoded within the DNA and how biochemical processes of the cell translate the genetic information into the phenotype. (781)

monoecious A term referring to plants in which male and female gametes are produced in the same individual. (83)

monohybrid cross A cross between two individuals that are both heterozygous for the same pair of alleles. (e.g., $Aa \times Aa$). By extension, the term also refers to crosses involving the pure-breeding parents that differ with respect to the alleles of one locus (e.g., $AA \times aa$). (41)

monoploidy An aberrant state in a normally diploid cell or organism in which only one complete set of chromosomes is present. (551)

monosomy An aberrant, aneuploid state in a normally diploid cell or organism in which one chromosome is missing, leaving one chromosome with no homolog. (551)

monozygotic twins *See* identical twins. (129)

multifactorial trait A trait influenced by multiple genes and environmental factors. (736)

multiple alleles Many alternative forms of a single gene. (101)

multiple crossovers More than one crossover occurring in a particular region of a chromosome in a meiosis. (158)

mutagen Any physical or chemical agent that significantly increases the frequency of mutational events above a spontaneous mutation rate. (519)

mutant allele Any nonwild-type allele of a gene. Mutant alleles may be dominant or recessive to wild-type alleles. (66)

mutation Any detectable and heritable change in the genetic material not caused by genetic recombination. (519)

mutational equilibrium A balance between a mutation occurring in one direction and a mutation occurring in the other direction in a population in Hardy-Weinberg equilibrium.

narrow-sense heritability The proportion of the phenotypic variance that results from additive genetic variance. (756)

natural selection Differential reproduction of genotypes. (760)

negative assortative mating A mating that occurs between dissimilar individuals more often than it does between randomly chosen individuals. (823)

neutral mutation A base-pair change in the gene that changes a codon in the mRNA such that there is no change in the function of the protein translated from that message. (521)

neutral-mutation hypothesis A hypothesis that replaced the classical model by acknowledging the presence of extensive genetic variation in proteins, but proposing that this variation is neutral with regard to natural selection. (797)

nick translation A process for labeling a double-stranded DNA molecule radioactively using DNase I and DNA polymerase to produce a labeled probe. (502)

nitrogenous base A nitrogen-containing base that, along with a pentose sugar and a phosphate, is one of the three parts of a nucleotide, the building block of RNA and DNA. (263)

non-autonomous development (c.f. **autonomous development**) When the development of a cell's phenotype is affected by the environment in which it develops. (448)

noncontributing alleles The alleles that do not have any effect on the phenotype of the quantitative trait. (751)

nonhistone A type of acidic or neutral protein found in chromatin. (297)

nonhomologous chromosomes The chromosomes containing dissimilar genes that do not pair during meiosis. (12)

non-Mendelian inheritance (cytoplasmic inheritance) The inheritance of characters determined by genes not located on the nuclear chromosomes but on mitochondrial or chloroplast chromosomes. Such genes show inheritance patterns distinctly different from those of nuclear genes. (715)

nonparental-ditype (NPD) One of three types of tetrads possible when two genes are segregating in a cross. The NPD tetrad contains four nuclei, all of which have recombinant (nonparental) genotypes, i.e., two of each possible type. (186)

nonsense codon *See* chain-terminating codon. (417)

nonsense mutation A gene mutation in which a base-pair change in the DNA causes a change in an mRNA codon from an amino acid–coding codon to a chain-terminating (nonsense) codon. As a result, polypeptide chain synthesis is terminated prematurely and is therefore either nonfunctional or, at best, partially functional. (521)

nonstructural gene A gene whose RNA transcripts are the final products of gene expression. They do not function as messenger molecules between DNA and protein. (355)

normal distribution A probability distribution in statistics, graphically displayed as a bell-shaped curve. (739)

northern blotting A similar technique to Southern blotting except that RNA rather than DNA is separated and transferred to a filter for hybridization with a probe. (507)

nuclear division (karyokinesis) The division of a cell nucleus in mitosis or meiosis. (14)

nuclease An enzyme that catalyzes the degradation of a nucleic acid by breaking phosphodiester bonds. Nucleases specific for DNA are termed deoxyribonucleases (DNases), and nucleases specific for RNA are termed ribonucleases (RNases). (258)

nucleofilament A fiber seen in chromatin. It is approximately 10 nm in diameter and consists of DNA wrapped around nucleosome cores. (298)

nucleoplasm The semisolid matrix of cell material in the nucleus. (63)

nucleosome The basic structural unit of eukaryotic nuclear chromosomes. It consists of about 140 base pairs of DNA wound almost two times around an octomer of histones. It is connected to adjacent nucleosomes by about 60 base pairs of DNA that is complexed with another histone. (298)

nucleotide A monomeric molecule of RNA and DNA that consists of three distinct parts: a pentose (ribose in RNA, deoxyribose in DNA), a nitrogenous base, and a phosphate group. (263)

nucleus A discrete structure within the cell that is bounded by a nuclear membrane. It contains most of the genetic material of the cell. (4)

nullisomy The aberrant, aneuploid state in a normally diploid cell or organism in which there is a loss of one pair of homologous chromosomes. (551)

nutritional mutation *See* auxotrophic mutation. (540)

Okazaki fragments The relatively short, single-stranded DNA fragments in discontinuous DNA replication which are synthesized during DNA replication and which are subsequently covalently joined to make a continuous strand. (330)

oligomers ("oligo" = few) Short DNA molecules. (271)

oncogenesis Tumor (cancer) initiation in an organism. (685)

one-gene–one-enzyme hypothesis The hypothesis, based on Beadle and Tatum's studies in biochemical genetics, that each gene controls the synthesis of one enzyme. (454)

oogenesis The development in the gonad of the female germ cell (egg cell) of animals. (27)

open reading frame In a segment of DNA, a potential protein-coding sequence identified by an initiator codon in frame with a chain-terminating codon. (509)

operon A cluster of genes whose expressions are regulated together by operator–regulator protein interactions, plus the operator region itself and the promoter. (607)

opposite polarity The polarity of two DNA strands in a double-helical molecule that are aligned in antiparallel fashion. One strand is oriented in the $5' \rightarrow 3'$ way, and the other is oriented $3' \rightarrow 5'$. (269)

ordered tetrads A structure resulting from meiosis in which the four meiotic products are in an order reflecting exactly the orientation of the four chromatids at the metaphase plate in meiosis I. (149)

origin A specific site on the chromosome at which the double helix denatures into single strands and continues to unwind as the replication fork(s) migrates. (226)

overdominance (heterozygote advantage) Condition in which the heterozygote has higher fitness than either of the homozygotes. (822)

ovum A mature egg cell. In the second meiotic division the secondary oocyte produces two haploid cells; the large cell rapidly matures into the ovum. (27)

P generation The parental generation, i.e., the immediate parents of an F_1. (41)

pachynema The stage meiosis during which the homologous pairs of chromosomes exchange chromosome regions. (21)

paracentric inversion An inversion in which the inverted segment occurs on one chromosome arm and does not include the centromere. (560)

parasexual system A system that achieves genetic recombination by means other than the regular alternation of meiosis and fertilization. (198)

parental-ditype (PD) One of three types of tetrads possible when two genes are segregating in a cross. The PD tetrad contains four nuclei, all of which are parental genotypes, with two of one parent and two of the other parent. (186)

parental genotypes (parental classes, parentals) Individuals among progeny of crosses that have combinations of genetic markers like one or other of the parents in P_1. (142)

partial dominance *See* incomplete dominance. (106)

partial linkage The linkage that occurs when homologous chromosomes exchange corresponding parts during meiosis through crossing-over; a linkage between two alleles that is not complete. (141)

particulate factors The term Mendel used to describe the factors that carried hereditary information and were transmitted from parents to progeny through the gametes. We now know these factors by the name *genes*. (43)

pedigree analysis A family tree investigation that involves the careful compilation of phenotypic records of the family over several generations. (83)

penetrance The frequency with which a dominant or homozygous recessive gene manifests itself in the phenotype of an individual. (123)

peptide bond A covalent bond in a polypeptide chain that joins the α-carboxyl group of one amino acid to the α-amino group of the adjacent amino acid. (408)

pericentric inversion An inversion in which the inverted segment includes the parts of both chromosome arms and therefore includes the centromere. (562)

phage lysate The progeny phages released following lysis of phage-infected bacteria. (234)

phage vector A phage that carries pieces of bacterial DNA between bacterial strains in the process of transduction. (238)

phenocopy An abnormal individual resulting from special environmental conditions. It mimics a similar phenotype caused by gene mutation. (128)

phenotype The physical manifestation of a genetic trait that results from a specific genotype and its interaction with the environment. (36)

phenotypic correlation An association between two traits. (763)

phenotypic variance (represented by V_p) A measure of a trait's variability. (754)

phosphate group A component, along with a pentose sugar and a nitrogenous base, of a nucleotide, the building block of RNA and DNA. Because phosphate groups are acidic in nature, DNA and RNA are called nucleic acids. (263)

phosphodiester bond A covalent bond in RNA and DNA between a sugar and a phosphate. Phosphodiester bonds form the repeating sugar-phosphate array of the backbone of DNA and RNA. (263)

photoreactivation (light repair) One way by which thymine dimers can be repaired. The dimers are reverted directly to the original form by exposure to visible light in the wavelength range 320–370 nm. (537)

plaque A round, clear area in a lawn of bacteria on solid medium that results from the lysis of cells by repeated cycles of phage lytic growth. (237)

plasmid An extrachromosomal genetic element consisting of double-stranded DNA that replicates autonomously from the host chromosome. (226)

pleiotropy The production, by a single-mutant allele, of multiple related and/or unrelated phenotypic effects. (107)

point mutation A mutation caused by a substitution of one base pair for another. (519)

polarity A term referring to a bacterial operon that codes for a polygenic mRNA. It is the phenomenon whereby certain nonsense mutations not only result in the loss of activity of the enzyme encoded by the gene in which they are located but also reduce significantly or abolish the synthesis of enzymes coded by structural genes on the operator-distal side of the mutation. The mutations are called polar mutations. (608)

poly(A) tail   A sequence of 50 to 200 adenine nucleotides that is added as a posttranscriptional modification at the 3′ ends of most eukaryotic mRNAs. (375)

polygene (multiple-gene) hypothesis for quantitative inheritance   The hypothesis that quantitative traits are controlled by many genes. (750)

polygenic mRNA (polycistronic mRNA)   A single mRNA transcript in prokaryotic operons of two or more adjacent structural genes that specifies the amino acid sequences of the corresponding polypeptides. (608)

polygenic traits   Traits encoded by many loci. (736)

polymorphic locus   Any locus that has more than one allele present within a population.

polynucleotide   A linear sequence of nucleotides in DNA or RNA. (264)

polypeptide   A polymeric, covalently bonded linear arrangement of amino acids joined by peptide bonds. (408)

polyploidy   The condition of a cell or organism that has more than its normal number of sets of chromosomes. (551)

polytene chromosome   A special type of chromosome representing a bundle of numerous chromatids that have arisen by repeated cycles of replication of single chromatids without nuclear division. This type of chromosome is characteristic of various tissues of Diptera. (558)

population   A group of interbreeding individuals that share a set of genes. (738)

population genetics   A branch of genetics that describes in mathematical terms the consequences of Mendelian inheritance on the population level. (781)

positive assortative mating   A mating that occurs more frequently between individuals who are phenotypically similar than it does among randomly chosen individuals. (723)

precursor RNA molecule (pre-RNA)   The initial transcript whose processing may involve the addition and/or removal of bases, the chemical modification of some bases, or the cleavage of sequences from the precursors. (371)

precursor mRNA (pre-mRNA)   The initial transcript of a gene that is modified and/or processed to produce the mature, functional mRNA molecule. In eukaryotes, for example, the transcript is modified at both the 5′ and the 3′ ends, and in a number of cases RNA sequences that do not code for amino acids are present and must be excised. (378)

precursor rRNA (pre-rRNA)   A primary transcript of adjacent rRNA genes (16S, 23S, and 5S rRNA genes in prokaryotes; 18S, 5.8S, and 28S rRNA genes in eukaryotes) plus flanking and spacer DNA that must be processed to release the mature rRNA molecules. (397, 401)

precursor tRNA (pre-tRNA)   A primary transcript of a tRNA gene whose bases must be extensively modified and that must be processed to remove extra RNA sequences in order to produce the mature tRNA molecule. In some cases the primary transcript may contain the sequences of two or more tRNA molecules. (388)

Pribnow box   A part of the promoter sequence in prokaryotic genomes that is located at about 10 base pairs upstream from the transcription starting point. The consensus sequence for the Pribnow box is TATAAT. The Pribnow box is often referred to as the TATA box. (363)

primary nondisjunction (nondisjunction)   A rare event in which sister chromatids (in mitosis) or chromosomes contained in pairing configurations (in meiosis) fail to be distributed to opposite poles. (73)

primase   The enzyme in DNA replication that catalyzes the synthesis of a short nucleic acid primer. (322)

primer   A preexisting polynucleotide chain in DNA replication to which new nucleotides can be added. (322)

principle of segregation   *See* Mendel's first law. (44)

principle of independent assortment   *See* Mendel's second law. (50)

primosome   A complex of *E. coli* primase and six or seven other polypeptides that together become functional in catalyzing the initiation of DNA synthesis. (322)

probability   The chance of occurrence of a particular event. (47)

processing control   The second level of control of gene expression in eukaryotes. This level involves regulating the production of mature RNA molecules from precursor-RNA molecules. (651)

product rule   The rule that the probability of two independent events occurring simultaneously is the product of each of their probabilities. (47)

prokaryote   A cellular organism whose genetic material is not located within a membrane-bound nucleus (*see* eukaryote). (4)

promoter site (promoter sequence, promoter)   A specific regulatory nucleotide sequence in the DNA to which RNA polymerase binds for the initiation of transcription. (360)

proofreading In DNA synthesis, the process of recognizing a base-pair error during the polymerization events and correcting it. Proofreading is a property of the DNA polymerase in prokaryotic cells. (330)

propeller twist In DNA, a rotation of the two bases of a base pair in opposite directions about their long axis. It serves to stabilize the helix. (273)

prophage A temperate bacteriophage integrated into the chromosome of a lysogenic bacterium. It replicates with the replication of the host cell's chromosome. (235)

prophase The first stage in mitosis or meiosis during which the chromosomes (already replicated) condense and become visible under the microscope. (16)

prophase I The first stage of meiosis. There are several stages of prophase I, including leptonema, zygonema, pachynema, diplonema, and diakinesis. (21)

proportion of polymorphic loci A ratio calculated by determining the number of polymorphic loci and dividing by the total number of loci examined. (797)

propositus (proband) The affected person, in human genetics, with whom the study of a character in a family begins. (83)

protein One of a group of high-molecular weight, nitrogen-containing organic compounds of complex shape and composition. (407)

protein degradation control Regulation of the protein degradation rate. (652)

proto-oncogenes A gene which, in normal cells, functions to control the normal proliferation of cells, and which, when mutated or changed in any other way, becomes an oncogene. (689)

prototroph A strain that is a wild type for all nutritional requirement genes and thus requires no supplements in its growth medium. (225)

Punnett square A matrix that describes all the possible gametic fusions that will give rise to the zygotes that will produce the next generation. (44)

pure-breeding *See* true-breeding. (40)

purine A type of nitrogenous base. In DNA and RNA the purines are adenine and guanine. (263)

pyrimidine A type of nitrogenous base. Cytosine is a pyrimidine in DNA and RNA; thymine is a pyrimidine in DNA; and uracil is a pyrimidine in RNA. (263)

Q banding A staining technique in which metaphase chromosomes are stained with quinacrine mustard to produce temporary fluorescent Q bands on the chromosomes. (292)

quantitative genetics Study of the inheritance of quantitative traits. (735)

quantitative (continuous) traits Traits that show a continuous variation in phenotype over a range. (735)

random mating Matings between genotypes occurring in proportion to the frequencies of the genotypes in the population. (789)

recessive An allele or phenotype that is expressed only in the homozygous state. (43)

recessive lethal An allele that causes lethality when it is homozygous. (121)

reciprocal cross A cross of males and females of one trait with males and females of another trait. In the garden pea example a reciprocal cross for smooth and wrinkled seeds is smooth female × wrinkled male and wrinkled female × smooth male. (42)

recombinant chromosome A chromosome that emerges from meiosis with a combination of genes different from a parental combination of genes. (21)

recombinant DNA molecule A new type of DNA sequence that has been constructed or engineered in the test tube from two or more distinct DNA sequences. (482)

recombinant DNA technology A collection of experimental procedures that allow molecular biologists to splice a DNA fragment from one organism into DNA from another organism and to clone the new recombinant DNA molecule. It includes the development and the application of particular molecular techniques, such as biotechnology or genetic engineering. This technology is important, for example, in the production of antibiotics, hormones, and other medical agents used in the diagnosis and treatment of certain genetic diseases. (482)

recombinants The individuals or cells that have nonparental combinations of genes as a result of the processes of genetic recombination. (142)

regression A statistical analysis assessing the association between two variables. (746)

regression line A mathematically computed line that represents the best fit of a line to the points. (746)

regulated gene A gene whose activity is controlled in response to the needs of a cell or organism. (604)

release factors *See* termination factors. (427)

replica plating The procedure for transferring the pattern of colonies from a master plate to a new plate. In this procedure a velveteen pad on a cylinder is pressed lightly onto the surface of the master plate, thereby picking up a few cells from each colony to inoculate onto the new plate. (540)

replication fork A Y-shaped structure formed when a double-stranded DNA molecule unwinds to expose the two single-stranded template strands for DNA replication. (330)

replicon (replication unit) The stretch of DNA in eukaryotes from the origin of replication to the two termini of replication on each side of the origin. (343)

repressor gene A regulatory gene whose product is a protein that controls the transcriptional activity of a particular operon. (610)

repressor molecule The protein product of a repressor gene. (610)

repulsion An arrangement in which each homologous chromosome carries the wild-type allele of one gene and the mutant allele of the other one. (154)

restriction endonucleases (restriction enzymes) Enzymes important for analyzing DNA and for constructing recombinant DNA molecules because of their ability to cleave double-stranded DNA molecules at specific nucleotide pair sequences. (482)

restriction enzyme cleavage site linker (linker) A relatively short, double-stranded oligodeoxyribonucleotide about 8 to 12 nucleotide pairs long which is synthesized by chemical means and which contains the cleavage site for a specific restriction enzyme within its sequence. (497)

restriction map A genetic map of DNA showing the relative positions of restriction enzyme cleavage sites. (507)

reverse mutation (reversion) A mutational change from a mutant allele back to a wild-type allele. (521)

ribonuclease (RNase) An enzyme that catalyzes the degradation of RNA to nucleotides. (258)

ribonucleic acid (RNA) A usually single-stranded polymeric molecule consisting of ribonucleotide building blocks. RNA is chemically very similar to DNA. The three major types of RNA in cells are ribosomal RNA (rRNA), transfer RNA (tRNA), and messenger RNA (mRNA), each of which performs an essential role in protein synthesis (translation). In some viruses, RNA is the genetic material. (2)

ribonucleotide A nucleotide that is a building block of RNA. (263)

ribose The pentose sugar component of the nucleotide building block of RNA. (263)

ribosomal DNA (rDNA) The regions of the DNA that contain the genes for the rRNAs in prokaryotes and eukaryotes. (355)

ribosomal proteins The proteins that, along with rRNA molecules, comprise the ribosomes of prokaryotes and eukaryotes. (396)

ribosomal RNA (rRNA) The RNA molecules of discrete sizes that, along with ribosomal proteins, comprise ribosomes of prokaryotes and eukaryotes. (396)

ribosome A complex cellular particle composed of ribosomal protein and rRNA molecules that is the site of amino acid polymerization during protein synthesis. (372)

ribosome-binding site The nucleotide sequence on an mRNA molecule on which the ribosome becomes oriented in the correct reading frame for the initiation of protein synthesis. (419)

R looping (R loops) A technique developed by M. Thomas, R. White, and R. Davis in which molecules of double-stranded DNA are incubated at temperatures below their denaturing temperature to open up short stretches of the DNA double helix so that single-stranded RNA molecules can begin to form DNA/RNA hybrids where the two are complementary. The DNA/RNA hybrid forms an R loop by displacing a single-stranded section of DNA. (374)

RNA *See* ribonucleic acid. (2)

RNA ligase An enzyme that splices together the RNA pieces once the intervening sequence is removed from the pre-tRNA. (392)

RNA polymerase An enzyme that catalyzes the synthesis of RNA molecules from a DNA template in a process called transcription. (357)

RNA polymerase I An enzyme in eukaryotes located in the nucleolus that catalyzes the transcription of the 18S, 5.8S, and 28S rRNA genes. (367)

RNA polymerase II An enzyme in eukaryotes found only in the nucleoplasm of the nucleus. It catalyzes the transcription of mRNA-coding genes. (367)

RNA polymerase III An enzyme in eukaryotes found only in the nucleoplasm. It catalyzes the transcription of the tRNA and 5S rRNA genes. (368)

RNA splicing A process whereby an intervening sequence between two coding sequences in an RNA molecule is excised and the coding sequences ligated (spliced) together. (383)

sample The subset used to give information about a population. It must be of reasonable size and it must be a random subset of the larger group in order to provide accurate information about the population. (738)

sampling error The phenomenon in which chance

deviations from expected proportions arise in small samples. (806)

satellite DNA   The DNA that forms a band in an equilibrium density band that is distinct from the band constituting the majority of the genomic DNA as a result of a different buoyant density. (311)

second law   *See* Mendel's second law. (50)

secondary nondisjunction   A nondisjunction of the X's in the progeny of females that were produced by a primary nondisjunction. (74)

secondary oocyte   A large cell produced by the primary oocyte. In the ovaries of female animals the diploid primary oocyte goes through meiosis I and unequal cytokinesis to produce two cells; the large cell is called the secondary oocyte. (27)

second-site mutation   *See* suppressor mutation. (521)

segmentation genes   One of the two categories of selector genes. (*See* selector genes.) (679)

selection coefficient   A measure of the relative intensity of selection against a genotype. (816)

selection differential   In natural and artificial selection, the difference between the mean phenotype of the selected parents and the mean phenotype of the unselected population. (761)

selection response   The amount that a phenotype changes in one generation when selection is applied to a group of individuals. (761)

selector genes   Special regulatory genes that play a major role in laying down an organism's basic body plan, starting early in embryonic development. (678)

self-assembly   The ability of certain complex biological structures (e.g., bacterial ribosomes) to assemble into a functional structure from the component molecules. (399)

self-fertilization (selfing)   The union of male and female gametes from the same individual. (39)

semiconservative replication model   A DNA replication scheme in which each daughter molecule retains one of the parental strands. (269)

semidiscontinuous   Concerning DNA replication, when one new strand is synthesized continuously and the other discontinuously. *See* discontinuous. (331)

sense codon   A sense codon, as opposed to a nonsense codon, in an mRNA molecule specifies an amino acid in the corresponding polypeptide. (417)

sense strand   One of the two DNA strands whose complementary base sequence is produced when an RNA molecule is produced by transcription. (356)

sex chromosome   A chromosome in eukaryotic organisms that is represented differently in the two sexes. In many organisms one sex possesses a pair of visibly different chromosomes. One is an X chromosome, and the other is a Y chromosome. Commonly, the XX sex is female and the XY sex is male. (12)

sexduction   *See F* duction. (229)

sex-influenced traits   The traits that appear in both sexes but either the frequency of occurrence in the two sexes is different or there is a different relationship between genotype and phenotype. (125)

sex-limited trait   A genetically controlled character that is phenotypically exhibited in only one of the two sexes. (125)

sex linkage   The linkage of genes located on the sex chromosomes. (72)

sex-linked inheritance   The pattern of hereditary transmission of sex-linked genes, i.e., located on the sex chromosomes. (72)

sexual reproduction   The reproduction involving the fusion of haploid gametes produced by meiosis. (13)

signal hypothesis   The hypothesis that the secretion of proteins from a cell occurs through the binding of a hydrophobic amino terminal extension to the membrane and the subsequent removal and degradation of the extension in the cisternal space of the endoplasmic reticulum. (429)

signal peptidase   An enzyme in the cisternal space of the ER that catalyzes removal of the signal sequence from the polypeptide. (430)

signal recognition particle (SRP)   In eukaryotes, a complex of a small RNA molecule with six proteins, which can temporarily halt protein synthesis by recognizing the signal sequence of a nascent polypeptide destined to be translocated through the ER, binding to it, and thereby blocking further translation of the mRNA. (429)

silencer element   In eukaryotes, a transcriptional regulatory element that decreases RNA transcription rather than stimulating it like other enhancer elements. (370)

silent mutation   A mutational change resulting in a protein with a wild-type function because of an unchanged amino-acid sequence. (521)

simple telomeric sequences   Simple, tandemly repeated DNA sequences at, or very close to, the extreme ends of the chromosomal DNA molecules. (306)

SINES (short interspersed repeated sequences)   One class of interspersed and highly repeated sequences that consists of dispersed families with unit lengths of fewer than 500 base pairs and repeated for as many as hundreds of thousands of copies in the genome. (311)

sister chromatid A chromatid derived from replication of one chromosome during interphase of the cell cycle. (15)

slope (regression coefficient) The change in one variable (y) associated with a unit increase in another variable (x). (746)

somatic cell hybridization The fusion of two genetically different somatic cells of the same or different species to generate a somatic hybrid for genetic analysis. (204)

Southern blot technique A technique invented by E. M. Southern and used in analyzing genes and gene transcripts, in which DNA fragments are transferred from a gel to a nitrocellulose filter. (505)

specialized transducing phage A temperate bacteriophage that can transduce only a certain section of the bacterial chromosome. (241)

specialized transduction A type of transduction in which only specific genes are transferred. (241)

spermatogenesis Development of the male animal germ cell within the male gonad. (27)

sperm cells (spermatozoa) The male gametes; the spermatozoa produced by the testes in male animals. (27)

spindle apparatus (spindle) A collection of fibers (spindle fibers), visible during mitosis and meiosis, which play a role in chromosome movement and the division of the cytoplasm. (16)

spontaneous mutations The mutations that occur without the use of chemical or physical mutagenic agents. (519)

sporophyte The generation, in the life cycle of plants, in pollenization that is initiated by the nucleus of one sperm cell fusing with the nucleus of the egg cell. (27)

stamen The male reproductive organ in a flowering plant that usually consists of a stalk, called a filament, bearing a pollen-producing anther. (28)

standard deviation The square root of the variance. It measures the extent to which each measurement in the data set differs from the mean value and is used as a measure of the extent of variability in a population. (742)

standard error of gene frequency A measure of the amount of variation among the gene frequencies of populations. It is the square root of the variance of gene frequency. (807)

stop codon *See* nonsense codon. (417)

stringent response (stringent control) A rapid shutdown of essential cellular activities. (630)

structural gene A gene that codes for an mRNA molecule and hence for a polypeptide chain. (355)

submetacentric chromosome A chromosome which has the centromere nearer one end than the other. Such chromosomes appear J-shaped at anaphase. (13)

sum rule The rule that the probability of either one of two mutually exclusive events occurring is the sum of their individual probabilities. (47)

suppressor mutation A mutation at a second site that totally or partially restores a function lost because of a primary mutation at another site. (521)

Svedberg units (S values) The conversions for sedimentation rates in sucrose density centrifugation. Svedberg units are used as a rough indication of relative sizes of the components being analyzed. (376)

synapsis The specific pairing of homologous chromosomes during the zygonema stage of meiosis. (21)

synaptinemal complex A complex structure spanning the region between meiotically paired (synapsed) chromosomes that is concerned with crossing-over rather than with chromosome pairing. (21)

synkaryon A fusion nucleus produced following the fusion of cells with genetically different nuclei. (205)

syntenic The genes that are localized to a particular chromosome by using an experimental approach (literally "together thread," i.e., another term for *linked*). (207)

TATA element *See* Goldberg-Hogness box. (369)

tautomeric shift The change in the chemical form of a DNA (or RNA) base. (526)

tautomers Alternate chemical forms in which DNA (or RNA) bases are able to exist. (526)

telocentric chromosome A chromosome that has the centromere more or less at one end. (13)

telomere-associated sequences Repeated, complex DNA sequences extending from the molecular gene of chromosomal DNA. Suspected to mediate many of the telomere-specific interactions. (305)

telophase A stage during which the migration of the daughter chromosomes to the two poles is completed. (20)

temperate bacteriophage The class of bacteriophages that, when they infect a bacterial cell, are capable of evoking the lysogenic response. That is, the phage genome integrates into the bacterial chromosome and replicates along with the bacterial chromosome in the integrated state. The phage genome is called the prophage. (583)

template strand The unwound single strand of DNA upon which new strands are made (following complementary-base-pairing rules). (322)

termination factors (or release factors, RF) The specific proteins in polypeptide synthesis (translation) that read the chain termination codons and then initiate a series of specific events to terminate polypeptide synthesis. (427)

terminator *See* transcription terminator sequence. (301)

testcross A cross of an individual of unknown genotype, usually expressing the dominant phenotype, with a homozygous recessive individual in order to determine the genotype of the individual. (48)

tetrad analysis Genetic analysis of all the products of a single meiotic event. Tetrad analysis is possible in those organisms in which the four products of a single nucleus that has undergone meiosis are grouped together in a single structure. (184)

tetrasomy The aberrant, aneuploid state in a normally diploid cell or organism in which an extra chromosome pair results in the presence of four copies of one chromosome type and two copies of every other chromosome type. (551)

tetratype (T) One of the three types of tetrads possible when two genes are segregating in a cross. The T tetrad contains two parental and two recombinant nuclei, one of each parental type and one of each recombinant type. (186)

three-point testcross A test involving three genes within a relatively short section of the chromosome. It is used to map genes for their order in the chromosome and for the distance between them. (160)

thymine (T) A pyrimidine base found in DNA but not in RNA. In double-stranded DNA thymine pairs with adenine. (263)

topoisomerases A class of enzymes that catalyze the supercoiling of DNA. (280)

totipotency The capacity of a nucleus to direct events through all the stages in development and therefore produce a normal adult. (513)

trailer sequence The sequence of the mRNA molecule beginning at the end of the amino acid–coding sequence and ending at the 3′ end of the mRNA. The trailer sequence is not translated and varies in length from molecule to molecule. (372)

transconjugants In bacteria, the recipients inheriting donor DNA in the process of conjugation. (222)

transcription The transfer of information from a double-stranded DNA molecule to a single-stranded RNA molecule. It is also called RNA synthesis. (354)

transcriptional control The first level of control of gene expression in eukaryotes. This level involves regulating whether or not a gene is to be transcribed and the rate at which transcripts are produced. (651)

transcription-controlling sequence The base-pair sequence found around the beginning and end of each gene that is involved in the regulation of gene expression. (651)

transcription terminator sequence A transcription regulatory sequence located at the distal end of a gene that signals the termination of transcription. (361)

transducing phage The phage that is the vehicle by which genetic material is shuttled between bacteria. (239)

transducing retroviruses Retroviruses that have picked up an oncogene from the cellular genome. (687)

transductants In bacteria, the recipients inheriting donor DNA in the process of transduction. (222)

transduction A process by which bacteriophages mediate the transfer of bacterial genetic information from one bacterium (the donor) to another (the recipient); a process whereby pieces of bacterial DNA are carried between bacterial strains by a phage. (222)

transfer RNA (tRNA) One of the three classes of RNA molecules, or transcripts, produced by transcription and involved in protein synthesis; molecules that bring amino acids to the ribosome, where they are matched to the transcribed message on the mRNA. (355)

transformant The genetic recombinant generated by the transformation process. (222)

transformation A process in which genetic information is transferred by means of extracellular pieces of DNA in bacteria. (222)

transition mutation A specific type of base-pair substitution mutation that involves a change in the DNA from one purine-pyrimidine base pair to the other purine-pyrimidine base pair at a particular site (e.g., AT to GC). (520)

translation (protein synthesis) The conversion in the cell of the mRNA base sequence information into an amino acid sequence of a polypeptide. (354)

translational control The regulation of protein synthesis by ribosome synthesis among mRNAs. (652)

translocation (transposition) A chromosome mutation involving a change in position of a chromosome segment (or segments) and the gene sequences it contains. (563) In polypeptide synthesis, translocation is the movement of the ribosome, one codon at a time, along the mRNA toward the 3′ end. (207)

transmission genetics (classical genetics) A subdivision of the science of genetics primarily dealing with how genes are passed from one individual to another. (781)

transport control Regulating the number of transcripts that exit the nucleus to the cytoplasm. (651)

transposable genetic element (TGE) A genetic element of chromosomes of both prokaryotes and eukaryotes that has the capacity to mobilize itself and move from one location to another in the genome. (574)

transposon (Tn) A mobile DNA segment that contains genes for the insertion of the DNA segment into the chromosome and for mobilization of the element to other locations on the chromosomes. (576)

transversion mutation A specific type of base-pair substitution mutation that involves a change in the DNA from a purine-pyrimidine base pair to a pyrimidine-purine base pair at the same site (e.g., AT to TA or GC to TA). (520)

trihybrid cross A cross between individuals of the same type that are heterozygous for three pairs of alleles at three different loci. (53)

trisomy An aberrant, aneuploid state in a normally diploid cell or organism in which there are three copies of a particular chromosome instead of two copies. (551)

trisomy-21 *See* Down syndrome. (552)

true-breeding or pure-breeding strain A strain allowed to self-fertilize for many generations to ensure that the traits to be studied are inherited and unchanging. (40)

Turner syndrome A human clinical syndrome that results from monosomy for the X chromosome in the female, which gives a 45,X female. These females fail to develop secondary sexual characteristics, tend to be short, have weblike necks, have poorly developed breasts, are usually infertile, and exhibit mental deficiencies. (78)

twin spots Two adjacent cell groups that differ in genotype and phenotype. They result from mitotic crossing-over within the somatic cells of a heterozygous individual. (195)

uniparental inheritance A phenomenon, usually exhibited by extranuclear genes, in which all progeny have the phenotype of only one parent. (715)

unique (single-copy) sequence A class of DNA sequences that has one to a few copies per genome. (309)

uracil (U) A pyrimidine base found in RNA but not in DNA. (263)

variance A statistical measure of how values vary from the mean. (741)

variance of gene frequency The variance in the frequency of an allele among a group of populations. (807)

vegetative reproduction *See* asexual reproduction. (13)

viral oncogene A viral gene that transforms a cell it infects to a cancerous state. *See* oncogenesis; cellular oncogene. (687)

virulent phage A phage like T4, which always follows the lytic cycle when it infects bacteria. (234)

virus A noncellular organism that can produce only within a host cell. It contains genetic material within a membrane or protein coat. Once a virus is within a cell, its genetic material causes the cellular machinery to produce progeny viruses. (2)

visible mutation A mutation that affects the morphology or physical appearance of an organism. (540)

wild type A strain, organism, or gene of the type predominating in the wild population. (66)

wild-type allele The allele found in the laboratory in the strain of the organism. (101)

X-linked dominant trait A trait due to a dominant mutant gene carried on the X chromosome. (87)

X-linked recessive trait A trait due to a recessive mutant gene carried on the X chromosome. (86)

X chromosome A sex chromosome present in two copies in the homogametic sex and in one copy in the heterogametic sex. (65)

X chromosome–autosome balance system A genotypic sex determination system. The main factor in sex determination is the ratio between the numbers of X chromosomes and autosomes. Sex is determined at the time of fertilization, and sex differences are assumed to be due to the action during development of two sets of genes located in the X chromosomes and in the autosomes. (75)

X chromosome nondisjunction An event occurring when the two X chromosomes fail to separate in meiosis so that eggs are produced either with two X chromosomes or with no X chromosomes, instead of the usual one X chromosome. (72)

X-linked   Referring to genes located on the X chromosome. *See also* sex linkage. (72)

X-linked dominant inheritance   A trait due to a dominant mutant gene carried on the X chromosome. Since females have twice the number of X chromosomes that males have, X-linked dominant traits are more frequent in females than in males. (87)

X-linked recessive inheritance   A trait due to a recessive mutant gene carried on the X chromosome. The most well-known X-linked recessive pedigree is hemophilia, in which affected males normally transmit the mutant gene to all their daughters but to none of their sons. (86)

Y chromosome   A sex chromosome that, when present, is found in one copy in the heterogametic sex, along with an X chromosome, and is not present in the homogametic sex. Not all organisms with sex chromosomes have a Y chromosome. (65)

Y-intercept   In a regression analysis, the value of Y when X is zero. (746)

Y-linked (holandric ["wholly male"]) trait   A trait due to a mutant gene carried on the Y chromosome but with no counterpart on the X. (89)

Y-linked (holandric) inheritance   A trait due to a mutant gene that is carried on the Y chromosome but has no counterpart on the X chromosome. (89)

zygonema   The stage during meiosis in prophase I at which homologous chromosomes begin to pair in a highly specific way and then to twist around one another. (21)

zygote   The cell produced by the fusion of the male and female gametes. (13)

Suggested Readings

Chapter 1

Brachet, J., and Mirsky, A. E., eds. 1961. *The cell,* vol. 3, *Meiosis and mitosis.* New York: Academic Press.

John, B., and Lewis, K. R. 1965. *The meiotic system.* New York: Springer-Verlag.

McLeish, J., and Snoad, B. 1958. *Looking at chromosomes.* New York: Macmillan.

Sturtevant, A. H. 1965. *A history of genetics.* New York: Harper & Row.

Wallace, R. A., King, J. L., and Sanders, G. P. 1981. *Biology, the science of life.* Santa Monica, CA: Goodyear.

Chapter 2

Bateson, W. 1909. *Mendel's principles of heredity.* Cambridge: Cambridge University Press.

Mendel, G. 1866. Experiments in plant hybridization (translation). In *Classic papers in genetics,* edited by J. A. Peters. Englewood Cliffs, NJ: Prentice-Hall.

Peters, J. A., ed. 1959. *Classic papers in genetics.* Englewood Cliffs, NJ: Prentice-Hall.

Srb, A. M., Owen, R. D., and Edgar, R. S. 1965. *General genetics.* San Francisco: Freeman.

Sturtevant, A. H. 1965. *A history of genetics.* New York: Harper & Row.

Tschermak-Seysenegg, E. von. 1951. The rediscovery of Mendel's work. *J. Hered.* 42:163–171.

Chapter 3

Bodmer, W. F., and Cavalli-Sforza, L. L. 1976. *Genetics, evolution, and man.* San Francisco: Freeman.

Bridges, C. B. 1916. Nondisjunction as a proof of the chromosome theory of heredity. *Genetics* 1:1–52, 107–163.

———. 1925. Sex in relation to chromosomes and genes. *Am. Naturalist* 59:127–137.

Dice, L. R. 1946. Symbols for human pedigree charts. *J. Hered.* 37:11–15.

Disteche, C. M., Casanova, M., Saal, H., Friedman, C., Sybert, V., Graham, J., Thuline, H., Page, D. C., and Fellous, M. 1986. Small deletions of the short arm of the Y chromosome in 46, XY females. *Proc. Natl. Acad. Sci. USA* 83:7841–7844.

Farabee, W. C. 1905. Inheritance of digital malformations in man. *Papers Peabody Museum Amer. Arch. Ethnol. (Harvard Univ.)* 3:65–78.

Levitan, M., and Montagu, A. 1971. *Textbook of human genetics.* New York: Oxford University Press.

McClung, C. E. 1902. The accessory chromosome—sex determinant? *Biol. Bull.* 3:43–84.

McKusick, V. A. 1965. The royal hemophilia. *Sci. Am.* 213:88–95.

Morgan, L. V. 1922. Non criss-cross inheritance in *Drosophila melanogaster. Biol. Bull.* 42:267–274.

Morgan, T. H. 1910. Sex-limited inheritance in *Drosophila. Science* 32:120–122.

———. 1911. An attempt to analyze the constitution of the chromosomes on the basis of sex-limited inheritance in *Drosophila. J. Exp. Zool.* 11:365–414.

Stern, C., Centerwall, W. P., and Sarkar, Q. S. 1964. New data on the problem of Y-linkage of hairy pinnae. *Am. J. Hum. Genet.* 16:455–471.

Sutton, W. S. 1903. The chromosomes in heredity. *Biol. Bull.* 4:231–251.

Wilson, E. B. 1905. The chromosomes in relation to the determination of sex in insects. *Science* 22:500–502.

Chapter 4

Allen, G. 1965. Twin research: Problems and prospects. *Prog. Med. Genet.* 4:242–269.

Ginsburg, V. 1972. Enzymatic basis for blood groups. *Methods Enzymol.* 36:131–149.

Landauer, W. 1948. Hereditary abnormalities and their chemically induced phenocopies. *Growth Symposium* 12:171–200.

Landsteiner, K., and Levine, P. 1927. Further observations on individual differences of human blood. *Proc. Soc. Exp. Biol. Med.* 24:941–942.

Reed, T. E., and Chandler, J. H. 1958. Huntington's chorea in Michigan. I. Demography and genetics. *Am. J. Hum. Genet.* 10:201–225.

Chapter 5

Barratt, R. W., Newmeyer, D., Perkins, D. D., and Garnjobst, L. 1954. Map construction in *Neurospora crassa. Adv. Genet.* 6:1–93.

Bateson, W., Saunders, E. R., and Punnett, R. G. 1905. Experimental studies in the physiology of heredity. *Rep. Evol. Committee R. Soc. II:*1–55 and 80–99.

Blixt, S. 1975. Why didn't Mendel find linkage? *Nature* 256:206.

Creighton, H. S., and McClintock, B. 1931. A correlation of cytological and genetical crossing-over in *Zea mays. Proc. Natl. Acad. Sci USA.* 17:492–497.

Fincham, J. R. S., Day, P. R., and Radford, A. 1979. *Fungal genetics.* 3rd ed. Oxford: Blackwell Scientific.

Fink, G. R. 1970. The biochemical genetics of yeast. In *Methods in enzymology,* edited by S. Colowick and N. O. Kaplan, vol. 17A, pp. 59–78. New York: Academic Press.

Morgan, T. H. 1910. Sex-limited inheritance in *Drosophila. Science* 32:120–122.

———. 1910. The method of inheritance of two sex-limited characters in the same animal. *Proc. Soc. Exp. Biol. Med.* 8:17.

———. 1911. An attempt to analyze the constitution of the chromosomes on the basis of sex-limited inheritance in *Drosophila. J. Exp. Zool.* 11:365–414.

———. 1911. Random segregation versus coupling in Mendelian inheritance. *Science* 34:384.

———, Sturtevant, A. H., Muller, H. J., and Bridges, C. B. 1915. *The mechanism of Mendelian heredity.* New York: Henry Holt.

Mortimer, R. K., and Hawthorne, D. C. 1966. Yeast genetics. *Annu. Rev. Microbiol.* 20:151–168.

Sturtevant, A. H. 1913. The linear arrangement of six sex-linked factors in *Drosophila,* as shown by their mode of association. *J. Exp. Zool.* 14:43–59.

Sutton, W. S. 1903. The chromosomes in heredity. *Biol. Bull.* 4:231–251.

Chapter 6

Chaleff, R. S., and Carlson, P. S. 1974. Somatic cell genetics of higher plants. *Annu. Rev. Genet.* 8:267–278.

Ephrussi, B., and Weiss, M. C. 1969. Hybrid somatic cells. *Sci. Am.* 220:26–35.

Kao, F., Jones, C., and Puck, T. T. 1976. Genetics of somatic mammalian cells: Genetic, immunologic, and biochemical analysis with Chinese hamster cell hybrids containing selected human chromosomes. *Proc. Natl. Acad. Sci. USA* 73:193–197.

McKusick, V. A. 1971. The mapping of human chromosomes. *Sci. Am.* 224:104–113.

———. 1978. Genetic nosology: Three approaches. *Am. J. Hum. Genet.* 30:105–122.

———, and Ruddle, F. H. 1977. The status of the gene map of the human chromosomes. *Science* 196:390–405.

Pontecorvo, G. 1956. The parasexual cycle in fungi. *Annu. Rev. Microbiol.* 20:151–168.

———, and Kafer, E. 1958. Genetic analysis based on mitotic recombination. *Adv. Genet.* 9:71–104.

Pritchard, R. H. 1955. The linear arrangement of a series of alleles of *Aspergillus nidulans. Heredity* 9:343–371.

Ruddle, F. H., and Creagan, R. P. 1975. Parasexual approaches to the genetics of man. *Annu. Rev. Genet.* 9:407–486.

———, and Kucherlapati, R. S. 1974. Hybrid cells and human genes. *Sci. Am.* 231:36–44.

Stern, C. 1936. Somatic crossing over and segregation in *Drosophila melanogaster. Genetics* 21:625–730.

Chapter 7

Archer, L. J. 1973. *Bacterial transformation.* New York: Academic Press.

Campbell, A. 1969. *Episomes.* New York: Harper & Row.

Curtiss, R. 1969. Bacterial conjugation. *Annu. Rev. Microbiol.* 23:69–136.

Delbruck, M. 1940. The growth of bacteriophage and lysis of the host. *J. Gen. Physiol.* 23:643–660.

Doermann, A. H. 1952. The intracellular growth of bacteriophages. I. Liberation of intracellular bacteriophage T4 by premature lysis with another phage or with cyanide. *J. Gen. Physiol.* 35:645–656.

Ellis, E. L., and Delbruck, M. 1939. The growth of bacteriophage. *J. Gen. Physiol.* 22:365–384.

Hayes, W. 1968. *The Genetics of bacteria and their viruses,* 2nd ed. New York: Wiley.

Hershey, A. D., and Rotman, R. 1949. Genetic recombination between host-range and plaque-type mutants of bacteriophage in single bacterial cells. *Genetics* 34:44–71.

Hotchkiss, R. D., and Gabor, M. 1970. Bacterial transformation with special reference to recombination processes. *Annu. Rev. Genet.* 4:193–224.

Jacob, F., and Wollman, E. L. 1961. *Sexuality and the genetics of bacteria.* New York: Academic Press.

Lennox, E. 1955. Transduction of linked characters of the host by bacteriophage P1. *Virology* 1:190–206.

Ravin, A. W. 1961. The genetics of transformation. *Adv. genet.* 10:61–163.

Susman, M. 1970. General bacterial genetics. *Annu. Rev. Genet.* 4:135–176.

Vielmetter, W., Bonhoeffer, F., and Schutte, A. 1968. Genetic evidence for transfer of a single DNA strand during bacterial conjugation. *J. Mol. Biol.* 37:81–86.

Wollman, E. L., Jacob, F., and Hayes, W. 1962. Conjugation and genetic recombination in *E. coli K–12. Cold Spring Harbor Symp. Quant. Biol.* 21:141–162.

Zinder, N., and Lederberg, J. L. 1952. Genetic exchange in *Salmonella. J. Bacteriol.* 64:679–699.

Chapter 8

Avery, O. T., MacLeod, C. M., and McCarty, M. 1944. Studies on the chemical nature of the substance inducing transformation of pneumococcal types. Induction of transformation by a deoxyribonucleic acid fraction isolated from pneumococcus type III. *J. Exp. Med.* 79:137–158.

Chargaff, E. 1951. Structure and function of nucleic acids as cell constituents. *Fed. Proc.* 10:654–659.

Davidson, J. N. 1972. *The biochemistry of the nucleic acids,* 7th ed. London: Chapman and Hall.

Dickerson, R. E. 1983. The DNA helix and how it is read. *Sci. Am.* 249 (December):94–111.

———. 1983. Base sequence and helix structure variation in B and A DNA. *J. Mol. Biol.* 166:419–441.

Fraenkel-Conrat, H., and Singer, B. 1957. Virus reconstitution: Combination of protein and nucleic acid from different strains. *Biochim. Biophys. Acta* 24:540–548.

Franklin, R. E., and Gosling, R. 1953. Molecular configuration of sodium thymonucleate. *Nature* 171:740–741.

Gierer, A., and Schramm, G. 1956. Infectivity of ribonucleic acid from tobacco mosaic virus. *Nature* 177:702–703.

Griffith, F. 1928. The significance of pneumococcal types. *J. Hyg. (Lond)* 27:113–159.

Hershey, A. D., and Chase, M. 1952. Independent functions of viral protein and nucleic acid in growth and bacteriophage. *J. Gen. Physiol.* 36:39–56.

Pauling, L., and Corey, R. B. 1956. Specific hydrogen-bond formation between pyrimidines and purines in deoxyribonucleic acids. *Arch. Biochem. Biophys.* 65:164–181.

Rich, A., Nordheim, A., and Wang, A. H.-J. 1984. The chemistry and biology of left-handed Z-DNA. *Annu. Rev. Biochem.* 53:791–846.

Saenger, W. 1984. *Principles of nucleic acid structure.* New York: Springer-Verlag.

"Structures of DNA," 1982. *Cold Spring Harbor Symp. Quant. Biol.* 47. New York: Cold Spring Harbor Laboratory.

Watson, J. D. 1968. *The double helix.* New York: Atheneum.

———, and Crick, F. H. C. 1953. Genetical implications of the structure of deoxyribonucleic acid. *Nature* 171–964–969.

———. 1953. Molecular structure of nucleic acids. A structure for deoxyribose nucleic acid. *Nature* 171:737–738.

———. 1953. The structure of DNA. *Cold Spring Harbor Symp. Quant. Biol.* 18:123–131.

Wilkins, M. H. F. 1963. The molecular configuration of nucleic acids. *Science* 140:941–950.

Wilkins, M. H. F., Stokes, A. R., and Wilson, H. R. 1953. Molecular structure of deoxypentose nucleic acids. *Nature* 171:738–740.

Wing, R. M., Drew, H. R., Takano, T., Broka, C., Tanaka, S., Itakura, K., and Dickerson, R. E. 1980. Crystal structure analysis of a complete turn of B-DNA. *Nature* 287:755–758.

Wu, H.-M., and Crothers, D. M. 1984. The locus of sequence-directed and protein-induced DNA bending. *Nature* 308:509–513.

Chapter 9

Blackburn, E. H. 1984. Telomeres: Do the ends justify the means? *Cell* 37:7–8.

———, and Challoner, P. B. 1984. Identification of a telomeric DNA sequence in *Trypanosoma brucei*. *Cell* 36:447–457.

———, and Szostak, J. W. 1984. The molecular structure of centromeres and telomeres. *Annu. Rev. Biochem.* 53:163–194.

Bloom, K. S., Amaya, E., Carbon, J., Clarke, L., Hill, A., and Yeh, E. 1984. Chromatin conformation of yeast centromeres. *J. Cell Biol.* 99:1559–1568.

Britten, R. J., and Kohne, D. E. 1968. Repeated sequences in DNA. *Science* 161:529–540.

Burlingame, R. W., Love, W. E., Wang, B.-C., Hamlin, R., Xuang, N.-H., and Moudranakis, E. N. 1985. Crystallographic structure of the octameric histone core of the nucleosome at a resolution of 3.3 Å. *Science* 228:546–553.

Carbon, J., 1984. Yeast centromeres: structure and function. *Cell* 37:351–353.

Clarke, L., Amstutz, H., Fishel, B., and Carbon, J. 1986. Analysis of centromeric DNA in the fission yeast *Schizosaccharomyces pombe*. *Proc. Natl. Acad. Sci. USA* 83:8253–8257.

Cold Spring Harbor Symposia on Quantitative Biology. 1973. *Chromosome structure and function,* vol. 38. New York: Cold Spring Harbor Laboratory.

Cold Spring Harbor Symposia on Quantitative Biology. 1978. *Chromatin,* vol. 42. New York: Cold Spring Harbor Laboratory.

Comings, D. 1978. Mechanisms of chromosome banding and implications for chromosome structure. *Annu. Rev. Genet.* 12:25–46.

D'Ambrosio, E., Waitzikin, S. D., Whitney, F. R., Salemme, A., and Furano, A. V. 1985. Structure of the highly repeated, long interspersed DNA family (LINE or L1Rn) of the rte. *Mol. Cell. Biol.* 6:411–424.

DuPraw, E. J. 1970. *DNA and chromosomes.* New York: Holt, Rinehart and Winston.

Eisenberg, J. C., Cartwright, I. L., Thomas, G. H., and Elgin, S. C. R. 1985. Selected topics in chromatin structure. *Annu. Rev. Genet.* 19:485–536.

Elgin, S. C. R., and Weintraub, H. 1975. Chromosomal proteins and chromatin structure. *Annu. Rev. Biochem.* 44:725–774.

Fitzgerald-Hayes, M., Clarke, L., and Carbon, J. 1982. Nucleotide sequence comparisons and functional analysis of yeast centromere DNAs. *Cell* 29:235–244.

Fredericq, E. 1982. Supramolecular aspects of chromatin structure. *Arch. Biol.* 93:127–142.

Freifelder, D. 1978. *The DNA molecule. Structure and properties.* San Francisco: Freeman.

Gellert, M. 1981. DNA topoisomerases. *Annu. Rev. Biochem.* 50:879–910.

Jabs, E. W., Wolf, S. F., and Migeon, B. R. 1984. Characterization of a cloned DNA sequence that is present at centromeres of all human autosomes and the X chromosome and shows polymorphic variation. *Proc. Natl. Acad. Sci. USA* 81:4884–4888.

Jelinek, W. R., and Schmid, C. W. 1982. Repetitive sequences in eukaryotic DNA and their expression. *Annu. Rev. Biochem.* 51:813–844.

Kornberg, R. D. 1977. Structure of chromatin. *Annu. Rev. Biochem.* 46:931–954.

———, and Klug, A. 1981. The nucleosome. *Sci. Am.* 244 (2):52–64.

Lee, A. S., Britten, R. J., and Davidson, E. H. 1977. Interspersion of short repetitive sequences studied in cloned sea urchin DNA fragments. *Science* 196:189–192.

Lewin, B. 1980. *Gene expression,* 2nd ed., vol. 2, *Eucaryotic chromosomes.* New York: Wiley.

Lilly, D., and Purdon, J. 1979. Structure and function of chromatin. *Annu. Rev. Genet.* 13:197–233.

Long, E. O., and Dawid, I. B. 1980. Repeated genes in eukaryotes. *Annu. Rev. Biochem.* 49:727–764.

MacHattie, L. A., Ritchie, D. A., and Thomas, C. A. 1967. Terminal repetition in permuted T2 bacteriophage DNA molecules. *J. Mol. Biol.* 23:355–363.

McGhee, J. D., and Felsenfeld, G. 1980. Nucleosome structure. *Annu. Rev. Biochem.* 49:1115–1156.

Marmur, J., Rownd, R., and Schildkraut, C. L. 1963. Denaturation and renaturation of deoxyribonucleic acid. *Prog. Nucleic Acid Res. Mol. Biol.* 1:231–300.

Mirkovich, J., Mirault, M.-E., and Laemmli, U. K. 1984. Organization of the higher-order chromatin loop: specific DNA attachment sites on nuclear scaffold. *Cell* 39:223–232.

Murray, A. W., and Szostak, J. W. 1983. Chromosome structure and behaviour. *Trends Biochem. Sci.* (Mar.):112–115.

Ng, R., Cumberledge, S., and Carbon, J. 1986. Structure and function of centromeres. UCLA Symposium: Yeast Cell Biology, pp. 225–239. New York: Alan R. Liss.

Olins, A. L., Carlson, R. D., and Olins, D. E. 1975. Visualization of chromatin substructure: 28 bodies. *J. Cell Biol.* 64:528–537.

Paris Conference. 1972. *Standardization in human cytogenetics.* New York: National Foundation, March of Dimes.

Richard, T. J., Finch, J. T., Rushton, B., Rhodes, D., and Klug, A. 1984. Structure of the nucleosome core particle at 7 Å resolution. *Nature* 311:532–537.

Singer, M. F. 1982. Highly repeated sequences in mammalian genomes. *Int. Rev. Cytol.* 76:67–112.

———. 1982. SINEs and LINEs: Highly repeated short and long interspersed sequences in mammalian genomes. *Cell* 28:433–434.

———, and Skowronski, J. 1985. Making sense out of LINES: Long interspersed repeat sequences in mammalian genomes. *Trends Biochem. Sci.* (Mar.):119–121.

Sinsheimer, R. L. 1959. A single-stranded deoxyribonucleic acid from bacteriophage ΦX174. *J. Mol. Biol.* 1:43–53.

Streisinger, G., Edgar, R. S., and Denhardt, G. H. 1964. Chromosome structure in phage T4, I. Circularity of the linkage map. *Proc. Natl. Acad. Sci. USA* 5:775–779.

Tartof, K. D. 1975. Redundant genes. *Annu. Rev. Genet.* 9:355–385.

Thomas, C. A., and MacHattie, L. A. 1967. The anatomy of viral DNA molecules. *Annu. Rev. Biochem.* 36:485–518.

Walmsley, R. W., Chan, C. S. M., Tye, B.-K., and Petes, T. D. 1984. Unusual DNA sequences associated with the ends of yeast chromosomes. *Nature* 310:157–160.

Woodcock, C. L. F., Frado, L.-L. Y., and Rattner, J. B. 1984. The higher-order structure of chromatin: Evidence for a helical ribbon arrangement. *J. Cell Biol.* 99:42–52.

Worcel, A. 1978. Molecular architecture of the chromatin fiber. *Cold Spring Harbor Symp. Quant. Biol.* 42:313–324.

———, and Benyajati, C. 1977. Higher order coiling of DNA in chromatin. *Cell* 12:83–100.

———, and Burgi, E. 1972. On the structure of the folded chromosome of *Escherichia coli. J. Mol. Biol.* 71:127–147.

———, Strogatz, S., and Riley, D. 1981. Structure of chromatin and the linking number of DNA. *Proc. Natl. Acad. Sci. USA* 78:1461–1465.

Yokoyama, R., and Yao, M.-C. 1986. Sequence characterization of *Tetrahymena* macronuclear ends. *Nucleic Acids Res.* 14:2109–2122.

Zimmerman, S. B. 1982. The three-dimensional structure of DNA. *Annu. Rev. Biochem.* 51:395–427.

Chapter 10

Bollum, F. J. 1975. Mammalian DNA polymerases. *Prog. Nucleic Acid Res. Mol. Biol.* 15:109–144.

Cairns, J. 1963. The bacterial chromosome and its manner of replication as seen by autoradiography. *J. Mol. Biol.* 6:208–213.

Cold Spring Harbor Symposia on Quantitative Biology. 1968. *Replication of DNA in microorganisms.* Vol. 33. New York: Cold Spring Harbor Laboratory.

Cozzarelli, N. R. 1980. DNA gyrase and the supercoiling of DNA. *Science* 207:953–960.

DeLucia, P., and Cairns, J. 1969. Isolation of an *E. coli* strain with a mutation affecting DNA polymerase. *Nature* 224:1164–1166.

De Pamphilis, M. L., and Wassarman, P. M. 1980. Replication of eukaryotic chromosomes: A close-up of the replication fork. *Annu. Rev. Biochem.* 49:627–666.

DuPraw, E. J. 1970. *DNA and chromosomes.* New York: Holt, Rinehart and Winston.

Edenburg, H. J., and Huberman, J. A. 1975. Eukaryotic chromosome replication. *Annu. Rev. Genet.* 9:254–284.

Gefter, M. L. 1975. DNA replication. *Annu. Rev. Biochem.* 44:45–78.

———, Hirota, Y., Kornberg, T., Wechsler, J. A., and Barnoux, C. 1971. Analysis of DNA polymerase II and III in mutants of *E. coli* thermosensitive for DNA synthesis. *Proc. Natl. Acad. Sci. USA* 68:3150–3153.

Gilbert, W., and Dressler, D. 1968. DNA replication: The rolling circle model. *Cold Spring Harbor Symp. Quant. Biol.* 33:473–484.

Groppi, V. E., and Coffino, P. 1980. G_1 and S phase mammalian cells synthesize histones at equivalent rates. *Cell* 21:195–204.

Hood, L. E., Wilson, J. H., and Wood, W. B. 1975. *Molecular biology of eucaryotic cells. A problems approach.* Menlo Park, CA: Benjamin.

Huberman, J. A. 1987. Eukaryotic DNA replication: A complex picture partially clarified. *Cell* 48:7–8.

———, and Riggs, A. D. 1968. On the mechanism of DNA replication in mammalian chromosomes. *J. Mol. Biol.* 32:327–341.

Klein, A., and Bonhoeffer, F. 1972. DNA replication. *Annu. Rev. Biochem.* 41:301–332.

Knippers, R. 1970. DNA polymerase II. *Nature* 228:1050–1053.

Korn, D., Fisher, P. A., Battey, J., and Wang, T. .S. F. 1979. Structural and enzymological properties of human DNA polymerases *a* and *b. Cold Spring Harbor Symp. Quant. Biol.* 43:613–624.

Kornberg, A. 1960. Biologic synthesis of deoxyribonucleic acid. *Science* 131:1503–1508.

———. 1980. *DNA replication.* San Francisco: Freeman.

———, Lehman, I. R., Bessman, M. J., and Simms, E. S. 1956. Enzymic synthesis of deoxyribonucleic acid. *Biochim. Biophys. Acta* 21:197–198.

Lehman, I. R. 1974. DNA ligase: Structure, mechanism, and function. *Science* 186:790–797.

Masters, M., and Broda, P. 1971. Evidence for the bidirectional replication of the *E. coli* chromosome. *Nature New Biol.* 232:137–140.

Mazia, D. 1974. The cell cycle. *Sci. Am.* 230:54–64.

Melli, M., Spinelli, G., and Arnold, E. 1977. Synthesis of histone messenger RNA of HeLa cells during the cell cycle. *Cell* 12:167–176.

Meselson, M., and Stahl, F. W. 1958. The replication of DNA in *Escherichia coli. Proc. Natl. Acad. Sci. USA* 44:671–682.

Nossal, N. G. 1983. Prokaryotic DNA replication systems. *Annu. Rev. Biochem.* 58:581–615.

Ogawa, T., Baker, T. A., van der Ende, A., and Kornberg, A. 1985. Initiation of enzymatic replication at the origin of the *Escherichia coli* chromosome: Contributions of RNA polymerase and primase. *Proc. Natl. Acad. Sci. USA* 82:3562–3566.

———, and Okazaki, T. 1980. Discontinuous DNA replication. *Annu. Rev. Biochem.* 49:424–457.

Okazaki, R. T., Okazaki, K., Sakobe, K., Sugimoto, K., and Sugino, A. 1968. Mechanism of DNA chain growth. I. Possible discontinuity and unusual secondary structure of newly synthesized chains. *Proc. Natl. Acad. Sci. USA* 59:598–605.

Pardee, A. B., Dubrow, R., Hamlin, J. L., and Kleitzien, R. F. 1978. Animal cell cycle. *Annu. Rev. Biochem.* 47:715–750.

Prescott, D. M. 1976. *Reproduction of eukaryotic cells.* New York: Academic Press.

Seale, R. L. 1977. Persistence of nucleosomes on DNA during chromatin replication. *Cold Spring Harbor Symp. Quant. Biol.* 42:433–438.

Sheinin, R., Humbert, J., and Pearlman, R. E. 1978. Some aspects of eukaryotic DNA replication. *Annu. Rev. Biochem.* 47:277–316.

Simchen, G. 1978. Cell cycle mutants. *Annu. Rev. Genet.* 12:161–191.

Taylor, J. H. 1970. The structure and duplication of chromosomes. In *Genetic organization,* edited by E. Caspari and A. Ravin, vol. 1, pp. 163–221. New York: Academic Press.

Van der Ende, A., Baker, T. A., Ogawa, T., and Kornberg, A. 1985. Initiation of enzymatic replication at the origin of the *Escherichia coli* chromosome: Primase as the sole priming enzyme. *Proc. Natl. Acad. Sci. USA* 82:3954–3958.

Wang, J. C., and Liu, L. F. 1979. DNA topoisomerases; enzymes that catalyze the concerted breakage and rejoining of DNA backbone bonds. In *Molecular Genetics,* edited by J. H. Taylor, part 3, pp. 65–88. New York: Academic Press.

Weissbach, A. 1977. Eukaryotic DNA polymerases. *Annu. Rev. Biochem.* 46:25–47.

Wickner, S. H. 1978. DNA replication proteins of *Escherichia coli. Annu. Rev. Biochem.* 47:1163–1191.

Williamson, D. H. 1985. The yeast ARS element, six years on: A progress report. *Yeast* 1:1–14.

Zyskind, J. W., and Smith, D. W. 1986. The bacterial origin of replication, *oriC. Cell* 46:489–490.

Chapter 11

Adhya, S., and Gottesman, M. 1978. Control of transcription termination. *Annu. Rev. Biochem.* 47:967–996.

Baker, S. M., and Platt, T. 1986. Pol I transcription: Which comes first, the end or the beginning? *Cell* 47:839–840.

Birnsteil, M. L., Busslinger, M., and Strub, K. 1985. Transcription termination and 3′ processing: The end is in site! *Cell* 41:349–359.

Brand, A. H., Breeden, L., Abraham, J., Sternglanz, R., and Nasmyth, K. 1987. Characterization of a "silencer" in yeast: A DNA sequence with properties opposite to those of a transcriptional enhancer. *Cell* 41:41–48.

Brennan, C. A., Dombroski, A. J., and Platt, T. 1987. Transcription termination factor rho is an RNA-DNA helicase. *Cell* 48:945–952.

Cold Spring Harbor Symposium on Quantitative Biology. 1970. *Transcription of genetic material.* Vol. 35. New York: Cold Spring Harbor Laboratory.

Corden, J., Wasylyk, B., Buchwalder, A., Sassone-Corsi, P., Kedinger, C., and Chambon, P. 1980. Promoter sequences of eukaryotic protein-coding genes. *Science* 209:1406–1414.

Hawley, D. K., and McClure, W. R. 1983. Compilation and analysis of *Escherichia coli* promoter DNA sequences. *Nucleic Acids Res.* 11:2237–2255.

Losick, R., and Chamberlin, M., eds. 1976. *RNA polymerase.* New York: Cold Spring Harbor Laboratory.

Marmur, J., Greenspan, C. M., Palecek, E., Kahan, F. M., Levine, J., and Mandel, M. 1963. Specificity of the complementary RNA formed by *Bacillus sub-*

tilis infected with bacteriophage SP8. *Cold Spring Harbor Symp. Quant. Biol.* 28:191–199.

McKnight, S. L., and Kingsbury, R. 1982. Transcriptional control signals of a eukaryotic protein-coding gene. *Science* 217:316–324.

Platt, T., and Bear, D. 1983. The role of RNA polymerase, rho factor, and ribosomes in transcription termination, in *Gene Function in Prokaryotes,* J. Beckwith, J. Davies, and J. Gallant, eds., pp. 123–162. New York: Cold Spring Harbor Laboratory.

Reznikoff, W. S., Siegele, D. A., Cowing, D. W., and Gross, C. A. 1985. The regulation of transcription initiation in bacteria. *Annu. Rev. Genet.* 19:355–387.

Rodriguez, R., and Chamberlin, M., eds. 1982. *Promoters: Structure and Function.* New York: Praeger.

Stewart, P. R., and Letham, D. S., eds. 1977. *The ribonucleic acids,* 2nd ed. New York: Springer-Verlag.

Struhl, K., 1987. Promoters, activator proteins, and the mechanism of transcriptional initiation in yeast. *Cell* 49:295–297.

Voss, S. D., Schlokat, U., and Gruss, P. 1986. The role of enhancers in the regulation of cell-type-specific transcriptional control. *Trends Biochem. Sci.* (July):287–289.

Chapter 12

Abelson, J. N., Brody, E. N., Cheng, S.-C., Clark, M. W., Green, P. R., Dalbadie-McFarland, G., Lin, R.-J., Newman, A. J., Phizicky, E. M., and Vijayraghavan, U. 1986. RNA splicing in yeast. *Chemica Scripta* 26B:127–137.

Apirion, D., ed. 1984. *Processing of RNA.* C.R.C. Press, Boca Raton, FL.

Baker, S. M., and Platt, T. Pol I transcription: Which comes first, the end or the beginning? *Cell* 47:839–840.

Banerjee, A. K. 1980. 5′-terminal cap structure in eucaryotic messenger ribonucleic acids. *Microbiol. Rev.* 44:175–205.

Black, D. L., Chabot, B., and Steitz, J. A. 1985. U2 as well as U1 small nuclear ribonucleoproteins are involved in pre-messenger RNA splicing. *Cell* 42:737–750.

Bogenhagen, D. F., Sakonju, S., and Brown, D. D. 1980. A control region in the center of the 5S RNA gene directs specific initiation of transcription II: The 3′ border of the region. *Cell* 19:27–35.

Brawerman, G. 1976. Characteristics and significance of the polyadenylate sequence in mammalian mes-*

senger RNA. *Prog. Nucleic Acid Res. Mol. Biol.* 17:117–148.

Breathnach, R., and Chambon, P. 1981. Organization and expression of eucaryotic split genes coding for proteins. *Annu. Rev. Biochem.* 50:349–383.

———, Mandel, J. L., and Chambon, P. 1977. Ovalbumin gene is split in chicken DNA. *Nature* 270:314–318.

Breitbart, R. E., Andreadis, A., and Nadal-Ginard, B. 1987. Alternative splicing: A ubiquitous mechanism for the generation of multiple protein isoforms from single genes. *Annu. Rev. Biochem.* 56:467–495.

Brimacombe, R., Stoffler, G., and Wittmann, H. G. 1978. Ribosome structure. *Annu. Rev. Biochem.* 47:217–249.

Brody, E., and Abelson, J. 1985. The "spliceosome"; Yeast premessenger RNA associates with a 40S complex in a splicing dependent reaction. *Science* 228:963–967.

Brown, D. D. 1981. Gene expression in eukaryotes. *Science* 211:667–674.

Cech, T. R. 1983. RNA splicing: Three themes with variations. *Cell* 34:713–716.

———. 1985. Self-splicing RNA: Implications for evolution. *Int. Rev. Cytol.* 93:3–22.

———. 1986. Ribosomal RNA gene expression in Tetrahymena: Transcription and RNA splicing. *Mol. biol. ciliated protozoa,* pp. 203–225. New York: Academic Press.

Chabot, B., Black, D. L., LeMaster, D. M., and Steitz, J. A. 1985. The 3′ splice site of pre-messenger RNA is recognized by a small nuclear ribonucleoprotein. *Science* 230:1344–1349.

———, and Steitz, J. A. 1987. Multiple interactions between the splicing substrate and small nuclear ribonucleoproteins in spliceosomes. *Mol. Cell. Biol.* 7:281–293.

Chambliss, G., Craven, G. R., Davies, J., Davis, K., Kahan, L., and Nomura, M., eds. 1980. *Ribosomes. Structure, function, and genetics.* Baltimore: University Park Press.

Chambon, P. 1975. Eukaryotic nuclear RNA polymerases. *Annu. Rev. Biochem.* 44:613–638.

Cheng, S.-C., and Abelson, J. 1986. Fractionation and characterization of a yeast mRNA splicing extract. *Proc. Natl. Acad. Sci. USA* 83:2387–2391.

Choi, Y. D., Grabowski, P. J., Sharp, P. A., and Dreyfuss, G. 1986. Heterogeneous nuclear ribonucleoproteins: Role in RNA splicing. *Science* 231:1534–1539.

Chu, F. K., Maley, G. F., West, D. K., Belfort, M., and Maley, F. 1986. Characterization of the intron in the phage T4 thymidylate synthase gene and evidence for its self-excision from the primary transcript. *Cell* 45:157–166.

Craig, N., Kass, S., and Sollner-Webb, B. 1987. Nucleotide sequence determining the first cleavage site in the processing of mouse precursor rRNA. *Proc. Natl. Acad. Sci. USA* 84:629–633.

Crick, F. H. C. 1979. Split genes and RNA splicing. *Science* 204:264–271.

Deutscher, M. 1984. Processing of tRNA in prokaryotes and eukaryotes. *Crit. Rev. Biochem.* 17:45–71.

Furuichi, Y., Morgan, M., Shatkin, A. J., Jelinek, W., Salditt-Georgieff, M., and Darnell, J. E. 1975. Methylated, blocked 5′ termini in HeLa cell mRNA. *Proc. Natl. Acad. Sci. USA* 72:1904–1908.

Gilbert, W. 1985. Genes-in-pieces revisited. *Science* 228:823–824.

Gluzman, Y., ed. 1985. Eukaryotic transcription: The role of cis- and trans-acting elements in initiation. Cold Spring Harbor Laboratory, Cold Spring Harbor, NY.

Grabowski, P. J., Padgett, R. A., and Sharp, P. A. 1984. Messenger RNA splicing in vitro: An excised intervening sequence and a potential intermediate. *Cell* 37:415–427.

———, Seiler, S. R., and Sharp, P. A. 1985. A multicomponent complex is involved in the splicing of messenger RNA precursors. *Cell* 42:355–367.

Green, M. R., 1986. Pre-mRNA splicing. *Annu. Rev. Genet.* 20:671–708.

Guarente, L. 1984. Yeast promoters: Positive and negative elements. *Cell* 36:799–800.

Guthrie, C. 1986. Finding functions for small nuclear RNAs in yeast. *Trends Biochem. Sci.* (Oct.): pp. 430–434.

Hall, B., Clarkson, S. G., and Tocchini-Valenti, G. 1982. Transcription initiation of eucaryotic transfer RNA genes. *Cell* 29:3–5.

Jacob, S. T. 1986. Transcription of eukaryotic ribosomal RNA genes. *Mol. Cell. Biochem.* 70:11–20.

Jeffreys, A. J., and Flavell, R. A. 1977. The rabbit beta-globin gene contains a large insert in the coding sequence. *Cell* 12:1097–1108.

Konarska, M. M., and Sharp, P. A. 1987. Interactions between small nuclear ribonucleoprotein particles in the formation of spliceosomes. *Cell* 49:763–774.

Kurland, C. G. 1977. Aspects of ribosome structure and function. In *Molecular mechanisms of protein biosynthesis,* edited by H. Weissbach and S. Pestka, pp. 81–116. New York: Academic Press.

Labhart, P., and Reeder, R. H. 1986. Characterization

of three sites of RNA 3′ end formation in the Xenopus ribosomal gene spacer. *Cell* 45:431–443.

———, and Reeder, R. H. 1987. Ribosomal precursor 3′ end formation requires a conserved element upstream of the promoter. *Cell* 50:51–57.

MacCumber, M., and Ornstein, R. L. 1984. Molecular model for messenger RNA splicing. *Science* 224:400–405.

Maden, B. E. H. 1976. Ribosomal precursor RNA and ribosome formation in eukaryotes. *Trends Biochem. Sci.* 1:196–199.

McStay, B., and Reeder, R. H. 1986. A termination site for Xenopus RNA polymerase I also acts as an element of an adjacent promoter. *Cell* 47:913–920.

Nazar, R. N. 1982. The eukaryotic 5.8 and 5 S ribosomal RNAs and related rDNAs. *Cell Nucleus* 11:1–28.

Nebins, J. R. 1983. The pathway of eukaryotic mRNA formation, *Annu. Rev. Biochem.* 52:441–466.

Nierhaus, K. H. 1980. The assembly of the prokaryotic ribosome. *Biosystems* 12:273–282.

Nishimura, S. 1974. Transfer RNA: Structure and biosynthesis. In *Biochemistry of nucleic acids,* edited by K. Burton, vol. 6 in *MTP international review of science,* pp. 289–322. London: Butterworths.

Nomura, M. 1973. Assembly of bacterial ribosomes. *Science* 179:864–873.

———, Morgan, E. A., and Jaskunas, S. R. 1977. Genetics of bacterial ribosomes. *Annu. Rev. Genet.* 11:297–347.

———,Tissieres, A., and Lengyel, P., eds. 1974. *Ribosomes.* New York: Cold Spring Harbor Laboratory.

O'Farrell, P. Z., Cordell, B., Valenzuela, P., Rutter, W. J., and Goodman, H. M. 1978. Structure and processing of yeast precursor tRNAs containing intervening sequences. *Nature* 274:438–445.

Padgett, R. A., Grabowski, P. J., Konarska, M. M., and Sharp, P. A. 1985. Splicing messenger RNA precursors: Branch sites and lariat RNAs. *Trends Biochem. Sci.* (April):154–157.

Perry, R. P. 1976. Processing of RNA. *Annu. Rev. Biochem.* 45:605–629.

Reed, R., and Maniatis, T. 1985. Intron sequences involved in lariat formation during pre-mRNA splicing. *Cell* 41:95–105.

Reeder, R. H. 1984. Enhancers and ribosomal gene spacers. *Cell* 38:349–351.

Rungger, D., and Crippa, M. 1977. Primary ribosomal DNA transcript in eukaryotes. *Prog. Biophys. Mol. Biol.* 31:247–269.

Ruskin, B., and Green, M. R. 1985. An RNA processing activity that debranches RNA lariats. *Science* 229:135–140.

Russell, P. J., and Wilkerson, W. M. 1980. The structure and biosynthesis of fungal cytoplasmic ribosomes. *Exp. Mycol.* 4:281–337.

Sakonju, S., Bogenhagen, D. F., and Brown, D. D. 1980. A control region in the center of the 5S RNA gene directs specific initiation of transcription I: The 5′ border of the region. *Cell* 19:13–25.

Schmidt, F. J. 1985. RNA splicing in prokaryotes: Bacteriophage T4 leads the way. *Cell* 41:339–340.

Sharp, P. A. 1987. Splicing of messenger RNA precursors. *Science* 235:766–771.

———, 1985. On the origin of RNA splicing and introns. *Cell* 42:397–400.

Sollner-Webb, B., and Tower, J. 1986. Transcription of cloned eukaryotic ribosomal RNA genes. *Annu. Rev. Biochem.* 55:801–830.

Starzyk, R. M. 1986. Prokaryotic RNA processing. *Trends Biochem. Sci.* (Feb.):60.

Sussman, J. L., and Kim, S. H. 1976. Three-dimensional structure of a transfer RNA in two crystal forms. *Science* 192:853–858.

Sylvester, J. E., Whiteman, D. A., Podolsky, R., Pozsgay, J. M., Respess, J., and Schmickel, R. D. 1986. The human ribosomal RNA genes: Structure and organization of the complete repeating unit. *Hum. Genet.* 73:193–198.

Tilghman, S. M., Curis, P. J., Tiemeier, D. C., Leder, P., and Weissman, C. 1978. The intervening sequence of a mouse β-globin gene is transcribed within the 15S β-globin mRNA precursor. *Proc. Natl. Acad. Sci. USA* 75:1309–1313.

———, Tiemeier, D. C., Seidman, J. G. Peterlin, B. M., Sullivan, M., Maizel, J. V., and Leder, P. 1978. Intervening sequence of DNA identified in the structural portion of a mouse beta-globin gene. *Proc. Natl. Acad. Sci. USA* 78:725–729.

Weinstock, R., Sweet, R., Weiss, M., Cedar, H., and Axel, R. 1978. Intragenic DNA spacers interrupt the ovalbumin gene. *Proc. Natl. Acad. Sci. USA* 75:1299–1303.

Zaug, A. J., and Cech, T. R. 1986. The intervening sequence RNA of Tetrahymena is an enzyme. *Science* 231:470–475.

Chapter 13

Bachmair, A., Finley, D., Varshavsky, A. 1986. In vivo half-life of a protein is a function of its amino-terminal residue. *Science* 234:179–186.

Blobel, G., and Dobberstein, B. 1975. Transfer of proteins across membranes. I. Presence of proteolytically processed and unprocessed nascent immunoglobulin light chains on membrane-bound ribosomes of murine myeloma. *J. Cell Biol.* 67:835–851.

Brenner, S., Jacob, F., and Meselson, M. 1961. An unstable intermediate carrying information from genes to ribosomes for protein synthesis. Nature 190:576–581.

Cold Spring Harbor Symposia for Quantitative Biology, Vol. 31, 1966: *The genetic code.* New York: Cold Spring Harbor Laboratory.

Colman, A., and Robinson, C. 1986. Protein import into organelles: hierarchical targeting signals. *Cell* 46:321–322.

Crick, F. H. C. 1966. Codon-anticodon pairing: The wobble hypothesis. *J. Mol. Biol.* 19:548–555.

———, Barnett, L., Brenner, S., and Watts-Tobin, R. J. 1961. General nature of the genetic code for proteins. *Nature* 192:1227–1232.

Dingwell, C. 1985. The accumulation of proteins in the nucleus. *Trends Biochem. Sci.* (Feb.):pp. 64–66.

Garen, A. 1968. Sense and nonsense in the genetic code. *Science* 160:149–159.

Griffiths, G., and Simons, K. 1986. The trans Golgi network: Sorting at the exit site of the Golgi complex. *Science* 234:438–442.

Gupta, N. K., Chatterjee, B., Chen, Y. C., and Majumdar, A. 1975. Protein synthesis in rabbit reticulocytes. A study of met-tRNA.fmet binding factor(s) and met-tRNA.fmet. *J. Biol. Chem.* 250:853–862.

Hardesty, B., Culp, W., and McKeehan, W. 1969. The sequence of reactions leading to the synthesis of a peptide bond on reticulocyte ribosomes. *Cold Spring Harbor Symp. Quant. Biol.* 34:331–344.

Haselkorn, R., and Rothman-Denes, L. B. 1973. Protein synthesis. *Annu. Rev. Biochem.* 43:397–438.

Horowitz, S., and M. A. Gorovsky. 1985. An unusual genetic code in nuclear genes of Tetrahymena. *Proc. Natl. Acad. Sci. USA* 82:2452–2455.

Khorana, H. G. 1966–67. Polynucleotide synthesis and the genetic code. *Harvey Lectures* 62:79–105.

Kozak, M. 1978. How do eucaryotic ribosomes select initiation regions in messenger RNA? *Cell* 15:1109–1123.

———. 1981. Mechanism of mRNA recognition by eukaryotic ribosomes during initiation of protein synthesis. *Curr. Top. Microbiol. Immunol.* 93:81–123.

———. 1983. Comparison of initiation of protein synthesis in procaryotes, eucaryotes, and organelles. *Microbiol. Rev.* 47:1–45.

———. 1984. Compilation and analysis of sequences upstream from the translation start site in eukaryotic mRNAs. *Nucleic Acids Res.* 12:857–872.

Kreibich, G., Czako-Graham, M., Grebenau, R. C., and Sabatini, D. D. 1980. Functional and structural characteristics of endoplasmic reticulum proteins associated with ribosome binding sites. *Ann. N.Y. Acad. Sci.* 343:17–33.

Maitra, U., Stringer, E. A., and Chaudhuri, A. 1982. Initiation factors in protein biosynthesis. *Annu. Rev. Biochem.* 51:869–900.

Meyer, D. I. 1982. The signal hypothesis—A working model. *Trends Biochem. Sci.* 7:320–321.

Moldave, K. 1985. Eukaryotic protein synthesis. *Annu. Rev. Biochem.* 54:1109–1150.

Morgan, A. R., Wells, R. D., and Khorana, H. G. 1966. Studies on polynucleotides. LIX. Further codon assignments from amino acid incorporation directed by ribopolynucleotides containing repeating trinucleotide sequences. *Proc. Natl. Acad. Sci. USA* 56:1899–1906.

Nirenberg, M., and Leder, P. 1964. RNA code words and protein synthesis. *Science* 145:1399–1407.

———, and Matthaei, J. H. 1961. The dependence of cell-free protein synthesis in *E. coli* upon naturally occurring or synthetic polyribonucleotides. *Proc. Natl. Acad. Sci. USA* 47:1588–1602.

Perara, E., Rothman, R. E., and Lingappa, V. R. 1986. Uncoupling translocation from translation: Implications for transport of proteins across membranes. *Science* 232:348–352.

Pestka, S. 1976. Insights into protein biosynthesis and ribosome function during inhibitors. *Prog. Nucleic Acid Res. Mol. Biol.* 17:217–245.

Pfeffer, S. R., and Rothman, J. E. 1987. Biosynthetic protein transport and sorting by the endoplasmic reticulum and Golgi. *Annu. Rev. Biochem.* 56:829–852.

Revel, M. 1977. Initiation of messenger RNA translation into proteins and some aspects of its regulation. In *Molecular mechanisms of protein biosynthesis,* edited by H. Weissbach and S. Pestka, pp. 246–321. New York: Academic Press.

Rogers, S., Wells, R., and Rechsteiner, M. 1986. Amino acid sequences common to rapidly degraded proteins: The PEST hypothesis. *Science* 234:364–368.

Schekman, R. 1985. Protein localization and membrane traffic in yeast. *Annu. Rev. Cell Biol.* 1:115–143.

Shatkin, A. J., Banerjee, A. K., Both, G. W., Furuichi, Y., and Muthukrishnan, S. 1976. Dependence of translation on 5'-terminal methylation of mRNA. *Fed. Proc.* 35:2214–2217.

Shine, J., and Delgarno, L. 1974. The 3'-terminal sequence of *Escherichia coli* 16S ribosomal RNA: Complementarity to nonsense triplet and ribosome-binding sites. *Proc. Natl. Acad. Sci. USA* 71:1342–1346.

Silhavy, T. J., Benson, S. A., and Emr, S. D. 1983. Mechanisms of protein localization. *Microbiol. Rev.* 47:313–344.

Smith, A. E. 1976. *Protein biosynthesis.* London: Chapman and Hall.

Tompkins, R. K., Scolnick, E. M., and Caskey, C. T. 1970. Peptide chain termination. VII. The ribosomal and release factor requirements for peptide release. *Proc. Natl. Acad. Sci. USA* 65:702–708.

Uy, R., and Wold, F. 1977. Posttranslational covalent modifications of proteins. *Science* 198:890–896.

Walter, P., Gilmore, R., and Blobel G. 1984. Protein translocation across the endoplasmic reticulum. *Cell* 38:5–8.

Watson, J. D. 1963. The involvement of RNA in the synthesis of proteins. *Science* 140:17–26.

Weissbach, H., and Ochoa, S. 1976. Soluble factors required for eukaryotic protein synthesis. *Annu. Rev. Biochem.* 45:191–216.

———, and Pestka, S., eds. 1977. *Molecular mechanisms of protein biosynthesis.* New York: Academic Press.

Chapter 14

Beadle, G. W., and Ephrussi, B. 1937. Development of eye colors in *Drosophila:* Diffusible substances and their interrelationships. *Genetics* 22:76–86.

———, and Tatum, E. L. 1942. Genetic control of biochemical reactions in *Neurospora. Proc. Natl. Acad. Sci. USA* 27:499–506.

Beet, E. A. 1949. The genetics of the sickle-cell trait in a Bantu tribe. *Ann. Eugen.* 14:279–284.

Bell, G. I., Selby, M. J., and Rutter, W. J. 1982. The highly polymorphic region near the human insulin gene is composed of simple tandemly repeating sequences. *Nature* 295:31–35.

Benzer, S. 1959. On the topology of the genetic fine structure. *Proc. Natl. Acad. Sci. USA* 45:1607–1620.

———. 1961. On the topography of the genetic fine structure. *Proc. Natl. Acad. Sci. USA* 47:403–415.

———. 1962. The fine structure of the gene. *Sci. Am.,* 206:70–84.

Fincham, J. 1966. *Genetic complementation.* Menlo Park, CA: Benjamin.

Garrod, A. E. 1909. *Inborn errors of metabolism.* New York: Oxford University Press.

Gilbert, F., Kucherlapati, R., Creagan, R. P., Murnane, M. J., Darlington, G. J., and Ruddle, F. H. 1975. Tay-Sachs' and Sandhoff's diseases: The assignment of genes for hexosaminidase A and B to individual human chromosomes. *Proc. Natl. Acad. Sci. USA* 72:263–267.

Gusella, J. F., Wexler, N. S., Conneally, P. M., Naylor, S. L., Anderson, M. A., Tanzi, R. E., Watkins, P. C., Ottina, K., Wallace, M. R., Sakaguchi, A. Y., Young, A. B., Shoulson, I., Bonilla, E., and Martin, J. B. 1983. A polymorphic DNA marker genetically linked to Huntington's disease. *Nature* 306:234–238.

Harris, H. 1974. *Prenatal diagnosis and selective abortion.* Cambridge: Harvard University Press.

———. 1975. *The principles of human biochemical genetics.* Amsterdam: North-Holland.

Ingram, V. M. 1961. Gene evolution and the haemoglobins. *Nature* 189:704–708.

———. 1963. *The hemoglobins in genetics and evolution.* New York: Columbia University Press.

Johnson, L. A., Gordon, R. B., and Emerson, B. T. 1976. Two populations of heterozygote erythrocytes in moderate hypoxanthine guanine phosphoribosyltransferase deficiency. *Nature* 264:172–174.

Kan, Y. W., and Dozy, A. M. 1978. Polymorphism of DNA sequence adjacent to human β-globin structural gene: Relationship to sickle mutation. *Proc. Natl. Acad. Sci. USA* 75:5631–5635.

Kolodny, E. H. 1976. Lysosomal storage diseases. *N. Engl. J. Med.* 294:1217–1220.

Lehmann, H., and Carrell, R. W. 1969. Variations in structure of human haemoglobin. *Br. Med. Bull.* 25:14–23.

———, and Huntsman, R. G. 1966. *Man's haemoglobins: Including the haemoglobinpathies and their investigation.* Amsterdam: North-Holland.

Livingstone, F. B. 1967. *Abnormal hemoglobins in human populations. A summary and interpretation.* Chicago: Aldine.

Loo, Y. H. 1967. Characterization of a new phenylalanine metabolite in phenylketonuria. *J. Neurochem.* 114:813–821.

Maniatis, T., Fritsch, E. F., Lauer, J., and Lawn, R. M. 1980. The molecular genetics of human hemoglobins. *Annu. Rev. Genet.* 14:145–178.

Milunsky, A. 1976. Prenatal diagnosis of genetic disorders. *N. Engl. J. Med.* 295:377–380.

Motulsky, A. G. 1964. Hereditary red cell traits and malaria. *Am. J. Trop. Med. Hyg.* 13:147–158.

———. 1973. Frequency of sickling disorders in U.S. blacks. *N. Engl. J. Med.* 288:31–33.

Murayama, M. 1966. Molecular mechanism of red cell "sickling." *Science* 153:145–149.

Neel, J. V. 1949. The inheritance of sickle-cell anemia. *Science* 110:64–66.

Pauling, L., Itano, H. A., Singer, S. J., and Wells, J. C. 1949. Sickle-cell anemia, a molecular disease. *Science* 110:543–548.

Perutz, M. F., and Mitchison, J. M. 1950. State of hae-moglobin in sickle-cell anemia. *Nature* 166:677–679.

Race, R. R., and Sanger, R. 1975. *Blood groups and man,* 6th ed. Oxford: Blackwell Scientific.

Scriver, C. R., and Clow, C. L. 1980. Phenylketonuria and other phenylalanine hydroxylation mutants in man. *Annu. Rev. Genet.* 14:179–202.

Srb, A. M., and Horowitz, N. H. 1944. The ornithine cycle in *Neurospora* and its genetic control. *J. Biol. Chem.* 154:129–139.

Stamatoyannopoulos, G. 1972. The molecular basis of hemoglobin disease. *Annu. Rev. Genet.* 6:47–70.

Weatherall, D. J., and Clegg, J. B. 1976. Molecular ge-netics of human hemoglobins. *Annu. Rev. Genet.* 10:157–178.

———. 1979. Recent developments in the molecular genetics of human hemoglobin. *Cell* 16:467–479.

Woo, S. L. C., Lidsky, A. S., Guttler, F., Chandra, T., and Robson, K. J. H. 1983. Cloned human phenylal-anine hydroxylase gene allows prenatal diagnosis and carrier detection of classical phenylketonuria. *Nature* 300:151–155.

Chapter 15

Aaij, C., and Borst, P. 1972. The gel electrophoresis of DNA. *Biochem. Biophys. Acta* 269:192–200.

Arber, W. 1965. Host-controlled modification of bacte-riophage. *Annu. Rev. Microbiol.* 19:365–378.

———. 1974. DNA modification and restriction. *Prog. Nucl. Acid Res. Mol. Biol.* 14:1–37.

———, and Dussoix, D. 1962. Host specificity of DNA produced by *Escherichia coli* I. Host controlled modification of bacteriophage lambda. *J. Mol. Biol.* 5:18–36.

———, and Linn, S. 1969. DNA modification and re-striction. *Annu. Rev. Biochem.* 38:467–500.

Baltimore, D. 1970. Viral RNA-dependent DNA poly-merase. *Nature* 226:1209–1211.

Blattner, F. R., Williams, B. G., Blechl, A. E., Dennis-ton-Thompson, K., Kiefer, D. O., Moore, D. D., Schumm, J. H. W. , Sheldon, E. L., and Smithies, O. 1977. Charon phages: safer derivatives of bacterio-phage lambda for DNA cloning. *Science* 196:161–169.

Boyer, H. W. 1971. DNA restriction and modification mechansms in bacteria. *Annu. Rev. Microbiol.* 25:153–176.

Chan, H. W., Israel, M. A., Garon, C. F., Rowe, W. P., and Martin, M. A. 1979. Molecular cloning of poly-oma virus DNA in *Escherichia coli*:plasmid vector system. *Science* 203:883–892.

Chang, L. M. S., and Bollum, F. J. 1971. Enzymatic synthesis of oligodeoxynucleotides. *Biochemistry* 10:536–542.

Curtiss, R. III. 1976. Genetic manipulation of microor-ganisms: Potential benefits and biohazards. *Annu. Rev. Microbiol.* 30:507–533.

Danna, K., and Nathans, D. 1971. Specific cleavage of simian virus 40 DNA by restriction endonuclease of *Haemophilus influenzae. Proc. Natl. Acad. Sci. USA* 68:2913–2917.

Danna, K. J., Sack, G. H. and Nathans, D. 1973. Stud-ies of simian virus 40 DNA. VII. A cleavage map of the SV40 genome. *J. Mol. Biol.* 78:363–376.

Donis-Keller, H., Barker, D., Knowlton, R., Schumm, J., and Braman, J. 1986. Applications of RFLP probes to genetic mapping and clinical diagnosis in humans. In *Current communications in molecular biology: DNA probes—Applications in genetic and infectious disease and cancer,* edited by L. S. Ler-man pp. 73–81. Cold Spring Harbor Laboratory, Cold Spring Harbor, NY.

Drayna, D., and White, R. 1985. The genetic linkage map of the human X chromosome. *Science* 230:753–758.

Dussoix, D., and Arber, W. 1962. Host specificity of DNA produced by *Escherichia coli.* II. Control over acceptance of DNA from infecting phage lambda. *J. Mol. Biol.* 5:37–49.

Freifelder, D. 1978. *Recombinant DNA: Readings from Scientific American,* San Francisco: W. H. Freeman.

Galjaard, H. 1986. Biochemical diagnosis of genetic diseases. *Experientia* 42:1075–1085.

Guttler, F., and Woo, S. L. C. 1986. Molecular genetics of PKU. *J. Inherited Metab. Dis.* 9 Suppl. 1:58–68.

Kelley, T. J., and Smith, H. O. 1970. A restriction en-zyme from *Haemophilus influenzae.* II. Base se-quence of the recognition site. *J. Mol. Biol.* 51:393–409.

Kessler, C., Neumaier, P. S., and Wolf, W. 1985. Recog-nition sequences of restriction endonucleases and methylases—A review. *Gene* 33:1–102.

Klee, H., Horsch, R., and Rogers, S. 1987. *Agrobacte-rium*-mediated plant transformation and its further applications to plant biology. *Annu. Rev. Plant Physiol.* 38:467–486.

Lee, A. S., and Sinsheimer, R. L. 1974. A cleavage map of bacteriophage ΦX174. *Proc. Natl. Acad. Sci. USA* 71:2882–2886.

Lobban, P. E., and Kaiser, A. D. 1973. Enzymatic end-to-end joining of DNA molecules. *J. Mol. Biol.* 78:453–471.

Luria, S. E. 1953. Host induced modification of vi-ruses. *Cold Spring Harbor Symp. Quant. Biol.* 18:237–244.

Maniatis, T., Fritsch, E. F., and Sambrook, J. 1982. *Molecular cloning: A laboratory manual.* Cold Spring Harbor Laboratory, Cold Spring Harbor, NY.

Maxam, A. M., and Gilbert, W. 1977. A new method for sequencing DNA. *Proc. Natl. Acad. Sci. USA* 74:560–564.

———, Tizard, R., Skryabin, K. G., and Gilbert, W. 1977. Promoter region for yeast 5S ribosomal RNA. *Nature* 267:643–645.

Meselson, M., Yuan, R., and Heywood, J. 1972. Restriction and modification of DNA. *Annu. Rev. Biochem.* 41:447–466.

Miller, W. L. 1981. Recombination DNA and the pediatrician. *J. Pediatr.* 99:1–15.

Moores, J. C. 1987. Current approaches to DNA sequencing. *Anal. Biochem.* 163:1–8.

Mulder, C., Arrand, J. R., Delius, H., Keller, W., Petterson, U., Roberts, R. J. and Sharp, P. A. 1974. Cleavage maps of DNA from adenovirus types 2 and 5 by restriction endonucleases EcoRI and HpaI. *Cold Spring Harbor Symp. Quant. Biol.* 39:397–400.

Peralta, E. G., and Ream, L. W. 1985. T-DNA border sequences required for crown gall tumorigenesis. *Proc. Natl. Acad. Sci. USA* 82:5112–5116.

Sanger, F., and Coulson, A. R. 1975. A rapid method for determining sequences in DNA by primed synthesis with DNA polymerase. *J. Mol. Biol.* 94:441–448.

Schell, J., and Van Montagu, M. 1983. The Ti plasmids as natural and practical gene vectors for plants. *Bio/Technology* 1:175–180.

Sharp, P. A., Sugden, B., and Sambrook, J. 1973. Detection of two restriction endonuclease activities in .ul Haemophilus parainfluenzae using analytical agarose-ethidium bromide electrophoresis. *Biochemistry* 12:3055–3063.

Sinsheimer, R. L. 1977. Recombinant DNA. *Annu. Rev. Biochem.* 46:415–438.

Smith, H. O., and Wilcox, K. W. 1970. A restriction enzyme from *Haemophilus influenzae.* I. Purification and general properties. *J. Mol. Biol.* 51:379–391.

Southern, E. M. 1975. Detection of specific sequences among DNA fragments separated by gel electrophoresis. *J. Mol. Biol.* 98:503–517.

Struhl, K., Cameron, J. R., and Davis, R. W. 1976. Functional genetic expression of eukaryotic DNA in *Escherichia coli. Proc. Natl. Acad. Sci. USA* 73:1471–1475.

Temin, H. M., and Mizutani, S. 1970. RNA-dependent DNA polymerase in virions of rous sarcoma virus. *Nature* 226:1211–1213.

Wu, R. 1978. DNA sequence analysis. *Annu. Rev. Biochem.* 47:607–634.

Chapter 16

Ames, B. N., Durston, W. E., Yamasaki, E., and Lee, F. D. 1973. Carcinogens are mutagens: A simple test system combining liver homogenates for activation and bacteria for detection. *Proc. Natl. Acad. Sci. USA* 70:2281–2285.

Auerbach, C. 1976. *Mutation Research.* London: Chapman and Hall.

———, and Kilbey, B. J. 1971. Mutation in eukaryotes. *Annu. Rev. Genet.* 5:163–218.

Boyce, R. P., and Howard-Flanders, P. 1964. Release of ultraviolet light-induced thymine dimers from DNA in *E. coli K12. Proc. Natl. Acad. Sci. USA* 51:293–300.

Collins, A., Johnson, R. T., and Boyle, J. M., eds. 1987. Molecular biology of DNA repair, Supplement 6, *J. Cell Sci.*

Devoret, R. 1979. Bacterial tests for potential carcinogens. *Sci. Am.* 241:40–49.

Drake, J. W. 1970. *The Molecular Basis of Mutation.* San Francisco: Holden-Day.

Haseltine, W. A. 1983. Ultraviolet light repair and mutagenesis revisited. *Cell* 33:13–17.

Lederberg, J., and Lederberg, E. M. 1952. Replica plating and indirect selection of bacterial mutants. *J. Bacteriol.* 63:399–406.

Moat, A. G., Peters, N., and Srb, A. M. 1959. Selection and isolation of auxotrophic yeast mutants with the aid of antibiotics. *J. Bacteriol.* 77:673–681.

Modrich, P. 1987. DNA mismatch correction. *Annu. Rev. Biochem.* 56:435–466.

Rädman, M., and Wagner, R. 1986. Mismatch repair in *Escherichia coli. Annu. Rev. Genet.* 20:523–538.

Russell, P. J., and Cohen, M. P. 1976. Enrichment for auxotrophic and heat-sensitive mutants of *Neurospora crassa* by tritum suicide. *Mutat. Res.* 34:359–366.

Setlow, R. B., and Carrier, W. L. 1964. The disappearance of thymine dimers from DNA: An error-correcting mechanism. *Proc. Natl. Acad. Sci. USA* 51:226–231.

Walker, G. C., Marsh, L., and Dodson, L. A. 1985. Genetic analyses of DNA repair: inference and extrapolation. *Annu. Rev. Genet.* 19:103–126.

Woodward, V. W.; DeZeeuw, J. R., and Srb, A. M. 1954. The separation and isolation of particular biochemical mutants of *Neurospora* by differential germination of conidia, followed by filtration and selective plating. *Proc. Natl. Acad. Sci. USA* 40:192–200.

Chapter 17

Auerbach, C. 1976. *Mutation research.* London: Chapman and Hall.

Barr, M. L., and Bertram E. G. 1949. A morphological distinction between neurones of the male and female, and the behavior of the nucleolar satellite during accelerated nucleoprotein synthesis. *Nature* 163:676–677.

Bloom, A. D. 1972. Induced chromosome aberrations in man. *Adv. Hum. Genet.* 3:99–153.

Dalla-Favera, R., Martinotti, S., Gallo, R., Erikson, J., and Croce, C. 1983. Translocation and rearrangements of the *c-myc* oncogene locus in human undifferentiated B-cell lymphomas. *Science* 219:963–997.

DeKlein, A., van Kessel, A. G., Grosveld, G., Bartram, C. R., Hagemeijer, A., Bootsma, D., Spurr, N. K., Heisterkamp, N., Groffen, J., and Stephenson, J. R. 1982. A cellular oncogene is translocated to the Philadelphia chromosome in chronic myelocytic leukemia. *Nature* 300:765–767.

DeVita, V., Hellman, S., and Rosenberg, S., eds. 1986. *Important advances in oncology.* Philadelphia: Lippincott.

Lyon, M. F. 1961. Gene action in the X-chromosomes of the mouse *(Mus. musculus L).* *Nature* 190:372–373.

Penrose, L. S., and Smith, G. F. 1966. *Down's anomaly.* Boston: Little, Brown.

Rowley, J. D. 1973. A new consistent chromosomal abnormality in chronic myelogenous leukemia identified by quinacrine fluorescence and Giemsa staining. *Nature* 243:290–293.

Shaw. M. W. 1962. Familial mongolism. *Cytogenetics* 1:141–179.

Chapter 18

Adams, S. E., Mellor, J., Gull, K., Sim, R. B., Tuite, M. F., Kingman, S. M., and Kingsman, A. J. 1987. The functions and relationships of Ty-VLP proteins in yeast reflect those of mammalian retroviral proteins. *Cell* 49:111–119.

Baltimore, D. 1985. Retroviruses and retrotransposons: The role of reverse transcription in shaping the eukaryotic genome. *Cell* 40:481–482.

Boeke, J. D., Garfinkel, D. J., Styles, C. A., and Fink, G. R. 1985. Ty elements transpose through an RNA intermediate. *Cell* 40:491–500.

Bukhari, A. I., Shapiro, J. A., and Adhya, S. L., eds. 1977. *DNA insertion elements, plasmids and episomes.* Cold Spring Harbor Laboratory, Cold Spring Harbor, NY.

Calos, M. P., and Miller, J. H. 1980. Transposable elements. *Cell* 20:579–595.

Cohen, S. N., and Shapiro, J. A. 1980. Transposable genetic elements. *Sci. Am.* 242:40–49.

Cold Spring Harbor Symposia for Quantitative Biology. 1980. Vol. 45. *Movable genetic elements.* Cold Spring Harbor Laboratory, Cold Spring Harbor, NY.

Engels, W. R. 1983. The P family of transposable elements in *Drosophila. Annu. Rev. Genet.* 17:315–344.

Finnegan, D. J. 1985. Transposable elements in eukaryotes: *Int. Rev. Cytol.* 93:281–326.

Foster, T. J., Davis, M. A., Roberts, D. E., Takashita, K., and Kleckner, N. 1981. Genetic organization of transposon Tn10. *Cell* 23:201–213.

Garfinkel, D. J., Boeke, J. D., and Fink, G. R. 1985. Ty element transposition: Reverse transcriptase and virus-like particles. *Cell* 42:507–517.

Jagadeeswaran, P., Forget, B. G., and Weissman, S. M. 1981. Short interspersed repetitive DNA elements in eucaryotes: Transposable DNA elements generated by reverse transcription of RNA pol III transcripts? *Cell* 26:141–142.

Kleckner, N. 1981. Transposable elements in prokaryotes. *Annu. Rev. Genet.* 15:341–404.

McClintock, B. 1961. Some parallels between gene control systems in maize and in bacteria. *Am. Naturalist* 95:265–277.

———. 1965. The control of gene action in maize. *Brookhaven Symp. Biol.* 18:162 ff.

Mount, S. M., and Rubin, G. M. 1985. Complete nucleotide sequence of the *Drosophila* transposable element copia: Homology between copia and retroviral proteins. *Mol. Cell. Biol.* 5:1630–1638.

Shapiro, J. A. 1977. DNA insertion elements and the evolution of chromosome primary structure. *Trends Biochem. Sci.,* Aug., 176–180.

———. ed. 1983. *Mobile genetic elements.* New York: Academic Press.

Spradling, A. C., and Rubin, G. M. 1981. *Drosophila* genome organization: Conserved and dynamic aspects. *Annu. Rev. Genet.* 15:219–264.

Varmus, H. E. 1982. Form and function of retroviral proviruses. *Science* 216:812–821.

Weiss, R. A., Teich, N., Varmus, H. E., and Coffin, J. M., eds. 1985. *Molecular Biology of Tumor Viruses: RNA Tumor Viruses,* 2nd ed. (2 vols.). Cold Spring Harbor Laboratory, Cold Spring Harbor, NY.

Chapter 19

Holliday, R. 1964. A mechanism for gene conversion in fungi. *Genet. Res.* 5:282–304.

Klar, A., and Strathern, J. N. 1986. *Mechanisms of yeast recombination.* Current communications in molecular biology, Cold Spring Harbor Laboratory, Cold Spring Harbor, NY.

Meselson, M., and Radding, C. M. 1975. A general model for genetic recombination. *Proc. Natl. Acad. Sci. USA* 72:358–361.

———, and Weigle, J. J. 1961. Chromosome breakage accompanying genetic reconstruction in bacteriophage. *Proc. Natl. Acad. Sci. USA* 47:857–868.

Orr-Weaver, T. L., and Szostak, J. W. 1985. Fungal recombination. *Microbiol. Rev.* 49:33–58.

Radding, C. 1982. Homologous pairing and strand exchange in genetic recombination. *Annu. Rev. Genet.* 16:405–437.

Recombination at the DNA level. 1984. Cold Spring Harbor Symposium on Quantitative Biology. vol. 49. Cold Spring Harbor Laboratory, Cold Spring Harbor, NY.

Roman, H. 1985. Gene conversion and crossing-over. *Environ. Mutagenesis* 7:923–932.

Stahl, F. 1979. *Genetic recombination: Thinking about it in phage and fungi.* San Francisco: W. H. Freeman.

Szostak, J., Orr-Weaver, T., Rothstein, R., and Stahl, F. 1983. The double-strand break repair model for recombination. *Cell* 33:25–35.

Whitehouse, H. L. K. 1973. *Towards an understanding of the mechanism of heredity.* 3rd ed. London: E. Arnold.

Chapter 20

Aloni, Y., and N. Hay. 1985. Attenuation may regulate gene expresson in animal viruses and cells. *CRC Crit. Rev. Biochem.* 18:327–383.

Beckwith, J. R., and Zipser, D. 1970. *The lactose operon.* Cold Spring Harbor Laboratory, Cold Spring Harbor, NY.

Bertrand, K., Korn, L., Lee, F., Platt, T., Squires, C. L., Squires, C., and Yanofsky, C. 1975. New features of the structure and regulation of the tryptophan operon of *Escherichia coli. Science* 189:22–26.

———, and Yanofsky, C. 1976. Regulation of transcription termination in the leader region of the tryptophan operon of *Escherichia coli* involves tryptophan as its metabolic product. *J. Mol. Biol.* 103:339–349.

Beyreuther, K. 1978. Revised sequence for the *lac* repressor. *Nature* 274:767.

Calos, M. P. 1978. DNA sequence for a low-level promoter of the *lac* repressor gene and an "up" promoter mutation. *Nature* 274:762–765.

Calvo, J. M., and Fink, G. R. 1971. Regulation of biosynthetic pathways in bacteria and fungi. *Annu. Rev. Biochem.* 40:943–968.

Cashel, M. 1975. Regulation of bacterial ppGpp and pppGpp. *Annu. Rev. Microbiol.* 29:301–318.

Culard, F., and Maurizot, J. C. 1982. Binding of *lac* repressor induces different conformational changes on operator and non-operator DNAs. *FEBS Lett.* 146:153–156.

Dickson, R. C., Abelson, J., Barnes, W. M., and Reznikoff, W. S. 1975. Genetic regulation: The *lac* control region. *Science* 187:27–35.

Echols, H., Conit, D., and Green, L. 1976. On the nature of cis-acting regulatory proteins and genetic organization in bacteriophage: The example of gene *Q* of bacteriophage λ. *Genetics* 83:5–10.

———, and Green, L. 1971. Establishment and maintenance of repression by bacteriophage lambda: The role of the *cI, cII,* and *cIII* proteins. *Proc. Natl. Acad. Sci. USA* 68:2190–2194.

Farabaugh, P. J. 1978. Sequence of the *lacI* gene. *Nature* 274:765.

Fisher, R. F., Das, A., Kolter, R., Winkler, M. E., and Yanofsky, C. 1985. Analysis of the requirements for transcription pausing in the tryptophan operon. *J. Mol. Biol.* 182:397–409.

Gilbert, W., Maizels, N., and Maxam, A. 1974. Sequences of controlling regions of the lactose operon. *Cold Spring Harbor Symp. Quant. Biol.* 38:845–855.

———, and Muller-Hill, B. 1966. Isolation of the *lac* repressor. *Proc. Natl. Acad. Sci. USA* 56:1891–1898.

Greenblatt, J. 1981. Regulation of transcription termination by the *N* gene protein of bacteriophage lambda. *Cell* 24:8–9.

Hershey, A. D., ed. 1971. *The bacteriophage lambda.* Cold Spring Harbor Laboratory, Cold Spring Harbor, NY.

Ho, Y-S., Wulff, D. L., and Rosenberg, M. 1983. Bacteriophage λ protein *cII* binds promoters on the opposite face of the DNA helix from RNA polymerase. *Nature* 304:703–708.

Jacob, F., and Monod, J. 1961. Genetic regulatory mechanisms in the synthesis of proteins. *J. Mol. Biol.* 3:318–356.

———. 1965. Genetic mapping of the elements of the lactose region of *Escherichia coli. Biochem. Biophys, Res. Commun.* 18:693–701.

Johnson, A., Mayer, B. J., and Ptashne, M. 1978.

Mechanism of action of the *cro* protein of bacteriophage λ. *Proc. Natl. Acad. Sci. USA* 75:1783–1787.

Kelley, R. L., and Yanofsky, C. 1982. *trp* aporepressor production is controlled by autogeneous regulation and inefficient translation. *Proc. Natl. Acad. Sci. USA* 79:3120–3124.

Lamond, A. I., and Travers, A. A. 1985. Stringent control of bacterial transcription. *Cell* 41:6–8.

Lee, F., and Yanofsky, C. 1977. Transcription termination at the *trp* operon attenuators of *Escherichia coli* and *Salmonella typhimurium:* RNA secondary structure and regulation of termination. *Proc. Natl. Acad. Sci. USA* 74:4365–4369.

Maizels, N. 1973. The nucleotide sequence of the lactose messenger ribonucleic acid transcribed from the UV5 promoter mutant of *Escherichia coli. Proc. Natl. Acad. Sci. USA* 70:3585–3589.

———. 1974. *E. coli* lactose operon ribosome binding site. *Nature (New Biol.)* 249:647–649.

Maniatis, T., Ptashne, M., Bachman, K., Kleid, D., Flashman, S., Jeffrey, A., and Maurer, R. 1975. Recognition sequences of repressor and polymerase in the operators of bacteriophage lambda. *Cell* 5:109–114.

Miller, J. H., and Reznikoff, W. S. 1978. *The operon.* Cold Spring Harbor Laboratory, Cold Spring Harbor, NY.

Pabo, C. O., Sauer, R. T., Sturtevant, J. M., and Ptashne, M. 1979. The λ repressor contains two domains. *Proc. Natl. Acad. Sci. USA* 76:1608–1612.

Platt, T. 1981. Termination of transcription and its regulation in the tryptophan operon of *E. coli. Cell* 24:10–23.

Ptashne, M. 1984. Repressors. *Trends Biochem. Sci.* 9:142–145.

———. 1986. *A genetic switch.* Oxford: Cell Press and Blackwell Scientific Publications.

———. 1967. Isolation of the λ phage repressor. *Proc. Natl. Acad. Sci. USA* 57:306–313.

———, Bachman, K., Humayun, M. Z., Jeffrey, A., Maurer, R., Meyer, B., and Sauer, R. T. 1976. Autoregulation and function of a repressor in bacteriophage lambda. *Science* 194:136–161.

———, and Gilbert, W. 1970. Genetic repressors. *Sci. Am.* 222:36–44.

Reichardt, L. F. 1975. Control of bacteriophage lambda repressor synthesis after phage infection. The role of the *N, cI, cII, cIII* and *cro* products. *J. Mol. Biol.* 93:267–288.

Reznikoff, W. S. 1972. The operon revisited. *Annu. Rev. Genet.* 6:133–156.

Rose, J. K., and Yanofsky, C. 1974. Interaction of the operator of the tryptophan operon with repressor. *Proc. Natl. Acad. Sci. USA* 71:3134–3138.

Squires, C. L., Lee, F., and Yanofsky, C. 1975. Interaction of the *trp* repressor and RNA polymerase with the *trp* operon. *J. Mol. Biol.* 92:93–111.

Ullmann, A. 1985. Catabolite repression 1985. *Biochimie* 67:29–34.

Winkler, M. E., and Yanofsky, C. 1981. Pausing of RNA polymerase during in vitro transcription of the tryptophan operon leader region. *Biochemistry* 20:3738–3744.

Yanofsky, C. 1981. Attenuation in the control of expression of bacterial operons. *Nature* 289:751–758.

———, and Kolter, R. 1982. Attenuation in amino acid biosynthetic operons. *Annu. Rev. Genet.* 16:113–134.

Chapter 21

Ashburner, M. 1970. Function and structure of polytene chromosomes during insect development. *Adv. Insect Physiol.* 7:1–95.

———, Chihara, C., Meltzer, P., and Richards, G. 1974. Temporal control of puffing activity in polytene chromosomes. *Cold Spring Harbor Symp. Quant. Biol.* 38:655–662.

Astrin, S. M., and Rothberg, P. G. 1983. Oncogenes and cancer. *Cancer Investigation* 1:355–364.

Baker, W. 1978. A genetic framework for *Drosophila.* development. *Annu. Rev. Genet.* 12:451–470.

Beerman, W., and Clever, U. 1964. Chromosome puffs. *Sci. Am.* 210:50–58.

Bishop, J. M. 1983. Cancer genes come of age. *Cell* 32:1018–1020.

———. 1983. Cellular oncogenes and retroviruses. *Annu. Rev. Biochem.* 52:301–354.

Bonner, J. J., and Pardue, M. L. 1976. Ecdysone-stimulated RNA synthesis in imaginal discs of *Drosophila melanogaster. Chromosoma* 58:87–99.

Brown, D. D. 1981. Gene expression in eukaryotes. *Science* 211:667–674.

Busch, H. 1984. *Onc* genes and other new targets for cancer chemotherapy. *J. Cancer Res. Clin. Oncol.* 107:1–14.

Cairns, J., and Logan, J. 1983. Step by step into carcinogenesis. *Nature* 304:582–583.

Case, M. E., and Giles, N. H. 1975. Genetic evidence on the organization of the *qa-1* gene product: A protein regulating the induction of three enzymes in quinate catabolism in *Neurospora crassa. Proc. Natl. Acad. Sci. USA* 72–553–557.

Charache, S., Dover, G., Smith, K., Talbot, C. C., Moyer, M., and Boyer, S. 1983. Treatment of sickle cell anemia with 5-azacytidine results in increased fetal hemoglobin production and is associated with nonrandom hypomethylation of DNA around the γ-δ-β-globin gene complex. *Proc. Natl. Acad. Sci. USA* 80:4842–4846.

Cooper, D. N. 1983. Eukaryotic DNA methylation. *Hum. Genet.* 64:315–333.

Cooper, G. M. 1982. Cellular transforming genes. *Science* 217:801–806.

Davidson, E. H. 1976. *Genetic activity in early development,* 2nd ed. New York: Academic Press.

De Robertis, E. M., and Gurdon, J. B. 1977. Gene activation in somatic nuclei after injection into amphibian oocytes. *Proc. Natl. Acad. Sci. USA* 74:2470–2474.

Doerfler, W. 1983. DNA methylation and gene activity. *Annu. Rev. Biochem.* 52:93–124.

Efstratiadis, A., Posakony, J. W., Maniatis, T., Lawn, R. M., O'Connell, C., Spritz, R. A., DeRiel, J. K., Forget, B. G., Weissman, S. M., Slighton, J. L., Blechl, A. E., Smithies, O., Baralle, F. E., Shoulders, C. C., and Proudfoot, N. J. 1980. The structure and evolution of the human β-globin gene family. *Cell* 21:653–668.

Eissenberg, J. C., Cartwright, I. L., Thomas, G. H., and Elgin, S. C. R. 1985. Selected topics in chromatin structure. *Annu. Rev. Genet.* 19:485–536.

Fraser, F. C. 1970. The genetics of cleft lip and cleft palate. *Am. J. Hum. Genet.* 22:336–352.

Gehring, W. 1979. Developmental genetics of *Drosophila. Annu. Rev. Genet.* 10:209–252.

Giles, N. H. 1978. The organization, function, and evolution of gene clusters in eucaryotes. *Am. Naturalist* 112:641–657.

Gordon, H. 1985. Oncogenes. *Mayo Clin. Proc.* 60:697–713.

Gross, D. S., and Garrard, W. T. 1987. Poising chromatin for transcription. *Trends Biochem. Sci.* (Aug.), pp. 293–297.

Gurdon, J. B. 1968. Transplanted nuclei and cell differentiation. *Sci. Am.* 219:24–35.

———. 1974. *The control of gene expression in animal development.* Oxford: Clarendon Press.

Hadorn, E. 1968. Transdetermination in cells. *Sci. Am.* 219:110–120.

Karlsson, S., and A. W. Nienhuis. 1985. Development regulation of human globin genes. *Annu. Rev. Biochem.* 54:1071–1078.

Kingston, R. E., Baldwin, A. S., and Sharp, P. A. 1985. Transcription control by oncogenes. *Cell* 41:3–5.

Krontiris, T. G. 1983. The emerging genetics of human cancer. *N. Engl. J. Med.* 309:404–409.

Loomis, W. 1970. *Papers on regulation of gene activity during development.* New York: Harper & Row.

Lucas, G. L., and Opitz, J. M. 1966. The nail-patella syndrome. *J. Pediatr.* 68:273–288.

McGinnis, W., Levine, M. S., Hafen, E., Kuroiwa, A., and Gehring, W. J. 1984. A conserved DNA sequence homeotic genes of the *Drosophila Antennapedia* and bithorax complexes. *Nature* 308:428–433.

Murdoch, J. L., Walker, B. A., Hall, J. G., Abbey, H., Smith, K. K., and McKusick, V. A. 1970. Achondroplasia—A genetic and statistical survey. *Ann. Hum. Genet.* 33:227–244.

O'Malley, B. W., and Schrader, W. T. 1976. The receptors of steroid hormones. *Sci. Am.* 234:32–43.

O'Malley, B., Towle, H., and Schwartz, R. 1977. Regulation of gene expression in eucaryotes. *Annu. Rev. Genet.* 11:239–275.

Postlethwait, J. H., and Schneiderman, H. A. 1973. Developmental genetics of *Drosophila* imaginal discs. *Annu. Rev. Genet.* 7:381–433.

Ratner, L., Gallo, R. C., and Wong-Staal, F. 1985. Cloning of human oncogenes. In *Recombinant DNA research and virus,* Y. Becker, (ed.) Boston: Martinus Nijhoff.

———, Josephs, S. F., and Wong-Staal, F. 1985. Oncogenes: Their role in neoplastic transformation. *Annu. Rev. Microbiol.* 39:419–449.

Reeves, R. 1984. Transcriptionally active chromatin. *Biochim. Biophys. Acta* 782:343–393.

Reinert, W. R., Patel, V. B., and Giles, N. H. 1981. Genetic regulation of the *qa* gene cluster of *Neurospora crassa:* Induction of *qa* messenger ribonucleic acid and dependency on *qa–1* function. *Mol. Cell. Biol.* 1:829–835.

Riggs, A. D., and Jones, P. A. 1983. 5-Methylcytosine, gene regulation, and cancer. *Adv. Cancer Res.* 40:1–30.

Ringold, G. M. 1985. Steroid hormone regulation of gene expression. *Annu. Rev. Pharmacol. Toxicol.* 25:529–566.

Scott, M. P., Weiner, A. J., Hazelrigg, T. I., Polisky, B. A., Pirotta, V., Scalenghe, F., and Kaufman, T. C. 1983. The molecular organization of the *Antennapedia* locus of *Drosophila. Cell.* 35:763–776.

Slamon, D. J., deKernion, J. B., Verma, I. M., and Cline, M. J. 1984. Expression of cellular oncogenes in human malignancies. *Science* 224:256–262.

Spelsberg, T. C., Littlefield, B. A., Seelke, R., Dani, G. M., Toyoda, H., Boyd-Leinen, P., Thrall, C., and Kon, O. I. 1983. Role of specific chromosomal proteins and DNA sequences in the nuclear binding sites for steroid receptors. *Recent Prog. Horm. Res.* 39:425–517.

Stein, G. S., Spelsberg, T. C., and Kleinsmith, L. J. 1974. Nonhistone chromosomal proteins and gene regulation. *Science* 183:817–824.

Tomkins, G., and Martin, D. W. 1970. Hormones and gene expression. *Annu. Rev. Genet.* 4:91–106.

Tyler, B. M., Geever, R. F., Case, M. E., and Giles, N. H. 1984. Cis-acting and transacting regulatory mutations define two types of promoters controlled by the *qa–1F* gene of *Neurospora. Cell* 36:493–502.

Vardimon, L., Renz, D., and Doerfler, W. 1983. Can DNA methylation regulate gene expression? *Recent Results Cancer Res.* 84:90–102.

Vesell, E. S. 1965. Genetic control of isozyme patterns in human tissues. *Prog. Med. Genet.* 4:128–175.

Wang, T. Y., Kostraba, N. C., and Newman, R. S. 1976. Selective transcription of DNA mediated by nonhistone proteins. *Prog. Nucleic Acid Res. Mol. Biol.* 19:447–462.

Weatherall, D., and Clegg, J. 1979. Recent developments in the molecular genetics of human hemoglobins. *Cell* 18:467–479.

Yamamoto, K. R., and Alberts, B. M. 1976. Steroid receptors: Elements for modulation of eukaryotic transcription. *Annu. Rev. Biochem.* 45:721–746.

Yunis, J. 1983. The chromosomal basis of human neoplasia. *Science* 221:227–236.

Chapter 22

Ashwell, M., and Work, T. S. 1970. The biogenesis of mitochondria. *Annu. Rev. Biochem.* 39:251–290.

Birky, C. W. 1978. Transmission genetics of mitochondria and chloroplasts. *Annu. Rev. Genet.* 12:471–512.

Boardman, N. K., Linnane, A. W., and Smillie, R. M. eds. 1971. *Autonomy and biogenesis of mitochondria and chloroplasts.* Amsterdam: North-Holland.

Borst, P., and Grivell, L. A. 1978. The mitochondrial genome of yeast. *Cell* 15:705–723.

Boynton, J. E., Gillham, N. W., and Lambowitz, A. M. 1979. Biogenesis of chloroplast and mitochondrial ribosomes. In *Ribosomes: Structure, function, and genetics,* edited by G. Chambliss, G. R. Craven, J. Davies, K. Davis, L. Kahan, and M. Nomura, pp. 903–950. Baltimore: University Park Press.

Clayton, D. A. 1982. Replication of animal mitochondrial DNA. *Cell* 28:693–705.

Collins, R. A., and Bertrand, H. 1978. Nuclear suppressors of the [*poky*] cytoplasmic mutant in *Neurospora crassa. Mol. Gen. Genet.* 161:267–273.

———, Bertrand, H., LaPolla, R. J., and Lambowitz, A. M. 1979. Mitochondrial ribosome assembly in *Neurospora crassa:* Mutants with defects in mitochondrial ribosome assembly. *Mol. Gen. Genet.* 177:73–84.

Ephrussi, B. 1953. *Nucleo-cytoplasmic relations in microorganisms.* New York: Oxford University Press.

Gillham, N. W. 1978. *Organelle heredity.* New York: Raven Press.

Green, M. R., Grimm, M. F., Goewert, R. R., Collins, R. A., Cole, M. D., Lambowitz, A. M., Heckman, J. E., Yin, S., and Raj-Bhandary, U. L. 1981. Transcripts and processing patterns for the ribosomal RNA and transfer RNA region of *Neurospora crassa* mitochondrial DNA. *J. Biol. Chem.* 256:2027–2034.

Jinks, J. L. 1964. *Extrachromosomal inheritance.* Englewood Cliffs, NJ: Prentice-Hall.

Kirk, J. T. O. 1971. Chloroplast structure and biogenesis. *Annu. Rev. Biochem.* 40:161–196.

Kuroiwa, T., Suzuki, T., Ogawa, K., and Kawano, S. 1981. The chloroplast nucleus: Distribution, number, size, and shape, and a model for the multiplication of the chloroplast genome during chloroplast development. *Plant Cell Physiol.* 22:381–396.

Lambowitz, A. M., LaPolla, R. J., and Collins, R. A. 1979. Mitochondrial ribosome assembly in *Neurospora.* Two-dimensional gel electrophoretic analysis of mitochondrial ribosomal proteins. *J. Cell Biol.* 82:17–31.

———, and Luck, D. J. L. 1976. Studies on the *poky* mutant of *Neurospora crassa. J. Biol. Chem.* 251:3081–3095.

LaPolla, R. J., and Lambowitz, A. M. 1982. Mitochondrial ribosome assembly in *Neurospora.* Structural analysis of mature and partially assembled ribosomal subunits by equilibrium centrifugation in CsCl gradients. *J. Cell Biol.* 95:267–277.

Levings, C. S. 1983. The plant mitochondrial genome and its mutants. *Cell* 32:659–661.

Linnane, A. W., and Nagley, P. 1978. Mitochondrial genetics in perspective: The derivation of a genetic and physical map of the yeast mitochondrial genome. *Plasmid* 1:324–345.

Rochaix, J. D. 1978. Restriction endonuclease map of the chloroplast DNA of *Chlamydomonas reinhardi. J. Mol. Biol.* 126:597–617.

Rush, M. G., and Misra, R. 1985. Extrachromosomal DNA in eukaryotes. *Plasmid* 14:177–191.

Sager, R. 1972. *Cytoplasmic genes and organelles.* New York: Academic Press.

Schmidt, R. J., Richardson, C. B., Gillham, N. W., and

Boynton, J. E. 1983. Sites of synthesis of chloroplast ribosomal proteins in *Chlamydomonas. J. Cell Biol.* 96:1451–1463.

Van Winkle-Swift, K. P., and Birky, C. W. 1978. The non-reciprocality of organelle gene recombination in *Chlamydomonas reinhardi* and *Saccharomyces cerevisae. Mol. Gen. Genet.* 166:193–209.

Chapter 23

Allard, R. W. 1960. *Principles of plant breeding.* New York: Wiley.

Brewbaker, J. L. 1964. *Agricultural genetics.* Englewood Cliffs, NJ: Prentice-Hall.

Davenport, C. B. 1913. *Heredity of skin color in negro-white crosses.* Washington, D.C.: Carnegie Institution of Washington, Pub. 554.

East, E. M. 1910. A Mendelian interpretation of variation that is apparently continuous. *Am. Naturalist* 44:65–82.

———. 1916. Studies on size inheritance in *Nicotiana. Genetics* 1:164–176.

———, and Jones, D. F. 1919. *Inbreeding and outbreeding.* Philadelphia: Lippincott.

Falconer, D. S. 1981. *Introduction to quantitative genetics,* 2nd ed. London: Longman.

Harrison, G. A., and Owen, J. J. T. 1964. Studies on the inheritance of human skin color. *Ann. Hum. Genet.* 28:27–37.

Mather, K. 1943. Polygenic inheritance and natural selection. *Biol. Rev.* 18:32–64.

———, and Jinks, J. L. 1971. *Biometrical genetics. The study of continuous variation.* Ithaca, NY: Cornell University Press.

Nilsson-Ehle, H. 1909. Kreuzungsuntersuchungen an Hafer und Weizen. *Lunds Univ. Aarskr. N. F. Atd.,* Ser. 2, 5 (2):1–122.

Thompson, J. N. 1975. Quantitative variation and gene number. *Nature* 258:665–668.

———, and Thoday, J. M., eds. 1979. *Quantitative genetic variation.* New York: Academic Press.

Wright, S. 1921. Systems of mating. *Genetics* 6:111–178.

Chapter 24

Ayala, F. J., ed. 1976. *Molecular evolution.* Sunderland, MA: Sinauer.

Cavalli-Sforza, L. L., and Bodmer, W. F. 1971. *Genetics of human populations.* San Francisco: Freeman.

Chu, E. H. Y., Brimer, P., Jacobson, K. B., and Merriam, E. V. 1969. Mammalian cell genetics. I. Selection and characterization of mutations auxotrophic for L-glutamine or resistant to 8-azaguanine in

Chinese hamster cells *in vitro. Genetics,* 62, 359–377.

Clarke, C. A., and Sheppard, P. M. 1966. A local survey of the distribution of industrial melanic forms in the moth *Biston betularia* and estimates of the selective values of these in an industrial environment. *Proc. R. Soc. Lond.* [*Biol.*] 165:424–439.

Crow, J. F., and Kimura, M. 1970. *An introduction to population genetics theory.* New York: Harper & Row.

Darwin, C. 1860. *On the origin of species by means of natural selection, or the preservation of favoured races in the struggle for life.* New York: Appleton.

Dobzhansky, T. 1951. *Genetics and the origin of species,* 3rd ed. New York: Columbia University Press.

———. 1955. A review of some fundamental concepts and problems of population genetics. *Cold Spring Harbor Symp. Quant. Biol.* 20:1–15.

———, Ayala, F. J., Stebbins, G. L., and Valentine, J. W. 1977. *Evolution.* San Francisco: Freeman.

Fisher, R. A. 1930. *The genetical theory of natural selection.* Oxford: Clarendon Press.

Ford, E. B. 1971. *Ecological genetics,* 3rd ed. London: Chapman and Hall.

Giles, N. H. 1951. Studies on the mechanism of reversion in biochemical mutants of *Neurospora crassa. Cold Spring Harbor Symp.,* 16, 283–313.

Glass, B., and Ritterhoff, R. K. 1956. Spontaneous-mutation rates at specific loci in *Drosophila* males and females. *Science,* 124, 314–315.

Haldane, J. B. S. 1931. *The causes of evolution.* New York: Harper.

Hardy, G. H. 1908. Mendelian proportions in a mixed population. *Science* 28:49–50.

Hartl, D. L. 1981. *A primer of population genetics.* Sunderland, MA: Sinauer.

Kettlewell, H. B. D. 1961. The phenomenon of industrial melanism in the Lepidoptera. *Annu. Rev. Entomol.* 6:245–262.

Kolmark, G., and Westergaard, M. 1949. Induced back-mutations in a specific gene of *Neurospora crassa. Hereditas,* 35, 490–506.

Lerner, I. M. 1958. *The genetic basis of selection.* New York: Wiley.

Lewontin, R. C. 1974. *The genetic basis of evolutionary change.* New York: Columbia University Press.

———, Moore, J. A., Provine, W. B., and Wallace, B. 1981. *Dobzhansky's genetics of natural populations I–XLIII.* New York: Columbia University Press.

Li, C. C. 1955. *Population genetics.* Chicago: University of Chicago Press.

Neel, J. V., Tiffany, T. O., and Anderson, N. G. 1973.

Approaches to monitoring human populations for mutation rate and genetic disease. In *Chemical Mutagens,* Vol. 3, A. Hollaender, ed., New York: Plenum Press. pp. 105–150.

Schlager, G., and Dickie, M. M. 1967. Spontaneous mutations and mutation rates in the house mouse. *Genetics,* 57, 319–330.

Speiss, E. 1977. *Genes in populations.* New York: Wiley.

Stadler, L. J., 1942. Some observations on gene variability and spontaneous mutation. Spragg Memorial Lectures (Third Series), Mich. State College.

Vogel, F., and Rathenberg, R. 1975. Spontaneous mu-

tation in man. *Adv. in Hum. Genet.,* 5, 223–318.

Wallace, B. 1981. *Basic population genetics.* New York: Columbia University Press.

———, and Srb, A. 1961. *Adaptation.* Englewood Cliffs, NJ: Prentice-Hall.

Wills, C. 1970. Genetic load. *Sci. Am.* 222:98–107.

Wright, S. 1931. Evolution in Mendelian populations. *Genetics* 16:97–159.

———. 1948. On the roles of directed and random changes in gene frequency in the genetics of populations. *Evolution* 2:279–294.

———. 1969. *Evolution and the genetics of populations.* 2 vols. Chicago: University of Chicago Press.

Solutions to Selected Questions and Problems

Chapter 1

1.3 c

1.5 a. Yes, if a sexual mating system exists in that species. In that case, two haploid cells can fuse to produce a diploid cell, which can then go through meiosis to produce haploid progeny. The fungi *Neurospora crassa* and *Saccharomyces cerevisiae* exemplify this positioning of meiosis in the life cycle.

b. No, because a diploid cell cannot be formed and meiosis only occurs starting with a diploid cell.

1.7 c

1.9 a. metaphase

b. anaphase

1.12 a. $\frac{1}{8}$. The probability of a gamete receiving A is $\frac{1}{2}$, because the choice is A or A'. Similarly, the probability of receiving B is $\frac{1}{2}$ and of receiving C is $\frac{1}{2}$. Therefore, the probability of receiving A, B, and C is $\frac{1}{2} \times \frac{1}{2} \times \frac{1}{2} = \frac{1}{8}$ (like the probability of tossing three heads in a row, the individual probabilities multiplied together).

b. $\frac{6}{8}$ ($=\frac{3}{4}$). The probability of gametes containing chromosomes from both maternal and paternal origin is 1 minus the probability that the gametes will contain either all maternal or all paternal chromosomes. The probability of a gamete containing all maternal chromosomes is $\frac{1}{8}$ (part a) and that of a gamete containing all paternal chromosomes is also $\frac{1}{8}$. Therefore the answer is $1 - \frac{1}{8} - \frac{1}{8} = \frac{6}{8}$.

Chapter 2

2.1 a. red

b. 3 red, 1 yellow

c. all red

d. $\frac{1}{2}$ red, $\frac{1}{2}$ yellow

2.4 The F_2 genotypic ratio (if *C* is colored, *c* is colorless) is $\frac{1}{4}$ *CC* : $\frac{1}{2}$ *Cc* : $\frac{1}{4}$ *cc*. If we consider just the colored plants, there is a 1:2 ratio of *CC* homozygotes to *Cc* heterozygotes. Therefore, if a colored plant is picked at random, the probability that it is *CC* is $\frac{1}{3}$ (i.e., $\frac{1}{3}$ of the *colored* plants are homozygous) and the probability that it is *Cc* is $\frac{2}{3}$. Only if a *Cc* plant is selfed will more than one phenotypic class be found among its progeny; therefore, the answer is $\frac{2}{3}$.

2.5 a. Parents are *Rr* (rough) and *rr* (smooth); F_1 are *Rr* (rough) and *rr* (smooth).

b. *Rr* × *Rr* → $\frac{3}{4}$ rough, $\frac{1}{4}$ smooth

2.7 Progeny ratio approximates 3:1, and so the parent is heterozygous. Of the dominant progeny there is a 1:2 ratio of homozygous to heterozygous, and so $\frac{1}{3}$ will breed true.

2.8 Black is dominant to brown. If *B* is the allele for black and *b* for brown, then female X is *Bb* and female Y is *BB*. The male is *bb*.

2.12 a. *WW Dd* × *ww dd*

b. *Ww dd* × *Ww dd*

c. *ww DD* × *WW dd*

d. *Ww Dd* × *ww dd*

e. *Ww Dd* × *Ww dd*

2.13 The cross is *Aa Bb Cc* × *Aa Bb Cc*.

a. Considering the *A* gene alone, the probability of an offspring showing the A trait from *Aa* × *Aa* is $\frac{3}{4}$. Similarly the probability of showing the B trait from *Bb* × *Bb* is $\frac{3}{4}$ and the C trait from *Cc* × *Cc* is $\frac{3}{4}$. Therefore, the probability of a given progeny being phenotypically ABC (using the product rule) is $\frac{3}{4} \times \frac{3}{4} \times \frac{3}{4} = \frac{27}{64}$.

b. Considering the *A* gene, the probability of an *AA* offspring from *Aa* × *Aa* is $\frac{1}{4}$. The same probability is the case for a *BB* offspring and for a *CC* offspring. Therefore, the probability of an *AA BB CC* offspring (from the product rule) is $\frac{1}{4} \times \frac{1}{4} \times \frac{1}{4} = \frac{1}{64}$.

2.16 a. The F_1 has the genotype *Aa Bb Cc^h* and is all agouti, black. The F_2 is $\frac{27}{64}$ agouti, black; $\frac{9}{64}$ agouti, black, Himalayan; $\frac{9}{64}$ agouti, brown; $\frac{9}{64}$ black; $\frac{3}{64}$ agouti, brown, Himalayan; $\frac{3}{64}$ black, Himalayan; $\frac{3}{64}$ brown; $\frac{1}{64}$ brown, Himalayan.

b. $\frac{4}{27}$

c. $\frac{1}{4}$

d. $\frac{3}{4}$

Chapter 3

3.1 The probability of a given homolog going to one particular pole is $\frac{1}{2}$. The probability of all five paternal chromosomes going to the same pole is $(\frac{1}{2})^5 = \frac{1}{32}$. The same answer applies for all maternal chromosomes going to one pole.

3.2

	Mother's		Father's	
	Mother	Father	Mother	Father
X chromosome	Yes	Yes	No	No
Y chromosome	No	No	No	Yes

3.4 The woman is heterozygous for the recessive color blindness allele, let us say c^+c. The man is not color-blind and thus has the genotype c^+Y. (It does not matter what phenotype his mother and father have because males are hemizygous for sex-linked genes; thus the genotype can be assigned directly from the phenotype.) All the female offspring will have normal color vision; half the male offspring will be color blind, and half will have normal color vision.

3.6 If *c* is the red-green color-blindness allele, and *a* is the albino allele, the parent's genotypes are c^+c^+ aa ♀ and cY a^+a^+ ♂. All children will be normal visioned (c^+c^+ ♀ and c^+Y ♂) and normally pigmented (a^+a, both sexes).

3.7 The parentals are $ww\ vg^+vg^+$ ♀ and w^+Y $vg\ vg$♂.

a. The F_1 males are all wY vg^+vg, white eyes, long wings. The F_1 females are all $w^+w\ vg^+vg$, red eyes, long wings.

b. The F_2 females are ⅜ red, long, females; ⅜ white, long, females; ⅛ red, vestigial, females; ⅛ white, vestigial, females; the same ratios of the respective phenotypes apply for the males.

c. The cross of F_1 male with the parental female is wY $vg^+vg \times ww\ vg^+vg^+$. All progeny have white eyes and long wings. The cross of an F_1 female with the parental male is $w^+w\ vg^+vg \times w^+$Y $vg\ vg$. Female progeny: All have red eyes, half have long wings and half have vestigial wings. Male progeny: ¼ red, long; ¼ red, vestigial; ¼ white, long; ¼ white, vestigial.

3.10 The simplest hypothesis is that brown-colored teeth are determined by a sex-linked dominant mutant allele. Man A was BY and his wife was bb. All sons will be bY normals and cannot pass on the trait. All the daughters receive the X chromosome from their father, and so they are Bb with brown enamel. Half their sons will have brown teeth because half their sons receive the X chromosome with the B mutant allele.

3.14 The answer is given in Figure 3.1.

3.16

	Pedigree A	Pedigree B	Pedigree C
Autosomal recessive	Yes	Yes	Yes
Autosomal dominant	Yes	Yes	No
X-linked recessive	Yes	Yes	No
X-linked dominant	No	Yes	No

3.18 a. X-linked recessive can be excluded, for example, because individual I.2 would be expected to pass on the trait to *all* sons and that is not the case. Also, the II.1, II.2 pairing would be expected to produce offspring all of whom express the trait.

Autosomal recessive can also be excluded because the II.1, II.2 pairing would be expected to produce offspring all of whom express the trait.

b. The remaining two mechanisms of inheritance are X-linked dominant and autosomal dominant. Genotypes can be written to satisfy both mechanisms of inheritance. X-linked dominance is perhaps more likely, for the II.6, II.7 pairing shows exclusively father-daughter inheritance; that is, all the daughters

$ww \times w^+$Y

Sperm

Eggs		w^+	Y
	ww	www^+	wwY
	O	w^+O	YO

Survivors are wwY (white ♀) and w^+O (sterile, red ♂)

Backcross ww Y $\times w^+$Y

Sperm

			w^+	Y
Eggs	Normal	w Y	$w^+\ w$Y Red ♀	w YY White ♂
		w	$w^+\ w$ Red ♀	w Y White ♂
	Secondary nondisjunction	ww	$w^+\ ww$ Triplo-X Usually dies	ww Y White ♀
		Y	w^+ Y Red ♂	Y Y Dies

Figure 3.1

and none of the sons exhibit the trait: this is a characteristic of the segregation of X-linked dominant alleles. If the trait were autosomal dominant, then half the sons and half the daughters would be expected to exhibit the trait.

3.21 a is false: the father passes on his X to his daughters, and so *none* of the sons will be affected.

b is false: because the disease is rare, the mother would be heterozygous and only 50% of her daughters would receive the X with the dominant mutant allele.

c is true: all daughters receive the father's X chromosome.

d is false: only 50% of her sons would receive the X with the dominant mutant allele.

3.23 a is true, that is, $aa \times aa$ produces only aa progeny.

b is false—because the disease is autosomal, the sex of the progeny makes no difference in the fact that all progeny will be affected if both parents are affected.

c is false—a child must receive one autosomal recessive allele from each parent to have the disease, but neither the parents nor the grandparents need have had the disease; all could theoretically be carriers.

d is false: if the unaffected parent is a carrier, then affected progeny can result.

3.25 Because hemophilia is an X-linked trait, the most likely explanation is that random inactivation of X chromosomes (lyonization, see p. 78) produces individuals with different proportions of cells with the normal allele. Thus, if some women had only 40% of their cells with an active h^+ allele, and 60% with the h (hemophilia) allele, but other women had 60% of their cells with an active h^+ allele, and 40% with the h allele, there would be a significant difference in the amount of clotting factor these two individuals would make.

Chapter 4

4.2 6 possible genotypes: *ww, ww1, ww2, w1w1, w1 w2, w2w2.*

4.6 The woman's genotype is $I^A I^B$ and the man's genotype is $I^A i$.

a. $\frac{1}{2} \times \frac{1}{2} = \frac{1}{4}$.

b. Zero. A blood group O baby is not possible.

c. $\frac{1}{2}$ (probability of male) $\times \frac{1}{4}$ (probability of AB) $\times \frac{1}{2}$ (probability of male) $\times \frac{1}{4}$ (probability of B) = $\frac{1}{64}$ probability that all four conditions will be fulfilled.

4.8 Blood type O, since genotype is *ii*.

4.10 Half will be *Rr* and hence will resemble the parents.

4.12 a. The cross is *FF oo* $\times$ *ff OO*. The F_1 is *Ff Oo,* which is fuzzy with round leaf glands.

b. Interbreeding the F_1 gives $\frac{3}{16}$ fuzzy, oval-glanded; $\frac{6}{16}$ fuzzy, round-glanded; $\frac{3}{16}$ fuzzy, no-glanded; $\frac{1}{16}$ smooth, oval-glanded; $\frac{2}{16}$ smooth, round-glanded; $\frac{1}{16}$ smooth, no-glanded.

c. Cross is *Ff Oo* $\times$ *ff OO.* The progeny are $\frac{1}{4}$ fuzzy, oval-glanded; $\frac{1}{4}$ fuzzy, round-glanded; $\frac{1}{4}$ smooth, oval-glanded; $\frac{1}{4}$ smooth, round-glanded.

4.18 To show no segregation among the progeny, the chosen plant must be homozygous. The genotypes comprising the $\frac{9}{16}$ colored plants are: 1 *AA BB* : 2 *Aa BB* : 2 *AA Bb* : 4 *Aa Bb.* Only one of these genotypes is homozygous; the answer is $\frac{1}{9}$.

4.20 a. If *Y* governs yellow and *y* governs agouti, then *YY* are lethal, *Yy* are yellow, and *yy* are agouti. Let *C* determine colored coat and *c* determine albino. The parental genotypes, then, are *Yy Cc* (yellow) and *Yy cc* (white).

b. The proportion is 2 yellow:1 agouti:1 albino. None of the yellows breed true because they are all heterozygous, with homozygous *YY* individuals being lethal.

4.22 a. *YY RR* (crimson) $\times$ *yy rr* (white) gives *Yy Rr* F_1 plants, which have magenta-rose flowers. Selfing the F_1 gives an F_2 as follows: $\frac{1}{16}$ crimson (*YY RR*), $\frac{2}{16}$ orange-red (*YY Rr*), $\frac{1}{16}$ yellow (*YY rr*), $\frac{2}{16}$ magenta (*Yy RR*), $\frac{4}{16}$ magenta-rose (*Yy Rr*), $\frac{2}{16}$ pale yellow (*Yy rr*), and $\frac{4}{16}$ white (*yy RR, yy Rr,* and *yy rr*). Progeny of the F_1 backcrossed to the crimson parent are $\frac{1}{4}$ crimson, $\frac{1}{4}$ orange-red, $\frac{1}{4}$ magenta, and $\frac{1}{4}$ magenta-rose.

b. *YY Rr* $\times$ *Yy rr* gives $\frac{1}{4}$ orange-red, $\frac{1}{4}$ yellow, $\frac{1}{4}$ magenta-rose, $\frac{1}{4}$ pale yellow.

c. *YY rr* (yellow) $\times$ *yy Rr* (white) gives $\frac{1}{2}$ magenta-rose (*Yy Rr*) and $\frac{1}{2}$ pale yellow (*Yy rr*).

4.24 a. The simplest approach is to calculate the proportion of progeny that will be black and then subtract that answer from 1. The black progeny have the genotype $A - B - C -$, and the proportion of these progeny is $(\frac{3}{4})^3$. Therefore the proportion of colorless progeny is $1 - (\frac{3}{4})^3 = 1 - \frac{27}{64} = \frac{37}{64}$.

b. Black is produced only when *cc* $A - B -$ results, and this offspring occurs with the frequency $(\frac{1}{4})(\frac{3}{4})(\frac{3}{4}) = \frac{9}{64}$ black, which gives $\frac{55}{64}$ colorless.

4.26 In males *HH* and *Hh* are horned, and *hh* is hornless; in females *HH* is horned, and *Hh* and *hh* are hornless. The cross is a *HH WW* male $\times$ *hh ww* female. The F_1 is *Hh Ww,* which gives horned white males and hornless white females. Interbreeding the F_1 gives the following F_2:

		male	female
$\frac{3}{16}$	*HH W–*	horned, white	horned, white
$\frac{6}{16}$	*Hh W–*	horned, white	hornless, white
$\frac{3}{16}$	*hh W–*	hornless, white	hornless, white
$\frac{1}{16}$	*HH ww*	horned, black	horned, black
$\frac{2}{16}$	*Hh ww*	horned, black	hornless, black
$\frac{1}{16}$	*hh ww*	hornless, black	hornless, black

In sum, the ratios are $\frac{9}{16}$ horned white : $\frac{3}{16}$ hornless white : $\frac{3}{16}$ horned black : $\frac{1}{16}$ hornless black males and $\frac{3}{16}$ horned white : $\frac{9}{16}$ hornless white : $\frac{1}{16}$ horned black : $\frac{3}{16}$ hornless black females.

4.28 Ewe A is *Hh ww*; B is *Hh Ww* or *hh Ww*; C is *HH ww*; D is *Hh Ww*; the ram is *Hh Ww*.

4.29 c

Chapter 5

5.3 From the χ^2 test, $\chi^2 = 16.10$; P is less than 0.01 at three degrees of freedom. This test reveals that the two genes do not fit a 1:1:1:1 ratio. It does not say why. Linkage might seem reasonable until it is realized that the minority classes are not reciprocal classes (both carry the *aa* phenotype). If the segregation at each locus is considered, however, the $B-:bb$ ratio is about 1:1 (203:197), but the $A-:aa$ ratio is not (240:160). The departure, then, specifically results from a deficiency of *aa* individuals. This departure should be confirmed in other crosses that test the segregation at locus *A*. In corn, further evidence would be that the *aa* deficiency might show up as a class of ungerminated seeds or of seedlings that die early.

5.6 There are 158 progeny rabbits. The recombinant classes are the English plus Angora and the non-English plus short-haired, and so the map distance between the genes is $(11 + 6)/158 \times 100\% = 10.8\% = 10.8$ mu.

5.8

```
  c     e        a              d          b
  +-----+--------+--------------+----------+
     <-7->  <-8->    <---38--->   <-13->
```

5.11 45% ab^+, 45% a^+b, 5% a^+b^+, 5% ab

5.12 a. $0.035 + 0.465 = 0.50$

b. All the daughters have *AB* phenotype.

5.15 a. The genotype of the F_1 is *AB/ab*. The gametes produced by the F_1 are 40% *AB*; 40% *ab*; 10% *Ab*; 10% *aB*. From a testcross of the F_1 with *ab/ab* we get 40% *AB/ab*; 40% *ab/ab*; 10% *Ab/ab*; 10% *aB/ab*.

b. The F_1 genotype is *Ab/aB*. The gametes produced by the F_1 are 40% *Ab*; 40% *aB*; 10% *AB*; 10% *ab*. From a testcross of the F_1 with *ab/ab* we get 40% *Ab/ab*; 40% *aB/ab* 10% *AB/ab*; 10% *ab/ab*. This question illustrates that map distance is computed between sites in the chromosome, and it does not matter whether the genes are in coupling or in repulsion, the percentage of recombinants will be the same, even though the phenotypic classes constituting the recombinants differ.

5.17 Each chromosome pair segregates independently. We can compute the relative proportions of gametes produced for each homologous pair of chromosomes separately from the known map distances (P = parental; R = recombinant):

P	R	P	R	P	R
AB 0.4	Ab 0.1	CD 0.45	Cd 0.05	EF 0.35	Ef 0.15
ab 0.4	aB 0.1	cd 0.45	cD 0.05	ef 0.35	eF 0.15

To answer the questions, simply multiply the probabilities of getting the particular gamete from the F_1 multiple heterozygote.

a. *ABCDEF* $= 0.4 \times 0.45 \times 0.35 = 0.063$ (6.3%)
b. *ABCdef* $= 0.4 \times 0.05 \times 0.35 = 0.007$ (0.7%)
c. *AbcDEf* $= 0.1 \times 0.05 \times 0.15 = 0.00075$ (0.075%)
d. *aBCdef* $= 0.1 \times 0.05 \times 0.35 = 0.00175$ (0.175%)
e. *abcDeF* $= 0.4 \times 0.05 \times 0.15 = 0.003$ (0.3%)

5.18 a. 47.5% each of *DPh* and *dph*; 2.5% each of *Dph* and *dPh*.

b. 23.75% each of *DpH*, *Dph*, *dPH*, and *dPh*; 1.75% each of *DPH*, *DPh*, *dpH*, and *dph*.

5.24 a. *FmW* and *fMw* will be the least frequent (because of double crossovers).

b. *MfW* and *mFw* will be the least frequent.

c. *MwF* and *mWf* will be the least frequent.

5.26 a. By doing a three-point mapping analysis as described in the chapter, we find that the order of genes in the chromosome is *dp-b-hk*, with 35.5 mu between *dp* and *b*, and 5.4 mu between *b* and *hk*.

b. (1) The frequency of observed double crossovers is 1.4%, and the frequency of expected double crossovers is 1.9%. The coefficient of coincidence, therefore, is 1.4/1.9 = 0.73. (2) The interference value is given by 1 - coefficient of coincidence, 0.27.

5.30 a. $a^+c\ b/a\ c^+b^+$

b. $a^+c^+b^+/Y$

c.

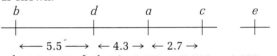
```
  a         c              b
  +---------+--------------+
     <-5->    <---10--->
```

5.33 a. *a*, *b*, *c*, and *d* are linked on the same chromosome, and *e* is on a separate chromosome. The map is as shown.

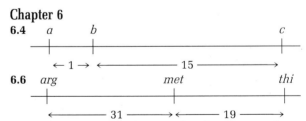
```
  b                  d       a          c           e
  +------------------+-------+----------+-----------+
     <----5.5---->    <-4.3->  <-2.7->
```

b. 1.6×10^{-5}; that is, $0.055 \times 0.043 \times 0.027 \times 0.5$ (for the half of the progeny produced by the triple crossover that have all the wild-type alleles) $\times 0.5$ (to give the proportion that are e^+)

Chapter 6

6.4
```
  a       b                              c
  +-------+------------------------------+
    <-1->   <----------15----------->
```

6.6
```
 arg                  met                thi
  +-------------------+------------------+
    <-------31------->  <------19------->
```

6.9 The two genes can be considered to assort independently because the number of parental ditype (20) is approximately equal to the number of nonparental ditype (18).

6.12
```
           b              c              a
 --0-------+--------------+--------------+--
    <-6->   <----11---->   <----14---->
```

6.13 Map distance is ½ × % of second division segregation asci, which in this case = 5 map units.

6.16 There are three linkage groups, as determined by which blocks of genes always stay together during haploidization: *sm-phe, pu-w-ad,* and *y-bi.*

Chapter 7

7.2 The whole chromosome would have to be transferred in order for the recipient to become a donor in an *Hfr* × *F⁻* cross; that is, the *F* factor in the *Hfr* strain is transferred to the *F⁻* cell last. This transfer takes approximately 100 min, and usually the conjugal unions break apart by then.

7.5 Strain A is *thr leu⁺* and B is *thr⁺ leu.* Transformed B should be *thr⁺ leu⁺,* and its presence should be detected on a medium containing neither threonine nor leucine.

7.7

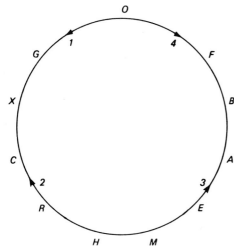

7.11 Two answers are compatible with the data:

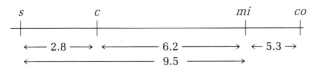

7.13 Only the volume of phage suspension plated is relevant to the calculation. 0.1 ml of the diluted phage suspension produced 20 plaques, meaning there were 200 phages/ml of the diluted suspension. The dilution factor was 10^8, and so the concentration in the original suspension was 200×10^8 or 2×10^{10}.

7.14 0.5% recombination (the numbers of plaques counted are so small that this value is a rather rough approximation)

7.17 The *trpB* marker cotransduced with the *cysB⁺* selected marker more frequently than did the *trpE* marker; thus *trpB* is closer to *cysB.*

7.20 Working from the principle that a relatively high cotransduction frequency means that the genes are relatively close, the order of the genes is: *supD − cheA − cheB − eda.*

Chapter 8

8.3 d

8.5 3′ and 5′ carbons are connected by a phosphodiester bond.

8.9 a. 3′ T C A A T G G A C T A G C A T 5′
b. 3′ A A G A G T C C T T A A G G T 5′

8.10 The adenine-thymine base pair is held together by two hydrogen bonds but the guanine-cytosine base pair is held together by three hydrogen bonds. Thus, the guanine-cytosine base pair requires more energy to break it and is the harder pair to break apart.

8.12 b, c, and d

8.17 a. 200,000
b. 10,000
c. 3.4×10^4 nm

Chapter 9

9.1 a.

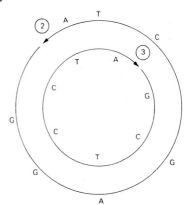

b. 3

9.5 Linkage number is the number of times the two chains of a double-stranded DNA molecule twist around each other.

9.8 See pp. 304–305. Each of the known *Saccharomyces cerevisiae* centromeres has a common core centromere region, which consists of three sequence domains. All three domains are important for centromere function. The centromeres of other eukaryotic organisms that have been characterized are quite different from the *Saccharomyces* centromeres.

9.10 Most protein-coding genes are found in unique-sequence DNA.

Chapter 10

10.3 Key: $^{15}N - ^{15}N$ DNA = HH; $^{15}N - ^{14}N$ DNA = HL; $^{14}N - ^{14}N$ DNA = LL.

a. Generation 1: all HL; 2: ½ HL, ½ LL; 3: ¼ HL, ¾ LL; 4: ⅛ HL, ⅞ LL; 6: ¹⁄₃₂ HL, ³¹⁄₃₂ LL; 8: ¹⁄₁₂₈ HL, ¹²⁷⁄₁₂₈ LL

b. Generation 1: ½ HH, ½ LL; 2: ¼ HH, ¾ LL; 3: ⅛ HH, ⅞ LL; 4: ¹⁄₁₆ HH, ¹⁵⁄₁₆ LL; 6: ¹⁄₆₄ HH, ⁶³⁄₆₄ LL; 8: ¹⁄₂₅₆ HH, ²⁵⁵⁄₂₅₆ LL.

10.5 A primer strand is a nucleic acid sequence that is extended by DNA synthesis activities. A template strand directs the base sequence of the DNA strand being made; for example, an A on the template causes a T to be inserted on the new chain, and so on.

10.6 a. One base pair is 0.34 nm, and the chromosome is 1100 μm. Thus the number of base pairs is $(1100/0.34) \times 1000 = 3.24 \times 10^6$.

b. There are 10 base pairs per turn in a normal DNA double helix; therefore it has a total of 3.24×10^5 turns.

c. 3.24×10^5 turns and 60 min for unidirectional synthesis; therefore $(3.24 \times 10^5/60)$ turns per minute = 5400 revolutions per minute.

d.

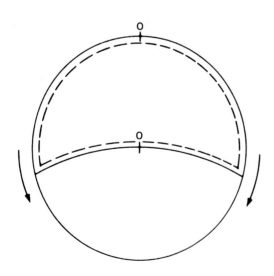

O   = origin
———   = parental DNA
— — —   = new DNA

10.9 See text, pp. 331–336.

10.11 See Tables 10.1 and 10.3.

10.15 See text, pp. 322–323. DNA polymerase; intact, high-molecular-weight DNA; dATP, dGTP, dTTP, and dCTP; and magnesium ions are needed.

10.16 Two principal lines of evidence show that the Kornberg enzyme is not the enzyme involved in *E. coli* chromosome duplication *in vivo*: First, a mutant that is deficient in the Kornberg enzyme nonetheless grows, replicates the DNA, and divides. Second, other mutants which are temperature-sensitive for DNA synthesis and which do not replicate the DNA or divide have Kornberg enzyme activity at the nonpermissive temperature.

Chapter 11

11.1 The DNA contains deoxyribose and thymine, whereas RNA contains ribose and uracil, respectively. Also, DNA is usually double-stranded, but RNA is usually single-stranded.

11.4 See text, pp. 357–364, 367–368.

11.5 See text, p. 361.

11.7 RNA polymerase I transcribes the major rRNA genes that code for 18S, 5.8S and 28S rRNAs; RNA polymerase II transcribes the protein-coding genes to produce mRNA molecules; and RNA polymerase III transcribes the 5S rRNA genes, the tRNA genes, and the genes for small nuclear RNA (snRNA) molecules.

In the cell the 18S, 5.8S, 28S, and 5S rRNAs are structural components of ribosomes. The mRNAs are translated to produce proteins, the tRNAs bring amino acids to the ribosome to donate to the growing polypeptide chain during protein synthesis, and at least some of the snRNAs are involved in RNA processing events.

11.9 An enhancer element is defined as a DNA sequence that somehow, without regard to its position relative to the gene or its orientation in the DNA, increases the amount of DNA synthesized from the gene it controls (discussed in the chapter on pp. 364–370).

Chapter 12

12.3 The three classes of RNA are mRNA (pp. 374–375), tRNA (pp. 388–389), and rRNA (pp. 396–401).

12.5 For mRNA: 5′ capping and 3′ polyadenylation in eukaryotes. For tRNA: modification of a number of the bases, removal of 5′ leader and 3′ trailer sequences (if present), and addition of a three-nucleotide sequence (5′-CCA-3′) at the 3′ end of the tRNA.

For rRNA in prokaryotes and pre-rRNA molecule is processed to the mature 16S, 23S, and 5S rRNAs. In eukaryotes a precursor molecule is processed to the 18S, 5.8S, and 28S rRNA molecules; 5S rRNA is made elsewhere. The large rRNAs in both types of organisms are methylated. Additionally, the same rRNAs in eukaryotes are pseudouridylated.

Chapter 13

13.1 c linear, folded chains of amino acids
13.2 b mRNA
13.10 a. 4 A:6 C

$$AAA = \left(\frac{4}{10}\right)\left(\frac{4}{10}\right)\left(\frac{4}{10}\right) = 0.064, \text{ or } 6.4\% \text{ lys}$$

$$AAC = \left(\frac{4}{10}\right)\left(\frac{4}{10}\right)\left(\frac{6}{10}\right) = 0.096, \text{ or } 9.6\% \text{ asn}$$

$$ACA = \left(\frac{4}{10}\right)\left(\frac{6}{10}\right)\left(\frac{4}{10}\right) = 0.096, \text{ or } 9.6\% \text{ thr}$$

$$CAA = \left(\frac{6}{10}\right)\left(\frac{4}{10}\right)\left(\frac{4}{10}\right) = 0.096, \text{ or } 9.6\% \text{ gln}$$

$$CCC = \left(\frac{6}{10}\right)\left(\frac{6}{10}\right)\left(\frac{6}{10}\right) = 0.216, \text{ or } 21.6\% \text{ pro}$$

$$CCA = \left(\frac{6}{10}\right)\left(\frac{6}{10}\right)\left(\frac{4}{10}\right) = 0.144, \text{ or } 14.4\% \text{ pro}$$

$$CAC = \left(\frac{6}{10}\right)\left(\frac{4}{10}\right)\left(\frac{6}{10}\right) = 0.144, \text{ or } 14.4\% \text{ his}$$

$$ACC = \left(\frac{4}{10}\right)\left(\frac{6}{10}\right)\left(\frac{6}{10}\right) = 0.144, \text{ or } 14.4\% \text{ thr}$$

In sum, 6.4% lys, 9.6% asn, 9.6% gln, 36.0% pro, 24.0% thr, and 14.4% his.

b. 4 G:1 C

$$GGG = \left(\frac{4}{5}\right)\left(\frac{4}{5}\right)\left(\frac{4}{5}\right) = 0.512, \text{ or } 51.2\% \text{ gly}$$

$$GGC = \left(\frac{4}{5}\right)\left(\frac{4}{5}\right)\left(\frac{1}{5}\right) = 0.128, \text{ or } 12.8\% \text{ gly}$$

$$CCG = \left(\frac{4}{5}\right)\left(\frac{1}{5}\right)\left(\frac{4}{5}\right) = 0.128, \text{ or } 12.8\% \text{ ala}$$

$$CGG = \left(\frac{1}{5}\right)\left(\frac{4}{5}\right)\left(\frac{4}{5}\right) = 0.128, \text{ or } 12.8\% \text{ arg}$$

$$CCC = \left(\frac{1}{5}\right)\left(\frac{1}{5}\right)\left(\frac{1}{5}\right) = 0.008, \text{ or } 0.8\% \text{ pro}$$

$$CCG = \left(\frac{1}{5}\right)\left(\frac{1}{5}\right)\left(\frac{4}{5}\right) = 0.032, \text{ or } 3.2\% \text{ pro}$$

$$CGC = \left(\frac{1}{5}\right)\left(\frac{4}{5}\right)\left(\frac{1}{5}\right) = 0.032, \text{ or } 3.2\% \text{ arg}$$

$$GCC = \left(\frac{4}{5}\right)\left(\frac{1}{5}\right)\left(\frac{1}{5}\right) = 0.032, \text{ or } 3.2\% \text{ ala}$$

In sum, 64.0% gly, 16.0% ala, 16.0% arg, and 4.0% pro.

c. 1 A:3 U:1 C; the same logic is followed here, using $\frac{1}{5}$ as the fraction for A, $\frac{3}{5}$ for U, and $\frac{1}{5}$ for C

AAA = 0.008, or 0.8% lys
AAU = 0.024, or 2.4% asn
AUA = 0.024, or 2.4% ile
UAA = 0.024, or 2.4% chain terminating
AUU = 0.072, or 7.2% ile
UAU = 0.072, or 7.2% tyr
UUA = 0.072, or 7.2% leu
UUU = 0.216, or 21.6% phe
AAC = 0.008, or 0.8% asn
ACA = 0.008, or 0.8% thr
CAA = 0.008, or 0.8% gln
ACC = 0.008, or 0.8% thr
CAC = 0.008, or 0.8% his
CCA = 0.008, or 0.8% pro
CCC = 0.008, or 0.8% pro
UUC = 0.072, or 7.2% phe
UCU = 0.072, or 7.2% ser
CUU = 0.072, or 7.2% leu
UCC = 0.024, or 2.4% ser
CUC = 0.024, or 2.4% leu
CCU = 0.024, or 2.4% pro
UCA = 0.024, or 2.4% ser
UAC = 0.024, or 2.4% tyr
CUA = 0.024, or 2.4% leu
CAU = 0.024, or 2.4% his
AUC = 0.024, or 2.4% ile
ACU = 0.024, or 2.4% thr

In sum, 0.8% lys, 3.2% asn, 12.0% ile, 2.4% chain terminating, 9.6% tyr, 19.2% leu, 28.8% phe, 4.0% thr, 0.8% gln, 3.2% his, 4.0% pro, and 12.0% ser. The likelihood is that the chain would not be long because of the chance of the chain-terminating codon.

d. 1 A:1 U:1 G:1 C; all 64 codons will be generated. The probability of each codon is $\frac{1}{64}$, so there is a $\frac{3}{64}$ chance of the codon being a chain-terminating codon. With those exceptions the relative proportion of amino acid incorporation is directly dependent on the codon degeneracy for each amino acid, and that can be determined by inspecting the code word dictionary.

13.11

	Word size	Number of combinations
a.	5	$2^5 = 32$
b.	3	$3^3 = 27$
c.	2	$5^2 = 25$

(The minimum word sizes must uniquely designate twenty amino acids.)

13.13 a. AAA ATA AAA ATA etc.
  b. TTT TAT TTT TAT etc.
  c. AAA for phe and AUA for tyr

13.14 A probable reason is that actinomycin D might block the transcription of a gene that codes for an inhibitor of an enzyme activity.

14.2

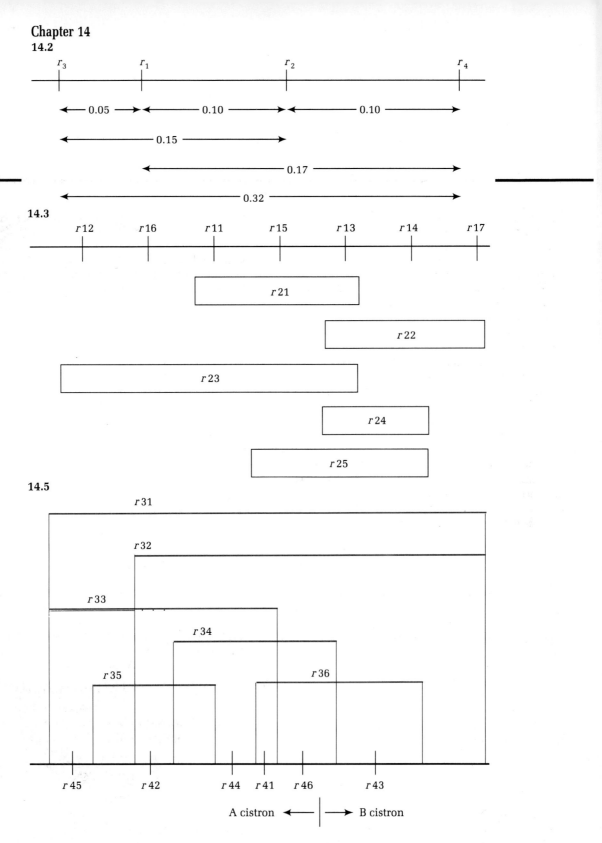

14.6 a. The simplest approach is to calculate the proportion of the F_2 that are colored and to subtract that answer from one. In this case the noncolorless are the brown or black progeny. The proportion of progeny that make at least the brown pigment is given by the probability of having the following genotype: $A- B- C-$ (*DD, Dd,* or *dd*). The answer is $\frac{3}{4} \times \frac{3}{4} \times \frac{3}{4} \times 1 = \frac{27}{64}$. Therefore the proportion of colorless is $1 - \frac{27}{64} = \frac{37}{64}$.

b. The brown progeny have the following genotype: $A- B- C- dd.$ The probability of getting individuals with this genotype is $\frac{3}{4} \times \frac{3}{4} \times \frac{3}{4} \times \frac{1}{4} = \frac{27}{256}$.

14.8 The colorless F_2's are *aa bb cc* in genotype, the probability for which is $\frac{1}{4} \times \frac{1}{4} \times \frac{1}{4} = \frac{1}{64}$.

a. The colorless F_2's are *aa bb cc* in genotype, the probability for which is $\frac{1}{4} \times \frac{1}{4} \times \frac{1}{4} = \frac{1}{64}$

b. $\frac{67}{256}$

14.10 a. Half have white eyes (the sons) and half have

fire red eyes (the daughters).

b. All have fire red eyes.

c. All have brown eyes.

d. All are w^+; a fourth are $bw^+ - st^+ -$, red; a fourth are $bw^+ - st\ st$, scarlet; a fourth are $bw\ bw\ st^+ -$, brown; a fourth are $bw\ bw\ st\ st$, the color of 3-hydroxykyrurenine plus the color of the precursor to biopterin, or colorless.

14.11 Wild-type T4 will produce progeny phages at all three temperatures. Let us suppose that model (1) is correct. If cells infected with the double mutant are first incubated at 17°C and then shifted to 42°C, then progeny phages will be produced and the cells will lyse. The explanation is as follows: The first step, A to B, is controlled by a gene whose product is heat-sensitive. At 17°C the enzyme works and A is converted to B, but B cannot then be converted to mature phages because that step is cold-sensitive. When the temperature is raised to 42°C, the A-to-B step is now blocked, but the accumulated B can be converted to mature phages for the enzyme involved with that step is cold-sensitive, and so the enzyme is functional at the high temperature. If model 2 is the correct pathway, then progeny phages should be produced in a 42°-to-17°C temperature shift, but not vice versa. In general, two gene product functions can be ordered by this method whenever one temperature shift allows phage production and the reciprocal shift does not, according to the following rules: (1) If a low-to-high temperature results in phages but a high-to-low temperature does not, then the hs step precedes the cs step (model 1); (2) If a high-to-low temperature results in phages but a low-to-high temperature does not, then the cs step precedes the hs step (model 2).

14.13 a. $c\ d^+$ and c^+d

b. The genes are not linked because parental ditype (PD) and nonparental ditype (NPD) tetrads occur in equal frequencies.

c. The pathway is Y to X to Z, with d blocking the synthesis of Y and c blocking the synthesis of X and Y.

14.16 He should testcross the male to a retinal atrophic female. Retinal atrophic pups would indicate that the male is heterozygous.

14.17 Baby Joan, whose blood type is O, must belong to the Smith family. Mrs. Jones, being of blood type AB, could not have an O baby. Therefore Baby Jane must be hers.

Chapter 15

15.1 At any one position in the DNA, there are four possibilities for the base pair: A-T, T-A, G-C, and C-G. Therefore the length of the base-pair sequence that the enzyme recognizes is given by the power to which four must be raised to equal (or approximately equal) the average size of the DNA fragment produced by enzyme digestion. The answer in this case is 6; that is, 4 to the power of 6 = 4096. If the enzyme instead recognized a four base-pair sequence, the average size of the DNA fragment would be 4 to the power of 4 = 256.

15.2

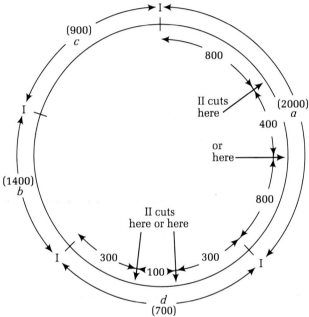

15.3 The sequencing gel banding pattern is shown in the following figure.

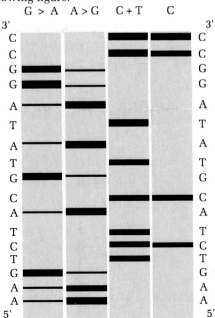

Chapter 16

16.1 b

16.3 e None of these (i.e., all are classes of mutations).

16.4 c induction of thymine dimers and their persistence or imperfect repair.

16.5 a Six.

b Three. The mutation results in replacement of the UGG codon in the mRNA by UAG, which is a chain termination codon.

16.6 a. UAG: CAG Gln; AAG Lys; GAG Glu; UUG Leu; UCG Ser; UGG Trp; UAU Tyr; UAC Tyr; UAA chain terminating.

b. UAA: CAA Gln; AAA Lys; GAA Glu; UUA Leu; UCA Ser; UGA chain terminating; UAU Tyr; UAC Tyr; UAG chain terminating

c. UGA: CGA Arg; AGA Arg; GGA Gly; UUA Leu; UCA Ser; UAA chain terminating; UGU Cys; UGC Cys; UGG Trp

16.8

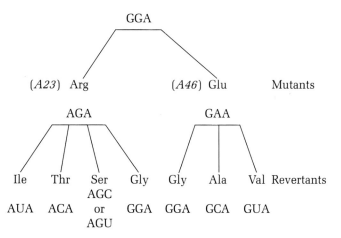

(wild type) Gly

GGA

(*A23*) Arg   (*A46*) Glu   Mutants

AGA   GAA

Ile   Thr   Ser   Gly   Gly   Ala   Val   Revertants
              AGC
AUA   ACA   or   GGA   GGA   GCA   GUA
              AGU

16.11 a. 5BU in its normal form is a T analog; in its rare form it resembles C. The mutation is a AT-to-GC transition.

$$A\text{-}T \rightarrow A\text{-}5BU \rightarrow G\text{-}5BU \rightarrow G\text{-}C$$

b. Nitrous acid can deaminate C to U, and so the reversion is to the original AT pair.

16.14 The reaction stops when it needs dGTP; only dGMP is present.

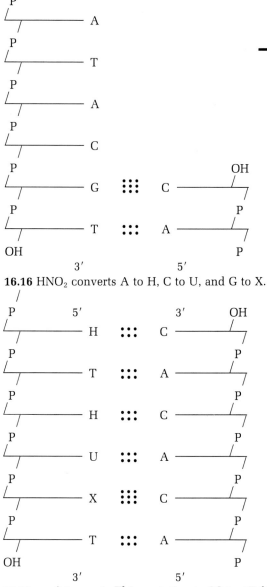

16.16 HNO_2 converts A to H, C to U, and G to X.

16.19 *ara*$^+$ to *ara*-1: This mutation is CG to AT for it is reverted by base analogs but not by HA or the frameshift mutagen. *ara*$^+$ to *ara*-2: This mutation is AT to GC because it is reverted by base analogs and HA but not by frameshift mutagen. *ara*$^+$ to *ara*-3: This mutation is AT to GC for the same reasons as given for the second mutant. Mutagen X causes transition mutations in both directions because mutants are revertible by base analogs, some are revertible by HA, and none (if this is a representative sample) by frameshift mutagens.

Chapter 17

17.3 a. 45

b. 47

c. 23

d. 69

e. 48

17.4 b An individual with three instead of two chromosomes is said to be *trisomic.*

17.5 a No Barr bodies in an XO Turner syndrome individual.

17.6 a Mother. The color-blind Turner syndrome child must have received her X chromosome from her father because the father carries the only color-blind allele. Therefore, the mother must have produced an egg with no X chromosomes.

b Father. Because the Turner syndrome child does not have colorblindness, she must have received her X chromosome from the homozygous normal mother. Therefore, the father must have produced a sperm with no X or Y chromosome.

17.7 Polyploids with even multiples of the chromosomes can better form chromosome pairs in meiosis than can polyploids with odd multiples. Triploids, for example, will generate an unpaired chromatid pair for each chromosome type in the genome, and so chromosome segregation to the gametes is irregular.

17.9 5*A*:1 *a*

17.12 The C plants are allotetraploids, containing a diploid set of chromosomes from both *A* and *B*. Fertile seed will be produced when they are crossed with other *C* plants because no abnormal chromosome pairing or unpaired chromosomes will result in that case.

17.14 See text, pp. 537–539.

17.17 An example of a two-strand double crossover and the resulting meiotic products is shown in the following figure:

In this case no dicentric bridge structure is formed, as with single crossovers, and all four meiotic products are viable.

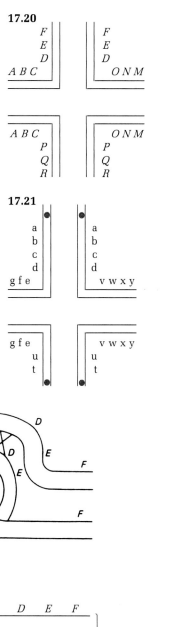

17.20

17.21

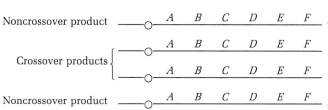

Noncrossover product   A   B   C   D   E   F

Crossover products { A   B   C   D   E   F

A   B   C   D   E   F

Noncrossover product   A   B   C   D   E   F

All four products have complete sets of genes; no duplications or deletions

Chapter 18

18.1 The four types are insertion sequences, transposons, plasmids, and certain temperate bacteriophages. See text pp. 575–591 for discussion.

Chapter 19

19.2 See text, pp. 598–600. Hybrid DNA resulting from branch migration may contain mismatched sequences. Mismatched sequences are recognized by special enzymes that remove a short segment of one DNA strand and then fill in the gap. The mismatch repair does not recognize which is the correct sequence.

19.3 In each case there is evidence that the segregation of one of the alleles in the tetrad has resulted from gene conversion caused by mismatch repair of heteroduplex DNA.

For the *a1 a2*⁺ × *a1*⁺ *a2* cross the tetrad shows 2:2 segregation of *a1*⁺:*a1* but 3:1 segregation of *a2*⁺:*a2,* indicating gene conversion of an *a2* allele to its wild-type counterpart. Similarly, the *a1 a3*⁺ × *a1*⁺ *a3* cross shows 2:2 segregation of *a3*⁺:*a3* and 3:1 segregation of *a1*⁺:*a1* resulting from gene conversion of an *a1* allele to *a1*⁺. In the *a2 a3*⁺ × *a2*⁺ *a3* cross, the *a2* allele segregates in a Mendelian fashion while the *a3* allele segregates 3:1 *a3*⁺:*a3* as a result of gene conversion of one *a3* allele to *a3*⁺.

19.4 The tetrad shows evidence of gene conversion of both alleles: a wild-type allele of *y1* has undergone gene conversion to a *y1* allele, and a wild-type allele of *y2* has undergone gene conversion to a *y2* allele.

Chapter 20

20.3 A constitutive phenotype can be the result of either an *lacI*⁻ or an *lacO*ᶜ mutation.

20.4 a. A missense mutation results in partial or complete loss of β-galactosidase activity, but there would be no loss of permeases and transacetylase activities. (b) A nonsense mutation is likely to have polar effects unless the mutation is very close to the normal chain-terminating codon for β-galactosidase. Thus permease and transacetylase activities would be lost in addition to the loss of β-galactosidase activity if nonsense mutation occurred near the 5′ end of the *lacZ* gene.

20.6 For the *E. coli lac* operon, write the partial-diploid genotype for a strain that will produce β-galactosidase constitutively and permease by induction.
lacI⁺ *lacO*ᶜ *lacZ*⁺ *lacY*⁻ /*lacI*⁺ *lacO*⁺ *lacZ*⁻ *lacY*⁺
(It cannot be ruled out that one of the repressor genes is *lacI*⁻)

20.8 Answer is in Figure 20.8 (p. 889)

20.11 The CAP, in a complex with cAMP, is required to facilitate RNA polymerase binding to the *lac* promoter. The RNA polymerase binding occurs only in the absence of glucose and only if the operator is not occupied by repressor (i.e., if lactose is absent). A mutation in the CAP gene, then, would render the *lac* operon incapable of expression because RNA polymerase would not be able to recognize the promoter.

20.14 See text, pp. 630–631. Stringent control is the regulatory mechanism that operates in bacteria to stop essential cellular biochemical activities when the cell is in a harsh nutritional condition, as when a carbon source runs out or when an auxotroph is starved for the nutrient involved. Characteristically, the unusual nucleotides ppGpp and pppGpp accumulate rapidly when the stringent-control system is in effect.

20.17 The *cI* gene codes for repressor protein that functions to keep the lytic functions of the phage repressed when lambda is in the lysogenic state. A *cI* mutant strain would be unable to establish lysogeny, and thus it would also follow the lytic pathway.

Chapter 21

21.3 An operon regulates transcription by using signals from the environment and provides for the coordinate production of proteins (e.g., enzymes) with related functions. The first property may or may not be found in a eukaryotic gene that encodes a multifunctional protein; that is, the gene may not be inducible or repressible. The existence, however, of a single multifunctional protein encoded by a single gene ensures the coordinate production of the related enzyme activities. Unlike the protein products of a prokaryotic operon (which are separate after synthesis), the multifunctional eukaryotic protein product has all the necessary functions at one location in the cell and thus is not susceptible to changes in the cellular microenvironment.

21.4 The final product of the rRNA genes is an rRNA molecule. Hence a large number of genes are required to produce the large number of rRNA molecules required for ribosome biosynthesis. Ribosomal proteins, in contrast, are the end products of the translation of mRNAs, which can be "read" over and over to produce the large number of ribosomal protein molecules required for ribosome biosynthesis.

21.11 four: A_3, A_2B, AB_2, B_3

21.16 Experiment A results in all the DNA becoming radioactively labeled, and so because DNA is a fundamental and major component of polytene chromosomes, radioactivity is evident throughout the chromosomes. Experiment B results in radioactive labeling of RNA molecules. Because radioactivity is first found

		Inducer absent		Inducer present	
Answer: Genotype		β-galacto- sidase	Permease	β-galacto- sidase	Permease
a. *lacI⁺ lacP⁺ lacO⁺ lacZ⁺ lacY⁺*		−	−	+	+
b. *lacI⁺ lacP⁺ lacO⁺ lacZ⁻ lacY⁺*		−	−	−	+
c. *lacI⁺ lacP⁺ lacO⁺ lacZ⁺ lacY⁻*		−	−	+	−
d. *lacI⁻ lacP⁺ lacO⁺ lacZ⁺ lacY⁺*		+	+	+	+
e. *lacIˢ lacP⁺ lacO⁺ lacZ⁺ lacY⁺*		−	−	−	−
f. *lacI⁺ lacP⁺ lacOᶜ lacZ⁺ lacY⁺*		+	+	+	+
g. *lacIˢ lacP⁺ lacOᶜ lacZ⁺ lacY⁺*		+	+	+	+
h. *lacI⁺ lacP⁺ lacOᶜ lacZ⁺ lacY⁻*		+	−	+	−
i. *lacI⁻ᵈ lacP⁺ lacO⁺ lacZ⁺ lacY⁺*		+	+	+	+
j. *lacI⁻ lacP⁺ lacO⁺ lacZ⁺ lacY⁺/* *lacI⁺ lacP⁺ lacO⁺ lacZ⁻ lacY⁻*		−	−	+	+
k. *lacIˢ lacP⁺ lacO⁺ lacZ⁺ lacY⁻/* *lacI⁺ lacP⁺ lacO⁺ lacZ⁻ lacY⁺*		−	−	−	−
l. *lacIˢ lacP⁺ lacO⁺ lacZ⁺ lacY⁻/* *lacI⁺ lacP⁺ lacO⁺ lacZ⁻ lacY⁺*		−	−	−	−
m. *lacI⁺ lacP⁺ lacOᶜ lacZ⁻ lacY⁺/* *lacI⁺ lacP⁺ lacO⁺ lacZ⁺ lacY⁻*		−	+	+	+
n. *lacI⁻ lacP⁺ lacOᶜ lacZ⁺ lacY⁻/* *lacI⁺ lacP⁺ lacO⁺ lacZ⁻ lacY⁺*		+	−	+	+
o. *lacIˢ lacP⁺ lacO⁺ lacZ⁺ lacY⁺/* *lacI⁺ lacP⁺ lacOᶜ lacZ⁺ lacY⁺*		+	+	+	+
p. *lacI⁻ᵈ lacP⁺ lacO⁺ lacZ⁺ lacY⁻/* *lacI⁺ lacP⁺ lacO⁺ lacZ⁻ lacY⁺*		+	+	+	+
q. *lacI⁺ lacP⁻ lacOᶜ lacZ⁺ lacY⁻/* *lacI⁺ lacP⁺ lacO⁺ lacZ⁻ lacY⁺*		−	−	−	+
r. *lacI⁺ lacP⁻ lacO⁺ lacZ⁺ lacY⁻/* *lacI⁺ lacP⁺ lacOᶜ lacZ⁻ lacY⁺*		−	+	−	+
s. *lacI⁻ lacP⁻ lacO⁺ lacZ⁺ lacY⁺/* *lacI⁺ lacP⁺ lacO⁺ lacZ⁻ lacY⁻*		−	−	−	−
t. *lacI⁻ lacP⁺ lacO⁺ lacZ⁺ lacY⁻/* *lacI⁺ lacP⁻ lacO⁺ lacZ⁻ lacY⁺*		−	−	+	−

Figure 20.8

only around puffs and later in the cytoplasm, we can hypothesize that the puffs are sites of transcriptional activity. Initially, radioactivity is found in RNA that is being synthesized, and the later appearance of radio-activity in the cytoplasm reflects the completed RNA molecules that have left the puffs and are being trans-lated in the cytoplasm. Experiment C provides addi-tional support for the hypothesis that transcriptional activity is associated with puffs. That is, the inhibi-tion of RNA transcription by actinomycin D blocks the appearance of ^{3}H-uridine (which would be in RNA) at puffs. In fact, puffs are much smaller, indicating that the puffing process is intimately associated with the onset of transcriptional activity for the gene(s) in that region of the chromosomes.

21.18 Preexisting mRNA that was by the mother and deposited in the egg prior to fertilization is translated up until the gastrula stage. After gastrulation new mRNA synthesis is necessary for production of proteins needed for subsequent embryo development.

21.19 The tissue taken from the blastula/gastrula has not yet been committed to its final differentiated state in terms of its genetic programming; that is, it is not yet determined. Thus when it is transplanted into the host, the determined tissues surrounding the transplant in the host communicate with the transplanted tissue and cause it to be determined in the same way as they are; for example, tissue transplanted to a future head area will differentiate into head material, and so on. In contrast, tissues in the neurula stage are now determined as to their final tissue state once development is complete. Thus tissue transplanted from a neurula to an older embryo cannot be influenced by the determined, surrounding tissues and will develop into the tissue type for which it is determined, in this case an eye.

Chapter 22

22.5 The two codes are different in codon designations. The mitochondrial code has more extensive wobble so that fewer tRNAs are needed to read all possible sense codons. As a consequence, fewer mitochondrial genes are necessary. The advantage is that fewer tRNAs are needed, and hence fewer tRNA genes need be present than for cytoplasmic tRNAs.

22.9 The features of extranuclear inheritance are differences in reciprocal-cross results (not related to sex), nonmappability to known nuclear chromosomes, Mendelian segregation not followed, the indifference to nuclear substitution.

22.12 The parental snails were *Dd* female and *d*-male. The F$_1$ snail is *dd*. (Given the F$_1$ genotype, the male can either be homozygous *d* or heterozygous, but the determination cannot be made from the data given.)

22.14 The first possibility is that the results are the consequence of a sex-linked lethal gene. The females would be homozygous for a dominant gene *L* that is lethal in males but not in females. In this case mating the F$_1$ females of an *L/L* × +/Y cross to +/Y males should give a sex ratio of 2 females:1 male in the progeny flies.

The second possibility is that the trait is cytoplasm-

ically transmitted via the egg and is lethal to males. In this case the same F$_1$ females should continue to have *only* female progeny when mated with +/Y males.

22.17 ½ *petite*, ½ wild-type *(grande)*

22.19 a. Parents: [*po*]*F*♀ × [*N*]+ ♂; progeny: [*po*]*F* and [*po*]+ in equal numbers. Because standard *poky* [*po*]+ is found among the offspring, gene *F* must not effect a permanent alteration of *poky* cytoplasm. The 1:1 ratio of *poky* to *fast-poky* indicates that all progeny have the *poky* mitochondrial genotype (by maternal inheritance) and that the *F* gene must be a nuclear gene segregating according to Mendelian principles. Thus the *poky* progeny are [*po*]+, and the *fast-poky* are [*po*]*F*.

b. Parents: [*N*]+ × [*po*]*F*; progeny: [*N*]+ and [*N*]*F* in equal numbers. These two types are phenotypically indistinguishable.

Chapter 23

23.1 a. Head width: Mean = 3.15, Standard deviation = 0.49 Wing length: Mean = 35.21, Standard deviation = 7.68.

b. Correlation coefficient = r = 0.973

c. Conclusions: Head width and wing length display a strong positive association, meaning that ducks with wider heads tend to have longer wings.

23.3 252/1024

23.4 Each pure-breeding parent is homozygous for the genes (however many there are) controlling the size character, and hence each parent is homogeneous in type. A cross of two pure-breeding strains will generate an F$_1$ heterozygous for those loci controlling the size trait. Since the F$_1$ is genetically homogeneous (all heterozygotes), it shows no greater variability than the parents.

23.7 a. 8 cm

b. 1 (2 cm) :6 (4 cm) :15 (6 cm) :20 (8 cm) :15 (10 cm) :6 (12 cm) :1 (14 cm)

c. Proportion of F$_2$ of *AA BB CC* is (¼)^3; proportion of F$_2$ of *aa bb cc* is (¼)^3; proportion of F$_2$ with heights equal to one or other is (¼)^3 + (¼)^3 = ^{2}⁄$_{64}$.

d. The F$_1$ height of 8 cm can be produced only by maintaining heterozygosity at no less than one gene pair (e.g., *AA Bb cc*). Thus although 20 of 64 F$_2$ plants are expected to be 8 cm tall, none of these will breed true for this height.

23.9 It is a quantitative trait. Variation appears to be continuous over a range rather than falling into three discrete classes. Also, the pattern in which the F$_1$ mean falls between the parental means and in which the F_2 individuals show a continuous range of variation from one parental extreme to the other is typical of quantitative inheritance.

23.12 The cross is *aa bb cc* (3 lb) × *AA BB CC* (6 lb), which gives an F_1 that is *Aa Bb Cc* and which weighs 4.5 lb. The distribution of phenotypes in the F_2 is, 1 (3 lb) :6 (3.5 lb) :15 (4 lb) :20 (4.5 lb) :15 (5 lb) :6 (5.5 lb) :1 (6 lb).

23.14 a. The progeny are *Aa Bb Cc Dd,* which are 18 dm high.

b. (1) The minimum number of capital-letter alleles is one and the maximum number is four, giving a height range of 12 to 18 dm. (2) The minimum number is one, and the maximum number is four, giving a height range of 12 to 18 dm. (3) The minimum number is four, and the maximum number is six, giving a height range of 18 to 22 dm. (4) The minimum number is zero, and the maximum number is seven, giving a height range of 10 to 24 dm.

23.16 a. Broad-sense heritability = V_G/V_P = $(V_A + V_D + V_I)/(V_A + V_D + V_I + V_E + V_{GE})$ = (4.2 + 1.6 + 0.3)/(4.2 + 1.6 + 0.3 + 2.7 + 0) = 6.1/8.8 = 0.69. Narrow-sense heritability = V_A/V_P = 4.2/8.8 = 0.48

b. About 69% of the phenotypic variation in leaf width observed in this population is due to genetic differences among individuals. About 48% of the phenotypic variation is due to additive genetic variation.

23.17 Heritability is specific to a particular population and to a specific environment and cannot be used to draw conclusions about the basis of populational differences. Because the environment of the farms in Kansas and in Poland differ, and because the two wheat varieties differ in their genetic makeup, the heritability of yield calculated in Kansas cannot be applied to the wheat grown in Poland. Furthermore, the yield of TK138 would most likely be different in Poland, and might even be less than the yield of the Russian variety when grown in Poland.

23.18 a. The narrow-sense heritability of number of triradii will equal the slope of the regression line of the mean offspring phenotype on the mean parental phenotype. For the data presented in this problem, the slope = 1.03.

b. All the observed variation in phenotype can be attributed to additive genetic variation among the individuals.

23.19 Selection differential = 14.3 − 9.7 = 4.6

Selection response = 13 − 9.7 = 3.3

Narrow-sense heritability = selection response/selection differential = 3.3/4.6 = 0.72

23.21 a. Fathers: Mean = 71.9, variance = 11.6

Sons: Mean = 70, variance = 10.25

b. Correlation = 0.49

c. Slope = b = 0.46

Narrow-sense heritability = 2(b) = 2(0.46) = 0.92

23.22 Selection response = narrow-sense heritability × selection differential.

Selection response = 0.60 × 10 g = 6 g

23.25 to obtain uniform lines to be used as the parents of hybrids

Chapter 24

24.1 Brown = $C^B C^B, C^B C^P, C^B C^Y$ = $p^2 + 2pq + 2pr$
$$= 236/500 = 0.472$$

Pink = $C^P C^P, C^P C^Y$ = $q^2 + 2qr$ = 231/500 = 0.462

Yellow = $C^Y C^Y$ = r^2 = 33/500 = 0.066

$f(C^Y C^Y) = r^2$

$\sqrt{f(C^Y C^Y)} = r^2$

$\sqrt{33/500} = r = \sqrt{0.066} = 0.26$

$f(C^P C^P + C^P C^Y + C^Y C^Y) = q^2 + 2qr + r^2$

$f(C^P C^P + C^P C^Y + C^Y C^Y) = (q + r)^2$

$$\frac{231 + 33}{500} = (q + r)^2$$

$0.528 = (q + r)^2$

$\sqrt{0.528} = q + r$

$0.727 = q + r$

$r = 0.26$

$0.727 - r = q$

$q = 0.467$

$p = 1 - q - r$

$= 1 - 0.467 - 0.26 = 0.273$

and so, $f(C^B) = p = 0.273$

$f(C^Y) = r = 0.26$

$f(C^P) = q = 0.467$

24.4 a. $\sqrt{0.16} = 0.4 = 40\%$ = frequency of recessive alleles; $1 - 0.4 = 0.6 = 60\%$ = frequency of dominant alleles; $2pq = (2)(0.4)(0.6) = 0.48$ = probability of heterozygous diploids. Then $(0.48)/[2 \times 0.16) + 0.48] = 0.48/0.80 = 60\%$ of recessive alleles are heterozygotes.

b. If $q^2 = 1\% = 0.01$, then $q = 0.1$, $p = 0.9$, and $2pq = 0.18$ heterozygous diploids. Therefore $(0.18)/[(2 \times 0.01) + 0.18] = 0.18/0.20 = 90\%$ of recessive alleles in heterozygotes.

24.5 $2pq/q^2 = 8$, and so $2p = 8q$; then $2(1 - q) = 8q$, and $2 = 10q$, or $q = 0.2$

24.8 a. Let p = the frequency of S and q = the frequency of s. Then

$$p = \frac{2(188) \, SS + 717 \, Ss}{2(3146)} = \frac{1093}{6292} = 0.1737$$

$$q = \frac{717 \, Ss + 2(2241) \, ss}{2(3146)} = \frac{5199}{6292} = 0.8263$$

b.

Class	Observed	Expected	d	d^2/e
SS	188	94.9	$+ 93.1$	91.235
Ss	717	903.1	$- 186.1$	38.361
ss	2241	2147.9	$+ 93.1$	4.032
	3146	3145.9	0	133.628

24.11 There is only one degree of freedom because the three genotypic classes are completely specified by two gene frequencies, namely, p and q. Thus the number of degrees of freedom = number of genes − 1. The X^2 value for this example is 133.628, which, for one degree of freedom, gives a P value less than 0.0001. Therefore the distribution of genotypes differs significantly from the Hardy-Weinberg equilibrum. Because the frequency of the trait is different in males and females, the character might be caused by a sex-linked recessive gene. If the frequency of this gene is q, females would occur with the character at a frequency of q^2, and males with the character would occur at a frequency of q. The frequency of males is $q = 0.4$, and thus we may predict that the frequency of females would be $(0.4)^2 = 0.16$ if this is a sex-linked gene. This result fits the observed data. Therefore the frequency of heterozygous females is $2pq =$

$2(0.6)(0.4) = 0.48$. For sex-linked genes no heterozygous males exist.

24.13 64/10,000 are color blind, i.e., $0.0064 = q^2$, and so $q = 0.08$ = probability of color-blind male.

24.16

$$q = \frac{u}{u + v} = \frac{6 \times 10^{-7}}{(6 \times 10^{-7}) + (6 \times 10^{-7})} =$$

$$\frac{6 \times 10^{-7}}{(6 \times 10^{-7}) + (0.6 + 10^{-7})} = \frac{6}{6.6} = 0.91$$

$$p = 1 - q = 1 - 0.91 = 0.09$$

Thus, the frequencies are 0.008 *AA,* 0.16 *Aa,* and 0.828 *aa*

24.17 a. 100

b. 96

c. 36

d. 7.8

24.20 $p'_{II} = mp_I + (1 - m)p_{II}$

p'_{II} II $(0.20)(0.5) + (1 - 0.20)(0.70) = 0.66$

24.23 mutation of A to a

24.25 $q = s/(s + t) = 0.12/(0.12 + 0.86) = 0.12$

24.27 $q = u/s = (5 \times 10^{-5})/0.8 = 0.0000625$

Credits

Text Credits

Figure 1.11: From Curtis and Barnes, *Biology,* 5th ed., Worth Publishers, New York, 1989. Reprinted by permission.

Figure 1.27: Reproduced from William T. Keeton and James L. Gould with Carol Grant Gould, *Biological science,* 4th ed. Illustrated by Michael Reingold. Copyright © 1986, 1980, 1979, 1978, 1972, 1967 by W. W. Norton and Company, Inc. Reprinted by permission of W. W. Norton and Company.

Figure 4.7: Explaining the forces of nature, in *The New York Times,* March 27, 1984. Copyright © 1984 by The New York Times Company. Reprinted by permission.

Table 4.4: From A. E. Mourant, A. C. Kopec, and K. Domaniewska-Sobczak, *The distribution of the human blood groups,* 2nd ed. (London: Oxford University Press, 1976). Reprinted by permission.

Table 4.5: Reprinted with permission of Macmillan Publishing Company from George W. Burns, *The science of genetics,* 5th ed., p. 44. Copyright © 1983 by George W. Burns.

Figure 5.5: From Ursula Goodenough, *Genetics,* 2nd ed., Fig. 12-1. Copyright © 1978 by Holt, Rinehart and Winston. Reprinted by permission of the publisher.

Table 5.2: Taken from Table IV of Fisher and Yates, *Statistical tables for biological, agricultural, and medical research,* 6th ed., published by Longman Group Ltd. London, 1974 (previously published by Oliver and Boyd Ltd. Edinburgh), by permission of the authors and publishers.

Figure 6.9: Source: Robert H. Tamarin, *Principles of genetics.* Boston: Prindle, Weber and Schmidt Publishers, 1982.

Figure 6.18: From Bruce Alberts et al., *Molecular biology of the cell.* Copyright © 1983 by Bruce Alberts, Dennis Bray, Julian Lewis, Martin Raff, Keith Roberts, and James D. Watson. Reprinted by permission of Garland Publishing Inc.

Figure 7.20: From W. H. Wood and H. R. Revel, The genome of bacteriophage T4, in *Bacteriological Review* 40(1976):847–68. Reprinted by permission.

Figures 8.19, 8.20: From R. E. Dickerson, The DNA helix and how it is read, in *Scientific American,* vol. 249, December 1983. Reprinted by permission.

Figure 8.21: From A. Rich, A. Nordheim, and A. H.-J. Wang, The chemistry and biology of left-handed Z-DNA. Reproduced, with permission, from *Annual Review of Biochemistry,* vol. 53. © 1984 by Annual Reviews Inc.

Figure 9.14: Adapted from R. J. Britten and E. Davidson, Distribution of genome size in animals, in *Quarterly Review of Biology* 46(1971):111. Reprinted by permission.

Figure 9.22: From Bruce Alberts et al., *Molecular biology of the cell,* Fig. 8-24. © 1983 by Bruce Alberts, Dennis Bray, Julian Lewis, Martin Raff, Keith Roberts, and James D. Watson. Reprinted by permission of Garland Publishing Inc.

Figures 9.25, 9.26, 9.27: From R. J. Britten and D. E. Kohne, Repeated sequences in DNA, in *Science,* vol. 161, August 9, 1978, pp. 529–40. Copyright 1978 by the AAAS. Reprinted by permission.

Table 9.4: From S. C. R. Elgin and H. Weintraub, Chromosomal proteins and chromatin structure. Reproduced, with permission, from *Annual Review of Biochemistry,* vol. 44. © 1975 by Annual Reviews Inc.

Figure 10.16: From J. D. Watson, N. H. Hopkins, J. W. Roberts, J. A. Steitz, and A. M. Weiner, *Molecular biology of the gene,* 4th ed., vol. 1, Fig. 10-22. Copyright © 1965, 1970, 1976, 1987 by The Benjamin/Cummings Publishing Company, Inc. Reprinted by permission.

Tables 10.4, 10.5: From R. Sheinin, J. Humbert, and R. E. Pearlman, Some aspects of eukarotic DNA replication. Reproduced, with permission, from *Annual Review of Biochemistry,* vol. 47. © 1978 by Annual Reviews Inc.

Figure 11.3: From J. Marmur et al., *Cold Spring Harbor Symposium of Quantitative Biology* 28(1963):191. Reprinted by permission.

Figure 11.8: From U. Siebenlist, R. Simpson, and W. Gilbert in *Cell* 20(1980):269–81. Copyright © 1980 by Cell Press. Reprinted by permission.

Figure 11.12: From Charles Yanofsky, Attenuation in the control of expression of bacterial operons. Reprinted by permission from *Nature,* vol. 289, p. 751. Copyright © 1981 Macmillan Journals Ltd.

Figure 12.12: Adapted from Proudfoot, Brownlee in *Nature,* vol. 263, September 16, 1976. Copyright © 1976 by Macmillan Magazines Ltd. Reprinted by permission.

Figure 12.15: From H. Busch, R. Reddy, and L. Rothblum in *Annual Review of Biochemistry,* vol. 51, 1982. Copyright © 1982 by Annual Reviews, Inc. Reprinted by permission.

Figure 12.25b: From M. Ellwood and M. Nomura in *Journal of Bacteriology* 149(1982):458. Reprinted by permission of the American Society for Microbiology.

Figure 12.31: Reprinted by permission from J. Sommerville in *Nature,* vol. 310, pp. 189–90. Copyright © 1984 Macmillan Magazines Ltd.

Table 13.2: From S. Michaelis and J. Beckwith in *Annual Review of Microbiology,* vol. 36. © 1982 by Annual Reviews Inc. Reproduced with permission.

Table 14.3: From A. Milunsky, Prenatal diagnosis of genetic disorders. Reprinted by permission of *The New England Journal of Medicine* 295(1976):337.

Figure 15.6: From Francisco Bolivar et al., in A. L. Bukhari, J. A. Shapiro, and S. L. Adhya, eds., *DNA insertion elements: Plasmids and episomes.* Copyright © 1977 by Cold Spring Harbor Laboratory. Reprinted by permission.

Figure 15.23: From Robert F. Weaver and Philip W. Hedrick, *Genetics.* Copyright © 1989 by Wm. C. Brown Publishers. Reprinted by permission.

Table 15.1: Adapted from R. J. Roberts in *CRC Critical Reviews in Biochemistry* 4(1976):123. Copyright © 1976 CRC Press, Inc., Boca Raton, Florida. Reprinted by permission.

Figure 16.5: Adapted from George W. Burns, *The science of genetics,* 5th ed. Copyright © 1983 by George W. Burns. Reprinted with permission of Macmillan Publishing Company.

Figure 16.6: From E. J. Gardner and D. P. Snustad, *Principles of genetics,* 5th ed. Copyright © 1984 by John Wiley and Sons. Reprinted by permission.

Figures 16.7, 16.8: From Bruce Alberts et al., *Molecular biology of the cell.* Copyright © 1983 by Bruce Alberts, Dennis Bray, Julian Lewis, Martin Raff, Keith Roberts, and James

D. Watson. Reprinted by permission of Garland Publishing Inc.

Figures 16.16, 16.17: From J. D. Watson, N. H. Hopkins, J. W. Roberts, J. A. Steitz, and A. M. Weiner, *Molecular biology of the gene,* 4th ed., vol. 1, Figs. 12-8, 12-9. Copyright © 1965, 1970, 1976, 1987 by The Benjamin/Cummings Publishing Company, Inc. Reprinted by permission.

Figure 18.11: A map of a Tyl-like genome, adapted by permission of Dr. Gerry Fink.

Figures 20.15, 20.17: From Charles Yanofsky, Attenuation in the control of expression of bacterial operons. Reprinted by permission from *Nature,* vol. 289, February 26, 1981. Copyright © 1981 Macmillan Journals Ltd.

Figure 21.11: From Bruce Alberts et al., *Molecular biology of the cell.* Copyright © 1983 by Bruce Alberts, Dennis Bray, Julian Lewis, Martin Raff, Keith Roberts, and James D. Watson. Reprinted by permission of Garland Publishing Inc.

Figure 21.17: From Weatherall and Clegg, *Thalassaemia syndromes,* 3rd ed., 1981. Reprinted by permission of Blackwell Scientific Publications Ltd.

Figure 21.23: From Y. Horta and S. Benzer, in F. H. Ruddle, ed., *Genetic mechanisms of development,* p. 129 (Orlando, Florida: Academic Press, 1973). Reprinted by permission.

Figure 21.25: Adapted from E. B. Lewis, *Embryonic development, Part A: Genetic aspects,* 1982, p. 272. Reprinted by permission of Allan Liss Inc.

Figure 21.29: From R. L. Garber, A. Kuroiwa, and W. J. Gehring, in *EMBO Journal* 2(1983):2027–36. Reprinted by permission or IRL Press Limited, Oxford, England.

Figure 21.38: Based on data from T. Takeya and H. Hanafusa in *Cell* 32(1983):881–90.

Table 21.2: From L. Chan and B. W. O'Malley in *Annals of Internal Medicine* 89(1978):649. Reprinted by permission.

Table 21.3: From R. Weiss et al., *RNA tumor viruses,* 2nd ed., 1985. Reprinted by permission.

Figure 22.10b: From R. Sager and D. Lane, Replication of chloroplast DNA in zygotes of *Chlamydomonas,* in *Federation Proceedings of the Federation of American Societies for Experimetnal Biology* 38(1969):347. Reprinted by permission.

Figures 22.18, 22.19: From Henry Allan Gleason et al., *New Britton and Brown, Illustrated flora of the northeastern U.S. and adjacent Canada.* Copyright 1952 by The New York Botanical Garden. Reprinted by permission.

Figure 22.25: From B. Singer, R. Sager, and Z. Ramanis in *Genetics* 83(1976):341. Reprinted by permission of the Genetics Society of America.

Figure 23.2: Adapted from C. M. Jackson, The physique of male students at the University of Minnesota: A study in constitutional anatomy and physiology, in *The American Journal of Anatomy,* vol. 40, p. 66, September, November, 1927; January 1928. Reprinted by permission of Allan Liss Inc.

Figure 23.3: From Harry Bakwin and Ruth Morris Bakwin, Body build in infants v. anthropometry in the new born, in *Human biology: A record of research,* vol. 6, 1934. Reprinted by permission of Johns Hopkins University Press.

Figure 23.5: From Sewall Wright, *Evolution and the genetics of populations,* vol. 1, pp. 113, 116, 118. Copyright © by The University of Chicago. All rights reserved. Reprinted by permission.

Figures 23.7, 23.9: From L. A. Synder, D. Freifelder, and D. L. Hartl, *General genetics,* 1985. Reprinted by permission of Jones and Bartlett Publishers, Inc.

Figure 23.12b: From A. H. Sturtevant and G. W. Beadle, *An introduction to genetics,* p. 265 (Philadelphia: W. B. Saunders Co., 1940). Reprinted by permission.

Figure 23.15: From Curtis and Barnes, *Biology,* 5th ed. Worth Publishers, New York, 1989. Reprinted by permission.

Figure 23.16: From D. S. Falconer, *Introduction to quantitative genetics,* 2nd ed., 1981. Reprinted by permission of Longman Group Ltd. London.

Figure 23.17: After Crane and Lawrence in *Journal of Pomology and Horticultural Science* 7(1929):290. Reprinted by permission of the *Journal of Horticultural Science.*

Figure 24.1: From E. B. Ford, *Ecological genetics.* Reprinted by permission of Chapman and Hall Ltd.

Figure 24.3: From R. K. Koehn et al. in *Evolution* 30(1976):6. Reprinted by permission of the Society for the Study of Evolution.

Figure 24.6: From Francisco J. Ayala and John A. Kiger, Jr., *Modern genetics,* 2nd ed. Copyright © 1984 and 1980 by The Benjamin/Cummings Publishing Company, Inc. Reprinted by permission.

Figure 24.8: From P. Buri in *Evolution* 10(1956):367. Reprinted by permission of the Society for the Study of Evolution.

Figure 24.9: From Philip W. Hedrick, *Genetics of populations,* 1983. Reprinted by permission of Jones and Bartlett Publishers, Inc.

Table 24.3: From R. K. Selander, Genetic variation in natural populations, in F. J. Ayala, *Molecular evolution,* 1976. Reprinted by permission of Sinauer Associates, Inc., Publishers.

Table 24.4: From Monroe W. Strickberger, *Genetics,* 3rd ed. Copyright © 1985 by Monroe W. Strickberger. Reprinted by permission of Macmillan Publishing Company.

Figure 26.1: Reprinted by permission from *Nature,* vol. 273, no. 5658, p. 104. Copyright © 1978 Macmillan Magazines Ltd.

Photo Credits

Chapter 1: p. 3 From Davis/Eisen/Ginsberg, *Microbiology,* Third Edition (Philadelphia: J.B. Lippincott Co., 1980): (a) Noyes, W.F.: *Virology* 23:65, 1964; (b) Finch, J.T.: *Journal of Molecular Biology* 8:872, 1964. Copyright by Academic Press, Inc., London; (c) Courtesy of P. Wildy; (d) Choppin, P.W., Stockenius, W.: *Virology* 22:482, 1964; (e) Courtesy of R.W. Horne; (f) A.N. Broers, B.J. Panessa, and J.F. Gennaro, Jr.; (g) S.G. Whitfield, Centers for Disease Control, Atlanta. p. 4 (a) © David Scharf/Peter Arnold, Inc.; (b) Lederle Laboratories/Photo Researchers, Inc.; (c) Omikron/Photo Researchers, Inc.; (d) J.J. Duda and J.M. Slack, West Virginia University/Biological Photo Service. p. 5 (top, a) P.W. Johnson and J. McN. Sieburth, University of Rhode Island/Biological Photo Service; (top, b) J.C. Burnham, Medical College of Ohio/Biological Photo Service; (bottom) T.J. Beveridge, University of Guelph/Biological Photo Service. p. 6 (a) Carolina Biological Supply Company; (b) © J.B. Woolsey Assoc.; (c) © J.B. Woolsey Assoc.; (d) Carolina Biological Supply Company; (e) J.B. Woolsey Assoc.; (f) Dr. Eric Schabtach and Dr. Ira Herskowitz, Department of Biochemistry and Biophysics, Division of Genetics, University of California, San Francisco; (g) Carolina Biological Supply Company; (h) PAR/NYC. p. 7 (top) John D. Cunningham © 1984; (bottom) M.C. Ledbetter, Brookhaven National Labroatory. p. 8 (top) R. Rodewald, University of Virginia/Biological Photo Service; (center) D.D. Kunkel, University of Washington/Biological Photo Service; (bottom) Courtesy of K.E. Carr, University of Glasgow, and P.G. Toner, Glasgow Royal Infirmary. From *Cell Structure,* p. 169. © 1982 Longman Group Limited. Used by permission of Churchill Livingstone, London. p. 9 H.S. Pankratz, Michigan State University/Biological Photo Service. p. 10 (top right) G.T. Cole, University of Texas, Austin/Biological Photo Service; (bottom left) P.W. Johnson and J.McN. Sieburth, University of Rhode Island/Biological Photo Service. p. 11 (top left) W.P. Wergin and E.H. Newcomb, University of Wisconsin, Madison/Biological Photo Service. p. 13 Courtesy of Dr. William C. Earnshaw, The Johns Hopkins University. p. 17 (a–f) Courtesy of Dr. Andrew Bajer. p. 19 (a) Produced from *Cytobios* 19:27 (1978), with permission of the Faculty Press, Cambridge, England; (b) Courtesy of Elton Stubblefield, Ph.D., produced from *Cytobios* 19:27 (1978), with permission of the Faculty Press, Cambridge, England; (c) © Dr. Don Fawcett/J.R. Paulson and U.K. Laemmli/Photo Researchers, Inc. p. 20 © David Scharf/Peter Arnold, Inc. pp. 22–23 A.D. Robinson, State of University of New York, Potsdam/Biological Photo Service. p. 24 Courtesy of M. Westergaard and D. von Wettstein.

Chapter 2: p. 35 (a–e) Louise Van der Meid, Salem, Oregon. p. 38 (top) Steve Woit/Picture Group; (center) Harvey Stein.

Chapter 3: p. 81 H.W. Pratt/Biological Photo Service. p. 82 (top) From *Principles of Genetics,* Seventh Edition, by Gardner and Snustad. © 1984 by John Wiley & Sons, Inc. Used by permission; (bottom left) Courtesy of Dr. Victor A. McKusick, Johns Hopkins Hospital; (bottom right) J.B. Woolsey Assoc. p. 84 From the *Journal of Heredity* 23 (September 1932), p. 345. p. 87 Harvey Stein. p. 90 Courtesy of Dr. Carl Witkop, University of Minnesota. p. 91 Courtesy of the American Society of Human Genetics. From Stern, Centerwall, and Sakar, "New Data on the Problem of Hairy Pemae," *American Journal of Human Genetics,* vol. 15 (1964), pp. 74–85.

Chapter 4: p. 107 (a) Courtesy of Carl L. Lewis, Jamestown, New York; (b–c) Bruce Iverson. p. 116 (a) Courtesy of the Development and Public Information Office, The Jackson

Laboratory, Bar Harbor, Maine; (b) © Russ Kinne/Comstock, Inc.; (c) © Dr. E.R. Degginger. p. 128 From the *Journal of Heredity* 24 (April 1933). p. 130 Peter J. Russell.

Chapter 5: p. 145 Courtesy of Jaroslav Cervenka, M.D., University of Minnesota.

Chapter 6: p. 185 Dr. Eric Schabtach and Dr. Ira Herskowitz, Department of Biochemistry and Biophysics, Division of Genetics, University of California, San Francisco. p. 187 (a–b) Carolina Biological Supply Company.

Chapter 7: p. 221 © David Scharf/Peter Arnold, Inc. p. 222 Dr. Edward Bottone/Mt. Sinai Hospital Department of Microbiology. p. 227 Micrograph courtesy of David P. Allison, Biology Division, Oak Ridge National Laboratory. p. 234 (a–b) Dr. Harold W. Fisher. p. 237 (a–b) Bruce Iverson. p. 238 Courtesy of Gunther S. Stent/University of California.

Chapter 8: p. 255 (top) From *Electron Microscopy in Human Medicine,* vol. 3, "Infectious Agents," by Jan Vincents Johannessen. © 1980 McGraw-Hill Book Company, Inc. Used by permission. (bottom) Dr. Edward Bottone/Mt. Sinai Hospital Department of Microbiology. p. 267 Courtesy of Professor M.H.F. Wilkins, Biophysics Department, King's College, London. p. 270 © Irving Geis. p. 271 (top) Courtesy of Dr. Richard Feldmann/NIH; (center) Courtesy of Richard Pastro—FDA/NIH; (bottom) Courtesy of Dr. Richard Feldmann/NIH. p. 272 © Irving Geis. p. 274 From Alexander Rich, "Right-handed and Left-handed DNA: Conformational Information in Genetic Material," *Cold Spring Harbor Symposia on Quantitative Biology,* vol. 47 (1983), p. 4. © 1983 Cold Spring Harbor Laboratory. Used by permission. p. 275 (a–b) From Alexander Rich, "Right-handed and Left-handed DNA: Conformational Information in Genetic Material." *Cold Spring Harbor Symposia on Quantitative Biology,* vol. 47 (1983). © 1983 Cold Spring Harbor Laboratory. Used by permission.

Chapter 9: p. 280 © K.G. Murti, © 1984/Visuals Unlimited. p. 281 (a–b) Dr. James C. Wany/Harvard University. p. 288 (top, a) R.W. Horne; (bottom, a) Courtesy of Harold W. Fisher; (b) From W.J. Thomas and R.W. Horne, "The Structure of Bacteriophage *Phi*X174," *Virology,* vol. 15 (1961), pp. 2–7 (Academic Press, N.Y.); (c) Courtesy of David Dressler and Kirston Koths, Harvard University. p. 290 A.P. Craig-Holmes/Biological Photo Service. p. 291 A.P. Craig-Holmes/Biological Photo Service. p. 292 Courtesy of Dr. C.C. Lin, University of Calgary. p. 294 Courtesy of Christine J. Harrison, Ph.D., Patterson Institute for Cancer Research, Christie Hospital and Holt Radium Institute, Manchester, England. p. 299 Victoria Foe. pp. 300–301 (top) Courtesy of Dr. Evangelos N. Moudrianakis, Johns Hopkins University. Reproduced from *Science,* 528 (3 May 1985), pp. 546–52. p. 301 (bottom) Barbara Hamakalo. p. 303 (a–b) A.P. Craig-Holmes/Biological Photo Service.

Chapter 10: p. 319 Courtesy of Dr. Matthew Messelson, Harvard University. p. 321 From J. Cairns, *Jopurnal of Molecular Biology,* vol. 6 (1963), pp. 208–13. With permission from the *Journal of Molecular Biology.* © 1963 by Academic Press, Inc. (London) Ltd. p. 323 Courtesy of N.P. Salzman,

NIH. p. 337 Courtesy of Professor T.A. Steitz, Yale University, Molecular Biophysics and Biochemistry. p. 342 From the *Proceeding of the National Academy of Science,* vol. 43, p. 122, article by J.H. Taylor et al. Autoradiograph provided by Dr. Taylor, Columbia Medical School. Used by permission of the National Academy of Science. p. 344 Courtesy of David Hogness, Department of Biochemistry, Stanford University School of Medicine, Stanford, California. From Kriegstein and Hogness, *Proceedings of the National Academy of Sciences,* U.S., vol. 71 (1974), p. 135.

Chapter 12: p. 373 Courtesy of O.L. Miller, Jr., Lewis and Clark Professor of Biology, University of Virginia. p. 379 Courtesy of Philip Leder, NIH. From Tilghman, Curtis, Tiemeier, Leder, and Weisman, *Proceedings of the National Academy of Sciences,* vol. 75 (1979), p. 1309. p. 381 Courtesy of Professor Pierre Chambon, Louis Pasteur University, Strasbourg, France. p. 389 (b) Courtesy of Evans and Sutherland. (c) Courtesy of Dr. Sung-Hou Kim, Duke University Medical Center. p. 397 Courtesy of Dr. James Lake, UCLA. p. 403 Courtesy of O.L. Miller, Jr., Lewis and Clark Professor of Biology, University of Virginia.

Chapter 13: p. 428 Courtesy of S.L. McKnight and O.L. Miller, Jr., University of Virginia.

Chapter 14: p. 439 Courtesy of Dr. D.P. Snustad, Department of Genetics and Cell Biology, College of Biological Sciences, University of Minnesota. p. 467 Courtesy of John A. Lond, Ph.D., University of California at San Francisco.

Chapter 15: p. 512 From "Coincidence of the Promoter and Capped 5' Terminus of RNA from the Adenovirus 2 Major Late Transcriptions Unit," by Edward B. Ziff and Ronald M. Evans of The Rockefeller University, New York. From *Cell,* vol. 15 (December 1978), pp. 1463–75 (image, p. 1468). Copyright © 1978 by MIT and 1988 by Cell Press, 50 Church Street, Cambridge, Massachusetts 01238.

Chapter 17: p. 554 Courtesy of U.S. Department of Health and Human Services. p. 555 Courtesy of Murray L. Barr, M.D. p. 556 Courtesy of the Clinical Cytogenetics Laboratory, Oregon Health Sciences University. p. 557 Courtesy of Dr. Victor A. McKusick, Johns Hopkins Hospital. p. 562 Courtesy of the Clinical Cytogenetics Laboratory, Oregon Health Sciences University.

Chapter 18: p. 587 From P.A. Peterson in *DNA Insertion Elements, Plasmids, and Episomes,* ed. A.I. Bukhari, J.A. Shapiro, and S.L. Adhya (Cold Spring Harbor, N.Y.: Cold Spring Harbor Laboratory, 1977). p. 590 Courtesy Harold W. Fisher. p. 593 From "The Functions and Relationships of Ty-VLP Proteins in Yeast Reflect Those of Mammalian Retroviral Proteins," by Salley E. Adams, Jane Mellorr, Susan M. Kingsman, Keith Gull, Mick F. Tuite, and Robert B. Sim.

From *Cell,* vol. 49 (10 April 1987), pp. 111–19. © 1987 by Cell Press, 50 Church Street, Cambridge, Massachusetts 02138.

Chapter 19: p. 597 Courtesy of R.G. Worton. p. 600 Huntington Potter and David Dressler, Department of Biochemistry and Molecular Biology, Harvard University. From page 4169, *Proceedings of the National Academy of Sciences USA,* vol. 74, no. 10 (October 1977), pp. 4168–72.

Chapter 21: p. 655 Courtesy of Dr. Harold Weintraub, Fred Hutchinson Cancer Research Center, Seattle. Used by permission. p. 671 (a–c) Courtesy of P. O'Farrell; (d–e) Courtesy of A. Ouelette. p. 674 Professor Bertil Daneholt, Karolinska Institutet, Stockholm. p. 680 (b–c) Courtesy of E.B. Lewis, California Institute of Technology. p. 681 (top left) From Suzuki et al., *Introduction to Genetic Analysis,* p. 485. © 1986 W.H. Freeman. Used by permission; (top right) F.R. Turner, Indiana University, Bloomington; (bottom) From David Suzuki et al., *Introduction to Genetic Analysis,* p. 485. © 1986 W.H. Freeman. Used by permission. p. 683 Courtesy of Grace Hospital, University of British Columbia Clinical, Genetics Unit, Vancouver, B.C. p. 684 Courtesy of Dr. John M. Opitz, Shodair Children's Hospital, Helena, Montana. p. 685 (left and right) National Medical Slide Bank/Graves Medical Audiovisual Library, London. p. 686 Courtesy of Dr. Harold W. Fisher and Dr. Robley C. Williams. p. 691 Courtesy of Dr. Robert A. Weinberg, Whitehead Institute, MIT.

Chapter 22: p. 701 Courtesy of Dr. Jack Griffith, Lineberger Cancer Research Center. p. 702 Courtesy of Professor David A. Clayton, Department of Pathology, Stanford University Medical School. p. 712 Fernandez-Moran/Omikron-Photo Researchers, Inc. p. 713 M. Murayama, Murayama Research Laboratory/Biological Photo Service. p. 716 Photo by John Schultz—PAR/NYC. Used by permission of the Department of Invertebrates, American Museum of Natural History, by kind cooperation of Harold Feinberg and Walter Sage, Senior Scientific Assistants. p. 718 © Derek Fell.

Chapter 23: p. 749 From E. W. Sinnott and L. C. Dunn, 1925. *Principles of Genetics* (New York: McGraw-Hill).

Chapter 24: p. 785 (top) Courtesy of F.M. Johnson, Ph.D., Research Geneticist, National Institutes of Health, National Institute of Environmental Health Sciences. p. 795 Field Museum of Natural History. p. 805 Courtesy of The University of Chicago Press. p. 813 (top) Richard Humbert/Biological Photo Service; (bottom) C.W. May/Biological Photo Service. p. 814 L.E. Gilbert, University of Texas, Austin/Biological Photo Service. p. 815 National Portrait Gallery, London. p. 816 (a–b) Courtesy of Dr. H.B.D. Kettlewell, Department of Zoology, Oxford, England.

Index